12th International Conference

High Magnetic Fields in the Physics of Semiconductors II

Volume 2

12th International Conference

High Magnetic Fields in the Physics of Semiconductors II

Volume 2

Würzburg, Germany
29 July–2 August 1996

Editors

G. Landwehr
W. Ossau

Physikalisches Institut der Universität Würzburg, Germany

World Scientific
Singapore • New Jersey • London • Hong Kong

Published by

World Scientific Publishing Co. Pte. Ltd.

P O Box 128, Farrer Road, Singapore 912805

USA office: Suite 1B, 1060 Main Street, River Edge, NJ 07661

UK office: 57 Shelton Street, Covent Garden, London WC2H 9HE

HIGH MAGNETIC FIELDS IN THE PHYSICS OF SEMICONDUCTORS II

ISBN 981-02-3076-1 (Set)
ISBN 981-02-3989-0 (Vol. 1)
ISBN 981-02-3990-4 (Vol. 2)

Printed in Singapore.

CONTENTS

VOLUME 1

I. Quantum Hall Effects

A. Integer Quantum Hall Effect

B. Fractional Quantum Hall Effect

C. Breakdown Phenomena

II. New Quasi-Particles

A. Composite Fermions

B. Skyrmions

III. Magnetotransport

IV. Tunneling

V. Small Quantum Systems

VOLUME 2

VI. Magneto-Optics

A. Quantum Well Structures

B. Heterostructures and 2D Systems

C. Dots and Microcavities

D. Coupled Systems

E. Excitons

F. Far Infrared Region

VII. Magnetic Semiconductors

VIII. Organic Conductors

IX. Si-Ge

X. Reports from Labs, Experimental Methods

XI. Post-Deadline Papers

INTERBAND OPTICAL STUDIES OF SEMICONDUCTOR HETEROSTRUCTURES IN HIGH MAGNETIC FIELDS

R. J. NICHOLAS, H. H. CHENG, A.N. PRIEST
Clarendon Laboratory, Department of Physics, University of Oxford, Parks Road, Oxford OX1 3PU, U. K.

Pulsed magnetic fields have been used to study a range of interband properties in doped and undoped quantum wells and heterojunctions. We will demonstrate how pulsed magnetic fields can be used for sensitive studies of photoluminescence, reflectivity and absorption in a wide range of semiconductor heterostructures. Typical examples of the systems studied include: Semimagnetic superlattices using materials such as ZnTe/ZnMnTe and CdTe/CdMnTe, high mobility 2-D electron systems in GaAs/AlGaAs at low temperatures, and wider gap materials such as AlGaInP/GaInP which show new band gap and effective mass renormalisation effects under high intensity excitation.

1. New excitonic signature of the magnetic field induced type I/type II crossover in semimagnetic superlattices

The field-induced band alignment change from type I to type II structure in semimagnetic semiconductors has been intensively investigated at low magnetic fields. Here, we describe photoluminescence (PL) and reflectivity measurements on a series of CdTe/$Cd_{1-x}Mn_xTe$ superlattices in both Faraday and Voigt configurations at high magnetic fields. In the Faraday configuration, the PL and reflectivity follow each other closely. In the Voigt configuration, the reflectivity also follows the transitions seen for Faraday. By contrast new behaviour is observed in the PL. It shows a sudden drop in emission energy (6 meV) at moderate fields and moves further below the transition energy measured in reflectivity. At high fields it falls even below the bulk CdTe 1s transition, and comes from a novel type II exciton.

The samples consist of a thick 2000 Å CdTe buffer layer and then 10 alternating layers of CdTe and $Cd_{1-x}Mn_xTe$ grown on InSb. Photoluminescence (PL) and reflectivity measurements were made at 4.2K in Faraday (B perpendicular to the SL plane) and Voigt configurations (B parallel to the SL plane) in pulsed magnetic fields up to 45 Tesla.. The emitted light was collected using a fibre bundle and analysed using a 0.25m spectrometer and nitrogen cooled CCD detector.

Typical sets of PL spectra for Faraday and Voigt

Table I Summary of sample structures.

Sample name	Mn %	1HH (PL)	well(b) width	$E_e(E_h)$ meV	B_c T	Bt T
H470	2	1611.9	77(180)	14.4(3.9)	2	
H215	5	1619.8	64(170)	24.6(6.5)	3	12
H473	6.4	1624.2	60(170)	29.7(7.7)	7	18
H360	7.3	1633.4	52(150)	36.3(9.7)	10	24
H484	11	1654.4	40(140)	58.3(15)	15	30
H311	15	1639.4	70(180)	40(9.4)	24	34

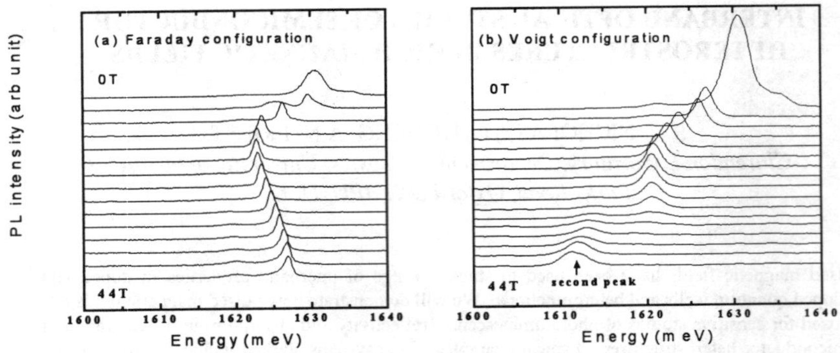

Figure 1: A family of PL spectra in Faraday and Voigt configurations for sample H360

geometry are shown in figure 1. At low field two emission lines are seen, one coming from the donor bound exciton (D_0X) and a higher energy feature from the free 1s heavy hole exciton (1HH). The peak energies at B=0 are listed in Table I. The precise Mn concentration is determined from the barrier transition energy. $E_e(E_h)$ is the calculated confinement energy of the first electron (hole) subband using the excitonic model of ref 1 and a band offset ratio[2] $\Delta Ec/\Delta Ev = 70/30$.
On applying small magnetic fields, the PL spectra show two changes. (1) The transition energy of both lines moves down due to the reduction of the confinment energy as the band edge modulation decreases. (2) The intensity of the low energy line (D_0X) decreases rapidly with field, and so the 1HH line dominates the spectra. This is due to the field-induced instability of the exciton binding to impurities[3].

We first discuss the behaviour in the Faraday geometry. The PL drops rapidly in energy, and by 10T has fallen by ~ 10meV. This is a little larger than the zero field hole confinement energy listed in Table I. Therefore a band alignment crossover from a type I to a type II structure is expected around this region where the electrons and holes are spatially separated in different layers. Based on the "conventional description" of a type I/type II transition [4], a plateau in the transition energy and a drop in oscillator strength is expected due to the change of excitonic binding energy and the spatial separation of the electron and hole wavefunctions. From the spectra, however, there is no clear signature to justify this picture, instead there is a smooth decrease in energy and emission intensity. This is due to the strong Coulomb attraction which keeps the electron-hole pair strongly bound close to the interface[5]. To provide evidence of a type II band alignment, the high field data are analysed by comparing the energy shift from different concentration samples. For low Mn content, the magnetic interaction saturates at low fields (10T), and above this the energy shift is dominated by the diamagnetic shift which is close to the values found for bulk CdTe. As the Mn^{2+} concentration increases a larger valence band gap modulation occurs, but only at high fields. Consequently the increasing band edge modulation balances out the diamagnetic shift. For a detailed discussion see reference 6.

We now move to discuss the reflectivity and PL in the Voigt configuration. At low fields the reflectivity shows a large anisotropy compared with the Faraday orientation, but for higher fields the two orientations give essentially similiar results (Figure 2). The origin of this behaviour is the reorientation of the quantization axis from the superlattice direction at low fields to the magnetic field direction at high fields. As a result, decoupling of the valence band states occurs, leading to a pure HH state.

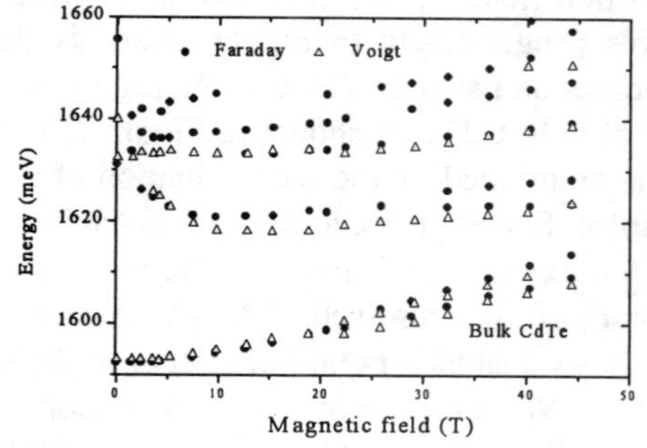

Figure 2: Peak positions from reflectivity for the Faraday and Voigt configurations with sample H360.

Turning now to the PL spectra, at low fields the transitions follow the lower branch of the Voigt reflectivity. Above 10 T, the behaviour of PL spectra fall into two categories: (a) low concentration samples: For samples with x< 7.3%, the main PL emission gradually weakens and is replaced by a second transition which

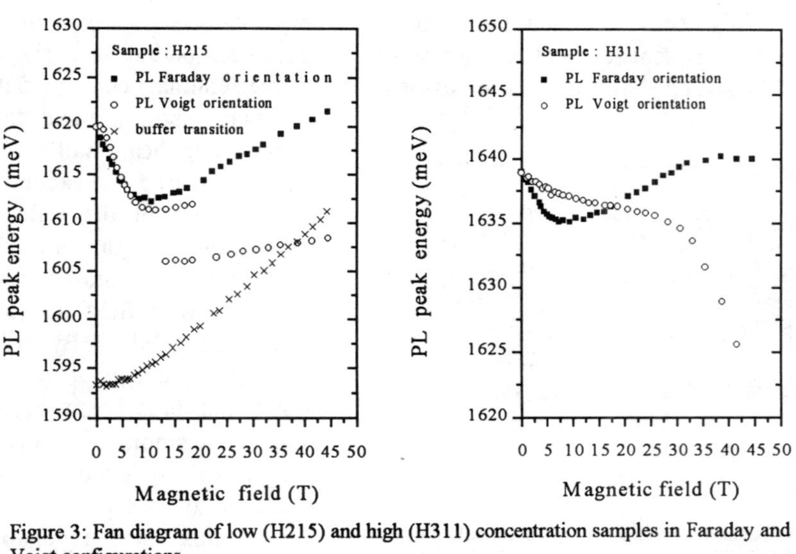

Figure 3: Fan diagram of low (H215) and high (H311) concentration samples in Faraday and Voigt configurations.

emerges at an energy 6 meV lower. This new transition dominates above a transition field B_t, and then moves down in energy more rapidly. This field value moves progressively to higher fields as the manganese content of the samples increases as listed in Table I.. At high fields, the PL transition moves even below that of bulk CdTe as shown in Figure 3, thus providing unequivocal evidence that PL is dominated by the recombination of a type II exciton. (b) High concentration samples: For x> 11%, a different behaviour is observed at high fields as shown in the fan diagram of Figure 3. The two separated transitions observed in the low x samples are not resolved. The PL trace initially follows the reflectivity as before, but shows a sudden rapid fall in energy above the threshhold field B_t.

No evidence of the sudden drop in PL transition energy is seen in the Faraday PL measurement. The origin of this difference almost certainly lies in the effect of magnetic field on the carrier motion. As the magnetic field is applied perpendicular to the SL direction in the Voigt configuration, it shrinks the carrier motion in the SL direction, introducing an additional confinement. This may allow the formation of a more strongly bound electron-hole pair, leading to an interface exciton with a large binding energy close to that found for the type I alignment.

2. Interband Magneto-optics of ZnTe/ZnMnTe superlattices

Changing to the ZnTe/ZnMnTe system, we can observe even larger magnetic field induced band alignment changes[1]. This is because the band gap difference produced by alloying from ZnTe to MnTe is only half that of the CdTe/CdMnTe system, while the sp-d exchange interaction is approximately 20% larger. This leads to much larger magnetic field induced changes in band alignment. A typical series of reflectivity spectra from a ten period SL is shown in fig. 4. The strong sharp features are associated with the semimagnetic superlattice, and again show a rapid fall in energy on application of the field. The bulk ZnTe 1s exciton can also be seen in the reflectivity spectra due to the underlying buffer layer. This moves upwards in energy with field due to the diamagnetic shift. By 6T the transition energy of the superlattice falls below even the bulk 1s transition, thus providing unequivocal evidence for the formation of a type II system[1]. Despite this there is again no dramatic decrease in either the

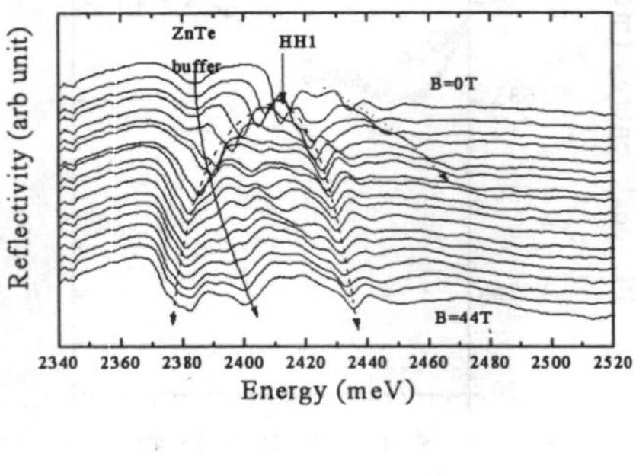

Figure 4: A family of reflectivity spectra in pulsed field for a 51/150 Å ZnTe/ZnMnTe superlattice with 7.5% Mn.

excitonic oscillator strength or the transition energy. This is further evidence for the importance of the excitonic Coulomb energy, which is sufficient to overcome the type II band alignment and still lead to the formation of a strongly bound exciton.

3. Photoluminescence in 2-D electron systems

Doped heterojunctions in GaAs/GaAlAs have been studied at temperatures down to 500mK. The luminescence shows a variety of features associated with the Landau level occupancy, including oscillating peak intensities[7-9], sudden energy shifts[9,10] and splittings occuring at ocupancies of $\nu = 1$, $\nu = 1/3$ and $\nu = 1/6$. Figures 5 and 6 show a series of PL traces taken on two heterojunctions with carrier densities of 1.2 and 2.3 x 10^{11} cm^{-2}. The data show dramatic changes in intensity of the PL peak related to level occupancy, as seen previously to be particularly strong for heterojunctions and wide quantum wells[9,10]. The peaks fall into two broad classifications, usually labelled E_0 and E_1, with the E_1 peaks only being strong in the region where the occupancy is nearly at an integer value and the system is in its incompressible state where screening effects are much reduced[11]. This upper peak is thought to be associated with the formation of a very weakly bound exciton which is almost free in character. The particularly unusual behaviour is the observation of

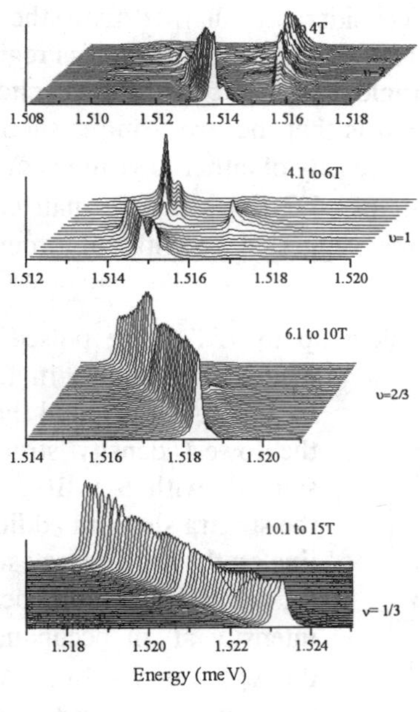

Figure 5: Families of PL peaks measured at 0.1T intervals at 0.5K in a GaAs/GaAlAs heterojunction with a carrier density of 1.2 x 10^{11} cm^{-2}.

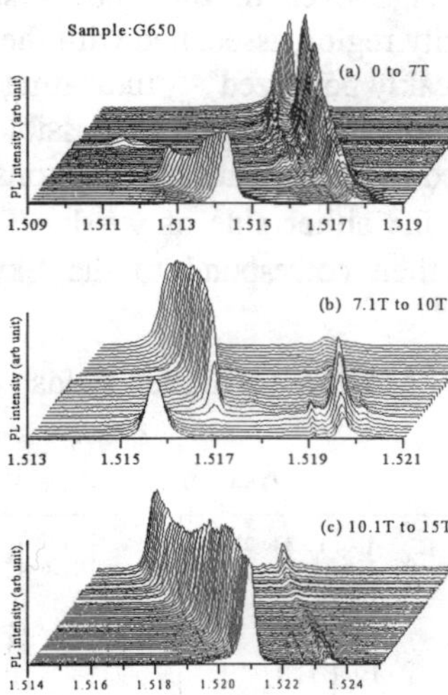

Figure 6: Families of PL peaks measured at 0.1T intervals at 0.5K in a GaAs/GaAlAs heterojunction with a carrier density of 2.3 x 10^{11} cm^{-2}.

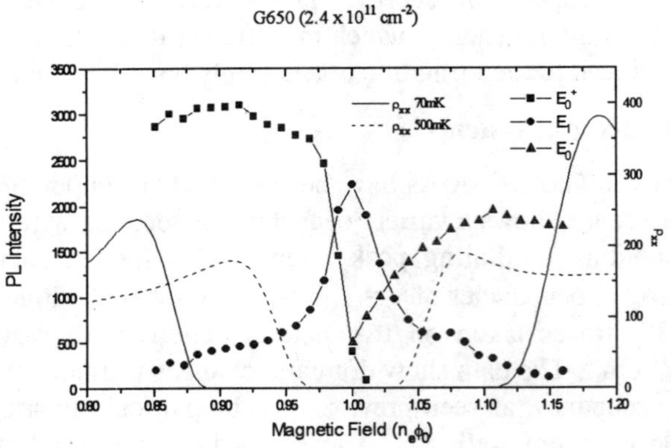

Figure 7: Plots of the intensities of the two transitions arising from E_0 either side of $\nu = 1$, and E_1 for the sample with a density of 2.3 x 10^{11} cm^{-2} ($n_e\Phi_0 = 9.4T$). Equivalent traces for the resistivity are shown at 500 and 70 mK.

a very sudden shift downwards (red-shift) in energy of the E_0 transition at an occupancy of $\nu = 1$ by several linewidths. This would seem to be best interpreted as a swap over in peak intensities between two transitions. This change over is very rapid, taking place over a change in occupancy of less than 5%. The change is shown in fig. 8, which shows a plot of the intensity of both the E_0 peaks and the E_1 peak, together with the associated resistivity traces at 500 and 70mK. It can be seen that the change over in the optical spectra is considerably sharper than the zero resistivity region associated with the quantum Hall plateau. The PL in this region is only weakly polarized[8,12], indicating that the single particle picture of the excitons is not appropriate, so that one possible explanation is that the two features should be assigned to the formation of excitons from a low density of either Skyrmions or anti-Skyrmions either side of $\nu = 1$. The red-shift of 6-7 times the Zeeman energy would then correspond to the Skyrmion sizes suggested recently of order 5-7 spins[13,14].

Moving to higher fields, data was taken up to 45T using pulsed field magnets, as shown in fig. 8. In this case it is found that for the lowest density structures studied, with 6 x 10^{10} cm^{-2}, the spectra show an additional line on the high energy side of E_0, which has a broad peak in intensity at an occupancy in the region of $\nu = 1/6$. As the occupancy is reduced to values as low as $\nu = 1/18$, the main resonance shows little extra structure, with only a

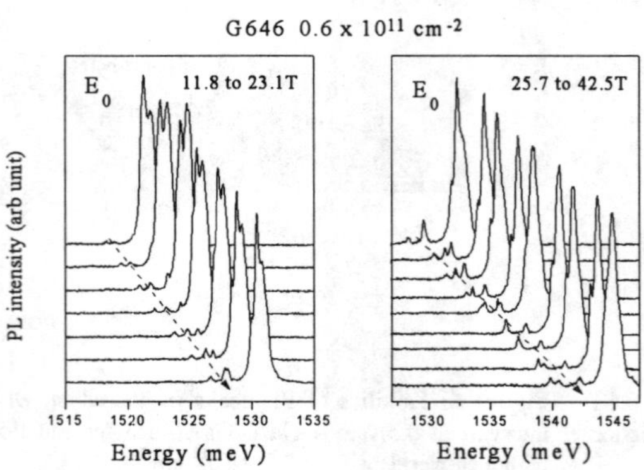

Figure 8: A family of PL spectra taken at 0.5K.

weak low energy shoulder appearing. At very high fields however a set of sharp new lines appears which is well separated from E_0 and is associated with the spin splitting of the valence bands.

4. Photoluminescence in dense electron-hole plasmas

The properties of dense electron-hole plasmas (EHP) has been somewhat controversial of late, with some reports of band gap renormalisation which is hardly dependent at all on carrier density[15], and others which describe a strong density dependence[16] as expected by theory[17]. To date the work has used relatively small gap materials such as GaInAs and GaAs. We have now studied EHP luminescence in GaInP/AlGaInP quantum wells in which carrier densities of order 10^{13} cm^{-2} have been photoexcited using pulsed excitation from a N_2 laser at 337nm. This enables a large number of Landau levels to be populated which are visible in the PL spectrum. A typical example is shown in fig. 9. It can be seen that in contrast to the results on GaAs[18], there is a continuous but slow shift in energy of the various Landau transitions with intensity. The effective mass is deduced from a fan diagram of the levels, as shown in fig. 10 for the highest intensity. This shows two distinct features. Firstly there remains a strong 1s like feature, which is almost unshifted from the low intensity results, and secondly there is a fan of Landau levels originating from an energy approximately one effective Rydberg below the 1s level. This corresponds to a band gap renormalisation of ~2Ry* for the maximum excited densities of ~1×10^{13} cm^{-2}, in good agreement with the results from GaAs[15]. In addition the effective masses are found to be much higher than expected from both conventional k.p perturbation theory and the values measured at low intesities from interband magneto-reflectivity. For the example shown, a 75Å MQW, the reduced effective mass for the high density results is ~40% larger. Such behaviour was seen in GaAs[15], but is completely opposite to that reported for the lighter mass GaInAs system[17].

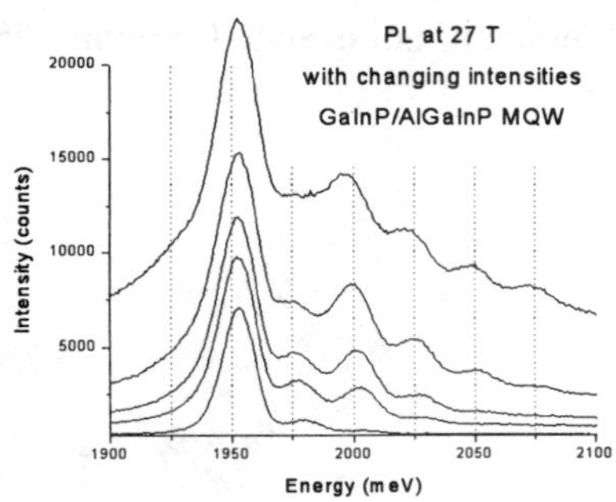

Figure 9: A series of PL spectra for high intensity excitation for a 120 Å well width MQW.

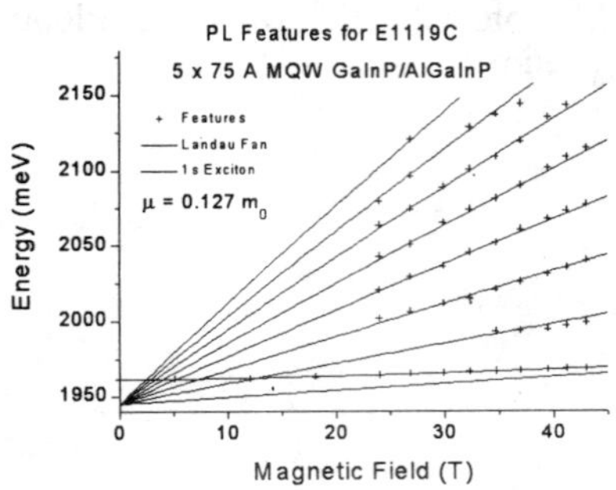

Figure 10: Landau levels for a 75 Å well width MQW with a photoexcited EHP density of ~1×10^{13} cm^{-2}.

References

1 H. H. Cheng, R. J. Nicholas, M. J. Lawless, D. E. Ashenford, and B. Lunn, Phys. Rev. B**52** 5269 (1995).

2 B. Kuhn-Heinrich, W. Ossau, T. Litz, A. Waag and G. Landwehr, J. Appl. Phys. **75**, 8046 (1994)

3 D. Heiman, P. Becla, R. Kershaw, D. Rifdgley, K. Dwight, A. Wold, and R. R. Galazka, Phys. Rev. B**34** 3961 (1986)

4 E. Delporte, J. M. Berroir, G. Bastard, C. Delalande, J. M. Hong and L. L. Chang, Phys. Rev. B**42** 5891 (1990)

5 J. Warnock, B. T. Jonker, A. Petrou, W. C. Chou, and X. Liu, Phys. Rev. B**48** 17321 (1993)

6 H. H. Cheng, R. J. Nicholas, D. E. Ashenford, and B. Lunn, to be published.

7. A.J. Turberfield et al, Phys. Rev. Lett. **65** 637 (1990)

8. A.J. Turberfield et al, Phys. Rev. B **47** 4794 (1993)

9. B.B Goldberg, D. Heiman, A. Pinczuk, L. Pfeiffer and K.W. West, Surf. Sci. **263** 9 (1992)

10. B.B Goldberg, D. Heiman, A. Pinczuk, L. Pfeiffer and K.W. West, in: High Magnetic Fields in Semiconductor Physics III, Wurzburg, Ed. G. Landwehr (Springer, Berlin 1990) p243

11. T. Uenoyama and L.J. Sham, Phys. Rev. B **39** 11044 (1989); S. Katayama and T. Ando, Solid State Commun. **70** 97 (1989)

12. F. Plentz et al, in High Magnetic Fields in the Physics of Semiconductors, ed. D. Heiman (World Scientific, Singapore, 1995) pp. 320-323

13. S.E. Barrett et al, Phys. Rev. Lett. **74** 5112 (1995)

14. E.H. Aifer, B.B. Goldberg and D.A. Broido, Phys. Rev. Lett. **76** 680 (1996)

15. R. Cingolani et al, Phys. Rev. B **43** 9662 (1991)

16. L.V. Butov, V.D. Kulakovskii, E. Lach, A. Forchel and D. Grutzmacher, Phys. Rev. B **44** 10680 (1991)

17. S. Schmitt-Rink, D.S. Chemla and D.A.B. Miller, Adv. Phys. **38** 89 (1989)

18. M. Potemski, J.C. Maan, K. Ploog and G. Weimann, Solid State Commun. **75** 185 (1990)

THE COUPLING OF COMPOSITE FERMIONS TO PHONONS MEASURED WITH MAGNETOTHERMOELECTRIC EXPERIMENTS

J.C. Maan, B. Tieke, A.K. Geim
Research Institute for Materials, High Field Magnet Laboratory, University of Nijmegen, Toernooiveld, 6525 ED, Nijmegen, The Netherlands.

R. Fletcher
Physics Department, Queens's University, Kingston, Ontario, Canada K7L 3N6

U. Zeitler, M. Henini
Department of Physics, University of Nottingham, Nottingham NG7 2RD, United Kingdom

Recent theoretical work shows that two dimensional electrons at high magnetic fields, where an even fraction of a Landau level is filled, may be described as new particles, named composite Fermions, in an effective zero magnetic field. This new concept has been quite succesful in describing experimental data of transport experiments. In this paper it will be shown that thermoelectric experiments at low temperatures and at high fields in high mobility GaAs/GaAlAs heterojunctions, probe directly the interaction of electrons with phonons, and the results are a strong support of the composite Fermion picture. In particular the e-ph interaction has the same value for even filling factors with the same denominator and only at these filling factors a T^4 temperature dependence is found, which is the same as electrons at zero field. Using this latter result, the phonon limited mobility for Composite Fermions can be obtained

1 Introduction

The fractional quantum Hall effect observed at low temperatures and at high magnetic fields where only the lowest Landau level is occupied, is a spectacular manifestation of the influence of electron-electron interaction. [1] This interaction is responsible for new ground states which appear at odd fractional filling factors. Recent theoretical understanding [2,3] describes these states as quantum oscillations of a new quasi-particle in an effective magnetic field given by $B_{eff} = |B - B_{1/2}|$, where $B_{1/2}$ is the field at half filling factor. This way minima in the resistance at magnetic fields with odd fractional filling factors $\nu = 1/3, 2/5, 3/7, 4/9..$ are now interpreted as Shubnikov-de Haas oscilations of the new particle in the effective field at new filling factors $\nu^* = 1, 2, 3, 4....$ [4,5]. Furthermore geometrical resonances of the cyclotron radius of this new particle with externally imposed length scales, like the wavelength of sound acoustic waves [6] or antidots [7], have convincingly been demonstrated.

Most of the previously mentioned experiments involve electrical transport and it is therefore interesting to study experimentally other interactions, like for instance that between CF and phonons. In this paper we will show that such effects can be studied by low temperature thermoelectric experiments [8]. The main results from this work are that experimentally the thermoelectric power at $n/2m$ filling factors, is shown to reflect the CF-phonon coupling for $n = 1, 3$ and $2m = 2, 4$, since it has the same value for the same m and different n and has the same temperature dependence as the zero field, 'normal' electron thermoelectric power. These data can be interpreted with an adapted zero field Bolzmann theory [9], and are shown to yield information about the CF mass. Finally the results can be formulated in terms of the CF-phonon limited mobility, which can than be shown to yield a T^{-5} dependence both for electrons as for CF, in contradiction with results inferred from transport data. [10]

2 Thermoelectric power

The thermoelectric coefficient (Seebeck coefficient) S at zero magnetic field is defined as (See figure 1):

$$S = \lim_{\Delta T \to 0} \frac{\Delta V}{\Delta T} = \frac{dV/dx}{dT/dx} \tag{1}$$

The microscopic origin of the effect is that carriers are driven by some mechanism from the hot to the cold side of the sample inducing a thermal current j_{th}, proportional to the temperature gradient dT/dx and a coefficient ϵ. Since no current is drawn out of the sample, this thermal current must be compensated by and electric current j_{el}, which leads to an electric field E, because $j_{el} = \sigma E$ (σ is conductivity). This compensating electrical field is measured as a voltage between contact probes [11]. Therefore:

$$j_{tot} = \sigma . E + \epsilon . \nabla T = 0 \Rightarrow E = -\frac{\sigma}{\epsilon} . \nabla T = S . \nabla T \tag{2}$$

The microscopic origin of ϵ (and thus of j_{th}) in semiconductors is either diffusion or phonon drag. Diffusion is related to the entropy of the 2 DEG[11,12] and its contribution is negligable in the experiment here. Phonon drag is caused by the flow of phonons from the hot to the cold side which through electron- phonon coupling drags electrons along. In the case of heterojunctions, the mechanism is conceptually very simple, because the phonon flow is almost exclusively determined by the substrate (bulk phonons) and is barely influenced by the very few electrons. Therefore electrons gain some momentum from the phonons, but the phonons loose momentum almost exclusively by other mechanisms. In fact in the present experiment, the phonon mean free path is determined by boundary scattering and is limited by the sample dimension, L (and thus independent of temperature).

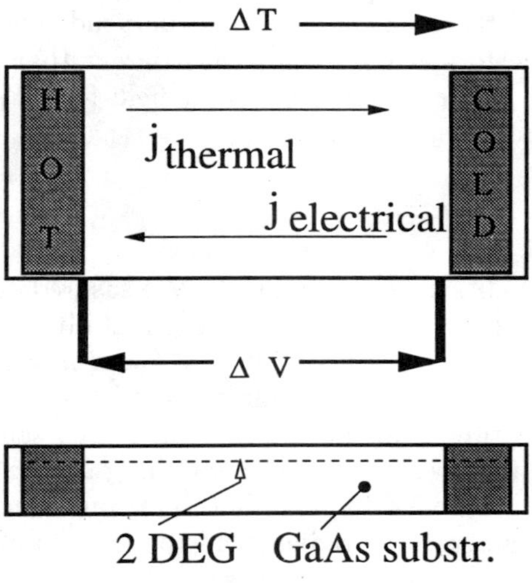

Figure 1: Schematic representation of the thermoelectric effect.

Within this framework an expression for the thermoelectric power S can be worked out [11], based on the thermodynamical relation $S = \Pi/T = \dot{Q}/Tj$ where Π is the Peltier coefficient defined as heat current $\dot{Q}$ associated with an electrical current j. Considering an electrical current flow driven by an electric field E; the electrons experience a force neE, of which a fraction f_{ep} is transferred to the phonons, leading to a net phonon flux P, which corresponds to a heat flow $\dot{Q} = v_s^2 P$ (v_s, the phonon sound velocity). Since on average there is no net current, electrical and phonon forces must balance each other and thus:

$$f_{ep} neE = P/\tau_p \tag{3}$$

with τ_p phonon relaxation time equal to L/v_s Using $j = \sigma E$ it follows,

$$S = \frac{1}{T}\frac{v_s L f_{ep} neE}{ne\mu E} = \frac{Lv_s}{\mu_p T} \tag{4}$$

$\mu_p = \mu/f_{ep}$, with μ_p is the phonon limited mobility, i.e. the fraction of the total mobility, including all other loss of momentum, determined by the phonons. This simple result arises because at low temperatures, the mobility is mainly determined by other mechanisms, such and impurity scattering and phonons contribute only a small fraction. Therefore an analysis of experimental data of S gives directly the phonon limited mobility.

The previously described analysis is only valid at zero field since both σ and ϵ, will become tensors in a magnetic field[12,11]. Therefore in principle two components S_{xx} (in the direction of the gradient) and S_{xy} (perpendicular to the gradient) can be measured, the former one being much larger than the second one. However, in the spirit of the effective magnetic field for CF the results should hold at even fractional filling factors.

3 Experimental results

The TEP of high mobility GaAs-GaAlAs heterostructures was measured down to 150 mK in magnetic fields up to 20T. The density and the mobility of the 2DEG were varied in the range of $n_e = (1.0 - 1.9) \times 10^{15} m^{-2}$ and $\mu = 60 - 100 m^2/Vs$, respectively, using illumination from an infrared diode. For TEP measurements, a calibrated temperature gradient was applied to the sample and the thermally created electric field was measured. For this purpose the substrate with the sample was mounted on a cold finger in the vacuum space of a dilution refrigerator. From the temperature gradient and the heat supplied the (known) heat conductivity of the substrate was measured, both with and without magnetic field, and the results used as a check on the thermometry in a magnetic field. The results showed no field dependence and the T^3 temperature dependence for the heat conductivity, expected for a boundary limited constant phonon mean free path.

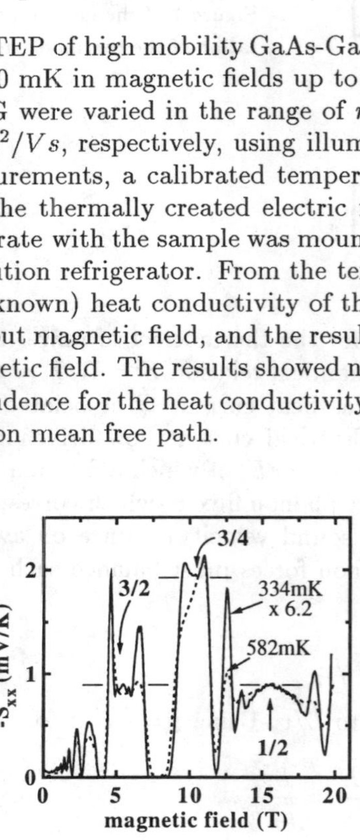

Figure 2: Magnetic field dependence of S_{xx} at two different temperatures. The data are normalized to the value at $\nu = 1/2$ to show that TEP at 1/2, 3/2 and 3/4 have a constant relative magnitude.

In figure 2 the results of S_{xx} are shown as a function of the magnetic field for two different temperatures. At first glance these curves resemble the well known resistivity measurement, showing an oscillatory magnetic field dependence with zeroes at integer and fractional filling factors.These zeroes are expected at these field values since the Fermi energy lies in an energy gap and no excitations can exist. More interesting however is the behaviour of S_{xx} as a function of temperature at half integer filling factors, shown in figure 3.

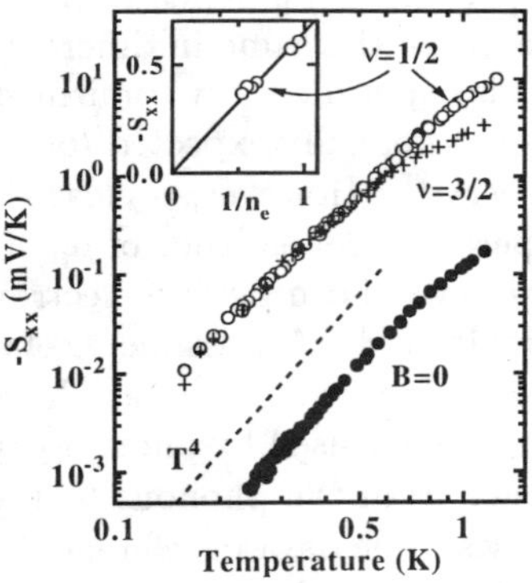

Figure 3: Temperature dependence of S_{xx} at $\nu = 1/2, 3/2$ and at $B = 0$. The inset show the density dependence at $\nu = 1/2$ and T= 440 mK.

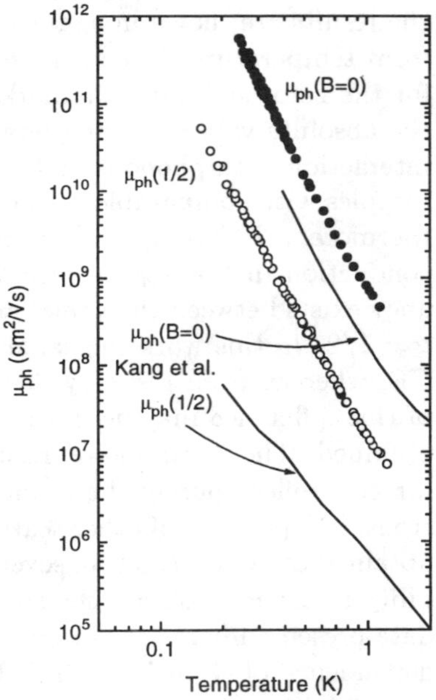

Figure 4: Phonon limited mobility at zero field and at $\nu = 1/2$, calculated from thermopower. The lines determined, from transport data, are taken from ref 10.

S_{xx} is found to have the same value for 1/2 and 3/2 over the whole temperature range where the FQHE is observable, even though the absolute magnitude varies a few orders of magnitude.The temperature dependence is proportional to $T^{4.0\pm0.5}$ both for $\nu = 1/2, 3/2$ and for $B = 0$. Furthermore S_{xx} at $\nu = 3/4$ is about 2.2 times that at $\nu = 1/2$ and has the same temperature dependence. Preliminary new data at higher fields (up to 30T using a hybrid magnet) have confirmed that also at $\nu = 1/4$ the same value is found as at 3/4, with roughly the same temperature dependence. These experimental facts are only true for even filling factors. At an arbitrary filling factor no particular relation with other filling factors is found, and the temperature dependence clearly deviates from a T^{-4} behaviour. Thermopower thus shows directly a unique "zero-field free electron like" behaviour at even filling factors, and thus strongly supports the CF theory at an effective zero magnetic field.

Using the simple theoretical model outlined in the beginning (equation 4), the phonon limited mobility can be calculated from the data in figure 3 and the results are shown in figure 4. Also shown in this figure are results obtained from temperature dependend resitivity measurements[10]. It can be seen that for the zero field data a remarkably good agreement is found for μ_p, both of the absolute values and the temperature dependence. μ_p depends only on the interaction with phonons and may expected to be the same in different GaAs samples with comparable densities as indeed is experimentally confirmed. Furthermore, a T^{-5} temperature dependence is theoretically expected for metallic conduction in this regime [11] and is also observed. However, a clear discrepancy exists between the values and the temperature dependence of μ_p of CF at $\nu = 1/2$. In this work the same T^{-5} law is obeyed both by free electrons and CF, whereas from resistivy data[10] a T^{-3} is derived. A possible explanation of this difference may lie in the different manner in which the data have been obtained. The thermopower results are exclusively caused by phonon-drag and directly reflect purely the momentum transfer from the phonons to the electrons, independent of other scattering processes. The data in reference [10], were obtained after substracting several other contributions from the measured mobility in order to obtain the small portion caused by phonon scattering. The data presented here are obtained more directly and show that the temperature dependence of el-ph or el-CF is the same.

The previous analysis show that TEP can directly be related to the phonon limited mobility of CF, but does not give any further insight in explaining relative magnitudes, temperature dependence and density dependence. It is therefore necessary to study in detail theoretical calculations in phonon drage thermopower. Cantrell and Butcher [9] have derived a theory, based on the Bolzmann equation for phonon drag thermopower in heterojunctions in the Drude model at zero magnetic field. Since σ is known in the Drude model, to calculate S only the thermoelectric coefficient ϵ must be obtained (equation 2). and this step contains three basic ingredients. i) A driving mechanism by which phonons (through deformation potential or piezoelectric coupling) transfer momentum to the electrons. ii) How long the electrons keep the acquired momentum (i.e. the momentum relaxation time). iii) How many phonons with a certain wavevector exist at a certain temperature which can couple to the electrons (i.e. matching the phonon statistical wavevector distribution to the electron c.q. CF distribution). Following ref. [9], the result can be written as:

$$S^g = -\frac{k^5 e E_1^2 L \tau(E_F)}{8\pi^3 \hbar^6 \rho v_s^3} T^4 \frac{m^*}{ne^2 \tau(E_F)} \times$$

$$\int_0^{2\xi_F} \int \frac{\xi^3(\xi^2+\xi_z^2)}{\sqrt{(\hbar k_F \xi/m^* v_s)^2 - (\sqrt{(\xi^2+\xi_z^2} - kT/2m^* v_s)^2}}$$

$$\times \frac{e^{\sqrt{\xi^2+\xi_z^2}}}{(e^{\sqrt{\xi^2+\xi_z^2}}-1)^2} d\xi_z d\xi \quad (5)$$

where all constants have their usual meaning, E_1 is the deformation potential, $\xi(_z) = hq_{(z)}/kT, \xi_F = hq_F/kT$, the normalized phonon wavevector $q_{(z)}$, and $\tau(E_F)$, the momentum relaxation time. This time cancels because the momentum loss of electrons after interaction with a phonon is almost exclusively determined by impurity scattering, just as the momentum lifetime for transport. This independence of S on the impurity lifetime makes phonon-drag thermopower such a meaningful experiment, since it measures directly the electron-phonon coupling. Eq. 5 is valid at zero magnetic field for free electrons and can as such not be used for CF, without some further adaptation. As it stands, equation 5 explains already the observed T^4 dependence of S_{xx} At temperatures where composite Fermions can be observed only very long wavelengths phonons exist and it can be shown that eq. 5 reduces to:

$$S \propto \frac{m_e}{n_e} \int_0^{2\xi_F} \int \frac{m_Q^* v_s}{\hbar k_F^Q} f(\xi, \xi_z) d\xi_z d\xi \quad (6)$$

In this equation the quasiparticle mass m_*^Q and wavevector k_F^Q are introduced in the integral expression in eq. 5, and descibes the matching of the phonon wavevector distribution function with that of the quasiparticles. Furthermore in the range of temperatures where CF exist, the first term $(\hbar k_F \xi/m^* v_s)$ is the most important one and may be taken outside the square root expression. $f(\xi, \xi_z)$ is the remainder of the integrand given explicitely in equation 5. Apart from the algebraic simplification eq. 6 constitutes an interpretation of eq. 5 for quasiparticles. The most intuitive way to arrive at this result is to retain all prefactors and consider only the coupling of the new particle wavevector k_F^Q to the normal phonons. One can also arrive at this result by an alternative route described by Tieke et al [8] where a careful analysis of both the thermal and the electrical current in terms of quasiparticle and electron current is used.

In any case, with this interpretation the TEP results can be explained straightforward as a measure of the CF mass. At $\nu = 1/2$ and at $\nu = 3/2$ TEP gives the same value since both m_*^Q and k_F^Q are dependent on $\sqrt{n_Q}$, and the ratio is thus constant. The different value of the $\nu = 1/4, 3/4$ family is than an indication of a higher mass for the CF states.

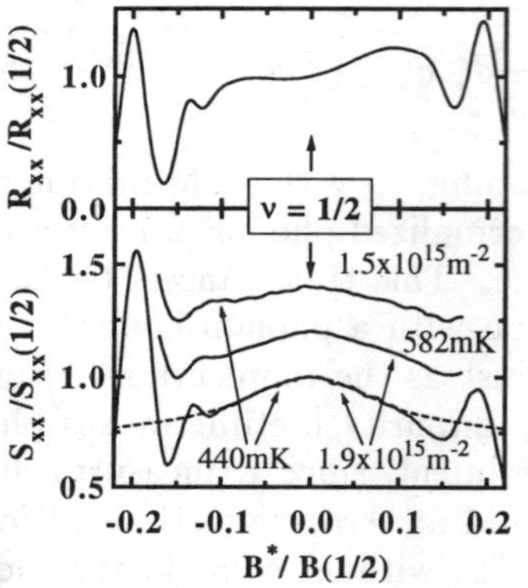

Figure 5: The shape of S_{xx} around $\nu = 1/2$, showing a clear symmetry for different densities and temperatures. For comparison a resistivity trace is also given.

That TEP is a measure of the CF mass is also very much suggested by the behaviour around $\nu = 1/2$, shown in figure 4. There is a striking symmetry around 1/2 which is not observed in magnetoresistance data (shown in the top trace). This symmetry reflects the fact that TEP does not depend on the impurity scattering time, which is often associated with a contribution to linear magnetoresistance in transport data. The results are indeed very suggestive for interpreting the $\nu = 1/2$ field value as an effective zero magnetic field. At different densities and temperatures the symmetry remains clearly visible at the same filling factor. In the CF theory[3] a weak logaritmic divergence of the effective mass is predictedaround filling factor 1/2. Using the logaritmic effective mass dependence the smooth background can be convincingly fitted with a simple distribution function for the broadening of the effective field, and the results are shown in the figure.

4 Conclusion

In conclusion it has been shown that phonon drag thermopower data provide a very strong support for the interpretation of quasiparticles in a 2DEG in high magnetic fields at even integer fractional filling factors. In this framework the phonon limited mobility at $\nu = 1/2$ is shown to have a T^{-5} dependence, just like normal electrons. Furthermore the same value of TEP at $\nu = 1/2$ and $3/2$, and $\nu = 1/4$ and $3/4$ can be explained by a different CF mass at the $1/2$ and $1/4$ states.

Acknowledgement This work is part of the research program of the "Stichting voor Fundamenteel Onderzoek der Materie (FOM)", which is financially supported by the "Nederlandse Organisatie voor Wetenschappelijk Onderzoek (NWO). Further support has been provided by the European Commission under the contract CHGT-CT93-0051.

References

1. H.L. Stormer, this proceedings
2. J.K. Jain, Adv. Phys. **41**, 105 (1992)
3. B.I. Halperin, P.A. Lee and M. Read, Phys. Rev. **B47**, 7312 (1993)
4. D.R. Leadley, R.J. Nicholas, C.T. Foxon and J.J. Harris, Phys. Rev. Lett **72**, 1906 (1994)
5. R.R. Du, H.L. Stormer, D.C. Tsui, L.N. Pfeiffer and K.W. West , Phys. Rev. Lett. **70**, 2944 (1993)
6. R.L. Willett, R.R. Ruel, K.W. West and L.N. Pfeiffer, Phys. Rev. Lett. **71**, 3846 (1993)
7. W. Kang, H.L. Stormer, L.N. Pfeiffer, W. Baldwin and K.W. West, Phys. Rev. Lett. **71**, 3850 (1993)
8. B. Tieke, U. Zeitler, R. Fletcher, S.A.J. Wiegers, A.K. Geim, J.C. Maan and M.Henini, Phys. Rev. Lett. **76**, 3630 (1996)
9. D.G. Cantrell and P.N. Butcher, J. Phys. C **20**, 1985, (1987); **20** 1993 (1987)
10. W. Kang, Song He, H.L. Stormer, L.N. Pfeiffer, W. Baldwin and K.W. West, Phys. Rev. Lett. **75**, 4106 (1995)
11. B.L. Gallagher and P.N. Butcher in *"Handbook of Semiconductors"*, Vol 1, 2nd edition, , p. 721, ed. T.S. Moss, North Holland, Amsterdam (1992)
12. R. Fletcher, J.C. Maan, K. Ploog and G. Weimann, Phys. Rev. **B33**, 7122 (1986)

ELECTRONIC AND OPTICAL PROPERTIES OF MAGNETOEXCITONS IN QUANTUM-WELL WIRES

M. GRAF

Walter Schottky Institut, TU München, D-85748 Garching, Germany

A.B. DZYUBENKO

General Physics Institute, Russian Academy of Sciences, Moscow 117942, Russia and Walter Schottky Institut, TU München, D-85748 Garching, Germany

We theoretically study binding energies, dispersion relations, and linear optical absorption of one-dimensional (1D) magnetoexcitons (MXs) in a quantum-well wire in strong magnetic fields. Results for the ground and excited MX states are presented. It is shown that 1D MXs possess a hydrogenic spectrum with bound states and a continuum of scattering states, in contrast to 2D MXs that are always bound. The appearance and shape of Fano resonances in the optical spectra are discussed.

The simultaneous effect of a magnetic field B and 1D lateral confinement in a quantum wire leads to strong modifications of the binding energies and oscillator strengths of excitons. This effect for the ground state of immobile 1D MXs has been studied theoretically before (see, e.g., [1]). In this work, we consider binding energies and dispersion relations for the ground and excited states of 1D MXs in a quantum-well wire with parabolic confinement for electrons (e) and holes (h). We also study optical absorption by 1D MXs in the region of the discrete spectrum, as well as in the continuum. In the latter region, due to the Coulombic coupling between the continuum and quasibound states belonging to lower-lying Landau level excitons, Fano-resonances appear. This effect has been studied for 3D MXs in the bulk.[2] Here we consider the evolution of Fano resonances with the lateral confinement in 1D quantum wires.

We take the 2D plane of the quantum well to be the $x_\perp$–$x_\parallel$–plane. The Hamiltonian in the perpendicular magnetic field B reads (we put $\hbar = 1$)

$$H = \frac{1}{2m_e}(\mathbf{p}_e+\frac{e}{c}\mathbf{A}_e)^2+\frac{1}{2m_h}(\mathbf{p}_h-\frac{e}{c}\mathbf{A}_h)^2+U_{eh}(\mathbf{r}_e-\mathbf{r}_h)+V_e(x_\perp^e)+V_h(x_\perp^h)\,, \quad (1)$$

U_{eh} denotes the Coulombic e–h interaction. The confining potentials are assumed parabolic: $V_j(x_\perp) = m_j\Omega_j x_\perp^2/2$ ($j = e, h$). The frequencies Ω_j are related to the width parameters of the quantum wire $L_j = (m_j\ \Omega_j)^{-1/2}$. 1D MXs are characterized by the conserved center-of-mass momentum $K_\parallel$. In high magnetic fields, $\ell = (\hbar c/eB)^{1/2} < a_{Bj} = \epsilon\hbar^2/m_j e^2$, we expand the eigenstates of 1D MXs with a given $K_\|$ in terms of unconfined 2D MXs in the high-field limit, $|nm\mathbf{K}\rangle$, which are characterized [3] by the conserved magnetic momentum $\mathbf{K}$ [here $\mathbf{K} = (K_\|, K_\perp)$] and by the indices n, m which label the electron and hole Landau levels, respectively:

$$|K_\|\rangle = \sum_{nm} \sum_{K_\perp} \Phi_{nmK_\|}(K_\perp)\,|nm\mathbf{K}\rangle \,. \tag{2}$$

Each pair of quantum numbers (n, m) defines a channel which contains an infinite number of states. The resulting equations for $\Phi_{nmK_\|}$ contain two types of channel coupling terms: originating in (i) the parabolic confinement (which couples the neighboring channels only), (ii) the Coulomb term $U_{n\,m}^{n'm'}(\mathbf{K}) = \langle n'm'\mathbf{K}|U_{eh}|nm\mathbf{K}\rangle$ (which couples *all* channels). The latter turns out to be the dominant one and leads to energy shifts and pronounced Fano resonances. Within one channel (n, m), we have the equation

$$-\frac{\Phi''_{nmK_\|}(K_\perp)}{2\mu_{\text{eff}}} + U_{nm}^{nm}(\mathbf{K})\,\Phi_{nmK_\|}(K_\perp) = E_{nmK_\|}\,\Phi_{nmK_\|}(K_\perp)\,, \tag{3}$$

where $E_{nmK_\|}$ is given relative to the ionization energy

$$I_{nmK_\|} = \left(\omega_e + \frac{\Omega_e^2}{2\omega_e}\right)\left(n + \frac{1}{2}\right) + \left(\omega_h + \frac{\Omega_h^2}{2\omega_h}\right)\left(m + \frac{1}{2}\right) + \frac{1 - s^2}{8\mu_{\text{eff}}} K_\|^2 \tag{4}$$

$s = \left(m_e\,\Omega_e^2 - m_h\,\Omega_h^2\right)/\mu\,\Omega^2$ measures the difference in confinement of e and h, and $\mu\,\Omega^2 \equiv m_e\,\Omega_e^2 + m_h\,\Omega_h^2$. Equation (3) appears as an effective 1D Schrödinger equation in the variable $K_\perp \ell^2$ (prime in Eq. (3) means the derivative in respect to $K_\perp \ell^2$). The parameter $\mu_{\text{eff}} = \mu\omega_c^2/\Omega^2$ plays the role of an effective mass, while the 2D MX dispersion $U_{nm}^{nm}(\mathbf{K})$ (with fixed $K_\|$) is the effective external potential. For the several lowest channels, $U_{nm}^{nm}(\mathbf{K})$ are shown in Fig. 1(a). Since $U_{nm}^{nm}(\mathbf{K}) \sim -K^{-1}$ for large K,[3] there should be an infinite number of bound states (labeled by a discrete quantum number ν) below the ionization energy and a continuum of scattering states above it. The eigenvalues for each bound state ν form bands with dispersion relations in $K_\|$. Coupling between channels does not change this result. Thus, we conclude that the lateral confinement of 2D MXs leads to the formation of ***unbound e–h*** states. This is in contrast with the 2D case where all MX states are bound.[3]

In Fig. 1(b) we show the numerically obtained dispersion relations for the lowest 15 MX states at $B = 10.7$ T in each of the channels $(0,0)$, $(0,1)$ and $(1,1)$. We used the parameters appropriate for the heavy-hole excitons in GaAs and take $L = L_e = L_h = 130$ Å for the confining potentials. The finite dispersion

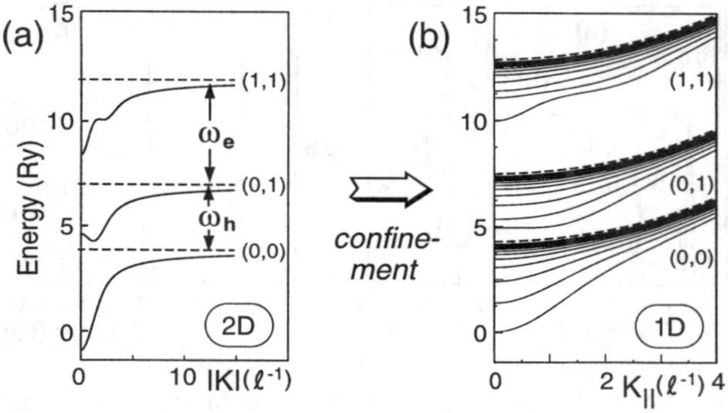

Figure 1: (a) Dispersions of 2D MXs in a GaAs quantum well at $B=10.7\,\mathrm{T}$ as a function of the total exciton momentum $|\mathbf{K}|$.[3] (b) Same but in a quantum wire of width L = 130 Å, as a function of momentum $K_{\|}$. The 1D states below the ionization energies (dashed lines) are quasi-bound states; above the dashed lines are continuum and quasi-bound states. The energy is in effective Rydbergs in GaAs $R=3.6\,\mathrm{meV}$.

of 1D MXs is a consequence of the interplay of both the lateral and magnetic field confinement, and the Coulomb e–h interaction. The resulting effective 1D MX masses are strongly enhanced in comparison with the $B=0$ case. For a given lateral confinement (Ω_e, Ω_h fixed) with increasing B the weak lateral quantization regime $\Omega \ll \omega_{ce(h)}$ is always realized. There are three different energy scales in this regime: (1) the cyclotron energy $\omega_{ce(h)} \sim B$ which determines the separation between the edges of different channels (n, m), (2) the Coulomb energy $E_0 \sim e^2/\epsilon\ell \sim \sqrt{B}$ which determines the binding energies of MXs and finite widths of bands in regard to different $K_{\|}$, and (3) the energy $\delta E \sim \left(E_0/\omega_{ce(h)}\right)^{1/2} \Omega \sim B^{-1/4}$ which determines fine structure — the energy separation between bound states within a given channel. δE is physically related[4] with the center-of-mass quantization of the 2D MX which occurs in the regime $\Omega \ll \omega_{ce(h)}$; thus the effective mass[3] of the 2D MX $2/E_0\ell^2 \sim \sqrt{B}$ enters the problem. Note also that, somewhat surprisingly, the MX ground state binding energy is *reduced* by lateral confinement. This is because the excitonic wave function gets not only contracted along the $x_\perp$-direction (which would increase the binding energy) but predominantly *elongated* along the $x_{\|}$-direction in high B. The inter-channel coupling, due to the Coulombic terms $U_{nm}^{n'm'}(\mathbf{K})$ [between bound and continuum states associated with different Landau level pairs], leads to the formation of pronounced Fano resonances. In this respect, 1D MXs resemble more closely the 3D case, whereas 0D MXs in quantum dots (see, e.g., [4]) only show discrete optical transitions again, analogously

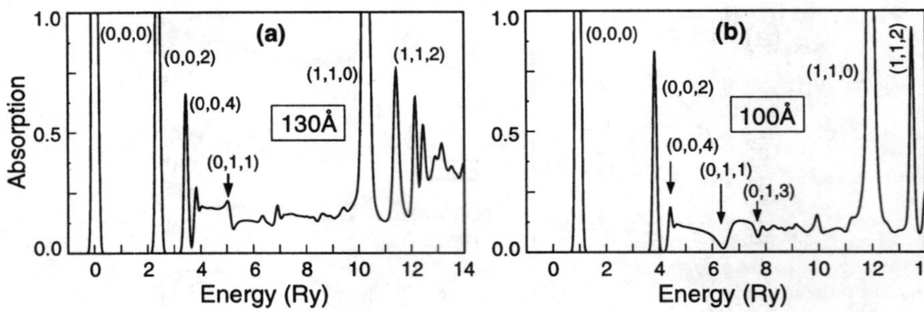

Figure 2: Predicted optical absorption spectrum of 1D MXs in GaAs quantum-well wires with different widths $L = 130\,\text{Å}$ (a), and $L = 100\,\text{Å}$ (b) at $B = 10.7\,\text{T}$. The bound states and Fano resonances in the continuum are labeled by the quantum numbers (n, m, ν).

to the situation in the 2D case. Some typical results for absorption in the low-lying channels $(0,0), (0,1), (1,0)$, and $(1,1)$ are shown in Fig. 2. All calculated spectra have been convoluted with a Gaussian of a width $\Gamma_G = 0.1\,R$. A noticeable feature is the suppression of the $1/\sqrt{E}$ type singularity at the ionization edges; this suppression is typical for 1D excitons.[5] Since only even bound and quasibound states of the optically active channels $(0,0)$ and $(1,1)$ contribute to the absorption, only every other of these states appears in the spectra (even quantum numbers ν). The pronounced Fano resonances result from the interaction of the continuum of the $(0,0)$ channel with odd quasibound states of the optically inactive channel $(0,1)$ and are labeled by odd ν. When the confinement becomes stronger, the Fano resonances are broadened and the absorption can even be strongly suppressed at the Fano resonances. This leads to the dip-shaped "window resonances", shown in Fig. 2(b). Thus, strong confinement would facilitate the resolution of Fano resonances.

We wish to thank P. Vogl for support and useful discussions. Financial support by the Deutsche Forschungsgemeinschaft (SFB 348), Bayerische Forschungsverbund (FOROPTO), Volkswagenstiftung, and the INTAS Grant 93-1495 is gratefully acknowledged.

1. T. Tanaka, Y. Arakawa, and G.E.W. Bauer, *Phys. Rev.* B **50**, 7719 (1994).
2. U. Siegner, M.-A. Mycek, S. Glutsch, and D.S. Chemla, *Phys. Rev.* B **51**, 4953 (1995) and the references herein.
3. I.V. Lerner and Yu.E. Lozovik, *JETP* **51**, 588 (1980).
4. A.B. Dzyubenko and A.Yu. Sivachenko, *J. Phys.* IV (Paris) **3**, 381 (1993).
5. T. Ogawa and T. Takagahara, *Phys. Rev.* B **44**, 8138 (1991).

MULTIPLE RESONANCES OF MAGNETO-EXCITONS IN $GaAs/Al_{0.30}Ga_{0.70}As$ QUANTUM WELLS

M. BAYER[a], T.L. REINECKE[a,b], S. WALCK[b], A. FORCHEL[a], V.B. TIMOFEEV[a]

[a] *Technische Physik, Universität Würzburg, Germany*
[b] *Naval Research Laboratory, Washington, USA*

A series of regularly spaced features has been observed for the excited heavy hole exciton states in photoluminescence excitation experiments on GaAs/Al0.30Ga0.70As quantum wells of high quality. These features occur in the energy dispersions of the excited states with magnetic field B superimposed on their Landau-level-like increase with B. We attribute these features to coupling of the transitions between the 1shh and the excited excitons with another excitation of the system, most likely an acoustic phonon at finite k.

Quasi-two-dimensional structures have been subject of intense optical investigations during the last decade. By applying a magnetic field normal to the heterostructure the evolution of the Landau fan has been studied in absorption and excitation spectroscopy. In the regime of rather high magnetic fields, in which the cyclotron energy $h\omega_c$ is larger than the excitonic Rydberg R_X, the transition energies increase almost linearly with magnetic field. To the best of our knowledge, up to now only smooth dependences of the exciton transition energies on B have been reported except states of the same symmetry which anticross when they become equal in energy.

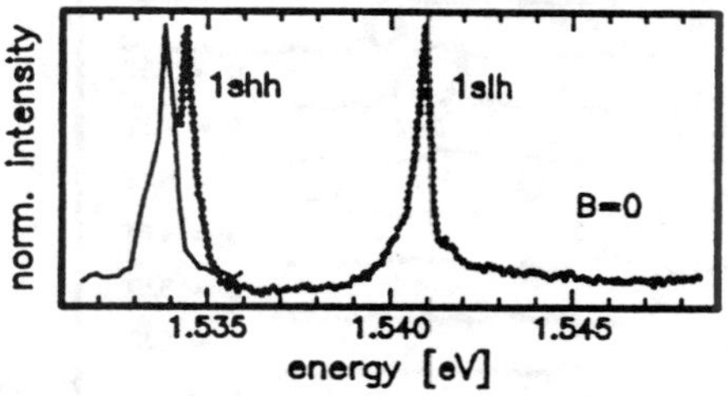

Fig. 1. Photoluminescence spectrum (solid line) and photoluminescence exciation spectrum (dotted line) of the 15 nm wide $GaAs/Al_{0.30}Ga_{0.70}As$ single quantum well at B = 0.

Here we report photoluminescence excitation studies of a 15 nm wide $Al_{0.30}Ga_{0.70}As/GaAs$ single quantum well of very high quality in magnetic fields up to 13 T. Surprisingly, we find in the magnetic field dispersions of the transition energies of excited exciton states that a regular, step-like dependence on B is superimposed on the linear increase of the Landau-level transition energies with B. We attribute this behaviour to the coupling of the transitions to another excitation in the system, most likely an acoustic phonon.

The samples were immersed in liquid helium (T = 4.2K) in a magnetocryostat with a superconducting solenoid ($B \leq 13$ T). The measurements were performed in Faraday-configuration. A Ti-sapphire laser pumped by an Ar^+-laser was used for optical excitation with an optical fiber of a diameter of 200 μm. The excitation powers were low being limited to 0.1 W cm^{-2}. The emission was dispersed by an double grating monochromator with a focal length of f = 1m and detected by an Peltier-cooled photomultiplier with a GaAs cathode which was interfaced by a photon counting system.

Figure 1 shows photoluminescence (solid lines) and photoluminescence excitation (dotted lines) spectra of the single quantum well at B = 0. In the photoluminescence spectrum only emission by the 1s ground state exciton of the heavy hole (hh) is observed. The full width at half maximum of this emission line is 500 μeV. In the excitation spectrum besides the 1shh exciton, the 1s exciton of the light hole (lh) is observed. In addition a weak shoulder on the high energy side of this exciton can be resolved which can be attributed to the 2shh exciton, as is confirmed in investigations in magnetic field. The Stokes-shift due to exciton localization is only about 0.6 meV. This, together with the narrow emission linewidth, indicates that the sample is of very high quality.

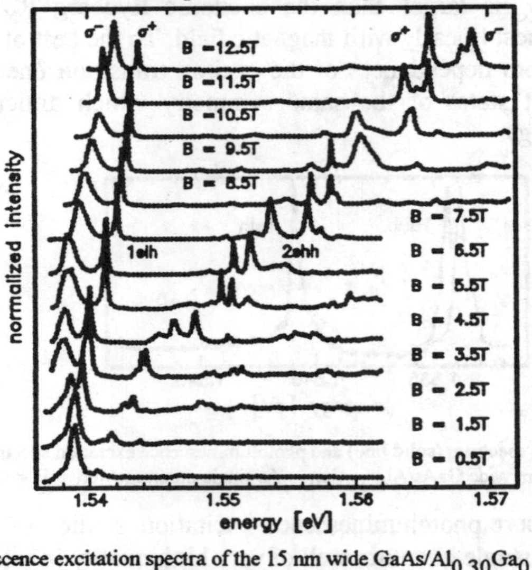

Fig. 2. Photoluminescence excitation spectra of the 15 nm wide GaAs/$Al_{0.30}Ga_{0.70}As$ single quantum well at various magnetic field up to 12.5 T.

Figure 2 shows photoluminescence excitation spectra of the single quantum well at varying magnetic fields up to 12.5 T. The detection wavelength was set to the energy of the 1shh exciton in each spectrum. In the figure we restrict ourselves to the region in energy of the 1slh and the 2shh excitons, because especially the 2shh exciton is of particular interest in the following. Both the 1slh exciton on the low energy side and the 2shh exciton on the high energy side split with increasing magnetic field, which arises from spin-splitting. Therefore the spectral lines are labeled with their corresponding circular polarizations.

The ns exciton states, n=1,2,... transform with increasing magnetic field into transitions between the N=(n-1) Landau levels. Therefore the energies of the 2shh excitons increase more strongly with field than do the energies of the 1slh excitons. Whereas the 1slh states show the expected smooth dependences on B, the transition energies of the 2shh excitons have non-linear features and show a step-like behaviour for both the σ^+ and the σ^- polarized lines. Therefore the energy splitting $\Delta_\pm$ between the σ^+ and σ^- transitions varies strongly. For example, the lines are close to each other at B = 4.5 T, and move apart up to 6.5 T. For increasing field $\Delta_\pm$ decreases (spectrum at 7.5 T), and at higher fields it increases again.

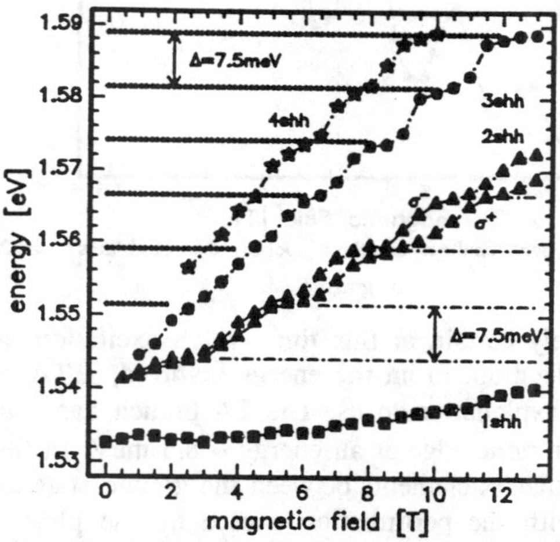

Fig. 3. Transition energies of the ground state 1s and excited heavy hole excitons as functions of the magnetic field. The dotted lines and the dashed dotted lines are guides to the eye.

The observed transition energies as functions of the magnetic field are shown in Figure 3, in which for clarity only the energies of the heavy hole excitons are plotted. We observe excitons up to n = 4. For all excited states n ≥ 2 the energies show the same quasi-oscillatory behaviour. We note that the oscillations are approximately equidistantly spaced in energy by about 7.5 meV. Because the transiton energies increase almost linearly with B the features are also approximately periodically in B with a period of about 3 T. The features in the transition energies of the two spin split 2shh transitions are not in phase. Therefore the splitting $\Delta_\pm$ is oscillatory in B in addition to the linear increase of the spin-splitting of a paramagnetic system, as can be seen from Fig. 4. At the highest B studied here the oscillations decrease in magnitude.

We propose that the step-like features in the transition energies of the excited exciton states shown in Fig. 3 result from coupling of the transition between the ground state 1shh exciton with each of the excited states, here 2s, 3s and 4s, with multiples of another excitation in the system. The energy of this excitation is approximately 7.5 meV, equal to the spacing in energy at which the steps in the transition energies occur.

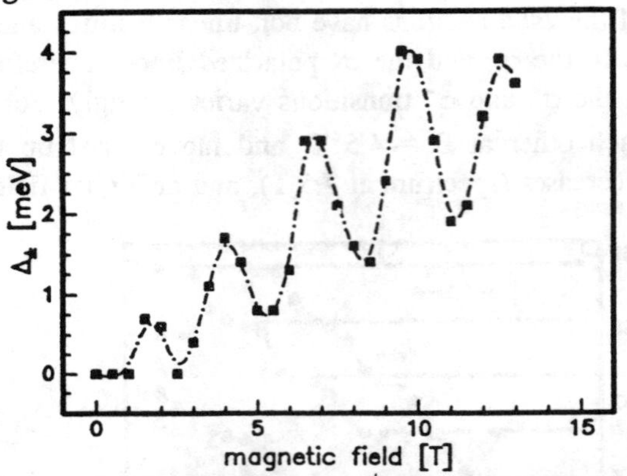

Fig. 4. Splitting in energy $\Delta_\pm$ between the σ^- and the σ^+ polarized spectral lines of the 2shh exciton versus magnetic field.

It is not entirely certain at this time which excitation of the system is involved in this hybridization. From the energy involved, a TA acoustic phonon seems the most likely explanation to us. The TA branch has a large density od states near the Brillouin zone edge at an energy ~ 8.5 meV. In the case here, the hybridization involves matrix elements between the ground state and each excited state of the exciton with the perturbation caused by the phonon. This matrix element has a maximum for non-zero wavevector away from the zone center, which could account for the energy being lower than the zone edge value.

The appearance of oscillations with magnetic field in the spin-splitting $\Delta_\pm$ shown in Figure 4 follows from the fact that the features in the transitions corresponding to the spin polarized states occur at approximately the same energy. Then the non-zero spin splitting causes the two states to have these plateau-like features at different magnetic fields, which results in oscillations of the spin splittings.

Acknowledgements

This work was supported by the State of Bavaria and by the US Office of Naval Research. TLR gratefully acknowledges the support of the Alexander von Humboldt Foundation.

MAGNETIC FIELD INDUCED EXCITON CONFINEMENT IN HEAVILY DOPED GaAs/$Al_{1-x}Ga_xAs$ QUANTUM WELLS

Jen-Y. JEN*, Yasuo OKA, James R. ANDERSON[a)], Malgorzata GORSKA[b)] and Iwao MOGI[c)]

Research Institute for Scientific Measurements, Tohoku University, Katahira 2-1-1, Aoba-ku, Sendai 980-77, Japan
e-mail: oka@rism.tohoku.ac.jp
[a)]*Department of Physics, University of Maryland, College Park, MD*
[b)]*Institute of Physics, Polish Academy of Science, Warsaw, Poland*
[c)]*Institute for Materials Research, Tohoku University, Sendai, Japan*
* *present address is a)*

The exciton confinement by high magnetic field is reported in a quantum well (QW) of the heavily doped GaAs/$Al_{1-x}Ga_xAs$ system. The QW sample doped with Si donors of 3×10^{18} cm^{-3} shows a broad luminescence band, at zero magnetic field, due to the QW sublevel transitions of the electron-hole pair. In the magnetic field above 10 T the QW exciton luminescence is discriminated from the Landau sublevel transitions. The result shows that the highly screened two-dimensional electron-hole pair state is confined by the high magnetic field to the QW exciton state having a finite binding energy.

1. Introduction

An exciton created in a narrow quantum well (QW) of semiconductor shows a markedly increased binding energy due to the confinement effect by the QW. If donor impurities are introduced in the QW, the electrons derived from the donors screen the Coulomb potential of the exciton and cause a decrease of the binding energy of the QW exciton.[1] It is interesting, therefore, to study the effect of external high magnetic field on the screened QW exciton state because the confinement of the exciton wave function by the magnetic field can recover the exciton binding which is decreased by the screening of the electrons.

In this paper the exciton photoluminescence (PL) of the heavily doped single QW (SQW) of the GaAs/$Al_{1-x}Ga_xAs$ system is studied in the magnetic field up to 21 T. The result shows that the highly screened 2-dimensional (2D) electron-hole pair state is confined, by the high magnetic field, to the "zero dimensional" QW exciton state having a finite binding energy .

2. Experimental

Heavily n-type-doped SQWs of GaAs/$Al_{1-x}Ga_xAs$ were grown by the molecular beam epitaxy method in Laboratory of Physical Science, University of Maryland. The width of the GaAs well (L_w) is 100 Å and that of the $Al_{1-x}Ga_xAs$ (x=0.22) barriers (L_b) is 300 Å, where the GaAs well layer was doped uniformly by Si-donors with doping level of

3×10^{18} cm^{-3}. The corresponding electron density in the well should be 3×10^{12} cm^{-2}. We refer this doped SQW sample as MB896. An undoped SQW with the same QW structure is also studied. Magneto-PL was measured by using a hybrid magnet in Institute for Material Research, Tohoku University. The magnetic field up to 21 T was applied perpendicular to the QW layer.

3. Results and Discussion

PL spectra of the doped SQW (MB896) are shown in Fig. 1. At 0 T (Fig. 1 (a)), the PL spectrum appears as a broad band peaking at 1.585 eV and extended to 1.610 eV in the high energy side.[2] The lower energy component of the broad band is extended to the region less than 1.53 eV and the sharp lines appear around 1.515 eV, which is the luminescence due to the bound excitons in the GaAs buffer layers. In the magnetic field of 18 T, the broad band splits into 4 components as shown in Fig. 1 (b), where we notice that the higher energy edge of the luminescence remains in the same energy position of 1.610 eV as shown by the arrow. The bound exciton lines of the GaAs buffer show multiple splitting. The lowest energy component at 1.535 eV in Fig. 1 (b) is related to the bound exciton in the QW. We do not discuss these bound-exciton-related luminescences in the present study.

Figure 2 shows the magnetic field dependence of the luminescence peak energies in MB896 (closed circles). The vertical bar in the figure displays the width of the decomposed-band-component around 1.55 eV. In the field below 7 T, however, the vertical bar displays the whole width of the luminescence band since the 1.55-eV component is not separated solely from the band. The peaks appeared above 1.56 eV for the field higher than 7 T show linear dependence to the magnetic field, which converge to 1.560 eV at 0 T (denoted as E_G). Therefore these luminescence peaks can be assigned to the recombination by the Landau level

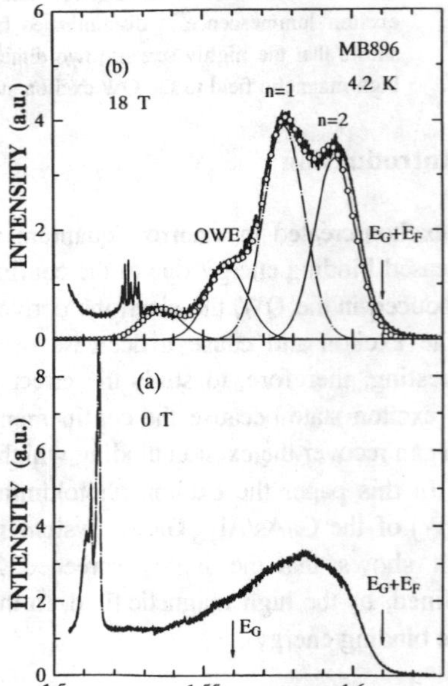

Fig. 1. Luminescence spectra of the QW of the GaAs/$Al_{1-x}Ga_xAs$ system doped with Si donors of 3×10^{18} cm^{-3} at the magnetic field of 0 T (a) and 18 T (b). L_w= 100 Å.

transitions with the Landau quantum number n = 1-6. Such Landau level luminescence can not be seen in the undoped QW sample. The observed Landau level luminescence in MB896 can be assigned as the recombination of the degenerated electrons in the conduction band QW with the photoexcited holes. The converging energy E_G corresponds to the lowest intersubband transition in the present SQW. In Fig.1 (a) the energy of E_G is shown by the arrow

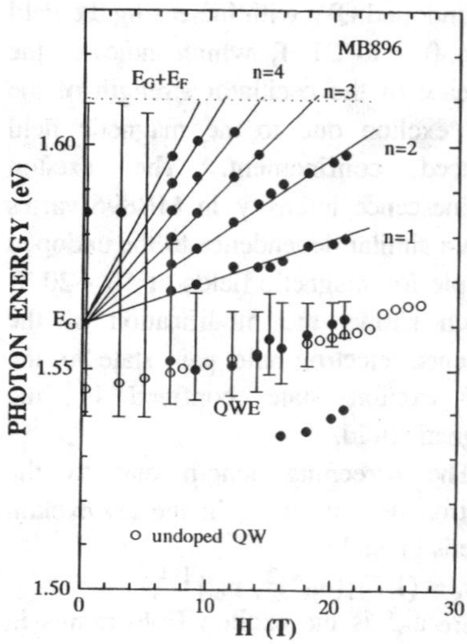

Fig. 2. Magnetic field dependence of the luminescence peaks of the QW exciton and Landau sublevel in MB896 (closed circles) and in the undoped sample (open circles).

The higher energy edge of the PL component always locates at 1.610 eV for the magnetic fields up to 21 T, indicating an existence of the Fermi level in the conduction-band Landau level. The energy difference between 1.610 eV and E_G is 50 meV which is the Fermi energy E_F of the electrons in the conduction band QWs. The Fermi energy for the electron in the 2D density of state is expressed as $E_F = \pi h^2 n_w / m^*$ (n_w and m^* are the density and the effective mass of the 2D-electrons, respectively). By using the obtained value of 50 meV for E_F and $m^* = 0.067\ m_0$, n_w is calculated as 1.4×10^{12} cm^{-2}. The doped electron density was 3×10^{12} cm^{-2}. Therefore 46 % of the doped electrons remains in the SQW and the rest of electrons might locate in the surface electronic states of in this QW specimen.

The luminescence component around 1.55 eV of MB896 (the closed circles in Fig. 2) shows a quadratic dependence to the magnetic field in the field region higher than 10 T, which agrees with that of the QW exciton (QWE) luminescence in the undoped sample shown by the open circles. This result indicates that the recombination of the QW exciton in the doped QW can be discriminated from the Landau sublevel transitions for the magnetic field higher than 10 T, while, for the field lower than 10 T, the QW excitonic transition is merged to the band-to-band recombination. Therefore the screening effect by the electrons in the QW smears the bound state of the QW exciton and the confinement of the exciton wavefunction due to the magnetic field causes the finite binding energy to the screened QW exciton state in case of the field higher than 10 T.

Figure 3 shows the luminescence intensity of the QW excitons in MB896 (closed circles) and the undoped sample (open circles) as a function of magnetic field. The QW exciton intensity increases by one order in the undoped QW with increasing the field from 0 T to 21 T, which indicates the increase of the oscillator strength of the QW exciton due to the magnetic field induced confinement. The exciton luminescence intensity in MB896 varies with a similar dependence to the undoped sample for magnetic fields of 10 - 20 T, which shows the modification of the screened electron-hole pair state to the QW exciton state confined by the magnetic field.

The screening length due to the electron density of n_w in the 2D exciton state is given by

$$r_s = (1/2)\{(a_B^*)^2 / n_w\}^{1/4},$$

where a_B^* is the exciton Bohr radius in the QW. The screening length r_s is calculated as 36Å for $n_w = 1.4\times10^{12}$ cm^{-2}. In the undoped QW at 0T, a_B^* = 63Å for the QW exciton with the binding energy of 15.7 meV. On the other hand the cyclotron orbit radius $a_c = (h/2eB)^{1/2}$ is 54 and 38 Å at 10 and 20 T, respectively. Therefore the Bohr radius of the screened QW exciton can be decreased to the value comparable with r_s by the magnetic field higher than 10 T. This magnetic field induced confinement perpendicular to the field causes the energy splitting of the screened QW exciton state from the Landau sublevel. Such a fully quantized screened-QW-exciton state is "the zero-dimensional exciton state", which has been reported in the case of undoped QWs.[3]

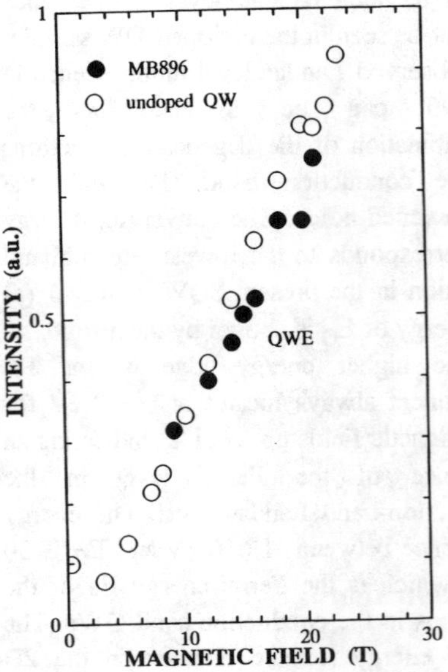

Fig. 3. Luminescence intensity of the QW exciton in MB896 and undoped sample as a function of the magnetic field.

References

1. D-S. Jiang et al. J. Appl. Phys. **53**, 999 (1982).
2. J. R. Anderson, M. Gorska, J.Y. Jen, Y. Oka et al., Proc. Nanostructure Conf., Sendai, (1995) to appear.
3. H. Sakaki, A Arakawa et al., Appl. Phys. Lett. **46**, 83 (1985).

OPTICAL AND MAGNETO-OPTICAL STUDIES OF MICROROUGHNESS IN QUANTUM WELLS

J.B.B. de OLIVEIRA[a,b], E.A.MENESES[a]
[a]*Instituto de Física Gleb Wataghin, Universidade Estadual de Campinas, Caixa Postal 6165,17083-970 Campinas, SP, Brazil*
[b]*Departamento de Física-FC, Universidade Estadual Paulista, Caixa Postal 473, 17033-360 Bauru, SP, Brazil*

E.C.F. da SILVA
Instituto de Física da Universidade de São Paulo, Caixa Postal 66318, 05389-970 São Paulo, SP, Brazil

In this work we estimated the lateral dimension of the interface microroughnes in GaAs/GaAlAs quantum wells (QW) and analized the full width at half maximum (FWHM) of the photoluminescence (PL) spectra using Singh's theory.

1 Introduction

Excitons are powerful sensors of quantum well interface roughness. Quantitative information can be obtained from photoluminescence (PL) line-shape analysis.[1] The PL emission lines are broadened due to final-state interactions (Lorentzian broadening) and to statistical fluctuations effects (Gaussian broadening). Shing *et al.* related the standard deviation of the Gaussian distribution to interface structural parameters, using thermodynamics arguments.[2] Based on this model, and using a infinitely high barrier approximation, Aksenov *et al.* described the PL-FWHM (σ) as:[3]

$$\sigma = 0.59\frac{\delta_1\delta_2}{\mu L_z^3 R_{ex}} \tag{1}$$

where δ_1 and δ_2 are the high and the lateral dimensions of the interface islands, $R_{ex} = \sqrt{\langle\Psi|x^2+y^2|\Psi\rangle}$ is the expectation value of the lateral exciton dimension, and μ is the reduced exciton mass in the direction perpendicular to the QW layer.

In this work we used PL and magneto-PL measurements to estimate δ_2 and to describe the interface microroughness effects on emission spectra, using Singh's model. The investigated sample was a $GaAs/Ga_{0.7}Al_{0.3}As$ multiple QW system grown by MBE. The barriers and wells were 300 and 80 Å wide, respectively.

The PL measurements were performed using an Ar-ion laser and conventional low-temperature and detection systems. The magneto-PL measurements were obtained using the Faraday configuration.

2 Experimental Results

A typical PL spectrum of the analyzed sample showed that, in addition to the heavy-hole emission, there is another peak, localized 1.4 meV below the free-exciton (FE) transition. The results of detailed PL and photoluminescence-excitation (PLE) measurements indicated that the low-energy emission is associated with excitons bound (BE) to defects localized at the GaAs/GaAlAs QW interfaces.

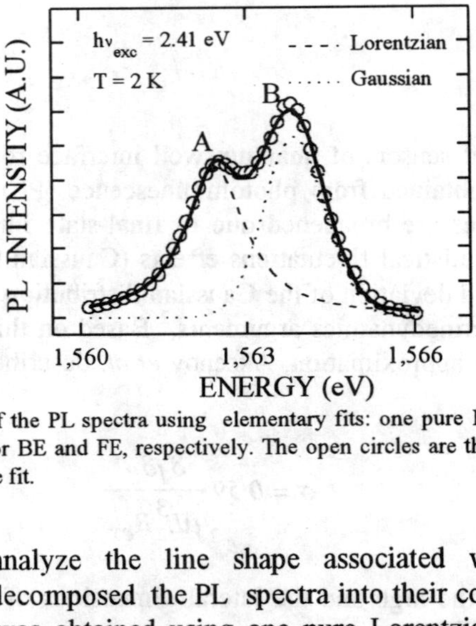

Fig. 1. Deconvolution of the PL spectra using elementary fits: one pure Lorentzian (dashed) and one pure Gaussian (dotted) for BE and FE, respectively. The open circles are the experimental data and the full line is the result of the fit.

In order to analyze the line shape associated with the free-exciton recombination, we decomposed the PL spectra into their constituents. The best fit (shown in Fig. 1) was obtained using one pure Lorentzian (BE) and one pure Gaussian (FE), indicating that statistical fluctuations effects dominate the FE emission and Lorentzian broadening can be neglected. On the other hand, alloy-composition fluctuations can also be neglected due to the negligible amplitude of the heavy-hole exciton wave function in the barrier.[4] Therefore, we attributed the observed Gaussian distribution of the FE emission as being predominantly determined by interface roughness.

In Fig. 2a the measured diamagnetic shift (ΔE) of the FE recombination has been plotted as a function of magnetic field up to 12 T. According to Ossau *et al.*, $\Delta E(1S) = \left(e^2 B^2 / 8\mu_{xy}\right) R_{ex}^2$, where all symbols have their usual meaning.[5] Using the results shown in Fig. 2a we calculated R_{ex} for each value of B, assuming an exciton in-plane mass $\mu_{xy} = 0.041 m_0$. The obtained values are plotted in Fig. 2b, where the full line represents a linear fit to the data. It is important to notice in this figure that R_{ex} decreases as a consequence of the additional exciton confinement induced by increasing magnetic field. The adjusted curve, extrapolated down to zero field, gives R_{ex} = 98 Å.

Using R_{ex} = 98Å, δ_1 = 2.83Å (one GaAs monolayer) and $\mu = 0.056\ m_0$ in EQ. 1, we obtained δ_2 = 37Å, a value in good agreement with the ones determined by Transmission Electronic Microscopy (TEM) for GaAs/GaAlAs QW.[6] Using EQ.1, δ_2 = 37Å and the R_{ex} values plotted in Fig 2b, we estimated σ for each B value. The results are plotted in Fig. 3, together with the values of σ measured directly from the PL spectra. It can be seen in this picture that the experimental values of σ and the ones estimated from Singh's theory are in a good agreement.

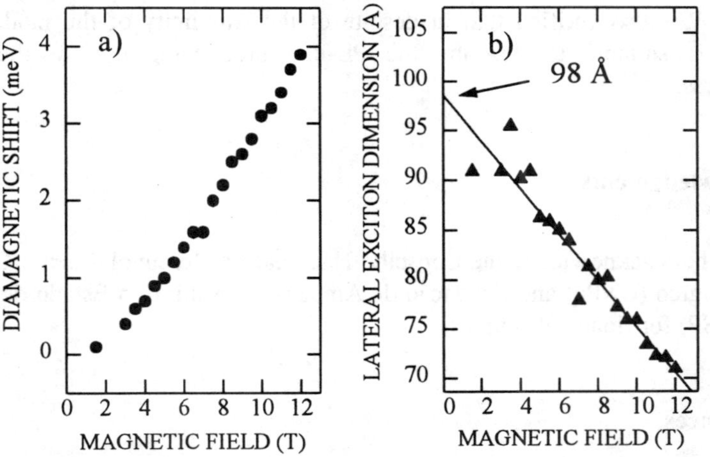

FIG 2 - a) Diamagnetic shift and b) lateral exciton dimension as a function of magnetic field. In (b) the line extrapolated back to zero-magnetic field gives R_{ex} = 98 Å.

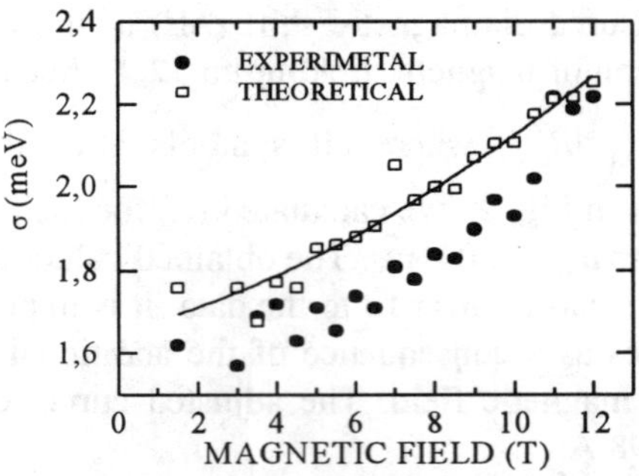

FIG. 3 - FWHM (σ) as a function of magnetic field: experimental values (filled circles) were obtained directly from the PL spectra. The theoretical results (open squares) come from Singh's theory. The full line is a guide for the eyes.

3 Conclusion

In this work we showed that the lateral dimension of interface microroughness in GaAs/GaAlAs QW were of the order of 37 Å, in good agreement with TEM results. We also verified that in despite of the simplicity of the model, Singh's theory is suitable to describe the PL-line broadening due to the interface roughness.

Acknowledgments

The authors acknowledge the Conselho Nacional de Desenvolvimento Científico e Tecnológico (CNPq) and Fundação de Amparo à Pesquisa do Estado de São Paulo (FAPESP) for financial support.

References

1. J. Christen and D. Bimberg, Phys. Rev. B **42**, 7213 (1990).
2. J. Singh, K. K. Bajaj and S. Chaudhuri, Appl. Phys. Lett. **44**, 805 (1984).
3. I. Aksenov, J. Kusano, Y. Aoyagi, T. Sugano, T. Yasuda, and Y. Segawa, Phys. Rev. B **51**, 4278 (1995).
4. R. L. Greene, K. K. Bajaj, and D. E. Phelps, Phys Rev. B **29**, 1807, (1984).
5. W. Ossau, B. Jakel, E. Bangert, and G. Landwehr, Surf. Sci. **174**, 188 (1986).
6. C. A. Warwick and R. P. Kopf, Appl. Phys. Lett. **60**, 386 (1992).

SPIN-FLIP RAMAN SCATTERING IN GaAs/AlAs MULTIPLE QUANTUM WELLS

A. A. SIRENKO [a], T. RUF, K. EBERL, and M. CARDONA
Max-Planck-Institut für Festkörperforschung, 70569 Stuttgart, Germany
sirenko@cardix.mpi-stuttgart.mpg.de

A. A. KISELEV and E. L. IVCHENKO
A. F. Ioffe Physical Technical Institute, 194021 St. Petersburg, Russia

K. PLOOG
Paul-Drude-Institut für Festkörperelektronik, 10117 Berlin, Germany

We report on resonant spin-flip Raman scattering in undoped GaAs/AlAs multiple quantum wells. Effective g factors of electrons and heavy-holes are directly measured as a function of layer thickness and the angle between the magnetic field and the structure growth axis. A large g factor anisotropy for conduction electrons in the Γ minimum is observed. To calculate the electron g factor we use the $\mathbf{k}\cdot\mathbf{p}$ perturbation theory within the Kane scheme. The g factor anisotropy appears because of the confinement-induced splitting of heavy- and light-hole states.

1 Introduction

Spin-flip Raman scattering (SFRS) in GaAs/AlGaAs multiple quantum wells (MQW) provides useful information on the exchange interaction and g factors of localized two-dimensional excitons and acceptor-bound holes.[1,2,3] The underlying theory has been developed recently.[1,3,4] In this work we investigate dependences of electron and heavy-hole g factors on the layer thickness (L) and compare their anisotropy with theoretical predictions,[4,5] both in type-*I* and type-*II* GaAs/AlAs MQWs.

2 Experimental results and theory

Undoped symmetric GaAs/AlAs MQWs with L between 12Å and 90Å were grown by molecular beam epitaxy on (001)-oriented GaAs substrates. Experiments were carried out in magnetic fields (**B**) up to 14 T at a temperature of 5 K. A tunable Ti-sapphire and dye lasers were used for resonant excitation of both the direct (Γ–Γ) and indirect (X–Γ) heavy-hole (*HH*) excitons.

In type-*I* MQWs ($L > 35$Å) we observe narrow Stokes and anti-Stokes SFRS lines with the Raman shifts proportional to **B**. Figure 1 shows typical

[a]on leave from *A. F. Ioffe Physical Technical Institute, 194021 St. Petersburg, Russia.*

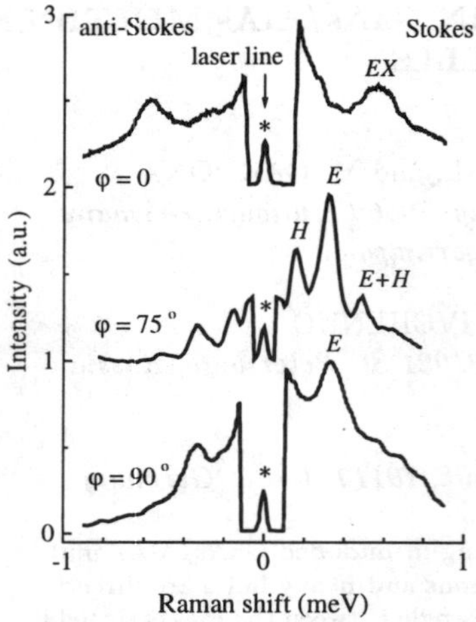

Figure 1: Spectra of SFRS for $\varphi = 0$, $\varphi = 75°$ and $\varphi = 90°$. B = 10T, L=40Å.

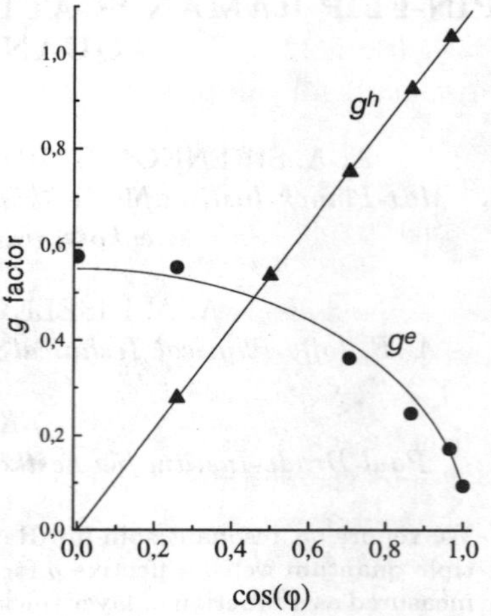

Figure 2: Electron (circles) and heavy-hole (triangles) g factors as a function of φ, L=40Å.

spectra for three different angles (φ) between **B** and the MQW growth axis (**z**) in a field of 10T. The lines are attributed to acoustic-phonon-assisted double resonant scattering with phonon energies equal to the Zeeman splittings of the *HH* exciton *1s* levels.[1] For $\mathbf{B}||\mathbf{z}$ ($\varphi = 0$) only one line (marked with the symbol *EX*) corresponding to the splitting between dipole-allowed *HH* exciton levels is observed. Both an electron and a hole change their spin in this process and the Raman shift is given by $\Delta = (g_{||}^h - g_{||}^e)\mu_0\mathbf{B}$, where $g_{||}^{h,e}$ are components of heavy-hole and electron g factors for $\mathbf{B}||\mathbf{z}$, respectively, and μ_0 is the Bohr magneton. A tilted magnetic field mixes the electron spin states $|+\frac{1}{2}\rangle$ and $|-\frac{1}{2}\rangle$ with respect to the **z** direction and transitions between all four *HH* exciton states can be observed. The lines marked with *H* and *E* in Fig. 1 ($\varphi = 75°$) correspond to Zeeman splittings of the heavy-hole and electron states in the *HH* exciton, respectively. For $\mathbf{B}\perp\mathbf{z}$ only the electron spin-flip line is observed (Fig.1, $\varphi = 90°$). Figure 2 shows the dependences of heavy-hole and electron g factors on φ for an MQW with L = 40Å. The lines represent fits using:

$$g^{h,e}(\varphi) = \sqrt{(g_{||}^{h,e} \cdot \cos\varphi)^2 + (g_{\perp}^{h,e} \cdot \sin\varphi)^2} \qquad (1)$$

with the parameters $g_{||}^e = 0.09\pm0.02$ and $g_{\perp}^e = 0.55\pm0.01$ for the electron and $g_{||}^h = 1.070\pm0.005$ and $g_{\perp}^h = 0\pm0.005$ for the heavy-hole. The signs of the g

factor components are determined based on the selection rules for a circular polarization of the Raman lines. The values of electron and heavy-hole g factor components for samples with different L are shown in Fig. 3. $g_{\parallel}^{h}$ decreases from 3.50 to 0.73 with increasing L from 12Å to 90Å whereas the absolute value of $g_{\perp}^{h}$ is close to zero: $|g_{\perp}^{h}| < 0.005$ (it is determined by the small valence band parameter q).[4]

In type-*II* MQWs ($L < 35$Å) we used the resonant excitation of both direct and indirect *HH* excitons. The electron spin-flip lines are observed only at the indirect exciton resonance giving small values for $g - 2$ and therefore close-to-isotropic electron g factors of about 1.98. For the direct exciton the electron spin-flip line is not detected because of the short electron life time as a consequence of effective transfer into the X valley. The heavy-hole spin-flip line is observed in both the indirect and direct exciton resonance regions giving the same values for the heavy-hole g factor which shows that the exchange splitting of the fourfold degenerate exciton levels at **B** = 0 is small compared to Zeeman splittings and can be neglected in our experiments.

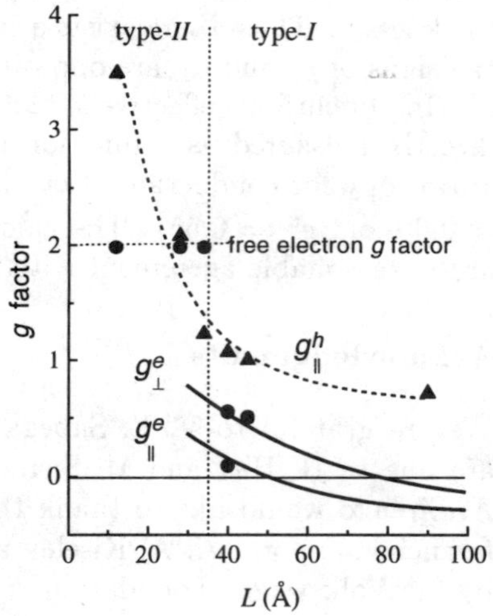

Figure 3: Dependence of electron (circles) and heavy-hole (triangles) g factor components on L. The dashed line is a guide to the eye for $g_{\parallel}^{h}$.

To calculate the electron g factor, we apply the Kane model: The $k_z p_z$ mixing between $c\Gamma_6, v\Gamma_8$ and $v\Gamma_7$ bands is taken into account exactly while the $k_x p_x + k_y p_y$ terms are treated by the second-order perturbation theory.[5] The electron g factor anisotropy is caused by the confinement-induced splitting of heavy- and light-hole states in the valence band. For the lowest conduction band state $e1$, the g factor components are given by[5]

$$g_{\parallel}^{e} = g_0 f_{\parallel} + m_0^{-1} \sum_{n} (| p_{+}^{e1,\frac{1}{2};n} |^2 - | p_{-}^{e1,\frac{1}{2};n} |^2)(E_0^{e1} - E_0^{n})^{-1} \tag{2}$$

$$g_{\perp}^{e} = g_0 f_{\perp} - 2m_0^{-1} \sum_{n} Re(p_{+}^{e1,\frac{1}{2};n} p_{z}^{e1,-\frac{1}{2};n})(E_0^{e1} - E_0^{n})^{-1} \tag{3}$$

Here m_0, g_0 are the free electron mass and Lande factor, $p_{\pm} = p_x \pm ip_y$, $p_{\alpha}^{n;m}$ ($\alpha = x, y, z$) is the momentum operator matrix element taken, in a single QW structure, between the subband states n, m at the extremum $k_x = k_y = 0$ and, in a superlattice, between the miniband states n, m at $k_x = k_y = Q = 0$, where Q is the wave vector component describing the electron motion along $\mathbf{z}$; E_0^n is the electron energy at this point. The coefficients $f_{\|}, f_{\perp}$ are close to unity and take into account that the average spin component for an electron in the state $|\, e1, \pm\frac{1}{2}\rangle$ differs slightly from $\pm\frac{1}{2}$ even at the extremum because of the $k_z p_z$ mixing with valence band states. The solid lines in Fig. 3 show calculated dependences of $g_{\perp}^e$ and $g_{\|}^e$ on L. Both values increase with decreasing L from the bulk GaAs value of -0.44 and invert their sign crossing zero at different thicknesses. Thus, there exists a range of L between about 50Å and 80Å where the signs of $g_{\|}$ and $g_{\perp}$ are opposite.

In conclusion, effective g factors of electrons and heavy-holes have been directly measured as a function of L and φ. For $L = 40$Å a large g factor anisotropy for conduction electrons in the Γ minimum has been observed ($g_{\|}^e = 0.09$ and $g_{\perp}^e = 0.55$). The calculated values (of $g_{\|}^e = 0.15$ and $g_{\perp}^e = 0.53$) are in reasonable agreement with the experiment.

Acknowledgments

We are grateful to V. F. Sapega and D. N. Mirlin for discussions. Thanks are due to H. Hirt and M. Siemers for their help with the experiments. A. A. Sirenko would like to thank the Alexander von Humboldt Foundation for financial support. A. A. Kiselev and E. L. Ivchenko were partially supported by the Volkswagen Foundation.

References

1. V. F. Sapega, M. Cardona, K. Ploog, E. L. Ivchenko, and D. N. Mirlin, *Phys. Rev.* B **45**, 4320 (1992).
2. D. N. Mirlin and A. A. Sirenko, *Fiz. Tverd. Tela Leningrad*, **34**, 205 (1992); [*Sov. Phys. Solid State*, **34**, 108 (1992)].
3. V. F. Sapega, T. Ruf, M. Cardona, K. Ploog, E. L. Ivchenko, and D. N. Mirlin *Phys. Rev.* B **50**, 2510 (1994).
4. E. L. Ivchenko, G. E. Pikus: *Superlattices and Other Heterostructures: Symmetry and Optical Phenomena*, Springer Ser. Solid-State Sci., Vol. 110 (Springer, Berlin, Heidelberg 1995).
5. E. L. Ivchenko, A. A. Kiselev, *Fiz. Tekh. Poluprovodn.* **26**, 1471 (1992); [*Sov. Phys. Semicond.*, **26**, 827 (1992)].

PHOTOCONDUCTIVITY STUDIES OF STRAINED InAsP/InP AND InAsP/GaInP STRUCTURES

D. KINDER, R.J. NICHOLAS
Clarendon Laboratory, Physics Department, Oxford University, Parks Road, Oxford OX1 3PU, U.K.

P.N. STAVRINOU
Department of Engineering Science, Oxford University, Oxford OX1 3PJ, U.K.

S.K. HAYWOOD
Department of Electronic and Electrical Engineering, University of Hull, Hull, U.K.

R. MOTTAHEDEH
Department of Electronic and Electrical Engineering, University College London, Torrington Place, London WC1E 7JE, U.K.

M. HOPKINSON, J.P.R. DAVID, G. HILL
EPSRC Semiconductor Growth Facility, University of Sheffield, Sheffield S1 3JD, U.K.

A series of InAsP/InP and InAsP/GaInP multiple quantum well structures with varying composition and strain relaxation have been studied with photoconductivity in both applied magnetic and electric fields. Both allowed and first forbidden transitions have been observed, indicating that InAsP/GaInP structures can be grown with sufficient quality for optical devices operating between 1.0μm and 1.55μm. Fitting the transitions seen in these spectra has led to an accurate set of parameters for modeling devices using this material system.

Introduction

The ternary semiconductor InAsP has great potential for opto-electronic devices operating between 1.0μm and 1.55μm, requiring less strain than more conventional GaInAs/GaAs structures. Furthermore, the large difference between the InAs and InP band gaps means that only a relatively small increase in the As content of the well material is required to move the operating wavelength out towards 1.55μm. The increased strain this composition change induces can be accommodated in the system by using a barrier material which is compressively strained relative to the InP substrate, such as GaInP. However, InAsP has received relatively little attention[1] compared to other ternary subsets of the quaternary GaInAsP system, with the result that there are few experimental reports on the material parameters of the $InAs_{1-x}P_x$ system. The samples studied are multiple quantum well diode structures, with nominal well compositions ranging from $InAs_{0.2}P_{0.8}$ to $InAs_{0.8}P_{0.2}$ and either InP or GaInP barriers. The samples were grown by conventional solid source molecular beam epitaxy (MBE), on a (100) n-type InP substrate, with a p-type InP capping layer.

Magnetic Field Data

Figure 1 shows photoconductivity spectra taken with magnetic fields for M737, nominally $InAs_{0.2}P_{0.8}/InP$ and M1217, nominally $InAs_{0.8}P_{0.2}/Ga_{0.08}In_{0.92}P$. For M737, the data was taken at 4K. M1217 has very deep quantum wells (>200meV) with the result that at 4K all the carriers are frozen in the wells, so the sample was heated to 200K while the data was taken.

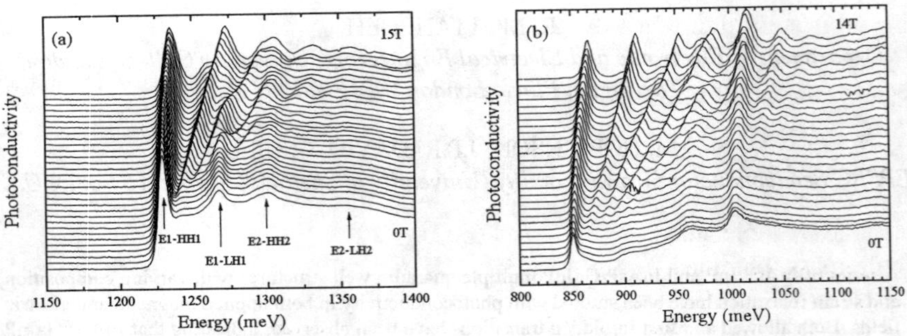

Figure 1: Photoconductivity spectra for (a) M737 and (b) M1217

In the presence of a magnetic field, each of the observed quantum well features will evolve into a Landau level fan, which for the E1-HH1 transition can be seen in the raw data, and can be made clearer by plotting the transition energies (i.e. the maxima in the photoconductivity spectra) as a function of magnetic field, as shown for another sample, M954 (nominally $InAs_{0.45}P_{0.55}/Ga_{0.06}In_{0.94}P$), in Figure 2.

Fitting this Landau fan to a numerical three dimensional exciton model[2], in which the electron and hole are treated as a hydrogenic system, gives information about the exciton mass and the binding energy of the exciton. The former of these is particularly important as mixing between light- and heavy-hole bands leads to a very complex valence band structure, from which it is difficult to calculate the in-plane heavy-hole mass. We assume that the electron mass is essentially the same as the bulk value, since the conduction band is far from any other band, allowing us to extract the in-plane heavy hole mass from the data.

To fit the energies of the different quantum well transition at zero field we have used a four-band **k.p** model. The most important input parameters of this model are well and barrier widths, the well and barrier band-gaps (which tell us the material composition), the strain relaxation and the band offset ratio between the well and barrier materials.

The superlattice period was taken from HRXRD measurements on the samples, with the well and barrier widths assumed to be in the same ratio as the nominal. Material parameters for InAsP and GaInP were obtained by linear interpolation of binary parameters taken from Ref. 3, except for the deformation potentials[4] and the InAsP band gap and spin orbit splitting[5]. A value of $0.55\Delta E_g$ was used for the conduction band offset ratio for all samples[6].

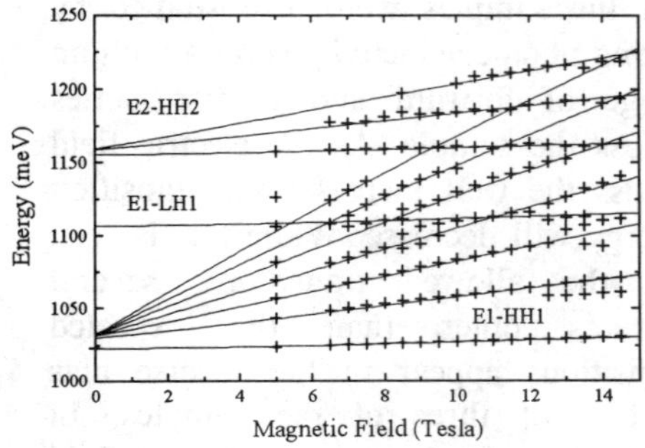

Figure 2: Fit to Landau Fan of M954

Using both the **k.p** and magnetic field models, the best fit to the E1-HH1 and E2-HH2 Landau fans plus the E1-LH1 transition in M954 is shown. Table 1 summarizes the fitted composition, the electron mass (calculated from the band-edge mass), the calculated heavy hole mass in the growth direction (||), and the fitted in-plane (⊥) heavy hole mass, obtained from the fitted reduced exciton mass by assuming that the effective electron mass is isotropic. The presence of strain increases the separation between the HH and LH bands, decreasing the mixing between them and thus decreasing the in-plane heavy hole mass. This is reflected in Table 1, with the most strained samples (those with the greatest HH1-LH1 splitting) having the lowest in-plane heavy hole masses.

Table 1: Calculated and fitted masses for all samples. The composition is given as the As content of the well followed by the Ga content of the barrier.

Sample	As Well/ Ga Barrier	HH1-LH1 Splitting	Transition	m_e^*	m_{hh}^* (‖)	m_{hh}^* (⊥)
M737	17%/0%	38meV	E1-HH1	$0.069m_0$	$0.57m_0$	$0.32m_0$
M738	18%/0%	43meV	E1-HH1	$0.069m_0$	$0.57m_0$	$0.38m_0$
M954	38%/6%	84meV	E1-HH1	$0.059m_0$	$0.53m_0$	$0.25m_0$
			E2-HH2	$0.066m_0$	$0.54m_0$	$0.55m_0$
M1216	50%/0%	108meV	E1-HH1	$0.052m_0$	$0.50m_0$	$0.15m_0$
M1217	55%/8%	113meV	E1-HH1	$0.049m_0$	$0.49m_0$	$0.15m_0$
			E2-HH2	$0.057m_0$	$0.50m_0$	$0.40m_0$

Electric Field Data

All the samples were also studied by taking photoconductivity spectra with an range of forward and reverse biases across the sample. As the electric field across the well increases the transition energy will decrease. While this is seen for the allowed transitions, several features other than the expected transitions appear in the reverse bias spectra of three of the samples. In particular, the increasing electric field causes a feature between E1-HH1 and E1-LH1 to become sharply defined (as seen in Figure 3), which calculations indicate is the E1-HH2 transition. With no electric field across the sample, we expect the transition strength to be zero, since the E1 wavefunction is symmetric and the HH2 wave-function anti-symmetric. However, when an electric field is applied the electrons and holes will tend to be spatially separated at different sides of the well, giving a non-zero overlap between the electron and hole wavefunctions.

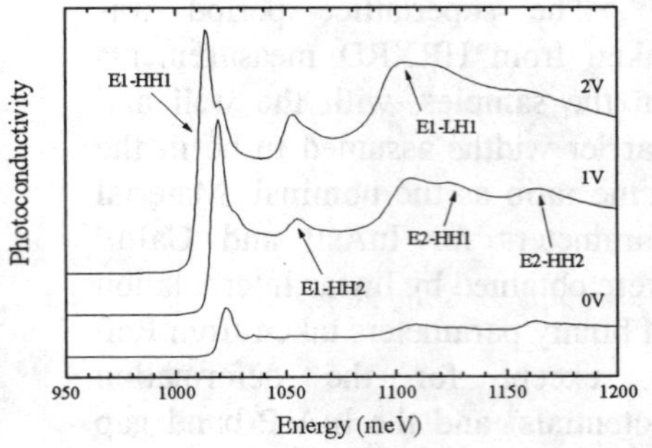

Figure 3: Reverse bias spectra for M954

Also observed is a splitting of the E1-HH1 transition peak, the origin of which is not clearly understood. Neither a non-linear electric field across the active region nor a monolayer difference in well width is sufficient to explain the observed splitting, but the two effects may both be contributing to the splitting.

In conclusion, we have shown that InAsP devices can be used for operation from 1.0μm up to 1.55μm. The introduction of Ga into the barriers allows longer wavelengths to be realized without loss of optical quality.

References

1. T.K. Woodward, T. Sizer II and T.H. Chiu, *Appl. Phys. Lett.* **58,** 1366 (1991).
2. P.C. Makado and N.C. McGill, *J. Phys. C.* **19,** 873 (1986).
3. *Numerical Data and Functional Relationships in Science and Technology*, edited by O. Madelung, Landolt-Börnstein, New Series, Group III, Vol. 17, Pt. a (Springer-Verlag, New York, 1982).
4. M.P.C.M. Krijn, *Semicond. Sci. Technol.* **6,** 27 (1991).
5. R.J. Nicholas, R.A. Stradling and J.C. Ramage, *J. Phys. C.* **12,** 1641 (1979).
6. D. Kinder, R.J. Nicholas, P.N. Stavrinou, S.K. Haywood, L. Hart, M. Hopkinson, J.P.R. David and G. Hill, *Semicond. Sci. Technol.* **11,** 34 (1996).

Exciton-LO Phonon Coupling and High Magnetic Fields Effects on $Zn_{1-x}Cd_xSe/ZnSe$ Multi Quantum Wells

T. Yasui, B. P. Zhang, T. Yasuda and Y. Segawa,
Photodynamics Research Center, RIKEN, 19-1399 Koeji, Nagamachi, Aoba, Sendai, 980, Japan

H. Kunimatsu, K. Uchida and N. Miura
Institute for Solid State Physics, University of Tokyo, Roppongi, Minato-ku, Tokyo, 106, Japan

Absorption spectra were measured in two $Zn_{1-x}Cd_xSe/ZnSe$ multi QWs under high magnetic fields up to 150 T at 6.5 K for circular polarizations ($\sigma+$ and $\sigma-$). In a sample with a well width of 43.5 Å, a sublinear field dependence of the HH-*1s* (HH-*2s*) exciton line was observed above 110 T(60 T). The transition from the quadratic to the sublinear occurred at a much lower field than expected from 2D magnetoexciton model. The cyclotoron energy exceeds the LO phonon energy at 110 T in this sample, and the effective mass change is induced at this field. In another sample, with a well width of 135 Å, a decrease of the field dependence was observed for the HH-*2s* exciton line at 110 T. Scattering of the magnetoexcitons to the Landau states via quasi-2D LO phonons in QW plays an important role in this phenomena.

1 Introduction

II-VI compounds show stronger exciton-longitudinal-optical (LO) phonon coupling as compared with III-V compounds due to the larger polarizability. The exciton binding energy of II-VI compounds is generally larger than that of III-V compounds, as well. In II-VI quantum well (QW) systems, we can decrease (increase) the exciton binding energy (Eb) below (above) LO phonon energy due to the quantum confinement effect. Therefore, in II-VI compounds, we can study the effects of the quantum confinement and the exciton-LO phonon coupling on the properties of excitons.[1] Reduction of the exciton-LO-phonon coupling was observed with increasing exciton confinements effects in II-VI QW systems.[1] When the exciton binding energy is increased beyond the LO phonon energy by the quantum confinement, the scattering probability between the exciton state and the continuum conduction band is decreased.

High magnetic fields quantize the conduction band and valence band[2] to the Landau levels. Not much has been known, however, on the scattering of the exciton states to the Landau levels by LO phonons in quantum wells. A dramatic change of the carrier-effective mass was observed in n-type ZnSe due to the undressing of free carriers when the cyclotron energy exceeds the 1-LO phonon energy of ZnSe (~30 meV) above 35 T.[3] In this paper, we study the exciton-LO-phonon coupling in II-VI QW under high magnetic fields. We discuss the relation between the magnetoexcitons confined in QW and the effective-mass change induced by high magnetic fields up to 150 T.

2 Experiment

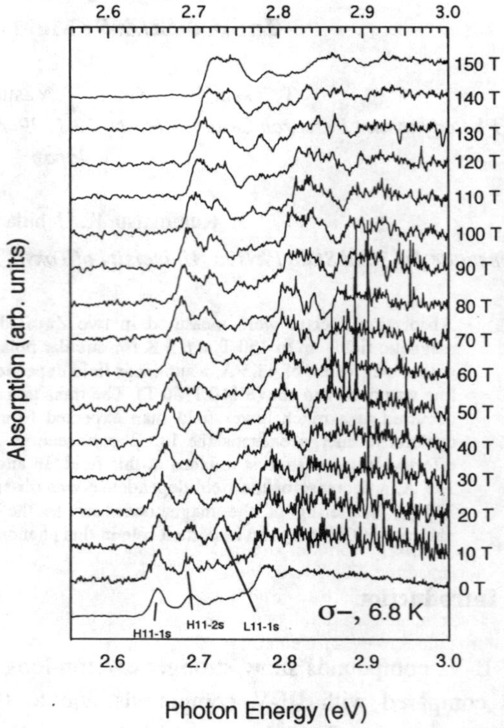

Fig. 1 Absorption spectra of the $Zn_{1-x}Cd_xSe/ZnSe$ MQW with σ− polarization at various magnetic field strengths in the photon energy region from 2.60 eV to 3.0 eV. The well width is 43.5 Å and Cd composition x is 0.15.

The samples were prepared by molecular beam epitaxy. The $Zn_{1-x}Cd_xSe/ZnSe$ multi-QWs with 20 periods were grown on a 1μm-ZnSe buffer layer on GaAs (100) substrate. We fabricated two samples for the transmission measurements: one with the well width (*Lw*) =43.5 Å and the Cd composition (x) =0.15 and the other with *Lw*=135 Å and *x*=0.16. The barrier widths were about 250 Å in all samples. After using a selective wet etching technique to remove the substrate, the samples for the transmission measurements were mounted on a quartz plate. The sample was mounted in a single-turn coil which generates pulsed high magnetic fields up to 150 T with about 7*μs* pulse duration. A Xe-flash lamp was used as a light source. The probing light was lead to the sample by an optical fiber and the transmitted light was sent to a spectrometer which was connected to an image converter camera and a cooled CCD camera. The light source and the detecting system were synchronized with the electrical controled system. The magnetic field strength was monitored by a calibrated pickup coil wound around the sample, with an accuracy of about 3 %.

3 Results and discussion

The transmission spectra of two $Zn_{1-x}Cd_xSe/ZnSe$ multi QWs were measured in high magnetic fields up to 150 T at 4.2 K for circular polarizations (σ+ and σ−). The well width and the Cd composition were *Lw*=43.5 Å, x=0.15 in sample A, *Lw*=135 Å, x=0.16 in sample B. The sample B shows the exciton-LO phonon coupling since the

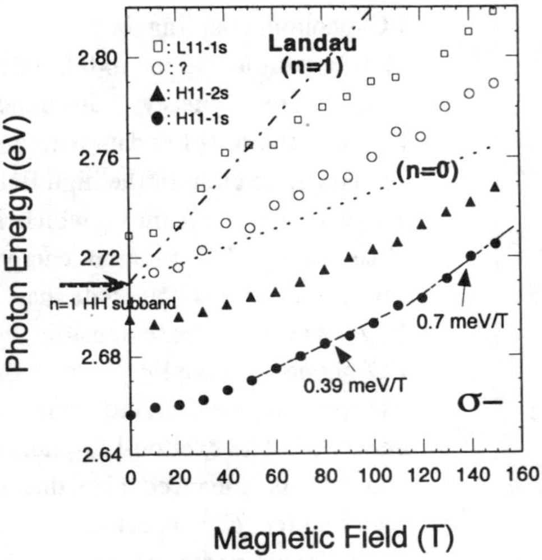

Fig. 2 Energy shifts of excitons up to 150 T with σ− polarization. The well width is 43.5 Å and Cd composition x is 0.15. The n=0 (n=1) Landau levels are indicated by the dotted (dashed) lines. Thin lines are guides to eyes for sublinear shifts.

Eb of the lowest *1s*-heavy hole (HH-*1s*) exciton is 35 meV, which is comparative with the 1-LO phonon energy of ZnSe, whereas the coupling is small in sample A whose *Eb* is 55meV.[1] Figure 1 shows the σ− absoption spectra for a sample B at various magnetic fields in the photon energy region from 2.60 eV to 3.0 eV. The absorption spectra were obtained by subtracting the background. The HH-*1s* and HH-*2s* excitons are clearly observed up to 150 T. Figure 2 shows the energy shifts of the exciton lines as a function of magnetic fields up to 150 T with σ− poalrization. We obtained *Eb*, by extrapolating the Landau levels linearly to the zero-magnetic field. For high fields the magnetic-field dependence becomes sublinear,[4] whereas for low fields the dependence shows a quadratic diamagnetc shifts. In sample A, the quadratic diamagnetic shift was observed below 70 T. The sublinear shift of the HH-*1s* line was observed above 70 T with 0.60 and 0.39 meV/T in the σ+ and σ− polarization, respectively, up to 110 T. In high magnetic fields above 110 T the gradiant becomes 0.70 meV/T, which is very close to the one for the n=1 Landau level. For the HH-*2s* exciton, the change from the low field behavior to the high field one occured at 60 T, with a high field gradiant of 0.42 meV/T above 60 T almost parallel to the n=0 Landau level. The cross over fields from 110 T for HH-*1s*, 60 T for HH-*2s*, are not in agreement with the values estimated from the 2D magnetoexciton model[4] in the high field regime. This is why the effective mass change is induced in the high magnetic field. The carrier-effective mass decreases due to the undressing of free carriers when the cyclotron energy exceeds the 1-LO phonon energy.[3] In sample A with *Lw*=43.5 Å, the change of the shift was observed at magnetic fields higher than in bulk n-type ZnSe. This is due to the scattering of the magnetoexcitons to the Landau states via quasi-2D LO phonons in QW.

In sample B with *Lw*=135 Å, the energy shift of the HH-*2s* exciton becomes smaller in the σ+ polarization above 110 T as shown in Fig. 3. Exceeding the 1-LO phonon energy above 110 T, the effective mass change is contrary to the sample A. To explain

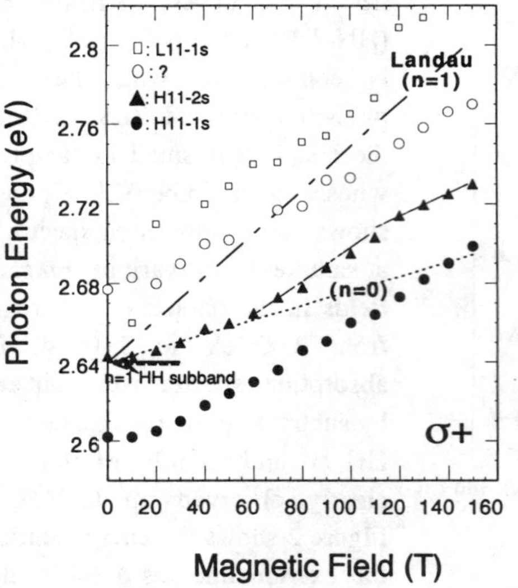

Fig. 3 Plots of the energy shifts of excitons up to 150 T with σ+ polarization. The well width is 135 Å and Cd composition x is 0.16. The n=0 (n=1) Landau levels are indicated by the dotted (dashed) lines. Thin lines are guides to eyes for sublinear shifts.

this, we should consider the large LO-phonon coupling between the HH-*2s* exciton and the n=0 Landau state. The energy difference between the n=0 Landau state and the HH-*2s* exciton in the high field range is about 30 meV, which is close to the LO phonon energy. Since the *2s* exciton state has a large wavefunction extension, the LO-phonon coupling of the excitons to the Landau states is enhanced. The exciton-LO-phonon coupling is enhanced also due to the smaller Eb^1 because of the larger well width of sample B. These two factors decrease the relative kinetic motion of HH-*2s* exciton, and thus reduce the enegy shift above 110 T.

4 Conclusion

The absorption spectra of two $Zn_{1-x}Cd_xSe$/ZnSe multi QWs were measured in high magnetic fields up to 150 T at 6.5 K for circular polarizations (σ+ and σ−). In a sample with 43.5 Å-well-width, the transition from the quadratic diamagnetic shift to the sublinear one of the HH-*1s* (HH-*2s*) exciton was observed at 110 T (60 T). This may be the reason why the cyclotoron energy exceeds the 1-LO phonon energy and the effective-mass change is induced at this field. The decrease of the HH-*2s* energy shift was observed in a sample with *Lw*=135 Å above 110 T. This is explained by the LO-phonon coupling between the HH-*2s* exciton state and n=0 Landau state.

References

1. N. T. Pelekanos, et. al., Phys. Rev. B 45, 6037, (1992).
2. J. C. Maan, G. Belle, A. Fasolino, M. Altarelli, and K. Ploog, Phys. Rev. B 30, 2253 (1984).
3. Y. Imanaka, N. Miura, H. Kukimoto, Phys. Rev. B 49, 16965, (1994).
4. O. Akimoto and H. Hasegawa, J. Phys. Soc. Jpn. 22, 181, (1967).

TIME-RESOLVED MAGNETO-LUMINESCENCE AND FINITE-WAVE-VECTOR RAMAN SPECTROSCOPY OF 2D-ELECTRONS IN EXTREME QUANTUM LIMIT

I.V. KUKUSHKIN[a,b], V.I. FALKO[b,c], R.J. HAUG[a,d], K.v. KLITZING[a], K. EBERL[a]

[a] *Max-Planck-Institut für Festkörperforschung, Heisenbergstr. 1, 70569 Stuttgart, Germany*

[b] *Institute of Solid State Physics, RAS, Chernogolovka, 142432 Russia*

[c] *School of Physics and Chemistry, Lancaster University, LA1 4YB, Lancaster, UK*

[d] *Inst. für Festkörperphysik, Univ. of Hannover, Appelstr. 2, 30167 Hannover, Germany*

We compare the results of transport and time-resolved magneto-luminescence measurements in disordered 2D electron systems in GaAs-AlGaAs heterostructures in the extreme quantum limit, in particular, in the vicinity of the metal-insulator transition (MIT). At filling factors $\nu < 1$, the optical signal has two components: a single-rate exponentially decaying part attributed to a uniform liquid and a power-law long-life-time tail characteristic for a microscopically inhomogeneous state of electrons. We interpret this result as a separation of the 2D electron system into a liquid and localized phases, especially because the MIT (detected in transport) occurs strikingly close to those filling factors where the liquid occupies ½ of the sample area (the percolation threshold condition in two-component media). We also demonstrated that in the best quality sample the transition at the critical temperature into a Wigner crystal is accompanied by an abrupt transition in the luminescence kinetics, whereas it shows no critical temperature behaviour in all other lower quality samples.

A Raman-like emission line shifted from the laser line by the electron cyclotron energy was found and investigated in GaAs/AlGaAs single heterojunctions. By using tilted magnetic fields we were able to separate signals of inelastic light scattering and of hot luminescence. It was shown that, together with a Raman signal from 2D-electrons, a line which has similar properties and which arises as a result of the recombination of hot bulk electrons and heavy holes predominates in the emission spectrum.

We used finite-wave-vector inelastic light scattering spectroscopy to study a spatial dispersion of magnetoplasmon mode of two-dimensional electrons in perpendicular, parallel and tilted magnetic fields. It is shown that in perpendicular magnetic-fields the Hall conductivity σ_{xy} can be extracted from the difference between magnetoplasmon energy and the cyclotron energy. We observe that a finite-wave-vector Hall conductivity $\sigma_{xy}(k)$ is quantized in the regimes of the Integer and Fractional Quantum Hall Effect and that the width of the plateau is almost independent of k.

It is demonstrated that an increase of k results in the nonlinear development of an additional plateau in σ_{xy} at half filling of the Landau level (at $\nu=1/2$) and the width of this plateau is approximately proportional to k.

1 Wigner Crystallization Versus Metal-Insulator Transition

The ground state of the two-dimensional (2D) electron system in the ultraquantum limit is formed by the interplay between electron-electron interaction and disorder. In a magnetic field both of them result in a frozen insulating state of the electronic system with different internal structures. In ideally pure systems, the formation of the Wigner solid is expected [1], whereas the disorder hinders crystallization and gives rise to a state with no well-defined (even local) order. In high quality semiconductor structures, the importance of interaction has been demonstrated experimentally by the observation of the series of incompressible fractional states [2] which is terminated in high magnetic fields by an insulating phase, as it was observed in different transport experiments [3-5]. However, transport method is rather insensitive to the internal structure of the electronic system, whereas the recombination kinetics of 2D-electrons was found to be a sensitive probe of the local spatial structure of the insulating phase of 2D-electrons [6]. One of the goal of the work was to compare the results of the transport and time-resolved magneto-luminescence measurements for wide variety of different quality samples, especially in the vicinity of the metal-insulator transition (MIT) and of the Wigner crystallization.

The metal-insulator transition in two-dimensional electron systems still remains to be a hot topic in the semiconductor physics, in particular, because of a variety of transitions between quantum Hall liquids and insulating states in the extreme quantum limit realized at high magnetic fields [7,8]. Although the single-particle scenario of localization of electrons at high magnetic fields was already thoroughly explored some years ago [9], the most recent theories [10] emphasize the influence of interactions on the MIT also in disordered systems, especially regarding the existence of a hierarchy of strongly correlated quantum Hall liquids. Being inspired by the success of scaling idea that was so productive in the single-particle localization models [9], the search of a global phase diagram of the MIT in the extreme quantum limit has been started [10] in order to describe the growing amount of experimental data. However, one should not ignore such a possibility that, in interacting electron systems, the transition between correlated liquids and disordered insulator is the Mott transition rather than the Anderson one, and that the mesoscopics-scale inhomogenuity in the strength of a random potential itself can result in the separation of the 2D system into phases with locally different properties (liquid versus localized insulator). In this scenario, the portion of a localized phase varies continuously under the variation of a magnetic field due to the sensitivity of the electron-electron correlations to the value of the filling factor, so that the MIT detected in the transport might manifest nothing but the destruction of the classical percolation through the liquid in a multicomponent media [11].

To find the truth, it is necessary to make a local measurement which would distinguish between different phases (liquid and insulator) and to evaluate the amount of each of them. We solve this problem by using the method based on the analysis of the time-resolved recombination of 2D electrons with a small number of acceptor bound holes in the δ-doped layer. We already know [6] that, in contrast to the quantum Hall effect state at the filling factor $\nu = 1$ that demonstrates the single-exponential decay $I(t) \propto \exp(-t/t_0)$, the magnetically frozen state manifests itself in a broad spectrum of recombination rates which results in the power-law kinetics of the magneto-luminescence. It is a peculiarity of the recombination process involving acceptor bound holes that the latter form a small neutral centers, and that the lifetime τ of each hole is determined by the local density of 2D electrons at the acceptor position, $\tau^{-1} \propto \rho_e$, so that the spectrum $f(\tau^{-1})$ of recombination rates τ^{-1} reflects the distribution of the local 2D electron density [12]. The latter can be found from the inverse Laplace transform of the measured temporal dependence I(t) of the luminescence intensity after a short illumination pulse, $I(t) \propto \int d\tau^{-1} f(\tau^{-1}) e^{-t/\tau}$. For example, the mentioned above single-exponential decay of the luminescence at $\nu = 1$ evidences the microscopic-scale homogeneity of the 2D electron density in this incompressible liquid state [13]m whereas the information about the distribution of local values ρ_e which we extract from the long-living tails of luminescence from any state at $\nu < 1$ may be used for identification of its intrinsic structure, like it has been done before in the Wigner solid regime [6].

We studied seven different quality (see Table 1) GaAs-AlGaAs single heterojunctions with a δ-doped monolayer of Be acceptors ($n_A = 1.3 \times 10^9$ cm^{-2}) located in the wide (1 μm) GaAs buffer layer at a distance about 30 nm from the interface [14]. The magnetoresistance tensor was measured on the Hall bar samples by use of a standard low-frequency (12 Hz) and low-current (10 nA) lock-in technique with a resistance 100MΩ connected in series with the sample. For photoexcitation, we used pulses from a tunable Ti-Sapphire laser (wavelength $\sim$ 800 nm) with duration 20 ns, peak power of 10^{-5} - 10^{-4} W/cm^2 and frequency of 10^5 - 10^3 Hz [14].

Fig. 1 shows the results of magneto-transport measurements (in the range of temperatures T = 50 - 150 mK) which were used for determining the MIT point in two samples B and E which have identical concentration but very different mobilities (see Table 1). The conductivity measured in samples of different qualities shows quite distinguishable behaviour, especially at $\nu < 1$. In particular, in better quality samples (A, B, C) one can find the inter-crossing between σ_{xx} (B) taken at different temperatures which can be used for determination of the MIT point [7,8]. However, such an inter-crossing does not appear in the magneto-conductivity data taken in the lowest mobility structures (D, E, F), so that another criterion of the MIT is desirable. The common feature of all studied samples near the MIT was an abrupt appearance of the out-of-phase signal both in the diagonal and Hall components of resistivity which is accompanied by a drastic decrease of a current and indicates the formation of a dielectric state. Fig. 1 (a, b) also shows the evolution of the magnetic field dependence of the current through the 2D system with the temperature. Under the decrease of a filling factor, the current shows a threshold at ν_c which become more pronounced at lower temperatures, so that ν_c will be treated below as the MIT point. Note that in the better quality samples (A, B, C) such a procedure gives the same result as the mentioned above conventional method.

In contrast to the transport which shows an abrupt change at the MIT point, the optical data taken in the same structures demonstrate only a continuous evolution both of the integral and time-resolved spectra. Insets to Fig. 1 (a) and (b) show the raw data of typical recombination kinetics measured in samples (B, E) in different magnetic fields at T = 50 mK and we see no critical changes occuring when $\nu = \nu_c$ is passed. the magneto-luminescence kinetics measured in lower- and higher-mobility samples has both common features and significant differences. The common feature of the recombination kinetics observed in all the samples is the purely exponential decay, i(t) $\propto$ exp $(-t/t_0)$, of the magneto-luminescence intensity at the filling factor $\nu = 1$. Since the density of the optically active holes is much smaller than the 2D electron density, the time dependence I(t) is controlled by the recombination rates of holes. In the homogeneous electron system, the latter are determined by the sheet electron density and the distance from the interface to the acceptor mono-layer and are the same for all of them, so that the single-rate ($\tau^{-1} = t^{-1}_0$ exponential decay manifests the microscopic scale homogeneity of the electron density. At $\nu = 1$, this is the well known property of a trial many-body wave-function corresponding to a completely filled Landau level [13]. For the lower fillings, the slow recombination tail develops, which indicates the formation of a short-range scale inhomogenuity of the local electron density (on the scale of $n_e^{-1/2}$). This behaviour is common for all samples we studied and evidences a partial localization of the system under an increase of a magnetic field which starts much earlier than the MIT takes place. Here, the difference between low and high mobility samples manifests itself: in the former, the long-living tails develop at considerably higher filling factors (lower magnetic fields). To make a comparison easier, in Fig. 2 we show the kinetics of magneto-luminescence I(t) for three groups of samples ({B, D}, {B, C, E}, {E, F}), each group has been studied at the same filling factor (ν = 0.19; 0.42; 0.58). Within each group of samples, the luminescence intensity presented in Fig. 2 is normalized to the same integral value (e.g., $\int I^{(B)}(t)dt = \int I^{(D)}(t)dt$ in the first group), which enables us to compare both the form of the long-living tails and their strength in samples with different parameter and at the same or different values of the filling factor.

First of all, after comparing the data taken in different samples at the same ν, we find that the strength of the long-living recombination tail is systematically more significant in lower quality samples. At the same time, the form of the tails at a long enough delay time is essentially the same for different samples studied at the same values of the filling factor. The presentation of the data in Fig. 2 using the log-log plot has been chosen in order to demonstrate that the recombination kinetics at the long-living tails obey the power law [6,15],

$$I(t) = A(t_0/t)^\alpha, \quad t > t_0 \tag{1}$$

and that the exponents α of the power law dependence coincide in different samples, if we make the measurement at the same filling factor, although the strength of the tail (i.e., parameter A) is higher when disorder is stronger. The power law kinetics has been predicted by the earlier theory of recombination in the extreme quantum limit [12], with the exponent α which can be approximated by $\alpha \equiv 1 + \nu \ll 1$. The fact that the exponent of the magnetoluminescence kinetics calculated in [12] depends only on the value of the filling factor was the result of the assumption that in the magnetically frozen state electrons are strongly localized at the interparticle distances - each sitting opposite to its own donor behind a narrow spacer. The spectrum of experimentally derived α's for all the available samples and for a broad range of filling factors, $0.75 > \nu > 0.15$, is presented in the inset to Fig. 2. Although these data correspond to the samples of a very different quality and concentration, the experimentally obtained values of α group near a single filling factor dependence $\alpha(\nu)$. This is what we call the *universality* of the power-law recombination kinetics of localized electrons. In a good approximation, the empirical α has the form $\alpha \approx (1 - \nu)^{-1}$, which is consistent with the theory [12] for $\nu \ll 1$.

As we discussed above, the kinetics of recombination of 2D electrons with acceptor-bound holes in a δ-doped layer is governed by the spectrum of the recombination rates of holes, $\tau^{-1} \propto \rho_e$, which, in its turn, reflects the character of the distribution $f(\rho_e)$ of the local 2D electron density, ρ_e. From this point of view, the observed power-law kinetics of recombination in Eq. (1) can be directly recalculated (using the inverse Laplace transform) into the power-law distribution of the local 2D electron density, $f(\rho_e) \propto \rho_e^{\beta}$ with $\beta = \alpha - 2 \approx (2\nu - 1)/(1 - \nu)$ and the observed universality of the luminescence kinetics has behind it the universality of the distribution of a local density in a localized disordered state of magnetically frozen electrons. Below, we use this fact for measuring the disorder-dependent portions of insulating and liquid phases in the sample. Fig. 3 shows that the extraction of the power-law tail from the total intensity gives the rest of it which has a typical single-exponential behaviour at the intermediate range of the delay time. The parameters (A and α) of the slow recombination tail $I_i(t) = A(t_0/t)^{\alpha}$ were determined from a fit of the power law asymptotics at $t \gg t_0$. The rest of the recombination intensity, $I_l = I(t) - I_i(t)$, is represented in Fig. 3 by solid circles and demonstrates an exponential decay, $I_l(t) \propto \exp(-t/t_0)$ which can be followed over the same number of decades as at $\nu = 1$. The single-rate decay is the property of a microscopically homogeneous density, so that we assign this part of the luminescence intensity to the liquid state of electrons and suggest that the 2D electron system represents a mixture of a liquid and disorder-induced localized phases.

The goal of our analysis is to construct a diagram illustrating the evolution of the portion of the sample area, $S_l(\nu) = [\int I_l dt] / [\int I(t)dt]$, occupied by a liquid under the decrease of the filling factor. The dependence $S_l(\nu)$ determined for six different samples is shown in Fig. 4. The relative amount of the liquid phase in all the structures tends to decrease under an increase of a magnetic field, which manifests the formation of the freeze-out regime. In the low-mobility samples (D,E,F), the localization is much more efficient and develops smoothly in the entire region of filling factors $\nu < 1$. In the better quality samples (A,B,C), the general tendency of the portion of the liquid to decrease with increasing of a magnetic field is accompanied by oscillations associated with the fractional quantum Hall effect states (see Fig. 4). The amount of the recombination events that we assign to the liquid has clear maxima at $\nu = 1/3$ in samples B and C, and at 1/3, 1/5, 2/5 in the sample A. This is a manifestation of the incompressibility of the Laughlin liquid, in which the interaction effects are strong enough (compared to disorder) to keep the electron density microscopically homogeneous [13].

Let us finally focus at the region near the critical filling factor ν_c of the MIT determined using the transport measurements (see Fig. 1). For each single sample, the optically characterized level of localization smoothly changes near ν_c and shows no striking features. However, been compared in a wide variety of structures, the MIT always occurs strikingly close to those filling factors where the continuously varying portion of a liquid fills one half of the sample area. The fundamental property of the percolation transition in macroscopically inhomogeneous 2D media is that it occurs when the insulating component occupies one half of the sample area, that's when the percolation path through the conducting phase is destroyed [11]. Our conclusion is that the magnetic field driven MIT which we observe at $\nu < 1$ is related to the *percolation threshold* through the liquid *in the systems with the phase separation into a liquid and disorder insulator*.

The kinetics of luminescence measured in our best quality sample (WC, see Table 1) indicates several specific features of Wigner crystallization, which were not detected in all other lower quality samples A-F. First of all, at low enough temperatures (T < 300 mK) the recombination kinetics corresponds not to the power law, but to a single exponent process with a very long recombination time, which exponentially depends on the filling factor [6]. Secondly, temperature behaviour of the recombination kinetics in the best quality sample shows an abrupt threshold at critical temperature, whereas no critical temperature could be extracted for samples A-F. In Fig. 5 we show the temperature dependence of the luminescence intensity, measured for two different quality samples (WC and A), at identical values of the filling factor and at similarly long time delay ($t/t_0 = 800$ in both cases). One can see from these data, that the measured temperature dependencies are qualitatively different: for the sample WC the it clearly demonstrates a threshold behaviour, whereas it is smooth for the sample A. This optical fingerprint of Wigner crystallization was used for determination of the phase diagram in our previous publications [16]. It is important to note, that the kinetics of luminescence, measured in the best quality sample WC considerably differs from the kinetics, measured for samples A-F only at very low temperatures [6], whereas at temperatures higher than the critical one they are similar and obey the power law (1). This fact gives us opportunity to estimate the magnetic field dependence of the portions of liquid in this sample (see Fig. 6) and to compare it with other dependencies, measured in other lower quality samples. One can see from Fig. 6 that at $T = 0.5$ K the relative portion of liquid measured at the same filling factor is considerably higher in sample WC than in all other samples, and due to this fact, the MIT in this sample is shifted to smaller values of filling factor. However, for such a comparison it is important to consider the influence of the temperature on the liquid portion, which we investigated in samples A-F. In Fig. 6 we also show the influence of the temperature on the magnetic field dependence of the portion of liquid phase, measured in the samples E and A. It is clear from this figure, that an increase of the temperature results in an increase of the portion of liquid phase and in the shift of the percolation threshold to the smaller filling factors, however, this delocalization effect is negligible in the temperature range $0.05\ K < T < 0.5\ K$, and becomes considerable only for $T > 1 - 2\ K$.

Therefore, we demonstrated, that in the best quality sample the transition into Wigner crystal at critical temperature is accompanied by an abrupt transition in luminescence kinetics, whereas it shows no critical temperature dependence in other lower quality samples in which the magnetic field driven MIT has a percolation nature.

2 Hot Luminescence Versus Inelastic Light Scattering

The great diversity of the fundamental effects in 2D electron systems makes it necessary to use different methods of measurements. In the last few years investigations of 2D systems by means of Raman spectroscopy have increased substantially. This method has been used to measure the characteristic energies of 2D electrons - intersubband splitting, cyclotron energy and energy of the roton minimum in the magnetoplasmon dispersion, spin splitting, and energy gap under the conditions of the Fractional Quantum Hall Effect [17, 18]. The most perfect GaAs/AlGaAs single heterojunctions or quantum wells were used in these works, and the Raman measurements are performed (as a rule) under resonance conditions, specifically, when the energy of the pump photon E_L or of the scattered photon E_S is equal to the energy of the real state. Under such conditions the question of separating the resonance Raman signal from the signal due to hot magnetoluminescence (HML) is nontrivial [19]. This question has still not been studied systematically, and the signals were separated on the basis of the assumptions that the spectral position of the HML lines in fixed magnetic field is also fixed and does not depend on E_L, whereas Raman signal follows E_L. It will be shown below that this simple criterion does not work in the case of high-quality heterostructures.

A series of spectra measured in high-quality GaAs/AlGaAs single heterojunction in different magnetic fields (oriented close to the normal direction) with resonance excitation is shown in Fig. 7a. All spectra contain two lines (the second line is shown only for $B = 2.37\ T$), shifted from the laser line by energy δ and 2δ. The energy δ is determined only by the magnitude of the magnetic field, and its dependence on B is shown in Fig. 7b. The dashed line on this figure corresponds to the magnetic field dependence of the electron cyclotron energy ($m_c = 0.067\ m_0$) and it is evident, that our data agree well with this dependence. The spectral position of the observed line measured as a function of E_L in $B = 6\ T$ is shown in Fig. 7c and it is clear that its spectral position follows the energy of the pump photon and shifted from E_L by the electron cyclotron energy. Note that in fixed B a change in E_L is accompanied by oscillation of the intensity of the line [20]. The properties of the line, shown in Fig. 7, at first glance indicate its Raman origin, however, three alternative possibilities can be discussed: (a) Raman scattering by 2D electrons, (b) Raman scattering by photoexcited electrons in the volume of GaAs, and (c) HML of photoexcited carriers. To choose between the 2D and 3D nature of the effect we employed a standard test - tilting of magnetic field with respect to the interface. If 2D electrons are important the spectral shift would follow the component of B, which is perpendicular to the interface. However, we found, that the properties of the dominant line and the value of δ are nearly the same for all values of the angle α. In Fig. 8a we show spectra measured in total magnetic field $B = 6\ T$, tilted on different angles with respect to the interface (at $\alpha = 0$, B is parallel to the interface). One can see, that dominant lines (HML) are almost insensitive to the tilting of B, however, there is also weaker line (MPL), whose position depend on the angle α and which is due to the Raman scattering by 2D electrons. Returning back to the line HML, we conclude that it has a 3D nature and further analysis of magneto-oscillations of the intensity of this line shows that it arises as a result of the recombination of hot volume photoexcited electrons and heavy holes [20]. Note that Raman-like HML lines dominate the spectrum only for the case when the energy of the pump photon does not exceed band gap of GaAs by the energy of two optical phonons, but a violation of this condition results in a fast energy relaxation of the photoexcited electrons and holes and, therefore, in considerable decrease of the HML intensity (see Fig. 8a). In contrast to the HMP lines, which are almost insensitive to a tilting of the magnetic field, the spectral position of the MPL line is defined by the orientation of B with respect to the interface. The angular dependence of the Raman shift of this line is presented in Fig. 8b. One can see that the Raman shift is maximal in perpendicular magnetic field (and it is very close to the cyclotron energy of 2D-electrons), but in parallel magnetic field it does not equal zero, but reaches minimal value. This minimal energy is due to the finite momentum k, which transferred from the light into 2D system and it is a measure of the plasmon energy $E_{pl} = (n_s e^2 k / 2m\epsilon\epsilon_0)^{1/2}$. One can see from Fig. 8b that a decrease of k result in a decrease of E_{pl} and this give us opportunity to measure the dispersion of the magnetoplasmon in perpendicular, tilted and parallel magnetic field [21]. The dependence of Raman shift on k is a specific property of the Raman process, in contrast to the HML line, which were found to be not sensitive to k. Therefore, this test also allows us to distinguish between Raman and HML lines.

3 Finite-Wave-Vector Raman Spectroscopy of 2D-Electrons

Up to now the studies of the spectra of elementary excitations of 2D electron systems were restricted by low values of the transferred momentum and only a small part of their dispersion curves was accesible experimentally. In the infra-red absorption, and microwave experiments only very long-wavelength modes could be excited. On the other hand, it is known from the existing theories that the dispersion of most of the collective modes in 2D electron systems is defined by the electron-electron interaction which dominates at the wavelengths comparable with the mean interparticle distances. Namely, the electron-electron interaction leads to the formation of spin-density-waves in the Integer Quantum Hall effect regime, of magneto-rotons in the incompressible Laughlin liquid, of magneto-phonons in the magnetically frozen Wigner crystal, and also renormalizes spectra of magnetoplasma and inter-subband modes. Therefore, a finite-wave-vector spectroscopy of 2D electron system in strong magnetic field is necessary. The first experiments on propagation of surface-acoustic-waves (with wavelength of about 300 nm) reveal new features of the electron system in the vicinity of $\nu = 1/2$, which were explained in terms of composite fermions. We developed the finite-wave-vector Raman spectroscopy with the values of transfered momentum, which is close to the values used in surface-acoustic-waves experiments with the aim to use this technique for the investigation of the excitation spectra under conditions of Integer and Fractional QHE.

In Fig. 9 we present typical finite-wave-vector Raman spectra, measured in a single quantum well in different perpendicular magnetic fields at $T = 0.4$ K. Top parts of figures 9a and 9b correspond to a rather low momentum resolution - $\delta k/k = 0.3$, whereas at the bottom parts of these figures we present finite-wave-vector Raman spectra, measured with high resolution of the momentum k, transferred into 2D electron system from the scattered light. The absolute value of the momentum k (we consider here only the component of the total momentum, which is parallel to 2D plane) was defined in our experiment by a relative geometrical configuration of the exciting and collecting fibers and the accuracy $\delta k/k$ was estimated from the size of the excited point and from the angles in which the light was emitted and collected [22]. It is clear from fig. 9 that an increase of the momentum resolution results in narrowing of the Raman line and that spectral position of the line is very sensitive to the absolute value of the momentum *k*. In Fig. 10a we show magnetic field dependence of the energy of magnetoplasmon mode, measured in perpendicular magnetic field at $T = 0.4$ *K* for different values of *k*. One can see from this figure, that at small magnetic fields, the energy of magnetoplasmon considerably deviates from the cyclotron energy E_c (shown by broken line) and the deviation is more pronounced for larger values of *k*. Such behaviour corresponds to the well known dispersion of the hybrid magnetoplasmon mode of 2D electrons in perpendicular magnetic field [23]:

$$E_{mpl}^2 = E_{pl}^2 + E_c^2; \quad E_{pl}^2 = n_s e^2 k / 2m\epsilon\epsilon_0. \tag{2}$$

A striking fact, also visible from Fig. 10a, is a presence of oscillations in the magnetic field dependence of magnetoplasmon energy. To enhance this effect we plot in Fig. 10b the magnetic field dependence of the difference $E^2_{mpl} - E_c^2 = E_{pl}^2$, which should be independent on magnetic field for fixed n_s and *k* (see (2)). However, it is clear from this figure, that the plasmon energy E_{pl} oscillates around average value as a function of the magnetic field. These oscillations were detected for all values of *k*, but they are better pronounced for lager momentum. We have checked that these oscillations are in phase with Shubnikov-de Haas oscillations, measured in σ_{xx}. The oscillations in E_{pl} could indicate the oscillation of concentration of 2D electrons, however, we was able to exclude this scenario by simultaneous investigation of the intersubband Raman scattering, which shows no oscillations of the intersubband energy. This fact is a good evidence of the constancy of the electron concentration since intersubband splitting is sensitive to n_s [24]. It follows from Fig. 10b that an extrapolation of the linear increase of the of E_{pl}^2 as a function of *B*, which is seen under the conditions of Integer Quantum Hall Effect, goes to zero for $B \to 0$. It means that the magnetic field dependence of the ratio E_{pl}/B is quantized as it is shown in Fig. 10c. This quantization is strikingly similiar to the quantization of σ_{xy}, which we measure in transport (see Fig. 10c and 10d). Therefore, we need to conclude that the concentration n_s standing in Eq. (2) corresponds not to the total value, but to the portion of delocalized electrons under the Fermi surface (similar portion of n_s is responsible for quantization of σ_{xy}). Note, that an increase of the temperature up to 2 - 3 K destroys the oscillations in the energy of magnetoplasmon mode, which indicates the importance of the localization effects in the observed phenomenon. Similar magnetic field dependence of E_{pl} was found in another low-density sample, however, in this case the oscillations were detected not only for Integer, but also for Fractional Quantum Hall Effect states. In Fig. 11 we compare the results measured in Raman scattering and in transport. For convenience the product $\sigma_{xy}{\cdot}B$ is also presented in Fig. 11b. One can see that the oscillations of E^2_{pl} derived from Raman scattering are similar to the features of $\sigma_{xy}{\cdot}B$, measured in transport both for Integer and Fractional QHE. However, there is one, but very essential discrepancy between these results - in a vicinity of $\nu = 1/2$ the transport coefficient $\sigma_{xy}{\cdot}B$ shows no anomaly, whereas it exists in E_{pl}. An important property of the anomaly observed in Raman scattering around $\nu = 1/2$ than for Integer and Fractional QHE states (see Fig. 11a). The dependence of the amplitude of the anomaly in plasmon energy ΔE^2_{pl} as a function of the momentum k, measured for different values of filling factor is presented in Fig. 12. It is clear from this Figure that for both Integer and Fractional QHE states the value of $\Delta E^2{}_{pl}$ is linearly increases with momentum k, whereas for $\nu = 1/2$ this dependence is superlinear and close to the parabolic one (dashed line in Fig. 12). It means that the width of the plateau measured by Raman analog of the Hall conductivity ($E^2{}_{pl}/B$) is independent on k and it is proportional to k at half filling of the Landau level. Note, that anomaly around $\nu = 1/2$ observed in the experiments on propagation of surface-acoustic-waves [25], which was explained in terms of geometrical resonance of composite fermions [26] also appears only for short wavelengths and develops nonlinear as a function of k. In our experiment no conditions of resonance is fulfilled and we believe that the anomaly of the spatial dispersion of magnetoplasmon mode in the electron system at half filling of the Landau level is responsible for the observed phenomenon. Therefore, we observed a quantization of Raman analog of the Hall conductivity and the width of the corresponding plateau is independent on momentum k for Integer and Fractional QHE, whereas it is approximately proportional to k at half filling of the Landau level.

References

[1] Yu. Lozovik and V. Yudson, JETP Lett. **22**, 11 (1975(; P. Lam, S. Girvin, Phys. Rev. B**30**, 473 (1984)

[2] D.C. Tsui et al., Phys. Rev. Lett. **48**, 1559 (1982)

[3] R.L. Willett et al., Phys. Rev. B**38**, 7881 (1988)

[4] H.W. Jiang et al., Phys. Rev. Lett. **65**, 633 (1990)

[5] V.J. Goldmann et al., Phys. Rev. Lett. **65**, 2189 (1990)

[6] I.V. Kukushkin et al., Phys. Rev. Lett. **72**, 3594 (1994)

[7] H.W. Jiang et al., Phys. Rev. Lett. **65**, 633 (1990); H.W. Jiang et al., Physica B **197**, 449 (1994)

[8] Glozman et al., Phys. Rev. Lett. **74**, 594 (1995); Shahar et al., ibid., 4511 (1995)

[9] H. Levine et al., Phys. Rev. Lett. **51**, 1915 (1983); D.E. Khmelnitskii, JETP Lett. **38**, 552 (1983)
[10] A. Karlhede et al., in Correlated Electron Systems, ed. V.J. Emery (World Scientific: 1992)
[11] A.L. Efros and B.L. Shklovskii, Electronic Properties of Doped Semiconductors (Springer: Berlin 1984)
[12] V.I. Fal'ko, Phys. Rev. B **49**, 2242 (1994)
[13] B. Janovici, Phys. Rev. Lett. **46**, 386 (1981)
[14] I.V. Kukushkin et al., Phys. Rev. B **40**, 7788 (1989); Europhys. Lett. **23**, 211 (1993)
[15] I.V. Kukushkin et al., Pis'ma ZhETF **63**, 118 (1996) (JETP Lett. **63**, 132 (1996))
[16] I.V. Kukushkin et al., Physica B **184**, 38 (1993); Europhys. Lett. **23**, 211 (1993)
[17] A. Pinczuk, Festkörperprobleme **32**, 45 (1992)
[18] A. Pinczuk et al., Semicond. Sci. Technol., **9**, 1865 (1994)
[19] M. Cardona, Light Scattering in Solids, Springer-Verlag, 1975
[20] V.E. Kirpichev et al., JETP Lett., **62**, 876 (1995)
[21] V.E. Kirpichev et al., JETP Lett., **63** (1996)
[22] I.V. Kukushkin et al., JETP Lett., **63** (1996)
[23] A.V. Chaplik, JETP, **35**, 395 (1972)
[24] S. Ernst et al., Phys. Rev. Lett. **72**, 4029 (1994)
[25] R. Willett et al., Phys. Rev. Lett. **71**, 3846 (1993)
[26] B.I. Halperin et al., Phys. Rev. **47**, 7312 (1993)

TABLE 1. Samples parameters: density n_e, mobility μ

sample	A	B	C	D	E	F	WC
n_e ($x10^{11}cm^{-2}$)	0.55	0.60	0.80	0.71	0.60	1.43	0.63
μ ($x10^5cm^2/Vs$)	12.1	3.85	0.94	0.43	0.21	0.11	30.5

Figure Captures

Fig. 1 Magnetic field dependence of the conductivity and current measured in samples B(a) and E(b) for different temperatures ((a) - T = 150; 100; 50

mK, and (b) - T = 100; 50 mK). The critical filling factor ν_c of the MIT is indicated. Insets represent the raw data of the time-dependent magneto-luminescence kinetics measured in the same samples at filling factors $\nu <$ 1 at T = 50 mK.

Fig. 2 Comparison of the recombination kinetics in groups of samples taken at three filling factors and at T = 50 mK. The integral luminescence intensity is normalized by the same value within each group. Dashed lines illustrate a draft fit of the long-living tail using a power law. Inset shows the dependence of α^{-1} on filling factor measured in all the studied samples.

Fig. 3 Example of a separation of the time-dependent magneto-luminescence intensity $I(t)$ into the power-law and exponential parts, $I = I_i + I_l$. Solid line correspond to the fit of the long-living tail of $I(t)$ by the power law $I_i(t) = A(t_0/t)^{\alpha(\nu)}$. Solid circles and dashed line are the dependence $I_l(t) = I_i(t)$ at an intermediate range of the delay time and a singe-exponential fit to it.

Fig. 4 The filling factor dependence of a portion of the sample area occupied by a liquid in all studied structures. Black squares indicate the points taken exactly at the filling factors of the MIT detected in the transport measurements. Dashed line and arrows illustrate the classical condition of a percolation threshold in a two-component systems mixed of a metallic and insulating phases.

Fig. 5 Temperature dependence of the luminescence intensity, measured for two different quality samples at close values of filling factor $\nu = 0.16$ and at similarly long time delay - $\Delta t = 800t_0$.

Fig. 6 Variation of the portion of liquid on filling factor, measured in different samples for various temperatures.

Fig. 7 (a) Emission spectra measured for different values of B and (b) magnetic field dependence of the energy shift of the line obtained for fixed pump photon energy $E_L = 1555.3\ meV$. (c) The dependence of the spectral position of the line as a function of E_L measured at $B = 6\ T$.

Fig. 8 (a) Variation of the emission spectra, measured for two values of E_L (1574.4 meV) in the fixed total magnetic field $B = 6\ T$ tilted on different angles α with respect to the interface. (b) Angular dependence of the Raman shift measured in the fixed total magnetic field $B = 6\ T$ for different values of k.

Fig. 9 Raman spectra, measured in a single quantum well (25 nm) at $T = 0.4\ K$ in perpendicular magnetic field $B = 2.5\ T$ (a) and $B = 10\ T$ (b) for different values of momentum k. $\delta k/k$ - is the relative momentum resolution.

Fig. 10 Magnetic field dependence of the Raman shift (a), of the plasmon energy $E^2_{pl} = E^2_{mpl} - E^2_c$ (b), of the value E^2_{pl}/B (c) and of the Hall conductivity (d), measured in a sample with electron concentration $4.6 \cdot 10^{11}\ cm^{-1}$ at T = 0.4 K. Optical data are presented for different values of the momentum k.

Fig. 11 Magnetic field dependence of the plasmon energy E^2_{pl} (a), of the Hall conductivity σ_{xy} and of the product $\sigma_{xy} \cdot B$ (b), measured in a sample with electron concentration $1.1 \cdot 10^{11}\ cm^{-1}$ at $T = 0.4\ K$.

Fig. 12 The dependence of the amplitude of the anomaly in plasmon energy ΔE^2_{pl} on the momentum k, measured for different values of filling factor.

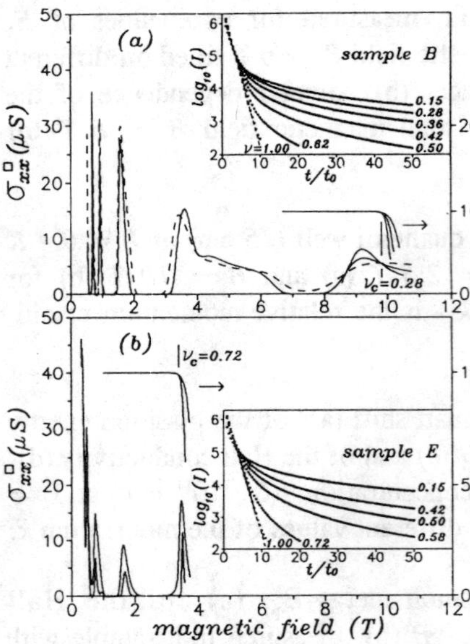

Fig.1

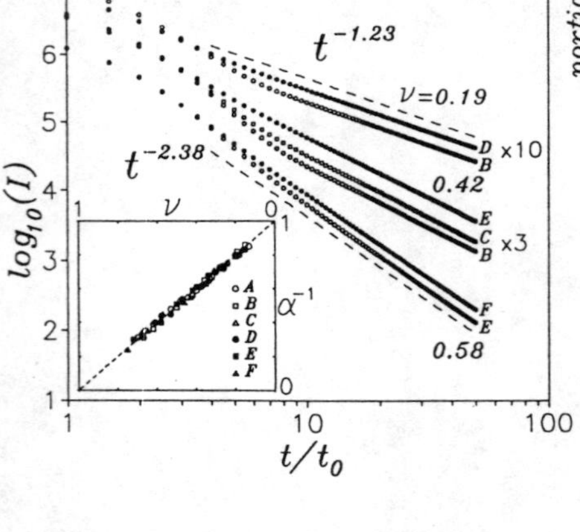

Fig.2

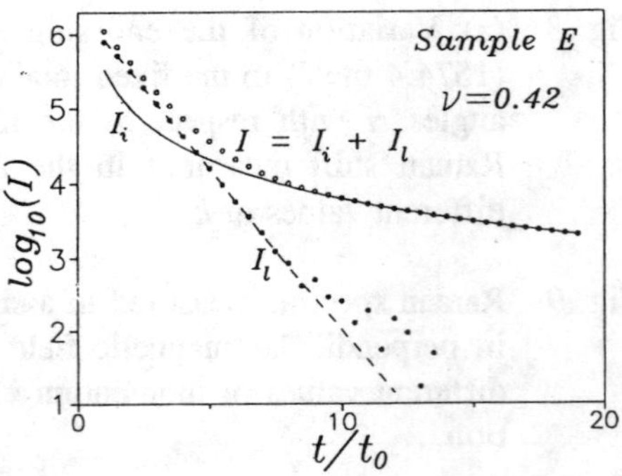

Fig.3

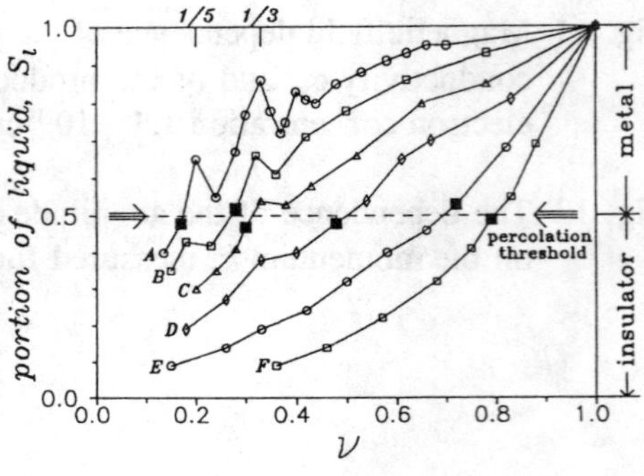

Fig.4

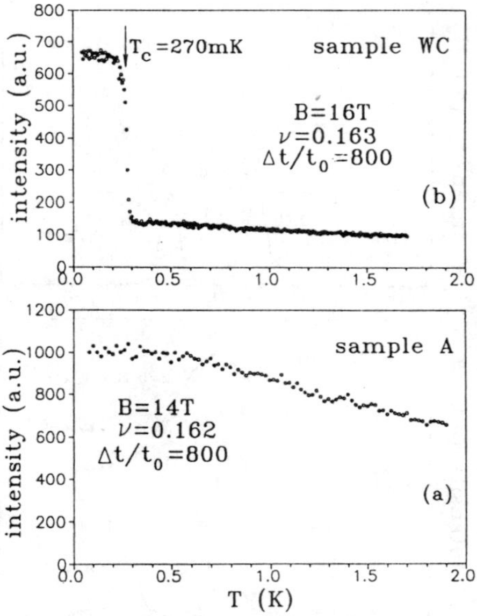

Fig.5

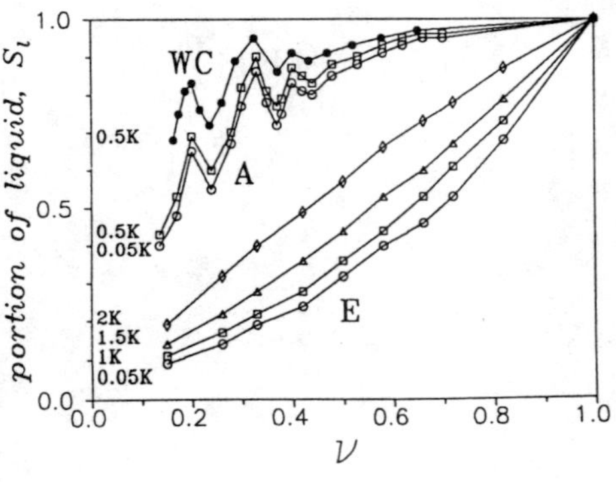

Fig.6

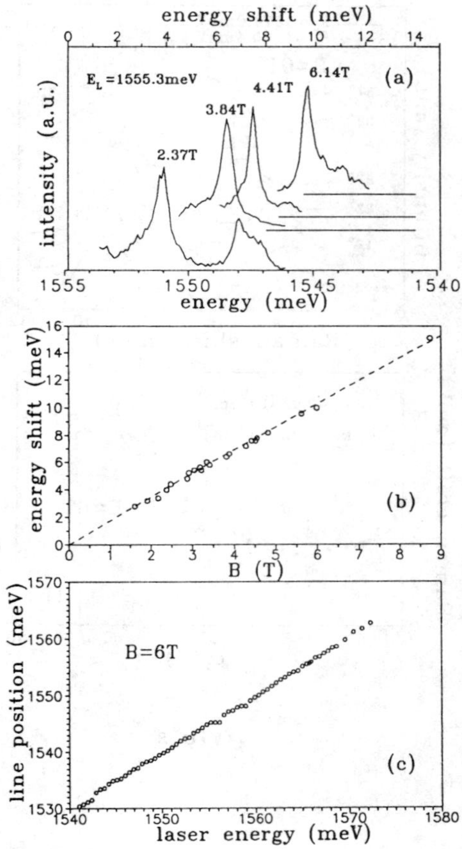

Fig.7

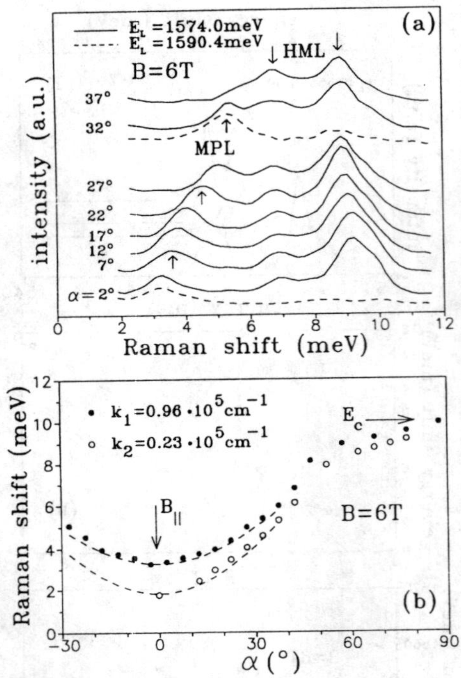

(a)
E_L =1574.0meV
E_L =1590.4meV
B=6T
HML
MPL
37°
32°
27°
22°
17°
12°
7°
α=2°
intensity (a.u.)
Raman shift (meV)
(b)
k1=0.96·10^5cm^-1
k2=0.23·10^5cm^-1
E_c
B_||
B=6T
Raman shift (meV)
α (°)

Fig. 8

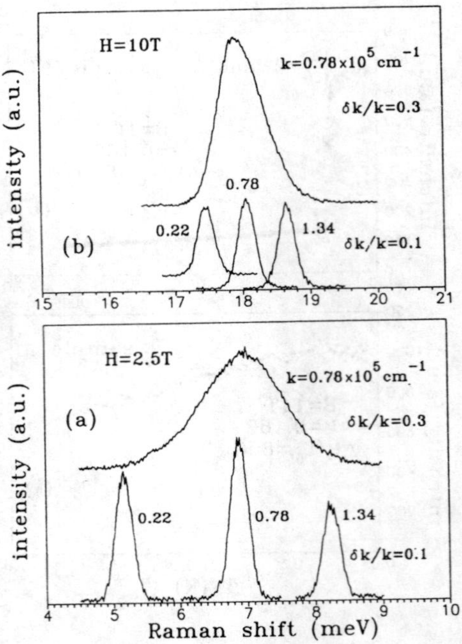

H=10T
k=0.78x10^5cm^-1
δk/k=0.3
(b)
0.22
0.78
1.34
δk/k=0.1
intensity (a.u.)
H=2.5T
k=0.78x10^5cm^-1
(a)
δk/k=0.3
0.22
0.78
1.34
δk/k=0.1
intensity (a.u.)
Raman shift (meV)

Fig. 9

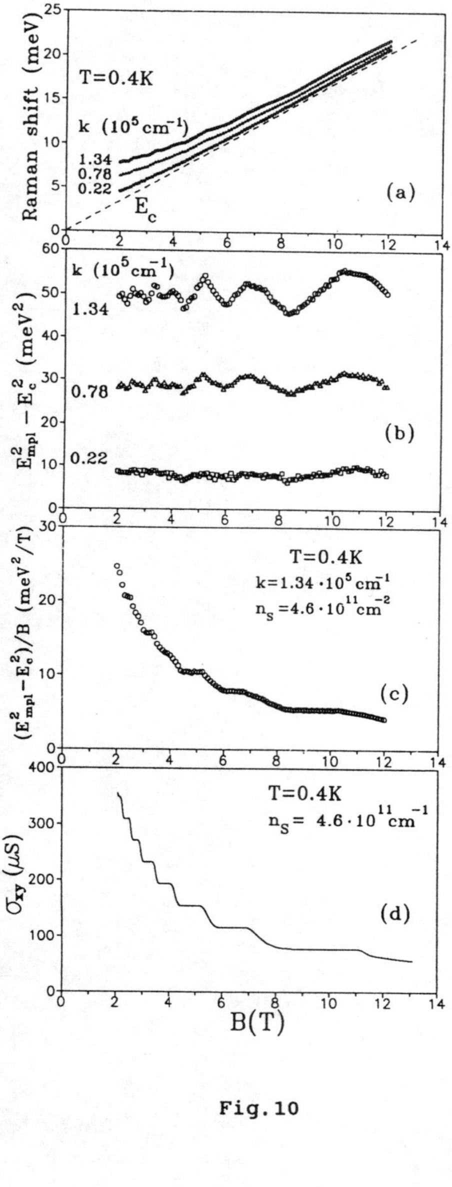

Fig.10

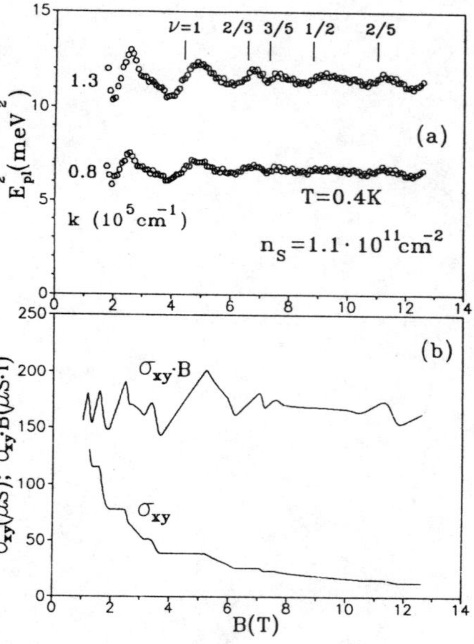

Fig.11

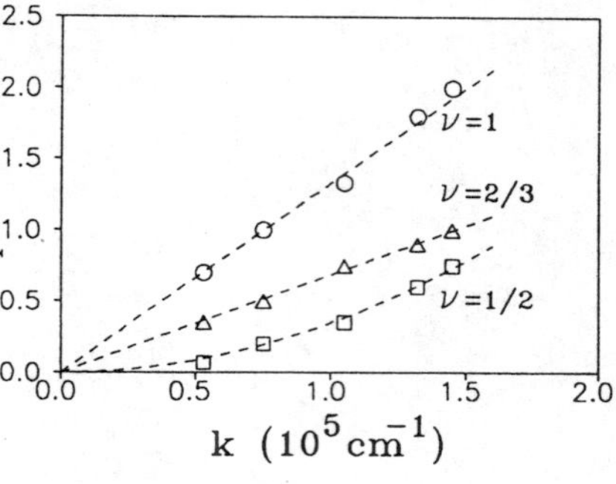

Fig.12

ANISOTROPY OF ZEEMAN SPLITTING IN ASYMMETRIC $GaAs/Al_{0.35}Ga_{0.65}As$ QUANTUM FILMS

S. Hallstein and M. Oestreich
Max-Planck-Institut für Festkörperforschung
Heisenbergstraße 1, D-70569 Stuttgart, Germany

W. W. Rühle
Fachbereich Physik der Philipps-Universität
Renthof 5, D-35032 Marburg, Germany

K. Köhler
Fraunhofer-Institut für Angewandte Festkörperforschung
Tullastraße 72, D-79108 Freiburg, Germany

We study by means of quantum beat spectroscopy the anisotropy of the electron Zeeman splitting for a magnetic field in the plane of a $GaAs/Al_{0.35}Ga_{0.65}As$ quantum film with triangular confinement potential. The combination of an asymmetric confinement potential and the lack of inversion symmetry in GaAs causes a different Zeeman splitting for the [110] and the [-110] crystal direction.

The Zeeman splitting in bulk semiconductors with zinc-blende structure is in a first approximation isotropic with a value of -0.44 for the electron Landé g factor of GaAs [1,2,3]. Reduction of the crystal symmetry by confinement of the electron in quantum films and quantum wires results in an obvious anisotropy of the electron g factor *parallel* and *perpendicular* to the confinement direction[4,5,6]. The interaction of the conduction band with the warped valence band leads to a small difference of the Zeeman splitting for a magnetic field $\vec{B} \parallel [100]$ and $\vec{B} \parallel [110]$ in bulk as well as in the plane of a quantum film. Our experiments, however, show that this anisotropy is smaller than expected [7,8]. We show for the first time experimentally, that an anisotropy of the electron g factor *in the plane* of a $GaAs/Al_{0.35}Ga_{0.65}As$ quantum film can be *induced*, if an electric field is applied in growth direction. This symmetry reduction together with the lack of inversion symmetry in GaAs leads to an electron g factor distinctly different for $\vec{B} \parallel [110]$ and $\vec{B} \parallel [-110]$ resulting in off-diagonal elements in the g-factor tensor [9].

The sample investigated consists of a 10 nm GaAs quantum film embedded in $Al_{0.35}Ga_{0.65}As$ barriers. Electric fields up to $9 * 10^4$ V/cm are applied perpendicular to the quantum-film plane, causing an asymmetric potential in growth direction (shown as inset in Fig.1). The sample can be rotated around the sample normal (the [001] direction) in a superconducting magnet with a maximum field of 16 Tesla. We use Voigt geometry with the light propagation direction perpendicular to the quantum film plane and to the magnetic field. The photoluminescence is analyzed by a circular polarizer, spectrally dispersed by a 0.32 m spectrometer, and temporally resolved by a synchroscan streak-camera with a time resolution of 7 ps.

Circularly polarized light pulses (FWHM 2 ps) of an 80 MHz Ti:sapphire laser create, according to the selection rules, a preferential spin orientation in the conduction band parallel and antiparallel to the excitation axis. The magnetic field forces the electron magnetic moment to precess with the Larmor frequency $\omega_L = g_e \mu_B B/\hbar$ around the magnetic field axis leading to a periodic change of the spin orientation parallel and antiparallel to the excitation direction. This change results in an oscillation between left- and righthanded cicular polarization of the emitted luminescence and is observed in a beating of time-resolved signal, if only one circular polarization is detected. The beating frequency yields the absolute value of the g factor.

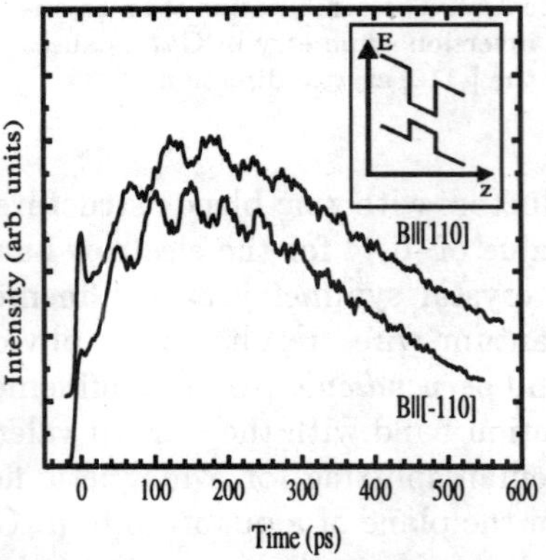

Figure 1: Beating in the circularly polarized luminescence for a magnetic field of 8 Tesla parallel to the [110] and [-110] crystal directions in the quantum-film plane. Inset: Asymmetric potential.

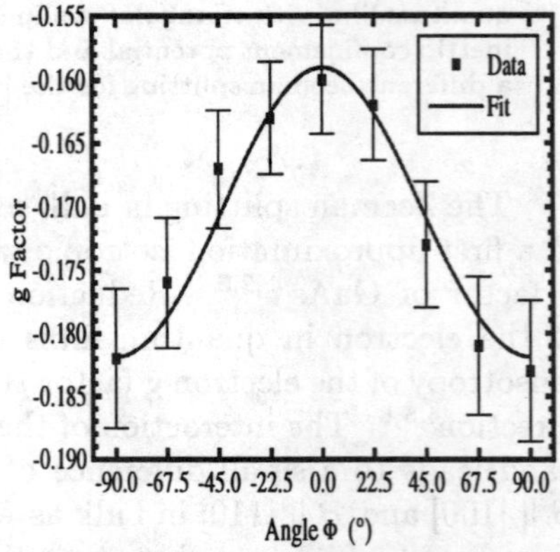

Figure 2: Anisotropy of the g factor in the quantum-film plane at an electric field of $5.6 * 10^4$ V/cm applied in growth direction. The angle of $\Phi = 0°$ corresponds to $\vec{B} \parallel [110]$.

Figure 1 shows the beating of the time-resolved excitonic luminescence for two orientations of a magnetic field of 8 Tesla in the plane of the quantum film ($\vec{B} \parallel [110]$ and $\vec{B} \parallel [-110]$) with an electric field of $5.6*10^4$V/cm applied in growth direction. The different beating frequencies clearly show that the [110] and the [-110] crystal directions are no longer equivalent, whereas no detectable anisotropy is found without electric field. The angular dependence of the g factor deduced from the spin-quantum-beat frequencies is shown in

Fig. 2. Note that the periodicity of the g factor is 180° compared to 90° for the non detectable anisotropy caused by valence band warping. A theoretical angular dependence

$$g_e(\phi) = \sqrt{g_{xx}^2 + g_{xy}^2 + 2g_{xx}g_{xy}\cos(2\phi)} \tag{1}$$

is expected according to the generalized expression for the Larmor frequency $\omega_L = \frac{\mu_B}{\hbar}|\hat{g} \cdot \vec{B}|$, where $\hat{g}$ is a tensor. ϕ is the angle between the magnetic field and the [110]-direction and $\vec{x} \parallel [100]$. The solid line is a least square fit according to Eq.1 and yields the tensor elements $g_{xx} = g_{yy} = -0.170 \pm 0.001$ and $g_{xy} = g_{yx} = 0.011 \pm 0.001$, where the errors are of statistical nature.

First order perturbation theory of the Hamiltonian discribing the spin splitting yields the following expression for the off-diagonal elements of the g-factor tensor [9]:

$$g_{xy} = \beta e(< z >< p_z^2 > - < p_z^2 z >) \tag{2}$$

with $\beta = 2\gamma_c/\hbar^3\mu_B$, where γ_c is the spin splitting constant of the momentum dependent spin splitting of an electron ($\propto p^3$) in semiconductors without inversion symmetry. Numerical calculations yield $g_{xy} = 0.045$ for the off-diagonal element of the g-factor tensor, if we use an electron wavefunction in a triangular potential corresponding to a quantum film with an applied electric field of $5.6 * 10^4$V/cm and a spin splitting constant [7,10] of $\gamma_c = 24.5$eV/Å^3 . The difference between the theoretical and experimental value might be caused by the calculation in first order perturbation theory and the neglection of excitonic effects: Localization of the electron and the hole in an exciton leads to an averaging over anisotropy.

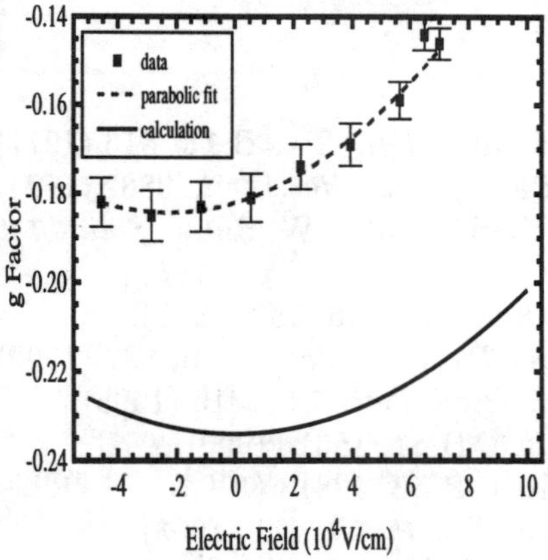

Figure 3: Dependence of the electron g factor on an electric field applied perpendicularly to the quantum-film plane ($\vec{B} \parallel [110]$).

The variation of the electron g factor with electric field strength is shown in Fig. 3. The experimental field dependence of the data is well reproduced by a parabola (dashed line) and caused by the energy dependence of the g factor. A simple model yields the solid line in Fig. 3: A linear approximation $g(\Delta E) = g_0 + a\Delta E$ is assumed for the energy dependence of the g factor, where $\Delta E = E_n - E_g^{'}(z)$, with E_n being the n-th quantized level and $E_g^{'}(z)$ the effective band gap $E_g^{'}(z) = E_g(z) + \frac{dV}{dz}z$. The terms V and E_g are the electric field potential and the band gap respectively. We use $g_0 = -0.44$ and $a = 6.3/eV$ for the quantum film and $g_0 = 0.50$ and $a = 2.0/eV$ for the barrier[5,7]. An expectation value of the electron g factor

$$< g >=< n|g_0|n > + < n|a(E_n - E_g)|n > -\frac{dV}{dz} < n|az|n > \qquad (3)$$

is obtained, where the first term is the expectation value of the g factor at the band gap, the second term describes the energy dependence without electric field, and the third term is introduced by the triangular potential. This simple model reflects well the *electric-field dependence* of the g factor although the absolute values we measure are smaller (see Fig.3).

In conclusion, we have shown that an electric field perpendicular to a quantum film plane causes an anisotropy of the Zeeman splitting in the film plane of a semiconductor without inversion symmetry. The g factor becomes a tensor with nonzero off-diagonal elements. The electron g factor exhibits a parabolic electric-field dependence, which is reasonably described by a simple model.

This work was partially supported by the "Deutsche Forschungsgemeinschaft" in the framework of the "Schwerpunkt Quantenkohärenz in Halbleitern".

References

1. Claude Weisbuch and Claudine Hermann, *Phys. Rev.* B **15**, 816 (1977).
2. A. P. Heberle, W. W. Rühle, and K. Ploog, *Phys. Rev. Lett.*, 3887 (1994).
3. R. M. Hannak, M. Oestreich, A. P. Heberle, and W. W. Rühle, *Solid State Commun.* **93**, 331 (1995).
4. V. K. Kalevich and V. L. Korenev, *JETP Lett.* **56**, 253 (1992).
5. E. L. Ivchenko and A. A. Kiselev, *Sov. Phys. Semicond.* **26**, 827 (1992).
6. M. Oestreich and W. W. Rühle, *Phys. Rev. Lett.* **74**, 2315 (1995).
7. P. Pfeffer and W. Zawadzki, *Phys. Rev.* B **41**, 1561 (1990).
8. P. Pfeffer and W. Zawadzki, *Phys. Rev.* B **53**, 12813 (1996).
9. V. K. Kalevich and V. L. Korenev, *JETP Lett.* **57**, 571 (1993).
10. D. Richards, B. Jusserand, H. Peric, and B. Etienne *Phys. Rev.* B **47**, 16028 (1993).

MAGNETO-LUMINESCENCE OSCILLATIONS AND QUANTUM HALL EFFECT IN ASYMMETRICAL MODULATION DOPED GaAs-GaAlAs QUANTUM WELLS

M. KAMAL-SAADI, P. VICENTE, B. COUZINET, A. RAYMOND
GES, URA CNRS 357-UM II, 34095 - Montpellier-Cédex, France

B. ETIENNE
L2M-CNRS, 98220 - Bagneux, France

We have investigated, both by transport and photoluminescence experiments, GaAs-GaAlAs modulation doped asymmetrical single quantum wells (QW) of 250 Å wide. In these high mobility structures we observed under continuous A_r^+ laser excitation the Shubnikov-de-Haas (SdH) effect and the Quantum Hall Effect (QHE) with well resolved plateaus. In the same conditions of excitation, the fan chart of the observed interband transition energies as a function of magnetic field (B) exhibits strong non linearities with quasi plateaus well correlated with the plateaus of the QHE. Assuming that under continuous illumination (CI) the degenerate 2D electron gas (2DEG) is not isolated, we analysed photoluminescence and transport experiments with the assumption that when B varied the Fermi level (E_F)remained constant in the whole structure even if N_s varied too.

It is commonly assumed that the QHE is a single-particle localization phenomenon of 2D electrons. The main question is then to know if the 2D electrons are localized inside or outside the 2D-system. The first hypothesis implies an isolated 2DEG with a density N_S magnetic field independent: due to disorder there exist 2D localized states which give rise to finite mobility gap between extended states in different Landau levels (LLs)[1]. The plateaus of the QHE are a consequence of the pinning of the Fermi level between two LLs by the localized 2D states, distributed in the tails of the LLs and in between them. In the second hypothesis, the 2DEG is non isolated, the pinning of the Fermi Level is due to existence of one or several reservoirs external to the 2DEG. The Hall plateaus are a direct consequence, when B varies, of the exchange of electrons between the reservoir(s) and the 2D system[2].

Last years magneto-photoluminescence experiments show non-linearities in the B dependence of interband transition energies[3,4]. For symmetrical modulation doped quantum wells (MDQW), of 250 Å width Uenoyama and Sham[5] as well as Katayama and Ando[6] interpreted this phenomenon in terms of many-body effects and particularly Coulomb electron-hole interactions:

they obtained non linearities which took place exactly at the even values of the filling factor ν.
In our study we used GaAs-$Ga_{0.67}Al_{0.33}As$ single QWs, used by P. Vicente et al.[7]. The fan chart of the observed interband energies as a function of B is shown in Fig.1. In Fig.2 we have reported SdH (ρ_{xx}) and QHE (ρ_{xy}) experimental curves obtained in the dark after illumination. We must notice that under CI the ρ_{xx} and ρ_{xy} structures are slightly distorted because of the conduction by photo-created holes; the 2DEG density N_s was also slightly larger

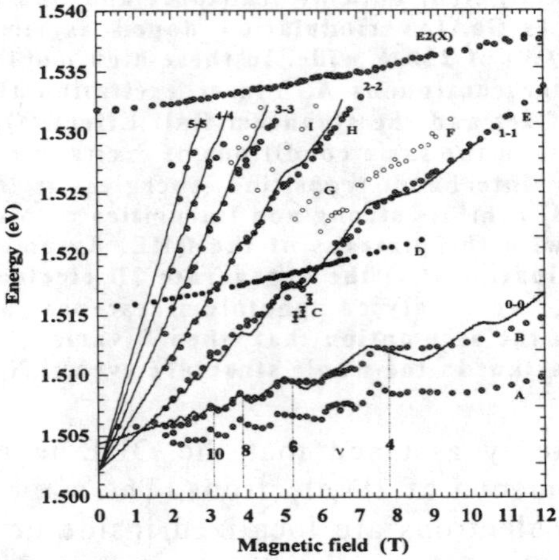

Figure 1: Magnetic Field dependence of the interband transition energies: points correspond to experimental results; lines correspond to theoretical results ($Ns(B=0T)=7.6x10^{11}$ cm^{-2}).

($N_s=7.6x10^{11}$ cm^{-2}). The nature of the different transitions reported in Fig. 1 has been recently analysed by Vicente et al[7]. We are here concerned by the non-linearities or quasi-plateaus observed on the free electron-hole interband transitions curves marked by the Landau number of the involved electron and hole LLs (n-n). In our asymmetrical QWs the outset of these quantum plateaus coincides with the outset of the QHE plateaus under CI.

To analyse magneto-transport and magneto-optic experiments under CI, we assumed that E_F is constant in the whole structure even if the 2DEG density N_s varied as a function of B (equivalent to thermo-dynamical equilibrium). This assumption agrees with the results of Kukuskin et al[8]and Hayne et al[9]. We assumed further that all the 2D states are not localized. As a consequence of the existence of gaps in the energy spectrum of the 2D states, the density of 2D electrons N_s oscillates when B varies.

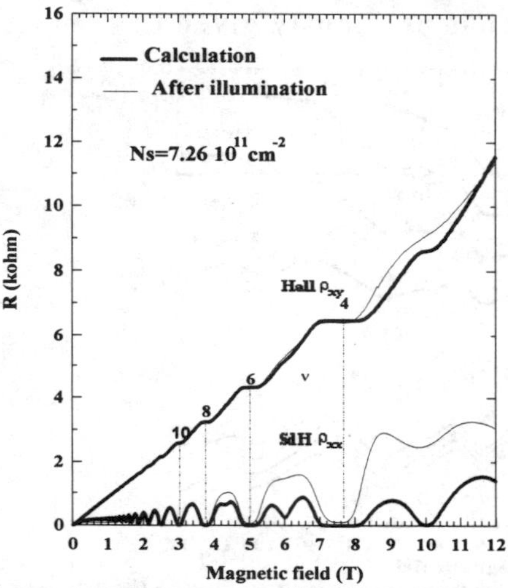

Figure 2: Transport coefficients ρ_{xx} and ρ_{xy} after illumination. The thin lines represent experimental results. The broad lines represents theoretical curves.

To calculate the energy profile and the subband energies, for a given value of N_s, we solved in a self-consistent way the effective Hamiltonian equation, in the parabolic approximation taking into account the electron-electron interaction, and the Poisson equation. In our calculation, we neglect the B dependance of electron-electron interaction (see refs 5, 6). The oscillations of N_s we previously discussed, induce variations of the energy profile and consequently of the electron and hole subband energies. To calculate the 2DEG concentration N_s, we used a density of state in the form of a sum of Gaussian peaks and we took into account the spin splitting enhancement by the exchange interaction of electrons to calculate the g factor. The transport coefficients σ_{xx}, σ_{xy}, ρ_{xx} and ρ_{xy} have been calculated as previously made by Raymond and Sibari[2]. On Fig. 2 we have reported the SdH (ρ_{xx}) and QHE (ρ_{xy}) curves obtained by the theoretical investigations in the dark after illumination (N_s= 7.26 10^{11} cm-^{2}). We can notice the excellent agreement for the Hall plateaus width and shape and for the SdH oscillations up to $\nu = 6$ (B = 5T). For larger values of B, our study shows that the disagreement is due to the B dependence of the broadening parameter Γ of the LLs and consequently of the calculation of the g factor enhancement: in order to simplify the calculations we took Γ equal to a constant directly given by the mobility, in the whole B range, when Ando and Uemura[10] give for short-range scattering $\Gamma \sim \sqrt{B}$ in the high field region.

Figure 3 shows the LLs of confined electrons and holes versus B, as obtained from the calculations assuming a pinned Fermi level. The

Δn = 0 energy difference gives us the interband transitions reported on Fig. 1. In this case too we obtained a very good agreement particularly up to 5T for the reasons previously given.

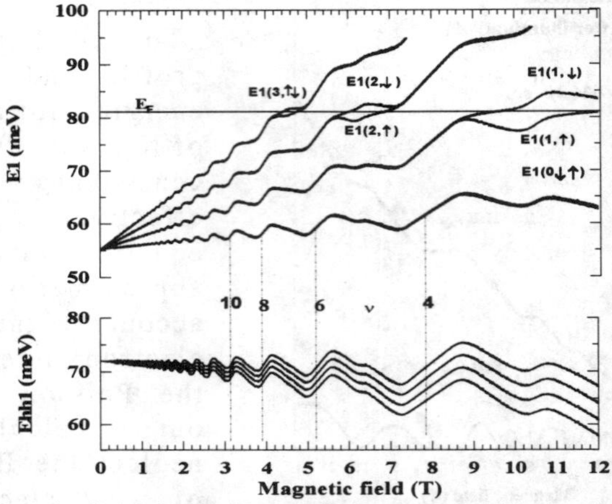

Figure 3: Magnetic field dependence of electrons and heavy holes LLs, in the non isolated QW assumption, for Ns(B=0T)=7.6 10^{11} cm^{-2} (under CI).

To conclude, This study shows that for asymmetrical wide MDQWs under continuous illumination at helium temperature, the assumption of a non isolated QW allows to account without fitting parameter for the magneto-luminescence oscillations as well as for the Hall plateaus width and shape and for the SdH oscillations without invoking the existence of 2D localized states.

REFERENCES

(1) R.E. Prange and S.M. Girvin (Ed).The Quantum Hall Effect Springer-Verlag, Berlin (Heidelberg/New-York 1987).
(2) A. Raymond and H. Sibari. Phys. Stat. Sol (b) 183, 159 (1994) and references there in.
(3) C.H. Perry et al. "High Magnetic Fields in Semiconductor Physics p. 202, Ed. G. Landwehr Springer Verlag, Berlin 1987
(4) M.S. Skolnik et al. Solid State Commun. 67, 637 (1988).
(5) T. Uenoyama and L.J. Sham, Phys. Rev. B39, 11044 (1989).
(6) S. Katayama and T. Ando, Solid State Commun. 70, 97 (1989).
(7) P. Vicente et al. Solid State Commun. 96, 901, (1995).
(8) I.V. Kukuskin et al. Phys. Rev. B40, 4179 (1989).
(9) M. Hayne et al. Phys. Rev. B.50 (1994).
(10) T. Ando and Y. Uemura, J. Phys. Soc. Japan, 36, 959 (1974).

CONDITIONS FOR OPTICAL DETECTION OF THE QUANTUM HALL EFFECT IN TWO-DIMENSIONAL HOLE SYSTEMS

Y.V. PONOMAREV, C.M. TOWNSLEY, A. USHER
Department of Physics, University of Exeter, Stocker Road, Exeter, EX4 4QL, UK

B.L. GALLAGHER, M. HENINI
Department of Physics, University of Nottingham, Nottingham, NG7 2RD, UK

G. HILL
Department of Electronic and Electrical Engineering, University of Sheffield, Mappin Street, Sheffield, S1 4DU, UK

Photoluminescence (PL) measurements have been performed on a high-mobility two-dimensional (2D) hole gas having a range of densities from 4×10^{10} to $2 \times 10^{11}\,cm^{-2}$, depending on illumination conditions. Within this range very significant changes occur in the system's optical response to the quantum Hall effect: At low densities, the PL is indicative of exciton recombination, and there is no optical signature of the quantum Hall effect. At higher densities, recombination of free carriers is observed, and pronounced minima are observed in the PL intensity at magnetic fields corresponding to both integer and fractional Landau-level occupancies, accompanied at the highest density by upward shifts in the PL energy. The trend from excitonic to free-carrier recombination, and the strengthening of the system's optical response to the quantum Hall effect as the 2D hole gas becomes more dense, is explained in terms of the strength of the interactions between the 2D holes and the photoexcited electrons. These interactions are affected both by screening and by the change in shape of the confining potential.

1 Introduction

Photoluminescence (PL) studies on two-dimensional electron systems (2DESs) have revealed various responses to the integer and fractional quantum Hall effects including upward[1] and downward[2] shifts in the recombination energy and large modulations in peak intensity.[3] Following the theoretical work of Apal'cov and Rashba,[4] the downward shifts in energy seen in δ-doped single heterojunctions, in which the carriers are held some 250 Å from the interface, were directly associated with the quantum Hall energy gaps. The upward shifts observed in quantum wells (QWs) result from the perturbation of the 2D system by the photoexcited carriers which, in these structures, are considerably closer to the interface.[5] The intensity modulations are a consequence of the weakened screening of the 2D system in the incompressible quantum Hall states.[5] In this paper we present the evolution of PL spectra in a two-dimensional hole system (2DHS) as a function of density (p_S) and compare our results with those obtained previously on 2DESs.

2 Experimental and sample details

A 150 Å single QW, with *p*-type one-side modulation doping and an undoped spacer width of 200 Å, provided the high mobility 2DHS for this experiment. The PL measurements were performed using excitation either above (ABG) or below (BBG) the $Al_{1-x}Ga_xAs$ band gap. Optical fibres delivered the excitation and collected the PL signal. The samples were cooled to a base temperature of 16 mK and under excitation the hole temperature was estimated to rise to approximately 100 mK. A magnetic field was applied perpendicular to the plane of the 2DHS, and simultaneous transport measurements were carried out, using a Hall-bar geometry ensuring uniform illumination of the 2DHS between the contacts.

3 Results

Figure 1 shows three PL spectra for a range of p_S using ABG excitation (the value of p_S being controlled by the excitation power), and a fourth spectrum obtained using BBG excitation which gives the highest value of p_S (p_S is independent of incident power for BBG excitation). At low p_S a twin-peak structure is observed. This structure is more pronounced in wider spacer samples (with even lower values of p_S) and has been attributed to recombination of neutral and charged excitons.[6] At higher values of p_S a single PL line, with a distinct low energy tail, is observed due to recombination of free carriers. The width of this PL structure (an approximate measure of the Fermi energy) increases as p_S increases.

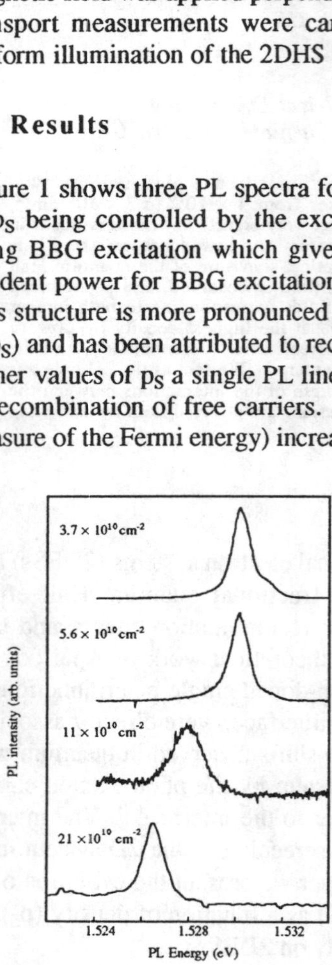

Figure 1: Top three spectra: 300 mK PL spectra over a range of p_s, obtained by ABG excitation of different powers. Bottom spectrum: 100 mK spectrum obtained using BBG excitation. p_s was determined by simultaneous magnetotransport.

The influence of p_S on the observation of the quantum Hall effect is demonstrated in figures 2 and 3. We present the magnetic field results for the low and high p_S limits respectively. In each case, the dependences of the integrated intensity and the normalized PL peak position on magnetic field are shown, together with simultaneous magnetotransport data. In both cases, at low temperature there are distinct minima in the integrated intensities at integer and fractional Landau-level filling factors, ν, which correlate well with the magnetotransport. These become progressively less distinct as the temperature is raised (representative 4.2 K data are given in figures 3a, 3b).

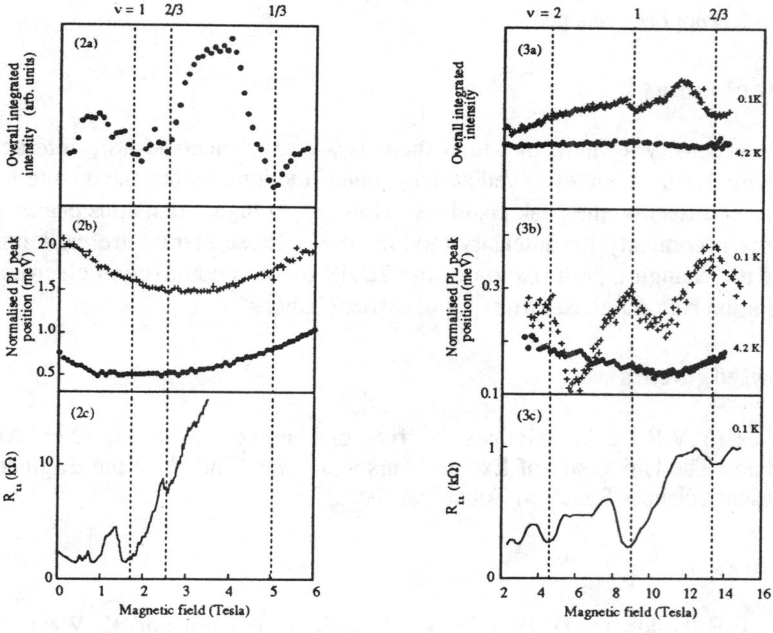

Figures 2 and 3: (a) the integrated PL intensity, (b) the PL peak positions (normalized by subtracting a linear field dependence from the data) and (c) a simultaneous measurement of the longitudinal resistance R_{xx} with $p_s = 4 \times 10^{10}$ cm^{-2} (figure 2) and $p_s = 2.1 \times 10^{11}$ cm^{-2} (figure 3). In figure 3 the integer and fractional quantum Hall effects influence both the intensity and peak position of the PL; BBG excitation was employed. In figure 2 no shifts in peak position are observed, but there is still clear evidence of the quantum Hall effects in the integrated intensity; ABG excitation was employed.

The behaviour of the PL energy at integer and fractional ν is strongly dependent on the strength of the electron-hole interaction.[1,5] In our structures, at low p_S the QW is almost symmetrical; the electron-hole separation in the growth direction is reduced almost to zero and no energy shifts are expected to occur, as shown in figure 2b. At higher p_S the QW becomes skew and the electrons and holes are significantly separated. The interaction becomes weaker and it is possible for upward shifts in the PL energy to occur, as seen in figure 3b, because of a decrease in the correlation self-energy of the 2D holes resulting from the perturbation caused by the photoexcited electrons. To observe the low interaction limit of downward shifts, higher p_S and larger well widths are required. When the photoexcited carriers lie in the same plane of the 2D system one does not expect intensity modulations. However in figure 2 very clear modulations are present, indicating that even at the lowest p_S this condition has not been reached.

4 Conclusions

In the low density system, in which there is a strong electron-hole interaction, intensity modulations are observed at integer and fractional ν, but the quantum Hall effect has no effect on the peak positions. However at higher densities peaks in the PL energy accompany the intensity modulations. These results are explained in terms of the changing perturbation of the 2DHS by the photoexcited electrons, as the separation between these two types of carrier changes.

Acknowledgments

One of us (Y.V.P.) acknowledges the financial support from the ORS Award Scheme and The University of Exeter. This work was funded by the Engineering and Physical Sciences Research Council of the UK.

References

1. B.B. Goldberg, D. Heiman, A. Pinczuk, L. Pfeiffer and K. West, *Phys. Rev.* B **44**, 4006 (1991).
2. For a review see I.V. Kukushkin and V.B. Timofeev, *Adv. in Phys.* **45**, 147 (1996).
3. A.J. Turberfield, S.R. Haynes, P.A. Wright, R.A. Ford, R.G. Clark, J.F. Ryan, J.J. Harris and C.T. Foxon, *Phys. Rev. Lett.* **65**, 637 (1990).
4 V.M. Apal'cov and E.I. Rashba, *JETP Lett.* **53**, 442 (1991).
5. A.H. MacDonald, E.H. Rezayi and D. Keller, *Phys. Rev. Lett.* **68**, 1939 (1992).
6. Y.V. Ponomarev, A. Usher, P.J. Rodgers, B.L. Gallagher, M. Henini and G. Hill, Submitted to *Phys. Rev.* B.

OPTICALLY DETECTED CYCLOTRON RESONANCE AND SHUBNIKOV DE HAAS OSCILLATIONS IN N- AND P-TYPE GAAS/GAALAS HETEROJUNCTIONS

J. X. SHEN, W. OSSAU, H. SCHÄFER, G. LANDWEHR
Physikalisches Institut der Universität Würzburg, Am Hubland, 97074, Würzburg

G. Z. ZHENG
Shanghai Institute of Technical Physics, Shanghai, 200083, China

G. WEIMANN
Walter Schottky Institut, TU München, 85748 Garching

The H band, an indirect transition in real space, is investigated by optically detected cyclotron resonance (ODCR) and optically detected Shubnikov de Haas oscillation (ODSdH) in n-and p-type GaAs/GaAlAs heterojunctions. The photoluminescence (PL) intensity of the H band and the PL intensity of excitons in the GaAs flat band region are correlated by the vertical transport of photo-excited carriers driven by the space charge potential. The H band emission arises from the recombination of indirect excitons located at different distances from the interface. The PL intensity oscillations (ODSdH) of the H band can be used to characterise the properties of carriers confined in the notch without contacts.

In order to investigate the electronic properties of carriers confined near the interface of a heterojunction, various methods have been applied where optical measurements seem to be the less developed because of the complicated nature of the indirect recombination in real space, referred to as the H band[1,2]. The H band was assumed to be the recombination between two dimensional electrons (holes) confined near the interface with three dimensional free holes (electrons) in the GaAs layer. In recent years, G. D. Gilliland[3] and D. C. Reynold[4] have found that besides the properties of an indirect transition, the H band has a strong excitonic behavior with binding energies between 0.4meV[4] and 2.8meV[5].

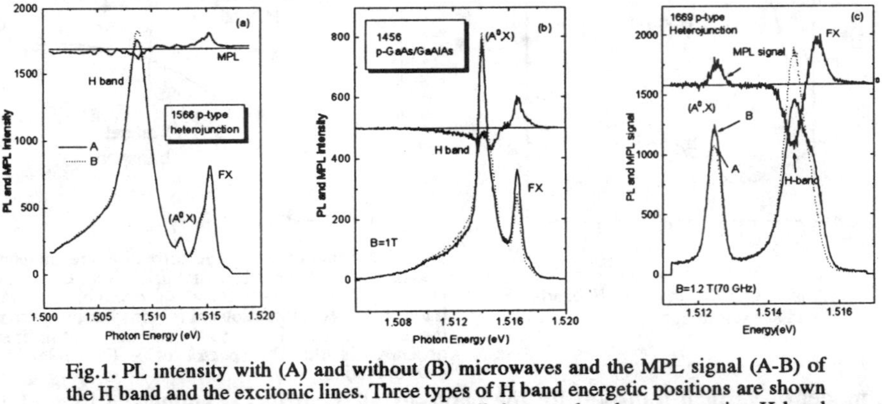

Fig.1. PL intensity with (A) and without (B) microwaves and the MPL signal (A-B) of the H band and the excitonic lines. Three types of H band energetic positions are shown (below, at and above the excitonic energy). All the samples show a negative H band MPL signal and a positive excitonic MPL signal. When the H band interferes with the excitonic lines, the MPL signal has a negative background with an upward peak (b).

Typical spectra of the H band are shown in Fig.1. The H band lies between the free exciton and the donor-acceptor pair band. The main characteristics of the H band are a rather broad linewidth and a pronounced blue shift with increasing laser excitation power[1,2]. With the application of microwaves, the two dimensional carriers in the notch are heated and their kinetic energy can be transferred to the photo-excited carriers near the interface by electron-electron interaction. The PL intensity of the H band decreases and the PL intensity of bound excitons and free excitons of the GaAs layer increases with the application of microwaves(Fig.1 (a), (c)). If the broad H band is mixed with bound exciton lines, the MPL spectrum exhibits a negative background with an upward peak at the free exciton line (Fig.1 (b)). The total integrated PL intensity changed by the microwaves equals zero.
Sweeping the magnetic field and adjusting the spectrometer to the luminescence peak, the PL intensity shows a resonant behaviour which is the optically detected cyclotron resonance (ODCR) (Fig.2). The H band ODCR signal is negative, while the excitonic ODCR signal is positive. From the resonance magnetic field strength, we obtain the effective mass values of electrons in n-type heterojunctions and those of holes in p-type heterojunctions. Although we have observed the ODCR signal of the confined carriers in the interface on the excitonic line, we do not believe that the normal exciton lines are caused by two dimensional carriers.

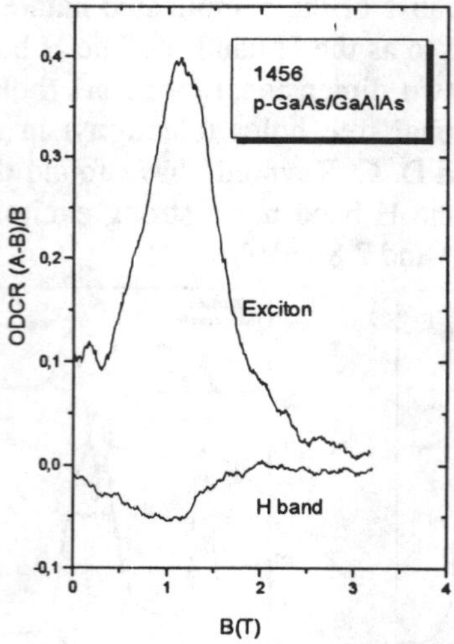

Fig.2. ODCR curves at H band and at exciton lines.

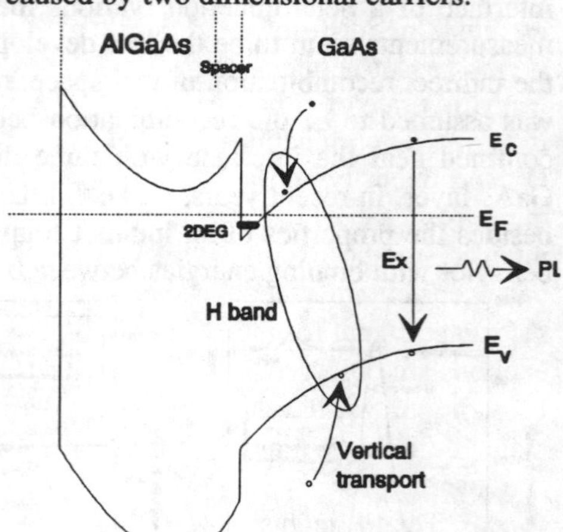

Fig.3 Diagram of the H-band formation mechanism. The photo-excited carriers first drift along the space charge potential, then form an exciton. The PL intensity of recombinations in regions far away from the interface, contributes to the normal exciton lines of GaAs in the PL spectra of a GaAs/GaAlAs heterojunction.

A model of vertical transport for the photo-excited carriers is suggested to explain the observed phenomena. For simplicity, we will discuss n-type heterostructures only (Fig. 3). However, the model can also be applied to the p-type structures. The laser excites electrons from the valence band to the conduction band. The photo-excited holes drift away from the interface and the photo-excited electrons drift towards the interface. Because of the strong interface electric field caused by the space charge, there are almost no photo-excited holes in the thin area close to the interface. The vertically drifted photo-excited holes capture electrons which drift towards the interface. The recombination of these excitons gives rise to the H band. The interface electric field lowers the energetic positions compared to the flat bandgap drastically and as well reduces the binding energy[3,4]. How strong the exciton binding energy will be reduced depends on the position of the exciton relative to the interface. The actual position of the H band exciton relative to the interface depends both on the vertical mobility of photo-excited carriers and the interface electric field. A strong interface electric field and a good vertical mobility enhances the distance of the H band exciton from the interface. This explains the different energetic positions of the H band excitons in different samples. Photo-excited carriers which are located far from the heterointerface do not feel the space charge potential. They recombine as normal three dimensional excitons.

Caused by the vertical transport properties photo-excited carriers, which are not recombining, drift towards the notch (electrons) or away from the interface (holes in n type structures) into the GaAs layer. With microwave application, the formation rate of H band excitons is reduced. Therefore, more photo-excited holes in n-type heterojunctions drift away from the notch resulting in a reduction of the H band intensity and an increase of the PL intensity of normal excitonic lines in the GaAs layer (Fig.3). That explains the fact that the increasing part of the PL intensity of the excitonic lines and decreasing part of the H band PL intensity balance each other. It also interprets why in ODCR experiments a negative ODCR resonance signal for the H band and a positive ODCR resonance signal for normal excitons are observed.

In addition, we have observed an oscillating PL intensity as a function of the magnetic field. We have detected the PL intensity oscillation both on the H band and on the free exciton luminescence in n-type GaAs/GaAlAs heterojunctions (Fig.4). The PL intensities of both lines oscillate but are shifted in phase by π. This again shows that the PL intensity of the H band and that of the excitons are correlated by carrier vertical transport.

The exact mechanism of the PL intensity oscillation of the H band in heterojunctions is not very clear yet. There are two possible reasons. First, the wave functions of the confined carriers near to the interface are extended far into the GaAs layer. The oscillatory properties of the screening effect in the 2DEG may influence the vertical transport of the photo-excited carriers. Second, when the H band exciton is composed of carriers lying very close to the states where the Fermi level is located, the H band exciton formation probability will also increase due to the Fermi level singularity[7]. The oscillation of the Fermi level energetic position therefore induces an oscillation of the H band PL intensity. When the photo-excited carriers recombine as the H band, they can not drift further to the GaAs layer and form normal excitons, so the PL intensity oscillates with a phase shift of π between the H band and the excitons in the GaAs layer.

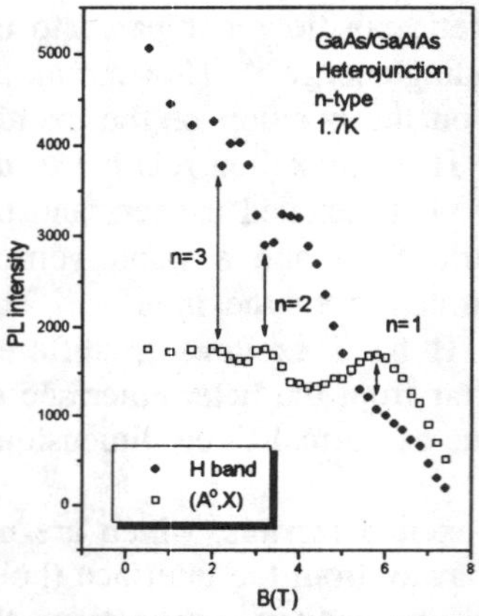

Fig.4 PL intensities of H band and normal (A^0,X) shows an oscillatory property. The oscillate in anti phase.

Our experimental results and the vertical transport model clearly show that the two dimensional properties observed for the PL intensity of normal excitonic lines do not mean that these excitons consist of two dimensional carriers. Although most of the H band excitons and all the bulk exciton lines may not originate directly from the 2DEG or 2DHG confined in the notch, many two dimensional properties of the 2DEG or 2DHG can still be detected by these transitions. However, consideration should always be paid that these lines may not always stem from confined particles in the notch.

As a conclusion, ODCR and ODSdH effects are discussed in n- and p-type heterojunctions. The indirect transition in real space (H band) arise from photo-excited carriers near the interface. The PL intensity of the H band excitons and the PL intensity of excitons in the GaAs flat band region are correlated by the vertical transport of photo-excited carriers. ODCR experiments on the H band and on bulk exciton lines provide values of the effective mass of carrier in the notch. The PL intensity oscillation of the H band is a very useful technique to study the properties of a 2DEG without contacts.

References:

1. Y.R.Yuan, M.A.A.Pudensli, et.al., J.Appl. Phys. 58, 397, (1985).
2. W.Ossau, E.Bangert, G.Weimann, Solid State Commun., 64, 711, (1987).
3. G.D.Gilliland, D.J.Wolford, et.al., Phys. Rev. B., 49, 8113, (1994).
4. D.C.Reynolds, D.C.Look, et.al., Phys. Rev. B, 50, 7461, (1994).
5. G.W.'t.Hooft, W.A.J.A.Van der Poel, et.al., Appl. Phys. Lett., 50, 1388, (1987).
6. W.M.Chen, B. Monemar, et.al., Semicond. Sci. Technol., 7, B253, (1992)
7. G. D. Mahan, Phys. Rev. 153, 882, (1966).

MAGNETOEXCITONS IN NEAR SURFACE QUANTUM WELLS: THEORY

A.L. YABLONSKII, A.B. DZYUBENKO, N.A. GIPPIUS, and S.G. TIKHODEEV
General Physics Institute, Russian Academy of Sciences,
Vavilova Street 38, Moscow 117942, Russia

We calculate the magnetoexciton transition and binding energies and wave functions in InGaAs/GaAs near-surface quantum well in quantizing magnetic fields. We take into account the modifications of the electron-hole Hamiltonian introduced by vacuum: a high potential barrier as well as the image-charge-caused modification of the Coulomb interaction, and trace in details the influence of a sharp, more than one order of magnitude, decrease of the dielectric constant at the semiconductor surface.

1 Introduction

It had been predicted[1,2] that an exciton in a thin semiconductor layer inside a dielectric should be enhanced due to the image-charge-caused modification of the electron-hole interaction (the so called dielectric enhancement of the exciton). Notwithstanding a lot of theoretical work (see, e.g., Ref. 3), the experimental proof of this effect has been only recently obtained[4] in InGaAs/GaAs near-surface quantum wells (NSQW), at zero and low magnetic field. An application of high magnetic fields to NSQWs opens new possibilities for detailed understanding of the dielectric confinement effect. In this work, we analyze the evolution of the magnetoexciton eigenstates and transition energies with the perpendicular quantizing magnetic field $B > 1$ T and with the thickness of the cap barrier layer, $L_{\rm cap}$, between the quantum well and vacuum.

The NSQW electron-hole Hamiltonian in the magnetic field $\mathbf{B} = (0, 0, B)$ takes the form

$$H = H_{ez} + H_{hz} + H_{\rm 2D} + U_{eh} \equiv H_0 + U_{eh} \,. \tag{1}$$

Here

$$H_{jz} = -\frac{\hbar^2}{2m_{jz}}\frac{\partial^2}{\partial z_j^2} + V_j(z_j) + V_{\rm self}(z_j),\ j = e, h. \tag{2}$$

The band-offset potentials are $V_j = \infty$ ($j = e, h$) in vacuum ($z < 0$), $V_e = -113$ meV and $V_h = -75$ meV inside a $\mathrm{In_{0.18}Ga_{0.82}As}$ QW ($L_{\rm cap} < z < L_{\rm cap} + L_{\rm QW}$) and $V_j = 0$ in the barrier GaAs layers. $L_{\rm QW} = 5$ nm is the QW width. Effective masses are $m_{ez} \equiv m_e = 0.067$ and $m_{hz} = 0.35$. The potential

$$V_{\rm self}(z) = \frac{e^2}{2\varepsilon}\left(\frac{\varepsilon - 1}{\varepsilon + 1}\right)\frac{1}{2\,|\,z\,|} \tag{3}$$

(where ε is the dielectric constant of the semiconductor) takes into account the repulsion of the charge from its self image. We neglect a small difference between ε in InGaAs and GaAs and take $\varepsilon = 12.5$. Hamiltonian H_{2D} describes the two-dimensional (2D) motion of a free e-h pair in the magnetic field,

$$H_{2D} = \frac{1}{2m_e}\left(-i\hbar\nabla_{\rho_e} + \frac{e}{c}\mathbf{A}_e\right)^2 + \frac{1}{2m_{h\|}}\left(-i\hbar\nabla_{\rho_h} - \frac{e}{c}\mathbf{A}_h\right)^2, \quad (4)$$

where $\mathbf{A}_j = \frac{1}{2}\mathbf{B}\times\boldsymbol{\rho}_j$ is the vector potential, $\boldsymbol{\rho} = \boldsymbol{\rho}_e - \boldsymbol{\rho}_h = (x, y)$, and $\mathbf{r} = (\boldsymbol{\rho}, z)$. We neglect also a small mass difference in the well and barrier layers and take $m_{h\|} = 0.2$. The potential

$$U_{eh}(\rho, z_e, z_h) = -\frac{e^2}{\varepsilon}\left[\frac{1}{\sqrt{\rho^2 + (z_e - z_h)^2}} + \frac{\varepsilon - 1}{\varepsilon + 1}\frac{1}{\sqrt{\rho^2 + (z_e + z_h)^2}}\right] \quad (5)$$

takes into account, in addition to the Coulomb interaction between the electron and hole, the attraction of the electron to the hole image, and of the hole to the electron image.

In order to solve the quantum-mechanical problem of the magnetoexciton in NSQW (1), we diagonalize numerically the matrix of the Coulomb potential (5) in the basis of noninteracting two-particle states of the Hamiltonian H_0. This approach has been also used in the problems of two-electron quasi-$2D^-$ centers [5] and of the excitons in double QW [6] in a magnetic field.

2 Results and Discussion

The introduction of a high vacuum potential barrier in a close vicinity of the quantum well causes an additional (to the QW) localization of electrons and holes, and thus an increase of free electron-hole transition energies with the decrease of cap layer thickness.[7] This effect has a tunneling origin (we will call it a "tunneling blue shift"). It depends on the cap layer thickness exponentially and influences the exciton binding energies only slightly.

A sharp decrease of the dielectric constant on the sample surface introduces the image charges that cause (*i*) an additional blue shift of free electron and hole (band gap) transitions and (*ii*) an enhancement of the exciton binding energies. The first effect (we will call it a "dielectric blue shift") is due to the electron and hole repulsion from their self images, Eq. (3), while the second effect (the dielectric enhancement) follows from the enhanced attraction between the electron and hole, the second term in Eq. (5).

The dielectric enhancement partly compensates the blue shift of the band gap transitions (which corresponds to the electro-neutrality of the exciton).

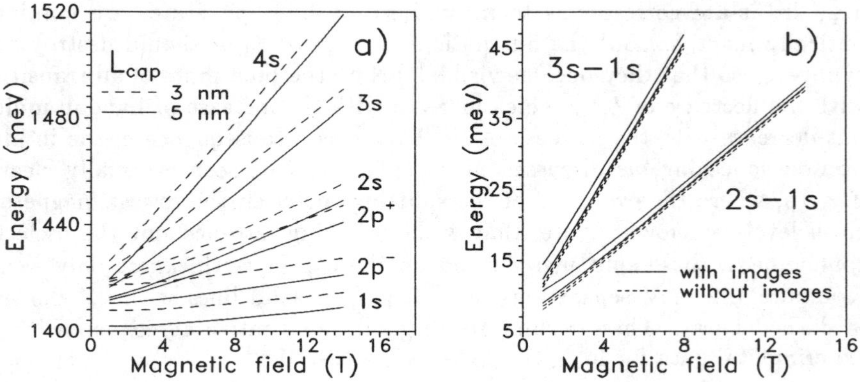

Figure 1: (a) Calculated Landau fan diagrams for magnetoexcitons in $In_{0.18}Ga_{0.82}As/GaAs$ NSQWs for L_{cap} = 3 and 5 nm. (b) Magnetoexciton $2s$–$1s$ and $3s$–$1s$ splittings in $In_{0.18}Ga_{0.82}As/GaAs$ NSQWs, for L_{cap} = 5 and 20 nm, calculated with and without accounting for dielectric confinement (solid and dashed lines, respectively). The upper line in each pair of lines corresponds to L_{cap} = 5 nm, and the lower one – to L_{cap} = 20 nm.

The smaller the in-plane radius of the exciton eigenstate, the larger the dielectric enhancement of the corresponding eigenenergy, and the better the compensation of the dielectric blue shift of the band gap transition energy. As a result, the net blue shift of the exciton ground state transition energy is mainly due to the tunneling blue shift. Hence, it is very difficult to separate the dielectric confinement effect from the dependence of the ***ground state*** transition energy on the cap layer thickness. The dielectric enhancement of the ***excited*** states binding energy is smaller, and the blue shift is larger. Thus, in order to trace the influence of the dielectric confinement, one has to investigate the inter-level splittings rather than the evolution of the ground state transition.

The perpendicular magnetic field causes the in-plane squeezing of the exciton wave functions. Thus, the dielectric enhancement of the magnetoexciton eigenstates increases with the decrease of L_{cap} and with the increase of B and, also, it depends on the Landau level indices of the magnetoexciton. The blue shifts of the corresponding transition energies decrease with increasing B.

We illustrate the above discussed evolution of the magnetoexciton transition energies with B and with decrease of L_{cap} in Fig. 1a, where the calculated Landau fan diagrams for the lowest ns ($n = 1, 2, \ldots 4$) and $2p^{\pm}$ magnetoexciton transition energies in $In_{0.18}Ga_{0.82}As/GaAs$ NSQWs are shown. In the limit of high magnetic fields, optically-active magnetoexciton states ns are bound states of the electron and hole with Landau level indices $n_e = n_h = n - 1$;

e.g., $2p^+$ state corresponds to $n_e = 1, n_h = 0$. [$2p^\pm$ states are nominally optically inactive. But the fluctuations of $L_{\rm cap}$ or $L_{\rm QW}$ should destroy axial symmetry so that they become visible.[8]] Note the blue shifts of all transitions with the decrease of $L_{\rm cap}$, which increase with the increase of Landau indices, but decrease with the increase of B. This is as a consequence of the in-plane exciton squeezing, as discussed above. This can be seen especially clear in Fig. 1b, where the evolution of the splittings between the lowest magnetoexciton levels is shown. Note, that without taking into account the dielectric confinement, these splittings depend on the cap layer thickness only weakly (compare 1–2 meV separations in each pair of solid lines with 0.4–0.5 meV in dashed ones). These results are in good quantitative agreement with the experimental data for $In_{0.18}Ga_{0.82}As/GaAs$ NSQWs.[8]

Acknowledgments

The authors are grateful to M. Bayer, A. Forchel, V.D. Kulakovskii, and L.V. Kulik for helpful discussions. This work was financially supported in part by INTAS grants 93-1495 and 94-2112, Russian Ministry of science program "Nanostructures" (grant 1-041) and RBFR grant 095-02-06062a.

References

1. N.S. Rytova, *DAN SSSR* **163**, 1118 (1965).
2. L.V. Keldysh, *Pis'ma Zh. Eksp. Teor. Fiz.* **29**, 716 (1979) [*JETP Lett.* **29**, 658 (1979)].
3. D.B. Tran Thoai, R. Zimmermann, M. Grundmann, D. Bimberg, *Phys. Rev.* B **42**, 5906 (1990); T. Takagahara, *Phys. Rev.* B **47**, 4569 (1993).
4. L.V. Kulik, V.D. Kulakovskii, M. Bayer, A. Forchel, N.A. Gippius, and S.G. Tikhodeev, *Phys. Rev.* B (1996) — in press.
5. A.B. Dzyubenko and A.Yu. Sivachenko, *Phys. Rev.* B **48**, 14 690 (1993).
6. A.B. Dzyubenko and A.L. Yablonskii, *Phys. Rev.* B **53**, June 15th (1996).
7. J. Dreybrodt, A. Forchel, J.P. Reithmaier *Phys. Rev.* B **48**, 14 741 (1993).
8. A.L. Yablonskii, A.B. Dzyubenko, S.G. Tikhodeev, L.V. Kulik, and V.D. Kulakovskii, *Pis'ma Zh. Eksp. Teor. Fiz.* **64**, 47 (1996) [*JETP Letters* (1996)]; V.D. Kulakovskii, L.V. Kulik, and A. Forchel, this Conference.

DIELECTRIC ENHANCEMENT OF MAGNETOEXCITONS IN NEAR SURFACE QUANTUM WELLS:EXPERIMENT

L.V. KULIK and V.D. KULAKOVSKII
Institute of Solid State Physics, Russian Academy of Sciences
142432 Chernogolovka, Russia

A. FORCHEL
Technische Physik, Universität Würzburg, Am Hubland, Würzburg, Germany

Magnetoexcitons in near-surface InGaAs/GaAs QWs have been investigated by photoluminescence excitation spectroscopy at 4.2 K. The effect of dielectric enhancement of excitons has been experimentally demonstrated by measuring the $1s - 2s$ exciton splitting. The maximum dielectric enhancement of the exciton binding energy occurs at B=0 T. It reaches 1.5 times for 5 nm wide quantum wells with cap layer thicknesses smaller than 3 nm. The effect decreases with magnetic field because of the reducing exciton size.

The spatial confinement of excitons in semiconductor quantum wells (QW) enhances strongly their binding energy (E_x). In the two dimensional (2D) limit it is four times larger than in bulk.[1] The enhancement has been predicted to be much stronger if the QW is surrounded by material of a smaller dielectric constant ε, which causes a redistribution of the electric field. This effect is referred to as dielectric confinement. It has been extensively discussed,[2–5] however, to our best knowledge, was not investigated experimentally for semiconductor nanostructures.

We have investigated the effect of the dielectric confinement in near surface quantum wells (NSQW), i. e. in the QWs located near the semiconductor-vacuum (SV) interface. In general, in the case of NSQW excitons there appear two important contributions in the exciton energy $\hbar\omega_x$, namely, from the change of the exciton localization and from the change of the dielectric environment. Indeed, an approaching high vacuum potential barrier (cf. Fig. 1) increases the exciton transition energy. This contribution depends exponentially on $L_{\rm cap}$ because of the tunnel origin and is referred to as tunnel blue shift. It is obvious that the shift is very similar for the ground and excited states.

The change of the dielectric environment because of a sharp decrease of ε at the sample surface results in the appearance at the SV interface of image charges. The repulsion of QW electrons and holes from self-images leads to an additional increase of the band-to-band transition energy in a NSQW, as shown in Fig. 1. This effect is referred to as dielectric blue shift. In excitons, there appear an additional attraction of the electron to the hole image and of the hole to the electron image which enhances the exciton binding energy and cancels partly the dielectric blue shift related to the band-to-band transition energy. The cancellation originates from the electroneutrality of excitons and hence is more effective for states with smaller in-QW-plane exciton radius a_x. In particular, theoretical estimates show that the effect of the dielectric confinement for the 1s exciton energy $\hbar\omega_{x,1s}$ in InGaAs/GaAs NSQWs is smaller than the tunnel blue shift.[6] Separation of these effects is only possible by measuring the energies of the excited exciton states.

We have investigated both the ground and excited states of magnetoexcitons confined in QWs with the different cap layer thicknesses. An external magnetic field normal to the QW plane changes a_x and allows us to control the dielectric enhancement of excitons at a fixed L_{cap}. Exciton transition energies have been determined from photoluminescence excitation (PLE) measurements at 4.2 K in a wide range of magnetic fields $B \leq 14\,\text{T}$. All measurements have been carried out by using the same $\text{In}_{0.18}\text{Ga}_{0.82}\text{As/GaAs}$ heterostructure with a QW thickness $L_{QW} = 5\,\text{nm}$. At this thickness, there is only one bound state in the conduction band and the splitting of light hole (lh) and heavy hole (hh) subbands exceeds 20 meV. The additional advantage of such QW structures are (i) a strong (one order of magnitude) decrease of ε at the SV interface, and (ii) a large exciton radius ($a_x \sim 60$ A). An as grown sample with $L_{\text{cap}} = 20\,nm$ was used as a reference. After measuring photoluminescence (PL) and PLE spectra the cap layer was thinned by dry etching. For the PLE studies we have used a Ti-sapphire laser and a double grating monochromator RAMANOR U1000.

Typical exciton PL and PLE spectra are shown in Fig. 1 for NSQWs with $L_{\text{cap}} = 20$, 5 and 3 nm for $B \leq 8\,\text{T}$. Only one line corresponding to an 1s exciton transition is observed in the PL spectra whereas the PLE spectra show peaks associated with 1s, 2s and 3s exciton transitions. A small red shift of the 1s exciton peak of about 1.5 meV in the PL spectrum is due to exciton localization by QW potential fluctuations.[7] The comparison of the PL and PLE spectra shows that the cap layer etching increases the exciton localization relatively weakly.

Figure 1 shows that all the exciton lines move to higher energies with decreasing L_{cap}. As discussed above, there are two important contributions in this shift, from an approaching high vacuum potential and from a change in the dielectric environment. The first effect is the same for all the exciton states whereas the latter increases with the radius of the exciton state. Comparison of the spectra in Figs. 1 and 2 shows a well pronounced increase in the line shift for excited $2s$ and $3s$ states at the fixed magnetic field which indicates the essential contribution in exciton energy from the dielectric confinement of

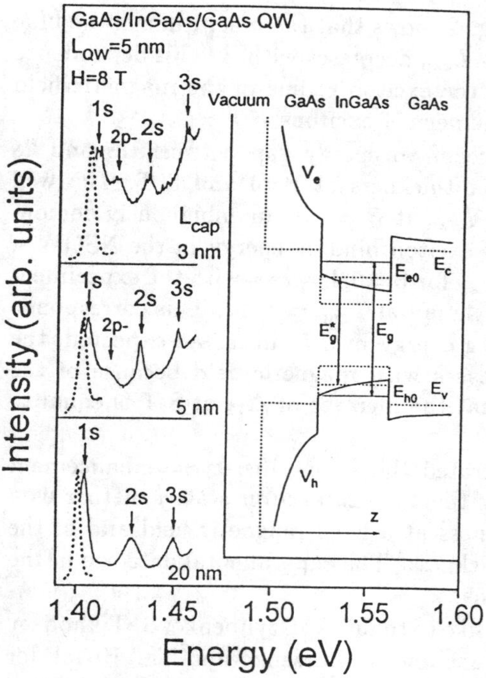

FIG. 1. PLE (solid lines) and PL (dashed lines) spectra for NSQWs with L_{cap} = 3, 5, and 20 nm at 8 T. The insert shows an energy scheme for the band-to-band QW transition in the NSQWs with (solid lines) and without (dotted lines) accounting of self-image charges.

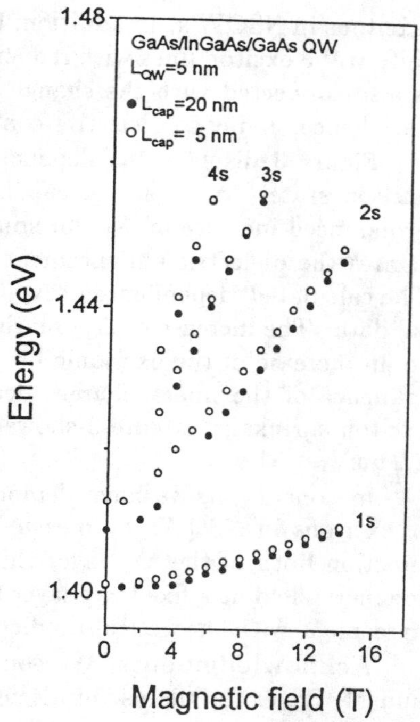

FIG. 2. Magnetoexciton transition energies as a function of magnetic field for L_{cap} = 5 and 20 nm.

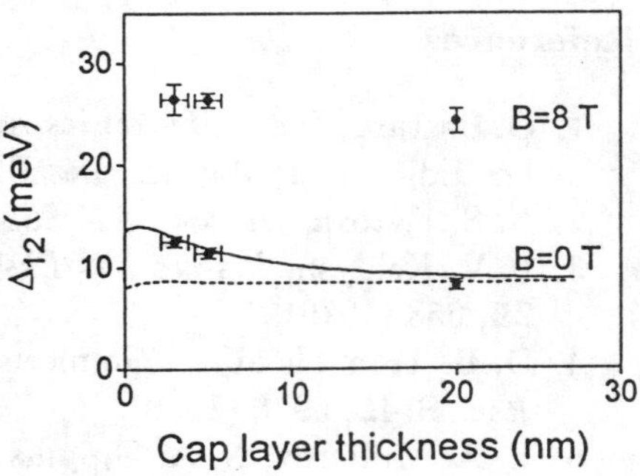

FIG. 3. Dependences of the energy gap between the 1s and 2s exciton states on the cap layer thickness in NSQWs for B = 0 and 8 T. The dependences calculated with and without accounting of self-image charges at B= 0 T are shown by solid and dashed lines, respectively.

excitons in NSQW's. In addition, Fig. 2 shows that the magnitude of the blue shift of the exciton lines with reducing L_{cap} decreases with B. This dependence is also connected with the change of the exciton radius in the magnetic field and, hence, is due to dielectric confinement of excitons.

Figure 3 displays the dependence of an energy gap between $1s$ and $2s$ exciton states, Δ_{12}, on the cap layer thickness at B=0 and 8 T. The well pronounced increase of Δ_{12} for small L_{cap} at $B = 0$ T unambiguously demonstrates the dielectric enhancement of exciton binding energy in the NSQW's. The calculated[6] dependence of $\Delta_{12}(L_{\text{cap}}$ for B= 0 T agrees with the experimental data. The increase of Δ_{12} reaches 4 meV at $L_{\text{cap}} = 3$ nm. This corresponds to an increase of the excitonic binding energy by 1.5 times. As expected, the influence of the image charges decreases with magnetic field because of the exciton shrinkage. Figure 3 shows that the increase of Δ_{12} at 8 T is equal to 2.2 meV.

In conclusion, we have demonstrated the strong dielectric enhancement of excitons in NSQWs by measuring the $1s - 2s$ exciton state splitting as a function both of the cap layer thickness at a fixed magnetic field and of the magnetic field at a fixed cap layer thickness. The experimental data are in the agreement with theoretical predictions.

Acknowledgments. We would like to thank A. Dzyubenko, S. Tikhodeev and M. Bayer for the useful discussions and T. Borzenko and Yu. Koval' for the assistance in the cap layer etching. The work was supported by INTAS grant 94-2112, Russian Foundation for Basic Research, grant 95-02-06062a, and Volkswagen Foundation.

References

1. G. Bastard, *Wave Mechanics Applied to Semiconductor Heterostructures*, Les Editions de Physics, Les Ulis, France, 1988, p.138.
2. N. S. Rytova, *Doklady AN SSSR* **163**, 1118 (1965).
3. L. V. Keldysh, *Pis'ma Zh. Eksp. Teor. Fiz.* **29**, 716 (1979) [*JETP Lett.* **29**, 658 (1979)].
4. D. B. Tran Thoai, R. Zimmermann, M. Grundmann, D. Bimberg, *Phys. Rev.* B **42**, 5906 (1990).
5. E. A. Muljarov, N. A. Gippius, S. G. Tikhodeev, and T. Ishihara, *Phys. Rev. B* **51**, 14370 (1995).
6. L. V. Kulik, V. D. Kulakovskii, M. Bayer, A. Forchel, N. A. Gippius, and S. G. Tikhodeev, *Phys. Rev.* B **54** (1996), in press.
7. R. Dingle, in *Festkorperprobleme XV (Advances in Solid State Physics)*, ed. by H. J. Queisser, Vieweg, Braunschweig (1975).

MAGNETOOPTICS OF TWO-DIMENSIONAL ELECTRONS AT A SEMICONDUCTOR SURFACE

P.D. ALTUKHOV
A.F. Ioffe Physical-Technical Institute, Politekhnicheskaya st. 26, St.-Petersburg 194021, Russia

Quantum oscillations of the circular polarization of the recombination radiation line of 2D electrons and nonequilibrium holes in a magnetic field are observed under formation of quantum interchanging electron and hole layers at [111] silicon surface in MOS structures. At low 2D electron density the surface luminescence line is attributed to localized charged excitons and biexcitons. The recombination radiation line of real excitons in a dense 2D electron gas is observed in luminescence spectra of [100] silicon MOS structures. It is shown, that the Kondo electron spin correlations of the exciton and 2D electrons contribute to disappearance of an anisotropic paramagnetic reduction of the luminescence line with decreasing 2D electron density.

The recombination radiation line (S-line) of 2D electrons and 2D nonequilibrium holes, bound to the [111] electron layer by the electron polarization attraction, is observed in luminescence spectra of silicon MOS structures [1]. At the 2D electron density $n_S > 10^{12}$ cm^{-2} and low surface electron-hole pair density n_P surface pairs exist as a 2D two-layer plasma and the additional hole layer is formed. Observation of the change of the circular polarization sign of the luminescence in a magnetic field at the high energy edge of the S-line with increasing n_P proves the existence of an additional quantum electron layer in an additional selforganized electron quantum well at $n_P \approx 2\cdot 10^{11}$ cm^{-2} [1]. The quantum oscillations of the circular polarization of the luminescence in a magnetic field at the high energy edge of the S-line (Fig.1) due to subsequent occupation of the electron and hole quantum levels in other additional electron and hole quantum wells are observed in this work.

The circular polarization degree of the S-line in a transverse magnetic field B in the Faraday geometry at $n_S > 10^{12}$ cm^{-2} and at low n_P is proportional to the 2D heavy hole average spin and is described by the formula

$$P_N = \sigma_N \mathrm{th}\left(3g_{h\perp}\mu_0 B / 2kT\right) \tag{1}$$

where $g_{h\perp}$=0.7 is the hole g-factor, μ_0 is the Bohr magneton, T is the temperature, σ_N=0.5 at the maximum of the TO-LO line. The σ_N decreases at the high energy edge of the S-line due to weak contribution of the LO-line to the luminescence [1]. At high density of holes in the second layer the circular polarization degree at the low energy part of the S-line in a weak magnetic field is equal to

$$P_N = 2\sigma_N\left[\left(3g_{h\perp}\mu_0 B / 4E_{Fh}\right) - \left(g\mu_0 B / 4E_{Fe}\right)\right] \tag{2}$$

where E_{Fh} is the Fermi energy of 2D holes in the second layer, E_{Fe} is the Fermi energy of 2D electrons in the first layer, $g \approx 2$ is the electron g-factor. Two

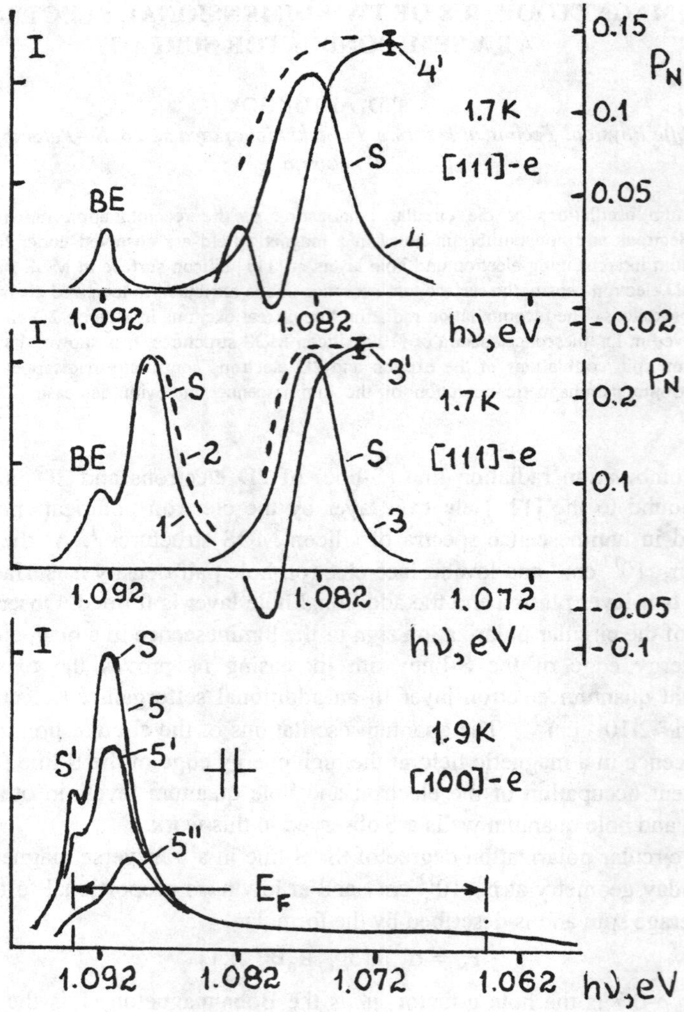

Fig. 1: The S-line spectra I (1-4, 5-5″) and the circular polarization degree P_N of the luminescence in the transverse magnetic field $B_\perp$ (3′, 4′) in [111] and [100] silicon MOS structures, TO-LO lines. The 2D electron density n_S, 10^{12} cm^{-2}: 1,2 - 0.46; 3,3′,4,4′ - 2.3; 5,5′,5″ - 5.1. The magnetic field $B_\perp$: 1,2,3,3′,4,4′ - 6 T; 5 - 0; 5′- 2.7 T; 5″ - 6.7 T. The excitation level, w·cm^{-2}: 1 - 10^{-3}; 2 - $3\cdot10^{-2}$; 3,3′ - 5; 4,4′ - 20; 5,5′,5″ - 10^{-3}. The dashed curves (3′,4′) represent the theory for two-layer model [1].

contributions in this formula represent the average hole spin and the average electron spin. At the beginning of a next layer formation the circular polarization at the high

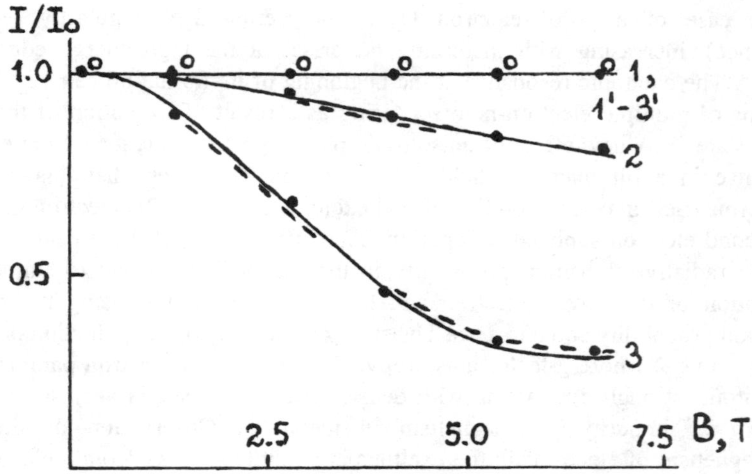

Fig. 2: The dependence of the intensity of the S-line maximum I/I_0 upon the magnetic field B at T=1.9 K, [100] 2D electron layer.
1-3 - $B=B_\perp$, 1'-3' - $B=B_\parallel$. The 2D electron density n_S, 10^{12} cm^{-2}: 1,1' - 3.2; 2,2' - 3.78; 3,3' - 5.1. The dashed curves represent the theory at T=1.9 K, $g_{h\perp}=0.5$, $g_{h\parallel}=0$: 3 - $T_K<T$ and $w_- \cdot w_+^{-1}=0.13$ (theory of ref. [3]); 2 - $T_K=5.4$ K and $w_- \cdot w_+^{-1}=0.53$.

energy edge of the S-line is defined by the average spin of carriers in the next layer due to low Fermi energy of the carriers. The electron and hole spin contributions to the circular polarization have opposite signs. So, the subsequent formation of the quantum interchanging electron and hole layers results in the quantum oscillations of the circular polarization sign. The maximum number of layers is estimated as five at $n_P \approx 5 \cdot 10^{11}$ cm^{-2}.

At $n_S \leq 4 \cdot 10^{11}$ cm^{-2} in the region of localization of 2D electrons the magnetic field dependence of the circular polarization of the S-line, described by the formula (1), is explained as a result of formation of localized charged excitons [2]. The ground state of two localized electrons in the charged exciton is the singlet state due to exchange interaction. The singlet state does not contribute to the circular polarization of the S-line, and the circular polarization is defined by the average hole spin. The exchange interaction of electrons in the surface localized biexcitons is not too strong, and the triplet electron state of the biexcitons at high excitation levels contributes to the circular polarization degree of the S-line at high magnetic field [2].

In the case of a [100] electron layer, the Fermi-edge singularity (excitonic resonance), increasing with increasing n_S, arises at the high energy edge of the S-line [3]. The excitonic resonance at the beginning of its formation can be explained in terms of multiple electron-hole scattering as a result of formation of the virtual bound state [4]. A new effect of anisotropic paramagnetic reduction of the excitonic resonance in a tilt magnetic field [3] (Figs 1 and 2) proves, that this resonance transforms itself into emission line of real excitons in a dense 2D electron gas before the second electron subband occupation. This effect is caused by a decrease of the average radiative transition probability in the magnetic field due to simultaneous occupation of the lowest electron and hole spin states, possessing the radiative transition probability equal to zero. The strong anisotropy of the reduction is induced by the strong 2D hole g-factor anisotropy. The problem of electron paramagnetism of excitons, strongly interacting with dense 2D electron gas, is very similar to the problem of impurity paramagnetism in metals [5]. Observation of the Curie paramagnetism of electrons in the excitons is possible at weak Kondo electron spin correlations of the exciton and 2D electrons. The electron paramagnetism of the excitons reveals the Curie behaviour at $n_S \approx (4 \div 6) \cdot 10^{12}$ cm^{-2} [3] (Fig.2, curve 3). A strong deviation of the electron paramagnetism of the excitons from the Curie paramagnetism due to Kondo electron spin correlations of the exciton and 2D electrons is observed at $n_S < 4 \cdot 10^{12}$ cm^{-2} as a result of a decrease of the distance between the excitons and 2D electrons (Fig.2, curve 2). In this case the paramagnetic reduction of the excitonic resonance intensity I at low temperatures is described by the formula

$$\frac{I}{I_0} \approx 1 + \frac{w_+ - w_-}{w_+ + w_-} 2\left\{-\left(g\mu_0 B_\perp / 4kT_K\right)\left[1 + \left(g\mu_0 B_\perp / 2kT_K\right)^2\right]^{-1/2}\right\} \tag{3}$$

where T_K is the Kondo temperature of the excitons, w_+ and w_- are the radiative transition probabilities for the upper and lower electron spin states [3], I_0 is the intensity of the resonance at B=0. The contents in the brackets represent the average electron spin for the Kondo singlet state [5].

References

1. P.D. Altukhov *at al, Fiz. Tverd. Tela* **34**, 3747 (1992).
2. P.D. Altukhov *at al, Pis'ma Zh. Tekhn. Fiz.* **19** (N24), 46 (1993).
3. P.D. Altukhov, *Workbook of the 11 Int. Conf. on the EP2DS*, p. 247. Nottingham, U.K. (1995).
4. G.D. Mahan, *Phys. Rev.* **153**, 882 (1967).
5. A.J. Heeger in *Solid State Physics,* Vol. **23**, Eds. F. Seitz, D. Turnbull, H. Ehrenreich (Academic Press, New York and London, 1969), p. 283.

RESONANCE EFFECT IN THE MAGNETO-OPTICAL PROPERTIES OF GaAs/InGaAs/AlGaAs MODULATION DOPED QUANTUM WELLS

F. IIKAWA, M. L. F. ABBADE, R. G. PEREIRA, AND J. A. BRUM
IFGW-Unicamp, CEP-13083.970, Campinas, SP, Brazil

A.A. BERNUSSI, AND A.L. GOBBI
CPqD-Telebras, CEP-13088-061, Campinas, SP, Brazil

G. BORGHS
IMEC, Kapeldreef 75, B-3001, Leuven, Belgiun

Magneto-optical properties of modulation doped $GaAs/In_{0.2}Ga_{0.8}As/Al_{0.25}Ga_{.75}As$ single quantum wells are studied for different 2D electron densities. A hydrogen passivation technique was used to modify the electron gas density in these structures. The remarkable feature observed in the magneto-luminescence spectra is the oscillation of the peak intensity at the Fermi energy when the second subband is in resonance with the Fermi level. The hydrogen-passivated sample exhibits a continuous decrease of the peak intensity as the Landau level empties. In contrast, the as-grown sample shows a peak intensity enhancement at even filling factor. Additionally, well defined resonant magneto-polaron effects were observed.

1. Introduction

Optical experiments have been extensively used in the last few years to investigate the properties of two-dimensional electron gas (2DEG) systems. Modulation doped quantum well (MDQW) is a suitable structure for this study because photocreated holes are spatially confined in the same region of the 2D electron gas. However, Coulomb interaction[1-3] between carriers leads to important modifications in the optical spectra. Enhancement of the Fermi-edge singularity (FES) effect and an oscillatory behavior of the luminescence peak intensity at the Fermi energy as a function of the magnetic field were observed in MDQW structures[3-4].

In this work we investigate the magneto-optical properties of GaAs/$In_{0.2}Ga_{0.8}As/Al_{0.25}Ga_{0.75}As$ MDQW structures with different 2D carrier concentrations. A nondestructive hydrogen passivation technique was used to control the 2DEG in a single sample. At increasing magnetic fields a distinct oscillatory behavior of photoluminescence (PL) peak intensity at the Fermi energy was observed for the as-grown and the passivated samples.

2. Experiment

The modulation doped single quantum well structure was obtained by Molecular-Beam Epitaxy. The structure consists of a semi-insulating (001) GaAs substrate, a thick undoped GaAs buffer layer, 13 nm of $In_{0.21}Ga_{0.79}As$ layer, 5 nm of $Al_{.22}Ga_{.78}As$ space layer, Si planar-doping (nominally $N_{si} = 5 \times 10^{12}$ cm^{-2}), 30 nm of $Al_{.22}Ga_{.78}As$ layer and finally 40 nm of *n*-type GaAs cap layer. The 2D electron density of the as-grown sample was estimated to be 2.1×10^{12} cm^{-2}. The hydrogen passivation was performed in a Reactive Ion Etching chamber. The estimated 2D electron concentration of *H*-passivated sample was 1.4×10^{12} cm^{-2}.

Magneto-luminescence measurements were performed at low temperature (T=2K) with magnetic fields (*B*) applied along the growth direction varying from zero to 13 Tesla.

3. Results and Discussion

Figure 1 shows the evolution of PL spectra with applied magnetic fields for the same MDQW sample but with two different electron densities. The luminescence spectrum of the as-grown sample (Fig. 1a) at B = 0 T presents two peaks and the *H*-passivated sample (Fig. 1b) only one. The low energy peaks exhibited in both spectra are attributed to recombination involving the first electron (e_1) and the heavy hole (h_1) subbands. The second luminescence peak shown in the spectra of the as-grown sample is attributed to the recombination between the second electron subband (e_2) and h_1. These peaks were identified based on self-consistent (Poisson and Schrödinger) calculations. In the case of the as-grown sample, the difference in the PL intensity strength between *e2-h1* emission and *e1-h1* is due to the stronger wave-function overlap involving *e2-h1* transition. As the magnetic field is increased several luminescence peaks are observed in the PL spectra. They are attributed to inter-Landau level transitions. These peaks involve dipole-allowed ($\Delta n=0$) and also forbidden recombination. They correspond to all occupied Landau levels of e_1 and e_2 and the Landau level of h_1 ($n_h = 0$). At low temperatures and high magnetic fields only the first Landau level of the valence band is populated by photocreated holes. Consequently transitions involving $\Delta n =$

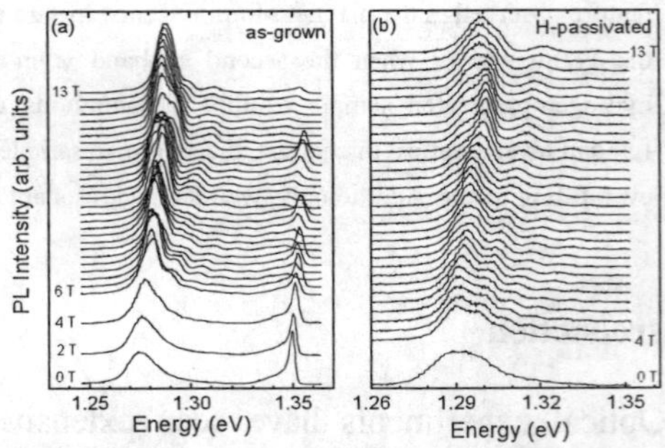

Fig. 1 - Magneto-luminescence spectra of (a) as-grown and (b) *H*-passivated samples. The magnetic field steps correspond to 0.25 T.

0 and $n_h > 0$ are expected to be unlikely. The dipole-forbidden transitions observed in Fig.1 are tentatively attributed to the scattering of electrons and holes with impurities or defects[5].

The striking feature of Fig. 1 is the PL peak intensity variation in the vicinity of the lowest and the highest energy peaks. The high energy peak corresponds to the recombination of the electron at the Fermi energy. The intensity of this peak oscillates as a function of magnetic field and its behavior is remarkably strongly different for two samples, as it is shown in Fig. 2. In the case of the as-grown sample, where the second electron subband is slightly occupied at B = 0 T, as the magnetic field is increased the peak intensity at the Fermi energy increases each time a Landau level is depopulated. The maxima occur at even filling factors. On the other hand, the *H*-passivated sample exhibits an opposite behavior (see Fig. 2b). The above results clearly show the strong influence of the second subband on the PL peak intensity at the Fermi energy. The Landau levels of the second subband become depopulated when the magnetic field is increased. At high magnetic fields this population becomes very small and the corresponding PL peak intensity oscillates due to the proximity of the Fermi level to the second subband[4]. In these conditions the exciton of the second subband becomes dominant compared to the free electron-hole pair. The small population of the second subband leads to an enhancement of the PL peak intensity at Fermi energy. In this situation the hybridization between the electron *sea* states and the second exciton subband takes place[4]. The decrease of PL peak intensity of *H*-passivated sample is attributed to the depopulation of the uppermost Landau level.

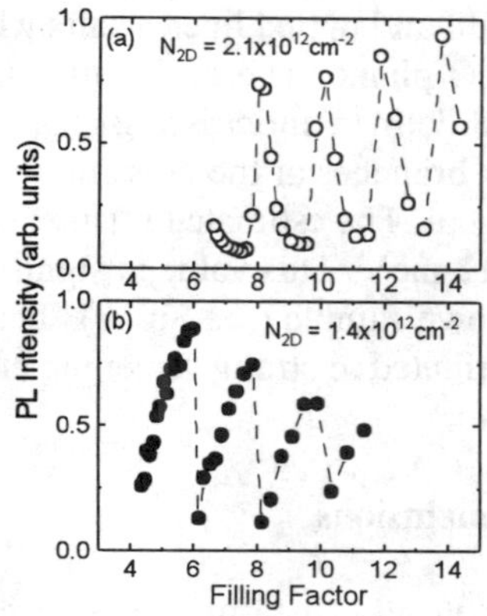

Fig. 2 - PL peak intensity at the Fermi energy *versus* filling factor for the (a) as-grown and (b) *H*-passivated samples.

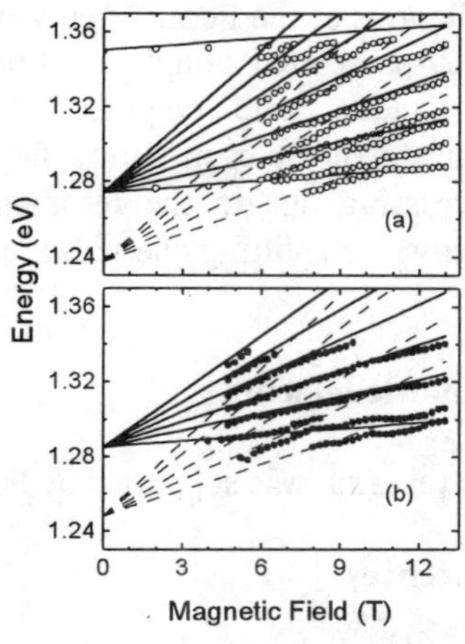

Fig. 3: PL peak energies *versus* magnetic field for the (a) as-grown and (b) *H*-passivated samples.

Figure 3 shows the PL peak energies for all transitions (see Fig.1) as a function of magnetic field for both samples. The straight lines correspond to inter-

Landau level transition energies and the dotted lines to inter-Landau level transition energies subtracted by one longitudinal optical (LO) phonon energy. The latter are associated to phonon assisted transitions[6]. The crossing point between straight and dotted lines occurs when an integer of the cyclotron energy is equal to one LO-phonon energy. In this case a resonant magneto-polaron effect[6] occurs and a well defined anticrossing is observed. The energy separation between upper and lower branches at the crossing region is related to the electron-phonon interaction potential. The estimated relative energy, using a pure 2D magneto-polaron model[7], is ~ 12 meV. This value is significantly higher than the ~ 5 meV obtained for the as-grown sample (see Fig. 3). The reduction of the anticrossing separation energy is attributed to strong screening effects of the electron-phonon interaction[6].

4. Conclusions

Magneto-PL measurements were performed on modulation doped GaAs /$In_{0.2}Ga_{0.8}As/Al_{0.25}Ga_{.75}As$ single quantum wells with two different 2D electron density. The density of carriers were controlled by using a hydrogen passivation technique. Magneto-PL spectra exhibited several structures that were attributed to interband Landau level and LO-phonon satellite transitions. Strong peak intensity oscillations at the Fermi energy were observed in the PL spectra of both samples. The *H*-passivated sample exhibits a continuous decrease of the peak intensity as the Landau level empties. The as-grown sample shows a peak intensity enhancement at even filling factors. The satellite emission lines show a clear anticrossing due to the resonant magneto-polaron effect. The small values of anticrossing splitting energies are attributed to screening effects.

Acknowledgments

This research was supported by Fapesp, Finep and CNPq.

References

1. T. Uenoyama and L.J. Sham, *Phys. Rev.* B**39**, 11044 (1989).
2. P. Hawrylak, N. Pulsford, and K. Ploog, *Phys. Rev.* B **46**, 15193 (1992).
3. M.S. Skolnick, *et al. Phys. Rev. Lett.*, **58**, 2130 (1987).
4. W. Chen, *et al. Phys. Rev. Lett.*, **64**, 2434 (1990).
5. S.K. Lyo, E.D. Jones, and J.F. Klem, *Phys. Rev. Lett.*, 61, 2265 (1988).
6. P.E. Simmonds, *et al. Phys. Rev.* B **45**, 9497 (1992); Swierkowski *et al. Phys. Rev.* B **51**, 9830 (1995).
7. F.M. Peteers and J.T. Devreese, *Phys. Rev.* B**31**, 3689 (1985).

ROLE OF HOLE MASS IN SHAKE-UP PROCESSES IN A 2DEG IN MAGNETIC FIELD

H.P.VAN DER MEULEN, J.RUBIO AND J.M.CALLEJA
Instituto Nicolas Cabrera, Universidad Autónoma de Madrid, Cantoblanco, 28049 Madrid, Spain

R.BERGMANN, F.SCHOLZ, V.HÄRLE AND H.SCHWEIZER
Physikalisches Institut der Universität Stuttgart, Pfaffenwaldring 57, 70550 Stuttgart, Germany

Hole mass has been found to play no significant role in the intensity of shake-up excitations observed in the photoluminescence spectra of a two-dimensional electron gas in a magnetic field. This result is in contrast with the observed clear dependence of the optical singularities at the Fermi edge on the hole mass, demonstrated by an increased singularity at increasing bandgap. Photoluminescence spectra at zero field of a non uniform modulation doped single quantum well of $In_{0.53}Ga_{0.47}As$ show correlated changes of the emission energy and the intensity at the Fermi-edge, which are attributed to the varying hole mass. This does not affect the shake-up emission peaks, in accordance to existing calculations

1 Introduction

Many-body processes are known to have a great influence on the optical properties of a degenerate electron gas[1,2]. In the photoluminescence spectra of a two-dimensional electron gas (2DEG) these processes manifest themselves in a strong emission at the Fermi level (Fermi-edge singularity or FES) and low energy contributions or sidebands in the emission spectra due to Auger-like processes (shake-up emission or SU). The occurrence of these processes is closely related to the Coulomb interaction between the electron gas and the photocreated hole. In a previous study[3] it has been demonstrated that the shake-up excitations show drastic changes, which are periodic in the filling factor, due to the magnetic field-dependent screening of the electron-hole Coulomb interaction.

In this work the dependence of these effects on the hole mass has been investigated by a series of experiments on the same sample, in which a continuous variation of the valence band structure exists due to non uniformity during growth.

2 Experiment

Photoluminescence spectra of a 2DEG have been measured on an 8 nm thick quantum well of nominally lattice-matched $In_{0.53}Ga_{0.47}As$, grown on a (100) InP substrate by low pressure metalorganic vapor phase epitaxy. The sample is modulation doped with 10 nm layers of InP containing 1.7×10^{18} S atoms per cm^3

in both barriers, separated from the well by 15 nm thick spacer layers, and covered by a 10 nm wide InP cap layer. The well has an electron density of about 8×10^{11} cm^{-2}, measured by Shubnikov-de Haas oscillations, which is in good agreement with the Fermi-energy of around 45 meV obtained from the emission spectra. The mobility at 4.2 K is about 175000 cm^2/Vs. Photoluminescence spectra have been obtained by illuminating with the blue-green lines of an Argon laser using 1 mW power, with the sample in a bath cryostat at a temperature of 2 K. Magnetic fields up to 13 T were applied in the Faraday configuration.

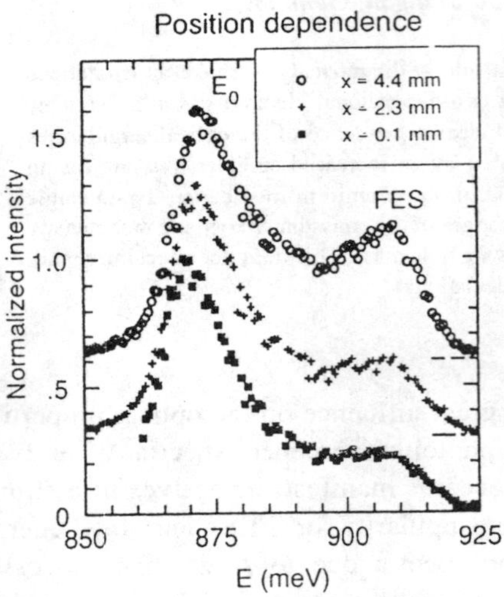

Figure 1 Photoluminescence spectra obtained by illuminating 3 different positions of the modulation doped quantum well. The increase in the bandgap is accompanied by an increase in the FES intensity.

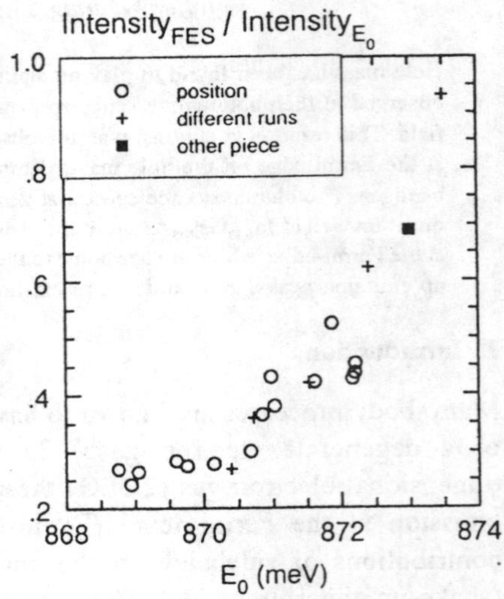

Figure 2 Relative intensity enhancement at the Fermi level vs the position of the bandgap E_0, in photoluminescence experiments.

3 Results

3.1 Fermi Edge Singularity without magnetic field

Luminescence spectra have been measured at different points of the sample (of dimensions 4 x 5 mm^2). Moving in one direction the bandgap shifts to higher energy and the intensity of the FES increases, whereas in the perpendicular direction no changes occur. Figure 1 presents a selection of three spectra showing this effect: a change in the position of the lowest emission line (from 868.7 to 872.2 meV) is interpreted as a change in the bandgap E_0, because the position of the Fermi level is constant over the whole sample. Simultaneously with the increase of E_0, the FES intensity relative to the intensity at E_0 increases.

Both facts are correlated, as shown in figure 2, where spectra taken at different sample points, or in different runs (between which the sample has been warmed up to room temperature), show the same indicated trend.

3.2 Magnetic Field Experiments

On applying magnetic field the electron states become quantized into Landau levels. The FES is seen now as an enhanced intensity of a Landau level when its energy is close to the Fermi level. Photoluminescence spectra taken in magnetic field, at different sample points, corresponding to the above mentioned runs, show also this variation in intensity at the Fermi level. Shake-up excitations are seen as peaks in the spectrum below the bandgap, which are separated roughly by the cyclotron energy[1].

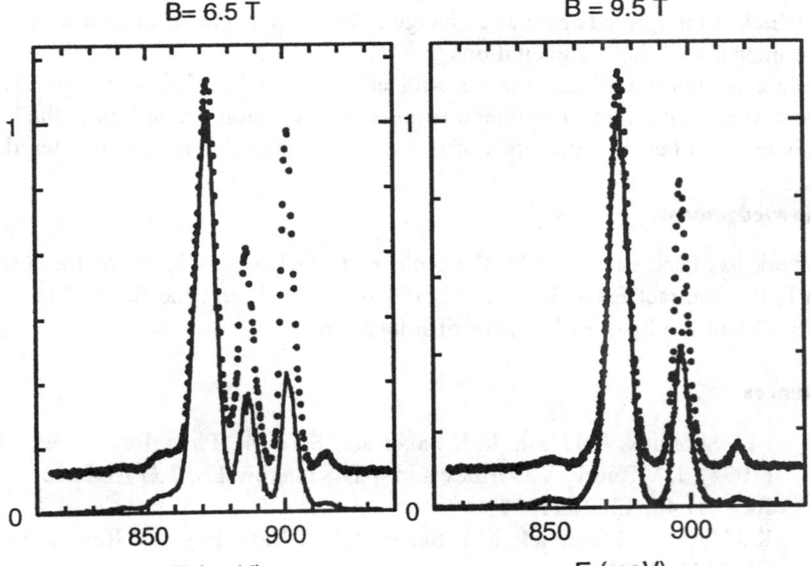

Figure 3 Photoluminescence in a magnetic field. For two different runs - which represent difference in hole mass - the intensity of the shake-up excitation remains unchanged, whereas the intensity near the Fermi edge varies strongly. This is presented for magnetic field of 6.5 and 9.5 T.

In figure 3 a comparison is made between spectra taken in two different runs, at two magnetic fields where the SU peak has its maximum (at even filling factor). A remarkable difference can be seen: the intensity of the FES (in the Landau level near the Fermi level) varies strongly, whereas the intensity of the shake-up excitations remains unaltered.

4 Discussion

The variation observed in the bandgap and in the FES intensity as function of the sample site can be explained by a gradually changing composition of the well, due to slightly inhomogeneous growing conditions. Experimental and theoretical studies of the composition variation in the InGaAs quantum well[4] have shown that a decrease of the In-concentration below the value of 53%, results in an increase of the hole mass (which gives rise to a higher FES intensity), and shifts the bandgap to higher energies. This is the correlation as presented in figure 2. Other explanations as a gradual change in the well thickness or the electron concentration cannot be completely ruled out, but would have hardly any effect on the valence band structure. The results in magnetic field show a difference in intensity at the Fermi level which is interpreted again as a change in hole mass. This change has no effect on the intensity of the SU excitations.

These results are in accordance with calculations by Uenoyama and Sham[5] demonstrating that on lowering the hole mass, the emission strength near the Fermi level is reduced but the intensities of the shake-up satellites remain unaltered.

Acknowledgements

This work has been supported by the project MAT94-0982-C02-01 of the Spanish CICYT, the contract ERBCHBICT9411193 of the E.U. and the fellowship SB95-A00789933 of the Spanish Ministry of Education.

References

1 M.S.Skolnick, K.J.Nash, M.K.Saker and S.J.Bass, Phys. Rev. B **50**, 11771 (1994); L.V.Butov, V.I.Grinev and V.D.Kulakowskii, T.G.Anderson, Phys. Rev. B **46**, 13627 (1992).
2 K.J.Nash, M.S.Skolnick, M.K.Saker and S.J.Bass, Physical Review Letters **70**, 3115 (1993).
3 H.P.van der Meulen, J.Rubio, J.M.Calleja, H.Schweizer, F.Scholz, V.Härle and R.Bergmann, to be published in Surf. Science.
4 V.Härle, thesis, University of Stuttgart, Germany (1994).
5 T.Uenoyama and L.J.Sham, Physical Review Letters **65**, 1048 (1990).

PHOTOLUMINESCENCE OF PERIODICALLY DELTA-DOPED InP IN HIGH MAGNETIC FIELDS

A.B. HENRIQUES, L.C.D. GONÇALVES, V. BINDILATTI, N.F. OLIVEIRA, JR.
Instituto de Física da Universidade de São Paulo
Caixa Postal 66318, 05389-970 São Paulo, Brazil

P.L. SOUZA, B. YAVICH
Centro de Estudos em Telecomunicações, Pontifícia Universidade Católica
22453 Rio de Janeiro, Brazil

The photoluminescence (PL) of InP with a periodical doping with Si was studied in magnetic fields of intensity 0–15 Teslas. The set of samples selected for the experiments had a doping period varying in the range 100–300 Å and a fixed sheet carrier density per doping period of approximately $4.3 \times 10^{12} cm^{-2}$. A very well defined Landau level fan diagram is obtained by measuring the PL intensity at a fixed wavelength as a function of applied magnetic field. The samples were characterized by C-V profiling, Hall, and Shubnikov–de Haas (SdH) measurements. On the basis of this characterization the theoretical Landau fan level diagram was calculated, which agrees with the experimental one with no adjustable parameters. The oscillator strength of the photoexcited holes with the individual minibands is discussed.

In n-type periodically delta-doped semiconductors equally spaced sheets of the host crystal are doped with shallow donors. In such systems the carriers released from the donor atoms are confined by a self-consistent periodic potential which gives rise to a miniband structure. The existence of a miniband structure has been demonstrated by PL [1,2] and by SdH measurements [3,4,5]. From the theoretical point of view it is predicted that under a magnetic field applied perpendicular to the dopant sheets each miniband will develop a single series of weakly defined Landau levels [6]. It should be possible to detect the formation of the Landau levels optically, however, optical experiments in high magnetic fields on periodically doped semiconductors have not been reported, which is probably a consequence of the strong electron-impurity interaction in these systems [7,8,9,10], which leads to a broad (unresolved) Landau structure.

The delta-doped InP structures were grown by LP-MOVPE at the Centro de Estudos em Telecomunicações of the Pontifícia Universidade Católica. The very high quality of the samples is demonstrated by their Hall and CV spectra, which showed that the donor atoms are spread over not more than two monolayers of the host crystal [11]. The doping periodicity was also extracted from CV measurements; Table 1 summarizes the results. PL was detected using an unfocused low power laser beam; the bandpass of the spectrometer was fixed at

Table 1: Data on the samples studied. Hall data was taken at room temperature. The sheet carrier density per period, n_{SdH}, was inferred from the Shubnikov-de Haas spectrum. Using the value of n_{SdH}, the position of the Fermi energy relative to the miniband E2, EF_2=EF-E2, is estimated from a self-consistent solution of Poisson and Schrodinger equations for the structure[13].

Sample	Period (Å)	$n_{Hall}(cm^{-2})$	$n_{SdH}(cm^{-2})$	EF	EF-E2
199	246	4.1×10^{12}	4.3×10^{12}	1.436	0.0373
200	176	4.2×10^{12}	4.2×10^{12}	1.445	0.0435

< 1 meV. Light was conveyed to the sample, and collected from the sample by using an optical fiber bundle. SdH measurements were made in the constant current mode, using the van der Pauw method, and the Joule heat dissipated on the sample was less than 10-20pW.

Figure 1 shows the near band-edge PL as a function of the magnetic field intensity for sample 200 at 4.2 K. The weak structure seen is due to the development of Landau levels in the E2 miniband. This is demonstrated by the concomitance of the oscillations in the PL intensity at a fixed photon energy as a function of applied magnetic field and the SdH oscillations, as Figure 2 shows. The SdH oscillations are dominated by the carriers occupying states in the E2 miniband, which have a higher mobility than carriers from the fundamental E1 miniband due to a smaller overlap of the E2 charge density with the impurity layers[10]. The Fourier transform of the SdH oscillations gives the fundamental field BF(E2)=32.3 T; using this value, and also the periodicity and donor layer width extracted from the CV profile, the electronic structure of the sample can be modeled[12], and the results of the theory are given in Table 1. In Figure 2 the photon energy was fixed at $h\nu = 1.445$ eV, which was chosen so as to have the oscillations in the PL intensity in phase with the SdH oscillations, which determines the position of the Fermi energy relative to the photoexcited holes to be 1.445 eV. Using the value of EF-E2 given in Table 1, the energy position of the electronic miniband relative to the photoexcited holes is obtained to be 1.4015. The Landau level diagram for the E2 miniband can then be drawn, as shown in Figure 3. In order to check this result, the position of the Landau levels as a function of the magnetic field intensity should be measured optically. To determine the field dependence of the Landau levels, the PL intensity was measured as a function of the field at a fixed photon energy; this is shown in Figure 4. The peaks shown obtained from Figure 4 are shown as solid circles in Figure 3, giving an excellent agreement with the theoretical lines.

Assuming that the holes have only a weak dispersion and that the electron-

Figure 1

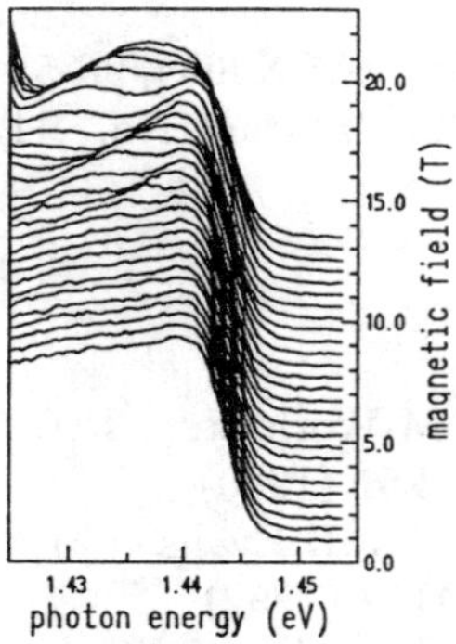

Figure 2

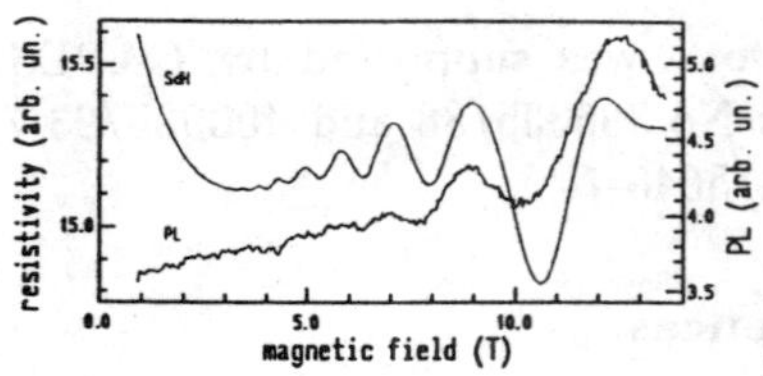

Figure 3

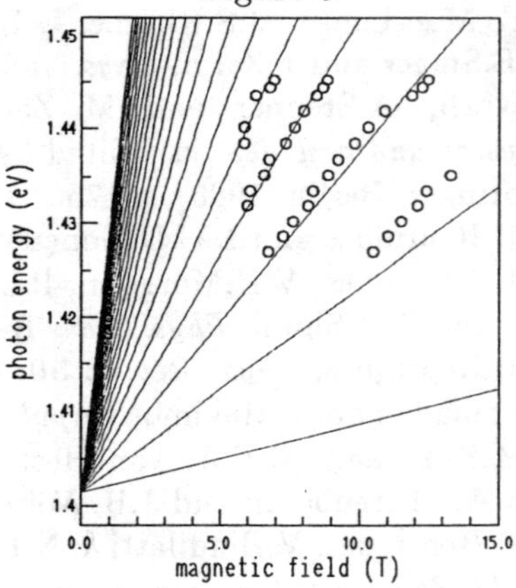

Figure 4

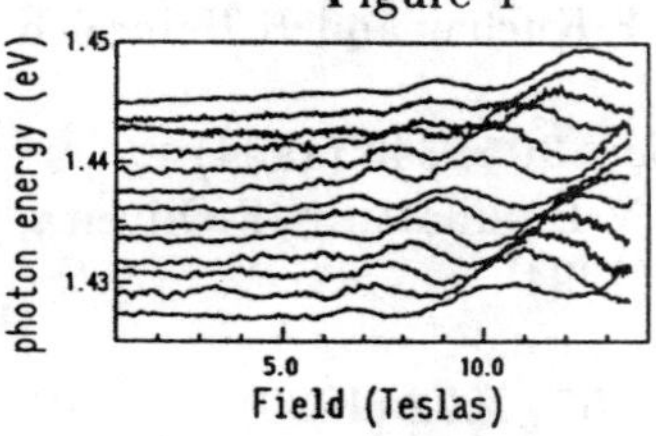

Figure 5

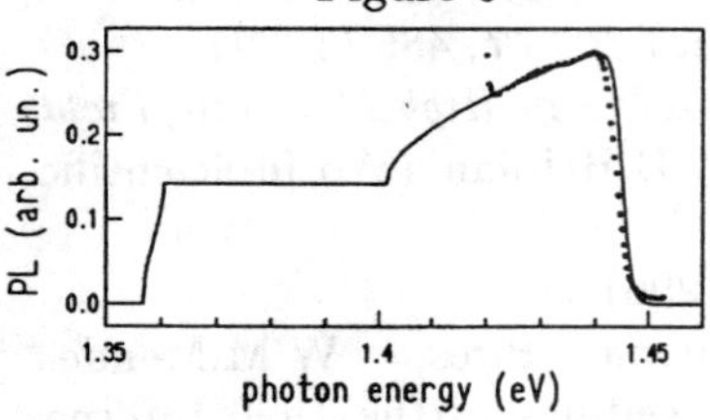

hole oscillator strength is energy independent, under weak excitation the zero-field PL lineshape will be given by

$$I(h\nu) = D(h\nu)f(h\nu, T) \tag{1}$$

where D(E) is the density of states (DOS) of the minibands and $f(E,T)$ is the Fermi-Dirac distribution. The DOS function was calculated using the parameters given above, and the results are shown in Figure 5. At lower energies, the DOS function overshoots the experimental curve, indicating that the overlap of the E1 miniband wave function with holes is weak.

Acknowledgments

This work was supported by CAPES, Grant No. AEX0654/96-5, CNPq, Grants No.306335/88 and 400935/93-7, and FAPESP, Grant No. 94/2155-9 and 94/5645-7.

References

1. A.C.Maciel, M.Tatham, J.F.Ryan, J.M.Worlock, R.R.Nahory, J.P.Harbison and L.T.Florez *Surf. Sci.* **228**, 251 (1990)
2. Ke Mao-Long, J.S.Rimmer, B.Hamilton, J.H.Evans, M.Missous, K.E.Singer and P.Zalm, *Phys. Rev.* B **45**, 14114 (1992)
3. F.Koch, A.Zrenner and M Zachau, in *Two-dimensional systems: physics and new devices*, edited by G.Bauer, F.Kuchar and H.Heinreich (Springer, Berlin, 1986), p.175.
4. A.B.Henriques and L.C.D.Gonçalves, *Surf. Sci.* **305**, 343 (1994)
5. A.B.Henriques, V.N.Morgoon, P.L.de Souza, V.Bindilatti, N.F.Oliveira, Jr., and S.M.Shibli, *Phys. Rev.* B **49**, 11248 (1994)
6. A.B.Henriques, *Phys. Rev.* B **50**, 8658 (1994)
7. S.Yamada and T.Makimoto *Appl. Phys. Lett.* **57**, 1022 (1990)
8. P.M.Koenraad, B.F.A. van Hest, F.A.P.Blom, R.van Dalen, M.Leys, J.A.A.J.Perenboom and J.H.Wolter, *Physica B* **177**, 485 (1992)
9. A.B.Henriques, V.Bindilatti & N.F. Oliveira Jr., in *High Magnetic Fields in the Physics of Semiconductors*, edited by D.Heiman (World Scientific, Singapore, 1995), p.604
10. A.B.Henriques, *Phys. Rev.* B , at press (1996)
11. B.Yavich, P.L. de Souza, M.Pamplona Pires, W.M.Mendes, A.B.Henriques and L.C.D.Gonçalves, Proceedings of the 1996 International Conference on MOVPE, Cardiff, U.K. (unpublished)
12. A.B.Henriques and L.C.D.Gonçalves, *Semicond. Sci. Technol.* **11**, 190 (1996)
13. A.B.Henriques, L.C.D.Gonçalves, P.L. de Souza and B.Yavich, *Semicond. Sci. Technol.* **8**, 585 (1993)

OPTICAL STUDY OF CdHgTe HETEROSTRUCTURES UNDER HIGH MAGNETIC FIELD

S. MORDENTI, M. ZIGONE, G. MARTINEZ
Grenoble High Magnetic Field Laboratory MPI-FKF/CNRS
BP. 166 F-38042 Grenoble Cedex 9, France

G. MULA, J. L. PAUTRAT, N. MAGNEA, J. BLEUSE
« Microsctructures de Semiconducteurs II-VI »
CEA/CNRS Joint Group, DRFMC, SP2M, LPS, BP. 85 X
F - 38041 Grenoble, Cedex, France

The luminescence properties of $Cd_xHg_{1-x}Te$ heterostructures are studied under high magnetic field. A quadratic blue shift of the luminescence peak is observed as field is increased. This field dependence is analysed in terms of an excitonic transition. This model reproduces quite well the data, but reveals some inconsistencies which favor more likely a donor- acceptor recombination or a localised electron-hole pair.

1 Introduction

CdHgTe heterostructures promise high-performance device for infrared detection and emission. However, although HgTe/CdTe binary material has been intensively studied[1], optical properties of $Cd_xHg_{1-x}Te$ based heterostructures are not well known[2-4]. The main optical transition observed in these heterostructures can be interpreted either as interband or excitonic or impurity related transitions. An unambiguous way to identify the origin of the transitions is to study them under high magnetic field B. In this work we present such an investigation, and show that the interpretation of spectra in terms of direct electron-hole recombination is not appropriate, because of a quadratic dependence of the peak of luminescence when the field is increased, which is more likely related to a donor-acceptor recombination or a localised electron-hole pair.

2 Samples and experimental technique

The two studied samples have been obtained by molecular beam epitaxy[4]. Both samples consist of five identical quantum wells of CdHgTe with a Cd concentration of 45 %, whereas in the barriers, built with the CdHgTe too, the Cd concentration reaches about 60 % and the width remains constant at about ≈ 100 Å. The wells are embedded in a GRINSCH structure which allows a better efficiency of the luminescence of the wells. The two samples differ by the thickness of the wells : 50 Å for sample A1 and 100 Å for sample A2. For both samples the energy of the interband transition e_1h_1 lies between 0.5 and 0.7 eV. Thus, the optical measurements are performed in the near infrared range. The luminescence spectra were excited with the 1.06 μm line of a continuous YAG laser, and performed at a temperature of about 5 K without polarisation, using a spectrometer equipped with an InSb detector. The excitation power on samples was kept at a level of about 10 W/cm^2, and the illuminated area was about 250 x 250 μm^2, which corresponds to the core diameter of the optical fibers. The focal length of the spectrometer is of 30 cm^{-1}, and the best resolution obtained with it is about 0.4 meV. A system of optical fibers, selected for infrared light propagation, provides the optical connections from the laser, the spectrometer and the sample which lies in a He-bath cryostat in the middle of a resistive magnet. This experimental set up prevents from water and CO_2 absorption band.

3 Experimental results and discussion

Fig. 1 depicts the luminescence spectra of samples A1 and A2 at a temperature of about 5 K and as a function of B applied perpendicularly to the layers. For both samples the general shape of spectra displays two important features when B is increased : i) no alteration of the line shape has been observed between 0 and 20 T. ii) only a weak blue shift of the peaks can be observed. For a quantitative study of the luminescence spectra we have calculated the moments[5] M_o, M_1 and M_2 of each spectrum.

The variations with B of M_0 and M_2 will not be discussed here. On the other hand, the variation of M_1 as a function of the field is shown in fig. 2 for both samples. It can be noticed that this variation is quadratic in both cases and moreover the quantitative shift is identical for the two thicknesses of the wells.

Between 0 and 20 T, the energy shift of the luminescence peak reaches about 10 meV. This kind of luminescence variation strongly indicates that the optical transition which is observed, does not correspond to a simple interband transition, but favors a recombination involving a binding energy. In the following we try to identify the origin of this binding energy. Different types of optical transitions are observed in an ideal 2D system of type I. Two of them can give a quadratic variation with the field : the transition between donors and acceptors and the recombination of the free electron-hole excitons[6, 7]. For the other mechanisms such as interband transitions the variation with field is expected to be linear at least at weak field. We have tentatively analysed our results in terms of an excitonic transition. We have calculated the behavior with B of a free exciton 1s-state using the method of the Two-point Padé Approximants developed by MacDonald and Richtie[6]. In this framework, assuming a strict 2D exciton in both samples, the energy dependence of the 1s state exciton with B can be written as follows :

$$E(B) = E_0 + R^*_{3D} \times \sum_{j=0}^{7} p_j z^j \Big/ \sum_{j=0}^{5} q_j z^j \qquad (1)$$

where $E_0 \approx 569.5$ meV for sample A1 and $E_0 \approx 511.3$ meV for sample A2. The effective Rydberg $R^*_{3D} = \mu e^4 / 2\hbar^2\varepsilon^2$ (μ is a reduced effective mass, e the absolute value of the electronic charge, ε is the dielectric function for the medium ≈ 14). The p_j and q_j coefficients are the Padé Approximants[8], which can be found in our case in ref. [6] (table V), and the variable $z = (\hbar\omega^*_\mu / 2R^*_{3D})^{1/2}$ where the cyclotron frequency $\omega^*_\mu = eB/\mu c$.

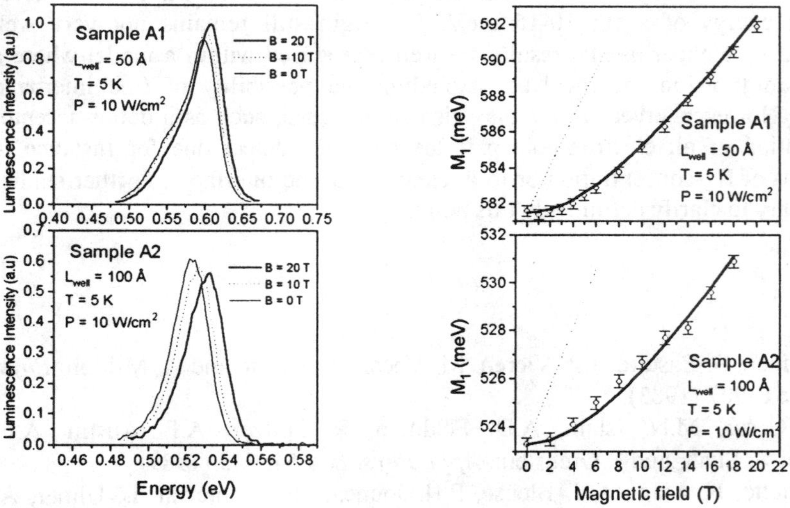

Fig. 1.Left Side : Luminescence line of sample A1 and A2 under high magnetic field.
Fig. 2. Right side : First moment M_1 (energy position) of the luminescence line for samples A1 and A2 as a function of the magnetic field. The open circles are the experimental points with error bars.
The solid line corresponds to the behavior expected from the excitonic model (see text).
The dotted line represents the calculated variation with field of a free electron-hole interband transition.

The results of this model are presented in Fig. 2 (continuous line). It is worthwhile to observe that the model reproduces quite well the experimental data. For both samples, we have found a reduce effective mass (as a fitting parameter) of $\mu^* = 0.044m_0$ (sample A1) and $\mu^* = 0.047m_0$ (sample A2) and therefore a value for $R^*_{3D} \approx 3$ meV. Indeed, with our assumptions of a strict 2D exciton, we can write the 2D binding energy : $R^*_{2D} = 4R^*_{3D}$. These results do not agree with the value $\mu^* = 0.026$ m_0 obtained by the dispersion relation. Also the full width at half maximum Γ is of about 40 meV, which means that $\Gamma >> R_{2D}^* \approx 12$ meV. These inconsistencies demonstrate the strong limitation of the model, and favor more likely the donor-acceptor or a localised electron-hole type of transition. This analysis contradicts the previous assignments of this optical transition, which was usually attributed to an electron-hole interband transition[4, 2].

4 Conclusion

We have studied the luminescence of CdHgTe heterostructures under magnetic field. A quadratic blue shift of the luminescence peak is found when the field is increased. This behavior reveals that the observed optical transition involves a binding energy of about 10-15 meV. Its origin still remains not very well understood. The experimental results are well reproduced within a model where a free exciton transition is involved, excluding the possibility of free interband transition. However, other origins may also be proposed, such as a donor-acceptor recombination or an electron-hole pair located on a defect due for instance to fluctuations of Hg-concentration or to fluctuations at the interfaces. Further studies are necessary to clarify definitively this point.

References

1. Y. Goldner, G. Bastard, J.P. Vieren, M. Voos, J.P. Faurie and A. Million, *Phys. Rev. Lett.* **51**, 907 (1983).
2. C. L. Cesar, M.N. Islam, A.D. Feldman, R. Spitzer, A.F. Austin, A.E. Digiovanni, J. Shah, and J. Orenstein, *Appl. Phys. Lett* **54**, 745 (1989).
3. H. Mariette, N. Magnea, J. Bleuse, P.H. Jouneau, E. Monterrat, L. Ulmer, A. Tardot, and J.P. Paulat , *Physica* **B 191**, 190 (1993).
4. E. Monterrat, L. Ulmer, R. Mallard, N. Magnea, J.L. Pautrat, *J. Appl. Phys.* **71**, 1774 (1992).
5. V.B. Timofeev in *Optical properties of semiconductors*, NATO ASI Series E, E= Applied Sciences, Vol. 228, edited by G. Martinez, *Kluiver Academic Publishers*, Dordrecht (1993), p 230.
6. A.H. MacDonald and D.S. Ritchie, *Phys. Rev* **B 33**, 8336 (1986).
7. M. Bugajski, W. Kuszko, K. Reginski, *Solid State Commun.* **60**, 669 (1986).
8. Georges A. Baker, *Essentials of Padé Approximants* (Academic Press, New York, 1975).

REFLECTIVITY, PHOTOLUMINESCENCE AND RESONANT RAMAN SCATTERING FROM SEMICONDUCTOR QUANTUM MICROCAVITIES

[1]M S Skolnick, [1]T A Fisher, [1]A Armitage, [1]A M Afshar, [1]D M Whittaker, [1]P Kinsler, [1]W R Tribe, [1]D J Mowbray, [1]D Baxter, [2]J S Roberts, [2]G. Hill and [2]M. A. Pate

[1]Department of Physics, University of Sheffield, Sheffield, S3 7RH, UK
[2]Department of Electronic and Electrical Engineering, University of Sheffield, Sheffield, S1 3DU, UK

The results of several new experiments on high quality semiconductor quantum microcavities are discussed. Particular topics addressed include the electro-optic tuning of the vacuum Rabi coupling, the effects of Zeeman splitting in high magnetic field, motional narrowing phenomena, the observation of a polariton relaxation bottleneck and anomalous Raman scattering from cavity polaritons.

1. INTRODUCTION

The ability to grow high quality semiconductor quantum microcavity structures (QMCs) has recently opened up a new area of study of exciton and polariton phenomena in semiconductors[1-9]. QMCs are Fabry-Perot cavities whose optical length is very close to the transition wavelengths of excitons in quantum wells (QWs) embedded in the cavity. The quantum wells in the QMC, and the cavity, provide separate control of the properties of the excitons and photons of the system, and of the interaction between them. A schematic diagram of one of our QMCs is shown in Fig 1.

In the resonance regime between the excitons and photons, the two modes couple together to form cavity polaritons[4]. The splitting between the coupled mode features on resonance is known as the vacuum Rabi splitting. It has an approximately square root dependence on exciton oscillator strength (f_{osc}). f_{osc} shows a marked variation with both applied electric and magnetic fields. As a result we show below that we are able to achieve controlled tuning of the vacuum Rabi coupling in external electric and magnetic fields.

The structures investigated here are of very high quality such that the strong coupling limit is easily achieved,[1,2] with coupled mode line widths on resonance very much less (by a factor of ~7) than the splitting of the two polariton modes. In the strong coupling limit, oscillations of the excitations of the system are expected between the two coupled modes (Rabi oscillations), in contrast to the weak coupling case where irreversible decay by spontaneous emission occurs.

Particular results which we emphasise include the modification of the vacuum Rabi splitting by controlled variation of the exciton oscillator strength, the effects of exciton Zeeman splitting, anomalous linewidth narrowing which we attribute to the occurrence of motional narrowing on resonance, the observation of a polariton relaxation bottleneck and Raman scattering phenomena. Experimental details are presented in section 2, followed by discussion of electric and magnetic field tuning in sections 3 and 4. The anomalous narrowing phenomena are described in section 5, followed by the photoluminescence and Raman scattering results in sections 6 and 7, followed by conclusions in section 8.

2. EXPERIMENTAL

The QMC structure investigated in most detail (see Fig 1) was grown by atmospheric pressure metal-organic vapour phase epitaxy. It consists of a λ GaAs cavity sandwiched between two distributed Bragg reflector mirrors which each consist of 20 periods of $\lambda/4$ AlAs and $Al_{0.13}Ga_{0.87}As$ layers. The mirrors are p and n doped so that the cavity forms the intrinsic region of a p-n junction. For further experimental details, see references 6 and 7.

3. ELECTRO-OPTIC TUNING OF THE VACUUM RABI COUPLING

Electric field tuning is achieved in the following way. The cavity-photon mode energy is unaffected by electric field, whereas the exciton energy is reduced by the Quantum Confined Stark Effect. Fig. 2 shows reflectivity spectra taken at a series of electric fields. At zero electric field (Fig 2(a)) the exciton feature (labelled X) is 7.7 meV to higher energy than the cavity peak. With increasing field the exciton moves closer to the cavity peak (Fig 2(b), 2(c)). At the same time the predominantly excitonic feature increases in intensity as coupling with the cavity increases, until at 37.8 kV/cm (Fig 2(d)) the exciton and photon are on-resonance and two peaks of equal intensity are seen. As the electric field is increased further (Fig. 2(e) and (f)) the exciton moves through the cavity mode and weakens again. The observed peak positions are plotted against electric field in Fig 3 (circles). The experimental points (circles) show clear anti-crossing behaviour in the resonance region (labelled Ω_{HHx}).

The experimental data were modelled using a Transfer Matrix Reflectivity (TMR) model [2,6,7]. The first step is to estimate the variation with electric field of the unperturbed exciton and cavity features. The cavity photon mode energy is constant with electric field and is estimated from its relatively unperturbed positions at 0 and 60kV/cm (Fig 3 - dashed line, labelled c). The variation of the unperturbed exciton energy was obtained from QW exciton calculations.

The variation of exciton oscillator strength with applied electric field is needed in order to carry out the fits to the data of Fig 3. At each electric field temperature is varied to bring the exciton and cavity photon modes on to resonance. We vary f_{osc} in the TMR calculations in order to fit the observed vacuum Rabi splitting (Ω_{HHx}) at resonance at each electric field. The variation of f_{osc} with elecric field is thus obtained. The splitting at resonance (Ω_{HHx}) is expected to vary approximately as $f_{osc}^{1/2}$, and thus is controlled by the variation of f_{osc} with electric field. The splitting was found to reduce from 5.1 meV at 6.9 kV/cm to 3.7 meV at 38.6 kV/cm and corresponds to a reduction in oscillator strength by nearly a factor of 2 from E=0 to 38.6 kV/cm. The polariton energies were then calculated using the unperturbed exciton and cavity positions and the deduced variation of f_{osc} with E. The fit to the experimental data is extremely good, as shown in Fig. 3 (solid lines)[6].

To summarise this section, we have demonstrated electro-optic control of the vacuum Rabi splitting by application of external electric fields. A reduction in the splitting by a factor of ~1.5 was found in a field of ~40kV/cm[6].

4. MAGNETIC FIELD ENHANCEMENT OF THE VACUUM RABI SPLITTING AND OBSERVATION OF ZEEMAN SPLITTING

With increasing magnetic field the exciton oscillator strength increases due to shrinkage of the exciton wave function, in contrast to the decrease of oscillator strength in electric field. In addition there is a diamagnetic upshift in the energy of the exciton, as opposed to the decrease in energy in applied electric field.

Reflectivity spectra as a function of temperature at a magnetic field of 14 Tesla are shown in Fig 4. At all magnetic fields the spectra are observed to be strongly polarised, with a clear Zeeman splitting observed between the σ^+ and σ^- components at each magnetic field. With increasing temperature, the exciton-like feature shifts to lower energy and passes through the resonance regime with the cavity mode at temperatures of 70 to 95K. The vacuum Rabi splitting (VRS) on resonance is 7.3 meV, as opposed to 5 meV at B=0, corresponding to the factor of ~2 increase in exciton oscillator strength expected from B=0 to 14T[7,9].

The Zeeman splittings for the two polariton branches for the σ^+ polarisation are plotted as a function of temperature in Fig 5. At low temperature the upper polariton branch shows large Zeeman splitting, whereas at high temperature its splitting is small. At high temperature the lower branch instead shows large Zeeman splitting. At resonance the two splittings are equal.

This behaviour can be understood qualitatively by noting that the exciton component of the polariton is split in magnetic field, whereas the cavity mode is unaffected. Away from resonance the splitting of the exciton-like component (the upper branch at low temperature, the lower branch at high temperature) will tend towards that of the uncoupled exciton. In the resonance regime, smaller splittings would be expected since the two modes are strongly mixed (we show in reference 7, using degenerate perturbation theory, that the on-resonance splitting should be one half the off-resonance value, in good agreement with the results of Fig 5). The splittings observed away from resonance (1.4meV at 14T) are reasonably close to those reported previously for InGaAs QWs of similar well width[10,11]. We also show in reference 7 that a very good fit to the spectra on resonance can be obtained using the TMR model, demonstrating that each exciton spin component interacts independently with each circularly polarised cavity mode.

5. MOTIONAL NARROWING

The term motional narrowing is applied to many situations in which the width of a spectral line in a disordered system is reduced by an averaging process. In this section we present experimental evidence for the observation of such motional narrowing in the reflectivity spectra of QMCs, the first report of such phenomena in the exciton spectra of semiconductors. The in-plane dispersion curves of the uncoupled exciton and cavity modes, and of the coupled cavity polaritons are shown in Fig 6. The cavity photon has strong dispersion in the near resonance regime, and can be described by a very small mass of $10^{-5}m_0$. As a result of the small photon 'mass', the polaritons also have very small in plane mass ($\sim 10^{-5}m_0$)[12], very much less than that of the bare exciton ($\sim 0.25m_0$). The polariton wavefuction is thus very extended in real space, and leads to marked averaging over in-plane potential fluctuations, and hence significant narrowing of the spectral lines in the resonance regime[13].

The experimental evidence for the motional narrowing is shown in Fig 7. The linewidths of the lower (upper) branch polariton features from Fig 5 are plotted as the filled (open) symbols on the figure, as a function of detuning from resonance. The horizontal axis is expressed as the exciton fraction of the polariton wavefunction ($|c_e|^2$). At resonance $|c_e|^2=1/2$. Linewidths from electric field tuning experiments are also plotted (the triangles) and are seen to be in very good agreement with the magnetic field results. This allows us to exclude the influence of temperature dependent effects on the linewidths from the magnetic field data. The most notable feature of Fig 7 is the decrease in the linewidth near resonance for the lower branch. This behaviour cannot be explained by simply averaging the homogeneous linewidths (the convolution of two Lorentzians). It also cannot be explained using a model which convolves a Gaussian exciton lineshape with the Lorentzian cavity function[14] (the dashed line on the figure). By contrast our motional narrowing model provides a very good description of the experimental data (the full line). This model[13] solves the Schrodinger equation for the centre of mass motion of the polariton in a white noise disorder potential having correlation length very much less than the length scale of the very extended polariton wavefunction.

The fit to the upper branch is much less good, with the observed linewidths being significantly greater than predicted on Fig 7 and observed for the lower branch. The greater linewidth of the upper branch has been ascribed previously at B=0 to coupling to exciton continuum states[2]. However, this cannot apply at high magnetic fields[7], where the nearest optically allowed states are at much higher energy. We speculate that the greater upper branch width may be due to coupling to higher energy optically forbidden states, mixed into the wavefunction by the disorder potential. We return to this discussion in section 7.

6. RELAXATION BOTTLENECK IN PHOTOLUMINESCENCE SPECTRA

Low temperature photoluminescence spectra at 1.7K as a function of incident laser power, are presented in Fig 8 for an on-resonance sample. The spectra are excited at 1.56eV, outside the high reflectivity region of the DBR mirrors. The light is absorbed strongly in the GaAs cavity, from where the excitons diffuse to the QWs, are captured and recombine as cavity polaritons[4,8]. The most striking feature of the spectra is the increasingly high relative intensity of the upper polariton peak as the laser power is reduced. A strongly non-thermal polariton population is observed (if thermal equilibrium were reached, then the population of the upper branch would be negligibly small at the lattice temperature of 1.7K, given that the Rabi splitting is 5meV), with a population inversion being found at the lowest powers.

This relaxation bottleneck for the cavity polaritons arises for the following reasons. Acoustic phonon scattering, already relatively slow in a QW (~100psec, compared to cavity polariton lifetimes of 10psec to 1nsec)[15] is expected to be further reduced in a QMC[12] because of the small mass and hence very low density of states of cavity polaritons, as discussed in the previous section. Indeed, Tassone et al[16] have recently predicted a population inversion inversion between the polariton branches close to k=0. In addition, two effects are likely to enhance further the bottleneck and population inversion in our work: the weakness of exciton-exciton scattering at low excitation levels where we observe the inverted population, and a reduction of the polariton lifetime due to tunnelling from the QWs under the residual p-i-n junction fields in our samples.

7. RESONANT RAMAN SCATTERING

Resonant Raman spectra are presented in the lower part of Fig 9 for the same on resonance sample employed for the PL spectra of Fig 8. The spectra are presented for the outgoing Raman channel, where the Raman-scattered photons, at the GaAs

LO phonon energy below the laser line, are resonant with the energies of the coupled polariton modes. The resonant Raman profile (RRP) is shown in the upper part of the figure. Two resonant features are observed, corresponding to the upper and lower polariton branches, with the higher energy feature being much weaker than the lower energy one. This intensity ratio is at first sight highly surprising since the two polariton states are expected to have equal excitonic character. However, from standard perturbation theory, the strength of the coherent Raman signal on resonance is proportional to (Γ_h^{-2}), where Γ_h is the homogeneous linewidth of the resonant state. It is likely that the upper state has shorter dephasing time and hence larger Γ_h, than the lower energy state, due to scattering into the excitonic continuum. As a result a much weaker Raman cross-section for the upper branch can thus be accounted for, as observed.

Finally we recall the remark we made at the end of section 5 on the possible influence of scattering into higher energy states leading to the greater linewidth of the upper polariton branch observed in reflectivity measurements. As discussed there whether this interpretation of the linewidths is correct is still an open question. However, Raman scattering being a coherent process is highly sensitive to dephasing processes. The results we present here provide further clear evidence that the two coupled polariton mode states are not fully equivalent; the present results show that dephasing processes are much more important for the higher energy polariton component.

8. CONCLUSION

To summarise this paper, we have presented the results of several of our recent experiments on semiconductor quantum microcavities. We have attempted to demonstrate the breadth of new phenomena which arise in the study of QMCs. Many of the these phenomena arise because of the controllability of the exciton-photon polariton interactions in QMCs. In addition we have shown how the very small in-plane mass of the cavity polaritons leads to several new interesting effects including the occurrence of 'motional narrowing', and a relaxation bottleneck for polaritons in microcavities. The discussion has focussed on our own results. However very notable contributions have of course been made by other workers, including the measurement of polariton dispersion curves in angle resolved PL[4], the demonstration of extremely strong enhancements of Raman cross-sections in QMCs[16], the observation of strongly reduced lifetimes for radiative recombination in the resonance regime[17], the observation of Rabi oscillations[18], the demonstration of Rabi coupling to quantised continuum states in magnetic field[9] etc. In these cases also the new advances have been achieved as result of the novel properties of exciton-polaritons in QMCs.

References

1 C Weisbuch, M Nishioka, A Ishikawa and Y Arakawa, Phys Rev Lett 69, 3314 (1992).
2 R Houdré et al Phys Rev B49, 16761 (1994).
3 R P Stanley et al, Appl. Phys. Lett, 65, 1833 (1994); ibid. 65 (2093) 1994
4 R Houdré, C Weisbuch, R P Stanley, and M Ilegems, Phys. Rev. Lett, 73, 2043 (1994)
5 T A Fisher, A M Afshar, D M Whittaker, M S Skolnick et al, Phys Rev B51, 2600 (1995)
6 T A Fisher, A M Afshar, M S Skolnick et al Solid State Electron 40, 493, 1996
7 T A Fisher, A M Afshar, D M Whittaker, M S Skolnick, J S Roberts, Phys Rev B53, R10469, 1996
8 R P Stanley, R Houdre, C Weisbuch, U Oesterle and M Ilegems, Phys Rev B53, 10995, 1996
9 J Tignon, P Voisin, C Delalande, M Voos et al Phys Rev Lett 74, 3967 (1995)
10 Th Wimbauer, K Oettinger, A L Efros and B K Meyer, Phys. Rev, B50, 889 (1994)
11 N J Traynor, R T Harley and R J Warburton, Phys Rev, B51, 7361 (1995)
12 S Pau, G Bjork, J Jacobson, H Cao and Y Yamomoto, Phys Rev B51, 7090, 1995
13 D M Whittaker, P Kinsler, T A Fisher, A Armitage, M S Skolnick, submitted
14 P Kinsler and D M Whittaker, Phys Rev B, 1996 in press.
15 J Bloch et al, Solid State Electron 40, 1996
16 F Tassone, C Piermarocchi, V Savona, A Quattropani et al Phys Rev B53, R76423, 1996
17 A Fainstein, B Jusserand and V Thierry-Mieg, Phys Rev Lett 75, 3764, 1996
18 T B Norris et al Phys Rev B50, 14663, 1994

Figure captions

Fig 1. Schematic diagram of high finesse quantum microcavity structure.

Fig 2. Electro-optic tuning of exciton-photon interaction, showing increase of intensity of the exciton feature as it approaches the cavity mode. At resonance the two modes are coupled and the features have equal intensity. With further increase of electric field, the exciton component moves further to lower energy and weakens again.

Fig 3. Experimental reflectivity peak positions as a function of electric field (solid circles). Dashed lines show unperturbed exciton (X) and cavity (C) energies. The full line is the TMR fit to the data.

Fig 4. Reflectivity spectra at 14 Tesla as a function of temperature for σ^+ (full line) and σ^- (dashed line) circular polarisations.

Fig 5. Zeeman splittings between the σ^+ and σ^- transitions of Fig 4 for the upper and lower polariton branches. At resonance the splittings of the two branches are equal.

Fig 6. In-plane dispersion curves in QMC. The dashed lines are the un-coupled modes, full lines - coupled polariton modes. The dotted curves are parabolic approximations to the coupled mode dispersion curves.

Fig 7. σ^+ linewidths against exciton fraction of the polariton $|c_e|^2$. Filled symbols - lower branch, open symbols - upper branch, squares - B field, triangles - E field data, full line - motional narrowing theory, dashed line - Gaussian-Lorentzian convolution.

Fig 8. PL measurements as a function of laser power.

Fig 9. Outgoing resonant Raman measurements - lower curves-series of resonant Raman spectra, upper part (RRP)- resonant Raman profile.

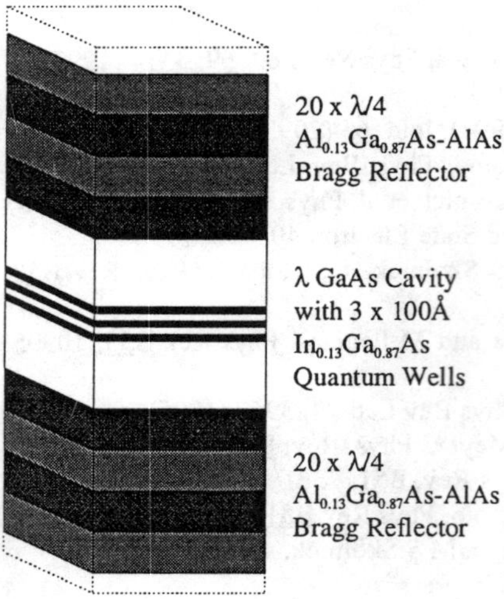
20 x λ/4
Al0.13Ga0.87As-AlAs
Bragg Reflector
λ GaAs Cavity
with 3 x 100Å
In0.13Ga0.87As
Quantum Wells
20 x λ/4
Al0.13Ga0.87As-AlAs
Bragg Reflector

1

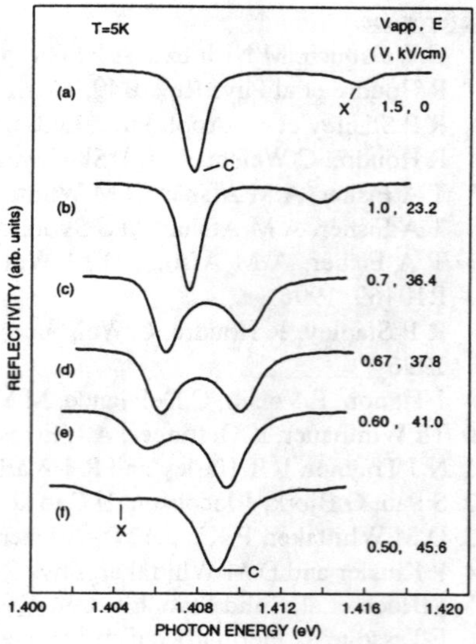
T=5K
Vapp, E
(V, kV/cm)
(a)
1.5, 0
(b)
1.0, 23.2
(c)
0.7, 36.4
(d)
0.67, 37.8
(e)
0.60, 41.0
(f)
0.50, 45.6
X
C
REFLECTIVITY (arb. units)
1.400
1.404
1.408
1.412
1.416
1.420
PHOTON ENERGY (eV)

2

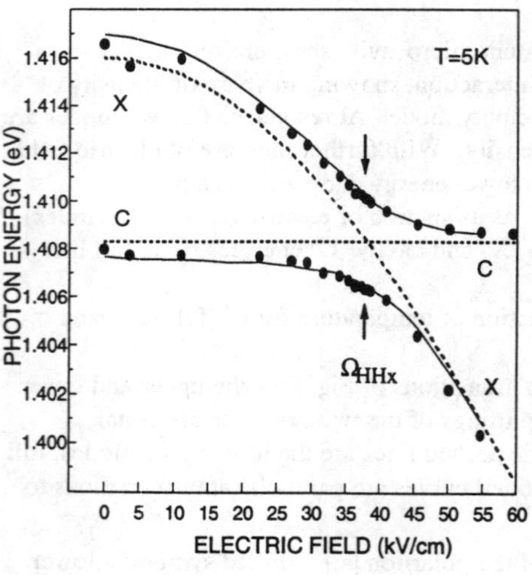
T=5K
X
C
C
X
ΩHHx
PHOTON ENERGY (eV)
1.416
1.414
1.412
1.410
1.408
1.406
1.404
1.402
1.400
0
5
10
15
20
25
30
35
40
45
50
55
60
ELECTRIC FIELD (kV/cm)

3

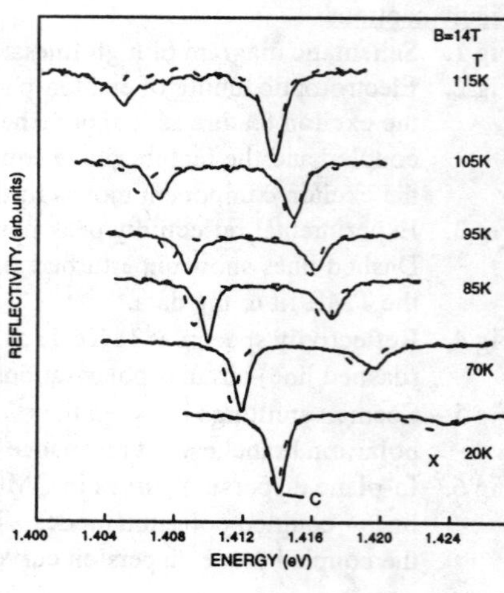
B=14T
T
115K
105K
95K
85K
70K
20K
X
C
REFLECTIVITY (arb.units)
1.400
1.404
1.408
1.412
1.416
1.420
1.424
ENERGY (eV)

4

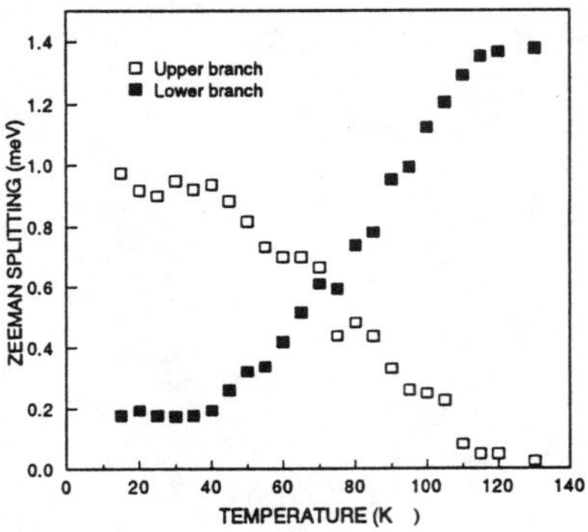

Upper branch
Lower branch
ZEEMAN SPLITTING (meV)
TEMPERATURE (K)

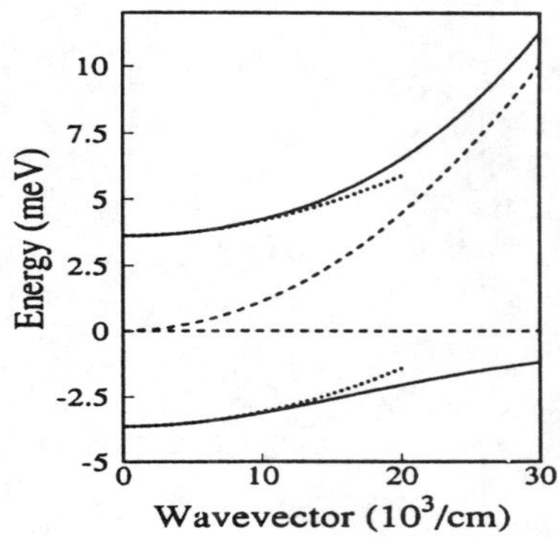

Energy (meV)
Wavevector (10^3/cm)

5

6

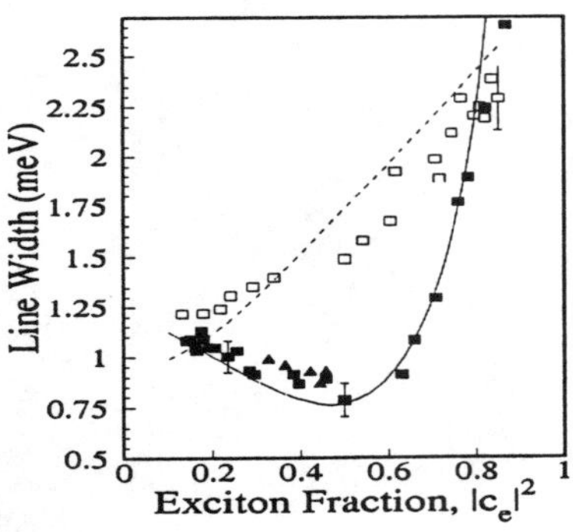

Line Width (meV)
Exciton Fraction, $|c_e|^2$

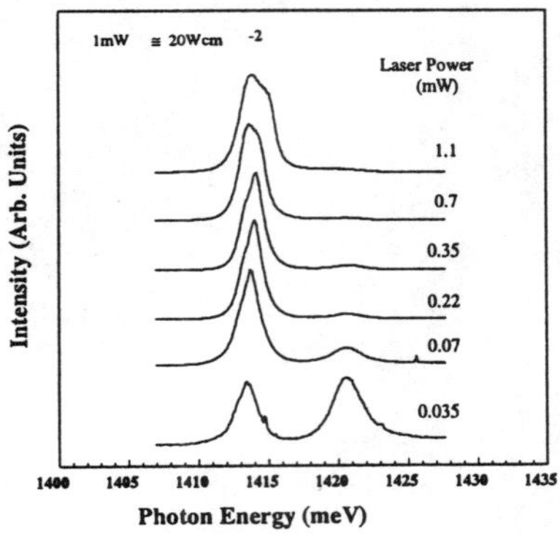

1mW ≡ 20Wcm^{-2}
Laser Power (mW)
1.1
0.7
0.35
0.22
0.07
0.035
Intensity (Arb. Units)
Photon Energy (meV)

7

8

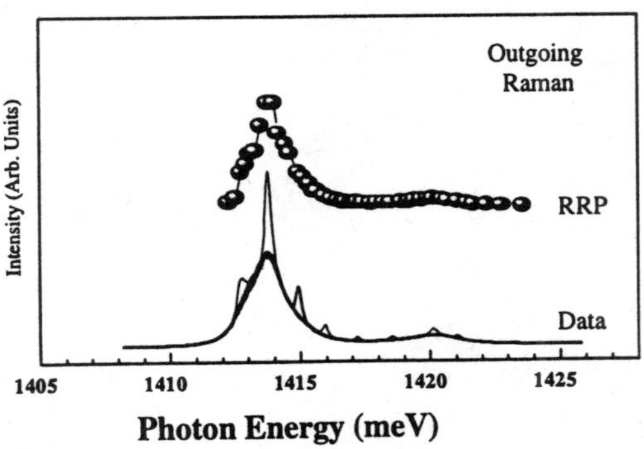

Outgoing Raman
RRP
Data
Intensity (Arb. Units)
Photon Energy (meV)

9

Diamagnetic shifts and magnetic field dependent excited states in coupled and uncoupled quantum dots

B. Kowalski, P. Omling, N. Carlsson, W. Seifert, M. Miller, S. Jeppesen, L. Samuelson
Department of Solid State Physics, University of Lund, Box 118, 221 00 Lund, Sweden

We studied the photoluminescence of the ground and excited states in selfassembled quantum dots under the influence of a high magnetic field *B*. The material systems were InP/GaAs, InP/GaInP and InAs/GaAs. The diamagnetic shift for *B* parallel and perpendicular to the growth direction were studied in uncoupled dots and in stacks of InAs dots, respectively. The differences between the diamagnetic shifts show a clear variation of the extension of the exciton wavefunctions for the different geometries sampled. For both the coupled InAs and uncoupled InP dots we were able to distinguish between one and three higher excited states employing bandfilling conditions, and could in this way study their evolution as a function of *B*.

Introduction

The Stranski-Krastanov growth mode to form quantum dots is used in a number of material combinations.[1-3] However, the electronic and excitonic properties related to the strong quantum confinement in all space coordinates in selfassembled dots is still fairly unexplored. Key features of these zero dimensional (0D) objects, compared to structures of higher dimensionality, are a larger spatial overlap of the electrons and holes, and a larger binding energy of the excitons. This stems from the fact that the excitons are more localized in space due to the stronger confinement. Here we present a study of the photoluminescence (PL) of the ground and excited states in both coupled and uncoupled dots subjected to a magnetic field *B*. The shift of the ground state energy with *B* is quadratic, i.e. diamagnetic, as long as the interaction of the "excitonic loop" with *B* is weak compared to the excitonic coulomb energy. The diamagnetic interaction for an hydrogenic exciton is given by[4]

$$\Delta E = \frac{e^2 B^2}{8\mu}\langle x^2 + y^2 \rangle = \beta B^2. \qquad (1)$$

$\langle x^2 + y^2 \rangle$ represents the extension of the squared exciton wavefunction perpendicular to B, μ is the reduced effective mass of the exciton and β is the diamagnetic coefficient. The interaction with B is smaller the more tightly bound the exciton is and the smaller its transverse extension is. This is dependent on dimensionality, e.g. in the theoretical limit for a 2D-exciton β_{2D} is about 5 times smaller than β_{3D}.[5] Moreover, the strength of the diamagnetic shift elucidates the shape of the exciton wavefunction, since it can be sampled by applying B in different directions relative to the quantum object.[5-7] In contrast to the diamagnetic shift of the ground state, it can be expected that the higher excited states of these quantum objects interact paramagnetically. Using as an approximation a rotational symmetry for the dots the characteristic features of the magnetic field evolution of the excited quantum dot states in the present experiments can in fact be explained by a simple calculation.

Experimental

Two different material systems will be described: i) uncoupled islands of InP on top of either a GaInP or a GaAs layer, grown by metalorganic chemical vapor deposition, or ii) InAs islands in GaAs, grown by chemical beam epitaxy. The substrate in both cases was [001] oriented GaAs. The InP dots are on the sample surface, i.e. they are not capped by GaInP or GaAs. The geometrical shape of the InP dots on top of GaInP has been determined previously by transmission electron microscopy (TEM).[8] The pieces used in that TEM- and the present magneto-PL study are adjacent bits of the same sample. The individual dots have the form of truncated pyramids, elongated along the [110] direction. The long and the short base diameter are within a range 55-65nm and 40-50nm, respectively, and the height is 12 to 18nm. The InP islands grown on top of the GaAs surfaces are about 12-14nm (18-20nm) high for a growth temperature of 580°C (610°C), as obtained by AFM. For InAs islands ten island layers were grown on top of each other with thin GaAs layers as tunneling barriers (5 nm thick). A TEM investigation on an adjacent piece of the same sample, as used in the present study, showed that the islands are well-aligned vertically to form complete stacks.[3] For the PL measurements as a function of B a 300 mK ^{3}He system equipped with an 18 T magnet was used. The sample was immersed in pumped ^{3}He at a temperature of $T \approx 1$K, and optically accessed by a fibre. The PL was excited with 488 nm laser light with typical powers of 0.5 to 5 mW before coupling into the fibre. The PL was dispersed with a 0.5 m grating spectrometer and detected with a Ge diode.

Results and Discussion

For B = 0T the PL from the InP islands on both GaInP or on GaAs (see Fig. 1a,b) consisted of features between 1.41 and 1.43 eV. With higher laser excitation power excited states of the InP dots are populated [9] (see Fig. 1b). The ground-excited states separation is 20meV and 12 meV for the smaller and larger InP/GaAs dots, respectively. The PL from stacked InAs dots in GaAs using low laser excitation power (Fig. 1c) peaks at 1.11 eV.

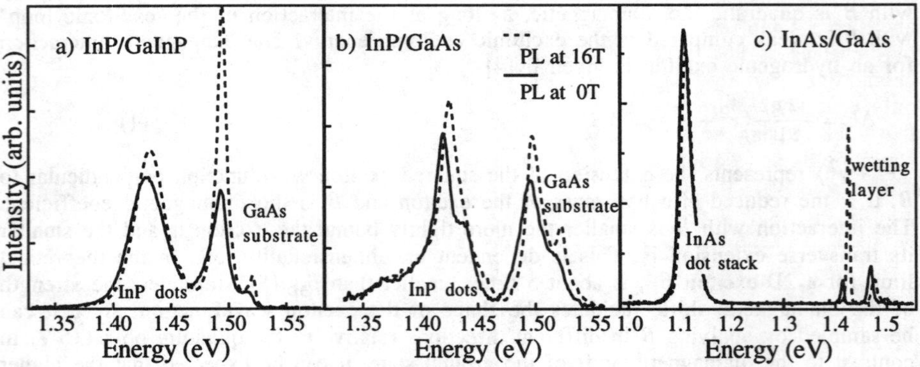

Fig. 1: Photoluminescence from uncovered dots of InP/GaInP and InP/GaAs (12-14nm height), as well as from InAs dot stacks in GaAs at B=0T and B=16T ($B \parallel [001]$).

When applying a magnetic field to the InP/GaInP dots the magnetic shift of the dot PL peak energy is purely quadratic up to B = 16T, i.e. diamagnetic. This behaviour was found for all three field directions considered in this study, $B \parallel [001]$, $B \parallel [-110]$ and $B \parallel [110]$ (see Fig. 2). However, the strength of the diamagnetic shift is different for the different directions, $\beta_{[001]} = 10\,\mu eV/T^2$ and $\beta_{[-110]} = 7.2\,\mu eV/T^2$, $\beta_{[110]} = 6.7\,\mu eV/T^2$. It has to be mentioned that the difference between the later two constants is smaller than the experimental error. The magnitude of these coefficients is smaller than the diamagnetic shift for InP in the theoretical 3D and 2D limit, $\beta_{3D} = e^2 a_B^2/4\mu = 58\ \mu eV/T^2$,

$\beta_{2D} = 12\ \mu eV/T^2$ ($\mu = 0.069 m_0$, $a_B^2 = 9.6\,nm$).[10] This is evidence for the enhanced excitonic localization in the InP dots. More interesting, however, is a comparison of the transverse excitonic extension seen in different directions with the geometrical shape of the InP dots given by TEM. From equ. (1) we can derive an effective exciton diameter d defined as

$$d = \sqrt{\langle x^2 + y^2 \rangle} = \frac{\sqrt{8\beta\mu}}{e}. \qquad (2)$$

For the exciton seen along the three directions, [001], [-110] and [110], the values of d are 5.6nm, 4.8nm and 4.6nm, respectively. In the 2D limit it is d_{2D} = 5.8 nm. These values can be compared with the geometrical dot shape obtained from TEM: the largest crossection of the dot is that perpendicular to the growth direction. Viewed along the [-110] and [110] crystallographic directions the dot exhibits a smaller crossection. Due to the elongation in the [110] direction[8] a slightly larger area is seen along [-110]. In the InAs/GaAs system the magnetic shift of the lowest energy PL peak of the dot stacks is also found to be purely quadratic. In this case, the diamagnetic shift was almost independent of field direction parallel or perpendicular to the axis of the stacks ($\beta_{\|} = 13\ \mu eV/T^2$ and $\beta_{\perp} = 12\ \mu eV/T^2$). This is in contrast to previous findings for individual or uncoupled InAs dots ($\beta_{\|} \approx 9\ \mu eV/T^2$ and $\beta_{\perp} \approx 4\ \mu eV/T^2$).[11] The point of interest here is the shape of the exciton wavefunction in the uncoupled dots as compared to the stacked dots. From equ. (2) it is evident that the transverse exciton extension in the uncoupled dots is 1.5 times larger viewed along [001] than it is viewed perpendicular to this axis, i.e. it is flattened. This is expected for the geometrical shape of the individual dots that is pyramidal with typical heights of 3-4 nm and a base width of 14-16nm.[3] For the InAs dot stack it is $d_{\|}$ = 3.5 nm and $d_{\perp}$ = 3.4 nm (for InAs $\mu = 0.022 m_0$).[10] This means that the exciton wavefunction of the stack is not flattened in the [001] direction, and hence does not resemble the geometrical shape of the individual dots. We conclude, that it is no longer confined to only one dot within the stack. This is an indication for a coupling of the dots within the stack. It can be induced quantum-mechanically, i.e. by tunneling. Also the large strain fields and variations in composition, which shift band edges and alignments, can play a role.

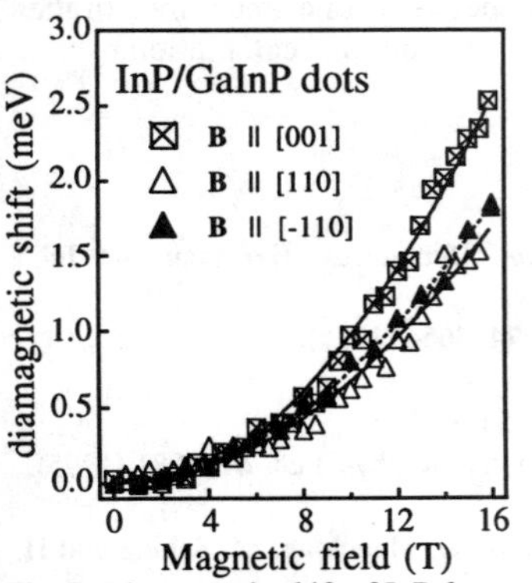

Fig. 2: Diamagnetic shift of InP dots on GaInP for the three crystallographic directions.

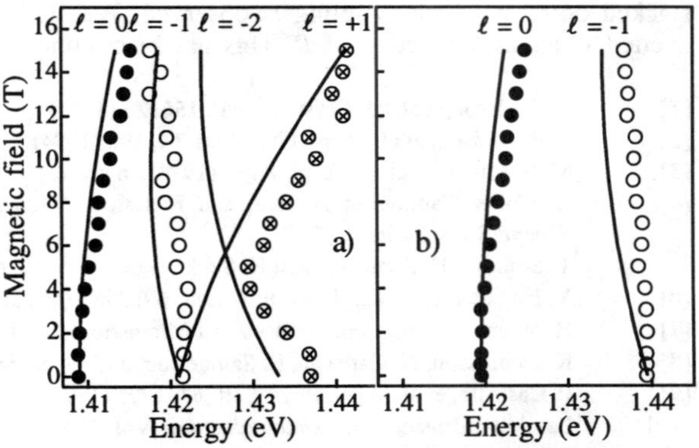

Fig.3: PL peak energies as a function of B for InP/GaAs dots of a height 18-20nm (a) and 12-14nm (b). The solid lines are calculated form equ. (3) for different angular momentum ℓ.

The PL of the InP dots on GaAs subjected to band-filling conditions has been studied as a function of B. In Fig 3a and b we present the field evolution of the PL of the smaller and the larger dots, respectively. The lowest energy PL peak of the dots with 12-14nm height showed a blue shift by more than 5 meV, which was not purely quadratic. Therefore we did not evaluate a diamagnetic coefficient. The second peak shifted by 3 meV to lower energy. For the dots with 18-20nm a third peak could be observed. An accurate theoretical treatment of these results is at present not possible, since the details of the strain and bandgap variations are not known. It is not clear, e.g., if both the electrons and holes are confined in the InP dots on GaAs. We therefore considered as a first approach the level spectrum of an electron subjected to a lateral parabolic confinement and a perpendicular magnetic field. This means that the confinement in the growth direction has been separated. The level spectrum is then given by[12]

$$E_{n,\ell} = (2n + |\ell| + 1)\sqrt{\hbar^2\Omega_0^2 + \tfrac{1}{4}\hbar^2\omega_c^2} + \tfrac{1}{2}\ell\,\hbar\omega_c \tag{3}$$

where $\omega_c = eB/m^*$ is the cyclotron frequency with $m^* = 0.077\,m_0$.[10] n is the radial quantum number, and $\ell = 0, \pm1, \pm2...$ represents the angular momentum. $\hbar\Omega_0$ is the harmonic oscillator energy at B=0, which we set to 12meV and 20meV for the smaller and the larger InP dots, respectively. The calculated energies are shown as solid lines in Fig.'s 5a and b, where a constant energy has been added to fit the lowest PL transition energy at B = 0T. The influence of B on the holes has been neglected, since their effective mass is about an order of magnitude larger. Despite the drastic simplifications made, it is interesting that this model can reproduce the essential features of the experimentally found variation of the PL emission energies with B. We therefore suggest that the main contribution to the B- dependent spectra stems from the electrons in the InP/GaAs dots. However, the details of the confining potential must still be regarded as unknown.

Conclusions

The PL of the ground and excited states in selfassembled quantum dots subjected to B has been studied. The material systems were InP/GaAs, InP/GaInP and InAs/GaAs. The diamagnetic shift resembles the geometrical shape of the quantum dots, except for the stacked dots, in which coupling is observed. We studied between one and three higher excited states as a function of B. This has been compared to a simple calculation.

[1] N. Carlsson,et al., J. Cryst. Growth **156**, 23 (1995).
[2] J. M. Moison, et al. Appl. Phys. Lett. **64**, 196 (1994).
[3] M. S. Miller, et al., J. Appl. Phys. (1996), in press.
[4] C. Cohen-Tannoudji, B. Diu, and F. Laloe, *Quantum Mechanics* (Hermann and John Wiley&Sons, Paris, 1977).
[5] T. Someya, H. Akiyama, and H. Skaki, Phys. Rev. Lett. **74**, 3664 (1995).
[6] Y. Nagamune, et al., Phys. Rev. Lett. **69**, 2963 (1992).
[7] H. Weman, et al., Superlattices and Microstructures **13**, 5 (1993).
[8] K. Georgsson, N. Carlsson, L. Samuelson, and W. Seifert, Appl. Phys. Lett. **67**, 2981 (1995).
[9] P. Castrillo, et al. Appl. Phys. Lett. **67**, 1905 (1995).
[10] Landolt-Börnstein, in *Semiconductors* ,Vol. 17a, edited by O. Madelung, M. Schulz and H. Weiss, (Springer Verlag, Berlin, 1982).
[11] H. Sigg, P. Christianen, R. Houdre, and A. Rudra, Physica Scripta **T54**, 81 (1994).
[12] V. Fock, Z. Physik **47**, 446 (1928).

BINDING ENERGY AND SPIN-SPLITTING OF EXCITONS IN $In_xGa_{1-x}As/GaAs$ QUANTUM DOTS

M. BAYER[a], A. FORCHEL[a], T.L. REINECKE[b], S. WALCK[b], V.B. TIMOFEEV[c]

a Technische Physik, Universität Würzburg, Germany
b Naval Research Laboratory, Washington, USA
c Institute of Solid State Physics, Chernogolovka, Russia

We have investigated the effects of confinement and dimensionality on excitons in $In_{0.10}Ga_{0.90}As/GaAs$ quantum dots with widely varying sizes. Measurements of the diamagnetic shifts of the exciton luminescence provide values of the exciton binding energies. The binding energies increase systematically with decreasing lateral sizes of the quantum dots. Detailed variational calculations of the exciton binding energies are found to be in good agreement with the experimental results. In addition, we find a significant enhancement of the spin-splitting with decreasing dot diameter.

Excitonic effects play a central role in the optical properties of semiconductors. In semiconductor nanostructures the electron-hole interaction is enhanced by spatial confinement and by decreased dimensionality, and thus in these structures excitonic effects are especially important. In case of quantum dots extensive theoretical studies of these effects on exciton binding energies have been made [1-3].

Nevertheless no systematic experimental studies of the binding energies in quantum dots have been reported. In the present work we investigate the effects of confinement on excitons in quantum dot systems. The diamagnetic shifts of the excitons are measured in low excitation magneto-luminescence experiments. These shifts are used to obtain the binding energies. These binding energies are shown to be in good agreement with detailed calculations using non-separable variational techniques. In addition, we investigate the spin-splitting of excitons and find a significant increase of the splitting with decreasing dot diameter.

Both buried and free standing $In_{0.10}Ga_{0.90}As/GaAs$ quantum dots of widely varying sizes have been studied in the experiments. These structures were fabricated by electron beam lithography and wet chemical etching. By using scanning electron microscopy we find that the dots are cylindrical in shape. The experiments were performed in Faraday-configuration in an optical split coil magnetocryostat ($B \leq 8$ T) at a temperature of $T = 2$ K. The samples were excited by a cw Ar^+ laser using very low excitation powers of 0.1 W cm^{-2} in order to avoid high density effects. The luminescence was dispersed by a 0.6 m monochromator and detected by a CCD camera.

Fig. 1 shows low excitation photoluminescence spectra of 52 nm and 106 nm wide, free standing $In_{0.10}Ga_{0.90}As/GaAs$ quantum dots at different magnetic fields up to 8 T. Both circular polarizations (σ^+ by solid and σ^- by dotted lines) are shown. Due to lateral quantization the spectral lines of the smaller quantum dots are shifted to higher energies by about 3 meV at $B = 0$. With increasing magnetic fields all features shift to higher energies. This shift is larger for the larger dots (more than 3.5 meV up to 8 T) and smaller for smaller dots (about 2.5 meV for the 52 nm wide dots). Moreover, with increasing magnetic field in both cases the spin splitting can be resolved, which is significantly larger for the 52 nm dots (~1.2 meV) than for the 106 nm dots (~ 0.5 meV).

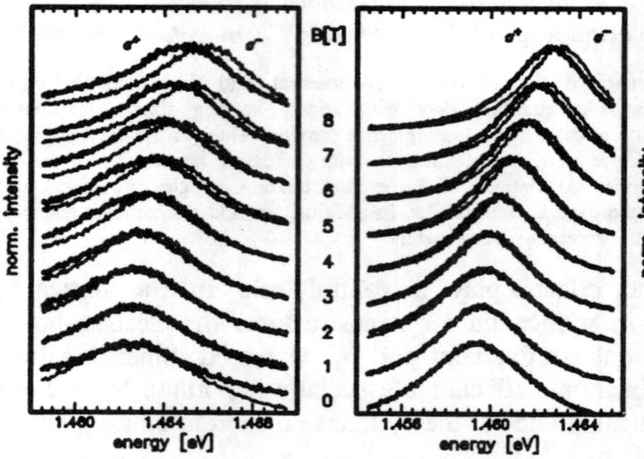

Figure 1: Low excitation photoluminescence spectra at varying magnetic fields up to B = 8 T for 52 nm and 106 nm wide free standing $In_{0.10}Ga_{0.90}As/GaAs$ dots. The σ^+ circularly polarized spectra are shown by the solid lines, the σ^- polarized ones by the dotted lines.

At low magnetic fields (B ≤ 4 T) the cyclotron energy $\hbar\omega_c$ and the lateral intersubband spacings ΔE_x are smaller than the exciton binding energy E_B. Then the shifts of the emission lines with increasing magnetic field are diamagnetic, $\Delta E = \gamma_2 B^2$, where γ_2 is the diamagnetic coefficient. We find a strong decrease of the diamagnetic coefficient with decreasing dot diameter. In case of the free standing dots which are laterally surrounded by vacuum γ_2 decreases from 70 μeV T^{-2} for large dots (which is identical to the quantum well value) down to less than 15 μeV T^{-2} for 30 nm dots. In case of the buried dots with their significantly weaker geometric confinement the decrease of γ_2 is smaller than for the free standing ones. For 30 nm dots γ_2 is about 25 μeV T^{-2}.The exciton binding energy can be evaluated by comparing the measured diamagnetic shifts with those for a bulk system having anisotropic masses [4]. The wavefunction is then given by a hydrogenic-like ellipsoid of revolution, which is characterized by direction dependent reduced exciton masses μ_i, i=x,y,z. For such an asymmetric hydrogenic wavefunction the diamagnetic shift is given by perturbation theory which gives

$$\gamma_2 = \frac{4\pi\hbar^2\varepsilon^2}{e^2\mu\mu_x\mu_y}, \qquad \frac{1}{\mu} = \frac{1}{3}\left(\frac{1}{\mu_x} + \frac{1}{\mu_y} + \frac{1}{\mu_z}\right). \qquad (1)$$

From the ratios of the experimentally observed diamagnetic coefficients for the dots to those of quantum wells we obtain from eqn. (1) values of the in-plane reduced masses $\mu x = \mu y$ for the quantum dots. Then the exciton binding energy E_B is given by

$$E_B = \frac{e^4 \mu}{32\pi^2 \hbar^2 \varepsilon^2}. \tag{2}$$

The resulting exciton binding energies (in units of the bulk Rydberg of 4.03 meV) for the quantum dots as functions of the dot diameter (in Bohr radii) are given in Figure 2. For the two-dimensional reference sample, eqn. (2) gives an exciton binding energy of 8.8 meV, which is identical to the value we obtain from the splitting of the 1s and the 2s exciton in photoluminescence excitation studies.

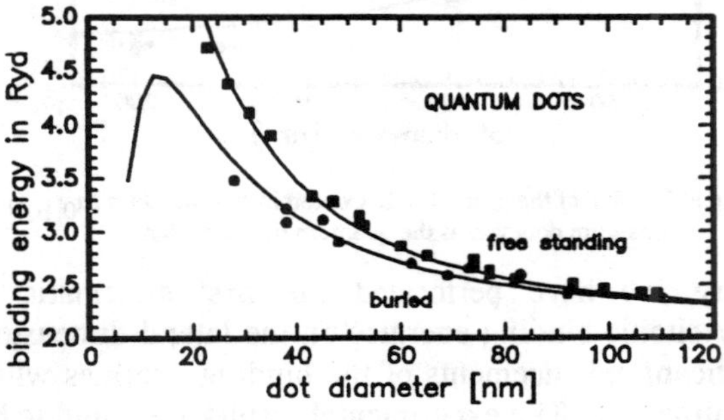

Figure 2: Exciton binding energies for the free standing and the buried $In_{0.10}Ga_{0.90}As$/GaAs quantum dots versus the dot diameter.

We have also made detailed calculations of the exciton binding energies for $In_{0.10}Ga_{0.90}As$/GaAs quantum dots based on the non-separable exciton wave functions

$$\Psi(\bar{r}_e, \bar{r}_h) = \zeta_e(\ldots)\zeta_h(\ldots)\exp(-\sqrt{\alpha\left((x_h - x_e)^2 + (y_h - y_e)^2\right) + \beta(z_h - z_e)^2}) \tag{3}$$

Here $\zeta_e(\ldots)$ and $\zeta_h(\ldots)$ are the envelop functions for the single particle confined subbands, which are functions of (z_h, r_h) and (z_e, r_e), where z is the coordinate along the quantum well growth direction and r is the radial coordinate in the cylindrical dot and the parameter α in eqn. (3) is a variational parameter. The results of these calculations are shown by lines in Fig. 2 [5]. The agreement with the experimental data is exceptionally good. From the results we see that for the dots the exciton binding energy is significantly enhanced for structures below ~100 nm size, where it increases smoothly up to slightly more than 3.5 Ryd for the buried quantum dots and almost up to 5 Ryd for the free standing dots, i.e. we find that the exciton binding energies in these dots are enhanced by almost a factor of five over the bulk value. These systematic results confirm the importance of excitonic effects for the optical properties of semiconductor nanostructures.

Fig. 3 shows the spin-splitting $\Delta E = E(\sigma^+) - E(\sigma^-)$ between the two circularly polarized lines at B = 8T versus dot diamater. A remarkable increase of ΔE with decreasing dot diameter is observed. In the quasi-two-dimensional sample the spin-splitting is only 0.2 meV but increases to 1.3 meV for 50 nm wide quantum dots.

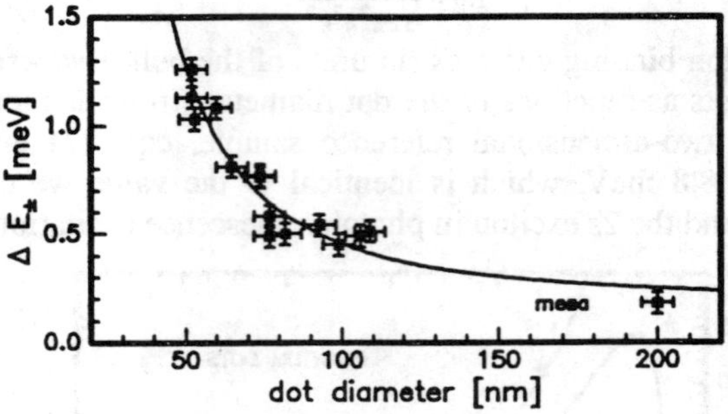

Figure 3: Spin splitting at B = 8 T of the ground state exciton in free standing $In_{0.10}Ga_{0.90}As$/GaAs quantum dots versus the quantum dot diamater.

In summary, we have performed the first systematic study of the dependence of the excitonic binding energies on the lateral dimension in quantum dots. We find significant enhancements of the binding energies which depend on the nature of the confinement. The experimental results are found to be in excellent agreement with the results of detailed calculations for the binding energies. We also find a significant increase of the spin-splitting with decreasing dot diameter.

Acknowledgements

We would like to thank the State of Bavaria and the U.S. Office of Naval Research for financial support of this work. T.L.R. gratefully acknowledges the financial support of the Alexander von Humboldt Foundation.

References

[1] G.W. Bryant, Phys. Rev. B **37**, 8763 (1988).
[2] W. Que, Phys. Rev. B **45**, 11036 (1992).
[3] S. LeGoff and B. Stébé, Phys. Rev. B **47**, 1383 (1993).
[4] S. Taguchi et al., J. Jap. Phys. Soc. **57**, 3256 (1988).
[5] The parameters used in the present calculations are: the electron mass m_e=0.062, the hole mass m_h=0.35, the conduction band offset between InGaAs and GaAs V_C = 69 meV, the valence band offset V_V = 37 meV, and the dielectric constant ε = 13.0.

Local spectroscopy of magnetoexcitons in different types of single quantum dots

W. Heller, U. Bockelmann and G. Abstreiter
Walter Schottky Institut, Technische Universität München, Am Coulombwall, 85748 Garching, Germany

We perform micro-photoluminescence experiments on two different types of single GaAs quantum dot structures in magnetic fields up to 5T at a temperature of 4K. In samples which were produced by laser induced thermal interdiffusion we find a Zeemann spin splitting depending on the size of the structure. In quantum dots formed by monolayer width fluctuations in a narrow quantum well we observe a ground state and two excited states with different spin splittings. The spin relaxation in this structure is quenched by a magnetic field.

Local photoluminescence (PL) spectroscopy of single artificial atoms in magnetic fields is an interesting experiment because it provides information on the confining potential. The interplay of lateral confinement, Coulomb interaction and magnetic confinement leads to a variation of the diamagnetic shift when the magnetic field is changed. The Zeemann spin splitting gives information on the exciton g-factor. In this article we focus on the spin splitting in quantum dots. Using a new micro-photoluminescence setup which allows to combine high spatial resolution and high magnetic fields (0-5T) we can address individual quantum dots and thus avoid inhomogeneous line broadening. We investigate a series of single GaAs/AlGaAs quantum dots of various size which was produced by laser induced thermal interdiffusion[1] and a narrow 3.5 nm wide quantum well where monolayer width fluctuations give rise to the localization of excitons. The exciting laser beam is focused by a microscope objective onto the sample which is mounted in a specially designed magnet cryostat. The PL is collected by the same objective and dispersed in a triple Raman spectrometer.

First we want to focus on the laser beam written samples. Fig. 1a shows a spectrum at B=5T of a structure with a lateral geometrical size of 450 nm. Using a quarter wave plate in front of the linearly polarizing spectrometer we measure the spin splitting of the three PL peaks[2] observed at low excitation power. We have also investigated other structures with a lateral geometrical size of 1 μm, 500 nm and 400 nm. The measured spin splittings are shown in Fig. 1b as a function of the peak energy. We find that the splitting increases with the peak energy. In a magnetic field the electron and hole states split up according to $\Delta E_{e,h} = g^*_{e,h}\mu_B B$ where μ_B is the Bohr magneton and $g_{e,h}$ are the effective g factors of electron and hole. The comparison of measured

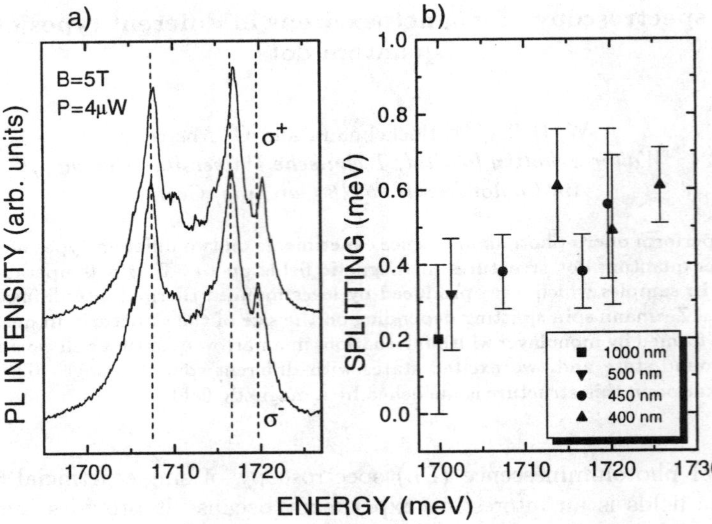

Figure 1: a) Micro-photoluminescence spectra of a single dot with a geometrical size of 450 nm, detected with opposite circular polarisations. b) Spin splitting at 5T as a function of the peak energy. A splitting of 0.5 meV corresponds to an exciton g-factor of 1.72.

electron and exciton g-factors shows that the spin splitting is dominated by the valence band. In quantum wells an enhancement of the exciton g-factor with decreasing well width was observed and attributed to the increase of the quantization energy[3]. An increase of g^* with the PL energy might also be expected for a quantum dot.

Let us now turn to the quantum dots formed by monolayer width fluctuations which are found in narrow quantum wells[4]. In the lowest curve of Fig. 2 a sharp exciton line about 12 meV below the dominant quantum well luminescence (PL) is shown. The line width of this 0D state with a lateral extension below the resolution limit is smaller than 0.1 meV. In photoluminescence excitation spectra (PLE) we find two excited levels corresponding to this exciton state with line widths of 50μeV and 120μeV, respectively[5]. When a magnetic field is applied we observe a spin splitting of these lines which is different for the ground state and the two excited levels. At B=5T we find a splitting of 0.4 meV for the ground state and about 0.7 meV for the excited states. For the second excited state, however, the sign of the g-factor is reversed. A similar behaviour has been predicted theoretically for magneto-excitons in a 10 nm wide quantum well by Bauer et al[6]. There it was attributed to valence band mixing. For a quantitative description of this effect calculations for excitons in

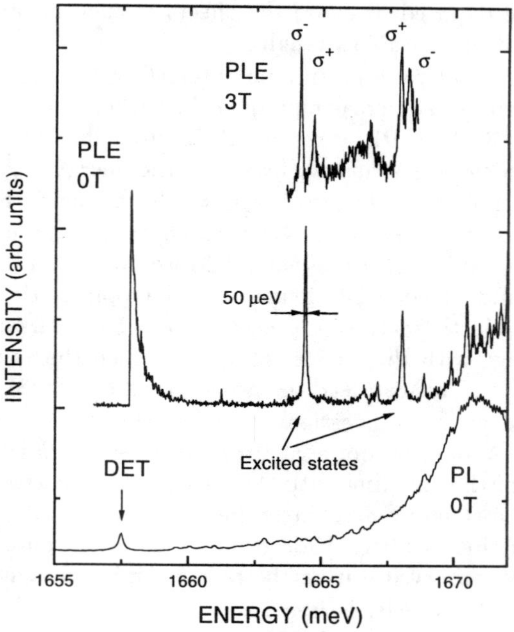

Figure 2: PL and PLE spectra of localized exciton states at T=4K. The excitation power is 4μW.

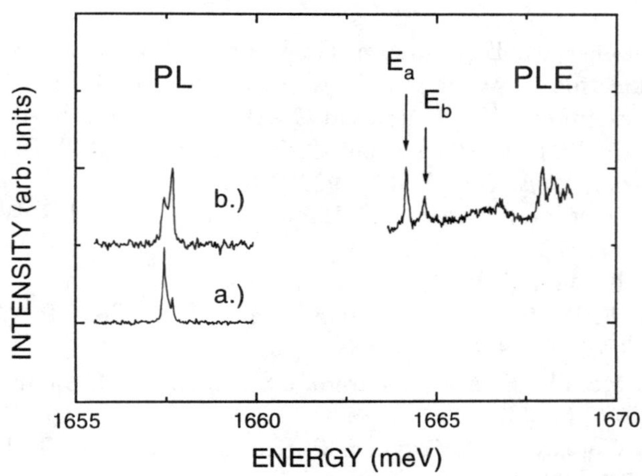

Figure 3: PL spectra (left side) which were recorded with resonant excitation into the spin split states of the first excited state at B=3T (marked by arrows).

quantum dots are required, because the effects of exciton binding and lateral confinement are of comparable strengh.

In a magnetic field we find another interesting feature, namely that the spin relaxation process is increasingly quenched when the field is raised. This is shown in Fig. 3. The PL is resonantly excited into the spin split states of the first excited level. When we excite at the energy E_a we find that the major part of the luminescence arises from the lower energy line of the ground state. This means that the spin is conserved to a large extent during the relaxation process. Without a magnetic field we do not find such a behaviour when we excite and detect with defined circular polarisation. While at zero magnetic field an elastic scattering process is possible between the spin states, an energy exchange with the lattice is required when the spin states are split by a magnetic field. This can lead to the observed decreased spin flip rate[7,8] .

In conclusion, we have investigated the spin splitting of exciton states in two types of single quantum dot structures. In laser beam written samples we find an increase of the splitting with the energy of the corresponding PL line. In quantum dots formed by monolayer fluctuations of a quantum well we find that the spin splitting is different for ground state and excited states and that the spin is widely conserved during the relaxation in a magnetic field.

We thank U. Rössler, R. Ferreira and G. Bastard for helpful discussions and K. Brunner, G. Böhm and G. Weimann for providing the samples. This work has been supported financially by the Deutsche Forschungsgemeinschaft (SFB 348).

References

1. K. Brunner, U. Bockelmann, G. Abstreiter, M. Walther, G. Böhm, G. Tränkle, and G. Weimann, *Phys. Rev. Lett.* **69**, 3216 (1992)
2. U. Bockelmann, W. Heller, and G. Abstreiter, submitted
3. M. J. Snelling, E. Blackwood, C. J. McDonagh, R.T. Harley and C. T. B. Foxon, *Phys. Rev.* B **45**, 3922 (1992)
4. K. Brunner, G. Abstreiter, G. Böhm, G. Tränkle, and G. Weimann, *Phys. Rev. Lett.* **73**, 1138 (1994)
5. W. Heller and U. Bockelmann, submitted
6. Gerrit E. W. Bauer, and Tsuneya Ando, *Phys. Rev.* B **37**, 3180 (1988), *Phys. Rev.* B **38**, 6015 (1988)
7. M. Z. Maialle, E. A. de Andrada e Silva, and L. J. Sham, *Phys. Rev.* B **47**, 15776 (1993)
8. R. E. Worsley, N. J. Traynor, T. Grevatt, and R. T. Harley, *Phys. Rev. Lett.* **76**, 3224 (1996)

RABI SPLITTING OF EXCITED MAGNETOEXCITONS IN A SEMICONDUCTOR MICROCAVITY: EXPERIMENTS AND CALCULATIONS

J. Tignon, J. Wainstain, R. Ferreira, P. Voisin, C. Delalande and M. Voos
Laboratoire de Physique de la Matière Condensée de l'Ecole Normale Supérieure
24, Rue Lhomond, F-75005 - Paris (France)

R. Houdré, R. Stanley and U. Oesterle
Institut de Micro- et Optoélectronique, E.P.F.L.
CH 1015 (Switzerland)

We present a new variational procedure to handle the magnetoexciton states in a quantum well. The theory is compared with reflectivity experiments on a semiconductor microcavity, a system which represents a convenient tool for the investigation of the magnetoexciton energies and oscillator strengths. The thresholds for the appearance of the reflectivity doublets are clearly observed.

As pointed out by Weisbuch et al[1], the microcavity effect is a good tool for the measurement of the exciton oscillator strength. Moreover, as discussed by Tignon et al[2], the variation of the excitonic oscillator strength under magnetic field in a semiconductor microcavity allows a tuning of the coupling between photons and excitons. A now well known semiclassical model[3 4], states that a non zero exciton linewidth leads to the existence of a threshold in the appareance of the doublet. This model predicts also that the Rabi splitting, observable in the absorption and reflectivity spectra in the strong coupling regime, is roughly proportional to the square root of the oscillator strength. In this work we discuss experimental and theoretical results concerning this field-dependent coupling, from vanishing to high (10 T) magnetic fields.

The sample[2] we used consists in three decoupled strained $In_{0.13}Ga_{0.87}As$-GaAs quantum wells embedded in a suitable Bragg-mirror cavity. Due to strain, the ligth and heavy holes are well separated ($\sim 70\ meV$) so that we can safely neglect the hole-band mixings[5]. Among the methods proposed to theoretically handle the magnetoexciton problem we quote : the diagonalisation of the envelope hamiltonian in a gaussian basis[6]; the exact numerical integration of the Schrödinger equation[7] and the two-point Padé approximant method of MacDonald and Ritchie[8]. We discuss now an original variational proposition to describe the magnetoexciton states in quantum wells. We assume a decoupling of the in-plane and confined (along the growth axis and B direction) motions[9]. At zero field the exciton levels are solutions of an effective two dimensional potential, $V_{eff}(\rho)$, obtained by averaging the three dimensional coulombic one over the confined electron and hole z-motions. By properly chosing the field gauge we have the well known $(K = 0)$ exciton hamiltonian [9]:

$$H_{exc}(\rho) = T_\rho + V_{eff}(\rho) + \frac{\mu}{2}\left(\frac{\omega_c}{2}\right)^2 \rho^2 \quad (1)$$

where the first term is the kinetic energy and the last one the magnetic field contribution. We take for the ground, $1S$-like exciton state the form:

$$\Phi_{1S}(\rho) = N_{1S}\, exp\left[-(\rho/\lambda_{1S})^{\alpha_{1S}}\right] \quad (2)$$

where λ_{1S} and α_{1S} are B dependent variational parameters and N_{1S} the normalisation constant. This envelope is expected to be flexible enough to smoothly interpolate between the $\alpha_{1S} \sim 1$ value at weak fields (when $B = 0$ the $\alpha_{1S} = 1$ value gives the exact result for the two dimensional case and is ofen used as a trial for excitons in actual quantum wells [9]) and the $\alpha_{1S} \sim 2$ value at very high fields. We have checked that this variational function reproduces with high accuracy the results of MacDonald and Ritchie [8] for the two dimensional case and the ones of Fisher et al [7] for an actual quantum well. Another parameter of interest is the exciton oscillator strength. According to eq. 2 it is simply given by $|N_{1S}(B)|^2$. We can easily generalise our approach to the excited exciton states. The variational envelope for the nS-like state reads:

$$\Phi_{nS}(\rho) = N_{nS} P_{nS}\left[\frac{\rho}{\lambda_{nS}}\right] exp\left[-(\rho/\lambda_{nS})^{\alpha_{nS}}\right] \quad (3)$$

$$P_{nS}[X] = \sum_{k=0}^{n-1} A_{nS,k}(X)^{k\alpha_{nS}} \quad (4)$$

where $A_{nS,0} = 1$ and the remaining n-1 coefficients $A_{nS,k}$ are determined by orthogonalisation with the $n'S$ envelopes: $< \Phi_{n'S}(\rho)|\Phi_{nS}(\rho) >= 0$ with $n' = 1, 2, ..., n-1$. Finally, λ_{nS} and α_{nS} are the only B dependent variational parameters and N_{nS} the normalisation constant. Exciton series with different angular dependences can be generated in a similar way. More details are presented in a forthcoming publication. The solid lines in fig.1 are the energies for the first four S-like magnetoexcitons ($n = 1$ to 4 in eq. 3 and eq. 4) in our structure. Note for a fixed n value the passage from a roughly quadratic variation at weak fields (diamagnetic shift) to a quasi-linear (Landau-like) behaviour at large fields. The evaluated oscillator strengths $OS_{nS} = |N_{nS}(B)|^2$ increase with B and show a roughly linear dependence at high fields, as expected (at high fields the extension of the in-plane wavefunctions decreases

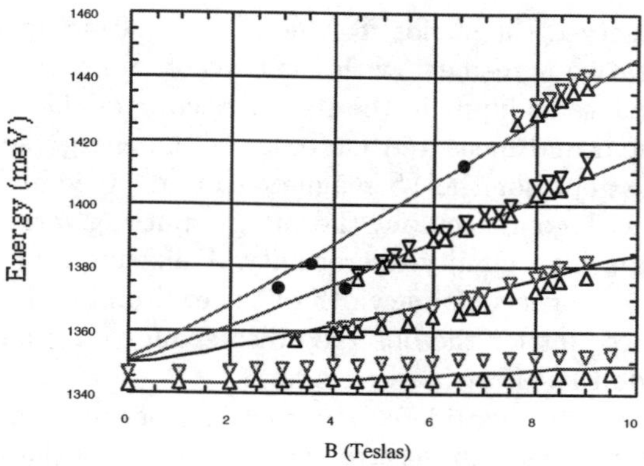

Figure 1: Energies of the reflectivity doublets at resonance, versus magnetic field (open symbols). Solid lines: calculated energies. Full symbols: energies of the magnetoexcitons below the thresholds, as deduced from the reflectivity width fluctuation.

like the cyclotron radius $\lambda_c \sim 1/\sqrt{B}$, which gives $OS \sim 1/\lambda_c^2 \sim B$). As shown above, $OS_{nS}(B)$ increases with B for a fixed n value and decreases with increasing n for a fixed B value. Due to the finite exciton linewidth ($5meV$), a vacuum Rabi splitting, resulting from the coupling between the cavity and the broadened nS-like magnetoexciton, is expected to be measured only above a critical field B_n, where B_n increases with n. The open symbols in fig.1 display the field variations of the observed energy positions of the reflectivity doublets at resonance. With respect to ref. (2), up to four series of doublets are now observed and the existence of different critical fields B_n is now clearly evidenced. Moreover, by varying B (before the strong coupling regime) with a fixed cavity energy, we observed for certain field values an increase in the width of the reflectivity signal. We associate these peaks to the field induced variation of the electronic density of states (and hence of the absorption coefficient). We can in this way access to the magnetoexciton energies before a splitting occurs (full symbols in fig.1). The latter results for the energies are in good quantitative agreement with the theoretical ones. Note finally that the measured energy splittings increase monotonously with increasing field $B > B_n$, as expected (they should reflect the enhancement by the field of the OS_{nS}). In order to be more quantitative, the reflectivity spectra were fitted by using an adaptation of the semiclassical linear dispersion model of Zhu et al [4] by Houdré et al [3]. Firstly, the observed thresholds for the $2S$, $3S$ and $4S$ magnetoexcitons are in good quantitative agreement with the ones obtained by using the theoretical $OS_{nS}(B)$ values as input in the semiclassical model. The comparison between the evaluated and measured increases of the energy splittings is, on the contrary, not so good for the $1S$ magnetoexciton. We checked, however, that our model fits with good accuracy the energy splitting variation measured by Fisher et al [7], in a high quality microcavity. Following ref.(10), we think that the effect of inhomogeneous linewidth of the exciton in our medium quality sample should account for the observed discrepancy. A study is presently under way to include this effect in our model.

In conclusion, we have considered the coupling between the cavity mode and excited magnetoexcitons. In the strong coupling regime the thresholds and the splittings were measured for the first four excitons. We have also discussed the variations of the microcavity linewidth, due to the field induced oscillations in the density of electronic states coupling to the photon mode in the intermediate field regime. In addition, a new variational procedure is presented to describe the magnetoexciton states in a single quantum well. Finally we showed that a semiclassical model correctly accounts for the observed splitting thresholds. A more complete study is needed to include the effects of the QW inhomogeneity.

Acknowledgments

The Laboratoire de Physique de la Matière Condensée is "Laboratoire Associé au CNRS (URA 1437) et aux Universités Paris 6 et Paris 7".

References

1. C. Weisbuch et al, *Phys. Rev. Lett.* **69**, 3314 (1992).
2. J. Tignon et al, *Phys. Rev. Lett.* **74**, 3967 (1995).
3. R. Houdré et al, *Phys. Rev.* B **49**, 16761 (1994).
4. Y. Zhu et al, *Phys. Rev. Lett.* **64**, 2499 (1990).
5. S.R. Eric Yang and L. Sham, Phys.Rev.Lett. **24**, 2598 (1987) and also G. E. W. Bauer and T. Ando, *Phys.Rev.*B **37**, 3130 (1988).
6. R. L. Greene and K. K. Bajaj *Phys. Rev.* B **31**, 6498 (1985).
7. T.A. Fisher et al, *Phys. Rev.* B **53**, 10469 (1996).
8. A.H. MacDonald and D. S. Ritchie, *Phys.Rev.*B **33**, 8336 (1986).
9. G. Bastard, Wave mechanics applied to semiconductor heterostructures (Les Editions de Physique, Les Ulis, 1988).
10. R. Houdré, R.P. Stanley, M. Ilegems, *Phys.Rev.* A **53**, 2711 (1996).

TUNING OF RABI-SPLITTING BY A MAGNETIC FIELD

T. GUTBROD[1], M. BAYER[1], A. FORCHEL[1], S.N. WALCK[2], T.L. REINECKE[1,2], M. RÖHNER[1], H. ZULL[1], J.P. REITHMAIER[1]

1 Technische Physik, Universität Würzburg, Germany
2 Naval Research Laboratory, Washington D.C., USA

We have studied the magnetic field dependence of the Rabi-splitting in semiconductor microcavities. We find an increase of the splitting which can be attributed to the increase of the exciton oscillator strength by the magnetic field. In addition we have investigated microcavities with full three-dimensional photon confinement. We observe magnetopolaritons composed of the ground state exciton and higher confined photon states.

In recent years the investigation of light matter interaction in semiconductor microcavities has been of central interest. The vacuum Rabi-splitting occurs in these systems in the strong coupling regime of exciton and photon where both excitations are discrete states. Here the photon and the exciton modes are no longer eigenstates of the system [1,2], but mix strongly and form cavity polaritons.

Here we report the increase of the vacuum Rabi-splitting in magnetic field, which can be attributed to the enhancement of the exciton oscillator strength by B and which is found to be in very good agreement with detailed calculations. In addition, we report the observation of polariton formation and Rabi-splitting between the ground state exciton X_0 and several confined photon states in semiconductor microcavities with full, three-dimensional photon confinement.

The MBE grown microcavity consists of a GaAs lambda-cavity, in the center of which a 6 nm wide $In_{0.14}Ga_{0.86}As$ quantum well is imbedded. This structure is sandwiched between two Bragg reflectors consisting of 17 quarter wave stacks of GaAs/AlAs layers on top and 21 at the bottom. In order to obtain three-dimensional photon confinement, rectangular columns were fabricated by patterning the sample in lateral direction.

The samples were mounted into the variable temperature insert of an optical split coil magnetocryostat ($B \leq 8$ T). An Ar^+ laser (514.5 nm) was used for optical exitation. The emitted light was dispersed by a monochromator with a focal length of 0.32 m and detected by a liquid nitrogen cooled Si - charge-coupled-devices camera.

In the as-grown sample without lateral patterning the photon states in the quantum well plane are quasi-continuous, and photon quantization occurs only along the quantum well growth direction. By the lateral patterning the cavity photon modes are discretized. Their eigenfrequencies are given by $\omega_C = c\,(k_x^2 + k_y^2 + k_z^2)^{1/2}$ with the wavenumbers $k_i = (\pi \cdot n_i)/L_i$. L_i is the length of the cavity along the i - direction and $k_z = \pi / \lambda$. The cavity modes are therefore characterized by three quantum numbers (n_x,n_y) which represent the lateral quantization and n_z=1.

Figure 1 shows photoluminescence spectra of the as-grown cavity sample for varying magnetic fields at the temperatures at which the two polariton branches are in resonance. At B=0 the splitting between the branches is about 2.7 meV and it is seen to increase continuously with increasing magnetic field. At 8 T the Rabi-splitting is about 3.6 meV. Figure 2 shows the observed Rabi-splittings (squares) as function of the magnetic field. The enhancement is seen to be almost linear.

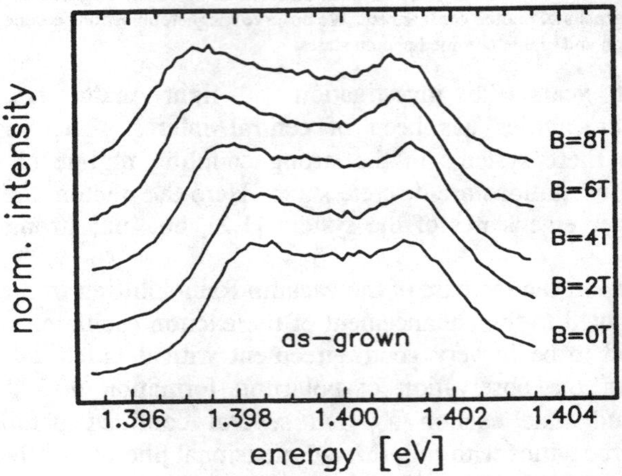

Fig. 1: Photoluminescence spectra of the as-grown microcavities for several magnetic fields at the corresponding polariton resonance temperature.

Such anticrossings as observed for the cavity polaritons occur if the dispersion relations of the states of a strongly coupled two-level-system come in resonance. In our case the two-level-system is formed by the microcavity photon modes and the 1shh exciton. Their dispersions are given by the temperature dependences of their energies. The two decoupled states are taken to be described by Lorentzian lineshapes characterized by their spectral halfwidths δ_C and δ_X for the cavity mode and the exciton, respectively. In case of resonance the splitting Ω between the two polariton modes is given by [3]:

$$\Omega^2 = \frac{F\alpha\ell}{\pi}\delta_X\delta_C - \frac{(\delta_X - \delta_C)^2}{4} \quad (1)$$

where $F = \pi R/(1-R)^{0.5}$ is the finesse of the cavity and ℓ is the quantum well width. α is the absorption coefficient in the center of X_0, which is given by $\alpha = (2 f N e^2)/(m\ \delta_x)$, where f is the oscillator strength of the exciton transition and N the density of the excitons.

The second term in Eq (1) is small, and then the Rabi-splitting is proportional to the square root of the exciton oscillator strength f(B). We have calculated the oscillator strength of the 1shh exciton as a function of magnetic field by solving for the exciton wavefunction exactly numerically. The square root of these results are scaled by the experimental results for the Rabi-splitting at B=0 and are shown by the dashed dotted lines in Figure 2. These theoretical results for the B dependence of the exciton oscillator strength are in excellent agreement with the experimental results.

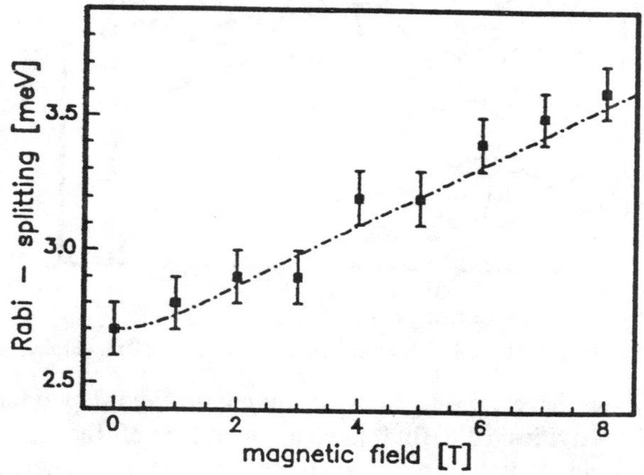

Fig. 2: Rabi-splitting as observed in the as-grown microcavity versus magnetic field. The dashed-dotted curve is obtained from theoretical calculations of the exciton oscillator strength.

Figure 3 shows photoluminescence spectra of a patterned cavity sample containing photonic quantum dots with a lateral size of 4.7 μm for varying temperatures (B=8T). Besides the 1s heavy hole exciton a number of additional transitions are observed, which can be attributed to three-dimensionally confined photon states. The different photon states are labeled by the quantum numbers (n_x, n_y).

In these laterally patterned microcavities the higher confined optical modes also are expected to form polaritons with the ground state exciton X_0 when their energies are comparable. To obtain resonances between excitons and photons the exciton energy was again varied with temperature. With increasing temperature the exciton mode comes into resonance with the fourth confined photon mode resulting in a clearly observable anticrossing. For higher temperatures also anticrossings with the lower lying photon states occur. This is the first time that the Rabi-splitting is observed in photonic quantum dots.

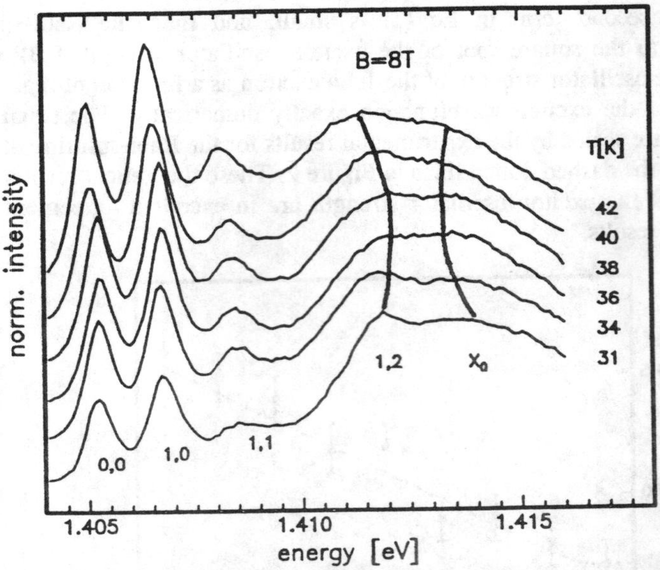

Fig. 3: Luminescence spectra of the 4.7 μm wide photonic quantum dots at B=8T for varying temperatures.

In summary, we have investigated the magnetic-field-dependence of the Rabi-splitting in microcavities. We find a clear increase of the splitting with increasing magnetic field, which can be attributed to the enhancement of the oscillator strength of the excitonic transition by the magnetic field. For the first time we observe Rabi-splitting between polaritons formed by the 1shh exciton and higher confined photon modes in microcavities with full three-dimensional photon confinement.

Acknowledgements.

This work was supported by the State of Bavaria and in part by the U.S. Office of Naval Research (S.N.W. and T.L.R.) and by the Alexander von Humboldt Foundation (T.L.R.).

References

[1] C. Weisbuch et al., Phys. Rev. Lett. 69, 3314 (1992).
[2] R. Houdré et al., Phys. Rev. Lett. 73, 2043 (1994).
[3] Y. Zhu et al., Phys. Rev. Lett. 64, 2499 (1990).

INTRA- AND INTERWELL MAGNETO-EXCITONS IN GaAs/AlGaAs DOUBLE QUANTUM WELLS AND SUPERLATTICES

V.B. TIMOFEEV
Institute of Solid State Physics, Russian Academy of Science, 142432 Chernogolovka, Moscow District, Russia

M. BAYER, A. FORCHEL
Technische Physik, Universität Würzburg, Am Hubland, D-97074 Würzburg, Germany

D. BIRKEDAL, J.M. HVAM
Microelectronic Centret, Technical University of Denmark, DK-2800 Lyngby, Denmark

M. POTEMSKI
Laboratoire des Champes Magnetiques Intenses, MPI/CNRS, 25 Avenue des Martyres BP 166, France

Abstract

Intra- and interwell excitons have been investigated in coupled double quantum wells (DQW's) and superlattices (SL's). In coupled DQW's in case of a flat band structure without applied bias spatially direct and indirect excitons are formed, if the width of the separating barrier is so large that the Coulomb interaction energy is larger than the single particle splittings (weak coupling regime). In contrast, in case of narrow barriers direct and indirect character are strongly mixed for all exciton states (strong coupling regime). While a magnetic field normal to the QW planes shows little effects on the coupling in case of narrow barriers, it causes a decoupling of the wells in case of wide barriers. Similar observations can be made for a symmetric superlattice, in which - beside the intrawell excitons - also interwell excitons can be formed. The localization of the carriers in adjacent quantum wells is only caused by the Coulomb interaction.

In case of a tilted band structure (application of a bias across the heterostructure) a direct-to-indirect crossover of the ground state exciton can be induced with increasing bias in coupled DQW's. The crossover depends on the magnetic field strength. The stronger the magnetic field is, the higher voltages have to be applied to induce the changeover. Vice versa, at a certain fixed voltage the indirect ground state exciton is stable only up to a certain magnetic field. For biased superlattices we observe the Wannier-Stark ladder of excitons. In addition, with increasing bias resonant tunneling between intrawell and interwell states is observed, which occurs when excited interwell states cross the intrawell ground state. This resonant tunneling leads to strong, anti-phase oscillations of the luminescence intensity of the intrawell and interwell excitons.

1. Introduction

The investigation of coupled double quantum wells (DQW's) and superlattices (SL's) in semiconductor heterostructures has gained considerable interest during the last decade [1-9]. In particular, this interest arises from the expectation that their electronic properties might be utilized in opto-electronic devices. On the other hand, the physical interest in these structures is strongly related to their excitonic properties, which have attracted attention both with respect to the single exciton problem as well as with respect to exciton-exciton interaction and to unusual collective phenomena which originate from this interaction [10-13].

Here the single exciton properties in DQW's and SL's in GaAs/AlGaAs heterostructures subject to high magnetic fields will be discussed. We will first give a short overview of the expected phenomena: If excitonic effects are neglected, each pair of electron/hole states (which are exactly degenerate without coupling) splits into a symmetric and an antisymmetric state in symmetric coupled DQW's. The splitting in energy between these states, $\Delta_{e/h}$, decreases exponentially with increasing barrier width L_B. For sufficiently large L_B the $\Delta_{e/h}$ are significantly smaller than the difference in the binding energies of an intrawell or direct (index D) and an interwell or indirect (index I) excitons: $\Delta_{e/h} \ll E_D - E_I$ (E_D = const for $L_b \Rightarrow \infty$, while $E_I \rightarrow 0$). Hence symmetric and antisymmetric single particle states are strongly mixed by the Coulomb interaction generating four different states in case of the $1s$ exciton, two states with predominantly direct and two with predominantly indirect spatial character. In the opposite limit (strong coupling) $\Delta_e, \Delta_h \gg E_D - E_I$ intrawell and interwell excitons are strongly mixed and the single particle picture can be used to describe the optical transitions. As a parameter for the strength of coupling and the resulting spatial characters of the excitonic states the following quantity can be defined $\alpha = \max(\Delta_e, \Delta_h)/(E_D - E_I)$. In general, the excitons can be distinguished by their symmetry with respect to the parity transformation $z \rightarrow -z$, where $z = z_e - z_h$ is the relative coordinate of electron and hole. Only the symmetric excitons are optically active. The splittings between the antisymmetric and the symmetric exciton states are usually small in comparison to the single particle spltittings, $\Delta_{A-S} \ll \Delta_{e/h}$ [8].

A magnetic field perpendicular to the DQW planes strongly modifies the excitonic states. In the high field limit ($l_c \ll a_{\text{ex}}$, l_c being the magnetic length) the asymptotic behaviours of the binding energies of purely direct and purely indirect excitons are given by: $E_D \sim e^2/\epsilon l_c \sim B^{1/2}$ and for $E_I \sim e^2/\epsilon$ $(l_c^2 + d^2)^{1/2} \Rightarrow e^2/\epsilon d$. Due to the different increase of the binding energies for direct and indirect excitons a magnetic field transforms the DQW structures to the wide barrier regime (weak coupling). This transformation is accompanied by drastic redistributions of the oscillator strengths of the I- and D- excitons.

If an electric field F is applied along the heterostructure growth direction its inversion symmetry is broken and pure spatially indirect (interwell) excitons are formed. With increasing F in DQW's hence a direct to indirect crossover can be obtained for the ground state exciton X_0, i.e. at low bias X_0 is still a intrawell state (index D), while at high bias it is an interwell state (index I^-). By applying a magnetic field this crossover with increasing electric field is significantly modified due to the enhancement of the Coulomb interaction by B.

Interwell excitons can also be formed in SL's without externally breaking the inversion symmetry: For SL's with narrow minibands the Coulomb interaction between electrons and holes in adjacent wells can be sufficient to localize the superlattice state and to form interwell excitons. As in case of the DQW's a magnetic field perpendicular to the plane of the SL's strongly modifies the intra- and interwell magnetoexciton states. By applying an electric field along the superlattice growth direction the electron and hole wave functions are localized in individual wells and the miniband structure is replaced by the Stark ladder.

2. Symmetric coupled DQW's

First we will consider how excitonic effects are modified in symmetric DQW's with varying barrier widths by a magnetic field. For this aim the hh and lh excitons in GaAs/AlGaAs DQW's separated by AlAs barriers of various widths have been investigated in magnetic field up to 20T. Photoluminescence (solid lines) and PLE (dotted lines) spectra recorded at $B = 0$ for DQW's with barriers 5, 3, 1 and 0 AlAs monolayers (ML) - one monolayer corresponds to a material thickness of 0.25 nm - are shown in Fig.1. Besides a large number of excited transitions the PLE spectra display four lines which can be attributed to transitions between the excitonic states of heavy and light holes which are split due to the coupling (the transitions are labeled n,m(hh/lh)$_{H/L}$ with n and m being the principal and angular quantum numbers and the indices H/I indicating the transition of higher (H) or lower (L) energy). With increasing barrier width the coupling between the two QW's is reduced. This is evidently reflected by the splitting in energy between the H and L states which continuously decreases with increasing L_B. By comparing the difference in the $2s - 1s$ splittings (which gives the exciton binding energies) of these exciton states with the calculated single particle splittings it becomes clear that the strong coupling regime is approached for AlAs barrier widths below 0.5 nm.

Fig. 2a shows a magnetic field series of PLE spectra of the sample containing 5 monolayers of AlAs as barrier (in this sample the intermediate to weak coupling regime is obtained). The transition energies are depicted in Fig. 2b. All spectral lines originating from the four $1s$ states of hh and lh, show prominent diamagnetic shifts to higher energies with increasing B, but for 1shh$_H$ and 1slh$_H$ states these shifts are significantly larger than that of the 1shh$_L$ and 1slh$_L$ states. This originates from the predominantly indirect spatial character of the $H = I$ states, while the $L = D$ states have predominantly direct character. Due to the spatial separation between electron and hole for the indirect exciton, the increase of its binding energy with B is significantly smaller than that of the direct exciton. Therefore the magnetic field results in an increase of the splitting in energy between the I- and D- excitons, Δ_{I-D}, and also in a reduction of the parameter α. However, this reduction of α decreases for excited states ($n > 1$) due to their larger in-plane extension.

Furthermore, the transformation to the weak coupling regime with increasing B causes a redistribution of the oscillator strength $f \sim | d |^2$ (d - dipole momentum) from the interwell (indirect) to the intrawell (direct) exciton states (Fig. 2c.), i.e. the reduction of the coupling constant α by B results in an increase of the ratio $R = f_D/f_I$ of the oscillator strengths of the direct and the indirect exciton. For example, in case of the 1shh exciton at $B = 13$T this ratio is almost 2.5 times larger than at $B = 0$. Therefore the magnetic field can be used to control the coupling between symmetric DQW's: B decreases the coupling significantly, if the splitting between A- and S- single particle states is comparable to the exciton binding energies.

In contrast, for strongly coupled systems as in case of the structures with AlAs barriers with widths of 0 or 1 monolayers, excitonic effects are negligible and the influence of the magnetic field is much weaker, i.e. B cannot reduce the coupling strongly. The splitting in energy between the H- and L- excitons as function of the magnetic field is then almost constant as is the ratio of oscillator strengths.

3. Biased coupled DQW's

Now we will discuss the behavior of magnetoexcitons in DQW's when a bias U is applied across the heterostructure. For this aim a n-i-n DQW structure with parameters corresponding to the weak couplin regime at U = 0 ($L_W = 100\AA$ and $L_B = 45\AA$) has been investigated. With applied electric field F the spectrum consists of direct - D and two types of indirect (I^+ and I^-) excitons both for the light and heavy hole excitons. The evolution of these magnetoexciton states with increasing bias at a fixed magnetic field strength (B = 12T) is shown in Fig. 3a, in which the transition energies as observed in PLE are plotted. By applying a bias the indirect magnetoexcitons split into states which move strongly either to lower (I^-) or to higher (I^+) energies with increasing U, while the energy of the direct excitons is almost constant. Therefore the I^- and the D excitons come close to each other, anticross and exchange their character, i.e. a direct-to-indircet crossover of the ground state exciton takes place. This crossover in accompanied by a sharp drop in the oscillator strength of X_0 with increasing bias, because the direct exciton has a significantly larger oscillator strength than the indirect one, the oscillator strength of which tends to zero (Fig. 3b). The crossover can therefore be characterized by an characteristic electric field strength F^*_{I-D}. Namely, at F^*_{I-D} the difference in energy between the single particle states due to the tilting of the band structure becomes equal to the difference in energy of the D and I exciton binding energies $eF^*_{I-D}d = E_D - E_I$. A magnetic field increases the difference in binding energies $E_D - E_I$ and, as consequence, increases F^*_{D-I}.

At a given electric field the diamagnetic shift of the indirect excitons is much larger than the shift of the direct excitons. The origin of this behaviour is the same as for $F = 0$: The binding energy of the I- excitons is less increased with increasing B as that of the D- excitons. At a given applied bias at which X_0 is an indirect state, the ground state exciton can be transformed back into a direct state due to the different diamagnetic shifts of the I- and D-states. This magnetic field induced crossover from I- to D- excitons determines the maximum magnetic field at which indirect excitons are stable for a given interwell tunneling.

4. Symmetric Superlattices

Next we will discuss the formation of spatially indirect excitons in symmetric SL's with narrow minibands. We have found an interwell excitonic state which lies energetically between the 1s exciton state and the onset of the miniband absorption (Fig. 4a). This state, with a hole confined in one well and an electron in a neighboring well, has considerable oscillator strength and is observed in PL, PLE and absorption spectra measurements.

The interwell exciton state is similar to the first Wannier Stark localized exciton in an electrically biased superlattice. But the carrier localization in the considered case of a symmetric SL (without applied bias) is mediated only by the Coulomb interaction between electron and hole. The condition for the formation of such interwell excitons is that the sum of the electron and hole miniband widths Δ is smaller than the exciton binding energies in the isolated QW's (E_{2D}) and smaller than the interwell Coulomb energy $E_I = e^2/\epsilon d$ (where d is the superlattice period): $\Delta < E_{2D}$, $e^2/\epsilon d$.

Here we present the experimental observation of interwell excitons in a GaAs/AlGaAs superlattice sample with barriers which are on one hand so thick that the interwell Coulomb energy is larger than the miniband width, but which are on the other hand so narrow that the oscillator strength of the I- exciton is large enough to be measured. We have found that this criterion is fulfilled in $GaAs/Al_{0.3}Ga_{0.7}As$ multiple QW structure with a well width of $80\mathring{A}$ and a barrier width of $50\mathring{A}$. For these parameters we find that $E_{2D} = 9.5$ meV, $E_I = 7$ meV and $\Delta = 2.6$ meV.

The interwell exciton state I_{hh} is clearly resolvable in the PLE spectra (this line is well separated from 2shh state even at $B = 0$T, see Fig.4a). This state demonstrates a stronger diamagnetic shift than the 1shh exciton. We have found that the ratio of diamagnetic shifts of the I_{hh} and 1shh is 1.85 clearly showing that the I_{hh} state is not related to the 2shh exciton. The oscillator strength of the interwell excitons continuously decreases with increasing magnetic field. This indicates that, as in case of symmetrical DQW's, the magnetic field transforms the system to the weak binding regime. Besides we have also observed interwell exciton states for light hole excitons (the corresponding lines are labeled as I_{lh} in Figs. 4a and b). These results are expected to be useful in the interpretation of recent four-wave mixing experiments which investigated Bloch-oscillations in superlattices [14,15].

5. Biased Superlattices

By applying an electric field along the SL growth direction, the electron and hole wave functions are localized in individual wells and this leads to the formation of Wannier Stark localized excitons [16,17]. In a biased GaAs/AlGaAs p-i-n 80/50/80 superlattice we have observed intrawell (D) and interwell (I_1^- and I_2^-) exciton states, which correspond to the Wannier-Stark ladder, in luminescence spectra (Fig.5a). These observations are confirmed by magneto-PL and PLE experiments. The interwell excitons exhibit much larger diamagnetic shifts in comparison to the shift of the intrawell exciton. Optical pumping of the interwell excitons can be obtained by selective resonant excitation of higher excited interwell exciton states.

In addition, resonant tunneling between intrawell and interwell magnetoexciton states can be observed by varying the magnetic field strength. B was again aligned along the SL growth direction. The tunneling occurs when excited states of the interwell excitons cross the ground state of the intrawell exciton. These crossings are accompanied by strong oscillations of the intensities of the luminescence of the inter- and intrawell magnetoexciton (Fig. 5a and 5b). For the D- and the I^- exciton states we find that these oscillations are antiphase and that the amplitude of the oscillations is increasing with increasing magnetic field.

Acknowledgements. V.B. Timofeev gratefully acknowledges the financial support of this work by the Alexander von Humboldt Foundation and by INTAS.

References

[1] Y.J. Chen, E.S. Koteles, B.S. Elman, and C.A. Arimento, Phys.Rev. B **64**, 4562 (1987).

[2] M.N. Islam, R.L. Hillman, D.A.B. Miller, D.S. Chemla, A.C. Gossard, and J.H. English, Appl.Phys.Lett. **50**, 1098, (1987).

[3] T. Kamizato and M. Matsuura, Phys.Rev.B **40**, 8378 (1989).

[4] M.M. Dignam and J.E. Sipe, Phys.Rev.B **43**, 4084 (1991); Phys.Rev.B **43**, 4097 (1991).

[5] G.W. Bryant, Phys.Rev.B **47**, 1683 (1993).

[6] T. Westgaard, Q.X. Zhao, B.O. Fimland, K. Johannessen, and L. Johnsen, Phys.Rev.B **45**, 1784 (1992).

[7] L.V. Butov, A. Zrenner, G. Abstreiter, A.V. Petinova and K. Eberl, Phys.Rev.B **52**, 12153 (1995).

[8] A.V. Dzyubenko and A.L. Yablonskii (unpublished).

[9] B. Deveaud, A. Chomette, F. Clerot, A. Regreny, J.C. Maan, R. Romestein, G. Bastard, H. Chu, and Y.-C. Chang, Phys.Rev.B **40**, 5802 (1989).

[10] Yu.E. Lozovik and V. Yudson, Sov.Phys.JETP Lett. **44**, 389 (1987).

[11] D. Yoshioka and A.H. MacDonald, J.Phys.Soc.Japan **59**, 4211, (1990).

[12] T. Fukuzawa, E.E. Mendez, and J.M. Wang, Phys.Rev.Lett. **64**, 3066 (1990).

[13] L.V. Butov, A. Zrenner, G. Abstreiter, G. Böhm, and G. Weimann, Phys.Rev.Lett. **73**, 304 (1994).

[14] K. Leo, P. Bolivar, F. Brüggemann, and R. Schwedier, Solid State Commun. **84**, 943 (1992).

[15] H. Roskos et al., Phys.Rev.Lett. **68**, 2216 (1992).

[16] E. Mendez, F. Aguilli-Rueda, and F. Hong, Phys.Rev.Lett. **60**, 2426 (1988).

[17] P. Voisin et al., Phys.Rev.Lett. **61**, 1639 (1988).

Figure Captions

Fig. 1: PL and PLE spectra at $B = 0$ of the GaAs/AlGaAs CDQW's separated by AlAs barriers with widths of 0, 1, 3 and 5 monolayers (ML).

Fig. 2: (a) PLE spectra of the CDQW sample containing the 5 ML wide AlAs barrier for different magnetic fields. (b) Positions in energy of the observed transitions as functions of the magnetic field. (c) Ratio of the oscillator strength of the 1$s$hh direct and indirect excitons versus magnetic field.

Fig. 3 (a) Observed transition energies for the GaAs/AlGaAs $n^+ - i - n^+$ CDQW at B=12T as functions of the applied voltage. (b) Ratio of the oscillator strengths of the I^- and the D states of the 1$s$hh exciton versus applied bias at $B = 6$ T und 12 T.

Fig. 4: (a) PLE spectra at different magnetic fields of the symmetric GaAs/AlGaAs SL. (b) PLE spectrum at $B = 0$ of the same SL.

Fig. 5 (a) PL spectra of the biased SL. (b) Relative intensity of the indirect I^- and the direct D hh exciton in the biased SL $\Delta = (I(I^-_{hh}) - I(D_{hh})/(I(I^-_{hh}) + I(D_{hh}))$.

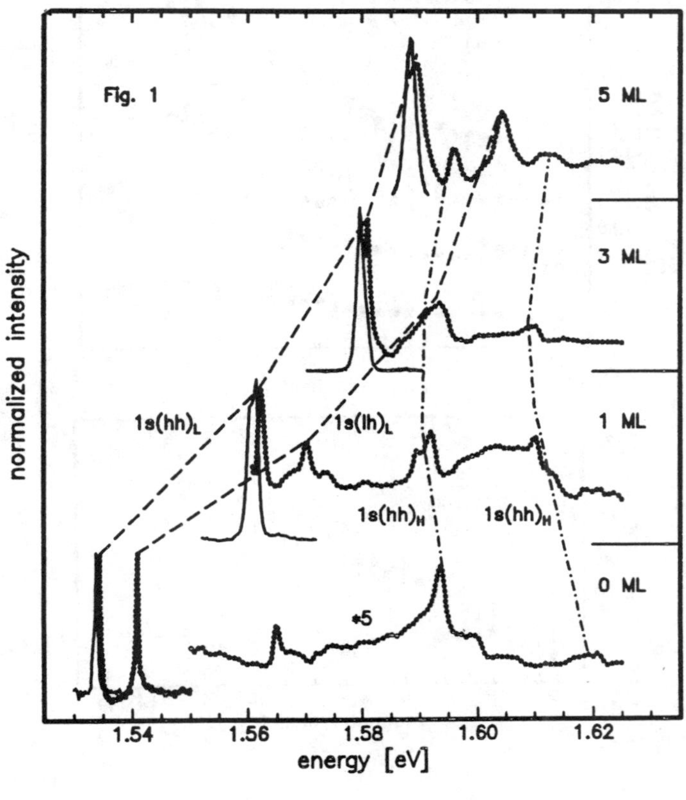

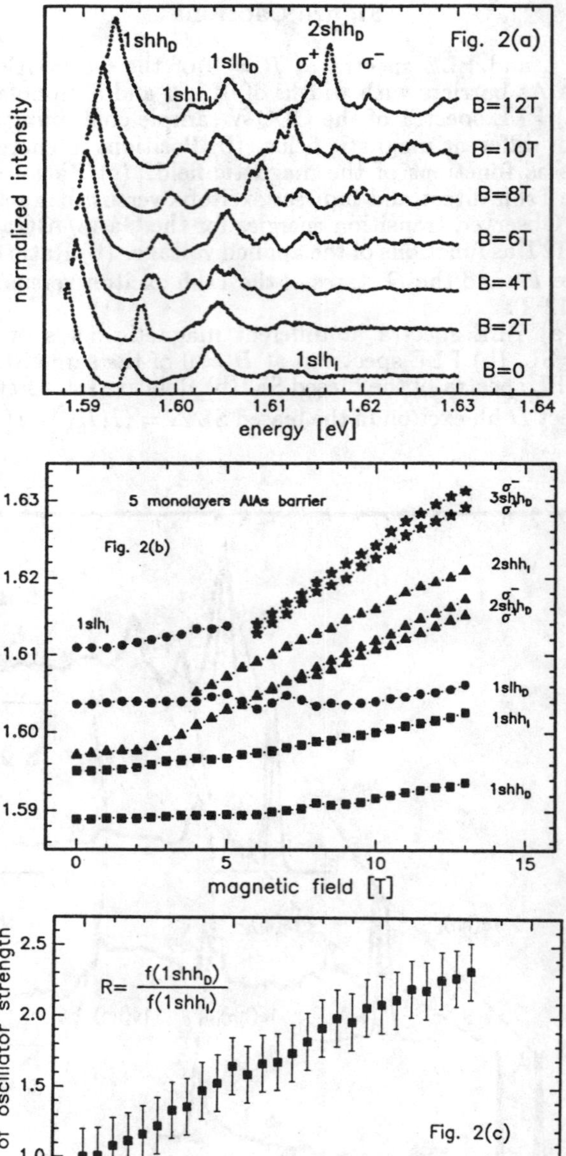
1shh$_D$
2shh$_D$
Fig. 2(a)
1slh$_D$
σ^+
σ^-
1shh$_I$
B=12T
B=10T
B=8T
B=6T
B=4T
B=2T
1slh$_I$
B=0
normalized intensity
1.59
1.60
1.61
1.62
1.63
1.64
energy [eV]
5 monolayers AlAs barrier
Fig. 2(b)
σ^-
3shh$_D$
σ^+
2shh$_I$
σ^-
2shh$_D$
σ^+
1slh$_I$
1slh$_D$
1shh$_I$
1shh$_D$
energy [eV]
1.63
1.62
1.61
1.60
1.59
0
5
10
15
magnetic field [T]
ratio of oscillator strength
2.5
2.0
1.5
1.0
R= f(1shh$_D$) / f(1shh$_I$)
Fig. 2(c)
0
5
10
15
magnetic field [T]

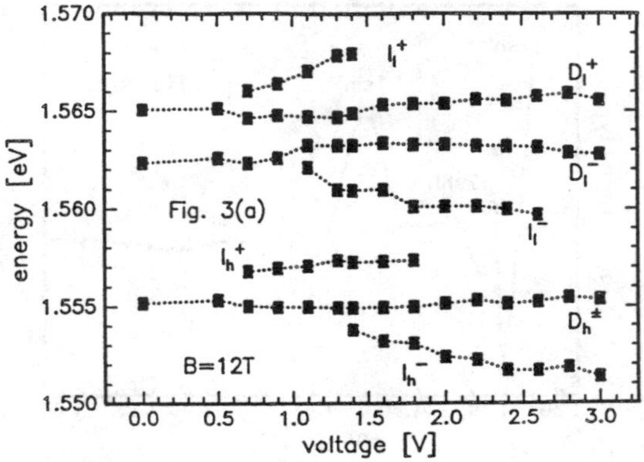

1.570
1.565
1.560
1.555
1.550
energy [eV]
Fig. 3(a)
B=12T
I_l^+
D_l^+
D_l^-
I_l^-
I_h^+
$D_h^\pm$
I_h^-
0.0 0.5 1.0 1.5 2.0 2.5 3.0
voltage [V]

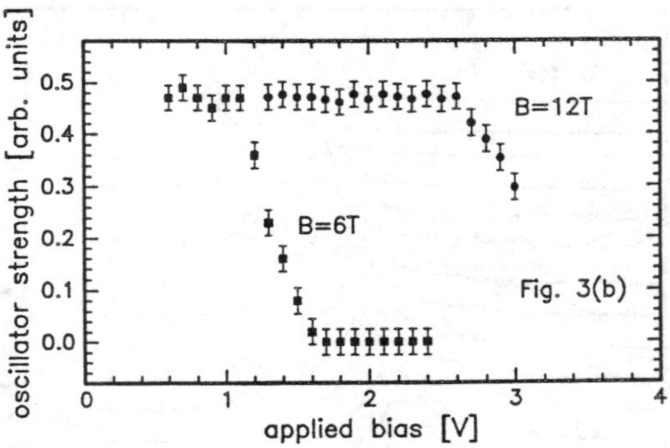

oscillator strength [arb. units]
0.5
0.4
0.3
0.2
0.1
0.0
B=12T
B=6T
Fig. 3(b)
0 1 2 3 4
applied bias [V]

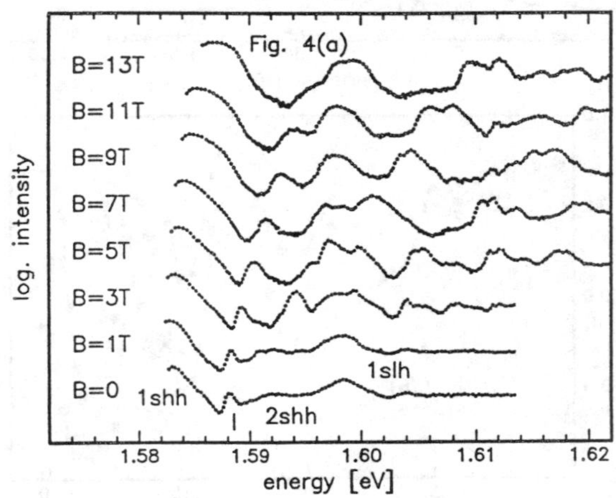

Fig. 4(a)
B=13T
B=11T
B=9T
B=7T
B=5T
B=3T
B=1T
B=0
1shh
2shh
1slh
log. intensity
1.58 1.59 1.60 1.61 1.62
energy [eV]

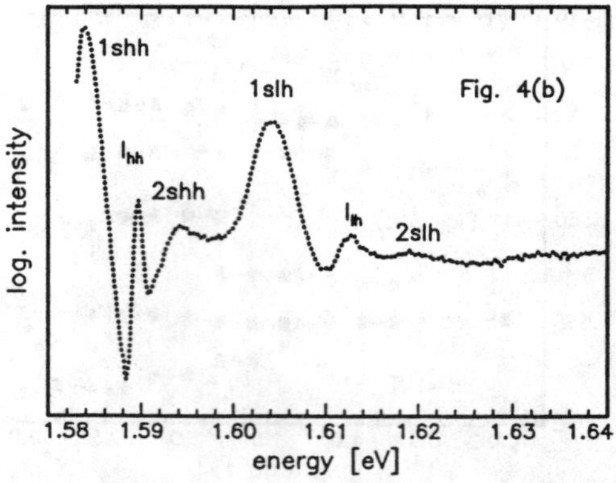
1shh
l_hh
2shh
1slh
l_lh
2slh
Fig. 4(b)
log. intensity
energy [eV]
1.58
1.59
1.60
1.61
1.62
1.63
1.64

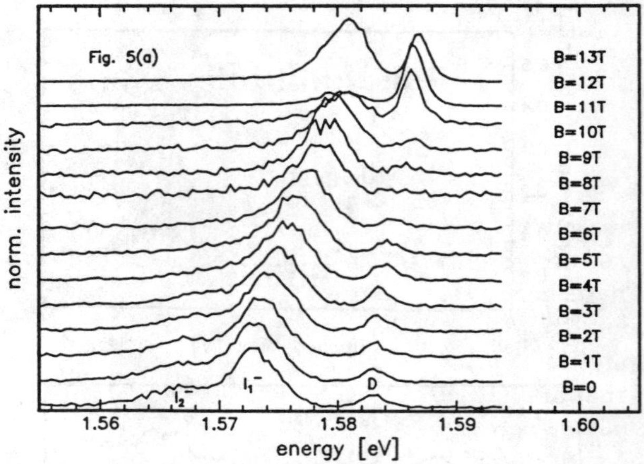
Fig. 5(a)
B=13T
B=12T
B=11T
B=10T
B=9T
B=8T
B=7T
B=6T
B=5T
B=4T
B=3T
B=2T
B=1T
B=0
l_2^-
l_1^-
D
norm. intensity
energy [eV]
1.56
1.57
1.58
1.59
1.60

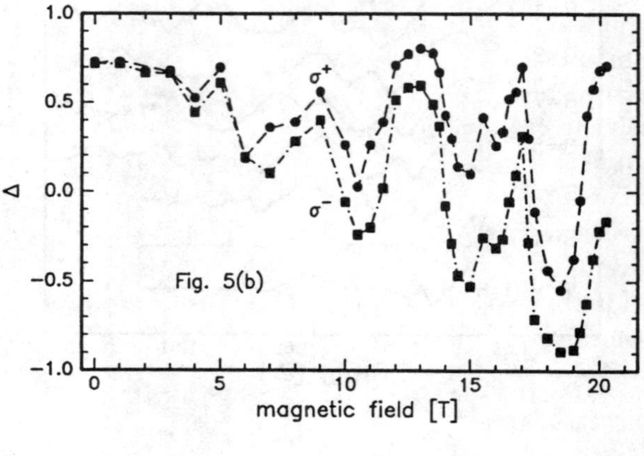
σ+
σ−
Fig. 5(b)
Δ
1.0
0.5
0.0
−0.5
−1.0
0
5
10
15
20
magnetic field [T]

NEW PLASMON MODE IN COUPLED QUANTUM WIRES IN HIGH MAGNETIC FIELDS

W. FRANK, W. HANSEN [a], A. GOVOROV [b], J. P. KOTTHAUS
Sektion Physik, Ludwig-Maximilians-Universität München, Geschwister-Scholl-Platz 1, D-80539 München, Germany

M. HOLLAND
Department of Electronics and Electrical Engineering, University of Glasgow, Glasgow G12 8QQ, United Kingdom

We present far-infrared (FIR) absorption measurements on field-effect induced quantum wire lattices with a unit cell that contains two closely spaced electron channels. We observe a plasmon spectrum that shows an additional resonance compared to the single-wire case. This mode is discussed in view of strong coupling between the electron channels. The coupling is found to be very sensitive to the influence of surface charges on the free GaAs surfaces between the gate electrodes.

In recent years, excitations in quantum wire lattices have been the object of intensive research.[1,2,3,4] The lattices have been defined with holographic lithography and, accordingly, only periodic gratings with one quantum wire per unit cell have been investigated. Here we examine for the first time a lattice with a unit cell that consists of two closely spaced quantum wires.

We generate the electron channels in metal-insulator-semiconductor heterostructures that are described in previous publications.[3,5] Fig. 1 shows schematically a unit cell of the lattice that is periodically repeated in the x-direction. The gate electrodes have been defined by e-beam lithography. The width of the stripes is approximately 150 nm, the smaller distance between the stripes is 125 nm and the larger one is 900 nm. By applying a positive voltage at the gate electrodes with respect to the back contact, electron channels are induced beneath the gates. With the gate voltage we control both the lateral confining potential and the electron density in the channels. All measurements were performed at 2.2 K with a constant magnetic field B in the z-direction.

In such wire arrays we observe for magnetic fields up to 4 T complex spectra with several modes similar to the spectra taken on single wire lattices.[5] At magnetic fields above 4 T we clearly resolve a new FIR mode with resonance position between the cyclotron resonance frequency and the fundamental mode

---

[a] present and permanent address: Institut für Angewandte Physik, Universität Hamburg, Jungiusstr. 11, D-20355 Hamburg, Germany

[b] permanent address: Institute of Semiconductor Physics, 630090 Novosibirsk-90, Russia

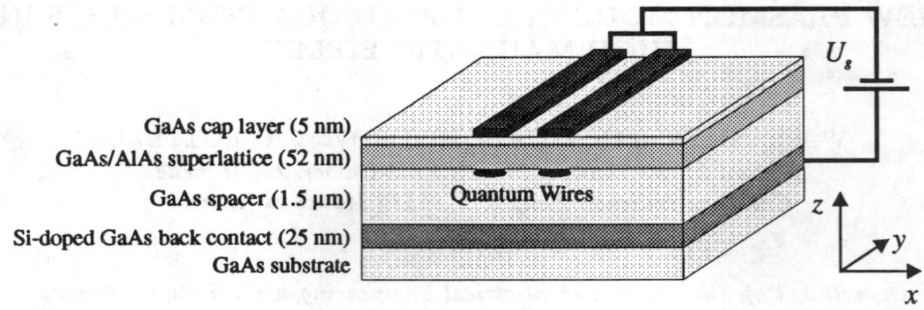

Figure 1: Schematic view of a unit cell of the sample lattice, showing the consecutive layers of the heterostructure (not to scale) and the gate layout.

of the single wires. In previous FIR experiments with single quantum wire lattices [1,5] such a mode has not been observed. Fig. 2 shows the behaviour of this mode with the gate voltage. The peak with the lowest energy is the cyclotron resonance that is excited in two-dimensional (2D) electron stripes that exist for technical reasons on the sample: They are induced beneath a coarse mesh of wider stripes that ensures homogeneous charging of the sample even if the grating of very fine gate stripes has some defects.

We first consider the spectra denoted by thick lines. At 0.95 V only the cyclotron resonance from the 2D regions and a weak broad peak at 115 cm^{-1} from the wire system are visible. With increasing gate voltage and hence electron density the oscillator strength (OS) of this resonance initially increases and then remains approximately constant. At 1.05 V a sharp peak at 111 cm^{-1} appears whose OS increases rapidly with the gate voltage. In the examined voltage range, no saturation of the OS of this resonance is visible.

The behaviour of the resonances for high magnetic fields can be understood qualitatively in a classical model based on the Drude approximation. We assume that with increasing gate voltage the channels are no longer completely separated, but connected by a region of lower electron density. The electron system is described by two parallel stripes of constant two-dimensional electron density n_1 that are connected by another stripe with lower electron density n_2. The lateral movement of the electrons in this system under the influence of a perpendicular electric field E_x is described by the following differential equation for the 2D current density $j(x)$:

$$\frac{\partial^2 j(x)}{\partial x^2} + \frac{m\epsilon\epsilon_0}{e^2 d}\frac{\omega^2 - \omega_c^2}{n_0(x)} j(x) = i\omega\frac{\epsilon\epsilon_0}{d}E_x. \qquad (1)$$

Here e is the elementary charge, m the effective electron mass, ω_c the cyclotron

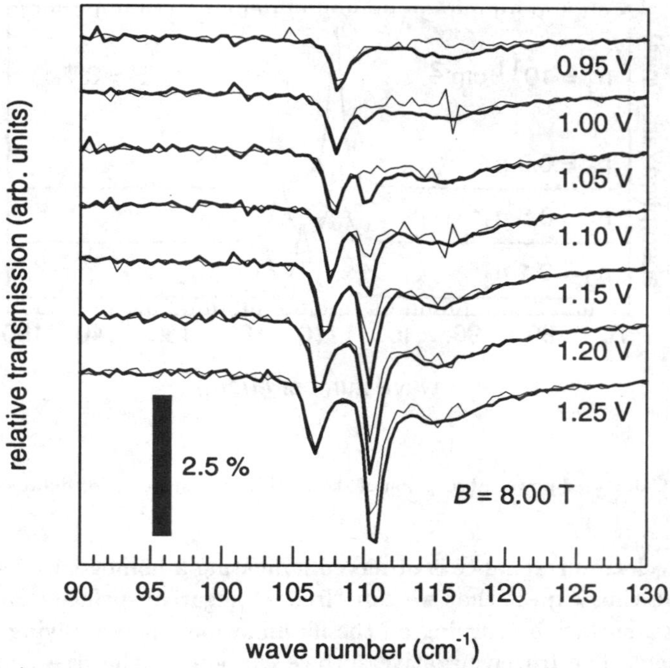

Figure 2: FIR transmission spectra of a typical sample for different gate voltages. Thick lines: Spectra taken without frozen surface charges, thin lines: Spectra taken with frozen negative surface charges (for discussion see text). The traces are offset for clarity.

frequency, $n_0(x)$ the equilibrium electron density in the channels (n_1 or n_2 depending on x) and d the distance from the channels to the gate electrodes. We solve this equation for $j(x)$ in order to calculate the power absorption that is depicted in Fig. 3. We find that with increasing n_2 the OS of the lower energy mode increases whereas the OS of the higher energy mode decreases.

The value of n_2 depends on the height of the potential barrier between the channels. Calculations of the bare electrostatic confinement potential show that this height depends critically on the surface charge density on the free GaAs surface between the gate stripes. Indeed, variations of the surface charge density result in a clearly visible change of the OS of the new FIR mode as demonstrated by the following experiment. Illumination with sub-bandgap IR radiation turns the free surface conducting. This is inferred from transmission spectra recorded on our samples under constant illumination, in which solely

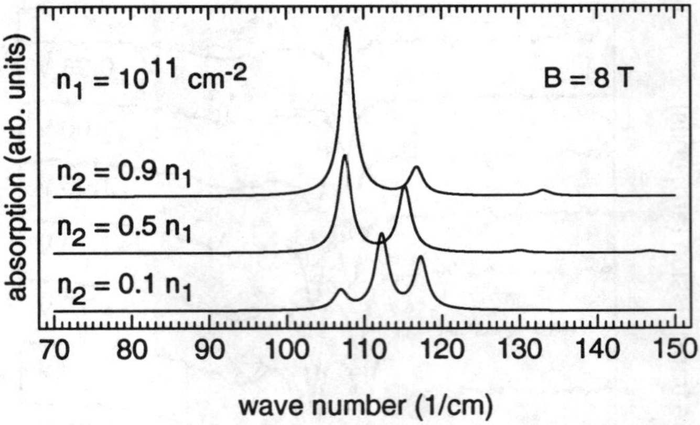

Figure 3: Calculated power absorption of the double wire system (for details see text).

a strong cyclotron resonance is observed indicating a homogenized surface potential. We thus expect that we can "freeze" negative surface charges at the open crystal surface by turning off the illumination while applying a negative gate voltage. The frozen negative surface charges on the free GaAs surface should, in turn, increase the potential barrier between the electron channels. As is seen in Fig. 2 (thin lines), with frozen negative surface charges the OS of the lower energy mode is clearly reduced, which agrees with our model.

Acknowledgments

This work was funded by Deutsche Forschungsgemeinschaft. A. G. gratefully acknowledges financial support from A. v. Humboldt-Stiftung.

References

1. W. Hansen *et al.*, Phys. Rev. Lett. **58**, 2586 (1987).
2. T. Demel, D. Heitmann, P. Grambow, und K. Ploog, Phys. Rev. Lett. **66**, 2657 (1991).
3. H. Drexler *et al.*, Phys. Rev. B **46**, 12849 (1992).
4. A. Schmeller *et al.*, Phys. Rev. B **49**, 14778 (1994).
5. G. Hertel *et al.*, Solid State Electronics **37**, 1289 (1994).

POLARIZED LUMINESCENCE STUDIES OF SINGLE AND BILAYER SYSTEMS IN THE INTEGER QUANTUM HALL REGIME

V. PELLEGRINI, A. PINCZUK, B.S. DENNIS, L.N. PFEIFFER, K.W. WEST
Bell Laboratories, Lucent Technologies, Murray Hill, NJ 07974, USA

We report the experimental investigation of the optical properties of GaAs single and coupled doubled quantum wells near the $\nu = 1$ quantum Hall state by polarized luminescence spectroscopy. The results uncover different manifestations of electron interaction effects in the two systems. In our analysis, polarized luminescence emerges as an easy and powerful technique to study electron screening, spin-polarization and inter-layer electron correlation effects in the quantum Hall regime.

1 Introduction

An intense research effort has been devoted to theoretical and experimental investigations of many-body effects in two-dimensional (2D) electron systems in the integer quantum Hall regime.[1,2] The competition between Zeeman and exchange energies is exemplified by recent evidence of spin textures,[3,4] and their impact on the spin-polarization of the 2D electron gas.[5]

In the coupled bilayer electron system of a double quantum well (DQW) a new degree of freedom is associated with the layer-index. The existence of such degree of freedom, that is described by a pseudospin quantum number, creates new manifestations of electron-electron interactions in 2D. When tunneling between the layers is allowed every spin-split Landau level splits further into symmetric and antisymmetric components separated by an energy gap Δ_{SAS}. This has interesting implications on the physics of the quantum Hall effect at $\nu = 1$.[6,7] Also, similar to the spin textures in single quantum wells (SQW), recent theoretical results predict the existence of pseudospin textures as the low-energy charged excitation at $\nu = 1$ in DQWs.[8] In analogy with spin textures, pseudospin textures modify the electron occupation of the symmetric and antisymmetric states of the the lowest Landau level.

In the following, we present the experimental investigation of many-body effects in single layer and coupled bilayer electron systems by polarized luminescence spectroscopy at ν close to unity. We show some remarkable effects associated with the intriguing electron-electron correlation at $\nu = 1$ in both single and bilayer systems. The results suggest that spin and pseudospin effects are revealed in polarized optical emission.

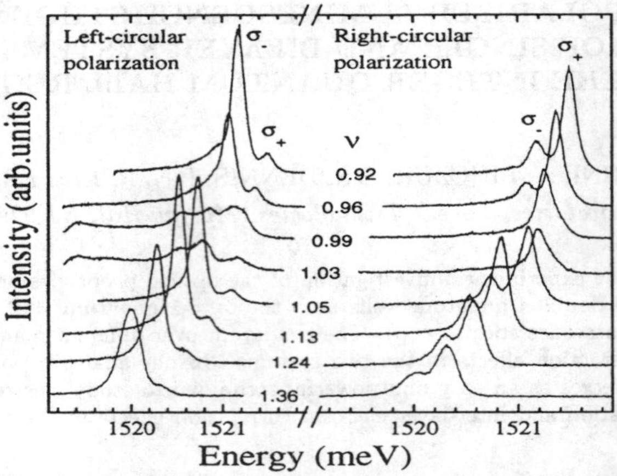

Figure 1: Left (I_-) and right (I_+) circularly polarized luminescence from single quantum well at T = 0.2 K and at different ν. ν = 1 corresponds to B = 5.45 T. The σ_- (σ_+) peak originates from recombination of spin-up (down) electrons in the lower Landau level.

2 Results and Discussion

We present results from two samples grown by molecular beam epitaxy. One is a one-side Si delta-doped 250 Å wide $Al_{0.1}Ga_{0.9}As/GaAs$ SQW, with electron density n = $1.34 \cdot 10^{11}$ cm^{-2} and mobility μ = $4 \cdot 10^6$ cm^2/Vs. The other is a modulation-doped DQW with n = $1.44 \cdot 10^{11}$ cm^{-2} and μ = $0.81 \cdot 10^6$ cm^2/Vs. The DQW consists of two 180 Å wide $GaAs$ wells separated by a 79 Å wide $Al_{0.1}Ga_{0.9}As$ undoped barrier. The samples were mounted in a $^3He/^4He$ dilution refrigerator with silica windows for optical access and inserted in a superconductive magnet cryostat. Spectra were excited by the cw linearly-polarized light of a dye-laser tuned to 7900 Å ($E_{ph} \approx 1570$ meV) and measured in a back-scattering geometry with multichannel detection. Accesible sample temperatures were in the range 0.2 K$\leq$ T$\leq$ 1.3 K. Power-densities were kept at 10^{-4} W/cm^{-2}. Figure 1 displays left (I_-) and right (I_+) circularly polarized luminescence at T= 0.2 K and at several filling factors around unity. The spectra show two different peaks with different preponderant polarization properties: the peak at lower energy arises from the left-circularly polarized σ_- optical recombination between the electron in the spin-up m = 1/2 state (m are the eigenvalues of the z-components of angular momentum) and the hole in the m = 3/2 state; the higher-energy peak from the right-circularly polarized σ_+ recombination of electrons in the spin-down m = -1/2 state with the hole in

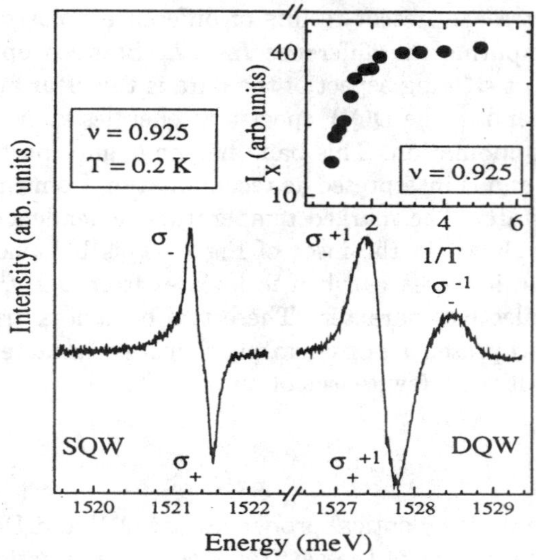

Figure 2: Difference polarized emission (DPE) $I_- - I_+$ spectra for the SQW (left) and the DQW (right) samples at $\nu = 0.925$ and T = 0.2 K. The superscripts ± 1 identify the symmetric or antisymmetric level. The inset displays the Arrhenius plot of the integrated intensity of the σ_-^{-1} band I_X at $\nu = 0.925$. The linear range has activation energy $E_a \approx 0.15$ meV.

the m = -3/2 state. σ_- spectra show a dramatic broadening due to enhancement of localization effects at $\nu = 1$ which are dominated by fluctuations in the long-range Coulomb potential from Si ionized donors. The selective manifestation of localization in the σ_- peak, and its very rapid suppression when ν departs from unity, points to the dominant role played by electron screening.[9] In addition, Fig.1 shows the intriguing behavior of σ_+ emission intensity: the σ_+ emission has a pronunced minimum at $\nu = 1$ and increases as ν departs from unity. This behavior can not be explained by changes in recombination and relaxation times.[10] The enhancement of σ_+ intensities are interpreted as arising from increases in occupation of the spin-down state of the lowest Landau level. The observed dependences of σ_+ intensities are thus consistent with the existence of spin textures for filling factors close to $\nu = 1$.

To understand the impact of spin and pseudospin textures in electron double-layers we compare in Fig.2 results from the SQW and DQW samples at ν = 0.925. To improve the contrast between the different structures in the optical spectra we present results of difference polarized emission (DPE) obtained by computing the difference $I_- - I_+$ between opposite polarization spectra. The most striking aspect of the data is the clear appearance of a new recombination band in the DQW spectra at energies above the doublet of the fundamental recombination. This band has no counterpart in polarized emission from SQW and is interpreted as recombination from the occupied spin-up antisymmetric state. The marked temperature dependence of the integrated intensity, I_X, is shown in the inset of Fig.2. This behavior and the striking dependence on ν, it increases when ν deviates from unity,[11]) are strong indications of its collective character. These two behaviors are suggestive of the recently predicted presence of pseudospin textures characterized by a binding-energy of the order of a few tenths of an meV. [8]

3 Conclusion

We have investigated the optical properties of SQW and DQW at $\nu = 1$. Our results dispay the impact of the electron-electron interactions in the interpretation of the polarized-luminescence data in the two systems. In thsi work, polarized luminescence emerges as a powerful technique to study pseudospin-interaction effects in bilayer systems.

References

1. S.L. Sondhi *et al.*, *Phys. Rev.* B **47**, 16419 (1993).
2. B.B. Goldberg *et al.*, *Phys. Rev. Lett.* **63**, 1102 (1989); I.V. Kukushkin *et al.*, *Europhysics Lett.* 222871993.
3. S.E. Barrett *et al.*, *Phys. Rev. Lett.* **74**, 5112 (1995).
4. A. Schmeller *et al.*, *Phys. Rev. Lett.* **75**, 4290 (1995).
5. E.H. Aifer *et al.*, *Phys. Rev. Lett.* **76**, 680 (1996).
6. G.S. Boebinger *et al.*, *Phys. Rev. Lett.* **64**, 1793 (1990); S.Q. Murphy *et al.*, *Phys. Rev. Lett.* **72**, 728 (1994).
7. E.H. Rezayi, and F.D.M. Haldane, *Bull. Am. Phys. Soc.* **32**, 892 (1987); T. Chakraborty, and P. Pietilainen, *Phys. Rev. Lett.* **59**, 2784 (1987); A.H. MacDonald *et al.*, *Phys. Rev. Lett.* **65**, 775 (1990); L.Brey, *Phys. Rev. Lett.* **65**, 903 (1990); H. Fertig, *Phys. Rev.* B **40**, 1087 (1989); K. Yang *et al.*, *Phys. Rev. Lett.* **72**, 732 (1994).
8. K. Moon *et al.*, *Phys. Rev.* B **51**, 5138 (1995).
9. B.B. Goldberg *et al.*, *Phys. Rev. Lett.* **65**, 641 (1990).
10. M. Dahl *et al.*, *Phys. Rev.* B **45**, 6957 (1992).
11. V. Pellegrini *et al.*, *to be published.*

INTERWELL MAGNETO-EXCITONS IN COUPLED DOUBLE QUANTUM WELLS

P.C.M. CHRISTIANEN AND J.C. MAAN
Research Institute for Materials, High Field Magnet Laboratory, University of Nijmegen, Toernooiveld, 6525 ED, Nijmegen, The Netherlands.

W. VAN DER VLEUTEN
COBRA Interuniversity Research Institute, Eindhoven University of Technology, P.O. Box 513, 5600 MB Eindhoven, The Netherlands

Interwell magneto-exciton transitions are observed in strongly asymmetric coupled double quantum well structures by photoluminescence excitation spectroscopy. The interwell transitions are absent at zero magnetic field and become only visible, at finite field, in resonance with excited states of intrawell magneto-excitons.

The excitonic spectrum of a coupled double quantum well (DQW) structure is strongly modified by Coulomb interaction across the barrier. For example, resonant tunneling occurs when single particle levels are close to resonance, but the actual tunneling event is the transfer of a spatially direct (intrawell) exciton to an indirect (interwell) exciton [1,2]. By application of an electric field the interwell exciton can be made the ground state, which, due to its long lifetime, is expected to be favorable for exciton condensation [3]. In a magnetic field Coulomb effects are larger, due to the enhanced exciton binding energy (E_b), and more pronounced, because of the extra quantization. Consequently we observe, at finite magnetic fields, interwell excitons in strongly asymmetric DQW's (ADQW's), which are coupled directly to excited states of intrawell excitons, without an accompanying single particle tunneling resonance.

The structures under study consist of two GaAs quantum wells (QW's), 50 Å (Narrow Well, NW) and 100 Å (Wide Well, WW) thick, separated by a (Ga,Al)As tunnel barrier of thickness d (d=30-80 Å). The DQW unit is sandwiched by 0.25μm (Ga,Al)As barriers and grown on top of a n^+-GaAs substrate and a GaAs/AlAs superlattice buffer layer. All experiments were repeated on a control sample, consisting of two QW's of 100 Å and 70 Å width decoupled by a 500 Å (Ga,Al)As barrier. The Al content of the (Ga,Al)As layers in all samples equals 35%. Polarized photoluminescence excitation (PLE) spectra were recorded at 4.2 K using a tunable Ti:sapphire laser, a single grating monochromator and a cooled GaAs detector. Magnetic fields up to 20 T were used, with the field directed along the growth axis.

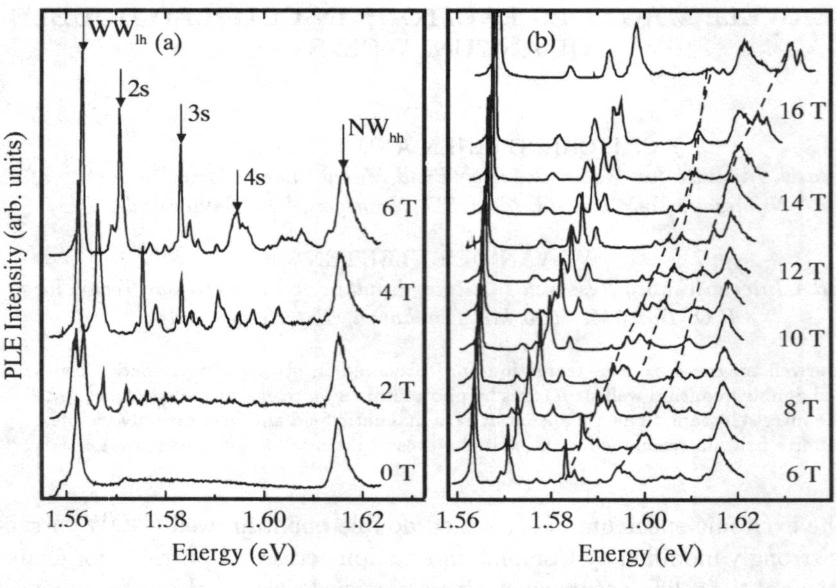

Figure 1: PLE spectra for an ADQW with 50 Å barrier, as function of magnetic field under $\sigma-$ excitation, for 0-6 T (a) and 6-18 T (b). Dashed lines were drawn to guide the eye.

Fig. 1 shows the resulting PLE spectra of the sample with d=50 Å for $\sigma-$ excitation and detection on the WW heavy hole (hh) exciton. At zero magnetic field the WW light hole (lh) (1.562 eV) and NW hh-exciton (1.615 eV) are visible. With increasing field the excited states of the WW hh- (labeled 2s, 3s, 4s in Fig. 1(a)) and lh magneto-excitons become apparent. For low fields (< 8 T) the fan diagram resembles the well known spectrum of an isolated 100 Å QW. However, for higher fields large deviations from this behavior appear. Instead of approaching more and more the behavior of free carrier Landau levels (LL's), i.e. linear with field, around 1.605 eV the magneto-exciton levels are bending, showing a clear anti-crossing with a new level, approximately 10 meV below the ground state NW exciton (dashed lines in Fig. 1(b)). This occurs for both $\sigma-$ and $\sigma+$ exciting polarizations, as is clear from Fig. 2(a), which plots the experimentally found energy positions for the 50 Å barrier sample. The region where the anti-crossing occurs is indicated by the solid line. It should be emphasized that the WW magneto-excitons do not couple to the NW ground state, which is unaffected. This indicates the importance of a new level, which is not visible at zero field, but which gains oscillator strength at finite field when it is resonant with excited WW states. This new level is clearly due to

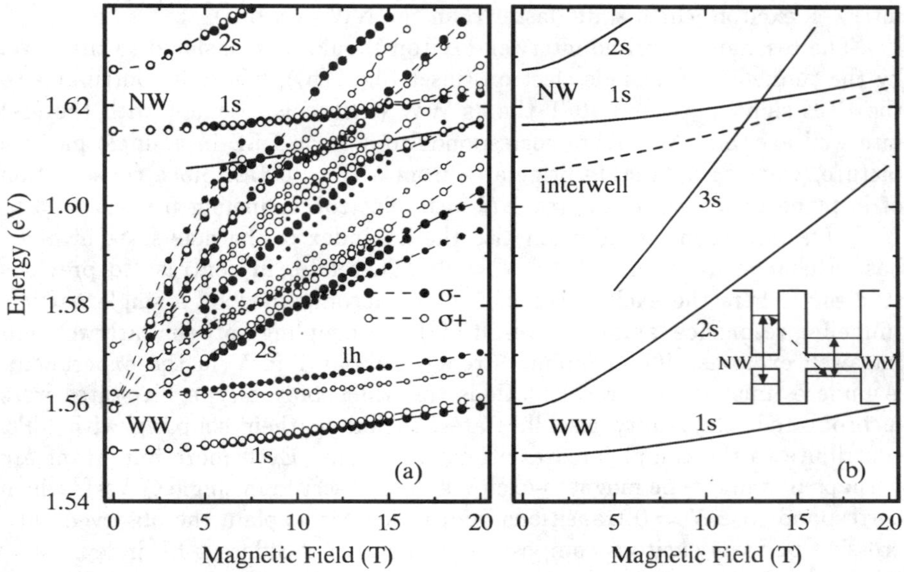

Figure 2: Experimental energy positions of the magneto-exciton transitions under $\sigma-$ and $\sigma+$ excitation (a). Calculated intrawell (solid curves) and interwell (dashed curve) magneto-exciton transitions (b). The inset shows schematically the relevant excitons.

the interwell coupling since it was observed for all samples with barrier widths 30-80 Å, but absent for the control sample with 500 Å decoupling barrier.

Fig. 2(b) shows theoretical magneto-exciton levels for the 50 Å barrier sample, from a variational calculation including Coulomb interaction, quantum confinement and magnetic field. In this calculation we compose the trial functions out of single particle wave functions for electrons and hh's of both wells, obtained numerically from the ADQW potential, and the sum of 13 Gaussian excitonic wave functions, which have been proven very successful in describing excitons and donors in single QW's[4].

The calculated 1s and 2s hh-exciton levels of both wells reproduce the experimental results very well, and yield (zero field) E_b values of 10.2 and 11.7 meV respectively for WW and NW. Striking is the appearance of an interwell exciton slightly below the NW exciton, formed by the electron in the NW and the hole in the WW (inset Fig. 2(b)), which explains the new level in the experimental data reasonably well. E_b of this exciton is calculated to be 4.1 meV, lower than that of the intrawell excitons due to the spatial electron-hole separation. This reduced E_b is also visible in the field dependence of the interwell exciton which shifts faster than the NW exciton [5,6].

The resonantly excited interwell exciton should be transfered to the WW, by the tunneling of a single electron (inset Fig. 2(b)), before it contributes to the PLE signal. Around 10 T the 4s WW exciton is resonant with the (1s) interwell exciton. A 4s-state corresponds to a $N = 3$ LL in a single particle picture, whereas a 1s state originates from a $N = 0$ LL. Since conservation of in-plane momentum requires $\Delta N = 0$, electron tunneling from $N = 3$ to $N = 0$ is forbidden. In addition also the LL index of the hole state involved has to change from $N = 3$ to $N = 0$. Therefore, in contrast to previous studies [1,2], here the exciton transfer is not accompanied by a single particle tunneling resonance, and is the result of direct coupling between interwell and intrawell excitons. The coupling strength, about 1 meV in the experiment, is underestimated in our calculations, in which only s-type hh states were incorporated. In reality also lh states, including their coupling with hh's, contribute to the complicated excitonic spectrum. Even more important are perhaps p- and d-type magneto-excitons, also observed in single QW's [7], which correspond to $\Delta N \neq 0$ transitions, and which can explain the observed anti-crossing between excitons composed out of states of different LL index.

In conclusion, interwell magneto-excitons were observed in the optical excitation spectra of strongly asymmetric DQW's, as a consequence of the direct coupling with excited states of intrawell excitons. The energy position of the interwell state is reproduced by a variational calculation for s-type hh-excitons, but the actual coupling mechanism is supposed to be due to the interaction of s-, p-, and d-type hh- and lh- magneto-excitons.

References

1. A.M. Fox, D.A.B. Miller, G. Livescu, J.E. Cunningham and W.Y. Jan, Phys. Rev. **B44**, 6231 (1991).
2. R. Ferreira, P. Rolland, Ph. Roussignol, C. Delalande, A. Vinattieri, L. Carraresi, M. Collocci, N. Roy, B. Sermage, J.F. Palmier and B. Etienne, Phys. Rev. **B45**, 11782 (1992).
3. Yu. E. Lozovik and V.I. Yudson, Sov. Phys. JETP **44**, 389 (1976).
4. Ronald L. Greene and K.K. Bajaj, Phys. Rev. **B31**, 913 (1985).
5. L.V. Butov, A. Zrenner, G. Abstreiter, A.V. Petinova and K. Eberl, Phys. Rev. **B52**, 12153 (1995).
6. A.B. Dzyubenko and A.L. Yablonskii, Phys. Rev. **B53**, 16355 (1996).
7. L. Viña, G.E.W. Bauer, M. Potemski, J.C. Maan, E.E. Mendez and W.I. Wang, Phys. Rev. **B41**, 10767 (1990).

Direct and indirect magnetoexcitons in $In_xGa_{1-x}As/GaAs$ coupled quantum wells: experiment and theory

L.V. Butov[1,2], A.B. Dzyubenko[2,3], A.L. Yablonskii[3], A. Zrenner[2], G. Abstreiter[2], A.V. Petinova[1], and K. Eberl[4]

[1] *Institute of Solid State Physics, Russian Academy of Sciences, 142432 Chernogolovka, Moscow District, Russia*

[2] *Walter Schottky Institut, Tecnische Universität München, D-85748, Germany*

[3] *General Physics Institute, Russian Academy of Sciences, 117942 Moscow, Russia*

[4] *Max-Planck-Institut für Festkörperforschung, D-70569 Stuttgart, Germany*

Abstract

We have performed experimental (PL and PLE spectroscopy at $B \leq 14$ T and $T = 350$ mK) and theoretical study of spatially direct (intrawell) and indirect (interwell) magnetoexcitons in InGaAs/GaAs coupled quantum wells. The evolution of the magnetoexciton states with the perpendicular magnetic field and electric field was studied. The peculiarities in the magnetoexciton spectra originating from the coupling between the QWs are analyzed.

A system of spatially separated electron (e) and hole (h) layers in high perpendicular magnetic fields has attracted considerable interest in the past. According to theoretical predictions, at low temperatures and in the limit of small disorder, a variety of ground states — exciton condensate, double Laughlin liquid or double Wigner solid state — can be realized in the system, depending on the ratio between the magnetic length, interparticle spacing and separation between e and h layers [1-3]. A spatially separated e–h system can be realized in coupled quantum wells (CQWs). The simplest e–h system in high magnetic fields in CQWs is a single e–h pair which forms a magnetoexciton. The objective of this work is experimental and theoretical study of magnetoexcitons in symmetric CQWs in the limit of low exciton densities (one-exciton problem). In order to be able to neglect the effects of mixing between heavy (hh) and light hole (lh) states, strained InGaAs/GaAs CQWs with a simple valence band structure were studied. In these CQWs the peculiarities in the exciton spectra originate entirely from the coupling between the QWs.

Due to the coupling between QWs, four magnetoexciton states correspond to each e and h Landau level pair (instead of one for a single QW). At zero electric field in the growth direction (F), the classification of the exciton states in symmetric CQWs (see [4,5] for the $B = 0$ case) depends on the ratio between the one-particle e and h symmetric-antisymmetric splittings and the difference between the direct (intrawell) and indirect (interwell) exciton binding energies, $\alpha = (\Delta_{SAS}^{(e)} + \Delta_{SAS}^{(h)})/(E_D - E_I)$. For $\alpha \ll 1$, exciton states are essentially direct (D) or indirect (I) in character. In order of increasing energy, the exciton states are: D symmetric, D antisymmetric, I antisymmetric and I symmetric. Antisymmetric exciton states are optically inactive. The oscillator strength (f_{osc}) of the I symmetric exciton is much smaller than that of the D symmetric exciton due to a smaller overlap between e and h. The energy splitting between the D and I exciton transitions $\simeq E_D - E_I$; the splittings between symmetric and antisymmetric excitons, $\Delta_X \ll \Delta_{SAS}^{(e)}, \Delta_{SAS}^{(h)}$ [4-6]. For $\alpha \geq 1$, the exciton states have mixed direct-indirect character. In the limit of zero Coulomb interaction, the exciton states can be classified as e-h single-particle pair states: $S_eS_h, S_eAS_h, AS_eS_h, AS_eAS_h$ (S and AS label the symmetric and antisymmetric single-particle states). S_eS_h and AS_eAS_h transitions have almost equal f_{osc}, while S_eAS_h, AS_eS_h are optically inactive. The energy splitting between the optical transitions $\simeq \Delta_{SAS}^{(e)} + \Delta_{SAS}^{(h)}$. For symmetric CQWs with arbitrary α, the first and the fourth exciton s–states in order of increasing energy are optically active, while the second and the third have zero f_{osc}. The ratio between f_{osc} of the first and the fourth exciton states increases with decreasing α.

At high electric fields F, the excitons become purely direct or indirect for arbitrary α. The transition from zero to high F regime, however, is expected to be dependent on α as it is strongly determined by the Coulomb interaction [4,5].

Electric field tunable n^+–i–p^+ CQW samples have been grown by MBE. The i-region consists of two 60 Å $In_{0.2}Ga_{0.8}As$ QWs separated by a 60 Å GaAs barrier and surrounded by 2000 Å GaAs barrier layers (Fig. 1). With increasing external gate voltage (V_g), the internal electric field in the structure is reduced and at $V_g \approx 1.52$ V the $F = 0$ regime is realized. PLE and PL spectra were recorded with the use of a *cw* Ti-Sapphire laser ($W_{ex} \approx 0.1$ W/cm^2) at $T = 350$ mK. Theoretically, exciton wave functions in $B > 2$ T are expanded in the basis of strictly 2D magnetoexcitons in the high-field limit and the wave functions describing the free perpendicular motion of e and h in a CQW in F. The details are presented elsewhere [6].

We first consider the $F = 0$ regime. At $B = 0$, two hh exciton transitions are observed in the PLE spectra (insert of Fig. 1). The splitting between them $\simeq 8.5$ meV slightly exceeds the calculated value of $\Delta_{SAS}^{(e)} + \Delta_{SAS}^{(h)}$=5.9+0.6=6.5 meV; f_{osc} of the transitions are approximately equal. Therefore, at $B = 0$ the excitons have mixed direct-indirect character, for assignment we label the transitions as D and I. The lh exciton is observed at ~ 1443 meV and is not considered in the paper.

At high B, the direct (D_N) and indirect (I_N) excitons composed from e and h with equal Landau level numbers (N-th) are optically active. The exciton energies are $\mathcal{E}_{D(I)_N} = (\hbar\omega_{ce} + \hbar\omega_{ch})(N + 1/2) - E_{D(I)_N}$, where $\hbar\omega_{ce}$ and $\hbar\omega_{ch}$ are e and h cyclotron energies, $E_{D(I)_N}$ is the binding energy of the D_N (I_N) exciton (a confinement energy is not included). Both D_N and I_N excitons from different Landau levels are observed in the spectra (see Fig. 1). At high B the exciton binding energies can be approximated by: $E_D \sim 1/l_B \sim B^{1/2}$ [7], $E_I \sim (l_B^2 + d^2)^{-1/2}$ (l_B is the magnetic length, d is effective interwell e–h separation). Due to the e–h separation for the I exciton, the increase of E_I with B is slower than that of E_D. Therefore, the magnetic field increases the energy splitting between the D and I excitons, Δ_{I-D}, and reduces the parameter α. The increase of Δ_{I-D} with B is observed in our experiments (see Figs. 1 and 2a) and is in an agreement with our calculations. At high magnetic fields, $\Delta_{I_0-D_0}$ markedly exceeds $\Delta_{SAS}^{(e)} + \Delta_{SAS}^{(h)}$ and is determined by the difference in the D and I exciton binding energies. The reduction of α with B results in an increase of the ratio between the oscillator strengths of the D and I excitons, f_D/f_I. While at B=0 $f_D \approx f_I$, at B=14 T f_D is considerably larger than f_I (insert of Fig. 1). The increase of f_D/f_I with B is in an agreement with our calculations (Fig. 2a).

The parameter α depends also on the Landau level number N because the exciton binding energy is reduced with N. The reduction of E_{I_N} with N is smaller than that of E_{D_N}, again due to the e–h separation for the I exciton. Due to this reason at high magnetic fields $\Delta_{I_1-D_1}$ is smaller than $\Delta_{I_0-D_0}$ (see Figs. 1 and 2a).

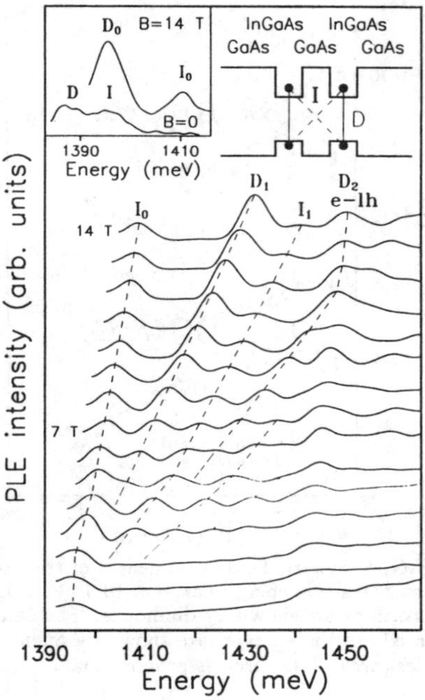

FIG. 1. PLE spectra recorded with detection on the maximum of the PL line at $F = 0$ for $B = 0 - 14$ T in 1 T steps. The dashed lines are guides for the eyes. Insert: PLE spectra recorded with detection on the low energy tail of the PL line at $F = 0$ for $B = 0$ and 14 T.

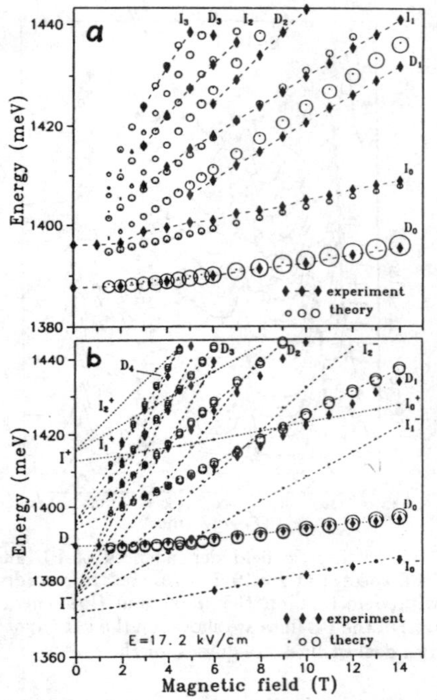

FIG. 2. Magnetic field dependence of the exciton transition energies for $F = 0$ (a) and for $F = 17.2$ kV/cm (b). PLE line positions are shown by diamonds. The calculated exciton energies are shown by open circles (area of the circle is proportional to f_{osc}).

At intermediate fields ($B = 3 - 6$ T) a complicated behavior of the exciton transitions is observed. Figure 1 shows that, under reduction of the magnetic field, f_{osc} of the D_N excitons start to drop quickly near $B \sim 6$ T; at the same time f_{osc} of the I_{N-1} excitons is increased ($N \geq 1$). This behavior is a consequence of the mixing between the D and I excitons from different Landau levels and can be seen also in the calculated dependences (Fig. 2a).

At high electric fields, due to low oscillator strengths of the I excitons, only the D excitons and I_0^- exciton (at high B) are observed in the PLE spectra (Fig. 2b). Similar to the $F = 0$ regime, at high F the diamagnetic shift of the I_0 exciton is higher than that of the D_0 exciton (see Fig. 2). The origin of this effect is the same as for the $F = 0$ regime: the increase of E_I with B is slower than that of E_D.

The shifts of the D_N and I_N excitons with electric field are essentially different (Figs. 3, 4). While the I_0^+ and I_0^- excitons experience a strong energy shift with F, the position and f_{osc} of D_0^- exciton are changed weakly at small electric field and start to drop quickly only at $F \approx 8.9$ kV/cm ($V_g \approx 1.15$ V). This electric field, F_{D-I}, corresponds to the anticrossing between the D and I excitons and, therefore, to the direct-indirect exciton crossover. At F_{D-I} the energy difference between the

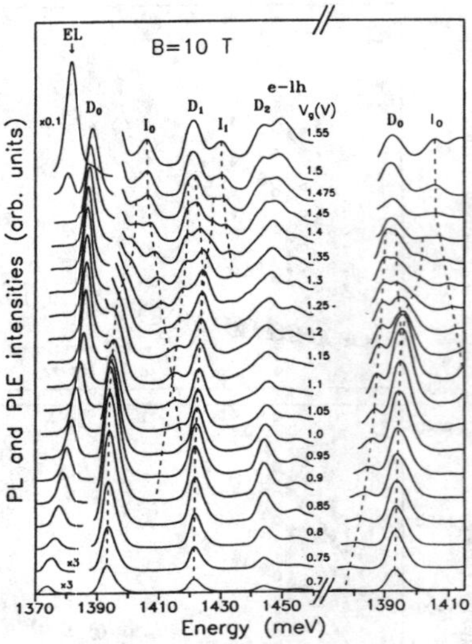

FIG. 3. Electric field dependences of PL and PLE spectra at B=10 T. PLE spectra recorded with detection on the maximum (low energy tail) of the PL lines are shown on the left (right). The dashed lines are guides for the eyes.

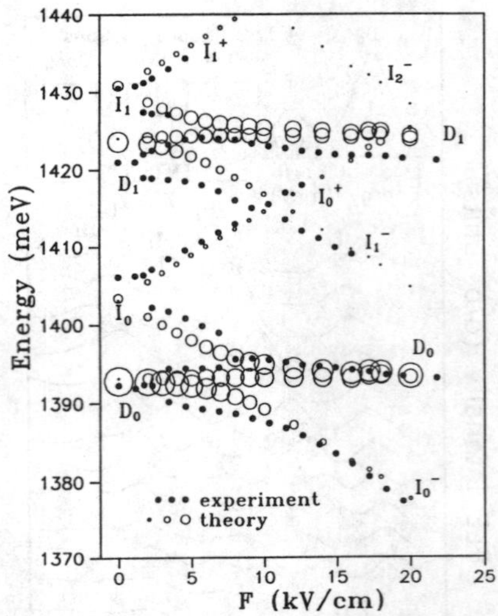

FIG. 4. Electric field dependences of the exciton transition energies at $B = 10$ T. PLE line positions are shown by diamonds. The calculated exciton energies are shown by open circles (area of the circle is proportional to f_{osc}).

single-particle direct and indirect pair states becomes equal to the difference in the D and I exciton binding energies: $eF_{D-I}d \approx E_D - E_I$. The increase of $E_D - E_I$ with increasing B results in an increase of F_{D-I}: at $B = 0$ the ground state exciton becomes essentially indirect already at $F \approx 4.1$ kV/cm ($V_g \approx 1.35$ V) [8]. Thus, the magnetic field modifies the direct-indirect exciton crossover due to the increase of the Coulomb interaction. In particular at $F \approx 7.7$ kV/cm ($V_g = 1.2$ V) the ground state exciton is indirect at $B = 0$ but direct at $B = 10$ T [6,8]. Similarly, F_{D-I} for the excitons from the first Landau level is smaller than F_{D-I} for the excitons from the 0th level (Figs. 3, 4), also due to the smaller Coulomb interaction for the former.

This work was supported by the Volkswagen Foundation and Russian Foundation of Fundamental Research.

[1] Y. Kuramoto and C. Horie, Solid State Commun. **25**, 713 (1978).
[2] D. Yoshioka and A.H. MacDonald, J. Phys. Soc. Jpn. **59**, 4211 (1990).
[3] X.M. Chen and J.J. Quinn, Phys. Rev. Lett. **67**, 895 (1991).
[4] M.M. Dignam and J.E. Sipe, Phys. Rev. B **43**, 4084 (1991).
[5] Garnett W. Bryant, Phys. Rev. B **47**, 1683 (1993).
[6] A.B. Dzyubenko and A.L. Yablonskii, Phys. Rev. B **53**, June 15th (1996).
[7] I.V. Lerner and Yu.E. Lozovik, Zh. Eksp. Teor. Fiz. **78**, 1167 (1980) [Sov. Phys. JETP **51**, 588 (1980)].
[8] L.V. Butov, A. Zrenner, G. Abstreiter, A.V. Petinova, and K. Eberl, Phys. Rev. B **52**, 12153 (1995).

THEORY OF INTRABAND MAGNETO-OPTICS OF EXCITONS IN COUPLED DOUBLE QUANTUM WELLS

A.B. DZYUBENKO[1,2] and A.L. YABLONSKII[1]
[1] *General Physics Institute, Russian Academy of Sciences, Moscow 117942, Russia*
[2] *Walter Schottky Institut, TU München, D-85748 Garching, Germany*

We study intraband magneto-optics of excitons in coupled double quantum wells (DQWs): $s \to p^{\pm}$ inter-exciton transitions in the far-infrared (FIR). Spatially direct (intrawell) and indirect (interwell) excitons in strained $In_xGa_{1-x}As/GaAs$ symmetric DQWs with simple valence band are considered. The evolution of transition energies and oscillator strengths with the perpendicular magnetic B and electric $\mathcal{E}$ fields that induce the direct-indirect crossover is studied.

Currently, there is considerable experimental interest in magnetoexcitons (MXs) in coupled double quantum wells (DQWs). In these systems, under the application of an external perpendicular electric field $\mathcal{E}$, the ground state becomes the indirect (interwell) exciton which has large radiative lifetimes. This opens a possibility to observe many-body effects in neutral e–h systems in high B at low excitonic temperatures. Spectroscopic identification of many-body effects requires detailed knowledge of magneto-optics of excitons. Theory of interband magneto-optics of MXs in DQWs developed by us[1] is in good agreement with experiment.[2] In this paper, we present the theory of intraband FIR magneto-optics of quasi-2D excitons in DQWs.

We consider symmetric strained $In_xGa_{1-x}As/GaAs$ DQWs. Due to the strain, the light-hole band is split off in such structures. As a result, the peculiarities in the excitonic spectra in a DQW are entirely due to the coupling through the barrier and not complicated by heavy- and light-hole anticrossings. Here we present the results for the particular values of the well and barrier widths $L_1 = L_2 = L_b = 60$ Å and $x = 0.2$. For such DQWs, at zero and low B, the intermediate barrier regime is realized. With increasing B, it gradually changes to a wide barrier regime.[1,2] This is because the difference in the Coulomb binding energies of the direct (D) and indirect (I) excitons, $\delta E_{DI} = E_D - E_I$, increases with B and becomes larger than the single-particle e and h symmetric-antisymmetric splittings Δ_e, Δ_h. For each $s, p^{\pm}, \ldots$ state, there are four excitonic terms in a DQW, instead of one for a single QW in the lowest electric subband (see [1,2] and references therein). Each D and I state splits to the symmetric (S) and antisymmetric (A) under inversion states. In the wide barrier regime, the excitonic symmetric-antisymmetric splittings[1] $\Delta_X \simeq \Delta_e \Delta_h / \delta E_{DI}$ are determined by the two-particle e-h co-tunneling through the barrier. The binding energies of the $2p^+$ D and I MXs

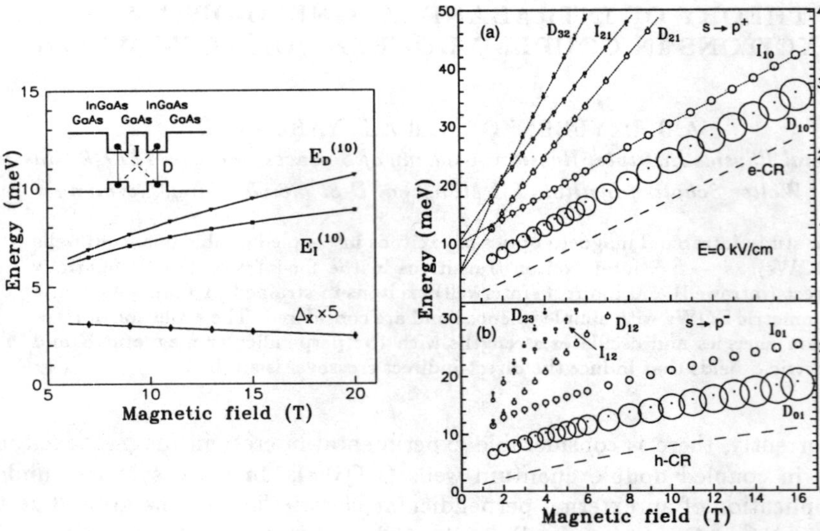

Figure 1: Left: The binding energies of the $2p^+$ direct, D_{10}, and indirect, I_{10}, excitons and the excitonic symmetric-antisymmetric splitting Δ_X vs B. Inset: the pairs of direct and indirect excitons in a DQW. Right: The evolution of the transitions from the ground symmetric $1s$ state $D_{00}S$ to excited p^+ states (a), and to p^- states (b) with B at $\mathcal{E} = 0$. The area of open circles is proportional to dipole transition matrix elements $|f|^2$. The lines with short dashes are guides to the eye showing positions of several weak transitions. Labeling shows the predominant character of the final states. Heavy-hole nonparabolicity results in a strong sublinear behavior of the $1s \rightarrow p^-$ transitions vs B.

are shown in Fig. 1. They are much smaller than these for $1s$ and $2s$ states.[1] As a result, the wide barrier regime for the $2p^{\pm}$ excitons is realized at comparatively higher B. The splitting Δ_X is rather small, suppressed by the excitonic effects ($\Delta_X \sim \delta E_{DI}^{-1}$) and, hence, decreases with B.

In the Faraday geometry (the light propagates along $\mathbf{B}$), the interaction of the exciton with the ac electric field (with the amplitude $\mathcal{F}_0$ and frequency ω) of the circularly-polarized FIR radiation is described by Hamiltonian

$$\delta\hat{V}^{\pm} = \frac{e\mathcal{F}_0}{\omega}\left(\frac{\pi_e^{\pm}}{m_e} - \frac{\pi_h^{\pm}}{m_h}\right)\exp(-i\omega t)\,. \qquad (1)$$

Here "+" is for the left ($\Delta l_z = 1$) and "−" is for the right ($\Delta l_z = -1$) circularly polarized light. $\pi_j^{\pm} = \pi_{jx} \pm i\pi_{jy}$ ($\boldsymbol{\pi}_j = -i\hbar\nabla_{\boldsymbol{\rho}_j} - \frac{e_j}{c}\mathbf{A}_j$, $j = e, h$) are the e and h inter-Landau raising (π_e^+, π_h^-) and lowering (π_e^-, π_h^+) operators. FIR intraband spectroscopy probes all populated excitonic states having fi-

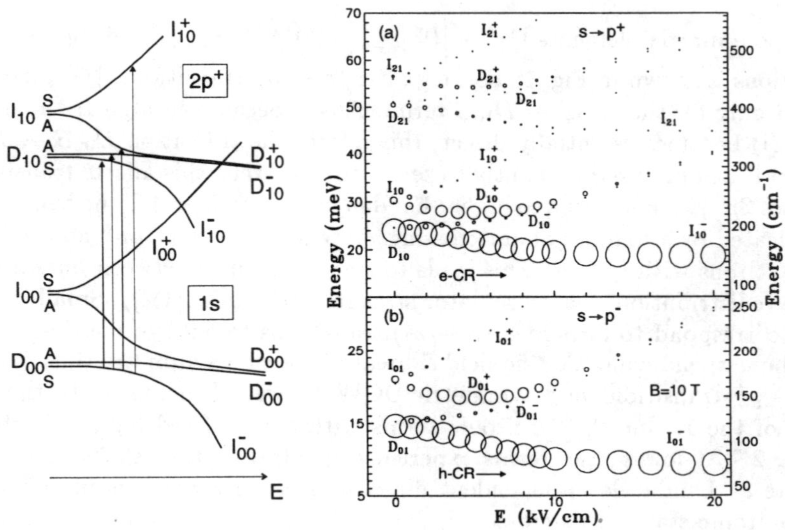

Figure 2: Left: A schematic diagram showing the dependencies on $\mathcal{E}$ of the excitonic $1s$ and $2p^+$ terms in a DQW at high B. Vertical arrows show the transitions from the *ground* $1s$ state. Right: The evolution of the transitions from the ground $1s$ state to excited p^+ states (a), and to p^- states (b) with $\mathcal{E}$ at $B = 10$ T.

nite momenta $\mathbf{K}$. Here, assuming low temperatures, we consider FIR-active transitions between excitons with $\mathbf{K} = 0$.

We find eigenfunctions of MXs, $\Psi_{\mathbf{K}=0,l_z}(\mathbf{r}_e, \mathbf{r}_h)$, using the expansion[1] in the basis formed by products of the electron, $\zeta_i(z_e)$, and hole, $\xi_j(z_h)$, wave functions of the free perpendicular motion in a DQW in $\mathcal{E}$ and wave functions of strictly-2D MXs in the high-field limit, $\phi_{nm}(\rho)$; here $\mathbf{r} = (\rho, z)$. For the MX with the factored wave function $\phi_{nm}(\rho)$, the quantum numbers n and m correspond to the e and h Landau levels, respectively; the angular momentum $l_z = n - m$. We label excitonic states by the high-field quantum numbers (D_{nm}, I_{nm}). At $\mathcal{E} = 0$ we also indicate the symmetry under inversion (S or A), in high $\mathcal{E}$ — the lower (D^-_{nm}, I^-_{nm}) or the upper (D^+_{nm}, I^+_{nm}) branches in the spectra (see Fig. 2).

In high B, the strong inter-exciton transitions $1s \to 2p^\pm$ can be thought of as the electron ($\phi_{00} \to \phi_{10}$) and hole ($\phi_{00} \to \phi_{01}$) cyclotron resonances (CR) modified by excitonic effects. Since the initial $1s$ state is more strongly bound than the final $2p^\pm$ states, the inter-exciton transition energies are *larger* than the bare CR energies $\hbar\omega_{ce(h)}$. The evolution with B at $\mathcal{E} = 0$ of energies and dipole matrix elements $|f|^2 \sim \left|\langle\Psi_{\mathbf{K}=0,p^\pm}|\delta\hat{V}^\pm|\Psi_{\mathbf{K}=0,s}\rangle\right|^2$ of the $1s \to p^\pm$ transitions is shown in Fig. 1. For, e.g., the $1s \to p^+$ transitions, $|f|^2$ increases with B only for the $D_{00}S \to D_{10}S$ term. This is because in high B the initial state: (i) becomes essentially direct; this affects the transition $D_{00}S \to I_{10}S$, and (ii) it belongs predominantly to zero Landau levels; this affects transitions to states $3p^+, \ldots$ belonging to higher Landau levels. At $B < 4$ T the behavior is complicated by a number of anticrossings between the direct np^+ and indirect $n'p^+$ excitons with $n' > n$. This leads to small splittings between lines and to visible redistribution of the oscillator strengths (cf. Refs. [1,2]). Small dots in Fig. 1 correspond to forbidden ($S \to A$) transitions to antisymmetric states.

The perpendicular electric field $\mathcal{E}$ breaks inversion symmetry thus making all $s \to p^\pm$ transitions in a symmetric DQW allowed. Schematically, the evolution of the $1s$ and $2p^+$ excitonic terms with $\mathcal{E}$ at a fixed high B is shown in Fig. 2. At low $\mathcal{E}$, the terms experience quadratic Stark shifts. At intermediate $\mathcal{E}$, Landau-level-dependent direct-indirect crossovers occur. For the $2p^+$ excitonic states, this crossover, due to diminished excitonic effects, takes place at lower fields than for the $1s$ states. This shows up in the behavior of FIR transitions. The transition to the first excited $2p^+$ state experiences a red shift which saturates in high $\mathcal{E}$. The net effect depends on B: the higher fixed B, the larger the red shift. It is due to the excitonic effects. Indeed, at low $\mathcal{E}$ both the initial D_{00} and final D_{10} states are spatially direct. At high $\mathcal{E}$, however, these become *indirect* MXs I_{00}^- and I_{10}^- with lower binding energies. As a result, the transition energy becomes closer in high $\mathcal{E}$ to the bare CR energies (shown by horizontal arrows in Fig. 2). Note also non-monotonous behavior of the transition to the third excited $2p^+$ state. (At $\mathcal{E}$=0 this transition is $D_{00}S \to I_{10}A$ and strictly forbidden.) It is due to the two consecutive direct-indirect crossovers — first for the final, next for the initial states. The first crossover (the third excited $2p^+$ state transforms from the indirect $I_{10}A$ to direct D_{10}^+ MX) occurs at lower $\mathcal{E}$, when the initial state is still predominantly direct D_{00}. Thus the gain of the oscillator strength and the red shift due to a larger Stark effect for the $2p^+$ state. Then, the direct-indirect crossover for the ground $1s$ states occurs (from $D_{00}S$ to I_{00}^-). After that, the transition becomes $I_{00}^- \to D_{10}^+$. As a result, it looses the oscillator strength and starts to experience the linear blue shift in $\mathcal{E}$ (due to the initial state Stark effect).

Financial support by Volkswagenstiftung (the Grant VW I/69 361) and the INTAS Grant 93-1495 is gratefully acknowledged.

1. A.B. Dzyubenko and A.L. Yablonskii, *Phys. Rev.* B **53**, June 15th (1996).
2. L.V. Butov, A. Zrenner, G. Abstreiter, A.V. Petinova, and K. Eberl, *Phys. Rev.* B **52**, 12153 (1995).

MAGNETOEXCITONIC OSCILLATIONS OF GROUND AND EXCITED STATES IN ASYMMETRIC QUANTUM WELL STRUCTURES

A.R. ALVES*, L.A. CURY, P.S.S. GUIMARÃES and M.V.B. MOREIRA
Departamento de Física, Instituto de Ciências Exatas, Universidade Federal de Minas Gerais, Caixa Postal 702, 30161-970, Belo Horizonte, MG, Brazil

We report on the oscillatory behaviour of the photoluminescence intensity from asymmetric $Al_{0.3}Ga_{0.7}As/In_{0.15}Ga_{0.85}As/GaAs$ quantum well structures in the presence of a perpendicular magnetic field. The opposite phase of the oscillations show clearly the competitive process between the transitions from the ground and first excited states. Electron transfer mechanisms can not explain the origin of these oscillations. The optical oscillations emerge from changes in the electron-hole interaction when a Landau level of the e_1 subband crosses the Fermi energy, which can be ascribed to screening effects and/or a Fermi edge singularity.

1. Introduction

The oscillatory behavior with magnetic field of the interband photoluminescence (PL) peak intensity in semiconductor asymmetric quantum wells has been widely investigated in the past few years[1-8]. The interband PL peak intensity is found to oscillate with a 1/B period when the magnetic field B is increased, which is close or identical to the period of the Shubnikov-de Haas (SDH) oscillations. Investigating GaAs quantum wells with just one conduction subband populated, Heiman et al.[1] and Goldberg et al.[4] observed minima in the PL intensity at Landau level (LL) integral filling factors ν. They suggest that the minima in the magnetophotoluminescence (MPL) intensity could be attributed to changes in the screening when the Fermi energy (E_F) is located at the localized states between LL's. Further results[2,5,7] have provided new insights to explain the origin of the MPL oscillations on the scope of Fermi edge singularity (FES). Another possible interpretation was given by Skolnick et al.[8] in terms of the crossing between the LL of the first subband e_1 with the second electron subband energy e_2. This crossing would lead to a maximum population in e_2, which in turn would cause the enhancement of the PL intensity by the greater wavefunction overlap.

In the present work we provide a clearer evidence of the competitive relation of the e_1-hh_1 and e_2-hh_1 recombination processes. An increase in the PL intensity from the e_1-hh_1 transition will necessarily lead to a decrease in the e_2-hh_1 transition (and vice-versa) since the intensity of both PL lines is determined by the

---

** Permanent address: Departamento de Física, Universidade Federal de Viçosa, 36570-000, Viçosa MG, Brazil*

relatively low number of photoexcited holes in the valence band. Our results do not corroborate to the interpretation by Skolnick et al.[8]. We show that the processes responsible for the MPL oscillations are due to an enhancement of the interaction between the holes in the valence band and the electrons in the first subband which occurs when the extended states of a LL of e_1 cross E_F. This enhancement could be interpreted in the context of a FES[2] but a different mechanism than the one proposed by Chen et al.[2] has to be considered.

2. Results and Discussion

The n-AlGaAs-InGaAs-GaAs asymmetrical quantum well used in this study was grown by MBE. The structure consists of the following layers: a 0.6 μm GaAs buffer layer; the 200 Å $In_{0.15}Ga_{0.85}As$ undoped quantum well; a 50 Å undoped $Al_{0.3}Ga_{0.7}As$ spacer layer; a 500 Å $Al_{0.3}Ga_{0.7}As$ Si-doped layer ($n = 4.8 \times 10^{18}$ cm^{-3}). The MPL measurements were performed at 2 K in a 17 T superconducting magnet. The Ar-laser line at 514.5 nm, with an intensity of 10 W/cm^2, was used as excitation light.

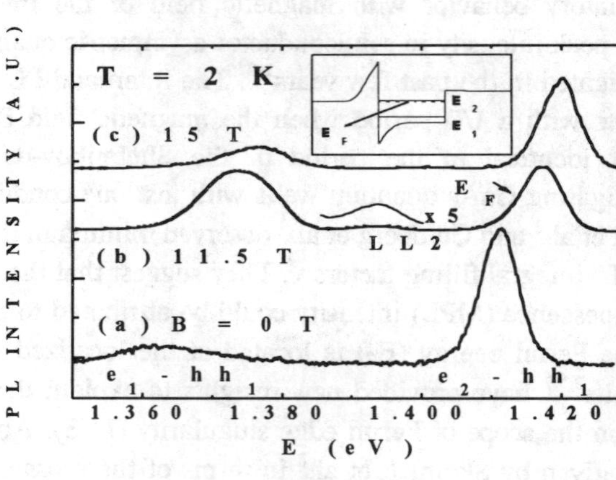

Figure 1 - Three representative photoluminescence spectra.

Figure 1 shows three representative PL spectra. At B=0 T, the e_1-hh_1 and e_2-hh_1 PL lines are identified. The e_2 subband is located 6 meV above E_F and 50 meV above the e_1, as shown in the inset of Fig. 1. The PL peak intensity for the e_2-hh_1 transition is about 6 times greater than that for the e_1-hh_1 transition at B=0 T. The overlap integral for the e_2-hh_1 is much stronger than the e_1-hh_1 case, which explains the large difference in PL intensity (I_{PL}). As the magnetic field increases a dramatic change in the relative I_{PL} of the two PL lines is observed. The

I_{PL} variation is apparent in the curves for 11.5 T and 15 T in Fig. 1. At B=11.5 T, the I_{PL}'s for the two PL lines are nearly the same. The peak labeled as LL2 comes from electron-hole (e-hh) recombination from the second Landau level of e_1, and its difference in energy to the e_1-hh_1 peak gives the value of the separation between Landau levels, $\hbar\omega_C$. The PL peak coming from recombination from LL3 of e_1, is located $2\hbar\omega_C$ above e_1-hh_1. This peak is not clearly seen in the figure due to its vicinity to the e_2-hh_1 transition. Reckoning from both the self-consistent calculations and by comparison with the position of the SDH maxima (top of Figure 2), the LL3 peak at B=11.5 T is in near resonance with E_F.

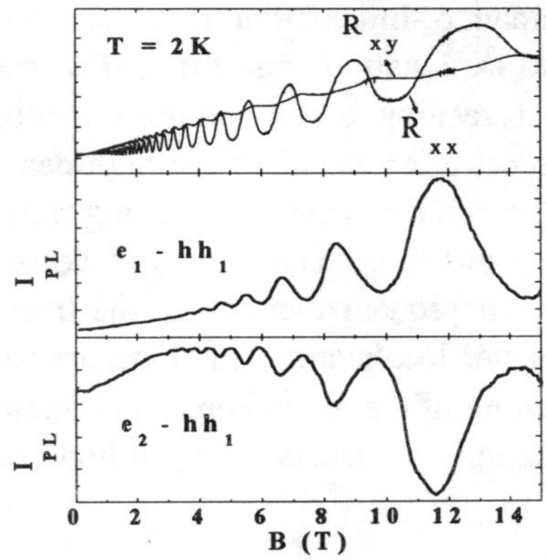

Figure 2- SDH oscillations, Hall plateaus and the MPL oscillations for e_1-hh_1 and e_2-hh_1 transitions.

The MPL oscillations are more clearly revealed in Fig. 2. Also shown in Fig. 2 are the SDH oscillations (R_{xx}) and the Hall plateaus (R_{xy}), measured simultaneously with the MPL. We observe a good correspondence between the SDH and the e_1-hh_1 MPL oscillations below 7 T. It is also striking to see that the oscillations for the e_2-hh_1 PL line, are strictly 180° out of phase with those oscillations in the e_1-hh_1, which is a direct evidence of the competitive process between them. Chen et al.[2] have also observed MPL oscillations in samples that are similar to ours. However, the oscillations in e_2-hh_1 seen by Chen et al. are *in-phase* with the oscillations in R_{xx}, opposite to the results shown in Fig. 2. Our results can not be explained by the electron transfer mechanisms suggested by Skolnick et al.[8]. These mechanisms could only explain an increase in the I_{PL} for the e_2-hh_1 transition as seen by Chen et al., and not a pronounced decrease as demonstrated in Figs 1 (b) and 2.

Our results imply that the MPL oscillations can only be caused by an enhancement of the e-hh interaction and not by electron transfer mechanisms[8]. The enhancement of the e-hh interaction could be explained on the FES scope[2,9]. The FES would occur at the magnetic fields for which the high density of extended states of a LL of the e_1 subband is resonant with E_F. However, this FES would have to lead to an enhanced interaction between all electrons in the e_1 subband and the holes in the valence band. It is not clear through which mechanism this should happen. It is worth mentioning that no detectable peak due to FES was observed in the PL spectra at E_F. It is possible though that at the relatively high incident powers used in this work the PL peak originating from FES will no longer be discernible due to the dominant e_2-hh_1 exciton transition[5].

Another possible explanation for the MPL oscillations resides in screening effects[1,4,6]. The screening of the excitons involving electrons in the second subband becomes more effective when the high density of extended states of a LL of e_1 crosses E_F. The enhancement of screening thus leads to an increased recombination time for electrons in e_2. The result is a decrease of the intensity of the e_2-hh_1 transition and a consequent increase in the intensity of the e_1-hh_1 PL line. On the other hand, carrier localization which occurs when the E_F is between two LL's reduces the screening of the e_2 excitons and therefore one should expect a smaller exciton recombination time and consequent higher I_{PL} for these excitons.

3. Conclusion

We show clearly that MPL oscillations from e_1-hh_1 and e_2-hh_1 transitions are competitive between them and that they could not be attributed to electron transfer mechanisms. The enhancement of electron-hole interaction responsible for these MPL oscillations could be due to screening effects and/or a Fermi edge singularity. However, a model based on a Fermi edge singularity will have to be able to explain the apparently conflicting results of Chen et al. and the ones presented here. We acknowledge FAPEMIG, CNPq and CAPES (Brazil).

References

1. D.Heiman et al., Phys. Rev. Lett., **61**, 605 (1988).
2. W.Chen et al., Phys. Rev. Lett., **64**, 2434 (1990).
3. A.J.Turberfield et al., Phys. Rev. Lett., **65**, 637 (1990).
4. B.B.Goldberg et al., Phys. Rev. Lett., **65**, 641`(1990).
5. W.Chen et al., Phys. Rev. B, **43**, 14738 (1991).
6. M.Dahl et al., Phys. Rev. B, **45**, 6957 (1992).
7. W.Chen et al., Phys. Rev. B, **45**, 8464 (1992).
8. M.S. Skolnick, P.E. Simmonds, T.A. Fisher, Phys. Rev. Lett., **66**, 963 (1991).
9. M.S. Skolnick et al., Phys. Rev. Lett., **58**, 2130 (1987).

TUNNELLING AND ENERGY TRANSFER IN ZnCdSe/ZnSe/ZnCdMnSe ADQW 's

W. HEIMBRODT, M. HAPP, F. HENNEBERGER
Humboldt-University, Institute of Physics, Invalidenstr.110, D-10115 Berlin, Germany

Tunnelling in novel semimagnetic ADQW's has been investigated in dependence on an external magnetic field. Evidence for a transfer of the exciton as an entity is given and reversal of the tunnelling direction is demonstrated beyond the magnetic-field-induced crossing of the exciton ground states in adjacent wells.

1 Introduction

Asymmetric double quantum wells (ADQW's) made of semimagnetic heterostructures are useful to study tunnelling processes[1,2]. The advantage of such combined diamagnetic/semimagnetic quantum well structures is the opportunity to shift well selected states with respect to others by means of the so called 'giant' Zeeman splitting in an external magnetic field, due to the s,p-d exchange interaction.

We report magnetooptical measurements on new semimagnetic ADQW's based on wide gap ZnSe. The structures contain a semimagnetic and a nonmagnetic well separated by a nonmagnetic barrier. We show in the present paper that magnetic field induced crossing of the exciton states leads to a reversal of the tunnelling direction. The underlying transfer mechanism for the exciton as an entity is discussed.

2 Samples and methods

The ADQW's were grown by MBE on (001) GaAs covered by a 1μm thick ZnSe buffer layer. The cladding layers as well as the barrier were made of ZnSe. Whereas a ZnCdSe layer forms the nonmagnetic well (NMW), the second well is composed of the semimagnetic ZnCdMnSe quaternary. The thickness of the ZnSe cap is 20 nm. Further details are given in table 1. We have designed ADQW structures as shown schematically in Fig.1.

Table 1: Parameters of the ADQW's

sample	No.	nonmagnetic well ZnCdSe; L_W= 6nm	6nm	semimagnetic well ZnCdMnSe; L_W= 6nm
a	2505	x_{Cd}=0.18	ZnSe	x_{Cd}=0.18, x_{Mn}=0.15
b	2805	x_{Cd}=0.15	ZnSe	x_{Cd}=0.15, x_{Mn}=0.09

Due to the additional incorporation of Mn, the semimagnetic well (SMW) exhibits the larger energy gap at zero magnetic field. Since most of the strain is adopted by the wells, the lh exciton is shifted to higher energies, for the SMW even beyond the barrier. We used photoluminescence (PL), photoluminescence excitation (PLE), and time-resolved PL spectroscopy in combination with a magnetooptical cryostat at T=1.8K accessible for fields up to 10 T in Faraday configuration. The ps-measurements were performed using a mode-locked synchronously pumped dye laser. The duration of the excitation pulses was 1-2 ps and the total resolution of the set up including a streak camera system was about 10 ps.

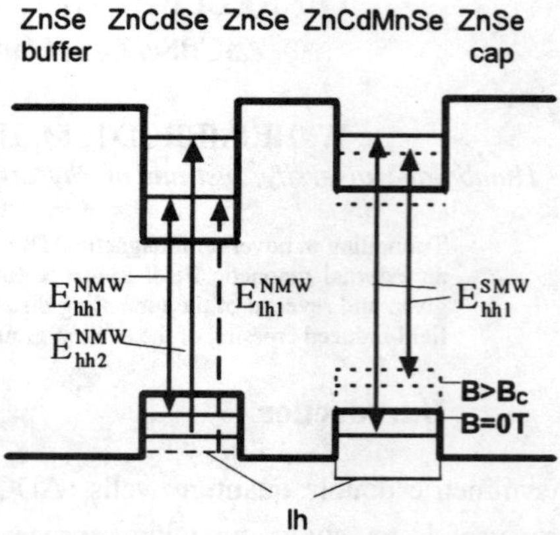

Fig.1 Schematic presentation of the band structure of the ADQW's

3 Results and discussion

In Fig.2 the PL spectra for the samples are shown at various magnetic fields with an above-barrier excitation. Due to the higher Cd concentration in the NMW of sample(a) the ZnCdSe PL at 2.592eV is energetically lower than that of sample(b), where the PL is dominated by a bound exciton (BE) line at about 2.622eV. The high-energy shoulder is due to the hh exciton. Further on the high-energy side another band is observable in both samples. This feature is due to the hh exciton of the SMW. It is less intensive and undergoes a strong shift to lower energies with increasing field, as a result of the 'giant' Zeeman effect. As the energy separation between the two well excitons in sample(a) is larger than 85 meV, no crossing occurs here up to highest fields, whereas this is accomplished in sample(b) because of the smaller separation (20 meV).

In Fig.3 the PLE spectra are shown in zero magnetic field. The spectra of the NMW exhibits four peaks for sample(a) and three peaks in case of sample(b). Besides the hh excitation band, the lh excitation is seen. The strong hh-lh splitting of about 60meV (sample(a)) and 50meV (sample(b)) is a result of the compressive in-plane strain. For both samples the excitation of the NMW-PL is observable also via the ground state of SMW, which yields direct evidence for a tunnelling or energy transfer from SMW to NMW. For comparison the PLE spectra of the SMW are given as dashed lines too.

As already mentioned, the SMW states shift in a magnetic field, but are still observable in the NMW-PLE spectra up to highest fields for sample(a) and up to the exciton crossing point of sample(b). The fourth excitation band of sample(a) is caused by the hh-e2 exciton state of the NMW, as we could conclude from the calculation of the ADQW states. The arrows in Fig.3 give the energy-level positions, which were calculated in the framework of a transfer matrix formalism for electron and hole states including an exciton binding-energy enumeration using the formula given by Leavitt and Little[3]. The commonly accepted value 70:30 is used as partition ratio for the valence and conduction band offset of ZnSe/ZnCdSe. The offset ratio for the ZnSe/ZnCdMnSe heterointerface was found to be in the range $\Delta V_{hh}/\Delta V_{e}= 20/80$ to 40/60, to fit the magnetic field induced shift of the SMW states. As the exchange integral $N_o\beta$=-1.31 eV for the valence band is five times larger than $N_o\alpha$=0.26 eV for the conduction band, the valence band bottom energy of the SMW (m_j=-3/2) shifts almost 50meV upwards B=7.5T, whereas the shift of electron state (m_j=-1/2) is only about 10meV. The resulting band alignment is given schematically in Fig.1 as dashed lines, too. Whereas the electron (m_j=−1/2) state in the SMW is still higher lying relative to NMW, the hh state (m_j=−3/2) of SMW represents the lowest hh state of the ADQW beyond a field strength B_{hh}. Thus at low fields, a tunnelling of electrons and holes from the SMW to NMW is possible, whereas beyond B_{hh} an independent hole tunnelling from the lowest SMW state becomes forbidden. Only a tunnelling or transfer of the exciton as a whole entity can explain the still observable energy transfer, which takes place as long as the exciton energy is larger in SMW than in NMW.

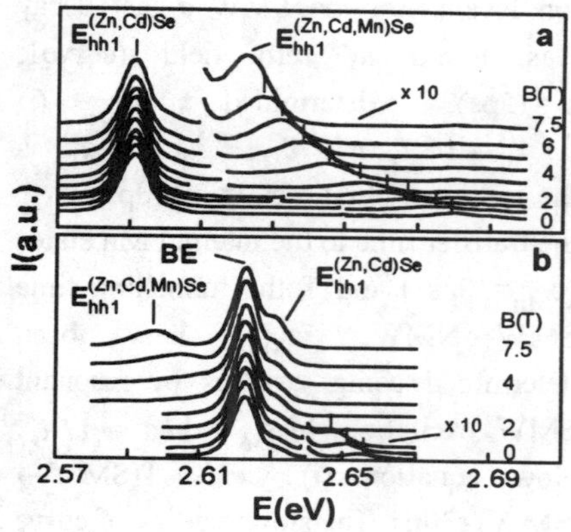

Fig.2 PL spectra of ZnCdSe/ZnSe/ZnCdMnSe ADQW structures at various magnetic field strengths

In Fig.4 the decay transients of sample(b) are given for both the hh PL of the SMW (curves 1-3) and hh PL of the NMW (curves 4,5), for a situation before and beyond crossing of the exci-

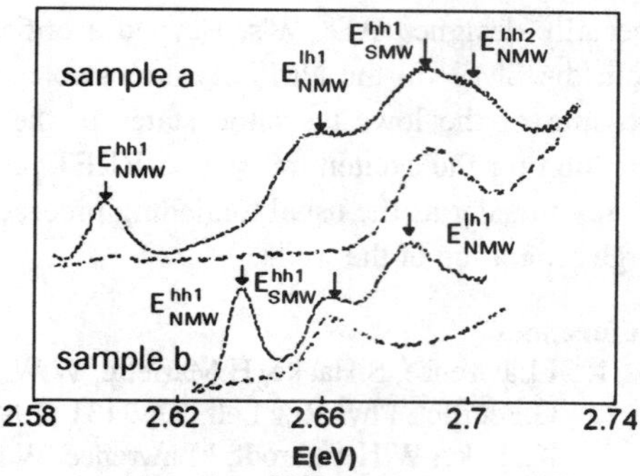

Fig.3 PLE spectra of hh^{NMW} (full line) and hh^{SMW} using the internal Mn luminescence as probe (dashed line).

ton levels. For the SMW a fast decay was found at zero field (curve1, τ_1=15ps), determined by (i) $1/\tau_1 = 1/\tau_r + 1/\tau_{Mn} + 1/T$. τ_r is the radiative lifetime (τ_r=150ps), τ_{Mn} the transfer time to the internal Mn states (τ_{Mn}=30ps) and T the tunnelling time SMW→NMW. τ_{Mn} has been determined using τ_2=30ps by resonant SMW excitation $1/\tau_{Mn} = 1/\tau_2 - 1/\tau_r$. Now equation (i) yields T(SMW→NMW)=30ps. The slower decay of curve 3 (τ_3=50ps) is due to the reversed tunnelling process, since now the NMW was resonantly excited. The corresponding tunnelling time T(NMW→SMW) lies in the range of 150...300 ps as follows from resonant NMW excitation at B=0T and 8T (curves 4 and 5) by $1/T = 1/\tau_4 - 1/\tau_5$, with τ_5=75ps and τ_4=50..60ps. The decrease of the decay time of NMW is only due to tunnelling, as the BE creation time was found to be independent of the field. The tunnelling times are surprisingly short for a barrier width of 6 nm and seem to be inconsistent with usual tunnelling mechanisms, especially since the energy difference between the involved states is smaller then the LO-phonon energy. To our opinion, an explanation can be given only in the framework of a radiationless resonance energy transfer (RRET) as was just recently discussed for III-V ADQW's[4].

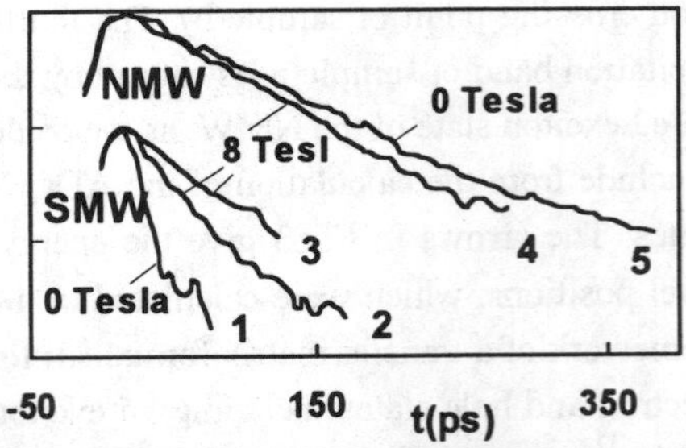

Fig.4 PL decay curves of SMW (1-3) and NMW (4,5) of sample(b) for B=0T and 8T

4 Summery

In summary we found evidence for a transfer process of excitons as whole entities in our specially designed ADQW's. Beyond a critical magnetic field, though hole tunnelling from the SMW to the NMW is impossible, an energy transfer is still observable. The crossing of the lowest exciton states of the SMW and NMW leads to a reversal of direction for the exciton transfer. A RRET process is the most likely explanation for the energy transfer as the usual tunnelling process is suppressed due to high barriers and the large separation of the wells.

References

1. I.Lawrence, S.Haake, H.Mariette, W.W.Rühle, H.Ulmer-Tuffigo, J.Cibert, G.Feuillet, Phys.Rev.Lett. **73**, 2131 (1994)
2. K.Hieke, W.Heimbrodt, I.Lawrence, W.Rühle, Th.Pier, H.E.Gumlich, D.E.Ashenford, B.Lunn, Solid State Commun. **93**, 257 (1995)
3. R.P.Leavitt, J.W.Little, Phys.Rev.B **42**, 11774 (1990)
4. A.Tomita, J.Shah, R.S.Knox, Phys Rev.B **53**, 10793 (1996)

ULTRAFAST DYNAMICS OF EXCITONS IN $Cd_{1-x}Mn_xTe/ZnTe$ (x=0-0.2) QUANTUM WELLS

Yasuo OKA, Kaisei MATSUI, Shuji TAKANO
Tomofumi HISATSUGU, Hiroshi OKAMOTO and Masaaki TAKAHASHI

Research Institute for Scientific Measurements, Tohoku University, Katahira 2-1-1, Aoba-ku, Sendai 980-77, Japan
e-mail: oka@rism.tohoku.ac.jp

The femtosecond and picosecond exciton dynamics is reported in the quantum well (QW) structure of the $Cd_{1-x}Mn_xTe/ZnTe$ system (x=0-0.2). The dephasing time T_2 of the free QW exciton is measured to be 360-450 fs, which shows the fast dephasing process influenced by the strains existing in the QWs. The longitudinal relaxation time of the free exciton is measured as 6.4-10 ps. In the asymmetric double QWs, the tunneling time of carriers between the QWs through the barrier is decreased by the external magnetic field due to the decrease of the magnetic polaron effect in the wells. The rapid dephasing of the free excitons and the magnetically affected tunneling of carriers are the characteristic dynamics in "the strained magnetic QWs".

1. Introduction

A quantum well (QW) composed of CdTe and ZnTe is a typical strained QW in which the CdTe well layer is compressed by the ZnTe barrier layers due to the mismatch of the lattice constants of these materials. The lattice constants of CdTe (6.48 Å) and ZnTe (6.10 Å) derive the lattice mismatch of 6 %. When Mn ions are introduced in the CdTe layer the magnetic interaction takes place between the electrons (holes) and the Mn ions in the QWs. Therefore the carriers and the excitons in the $Cd_{1-x}Mn_xTe/ZnTe$ QW system are affected by both the strain and magnetic effects. In this paper we report the femtosecond and picosecond dynamics of excitons and carriers in the multiquantum wells (MQW) and the asymmetric double quantum wells (ADQW) of the $Cd_{1-x}Mn_xTe/ZnTe$ system.

2. Experimental

The MQW and ADQW of $Cd_{1-x}Mn_xTe/ZnTe$ (x=0-0.2) were prepared by the hot wall epitaxy method. Two types of CdTe/ZnTe MQWs were studied. One has the QW structure in which the width of the CdTe well L_w is 124 Å and that of the ZnTe barrier L_b is 8 Å with the periods P of 35 (we denote the sample as CZ1), while the other (CZ2) has a structure of L_w = 19 Å, L_b = 35 Å and P = 100. In CZ1, the effect of the strain in the CdTe well layer is small because much of the strain is absorbed by the elastic deformation of the thin ZnTe layers. The ADQW samples of $Cd_{1-x}Mn_xTe/ZnTe$ (x = 0.1) were fabricated with the width of the narrow well (NW) of $Cd_{1-x}Mn_xTe$, L_{NW} = 20 Å and that of the wide well (WW), L_{WW} = 30 Å, which are separated by the ZnTe barrier of 40-320 Å thickness.

The exciton dynamics was studied by the transient degenerate four wave mixing (DFWM), the transient absorption spectroscopy by pump and probe method (TAS) and the time resolved photoluminescence (TRPL). These optical measurements were made by using 200 fs pulses of the mode locked Ti-sapphire laser system.

3. Experimental Results and Discussion

The DFWM signals in CZ1 at the resonance excitation on the exciton state is shown in Fig. 1, with parallel and perpendicular polarization configurations for the two incident beams. In the perpendicular polarization configuration, the DFWM signal is mostly decayed with T_2 of 360 fs. A doubly-decayed signal is observed in the parallel configuration, where the fast decay component corresponds to the dephasing time T_2 = 450 fs. This fast dephasing component of the DFWM signal is caused by the free exciton in the QWs, which is confirmed by the resonance behavior of this component at the exciton absorption peak. The signal intensity ratio for the perpendicular to parallel configurations is 1/5, which is much higher than that observed in GaAs/GaAlAs QWs (1/100 in this case). Therefore the origin of the signal in the perpendicular configuration is different from the biexciton model of DFWM.[1,2] The slow decay component in the parallel configuration partly involves the longitudinal relaxation rate of the exciton in the QW by the effective pump and probe measurement, where the longitudinal relaxation time T_1 is 6.4 ps.

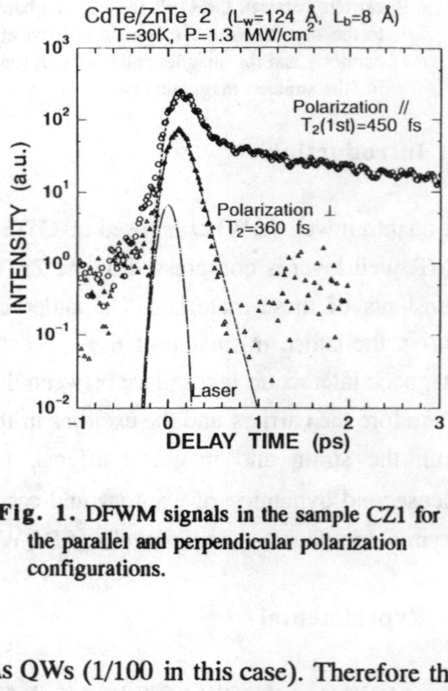

Fig. 1. DFWM signals in the sample CZ1 for the parallel and perpendicular polarization configurations.

The observed T_2 of the QW exciton is independent of the lattice temperature, which shows a less affection of acoustic and optical phonon relaxations than in the GaAs system. The fast dephasing of the exciton in the CdTe/ZnTe system is caused by the relaxation due to the lattice inhomogeneity involved in the present strained QWs because the free exciton is much affected by the inhomogeneous relaxation of the lattice. The DFWM signal in the perpendicular polarization is also induced by the lattice inhomogeneity by partial breaking of the polarization selection rule for the QW exciton transition. Present experimental results display the dynamics and the dephasing process of the excitons in the strained quantum wells.

Fig. 2 shows the TAS spectra of CZ2, which has parameters of L_w = 19 Å and L_b = 33 Å with P = 100. The spectra in the top are the exciton absorption and luminescence spectra in the steady state. The exciton in this MQWs locates at 2.09 eV, while the exciton luminescence peak is shifted by 40 meV to the lower energy side. The large Stokes shift of the exciton luminescence shows the existence of the fluctuation of the well thickness in the MQWs. The transient absorption spectra in the lower part of the Fig. 2 show the variation of the saturated absorption dip energy from 2.090 to 2.075 eV within 22 ps after the pump excitation and also show the further decrease of the dip energy to 2.06 eV by 178 ps after the pumping. From the analysis using the rate equations for the MQW sublevels with the well-width fluctuation, the energy relaxation time T_1 of the exciton in the MQWs is 10 ps, while the lifetime decaying from the MQWs is 160 ps.[3] For the excitation higher than 95 MW/cm^2 (which corresponds to 6.8×10^{12} photons) the relaxations with the time constant of 30-80 ps newly appear. This new relaxation in the high excitation case is induced by both the exciton-exciton scattering and the state filling effect in the MQWs. The measured T_1 of the exciton in the MQWs, 10 ps, almost agrees with T_1 obtained by DFWM in the sample CZ1. The relaxations by the exciton-exciton scattering and by the state filling effect in the QWs are also observed in the high density exciton case.

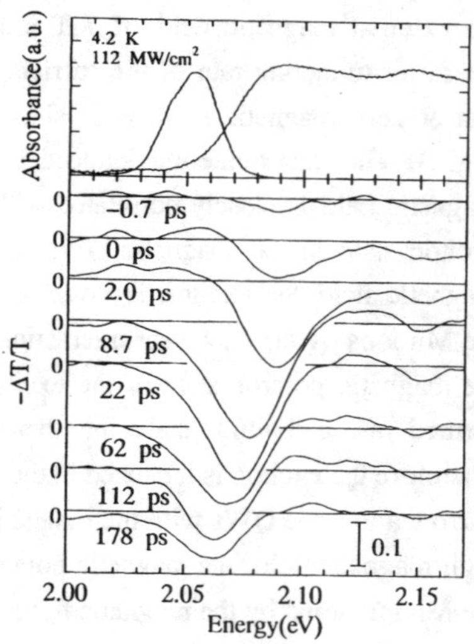

Fig. 2. Transient absorption spectra in the sample CZ2. The top spectra are the absorption and luminescence spectra of the QW exciton in steady state.

The result of TRPL in the ADQWs of $Cd_{1-x}Mn_xTe/ZnTe$ is shown in Fig. 3. The exciton luminescence in NW and WW of the ADQWs shows the tunneling effect of the carriers from NW to WW with the tunneling time of 1-1000 ps for the variation of the L_b of 40-320 Å. In the analysis to derive the tunneling time, it is necessary to take into account the exciton localization effect in the wells, which might be due to the alloy fluctuation of Mn ions existing in the well. An external magnetic field of 7 T induces the faster tunneling rate of the carriers than that at zero magnetic field as is shown in Fig. 3. This fast tunneling induced by the magnetic field is closely correlated with the exciton state in the magnetic QW: At zero magnetic field the exciton is coupled with the Mn ions by the exchange interaction and the magnetic polaron state of the exciton is formed as a bound state in the QW. Therefore the exciton is localized in the narrow and wide QWs with the magnetic polaron binding energy. On the other hand, in the high magnetic field, the magnetic polaron effect decreases due to the external alignment of the Mn ion spins by the magnetic field. In this case the exciton can tunnel more easily from NW to WW owing to the decrease of the localization effect of the magnetic polaron. The tunneling process is analyzed by using the semiclassical carrier tunneling model. The calculated tunneling time of an electron through the barriers shows very steep L_b dependence in case of the valence band offset of 20 % in the ADQWs, which suggests a dominance of the tunneling by the holes with the small valence band offset of 10 meV.[4]

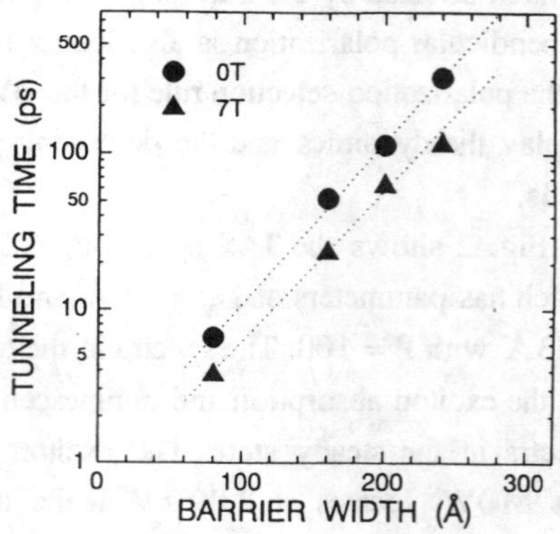

Fig. 3. Tunneling time of carriers in the $Cd_{1-x}Mn_xTe/ZnTe$ (x = 0.1) ADQWs as a function of the barrier width for 0 and 7 T.

In conclusion the present transient spectroscopic studies clarified the longitudinal and dephasing relaxations of the excitons as well as the carrier tunneling which is affected by both the lattice inhomogeneity and the exchange interaction in the strained magnetic QWs.

References

1. H.H. Yaffe, Y. Prior, J.P. Harbison and L.T. Florez, J. Opt. Soc. Am.**10**, 578 (1993)
2. J.P. Doran, PR. Stanley, J. Hegarty, R.D. Feldman, R.F. Austin, J. Crys. Growth **138**, 826 (1994).
3. A. Schulzgen, F. Kreller, F. Henneberger, M. Lowisch and J. Puls, J. Cryst. Growth **138**, 575 (1994).
4. S. Ten, F. Henneberger, M. Rabe and N. Peyghambarian, Phys. Rev. B**53**, 12637 (1996).

CHARACTERIZATION OF DEFECTS IN II-VI-MATERIALS BY MAGNETOLUMINESCENCE

L. WORSCHECH, W. OSSAU, H. SCHÄFER, G. LANDWEHR
Physikalisches Institut der Universität Würzburg, Am Hubland, 97074 Würzburg

ABSTRACT

Defects in ZnSe and CdTe and their dependence on high magnetic fields are reviewed. The anisotropy of the Zeeman-splitting is investigated. The role of reduced site symmetry of the impurities is discussed and compared with the experimental findings. An analysis of a prominent luminescence observed in indium-doped CdTe is given. The local structure of an isoelectronic defect complex and the associated strain directions are identified. The strain-inducing complex has tetrahedral structure consisting of two substitutional indium atoms, one cadmium vacancy and an additional cadmium atom on a lattice site. It is shown that the symmetry of the exciton binding trap is lower than trigonal. Internal strain fields in MBE grown ZnSe epilayers are identified by linearly polarized luminescence. A series of ZnSe epilayers is investigated, in which the layer thickness has been varied from 165 to 4300 nm. The photoluminescence of the Y_0 line at 2.61 eV shows a degree of linear polarization up to 80 % with respect to the <-110>-axis indicating a strong internal distortion preferentially aligned collinear to one of the four <110>-axes. The Zeeman-effect exhibits a <110>-symmetry identical with the symmetry axis of the associated structural defect. A dislocation related exciton model is proposed to explain the origin of the Y_0 line.

1. Introduction

In order to control the optical and electrical properties in compound semiconductors a detailed understanding of the defect structure is required. Recently defects in CdTe and ZnSe have been intensely investigated.[1-3] Whereas some defects can be attributed to chemical impurities on a lattice site, a large number of them have been connected with defect complexes. Magnetoluminescence spectroscopy has been employed to identify the symmetry, microscopic structure and chemical origin of these defects. However, the observed lines are frequently broadened due to sample inhomogeneity or internal strain gradients to such an extent that the potential of the magnetoluminescence spectroscopy can not fully be exploited. The quality of the epitaxial layers grown in our laboratory by **m**olecular **b**eam **e**xpitaxy (**MBE**) has reached such a level that detailed studies have become possible.

Some long-standing problems hampering the fabrication of light emitting optoelectronic devices have been attributed to structural defects. Compensation can occur limiting the doping efficiency in various MBE grown binary II-VI semiconductors. It is well known, that an increase of dopants can in fact cause a decrease of the free carrier concentration. The origin of particular compensation mechanisms have been assigned to defect complexes consisting of chemical dopants in conjunction with vacancies.[4]

In addition to that a main problem which has come up in epitaxially grown ZnSe layers on GaAs substrates is related to the formation of extended defects. The reduction of the number of extended defects is a key issue in the attempt to improve the lifetime of laser diodes emitting blue-green light. However, the low density of extended defects in state of the art devices it difficult to measure by conventional methods.

In this contribution we present results of magnetooptical investigations revealing different features in the photoluminescence spectra of CdTe and ZnSe. First the anisotropy of the Zeeman splitting of a prominent line detected in indium doped CdTe is analyzed. We will show that the defect complex induces an asymmetric displacement of neighbouring ions by distortion. The strain-inducing complex has tetrahedral structure consisting of two substitutional indium atoms, one cadmium vacancy and another cadmium atom.

We observed linearly polarized luminescence for different features in the photoluminescence spectra of ZnSe epitaxial layers. Linearly polarized luminescence is a fingerprint of structural defects. This additional polarization can be observed easily and used as a powerful tool to identify intrinsic structural defects.

2. Experimental Setup

The photoluminescence spectra were measured with a single grating spectrometer with 1 m focal length and a cooled photomultiplier. The sample temperature was 1.7 K. The linearly polarized luminescence intensities as a function of the polarization axis $\vec{E}_{pol}$ were determined by rotating samples in a plane perpendicular to the wavevector of the luminescence $\vec{k}_{lum}$. The degree of linear polarization with respect to a crystal axis is defined as the ratio $P = (I_{//} - I_{\perp}) / (I_{//} + I_{\perp})$, where $I_{//}$ is the integrated photoluminescence intensity parallel and $I_{\perp}$ perpendicular to the <110>-axis. To investigate the Zeeman effect the samples have been mounted in a superconducting split coil magnet with a maximum field of 8 T. The anisotropy of the Zeeman splitting has been determined by rotating the samples in small angular steps with respect to the magnetic field direction. A series of ZnSe samples has been grown on (001) GaAs substrates by **m**olecular **b**eam **e**xpitaxy (MBE). The layer thickness has been varied from 165 nm up to 4300 nm.

3. Theory

The spin-Hamilton formalism is a theoretical tool to describe electronic states associated with excitons bound to defects. As a result of symmetry properties donors, acceptors and isoelectronic traps exhibit different fine structures of the bound exciton. Therefore one can identify the electronic structure of the associated defect by an analysis of the electronic structure of the bound exciton complex. The origin of the fine structure of bound excitons is attributed to exchange interactions between the bound particles. In case of an exciton bound to a donor the resulting excited state is related to the angular momentum j = 3/2. However, a complex term scheme is expected for an exciton bound to an acceptor. The two Γ_8 holes with the angular momentum j = 3/2 are combined to form J = 0 and J = 2 states. The additional coupling to a Γ_6 electron with angular momentum s = 1/2 would lead to four bound exciton states, which can be described by the following effective Hamiltonian H_{BE}:[5]

$$H_{BE} = a\,\vec{j}_1\,\vec{j}_2 + b\,\vec{J}\,\vec{s} + \frac{c}{6}\left[\tfrac{5}{4}\left(J_x\,J_y\,J_z + J_z\,J_y\,J_x\right)^2 - J\left(J+1\right)\right], \tag{1}$$

where $\vec{J} = \vec{j}_1 + \vec{j}_2$, s = 1/2, j_1 = 3/2 and j_2 = 3/2. The parameters a and b describe the strength of the hole-hole interaction and the electron-hole interaction, respectively. A cubic field splits the J = 2 hole-hole state in a Γ_3 doublet and a Γ_5 triplet. The energy splitting is equal to 5/2 c. The ground state of an isoelectronic trap is associated with the angular momentum j = 0. For the bound exciton states the angular momenta are J = 1 and J = 2 coupled by the angular momenta of the isoelectronic trap j = 0, the electron s = 1/2 and the hole j=3/2. Fig.1 illustrates that in the case of a small crystal field dipole transitions from the initial J = 2 state to the final J = 0 state are forbidden. But for a sufficiently strong, uniaxial field the distortion of the lattice removes the degeneracy of the j = 3/2 hole state and the level splits into twofold degenerate branches. The two $m_j = \pm 1$ states of J = 1 and J = 2 become mixed due to the transformation behaviour of their wavefunctions and then $m_j = \pm 1$ for J = 2 becomes an allowed transition.

In the frame of a spin-Hamilton formalism for tetrahedral symmetry the magnetic coupling is best treated for Γ_8 holes and Γ_6 electrons by using the method of group invariants:[6,7]

$$\begin{aligned} H_B = g_e\,\mu_B\,\vec{s}\,\vec{B} + \mu_B\,\left[K\,\vec{j}\,\vec{B} + L\left(B_x\,j_x^3 + B_y\,j_y^3 + B_y\,j_y^3\right)\right] \\ c_1\,B^2 + c_2\left(\vec{j}\,\vec{B}\right)^2 + c_3\left(B_x\,B_y\,\tfrac{1}{2}\left(j_x\,j_y + j_y\,j_x\right) + CP\right) \end{aligned}, \tag{2}$$

where B is the magnetic field, j =3/2, s = 1/2 and μ_B the Bohr magneton. *K* and *L* are the isotropic and cubic anisotropic g values of the hole, respectively, which together with the electron g_e value determine the linear Zeeman-splitting. c_1 and c_2 describe the isotropic and cubic anisotropic diamagnetism, respectively.

A perturbation H_S induced by a distortion can be classified according to the irreducible representations of the crystal point group T_d. By using only linear combinations of the matrices associated with the irreducible representations Γ_3 and Γ_5 of T_d one can write H_S with j = 3/2:[1]

$$H_S = -D\left[j_\varsigma^2 - \tfrac{1}{3} j(j+1)\right] - W\left[l\,m\{j_x, j_y\} + l\,n\{j_x, j_z\} + n\,m\{j_y, j_z\}\right] \tag{3}$$

$$\text{with } j_\varsigma = \left(l\, j_x + m\, j_y + n\, j_z\right)$$

where D and W are fitting constants and $\vec{n} = (l, m, n)$ is a vector collinear to the axis of distortion.

4. Indium related defect complexes in CdTe

The doping of CdTe samples with indium usually results in a material with high resistivity contrary to the expected n-type conductivity connected with the appearance of the C line in the photoluminescence spectrum.[6-9] The strong correlation of the C line with Cd vacancies has been shown by variation of the Cd flux in MBE grown indium doped samples.[10]

For bulk materials we observed that the C line exhibits an increase of the halfwidth and a decrease in intensity by a factor of 5 in a period of half a year. However, etching these samples with a bromine-methanol solution and thus removing the surface layers restores the sharpness and intensity of the line. Obviously only defects near the surface are affected by this ageing-effect, whereas bulk defects are not disturbed with time. Similar findings have been observed for MBE grown CdTe: In. Two PL-spectra of a MBE grown CdTe sample are depicted in Fig.2. Spectrum (a) was measured immediately after the growth of the sample. The intensity of the acoustical phonon side band with a maximum in the TA(X) phonon is more than 50 percent of the C line intensity. For the same sample the C line was not detectable in the PL spectrum (b) after one month storage at room temperature.

The assignment of the C line luminescence to the isoelectronic complex In_{Cd}-V_{Cd}-In_{Cd} is confirmed by the observed strong anticorrelation of the C line intensity with the Cd-flux in MBE grown CdTe layers.[10] With increasing Cd flux the oscillator strength of the C line decreases and the mobility and concentration of the charged carriers is enhanced strikingly. No additional line in the PL-spectra is related to the C line. The absence of two particle transitions, for instance the donor-acceptor pair recombination supports the concept of an isoelectronic state, correlated with the C line.

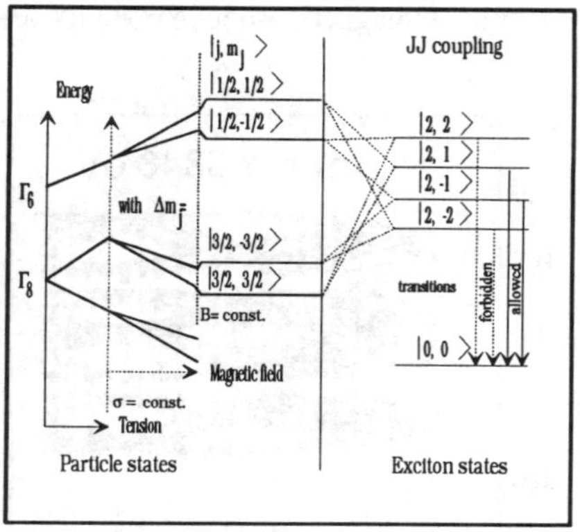

Fig.1: Transition scheme of the electron and the hole states in a strained cubic crystal. The perturbation due to the strain is much larger compared to the JJ coupling of the bound exciton.

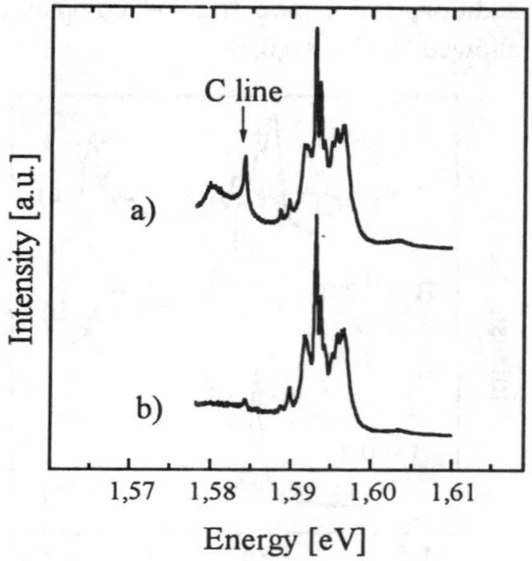

Fig.2: Photoluminescence spectra of a MBE grown CdTe film. Spectrum a) was detected immediately, and b) one month after the growth of the sample.

In Fig.3 we show PL spectra of the C line for B=0 T (c) and the Zeeman pattern for the C line in Voigt geometry for a polarization parallel (a) and perpendicular (b) to a (1,1,0)-plane. The magnetic field has been rotated by 15° in the (1,1,0)-plane from the <1,1,0>-axis. The observed manifold of this Zeeman pattern with 28 magnetic subcomponents could not be explained with the common theory of bound excitons. For a detailed investigation the positions of the observed peaks as a function of angle of rotation have been plotted in Fig.4. The following conclusions can be drawn from these spectra. It can be seen in Fig.3 that there is a mirror symmetry of the Zeeman pattern with respect to the centre of gravity of the line. Therefore, the electron-hole exchange interaction $H_{EX} = a\,\vec{s}\,\vec{j}$ must be negligibly small and not responsible for the unusually high number of magnetic subcomponents resolvable in the PL-spectra (see (a) and (b) Fig.3). In addition, the diamagnetic shift of the centre of gravity ($1.143 \cdot 10^{-2}$ meV/T^2) allows the conclusion that the electron is weakly bound, the local potential of this defect is predominantly attractive for holes. From the peak positions one can conclude, that the orientation of the defect complex is close to the trigonal <1,1,1> axis, because B || <1,1,1> led to the maximum Zeeman splitting. However the trigonal site symmetry is broken. On the one hand one can recognize a large number of lines and on the other hand a clear anticrossing of these lines for a range of energies labelled as K1 and K2 in Fig.4 is obvious. This can be understood only if we assume a symmetry lower than trigonal for the defect complex. In addition, for a pure trigonal complex symmetry only 4 magnetic subcomponents are allowed for B || <0,0,1>.

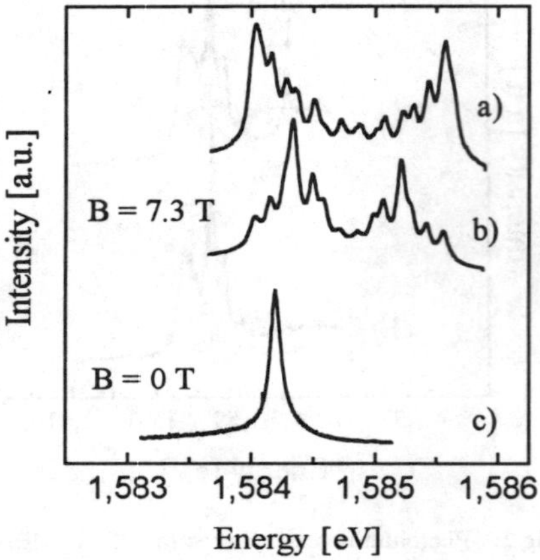

Fig.3: Photoluminescence spectra of an indium doped CdTe sample with magnetic field in Voigt geometry for polarization parallel a) and perpendicular to the magnetic field.

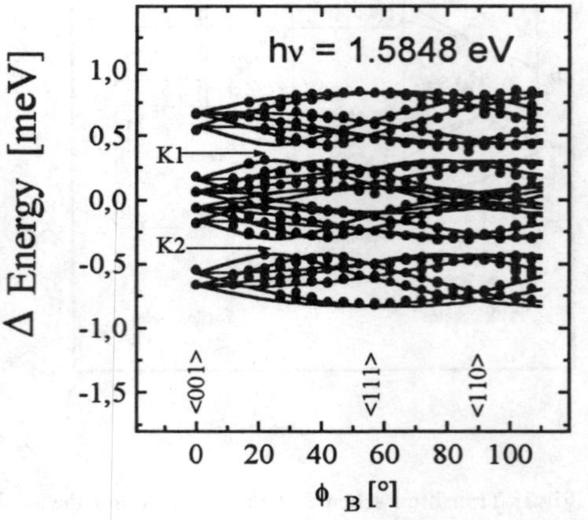

Fig.4: Angular dependence of the Zeeman-splitting of the C line subcomponents at B = 7.3 T. The crystal was rotated in a (110)-plane. B || <001> is denoted by 0°.

The experimental data ($\rightarrow$ *Fig.* 3 *and Fig.* 4) are in excellent agreement with the numerical data obtained in the frame of a perturbation Hamiltonian $H_B + H_S$ ($\rightarrow$ *equ.*2 *and* 3) using the parameters listed in table 1. The spin-Hamiltonian is applicable for arbitrary directions of the strain. The defect complex is formed of a strained planar triangle In_{Cd}-V_{Cd}-In_{Cd} where the strain axis is tilted by a small angle of 2° with respect to the <1,1,1> axes in the direction of the cartesian axes.[1] Due to the fact that the defect complex In_{Cd}-V_{Cd}-In_{Cd} does not fit into the host lattice structure the intrinsic defect is not oriented along the trigonal <1,1,1> axis but slightly inclined. Close to the sample surface no balance exists between the strain forces caused by the defect and the lattice. As a consequence the defect axis will eventually relax resulting in a broadening of the halfwidth of the C line. This relaxation starts at the surface and migrates into the sample with time. This explains why the C line luminescence broadens by a factor of 5 within 6 months. If the surface of the sample is etched the C line becomes narrow again.

Table 1: Parameters describing the Zeeman-splitting of the C line

g	K	L	c_1	W/D	l	m	n
-1.73	0.75	0.0	1.143×10^{-2}	-1.10	0.56	0.56	0.61

5. Extended Defects in ZnSe

One property of structural defects is that they may induce a sufficiently large local stress. The deformation can reach values high enough that new types of dislocations are induced.[11] The magnitude of the stress depends on the distance between the dislocations and therefore on their density in the ZnSe samples. Charged carriers are affected by stress in this way. The electronic structures of electrons and holes are shifted and especially for holes the lh-hh-degeneracy (**h**eavy **h**ole = **hh** and **l**ight **h**ole = **lh**) is lifted. Assuming an uniaxial tension preferentially aligned in one <110> direction a linear polarization results for the associated transitions. According to symmetry considerations transitions from an initial electron state into a final hh-state polarized parallel to the <110>-axis are forbidden, but transitions polarized perpendicular to <110> are allowed. In the case of a final lh-state the intensity ratio is 1/4 for transitions polarized in <-110> and <110>, respectively.[12]

The correlation of various structural defects such as misfit dislocations and even large crystallographic defects like grain boundaries or twins with the Y_o line has been suggested by several authors.[11,13-16] This has been concluded from transmission electron microscopy (**TEM**) data, X-ray diffraction analysis and the dependence of the intensity of the Y_o line on the epilayer thickness. It has been observed that the Y_o transitions and the I_v^o line at an energy of 2.774 eV are strong in those samples which have very low concentration of background impurities and a high density of structural defects.[16] The relative intensities of these lines are inversely related to the concentration of donors.

It is shown in Fig.5 that the photoluminescence intensities associated with the Y_o line and the corresponding LO-phonon replicas differ significantly depending wether the polarization of the luminescence is measured parallel or perpendicular to a <-110> direction with $\vec{E}\,//<-110>$ (solid curve) or <110> (dotted curve), respectively. In contrast to this observation the features in the photoluminescence spectra associated with a donor-acceptor- (D°, A°) and free electron-acceptor-transitions (e, A°) exhibit less polarization effects. This demonstrates that the biaxial strain caused by the lattice mismatch cannot be the origin of the polarization of the Y_o line.

In addition linear polarization of the donor-bound exciton I_2 emission in ZnSe epitaxial layers on (001) GaAs substrate has been reported by Nakano *et al.*.[17] The anisotropic lattice relaxation observed in X-ray diffraction has been discussed as a possible reason for this polarization resulting from different dislocation propagations along <110> and <-110> directions (α and β dislocations) which is mediated by the orientation of the substrate surface. In case of the I_2 emission the degree of polarization of the I_2 emission increases up to a layer thickness of 2 μm and then starts to decrease. In Fig.6 the degree of linearly polarized Y_o luminescence is plotted as a function of the layer thickness. The maximum value of the degree of polarization of 80% has been observed for an undoped ZnSe layer with a thickness of 165 nm, which reveals a strong contribution of the hh-states which are mainly involved in the electronic structure of the Y_o component with the largest intensity.

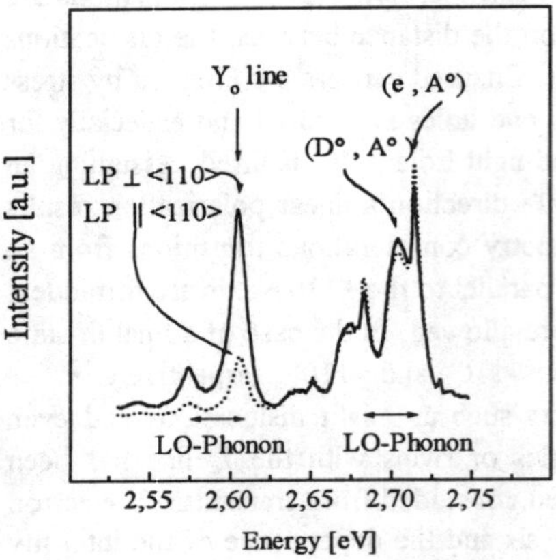

Fig.5: Photoluminescence spectra of ZnSe/GaAs film with linear polarization ∥ <110> (solid curve and linear polarization ⊥ <110> (dotted curve)

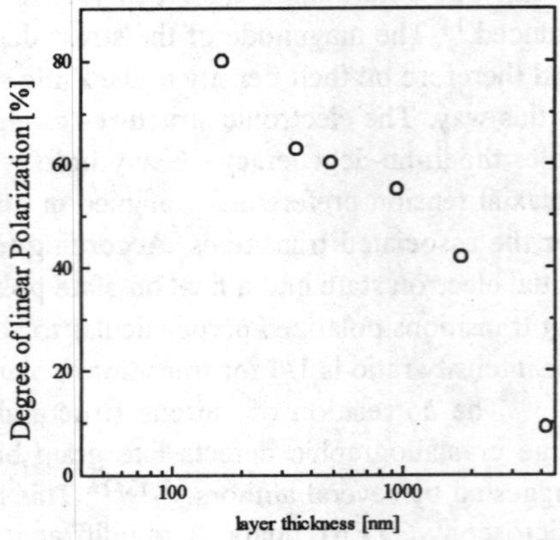

Fig.6: The degree of linear polarization of various layer thicknesses.

The degree of polarization decreases with increasing layer thickness. This behaviour differs from that of the I_2 emission mentioned above. The structural defects associated with the Y_o line are suggested to be strongly correlated with the process of lattice relaxation even for high quality samples. The intensity of the Y_o line in undoped ZnSe is associated with the number of misfit dislocations while the degree of polarization reflects the orientation of the stress, the ratio of misfit dislocations in the different <110>-directions. Also, other mechanisms reducing the polarization degree such as reabsorption, strain gradients and scattering processes will influence the intensity of the Y_o line.

For samples with a large degree of polarization the total luminescence as well as the free exciton transition have been found to be linearly polarized. We suggest that the localized stress of the structural defects is spread out over the whole sample by an increase of the number of strain inducing structural defects.

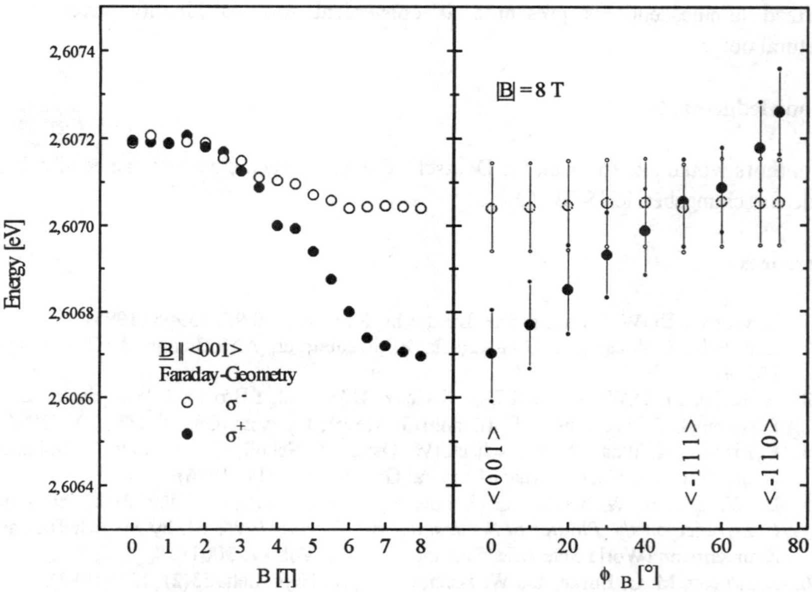

Fig.7: The Zeeman-splitting of the component of the Y_o line with the largest intensity shifts towards lower energies in Faraday configuration $\vec{B} \parallel$ <001> with increasing strength of the magnetic field. The angular dependence of the Zeeman-splitting is shown for a magnetic field in a (110)-plane. $\phi_B = 0°$ indicates $\vec{B} \parallel$ <001> and for $\phi_B = 54.3°$ is $\vec{B} \parallel$ <111>

A suitable tool to investigate the symmetry axis associated with the Y_o line are magnetooptical experiments. The Zeeman-splitting of the component of the Y_o line with the largest intensity is shown in Fig.7. The magnetic subcomponents shift towards lower energies in Faraday configuration $\vec{B} \parallel$ <001> with increasing strength of the magnetic field, indicating a completely different exciton binding system compared to bound exciton recombinations associated with point defects in ZnSe.[3]

The angular dependence of the Zeeman-splitting is depicted for a magnetic field in a (110)-plane. For a magnetic field $\vec{B} \parallel$ <001> a maximum of the Zeeman-splitting is detected. The effective g-factor $g_{eff} = 0.7\pm0.2$ has been determined. On the contrary a directional degeneracy of the magnetic subcomponent is found for $\vec{B} \parallel$ <111>. In Voigt geometry for $\vec{B} \parallel$ <110> no splitting is observed, again confirming the <110> direction as a symmetry axis for the associated defects.

In conclusion we have shown that the defects associated with the Y_0 line in ZnSe produce a local uniaxial strain which is not caused by the lattice mismatch. The data demonstrate, that the changes in the intensity must be caused by local strain. Investigation of linearly polarized luminescence is presented as convenient tool to identify strain inducing structural defects.

Acknowledgements

The authors would like to thank the Deutsche Forschungsgemeinschaft, especially through Sonderforschungsbereich SFB410.

References

1. L. Worschech, W. Ossau, and G. Landwehr, Phys. Rev. B **95**, 13965 (1995).
2. U.W.Pohl, D.Wiesmann, G.H.Kudlek, B.Litzenburger, A.Hoffmann, J. Crystal Growth **159**, 414 (1996).
3. A.Hoffmann, D.Wiesmann, I.Loa, R.Heitz, U.W.Pohl, I.Broser, L.Worschech, E.Kurtz, D.Hommel, G.Landwehr, D.Hoffman, B.K.Meyer, J. Crystal Growth **159**, 302 (1996).
4. F. Fischer, A. Waag, L. Worschech, W. Ossau, S. Scholl, G. Landwehr, J. Mäkinen, P. Hautojärvi, C. Corbel, Journal of Crystal Growth **161**, 214 (1996).
5. L. Worschech, W. Ossau, G. Landwehr, in *Proceedings of the 23rd International Conference on the Physics of Semiconductors, Berlin, 1996,* ed. by M. Scheffler and R. Zimmermann (World Scientific, Singapore, 1996), Vol. IV, 3001.
6. Z.C. Feng, M. G. Burke, and W. J. Choyke, Appl. Phys. Lett. **53(2)**, 128 (1988).
7. C. E. Barnes, and K. Zanio, J. Appl. Phys. **46(9)**, 3959 (1975).
8. N. C. Giles, R. N. Bicknell, R.L. Harper, S. Hwank, K. A. Harris, and J. F. Schetzina, J. Crystal Growth **86**, 348 (1988).
9. H. Zimmermann, R. Boyn, M. U. Lehr, H.-J. Schulz, P. Rudolph, and J.-Th. Kornack, Semicond. Sci. Technol. **9**, 1598 (1994).
10. W. Ossau, T. A. Kuhn, A. Hauck, D. Hommel, A. Waag, S. Scholl, G. Landwehr,and D. Sinerius, *Proc. of the 21st International Conference on the Physics of Semiconductors*, Beijing, 1992, ed. by P. Jiang and H.-Z. Zheng, World Scientific Singapore, (1992), Vol. II, 1633.
11. L.H. Kuo, L. Salamanca-Riba, J.M. DePuydt, H. Cheng, and J. Qui, Appl. Phys. Lett. **63** (23) (1993), 3197.
12. F. H. Pollak and M. Cardona, Phys. Rev. **172** (1968) 816.
13. E. Krause, H. Hartmann, J. Menninger, A. Hoffmann, Ch. Fricke, R. Heitz, B. Lummer, V. Kutzer and I.Broser, J. Crystal Growth **128** (1994) 441.
14. P. J. Dean, phys. stat. sol. (a) **81** (1984), 625.
15. K. Wolf, S. Jilka, H. Sahin, H. Stanzl, T. Reisinger, A. Naumov, W. Gebhardt, J. of Crystal Growth **152** (1995), 34.
16. K. Shahzad, J. Petruzzello, D. J. Olego, D. A. Cammack, and J. M. Gaines Appl.Phys.Lett.**57** (23) (1995), 2452.
17. K. Nakano, H. Okuyama, T. Miyajima and K. Akimoto, J. of Crystal Growth **117** (1992) 797.

OPTICALLY DETECTED RESONANCE SPECTROSCOPY OF SEMICONDUCTOR QUANTUM WELLS IN HIGH MAGNETIC FIELDS

G. S. HEROLD, M. S. SALIB, H. A. NICKEL, J. KONO*, S. T. LEE, A. PETROU, AND B.D. McCOMBE

Department of Physics, State University of New York at Buffalo, Buffalo, NY 14260, U.S.A.

R. CHEN and K. K. BAJAJ

Department of Physics, Emory University, Atlanta, GA 30322, U.S.A.

A sensitive technique that combines visible/near infrared (photoluminescence) and far infrared (FIR) methods, optically detected resonance (ODR) spectroscopy, has been used to study electronic states in GaAs/AlGaAs quantum-well-structures in magnetic fields up to 15T. For wide-well (~ 200 Å) samples doped in the well centers with Si donors (7×10^9 cm^{-2}) the $1s$ - $2p_+$ transition of neutral donors, and singlet and triplet transitions of negative donor ions (D^-) were observed in addition to electron cyclotron resonance. Studies of both doped and undoped narrower well samples (80 - 125 Å) have revealed $1s$ - np internal transitions of confined hydrogenic excitons, as well as hole cyclotron resonance. The measured energies of the excitonic $1s \rightarrow 2p_+$ transition vs. magnetic field are in good agreement with variational calculations. Mechanisms leading to observed positive and negative ODR signals are discussed.

1. Background

A new type of detection scheme has been recently developed for investigating cyclotron resonance in semiconductors. This technique, optically detected cyclotron resonance (ODCR), originally developed for the microwave region, was extended to the far infrared (FIR) by Wright *et al.*[1] in studies of epitaxial GaAs, and subsequently by others.[2-7] Most of this work has concentrated on CR; there has been only one detailed study of impurities (in epitaxial layers)[7] until very recently,[8] and the technique has not been extensively utilized to study low dimensional systems.[9] In ODCR resonant absorption of FIR laser radiation is monitored through induced changes in the intensity of photoluminescence (PL), simultaneously excited by a visible laser. This technique possesses several significant advantages over conventional FIR methods, including spectral specificity of the PL and the FIR resonances and sensitivity. With this technique FIR spectroscopy and magnetospectroscopy can be carried out on "clean" systems, with increased scattering time, τ, and in materials for which doping is difficult. However, details of underlying mechanisms leading to the ODR signals are not well understood for many situations.

Applying this technique to well-center-doped GaAs quantum wells (QW's), we have observed, in addition to electron CR, the $1s \rightarrow 2p_+$ transition of neutral donors (D^0), and singlet and triplet transitions of negative donor ions (D^-).[8] We and others have recently extended these studies to undoped GaAs/AlGaAs QWs and have observed $1s \rightarrow np$ (in the hydrogen atom notation) internal transitions of free excitons.[10] To emphasize its generality we use the term Optically Detected Resonance (ODR) spectroscopy. Our results demonstrate the power and utility of ODR spectroscopy for investigating free carriers, donors and excitons in semiconductor quantum wells.

2. Experimental

With the ODR method a monochromatic FIR laser beam excites electronic transitions whose energies are tuned into resonance with an applied magnetic field. The FIR absorption induces changes in the intensity of the band-edge luminescence, which is excited by a separate laser whose photon energy is larger than the effective bandgap of the heterostructure under investigation. The present apparatus employs visible and FIR lasers, modulation of the FIR laser output, and synchronous detection of changes in the near IR (~800 nm) PL. The FIR laser light is guided and focused by a light-pipe/condensing-cone system onto the sample, which is maintained near 4.2 K in helium exchange gas at the center of a 17 T superconducting magnet. Photoluminescence (PL) is excited with the 6328 Å line of a He-Ne laser via a 600 μm diameter optical fiber, and the PL signal is collected with a second 600 μm fiber and analyzed with a single grating spectrometer equipped with a Si-diode detector. A CO_2-pumped FIR laser generates laser lines at 393.4μm, 163.0 μm, 118.8 μm, 96.5 μm, and 70.5 μm. Changes in the PL intensity induced by absorption of the FIR beam are synchronously detected with a lock-in amplifier referenced to the chopped output of the FIR laser. A dedicated computer is used to record simultaneously the magnetic field values and the corresponding detected changes in the PL, and to control the monochromator drive to track the center of the desired PL feature as it shifts with magnetic field.

The samples used in this study were MBE-grown $GaAs/Al_{0.3}Ga_{0.7}As$ MQW structures having the following parameters: Sample 1-125 Å wells/150 Å barriers, 30 repetitions, undoped; Sample 2 - 125 Å wells/150 Å barriers, 30 repetitions, doped over the central one-third of the wells; Sample 3 - 80 Å wells/150 Å barriers, 45 repetitions, doped over the central one-third of the wells; Sample 4 - 210 Å wells/150 Å barriers, 6 repetitions, doped over the central 1/3 of the wells; and Sample 5 - 200 Å wells/600 Å barriers, 20 repetitions, undoped. In all cases the dopant was Si at 1×10^{16} cm^{-3}. The PL and FIR magneto-transmission from samples 1-4 have been previously studied with emphasis on the donor-related features.[11]

3. Donor-Related Transitions

Intra-donor transitions in semiconductor quantum wells in the FIR have been studied for a number of years, and the effects of confinement and location of the impurity ions along the growth direction in the well are understood for neutral donors.[12] In general, for well-center impurities confinement increases both the binding and transition energies for fixed position of the impurity, while displacement of the impurity from the center of the well along the growth direction toward the edge and into the barrier at constant well width decreases the binding and transition energies. Recently, stable negative donor ions (D^-) have been identified in GaAs/AlGaAs quantum-well structures.[13,14] Both singlet and triplet transitions have been observed.[15] These two-electron ions are stable at low temperatures in QW structures that are doped in both the wells and barriers with the singlet state lowest. However, even in samples doped only in the wells it is possible to observe some significant fraction of donors binding two electrons at elevated temperatures.[15]

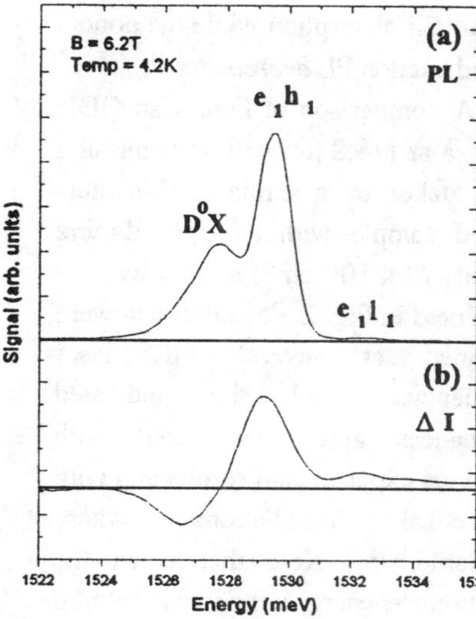

Fig. 1. (a) Pl spectrum, and (b) change (ΔI) in PL intensity due to FIR absorption at the peak of electron cyclotron resonance. The magnitude of (ΔI) at the e_1h_1 peak is about 1% of the total PL intensity.

ODR spectroscopy has been applied to a MQW sample doped over the central 1/3 of the wells with Si donors (Sample 4) to explore the sensitivity of the technique for donor studies, and to investigate mechanisms that give rise to the ODR signals. Some results are presented in Fig. 1. Figure 1 (a) shows the PL spectrum; the PL peak e_1h_1 is the heavy-hole free exciton associated with the ground state subbands, and the peak D^0X is associated with donor recombination (primarily neutral-donor-bound excitons). The weak feature e_1l_1 is free exciton recombination associated with the ground state light-hole subband. Fig. 1 (b) shows the changes in PL (Δ I) at fixed magnetic field (energy scan) induced by electron cyclotron resonance absorption of FIR radiation (resonant with the

118.8 μ photons at 6.2T). These data show that the free exciton PL *increases* due to CR absorption while the donor-bound exciton PL *decreases*.

A comparison of field-scan ODR spectra at 118.8 μm with transmission data taken on a similar well-center-doped sample with a larger doping density (2 x 10^{10} cm^{-2}) and 50 wells is presented in Fig. 2. Similar data were obtained at several FIR laser frequencies, and the indicated resonances agree very well with previous experimental results and with theoretical calculations where available.[11,16] Note that when the spectrometer energy window is centered on the e_1h_1 free exciton the entire spectrum is positive (increasing PL signal), while it is negative when the spectrometer is centered on the bound exciton recombination line.

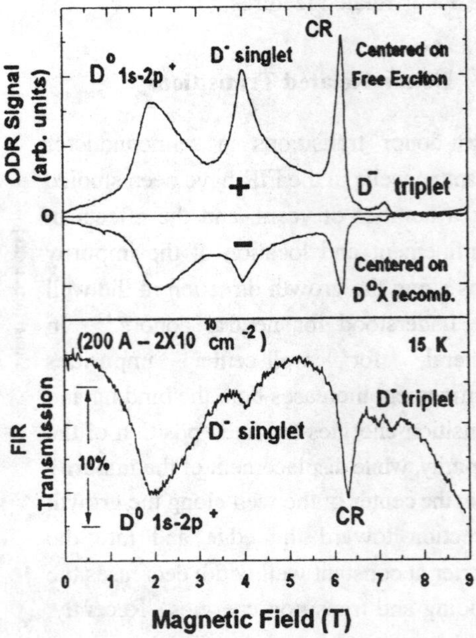

Fig. 2. Transmission (lower panel) and ODR signal (upper panel) vs. magnetic field at 118.8 μm. The 200 Å well-width sample for transmission measurements consists of 50 wells.

This comparison of ODR data with transmission data at the same wavelength (taken at 15K) demonstrates clearly the very high sensitivity of the ODR technique (The effective doping density is 25 times larger in the lower panel). These results also show, from the fact that the triplet transition is clearly observed, that the carrier system is heated significantly at this excitation intensity (~ 0.7 W/cm^2). The triplet ground state is an excited state of the system and is not observable in the transmission spectra at temperatures below about 15 K.[15] The D^--singlet transition in the ODR spectra is very strong, comparable to CR, in contrast to the rather weak singlet transition observed in the laser transmission data for a different well-only-doped sample. Ideally in such samples no D^- ions can exist in the absence of optical or thermal excitation; and strong D^- transitions are usually observed only in samples that are also doped in the barriers. It is possible that there is accidental dopant in the barriers of this sample, which gives rise to additional stable D^- -ions, or alternatively the optical pumping creates a steady state situation in which a much larger fraction donors is occupied by two electrons than in equilibrium. Another possibility that cannot be excluded at present is that D^--singlet absorption is much more efficient in inducing changes in the PL.

4. Internal Transitions of Excitons

Free excitons in GaAs have an internal energy spectrum similar to that of hydrogenic donors except that the hole mass is finite, leading to smaller binding ($E_B = m_r e^4/2\hbar^2\epsilon_0^2$, with m_r the reduced mass) and transition energies. Thus confinement of excitons in a Type-I quantum-well-structure results in increases of binding and transition energies similar to those described above for well-center impurities. In near-band-edge spectroscopic studies a series of excitonic lines appears corresponding to the various interband transitions among confined electron and hole subbands. These confined excitons have been actively studied in the GaAs/AlGaAs system for many years. Information about the various *s*-like excited states of the lowest confined exciton in the GaAs/AlGaAs system has been obtained from photoluminescence excitation (PLE)[17] and magneto-reflectance[18] experiments; however, the $1s \to np$ (n=2,3,...) transitions cannot be investigated by these spectroscopies.

We have observed several $1s \to np_+$ internal transitions of the e_1h_1 exciton in GaAs/AlGaAs multiple-quantum-well (MQW) samples.[10] Internal excitonic transitions have been previously reported for type-II GaAs/AlAs quantum wells in photoinduced absorption studies.[19] The study of internal transitions of excitons by ODR spectroscopy provides information complementary to conventional PL, PLE, or absorption/reflectivity studies of excitons, and permits a determination of the exciton binding energy that is not dependent on separate measurements of the free electron-hole energy and associated complications, such as band gap renormalization. Thus it should be possible to investigate, and develop unique information on, the energy spectrum of excitons under a variety of conditions.

The zero magnetic field PL spectrum (I_{PL}) of sample 1 (undoped) in the vicinity of the gap exhibits one dominant feature, the ground state e_1h_1 excitonic transition at 1.5411 eV. Recombination associated with excitons bound on residual impurities leads to a very weak asymmetric low energy tail. An energy scan of the ODR signal ΔI_{PL}, ($\lambda_{FIR} = 118.8$ μm) at a magnetic field of 2.4 T, the resonant field for the strongest feature attributed to intraexcitonic transitions (see Fig. 3), shows an entirely *negative* signal, while a similar ODR scan at 6.2 T, the resonant field for electron CR, shows a *positive* signal.

In Fig. 3 the change in the luminescence intensity, ΔI_{PL}, of the e_1h_1 feature from Sample 1 is plotted as a function of magnetic field for four FIR wavelengths. At λ_{FIR} =

118.8 μm the changes in the electron-CR correspond to approximately 1% of the peak PL intensity. A variety of resonances (labeled by the letters) is present in these scans. Features (d) and (e) are due to cyclotron resonance of electrons and holes, respectively. Hole cyclotron resonance in this case corresponds to transitions originating on the highest hole ($m_j = \pm 3/2$) Landau levels.[20] Resonances (a) and (c) are attributed to the $1s \rightarrow 3p_+$ and $1s \rightarrow 2p_+$ intra-excitonic transitions of the e_1h_1 confined exciton, respectively. The origin of features (b) and (f) is uncertain at present. We concentrate on the internal transitions of the e_1h_1 exciton, and reserve discussion of the hole CR and other transitions for a later paper.

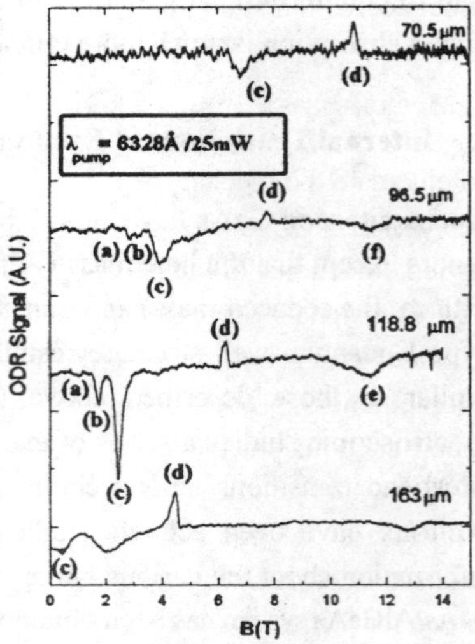

Fig. 3. ODR signals (ΔI_{PL}) as a function of magnetic field for Sample 1 for the indicated FIR wave-lengths at 4.2 K.

We have used a variational approach to calculate the energies of the $1s$ and $2p_\pm$ states of an $m_j = \pm 3/2$ (e_1h_1) exciton in $GaAs/Al_{0.3}Ga_{0.7}As$ quantum-well structures as functions of the well width and magnetic field along the growth direction. Trial wave functions were taken to be products of the electron and hole subband wave functions in a finite quantum well and those which describe the internal motion of the exciton in the presence of a magnetic field. Effects due to hole subband mixing and image charges were not included in these calculations. The energies of the various ODR features for Samples 1 and 3 are plotted vs. magnetic field in Fig. 4a) and 4b), respectively. The bold solid lines are the variationally calculated values for the $1s \rightarrow 2p_+$ intra-excitonic transition for these structures; they agree very well with feature (c) and lend strong support to the assignment. The attribution of the weak $1s \rightarrow 3p_+$ transition (feature (a) in Figs 4a) and 4b)) is based on the energy separation between features (a) and (c), and the assignment of the very weak feature (a′) (Fig. 4b)) to the $1s \rightarrow 4p_+$ transition is based on similar reasoning.

We have also carried out measurements on a well-center-doped MQW structure (Sample 2) with dimensions identical to those of sample 1 (grown in the same growth run). The ODR spectra from Sample 2 contain negative-going features (a), (b), and (c), as well as the donor-related transitions which: (i) are dominant in the ODR spectra, (ii) are clearly well-separated in energy, and (iii) give rise to a positive ODR signal ($\Delta I_{PL} > 0$). These features argue strongly against the possibility that transitions (a), (b), and (c) are associated with residual donors.

5. Mechanisms

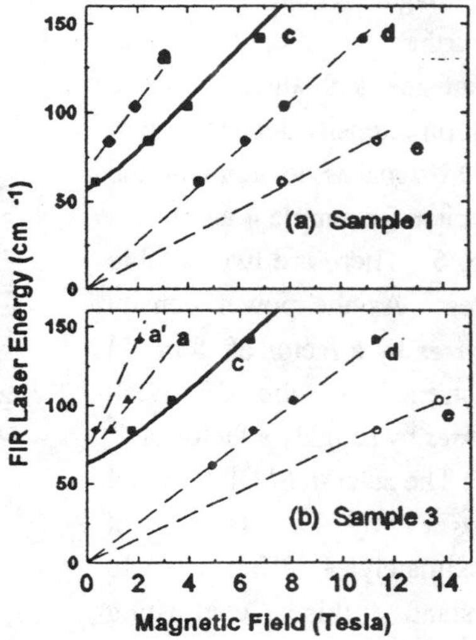

Fig. 4. Energies of the ODR resonances vs. magnetic field at 4.2 K: (a) Sample 1; (b) Sample 3. Solid lines - variational calculation. Dashed lines - guides to the eye.

To elucidate the mechanisms leading to the various ODR signals, we have carried out studies of the dependence of the ODR signals on the particular FIR resonant absorption line, on the PL recombination process (free exciton, bound exciton), on the excitation laser intensity, and studies of the lattice temperature dependence of the PL spectra. The main features of the ODR spectra for Sample 4 (see Figs. 1 and 2), namely a positive ODR signal for all FIR resonant absorption processes when centered on the e_1h_1 free exciton and a "mirror image" negative ODR signal when centered on the donor- related PL, can be qualitatively explained by resonant heating coupled with two main competing PL processes, free excitons and donor-bound excitons. As the effective carrier temperature is increased through CR absorption, bound excitons dissociate into free excitons and donors which decreases the bound exciton channel and increases the free exciton channel in the steady state. In the case of resonant impurity absorption a similar heating mechanism is operable since the final states of the transitions are associated with the N = 1 Landau level, and this excess energy can be rapidly distributed throughout the carrier gas. Any FIR process that heats the carriers at low lattice temperature and low visible laser excitation leads to a decrease in the number of donor-bound excitons and an increase in the number of free excitons, resulting in a *negative* ODR signal in the former case and a *positive* signal in the latter case. We have compared high resolution ODR energy scans at different magnetic fields with differences between PL spectra at different lattice temperatures, and find that most of the qualitative features of the ODR spectra are reproduced by PL difference spectra, $I_{T1}(\nu)$ - $I_{T2}(\nu)$, where $I_T(\nu)$ represents the PL spectrum at temperature T, $T_{1,2}$ are lattice temperatures, and $T_1 > T_2$. The strongest discrepancies between the ODR spectra and the PL temperature difference spectra occur at zero magnetic field.

Further insight into mechanisms is obtained from the excitation intensity dependence of the ODR signal associated with the e_1h_1 exciton for sample 4 as shown in Fig. 5. There are two notable features: As the power density *decreases* by a factor of four: 1) The strength of the CR peak *decreases* by roughly a factor of 4; and 2) The *zero field* ODR signal *increases* by a factor of approximately 4. In order to understand this contrasting behavior we recall that there are always two dominant competing channels for recombination, free excitons and donor-bound excitons. At CR, FIR power is absorbed in exciting electrons from the N = 0 to the N = 1 Landau level, further heating the electron system which is already at an elevated temperature due to the He-Ne excitation. This leads to an increase in the effective carrier temperature and a concomitant *increase* in the free exciton (e_1h_1) recombination, because the competing donor-bound-exciton channel will *decrease* due to the reduced probability of binding excitons. Since the power absorbed from the FIR laser is proportional to the density of free electrons, and a larger visible laser excitation intensity will increase the steady-state electron density, the ODR signal at CR should *increase* with increasing He-Ne laser intensity, as observed. At B = 0 there are two processes whereby power can be absorbed from the FIR beam at 118.8 μm: a) free carrier absorption and b) direct photoionization of donors. As the excitation laser intensity is *increased* the density of free carriers *increases* and the density of neutral donors *decreases*. Thus if free carrier absorption and concomitant heating of the carriers were the dominant process, the B = 0 ODR signal should *decrease* with *decreasing* He-Ne intensity, opposite to the observed dramatic *increase*. We conclude that the dominant process at B = 0 is photoionization of D^0, since the absolute ODR signal is a factor of 4 stronger at the lowest excitation intensity. Thus there are two competing mechanisms leading to the ODR signals in this

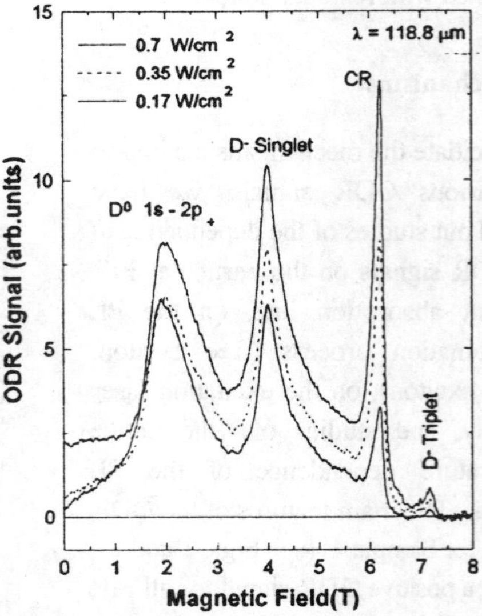

Fig. 5. Dependence of the ODR spectra at 118.8 μm on the He-Ne laser intensity. The monochromator energy window was centered on the e_1h_1 free exciton PL.

case, carrier heating, and photoionization of impurities; the latter dominates a B = 0.

In the samples in which internal transitions of excitons were observed, the ***positive*** ODR signal at e-CR and the *negative* ODR signal for the intra-excitonic transition require closer examination. Carrier heating can explain the negative ODR signal for internal transitions of excitons under certain conditions; however, the observed positive ODR signal at e-CR under the same conditions is not compatible with this mechanism. Carrier heating will dominate if there is no competing recombination process (bound excitons), or at high enough carrier temperatures for which the competing bound exciton channel is ineffective. However, in this case negative ODR signals are expected for *both* resonant FIR absorption processes. Competing PL mechanisms provide an alternative explanation. In this interpretation excitons absorb FIR laser photons and are promoted to an internal excited state from which they are easily thermally ionized. This reduces the population of free excitons available for recombination, and increases the probability of electrons and holes forming bound excitons, thus enhancing the probability of recombination via this channel, and decreasing the ODR signal from the e_1h_1 free exciton. The *positive sign* of the ODR signal at electron CR is attributed to resonant carrier heating. The effective temperature of the electrons is increased by resonant absorption of FIR radiation, and this energy is shared among all the carriers and excitons, leading to dissociation of bound excitons into free excitons and donors. This increases the free-exciton population, decreases the competing bound exciton recombination channel, and hence leads to an increase in the free-exciton PL intensity (ΔI_{PL}).

6. Summary and Conclusions

An optically detected resonance technique has been used to study various electronic transitions in a series of GaAs/AlGaAs quantum-wells samples. Various donor-related features have been observed and investigated in doped QWs, and several electric-dipole-allowed internal transitions of e_1h_1 excitons ($1s \rightarrow np$) have been identified. These results demonstrate the sensitivity of the technique, and show that ODR can be used to study excitons and other electronic states in semiconductor nanostructures under a variety of conditions for which they are not accessible by other means.

Acknowledgements

Work at SUNY at Buffalo was supported in part by ONR/MFEL Grant No. N00014-91-J-1939.

References

* Present Address *Quantum Institute, University of California, Santa Barbara CA 93106*

1. M. G. Wright *et al.*, *Semicond. Sci. Technol.* **5**, 438 (1990).
2. S. I. Gubarev *et al.*, *Pis'ma Zh. Teor. Fiz.* **54**, 361 (1991) [JETP Lett. **54**, 355 (1991)].
3. A. Moll *et al.* Phys. Rev. B **45**, 1504 (1992).
4. P. Omling, C. Wetzel, Al. L. Efros, and B. K. Meyer, *Physica* B **184**, 164 (1993).
5. R. J. Warburton *et al.*, *Phys. Rev.* B **46**, 13394 (1992).
6. J. G. Michels *et al. Physica* B **184**, 159 (1993).
7. J. G. Michels, R. J. Warburton, R. J. Nicholas and C. R. Stanley, *Semicond. Sci. Technol.* **9**, 198 (1994).
8. J. Kono *et al.*, *Phys. Rev.* B **52**, R8654 (1995b)
9. For recent studies of modulation-doped single heterostructures see A. A. Dremin *et al.*, *22nd Internation Conference on the Physics of Semiconductors*, ed. by D. J. Lockwood (World Scientific, Singapore, 1995) p 1157.
10. M. S. Salib *et al.*, *Phys. Rev. Lett.*, July 22 issue, to be published; J. Cerné *et al.*, preprint, submitted to *Phys. Rev. Lett.*
11. N.C. Jarosik *et al.*, *Phys. Rev. Lett.* **54**, 1283 (1985); X. Liu *et al.*, *Phys. Rev.* B **38**, 8522, (1988).
12. See e.g., B. D. McCombe and A. Petrou, Chapter 6 in *Handbook of Semiconductors Vol. 2 Optical Properties*, ed. by M. Balkanski (Elsevier Science Publishers, Amsterdam, 1994) pp 285-384, and references therein.
13. S. Huant, S. P. Najda and B. Etienne, *Phys. Rev. Lett.* **65**, 1486 (1990).
14. S. Holmes, J.-P. Cheng, B. D. McCombe and W. Schaff, *Phys. Rev. Lett.* **69**, 2571 (1992).
15. S. R. Ryu *et al.*, *Phys. Rev. B*, Rapid Communications, to be published.
16. R. L. Greene and K. K. Bajaj, *Phys. Rev.* B**31**, 913 (1985); R. L. Greene and P. Lane, *ibid.* **34**, 8639 (1986).
17. P. Dawson, *et al.*, *Phys. Rev.* B **34**, 6007 (1986).
18. G. Duggan, *Phys. Rev.* B **37**, 2759 (1988).
19. C.C. Hodge, *et al.*, *Phys. Rev.* B **41**, 12319, (1990).
20. C. Delalande, *et al.*, *Superlattices and Microstructures,* **3**, 29 (1987).

CLASSICAL AND QUANTUM MAGNETO-TRANSPORT OF EXCITONS IN TWO-DIMENSIONAL SYSTEMS

P.I. ARSEYEV

Lebedev Physical Institute, Russian Academy of Sciences, Moscow 117924, Russia

A.B. DZYUBENKO

General Physics Institute, Russian Academy of Sciences, Moscow 117942, Russia and Walter Schottky Institut, TU München, D-85748 Garching, Germany

G.E.W. BAUER

Faculty of Applied Physics/DIMES, TU Delft, 2628 CJ Delft, The Netherlands

We study the effect of a magnetic field B on transport of composite neutral particles — excitons in two dimensions (2D). We show that for classical transport (interference between paths disregarded), B monotonously suppresses the exciton static diffusion constant $D(B)$. Quantum corrections to transport lead to localization of excitons at $B = 0$. At finite $B \neq 0$, suppression of weak localization results in a positive magneto-diffusion effect. In the limit of high magnetic fields, however, a crossover to weak localization takes place.

Recently, interesting low-temperature anomalies in transport of quasi-2D excitons in a magnetic field B have been reported [1]. In particular, the exciton diffusion constant D was demonstrated to be a non-monotonous function of B with an increase at intermediate fields. This has been interpreted as an indication of Bose-Einstein condensation of excitons and onset of exciton superfluidity. In this work, we theoretically consider classical and quantum magneto-transport of 2D excitons in the normal phase. We investigate whether effects associated with the suppression of weak localization of neutral 2D excitons in B exist and can lead to anomalies in transport.

We consider elastic scattering by disorder which is assumed weak. We adopt the "exciton" approximation in which the exciton, being in the ground state of the relative e–h motion, is scattered as a whole (see [2,3]); scattering to higher-lying excited states is neglected. In concrete calculations we consider interface roughness (IFR) which is the dominant scattering mechanism at low temperatures. To calculate exciton transport properties, we exploit diagram technique which involves disorder-averaged exciton propagators in the magnetic field [3]. The two limiting cases of low magnetic fields $\ell_B = (\hbar c/eB)^{1/2} \gg a_B = \epsilon\hbar^2/(m_e + m_h)e^2$ and the magnetic quantum limit $\ell_B \ll a_B$ are studied in detail; we make plausible conjectures for intermediate fields $\ell_B \sim a_B$. The partial diffusion constant of excitons (with conserved energy E) at frequency ω, $D(\omega, E)$, can be obtained from the expression for

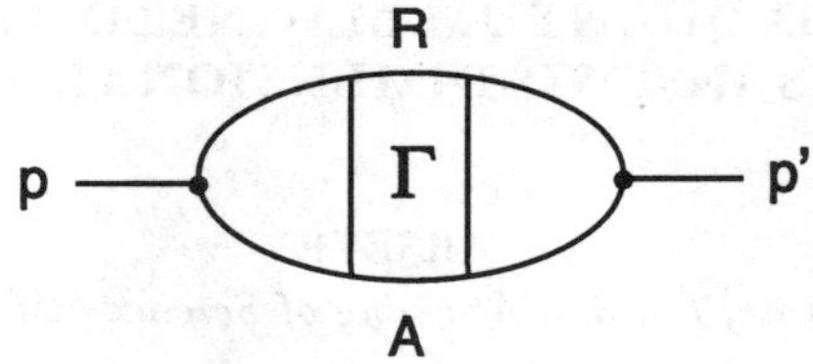

Figure 1: Diagrammatic representation of Eq. (1). Γ is irreducible impurity vertex, lines labeled A and R represent disorder-averaged magnetoexciton propagators G^A, G^R, current vertices correspond to the exciton center-of-mass velocity.

generalized conductivity (Fig. 1) $\sigma(\omega, E) = D(\omega, E)\mathcal{N}(E)$,

$$\sigma(\omega, E) = \frac{1}{2\pi}\int d\mathbf{p} \int d\mathbf{p}' \left\langle V_x(\mathbf{p})\, G^R(\mathbf{p}, \mathbf{p}', E+\omega) G^A(\mathbf{p}', \mathbf{p}, E) V_x(\mathbf{p}')\right\rangle , \quad (1)$$

where $\mathbf{V}$ is the exciton center-of-mass velocity, $\mathcal{N}(E)$, the exciton density of states. We are interested is the static limit $D(E) = D(\omega \to 0, E)$. The *total* static diffusion constant $D = D(T)$ (corresponding to all exciton density fluctuations) can be determined from the partial microscopic constants $D(E)$ using generalized Einstein relation

$$D = \int dE\, \mathcal{N}(E) \left[-\frac{\partial f}{\partial E}\right] D(E) \left(\int dE\, \mathcal{N}(E) \left[-\frac{\partial f}{\partial E}\right]\right)^{-1} , \quad (2)$$

where $f = f(\mu, T)$ is the exciton distribution function.

Classical transport (interference effects disregarded) is described by the approximation in which the impurity vertex Γ only includes ladder diagrams. In this approximation B monotonously suppresses transport of excitons. At low B the static diffusion constant $D(B) \sim D_0\,[1 - B^2/B_0^2]$, where $B_0 a_B^2 \sim \phi_0 = hc/e$. At high fields, we obtain $D \sim B^{-1}$, consistent with the previous consideration [2] of magnetoexciton transport relaxation times. Physically, this reflects the fact that the exciton effective translational mass and density of states $\mathcal{N}(E)$ increase with increasing B. As an example, in Fig. 2 we show the dependence of the total diffusion constant $D(T)$ on B for type-II GaAs/AlAs quantum wells with different widths of the AlAs layer d_e (at a fixed width of the GaAs layer d_h). In calculations, we assumed a Maxwell-Boltzmann distribution function f and some typical parameters of IFR [2].

Quantum corrections to transport (weak localization effects) correspond to the inclusion of maximally crossed diagrams to Γ. At $B = 0$ they lead to localization of excitons, i.e., $D(B = 0) = 0$. When $B \neq 0$, it is known

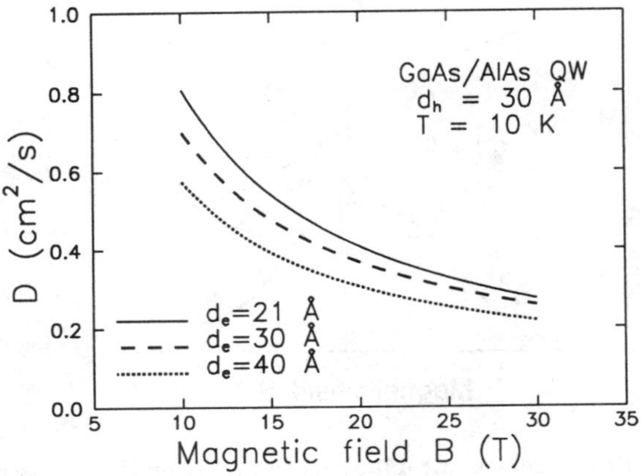

Figure 2: Dependence of the total exciton diffusion constant D on the magnetic field B for type-II GaAs/AlAs quantum wells determined by IFR for classical transport.

for *charged* electrons that broken time-reversal symmetry leads to negative magneto-resistance. For charge-neutral excitons, the important question is whether $t \to -t$ symmetry is broken in B. Consider the Hamiltonian

$$H = \frac{1}{2m_e}\left(-i\hbar\nabla_{\rho_e} + \frac{e}{c}\mathbf{A}_e\right)^2 + \frac{1}{2m_h}\left(-i\hbar\nabla_{\rho_h} - \frac{e}{c}\mathbf{A}_h\right)^2 + U_{eh} + V_e + V_h \quad (3)$$

which describes the motion of an interacting neutral e–h pair in a uniform B and in (random scattering) potentials $V_e(\mathbf{r}_e)$ and $V_h(\mathbf{r}_h)$. When $m_e = m_h$ and $V_e \equiv V_h$, the spin-$\frac{1}{2}$ e and h are exact time-reversed counterparts. In this case Hamiltonian (3) commutes with the operator of time reversal: $[H, \hat{T}] = 0$. Thus, time-reversal symmetry is preserved for neutral excitons in B, and they should be localized in 2D despite the presence of B.

In general, however, time reversal symmetry is broken. As a consequence, the diffusion constant D becomes finite in B even without inelastic phase-breaking processes. This is because the exciton, being scattered at an impurity, acquires some random phase shift. After many scattering events, the accumulated phase shift ~ 1, and the coherence is broken. Formally, this leads to a finite phase-breaking time $\tau_B^{-1} \neq 0$ which cuts off the singularity in the Cooperon [3]. However, in both limits $\ell_B \gg a_B$ and $\ell_B \ll a_B$ quantum corrections to diffusion constant diverge logarithmically with B. To obtain the dependence $D(B)$, we apply a self-consistent treatment (see [4]). In this approximation, in high B, weak localization corrections strongly renormalize the diffusion constant behavior to $D \sim B^{-2}$ (from $D \sim B^{-1}$ for classical

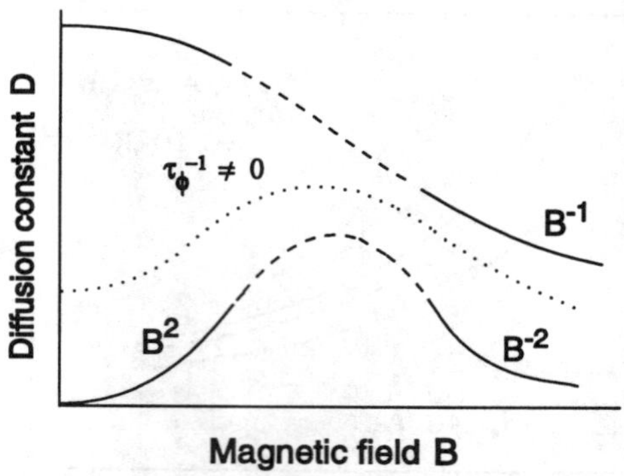

Figure 3: Dependence of the exciton static diffusion constant D on B for: classical transport (upper curve), with quantum corrections (lower curve), when also small inelastic scattering is included (dotted curve).

transport), i.e., diminish D. Physically, this is because the magnetoexciton is squeezed in high B and increasingly appears like a structureless neutral boson. At low fields, however, we obtain a positive magneto-diffusion effect: $D \sim B^2$. Qualitatively, the dependence of the static diffusion constant on the magnetic field B is shown in Fig. 3. Dashed parts of the curves correspond to intermediate fields $\ell_B \sim a_B$, where we conjecture that the maximum of $D(B)$ should occur. Weak inelastic scattering characterized by phase-breaking time τ_ϕ can be also incorporated phenomenologically into our model. The following dependence $D(B)$ is shown in Fig. 3 by dotted line. Though non-monotonous in B, the theoretical field dependence differs from the measured one [1]. We conclude, in particular, that weak localization effects alone cannot explain the low-temperature anomalies observed in [1].

We wish to thank L.V. Butov for useful discussions. This work was financially supported by the Volkswagen foundation (the Grant No. VW I/69 361), the INTAS Grant 93-1495, and the Dutch Science Foundation NWO.

1. L.V. Butov, A. Zrenner, M. Hagn, G. Abstreiter, G. Böhm, and G. Weimann, to be published in *Surface Science*. see also L.V. Butov, A. Zrenner, G. Abstreiter, G. Böhm, and G. Weimann, *Phys. Rev. Lett.* **73**, 304 (1994).
2. A.B. Dzyubenko and G.E.W. Bauer, *Phys. Rev.* B **51**, 14524 (1995).
3. P.I. Arseyev and A.B. Dzyubenko, *Phys. Rev.* B **52**, R2261 (1995).
4. D.Vollhardt and P. Wölfle, *Phys. Rev.* B **22**, 4666 (1980).

SPECTROSCOPIC STUDY OF THE TRION/EXCITON PHASE IN THE TWO-DIMENSIONAL ELECTRON LIQUID AT LOW FILLING FACTORS

D. GEKHTMAN, E. COHEN, ARZA RON
Solid State Institute, Technion-Israel Institute of Technology, Haifa 32000, ISRAEL
and L.N. PFEIFFER
Bell Laboratories, Murray Hill, NJ 07974, USA

We studied the circularly polarized photoluminescence (PL) and its excitation (PLE) spectra of the trion singlet (X_s^-) and exciton (X) transitions in the two-dimensional electron system (2DES) with a variable density, formed in $GaAs/Al_{0.1}Ga_{0.9}As$ quantum wells at T=2K. A perpendicular magnetic field is applied and the filling factor varies in the range $0<\nu<1$. The results indicate a significant effect of the interacting 2DES on the optically created X_s^- spin states: their Zeeman splitting shows a strong nearly symmetric enhancement around $\nu \simeq 1/2$. In the vicinity of $\nu=1$ a new low energy feature appears in σ_- polarized PL spectra, that is attributed to a cooperative transition of X_s^- and the 2DES spinwave. We also obtained the chemical potential (μ) of the 2DES by measuring the X_s^-/X PL and PLE intensity ratio at $\nu \leq 1/2$. It is found that $d\mu/d\nu$ becomes negative in the range of filling factors corresponding to the increase of X_s^- Zeeman splitting.

Recent spectroscopic investigations[1] in $GaAs/Al_xGa_{1-x}As$ modulation n-doped quantum wells (MDQW) have revealed that as the density (n_e) of the two-dimensional electron system (2DES) decreases, the exciton (X) appearance is preceded by the formation of a charged exciton (trion, X^-). Under the application of a perpendicular magnetic field the trion singlet (X_s^-) and triplet (X_t^-) transitions were analyzed in circularly polarized optical spectra.

The purpose of the present work is to study the effect of the interacting 2DES on the X_s^- spin states in a strongly quantizing magnetic field. We studied the interband photoluminescense (PL) and its excitation (PLE) spectra in a $GaAs/Al_{0.1}Ga_{0.9}As$ MDQW at T=2K and under magnetic fields ($0 \leq B \leq 5T$), applied perpendicularly to the quantum well (QW) plane (and parallel to the light direction). The structure[2] under study has ten 300Å-thick GaAs wells with 3500Å-wide $Al_{0.1}Ga_{0.9}As$ barriers. It is symmetrically δ-doped in the middle of each barrier to produce a QW confined 2DES with a density $7.1\times10^{10}cm^{-2}$ and high mobility ($\sim4\times10^6cm^2/Vs$). The emitted light was analyzed with circular polarizers and dispersed through a double spectrometer (resolution of $\simeq 10\mu eV$). The MDQW is resonantly excited with a laser energy $E_{L1} \simeq 1.52eV$ and intensity $\simeq 1mW$, modulated at a frequency of $\sim 1kHz$. The 2DES density is reduced optically[3] in the range $7.1 \geq n_e[10^{10}cm^{-2}] \geq 0.8$

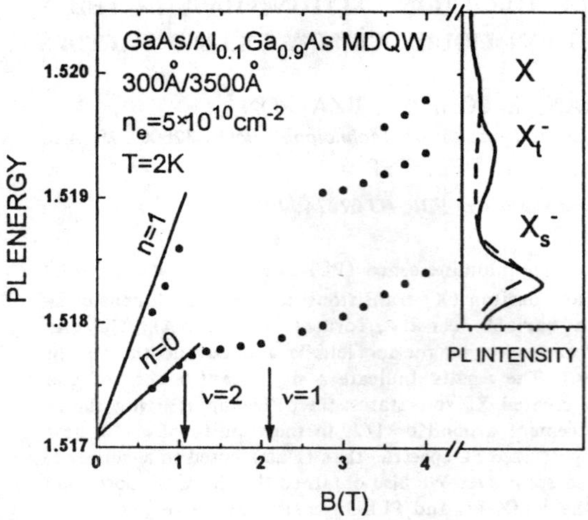

Figure 1: The PL transition energies (solid dots) measured as a function of magnetic field B. The values of B corresponding to filling factors $\nu=2$ and $\nu=1$ are shown by arrows. The solid lines are fittings to the n=1 and n=0 Landau transition dependence. The inset shows the PL spectra measured at B=4Tfor σ_+ (dashed line) and σ_- (solid line) polarized light.

by above barrier bandgap excitation with another laser of energy $E_{L2} \simeq 1.7$eV and various intensities.

Fig. 1 shows the B-dependence of the observed PL transitions energy for a 2DES with $n_e=5\times10^{10}cm^{-2}$. For filling factor $\nu \equiv \frac{hcn_e}{eB} > 2$ the n=0 and the n=1 Landau transitions are observed, while for $\nu<2$ the slope of the PL energy (with respect to B) discontinuously changes. This is due to the electron-hole interaction "freeze-out" resulting from the depletion of the n=1 Landau level. It results in the formation of a many electron-bound hole complex for $1<\nu<2$, that transforms at $\nu=1$ into the X_s^- ground state with the subsequent appearance of the higher energy X_t^- and X excited states (their polarized spectra observed at B=4T are shown in the inset of Fig. 1). The transition into the X^-/X phase at $\nu=1$ is interpreted[4] as being induced by spatial localization of the 2DES which is occupying the lowest spin-polarized (n=0) Landau band. As a result, the X^- states are formed inside large QW areas filled with electrons in accordance with the value of ν. These areas are determined by potential fluctuations due to the charged donor (in-barrier) distribution. In order to elucidate how strongly the surrounding electrons interact with the optically created X^- we study the polarized PL spectra from the X_s^- transition as a function of $0<\nu<1$ and at different values of B. For high B (Fig. 2a) the observed Zeeman splitting between σ_+ and σ_- polarized peak energies changes appreciably with ν. However, for low B (Fig. 2b) the Zeeman splitting is negligibly small, but a new low energy transition appears with increasing ν in the σ_- polarization (it is also seen in Fig.2a as a broadening of the low energy tail).

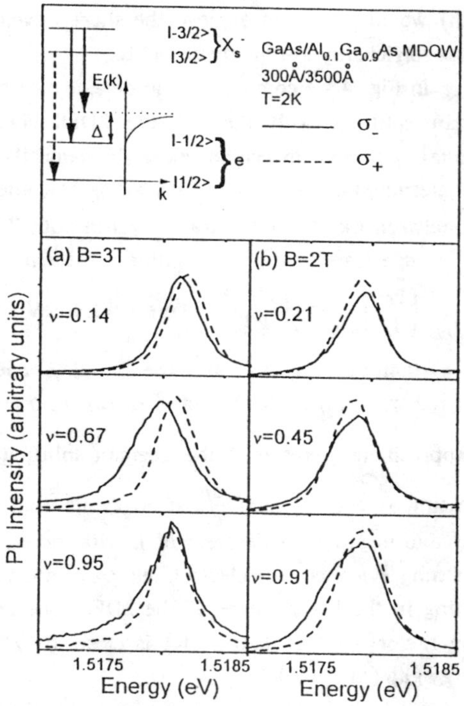

Figure 2: The X_s^- polarized PL spectra measured at different filling factors ν for (a) B=3T and (b) B=2T. The possible transitions between spin projection states $|s_z\rangle$ are shown schematically in the inset.

We interpret these observations as an evidence for the correlation between the X_s^- spin states and those of the 2DES. The ν-dependent interaction between the electrons within the lowest, partially filled spin-band couples the optically induced electron spin-flip (see the inset in Fig. 2). In Fig. 3 we show the measured Zeeman splitting extracted from the spectra in Fig. 2a. We notice that it strongly enhances nearly symmetrically around $\nu=1/2$, which is in agreement with the electron-(quasi)hole symmetry. The spin splitting at $\nu=0.95$ is very close to that of excitons measured in intrinsic QW's of a comparable width[5]. At $\nu\to 1$ the inter-spin-band 2DES excitation (spinwave) with a short wave-length is a plausible explanation for the appearance of the low energy mode in σ_- polarized spectra. It is due to the well-known exchange enhancement[6] of the electron effective g-factor at $\nu=1$. The observed energy seperation between the two σ_- modes, $\Delta\simeq 0.3$meV at B=2T, agrees with recently reported transport measurments[7]. In order to have such an interband transition, some mechanism of momentum relaxation must be involved (for instance, due to short range scattering). The resulting cut-off in spinwave k-space (scaled by the inverse magnetic

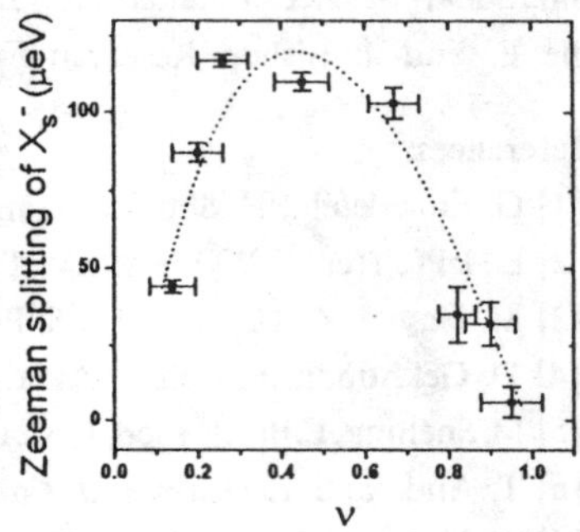

Figure 3: Zeeman splitting of the X_s^- transition measured at B=3T as a function of filling factor ν. The dotted the guide to the eye.

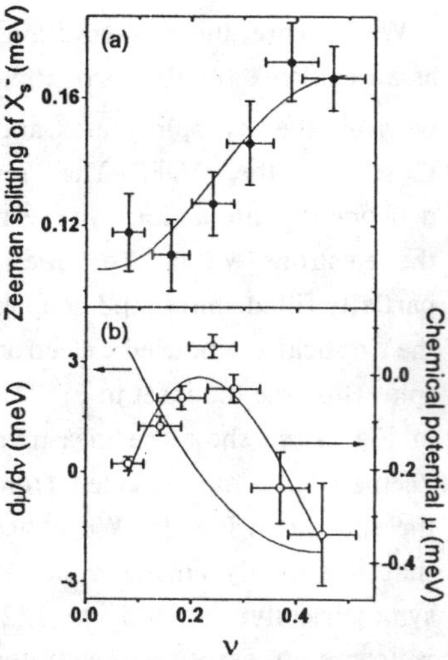

Figure 4: Comparison between the values of (a) X_s^- Zeeman splitting and (b) 2DES chemical potential measured at B=4.5T as a function of ν. The solid line denotes the fitting to a continuos function.

length) would permit to achieve the short wavelength limit at sufficiently low magnetic fields.

In Fig. 4 we compare the ν-dependence of the Zeeman splitting with that of the 2DES chemical potential (μ), both measured under the same B=4.5T. The determination of μ is by measuring the intensity ratio between the X^- and X transitions in both PL and PLE spectra and relating them by $\left(\frac{I_{X^-}}{I_X}\right)_{PL} = \left(\frac{I_{X^-}}{I_X}\right)_{PLE} exp\left(\frac{\mu+\varepsilon}{kT}\right)$, where $\varepsilon \simeq 1.4$meV for X_s^- and $\varepsilon \simeq 0.4$meV for X_t^- are the binding energies measured with respect to that of X. It can be seen that the appreciable increase of the Zeeman splitting with increasing ν occurs when $\frac{d\mu}{d\nu}$ changes its sign from positive to negative. A decrease of μ with increasing ν is a strong evidence[8] for electron-electron interactions resulting in the liquid phase of the 2DES. The critical filling factor $\nu_c \simeq 0.25$ (at T=2K) is close to 1/3 that corresponds to the largest fractional energy gap (at $T \rightarrow 0$.)

Acknowledgements: The work at Technion was done in the Barbara and Norman Seiden Center for Advanced Optoelectronics and was supported by the US-Israel Binational Science Foundation (BSF), Jerusalem. A.R. acknowledges the support of the E. and J. Bishop Research Fund.

References:

[1] G. Finkelstein, H. Shtrikman and I. Bar-Joseph, *Phys. Rev.* **B53**, R1709 (1996)
[2] L.N. Pfeiffer, K.W. West, J.W. Baldwin, P. Gammel, *Appl. Phys. Lett.* **61**, 1211 (1992)
[3] M. Hayne, A. Usher and A.S. Plaut, *Phys. Rev.* **B50**, 17208(1994)
[4] D. Gekhtman, E. Cohen, Arza Ron and L.N Pfeiffer, submitted to *Phys. Rev.* **B**
[5] M. Snelling, E. Blackwood, C. McDonagh, R. Harley, C. Foxon, *Phys. Rev.* **B45**, 3922(1992)
[6] T. Ando and Y. Uemura, *J. Phys. Soc. Jpn.* **37**, 1044 (1974)
[7] A. Usher, R.J. Nicholas, J.J. Harris and C.T. Foxon, *Phys. Rev.* **B41**, 1129(1990)
[8] S.V. Kravchenko, D.A. Rinberg, S. G. Semenchinsky, V.M. Pudalov, *Phys. Rev.* **B42**, 3741 (1990)

BINDING ENERGY OF NEGATIVELY CHARGED MAGNETO-EXCITONS IN QUANTUM WELLS

A J SHIELDS AND M PEPPER
Toshiba Cambridge Research Centre,
260, Science Park, Milton Rd, Cambridge CB4 4WE. U.K.

P C M CHRISTIANEN AND J C MAAN
High Field Magnet Laboratory, Univ. of Nijmegen, 6525 ED Nijmegen, Netherlands

M Y SIMMONS AND D A RITCHIE
Cavendish Laboratory, Univ. of Cambridge, Madingley Road, Cambridge. U.K.

We study the second electron binding energy of negatively charged excitons in remotely doped GaAs quantum wells. The repulsion of the two electrons produces a very different dependence on magnetic field than for neutral excitons.

A magnetic field applied perpendicular to the plane of a quantum well (QW) exerts an additional confinement potential upon the excitonic wave-function. At the magnetic fields where the cyclotron radius becomes smaller than the exciton Bohr radius, the electron and hole are forced closer together, enhancing their mutual Coulomb interaction and thereby their binding energy. This produces a monotonic increase in the binding energy of neutral excitons (X) with magnetic field.[1] We demonstrate here that the negatively-charged exciton[2,3,4,5,6,7] (X^-) behaves in a fundamentally different manner. Since X^- consists of two electrons and a hole, the binding energy does not necessarily always increase with magnetic field. Indeed at the magnetic fields where the electron-electron separation is smaller than electron-hole, the binding energy could actually decrease with field.

We studied an n-type remotely doped 300 Å GaAs/$Al_{0.33}Ga_{0.67}As$ QW. A voltage applied between a NiCr Schottky contact evaporated on the sample surface and an Ohmic contact to the QW layer, allowed the electron density to be varied. Magnetic fields were applied normal to the QW using either a split-coil superconducting magnet, for measurements up to 8T, or a Bitter magnet (up to 20 T). Photoluminescence (PL) was measured for both circular polarisations with respect to the field axis.

Figure 1a plots PL spectra recorded with magnetic fields up to 8 Tesla, emitted in σ^- polarisation. The Schottky bias was fixed for these measurements to produce a small electron density in the QW. The two lines observed at zero field are assigned to the neutral (X) and ground state (spin-singlet) of the negatively charged (X_s^-) exciton, by the dependence of their intensity on

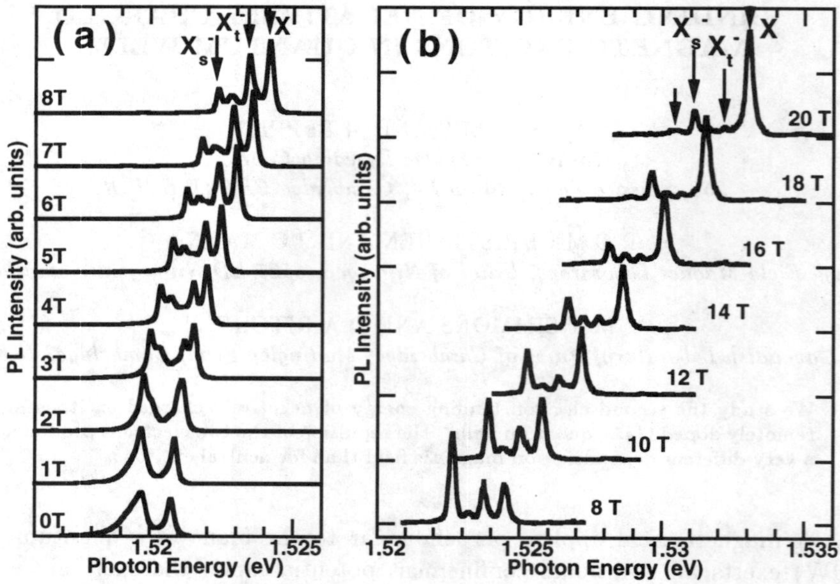

Figure 1: PL spectra, emitted in σ^- polarisation, recorded at $\sim$ 2 K and with different magnetic field applied normal to the well of between (a) 0-8 T and (b) 8-20 T.

the electron density in the QW.[3] Around 3 T another peak emerges below X, which is due to an excited state of X^- for which the spin wavefunction of the two electrons is symmetric upon interchange, i.e. a spin-triplet state (X_t^-).[5,6] The weak peak between X_s^- and X_t^- at 8T, which is strongest at low electron density and is not seen in excitation spectra, is due to a donor bound X.

Spectra recorded at higher magnetic fields are plotted in Fig. 1b. These spectra were recorded with a larger QW electron density, so as to compensate for the apparent weakening of X^- with field. Notice that the 8 T spectrum shows the same three lines due to X, X_t^- and X_s^-, as can be seen in Fig. 1a. (The small shift in energy of the three peaks is explained by the measurements being performed on different mesas.) Beyond 8T, the excitonic lines shift to higher energy, in almost linear fashion, with field.

Notice in both Figs. 1a and 1b that the X_s^- peaks weakens, while X strengthens, with increasing field. This phenomenon is also observed in electroreflectance spectra taken on this sample,[6] and can be explained using the real space filling model discussed in Ref.2. The magnetic field shrinks the spatial extent of the exciton and excess electron wavefunctions, thus making the formation of neutral excitons more likely at a fixed electron density.

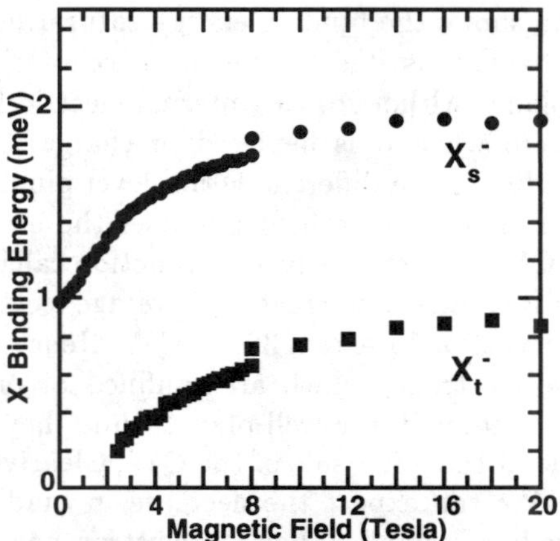

Figure 2: Magnetic field dependence of the second electron binding energy for both the singlet, X_s^- (circles) and triplet, X_t^- (squares) states.

The field dependence of the second electron binding energy, as determined from the spectral separation of the X and X^- PL peaks, is plotted in Fig. 2. Notice that the binding energy of the singlet increases very rapidly at low field and appears to saturate around 10 T at roughly twice its zero field value. The binding energy of the triplet state, which becomes finite around 2 T, also increases rapidly before saturating at high field.

The large increase in the second electron binding energy of X^- at fields less than 10 T can be understood by regarding its spatial wavefunction as a combination of two one-electron orbitals. Variational calculations for D^- with this approach have shown that at zero magnetic field, the wavefunction consists of an inner orbital which is similar to that of D^o and an extended outer orbital which has roughly twice the diameter of the inner one.[8] In this description we can picture X^- as a neutral exciton being orbited by a second, weakly bound electron. A magnetic field applied perpendicular to the QW squeezes the extended outer electron orbital towards the core, thereby increasing their mutual Coulomb interaction and enhancing the second electron binding energy of X^-. The second electron binding energy can be expected to increase for fields where the Bohr radius becomes smaller than the Bohr radius of the outer orbital. Taking this to be twice that of the neutral exciton (i.e. ~300 Å) yields a field of about 0.7 Tesla in good agreement with the observed behaviour.

For an ideal system the X^- binding energy would continue to increase with magnetic field, although less rapidly than for a neutral exciton due to the additional electron-electron repulsion discussed above. On the other hand the experimental data shows the binding energy saturating at high fields. We suggest that this behaviour is due to the presence of a small electric field normal to the QW plane. Although, the internal electric field across the well should decrease to zero when it is depleted of charge, there will still be a residual electric field due to the different Fermi level pinning at the front and back of the epilayers. This electric field polarises the electrons and holes to opposite faces of the QW. In fact an envelope function calculation shows that a field of just 5 kV/cm is sufficient to create an average electron-hole separation along the growth direction of 75 Å in a 300 Å QW. Hence under an increasing electric field, the two electrons, which are confined at one side of the well, will be forced closer together in the well plane, while they will remain a fixed distance from the hole at the other side of the QW. Clearly this will reduce the X^- binding energy. We can expect the decrease in binding energy to occur when the cyclotron radius is smaller than the electron-hole separation. Taking the latter to be 75 Å yields a field of about 12 T, around the field where we see the binding energy saturate.

Zhu and Xu[9] calculated the second electron binding energy of D^- for the case where the donor ion is displaced away from the centre plane of the QW. Since this again involves displacing the positive charge away from the two electrons, it is a somewhat analogous case to the situation discussed above for X^- in the presence of an electric field. They find that displacing the donor ion causes a large reduction in the D^- binding energy at high field. The binding energy can then level off, or even decrease, with increasing magnetic field. These calculations are highly supportive of our interpretation of the X^- data.

In conclusion, the X^- second electron binding energy saturates at high magnetic field. The magnetic field induced enhancement is less than for the first electron due to the extra repulsion of the two electrons. The X^- binding is weakened further by a small residual electric field normal to the QW.

References

1. D M Whittaker and R J Elliott, Solid State Comm. **68**, 1 (1988).
2. K Kheng et al, Phys Rev Lett **71**, 1752 (1993).
3. A J Shields et al, Phys Rev B **51**, 19049 (1995).
4. G Finkelstein et al, Phys Rev Lett **74**, 976 (1995).
5. A J Shields et al, Phys. Rev B. **52**, 7841 (1995).
6. A J Shields et al, Advances in Physics **44**, 47 (1996).
7. G Finkelstein et al, Phys Rev B **53**, R1709 (1996).
8. N P Sandler and C R Proetto, Phys. Rev. B **46**, 7707 (1992).
9. Zhu and Xu, J. Phys: Condens. Matter **6**, L299 (1994).

EFFECTS OF EXCITON - ELECTRON INTERACTION IN CdTe/CdMgTe QUANTUM WELL STRUCTURES

J.X.SHEN[1], D.R.YAKOVLEV[1,2], V.P.KOCHERESHKO[2], H.SCHENK[1], W.OSSAU[1], A.WAAG[1], and G.LANDWEHR[1]

[1]*Physikalisches Institut der Universität Würzburg, D-97074 Würzburg, Germany*
[2]*A.F.Ioffe Physico-Technical Institute RAS, 194021 St.Petersburg, Russia*

Optical tuning of electron concentration in a quantum well is realized in CdTe/(Cd,Mg)Te heterostructures. Effects governed by the exciton-electron interaction under low electron density are studied by use of photoluminescence in magnetic fields and under microwave radiation.

Effects resulting from the exciton-electron interaction in the presence of a two-dimensional electron gas of low density at $n_e a_B << 1$, where n_e is the electron concentration and a_B is the exciton Bohr radius, became subject of intensive investigations very recently. This interest has been stimulated by the paper of K.Kheng *et al.*, where the observation of a negatively charged exciton (X^-) has been reported in CdTe/(Cd,Zn)Te modulation-doped QW's[1]. After that, negatively and positively charged excitons have been observed in GaAs/(Al,Ga)As modulation-doped QW's with the free carrier concentration tuned by the variation of the gate voltage[2]. Recently it has been reported about the enhancement of energy- and phase relaxation rates for excitons in the presence of electrons[3,4].

In this paper effects determined by the exciton-electron interaction are studied in a type I CdTe/$Cd_{0.74}Mg_{0.26}Te$ heterostructure consisting of a 75-Å-thick QW sandwiched between 1000-Å-thick short-period superlattices (SL's) (30Å/30Å, 16 periods). The QW is separated from the SL's by 200-Å-thick barriers. Tunneling probabilities for electrons and holes from the SL's into the QW differ by about three orders of magnitude. This allows to tune the concentration of electrons in the QW by varying the intensity of photoexcitaion at energies exceeding the band gap of the SL. The structure has been grown by molecular-beam epitaxy on an (100)-oriented $Cd_{0.97}Zn_{0.03}Te$ substrate. Photoluminescence (PL) spectra have been measured at 1.6 K and in magnetic fields up to 7.5 T applied perpendicular to QW layers (Faraday geometry). Time-resolved measurements have been performed by the use of an acousto-optical modulator with time resolution of 100 ns. A dye-laser (Pyridine) with the photon energy below the gap of the SL has been used for a direct excitation of excitons in the QW. An Ar-ion laser, which energy exceeds the SL band gap, has been used for variation of the electron concentration in the QW. Excitation density of the indirect excitation does not exceeds 3 W/cm^2 and PL intensity from QW under indirect excitation was very small in comparison with the PL intensity taken under direct excitation (compare spectra 3 and 1 in Fig.1). The PL spectrum 1 consists of two lines. The X-line is corresponded to the

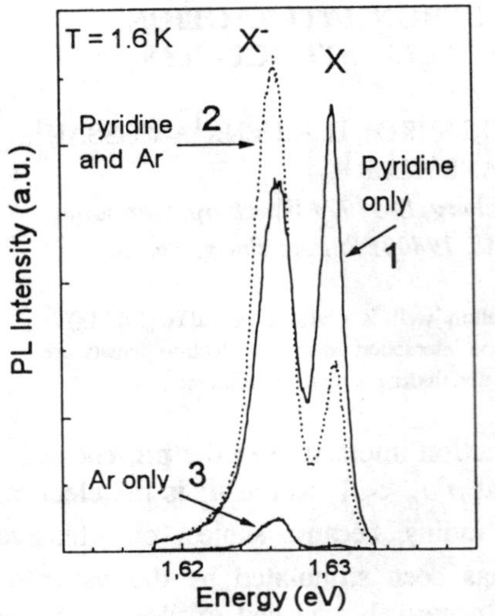

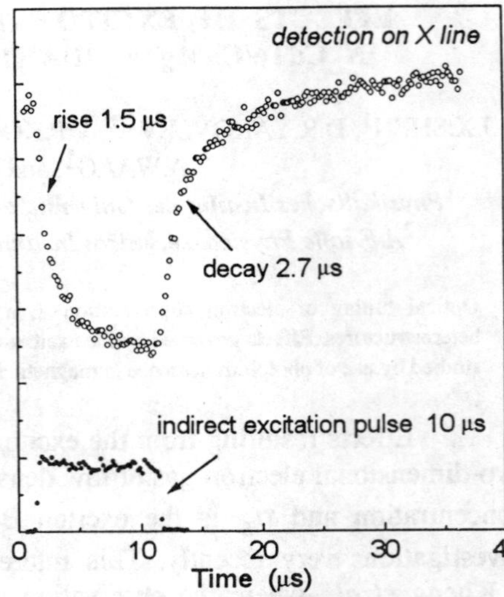

Fig.1 PL spectra of a 75-Å-thick CdTe/$Cd_{0.74}Mg_{0.26}Te$ QW taken under different excitation conditions.

Fig.2 Temporal changes of X-line intensity measured under pulsed indirect excitation (open circles).

recombination of excitons localized on monolayer fluctuations of QW width. Its linewidth of 2 meV and Stokes shift of 1 meV reflect high interface quality of the studied structure. A low energy line (X^--line) at 1.627 eV is observed 3.5 meV below the X-line and corresponds to the exciton bound state. It was recently shown that this binding energy is characteristic for the negatively charged exciton state in CdTe-based QW's and is smaller than the typical binding energy of an exciton to a neutral donor[1]. In addition, the indirect excitation causes a strong redistribution of the PL intensity in favour of the X^--line (compare spectra 1 and 2 in Fig.1). These changes are understandable, as indirect excitation increases the number of electrons in the QW and increases, respectively, the probability of the X^- formation.

A temporal characteristics of this redistribution are shown in Fig.2. 10 μs pulses with 30 μs intervals have been used for the indirect excitation. PL intensity has been detected at the maximum of the X-line, where the additional electron concentration causes a decrease of the intensity. The rise time of the effect is 1.5 μs and increases with the decrease of the density of indirect excitation. It is determined by the number of electrons collected into the QW. The decay of the effect within 2.7 μs is independent of the excitation density and is due to the indirect recombination of the QW electrons with holes located in the SL's, which are separated by the 200-Å-thick barriers. The recombination dynamics of excitons in the QW amounts to 200 ps at 1.6 K, which is much shorter than the dynamical processes measured in the μs-range.

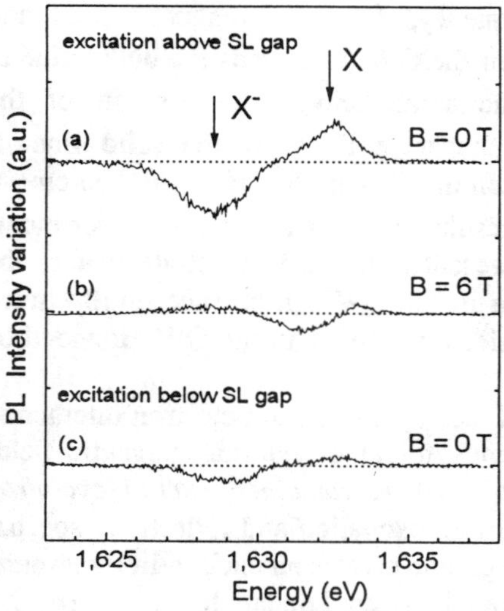

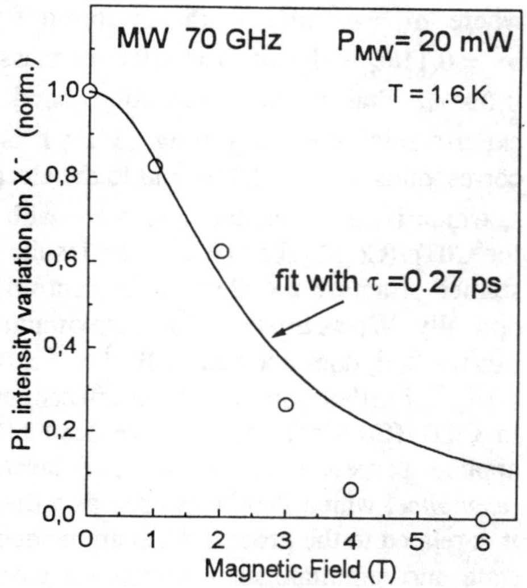

Fig.3 Microwave radiation effect on a PL intensity of a 75-Å-thick $CdTe/Cd_{0.74}Mg_{0.26}Te$ QW.
Fig.4 Magnetic field dependence of microwave radiation effect on a PL intensity of X^- line.

Fig.3 depicts the effect of microwave (MW) radiation on the PL intensity of the QW. MW radiation with a frequency of 70 GHz and a power on the sample of about 20 mW has been used in this experiment. Being effectively absorbed by charged carriers the MW radiation heats the electrons and reduces the probability of the X^- formation. Therefore, under MW radiation the intensity of the X^--line should decrease and more excitons should recombine as the free excitons increasing intensity of the X-line. This behavior is shown in Fig.3 by curve *a* where the change of the PL intensity is plotted. The modulation of the signal raises up to 6% of the PL intensity under the excitation above the SL gap and is strongly reduced for excitation below the SL gap when no electrons are collected from the SL's into the QW (compare curves *a* and *c*). The MW radiation effect on the X^--line is reduced by the increase of a magnetic field (compare curves *a* and *b*). Fig.4 shows a magnetic field dependence of the MW radiation effect on the X^--line. The dependence shows monotonic decrease without pronounced feature corresponding to the optically detected cyclotron resonance. Such behavior is typical for structures with $\omega\tau << 1$, where ω is the MW frequency and τ is the momentum relaxation time[5]. Under this condition the power absorbed by the electrons is given by[5]:

$$\frac{P(B)}{P(B=0)} = \frac{1}{1+(\omega_c\tau)^2} \qquad (1),$$

where $\omega_c = eB/m^*c$ is the cyclotron frequency, B is the magnetic field and $m^* = 0.11m_0$ is the electron effective mass in the QW (it exceeds the bulk value of $0.096m_0$ due to the conduction band nonparabolicity). The best fit of the experimental data according to Eq.1 is shown in Fig.4 by the solid line. It corresponds to $\tau = 0.27\,ps$ and to the electron mobility in the QW $\mu = 4000$ cm^2/V s, which is in reasonable agreement with results of magnetotransport experiments for CdTe/(Cd,Mg)Te QW's[6]. The results presented above demonstrate that in the studied structure the electron concentration in the QW can be successfully tuned optically. We estimate that the maximum electron density in the QW, provided by this method, does not exceed 10^{10} cm^{-2}.

Furthermore, two other effects governed by the exciton-electron interaction in CdTe/(Cd,Mg)Te QW's have been recently found in external magnetic fields applied perpendicular to the QW layers: (i) A *combined exciton-cyclotron resonance*, which manifests itself as a line in PL excitation and reflectivity spectra. It is related to the process when an incident photon creates an exciton in the ground state and simultaneously excites an electron between Landau levels[7]. (ii) An *exchange exciton-electron scattering*, which is an inelastic and spin-dependent process. In this process the photogenerated exciton loses its energy by scattering to an orthoexciton state and the electron is simultaneously excited to the upper Zeeman sublevel[8].

Acknowledgments

This work has been supported in part by the Deutsche Forschungsgemeinschaft through SFB 410, the Volkswagen Foundation and the Russian Foundation for Basic Research through grant No.95-02-04061a.

References

1. K.Kheng *et al*, Phys.Rev.Lett. **71**, 1752 (1993)
2. G.Finkelstein *et al*, Phys.Rev.B **53**, R1709 (1996).
3. B.M.Ashkinadze *et al*, Phys.Rev.B **51**, 1938 (1995)
4. M.Koch *et al*, Phys.Rev.B **51**, 13887 (1995).
5. K.Seeger, *Semiconductor Physics* (Springer-Verlag, Wien 1973), Chap.11.
6. S.Scholl *et al*, Solid State Commun. **94**, 935 (1995); and J.Gerschütz et al., this volume.
7. D.R.Yakovlev *et al*, *Proc. 23-d Int. Conf. on Physics of Semiconductors*, Berlin 1996.
8. V.P.Kochereshko *et al*, *Proc. 23-d Int. Conf. on Physics of Semiconductors*, Berlin 1996.

TRANSPORT AND FAR INFRARED SPECTROSCOPY OF QUANTUM WIRES

E. GORNIK, G. PLONER, V.ROßKOPF, G. STRASSER,
Institut für Festkörperelektronik,Mikrostrukturzentrum der TU Wien
Floragasse 7, A-1040 Wien, Austria

C.M. ENGELHARDT

Walter Schottky Institut, TU München, Am Coulombwall,
D- 85748 Garching, Germany

Systematic studies of the electron transport in quantum wires are reported. Magnetophonon as well as magnetic depopulation experiments are applied for a systematic analysis of subband spacings as a function of carrier density. It is found that the subband energy spacings and the polaron mass increase systematically with decreasing 1D electron density. Both effect reflect a decreasing screening and an enhancement of the electron-LO phonon coupling with decreasing carrier density.

These experiments are complimented by far infrared transmission experiments. The transmission spectra reveal a dominant magnetoplasmon peak which allows the determination of the subband energies in the bare potential.Deviations from the plasmon dispersion are observed at high magnetic fields.In the quantum limit a deformation of the resonance showing an onset of spin splitting is observed, also indicating a reduced Coulomb interaction in wires. In addition photoconductivity measurements were performed on samples consisting of 10 to 100 wires. In shallow etched wires a narrow single photoconductive peak with a linewidth below 1 cm^{-1} is observed which shifts with the magnetic field according to the magnetoplasmon dispersion.

1 Introduction

Transport properties of one-dimensional electrongases (1DEG) are of great interest due to theoretical predictions of enhanced mobilities as compared to a two-dimensional electron gases (2EDG)[1]. Very little experimental work has been devoted to transport studies on etched wires. Only very recently Wirner et.al.[2] have reported a systematic study of side gated etched wires, which show an enhanced mobility. With improving wire quality techniques for the analysis of the wire potential become more and more important. Conventional magnetotransport measurements at low temperatures fail, when only a very few subbands are occupied. The study of magnetophonon resonances (MPR) in quantum wires is an valuable tool particularly due to the fact that they can also be used to estimate quantization energies in the lateral confinement potential of the wires. Since one can investigate sublevel spacings by MPR even when only low numbers of subbands are occupied in the wire, it is considered to be the only transport technique for the characterization of wires near the quantum limit. In the case of MPR in 2D electron systems there is a wealth of theoretical and experimental work which show very complete understanding of this effect [3-5]. However, to our knowledge, only very little experimental work is reported on MPR in 1D systems even though this subject has raised considerable interest to theoreticians[6-9] which predicted several interesting effects worth while being studied experimentally. It is therefore of interest under which conditions MPR in shallow etched quantum wires give reliable information on quantization energies and polaron masses. A systematic investigation on carrier density and temperature has been performed on wires with considerable different mobility's due to different preparation techniques.

Additional insight into the electronic properties of low-dimensional electron systems is gained from investigations of their FIR excitations. It has been demonstrated that in etched[10] as well as field-effect[11] induced low-dimensional electron systems the bare confinement potential is of parabolic shape to a very good approximation. Taking the generalised Kohn theorem[12] into account, the FIR conductivity exhibits a single dimensional resonance at a frequency that corresponds to the characteristic frequency of the bare potential. The meaning of "bare potential" here is, the potential is induced by all external charges contributing to the confinement as e.g. charges on a gate pattern, in surface states or image charges on the semiconductor interfaces, but not the electrons occupying the quantum wire. Experiments[13] as well as model calculations demonstrate that the electron systems usually are smaller than the lithographically defined patters on the surface due to lateral depletion. In the present paper FIR transmission measurements are used to investigate 1D subband spacings and Coulomb interactions on GaAs-GaAlAs heterostructures.

2 Samples

The samples used for our experiments were conventional modulation doped MBE (molecular beam epitaxy) grown GaAs/$Al_xGa_{1-x}As$ heterostructures. Quantum wires were fabricated by laser holography and subsequent wet chemical etching.

The samples used for transport studies consisted of arrays of 40 quantum channels. The samples, referred to as g166 have the following structure. A 10nm undoped GaAs cap layer is followed by a 30nm AlGaAs with $n_D=2x10^{18}$ cm^{-3}. A 60nm AlGaAs spacer separates the 2DEG from the doped region. The electron mobility and electron density at 4.2 K of the unstructured sample are 1.6 $x10^6$ cm^2/Vs and 1.1 $x10^{11}$ cm^{-2}, respectively.
The variation of the electron density was achieved by illuminating the wires with a light emitting diode. The wires are fully depleted immediately after the etching.

After illumination magnetic depopulation (MD) experiment are performed at 2 K, followed by slowly heating the sample up to temperatures between 100 K and 160 K awaiting a stable sample resistance. Subsequent to the high temperature measurement the sample was cooled down slowly and again a MD experiment is performed to obtain the electron density. The density of charged carriers which has been generated by illumination is reduced during the heating process, but we assume that the cooling subsequent to the high temperature measurement freezes the electron density which is therefore the same in the last two steps. The persistent electron density obtained after illumination turned out to be extremely stable.

For the transmission experiments the same samples as for transport were used. The laser holographic gratings have periods between 475 nm and 630 nm. The geometrical dimensions of the structured sample are 3 x 3 mm^2. Etching times varying from 10 s to 50 s were used to increase the subband spacing. By consecutive bandgap illumination with a red LED the filling of the wires could be varied. For shallow etched wires illumination changed the electronic system from electrically isolated wires to a density modulated system and back to an almost perfect two dimensional system.

3 Magnetophonon Results

A strong dependence of the Magnetophonon amplitude (magnetoresistance oscillations) on sample structure and preparation parameters is observed. The effect is most pronounced in samples where the impurity and remote impurity scattering is reduced , which can be achieved by a low doping and large spacer layer. Since the wet chemical etching defining the wires induces additional sources of scattering (surface roughness scattering) the sample has to be prepared in such a way that the etching has no additional effect. This restricts etching to very shallow etching and thus to weak confinement potentials.

In Fig. 1 the oscillatory part R_{osc} of the magnetoresistance obtained after subtraction of the background resistance is shown for two samples g166/1 and g166/8. These two samples are from the same waver but differ by the depth of the etch pits, g166/1 being etched slightly longer than g166/8. As can be seen from the figure the amplitude of the resonance's is smaller by almost an order of magnitude for sample g166/1, whereas no striking broadening of the resonance's is observed.

This dependence of the strength of the MPR on the preparation process of the sample can be understood partly in terms of the broadening of the magnetoelectric hybrid levels induced by additional scattering processes due to the quantum wire fabrication process. A broadened density of states weakens the MPR and smears out the weak low field oscillations. This is a drastic effect which shows that the different etching times change the resonance amplitudes by an order of magnitude.

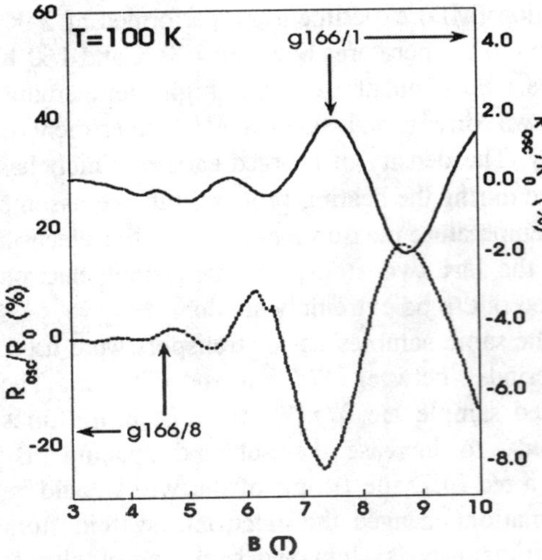

Fig.1: Oscillatory part of the magnetoresistance for two different samples

The subband spacing E_0 determined by MPR is in all cases larger by 30-50% than the values obtained from MD experiments. This difference can be explained as follows: In shallow etched samples the confinement potential is closer to a cosine than to a parabolic potential. Therefore the subband spacing of levels near the upper edge of the confining potential is smaller than it is near its bottom. In the MD experiments only the subbands at the Fermi level in the upper part of the potential are probed at small magnetic fields[9]. When MPR are investigated, however, the higher temperature broadens the Fermi distribution and also electrons at the bottom of the confining potential can absorb LO phonons and scatter into empty states 36 meV higher in energy. In addition MPR are observed at relatively high magnetic fields. Therefore the parabolic confinement at the bottom of the electrostatic potential will contribute appreciably to the magnetoelectric hybrid levels leading to a negligible influence of its nonparabolic parts on the probed subbands. Hence the subband spacings determined from MPR resemble those near the bottom of the confining cosine potential which are larger than those estimated from the MD experiments.

Whenever the resonance condition

$N\hbar\omega_{eff} = \hbar\omega_{LO}, \quad \omega_{eff}^2 = \omega_{cycl}^2 + E_0^2/\hbar^2$ (E_0: subband spacing) is satisfied, minima occur in R_{osc}. From the magnetic field positions of these minima, the confinement energy E_0 and the polaron mass can be estimated according to

$$B^2 = \left(\frac{m^*_{Pol}}{e\hbar}\right)^2 \frac{E_{LO}^2}{N^2} - \left(\frac{m^*_{Pol}}{e\hbar}\right)^2 E_o^2$$

From the data shown in Fig.1 we find E_0=1.6±0.3 meV and a relative polaron mass m_{pol}/m_e=0.069 for the better sample. A MD experiment performed after the high temperature measurement yields E_0=1.1±0.2 meV. This difference is observed for all samples and consistent with the above explanation. One basic problem which are faced by both techniques is the asumtion of a parabolic potential for the wire. This is most likely not a good approximation at low temperatures. At temperatures where the MPR are measured it might be allready valid due to weaker screening.

Using the illumination and temperature cycle procedure described in the experimental part the electron density in the wires was varied to investigate the dependence of E_0 and m_{pol} on the carrier density. One thereby eventually reaches densities so low that only 2-3 subbands are occupied. The only information that can then be extracted from MD measurements is the 1D electron density. Using the E_0 values obtained from MPR, n_{1D} can be estimated from the slope of the (linear) N(1/B) plot for the high B minima in R_{xx} at 2 K [9]. Fig. 2 shows the sublevel spacing extracted from the MPR data of sample g166/8 as well as from MD experiments (solid circles) as a function of the 1D carrier density in the wire. The observed increase of E_0 with decreasing n_{1D} is a consequence of the reduced screening by the free electrons of the bare confinement potential.

Fig. 3 shows the temperature dependence of subband spacing and polaron mass for sample g166/1 a sample which has been deeper etched. The electron density for this sample is in the same range as for the other sample shown in Fig.2, the confinement is somewhat stronger (cf. Fig.1). The subband spacing is found to decrease with increasing temperature. Since E_0 turned out to depend strongly on the electron density the observed increase is due to the thermal generation of charge carriers at elevated temperatures. The polaron mass is found to depend on the sample structure as well as on the strength of the confinement and is particularly high for sample g166/1. The observed slight reduction of m_{pol} with increasing temperature (Fig. 3 b) is probably due to changes in the band structure.

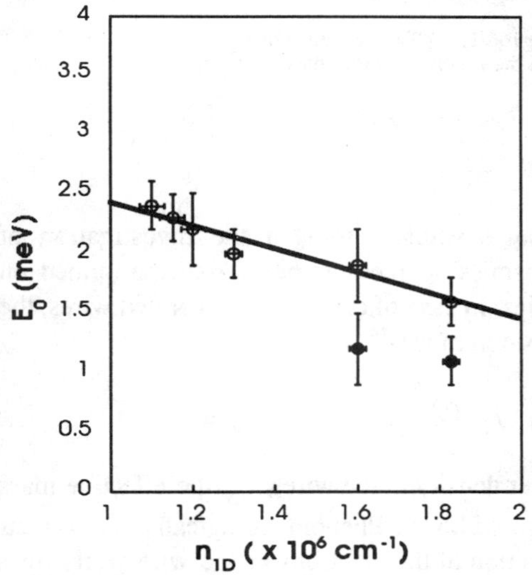

Fig.2: Subband spacing vs. 1D electron density for sample g166/8. Values obtained from MD experiments are indicated by solid circles

From this we can conclude that the mass is a sensitive tool for potential fluctuations. While the two samples have the same subband splitting they show a strong difference in the size of the MPR and the polaron mass. This means that the study of subband splittings alone does not provide enough information for a quality quality analysis of quantum wires. While the subband splitting gives a measure for the form of the potential, the polaron mass and the size of the MPR oscillations are related to defect numbers and thus the size of the fluctuations in the lateral potential.

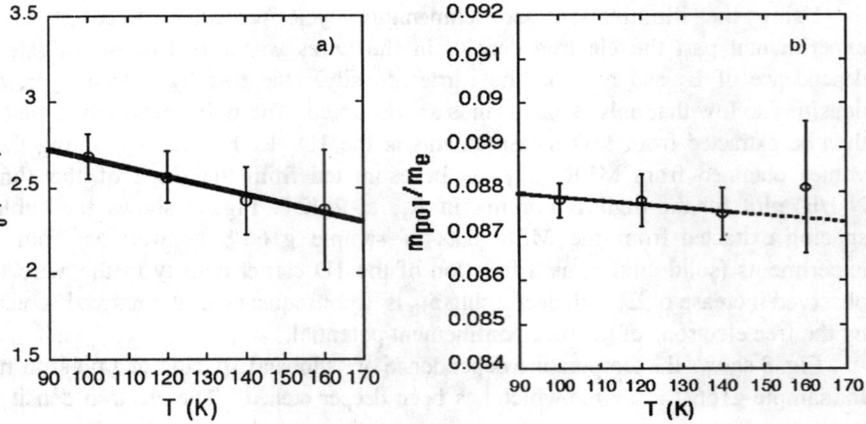

Fig.3: results for sample g166/1: (a) T-dependence of E_0.
(b) T-dependence of the relative polaron mass.

4 Far Infrared Spectroscopy

Far infrared (FIR) spectroscopy provides a valuable tool in the investigation of mesoscopic structures. Far infrared transmission measurements were performed on the same structures as the transport studies. In case of electrically separated wires, the FIR excitation excites odd confined plasmon modes[14]

$$\omega_p^{\ 2} = \frac{e^2}{2\varepsilon\varepsilon_0 m^*} n_{2D} (j+\alpha)\frac{\pi}{w}, \qquad j = 1, \quad 3, \quad 5, \quad \ldots$$

with the effective two dimensional carrier density in the wires n_{2D}, the effective mass of the carriers m* and the electrical wire width w. Phenomenologically α corrects for the phase shift occurring due to reflection of the electrons at the wire walls. In a magnetic field the plasmon should couple quadratically $\omega^2=\omega_c^{\ 2} + \omega_p^{\ 2}$ to the cyclotron resonance (CR) $\omega_c = eB/m^*$.

FIR transmission spectroscopy has been successfully used to detect 2D plasmons dispersions, localised plasmons and depolarisation shifted 1D intersubband transitions[15]. We report here measurements of electronic excitations in the confinement regime between zero and two dimensions. The far infrared response of arrays of periodic quantum wires has been investigated by cyclotron resonance transmission measurements. Due to narrow geometrical dimensions (300 nm) quantum confinement arises and leads to the formation of one-dimensional electronic subbands with a typical energy separation of 1-3 meV in the case of the heterostructures and up to 9 meV in the case of the quantum wells[16]. The far infrared transmission spectra of the quantum wire structures show one strong resonance, which can be described by an harmonic oscillator model, assuming that the confining potential is of parabolic shape. Depending on the intensity of bandgap illumination a well pronounced transition from an one-dimensional electronic system behaviour to a modulated 1D system and finally to a pure 2 dimensional system can be achieved for the heterostructure samples. From the frequency of this localised plasmons the subband quantisation and the electrical width of the quantum wires is derived.

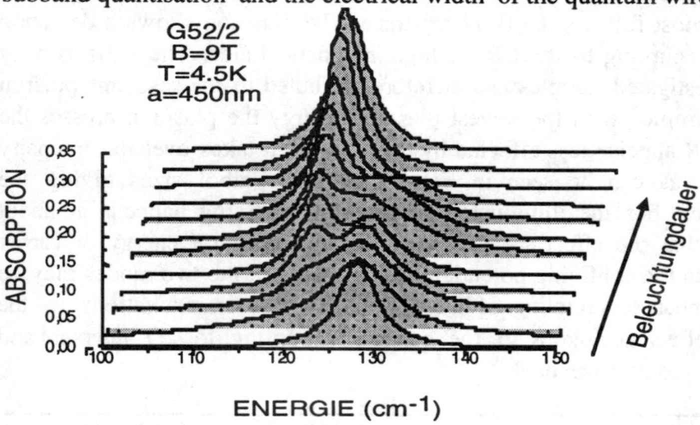

Fig.4: FIR absorption spectra of a quantum wire at 9T at different illumination times. The illumination times differ by 1 ms (bottom to top)

In figure 4 FIR absorption spectra of shallow etched quantum wire structures at 9T for different band gap illumination times are shown. One can clearly see the transition from a well separated wire system with a single resonance to a modulated 2D system, where two lines are present. At the highest illumination time the system becomes purely 2 dimensional with a single line at the cyclotron frequency. The experiment clearly identifies the plasmon shifted CR with the first odd mode j=1.

In a density modulated system, one expects to find two resonances[13] the unperturbed CR and an extended plasmon, which is a better description, when coupling between the electrons in the different parallel wires becomes important. Then ω_p is dominated by the grating period a[17] the plasmon frequency is given by

$$\omega_p^2 = \frac{e^2}{2\varepsilon\varepsilon_0 m^*} n_{2D} \frac{2\pi}{a}$$

By increasing the illumination time, the potential modulation disappears at all, and one finds the pure two dimensional behaviour of the system. By subsequent etching the resonance positions shift to higher energies. Assuming an harmonic oscillator model for the confining potential we derive a subband spacing ω_o = 3 meV and an electrical width w = 380 nm (dark) and w = 510 nm (illuminated).

An anomaly appears for weak confinement potentials in high magnetic fields. Fig. 5 shows spectra for a situation when two resonance's are present. The CR line is weak, the dominant peak at low magnetic fields corresponds to the plasmon mode. Spectra as obtained from transmission measurements with the Fourier spectrometer and fits with 3 lines are shown for different fields. Fig.6 shows the dependence of the resonance energies as function of magnetic field. The CR energy for $m^* = 0.07\ m_o$ has been subtracted to enhance visibility of small variations. The unperturbed CR almost follows the 0.07 m_o trace. The plasmon shows a deviation from the quadratic coupling to the CR at high magnetic fields. The shift is very similar for all investigated samples and therefore attributed to the resonant polaron effect [18]. For the sample with the lowest plasmon energy the plasmon crosses the CR at 14 T. The CR appears less affected by the polaron, it takes over the intensity from the plasmon, as can be seen in Fig. 6 by the symbol sizes giving the resonance intensities. For the illuminated sample the crossing happens at about 15 T indicating that the effect scales with either the plasmon energy or carrier density and not with the confining potential. This crossing of the two modes may be explained by an enhanced coupling of the collective plasmon excitations to the resonant polaron effect analogous to the enhanced coupling to 2D intersubband transitions as observed by Kern et al.[19].

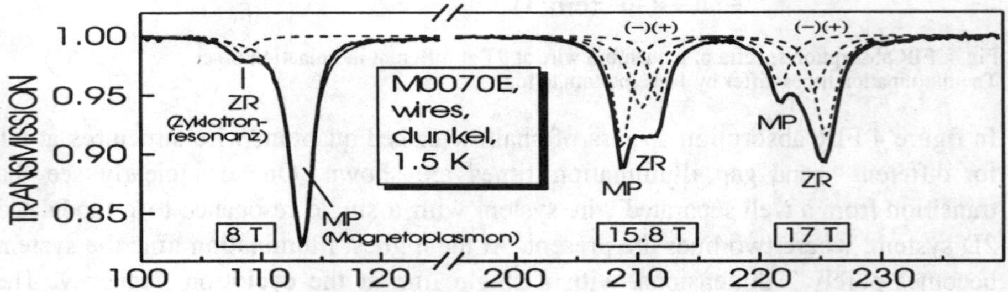

Fig.5: Transmission spectra for 3 different magnetic fields. The dashed lines show the convolution of the measured spectra with a combination of 3 lines indicating spin splitting.

In this regime the CR appears to be deformed. It cannot be fitted by a single Lorentzian. A good fit is obtained with two lines. The energy separation of the two resonances comes out to be close to the spin splitting. The spin splitting is an effect of nonparabolicity and has been observed in 2D systems only in the extreme quantum limit[20,21] when Coulomb coupling of the resonances is weakened [22]. The temperature dependence of the two constituents of the CR show a very similar

behavior as in 2D. With increasing temperature up to 12 K an exchange of intensities between the two lines is observed with an activation energy very close to the spin splitting. This splitting is in the order of 5K in agreement with the Zeeman splitting in the magnetic field regime under consideration.
This might be an effect very specific to wire systems which show a reduced Coulomb coupling due to a reduced lateral electron density. The observation of spin splitting happens in 1D systems at filling factors (here the filling factor is little below 1) much above the critical filling factors of the 2D system, which are about 1/6. To answer this question more systematic studies in this regime have to follow. As long as the origin of the crossing of the plasmon and CR is not understood in detail, the observed deformation can be part of the effect.

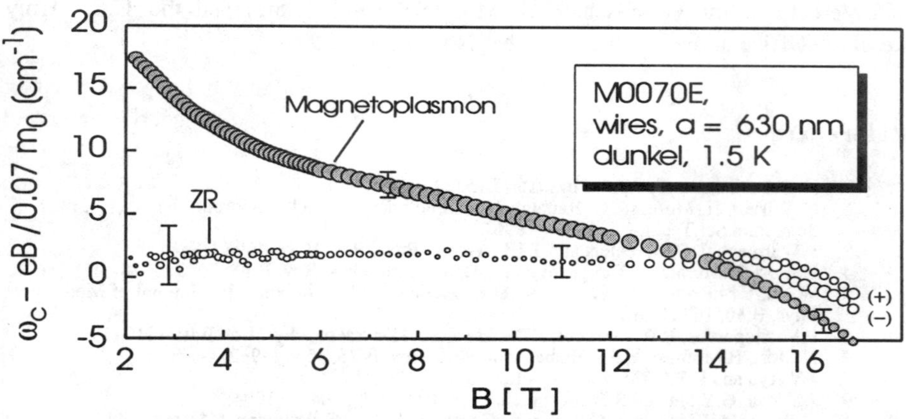

Fig.6: Energy shift of the resonances with respect to the cyclotron resonance (CR) with $m^* = 0.07\ m_o$. The intensity is given by the symbol size. The crossing of plasmon (filled circles) and CR (open circles) is illustrated by two example spectra. The development of the asymmetry in the CR is described by two constituents (dashed lines).

However the activation behaviour of the two cyclotron lines is a strong indication for the appearance of spin splitting showing that it is easyer to reduce screening in 1D than in 2D systems.

4 Summary

Magnetophonon and FIR transmission experiments are complementary techniques to study energy levels, transport phenomena and correlation effects in quantum wire systems. In the course of these experiments it turned out that only shallow etched wires in heterostructures resemble high enough quality to study magnetophonon effects and electron-electron interaction effects. Very large differences in magnetophonon oscillation amplitude are observed in wires depending on the etching depth. This indicates a strong influence of even the wet etching process on transport properties. The plasmon energy is reduced at high magnetic fields due to the resonant polaron effect. In the quantum limit the plasmon mode crosses the cyclotron mode. The cyclotron line splits into 2 line indicating the onset of spin splitting due to a reduced Coulomb ineraction in the wire case.

Acknowledgements

This work was suported in part by the Bundesministerium für Wissenschaft, Kunst und Verkehr , the Gesellschaft für Mikroelektronik (Gme) and the U.S. Army Research Office under contract number N68171-96-C 9014.

References

1. H. Sakaki, Japan. J. Appl. Phys. **19** , L753 (1980).
2. C. Wirner, H. Momose, C. Hamaguchi, J. Smoloner, A. Köck, E.Gornik, Semicond.Sci.Techn.**11** , 1065 (1996
3. P.Warmenbol, F.M.Peeters and J.T.Devreese, Phys. Rev. B **37**, 4694 (1988)
4. N.Mori, H.Murata, K.Taniguchi and C.Hamaguchi, Phys. Rev. B 38, 7622 (1988)
5. W.Xu, F.M.Peeters, J.T.Devreese, D.R.Leadley and R.J.Nicholas, Int. Journal of Mod. Phys. B **10**, 169 (1996).
6. P.Vasilopoulos, P.Warmenbol, F.Peeters and J.TDevreese, Phys. Rev.**B40**, 1810 (1989).
7. N.Mori, H.Momose and C.Hamaguchi, Phys. Rev. B **45**, 4536 (1992).
8. J.Y.Ryu and R.F.O´Connell, Phys.Rev. B **48**, 9126 (1993).
9. J.Y.Ryu, G.Y.Hu and R.F.O´Connell, Phys. Rev. B **49**, 10437 (1994).
10. G.Berthold, J.Smoliner, C.Wirner, E.Gornik, G.Böhm, G.Weimann, C.Hamaguchi, N.Mori and H.Momose, Semicond.Sci.Technol. **8**, 735 (1993).
11. K.-F.Berggren, G.Roos and H.van Houten, Phys. Rev. B **37**, 10118 (1988).
12 S.K.Yip, Phys.Rev. **B43**, 1707 (1991)
!3 R.Strenz, V. Roßkopf, F. Hirler, G. Abstreiter, G. Böhm. G. Tränkle, G. Weimann, Semicond. Sci.Techn. **79**, 399 (19994)
14 W.L.Schaich, A.H. MacDonald, Solid State Commun. **83** , 779 (1992)
15 W. Hansen, J.P. Kotthaus, U. Merkt, Semicond. and Semimetals, **35** ,279 (1990)
16 V. Roßkopf, P.Auer, E. Gornik, R. Strenz, G. Abstreiter, G. Böhm, G. Wemann, Solid State Electronics, **40** ,333 (1996)
17 T. Demel, D. Heitmann, P. Grambov, K. Ploog, Phys.Rev. **B38** , 12732 (1988)
18 H. Sigg, H.J.A. Bluyssen, P. Wyder, Solid stae commun. 48 , 897 (1983)
19 K.Kern, D. Heitmann, R.R. Gerhardts, P. Grambov, Y.H. Zhang, K.Ploog, Phys.Rev. **B44**, 1139 (1991)
20 G.M. Summers, R.J. Warburton, J.G. Michels, R.J. Nicholas, J.J. Harris, C.T. Foxon, Phys.Rev.Lett. **70** ,2150 (1993)
21 C.M. Engelhart, E. gornik, M. Besson, G.Böhm, G. Weimann, Surf.Sci. , **305** , 23, (1994)
22 N.R. Cooper, J.T. Chalker, Phys.Rev. **B72** , 2057 (1994)

SEMIRELATIVITY IN SEMICONDUCTORS

W. ZAWADZKI

Institute of Physics, Polish Academy of Sciences
Al. Lotników 32/46, 02-668 Warsaw, Poland

A theoretical and experimental analogy is traced between behavior of electrons in narrow-gap semiconductors and that of relativistic electrons in vacuum. Beginning with the similarity of energy-momentum relations for both systems, we consider various effects in the presence of a magnetic field and in crossed electric and magnetic fields. Finally, the relativistic Zitterbewegung of electrons in semiconductors is described and its possible consequences are discussed.

1 Relativistic Analogy

It is well known that electrons in a crystalline solid can be described to a good approximation as free particles with an effective mass. The purpose of this presentation is to show that the analogy of behavior in a solid and in vacuum can be extended for some semiconductor materials into relativistic domain.

We begin with the **kp** theory, as formulated by Luttinger and Kohn [1] (cf. also Zak and Zawadzki [2]). The dispersion relations $\epsilon(k)$ for the energy bands in a crystaline semiconductor can be determined from the set of equations

$$\sum_l \left[\left(\epsilon_{l0} + \frac{\hbar^2 k^2}{2m_0} - \epsilon \right) \delta_{l'l} + \frac{\hbar}{m_0} \mathbf{k} \cdot \mathbf{p}_{l'l} \right] c_l(\mathbf{k}) = 0. \tag{1}$$

where the sum is over all bands, and $l' = 1, 2, 3, ...$ runs also over all bands. Here m_0 is the free electron mass, $\hbar\mathbf{k}$ is the pseudomomentum, ϵ_{l0} are the band-edge energies, $\mathbf{p}_{l'l}$ are the interband matrix elements of momentum operator, and $c_l(\mathbf{k})$ are coefficients to be determined.

Kane[3] solved the set (1) for the narrow-gap material InSb by taking exactly into account three bands and treating all other (far) bands by the perturbation theory. We will follow Kane's method for narrow-gap materials, assuming for simplicity that far bands can be totally neglected. We will also omit the free electron term in Eq.(1), as it is negligible in materials of our interest. Taking into account only the conduction and the valence bands, one has

$$\begin{pmatrix} -\epsilon + \epsilon_g/2 & (\hbar/m_0)\mathbf{k} \cdot \mathbf{p}_{12} \\ (\hbar/m_0)\mathbf{k} \cdot \mathbf{p}_{21} & -\epsilon - \epsilon_g/2 \end{pmatrix} \begin{pmatrix} c_1 \\ c_2 \end{pmatrix} = 0, \tag{2}$$

This can be solved for the energies. The result is

$$\epsilon = \pm \left[\left(\frac{\epsilon_g}{2} \right)^2 + \epsilon_g \frac{\hbar^2 k^2}{2m_0^*} \right]^{1/2} \tag{3}$$

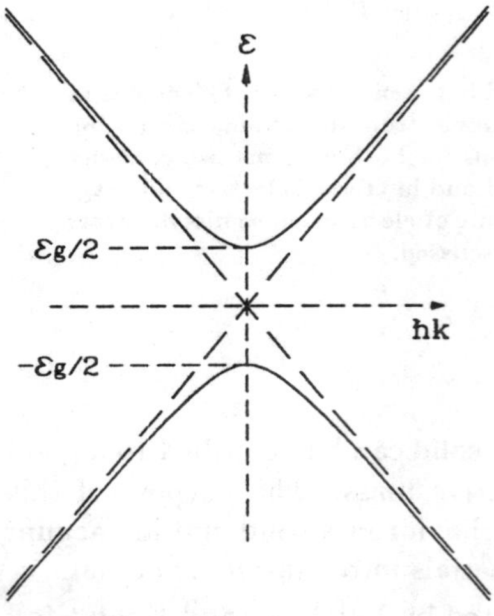

Figure 1: Energy-wave vector relations for the conduction and light-hole bands, as given by the two-band model (schematically).

where the effective mass at the band edge has been defined as $1/m_0^* = 2 \mid p_{12}^x \mid^2 / m_0^2 \epsilon_g$. It has been assumed that $p_{12}^x = p_{12}^y = p_{12}^z$, so that the resulting bands are spherical.

The valence band is usually degenerate as a result of the p-type symmetry, but Eq.(3) correctly describes the conduction and the light hole bands in InSb, InAs, GaSb, HgTe, HgCdTe, HgSe, β-HgS. In lead salts PbTe, PbSe, PbS the band extrema occur at the L point of the Brillouin zone and the resulting bands are spheroidal. The above Hg compounds are zero-gap materials, but the energy interval between p-type and s-type levels is not zero, so that relation (3) is still valid.

It may appear that Eq.(3) is limited to the description of narrow-gap materials. However, expanding quite generally the $k^2(\epsilon)$ relation near its extremum, one obtains[4]

$$k^2(\epsilon) = \lambda_0 + \lambda_1 \epsilon + \lambda_2 \epsilon^2 + \lambda_3 \epsilon^3 + \ldots \tag{4}$$

The first two terms describe a parabolic standard band, while the third (first nonparabolic correction) leads to Eq.(3). Thus, its importance goes beyond narrow-gap materials.

The two bands described by Eq.(3) are shown schematically in Fig.1. For small k values the relations $\epsilon(k)$ are parabolic, for high k they become linear. At the Brillouin zone boundaries the $\epsilon(k)$ dispersion must be horizontal, it is then clear that the two-band description is realistic only until the inflection points on the $\epsilon(k)$ curve.

The dispersion (3) looks similar to the energy-momentum relation for free relativistic electrons:

$$\mathrm{E} = \pm \left[m_0 c^2 + c^2 p^2 \right]^{1/2} = \pm \left[\left(\frac{2m_0 c^2}{2} \right)^2 + 2m_0 c^2 \frac{p^2}{2m_0} \right]^{1/2} \tag{5}$$

In fact, the two formulas are identical, with the correspondence [5]

$$
\begin{aligned}
2m_0c^2 &\to \epsilon_g \\
m_0 &\to m_0^*
\end{aligned}
\tag{6}
$$

According to the initial Dirac interpretation, the negative energies in Eq.(5) represent completely filled "electron sea", which can not be observed because of the Pauli exclusion principle for fermions. In this picture the electron-positron pair creation by a γ photon, having the energy $\hbar\omega > 2m_0c^2$, is simply an excitation of an electron from negative to positive energies, the hole in the Dirac sea being equivalent to the positron. The identical picture is used in semiconductor physics for an optical excitation between valence and conduction bands.

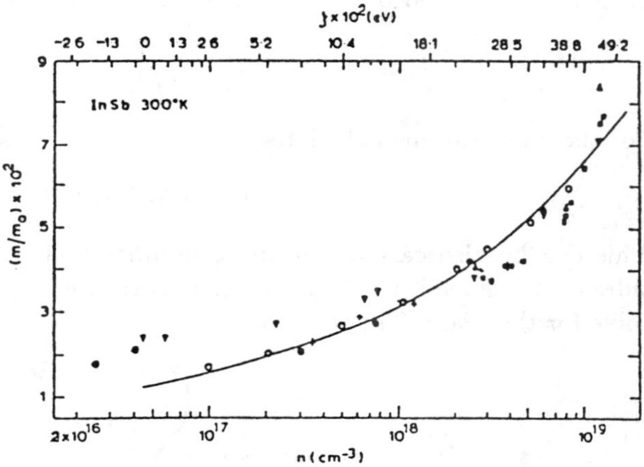

Figure 2: Semirelativistic increase of the effective mass of electrons in InSb vs Fermi energy (upper abscissa). Corresponding electron densities are indicated on the lower abscissa. The solid line is calculated. After Zawadzki, Ref.4.

It is easy to see that the electron velocity in the conduction band described by Eq.(3) reaches a saturation value as k increases. One has

$$
v_i = \frac{1}{\hbar}\frac{\partial\epsilon}{\partial k_i} = \frac{1}{\hbar}\frac{d\epsilon}{dk}\frac{k_i}{k} \to \left(\frac{\epsilon_g}{2m_0^*}\right)^{1/2} = u, \tag{7}
$$

where the limit is taken for large values of k_i. The same result is obtained using the relativistic analogy (6). The maximum velocity is $c = (2m_0c^2/2m_0)^{1/2} \to (\epsilon_g/2m_0^*)^{1/2}$. According to the two-band model there is $u = \mid p_{12}^x \mid / m_0$. This value is almost the same for various materials since $\mid p_{12}^x \mid$ depends only on the symmetry of the band-edge functions. Using experimental values of ϵ_g and m_0^* one obtains on average: $u \approx 1.3 * 10^8 cm/sec$. This is a reasonable value, since typical electron velocities in semiconductors are $10^7 cm/sec$. Still, u is roughly

300 times smaller than c, so that one can expect "semirelativistic" phenomena in semiconductors to occur at correspondingly lower velocities. This is indeed the case.

We define an effective mass tensor relating the electron velocity to its pseudomomentum: $\hat{m}^*\mathbf{v} = \hbar\mathbf{k}$. This is not the only possible definition of the mass, but it is the most useful one. For a spherical energy band the above mass is a scalar and it is given in general by the formula

$$\frac{1}{m^*} = \frac{1}{\hbar^2 k}\frac{d\epsilon}{dk} \tag{8}$$

For the band (3) one calculates

$$m^* = (2m_0^*/\epsilon_g)\epsilon \tag{9}$$

This can be also cast in the more familiar relativistic form. Squaring both sides of (9) and substituting for ϵ the expression (3), where $\hbar k = m^* v$, one can solve for the mass. The result is

$$m^* = m_0^*\left(1 - \frac{v^2}{u^2}\right)^{-1/2} \tag{10}$$

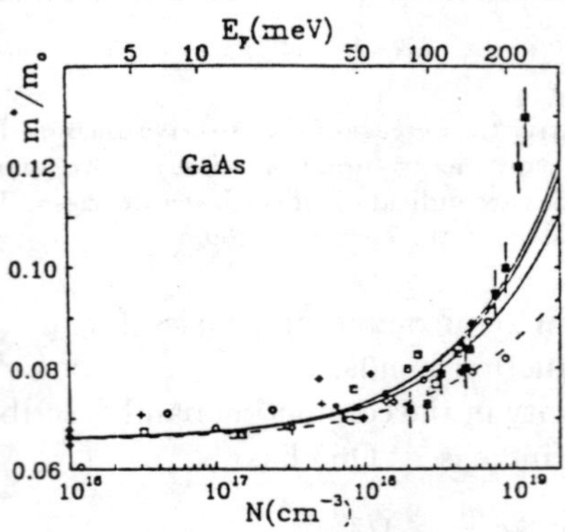

Figure 3: The same as in Fig.2, but for GaAs. After Pfeffer and Zawadzki, Ref.6.

This is the well known formula for the Lorentz mass, in which c is replaced by u, as defined in Eq.(7).

In Figs 2 and 3 we show the mass dependence on the energy in InSb and GaAs, as described by Eq.(9). The mass is measured by various experiments at the Fermi energy. The latter is increased by filling the band with electrons. GaAs is not a narrow-gap material, so that its conduction band is described by an effective two-band formula (3), in which the real gap is replaced by an effective gap $\epsilon_g^* \approx 1eV$, in agreement with the more general formulation of Eq.(4) (cf. Pfeffer and Zawadzki [6]).

Finally, Eq.(9) can be rewritten in the form

$$\epsilon = m^* u^2 \tag{11}$$

This formula is analogous to the famous Einstein relation $E = mc^2$, in which c is replaced by u. The rest energy is obtained for $m^* = m_0^*$, which gives

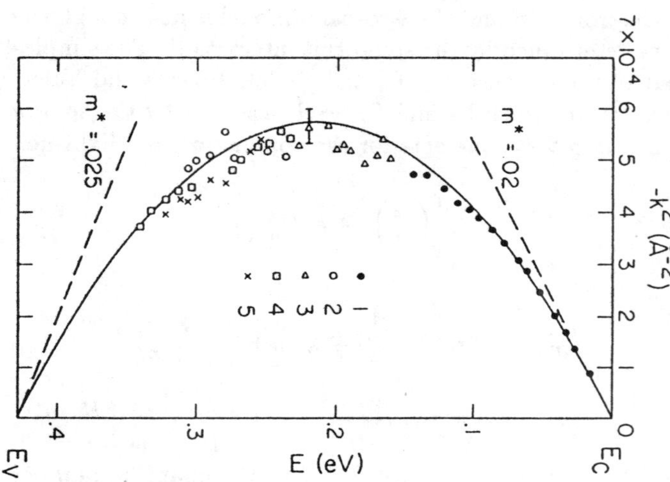

Figure 4: Energy vs wavevector (squared) in the gap of InAs. The solid line is calculated. After Parker and Mead, Ref.7.

$\epsilon_r = m_0^* u^2 = \epsilon_g/2$. This again corresponds to the free electron rest energy $E_r = m_0 c^2$, cf. Eq.(5).

The dispersion relation (3) can also be used for the gap. This region is described by imaginary values of $k = i\kappa$, so that $k^2 < 0$. Figure 4 shows the $\epsilon(k)$ relation in the gap of InAs, as measured by Schottky-barrier experiments. Similar data were obtained on GaAs, but the results are not of equally good quality (cf. Ref.6).

2 Electrons in a Magnetic Field

One can generalize the **kp** theory to account for the presence of an external magnetic field. The result is (cf. Zawadzki[8]).

$$\sum_{l} \left[\left(-\epsilon + \frac{P^2}{2m_0} + \epsilon_{l0} \right) \delta_{l'l} + \frac{\mathbf{p}_{l'l} \cdot \mathbf{P}}{m_0} + \mu_B \mathbf{H} \cdot \sigma_{l'l} \right] f_l = 0 \ , \qquad (12)$$

where l'=1,2,3... Here $\boldsymbol{\sigma}$ is the Pauli spin operator, $\mu_B = e\hbar/2m_0c$ is the Bohr magneton. This form is similar to **kp** theory of Eq.(1), with $\hbar\mathbf{k}$ replaced by $\mathbf{p} + (e/c)\mathbf{A}$, in which $\mathbf{A}$ is the vector potential of magnetic field. Eq.(12) represents the set of coupled differential equations for the envelope functions f_l. For spherical energy bands the set (12) is soluble in terms of harmonic oscillator functions. In the presence of a magnetic field one must account also for the electron spin and the two-band model is not enough since the spin properties are determined by the spin-orbit interaction. The simplest adequate model is that of three levels: Γ_6, Γ_8 and Γ_7 (cf. Bowers and Yafet [9], Kacman and Zawadzki [10]), in which Γ_8 and Γ_7 levels are split by the spin-orbit energy Δ. In the limit $\Delta >> \epsilon_g$, one gets for the conduction and light-hole bands

$$\epsilon = \pm \left[\left(\frac{\epsilon_g}{2} \right)^2 + \epsilon_g D_{nk_z}^{\pm} \right]^{1/2} , \tag{13}$$

where

$$D_{nk_z}^{\pm} = \hbar\omega_c (n + \frac{1}{2}) \pm \frac{1}{2} g_0^* \mu_B H + \frac{\hbar^2 k_z^2}{2m_0^*} . \tag{14}$$

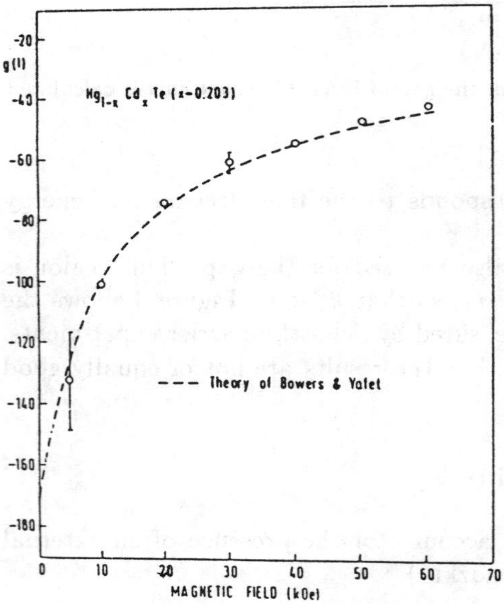

Figure 5: The spin g-factor in HgCdTe vs magnetic field. The dashed line is calculated. After McCombe, Wagner, and Prinz, Ref.12.

Here $\omega_c = eH/m_0^* c$ is the cyclotron frequency, n=0, 1, 2... is the Landau quantum number, and g_0^* is the spin g-factor at the band edge.

Because of the square root in Eq.(13) both the orbital and the spin splittings diminish as n increases. If one defines the cyclotron mass m_c^* from the relation $\epsilon_{n+1}^{\pm} - \epsilon_n^{\pm} = \hbar eH/m_c^* c$, then the decrease of the orbital splitting as a function of energy can be interpreted as a relativistic increase of the mass, described by Eq.(9). One can, in fact, show that the "momentum" mass, given by Eq.(8) is also the cyclotron mass. Similarly, one can define the effective spin g-value by the relation $\epsilon_n^+ - \epsilon_n^- = g_s^* \mu_B H$. This g-value decreases as a function of energy, as illustrated experimentally in Fig.5.

Using Eqs.(13) and (14) one can show that in the limit of small magnetic fields there is: $g_s^*(\epsilon) \sim 1/m^*(\epsilon)$ (cf.Zawadzki [11]). Thus the decrease of g-value is directly related to the increase of m^*. The diminishing of g_s^* is equivalent to the decrease of the electron spin magnetic moment.

Formula (13) is again in direct correspondence to that describing the energies of relativistic electrons in a magnetic field. They are given by solutions of the Dirac equation [13],

$$E^{\pm}_{nk_z} = \pm\left[\left(\frac{2m_0c^2}{2}\right)^2 + 2m_0c^2 G^{\pm}_{nk_z}\right]^{1/2} \tag{15}$$

where

$$G^{\pm}_{nk_z} = \hbar\omega_c(n+\frac{1}{2}) \pm \frac{1}{2}g_0\mu_B H + \frac{\hbar^2 k_z^2}{2m_0}, \tag{16}$$

Here $\omega_c = eH/m_0c$ and $g_0 = 2$ stands for the g-factor of the free electron. The structure of Eqs.(13)-(14) and (15)-(16) is the same, with the correspondence given in Eq.(6). In addition one has

$$g_0 \to g_0^* \tag{17}$$

The main difference between the energies of electrons in a semiconductor and free electrons is that for the free electron the spin splitting is equal to the orbital splitting, while in semiconductors they are not the same since the spin splitting is strongly influenced by the spin-orbit interaction (of the atomic origin), which does not enter into play in vacuum.

One can now ask: what happens to the spin magnetic moment of the relativistic electron when it is accelerated to high energies? A qualitative answer is already given by Eqs.(15) and (16) - the spin splitting decreases with energy, which means that the spin magnetic moment becomes smaller. The quantitative answer (cf.Zawadzki [14]) is obtained by keeping $g_0 = 2$ and defining the spin magnetic moment as (cf. Eqs.(15) and (16)),

$$\mu_s = \frac{E^+_n - E^-_n}{2H} \to \frac{e\hbar}{2m(E)c} \tag{18}$$

The limit is taken for $H \to 0$, which, for a given energy, corresponds to high values of n. The mass $m(E)$ in Eq.(18) is just the Lorentz energy-dependent relativistic mass! Thus the spin magnetic moment is given by the Bohr magneton, in which the rest mass is replaced by the Lorentz mass. Clearly, this result is in close correspondence to the behavior of semiconductor electrons described above, in that the magnetic moment is inversely proportional to the energy dependent mass. To our knowledge, this energy dependence of the spin magnetic moment has never been observed experimentally for free electrons.

3 Electrons in Crossed Fields

The semirelativistic behavior of semiconductor electrons is spectacularly manifested in crossed magnetic and electric fields. The classical equation of motion reads,

$$m\ddot{\mathbf{r}} = e\mathbf{E} + \frac{e}{c}(\mathbf{v} \times \mathbf{H}) \tag{19}$$

For **H** || **z** and **E** || **y** the classical trajectory is shown in Fig.6. It represents a superposition of the cyclotron circle and a drift along the x direction with the constant velocity $v_{dr} = cE/H$. The motion is of magnetic type: one deals with the cyclotron oscillations and the electron does not run away in the **y** direction (which characterizes the electric type of motion).

In the one-band effective mass approximation the Schroedinger equation reads

$$\left[\frac{1}{2m_0^*}\left(\mathbf{p} + \frac{\mathbf{e}}{\mathbf{c}}\mathbf{A}\right)^2 + eEy\right]\Psi = \epsilon\Psi, \tag{20}$$

Using the gauge $\mathbf{A} = [-Hy, 0, 0]$ one separates the variables by looking for solutions in the form: $\Psi = exp(ik_x x + ik_z z)\Phi(y)$. Equation (20) can be then reduced to that for the harmonic oscillator and the resulting eigenenergies are

$$\epsilon = \hbar\omega_c(n + \frac{1}{2}) + \frac{\hbar^2 k_z^2}{2m_0^*} + eEk_x L^2 - \frac{1}{2}m_0^* c^2 \frac{E^2}{H^2}, \tag{21}$$

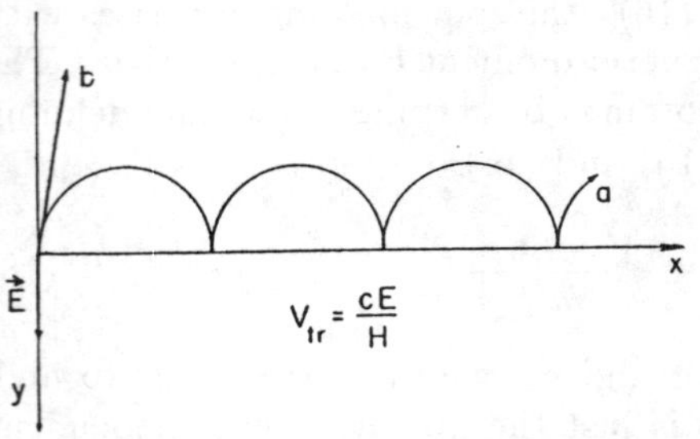

Figure 6: Classical electron trajectory in crossed fields: a) magnetic motion, b) electric motion (schematically).

where $L = (\hbar c/eH)^{1/2}$ is the magnetic radius. In agreement with the classical picture the quantum description represents the magnetic type of motion, characterized by the Landau quantization.

For electrons in the valence band one has to change the sign of mass in Eq.(21). Thus the valence Landau levels shift upwards due to electric field. It is then possible to observe the shift of levels in crossed fields in interband magnetooptical experiments (Aronov [15]). Figure 7 shows the decrease of interband energies, as observed in Ge [16]. They are reduced by $(m_e^* + m_{lh}^*)c^2 E^2/H^2$, in agreement with the theory.

It can be seen from Eq.(21) that the above theory leads to difficulties when the magnetic field tends to zero. It is also not physically reasonable that the classical motion described by Eq.(19) is always of magnetic type, even for large E and weak H. The answer to these problems is the relativity for free electrons and the two-band model for semiconductor electrons. When the drift velocity

$v_{dr} = cE/H$ becomes comparable to the maximum velocity in the system (c or u) the nonrelativistic equation of motion is given by Eq.(19), in which LHS is $d(m\mathbf{v})dt$, where m is Lorentz mass. The resulting motion is of magnetic type for $E/H < 1$ and the drift velocity is $v_{dr} = cE/H$. The motion becomes of electric type for $E/H > 1$ and the drift velocity is then $v_{dr} = cH/E$, cf. Fig.6. The two types of motion can be deduced quantum mechanically from the relativistic Klein-Gordon equation:

$$\left[(m_0c^2)^2 + c^2\left(\mathbf{p} + \frac{e}{c}\mathbf{A}\right)^2 - (\mathrm{E} - eEy)^2\right]\Psi = 0, \tag{22}$$

The leading term y^2, decisive for the character of the motion, are $e^2(H^2 - E^2)y^2$. Thus for $E/H < 1$ one deals with a parabolic quantum well, the resulting motion is confined and the energy is quantized (magnetic regime). For $E/H > 1$ a well becomes a barrier, the motion is not confined and the energy is not quantized (electric regime). The main difference between the nonrelativistic and the relativistic treatment is that in the former only U term appears, while in the latter one has both U and U^2 terms.

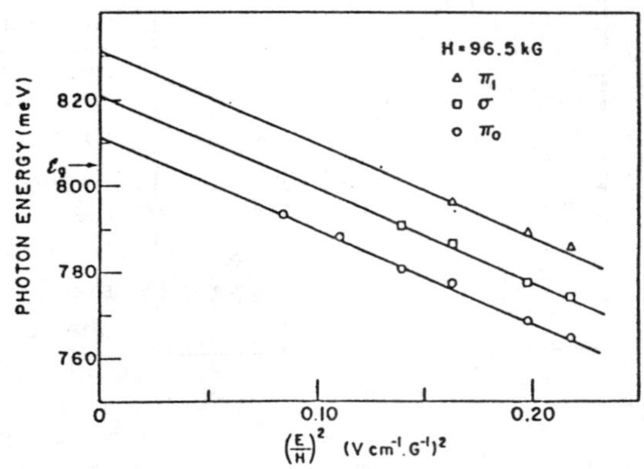

Figure 7: Shift of interband magnetooptical transitions in Ge in crossed fields vs E^2/H^2 (magnetic regime). After Vrehen, Zawadzki and Reine, Ref.16.

For $E/H < 1$ the leading terms can be written as $e^2H^2(1-\delta^2)y^2$, where $\delta^2 = v_{dr}^2/c^2$ with $v_{dr} = cE/H$. On the other hand, for $H/E < 1$ one can write $-e^2E^2(1-\delta^2)y^2$, where $\delta^2 = v_{dr}^2/c^2$ with $v_{dr} = cH/E$. According to the relativistic interpretation in the first case there exists the Lorentz transformation to the moving system, in which the magnetic field is $H' = H(1-\delta^2)^{1/2}$ and $E' = 0$. In the second case one transforms to the system, in which $E' = E(1-\delta^2)^{1/2}$ and $H' = 0$.

For semiconductor electrons one has to solve the two-band model. The set (12) has now in addition the potential U on the diagonal. Neglecting again the free electron terms, one has

$$\begin{pmatrix} -\epsilon + \epsilon_g/2 + eEy & (1/m_0)\mathbf{p}_{12}\cdot\mathbf{P} \\ (1/m_0)\mathbf{p}_{21}\cdot\mathbf{P} & -\epsilon - \epsilon_g/2 + eEy \end{pmatrix} \begin{pmatrix} f_1 \\ f_2 \end{pmatrix} = 0, \tag{23}$$

Eliminating f_2 one obtains (neglecting a small term),

$$\left[\frac{1}{2m_0^*}P^2 - \frac{1}{\epsilon_g}(\epsilon - eEy)^2 + \frac{\epsilon_g}{4}\right] f_1 = 0 \ , \tag{24}$$

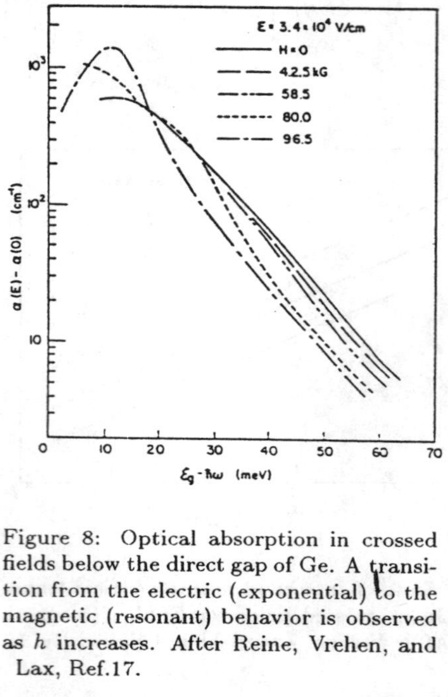

Figure 8: Optical absorption in crossed fields below the direct gap of Ge. A transition from the electric (exponential) to the magnetic (resonant) behavior is observed as h increases. After Reine, Vrehen, and Lax, Ref.17.

The leading terms are $[(e^2H^2/2m_0^*c^2) - (e^2E^2/\epsilon_g)]y^2$. The transition between magnetic and electric type of motion occurs now for $v_{dr} = cE/H = u$. The effective cyclotron frequency in the magnetic regime is $(eH/m_0^*c)(1-\delta^2)^{1/2}$, where $\delta = v_{dr}/u$. At $v_{dr} = cE/H = u$ the Landau quantization disappears. For $cE/H > u$ the motion becomes electric-like and $v_{dr} = cH/E$. Fig.8 illustrates the transition from the electric (exponential) to the magnetic (resonant) type of interband absorption as H increases. Figure 9 shows the cyclotron mass in InSb, measured in crossed fields on MOS structures. The CR mass is traced as a function of B for $E = 0$ and for two non-vanishing electric fields. For $E = 0$ the mass m_c^* reaches the band-edge value m_0^* at $H = 0$. For $E \neq 0$ the CR mass is strongly enhanced as $H \to 0$, which can be interpreted as the relativistic increase when v_{dr} approaches u. In order to describe these results quantitatively one needs a three-level model since the spin splitting in InSb is comparable to the orbital splitting [18]. The final result for the energies in the $\Delta \gg \epsilon_g$ limit is

$$\epsilon = \frac{eE}{H}k_x + (1-\delta^2)^{1/2}\left(\frac{\epsilon_g^2}{4} + \epsilon_g D_n^{\pm}\right)^{1/2} , \tag{25}$$

where

$$D_n^{\pm} = \hbar\omega_c(1-\delta^2)^{1/2}(n+\frac{1}{2}) \pm \frac{1}{2}g_0^*\mu_B H \tag{26}$$

These formulas describe the data of Fig.8 (solid lines) with realistic band parameters of InSb (cf. Ref.18).

The difference $\epsilon^{\pm}_{n+1} - \epsilon^{\pm}_{n} = \hbar\omega$ can be interpreted as the Transverse Doppler Shift (TDS) of the radiation frequency ω_0 emitted by the moving source. In relativity TDS is described by $\omega = \omega_0(1 - v^2/c^2)^{1/2}$, which corresponds to $(1 - \delta^2)^{1/2}$ term in front of the root in Eq.(25). The truly relativistic TDS is considered to be a direct manifestation of the time dilatation [19]. Thus, the relativistic analogy contains also time.

It should be mentioned that we do not deal in reality with truly relativistic drift velocities of conduction electrons: $v_{dr} = 0.85 \times 10^8 cm/sec$ for the highest point on the upper curve of Fig.9. The semiralativistic increase of the mass is well observable because u^2 in the semiconductor is almost 10^5 times smaller than c^2.

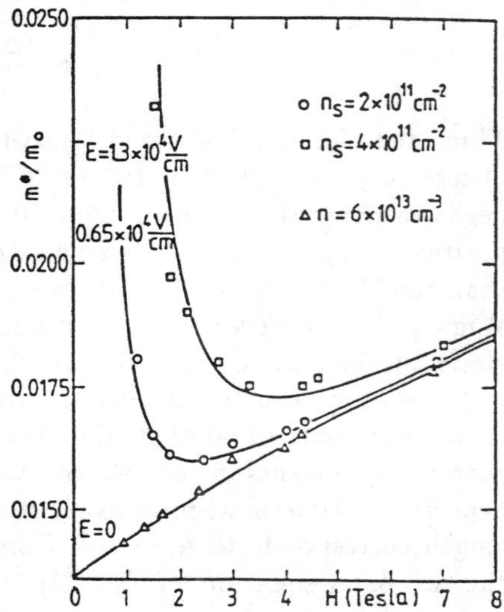

Figure 9: Cyclotron masses in InSb in crossed fields vs magnetic field for two values of electric field. As E/H increases the mass is semirelativistically enhanced. The results for $E = 0$ are shown for comparison. After Zawadzki, Klahn and Merkt [18].

4 Zitterbewegung

It was first shown by Schroedinger [20] that the quantum motion of a free relativistic electron contains a quickly oscillating component, called the Zitterbewegung (the trembling motion). The amplitude of this motion is roughly the Compton wavelength $\lambda_C = \hbar/m_0c$ and its frequency $\omega = 2m_0c^2/\hbar$. The Zitterbewegung is caused by an interference of positive and negative energy components in the electron wavefunction. Foldy and Wouthuysen [21] showed how to eliminate the negative (say) energy components by an expansion in powers of $p/2m_0c$. Writing $p = \hbar k$ the FW transformation is an expansion in $k\lambda_C/2$. One can also show using either the Klein-Gordon [22] or the Dirac [23] equation that, if a relativistic electron represents a pure positive (say) energy state, its spacial extension is of the order of λ_C. This can be easily seen from the following simple consideration. Suppose that $\Delta x \leq \hbar/2m_0c$. The uncertainty relation gives $\Delta p \geq 2m_0c$, which leads to the energy uncertainty

$$\Delta \mathrm{E} \approx \frac{(\Delta p)^2}{2m_0} \geq 2m_0c^2 \tag{27}$$

Thus, the electron "squeezed" into the region $\Delta x \leq \hbar/2m_0c$ has the energy uncertainty so large that it doesn't "know" whether it belongs to positive or negative energies. For this reason "It is often stated that a measurement of the position of a particle, such as an electron, if carried out with a greater precision than the Compton wavelenght, would lead to pair production..." (Newton and Wigner [24]). However, such a situation is not a practical possibility for free electrons since $\hbar/m_0c = 3.87 \times 10^{-3}\mathring{A}$.

It is of interest to consider the corresponding situation in semiconductors. It is clear that, because of the relativistic analogy and the mixing of both energy components in the electron wavefunction (cf. Kane [3], Zawadzki [4]), one expects the Zitterbewegung to occur also in this system. The question is: what length corresponds to $\hbar/m_0c$? Comparing the Klein-Gordon equation with the two-band equation [cf. Eq.(3)],

$$\left(\nabla^2 + \frac{1}{\hbar^2c^2}\mathrm{E}^2\right)\Psi = \left(\frac{m_0c}{\hbar}\right)^2\Psi \tag{28}$$

$$\left(\nabla^2 + \frac{1}{\hbar^2u^2}\epsilon^2\right)\Psi = \left(\frac{m_0^*u}{\hbar}\right)^2\Psi \tag{29}$$

one can see that the Zitterbewegung length λ_Z is

$$\lambda_Z = \frac{\hbar}{m_0^*u} \tag{30}$$

Thus λ_Z can be obtained by using the correspondence $m_0 \to m_0^*$ and $c \to u$.

It should be emphasized that λ_Z is not related to the Compton-like scattering of light in solids, since the light velocity in a solid is not u but c/n, where n is the refractive index.

The Luttinger-Kohn [1] scheme to decouple bands in a semiconductor uses an expansion in powers of $\hbar\mathbf{k}\cdot\mathbf{p}_{12}/m_0\epsilon_g$. Using our definition of the effective mass this can be expressed as $k\lambda_Z/2$. Thus the Luttinger-Kohn decoupling scheme for semiconductors is equivalent to that of Foldy-Wouthuysen for relativistic electrons.

If an electron in a semiconductor is confined to $\Delta x \leq \lambda_Z/2$, then $\Delta p \geq 2\hbar/\lambda_Z$, so that

$$\Delta\epsilon \approx \frac{(\Delta p)^2}{2m_0^*} \geq \frac{4\hbar^2}{\lambda_Z^2 2m_0^*} = 2m_0^*u^2 = \epsilon_g \quad , \tag{31}$$

which corresponds to Eq.(27). Thus, an electron confined to $\Delta x \leq \lambda_Z/2$ has the energy uncertainty larger than the gap.

Clearly, λ_Z is much larger than λ_C. For GaAs ($m_0^* = 0.066m_0$) we calculate $\lambda_Z = 13.5\mathring{A}$, for InSb ($m_0^* = 0.0136m_0$) $\lambda_Z = 67\mathring{A}$. This value can be made even higher by taking $Hg_{1-x}Cd_xTe$ alloys with a smaller effective mass. The modern technology makes it possible to confine electrons to such dimensions. Also an external magnetic field can be used for such purpose. The magnetic radius is $L = (\hbar c/eH)^{1/2} \approx 80\mathring{A}$ for H =10T.

According to the above considerations, one can expect in such conditions a spontaneous creation of electron-hole pairs. This means for example that in a confined system one should always have some intrinsic electrons in the conduction band at T=0K, even if in the bulk material the conduction band is completely empty. This should also happen in a magnetic field. In fact, one should be able to observe an insulator-metal transition as a function of the field in the region $L \approx \lambda_Z/2$.

5 Conclusion

There exists a close analogy between the behavior of electrons in narrow-gap semiconductors and that of relativistic electrons in vacuum. The main difference between the two systems is the fact that in vacuum the velocity of light is also the maximum velocity of electrons, whereas in semiconductors the two velocities are different. In some cases it is much easier to observe relativistic effects in semiconductors than in vacuum. In particular, it is predicted that a confinement of electrons to dimensions smaller than the Zitterbewegung length $\lambda_Z = \hbar/m_0^* u$ should lead to an observable creation of electron-hole pairs.

All the above effects can be made even more spectacular by the use of minibands in superlattices.

Acknowledgments

I am grateful to Dr P.Pfeffer for elucidating discussions and help in the preparation of manuscript. This work was supported in part by the Polish Committee for Scientific Research under grant No. 2P03B13911.

References

1. J.M.Luttinger and W.Kohn, *Phys. Rev.* **97**, 869 (1955).
2. J.Zak and W.Zawadzki, *Phys. Rev.* **154**, 536 (1966).
3. E.O.Kane, *J.Phys. Chem. Solids* **1**, 249 (1957).

4. W.Zawadzki, *Adv. Physics* **23**, 435 (1974).
5. W.Zawadzki, in *Optical Properties of Solids*, edited by E.D.Haidemenakis, (Gordon and Breach 1970), p. 179.
6. P.Pfeffer and W.Zawadzki, *Phys. Rev.* B **41**, 1561 (1990).
7. G.H.Parker and C.A.Mead, *Phys. Rev. Lett.* **21**, 605 (1968).
8. W.Zawadzki, in *Narrow Gap Semiconductors. Physics and Applications*, edited by W.Zawadzki, Lecture Notes in Physics, Vol. **133**, (Springer 1980), p. 85.
9. R.Bowers and Y.Yafet, *Phys. Rev.* **115**, 1165 (1959).
10. P.Kacman and W.Zawadzki, *Phys. Stat. Sol.*(b) **47**, 629 (1971).
11. W.Zawadzki, *Phys. Letters* **4**, 190 (1963).
12. B.D.McCombe, R.J.Wagner, and G.A.Prinz, *Sol. State Commun.* **8**, 1687 (1970).
13. I.I.Rabii, *Z. Physik* **49**, 7 (1928).
14. W.Zawadzki, *Phys. Rev.* D **3**, 1728 (1971).
15. A.G.Aronov, *Sov. Phys.-Solid State* **5**, 402 (1963).
16. Q.H.Vrehen, W.Zawadzki, and M.Reine, *Phys. Rev.* **158**, 702 (1967).
17. M.Reine, Q.H.F.Vrehen, and B.Lax, *Phys. Rev.* **163**, 726 (1967).
18. W.Zawadzki, S.Klahn, and U.Merkt, *Phys. Rev. Lett.* **55**, 983 (1985).
19. H.E.Ives and G.R.Stilwell, *J. Opt. Soc. Am.* **28**, 215 (1938) ; G.Otting, *Phys. Z.* **40**, 681 (1939).
20. E.Schroedinger, *Sitzber. Preuss. Akad. Wiss. Physik-Math.* **24**, 418 (1930).
21. L.L.Foldy and S.A.Wouthuysen, *Phys. Rev.* **78**, 29 (1950).
22. W.Greiner, *Relativistic Quantum Mechanics* (Springer 1987), p. 59.
23. M.E.Rose, *Relativistic Electron Theory* (Wiley, New York 1961), p. 90.
24. T.D.Newton and E.P.Wigner, *Rev. Mod. Phys.* **21**, 400 (1949).

MAGNETIC LEVELS IN QUANTUM DOTS: SINGLE AND MULTI-PARTICLE EFFECTS

Ph. LELONG, G. BASTARD
Laboratoire de Physique de la Matière Condensée ENS
24, rue Lhomond F-75005, Paris, France

We report on perturbative and variational calculations of the binding energies of excitonic complexes under magnetic field in $In_xGa_{(1-x)}As$ / GaAs quantum dots. We analyse two kinds of systems, self-assembled boxes modeled by cones or hemispherical caps floating on a 2D plane, or a cylindrical box fabricated by lithographic etching.

One way to study the properties of quantum dots is to consider the effects of a magnetic field on their eigenstates. Actually, experiments [1] or theoretical studies [2-4] have already been undertaken. Our aim is to take into account the effects of a magnetic field on the single and multi-particle states for various shapes of quantum boxes [5]. We have neglected stress and piezo-electric field effects [6] in our calculations and taken position independent effective masses and dielectric constant.

Single particle in a magnetic field

For any kind of dot with cylindrical symmetry the single particle Hamiltonian in the presence of a magnetic field can be written :

$$H(B) = \frac{P_z^2}{2m_\perp} + \frac{P_\rho^2}{2m_\parallel} + \frac{m_\parallel}{2}\left(\frac{\omega_c}{2}\right)^2 \rho^2 - \frac{q}{|q|}\frac{\hbar\omega_c}{2} n + V_{conf}(z,\rho) \qquad (1)$$

where $\hbar n$ is the eigenvalue of the angular momentum L_z, $\omega_c = \frac{eB}{m_\parallel}$ is the in-plane cyclotron frequency, q the charge of the particle. The corresponding eigenstates are calculated by diagonalization of H in the basis of Fourier-Bessel functions which are the exact free particle solutions for a large cylinder (height Z, radius R): the wave function $\Psi^{kn}(\vec{r})$ is given by the following decomposition

$$\Psi^{kn}(\vec{r}) = \langle \vec{r} \mid k, n \rangle = e^{in\theta} \sum_{k(i,j)=1}^{\infty} \eta_{i,j}^n \sin\left(\frac{\pi j z}{Z}\right) J_n\left(\frac{\lambda_i^n \rho}{R}\right) \qquad (2)$$

In eq. 2, the k index of the wave function classification is determined by the eigenenergy relation $(\varepsilon_{n,k} \leqslant \varepsilon_{n,k+1})$. In the case of a conical cylindrical quantum dot, the extended electronic levels of the underlying wetting layer are given by a 2D Landau regime with an energy spacing equal to $\hbar\omega_c$ at large B.

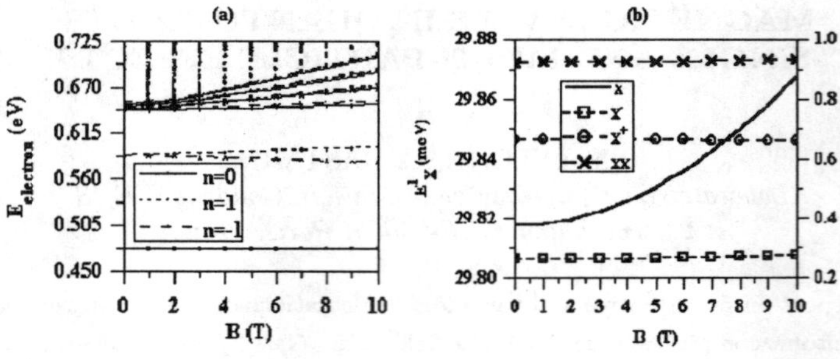

Figure 1: Electron level and excitonic complexes energies in a conical box

We find this regime (fig. 1. (a)) for the electronic spectrum, while a quadratic regime prevails at small B due to the finite radial size of the confining cylinder. The n dependency of ladders energies at large B corresponds to :

$$\varepsilon_n = \varepsilon^z_{j_l} + \hbar\omega_c(i_l + \tfrac{|n|-n+1}{2} + \tfrac{1}{2}) \tag{3}$$

with (i_l,j_l) is the reference of the separable states in the ideal limit. There is no change for n=1 and a step of $\hbar\omega_c$ for n=-1 respectively compared to n=0 ladders.

Complexes in self assembled boxes

The shape of self assembled dots is modelled by cones or hemispherical caps floating on a 2D wetting layer [5]. Moreover, the large separation between the state confined in the dots suggests to treat coulombic effects by perturbation theory. Typically, there are three levels confined in the dot, which are the ground states corresponding to the quantum number n=-1,0,1. The second order of perturbation theory applied to a biexcitonic complex leads to

$$\begin{aligned} E_{XX} &= 2(E_X+E_g)+\langle (00)_e;(00)_h|\,V_{ee}+4V_{eh}+V_{hh}\,|(00)_e;(00)_h\rangle \\ &\quad +\sum_{n\neq 0}\frac{\left|\langle (00)_e;(00)_h\left|4V_{eh}\right|(0n)_e;(0-n)_h\rangle\right|^2}{E_{e0}-E_{en}+E_{h0}-E_{hn}}+\frac{\left|\langle (00)_e;(00)_e\left|V_{ee}\right|(0n)_e;(0-n)_e\rangle\right|^2}{2(E_{e0}-E_{en})} \\ &\quad +\frac{\left|\langle (00)_h;(00)_h\left|V_{hh}\right|(0n)_h;(0-n)_h\rangle\right|^2}{2(E_{h0}-E_{hn})} \end{aligned} \tag{4}$$

$$E^l_{XX} = 2(E_X+E_g) - E_{XX} \tag{5}$$

E_X is the excitonic energy given by a similar expansion. For a conical box (InAs/GaAs, radius 8 nm, height 1.7 nm), we find very weak effects for the B variations of the excitonic complexes (fig. 1. (b)). The exciton binding energy increases by 0.1 meV at most, and the biexciton E^l_{XX} or charged exciton binding energies are practically magnetic field independent. In another kind of dots, with a hemispherical shape ($InAs_{0.3}GaAs_{0.7}/GaAs$, radius 10 nm, height 1.5 nm), only exciton (X) and charged - exciton (X^-) are found bound relative to single particle and exciton respectively (E^l_X = 14.4 meV and $E^l_{X^-}$ = 2.6 meV at B=0). The binding energy increases by 1.5 meV for (X) and 0.1 meV for (X^-) at B=10 T. This is because the one particle states are much less bound at B=0 than in the previous dot.

Exciton in a cylindrical quantum box

In this study, we are interested in the consequence of wire-dot-well transition for the 1S exciton [7]. The model simulates cylindrical dots obtained by lithography process on a quantum well structure (material $In_{0.1}Ga_{0.9}As/GaAs$, height L =5 nm). The calculation is based on a variational method using the "exact" electron and hole wave functions $\Psi^{00}_{e,h}(\vec{r}_{e,h}; B)$ under magnetic field given in eq. 2, corresponding to the energy $\varepsilon^0_{e,h}(B)$, i.e. we assume that the particle cannot leak out of the wire. The excitonic trial wave function is

$$\Phi_X(\vec{r}_e, \vec{r}_h; \mu, B) = \Psi^{00}_e(\vec{r}_e; B)\Psi^{00}_h(\vec{r}_h; B)e^{-\frac{\mu}{2}\|\vec{r}_e - \vec{r}_h\|} \tag{6}$$

The confinement potential is given by $V_{conf} = V_{InGaAs/GaAs}\ \theta\left(z^2 - (L/2)^2\right)$ $\theta(R - \rho)$. We obtain for the functional of the excitonic energy :

$$\begin{aligned} E_X(B,\mu) &= \varepsilon^0_e(B) + \varepsilon^0_h(B) + E_g + \frac{\hbar^2}{2\eta}\left(\frac{\mu}{2}\right)^2 + \frac{e^2}{\kappa}\frac{F_\mu(B)}{\partial_\mu F_\mu(B)} \qquad (7) \\ F_\mu(B) &= \iint_{S_e S_h} d^3\vec{r}_e d^3\vec{r}_h \frac{|\Phi_X(\vec{r}_e, \vec{r}_h; \mu, B)|^2}{\|\vec{r}_e - \vec{r}_h\|} \end{aligned}$$

η is the reduced mass of the complex, E_g is the gap energy. We can observe in (fig. 2. (a)) the decrease of the excitonic E_X and exciton binding E^l_X energies down to the quantum well limit when R increases. The E_X behaviour for small radius is dominated by the electron and hole energies :

$$E_X(0,\mu^*) \underset{R\to 0+}{\sim} \frac{1}{R^2} \tag{8}$$

$$E^l_X(0,\mu^*) \underset{R\to 0+}{\sim} \frac{1}{R} \tag{9}$$

This dependency in log-scale is linear with a slope equal to -2 (eq. 8, fig. 2. (a)). For the asymptotic study of the exciton binding energy E^l_X, we have taken delta functions for the probability densities along the z direction and found a $\frac{1}{R}$ behaviour (eq. 9, fig. 2. (a)).

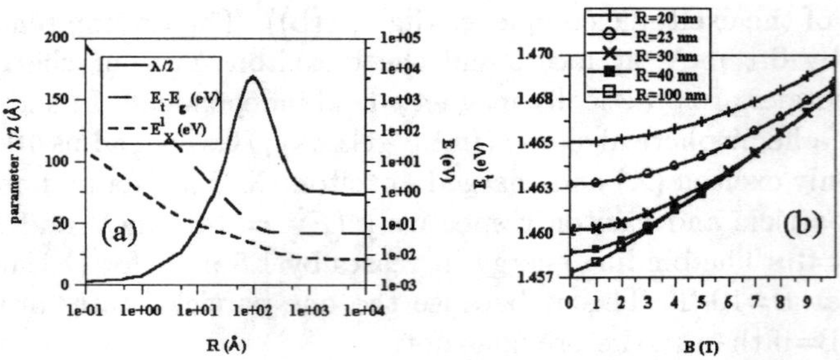

Figure 2: Excitonic energies and excitonic Bohr radius

The parameter $\lambda(=\frac{1}{\mu^*})$ (exciton Bohr radius) displays three kinds of variations. For small radius, the wire limit, λ decreases down to 0 since the excitonic effect grows indefinitely, eq. 9. In the second (dot) regime, the excitonic energy depends very weakly on the excitonic coupling term. The latter could be treated perturbatively. The third regime corresponds to wave functions which are increasingly delocalized in the xy-plane, and leads to an important excitonic coupling term (λ is small and constant): the structure is in fact equivalent to a quantum well in the limit of large R. Under magnetic effect (fig. 2. (b)), we remark that the excitonic energy tends to the quantum well limit for high magnetic fields. The convergence is the quicker for the larger R.

Acknowledgments

We gratefully acknowledge discussions with O. Heller and R. Ferreira. The Laboratoire de Physique de la Matière Condensée de l'Ecole Normale Supérieure is Unité Associée au CNRS (UA 1437) et aux Universités Paris 6 et Paris 7.

References

1. M. Fricke *et al*, to appear in *Phys. Rev.* B (1996)
2. A. Wojs, P. Hawrylak, *Phys. Rev.* B**51**, 10880 (1995)
3. A. Barenco & M.A. Dupertuis, *Phys. Rev.* B**52**, 2766 (1995)
4. A. Wojs, P. Hawrylak, S. Fafard & L. Jacak, submit to *Phys Rev* B
5. Ph. Lelong & G. Bastard, *Solid State Commun.* **98**, 819 (1996)
6. M. Grundmann *et al*, *Phys. Rev. Lett.* **74**, 4043 (1995)
7. A. Forchel, 1995, *Private Communication*

FAR-INFRARED RESPONSE AND COLLECTIVE EXCITATIONS OF A SQUARE LATTICE OF ANTIDOTS

S. A. MIKHAILOV
Institut für Theoretische Physik, Universität Regensburg,
D-93040 Regensburg, Germany

An analytic theory of electromagnetic response and collective excitations of a square lattice of antidots is developed. The response functions (the macroscopic conductivity, the absorption coefficient) and the excitation spectra are calculated as functions of a magnetic field and antidot lattice parameters. New collective modes of the system are predicted. Two-dimensional electron systems (2DES) both with and without the screening metal electrodes are considered. The theory quantitatively agrees with experimental data.

1 Introduction

FIR transmission experiments on antidot superlattices [1,2,3] demonstrate the behavior of excitation modes, which is quite different from that of dot systems and is not yet completely understood. We present here an analytic theory of electromagnetic response and collective excitations of antidot superlattices. Contrary to the theoretical papers published so far (see [4,5] and references in [5]) we take fully into account the real spatial symmetry of the structure and obtain explicit functional dependencies of the response functions and the excitation spectrum on lattice parameters. The theory correctly describes the behavior of all observed excitation modes and predicts the new ones.

2 Formulation of the problem and results

2.1 Local theory

Let the 2DES be placed at the plane $z = 0$ between two metal electrodes (at $z = d_1$, $z = -d_2$), the background dielectric constants equal to ϵ_1 at $0 < z < d_1$ and ϵ_2 at $-d_2 < z < 0$. In the local approach we assume that the conductivity of 2D electrons $\sigma_{\alpha\beta}(\omega)$ is proportional to an equilibrium electron density

$$n_0(\mathbf{r}) = n_0\{1 - \Theta(\mathbf{r})\}, \quad \Theta(\mathbf{r}) = \sum_{i,j} \vartheta(\mathbf{r} - \mathbf{a}_{i,j}), \tag{1}$$

where n_0 is the background density of a homogeneous 2DES and $\mathbf{a}_{i,j} = a(i,j)$, i, j integer, are the lattice vectors. The function $\vartheta(\mathbf{r}) \equiv \vartheta(r)$ describes the density profile near an antidot; it changes from $\vartheta(r) = 1$ inside of the antidots to $\vartheta(r) = 0$ outside of the antidots on a distance small compared to both a and R. An external ac electric field $E^0_\pm$ with + and − circular polarization is assumed to be uniform and parallel to the plane $z = 0$.

Solving the system of Poisson, continuity equations, and the local Ohm's law we obtain the macroscopic conductivity $\sigma^{\rm macro}_\pm(\omega) = \sigma_\pm(\omega)\{1-\langle\vartheta\rangle/\zeta_\pm(\omega)\}$ and the absorption coefficients $\gamma_\pm(\omega)$ of the structure,

$$\gamma_\pm(\omega) = {\rm Re}\sigma_\pm - \frac{\langle\vartheta\rangle{\rm Re}\sigma_\pm}{|\zeta_\pm(\omega)|^2} + \frac{{\rm Re}\sigma_{xx}}{2}\left|\frac{2\pi\langle\vartheta\rangle\sigma_\pm}{\omega\zeta_\pm(\omega)}\right|^2 \sum_{\mathbf{G}\neq\mathbf{0}} \frac{G^2\alpha(G)}{|\epsilon(G,\omega)|^2}. \qquad (2)$$

Here $\sigma_\pm(\omega) = \sigma_{xx} \pm i\sigma_{xy}$, $\langle\vartheta\rangle$ is the area filling factor (the angular brackets mean the average over the area of an elementary cell), $\mathbf{G} = \mathbf{G}_{m,n} = (2\pi/a)(m,n)$ are reciprocal lattice vectors, the function

$$\zeta_\pm(\omega, B) = 1 - \frac{\pi i\langle\vartheta\rangle\sigma_\pm(\omega)}{\omega} \sum_{\mathbf{G}\neq\mathbf{0}} \frac{G\alpha(G)}{\epsilon(G,\omega)} \qquad (3)$$

has the meaning of the response function of one antidot in the lattice, $\alpha(G) = |\langle\vartheta e^{i\mathbf{Gr}}\rangle/\langle\vartheta\rangle|^2$ is a form-factor determined by the Fourier components of equilibrium electron density, the dielectric function of the system "metal - dielectric - uniform 2DES with the density n_0 - dielectric - metal",

$$\epsilon(G,\omega) = [\epsilon_1 \coth(Gd_1) + \epsilon_2 \coth(Gd_2)]/2 + 2\pi i\sigma_{xx}(G,\omega)G/\omega, \qquad (4)$$

describes the screening of the charge density fluctuations in the system.

Figure 1a demonstrates the frequency dependence of the absorption coefficients $\gamma_+(\omega)$ (solid) and $\gamma_-(\omega)$ (dashed curves) at different magnetic fields, for $d_1 = d_2 = \infty$, $\langle\vartheta\rangle = 0.5$ and $\omega_0\tau = 20$, where $\omega_0 \equiv \omega_p(G_{1,0})$ is the frequency of 2D bulk plasmon with the wavevector $q = G_{1,0}$, and τ is the momentum relaxation time. In addition to two main resonances which were found experimentally one can see the weak − -polarized mode at $\omega \simeq \omega_0$ and a set of high-frequency + and − -polarized modes at $\omega/\omega_0 > 1$ with smaller oscillator strengths. The position of resonances (Fig. 1b, where $\langle\vartheta\rangle = 0.25$) is determined by the poles of $\sigma_\pm(\omega)$ (the ***single particle*** contribution $\omega = \omega_c$) and the zeros of $\zeta_\pm(\omega)$ (*collective* modes). At $\omega < \omega_c$ we find the + -polarized edge magnetoplasmon (EMP) mode. At $\omega > \omega_c$ the functions $\zeta_\pm(\omega)$ have an infinite set of poles at the frequencies of 2D bulk magnetoplasmons $\omega_{\rm mp}(\mathbf{G}_{m,n}, B)$ (dotted curves) and an infinite set of associated zeros - the modes $\omega^{(\pm)}_{m,n}$ [solid (dashed) curve for + (−) -polarization]. In Fig. 2 the results are compared with experimental data from Refs. [3] (a) and [1] (b). Fig. 2c shows the magnetic field dependence of oscillator strengths $S^{(\pm)}_{m,n}(B)$ of the modes $\omega^{(\pm)}_{m,n}$. The dependence of zero-B oscillator strengths $S_{m,n}(0)$ on $\langle\vartheta\rangle$ is shown in the inset.

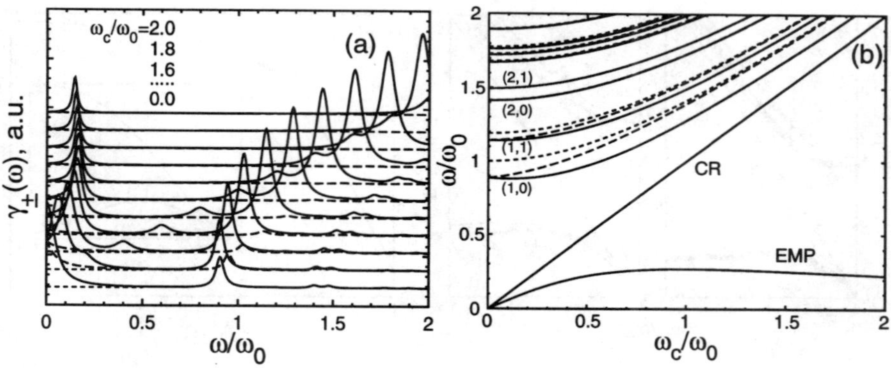

Figure 1: The absorption (a) and excitation (b) spectra of the square lattice of antidots.

2.2 Nonlocal effects

As seen from Fig. 2a,b, the local theory quantitatively explains the observed excitation spectra of antidots, except the anticrossing at $\omega_{1,0}^{(+)}(B) \simeq 2\omega_c$, Fig. 2b. To describe this anticrossing, we take into account nonlocal corrections [6] to the conductivity $\sigma_{xx}(G,\omega)$ of the uniform 2DES in Eq. (4). The dispersion equation near the intersection point $\omega_{1,0}^{(+)} = 2\omega_c$ then assumes the form

$$(\omega - \omega_{1,0}^{(+)})(\omega - 2\omega_c) = \Gamma^2, \qquad \Gamma = \sqrt{3}\pi\omega_p(G_{1,0})r_c/4a, \tag{5}$$

and describes the mode splitting ($r_c \ll a$, r_c is the cyclotron radius).

2.3 Quantum oscillations

The quantum oscillations of the EMP frequency at large B, observed in [3] (Fig. 2a), can be qualitatively understood as follows. As seen from (2), (3), all observable values are functionals of the equilibrium electron density $\vartheta(r)$ in antidots. Effects of nonlinear screening [7] result in the B-dependent oscillations of $\vartheta(r,B)$ and hence $\alpha(G,B)$. This should result in the B-dependent oscillations of the frequency and the linewidth of *all* collective modes in the system.

3 Conclusions

The full dipole excitation spectrum of antidots consists of three types of modes: the single particle CR, the EMP mode, and an infinite set of high-frequency

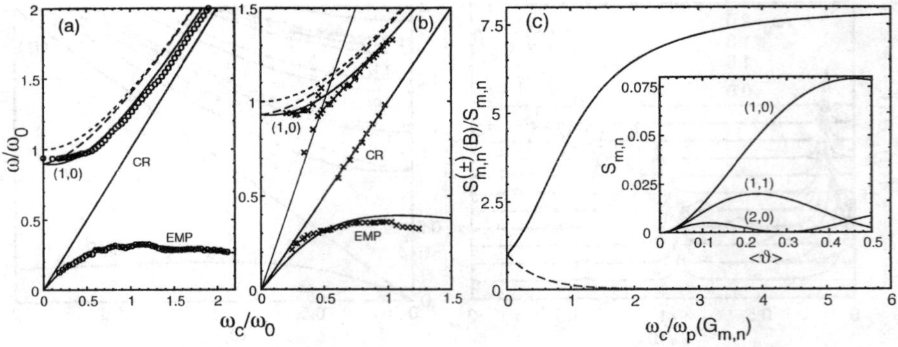

Figure 2: Comparison with the data from[3] [(a), $\langle\vartheta\rangle = 0.16$] and[1] [(b), $\langle\vartheta\rangle = 0.087$]. c) The oscillator strengths of the modes $\omega_{m,n}^{(+)}$ (solid) and $ga_{m,n}^{(-)}$ (dashed curve) vs B and $\langle\vartheta\rangle$.

modes $\omega_{m,n}^{(\pm)}(B)$. The modes $\omega_{m,n}^{(\pm)}$ are excited in the system due to the antidot lattice, which can be considered as a grating coupler imposed on the uniform 2DES. Our results quantitatively agrees with experimental data obtained in different semiconductor structures with different parameters.

Acknowledgments

The work was supported by the Alexander von Humboldt Foundation and the NATO Science Program.

References

1. K. Kern, D. Heitmann, P. Grambow, Y. H. Zhang, and K. Ploog, Phys. Rev. Lett. **66**, 1618 (1991).
2. Y. Zhao, D. C. Tsui, M. Santos, M. Shayegan, R. A. Ghanbari, D. A. Antoniadis, and H. I. Smith, Appl. Phys. Lett. **60**, 1510 (1992).
3. K. Bollweg, T. Kurth, D. Heitmann, V. Gudmundsson, E. Vasiliadou, P. Grambow, and K. Eberl, Phys. Rev. Lett. **76**, 2774 (1996).
4. G. Y. Wu and Y. Zhao, Phys. Rev. Lett. **71**, 2114 (1993).
5. S. A. Mikhailov and V. A. Volkov, Phys. Low-Dim. Struct. **1**, 31 (1994); Phys. Rev. B **52**, 17260 (1995).
6. K. W. Chiu and J. J. Quinn, Phys. Rev. **B 9**, 4724 (1974).
7. V. G. Burnett, A. L. Efros, and F. G. Pikus, Phys. Rev. **B 48**, 14365 (1993).

HIGH FIELD MILLIMETER WAVE SPECTROSCOPY

M.E.J. BOONMAN, J.A.A.J. PERENBOOM, P.J.M. VAN BENTUM,
A. WITTLIN, J.C. MAAN
Research Institute for Materials & High Field Magnet Laboratory, University of Nijmegen, Toernooiveld, 6525 ED Nijmegen, the Netherlands

W. MAC, A. TWARDOWSKI
Institute of Experimental Physics, Warsaw University, Hoża 69, PL-00-681 Warsaw, Poland

S.O. HILL
National High Magnetic Field Laboratory, Florida State University, 1800 E. Paul Dirac Drive, Tallahassee Fl 32310, U.S.A.

R. KATO, H. SAWA, S AONUMA
Institute for Solid State Physics, University of Tokyo, 7-22-1 Roppongi, Minato-ku, Tokyo 106, Japan

The combination of high magnetic fields and a tunable millimeter wave source with phase sensitive heterodyne detection enables us to perform high field millimeter wave spectroscopy in the energy range of 1 to 7 cm^{-1} , making many interesting fundamental low energy excitations like EPR, plasmons, magnons, spin density waves, etc., experimentally accessible. High field EPR measurements of Jahn-Teller distorted Cr^{2+} centers in ZnTe and a cyclotron resonance experiment on an organic conductor $(DMe\text{-}DCNQI)_2\text{-}Cu$ with a high effective mass will demonstrate the potential of this instrumental technique.

1 Introduction

Far infrared magneto-optics and electron paramagnetic resonance techniques have proven to be very successful methods to study excitations in solids. With a recently developed new type of millimeter wave vector network analyzer [1] (MVNA) it has become possible to conveniently perform phase sensitive spectroscopy on solids in the frequency range between 30 and 210 GHz. [a] The amplitude and phase signals allow to observe directly the absorptive and dispersive part of a resonance, without using any bridge technique. The energy range from 1 to 7 cm^{-1} bridges the gap between the conventional EPR and far infrared spectroscopy. Therefore this system is especially useful to perform high field EPR experiments ($\omega_L = g\mu_B B/\hbar$; g=2 implies B=3.57 T for f=100 GHz) and cyclotron resonance experiments involving charge carriers with a high effective mass ($\omega_c = eB/m^*$; m^*=1 implies B=3.57 T for f=100 GHz).

[a] f=30 GHz $\Rightarrow k$=1 $cm^{-1} \Rightarrow E$=0.12 meV $\Rightarrow \lambda$ =10 mm

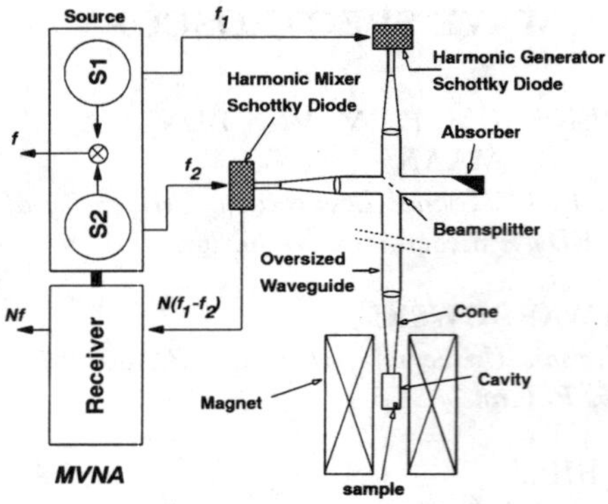

Figure 1: Schematic overview of the setup used for millimeter wave spectroscopy

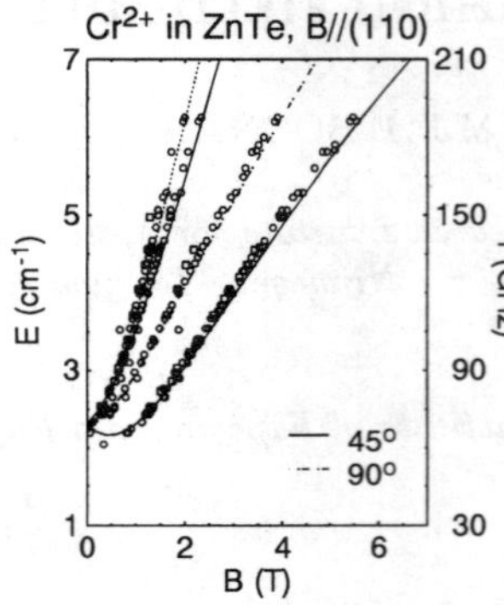

Figure 2: Cr^{2+} Transition energy versus magnetic field for $B \parallel (110)$ axis. (○, □ = data from 2 samples, lines = calculations)

2 Experimental Technique

The MVNA is a solid state based vector analyzer measuring the amplitude and phase of a signal using heterodyne detection. The continuously variable YiG frequency between 8 and 18 GHz is guided to frequency multiplying Schottky diodes via coaxial cables allowing a large flexibility of the system, see figure 1. The generated millimeter wave radiation is quasi-optically guided with oversized waveguides (light pipes) to the magnetic field center, where the vertical waveguide section is terminated by an oversized resonant cavity. With a mylar sheet the detector in the horizontal section is coupled to the vertical section. Since we use an oversized copper cavity (l=20 mm, d=12.5 mm, $f_{\mathrm{TE110}} \approx$ 37 GHz) and have the advantage of tunability of the millimeter wave source, which allows to tune the frequency in resonance with one of the many resonant cavity modes, we are not restricted in working frequencies. This frequency tunability is important for studying excitations with a zero field offset in the energy versus field diagram or with strong non-linear behaviour in which case a measurement at a single frequency is not conclusive, see for example figure 2. The observed signal consists of a resonant part and a non-resonant background signal (complex leak signal). If we fit an observed resonance with a circle in the polar plane, not defining any lineshape, we obtain the value of this complex leak signal, which we can subtract from the measured signal, and also find a complex normalization factor. Assuming the leak signal is constant over the whole measurement, we find after amplitude and phase transformation two signals corresponding to the absorption and dispersion of the resonance. If we use a small sample compared to the cavity volume and assume the fields to penetrate throughout the whole sample [2], the transformed amplitude signal

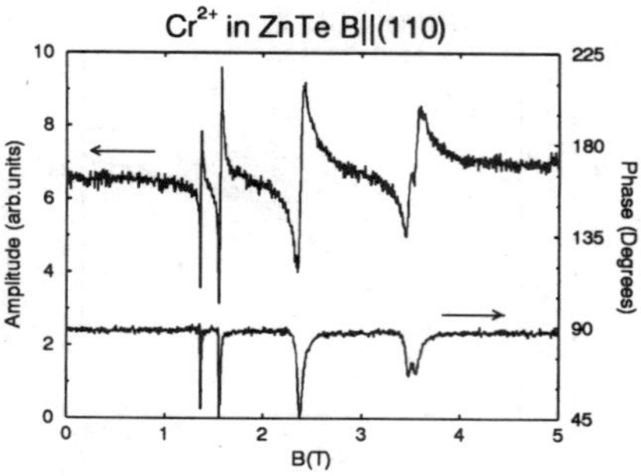

Figure 3: Measured complex EPR signal at f=137.875 GHz, T=1.2 K and $B \parallel (110)$

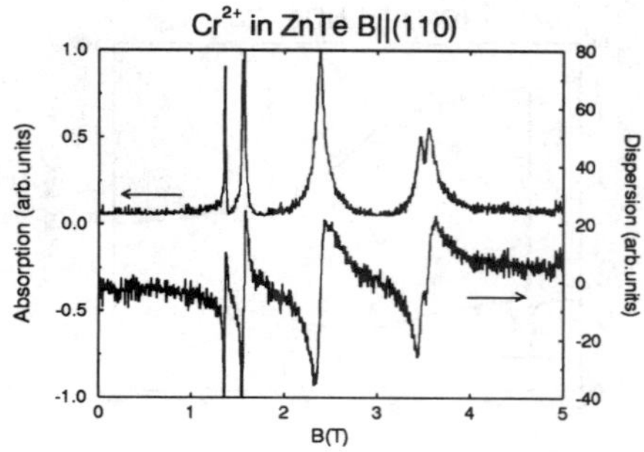

Figure 4: Transformed EPR signal after leak subtraction and normalization

will be proportional to ϵ_2 and the transformed phase signal to ϵ_1.

3 Diluted Magnetic Semiconductors

High magnetic field millimeter wave EPR experiments have been performed on diluted magnetic semiconductors (DMS) with chromium [3]. The Cr^{2+} center in a tetrahedral crystal field undergoes a static Jahn-Teller distortion along one of the (100) axes, which results in a strong anisotropy of the Cr^{2+} ion energy level structure. We studied the low energy levels of the Cr^{2+} ion in a ZnTe host lattice in magnetic fields oriented along the (110) axis. Figure 2 displays the energy level diagram of two orientationally different Cr^{2+} centers observed, showing a zero field offset of 2.2 cm^{-1} and strong non-linearity. The measurements at fixed frequency show resonances corresponding to low energy level transitions of the Cr^{2+} ion, see figure 3. The fact that the last resonance in figures 3 and 4 is a double resonance, due to a slight misalignment of the magnetic field along the (110) axis and corresponding to two Cr^{2+} levels distorted at 45°, demonstrates the existence of three types of Jahn-Teller distorted Cr centers. Fitting one of the four resonances with a circle in the polar plane and correcting the measured data by subtraction of the leak signal and complex normalization, we find the absorption and dispersion curves of these resonances, see figure 4.

4 Organic Conductors

Cyclotron resonance experiments have been performed on the novel molecular conductor $(DMe\text{-}DCNQI)_2$-Cu $d_2[1,1;0]$. The $p\pi$ orbitals of the conducting stack of (DMe-DCNQI) molecules overlap with the 3d orbitals of the interconnecting Cu ions. This orbital overlap results in complicated characteristic 1D and 3D Fermi surfaces. Applying a magnetic field along the crystallographic c-axis, parallel to the (DMe-DCNQI) stacks, one particular cyclotron orbit of

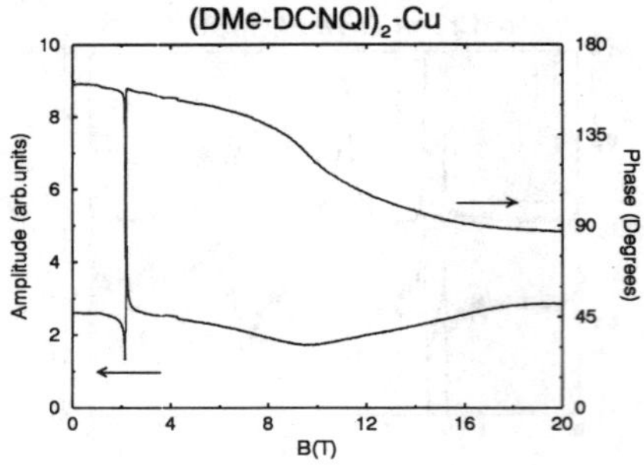

Figure 5: Measured complex CR (and EPR) signal at f=78.59 GHz and T=1.2 K

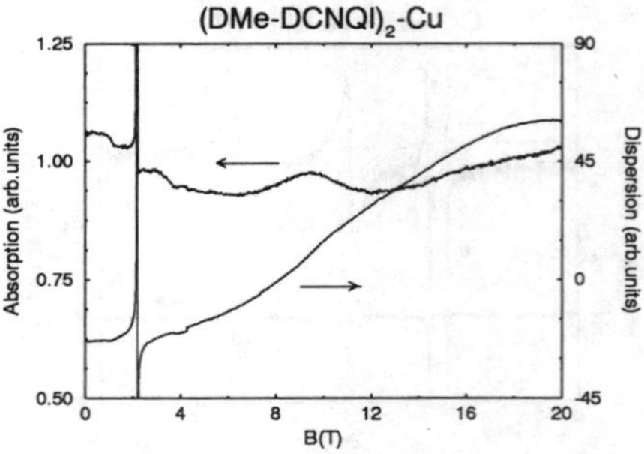

Figure 6: Corrected CR signal after leak subtraction and normalization

the Fermi surface can be probed with a dHvA effective mass of $m^* = 3.5\ m_e$ [4]. After subtraction of the background signal and normalization our measurements clearly show a very broad cyclotron resonance corresponding with a CR effective carrier mass of $m^* \approx 3.4\ m_e$, see figure 6. The mobility of the carriers can be estimated from the width to be $\mu \approx 10^3$ cm^2/Vs. At lower fields a very strong EPR signal is observed, see figure 5 with g=2.32 and a zero field splitting of 11 GHz. This corresponds to Cu^{2+} ions which are not contributing to the inter (DMe-DCNQI) column conductivity. The low mobility and the presence of Cu^{2+} ions can possibly be explained by a hopping type of conductivity between the (DMe-DCNQI) molecules and the Cu^{2+} ions.

Acknowledgments

This work is part of the research program of the "Stichting voor Fundamenteel Onderzoek der Materie (FOM)", which is financially supported by the "Nederlandse Organisatie voor Wetenschappelijk Onderzoek (NWO)", and has been supported by the European Commission under the contract no. CIPD-CT94-0767 and CHGT-CT93-0051.

References

1. P. Goy, M. Gross and J.M. Raimond, in: *Proc.15th Int.Conf.on IR and mm Waves, Orlando, Florida (1990)*, Ed. R.J. Temkin (Plenum Press, New York, London, 1990), p. 172
2. O. Klein, S. Donovan, M. Dressel and G. Grüner, *Int. J. Infrared Millimeter Waves* **14**, 2423 (1993)
3. M.E.J. Boonman, W. Mac, A. Twardowski, A. Wittlin and M. Demianiuk, *Acta Physica Polonica A* **88**, 829 (1995)
4. S. Uji, T. Terashima, H. Aoki, J.S. Brooks, R. Kato, H. Sawa, S. Aonuma, M. Tamura and M. Kinoshita, *Phys. Rev. B* **50**, 15597 (1994)

FAR-IR SPECTROSCOPY OF DOUBLE LAYER TWO-DIMENSIONAL ELECTRON SYSTEMS IN HIGH MAGNETIC FIELDS

YU.B.VASILYEV
Ioffe Physical Technical Institute, 194021 St.Petersburg, Russia and Max-Planck-Institut für Festkörperforschung, D-70569, Stuttgart, Germany

D.BERTRAM, M.DILGER, K.v.KLITZING, and K.EBERL
Max-Planck-Institut für Festkörperforschung,
Heisenbergstraße 1, D-70569 Stuttgart, Germany

We present cyclotron resonance (CR) experiments on GaAs-AlGaAs coupled double layer electron systems. A nonparabolicity-induced CR splitting from the two electron layers in the individual wells is analysed as a function of the carrier densities in the wells and the strength and direction of the magnetic field. It is found that the splitting disappears with the increasing coupling between the wells which takes place at strong magnetic fields or in the presence of a small parallel component of magnetic field.

1 Introduction

There has been recent considerable interest in the physics of two parallel high mobility two-dimensional electron gases (2DEG's). For such a system, a variety of new effects are expected to occur which are not accessible in a single 2DEG.[1] Among them is a subtle physics of electron-electron interaction which dominates at high magnetic fields when the electron spectrum is completely quantized. Until now, different interaction phenomena in double quantum well (DQW) systems were studied, primarily via transport[2] or capacitive[3] experiments, and there have been very few far-infrared transmission studies of these systems.[4,5] We present preliminary results of far-infrared magneto-optical spectroscopy on DQW structures. Measurements both in perpendicular and tilted magnetic fields (B) reveal a number of novel features. Here we concentrate on cyclotron resonance (CR) splitting behaviour in an asymmetric DQW sample.

2 Experiment

The DQW sample was grown by molecular-beam epitaxy and consists of two GaAs quantum wells with the upper well (top layer) width of 30nm and lower well (back layer) of 25nm separated by an 5nm $Al_{0.33}Ga_{0.67}As$ barrier. The electron density in the upper well is $1.2\text{x}10^{11} cm^{-2}$ and in the lower one is

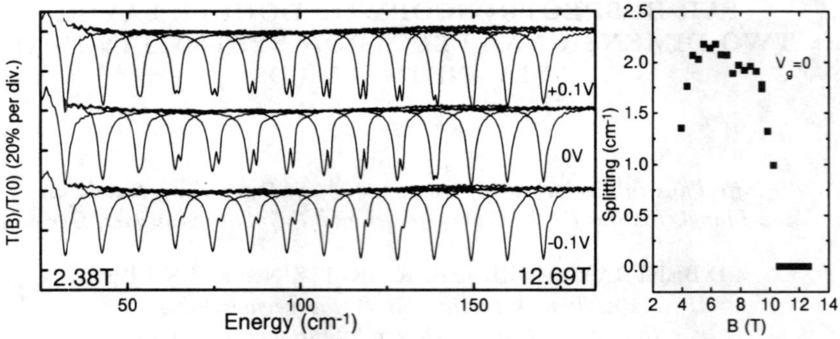

Figure 1: A typical series of CR spectra from 2.38 T to 12.69 T for three voltages V_g. The CR splitting energy as a function of magnetic field for V_g=0.

$4.2x10^{11} cm^{-2}$ as defined from Shubnikov-de Haas oscillations. Additionally a front gate was evaporated that allows the variation of the electron density in the upper well. CR measurements were performed using Fourier transform spectroscopy with a resolution of 0.25 cm^{-1} at different magnetic fields with tilt angles up to 45°. The sample transmission was always normalized to a reference spectrum taken at B=0T. All experiments are done at a device temperature of T=2.2 K.

3 Results and Discussions

First we turn to our data in magnetic fields directed perpendicular to the plane of the structure. A representative series of CR traces is shown in fig.1 (left plot) as a function of wave number for different magnetic fields and gate voltages. At low (B $<$ 4 T) and high (B $>$ 10 T) magnetic fields only one CR peak is seen. The position of the peak linearly depends on magnetic field with the corresponding effective mass of $0.0695 m_o$. In a certain interval of both magnetic field and front gate voltage an additional CR peak is observed on the high-energy side of the main peak. The intensity of this additional peak increases proportionally with the front gate voltage from V_g=-0.3V when it appears to V_g=0.3V when it merges with the main peak. The splitting between these two peaks accordingly decreases smoothly from 4.3 cm^{-1} at V_g=-0.3V to 0 at V_g=0.3V. The appearance of the second peak is attributed to the presence of two electron layers. Most of the observed characteristics of the CR splitting are in agreement with what is expected for absorption in

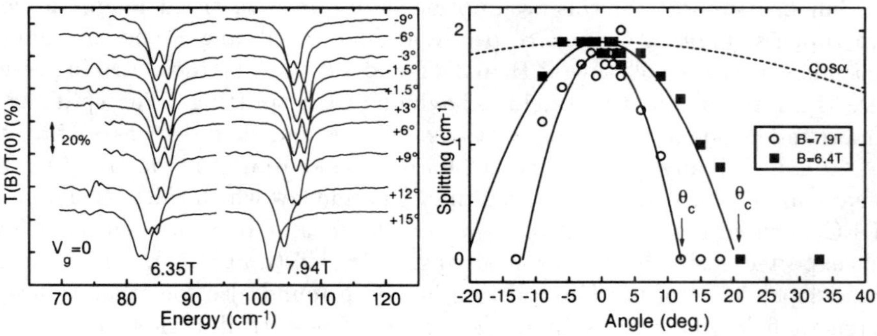

Figure 2: A series of experimental CR traces taken at V_g=0 at different tilt angles around 0° for two magnetic fields 6.35 T and 7.94 T. CR splitting as a function of tilt angle. Cosine dependence is also shown (dashed line).

two independent 2D layers. Indeed, the asymmetry of the structure results in different subband energies for the two wells so that no resonant coupling exists. In this case the electron wave functions are localized in the individual wells and subband densities are equal to the electron density in the wells. Because of the different electron densities the CR peak, arising from the larger carrier density in the lower layer, has a smaller energy in accordance with nonparabolicity of the conduction band. It is found that the amplitude of the second peak increases with positive gate voltage and the splitting decreases as expected for a decoupled DQW structure.

The most striking result here is the dependence of the CR splitting on the magnetic field which is shown in fig.1 (right plot) for V_g=0. Instead of the expected monotonic behaviour as a function of magnetic field, the splitting disappears abruptly at magnetic field larger than the critical field of B≈10 T. Similar dependences of the CR splitting were observed for other gate voltages. The disappearance of the additional peak at high magnetic fields does not mean that the upper well is completely depleted. We believe that the splitting behaviour reflects the occurrence of different regimes of Landau level coupling. Appearance and disappearance of the additional CR peak corresponds to regimes of strong and weak interlayer coupling, respectively. Indeed, it is known that interlayer electron-electron interactions dominate over kinetic effects at high magnetic fields. Thus the collapse of the CR splitting is the result of a transition from independent CR absorption in individual layers to a single mode dominated by Coulomb interaction at high magnetic fields.

Further support for this assumption is obtained in tilted magnetic field experiments. Figure 2 shows CR spectra at different tilt angles and the angular dependences of the splitting at B=6.35 T and 7.94 T. The most striking result here is an anomalous tilted field behaviour of the splitting. The splitting is almost independent on the angle at small angles and then decreases abruptly to 0 at a critical angle θ_c (around 20° for 6.4 T and around 11° for 7.9 T). This behaviour is inconsistent with the ordinary cosine law which rules behaviour of 2DEG in tilted magnetic fields. However, the mean CR peak position follows this expected cosine behaviour at all angles studied (up to 45°). Such unusual angle dependence of the CR splitting has been found also for other values of magnetic field where the splitting is present. The critical angle θ_c is found to decrease roughly linearly with the magnetic field. The parallel field component, which is small at small angles, is found to reduce the splitting but does not change the electron densities in the wells, i.e. it increases the electron-electron interaction between the layers, resulting in hybridisation. Thus the parallel magnetic field transfers the system into a new phase where states from different layers are coupled. For the case of weakly coupled DQW systems no anomalous behaviour in angle dependences has been found.

Conclusion

In conclusion, we have carried out FIR Fourier-transform magnetospectroscopy on DQW structures in magnetic fields up to 13 T. Two regimes of strong and weak interlayer coupling have been observed on asymmetric DQW samples. We have also found that a small parallel component of the magnetic field drastically suppresses the weak coupling regime.

Acknowledgments

We would like to thank M.Riek for the help with the sample preparation. Yu.Vasilyev thanks the MPI for financial support and the hospitality during his stay in Stuttgart.

References

1. J.P.Eisenstein, *Superlatt. Microstruct.* **12**, 107 (1992)
2. X.Ying *et al*, *Phys. Rev.* B **52**, 11611 (1995).
3. J.P.Eisenstein *et al*, *Phys. Rev.* B **50**, 1760 (1994).
4. K.Ensslin *et al*, *Phys. Rev.* B **39**, 11179 (1989).
5. A.Lorke *et al*, *Phys. Rev.* B **42**, 1321 (1990).

Resonant Four-Wave Mixing of Near-Infrared and Far-Infrared Radiation in GaAs Quantum Wells at High Magnetic Fields

J. KONO,* T. INOSHITA, H. SAKAKI
Quantum Transition Project, Research Development Corporation of Japan, and Research Center for Advanced Science and Technology, University of Tokyo, Tokyo 153, Japan

J. CERNE, M.S. SHERWIN
Center for Free-Electron Laser Studies, University of California, Santa Barbara, CA 93106, U.S.A.

M. SUNDARAM,# A.C. GOSSARD
Materials Department, University of California, Santa Barbara, CA 93106, U.S.A.

(Received 1 July 1996)

We have carried out experimental and theoretical studies of the nonlinear optical response of GaAs/AlGaAs quantum wells in high magnetic fields to combined near-infrared (NIR) and far-infrared (FIR) radiation. Bright emission lines, or *sidebands*, occur at frequencies $\omega_{sideband} = \omega_{NIR} \pm 2\omega_{FIR}$. The sideband emission intensity exhibits resonances as the magnetic field or ω_{NIR} is tuned, leading to a novel observation of *non-linear* optically-detected resonances. A theoretical model of the results, based on a four-wave mixing process involving 1 NIR photon and 2 FIR photons, exhibits rich structure and predicts a variety of interesting phenomena arising from the non-linear dynamics of terahertz-driven carriers in quantizing high magnetic fields.

1 Introduction

Recently, the simultaneous use of far-infrared (FIR) and near-infrared (NIR) / visible radiation has successfully led to optical detection of FIR resonances (ODR) in many semiconductor systems [1]. This technique has made it possible to observe *internal transitions* of confined hydrogenic excitons in quantum wells (QW's) [2]. However, experiments with intense FIR radiation to probe optical non-linearities have not been well explored. Here we present results of a non-linear optical experiment employing a two-beam set-up similar to ODR. Undoped GaAs/AlGaAs QW's mediate four-wave mixing (FWM) of FIR and NIR radiation, resulting in bright and narrow emission lines, or *sidebands*.

2 Samples and Experimental Methods

We used undoped GaAs/$Al_{0.3}Ga_{0.7}As$ (10nm/15nm×50) multiple QW's. NIR radiation from a Ti:Sapphire laser excited the QW's, which were simultaneously driven by intense FIR radiation from the UCSB Free-Electron Lasers (FEL's). The FEL's are tunable between 4 and 160 cm^{-1} and can attain peak powers of order 1 MW/cm^2. Magnetic fields up to 9 T were applied in the growth direction, parallel to both laser beams. While the FIR radiation was linearly polarized, the NIR radiation was unpolarized.

3 Experimental Results

The sidebands were observed at frequencies $\omega_{sideband} = \omega_{NIR} \pm 2\omega_{FIR}$, where ω_{NIR} and ω_{FIR} are the frequency of the NIR and FIR radiation respectively. Figure 1(a) shows emission spectra for the QW's when pumped with NIR radiation with a frequency of 12736 cm^{-1} in the absence (dashed line) and the presence (solid line) of FIR radiation with a frequency of 84 cm^{-1}. Within the heavy-hole (HH) exciton PL line at 12560 cm^{-1}, a strikingly narrow (width < 2 cm^{-1}) line is seen at exactly 12568 cm^{-1} (= 12736 - 2 × 84 cm^{-1}) only in the presence of FIR radiation.

The sideband intensity was resonantly enhanced when NIR radiation resonantly created HH excitons ($\omega_{NIR} = \omega_{HH}$) and the FIR frequency was near electron cyclotron resonance ($\omega_{FIR} \sim \omega_c$). In Fig. 1(b) the $+2\omega_{FIR}$ sideband intensity is plotted vs. B for ω_{FIR} = 103 cm^{-1} with $\omega_{NIR} = \omega_{HH}$ at all B. The data demonstrates a novel *non-linear* optically-detected resonance. More detailed descriptions of experimental results can be found in a separate publication [3].

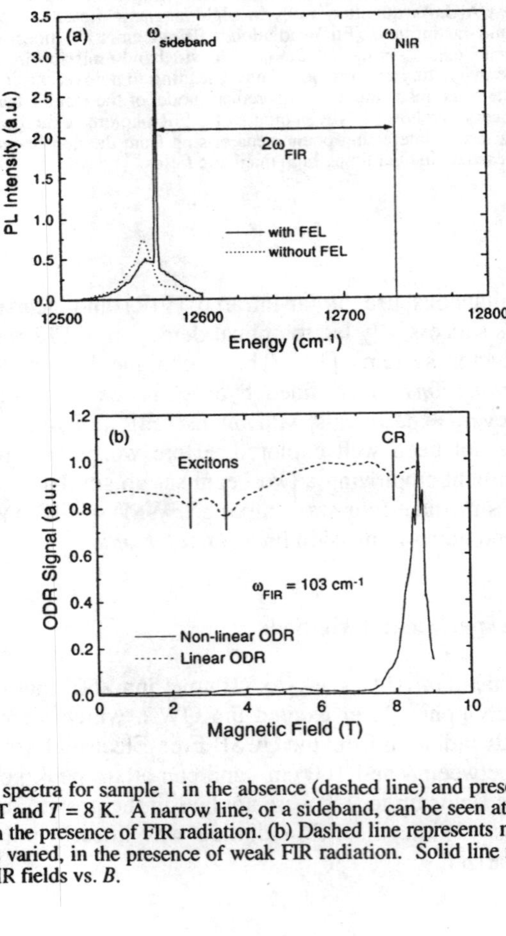

Fig. 1 (a) Emission spectra for sample 1 in the absence (dashed line) and presence (solid line) of FIR radiation at B = 8.5 T and T = 8 K. A narrow line, or a sideband, can be seen at 12568 cm^{-1} ($\omega_{sideband} = \omega_{NIR} - 2\omega_{FIR}$) only in the presence of FIR radiation. (b) Dashed line represents modulation of the PL as the magnetic field is varied, in the presence of weak FIR radiation. Solid line represents the sideband intensity in strong FIR fields vs. B.

4 Discussion

To the lowest order in electron-photon interactions, the generation of the $+2\omega_{FIR}$ sideband can be considered to be a process in which one ω_{NIR} photon along with two ω_{FIR} photons are destroyed and one sideband photon $\omega_{NIR} + 2\omega_{FIR}$ is created; in the case of the $-2\omega_{FIR}$ sideband, two ω_{FIR} photons are created rather than destroyed. These processes may be conveniently modeled by a third-order nonlinear susceptibility $\chi^{(3)}(\omega_{NIR} \pm 2\omega_{FIR}; \omega_{FIR}, \omega_{FIR}, \omega_{NIR})$.

We have calculated $\chi^{(3)}$ assuming that (1) both the conduction and valence bands are isotropic, parabolic and nondegenerate, and (2) the NIR and FIR photons induce only interband and intraband transitions respectively. Exciton effects are neglected. The resulting expression consists of both resonant and nonresonant parts. In the case in which all the four photons have the same linear in-plane polarization, the resonant part for the $-2\omega_{FIR}$ sideband is of the form:

$$\chi^{(3)} \propto \frac{B}{\omega_{NIR}\omega_{FIR}^2}\sum_n \frac{1}{\omega_{n\tilde{n}}}\sum_{n'=n\pm1\geq0}\frac{\max(n,n')}{\omega_{NIR}-\omega_{n\tilde{n}}+i\Gamma_{n\tilde{n}}}$$
$$\times\frac{1}{(\omega_{NIR}-\omega_{FIR}-\omega_{n'\tilde{n}}+i\Gamma_{n'\tilde{n}})(\omega_{NIR}-2\omega_{FIR}-\omega_{n\tilde{n}}+i\Gamma_{n\tilde{n}})} \quad (1)$$

Here n and $\tilde{n}$ (= 0,1,2,...) are the Landau level (LL) indices for the conduction and valence bands, respectively, $\omega_{ij} = (E_i - E_j)/\hbar$ where E_i is the energy of the i-th LL, and Γ_{ij} is a damping factor. In Eq. (1), FIR-induced transitions between the valence band LLs were neglected for simplicity, and only those between the conduction band LLs were taken into account. Diagrammatic representations of the n'=n+1 and n'=n-1 terms in Eq. (1) are given in Figs. 2(a) and (b), respectively.

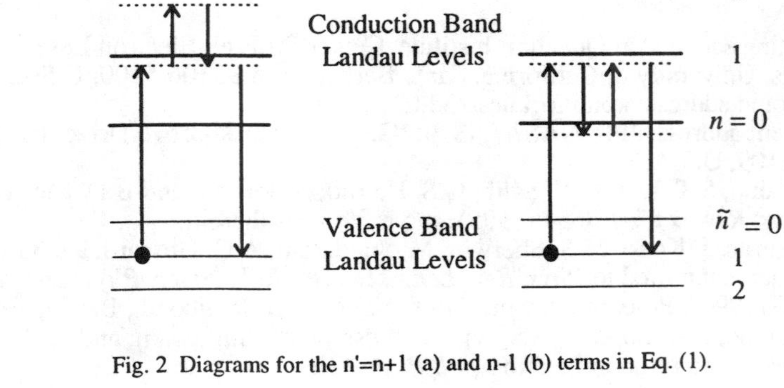

Fig. 2 Diagrams for the n'=n+1 (a) and n-1 (b) terms in Eq. (1).

Equation (1) indicates that the sideband is resonantly enhanced when (A) $\omega_{NIR} = \omega_{nn}$ (for some n), (B) $\omega_{NIR} = \omega_{nn} \pm \omega_c - \omega_{FIR}$, or (C) $\omega_{NIR} = \omega_{nn} + 2\omega_{FIR}$. Note that (A) and (B) may be simultaneously satisfied, giving a stronger resonance. This is consistent with the resonance conditions found in the experiment, i.e., $\omega_{NIR} = \omega_{HH}$ and $\omega_{FIR} \sim \omega_c$, if we identify ω_{HH} with ω_{00}. Note that if ω_{NIR} is detuned from ω_{00}, the value of ω_{FIR} giving maximum sideband intensity is also shifted from ω_c. This prediction and the resonance condition (C) are yet to be tested experimentally. The lowest-order perturbation approach described above is obviously inadequate when very close to a resonance or when the light fields are very strong. A more rigorous calculation based on an exact solution of the Schrödinger equation for a LL electron driven by a sinusoidal ac field is in progress and will be reported in our forthcoming article.

In our experimental set-up, the emitted light was collected in the reflection geometry, in which the total wave vector change multiplied by the interaction distance is of order unity (large phase mismatch). The use of the transmission geometry will significantly enhance sideband generation through improved phase matching.

Finally, it is interesting to point out that the FWM observed in our Landau-quantized electron system may be observed also in an array of quantum dots (such as InAs self-organized dots) without a magnetic field. We expect to see similar resonant enhancement of sideband intensities when $\omega_{NIR} = \omega_{HH}$ and ω_{FIR} is close to the energy separation between adjacent dot levels.

We conclude that the NIR-FIR Four-Wave-Mixing we discovered and explained in terms of a simple model is a sensitive and powerful tool for probing the nonlinear dynamics of quantum-confined carriers driven by strong ac electric fields in high dc magnetic fields.

Acknowledgments

The authors are grateful to Profs. S.J. Allen, Jr., T. Ando, A. Imamoglu, and L.J. Sham for useful and stimulating discussions, and D.P. Enyeart and J.R. Allen for technical support. This work was supported by QUEST, ONR, and JRDC.

References

* Visiting scientist at Quantum Institute, Center for Free-Electron Laser Studies, University of California, Santa Barbara, CA 93106-5100, U.S.A. Electronic address: kono@qi.ucsb.edu.
# Present address: JPL, NASA, NS 302-231, 4800 Oak Grove Drive, Pasadena, CA 91109, U.S.A.

1. J. Kono, S.T. Lee, M.S. Salib, G.S. Herold, A. Petrou, and B.D. McCombe, *Phys. Rev.* B **52**, R8654 (1995), and references therein.
2. J. Cerne, J. Kono, M.S. Sherwin, M. Sundaram, A.C. Gossard, and G.E.W. Bauer, submitted to *Phys. Rev. Lett.*; see also, *Bull. Amer. Phys. Soc.* **41**, 240 (1996); Proc. of 23rd Int. Conf. on Phys. of Semicond., Berlin, 1996.
3. J. Cerne, J. Kono, M.S. Sherwin, T. Inoshita, M. Sundaram, and A.C. Gossard, to be submitted to *Appl. Phys. Lett.*

COUPLED CYCLOTRON MODES IN GaAs/(Al,Ga)As -2D HOLE SYSTEMS MEASURED IN THE MAGNETIC ULTRA-QUANTUM LIMIT

B.E. COLE, J.M. CHAMBERLAIN, M. HENINI, T. CHENG
Department of Physics, The University of Nottingham, Nottingham NG7 2RD, United Kingdom

W. BATTY
School of Electronic Engineering Science, University of Wales, Bangor, Gwynedd LL57 1UT, United Kingdom

L. LI, L. VAN BOCKSTAL
Lab. voor Vaste-stoffysica en Magnetisme, Katholieke Universiteit Leuven, Celestijnenlaan 200 D, B-3001 Leuven, Belgium

Y. IMANAKA, Y. SHIMAMOTO, N. MIURA
Institute for Solid State Physics, The University of Tokyo, Minato-ku, Tokyo 106, Japan

J. SINGLETON
Department of Physics, University of Oxford, Clarendon Laboratory, Oxford OX1 3PN, United Kingdom

The cyclotron resonance (CR) of high mobility holes in GaAs/(Ga,Al)As heterojunctions has been studied in pulsed magnetic fields of up to 50T in the submillimetre spectral range. The CR is found to shift to lower magnetic fields as the sample temperature is raised from 4K to 20K; the size of this shift varies with the crystallographic growth direction of the sample and the CR energy. The influence of the Coulomb interaction between holes is explored as an explanation for the observed temperature dependence in terms of the electrostatic coupling of multiple single-particle CR transitions of differing energy. The Landau level structure has been modelled for (100), (311) and (111) oriented structures in the axial approximation using a rotated Luttinger Hamiltonian. Evidence for the coupled mode model is discussed with respect to these calculations.

1 Introduction

In a severely non-parabolic system such as that formed by the 2D holes system (2DHS) at a GaAs/(Al,Ga)As heterojunction[1, 2], the cyclotron resonance (CR) may be strongly influenced by hole-hole interactions. A theory for CR in 2D electron systems (2DES) in the ultra-quantum limit, including the coulomb interactions between electron in different spin states, has been put forward by Cooper and Chalker[3] in order to explain some previous studies of CR in 2DESs[4]. It was found that in the limit of strong electron-electron interactions, the CR consists of a single resonance at the weighted mean of the single-particle transitions. In the 2DHS holes in different Landau levels can undergo cyclotron motion at quite different frequencies. If strong hole-hole coupling is assumed then the coupled CR can be expected to depend sensitively on the strength of the composing single-particle transitions leading, for example, to a pronounced temperature dependence of the CR mass.

In this study we have measured the temperature dependence, from 4K to 20K, of the CR in high mobility, p-type GaAs/(Al,Ga)As heterojunctions grown on both (100), (111) and (311) orientation substrates at far-infrared energies and magnetic fields up to 50T. Under these conditions the samples are firmly in the quantum limit. The results are interpreted in terms of a coupled CR with a temperature dependent contribution from higher energy levels.

2 Experimental details

The far infrared (FIR) magneto-transmission of the samples was measured using 7.6meV and 10.4meV radiation from an optically pumped molecular gas laser. The pulsed magnetic fields were obtained at the Departement Natuurkunde, KU Leuven, Belgium and the Institute for Solid State Physics, Tokyo; details of these facilities are given elsewhere[5,6]. The samples were grown by MBE and had the following layer structures: 170Å GaAs cap layer, 400Å doped (10^{18} cm^{-3}) ($Al_{0.33}$,$Ga_{0.67}$)As, undoped ($Al_{0.33}$,$Ga_{0.67}$)As spacer , 5000Å undoped GaAs, superlattice buffer, semi-insulating substrate. Sample 1 was (111) orientation, Be doped and had a 500Å spacer layer, p_s=3.2x10^{11} cm^{-2} and μ_{4K}=40,000 $cm^2V^{-1}s^{-1}$. Sample 2 was grown on a (100) orientation substrate, Be doped and had a 407Å spacer layer , hole density (p_s) of 2.8x10^{11} cm^{-2} and a 4K mobility (μ) of 120,000 $cm^2V^{-1}s^{-1}$. Sample 3 was (311)A orientation, Si doped and had a 200Å spacer layer, p_s=3.3x10^{11} cm^{-2} and μ_{4K}=80,000$cm^2V^{-1}s^{-1}$. The samples were wedged to avoid interference effects and were mounted in exchange gas in a gas-flow cryostat with FIR optical access down 6mm light-pipes; the sample temperature could be varied from 4K to 30K and was measured using either a Si diode or thermocouple adjacent to the sample. The magnetic field was measured using a calibrated pick-up coil at the sample position. The FIR transmission was measured using an external He cooled Ge(Ga) photoconducting detector.

3 Results and Discussion

CR magnetotransmission spectra for samples 1 and 2 are shown in Figs. 1a and 1b for a range of temperatures and a FIR energy of 10.4meV. A clear shift to *lower* fields is observed in both samples as the temperature is raised from 4K to 20K although the shift appears somewhat less in the (111) case (sample 1). At 7.6meV, a similar shift to lower fields was obtained for sample 3 but *not* for sample 1. Lorenzian lineshapes are obtained whose linewidth increases sharply with temperature, consistent with the accoustic phonon scattering limited lifetimes expected in these samples[7]. We suggest the shift to lower field, or decrease in CR mass eB/ω_c, should be interpreted as being the result of thermal population of Landau levels of lower CR mass. These additional transitions are coupled to the dominant CR by strong hole-hole interactions and weight the observed CR to a lower mass. A simple two-mass scheme is illustrated in Fig. 1c.

The Landau level fan, calculated in the axial approximation for the (100)

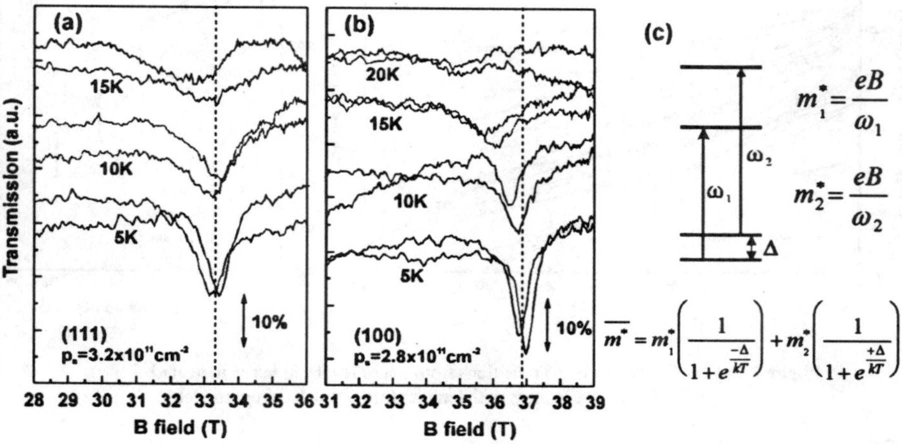

Figure 1: (a)CR spectra at 10.4meV for samples 1 (b) CR spectra at 10.4meV for sample 2. (c) Two mass scheme for CR temperature dependence.

accumulation layer is shown in Fig. 2a. Details of the model are given elsewhere[8]. Further calculations (not shown) were made for (100) inversion layers and also for (111) and (311) accumulation and inversion layers. The selection rule for CR transitions in the axial scheme is $\Delta n=\pm 1$, where n is the harmonic oscillator wavefunction index. The possible transitions from states above the ground Landau level are indicated on Fig. 2a. The data for the (100) sample can be fitted by the simple two-mass scheme of Fig. 1c and yields an activation energy, Δ, of 4meV, in excellent agreement with the ground-to first Landau level separation shown in Fig. 2a. In Fig. 2b, we estimate the temperature dependence for the (111) and (100) samples, considering transitions from three closely grouped levels ~4meV above the ground Landau level, assuming all allowed transitions to have equal oscillator strength. The experimental data is plotted for comparison. The model suggests a temperature dependence of similar order to that observed however a greater shift is predicted for (111) over (100), in contrast to the experiment.

Finally, we have considered the slight modification to the self-consistent confinement potential due to thermal population of higher levels as a mechanism for the observed temperature dependence; we find the $n=1\rightarrow n=2$ CR transition to be very insensitive to changes in potential shape; comparing accumulation and inversion layers, less than 3% difference is found in the energy of this transition thus any shift due to the far smaller thermal change in potential shape would not be expected to yield a temperature dependence of the magnitude observed here.

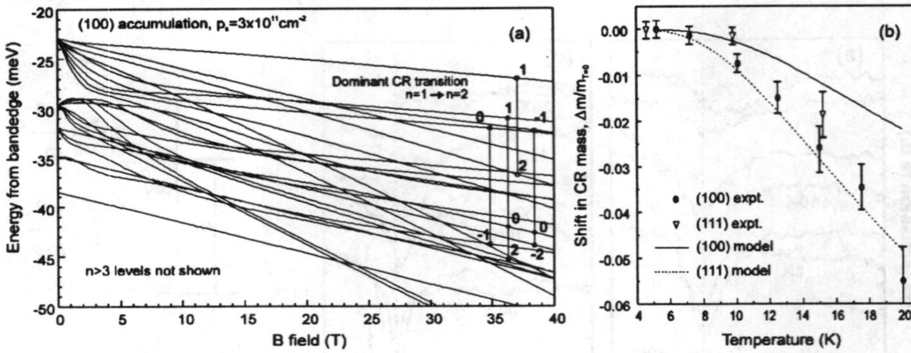

Figure 2: (a) (100) Landau fan. Transitions from upper levels, index *n*, indicated.
(b) Variation in CR mass with temperature estimated from model.

4 Summary and Conclusions

To conclude, we have observed the CR in 2DHSs to shift to lower mass as the temperature is raised from 4K to 20K under conditions of extreme magnetic quantisation. Where measured, the shift is characterised by an activation energy close to the energy gap between the ground and higher Landau levels. A coupled picture of the CR is assumed and the temperature dependence is estimated using single-particle Landau level calculations.

References

1. S.J. Hawksworth, S.O. Hill et al. *Semicond. Sci. Tech.* **8** 1465 (1993).
2. U. Ekenberg and M. Altarelli *Phys. Rev. B* **32**(6) 3712 (1985).
3. N.R. Cooper and J.T. Chalker *Phys. Rev. Lett.* **72**(13) 2057 (1994).
4. J.G. Michels, S.O. Hill, R.J. Warburton et al. *Surf. Sci.* **305** 33 (1994).
5. F. Herlach, R. Bogaerts, I. Deckers et al. *Physica B* **201** 542 (1994).
6. N. Miura, H. Nojiri and Y. Imanaka *Proc. 22nd Int. Conf. on Physics of Semiconductors* D. Lockwood ed., World Scientific, Singapore (1995).
7. M. Henini, P.J. Rodgers et al. *Appl. Phys. Lett.* **65**(16) 2054 (1994).
8. W. Batty, B.E. Cole, J. Singleton et al. *Proc. 23rd Int. Conf. on the Physics of Semiconductors* pub. World Scientific, Singapore (1996).

CYCLOTRON RESONANCE OF TYPE-I $(GaAs)_n/(AlAs)_m$ SUPERLATTICES UNDER TILTED MAGNETIC FIELDS

K. Yamanaka, Y. Suzuki, H. Momose, N. Mori, and C. Hamaguchi
Department of Electronic Engineering, Osaka University
Suita City, Osaka 565, Japan

H. Arimoto, Y. Imanaka, Y. Shimamoto, and N. Miura
Institute for Solid State Physics, University of Tokyo
Roppongi, Minato-ku, Tokyo 106, Japan

Cyclotron resonance (CR) measurements in short period $(GaAs)_n/(AlAs)_m$ superlattices (SLs) of $(n,m) = (8,4)$, $(16,16)$ and $(20,20)$ have been performed under tilted magnetic fields for studying Landau level mixing between different subbands. The CR peak positions for thicker AlAs barriers of 16 ML or 20 ML are found to deviate to higher magnetic field from a simple $\cos\theta$-dependence, while the CR peak positions for thinner AlAs barrier of 8 ML deviate to lower magnetic field. This lower shift is attributed to miniband formation in $(GaAs)_8/(AlAs)_4$ SL.

1 Introduction

Cyclotron resonance (CR) measurement under high magnetic fields has played an important role not only in determining effective masses but also in investigating the energy band structure of two-dimensional electron gas (2DEG) in single heterostructures, quantum wells and superlattices (SLs). Although extensive study of CR measurement for a tilted magnetic field has been performed on 2DEG in Si inversion layers [1,2,3], GaInAs/AlInAs heterostructures [4] and GaAs/AlGaAs multi quantum well structures [5], short period SLs with thinner barriers have been studied very little. In the previous study [6], we investigated the Landau level mixing, which is caused by a magnetic field component parallel to the hetero-interface, in type-II $(GaAs)_n/(AlAs)_n$ SLs, and found that the longitudinal effective mass of X-like states in AlAs layer can be deduced by analyzing a deviation of CR position from a simple $\cos\theta$-dependence. In the present study, we report CR measurements in type-I $(GaAs)_n/(AlAs)_m$ SLs under titled magnetic fields.

2 Experimental Procedures

2.1 Sample Structures

The SL samples used in the present study are type-I $(GaAs)_n/(AlAs)_m$ SLs with $(n,m) = (8,4)$, $(16,16)$ and $(20,20)$. All the samples were grown by molecular beam epitaxy (MBE) on semi-insulating GaAs (001)-oriented substrates. 80 periods of n-GaAs/AlAs SLs are sandwiched with 10 periods undoped GaAs/AlAs SLs (total 100 periods). The number of GaAs and AlAs layers were accurately controlled by counting the periods of the intensity oscillations of a specularly reflected beam in a reflection high-energy electron diffraction (RHEED).

2.2 Experimental Apparatus

A block-diagram of the experimental apparatus is shown in Fig. 1. Pulsed high magnetic fields up to 150 T were produced by a single-turn coil technique [7]. The pulsed magnetic field was applied to the samples changing the angle θ between the magnetic field B and the normal to the SL layers. The magnetic filed was measured by a pick-up coil wound on the sample. CR signals were observed by using an externally chopped cw CO_2 laser (10.6 μm and 9.60 μm).

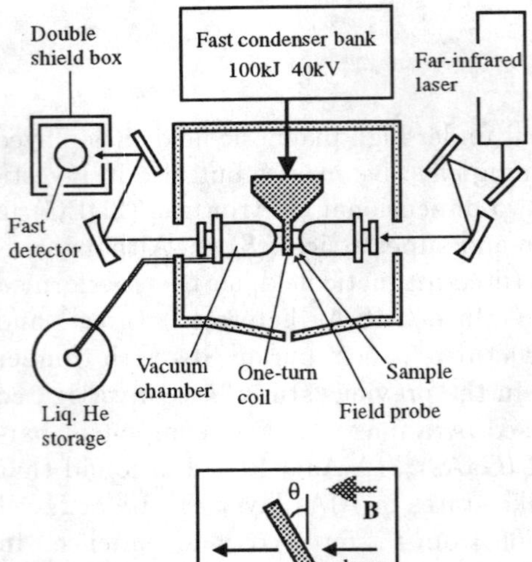

Figure 1: Block-diagram of the experimental apparatus for CR measurement. The magnetic field is applied changing the angle θ between the magnetic field B and the normal to the SL layers.

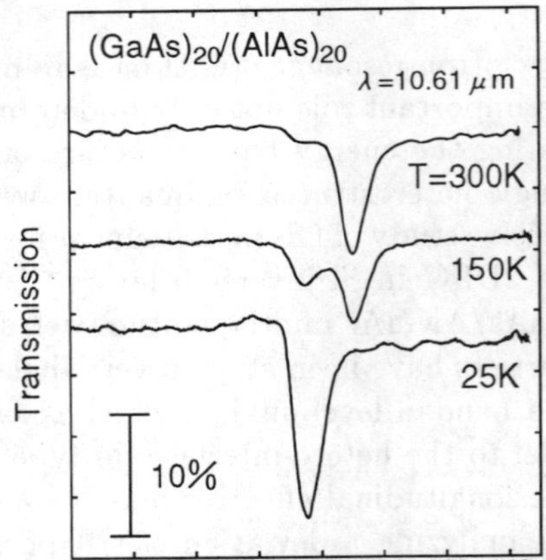

Figure 2: CR for $(GaAs)_{20}/(AlAs)_{20}$ SLs in the transmission of 10.6 μm radiation at 25, 150 and 300K.

3 Experimental Results

3.1 Temperature Dependence of Cyclotron Resonance

At low temperatures, electrons freeze out onto the shallow donor states and 1s-$2p^+$ transition of shallow donors is observed in CR measurement. We measured CR peaks in $(GaAs)_n/(AlAs)_n$ SLs of $n = 16$ and 20 at various temperatures to estimate impurity binding energy. Figure 2 shows the temperature dependence of CR for $n = 20$ SL in the transmission of 10.6 μm at 25, 150 and 300 K. The resonance position of 75 T at 25 K shifts to 99 T at 300 K. We estimate the $1s$-$2p^+$ binding energy difference $\Delta E_B = 17$ meV at 75 T for $(GaAs)_{20}/(AlAs)_{20}$ SL.

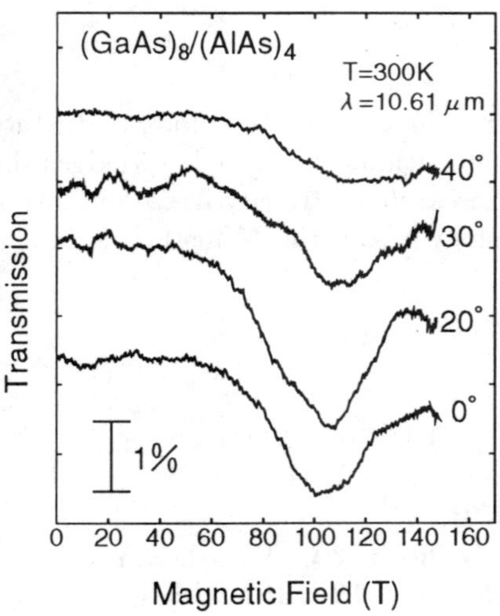

Figure 3: CR signals of $(GaAs)_8/(AlAs)_4$ SL for $\theta = 0°$, $20°$, $30°$ and $40°$ at $T =$ 300 K.

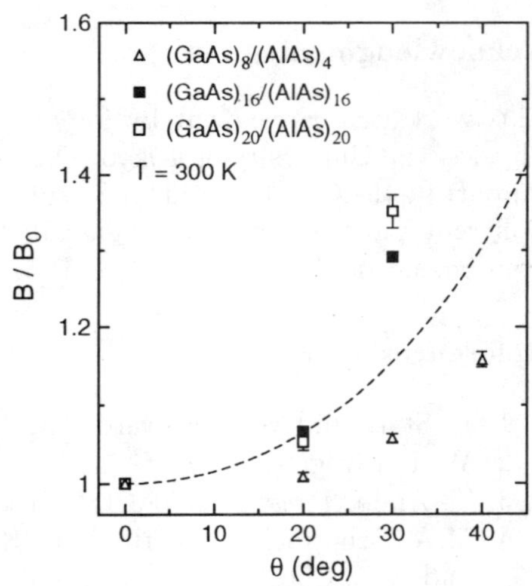

Figure 4: CR peak positions normalized by B_0 (CR peak position for $\theta = 0$) compared with a simple $\cos\theta$-dependence (dashed line).

3.2 Angular Dependent Cyclotron Resonance

Figure 3 shows the experimental CR signals of $(GaAs)_8/(AlAs)_4$ SL for several angles θ. The CR peak positions are plotted in Fig. 4 as a function of θ. In Fig. 4, we also plot the experimental CR positions for $(GaAs)_n/(AlAs)_n$ SLs of $n = 16$ and 20. We found that the CR peak positions for thicker AlAs barriers of 16 ML or 20 ML show deviation to higher magnetic field from a simple $\cos\theta$-dependence, while the CR peak positions for thinner AlAs barrier of 8 ML deviate to lower magnetic field. The higher shift for $n = 16$ and 20 is considered to be caused by the Landau level mixing observed in type-II SLs[6], although the deviation is much larger than a theoretical value in a simple effective mass theory[8]. This disagreement may be due to the $\Gamma - X$ mixing at around 100 T. The lower shift observed in $(GaAs)_8/(AlAs)_4$ SL is attributed to miniband formation because of thin barrier width of ~ 11 Å. We can fit the CR peak shifts in the ellipsoidal effective mass model for three dimensional systems by assuming $m_z/m_x = 2.3$ with m_z being the effective mass along the growth direction and m_x being the effective mass along the hetero-interface.

Acknowledgments

This work was carried out by joint research in the Institute for Solid State Physics, the University of Tokyo. One of the authors (C.H.) acknowledges the support by the Grant-in-Aid for Scientific Research on Priority Area "Quantum Coherent Electronics, Physics and Technology" from the Ministry of Education, Science and Technology.

References

1. F. Stern and W.E. Howard, *Phys. Rev.* **163**, 816 (1967).
2. W. Beinvogl and J.F. Koch, *Phys. Rev. Let.* **40**, 1736 (1978).
3. T. Ando, *Phys. Rev.* B **19**, 2106 (1979).
4. M.A. Brummell, M.A. Hopkins, R.J. Nicholas, J.C. Portal, K.Y. Cheng, and A.Y. Cho, *J. Phys. C: Solid State Phys.* **19**, L107 (1986).
5. B.C. Cavenett and E.J. Pakulis, *Phys. Rev.* B **32**, 8449 (1985).
6. K. Yamanaka, H. Momose, N. Mori, C. Hamaguchi, H. Arimoto, Y. Imanaka, Y. Shimamoto, and N. Miura, in *Proc. 3nd Int. Workshop on Mesoscopic Systems*, Maui 1995, to be published in *Physica B.*
7. M. Nakayama, I. Tanaka, I. Kimura, and H. Nishimura, *Jpn. J. Appl. Phys.* **29**, 41 (1990).
8. K. Yamanaka, unpublished.

RESONANT MAGNETOPOLARON EFFECTS IN GaAs/AlGaAs MQWs AT HIGH MAGNETIC FIELDS

Y. J. WANG
National High Magnetic Field Laboratory at Florida State University
Tallahassee, Florida 32306, U.S.A.

H. A. NICKEL and B. D. McCOMBE
Department of Physics, State University of New York at Buffalo
Buffalo, New York 14260, U.S.A.

F. M. PEETERS, G.Q. HAI, J.M. SHI and J.T. DEVREESE
Department of Physics, Universiteit Antwerpen (UIA)
B-2610 Antwerpen, Belgium

X.G. Wu
National Laboratory for Superlattices and Microstructures
Institute of Semiconductors, Chinese Academy of Science, Beijing, China

A study of polaron cyclotron resonance (CR) has been carried out both experimentally and theoretically in two modulation-doped $GaAs/Al_{0.3}Ga_{0.7}As$ multiple quantum well samples at fields up to 30 T. Resonant polaron effects due to the interaction between electrons and AlAs-like interface optical phonon modes have been observed for the first time. Large CR splittings due to resonant polaron effects are also observed near the GaAs reststrahlen region. The results of the theoretical calculations which include the effects of confined slab phonons and interface phonons on the polaron CR frequency are in good agreement with the experimental observations.

The interaction of electrons with optical phonons in quasi-two-dimensional systems has been of considerable interest for several years[1], since the electronic properties of semiconductors are strongly affected by the interaction between charge carriers and optical phonons. In order to explore the higher frequency region around the barrier AlAs-like and interface AlAs-like optical modes, we have carried out a magnetotransmission study of electron CR versus magnetic field (up to 30 T) in two modulation-doped $GaAs/Al_{0.3}Ga_{0.7}As$ multiple quantum well samples; the transistion energies including interactions of the elctrons with appropriate optical phonons were calculated for our specific sample structures within the framework of degenerate second-order perturbation theory[2].

The transmission of far infrared light through the sample, maintained at 4.2 K, was measured at magnetic fields up to 30 T in a resisitive magnet with a Bruker 113v interferometer, and detected by a 4.2 K silicon bolometer. The two samples studied are grown by MBE and they have the same well and barrier width, 240 Å by 240 Å. Both samples are doped with silicon over central 1/3 of the barriers; the doping concentrations are 1.5 x $10^{11}cm^{-2}$ and 3.0 x $10^{11}cm^{-2}$ for samples A and B, respectively.

The transmission results show large CR splittings in the GaAs reststrahlen region. At magnetic fields corresponding to CR above this region, a much weaker splitting of the CR into two branches is observed at energies in the vicinity of the AlAs-like optical phonon frequencies for sample A. Fig. 1 shows examples of the transmission spectra for sample A at different magnetic fields from 25 T to 29 T. At both 25 T and 29 T, there is only one resonance minimum. But when the field is tuned through the AlAs-like reststrahlen region, the resonance is clearly split into two branches. The lower energy branch loses intensity, and the higher energy branch gains intensity gradually when the field is increased. Both branches can be seen clearly from 26 T to 28 T.

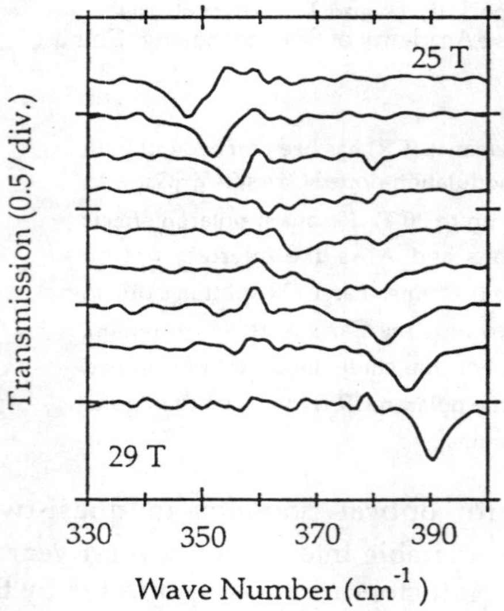

Fig. 1 Transmission spectra at different field divided by zero field spectra for sample A. The traces are spaced every half tesla.

The integrated amplitude of the electron wavefunctions that penetrate into the $Al_{0.3}Ga_{0.7}As$ barrier is estimated to be less than 0.5 % of the total for these wide well samples. Therefore the observed splitting can not come from the interaction of the electrons in the wells and the AlAs-like LO phonons in the barriers. The interaction between electrons and the symmetric AlAs interface optical (IO) phonon modes (with significant amplitude in the GaAs wells) was included in the calculations to compare with the present experimental results. The theoretical calculation gives a splitting near 375 cm^{-1}. Based upon these facts, the splitting observed in the AlAs reststrahlen region in our measurements is attributed to this interaction.

Both samples have been previously studied at low magnetic fields[3], and the effective mass obtained at low magnetic fields is about $0.069m_0$, with m_0 the free electron mass. Fig. 2 shows the normalized (to the band mass, m_b) spectroscopic effective mass ($m^*/m_0 = eB/m^*\omega_{res}c$) vs. field at high magnetic fields for the two samples. The points are the measured data and the curves are the calculated. The straight dashed line in Fig.2 is the "expected" effective mass at high fields due to the non-parabolicity. The measured effective masses of both energy branches start to deviate from the line far away from the GaAs-like optical phonon energies, and there is a large discontinuity of the effective masses in the GaAs reststrahlen region. The difference between the two effective mass branches approaches a minimum when the unperturbed cyclotron energy ($\propto$ B) resonance with the GaAs LO phonon frequency.

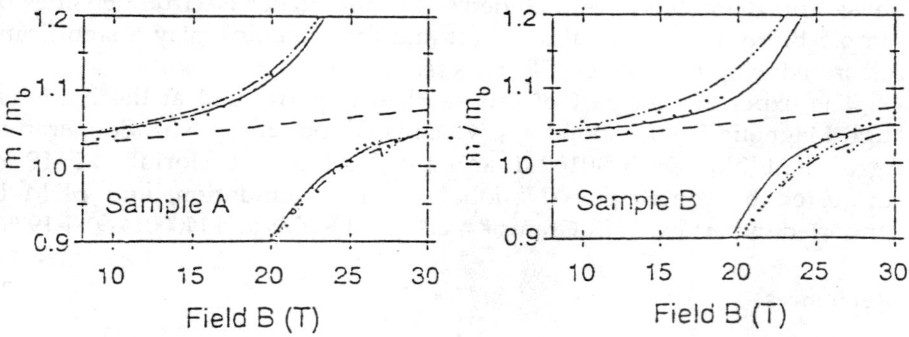

Fig. 2 The normalized effective mass m^*/m_b vs. B. The points are measured data, and all curves are calculated. The straight dashed line is the "expected" m^* due to the non-parabolicity. The dashed curves are for coupling with only bulk GaAs optical phonons. The dotted curves consider the effects of IO-phonon and slab LO-optical phonon modes. The solid lcurves include the screening effects in addition to the coupling with the GaAs bulk phonons.

Since so-called dielectric artifacts can also give rise to splittings[4] in the reststrahlen region due to the resonant dielectric function of the material at the TO frequency (ies), a computer simulation[5] of the classical dielectric effects for our samples has been performed to check the contribution from the dielectric artifacts to the observed resonant splitting. For the present sample parameters, there is no measurable CR splitting in the GaAs reststrahlen region from the simulation. The observed large discontinuity must be due to resonant polaron effects.

Theoretical calculations have been done to compare with the experimental results. Since our samples are multiple quantum well structures, the phonon modes have to be modified due to confinement and the presence of the interfaces. The dashed line in Fig. 2 is obtained by considering the effects of IO-phonon and confined slab LO-phonon modes without including many body effects. As one can see, the calculations agree very well with the experimental results over the entire resonant region. The calculation has also been done for coupling with only bulk GaAs 3D phonons as shown by the dotted line. The results are nearly the same, except close to the AlAs like LO-phonon frequency, as expected for such wide quantum wells.

The doping concentrations of both samples are rather high; therefore screening effects are expected to play a role in the strength of the interaction and thus in the magnitude of the resonant splitting. The solid curves in Fig. 2 are the calculated results in which the occupation effect resulting from the Pauli principle and the screening, within the random phase approximation, are included[6]. For the higher electron densities of sample B, both the occupation effect and the screening play a significant role in reducing the polaron CR-mass renormalization.

The experimental part of this work was performed at the National High Magnetic Field Laboratory, which is supported by NSF Cooperative Agreement No. DMR-9016241 and by the State of Florida. FMP is supported by the Belgian National Science Foundation, and BDM is supported in part by ONR Grants # N00014-89-J-1673 and N00014-91-J-1939.

References

1. For a review, see e.g., B.D. McCombe and A. Petrou, Handbook on Semiconductors, Vol. 2, Optical Properties of Semiconductors, edited by M. Balkanski, pp 285-384 (Elsevier, New York, 1994).
2. G.Q. Hai, F.M. Peeters, and J.T. Devreese, Phys. Rev. B **47**, 10358 (1993).
3. Y.J. Wang, B.D. McCombe, R. Meisels, F. Kuchar, and W. Schaff, Phys. Rev. Lett. **75**, 906 (1995).
4. M. Ziesmann, D. Heitmann, and L.L. Chang, Phys. Rev. B 35, 4541 (1987).
5. J.P. Cheng and B.D. McCombe, Phys. Rev. Lett. **62**, 1925 (1989).
6. F.M. Peeters and J.T. Devreese, Phys. Rev. B **31**, 3689 (1985).

TILTED FIELD CYCLOTRON RESONANCE IN MEGAGAUSS FIELDS USING THE SEMIMETALLIC SUPERLATTICE SYSTEM InAs/GaSb

G.M. SUMMERS, R.J. NICHOLAS, N.J. MASON, P.J. WALKER
Physics Department, Oxford University, Clarendon Laboratory, Parks Road, Oxford. OX1 3PU UK

Y. IMANAKA, H. ARIMOTO, N. MIURA,
I.S.S.P., Tokyo University, Roppongi, Minato-ku, Tokyo, Japan

High energy (130 meV) cyclotron resonance (CR) has been studied in tilted magnetic fields up to 140 Tesla in a series of InAs/GaSb superlattices. This shows two effects: (i) Samples with wide InAs layers (>100Å) exhibit a decrease in resonance position with increasing tilt angle, and (ii) Samples with thin InAs layers (<100Å) show a rapidly increasing CR position on tilting. These two effects are explained by the resonant interaction of the cyclotron resonance with intersubband transitions, which couple strongly in tilted magnetic fields.

Introduction

We report a study of cyclotron resonance in very high tilted magnetic fields in superlattices of the semimetallic system InAs/GaSb using the megagauss facility of the University of Tokyo. InAs/GaSb systems are of interest due to the crossed alignment of the bands where the conduction band edge of the InAs lies 140 meV above the top of the valence band of the GaSb. The application of strong magnetic fields on this system has substantial effects on the band structure, allowing both semimetallic and semiconducting behaviour[1,2] to be exhibited. Previous CR experiments in long period superlattices have shown in the region of 40-50T a semimetallic to semiconductor transition due to the additional magnetic confinement[3].

The experiments presented here consist of a series of measurements of the transmission of infrared radiation in the region of 10μm, from a CO_2 laser, as a function of magnetic field generated using a single turn coil pulsed magnetic field system. Fields up to 140T were used with a typical pulse duration of 5μsec. Four superlattices were examined with InAs layer thickness' from 66 to 220Å, grown by MOVPE on a semi-insulating GaAs substrate. The experiments were performed at room temperature to produce sufficient electron carrier densities, the CR linewidths were typically 10-20T.

Results

Figure 1 shows the experimental traces obtained in two superlattice samples with InAs thickness' 220Å and 75Å. The plot shows series of CR spectra against magnetic field for tilt angles 0-60° at an incident laser wavelength of 9.2μm.

The 220Å InAs layer sample, upper plot, exhibits a split CR peak which moves to lower magnetic field on titling. The splitting of the CR is due to the resolution of the two possible spin transitions from the lowest Landau level. The behaviour of the CR moving to lower fields is unexpected as a 2D electron system is expected to show an increase in field position with a $(\cos\theta)^{-1}$ dependence with tilt angle θ, from:

$$B = \frac{\hbar\omega m^*}{e\cos\theta} \qquad (1)$$

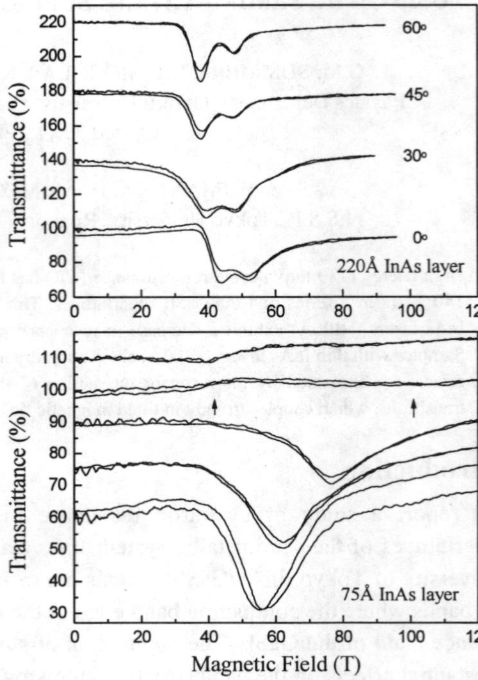

Figure 1. Cyclotron resonance spectra are shown for various tilt angles in two samples with InAs layer thickness' 220 and 75Å, at an incident wavelength of 9.2μm.

and in a 3D system the resonance position would be expected to be dependent on the total applied field B and independent of tilt angle. The scale of the cyclotron motion (cyclotron radius l_c=36Å at 50T) is significantly less then the InAs thickness and essentially 3D behaviour would be expected. The movement to lower field is still unusual and has been explained previously in longer period samples by a mixing of the cyclotron motion and the confinement, leading to an effective mass decreased through reduced non-parabolicity thus and a shift to lower field[4].

In complete contrast the sample with the 75Å InAs layer, lower plot of Figure 1, shows a rapidly increasing CR position on tilting, considerably faster than $(\cos\theta)^{-1}$. Above 30° tilt the main CR feature is significantly reduced in intensity with possibly an additional branch of the CR appearing at a lower field position. It would appear the behaviour is more two dimensional in nature, as might be expected as the cyclotron radius approaches the superlattice layer thickness, but the rapid rise to higher fields and the additional CR branch are unexpected.

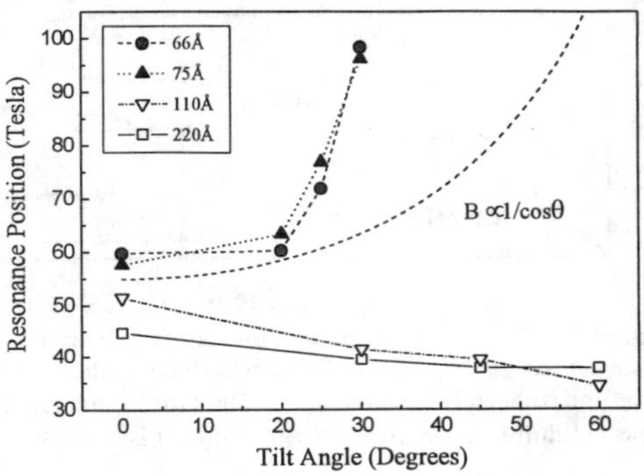

Figure 2. Summary of the trend in cyclotron resonance position on applying a titled magnetic field up to 60°. Data from four samples with InAs thickness' from 66-220Å are shown. The dashed line shows the expected two dimensional behaviour.

Fig. 2 summarises the observed behaviour in four different samples with InAs layer thickness' from 66Å to 220Å. The cyclotron resonance position is displayed against tilt angle, compared with the expected two dimensional behaviour shown by the dashed line. The division can be clearly seen between wide and narrow layer samples. Also the data from the 110Å sample shifts to lower fields more rapidly then the wider 220Å sample. The 'flatter' data from the wider sample is consistent with the dependence tending towards three dimensional characteristics, however the 110Å data is tending away from the 2D curve. This would lead to the conclusion that there is some critical divergent behaviour occurring between the 75Å and 110Å superlattice samples.

Resonant Subband Landau Level Coupling

The data is consistent with a picture of resonant subband Landau level coupling, where the CR is strongly coupled to the intersubband transitions. The coupling strength increases with tilt angle and with the proximity of the CR energy to the intersubband transitions, leading to an anticrossing and pinning of the CR[4,5].

Figure 3 illustrates the expected CR behaviour in the vicinity of the subband energy spacing of the zeroth and first subbands (E_{01}). Two distinct regions exhibit themselves:

(a) With E_{10} well below the CR energy, three dimensional behaviour is expected with the resonance position dependent on the total applied magnetic field. However, close to E_{01} the CR would be expected to be pushed to lower fields on increasing θ due to increased resonant coupling.

(b) When E_{01} is well above the CR energy the peak position should behave in a 2D manner and follow $(\cos\theta)^{-1}$. It will also be pushed more dramatically to higher fields on tilting in the vicinity of E_{01}.

As we go to samples with progressively narrower InAs layers the enhanced confinement increases the value of E_{01}, while in this experiment the CR energy is fixed at 133 meV (9.2μm). The relative trends in CR positions are shown schematically by the four arrows in Figure 3. The closer the CR energy is to E_{01} the larger the expected shift and this is consistent with the data in Fig. 2 which shows stronger responses in the 75Å and 110Å samples.

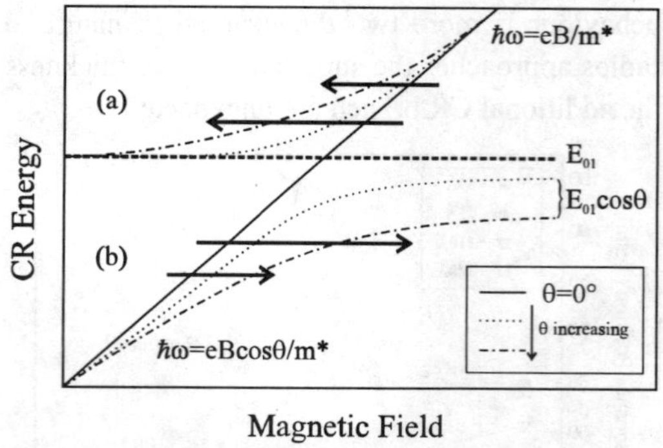

Figure 3. Schematic diagram showing CR anticrossing behaviour on applying a tilted magnetic field in the vicinity of an intersubband transition (E_{01}). The arrows indicate the expected shift in CR at fixed energy as the tilt is increased.

Energy level calculations

Table 1. Calculated intersubband spacing

Sample	InAs/GaSb	E_{01} (meV)
(a)	220/180 Å	67.2
(b)	110/90 Å	248.0
(c)	75/71 Å	433.7
(d)	66/47 Å	531.9

Table 1 shows the results of a self consistent single band calculation of the electron energy levels in the samples examined, with non-parabolicity taken into account. It can be seen that sample (a) does indeed meet the condition of $E_{cr}>E_{01}$, E_{cr} being 133meV in this case. And the data from samples (c) and (d) are also in agreement with the observed data with $E_{cr}<E_{01}$. However, the calculated energy for sample (b) appears to be the wrong size of E_{cr}. The nature of the calculation indicates that the values of E_{10} are likely to be overestimates. Additional errors may be incurred by inaccuracies in the measured well widths.

Further work is in progress comparing the experimental CR data with more sophisticated ***k.p*** calculations.

References

1. G.A Sai-Halasz, L. Esaki and W.L. Harrison, *Appl. Phys. Lett.* **30** (1977) 651
2. R.J. Nicholas *et al*, *Physica B* **184** (1993) 268
3. D.J. Barnes, R.J. Nicholas, R.J. Warburton, N.J. Mason, P.J. Walker and N. Miura, *Phys. Rev.* **B49** (1994) 10474
4. S. Huant, M.Grynberg, G. Martinez and B. Etienne, *Sol. State Commun.* **65** (1988) 457
5. R. Merlin, *Sol. State Commun.* **64** (1987) 99

ANOMALOUS CYCLOTRON RESONANCE IN N-TYPE CdS AT ULTRA-HIGH MAGNETIC FIELDS

Y.IMANAKA, N.MIURA
Institute for Solid State Physics, University of Tokyo,
7-22-1 Roppongi, Minato-ku, Tokyo, 106, Japan

S.LUTHER, M. VON ORTENBERG
Institut für Physik der Humboldt-Universität
zu Berlin Invalidenstrasse 110, D-10115 Berlin, Germany

We performed cyclotron resonance experiments in the low magnetic field region to investigate the ground state of n-type CdS. The resonance was observed down to 4.2K without a freeze out effect. The extrapolation of the resonant energy vs. field line to the zero field does not intersect the origin. These facts indicate that there is a very shallow trap and electrons are confined or localized at low temperatures in this area. If we assume the existence of a harmonic oscillator potential, its characteristic frequency and the polaron mass are obtained $\hbar\omega_0 \simeq 1$meV and $m_p^* = 0.185 m_0$, respectively, from the fitting with the theory of quantum dots. The characteristic length of this trap is estimated to be about r≃200Å.

In n-type CdS, a number of cyclotron resonance experiments in the low energy region were performed, in contrast with other II-VI semiconductors, to observe the effect of the piezoelectric polaron in 1970's[1-3]. However, the temperature dependence of the resonant magnetic fields was not in good agreement with the several theoretical prediction of the piezoelectric polaron[4-6]. Moreover, the origin of various anomalies, such as the line shape of the resonance and the problem of residual carrier at low temperatures, have not been understood yet. Since II-VI semiconductors have a significantly large Fröhlich electron-phonon coupling constant α, II-VI materials are one of the most suitable materials to study the polaron effect[7,8]. In addition to the polaron effect with the LO phonon interaction, n-type CdS has a possibility of piezoelectric polarons.

Recent advance of high field magnet technology has enabled us to study cyclotron resonance in ultra-high magnetic fields above 100T (megagauss fields). We observed anomalous cyclotron resonance in n-type CdS above LO phonon energy ($\hbar\omega_{LO}$=37.8meV) at a low temperature region. The origin of this anomalous behabior is not clear yet. Many unidentified absorption peaks were observed in this material at low temperatures and at ultra-high magnetic fields[9].

In the present paper, we focus on far infrared transmission experiments in n-type CdS to investigate the ground state of this materials at low temperatures by using several kinds of molecular gas lasers (λ=70.5μm~1.90mm), especially in the relatively low field region up to 33T. The high magnetic fields were generated by using a long pulse magnet and a superconducting magnet. The sample temperature was controlled from room temperature to 4.2K. The transmission signal was detected by using several kind of detectors, which have a sufficiently fast response time for the measurements in very short pulse magnetic fields.

The bulk n-type CdS samples were supplied from Eagle-Picher Research Lab. and Cleaveland Crystals Inc., which were grown by the vapor phase transport method. These samples were not intentionally doped and the carriers probably come from vacancies. The mobility and the resistivity of the sample are approximately 200cm^2/V·s and 20Ω·cm at room temperature, respectively. The typical size of the sample for these measurements was 2.5×2.5×0.2mm^3. The magnetic field was applied parallel to the c-axis in the present works.

Figure 1 shows a temperature dependence of the transmission spectra at a wavelength of 119μm, measured using a long pulse magnet up to 31T. We can see several anomalous behaviors. First, it is an astonishing result that the resonance is observed down to very low temperatures(5K) without a excitation light. In other materials, we can not observe cyclotron resonance in this field region because almost all carriers are bound the impurity state (freeze out effect) at low temperatures. Moreover, the line shape and the temperature dependence of the resonant field are anomalous. The mobility of this resonance is estimated about 30000cm^2/V·s, which is much higher than other II-VI semiconductors.

The wavelength dependence of transmission spectra at 4.2K in n-type CdS is shown in Figure 2. The width of the resonance is broadened as the wavelength increases. The condition ($\omega_c\tau\geq 1$) is no more satisfied above λ~1mm. This fact indicates that there is a large energy (or magnetic field) dependence of the mobility at 4.2K. Such a drastic change of the mobility is unusual feature in normal cyclotron resonance.

The solid circles in figure 3 denote the resonant position at 4.2K for each wavelength (photon energy). The extrapolated line of these data points to the zero field does not intercept the origin. These facts indicate that the very sharp resonance near liquid helium temperature is not due to the response of the carriers in the conduction band, but of the carriers affected by some perturbation. In the resistivity measurement, we observed a behavior like normal semiconductor, namely, the carriers which appear in the cyclotron resonance at 4.2K do not contribute to the DC current. This means that the wavefunction

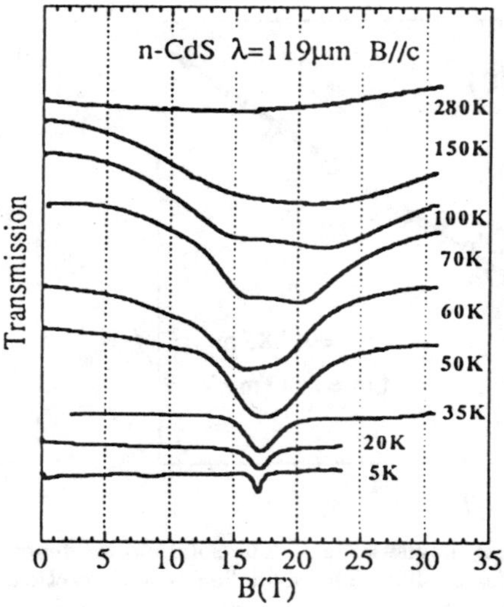

Figure 1: Temperature dependence of the magneto-transmission spectra in n-type CdS at 119μm.

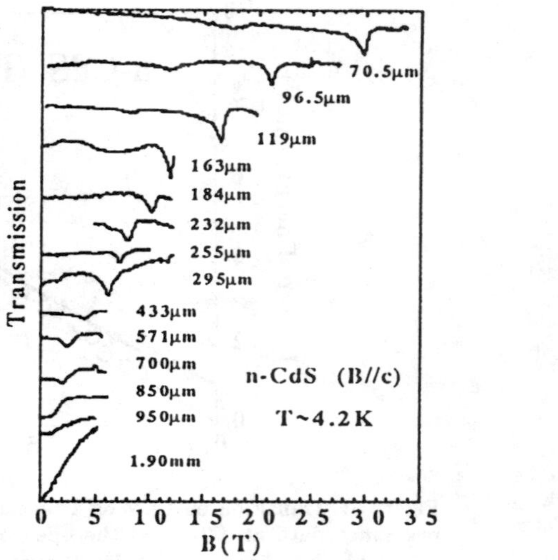

Figure 2: Wavelength dependence of the magneto-transmission spectra in n-type CdS at 4.2K.

is not extended in the system at 4.2K. If we assume the existence of the very shallow trap in this material and the confinement or localization of electrons at low temperatures in this area, we can explain these anomalies in this system. When the cyclotron radius is smaller than the characteristic length of the trap in high fields, the local mobility becomes higher. However, as the magnetic field decreases, the cyclotron radius becomes larger than the characteristic length of the trap and the mobility decreases drastically due to the scattering at the boundary of the trap. The origin of this trap is not clear yet, but it may exist due to the fluctuation of the residual impurity (contamination). The solid line in figure 3 is a theoretical fitting based on a quantum dot's formula by employing two parameters, a polaron mass $m_{\rm p}^*$=0.185m_0 and a harmonic oscillator energy $\hbar\omega_0$=1.03meV, respectively. The characteristic length of this trap is estimated to be about r$\simeq$200Å from a harmonic oscillator frequency ω_0. The dashed line is a calculated line of the cyclotron resonance when we employed the effective mass $m_{\rm p}^*$=0.185m_0. The dashed line explains the open circles well. The open circles are the resonant position at 40K. Thus it seems reasonable to assume that the existence of very shallow trap and the shift of the resonant position between 5K and 50K in figure 1 reflects on the excitation of the carrier from the trapping state to the free state in the conduction band from the present results.

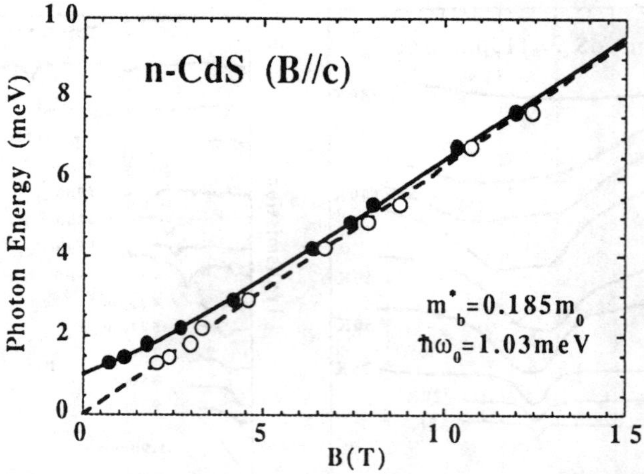

Figure 3: Transition energies as a function of magnetic field. The solid circles denote the resonance data at 4.2K and the open circles at 40K. The solid line is a theoretical line calculated by the quantum dot theory. The dashed line represents cyclotron resonance energy against magnetic field by using the effective mass, m_p^*=0.185m_0.

Acknowledgments

This work was partly supported by a Grant-in-Aid for JSPS Research Fellowships for Young Scientists.

References

1. K. J. Button, B. Lax and D. R. Cohn, Phys. Rev. Lett. **24**, 375 (1970).
2. D. R. Cohn, B. Lax, K. J. Button and W. Dreybrodt, Solid State Commun. **9A**, 441 (1971).
3. K.Nagasaka, Phys. Rev. B **15**, 2273 (1977).
4. G.D.Mahan and J.J.Hopfield, Phys. Rev. Lett. **12**, 241 (1964).
5. M.Saitoh and A.Kawabata, J. Phys. Soc. Japan **23**, 1006 (1967).
6. S. J. Miyake, Phys. Rev. **170**, 726 (1968).
7. Y. Imanaka, N. Miura and H. Kukimoto, Phys. Rev. B **49**, 16965 (1994).
8. Y. Imanaka and N. Miura, Phys. Rev. B **50**, 14065 (1994).
9. Y. Imanaka and N. Miura, Phisica B **201**, 284 (1994).
10. F. M. Peeters and J. T. devreese, Phys. Rev B **34**,7246 (1986).

A CYCLOTRON RESONANCE STUDY OF THE RESONANT POLARON COUPLING IN CdTe/CdMgTe QUANTUM WELLS

C. C. CHANG, J. G. MICHELS, H. H. CHENG, R. J. NICHOLAS,
Clarendon Laboratory, Department of Physics, University of Oxford, Parks Road, Oxford OX1 3PU, U. K.

F. M. PEETERS,
Department of Physics, University of Antwerp(UIA), B-2610 Antwerp, Belgium

X. G. WU,
Chinese Academy of Sciences, P. O. Box 912, Beijing, China

W. OSSAU, A. WAAG, G. LANDWEHR
Physikalisches Institut der Universitat Wurzburg, Am Hubland, D97074, Wurzburg, Germany

We describe a study of the resonant polaron effect by cyclotron resonance for the first time in II-VI quantum wells. Several doped CdTe/CdMgTe quantum wells have been studied both above and below the resonant polaron coupling regime using both steady and pulsed magnetic fields. The results are compared with calculations using the memory function approach, which demonstrate that there is a large reduction in the resonant coupling due to level occupancy effects.

The polaron effect modifies the electronic properties of quasi-two dimensional semiconductors made of polar materials [1]. CdTe has been recognised for several years as one of the best possible materials for testing experimentally the predictions of the polaron effect. The main reason is CdTe is one of the few polar semiconductors which has a relatively large electron-phonon interaction with a small effective mass. MBE growth has recently made it possible to achieve a mobility high enough to permit the measurement of sharp resonances in quantum wells. The three dimensional polaron effect in the CdTe bulk has been studied for some time [2]. However, this is the first time that the two dimensional properties in quantum wells are reported.

Four n-type based CdTe samples were studied, which were grown by MBE on undoped CdZnTe substrates with a 10nm undoped CdTe buffer layer followed by 300nm of CdMgTe. The quantum wells were typically 10nm wide and modulation doped in the CdMgTe barriers, using a spacer layer. The Mg concentration in the

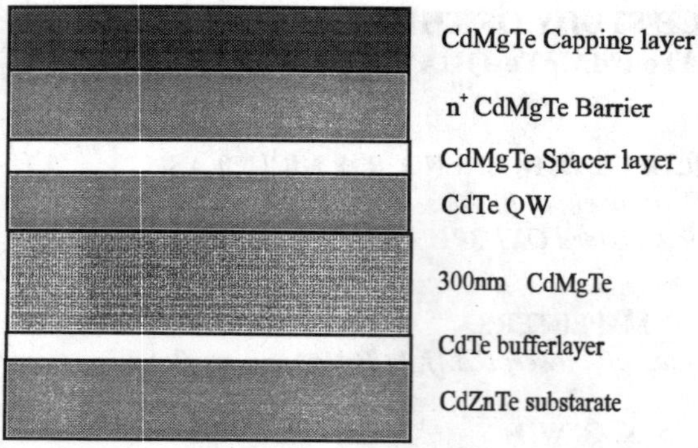

Fig. 1 A schematic diagram of the samples investigated

CdMgTe layers is around 30%. The resulting two dimensional carrier densities were in the range 7 - 8.5 x 10^{11} cm^{-2}. A schematic diagram is shown in Fig.1.

Cyclotron resonance at 4.2 Kelvin was measured with an optically pumped far infrared laser. For wavelengths longer than 77 μm, the direct transmission signals were measured in steady magnetic fields. A long (10msec) pulsed field magnet was used for shorter wavelengths due to the high field required (beyond 17 Tesla). Transmission measurements become impossible in the reststrahlen band of the CdTe and CdZnTe substrate (between 77 μm and 55 μm). Reflectivity was studied instead in this region. The traces for several different wavelengths are illustrated in Fig. 2. Clear resonances were seen in transmission both above and below the Reststrahlen band. One weak resonance was also detected in reflectivity, using a wavelength of 57.2 μm, corresponding to an energy just above the CdTe LO phonon. The reflectivity is known to give both maxima and minima at resonance, depending on the carrier density and the frequency, with minima occurring for very low densities or for frequencies within the Reststrahlen band [3]. As the density is high and we are just beyond the Reststrahl for the active layer we assume that the broad peak around 23 T is the resonance, although this point should be taken as having considerable error.

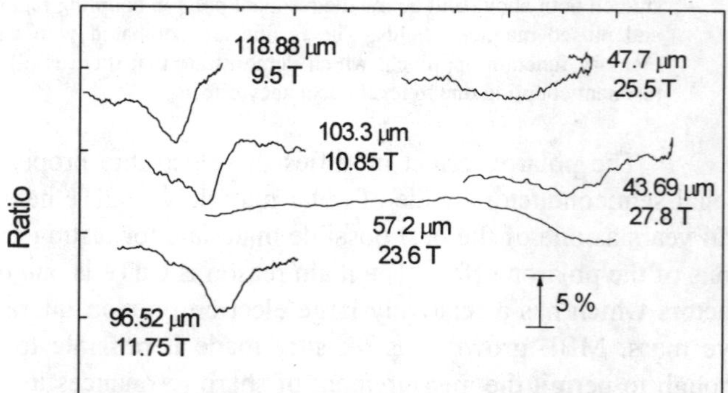

Fig.2 Normalized resonance traces in pulsed magnetic fields at different cyclotron energies.

The effective masses deduced from the experimental data are shown as a function of magnetic field in Fig. 3. The quantum well effective masses show a gentle background increase as a function of magnetic field, due to the band non-parabolicity, upon which the resonant polaron contribution to the effective mass is superimposed. There is a very small discontinuity of around 5% in the mass as the energy passed through the LO phonon frequency in the region of 20 Tesla. A small overall (~6 %) increase relative to the bulk value (0.096 m_0) is also observed resulting from the confinement potential of quantum wells.

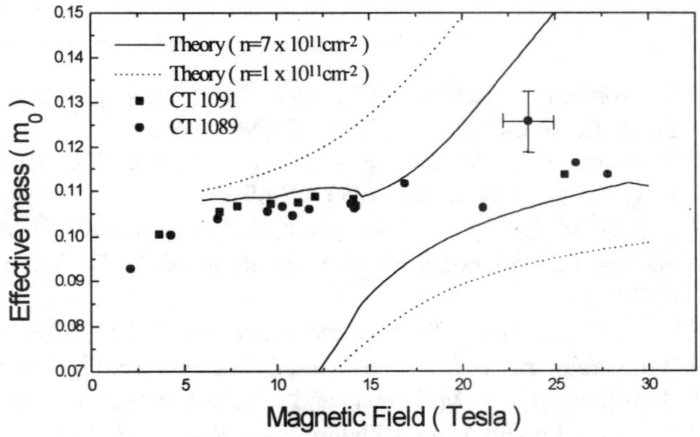

Fig. 3 Effective masses of both the samples and the theoretical prediction

Theoretical calculations were performed to second-order perturbation in magnetic fields using the memory function approach [4,5]. The results of this calculation are shown in Fig. 3, together with the experimental data for two different carrier densities. For the lower density ($1x10^{11}$ cm^{-2}) there is a large resonant polaron coupling which leads to a very large divergence in the effective mass, as would be expected. By contrast an increase in the density to $7x10^{11}$ cm^{-2} causes a very dramatic suppression of this resonant coupling, in excellent agreement with experiment. The main cause of this is the essentially full occupany of the Landau levels at low field, which blocks possible coupling between levels. It can be seen in Fig. 3 that there is in fact a small oscillatory term in the mass due to these 'Shubnikov-de Haas' oscillations of the polaron effect. Such behaviour was first observed in GaAs/AlGaAs [3] heterojunctions, following the predictions of Xiaoguang Wu et al [5], although only the lower polaron branch was observed in this less polar material. The present results allow a much clearer comparison with theory due to the larger coupling strength, and the observation of the resonances from the upper branch. Our conclusion from the comparison of the fits to the experimental data is that if anything the resonant polaron coupling has been suppressed even more than predicted by theory, which includes a treatment of only the bulk phonon modes but not interface optical phonon and confined slab LO phonon modes [6,7].

In summary, we have investigated the resonant polaron cyclotron resonance in several doped CdTe quantum wells. None of them show obviously discontinuities in the effective masses near the LO phonon frequency, which is a sign of strong coupling. This is due to the level occupancy effects resulting from the high carrier densities of the samples. A theoretical treatment including the effects of both level degeneracy and static screening is in good agreement with the experiment data.

---

[1] H. Frohlich, Advances in Physics **3**, 325 (1954); D. H. Dickey, E. J. Johnson and D. M. Larson, Phys. Rev. Lett. **18**, 599 (1967).
[2] J. Waldman, D. M. Larson, P. E. Tannenwald, C. C. Bradley, D. R. Cohn and B. Lax, Phys. Rev. Lett. **23**, 1033 (1969).
[3] C. J. G. M. Langerak, J. Singleton, P. J. Van der Wel, J. A. A. J. Perenboom, D. J. Barnes, R. J. Nicholas, M. A. Hopkins and C. T. B. Foxon, Phys. Rev. B **38**, 13133 (1988).
[4] F. M. Peeters and J. T. Devreese, Phys. Rev. B **31**, 3689 (1985).
[5] Wu Xiaoguang, F. M. Peeters and J. T. Devreese, Phys. Rev. B **34**, 8800 (1986).
[6] Guo-qiang Hai, F. M. Peeters and J. T. Devreese, Phys. Rev. B **42**, 11063 (1990).
[7] Tianquan Lu and Yisong Zheng, Phys. Rev. B **53**, 1438 (1996).

MAGNETO-OPTICAL STUDIES OF INVERTED SEMICONDUCTING HgTe/(HgCd)Te SUPERLATTICES GROWN ON CdZnTe AND GaAs SUBSTRATES

M. von TRUCHSESS, A. PFEUFFER-JESCHKE, V. LATUSSEK, C. R. BECKER, AND E. BATKE

Physikalisches Institut der Universität Würzburg, Am Hubland, D-97074 Würzburg, Germany

We report on far-infrared magneto-optical investigations of inverted semiconducting HgTe/$Hg_{1-x}Cd_xTe$ superlattices. Cyclotron resonances and various interband transitions were observed and compared to the predictions of an 8×8 $\mathbf{k} \cdot \mathbf{p}$ band structure calculation including an external magnetic field. We will discuss two characteristic transitions that allow us to distinguish an inverted from a semimetallic or normal semiconducting superlattice.

1 Introduction and Experimental Details

Typ-III superlattices composed of the semimetal HgTe and the wide-gap semiconductor $Hg_{1-x}Cd_xTe$ have the unique feature that they can exist either in the normal semiconducting, the semimetallic or the inverted semiconducting regime.[1] Here we present magneto-optical investigations of inverted semiconducting HgTe/$Hg_{1-x}Cd_xTe$ superlattices. Many different transitions that can be classified as cyclotron resonances (CR) and interband transitions (IB) were observed and compared to the predictions of an 8×8 $\mathbf{k} \cdot \mathbf{p}$ band structure calculation including an external magnetic field B. We found two transitions that are characteristic for an inverted semiconducting superlattice. One is an IB that exhibits a negative magnetic field dispersion at small magnetic fields, the other is a hole cyclotron resonance that can occur only at a magnetic field strength $B > B_C$, where B_C is the critical field for the magnetic field induced

Table 1: Sample parameters for HgTe/$Hg_{1-x}Cd_xTe$ superlattices. d_W and d_B indicate well and barrier width, respectively.

sample	d_W(nm)	d_B(nm)	x	periods	substrate
A	13.8	7.2	0.68	90	(100) $Cd_{0.96}Zn_{0.04}Te$
B	11.9	7.6	0.68	80	(100) GaAs
C	14.0	7.5	0.90	80	(211)B $Cd_{0.96}Zn_{0.04}Te$

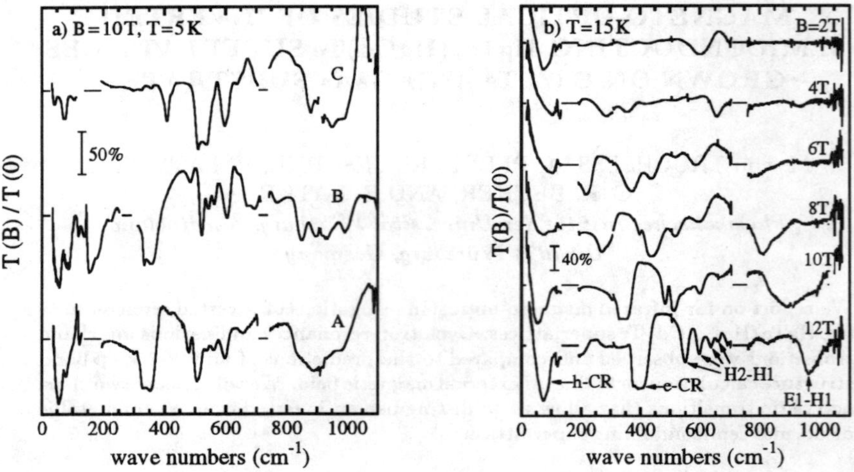

Figure 1: Normalized transmission for a) the samples of Tab. 1 and b) sample A for different magnetic field strengths.

transition from the inverted to the normal semiconducting regime. Our superlattices were grown by molecular beam epitaxy on different types of substrate.[2] The sample parameters are listed in Tab. 1. Frequency dependent transmission measurements were performed with unpolarized radiation in Faraday geometry with magnetic field up to 12.5 T in the growth direction. Resonances were obtained from the normalized transmission $T(B)/T(0)$.

2 Results and Discussions

In Fig. 1a) we compare transmission spectra for our samples at temperature T = 5 K and magnetic field strength B = 10 T. Independent of the substrates, we observe for all samples qualitatively similar spectral structures. All samples showed n-type behaviour. For samples A and B the resonance with the rectangular shaped line at about 350 cm^{-1} is an electron CR. Due to the high total electron density this line is saturated, i.e. the CR amplitude is at the theoretical limit of 50% which applies for unpolarized radiation. Because of the smaller total electron density in sample C, the CR line is not saturated. Thus we observe a nearly symmetric line shape at about 400 cm^{-1} . Transitions with frequencies above the cyclotron frequency are IB, and those with energies lower than 120 cm^{-1} are hole CR as will be shown below from an analysis of magnetic field dependent measurements for sample A (see Fig. 1b). In Fig. 1b)

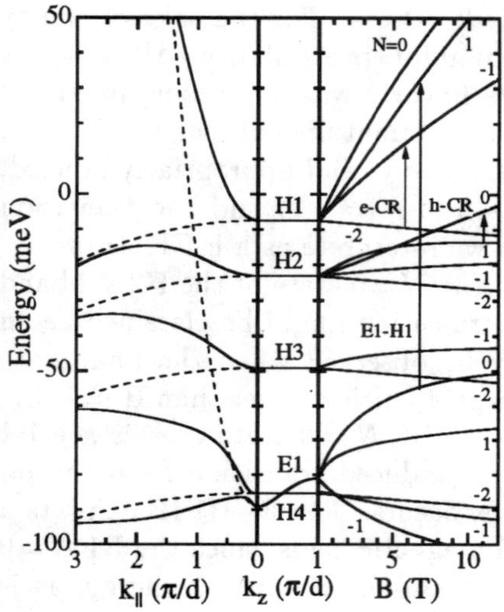

Figure 2: Calculated band structure for sample A at $T = 15$ K.

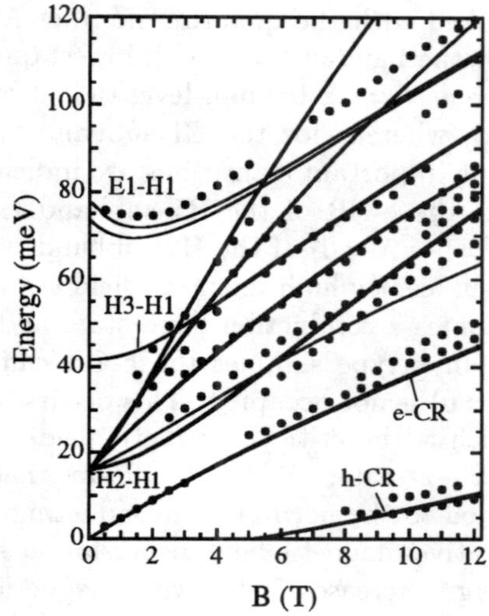

Figure 3: Comparsion between theory and experiment for sample A at $T = 15$ K.

for 12 T electron (e) and hole (h) CR, H2-H1 and E1-H1 IB are marked. All resonances shift with decreasing magnetic field to smaller energies. The h-CR disappears for B < 8T, i.e. can not be separated from the Drude background absorption which induces a frequency dependence of $T(0)$ and increases the normalized transmission in the frequency regime $\bar{\nu} < 100\,\mathrm{cm}^{-1}$. In Fig. 2 we show the calculated band structure for sample A and in Fig. 3 the experimental resonance positions for sample A are compared with the theoretically predicted transition energies. The numerical calculation takes into account the full 8×8 $\mathbf{k} \cdot \mathbf{p}$ Hamiltonian including all second-order terms representing the far-band contributions, an external magnetic field and the effects of strain. The calculation is based on a recently established set of band parameters.[3] The wave vektor $\mathbf{k_z}$ and $\mathbf{k}_{||}$ indicate the dispersion parallel and perpendicular to the growth direction, respectively. The dotted lines show the $\mathbf{k}_{||}$-dispersion if we neglect coupling of the subbands at finite $\mathbf{k}_{||}$. At $\mathbf{k}_{||}$=0 the dispersionless subbands H_j ($j = 1, \ldots, 4$) are Γ_8 heavy-hole states, whereas E1 crossing the H4 subband is essentially a mixed Γ_6-Γ_8 light-hole state. Such a band structure is referred to as inverted semiconducting, since H1 forms the conduction band and E1 a valence band. The Fermi energy for our n-type samples lies in the H1 subband. At finite magnetic fields all bands split into Landau-levels, marked with the quantum number $N = -2, -1, \ldots$. For the sake of clarity only the Landau-levels with lowest quantum numbers are shown. All H_j states have a $N = -2$ Landau-level that decreases linearly with increasing magnetic field, whereas for the E1 subband the lowest quantum number is $N = -1$. Some important transitions are indicated by arrows and appropriatly labeled, e.g. the e-CR of the H1 subband ($N = -2 \rightarrow N = -1$) and the transition ($N = 1 \rightarrow N = 0$) of the H2 subband, which we interprete as a h-CR. However, at sufficently high magnetic field strengths the $N = 0$ state of the H2 subband becomes a conduction band state and the transition might be classified as an IB. In n-type samples the h-CR can only be observed when the final state ($N = 0$) is not occupied. This occurs for magnetic fields larger than B_C=7.8 T, which is the critical magnetic field where the H2 $N = 0$ state crosses the H1 $N = -2$ state. At B=B_C the magnetic field induced transition from the inverted to the normal semiconducting regime occurs. The E1-H1 IB exhibits a negative magnetic field dispersion at small magnetic fields, since the E1 $N = 0$ energy increases faster with magnetic field than the H1 $N = 1$ energy, as is observed (see Fig. 3). At higher magnetic fields, E1 $N = 0$ is essentially independent of B such that the transition energy is dominated by the magnetic field dependence of the H1 $N = 1$ state. In the limit of vanishing magnetic fields, only the H2-H1 IB can be nearly linearly extrapolated to the H2-H1 subband separation. In conclusion, CR and IB were investigated from magneto-optical measurements on inverted semiconducting $HgTe/Hg_{1-x}Cd_xTe$ superlattices. Good agreement between the experiment and a $\mathbf{k}\cdot\mathbf{p}$ band structure calculation was obtained. Two characteristic transitions for the inverted semiconducting regime were discussed.

Acknowledgements

We gratefully acknowlegde financial support by the Deutsche Forschungsgemeinschaft and the Bundesministerium für Forschung und Technologie.

References

1. N. F. Johnson, P. M. Hui, and H. Ehrenreich, *Phys. Rev. Lett.* **61**, 1993 (1988).
2. C. R. Becker, L. He, M. M. Regnet, M. M. Kraus, Y. S. Wu, G. Landwehr, X. F. Zhang, and H. Zang, *J. Appl. Phys.* **74**, 2486 (1993).
3. M. von Truchsess, V. Latussek, R. Sizmann, P. Helgesen, F. Goschenhofer, C. R. Becker, G. Landwehr, and E. Batke, *Phys. Rev.* B **51**, 17618 (1995).

MAGNETOOPTICAL PROPERTIES OF GATED HgTe/CdTe HETEROSTRUCTURES

M. SCHULTZ, F. HEINRICHS, A. ECKSTEIN, U. MERKT

Institut für Angewandte Physik und Zentrum für Mikrostrukturforschung, Universität Hamburg, Jungiusstraße 11, 20355 Hamburg, Germany

T. COLIN, P. HELGESEN, T. SKAULI, S. LØVOLD

Norwegian Defence Research Establishment, PO Box 25, Kjeller, Norway

We study a metal-oxide-semiconductor (MOS) structure with a HgTe quantum well by far-infrared Fourier spectroscopy. In particular, this allows us to investigate filling factor dependent splittings of the cyclotron resonance in strong magnetic fields. We also find evidence for a new resonance which indicates the transition from the inverted to the normal semiconducting band structure in a magnetic field.

1 Introduction and experimental details

Narrow gap HgTe/CdTe heterostructures can have several interesting properties, like small effective electron mass, inverted band structure, large effective Landé g^* factor and large optical nonlinearity, some of which could be interesting for future device applications. Here we report on a single-quantum well in which the electron density can be varied via the electric field effect.

The HgTe quantum well (115Å±10Å) was grown between two CdTe layers by MBE on a CdZnTe(211)B substrate. On this sample, a MOS capacitor was fabricated via a 1000Å thin plasma enhanced chemical vapour deposited (PECVD) SiO_2 gate insulator and a semitransparent 50Å NiCr gate. [1] To eliminate interference effects the back of the sample was wedged at an angle of about 3°. The electron gas was contacted at the edges of the sample to enable Shubnikov-de Haas (SdH) measurements in van der Pauw geometry. In the spectroscopic experiments the relative change of transmittance $-\Delta T/T = -(T(V_g) - T(V_g = 0))/T(V_g = 0)$ between the actual gate voltage V_g and $V_g = 0$ is recorded with a Fourier transform spectrometer. At the threshold voltage $V_{th} = 4V$, the conduction band of the quantum well is depleted of electrons. By varying the gate voltage the electron density can be tuned between $n_s = 0$ and $n_s = 1.3 \cdot 10^{12} cm^{-2}$. A typical mobility $\mu = 7.5 \cdot 10^4 cm^2V^{-1}s^{-1}$ is determined for a density of $1 \cdot 10^{12} cm^{-2}$ from the linewidth of the cyclotron resonance (CR) at $B = 3T$.

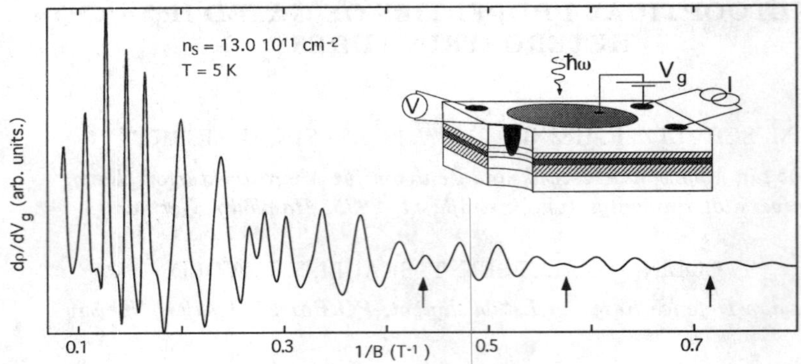

Figure 1: Shubnikov-de Haas (SdH) oscillations measured as the derivative of the magnetoresistance with respect to the gate voltage. The arrows mark the beat nodes due to the Rashba effect. At higher magnetic fields ($B \geq 5T$) the frequency of the oscillations doubles as a result of Zeeman spin splitting. The inset shows a sketch of the sample.

2 Results and discussion

Figure 1 depicts SdH oscillations for a carrier density $n_s = 1.26 \cdot 10^{12} cm^{-2}$. At lower magnetic fields a beating pattern is apparent which is caused by the spin splitting of the ground electric subband due to the Rashba effect. At higher fields ($B^{-1} \leq 0.2T^{-1}$) the frequency of the oscillations doubles. In this high-field regime the Zeeman spin splitting of the Landau levels becomes more important and leads to the doubling of the frequency. The carrier densities deduced from the SdH measurements are used for the analysis of the CR measurements in the high-field regime. In Fig. 2 (a) we show CR measurements for a fixed carrier density $n_s = 1.3 \cdot 10^{12} cm^{-2}$ for various magnetic field strengths B. The CR shows a strong splitting for $B = 12T$. As the magnetic field is decreased, the split resonances exchange their oscillator strengths. We observe the onset of new resonances roughly at integer filling factors $\nu = \frac{n_s h}{eB}$ as can be seen in the inset of Fig. 2 (a). This scaling of the onset with the filling factor is observed for all carrier densities. Filling factor dependent splittings of the CR in narrow-gap semiconductors have been observed previously by Truchsess *et al.* in HgTe/CdTe multiple-quantum wells [2] and Scriba *et al.* in InAs/AlSb single-quantum wells [3].

Figure 3 (a) shows measurements for a magnetic field of $B = 9T$ and carrier densities $n_s = 2.9$ to $13.0 \cdot 10^{11} cm^{-2}$. The cyclotron masses that correspond to the split resonances are plotted versus density in Fig. 3 (b). Again, the onsets of the splittings appear at integer filling factors ν. Note that a resonance with a

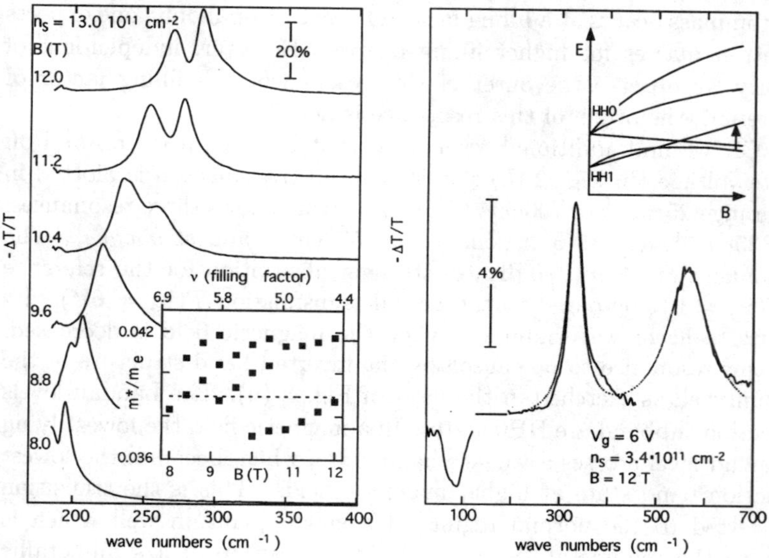

Figure 2: (a) Cyclotron resonance (CR) spectra for various magnetic fields and a fixed carrier density of $n_s = 1.3 \cdot 10^{12} cm^{-2}$ at $T = 5K$. The inset shows the cyclotron masses versus magnetic field strength. The dashed line indicates the effective mass observed in low magnetic fields ($B \leq 3T$). (b) Normalized transmission spectra for a fixed carrier density of $n_s = 3.4 \cdot 10^{11} cm^{-2}$ and a magnetic field of $B = 12T$ at $T = 5K$. The inset shows schematically the inverted band structure of the HgTe quantum well as a function of magnetic field strength. The dotted line is a two-Lorentzian fit.

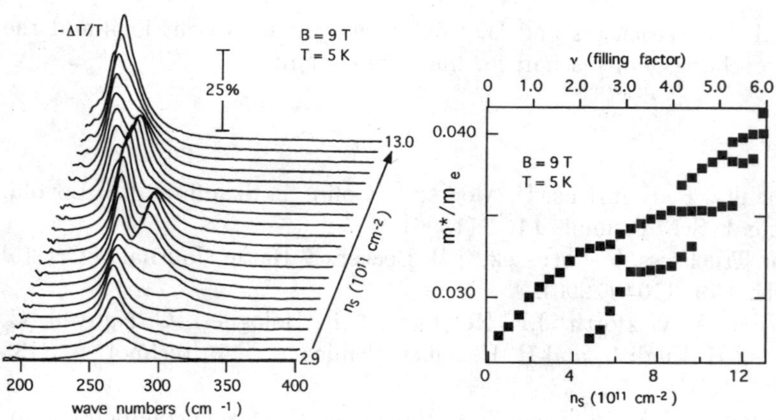

Figure 3: (a) CR for a fixed magnetic field of $B = 9T$ and various carrier densities $n_s = 2.9 - 13.0 \cdot 10^{11} cm^{-2}$. (b) Resonance positions interpreted in terms of effective masses.

lower cyclotron mass occurs at a filling factor of $\nu \simeq 2$. Commonly, one expects higher cyclotron masses for higher filling factors. However, independent of carrier density we observe the onset of this resonance at a filling factor of $\nu \simeq 2$. At present, the origin of this resonance is not clear.

Besides CR we find additional resonances that can be turned on and off with the gate voltage. In Fig. 2 (b) the normalized transmission is plotted in the spectral range $25cm^{-1}$ to $700cm^{-1}$. The spectrum shows three resonances: the CR at $335cm^{-1}$, as well as resonances at $570cm^{-1}$ and at $95cm^{-1}$. The inverted resonance at $95cm^{-1}$ indicates stronger absorption for the reference spectrum $T(V_g = 0)$ compared to the actual transmission $T(V_g = 6V)$. Its position shifts to lower wave numbers when the magnetic field is decreased. We assume this resonance to be caused by the inverted band structure of the HgTe quantum well as sketched in the inset of Fig. 2 (b). The Landau levels of the conduction subband are HH0 states. In a magnetic field the lowest lying conduction band level crosses a valence band state, which becomes the lowest lying conduction band state at higher magnetic fields. This is the transition from the inverted to the normal regime of a single-quantum well which is very similar to the semimetal-to-semiconductor transition of a semimetallic superlattice[4]. At zero gate voltage this level is depleted and transitions which have a much lower energy than the CR become allowed (see arrow). We identify the resonance at $95cm^{-1}$ for $B = 12T$ with these transitions. The resonance at $530cm^{-1}$ presumably is a $HH1 \rightarrow HH0$ intersubband transition.

Acknowledgments

We thank M. von Truchsess and U. Rößler for valuable discussions and the Deutsche Forschungsgemeinschaft for financial support.

References

1. M. Schultz, F. Heinrichs, U. Merkt, T. Colin, T. Skauli, and S. Løvold, Semicond. Sci. Technol. **11**, 1 (1996).
2. M. von Truchsess, V. Latussek, C.R. Becker, E. Batke, Journal of Crystal Growth **159**, 1104 (1996).
3. J. Scriba, A. Wixforth, J.P. Kotthaus, C.R. Bolognesi, C. Nguyen, G. Tuttle, J.H. English, and H. Kroemer, Semicond. Sci. Technol. **8**, S133 (1993).
4. J.R. Meyer, C.A. Hoffman, F.J. Bartoli in II-VI semiconductor compounds, ed. M. Jain (World Scientific, Singapore, 1993)

DETERMINATION OF $Hg_{1-x}Mg_xTe$ BAND PARAMETERS

B. KÖNIG, A. PFEUFFER-JESCHKE, S. OEHLING, C. R. BECKER, AND E. BATKE

Physikalisches Institut der Universität Würzburg, D-97074 Würzburg, Germany

The band structure of $Hg_{1-x}Mg_xTe$ ($0.04 \leq x \leq 0.20$) epitaxial layers was investigated by temperature dependent magneto-optical measurements. Electron cyclotron resonances and interband transitions were analysed in the framework of the quasi-Germanium model. A set of band parameters is obtained that excellently describes the experiment over a wide regime of temperatures.

1 Introduction and Experimental Details

We report here on temperature dependent magneto-optical investigations of the novel alloy $Hg_{1-x}Mg_xTe$, and the resulting parameters for $\mathbf{k} \cdot \mathbf{p}$ band structure calculations. In the framework of the quasi-Germanium model [1] an universal set was obtained that excellently describes the experiment with a temperature and composition dependent gap between the bands of Γ_6 and Γ_8 symmetry in a form according to Laurenti.[2] Although the quasi-Germanium model is strictly valid only for systems with inversion symmetry, contributions arising from the lack of inversion are extremely small and can be neglected here. It will be demonstrated that for our composition ($0.04 \leq x \leq 0.20$) a semimetal with inverted band structure and a narrow-gap semiconductor can be realized.

Our $Hg_{1-x}Mg_xTe$ samples were epitaxial layers with thicknesses d of about 1 μm grown at 180°C on (100) $Cd_{0.96}Zn_{0.04}Te$ substrates by molecular beam epitaxy.[3] To improve the material quality a thin CdTe buffer was deposited initially. The growth conditions were chosen such that crystallization in zinc blende structure occured. Frequency and temperature (5 K to 200 K) dependent transmission experiments were performed with unpolarized radiation in the Faraday geometry with magnetic field direction parallel to the [100] crystal axis. Electron cyclotron resonances (CR) and interband transitions (IB) were obtained from normalized transmission spectra $T(B)/T(0)$.

2 Results and Discussion

Depending on the composition one can expect a zero-gap (semimetal), narrow-gap or wide-gap semiconductor for $Hg_{1-x}Mg_xTe$ as is well known for the alloy

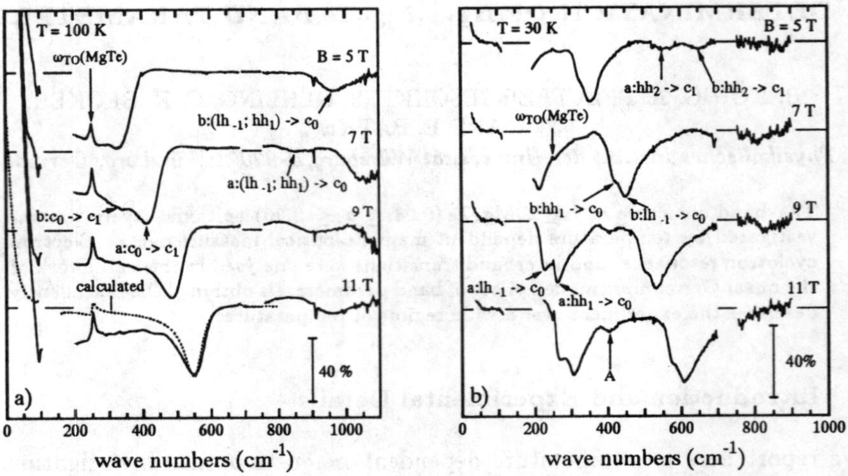

Figure 1: Magnetic field dependent spectra for a) $Hg_{0.90}Mg_{0.10}Te$ and b) $Hg_{0.94}Mg_{0.06}Te$.

$Hg_{1-x}Cd_xTe$. However, due to the lower mass of Mg compared to Cd the gap between the bands of Γ_6 and Γ_8 symmetry can be extended to a value more than twice as large for $Hg_{1-x}Mg_xTe$ ($\approx$ 3.5 eV) than for $Hg_{1-x}Cd_xTe$ ($\approx$ 1.6 eV).[3] Magnetic field dependent spectra $T(B)/T(0)$ are shown in Fig.1 for a a) semiconducting and b) semimetallic layer. CR and IB are observed which shift with increasing magnetic field strength B to higher energies. In the frequency regimes of $100 \leq \bar{\nu} \leq 200\,\mathrm{cm}^{-1}$ and $700 \leq \bar{\nu} \leq 750\,\mathrm{cm}^{-1}$, the radiation is blocked due to the reststrahlen absorption in the substrate and absorption by an optical window. The sharp peak at $250\,\mathrm{cm}^{-1}$ is due to influences of the transverse optical MgTe-like phonon mode, and the resonance marked A in Fig.1b) arises from an acceptor in the substrate. The increase in the normalized transition for energies smaller than about $100\,\mathrm{cm}^{-1}$ arises from Drude absorption of quasi-free carriers, which influence $T(0)$. The CR and IB are marked according to the notation of Weiler [1] assuming the quasi-Germanium model. At zero wave vector k_z parallel to the magnetic field direction, the electron and hole Landau-levels can be classified in decoupled sets a and b. The abbreviations c, lh and hh indicate the conduction, light-hole and heavy-hole bands, respectively, and the lower index denotes the Landau-level quantum number N. CR and IB can easily be identified from their characteristic line shapes. CR exhibits always a lower, IB a higher energy tail. This can easily be understood from the k_z-dispersion of the bands for motion parallel to the magnetic field

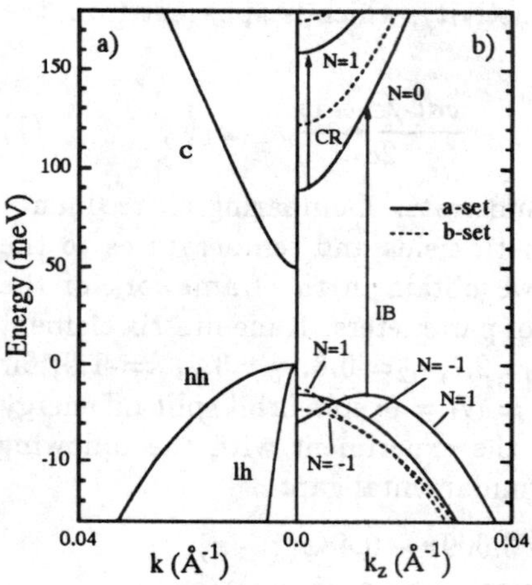

Figure 2: Band dispersion in [100] direction according to the quasi-Germanium model for $Hg_{1-x}Mg_xTe$ with $\Gamma_6 - \Gamma_8$ band gap of 50 meV for a) B=0 and b) B=10 T.

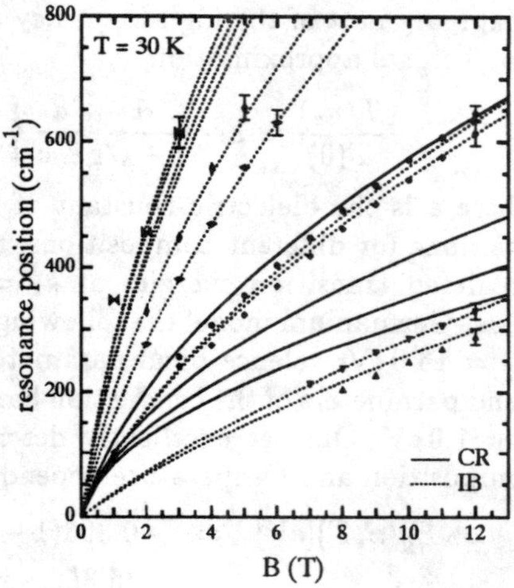

Figure 3: Experimental and calculated transition energies for $Hg_{0.94}Mg_{0.06}Te$.

direction. In Fig.2a) the dispersion of the c, hh and lh bands for the [100] direction is shown for a semiconducting sample with a fundamental gap of 50 meV. In a magnetic field the bands split into Landau-levels as shown in Fig.2b) for the c ($N = 0, 1$), hh ($N = 1$) and lh ($N = -1$) bands. For a better overview different scales for postive and negative energies are chosen. Please note the 'camel-back' structure for the hole Landau-bands of the a-set. To our knowledge no direct verification has been provided yet for this interesting feature. In our experiment the mobility is not sufficient to allow its resolution. CR and IB show a different asymmetry of the line shapes, since with increasing k_z the separation of the c band Landau-levels decrease, whereas the interband separation increase. However, the asymmetry of the CR is not caused by the k_z-dispersion of the conduction band alone. A further contribution to the form of the line shape arises from an optical interference effect in the epitaxial layer. This is apparent if we compare a calculated $T(B)/T(0)$ assuming a Drude-type high-frequency conductivity $\sigma_+(\omega) = \sigma'_+(\omega) - i\sigma''_+(\omega) = [n_e e^2 \tau]/\{m^*[1+i(\omega-\omega_c)\tau]\}$ for a parabolic band as shown in Fig.1a) with a dotted line. From this calculation we obtain an electron density $n_e = 6.5 \cdot 10^{17}\,\text{cm}^{-3}$ and a scattering time $\tau = 0.2\,\text{ps}$. The influence of interference induces a coupling of the real and imaginary part of the high-frequency conductivity, which is apparent from the small signal approximation

$$\frac{T(n_e)}{T(0)} \approx 1 - \frac{1}{1+\sqrt{\varepsilon}\,\varepsilon_0 c}\frac{d}{\ }\left\{\sigma'_+(\omega) - \frac{\omega d(\sqrt{\varepsilon}-1)}{2c}\sigma''_+(\omega)\right\}, \qquad (1)$$

where ε is the dielectric constant of the substrate. Comparing all resonance positions for different compositions, magnetic fields and temperatures to the predicted transition energies at $k_z = 0$ we obtain in the framework of the quasi-Germanium model the following set of parameters: Kane matrix element $E_p = 16.0\,\text{eV}$; valence band parameters γ_1=2.7, γ_2=-0.6, γ_3=0.3, κ=-1.9; far band parameters of the conduction band $F = N_1 = 0$; spin orbit split-off energy Δ_0=1.0 eV. Our set excellently describes the experiment with the following composition and temperature dependent fundamental gap:

$$E_g(x,T)[eV] = -0.303(1-x) + 3.669x - 0.48x(1-x) + \frac{[6.3(1-x) - 8.99x + 13.0x(1-x)]10^{-4}T^2}{11.0(1-x) + 140.5x + T}. \qquad (2)$$

In Fig.3 the good aggreement between experiment and theory is demonstrated. At fixed T=5 K, a transition from a semimetallic to a normal semiconducting layer occurs for $x \approx 0.086$. We have experimental evidence that for $x = 0.06$ there is also a temperature induced transition at $T \approx$170 K.

In conclusion, band parameters for $Hg_{1-x}Mg_xTe$ ($0.04 \leq x \leq 0.20$) epitaxial layers were determined from magneto-optical investigations in the framework of the quasi-Germanium model.

Acknowledgements

We would like to gratefully acknowlegde financial support by the Deutsche Forschungsgemeinschaft and the Bundesministerium für Forschung und Technologie.

References

1. M. H. Weiler, in *Defects*, (HgCd)Se, (HgCd)Te, edited by R. K. Willardson and A. C. Beer, *Semiconductors and Semimetals* Vol. **16** (Academic Press, New York. 1981), p.119.
2. J. P. Laurenti, J. Camassel, A. Bouhemadou, B. Toulouse, R. Legros and A. Lussau *J. Appl. Phys.* **67**, 6454 (1990).
3. S. Oehling, U. Lunz, H. Heinke, G. Plahl, C. R. Becker and G. Landwehr, *J. Cryst. Growth* **159**, 1157 (1996).

ZEEMAN EFFECT IN DILUTED MAGNETIC SEMICONDUCTORS

S. Rodriguez[1], *E. Kartheuser*[2], *and M. Villeret*[3]

[1]Department of Physics, Purdue University, West Lafayette, IN 47907, U.S.A.
[2]Institut de Physique, Université de Liège, B-4000 Liège, Belgium
[3]Department of Mathematics, City University, London EC1V 0HB, U.K.

We present an analysis of the splitting of the energy levels of the magnetic ions in diluted magnetic semiconductors (DMS's) in the presence of a magnetic field $\boldsymbol{B}$. The g-factors of the energy levels in the lowest terms of Fe, Ni, Co and Cu in zinc-blende DMS's are obtained. The magnetic-field-dependence of the far infrared Raman lines of an iron-based wurtzite DMS is described. The effect of the strong sp-d exchange interaction between the conduction and valence electrons on the one hand and the 3d localized states of the magnetic ion on the other gives rise to large splittings of the exciton levels of the DMS's. As an application we display the energy levels of excitons in $Cd_{1-x}Fe_xTe$ as functions of $|\boldsymbol{B}|$ for two directions of $\boldsymbol{B}$ with respect to the crystallographic axes.

I. Introduction

Diluted Magnetic Semiconductors (DMS's) [1] are materials obtained by alloying a compound AB, where A and B are elements of the groups II and VI of the periodic table of the elements, respectively, with MB where M is a transition metal ion. They are usually designated by the chemical formula $A_{1-x}M_xB$ where x is the concentration of the magnetic ion. These II-VI semiconductors crystallize in the zinc-blende or the wurtzite structure and the element M enters the crystal substitutionally at cation sites in the compound AB. DMS's exhibit a rich variety of novel physical phenomena and provide a unique combination of semiconducting and magnetic properties. They are of great fundamental and technological interest because of the tunability of their properties by control of the concentration of magnetic impurities. Recently, quaternary compounds containing two different transition metal ions, e.g., $Cd_{1-x_1-x_2}Mn_{x_1}Fe_{x_2}Te$, have also received attention [2].

The physical properties of the transition metal ions in DMS's are strongly influenced by the crystalline environment. As a consequence, DMS's display a wide range of unusual responses to external magnetic fields. This paper will focus on some of the manifestations of an externally applied magnetic field on the properties of diluted magnetic semiconductors.

Depending on the number of electrons in the partially filled 3d-shell of the magnetic ion DMS's behave as paramagnets or display a magnetic-field-induced temperature independent paramagnetism (Van Vleck paramagnetism). DMS's containing

Sc^{2+}, Ti^{2+}, V^{2+}, Mn^{2+}, Co^{2+}, or Cu^{2+} have degenerate ground states and behave as paramagnets while the ground states of Fe^{2+}, Cr^{2+} and Ni^{2+} are singlets and, hence, possess no permanent magnetic moment and exhibit Van Vleck paramagnetism.

In Sec. II, we present a systematic study of the energy level spectra of transition metal ions in DMS's in the absence of an external magnetic field. The lowest terms of the magnetic ions are split by the crystal potential. In zinc-blende crystals the symmetry of the site of the impurity is tetrahedral (point group T_d) whereas in wurtzite-structure crystals a distortion along a [111] direction results in trigonal symmetry (point group C_{3v}). The spin-orbit interaction partially removes the degeneracies of the crystal-field-split states.

Section III is devoted to the effect of an external magnetic field on the energy levels of selected transition metal ions in DMS's. These effects are reflected, e.g., in the g-factors that can be measured by paramagnetic resonance experiments [3, 4]; in the striking anisotropy of the magnetization of iron-based DMS's having the zinc-blende structure [5]; and in the magnetic field dependence of the Raman lines of $Cd_{1-x}Fe_xSe$ [6].

The exchange interaction between s- or p-like band electrons and the localized electrons from the partially filled 3d shells of the transition metal ion also plays an important role on the properties of DMS's [1]. This strong exchange interaction manifests itself for example in the large splitting of the exciton lines in an externally applied magnetic field [7]. Section IV will be devoted to a discussion of this effect.

II. Energy levels of magnetic ions in II-VI compounds

The properties of the magnetic impurity when embedded in a semiconductor differ from those of the free ion. In the crystal, the magnetic ion is surrounded by other ions giving rise to an electrostatic potential called the crystalline potential (or crystal field). The total Hamiltonian of the ion in the absence of a magnetic field is

$$H = H_0 + V_c + \lambda \boldsymbol{L} \cdot \boldsymbol{S} \tag{1}$$

where H_0 is the Hamiltonian of the free ion excluding the spin-orbit interaction, $\lambda \boldsymbol{L} \cdot \boldsymbol{S}$, and V_c is the crystal potential.

Besides Mn^{2+}, all the doubly ionized iron-group ions have ground states that are either D or F terms. Symmetry arguments show that while a D term splits into a doublet Γ_3 and a triplet Γ_5 in a tetrahedral field, F terms split into $\Gamma_2 + \Gamma_4 + \Gamma_5$. The spin-orbit interaction partially removes the degeneracies of these multiplets. The lowest electronic configurations of Fe^{2+}, Co^{2+}, Ni^{2+}, and Cu^{2+} are $(3d)^6$, $(3d)^7$, $(3d)^8$, and $(3d)^9$, respectively, and the energy level spectra of these ions in a tetrahedral crystal field are displayed in Fig. 1. The spectra for Cr^{2+}, V^{2+}, Ti^{2+}, and Sc^{2+} are obtained by reversing the order of the crystal-field-split levels of Fe, Co, Ni, and Cu, respectively and changing the sign of the spin-orbit coupling constant. Cr^{2+} is a special case in that its orbital ground state is split by a static Jahn-Teller distortion [8]. The case of Mn^{2+} does not appear in Fig. 1 and must be considered separately. Mn^{2+} has an exactly half-filled 3d shell and its ground state is $^6S_{5/2}$; the crystal field has a negligible effect on its properties and isolated Mn^{2+} ions in DMS's behave as if they were free.

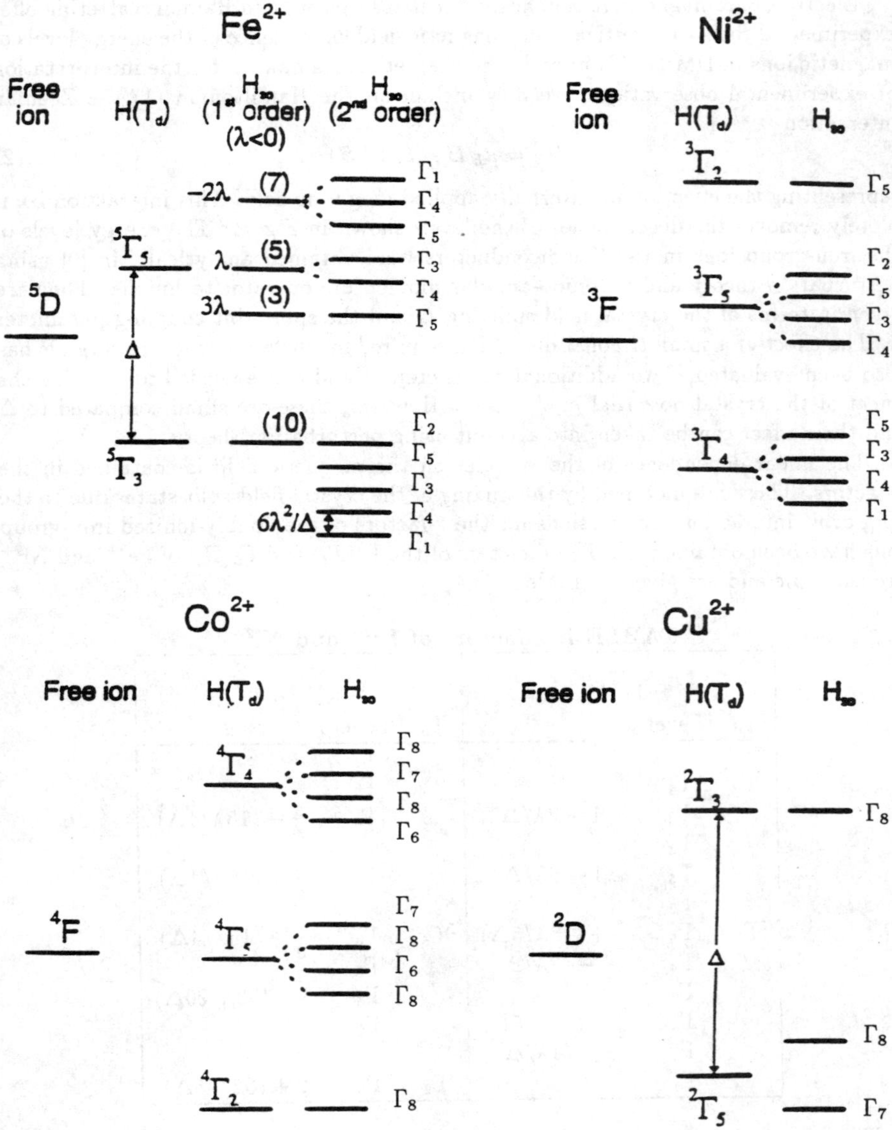

Figure 1. Energy level spectra of iron-group ions in a tetrahedral potential. From left to right we show the splittings of the D and F terms of the free ion caused by the crystal field and the spin-orbit interaction.

III. Zeeman energy levels

Electron Paramagnetic Resonance, optical absorption, and Raman scattering offer experimental tools to investigate the magnetic field dependence of the energy levels of magnetic ions in DMS's. To provide the theoretical framework for the interpretation of experimental observations, we now include in the Hamiltonian (1) the Zeeman interaction

$$H_Z = \mu_B \boldsymbol{B} \cdot (\boldsymbol{L} + 2\boldsymbol{S}) \tag{2}$$

representing the effect of an externally applied magnetic field. This interaction completely removes the degeneracies of the levels shown in Fig. 1. The energy levels of the iron-group ions in II-VI semiconductors were obtained analytically in [9] using perturbation theory and a pseudo-angular momentum operator technique. They are given in terms of the crystal field splitting Δ and the spin-orbit coupling parameter λ. The effect of a small trigonal distortion required in wurtzite-structure crystals has also been evaluated. Two additional parameters, b and c, are needed to describe the effect of the crystal potential in this case. However, these are small compared to Δ and their effect can be taken into account using perturbation theory.

The linear dependence of the energies on the magnetic field is contained in the g-factors. These are modified by the mixing of the crystal-field-split states due to the spin-orbit interaction. Expressions for the g-factors of the doubly-ionized iron-group ions have been obtained [9]. The g-factors of the $\Gamma_4(T_d)$ and $\Gamma_5(T_d)$ of Fe^{2+} and Ni^{2+} are isotropic and are given in Table I.

TABLE I. g-factors of Fe^{2+} and Ni^{2+}

Fe^{2+}			Ni^{2+}		
T_d	Level	g	T_d	Level	g
$^5\Gamma_3$	Γ_1		$^3\Gamma_4$	Γ_1	
	Γ_4	$1+2\lambda/\Delta$		Γ_4	$\frac{1}{4}-(15\lambda/4\Delta)$
	Γ_3			Γ_3	
	Γ_5	$1-6\lambda/\Delta$		Γ_5	$\frac{1}{4}+(15\lambda/4\Delta)$
	Γ_2				
$^5\Gamma_5$	Γ_5	$\frac{7}{2}+(18\lambda/5\Delta)$	$^3\Gamma_5$	Γ_4	$\frac{5}{4}-(15\lambda/4\Delta)$
	Γ_4	$\frac{3}{2}-(2\lambda/\Delta)$		Γ_3	
	Γ_3			Γ_5	$\frac{5}{4}-(53\lambda/20\Delta)$
	Γ_5	$\frac{3}{2}-(12\lambda/5\Delta)$		Γ_2	
	Γ_4	$\frac{1}{2}-(4\lambda/\Delta)$			
	Γ_1		$^3\Gamma_2$	Γ_5	$2+(32\lambda/5\Delta)$

The states of Co^{2+} and Cu^{2+} shown in Fig. 1 belong to the irreducible representations Γ_6, Γ_7 or Γ_8 of the group T_d. The matrix representation of the Zeeman

Hamiltonian can be written as

$$H_Z^{(6,7)} = \frac{1}{2} g^{(6,7)} \mu_B \boldsymbol{B} \cdot \boldsymbol{\sigma} \tag{3}$$

for the Γ_6 and Γ_7 levels, and

$$H_Z^{(8)} = g_1 \mu_B \boldsymbol{B} \cdot \boldsymbol{S} + g_2 \mu_B \sum_i B_i S_i^3 \tag{4}$$

for the Γ_8 states where $\boldsymbol{\sigma}$ is the Pauli matrix and $\boldsymbol{S}$ the angular momentum operator for $S = 3/2$. These forms, demanded by symmetry, are the only invariants of T_d linear in $\boldsymbol{B}$. The quantities $g^{(6,7)}$, g_1, and g_2 are listed in Tables II and III for Co^{2+} and Cu^{2+}.

TABLE II. g-factors of Co^{2+}.

		Co^{2+}		
T_d	Level	$g^{(6,7)}$	g_1	g_2
$^4\Gamma_4$	Γ_6	$\frac{13}{3} + (75\lambda/8\Delta)$		
	Γ_8		$\frac{16}{15} + (65\lambda/4\Delta)$	$-5\lambda/\Delta$
	Γ_7	$1 - (75\lambda/8\Delta)$		
	Γ_8		$\frac{5}{2} + (435\lambda/16\Delta)$	$\frac{6}{5} - (45\lambda/4\Delta)$
$^4\Gamma_5$	Γ_8		$\frac{35}{6} - (435\lambda/16\Delta)$	$-\frac{14}{5} + (45\lambda/4\Delta)$
	Γ_6	$\frac{7}{3} + (75\lambda/8\Delta)$		
	Γ_8		$\frac{8}{5} - (33\lambda/4\Delta)$	$5\lambda/\Delta$
	Γ_7	$3 - (75\lambda/8\Delta)$		
$^4\Gamma_2$	Γ_8		$2 - (8\lambda/\Delta)$	0

TABLE III. g-factors of Cu^{2+}.

		Cu^{2+}		
T_d	Level	$g^{(6,7)}$	g_1	g_2
$^2\Gamma_3$	Γ_8		$\frac{7}{3} + (9\lambda/\Delta)$	$-\frac{4}{3} - (4\lambda/\Delta)$
$^2\Gamma_5$	Γ_7	-2		
	Γ_8		$-9\lambda/\Delta$	$4\lambda/\Delta$

A striking example of the effect of a magnetic field on DMS's is provided by the unusual response to external magnetic fields of iron-based DMS's having the cubic zinc-blende structure. For low fields. $\boldsymbol{B}$, the magnetic susceptibility is isotropic as expected for cubic crystals. However, when the magnetic field is sufficiently high so that the Zeeman energy cannot be neglected in comparison to the spacing of the low-lying levels of Fe^{2+} (owing to the spin-orbit interaction), the nonlinear magnetization exhibits a striking anisotropy. This phenomenon is related to the Van Vleck paramagnetism primarily due to the nature of the ground state of Fe^{2+}, namely a singlet in the crystal potential of tetrahedrally coordinated semiconductors. The high-field anisotropy of the magnetization has been observed by Testelin *et al.* [10] in $Cd_{1-x}Fe_xTe$. To understand the physical origin of the anisotropy it is best to use a perturbation picture [11]. The Zeeman interaction (2), being a pseudo-vector operator, can only connect the ground state Γ_1 of Fe^{2+} (see Fig. 1) to a state of symmetry Γ_4 so that, in second order of perturbation theory, the induced magnetic moment,

$$\mu(\hat{n}) = -\frac{\partial E_0^{(2)}}{\partial B} = 2\mu_B^2 B \sum_{\nu} \frac{|< \nu \mid (\boldsymbol{L}+2\boldsymbol{S})\cdot\hat{n} \mid 0 >|^2}{E_\nu - E_0} \tag{5}$$

is independent of the orientation of B. In Eq. (5), the summation over ν is over all excited states, μ_B is the Bohr magneton and $\hat{n}$ is a unit vector in the direction of the applied magnetic field. Γ_1 is connected directly to Γ_4 only but, in turn, the matrix elements of H_Z between Γ_4 and Γ_3 are different from zero when $\boldsymbol{B} \parallel < 001 >$ but vanish when $\boldsymbol{B}$ is along $< 111 >$ or $< 110 >$. Therefore in fourth-order perturbation theory, the contribution to $\mu(\hat{n})$ cubic in B depends on the orientation of the magnetic field with respect to the cubic axes. Sixth- and eigth-order perturbation terms also contribute to the anisotropy in the magnetic moment and introduce the anisotropy between the $< 111 >$ and $< 110 >$ directions. The anisotropy of the magnetization is also reflected in the magnetic field dependence of the exciton energies as discussed in Sec. IV.

As a final illustration of the effect of a magnetic field on the energy level spectra of transition metal ions in DMS's we discuss the magnetic field dependence of the Raman lines of Fe^{2+} in CdSe [12]. Studies of the electronic Raman spectrum of $Cd_{1-x}Fe_xSe$ as a function of magnetic field were carried out by Mauger *et al.* [6]. CdSe crystallizes in the wurtzite structure and the site symmetry of Fe^{2+} is then C_{3v}. The first excited Γ_4 state of the energy level spectrum of Fe^{2+} given in Fig. 1 splits into $\Gamma_2(C_{3v})$ and $\Gamma_3(C_{3v})$. Mauger *et al.* [6] observed the Raman transitions between the ground state Γ_1 and these first two excited states. These electronic transitions occur in the far infrared region of the electromagnetic spectrum. Absorption studies of $Cd_{1-x}Fe_xSe$ in the near infrared in zero-magnetic field have been carried out by Udo *et al.* [13]. To give a theoretical interpretation of the magnetic field dependence of the Raman lines observed by Mauger *et al.* [6] we first need to determine the constants Δ, λ, b, c taken as phenomenological parameters. Since the precision of the Fourier-transform absorption measurements is greater than that of Raman scattering data we use the optical absorption lines to determine these parameters.

Figure 2 shows the Raman lines as functions of the magnetic field $\boldsymbol{B}$ parallel to and perpendicular to $\hat{\zeta}$, the trigonal axis. The solid lines represent the calculated results obtained using the parameters $\Delta = 2550.6$ cm^{-1}, $\lambda = -94$ cm^{-1}, $b = 28$ cm^{-1}, $c = -22$ cm^{-1} determined from a fit to the absorption data of [13]. The points

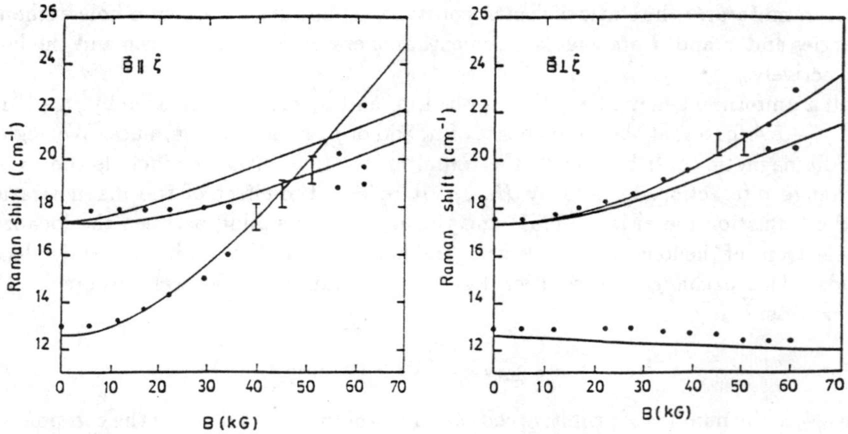

Figure 2. Raman lines as functions of the magnetic field $\boldsymbol{B}$ parallel to and perpendicular to $\hat{\zeta}$, the trigonal axis.

are the experimental data of Mauger *et al.* [6].

IV. Zeeman effect of the exciton states

We consider the exciton levels in DMS's and quaternary alloys. The exciton levels can be described by electron-hole wave functions which modulate an envelope function $\Phi(\boldsymbol{r}_e - \boldsymbol{r}_h)$ where $\boldsymbol{r}_e$ and $\boldsymbol{r}_h$ are the electron and hole coordinates, respectively. The envelope function varies slowly over the primitive cell of the crystal and, in the absence of a magnetic field and of the interaction with the magnetic ions, satisfies a Schrödinger equation of the form

$$\left[H_c - H_v - \frac{e^2}{\epsilon \mid \boldsymbol{r}_e - \boldsymbol{r}_h \mid} + \Delta_1 \boldsymbol{S}\cdot\boldsymbol{J} + \Delta_2(S_x J_x^3 + S_y J_y^3 + S_z J_z^3)\right]\Phi = E\Phi. \qquad (6)$$

In a crystal of tetrahedral symmetry in which the minimum of the conduction band occurs at the center of the Brillouin zone,

$$H_c = \frac{p_e^2}{2m^*} \qquad (7)$$

where $\boldsymbol{p}_e$ is the crystal momentum of the electron and m^* its effective mass. The valence band Hamiltonian, to second order in the crystal momentum is [14]

$$H_v = -\frac{1}{m}\left[\frac{1}{2}(\gamma_1 + \frac{5}{2}\gamma_2)p_h^2 - \gamma_2 \sum_{i=1}^{3} p_{hi}^2 J_i^2 - \frac{\gamma_3}{2}\sum_{i<j}\{p_{hi}, p_{hj}\}\{J_i, J_j\}\right]. \qquad (8)$$

The quantities γ_1, γ_2 and γ_3 are known as the Luttinger parameters, the indices i, j range over x, y, and z, the directions of the cubic axes, $\{u, v\} = uv + vu$ is the anticommutator, ϵ the static dielectric constant, Δ_1 and Δ_2 are electron-hole exchange energies and $\boldsymbol{S}$ and $\boldsymbol{J}$ are angular momentum operators for the electron and the hole, respectively.

If a uniform magnetic field $\boldsymbol{B}$ is applied, $\boldsymbol{p}_e$ and $\boldsymbol{p}_h$ must be replaced by $\boldsymbol{p}_e + e\boldsymbol{A}/c$ and $\boldsymbol{p}_h - e\boldsymbol{A}/c$ so that the components of $\boldsymbol{p}_e$ and of $\boldsymbol{p}_h$ no longer commute. We neglect the diamagnetic contributions to the Hamiltonian. In addition we include the $sp-d$ exchange interaction denoted by H_{sp-d}. It reflects the effect of the magnetization of the transition metal ions (TMI) and the exchange coupling between the localized 3d-electrons of the ions and the electron and hole states in the conduction and valence bands. This exchange contribution has been introduced by several authors [1, 15]. Its expression is

$$H_{sp-d} = N_0 \textstyle\sum_i x_i < S_{mz}^{(i)} > (\alpha_i S_z - \frac{1}{3}\beta_i J_z). \tag{9}$$

Here N_0 is the number of primitive cells per unit volume, S_z and J_z are the components of the $\boldsymbol{S}$ and $\boldsymbol{J}$ operators along the direction of the magnetic field $\boldsymbol{B}$, α_i and β_i are exchange constants depending on the host and the TMI considered and $< S_{mz}^{(i)} >$ is the average spin of the ith TMI in the material with atomic concentration x_i.

The states of the exciton in the absence of the electron-hole exchange interaction ($\Delta_1 = \Delta_2 = 0$) and of an external magnetic field have the symmetry of the product of the Bloch states needed to form their wave functions. States at the center of the Brillouin zone belong to $\Gamma_6 \times \Gamma_8 = \Gamma_3 + \Gamma_4 + \Gamma_5$ and are eight-fold degenerate. This degeneracy is removed by the exchange interaction into a doublet Γ_3 and two triplets (Γ_4 and Γ_5). This splitting is, however, small and has not been detected experimentally. Thus, we neglect the terms in Δ_1 and Δ_2. Furthermore, H_v can be separated into a spherically symmetric part and a cubic contribution. Keeping only the former, transforming to the relative coordinate $\boldsymbol{r} = \boldsymbol{r}_e - \boldsymbol{r}_h$ and concentrating on the case in which the exciton is at rest [16], we rewrite its Hamiltonian as

$$H_{ex}^{(S)} = \frac{p^2}{2\mu} - \frac{e^2}{\epsilon r} + \frac{\overline{\gamma}}{m}\left(\frac{5}{4}p^2 - (\boldsymbol{p}\cdot\boldsymbol{J})^2\right) + H'. \tag{10}$$

Here

$$\frac{1}{\mu} = \frac{1}{m^*} + \frac{\gamma_1}{m}, \tag{11}$$

$$\overline{\gamma} = (2\gamma_2 + 3\gamma_3)/5 \tag{12}$$

and

$$H' = N_0 \sum_i x_i < S_{mz}^{(i)} > (\alpha_i S_z - \frac{1}{3}\beta_i J_z) + \mu_B B(g_e S_z - K J_z). \tag{13}$$

Here g_e and K are appropriate g-factors for the conduction and valence bands, respectively. We note that because $H_{ex}^{(S)} - H'$ is spherically symmetric we need not specify the direction of $\boldsymbol{B}$ with respect to the crystallographic axes of the material.

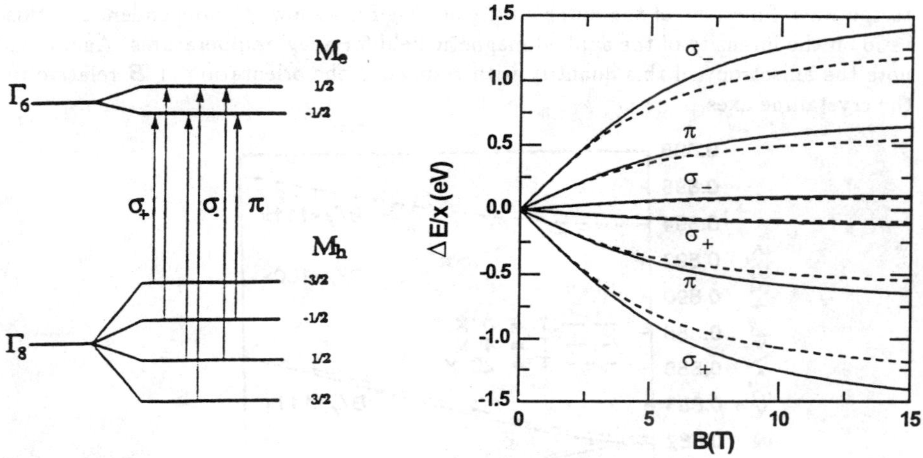

Figure 3. Selection rules and exciton energies for $Cd_{1-x}Fe_xTe$ in a magnetic field. The solid lines correspond to $\boldsymbol{B} \parallel < 100 >$ and the dashed lines to $\boldsymbol{B} \parallel < 111 >$.

The energy eigenvectors of $H_{ex}^{(S)}$ can be written in terms of the quantum numbers ν, M_e and M_h. The first describes the states of $H_{ex}^{(S)} - H'$ without regard for the mixing caused by the contribution $(\boldsymbol{p} \cdot \boldsymbol{J})^2$ which will be treated in perturbation theory. The quantum numbers $M_e = -1/2, 1/2$ and $M_h = -3/2, -1/2, 1/2, 3/2$ are the eigenvalues of S_z and J_z, respectively. For example the ground state of the first two terms in Eq. (10), namely $(\pi a^3)^{-1/2} \exp(-r/a)$ with $a = \epsilon\hbar^2/\mu e^2$, yields a zero expectation value for the third term. For such a state the exciton levels are specified by M_e and M_h with energies

$$\Delta E = M_e(N_0 \sum_i \alpha_i x_i < S_{mz}^{(i)} > + g_e \mu_B B) - M_h(\frac{N_0}{3} \sum_i \beta_i x_i < S_{mz}^{(i)} > + K \mu_B B) \quad (14)$$

in excess of the zero-magnetic field value.

The selection rules for exciton creation are identical to those in $Cd_{1-x}Mn_xTe$ [17] and are best described by means of the diagram in Fig. 3a. Figure 3b shows the dependence of the energies of the exciton lines as functions of the magnetic field B. Here we have neglected the terms in g_e and K in Eq. (14) in comparison with the effect of the $sp-d$ exchange interaction. The graph shows $\Delta E/x$ for $Cd_{1-x}Fe_xTe$ for which $N_0\alpha = 300$ meV and $N_0\beta = -1270$ meV [18]. The values of $< S_{mz} >$ are obtained using the following material parameters: $\Delta = 2470$ cm^{-1}, $\lambda = -91$ cm^{-1}, and calculated at 4.2 K for $\boldsymbol{B} \parallel < 100 >$ (solid lines) and $< 111 >$ (dashed lines).

We emphasize that $< S_{mz} >$ in Eqs. (9), (13) and (14) contains only the spin magnetization and differs from that discussed in Ref. [11]. The ratio $2 < S_z > /(< L_z + 2S_z >)$ would be unity if we neglected the admixture of the $^5\Gamma_5$ multiplet into the $^5\Gamma_3$ states. The spin-orbit interaction gives a contribution proportional to (λ/Δ)

to the wave functions of the lower multiplet. Figure 4 shows the dependence of this ratio on the intensity of the applied magnetic field for a few temperatures. Again, we note the anisotropy of this quantity with respect to the orientation of $\boldsymbol{B}$ relative to the crystalline axes.

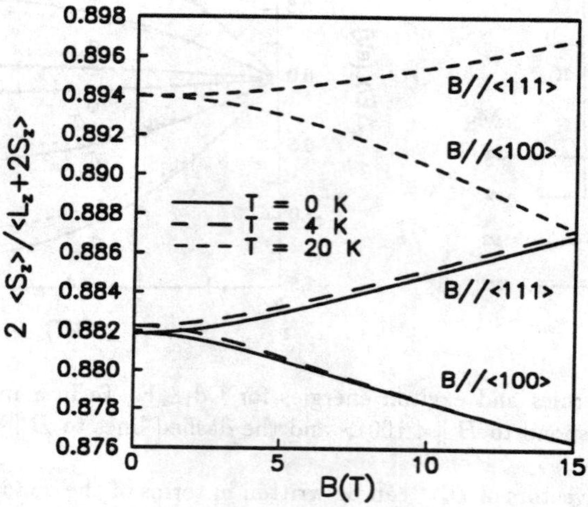

Figure 4. The ratio $2 < S_z > /(< L_z + 2S_z >)$ as a function of the applied field B, for several temperatures and two orientations of $\boldsymbol{B}$.

Aknowledgments

The authors thank Professor A. K. Ramdas for stimulating discussions. This work was supported by the U.S. National Science Foundation (MRG Grant No. DMR92-21390), by the Fonds National de la Recherche Scientifique (Belgium) and by the Royal Society of London, U.K.

References

[1] See, for example, *Diluted Magnetic Semiconductors*, Semiconductors and Semimetals, Vol. 25, edited by J. K. Furdyna and J. Kossut (Academic, Boston, 1988).

[2] E. Oh, C. Parks, I. Miotkowski, M. D. Sciacca, A. J. Mayur, and A. K. Ramdas, Phys. Rev. B **48**, 15040 (1993).

[3] T. Hoshina, J. Phys. Soc. Jpn **21**, 1608 (1966).

[4] D. U. Bartholomew, E. K. Suh, A. K. Ramdas, S. Rodriguez, U. Debska, and J. K. Furdyna, Phys. Rev. B **39**, 5865 (1989).

[5] See, for instance, E. Kartheuser, S. Rodriguez, and M. Villeret, Phys. Rev. B **48**, 14127 (1993).

[6] A. Mauger, D. Scalbert, J. A. Gaj, J. Cernogora, and C. Benoit à la Guillaume, Phys. Rev. B **43**, 7102 (1991).

[7] R. L. Aggarwal, S. N. Jasperson, P. Becla, and R. R. Galazka, Phys. Rev. B **32**, 5132 (1985).

[8] J. T. Vallin and G. D. Watkins, Phys. Rev. B **9**, 2051 (1974).

[9] M. Villeret, S. Rodriguez, and E. Kartheuser, Phys. Rev. B **41**, 10028 (1990); **42**, 11375(E) (1990).

[10] C. Testelin, A. Mauger, C. Rigaux, M. Guillot, and A. Mycielski, Solid State Commun. **71**, 923 (1989).

[11] S. Rodriguez, M. Villeret, and E. Kartheuser, Physica Scripta T **39**, 131 (1991).

[12] M. Villeret, S. Rodriguez, and E. Kartheuser, Phys. Rev. B **47**, 1228 (1993).

[13] M. K. Udo, M. Villeret, I. Miotkowski, A. J. Mayur, A. K. Ramdas, and S. Rodriguez, Phys. Rev. B **46**, 7459 (1992).

[14] J. M. Luttinger, Phys. Rev. **102**, 1030 (1956).

[15] G. Bastard, C. Rigaux, and A. Mycielski, Phys. Stat. Sol. B **79**, 585 (1977).

[16] K. Cho, S. Suga, W. Dreybrodt, and F. Willmann, Phys. Rev. B **11**, 1512 (1975).

[17] A. K. Ramdas, in *High Magnetic Fields in Semiconductor Physics II*, ed. G. Landwehr (Springer-Verlag, Berlin, 1989), p. 464.

[18] C. Testelin, C. Rigaux, A. Mycielski, M. Menant, and M. Guillot, Solid State Commun. **78**, 659 (1991).

MAGNETO-OPTICS OF TERNARY AND QUATERNARY II-VI DILUTED MAGNETIC SEMICONDUCTORS & THEIR HETEROSTRUCTURES

A. K. RAMDAS, H. ALAWADHI, C. PARKS, EUNSOON OH,* I. MIOTKOWSKI and S. RODRIGUEZ
Department of Physics, Purdue University
West Lafayette, IN 47907, U.S.A.

The large **excitonic Zeeman effect**, characteristic of II-VI diluted magnetic semiconductors (DMSs), and responsible for the **giant Faraday effect** they display, is investigated with the special advantages of modulation spectroscopy. The large linear birefringence in DMSs, induced by an external magnetic field (the **Voigt effect**) is compared with the circular birefringence (the **Faraday effect**). The transformation of the **Raman-EPR** of the magnetic ions in the paramagnetic phase of a DMS into the high frequency component of the magnon in an external magnetic field in its magnetically ordered phase is illustrated in the quaternary $Cd_{1-x-y}Zn_xMn_yTe$.

1 Introduction

A tetrahedrally coordinated II-VI semiconductor with a fraction of its group-II ions randomly replaced with one of the transition metal ions of the iron group — a **diluted magnetic semiconductor (DMS)** — displays spectacular magnetic phenomena.[1] Originating in the spin-spin exchange interaction between the 3d electrons of the transition metal ions on the one hand and the band electrons on the other hand, *i.e.*, the so-called sp-d exchange interaction, it exhibits striking magnetic-field-induced circular birefringence (the Faraday effect)[2]; huge red-shift of the excitonic recombination radiation, and and enormous linear birefringence (the Voigt effect).[3] All these phenomena show a characteristic saturation at high magnetic fields and low temperatures. In addition, the magnetic interaction *between* the transition metal ions is known to occur through an antiferromagnetic coupling. This, too, leads to magnetic phenomena unique to the DMSs: magnetic phases due to long-range magnetic ordering along with the associated collective excitations.[4]

We note here that DMS single crystals have been grown as ternaries (*e.g.*, $Cd_{1-x}Mn_xTe$) and as quaternaries (*e.g.*, $Cd_{1-x-y}Zn_xMn_yTe$).[5] The latter offer the additional flexibility accessible with the control of material parameters, for example, a fixed band gap with variable magnetic property or *vice versa*.

The tetrahedrally coordinated II-VI compound semiconductors represent an important class of semiconductors in the context of the physics of quantum well structures and their opto-electronic applications. They offer a wide range of total band off-sets. A judicious use of molecular beam epitaxy (MBE) also allows the fabrication of strain-free as well as strained layer superlattices or quantum well structures. MBE growth techniques have significantly increased the composition range of the II-VI alloys including DMSs; it has enabled the growth of epilayers and quantum well structures not accessible in bulk growth.[6,7]

In this paper the focus is on the excitonic Zeeman effect of DMSs in bulk crystals as well as in quantum well structures observed with modulated reflectivity techniques and photoluminescence; Faraday and Voigt effects driven by the excitonic transitions; and Raman spectroscopy of the localized and collective magnetic excitations of the magnetic ions, illustrative examples being selected from the investigations of the authors.

2 Excitonic Zeeman Effect

A tetrahedrally coordinated, direct gap, zinc-blende DMS is characterized by a zone center Γ_6 conduction band minimum and a Γ_8 valence band maximum. In the presence of an external magnetic field $\boldsymbol{H} \parallel \hat{z}$ (referred to an orthogonal right handed coordinate system xyz), the Γ_6 level splits into two levels, $\Gamma_6(1/2)$ and $\Gamma_6(-1/2)$, with energies $E_g + 3A$ and $E_g - 3A$, respectively; the Γ_8 level splits into four levels given by $\pm B$ and $\pm 3B$ corresponding to $\mp 1/2$, $\mp 3/2$ spins, respectively.[1] Here the zero of the energy scale is defined by the valence band maximum at $H = 0$; E_g is the energy gap at $k = 0$; A and B are proportional to α and β, the 's-d' and 'p-d' exchange integrals for the conduction and valence band, respectively, and to M, the magnetization per unit volume. M for small x is described by a Brillouin function appropriate for the transition metal ion involved. In the Faraday geometry, *i.e.*, for light propagation along $\boldsymbol{H}$, the excitonic transition at E_o splits into four Zeeman components expected at $E_a(-3/2 \rightarrow -1/2) = E_o + 3B - 3A$; $E_b(-1/2 \rightarrow 1/2) = E_o + B + 3A$; $E_c(1/2 \rightarrow -1/2) = E_o - B - 3A$; and $E_d(3/2 \rightarrow 1/2) = E_o - 3B + 3A$; the a and b transitions are allowed for $\hat{\sigma}_+ = (1/\sqrt{2})(\hat{x} + i\hat{y})$ polarization whereas the c and d occur for $\hat{\sigma}_- = (1/\sqrt{2})(\hat{x} - i\hat{y})$. ($E_o$ differs from E_g by the excitonic binding energy.) In the Voigt geometry, *i.e.*, for light propagation normal to H, the a, b, c and d components are all polarized $\perp H$ whereas two additional Zeeman components e and f at $E_e(-1/2 \to -1/2) = E_o + B - 3A$ and $E_f(1/2 \to 1/2) = E_o - B + 3A$, polarized along H, are expected.

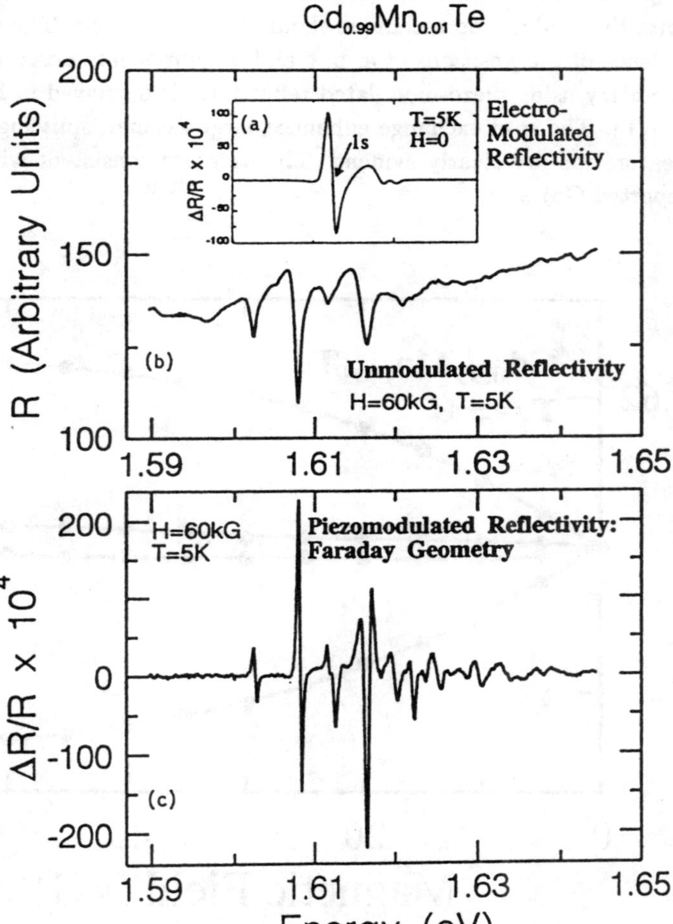

Fig. 1. Excitonic Zeeman effect in $Cd_{1-x}Mn_xTe$. (Ref. 8).

In Fig. 1 the electromodulated reflectivity spectrum at $H = 0$ (a), the unmodulated reflectivity spectrum at $H = 60$ kG (b), and the piezo-modulated reflectivity spectrum $H = 0$ at 60 kG, (c) all at 5K, are displayed for $Cd_{0.99}Mn_{0.01}Te$.[8] The spectra for $H \neq 0$ are recorded in the Faraday geometry with unpolarized light. The striking enhancement of the spectral features achieved with the modulation technique should be underscored. The magnetic field dependence of the positions of a, b, c and d components observed in the Faraday geometry using piezo-modulated reflectivity is displayed in Fig. 2 for $Cd_{0.99}Mn_{0.01}Te$. The sp-d exchange enhanced large Zeeman splitting and the high field saturation are clearly evident. The data are consistent with $(\beta/\alpha) = -4.0$ reported Gaj *et al.*.[9]

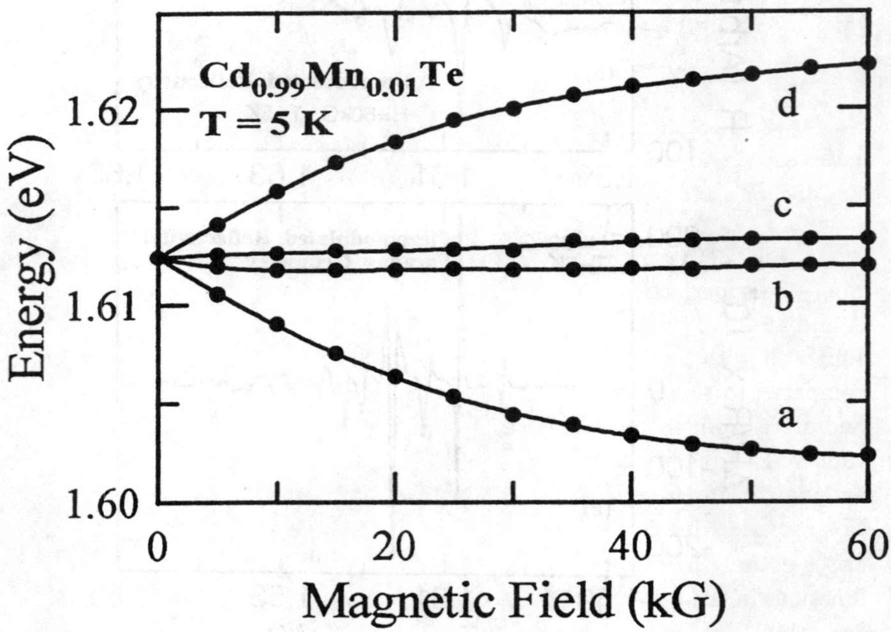

Fig. 2. Energies of the excitonic Zeeman components of $Cd_{0.99}Mn_{0.01}Te$ as a function of magnetic field in the Faraday geometry. (Ref. 8).

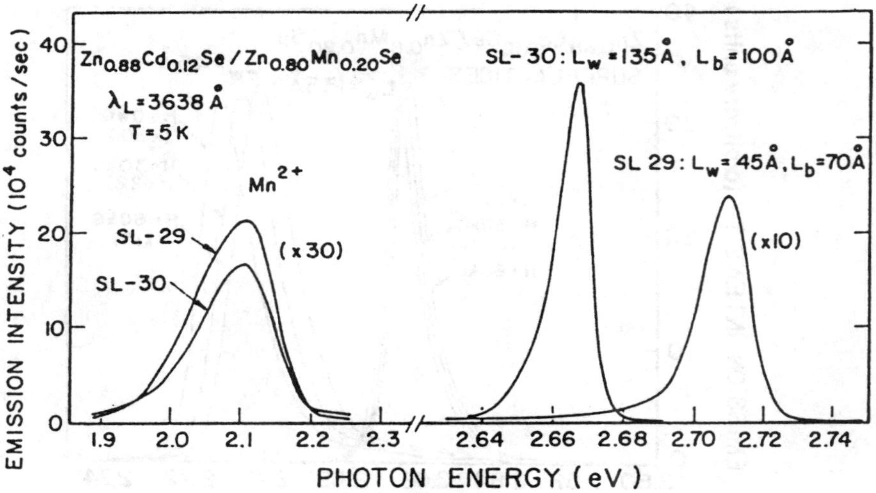

Fig. 3. Photoluminescence spectra of $Zn_{0.88}Cd_{0.12}Se/Zn_{0.80}Mn_{0.20}Se$ superlattaices showing luminescence from quantum-confined excitonic transitions from the ZnCdSe wells and Mn^{2+} emission from the ZnMnSe barriers. (Ref. 10).

In Fig. 3 the photoluminescence spectra of two $Zn_{0.88}$ $Cd_{0.12}Se/Zn_{0.80}$ $Mn_{0.20}Se$ superlattices are shown for the $Zn_{0.88}Cd_{0.12}Se$ wells of widths 135 Å and 45 Å.[9] The *blue shift* of the recombination radiation for the 45 Å quantum well with respect to that from the 135 Å quantum well and for both compared to that of the bulk $Zn_{0.80}Cd_{0.12}Se$ at 2.656 eV are clearly evident and arises from the quantum confinement of the electronic levels in the quantum well. Equally noteworthy is the simultaneous observation of the 2.2 eV emission arising from the Mn^{2+} in the barriers. The *red shift* of the recombination radiation due to the excitonic Zeeman effect discussed for a bulk DMS now arises entirely from the penetration of the quantum-confined electronic wave functions in the *non-magnetic* $Zn_{0.88}Cd_{0.12}Se$ quantum wells into the *magnetic* $Zn_{0.80}Mn_{0.20}Se$ barriers. (See Fig. 4.) The effect (including the difference arising from quantum well widths) can be quantitatively understood on the basis of the transfer matrix method based on the $\boldsymbol{k}\cdot\boldsymbol{p}$ formalism. This is an example of 'wave function' engineering revealed through effects unique to DMSs.

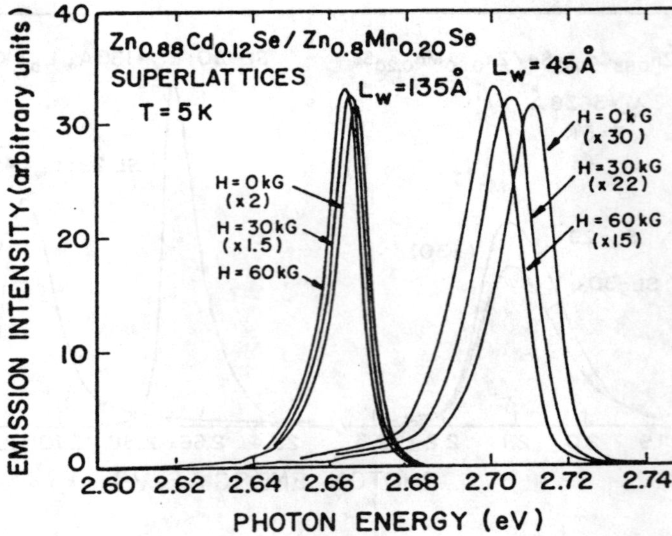

Fig. 4. Photoluminescence spectra associated with the magnetic-field-induced, 'red shifted' recombination radiation from the non-magnetic $Zn_{0.8}Cd_{0.12}Se$ quantum wells illustrating the penetration of the 'quantum-confined' electronic wave functions into the magnetic $Zn_{0.8}Mn_{0.20}Se$ barriers. (Ref. 10.)

3 Faraday and Voigt Effects in DMSs

In the presence of an external magnetic field, the dispersive effects of the excitonic Zeeman components of the zinc-blende DMS observed in σ_+ and σ_- polarizations for observation along H, produce circular birefringence and hence a rotation of the plane of polarization for light propagating along H. This magnetic-field-induced rotation, *i.e.*, the Faraday rotation, Θ_F, is $(E\ell/2\hbar c)(n_- - n_+)$ where $n_\pm$ correspond to refractive indices of light with $\sigma_\pm$ polarizations; $E = \hbar\omega$, the photon energy; and $\hbar$ and c have the usual meaning and ℓ is the distance travelled in the medium.

In Fig. 5, Θ_F as a function of photon energy is displayed by observing the intensity oscillations in the white light transmitted along H through $Cd_{1-x}Mn_xTe$ held between two polarizers at 45° and spectrally analyzed.[11] Successive transmission maxima correspond to Faraday rotations of $(m\pi+45°)$, $m = 0, 1, 2, 3, \ldots$. The enormous Faraday rotation and its dispersion as $\hbar\omega$ approaches the excitonic transition energies are the striking features generated by the large sp-d exchange interaction; Θ_F at a specific $\hbar\omega$ shows the saturation at high values of (H/T) since it is proportional to M.

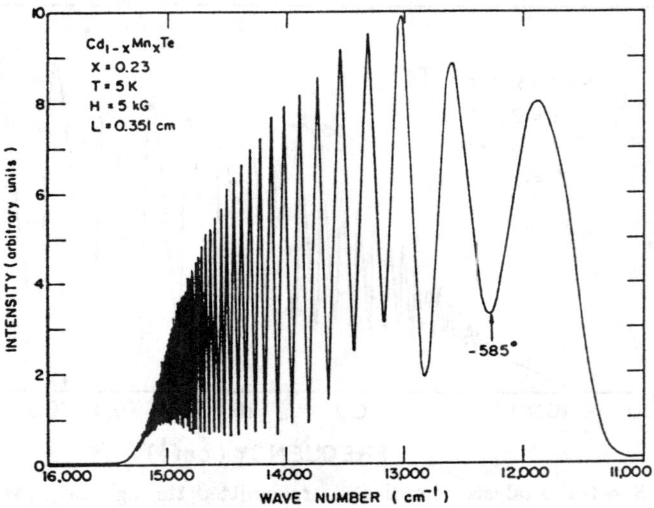

Fig. 5. Transmission through $Cd_{1-x}Mn_xTe$ ($x = 0.23$) in the Faraday geometry displaying giant Faraday rotation and its dispersion. (Ref. 11.)

The double refraction induced in a medium by an external magnetic field, (H), the direction of light propagation being normal to it, is called the Voigt effect. The same factors which enhance the Faraday effect in a II-VI DMS, in contrast to that observed in a non-magnetic counterpart, also lead to a huge Voigt effect in the DMSs as demonstrated by Oh *et al.*[3] The spectral analysis of white light transmitted through a specimen of $Cd_{0.65}Mn_{0.35}Te$ kept between crossed linear polarizers, at ±45° with respect to H, and propagating normal to H, shows oscillations in intensity as a function of photon energy (Fig. 6). They arise from the variations in $\phi = (E\ell/\hbar c)(n_\perp - n_\parallel)$, where $E =$ photon energy, ℓ is the thickness and $n_\perp$ and $n_\parallel$, the refractive indices for electric vector perpendicular and parallel to H, respectively. The transmitted intensity displays maxima for $\phi = (2p+1)\pi$, p being an integer. The very large Voigt effect, its dispersion and the saturation it exhibits at high H/T... all of these can be accounted for in terms of the large excitonic Zeeman effect discussed in Sec. 2. The $\hat{\sigma}_+$ (a,b) and $\hat{\sigma}_-$ (c,d) transitions which produce the circular double refraction in the Faraday effect, now contribute to $n_{\perp}$ whereas those polarized along $\boldsymbol{H}$, *viz.*, the e and f transitions, determine $n_{\parallel}$.

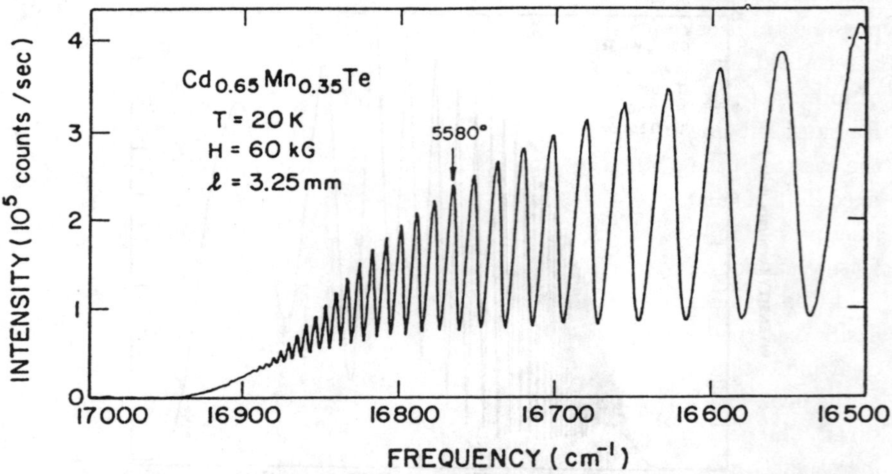

Fig. 6. Spectral analysis of radiation transmitted through $Cd_{1-x}Mn_xTe$ ($x = 0.35$) in the Voigt geometry. The sample is held between two crossed polarizers $\pm 45°$ with respect to H, radiation propagating normal to it. Maxima correspond to $\phi = (E\ell/\hbar c)(n_{\perp} - n_{\parallel}) = (2p+1)\pi$, p being integers. (Ref. 3.)

4 Magnetic Raman Scattering from II-VI DMSs and their Quantum Well Structures

DMSs display fascinating magnetic phases[1] associated with the magnetic coupling between the magnetic ions. In the best studied case of $Cd_{1-x}Mn_xTe$, a new line appears at ω_M below the Néel temperature (T_N) as a paramagnetic → antiferromagnetic phase transition occurs.[4] This Raman feature softens in frequency as the temperature approaches T_N from below, has polarization features expected for a one-magnon Raman line, and splits into a doublet predicted for one-magnon excitation.[11] The Mn^{2+} ions precess independent of one another for $T > T_N$, and Raman lines corresponding to the $\Delta M_S = \pm 1, \pm 2$ transitions of the Mn^{2+} ground state, *i.e.*, the Raman-EPR lines at a ω_{PM} and $2\omega_{PM}$ are observed; it evolves smoothly into the high-frequency component of ω_M observed in an external magnetic field.[11] With decreasing Mn concentration, ω_M decreases, and below $\sim 40\%$ it is difficult to observe because of its close proximity to the laser line and its significantly larger width compared to that of the Raman-EPR line. However, the application of an external magnetic field can expose the high-frequency component of the magnon, distinguished from the Raman-EPR line by its larger frequency and line width. Thus it is possible to discover the occurrence of an antiferromagnetic state even when a zero-field magnon is difficult to observe. With these considerations it is of interest to observe the effect of Mn concentration and temperature in the Raman-EPR line in the quaternary, $Cd_{1-x-y}Zn_xMn_yTe$.[5] Fig. 7(a) displays the change in the peak position of the Raman-EPR line with increasing y for a fixed T (5 K) and x (0.2). All the spectra were taken with the same slit width, laser power, and scattering geometry. As mentioned above, although a magnon feature cannot be observed with $y \leq 0.4$ in the absence of an external magnetic field, the significant shift of the peak position and the line broadening can be easily discerned as the Mn concentration is increased from $y = 0.1$ to 0.4, a clear indication of a magnon excitation. In Fig. 7(b) we show the *sharpening* of the peak for $Cd_{1-x-y}Zn_xMn_yTe$, $x = 0.2$, $y = 0.4$ with increasing temperatures. The peak position at 7.1 cm^{-1} for $T = 20$ K approaches 5.6 cm^{-1} with increasing temperature as the Mn^{2+} ions start precessing independent of one another.

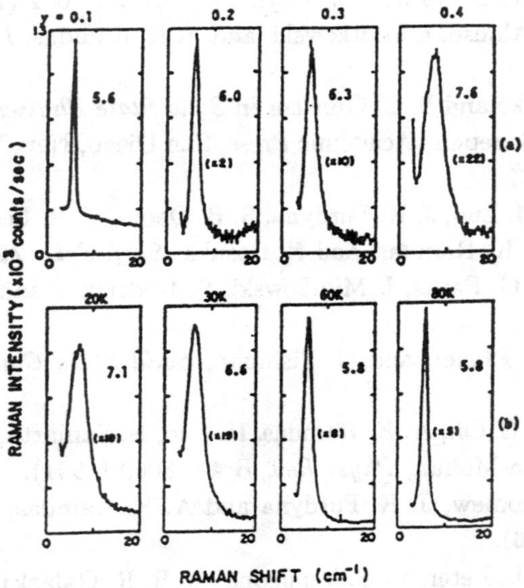

Fig. 7. The evolution of the Raman-EPR line (ω_{PM}) of $Cd_{1-x-y}Zn_xMn_yTe$ into the high frequency component of the magnon doublet in an exernal magnetic field: (a) as a function of y at fixed T = 5 K and x = 0.2 and (b) as a function of T at fixed x = 0.2 and y = 0.4. The peak positions in cm^{-1} are indicated in each panel. (Ref. 5.).

Acknowledgments

The work reported in this paper received support from the U.S. National Science Foundation (DMR 93-03186 and DMR 92-21390).

References

1. See articles in: *Diluted Magnetic Semiconductors*, eds J. K. Furdyna and J. Kossut, volume 25 of the series *Semiconductors and Semimetals*, eds. R. K. Willardson and A. C. Beer (Academic Press, San Diego, 1988).
2. E. Oh, A. K. Ramdas and J. K. Furdyna, *J. Lum.* **52**, 183 (1992). This review lists references to the earlier studies on the Faraday effect in DMSs including to the first reports by Komarov *et al.*, *Sov. Phys. JETP* **46** 318 (1977) and Gaj *et al.*, *Solid State Commun.* , **25**, 193 (1978).
3. Eunsoon Oh, D. U. Bartholomew, A. K. Ramdas, J. K. Furdyna and U. Debska, *Phys. Rev.* B **44**, 10551 (1991).
4. S. Venugopalan, A. Petrou, R. R. Galazka and A. K. Ramdas *Solid State Commun.* **38**, 365 (1981); S. Venugopalan, A. Petrou, R. R. Galazka, A. K. Ramdas and S. Rodriguez, *Phys. Rev.* B **25**, 2681 (1982).
5. E. Oh, R. G. Alonso, I. Miotkowski, and A. K. Ramdas, *Phys. Rev.* B**45**, 10934 (1992).
6. A. V. Nurmikko and R. L. Gunshor in *Solid State Physics* eds. H. Ehrenreich and F. Spaepen (Academic Press, San Diego, New York, 1995) Vol. 49. pp 205–282.
7. N. Samarth, H. Luo, J. K. Furdyna, S. B. Qadri, Y. R. Lee, R. G. Alonso, E.-K. Suh, A. K. Ramdas, and N. Otsuka, *Surf. Sci.* **228**, 226 (1990).
8. H. Alawadhi, C. Parks, I. Miotkowski, S. Rodriguez and A. K. Ramdas (unpublished).
9. J. A. Gaj, R. Planel, and G. Fishman, *Solid State Commun.* **29**, 439 (1979).
10. R. G. Alonso, E. Oh, A. K. Ramdas, H. Luo, N. Samarth, J. K. Furdyna, and L. R. Ram-Mohan, *Phys. Rev.* B **44**, 8009 (1991).
11. D. U. Bartholomew, J. K. Furdyna and A. K. Ramdas, *Phys. Rev.* B **34**, 6943 (1986).
12. A. Petrou, D. L. Peterson, S. Venugopalan, R. R. Galazka, A. K. Ramdas and S. Rodriguez, *Phys. Rev.* B **27**, 3471 (1983).

SPECTROSCOPY OF MAGNETIC-IONS IN HIGH MAGNETIC FIELDS

D. HEIMAN
MIT Francis Bitter Magnet Lab, Cambridge, MA 02139 USA

Experiments in high magnetic fields provide a valuable means to measure the energetics of magnetic ions. In materials containing dilute transition metal ions, the *d-d* exchange between magnetic ions can be accurately measured by the method of "magnetization steps". This method, often ranging up to fields of 60 T, has been applied to II-VI magnetic semiconductors. High fields have also been used to determine further-neighbor exchange, anisotropic exchange, and intra-ion energy levels. Experiments to study the *sp-d* exchange between magnetic ions and semiconductor carriers require the measurement of both magnetization and optical Zeeman splitting. This has been applied to materials containing Mn, Fe, Co, and Cr. Recent measurements, theoretical models, and global trends of the exchanges are discussed.

1. Introduction

High magnetic fields have been indispensable in gathering information on materials containing magnetic ions. In no place has this been more evident than in the study of exchange effects in magnetic semiconductors (MS). In the subclass of diluted magnetic semiconductors (DMS) — exemplified by the "fruit-fly" $Cd_{1-x}Mn_xTe$ [1-3] — experiments in high magnetic fields have been essential in studying:

(i) isotropic d-d exchange constants between magnetic ions;
(ii) Dzyaloshinski-Moriya anisotropic exchange;
(iii) internal energy level structure of single ions;
(iv) *s-d* and *p-d* exchange energies between localized ions and mobile carriers.

In the following I briefly review pertinent experimental methods, describe recent high magnetic field experiments, and discuss relevant models.

2. *d-d* Exchange and Magnetization Steps of Clusters

In 1984 an important method was discovered for studying and accurately determining the *d-d* exchange constant between localized magnetic ions in dilute materials. Simultaneously, Shapira working at the MIT Magnet Lab, [4] and Aggarwal working at the ISSP High Field Lab in Tokyo, [5] were studying the magnetization (M) of $Cd_{1-x}Mn_xSe$. They found that at high fields M(B) rises above the expected saturation appropriate to a paramagnetic Brillouin function, and discovered additional structure in M(B). With a background of many years of experience in magnetization studies Shapira realized that they were observing "magnetization steps" resulting from pairs of Mn ions that were undergoing a change in magnetic moment. [4] The width of the steps were found by Aggarwal to be related to temperature and further-neighbor interactions.

At low fields and low temperatures these pairs are antiferromagnetically locked into a total spin $S_T = S_1 + S_2 = 0$ state. At a material-dependent transition field B_1 the magnetic field overcomes the internal exchange field forcing the pair to be in the $S_T = -1$, $S_Z = -1$ state. This is equivalently viewed as an energy level crossing where the $S_Z = -1$ state becomes lower in energy than the $S_Z = 0$ state. Since the pair has transformed from a nonmagnetic state to a state having a magnetic moment, a step-like increase is generated in M(B). These crossings and steps are diagrammed for S = 5/2 Mn^{2+} ions in Fig. 1. Five energy level crossings in the upper part of the figure give rise to five magnetization steps shown in the lower part. The steps appear after the single spins saturate at low fields. The magnitude of the step is optimized at x = 5 % where the number of pairs is maximum. All five steps were observed in $Cd_{1-x}Mn_xTe$ by Foner and coworkers in experiments up to 60 T. [6] In these pulsed-field experiments the sweeping field naturally gives the derivative dM/dB. Figure 2 shows five peaks in dM/dB, corresponding to the five magnetization steps. These are essentially *spectral* peaks in magnetic field space.

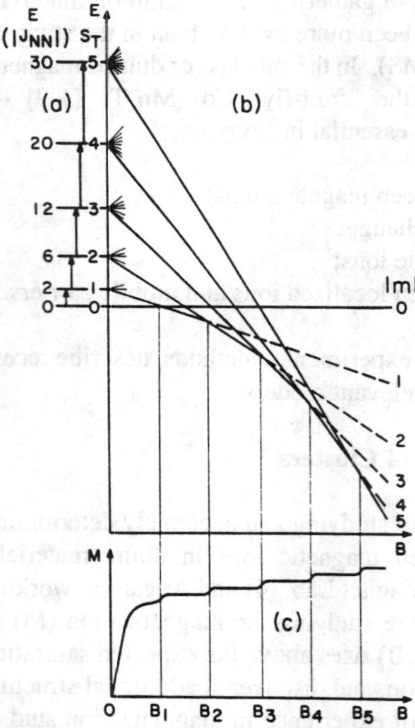

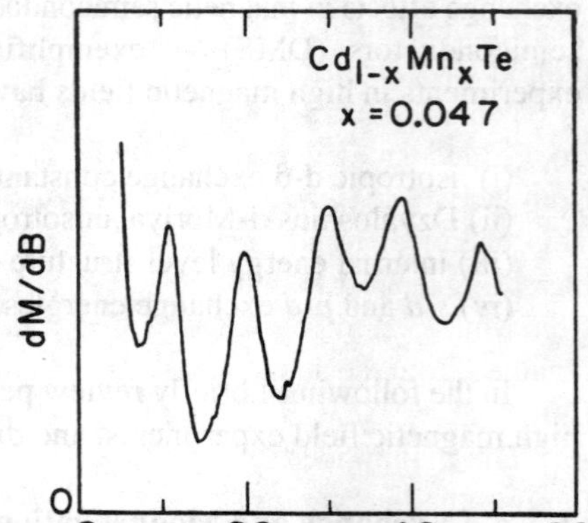

Figure 2: Derivative of magnetization versus magnetic field for $Cd_{1-x}Mn_xTe$, x=0.047, taken in a 5 ms pulsed magnetic field. The temperature at the start of the pulse was 1.4 K. The 5 peaks correspond to the 5 magnetization steps. (From Ref. 6)

Figure 1: Diagram showing energy level crossings and the resulting magnetization (M) steps for pairs of Mn^{2+}. In (a) is the energy level diagram in units of $|J_1|$ for pairs having total spin S_T. In (b) are the energies versus B for the lowest m-state (S_Z) of each S_T state. In (c) is the total M versus B, showing the five M steps, in addition to the modified Brillouin function saturation of singles at low field. (From Ref. 6)

The great utility of these measurements is that the fields at which the steps occur give an accurate value for the exchange energy J_1 for nearest neighbors. [7,8] The n-*th* step occurs at B_n according to $g\mu_B B_n = 2\,n\,|J_1| + \Delta$. The additional parameter Δ is due to further-neighbor exchange and was introduced by Barilero. [9] In general Δ is only weakly-dependent on B (or n), so values for J_1 are best extracted from the difference between the first two steps. The accuracy for measuring J_1 by the magnetization step method is typically 2-5%, which is more accurate than estimating J_1 from high-temperature magnetic susceptibility. The magnetization step method has also been applied to further-neighbor pairs [10] and triplets [11].

The steps can be measured by a variety of methods, including direct magnetization measurements in *dc* or pulsed fields, or optical experiments that measure the Zeeman splitting of the bands which is dominated by M(B). In addition to the optical measurements of bandgap reflection and absorption, Faraday rotation [12,13] and spin-flip Raman scattering [3,14] measurements have been useful. The Faraday rotation approach was used in a pulsed field experiment to 60 T in order to observe the first step in CdS:Co, which occurs at B = 41 T due to the large value $J_1/k_B = -30$ K. [13] Applications of these techniques to the measurement of magnetization steps are discussed in several review articles. [3,7,8,15] For some appropriate materials, direct measurements of J_1 have also been obtained from inelastic neutron [16] and inelastic light [17,18] scattering experiments.

Table I shows a compilation of J_1 values for various II-VI materials containing Mn, Co, Fe, and Cr. The general trend for a given magnetic ion is for J_1 to increase going from heavier to lighter ions — from Cd to Zn cations, and from Te to Se to S anions. This chemical trend is primarily due to the exponential decrease in exchange for increasing nearest-neighbor distance. [19] There is also a trend of increasing J_1 going from Mn(S=5/2) to Fe(S=2) to Co(S=3/2). This results from the dependence $J_1 \propto (2S)^{-1}$, obtained from the sum over intermediate states in the fourth-order perturbation theory. [20]

The magnitude of the exchange is dominated by superexchange that is mediated through the intervening Te, Se, and S anions. The exchange mechanism results from the hybridization of the *d*-levels of the magnetic ions with the *p*-levels of the anions. In a real space picture, the *d*-orbitals spread out and lower their energy for antiparallel adjacent spins. Quasiparticles can hop between adjacent Mn sites through virtual double occupation of neighboring sites which reduce the kinetic energy. For parallel spins, Pauli exclusion does not permit this reduction. Larson and coworkers have used a three-level model of kinetic superexchange and successfully applied it to CdTe:Mn. [19] In this model for Mn^{2+}, and extensions to other ions, [20] both the *d-d* exchange and *p-d* exchange depend on the hybridization of *p*- and *d*-orbitals. The contribution from the Bloembergen-Rowland mechanism (RKKY effect in insulators) is an order of magnitude smaller. [19] It is interesting to note that the *d-d* exchange for Cr^{2+} could be ferromagnetic instead of antiferromagnetic, however, this remains to be determined.

Table I -- Values for the d-d exchange constant for nearest-neighbors (J_1) and next-nearest neighbors (J_2); and the sp-d exchange energies αN_O and βN_O. See Ref. 2, 7, and 15 for additional references to original work.

Material	$-J_1/k_B$ (K)	$-J_2/k_B$ (K)	Ref.	αN_o (meV)	$-\beta N_o$ (meV)	Ref.
HgTe:Mn	4.7 ± 0.4	1	15	400	600	15
HgSe:Mn	5.6 ± 0.4		15	400	700	15
CdTe:Mn	6.1 ± 0.2	1.1 ± 0.2	7	220 ± 10	880 ± 40	29
CdSe:Mn	7.6 ± 0.2	1.6 ± 1.5	7, 22	258 ± 5	1110 ± 70	3, 15
CdS:Mn	9.6 ± 0.2	5.3 ± 0.3	7	217 ± 11	(1800)	3, 15
	11.0 ± 0.2		7			
ZnTe:Mn	9.0 ± 0.2	0.6	7	200 ± 40	1120 ± 240	35
ZnSe:Mn	12.2 ± 0.2	2.7 ± 2	7	243 ± 10	1220 ± 100	15, 36
ZnS:Mn	16.9	0.6	7			
HgSe:Fe	17 ± 2		27			
CdTe:Fe					1270	15
CdSe:Fe	18 ± 2		27	225 ± 5	1520	24, 27
CdS:Fe				200 ± 30		37
ZnTe:Fe					1900 ± 400	
ZnSe:Fe	22 ± 2		27		1740	27
CdTe:Co				(250)	1870 ±100	38
CdSe:Co	31.0 ± 2		8	275 ± 6	2120 ± 50	39
CdS:Co	30.4 ± 1.1		12	179 ± 4	1873 ± 42	39
ZnTe:Co	38 ± 2	5.7 ± 0.6	16			
ZnSe:Co	49.5 ± 1	3.0 ± 0.1	16		2200	40
ZnS:Co	47.5 ± 0.6	2.3 ± 0.2	16			
CdS:Cr				220 ± 20		41
ZnSe:Cr					850 ± 100	31
ZnS:Cr					570 ± 100	40

3. Dzyaloshinski-Moriya Anisotropic Exchange

High field experiments have been crucial in studying the Dzyaloshinski-Moriya (DM) anisotropic exchange $H = -2\mathbf{D} \cdot \mathbf{S}_1 \times \mathbf{S}_2$. The magnitude of **D** has been measured in magnetization measurements; its effect on spin relaxation has been modeled and quantified; and it contributes to narrowing of magnetization steps in pulsed-field experiments.

It has been shown that D can be determined by analyzing the lineshape of magnetization steps in CdSe:Mn. [21] The linewidth of a step is usually dominated by thermal broadening to a full-width at half-maximum δB_T given by $g\mu_B\delta B_T = 3.53\ k_BT$. (Some inhomogeneous broadening arises from distant-neighbor ions.) There is also step broadening from the DM interaction given by $g\mu_B\ \delta B_{DM} = 3.07\ C_n\ D_\perp$, where $C_1 = (6/35)^{1/2}$ for the first step, and $D_\perp$ is the projection of **D** on the plane perpendicular to the B-field. In the wurtzite structure each magnetization step is split into two components due to the presence two types of nearest-neighbor ion pairs. There are six pair configurations where the ions lie in the plane perpendicular to the hexagonal c-axis and six lying out of the plane. Because $D_\perp$ for the two configurations are not equal, the DM interaction causes the component at higher field to be narrower and larger. At T=0 the upper component is twice as narrow. In addition, the difference in linewidth of the two components depends on the crystal orientation in the field. Measurements of the first step at B=12 T and T=0.08 K, and of the second step at 24 T and T=0.6 K, give $D/k_B = 0.21 \pm 0.1$ K. This is comparable to the theoretical value of 0.16 K. [19]

Spin-lattice relaxation has been studied for Mn pairs in CdSe:Mn, and attributed to the DM interaction. [22] In general, spin-lattice relaxation has two essential ingredients: relaxation of the energy, and angular momentum selection rule requirements. Spins normally relax by emitting phonons via strain-induced modulation of J through its distance dependence, dJ/dr. However, relaxation cannot occur solely through the isotropic exchange J because the interaction conserves total spin S_T. On the other hand, the DM interaction is both the strongest non-spin conserving interaction, and the only one that has the selection rule $\Delta S = \pm 1$. The DM interaction makes the spin transition allowed by mixing pair-states differing by $\Delta S = \pm 1$. Modulated Faraday rotation experiments were used to measure spin relaxation times as a function of frequency. [22] At the first pair step at B=11.75 T two relaxation times were found, 3×10^{-5} and 1.8×10^{-4} s, differing by a factor six. These two times are attributed to the two sets of paired ions discussed in the previous paragraph. For these sets, the direction of the D-vector with respect to the B-field is different, making the strength of the mixing and hence the relaxation times different. Analyzing the times gives a value of $D/k_B = 0.24 \pm 0.1$ K, similar to the theory, [19] and that obtained from the step lineshape [21]. This cross relaxation model can also be effective for relaxing single spins through coupling to further neighbors.

Narrowing of magnetization steps has been observed in 10^{-2} s pulsed-field experiments. [23] The width of the steps can be a factor of two or more narrower than expected for thermal broadening. This is due to the inability of the spin system to stay in thermal equilibrium as the field sweeps past the level crossings. The narrowing is greatest for CdS, smaller for CdSe, and no narrowing is observed in CdTe. For these compounds, theory finds $D/k_B = 0.055$, 0.16, and 0.30 K, respectively. [19] The narrowing appears to be inversely related to the magnitude of **D**, possibly through the relaxation time. Two mechanisms have been proposed, including a sample-bath bottleneck, but a detailed model for the narrowing has not been fully developed.

4. Internal Level Structure of Single Ions

Both Fe^{2+} (S=2, L=2) and Co^{2+} (S=3/2, L=3) — unlike the spin-only ion Mn^{2+} (S=5/2, L=0) — have nonzero orbital moments which produce complex structures in the energy levels of single ions. High magnetic fields have been essential for studying these levels which ultimately determine their magnetism.

The single-ion energy levels of many II-VI materials containing Fe have been studied using *inter*level Raman scattering experiments. The ground state of Fe^{2+} is a singlet A_1 and has no magnetic moment, only a field-induced moment. [24] Transitions between the ground and higher levels range between 1–10 meV. Several of the lower transitions were first observed by Heiman [25] in high-field Raman scattering experiments, and their B-dependence [25, 26] has been crucial to modeling the ion levels. The bulk of this work has been carried out by Twardowski and co-workers, who have also studied the complexities of Fe-Fe pairs, and also found it necessary to incorporate Jahn-Teller coupling in modeling several compounds. [27]

An isolated Co^{2+} ion normally functions as a spin-only S = 3/2 ion, while the nonzero angular momentum is responsible for g = 2.3 being different from the free-space value. The crystal field in wurtzite CdS produces a splitting of the four-fold S = 3/2 level into two spin doublets, $S_z = \pm 3/2$ and $S_z = \pm 1/2$. Figure 3 shows that at B = 0 the ± 1/2 levels are below the ± 3/2 levels by the single-ion anisotropy energy 2δ. This ordering gives rise to a magnetization step at B_o=1.1 T corresponding to the −1/2 → −3/2 level crossing. At low temperatures, the ions have a magnetic moment appropriate to spin-1/2 for $B < B_o$, while spin-3/2 is appropriate for $B > B_o$. The increase in magnetic moment produces a step where M increases by a factor of three. This step has been observed in Faraday rotation experiments at T=0.1 K in CdS:Co.[12]

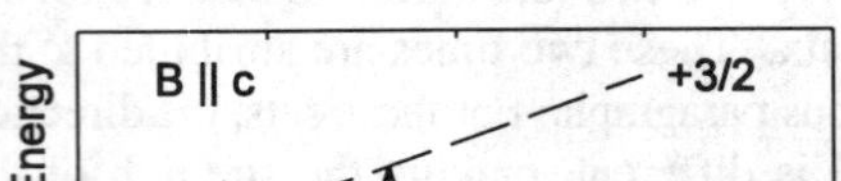

Figure 3: Energy versus B for the four Co^{2+} levels ±3/2 and ±1/3. The ±1/2 and ±3/2 levels are split by 2δ at B=0. At B=B_O the lowest level switches from −1/2 to −3/2. The arrows labeled a, b, c, d refer to transitions between levels. (From Ref. 28.)

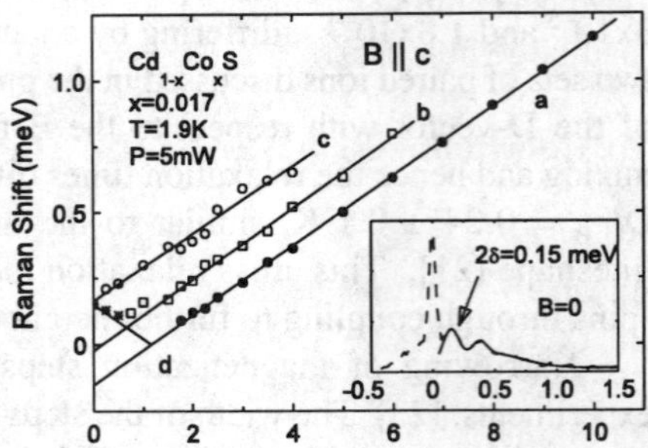

Figure 4: Transition energies versus B for the four transitions (a, b, c, d) in CdS:Co from Raman scattering. Inset displays Raman spectrum at B=0, and the arrow shows the peak due to the transition from the ±1/2 to the ±3/2 level. (From Ref. 28.)

Several transitions in CdS:Co have been studied by Raman scattering experiments. [28] The inset of Fig. 4 shows the B=0 spectrum. The peak at 0.15 meV is the splitting 2δ. Figure 4 shows data for the four transitions (a, b, c, d) as a function of magnetic field. From these experiments the value $\delta/k_B = 0.93 \pm 0.01$ K is obtained.

5. *sp-d* Exchange

The *sp-d* exchange interaction acts between the localized magnetic ions and mobile semiconductor carriers. Table I lists the *s-d* and *p-d* exchange energies, αN_o and βN_o, respectively. These values can be obtained by combining magnetization measurements with optical Zeeman splitting measurements. [29] Accurate values for αN_o in wide bandgap materials are usually obtained from spin-flip Raman scattering experiments, [3] and are approximately +0.2 eV. This small and nearly constant value for all the materials arises from the negligible hybridization of the *d*-states with the conduction band minima. [30] On the other hand, there is strong hybridization of the *d*-states with the valence band. [19, 30] This is the same hybridization that gives rises to the *d-d* exchange discussed in section 2. In essence, the *d-d* exchange is the *p-d* exchange acting twice, *d-d* = *d-p* + *p-d*. βN_o is negative (antiferromagnetic) and larger than αN_o. Values for βN_o are usually obtained from magneto-reflection experiments and are typically –1 to –2 eV. [29] The tabulated values demonstrate that $|\beta N_o|$ increases going from Cd to Zn and from Te to Se to S, in a similar manner to J. However, there are a few exceptions. The *p-d* exchange for Cr^{2+} ions is positive (ferromagnetic). [31] This change of sign arises because the spin-up t-type orbital t^3_+ is only partially occupied instead of fully occupied as in Mn^{2+}, Fe^{2+}, and Co^{2+}. Another exception is the apparent strong dependence of βN_o on Mn concentration in $Cd_{1-x}Mn_xS$, where the Zeeman splitting of the valence band is not linear in $(x\beta N_o)$ at small x. The large βN_o and large hole mass conspire to nearly localize holes at the Mn sites. [32]

6. Concluding Remarks

Both the *sp-d* and *d-d* exchange is likely to impact on future magneto-optical and magneto-electronic device applications. Recently, the Japanese company Tokin began marketing Faraday isolator devices utilizing HgCdMnTe as the active material. The devices operate at room temperature and are tailored for near-IR lasers for optical communications. This is the first example of an application for magnetic semiconductor materials and it places a renewed emphasis on investigations of the exchange in magnetic semiconductors.

For device applications it will be especially important to maximize the magnetization. Thus we would like to have a ferromagnetic (FM) response of the magnetic ion system instead of paramagnetic or antiferromagnetic (AF) response. This might be accomplished with a *ferri*magnetic material, an intrinsic FM *d-d* exchange, or a FM RKKY-type *sp-d* exchange that overcomes the AF *d-d* exchange.

Magnetic ions in III-V semiconductors are beginning to be studied. There have been a few efforts to investigate the *d-d* and *sp-d* exchange in GaAs:Mn [33] and InAs:Mn [34], but more work remains to be done in this area.

In summary, Figs. 5 and 6 show the donor ($M^{2+} \rightarrow M^{3+} + e$) and acceptor ($M^{2+} \rightarrow M^{+} + h$) levels of the various transition metal ions, along with the relative band edges for II-VI and III-V semiconductors. The approximate band lineups were determined from the positions of the transition metal levels.

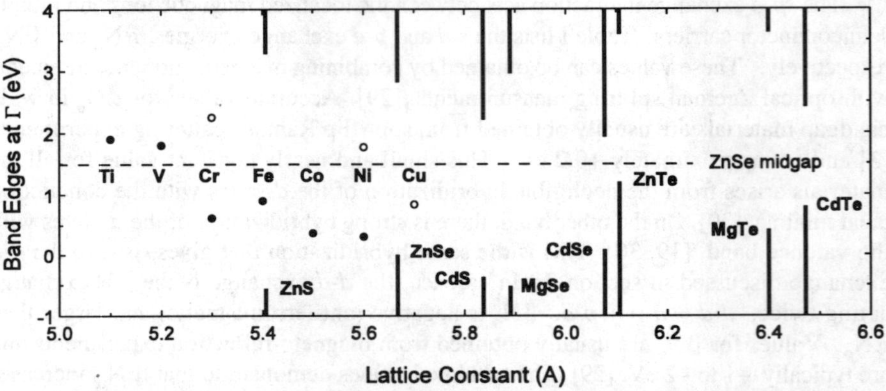

Figure 5: Relative band edges at Γ-point for II-VI group semiconductors versus lattice constant. The materials are designated next to the top of their respective valence band. Donor (solid points) and acceptor (open points) ground state energies are listed for various transition metals. The dashed horizontal line represents the midpoint of the ZnSe bandgap. (After Ref. 3.)

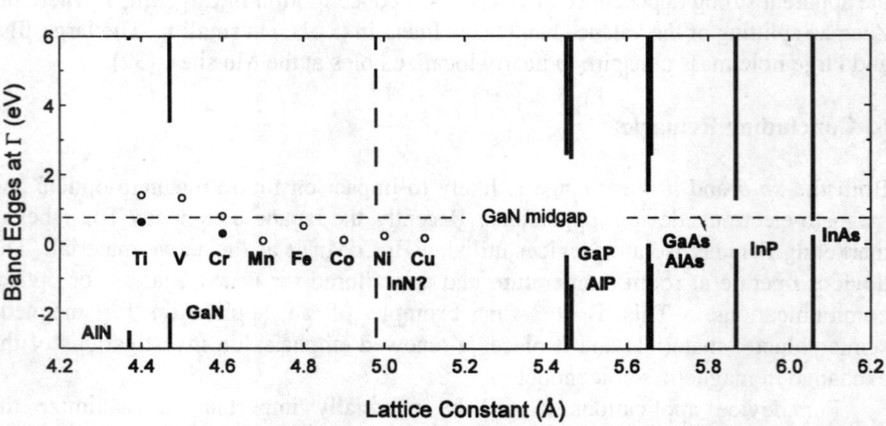

Figure 6: Relative band edges at Γ-point for III-V group semiconductors versus lattice constant. The materials are designated next to the top of their respective valence band. Donor (solid points) and acceptor (open points) ground state energies are listed for various transition metals. The dashed horizontal line represents the midpoint of the GaN bandgap.

Acknowledgements

I wish to thank S. Foner, Y. Shapira, and P.A. Wolff for many useful and important conversations over the years, and thank collaborators at MIT: P. Becla, M. Dahl, U. Gennser, M.J. Graf, X.C. Liu, T.Q. Vu, and X. Wang; and T. Dietl and A. Twardowski.

References

1. *II-VI Compounds and Semimagnetic Semiconductors*, eds. H. Heinrich and J.B. Mullin, *Mat. Sci. Forum* **182-184** (Trans Tech Publications, 1995).
2. J. Kossut and W. Dobrowolski, in *Handbook of Magnetic Materials*, ed. K.H.J. Buschow, vol. 7, (North Holland, Amsterdam, 1993), p. 231.
3. D. Heiman, Ch. 1, in *Laser Spectroscopy of Semiconductors*, eds. D.G. Seiler and C.R. Littler, Semiconductors and Semimetals, vol. 36, (Academic Press, 1992), p. 1.
4. Y. Shapira, S. Foner, D.H. Ridgley, K. Dwight, and A. Wold, *Phys. Rev.* B **30**, 4021 (1984).
5. R.L. Aggarwal, S.N. Jasperson, Y. Shapira, S. Foner, T. Sakakibara, T. Goto, N. Miura, K. Dwight, and A. Wold, *Proc. 17th Int. Conf. on the Physics of Semiconductors*, eds. J.D. Chadi and W.A. Harrison (Springer, 1985), p. 1419.
6. S. Foner, Y. Shapira, D. Heiman, and P. Becla, *Phys. Rev.* B **39**, 11793 (1989).
7. Y. Shapira, *J. Appl. Phys.* **67**, 5090 (1990).
8. S. Foner, *Physica* B **201**, 13 (1994).
9. G. Barilaro, C. Rigeaux, N.H. Hau, J.C. Picoche, and W. Giriat, *Sol. State Commun.* **62**, 345 (1987).
10. T.Q. Vu, V. Bindilatti, Y. Shapira, E.J. McNiff, C.C. Agosta, J. Papp, R. Kershaw, K. Dwight, and A. Wold, *Phys. Rev.* B **46**, 11617 (1992).
11. X. Wang, D. Heiman, S. Foner, and P. Becla, *Phys. Rev.* B **41**, 1135 (1990).
12. T.Q. Vu, D. Heiman, Y. Shapira, R. Kershaw, K. Dwight, and A. Wold, *Sol. State Commun.* **94**, 197 (1995).
13. M. Dahl, D. Heiman, S. Foner, Y. Shapira, T.Q. Vu, R. Kershaw, K. Dwight, and A. Wold, *Phys. Rev.* B **51**, 17561 (1995).
14. E.D. Isaacs, D. Heiman, P. Becla, Y. Shapira, R. Kershaw, K. Dwight, and A. Wold, *Phys. Rev.* B **38**, 8412 (1988).
15. W.J.M. de Jonge and H.J.M. Swagten, *J. Magn. Magn. Mater.* **100**, 322 (1991).
16. T.M. Giebultowicz, J.J. Rhyne, J.K. Furdyna, and P. Klosowski, *J. Appl. Phys.* **67**, 5096 (1990).
17. D.U. Bartholomew, E.-K.Suh, S. Rodriguez, A.K. Ramdas, and R.L. Aggarwal, *Sol. State Commun.* **62**, 235 (1987).
18. J. Stuhler, G. Schaak, M. Dahl, A. Waag, G. Landwehr, K.V. Kavokin, I.A. Merkulov, in Ref. 1, p. 781; and *J. Raman Spectroscopy* **27**, 281 (1996).
19. B.E. Larson and H. Ehrenreich, *J. Appl. Phys.* **67**, 5084 (1990); *Phys. Rev.* B **39**, 1747 (1989); and K.C. Hass, in *Diluted Magnetic Semiconductors*, eds. M. Balkanski

and M. Averous, (Plenum, 1991) p. 59.
20. J. Blinowski, P. Kacman, and J.A. Majewski, in Ref. 1, p. 779; and J. Blinowski and P. Kacman, *Phys. Rev.* B **46**, 12298 (1992).
21. V. Bindilatti, T.Q. Vu, Y. Shapira, C.C. Agosta, E.J. McNiff, R. Kershaw, K. Dwight, and A. Wold, *Phys. Rev.* B **45**, 5328 (1992).
22. X. Wang, M. Dahl, D. Heiman, P.A. Wolff, and P. Becla, *Phys. Rev.* B **46**, R11216 (1992).
23. Y. Shapira, S. Foner, D. Heiman, P.A. Wolff, and C.R. McIntyre, *Sol. State Commun.* **71**, 355 (1989); and V. Bindilatti et al., *Sol. State Commun.* **77**, 423 (1991).
24. D. Heiman, A. Petrou, S.H. Bloom, Y. Shapira, E.D. Isaacs, and W. Giriat, *Phys. Rev. Lett.* **60**, 1876 (1988).
25. D. Heiman, E.D. Isaacs, P. Becla, A. Petrou, K. Smith, J. Marsella, K. Dwight, and A. Wold, in *Proc. 19th Int. Conf. on the Physics of Semiconductors*, ed. W. Zawadzki (Polish Academy of Science, Warsaw, 1988), p. 1539.
26. D. Scalbert, J. Cernogora, A. Mauger, C. Benoit a la Guillaume, and A. Mycielski, *Sol. State Commun.* **69**, 453 (1989).
27. A. Twardowski, in *Dilute Magnetic Semiconductors*, eds. M. Balkanski and M. Averous (Plenum, 1991), p. 275; and *J. Appl. Phys.* **67** 5108 (1990).
28. X. Liu, U. Gennser, T.Q. Vu, D. Heiman, A. Twardowski, and M. Demianiuk, *Phys. Rev.* B **52**, 17199 (1995).
29. J.A. Gaj, in *Diluted Magnetic Semiconductors*, eds. J.K. Furdyna and J. Kossut, Semiconductors and Semimetals, vol. 25, (Academic Press, 1988) p. 276.
30. A.K. Bhattacharjee, G. Fishman, and B. Coqblin, *Physica* **117B&118B**, 449 (1983); A. K. Bhattacharjee, *Phys. Rev.* B **46** 5266 (1992); and Ref. 1, p. 681.
31. W. Mac, N.T. Khoi, A. Twardowski, J.A. Gaj, and M. Demianiuk, *Phys. Rev. Lett.* **71**, 2327 (1993).
32. C. Benoit a la Guillaume, D. Scalbert, and T. Dietl, *Phys. Rev.* B **46**, 9853 (1992).
33. X.C. Liu, D. Heiman, J. Hao, and K.C. Hsieh, *High Magnetic Fields in the Physics of Semiconductor*, ed. D. Heiman (World Scientific, Singapore, 1995), p. 658; and Ref. 1, p. 627.
34. H. Ohno, Ref. 1, p. 443; and *Phys. Rev. Lett.* **68**, 2664 (1992).
35. A. Twardowski, *Phys. Lett.* **94A**, 103 (1983).
36. D. Heiman, Y. Shapira, and S. Foner, *Sol. State Commun.* **51**, 603 (1984).
37. A. Twardowski, D. Heiman, Y. Shapira, T.Q. Vu, and M. Demianiuk, *Sol. State Commun.* **82**, 229 (1992).
38. O. Shih, R.L. Aggarwal, T.Q. Vu, and P. Becla, *Sol. State Commu.* **81**, 245 (1992).
39. U. Gennser, X.C. Liu, T.Q. Vu, D. Heiman, T. Fries, Y. Shapira, M. Demaniuk, and A. Twardowski, *Phys. Rev.* B **51**, 9606 (1995).
40. A. Twardowski, Ref. 1, p. 599.
41. A. Twardowski, D. Heiman, M.T. Liu, Y. Shapira, and M. Demaniuk, *Phys. Rev.* B **53**, 10728 (1996).

MAGNETIZATION OF DILUTED MAGNETIC SEMICONDUCTOR EPILAYERS IN HIGH MAGNETIC FIELDS

H. BEDNARSKI[a,b], J.C. PORTAL[a] and J. CISOWSKI[b,c]
[a]*Laboratoire des Champs Magnétiques Intenses, CNRS, 25 Avenue des Martyrs, 38042 Grenoble, France*
[b]*Department of Solid State Physics, Polish Academy of Sciences, ul.Wandy 3, 41-800 Zabrze, Poland*
[c]*Institute of Physics, Silesian Technical University, Katowice, Poland*

Analysing the *d-d* interaction in thin diluted magnetic semiconductor (DMS) epilayers, we have used our generalized pair approximation which permits to take into account the specific cation arrangement in such structures. We have subsequently applied our approach to calculated the magnetization (or, alternatively the mean spin projection of Mn^{2+} ions) in a series of $Cd_{1-x}Mn_xTe$ epilayers with moderate magnetic ion concentration ($x < 10\%$).

The magnetization of the first few monolayers of a dilute magnetic semiconductor (DMS) adjacent to the interface with a nonmagnetic semiconductor is significantly different from that of the bulk. This phenomenon is due to the difference in the number of neighbouring cation sites which may be occupied by magnetic ions, depending on whether a given ion is inside the DMS layer or it lies near the interface. The magnetization of DMS quantum structures (QS) has been up to now approximated by $M = M_0(x)B_S(B,T_{eff}(x))$[1], where x is the magnetic ion concentration, B is the external magnetic field, B_S is a Brillouin function for the spin value S, and $M_0(x)$ and $T_{eff}(x)$ are adjustable parameters. As has been proposed recently[2], in the limit of high magnetic field, a saturation value has to be replaced by a term $M_0(x) + \delta M_0(x)$ with $\delta M_0(x) = (Sg\mu_B - M_0(x))B_S(B,T'_{eff}(x))$, where g is the Landé factor, μ_B is the Bohr magneton and $T'_{eff}(x)$ is an additional effective temperature that describes the field-dependent increase of the saturation value. However, such an approximation, reproducing satisfactorilly the experimental data, is justified only for bulk materials, in which the random and homogeneous spatial distribution of the magnetic ions may be assumed[3]. Moreover, this approach is not very useful when studying the exchange interaction between the magnetic ions at very low temperatures, where a step-like behaviour of magnetization occurs and values of the adjustable parameters should depend on the number of monolayers of an epilayer.

The purpose of this work is to present a theoretical model for a description of the magnetization in DMS epilayers for moderate magnetic ion concentration ($x < 10\%$). Our method is based on the generalized pair approximation (GPA)[4], originally applicable to bulk DMS in which the magnetic ions may occupy a few kinds of inequivalent sites of the crystal lattice.

Within the GPA, the total free energy of DMS is given by a sum of free energies of interacting spins multiplied by appropriate probabilities[4]. A thin DMS epilayer, from the point of view of the GPA, is a solid with inequivalent planes perpendicular to the growth direction. It means that all ions belonging to a given plane has the same numbers of first (1N), second (2N), third (3N), etc., neighbours, so they are all equivalent but the ion sites belonging to different planes may be inequivalent. Thus, the magnetization M or, alternatively, the averaged spin of magnetic ions $<S_z>$, can be expresed as

$$M = < S_z > g\mu_B = -\sum_{\alpha=1}^{t} \frac{n_\alpha}{2} \sum_{\nu}^{\infty} P_\nu^\alpha(x) \frac{\partial F_\nu(J(R_\nu))}{\partial B}, \qquad (1)$$

where n_α is the number of spins in the α plane, F_ν is the free energy of a pair of interacting spins, $J(R_\nu)$ is an exchange constant between these spins separated by a distance R_ν and $P_\nu^\alpha(x)$ is the probability of finding such a pair.

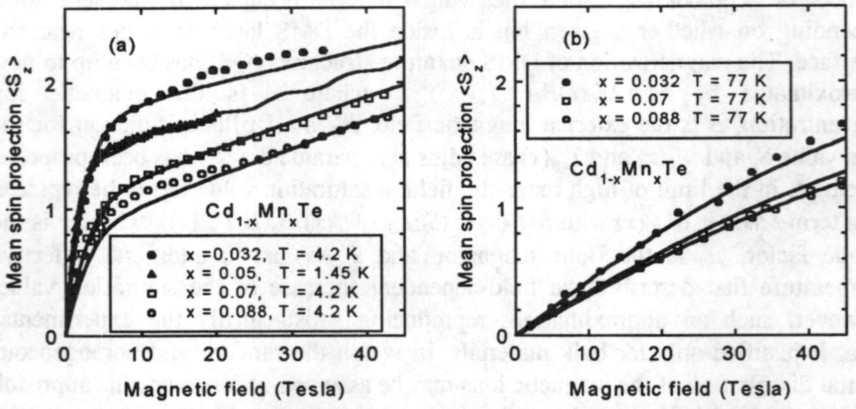

Figure 1: Comparison of the calculated mean spin projection (solid lines) with experiment for $Cd_{1-x}Mn_xTe$ as a function of magnetic field, for a series of different Mn^{2+} concentrations. Experimental data at 4.2 K (a) and 77 K (b) are taken from Ref. 2 and those at 1.45 K (a) from Ref. 6.

An example of application of our model to a series of $Cd_{1-x}Mn_xTe$ epilayers, grown by MBE on [001] InSb substrates is shown in Fig. 1, where, for the purpose of comparison, the data obtained for the bulk $Cd_{0.95}Mn_{0.05}Te$ are also presented. The calculations were performed assuming the *d-d* exchange interaction of the form

$J(R_v) = J_1 R^{-6.8}$ with the value of 1N exchange interaction $J_1 = -6$ K as found from inelastic neutron scattering in $Cd_{0.35}Mn_{0.65}Te$[5] and with an effective concentration, lowered by 10% ($x_{eff} = 0.9x$). As can be seen, our curves reproduce satisfactorilly all the experimental data for x up to 7%.

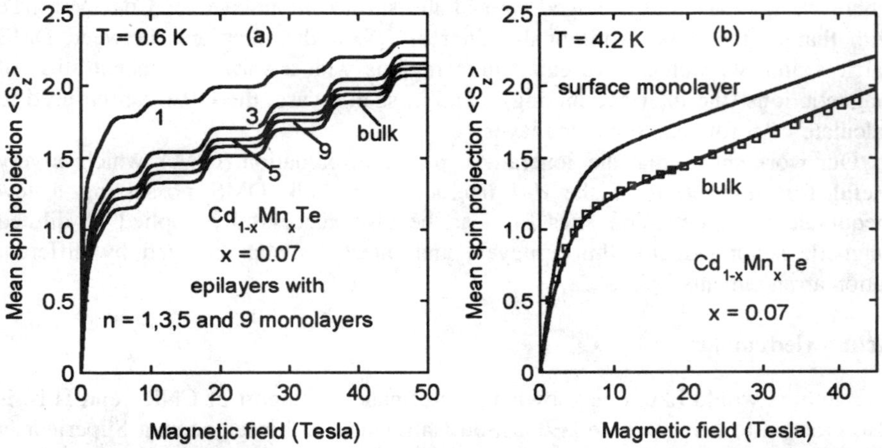

Figure 2: Calculated magnetic field dependence of the averaged spin of Mn^{2+} ions in $Cd_{0.93}Mn_{0.07}Te$: (a) thin epilayers consisting of 1, 3, 5 and 9 monolayers as compared with the bulk at T = 0.6 K; (b) the surface monolayer of the bulk DMS as compared with the bulk at T = 4.2 K (experimental data of Ref. 2).

We can now apply our method to calculate the magnetization in a number of very thin epilayers. The main difference in the magnetic properties of such quantum structures has only a statistical origin and the appropriate probabilities in Eq. 1 must be calculated for each specific epilayer. In Fig. 2 the calculated mean spin projection for the $Cd_{0.93}Mn_{0.07}Te$ epilayers consisting of odd numbers of monolayers is shown together with that of the bulk. Generally, the averaged spin in these thin DMS structures increases systematically as the number of layers decreases. This takes place also in low magnetic fields, where only pairs separated by distances greater than the 1N distance contribute to the averaged spin and saturate at about 8 T. Behaviour of the one-monolayer structure is characterized by a giant enhancement in low magnetic fields, which is connected with completely different cation arrangement than in a thicker epilayer. At higher fields and at sufficiently low temperatures, $\langle S_z \rangle$ exhibits a temperature-broadened, step-like increase due to the magnetic-field-induced alignments of the antiferromagnetically coupled 1N pairs and the magnitude of the steps does not depend of the number of monolayers except for the one-monolayer for which the steps are smaller.

Finally, we would like to discuss shortly a possibility of application of the presented method to describe the confined states in DMS quantum wells or superlattices. As is known, a barrier potential is proportional to $<S_z>$, so even in an ideal DMS quantum well with the random and homogenous distribution of magnetic ions in the barriers, the wells are not rectangular. This can be deduced from Fig. 2b, where we compare the averaged spin of the surface monolayer of $Cd_{0.93}Mn_{0.07}Te$ with that of the bulk. Additionally, in real QS at the interfaces between DMS barriers and SC wells, there exist finite regions with a variable concentration of magnetic ions (the interface mixing)[7,8] and in such a case, the GPA can be used to calculate $<S_z>$ for each separated layer.

Our work shows that the generalized pair approximation (GPA), which is very useful for description of the *d-d* interaction in bulk DMS possessing a few inequivalent magnetic ion sites [4], can be also successfully applied to diluted magnetic semiconductor thin epilayers and interfaces characterized by different cation arrangements.

Acknowledgments

The authors would like to acknowledge the financial support of CNRS and H.B. is also grateful to Ministére de l'Education Nationale, de l'Enseignement Superieur et de la Recherche of France for granting the post-doctoral fellowship.

References

1. J.A. Gaj, R. Planel and G. Fishman, *Solid State Commun.* **29**, 435 (1979).
2. R. J. Nicholas, M. J. Lawless, H. H. Cheng, D.E. Ashenford and B. Lunn, *Semicond. Sci. Technol.* **10**, 791 (1995).
3 J. M. Fatah, T. Piorek, P. Harrison, T. Stirner and W.E. Hagston, *Phys. Rev.* B **49**, 10341 (1994).
4. H. Bednarski and J. Cisowski, *Phys. Rev.* B **48**, 5113 (1993).
5. W.J.M. de Jonge and H.J.M. Swagten, *J. Magn. Magn. Mat.* **100**, 322 (1991).
6. R.L. Aggarwal, S.N. Jasperson, P. Becla and R.R. Galazka, *Phys. Rev.* B **32**, 5132 (1985).
7. J.A. Gaj,W. Grieshaber, C. Bodin-Deshayes, J. Cibert, G. Feuillet, Y. Merle d'Aubigné and A. Wasiela, *Phys. Rev.* B **50**, 5512 (1994).
8. T. Stirner, P. Harrison, W.E. Hagston and J.P. Goodwin, *Surface Science* **313**, 417 (1994).

ION-ION INTERACTIONS IN MIXED-VALENCY DILUTED MAGNETIC SEMICONDUCTORS

J. BLINOWSKI

Institute of Theoretical Physics, Warsaw University, Hoża 69, 00-681 Warszawa, Poland.

P. KACMAN

Institute of Physics, PAS, Al. Lotników 32/46, 02-668 Warszawa, Poland

J.A. MAJEWSKI

Walter Schottky Institut, Technische Universität München, Garching, Germany

We study the non-standard spin-spin interactions possible in mixed-valency diluted magnetic semiconductors . We consider the superexchange between d^4 and d^5 (e.g., Mn^{3+}- Mn^{2+}) ions and the double exchange - a mechanism which correlates the spin of magnetic ions in different charge states by virtual hopping of a d-electron from one ion to the other. We derive simultaneously the effective Hamiltonians for both the double exchange and the superexchange and calculate the energy spectrum for the pair of magnetic ions in different charge states.

By the expression "mixed-valency diluted magnetic semiconductors" we term semiconductors in which one can find coexisting transition metal (TM) ions of the same chemical nature but different charge state. This situation is by no means unusual and can be encountered in most $A^{II}B^{VI}$ DMS with TM ions, other than Mn, partially compensated by some shallow impurities. In $A^{III}B^{V}$ materials the state of mixed valency is possible even for Mn ions. Coexisting different charge states of the magnetic ions were observed in $A^{III}B^{V}$ semiconductors ($Ga_{1-x}Mn_xAs$,[1] and $In_{1-x}Mn_xAs$,[2]), though the precise nature of the Mn-related 3+ centers in these materials is not finally determined. Some authors suggest that these centers are complexes of Mn^{2+} ions with a tightly bound hole. Furthermore, the mixed valency occurs, even in the absence of any extra impurities, in DMS with resonant states of TM ions. The well known materials with this property are the zero-gap $A^{II}B^{VI}$ zinc-blende type semiconductors containing magnetic ions with the 2+/3+ donor level degenerated with the conduction band. HgSe and HgS with Fe ions are the best studied examples of the above category [3], but possible candidates are also mercury chalcogenides with chromium and vanadium magnetic ions.

The coexistence of the magnetic ions in different charge states opens a new channel of ion-ion exchange interactions resulting from the virtual hopping of the "extra" d-electron from one ion to the other. Such an interaction, termed the double exchange, was proposed by P.W. Anderson and H. Hasegawa [4] for manganites of perovskite structure. These authors considered a pair of Mn^{2+} - Mn^{3+} ions with one d-electron hopping virtually from one ion to the other via the p-orbitals of the neighboring anions.

In this paper we consider a model d^4-d^5 pair of ions in a semiconductor crystal. In this case the internal exchange integral responsible for the intrashell correlations leading to the Hund's rule is much larger than the anion-cation sp-d hybridization integral, so that the perturbation theory can be used to derive the effective Hamiltonian for the low energy dynamics of the system. One has to account, however, for the fact that the virtual electronic states in a semiconductor are the band states rather than the localized anion levels. Using the k-space perturbation theory with respect to the sp-d hybridization Hamiltonian we derive the effective Hamiltonian describing simultaneously the double exchange and the superexchange for a d^4 -d^5 pair:

$$\hat{H}_{eff} = -2J\hat{S}_1\hat{S}_2 + \left[b\left(\hat{a}^+_{2l\sigma}\hat{a}_{1l\sigma} + \hat{a}^+_{2l(-\sigma)}\hat{a}_{1l(-\sigma)} \right) + h.c. \right] \quad (1)$$

$\hat{S}_1$ and $\hat{S}_2$ are the spin operators for d-electrons in the ion 1 and 2, respectively. $\hat{a}^+_{il\sigma}\hat{a}_{jl\sigma}$ are hopping operators transferring the electron with the spin σ from the d-orbital l of the j-th ion to the orbital l of the i-th ion. The superexchange constant J results from the fourth order perturbation theory. The double exchange constant b comes already from the second order. Both constants depend on the excitation energies of the system and involve summations over the valence bands in the entire Brillouin zone. They also differ in their dependencies on the interionic distance $\vec{R}_1 - \vec{R}_2$.

We start with the discussion of the case when the effective Hamiltonian (1) acts in the space of degenerate states, in which the unperturbed states $1d^4 2d^5$ and $1d^5 2d^4$ (i.e., the state with 4 d-electrons on the ion 1 and 5 on the ion 2, and vice versa) have the same energy. The diagonalization of the Hamiltonian (1) gives the energy spectrum of the pair in the form:

$$E_{\pm}(S) = -JS(S+1) \pm \frac{|b|}{5}\sqrt{S(S+1) + 1/4} + c \quad (2)$$

where S = 5/2 +2 = 9/2, 7/2, ... , | 5/2 - 2 | =1/2, is the total spin of the pair.

The states with lower energy $E\_(S)$ are, depending on the sign of the double exchange constant b, symmetric ($b<0$) or antisymmetric ($b>0$) combinations of the states $1d^4 2d^5$ and $1d^5 2d^4$. Regardless the sign of b, the second order double exchange term favors the large values of the total spin of the pair. It has, therefore, essentially ferromagnetic, though non-heisenberg character. The superexchange term has a standard heisenberg-like form, which, depending on the sign of the superexchange constant J, can be ferromagnetic ($J>0$) or ferrimagnetic ($J<0$). For $J>0$ the lowest energy state of the pair is the state of parallel spins of the ions. .For $J<0$ there is a competition between the heisenberg-like superexchange and non-heisenberg double exchange terms. Minimizing the energy (2) with respect to the argument S(S+1)+1/4 we can distinguish three possible spin alignments corresponding to the lowest energy state, depending on the ratio $\Re = b/(10J)$:

a) $|\Re| > 9/2$ - the lowest energy is given by the maximum value of the total spin S=9/2 and, like in the $J>0$ case, the ferromagnetic alignment is privileged;

b) $|\Re| < 3/2$ - the lowest energy state corresponds to the minimum value of the total spin S=1/2 (ferrimagnetic alignment);

c) $3/2 < |\Re| < 9/2$ - the lowest energy state is a state with the intermediate values of the total spin $1/2 < S < 9/2$ (canted alignment).

The ratio $\Re$ determines the magnetic properties of the pair, since the effect of the external magnetic field on the energy spectrum (2) reduces, in the first approximation, to an additive Zeeman-like term $g\mu_B S_z B$. The numerical calculations of the constants J and b we performed with the parameters taken for ZnSe with Cr (d^4 and d^5) ions. It is not clear yet whether the conclusions apply also to other systems. The calculated constants J are negative (antiferromagnetic interaction, in contrast to the ferromagnetic Cr^{2+}-Cr^{2+} superexchange [5]) but much smaller than for typical DMS with d^5 (Mn^{2+}) ions ($|J|$<1K, even for the n.n. ions). The constants b turned out to be quite large for this particular system (of the order of 400 K for the nearest, 60 K for second and 20 K for the third and fourth neighbors), so that the ferromagnetic alignment of the case a) can be expected. It is interesting to observe, however, that the double exchange, even in this case, has no effect on the pair contribution to the high temperature susceptibility of the system, which is completely determined by the superexchange. This is due to the cancellation of the terms with plus and minus sign in Eqn (2) in the high temperature expansion.

We pass now to the case in which the local potential energy fluctuations remove the degeneracy between the states $1d^4 2d^5$ and $1d^5 2d^4$. The energies of these states we denote by E_1 and $E_2 > E_1$, respectively. In this case the energy spectrum of the pair has a more complex form:

$$E_\pm(S) = -JS(S+1) + \frac{E_1 + E_2}{2} \pm \sqrt{\left(\frac{E_2 - E_1}{2}\right)^2 + \frac{b^2}{25}[S(S+1) + 1/4]} + c \qquad (3)$$

In the limiting case when the energy splitting (E_2-E_1) is much larger than the double exchange term the square root may be expanded and we get the purely heisenberg form of the spin-spin interaction. In this limit the double exchange is still ferromagnetic, but strongly suppressed as compared to the degenerate case.

The description of the macroscopic magnetic properties of semiconductors with coexisting magnetic ions in different charge states represents a difficult theoretical problem. The spatial distribution of the ions carrying the extra charge is not random. It depends very much on the character of the compensating charge. In this paper we have presented the discussion of the spin-spin interactions within a d^4-d^5 pair of magnetic ions in a semiconductor. Such a pair model represents only a first step towards understanding the magnetic properties of mixed valency DMS's, shining light on the microscopic mechanisms of spin-spin interactions in these materials. We believe, however, that there are some specific situations, in which the pair model might be of more direct use:

1. in zero-gap materials the extra charge of magnetic donors is compensated by the free carriers in the conduction band, quasi-uniformly smeared out over the entire crystal. As shown by Wilamowski et al.[6], the charge-charge correlations keep the charge donors as far as possible one from each other. The position of each extra charge is not completely fixed, however. When two (or more) ions are situated in the vicinity of the local potential energy minimum, the extra charge is bound by a pair (or larger cluster) rather than by an individual ion. One may expect the properties of such small clusters to be properly described by the discussed in this paper pair model or by its simple generalization for slightly larger clusters.
2. in $A^{III}B^{V}$ materials with Mn ions the Mn^{2+} centers are negatively charged compared to the other cations. In n-type materials this extra charge is compensated by ionized donors. In this situation negatively charged are those manganese ions which are nearest to the donors. Here again the pair model might be useful for the description of the magnetic properties of pairs formed by the Mn^{2+} ions with the nearby Mn^{3+} centers. The conclusions of the d^4-d^5 pair model should remain qualitatively correct, in our opinion, even in the case when the Mn^{3+} centers represent complexes "Mn^{2+} + hole", provided the spin of the hole is strongly coupled to the spin of the ion.

Acknowledgments

This work was partially supported by the Polish Scientific Committee (KBN), grant No. 2 P03B 025 11.

References

1. X.C. Liu *et al*, *Mat. Science Forum* **182-184,** 627 (1995)
2. H. Munekata *et al*, *Phys. Rev. Lett.* **63**, 1849 (1989).
3. Z. Wilamowski, *Acta Phys.Polon.* **A 77,** 133 (1990) and the references therein.
4. P.W. Anderson and H. Hasegawa, *Phys. Rev.* **100,** 675 (1955).
5. J. Blinowski *et al, Phys. Rev.* **B 53,** 9524 (1996-I).
6. Z. Wilamowski *et al*, *Solid State Commun.*, **74**, 833 (1990).

THEORETICAL MODEL FOR THE DESCRIPTION OF LOW DIMENSIONAL STRUCTURES IN TILTED MAGNETIC FIELDS

D. SUISKY, F. NEUGEBAUER, W. HEIMBRODT
Humboldt-Universität zu Berlin, Institut für Physik, Invalidenstr. 110, D-10115 Berlin, Germany

P. ENDERS
Max-Born-Institut für Nichtlineare Optik und Kurzzeitspektroskopie, Rudower Chausse 6, D-12489 Berlin, Germany

We calculate the Zeeman splitting pattern in dimension reduced structures for magnetic fields of an arbitrary orientation relative to the sample symmetry axis. We discuss the coupling mechanism of the valence band states induced by the magnetic field and explain typical features of the magnetic field dependence observed on CdTe/CdMnTe and CdMnTe/CdMgTe quantum well structures.

1 Introduction

Dimension reduced structures like superlattices (SL's), double quantum wells (DQW's) and quantum wires (QWR's) have been achieved a great deal of interest due to their tunable sample parameters which are obtained by changing the well widths and depths combining materials with different bulk properties. The special features of semimagnetic materials allow due to the s,p-d interaction between the localized spins of the Mn ions and the spins of the band electrons of the host material [1,2] for an additional tuning of the parameters by an external magnetic field. Typical Zeeman splitting pattern have been observed experimentally in the energy range of the excitonic transitions where in the saturation range of the Zeeman splitting a heavy hole splitting of about 80 to 100 meV is obtained for field strengths of 6 to 8 Tesla and for Mn concentration of $x_{Mn} = 0.2$ where a saturation of the alignment of the spins of the Mn ions is observed. Beside the saturation which is independent of the orientation of the magnetic field orientation dependent effects have been observed experimentally [3–7] and investigated theoretically [4,5,7]. The majority of experimental data reported in literature are obtained either in Faraday or in Voigt configuration where the magnetic field is oriented parallel or perpendicular to the growth direction, respectively. Intermediate tilted field configurations are rarely investigated [3].

The theoretical model for the explanation of the field dependence considers the complex interaction of different perturbation on the band structure like the confinement, the misfit induced strain and the Coulomb interaction between electrons and holes. The explanation of the experimentally observed different Zeeman splitting pattern in Faraday and Voigt configurations has been given in terms either of a magnetic field induced [4] or a magnetic field independent coupling [5,7] of the valence band states.

Thus, it seems to be quite naturally to study intermediate tilted field orientations in order to link the two limiting cases and to obtain an insight in the coupling mechanism as well as the interplay of the magnetic field with other perturbations. This will be done in the present paper.

2 Theory

The model is based on the bulk material characterization of each layer of the SL or the QW structure by an Kohn-Luttinger Hamiltonian

$$H = H_{kp} + H_{strain} + H_{ex} \tag{1}$$

where H_{kp} describes the bandstructure, H_{strain} the misfit induced deformation and H_{ex} the exchange interaction present either in the well or in the barrier material. The magnetic field dependence is caused only by the third term. The whole structure is described by the transfer matrix formalism [8] where the confinement enters as constant potential added to the corresponding parts of the Hamiltonian which describe the material of different layers. Assuming the z-direction parallel to the symmetry axis, the equivalence of the x- and y- inplane directions in the homogeneously deformed cubic material will be expressed by the Kohn-Luttinger representations of H_{kp} and H_{ex} which read

$$H^v_{cryst} = \begin{pmatrix} F & K & J & 0 \\ K^* & G & 0 & J \\ J^* & 0 & G & -K \\ 0 & J^* & -K^* & F \end{pmatrix}, \quad H^v_{ex} = \begin{pmatrix} \frac{3}{2}\delta & \gamma & 0 & 0 \\ \gamma^* & \frac{1}{2}\delta & \alpha & \\ 0 & \alpha^* & -\frac{1}{2}\delta & \gamma \\ 0 & 0 & \gamma^* & -\frac{3}{2}\delta \end{pmatrix}, \tag{2}$$

where F, G, J, and K depend on the Kohn-Luttinger parameters γ_1, γ_2 and γ_3 and the $\vec{\mathbf{k}}$ vector. The typical magnetic field dependence is described by H_{ex} where α, γ, and δ depend on the effective field components B_x^{eff}, B_y^{eff}, and B_z^{eff}, respectively. Examining the nondiagonal matrix elements K and J of H_{kp} we find that the isotropy of the bulk material properties parallel to the layers is well described because of the equivalence of the x- and y-directions. $J = \sqrt{3}\gamma_2(-k_x^2 + k_y^2) + 2\iota\sqrt{3}\gamma_3 k_x k_y$ depends only on k_x and k_y and the coupling between the directions parallel and perpendicular to the growth direction is described by K which depends on the products $k_x k_z$ and $k_y k_z$. Studying only

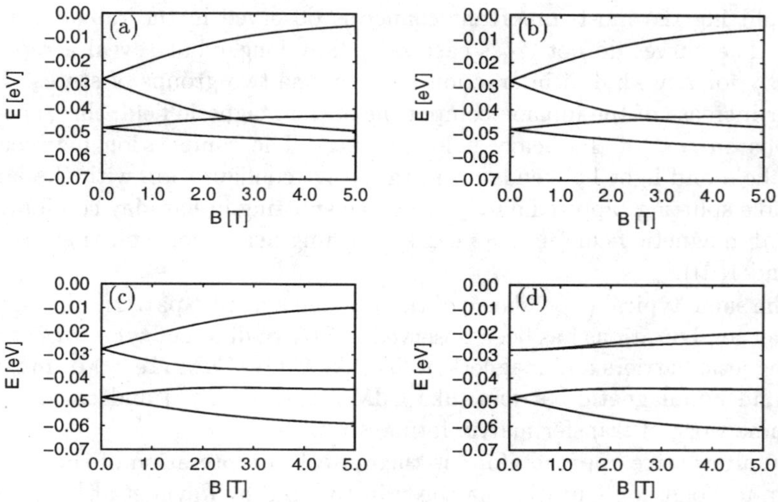

Figure 1: Zeeman splitting pattern for a semimagnetic single quantum well in tilted field configuration. Parameters: Well width $L_w = 30$Å, well depth $E_v = 80meV$, Mn concentration $x_{Mn} = 0.2$. Ratio of the magnetic field components $R = B_z/B_x$. (a) Faraday configuration: $B_z = B$, $B_x = 0$, (b) $R = 1/3$, (c) $R = 1$, (d) Voigt configuration: $B_z = 0$, $B_x = B$.

zone center states for optical excitonic transition we can set $k_x = k_y = 0$ and the matrix (2) becomes diagonal and is finally introduced in the transfer matrix formalism replacing k_z by $-i\partial/\partial z$. Then, the only remaining nondiagonal terms in the Hamiltonian are found in the magnetic field dependent H_{ex} part if the magnetic field is applied either in Voigt configuration or in any tilted field configuration.

The quantitative study of the dependence on the tilting angle will be given in terms of the Zeeman splitting pattern in the next section.

3 Results and Discussion

The results of the calculations for a single magnetic quantum well embedded between nonmagnetic layers are shown in Fig 1. In Faraday configuration, we observe the typical Zeeman splitting (see Fig. 1(a)). The heavy hole splitting is larger than the light hole splitting.

In Fig. 1(b) the splitting pattern in the presence of an inplane component of the magnetic field B_x accompanied by a reduction of the B_z component is shown. Beside the reduction of the splitting ratio of heavy hole and light hole splitting the most important change is observed in the range around 3 Tesla. The curves do not cross each other any longer but reveal a repulsion typically for any kind of interaction between the two groups of states. For a further increase of the inplane component of the magnetic field, this repulsion develops on cost of the heavy hole splitting. The competition between the heavy hole and light hole ends up in the Voigt configuration with the largest light hole splitting (approximately twice the splitting in Faraday configuration for small magnetic fields) and a nearly vanishing heavy hole splitting (see Fig. 1(c) and 1(d)).

The same typical dependence of the Zeeman splitting pattern in Voigt and Faraday configurations has been observed in SL's of different type consisting of nonmagnetic barriers and magnetic wells (like CdTe/CdMnTe [4,5]) or magnetic wells and nonmagnetic barriers (like CdMnTe/CdMgTe [7]) and calculated in the frame work of transfer matrix formalism [4].

An alternative approach for the theoretical interpretation of the splitting pattern in Voigt configuration has been introduced by Peyla et al. [5] which has been used for the interpretation of the magnetic field dependence observed on CdMnTe/CdMgTe structures by Kuhn-Heinrich et al. [7]. The H_{kp} Hamiltonian is decomposed into two blocks which belong to the two independent subspaces [3/2, -1/2] and [-3/2, 1/2] neither coupled by the strain nor by the magnetic field nor by another perturbation. Obviously, a tilted field configuration cannot be studied in that model and, furthermore, it is impossible to describe the magnetic field dependence in Voigt and Faraday configurations in the same reference system. As a consequence, the agreement between experiment and theory is at least only qualitatively [7].

Summarizing we can state our model allows to interpret the typical features of tilted field configurations in terms of the coupling of heavy hole and light hole states including the quantitative agreement with the experimental data in the limiting cases of Faraday and Voigt configurations.

References

1. R. R. Gałązka and J. Kossut, *Lecture Notes in Physics* **133**, 245 (1980)
2. O. Goede and W. Heimbrodt, *phys. stat. sol.* (b)**146**, 11 (1988)
3. E. Vornberger et al., *Proc. 21st Int. Conf. Phys. Semicond.*, 1990
4. F. Neugebauer et al., *Proc. 22nd Int. Conf. Phys. Semicond.*, 1992
5. P. Peyla et al., *Phys. Rev.* **B47**, 3783 (1993)
6. W. Heimbrodt et al., *J. Luminescence* **60/61**, 344 (1994)
7. B. Kuhn-Heinrich et al., *Solid State Commun.* **91**, 413 (1994)
8. B. Chen et al., *Phys. Rev.* **B45**, 1204 (1992)

EXCITON STATES IN CdTe/$Cd_{1-x}Mn_xTe$ MULTIPLE QUANTUM WELLS IN THE MEGAGAUSS FIELD RANGE

Shinji KURODA, Kazutoshi KOJIMA and Kôki TAKITA
Institute of Materials Science, University of Tsukuba, Tsukuba, Ibaraki 305, JAPAN

Kazuhito UCHIDA and Noboru MIURA
Institute for Solid State Physics, University of Tokyo, Roppongi, Minato-ku, Tokyo 106, JAPAN

Exciton states of CdTe/$Cd_{1-x}Mn_xTe$ multiple quantum wells (MQW's) with Mn-content of $x \simeq 30\%$ were studied in the megagauss field range. The field dependence of observed exciton energies is compared with an effective-mass calculation taking the valenceband mixing effect into account and features which appear associated with the type I – type II transition are discussed.

1 Introduction

In CdTe/$Cd_{1-x}Mn_xTe$ multiple quantum wells (MQW's), the effective band offset can be tuned due to the enhanced Zeeman splitting in the semimagnetic $Cd_{1-x}Mn_xTe$ layer. Especially, applying high magnetic fields in the megagauss field range on MQW's with a relatively large Mn concentration enables us to change the superlattice energy structure drastically, including the field-induced type I – type II transition. In a previous paper[1], we reported evident features of the type I – type II transition which were observed in the field dependence of the heavy-hole exciton of the first subband (E_1H_1). In this work, the exciton states under high magnetic field including those other than the E_1H_1 exciton are studied, with a special attention to new features associated with the type I – type II transition.

2 Experimental Methods

MQW samples were grown by the conventional MBE method described in Ref.[2]. The magneto-absorption measurement in the megagauss field range was performed using the single-turn coil technique in a similar way to those described in Ref.[1]. Magnetic fields were applied along the growth direction of samples and light from a Xe flash lamp was transmitted along this direction (Faraday configuration). The temperature at the measurement was typically $16 \sim 18\,K$.

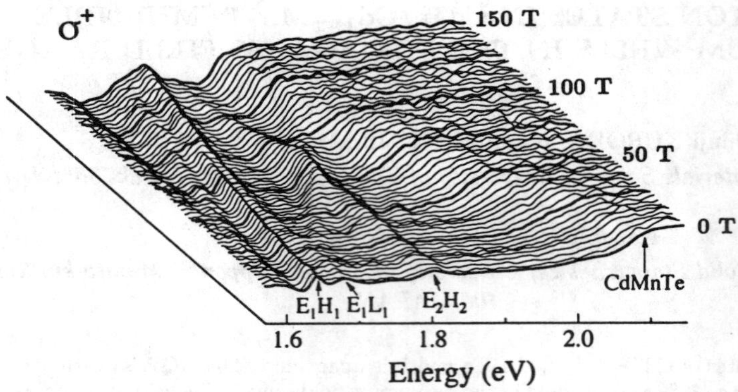

Figure 1: Absorption spectra of a sample with $L_W = 67$ Å, $L_B = 163$ Å and $x = 0.32$ for the σ^+ polarization under magnetic fields from 0 to 150 T at an interval of 2.5 T.

3 Experimental Results and Discussion

Figure 1 shows a typical example of absorption spectra under magnetic fields up to 150T for the σ^+ polarization. This is the result of a sample with well and barrier widths of 67 Å and 163 Å and Mn-content $x = 0.32$. The assignment of absorption peaks at zero field is indicated in the figure by the following notation; EH(EL) means the heavy-hole (light-hole) exciton and the subscript is the subband in the conduction and the valenceband. When the magnetic field is increased, the E_1H_1 peak becomes weak and disappears at about 100T[1]. This phenomenon is interpreted to be associated with the field-induced type I – type II transition; the oscillator strength is reduced as the heavy-hole wavefunction becomes confined in the $Cd_{1-x}Mn_xTe$ layer under the type II band alignment.

In Fig.2, peak energies in the absorption spectra for the σ^+ polarization are plotted against magnetic field. In the figure, circles and triangles denote peaks which are assigned to the heavy-hole (E_1H_1, E_2H_2) and the light-hole exciton (E_1L_1), respectively. Crosses denote the absorption edge of the $Cd_{1-x}Mn_xTe$ cladding and capping layers and small circles do the absorption peak of the CdTe buffer layer. As already seen in Fig.1, the E_1H_1 peak dissappears beyond 100 T for the σ^+ polarization. It is to be noted that the E_2H_2 peak begins to shift rapidly to the lower energy when field is increased beyond 100 T. This rapid decrease of the E_2H_2 exciton energy is also interpreted to be an evidence of the type transition; that is, in the field region of the type II band alignment, the heavy hole with $m_j = -3/2$ is confined in the $Cd_{1-x}Mn_xTe$ layer and its energy decrease rapidly with the increase of magnetic field following the giant

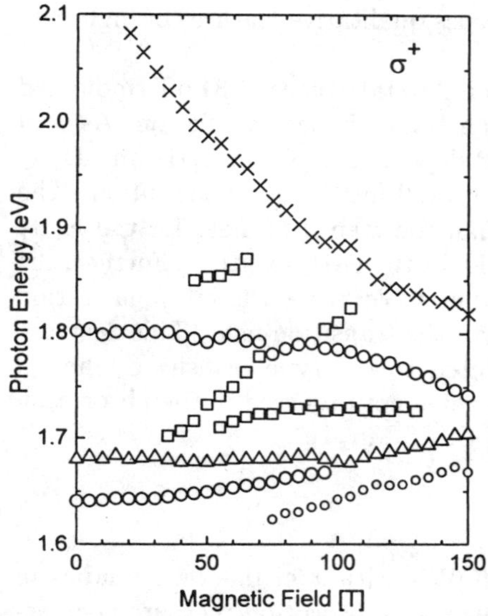

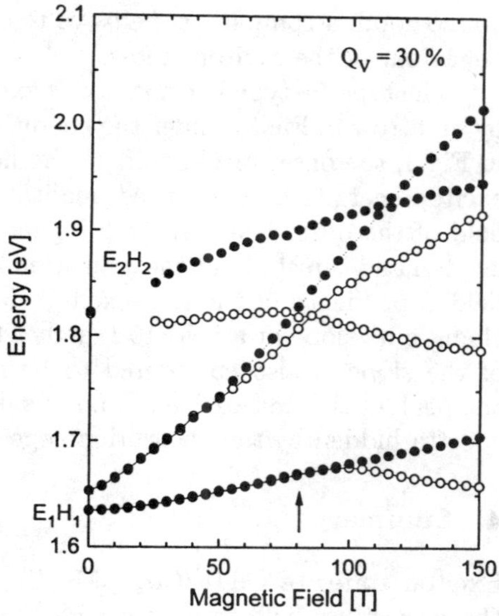

Figure 2 (left): Peak energies in the absorption spectra shown in Fig.1 are plotted against magnetic field. See text for the notation of symbols.

Figure 3 (right): The calculated exciton energies as a function of magnetic field. The E_2H_2 exciton energies in intermediate field range below 20 T could not be calculated accurately due to the limitation of the basis function and are not plotted here.

Zeeman shift in $Cd_{1-x}Mn_xTe$. On the other hand, in the field dependence of the E_1H_1 exciton energy, a feature associated with the type of transition is not observed because the E_1H_1 peak dissappears totally in the type II field region. In other samples with about the same Mn contents, however, the E_1H_1 peak survives in the type II region and the onset of a red shift is observed[1].

We have calculated the exciton energies under magnetic fields on the basis of the magneto-exciton model, taking full acount of the valenceband mixing after Ref.[3]. In the calculation, the alignment of Mn^{2+} spins was estimated based on the mean-field approach[4]. The result of the calculation shows a complicated energy structure consisting of many levels with various quantum indices of the subbands in the quantum well and of the 2D hydrogenic orbitals. Among them, the ground state and the first-excited Landau sublevel of the heavy-hole exciton are extracted and plotted in Fig.3. Here the valenceband offset ratio $Q_V = 30\,\%$ is assumed. In the figure, open (closed) circles denote exciton states the transition of which is allowed for the σ^+ (σ^-) polarization. Though quantitative agreement between the calculation and the experiment is not so good, a comparison helps us to discuss qualitative features of high-field behaviour of the exciton states.

The type I – type II transition is expected to take place at 81.6 T (indicated by an arrow in Fig.3) under the condition assumed in the calculation. As seen in Fig.3, the onset of red shift in the field dependence of the E_1H_1 and E_2H_2 exciton energies is reproduced qualitatively well for the σ^+ polarization. The field of the onset is about 20 T higher than the transition field for the E_1H_1 exciton and is just about the transition field for the E_2H_2 excition. Further, the field dependence of the first-excited Landau sublevel for the σ^+ polarization changes its slope at a field 10 T higher than the transition field. This decrease of the slope is also considered to be related to the type transition, though the peak of the first-excited Landau sublevel is not observed in the absorption spectra hidden by the absorption edge of $Cd_{1-x}Mn_xTe$.

4 Summary

Exciton states of $CdTe/Cd_{1-x}Mn_xTe$ MQW's with $x \simeq 0.3$ were studies in the megagauss field range through the magneto-absorption measurement using the single-turn coil technique. In the field dependence of the E_1H_1 and E_2H_2 exciton, the onset of red shift associated with the type I – type II transition was observed and was reproduced qualitatively well by the effective-mass calculation. The result of the calculation also suggests that the energy of the first-excited Landau sublevel of the heavy-hole exciton should change its slope against magnetic field due to the type transition.

Acknowledgment

This work is supported in part by Grant-in-Aid for Scientific Research from the Ministry of Education, Science and Culture of Japan.

References

1. S. Kuroda, K. Kojima, K. Takita, K. Uchida and N. Miura, J. Crystal Growth **159** 967 (1996).
2. S. Kuroda, K. Kojima, K. Kobayashi, Y. Shirai, K. Takita, K. Uchida and N. Miura, Physica B **201** 415 (1994).
3. G. E. W. Bauer and T. Ando, Phys. Rev. B **38** 6015 (1988).
4. E. D. Isaacs, D. Heiman, X. Wang, P. Becla, K. Nakao, S. Takeyama and N. Miura, Phys. Rev. B **43** 3351 (1991).

MAGNETO-RAMAN SCATTERING AND MAGNETO-OPTIC KERR EFFECT OF THE DILUTED MAGNETIC SEMICONDUCTOR (Cd,Fe)Te

J. Stühler, R. Meyer, M. Dahl, and G. Schaack
Physikalisches Institut der Universität Würzburg, Am Hubland, 97074 Würzburg, Germany

The exchange integrals of the diluted magnetic semiconductor $Cd_{0.985}Fe_{0.015}Te$ were investigated by means of resonant Raman spectroscopy and polar magneto-optic Kerr effect in magnetic fields to B=14T and B=7T respectively. The ratio $\frac{\alpha}{\beta}$ of the electron and hole exchange is derived by combining spin-flip Raman scattering and Raman resonance data or results from magneto-optic Kerr experiments. Analysis of the phase relation of Kerr rotation and ellipticity reveals the existence of a magneto-optically inactive surface layer.

1 Introduction

The unique properties of diluted magnetic (semimagnetic) semiconductors (DMSs) result from the strong exchange interaction between the spins of the magnetic ions and the free carriers (conduction band (CB) electrons and valence band holes). Contrary to the very intensively studied Mn-based DMSs the knowledge about systems based on other magnetic ions is still fragmentary. In the following we present results for a $Cd_{1-x}Fe_xTe$ sample. The magnetic behavior of this material system is rather different from the Mn-based DMSs. The cubic crystal field and the spin-orbit interaction lead to a splitting of the atomic ^5D multiplet of the Fe^{2+} ion. The lowest manifold of states consists of two singlets, two triplets and a doublet separated by a few meV. The ground state is a Γ_1 singlet with quenched angular momentum. The paramagnetic properties of the Fe^{2+} system are thus due to magnetic field induced magnetic moments (Van Vleck paramagnet) [1].

2 Experimental

We used Raman scattering in Voigt geometry and polar magneto-optic Kerr effect (MOKE) to probe the CB spin splitting and the Zeeman splitting of the heavy-hole (HH) interband transitions in an external magnetic field. The Fe concentration x=0.015 of the sample was determined using the energy dependence of the HH exciton transition [2] and the position determined from the MOKE data.

Figure 1.a shows the energies of the spin-flip (SF) and two intra-iron tran-

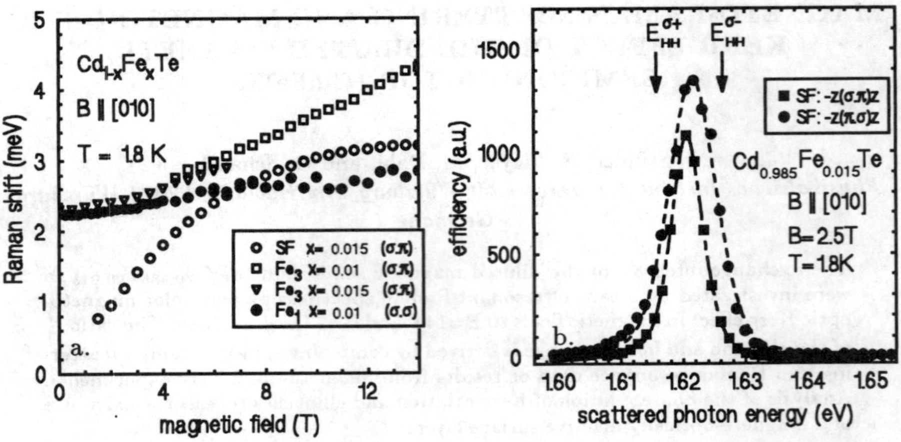

Figure 1: a. Magnetic field dependence of the spin-flip and intra-iron transitions of Fe^{2+} in CdTe. b. Resonance behavior of the of the spin-flip Raman transitions for $(\pi\sigma)$ and $(\sigma\pi)$ polarizations. The arrows indicate the positions of the heavy-hole transitions determined from the MOKE data.

sitions in magnetic fields. The nonlinearity of the SF energy for fields above 4T is a consequence of the small energy difference between the 10 lowest iron levels which allows a strong admixture of excited Fe states to the ground state. The interaction between the SF and the lowest intra iron transition of the same symmetry causing an anticrossing in (Cd,Fe)Se [3] and (Cd,Fe)S [4] is not important here since for the low iron concentration of our sample even the interaction-free levels do not cross due to the nonlinearity of the magnetization. The SF energy E_{SF} which provides a direct measurement of the CB splitting is therefore, like in Mn-based materials, given by $\Delta E_{CB} = N_0 \alpha x \langle S_z \rangle$.

The scattering process of the SF Raman transition shows sharp resonances in the energy range just below the E_0 band gap, an example is given in figure 1.b. The two resonance peaks can be assigned to scattering processes involving light-hole (LH) excitations as intermediate states, scattering processes involving HH transitions are forbidden by selection rules. The separation $\Delta E_{res} = -1.0 \pm 0.2\, meV$ at $B = 2.5T$ of the resonance maxima is given by the LH splitting $\Delta E_{LH} = \frac{1}{3} N_0 \beta x \langle S_z \rangle$. The sign of ΔE_{res} follows by means of the selection rules from the observation of the $\sigma\pi$ resonance at lower energy than the $\pi\sigma$ resonance. The arrows mark the energies of the HH transitions as determined by MOKE. Using the SF energy and the splitting of the resonances allows a determination of the ratio of the exchange integrals $\frac{\alpha}{\beta} = \frac{\Delta E_{SF}}{3 \Delta E_{res}} = -0.4 \pm 0.1$ for Fe^{2+} in CdTe.

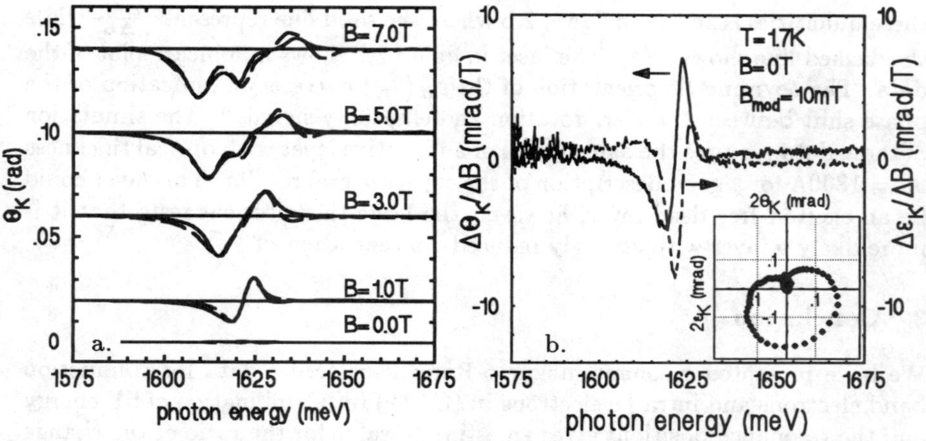

Figure 2: a. Dispersion of the magneto-optic Kerr angle Θ_K for a $Cd_{1-x}Fe_xTe$ sample at T=1.7K for different magnetic fields. Dashed line: experimental data; solid line: fit result. b. Dispersion of the differential magneto-optic Kerr angle (solid line) and Kerr ellipticity (dashed line) for a $Cd_{1-x}Fe_xTe$ sample. Inset: Poincaré-type plot of the data.

Figure 2.a shows the Kerr angle for the $Cd_{0.985}Fe_{0.015}Te$ sample in static magnetic fields to B=7T. The MOKE of DMS materials is determined by the excitonic contribution to the dielectric function and can be analyzed in terms of an oscillator model for the dielectric function[5]. The solid lines in figure 2.a are the results of a fitting procedure where the valence band exchange $N_0\beta x \langle S_z \rangle$, the longitudinal-transverse splitting ΔE_{LT}, damping constants for light and heavy hole, and a background dielectric constant were treated as adjustable parameters. The values for the conduction band exchange $N_0\alpha x \langle S_z \rangle$ were kept constant at values taken from the SF Raman data. The ratio of the exchange integrals $\frac{\alpha}{\beta} = -0.15 \pm 0.05$ derived from the fit results is much smaller than the value obtained above. Both ratios differ significantly from the value $\frac{\alpha}{\beta} = -0.24$ given by Testelin et al.[2] and determined from magnetoabsorption experiments.

Possible explanations for these discrepancies could be inhomogeneities of the iron distribution in the sample since the Raman data were not taken at exactly the same sample position as the MOKE data. Also of importance could be the different absorption length for the Raman and MOKE experiments, i.e. the depth below the sample surface, where the information is collected from, is different. The absorption of the LH transitions is much weaker, the Raman signal originates therefore from deeper inside the sample. The existence of an inactive surface layer can be deduced from the phase relation between the Kerr angle θ_K and ellipticity ε_K. An example for the dispersion of the derivative of these quantities is shown in figure 2.b where the solid line represents $\frac{\Delta\theta_K}{\Delta B}$ while the dashed line shows $\frac{\Delta\theta_K}{\Delta B}$. The inset in figure 2.b shows a Poincaré plot of the data. The asymmetric orientation of the $\varepsilon_K(\theta_K)$ curve is an indication of the phase shift between the Kerr rotation and ellipticity signals [6]. The simulation of these data requires the assumption of an inactive layer with optical thickness nd = 1800Å for a good description of the experimental results. This layer could be an exciton free dead layer, however, the large thickness suggests that it is more likely a layer with strongly reduced concentration of Fe^{2+}.

3 Conclusion

We have presented resonant magneto-Raman scattering data for conduction band electrons and intra Fe electrons in (Cd,Fe)Te. Combination of SF energy and the resonance positions gives an estimate value for the ratio of the change constants α and β. The combination of SF Raman data with results from MOKE provides a second possibility to derive $\frac{\alpha}{\beta}$. Analysis of Kerr rotation and ellipticity reveals the existence of an inactive surface layer.

Acknowledgments

We are grateful to Dr. Kejun Ma for providing the sample.

References

1. C. Benoit à la Guillaume in *Semimagnetic Semiconductors and Diluted Magnetic Semiconductors* ed. by M. Averous and C. Balkanski (Plenum, New York, 1991)
2. C. Testelin, C. Rigaux, A. Mycielski, M. Menant, and M. Guillot, Solid State Commun. **78**, 659 (1991)
3. D. Scalbert, A. Mauger, J. A. Gaj, J. Cernogora, M. Nawrocki, and C. Benoit à la Guillaume, Phys. Rev. B**43**, 7109 (1991)
4. A. Twardowski, D. Heiman, Y. Shapira, T.Q. Vu, H. Demianiuk, Solid State Comm. **82**, 229 (1989)
5. B. Buda, M. Dahl, N.v. Truchsess, and A. Waag, J. Cryst. Growth **138**, 652 (1994)
6. M. Dahl, to be published

MAGNETOOPTICAL INVESTIGATIONS OF $Cd_{1-X}Mn_XS$, A DILUTED MAGNETIC SEMICONDUCTOR, IN B-EXCITON REGION

Y.G. SEMENOV

Groupe d'Etude des Semiconducteurs, Université Montpellier 2, Place Eugène Bataillon, 34095 Montpellier Cedex, France

Permanent address: *Institute of Semiconductor Physics, National Academy of Sciences of Ukrainian Prospekt Nauki 46, Kiev 252028, Ukraine*

V.G. ABRAMISHVILI, A.V. KOMAROV and S.M. RYABCHENKO

Institute of Physics, National Academy of Sciences of Ukraine, Prospekt Nauki 46, Kiev 252028, Ukraine

An approach to determine the band parameters Δ_1, Δ_2, Δ_3 and the hole-ion exchange constant J_h of hexagonal diluted magnetic semiconductors is suggested. It is based on the description of magnetoreflection in the region of B-excitonic spectrum in a mean field approximation. We apply our method for the treatment of exciton reflectivity spectra obtained for $Cd_{1-x}Mn_xS$ crystals as a function of x. In the limiting case of small x, we determine following band parameter values Δ_1= 14.5±2 meV, Δ_2= 26.5±0.7 meV, Δ_3= 21.5±0.3 meV. They differ from those known in the literature for pure CdS. The value J_h= -1.2 eV turns out to be several times smaller than that obtained from magnetooptical measurements of A-excitons in a mean field approximation. However, this value is in good agreement with the value obtained earlier from A-excitonic spectra treatment beyond the mean field approximation.

The most widely discussed anomaly of the giant spin splitting (GSS) of exciton reflection spectra in $Cd_{1-x}Mn_xS$ crystals is the sublinear concentration dependence of A-excitons GSS, which was interpreted in Ref 1 as a consequence of the effective dependence of hole-ion exchange constant $J_h = J_h(x)$ on magnetic component fracture x; thus a J_h value of 3-4 eV, several times larger than its value in the other diluted magnetic semiconductors, was deduced.

The peculiarities of A-exciton GSS in $Cd_{1-x}Mn_xS$ was the qualitative understanding as a manifestation of multiple scattering of hole, which influence GSS due to large effective mass of A-valence electrons. A quantitative analysis was carried out in Ref 2 on the basis of simplified band structure in the isotropic effective mass approximation with m^*_h =1.3 m_0 for the A subband. The bare exchange constant J_h=-1.4 eV, obtained in the latter approach, is in agreement with the data for other $A^2{}_{1-x}Mn_xB^6$ and appears reasonable.

Much less attention has been paid to magnetooptical investigations of B-exciton band in $Cd_{1-X}Mn_XS$, where substantial splitting was not observed in $\sigma^{\pm}$ polarization with magnetic field **H**| |**c** (**c**-is hexagonal axis of crystal) (Ref 3) where a reconsideration of the well-known band parameters Δ_1, Δ_2, Δ_3 in the literature showed Δ_1-Δ_2 turned out to have a different sign.

Taking into account, that the effective masses of B - valence electrons $m^*_B ||=0.51m_0$, $m^*_{B\perp}=0.67m_0$ are substantially less than $m^*_A|| =3.03\ m_0$ and have the same order of magnitude as those for other DMS, we assume the multiple scattering of B-hole is not of importance in this case, and the B-exciton GSS can be described by mean field approximation. This assumption is a basis of our approach to the determination of parameters Δ_1, Δ_2, Δ_3 and J_h using the spin splitting of the B-exciton, measured with **H**||**c** and **H**⊥**c** as well as the energy differences between the A-, B- and C- exciton components measured at H=0.

$E_{A,B,C}$,the energies of the tops of the A, B or C valence subbands, are:

$$E_A = \Delta_1 + \Delta_2, \tag{1}$$

$$E_A - E_{B,C} = \frac{1}{2}\left\{\Delta_1 \mp 3\Delta_2 \sqrt{(\Delta_1 - \Delta_2)^2 + 8\Delta_3^2}\right\} \quad . \tag{2}$$

It is convenient introduce the ratio $P=(E_A - E_B)/(E_A - E_C)$, which can be easily expressed in terms of $\delta=(\Delta_1-\Delta_2)/(\Delta_1+\Delta_2)$ and $(\Delta_3-\Delta_2)/(1+\delta)\Delta_2$.

The exchange interaction of electrons with magnetic ions is described by the Hamiltonian

$$H_{ex} = -\sum_{R_i} J_{e(h)}(\mathbf{r} - \mathbf{R}_i)\,\mathbf{S}_{l\,i}\,\mathbf{S}_{e(h)}\,. \tag{3}$$

This Hamiltonian can be reduced to Zeeman one with mean (or exchange) field $\mathbf{G}_{e(h)}$ after the averaging over magnetic ion coordinates $\mathbf{R}_i$ and electron wave function:

$$H_{ex} = \mathbf{G}_{e(hi)}\mathbf{S}_{e(h)} = -J_{e(hi)}\, x'\langle \mathbf{S}_l\rangle\, \mathbf{S}_{e(hi)}, \tag{4}$$

where $\langle \mathbf{S}_l \rangle$ are thermodynamic averages of spin components, x is active content of magnetic ions, hi=hA or hB for different valence subbands.

Evidently, the effective fields are different for A or B valence electrons due to the differences of multiple scattering in subbands with different effective masses. Thus, the B -valence electron is affected by an exchange field $\mathbf{G}_{h(B)}$ with the bare exchange constant while the $\mathbf{G}_{h(A)}$ can be described by the modified exchange constant. In the framework of our approach $G_{hB} \neq G_{hA}$.

In the B-subband eigenvectors basis, the Hamiltonian (4) still has the Zeeman form but with an anisotropic g-tensor

$$H_{vB} = -\xi_0 G_{hBZ} S_{BZ} + \sqrt{(1-\xi_0)/2}\,(G_{hBX}S_{BX} + G_{hBY}S_{BY}) \tag{5}$$

where $\xi_0=(\Delta_1 - \Delta_2) / \sqrt{(\Delta_1-\Delta_2)^2+8\Delta_3^2}$, S_B is operator of pseudo spin 1/2, introduced with $|B^+\rangle$ and $|B^-\rangle$ states. The equation (5) describes the B-band valence electrons GSS, $\Delta E^{(B)}$, linear in exchange field G_h approximation. This is the case for small x as well as we attemp to describe the linear part of the B exciton GSS extracted from the experiment. It is convenient to relate the ratios of B valence band splitting for different **H** directions, $\Delta E^{(B)}|_{H||c}$ and $\Delta E^{(B)}|_{H\perp c}$, to the splitting of the conduction band $\Delta E^{(e)}$:

$$Q = \Delta E^{(B)}|_{H||c} / \Delta E^{(e)} = -\eta\xi_0 \; ; \qquad (6)$$

$$R = \Delta E^{(B)}|_{H\perp c} / \Delta E^{(e)} = |Q/\xi_0| \sqrt{(1-\xi_0)/2} \; . \qquad (7)$$

The experimental determination of P, Q and R together with E_A -E_B allows one to find the band parameters Δ_1, Δ_2, Δ_3 and ratio $\eta= J_{hB}/J_e$.

The reflection spectra in the region of A, B and C excitons of the $Cd_{1-x}Mn_xS$ crystals with Mn content x_1=0.0002, x_2=0.0013 and x_3=0.0056 were obtained at T=2K in magnetic fields H up to 3.5T for **H**||**c** and **H**⊥**c**. The energy position of A,B and C excitons and the value of P- parameters are shown in Table 1.

Table 1: Measured positions of A, B and C-excitons bands in zero magnetic field for the three crystals investigated and the ratios $P=(E_A - E_B)/(E_A - E_C)$, calculated from our data.

x	E_A (eV)	E_B (eV)	E_C (eV)	P
0,013[a]	2.543[a]	2.575[a]	2.650[a]	0,30
0,0056	2.542	2.568	2.640	0,26
0,0013	2.550	2.571	2.635	0,24
0,0002	2.554	2.570	2.632	0,21

[a] the data of Ref. 4

The reflection spectra in B-exciton region show the absence of visible splitting in σ^+ and σ^- circular polarizations for **H**||**c** for all of samples investigated. It was interpreted as $\Delta E^{(B)}|_{H||c} \cong \Delta E^{(e)}$. The results of B-excitons GSS for **H**⊥**c** at several magnetic fields as well as the calculated values of conduction band GSS with J_e =0.23 eV [3] are presented in Table 2. In terms of these data, we have found the value of the exchange constant ratio η= - 5.2.

The results of such analysises of experimental data are presented in Table 3.

Table 2: Measured B-exciton spin splitting $\Delta E^{(B)}|_{H\perp c}$ for crystal with x=0.056 at several magnetic field values for **H⊥c** as well as conduction band GSS's and their ratio, calculated from our data.

H (KOe)	$\Delta E^{(e)} = G_e$ (meV)	$\Delta E^{(B)}|_{H\perp c}$ (meV)	$R = \Delta E^{(B)}|_{H\perp c} / \Delta E^{(e)}$
10	1,90	8,05±1	4,24±0.5
15	2,36	10,6±1	4,50±0.5
20	2,60	11,8±1	4,52±0.5
25	2,76	12,55±1	4,55±0.5

Table 3. $Cd_{1-x}Mn_xS$ band parameters values, calculated from the equations (1),(2), (6),(7) at Q=1.0±0.2, R=4.0 ±0.5 (for x=0.0056) and P, taken from Table 1.

x	Δ_1 (meV)	Δ_2 (meV)	Δ_3 (meV)
0,013	23.9±2	38.4±0,7	26.0±0,3
0,0056	20.3±2	34.2±0,7	25.0±0,3
0,0013	17.1±2	29,5±0,7	22,3±0.3
0,0002	14.5±2	26.5±0,7	21.5±0.3

One more result seems to be important : the valence band exchange constant $J_{hB} = \eta J_e = -1.2$ eV turned out to be substantially smaller than the J_h value obtained by the application of mean field approximation to the A-exciton GSS description. Nevertheless, the J_{hB} value is almost similar to J_h value obtained in the model [2] of the valence electron multiple scattering by the magnetic impurities. This agreement perhaps gives an additional evidence that the approximation used here is reasonable.

This work was supported by INTAS grant.

References

1. S.I.Gubarev, *Sov. Phys. JETP* **53**, 601 (1981)]; V.G.Abramishvili et al,, *Sov. Phys. Solid State* **26**, 666 (1984).
2. C. Benoit a la Guillaume et al, *Phys. Rev.* **46**, 9853 (1993); D.Scalbert et al, *Phys. Rev.* **48**, 17752 (1993);J. Tworzydlo, *Phys. Rev.* **50**, 14591 (1994); *Solid State Commun.* **94**, 821 (1995).
3. V.G.Abramishvili et al, *Proceedings 1th Russian Conf. Defects Semicond.* (Word Scientific, St-Petersburg,1992) p. 57 .
4. M.Nawrocki et al *MRS Symposia Proceedings* No. 89, 65 (1987).

NON-SATURATING GIANT ZEEMAN SHIFT AND ANOMALOUS PL INTENSITY IN $Zn_{1-x}Mn_xSe$ AT HIGH MAGNETIC FIELDS

T.SCHMIEDEL
National High Magnetic Field Laboratory, FSU, Tallahassee,
FL 32308, U.S.A.
A.PAREEK, S.LEE, M.DOBROWOLSKA, J.K.FURDYNA
Notre Dame University, Notre Dame,
IN 46556, U.S.A.

Magnetophotoluminescence was studied in a series of MBE-grown Zn_{1-x} Mn_xSe epilayers (0<x<0.25) and $Zn_{1-x}Mn_xSe$/ZnSe superlattices. In the presence of a magnetic field, the photoluminescence (PL) peak exhibits a giant red shift, as well as a large increase in intensity. The Zeeman shift of the PL peak is caused by the interaction of the carriers with the thermally averaged Mn spins in the system. We will explain different scenarios of saturating, steplike and non-saturating magnetization behavior. We will also attempt to propose different models for the PL peak intensity increase in these structures.

1 Introduction

Studies of Mn based diluted magnetic semiconductors (DMSs) have been performed for more than 15 years [1] revealing a wide variety of effects present in this system. With the primary emphasis on investigating effects in quantum structures, bulk materials of this group are to date not fully understood. We report about effects seen in photoluminescence spectra taken at 1.8K in magnetic fields of up to 30T. The luminescence was excited by the 364nm Argon laser line using fiber optics for (unpolarized) transmission of light to and from the sample.

Figures 1 and 2 show the dependence of the excitonic luminescence peak energy shift on the magnetic field for various Mn concentrations x. We have normalized the spectra (Fig.2) to coincide in their energy at B=0. The effects seen here can be divided into 3 categories: At concentrations below x=0.05 a clear saturation behavior of the Zeeman shift can be seen at fields higher than about 5T, disregarding effects due to diamagnetic shift. For 0.05 <x<0.16, steps start to occur at a critical

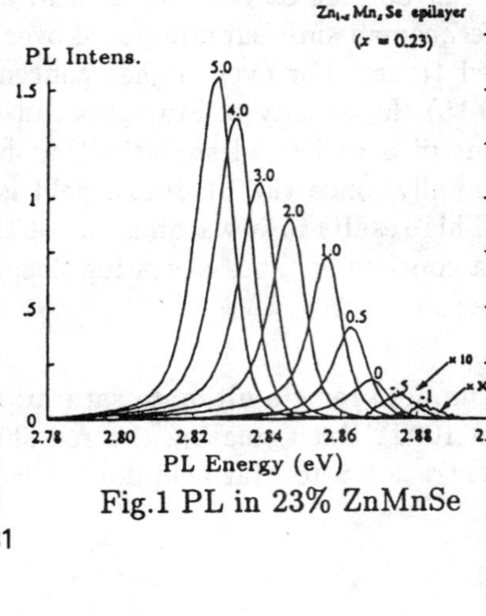

Fig.1 PL in 23% ZnMnSe

field which indicate a sudden increase of magnetization. At concentrations x>0.16, the magnetization increases continuously with field, even above 5T. For all concentrations, the intensity of the PL peak increases by a factor of about 10 upon applying a magnetic field (see Fig.3).

In the following, we will describe the different scenarios in the red-shift in terms of correlation within the collective of Mn spins in the host material. Those spins make up the value of the thermally-averaged spin $<S_z>$ of the Mn ions and thus are responsible for the giant red shift, observed in this system. We will also attempt to explain the dramatic increase in the intensity of the emission by giving several scenarios of excitonic and band to band recombination.

2 Discussion of energy shifts

Let us first consider a distribution of Mn ions with a spin between +5/2 and -5/2 in a ZnSe crystal lattice. For x<0.05 these spins can be considered isolated (uncorrelated) spins at random orientation at zero field. Some of the spins might find a nearest neighbor and form a pair which will align antiferromagnetically thus not being able to contribute to the overall magnetization. Applying a magnetic field to such a system will quickly align all spins so that the magnetization is saturated for fields greater than 5T.

With a concentration of Mn ions above x=0.05, more spins will have nearest neighbors and correlate in clusters. In the 6 level d-spin system of Mn, the clusters are initially spin compensated <-5/2,+5/2>. As the magnetic field increases, a gradual alignment of the quantized spins turn them into a 3/2 and 1/2 position causing a net magnetization of the clusters at higher fields. This can be seen at the critical field as a step in the Zeeman shift superimposed over the saturated trace. For even higher concentrations (x>0.15) the density of Mn spins causes interactions of a variety of stengths that break up continually, once the magnetic field is turned on. This results in a washing-out of the steps into a continuous, non-saturating Zeeman shift as seen in Fig 2.

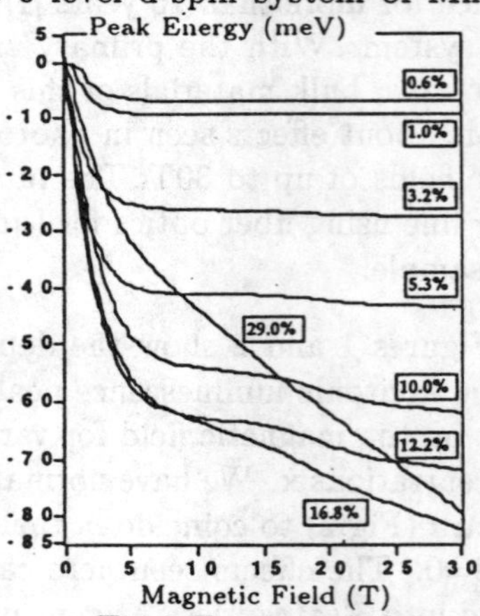

Fig.2 Zeeman shift in $Zn_{1-x}Mn_xSe$.

This model predicts ultimate saturation of the Zeeman shift to occur at just above 100T. Extrapolating the red shift observed on samples with high Mn concentration x to that field does in fact yield a value of the Zeeman splitting expected when all x spins are aligned with the applied field. The high magnetic field measurements on samples with low Mn concentrations have, upon saturation of the Zeeman splitting due to the sp-d exchange interaction, yielded clear data on the diamagnetic shift of the exciton line. The diamagnetic shifts observed in $Zn_{1-x}Mn_xSe$ layers with x$\leq$ 0.05 is consistent with that observed in high-quality ZnSe epilayers measured for comparison. Zeeman splittings observed in Zn_{1-x} Mn_xSe/ZnSe superlattices (including the ZnSe/MnSe system), also show extremely large non-saturating Zeeman shifts. A comparison of the superlattice and epilayer data provides valuable information on the alignment of spins at the ZnMnSe/ZnSe interfaces as a function of the applied field.

3 Discussion of PL intensity

A very striking feature of the PL, observed in $Zn_{1-x}Mn_xSe$ is the dramatic increase of its intensity with increasing magnetic field. This is illustrated in Fig.1 and 3, where the relative PL intensity is plotted versus the magnetic field. The increase by more than an order of magnitude over its zero field value is much beyond what one would expect for a PL intensity increase in other materials in a magnetic field. The behavior shown in this figure is typical for all the $Zn_{1-x}Mn_xSe$ epilayers investigated, including specimens of very low Mn concentrations (e.g. x=0.006). Here, we propose the several explanation for this dramatic behavior and discuss their nearing on this system.

Photo-excited carriers can recombine via a radiative channel (Rad), resulting in PL emission, or non-radiative processes (NRad). Furthermore, the PL intensity will be proportional to the overlap integral between the electron and the hole wavefunction and of course the number of excitons relating to the density of states. We see here that processes increasing the number of excitons in the Rad channel over the NRad channel will result in increased PL intensity.

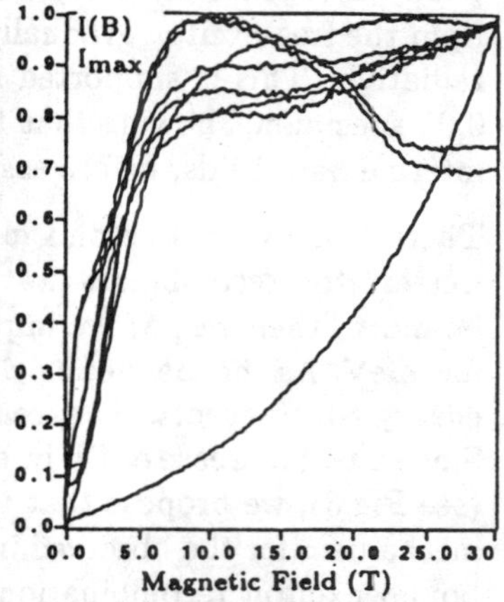

Fig.3 Rel. PL intensity in $Zn_{1-x}Mn_xSe$.

We first consider deep levels, being NRad traps for excitons. Since a deep level would experience a smaller Zeeman splitting than a free exciton (considered to be primarily Rad), the energy of the free exciton will eventually move below the deep level making it the ground state. Emission from the free exciton recombination will then become dominant, resulting in much stronger PL intensity. We rule out this interpretation, because the effect that we observe is common to all samples, even those with low Mn concentrations. For example, we observe a ten-fold increase in PL intensity in $Zn_{1-x}Mn_xSe$ with x= 0.006, where the maximun Zeeman spliting is only 6meV. It does not appear likely that such a small shift would suffice to substantially change the role of a deep level as a channel for non-radiative recombination. We also note that we would expect a drastically different behavior for higher compared to lower concentration samples which is not seen here.

Considering the universality of the effect (we also see qualitatively similar increases in PL intensity in ZnMnTe, CdMnSe, and CdMnTe), it is reasonable to look for the cause of this behavior in the intrinsic properties of interband transitions, together with the role played by naturally occuring centers. Inspection of a typical PL spectrum at B=0 (see Fig. 4) shows two prominent peaks, which we interpret as a free and a bound exciton. Two mechanisms suggest themselves when we compare bound to free excitons.

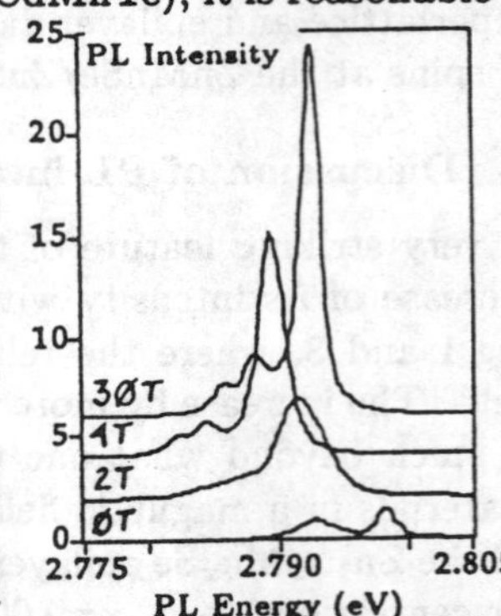

Fig.4 PL in 1% ZnMnSe

First, it has been shown by Pareek [2] that the red shift of free excitons is substantially larger than either that of the donor- or the acceptor-bound excitons. Thus, as the field increases, the free exciton can eventually obtain the ground-state status. As the recombination from the free-exciton gradually becomes dominant, it will become increasingly radiative. This is supported in Fig.4 with a sequence of spectra for the x = 0.01 specimen, showing that the well-resolved features of 0-field PL merge at intermediate fields, before assuming their high-field character.

There also exists a second mechanism which leads to the dominance of Rad free-exciton recombination at high fields. The binding energy of excitons are bound to their respective impurities by weak Coulomb force [3], is typically a few meV. As the magnetic field increases, the bound exciton acquires magnetic energy which eventually exceeds it's binding energy rendering it a free exciton. Since a fast increase of PL intensity with magnetic field is present in all samples (see Fig.3), we propose that the breaking up of bound excitons is the primary mechanism for the observed increase in PL intensity. It would then follow that bound exciton recombination is primarily non-radiative, by a ratio of about 10:1.

References

1 J.K.Furdyna, J.Appl.Phys., **64**, (1988), R29 and references therein.

2 A. Pareek, Ph.D. Dissertation, University of Notre Dame 1995 (unpubl.).

3 J. Mycielski, in Semiconductors and Semimetals Vo.25, edited by J.K. Furdyna and J. Kossut, (Academic, New York, 1988), Ch.8.

ELECTRICAL AND MAGNETIC PROPERTIES OF (In,Mn)As/(Al,Ga)Sb HETEROSTRUCTURES AND BULK (Ga,Mn)As

A. OIWA, Y. IYE, S. KATSUMOTO, A. ENDO, M. HIRASAWA
Institute for Solid State Physics, University of Tokyo Tokyo 106, Japan

H. OHNO[a], F. MATSUKURA, A. SHEN
Research Institute of Electrical Communication,
Tohoku University Sendai 980-77, Japan

H. MUNEKATA
Imaging Science and Engineering Lab.,
Tokyo Institute of Technology Yokohama 227, Japan

III-V based diluted magnetic semiconductors are grown by molecular beam epitaxy. The interesting phenomena depending on spins occur in magnetic and transport properties. We have investigated the two systems of III-V based diluted magnetic semiconductors. The first is *p*-type (In,Mn)As/(Al,Ga)Sb heterostuructures. The second is a bulk (Ga,Mn)As, which a ferromagnetic order with $T_C \sim 70$ K are observed in the magnetization measured by SQUID.

1 Introduction

We have recently succeeded by a low temperature molecular beam epitaxy technique, in growth of III-V semiconductors, InAs, with Mn up to high concentrations ($10^{20} - 10^{21}\mathrm{cm}^{-3}$) maintaining high quality of the hetero-interface with a (Al,Ga)Sb buffer layer. Previous studies on bulk thick film (In,Mn)As[1] and (In,Mn)As/(Al,Ga)Sb heterostructure[2,3,4] revealed a carrier-induced ferromagnetic transition at ~ 7 K and ~ 35 K, respectively. Transport in the low temperature magnetic phase exhibits conspicuous spin-dependent behavior characterized by the anomalous Hall effect and large magnetoresistance. Hall resistivity in the low field range is dominated by the so-called anomalous Hall term proportional to the magnetization M,

$$\rho_{xy} = R_0 B + R_S M, \tag{1}$$

where R_0 and R_S are the normal and anomalous Hall coefficients. Recently, high concentration Mn doping to GaAs have been achieved by MBE[5]. In this paper, we report on transport and magnetic properties of (In,Mn)As/(Al,Ga)Sb heterostructures and a newly prepared (Ga,Mn)As thick film sample.

[a] also at *Research and Development Corporation of Japan (JRDC)*

Table 1: Samples of (In,Mn)As heterostructures and bulk (Ga,Mn)As.

Sample	Materials	thickness of magnetic layer (nm)
#1	$In_{0.82}Mn_{0.18}As/Al_{0.3}Ga_{0.7}Sb$	20
#2	$In_{0.93}Mn_{0.07}As/GaSb$	9
#3	$In_{0.93}Mn_{0.07}As/AlSb$	9
#4	$Ga_{0.965}Mn_{0.035}As$	150

2 Experimental

Three samples of (In,Mn)As/(Al,Ga)Sb heterostructures and one sample of (Ga,Mn)As thick film listed in Table 1 were investigated. They were grown by molecular beam epitaxty on the (100) surface of GaAs substrates with low ($T_S = 200 \sim 250$ °C) growth temperature. Magnetotransport properties were measured from 0.4 to 200 K under magnetic fields up to 15 T by a standard DC four probe method. For #3, transport measurements were carried out down to the millikelvin region using a dilution refrigerator. Magnetization were studied using a commercial SQUID magnetometer.

3 Results and discussions

Figure 1 shows the Hall resistance and magnetoresistance of #3. Both #1 and #2 exhibit similar overall behavior. The carrier (hole) density as estimated from the slope of the Hall resistivity is on the order of 10^{19}cm^{-3} and the mobility is ~100 cm^2/Vs at 4.2 K. A ferromagnetic order of Mn spins develops below ~ 35 K. Magnetotransport in the low temperature ordered phase is characterized by an anomalous Hall effect with a hysteretic behavior and a large negative magnetoresistance. Figure 2 compares hysteresis loops of the Hall resistance with magnetic fields applied perpendicular to the

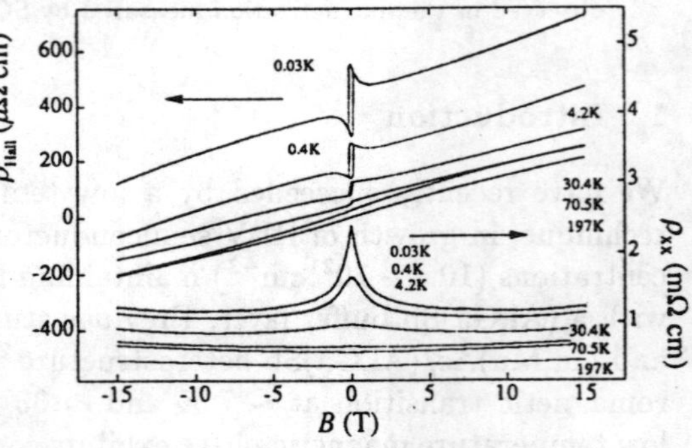

Fig. 1: Hall resistance and magnetoresistance of #3,(In,Mn)As/AlSb, at each temperatures. Note that the Hall resistance curves are shifted vertically for clarity.

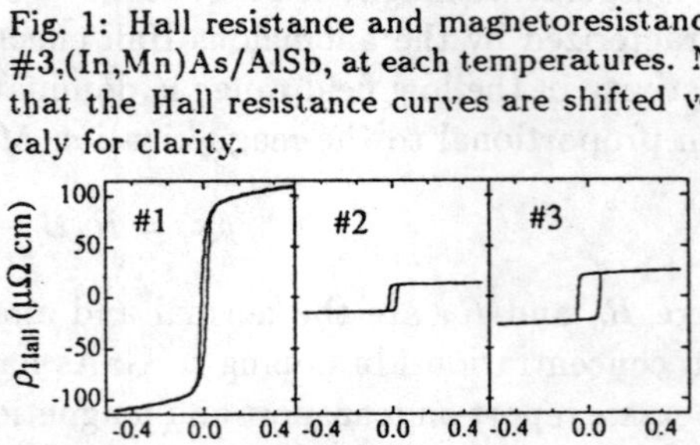

Fig. 2: Low-field Hall resistance of three (In,Mn)As/(Al,Ga)Sb heterostructures at 4.2 K.

sample plane. The square hysteresis loops observed in #2 and #3 suggest that they have a relatively strong magnetic anisotropy with the magnetic easy axis perpendicular to the film plane. By contract, the rounded loop and small coercive field of #1 indicates that its magnetic anisotropy is small. Measurements of the field-angle dependence of the Hall resistance shown in Fig. 3 also lead to the same conclusion. The origin of perpendicular magnetic anisotropy is attributed to the tensile strain caused by lattice mismatch at the heterojunction. When the thickness of the (In,Mn)As layer exceeds beyond the critical thickness for lattice relaxation, the strain and thus the anisotropy is lost.

As regards the mechanism for the anomalous Hall effect, two models, "skew scattering" and "side jump", are often invoked. The relation between the anomalous Hall coefficient R_S and resistivity ρ is $R_S \propto \rho$ for the former and $R_S \propto \rho^2$ for the latter model, respectively. For #1, the temperature dependence of ρ_{xy}/ρ and the saturation magnetization as measured by the SQUID magnetometer, are in good agreement[3]. For #3, however, ρ_{xy}/ρ^2 seems to yield a better agreement with the SQUID data as shown in Fig. 4 than ρ_{xy}/ρ dose. The saturation magnetization $M_S \sim 0.04$ T at 2 K is roughly a half of the value 0.079 T which is calculated from

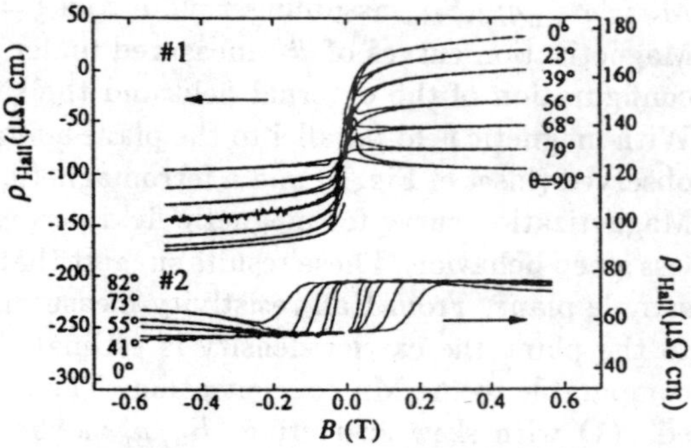

Fig. 3: Field-angle dependences of the Hall resistance of #1 and #2. The angle θ is defined as the angle between the direction of the applied field and the normal of the sample plane.

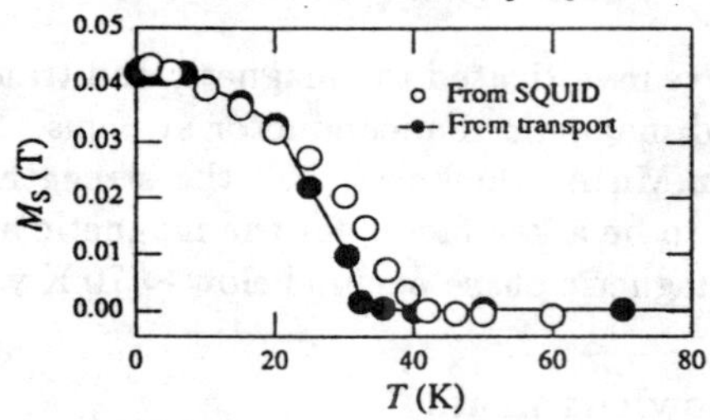

Fig. 4: Temperature dependence of #3, (In,Mn)As/AlSb. The open circls are measured by the SQUID. the solid circles are ρ_{xy}/ρ^2 deduced from transport measurements.

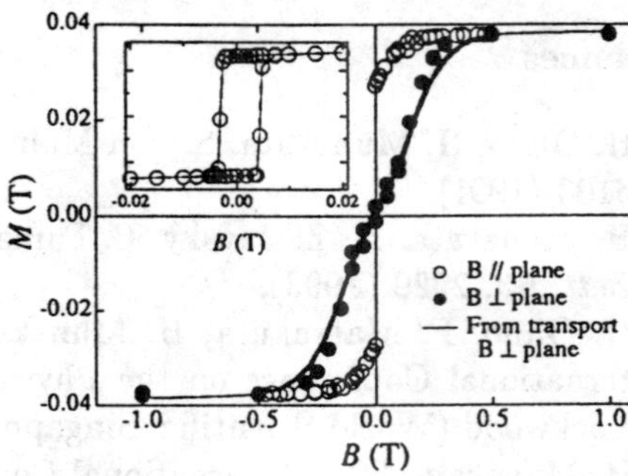

Fig. 5: Magnetizaton curves of Smaple #4, (Ga,Mn)As, at 5 K with the two configuration of magnetic field, $B \parallel$ plane (open circle), and $B \perp$ plane (solid). The solid line is deduced from transport using eq. (1) and assuming ρ_{xy}/ρ. Inset shows the low field behavior for $B \parallel$ plane.

$M_S = N_{Mn} g\mu_B S_{Mn}$ assuming x=0.07 and g=2, S_{Mn} = 5/2. Magnetization curves of #4 measured under magnetic field with two different configuration of the external field and the sample plane are shown in Fig. 5. With magnetic field parallel to the plane a rectangular hysteresis loop is clearly observed (inset of Fig. 5) and a ferromagnetic transition is observed at $\sim$ 70 K. Magnetization curve for magnetic field perpendicular to the plane has a much less steep behavior. These results suggest that the magnetic easy axis lies in the sample plane. From Hall resistivity measurement with the field perpendicular to the plane the carrier density is estimated as $p \sim 1 \times 10^{21}\text{cm}^{-3}$, which is comparable to the Mn concentration. The magnetization curve deduced using eq. (1) with skew scattering, R_S/ρ, as shown in Fig. 5 has a rounded field dependence similar to that measured by SQUID.

4 Summary

We have investigated the magnetic and transport properties of two III-V based diluted magnetic semiconductor systems. In the (In,Mn)As heterostructures, the (In,Mn)As thickness, i.e. the strength of the lattice mismatch strain, is found to be a key factor for the magnetic anisotropy. In bulk (Ga,Mn)As, the ferromagnetic phase occurs below $\sim$ 70 K with the easy axis in the layer plane.

Acknowledgments

The authors thank Prof. H. Takagi for the collaboration with SQUID magnetometer measurements.

References

1. H. Ohno, H. Munekata, S. von Molnar, L.L. Chang, *J. Appl. Phys.* **69**, 6103 (1991).
2. H. Munekata, A. Zaslavsky, P. Fumagalli, R. J. Gambino, *Appl. Phys. Lett.* **63**, 2929 (1993).
3. H. Ohno, F. Matsukura, H. Munekata, Y. Iye, J. Nakahara, 22nd International Conference on the Physics of Semiconductors, ed. by D. J. Lockwood (World Scientific, Singapore, 1994), p. 2605.
4. H. Munekata, 22nd International Conference on the Physics of Semiconductors, ed. by D. J. Lockwood (World Scientific, Singapore, 1994), p. 2517.
5. H. Ohno, A. Shen, F. Matsukura, A. Oiwa, A. Endo, S. Katsumoto, Y. Iye, *Appl. Phys. Lett.* **69**, 363 (1996).

MAGNETOTRANSPORT PROPERTIES OF $Pb_{1-x}Eu_xSe$ EPILAYERS GROWN BY MOLECULAR BEAM EPITAXY

W. DOBROWOLSKI, E. GRODZICKA, T. STORY
Institute of Physics, Polish Academy of Sciences
Al. Lotników 32/46, 02-668 Warsaw, Poland

A. LAMBRECHT, H. BÖTTNER, M. TACKE
Fraunhofer Institut für Physikalishe Messtechnik
Heidenhofstrasse 8, D-79110 Freiburg, Germany

Hall effect and electron conductivity of $Pb_{1-x}Eu_xSe$ ($0 \leq x \leq 0.06$) MBE grown epilayers were investigated as a function of temperature and magnetic field. The Hall coefficient depends strongly on magnetic field for *p*-type samples grown with Se excess. The hole-Eu ion exchange interaction seems to be very weak.

Lead chalcogenides mixed crystals containing europium have found wide applications as confinement layers and active layers in infrared diode lasers and as barriers in quantum well structures [1]. Several reasons contributed to this situation. Rare earth monochalcogenides share the rock salt structure with IV-VI compounds and they have similar crystallographic lattice constants. Because their band gaps are much larger than those of IV-VI compounds even a moderate addition of a rare earth monochalcogenide substantially increases the band gap of a resulting mixed crystal. Among rare earth elements Eu is the most stable in 2^+ valence state. Moreover, it has a small diffusion coefficient which enables obtaining abrupt heterojunctions.

While optical and magnetooptical properties of $Pb_{1-x}Eu_xSe$ were rather extensively studied (see e.g.[2,3]) transport properties are relatively poorly known. In this work we present the study of *p*- and *n*- type $Pb_{1-x}Eu_xSe$ ($0 \leq x \leq 0.06$) epilayers. Investigated layers were grown by the molecular beam epitaxy (MBE) on (111) BaF_2 substrates. Thicknesses of layers varied between 2.5 and 5.5 μm. Measurements of dependencies of the electron conductivity σ and the Hall coefficient R_H (at $B = 1$ T) on temperature were performed in a continuous flow cryostat at the temperature range 2 - 300 K. High field measurements were performed in 13 T superconducting magnet at 5 to 10 different temperatures from the range 1.8 to 270 K. The list of investigated samples and their electrical parameters are presented in Table 1.

Interesting properties revealed the group of samples enriched with Se. Figure 1 shows the dependence of the Hall coefficient on the temperature for three such samples. With the increasing Eu content the character of the curves changes exhibiting in the middle range of temperatures a feature increasing

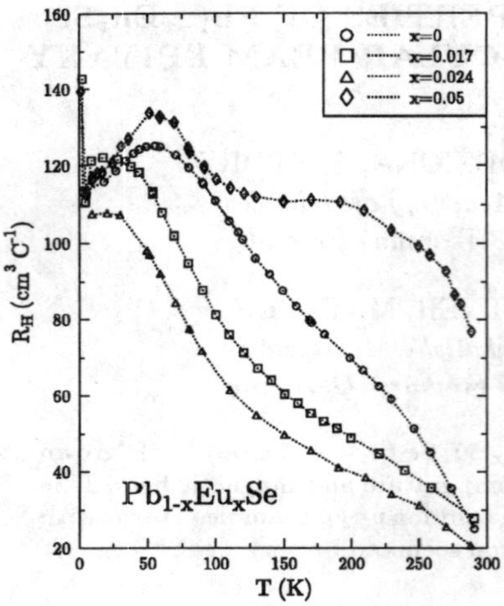

Figure 1: Temperature dependence of the Hall coefficient R_H determined at $B = 1$ T in $Pb_{1-x}Eu_xSe$ epilayers.

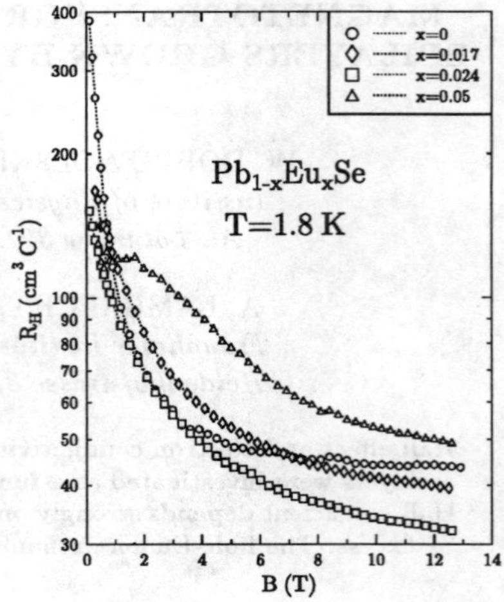

Figure 2: The magnetic field dependence of the Hall coefficient R_H in $Pb_{1-x}Eu_xSe$ epilayers at 1.8 K.

with increasing x. Our computer simulation shows that similar shapes may be generated numerically when the presence of an acceptor level in the vicinity of the valence band edge is assumed.

The Hall mobility is a rapid function of the Eu content. For example, for $x = 0.017$ the mobility is lowered 3 times in comparison to that in the layer of undoped PbSe with the comparable carrier concentration (cf. Table 1). Among reasons leading to the mobility suppressing a rapid increase of the energy gap with the growing Eu content may be distinguished. As a result the effective mass of holes increases. Additionally, in the mixed crystals the alloy scattering is expected.

Magnetoconductivity and Hall coefficient data obtained for p-type samples with the small carrier concentration, $p \approx 5 \cdot 10^{16}$ cm^{-3} (see Table 1), exhibit strong temperature and magnetic field dependencies. In $Pb_{1-x}Eu_xSe$, $x = 0.05$, the value of R_H diminishes more than two times in low temperatures when magnetic field rises to 13 T, but the effects are even more pronounced for samples containing less europium. In particular a significant decrease of the Hall coefficient R_H with an increasing magnetic field may be observed for sample containing no Eu (see Fig. 2). The effect is strongest at low temperatures. The drop of R_H at $T = 1.8$ K for the sample shown in Fig. 2 equals almost an order of magnitude. For the same sample even at 160 K the

30% change of R_H is observed. Because the effect does not disappear even at the room temperature an immediate conclusion may be drawn. A value of the Hall coefficient is of practical importance only when it is considered together with the value of the magnetic field in which it has been measured.

The fact that the Hall coefficient changes with the temperature even at the lowest temperatures may evidence that $Pb_{1-x}Eu_xSe$ epilayers, having metallic properties, approach to the metal-insulator transition (MIT). A similar picture was observed in PbTe epilayers on the metallic side of the MIT[4].

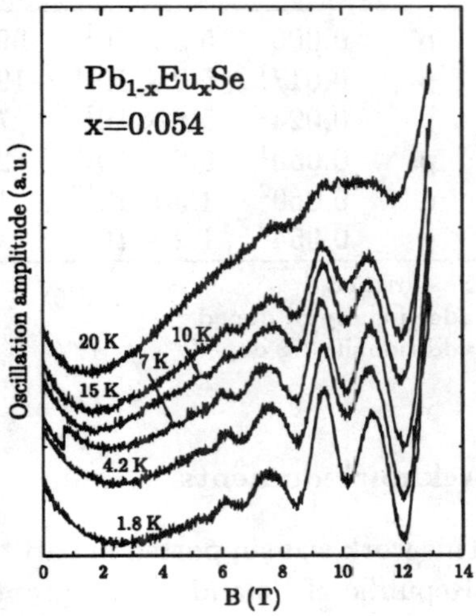

Figure 3: Shubnikov–de Haas oscillations in the $Pb_{1-x}Eu_xSe$ epilayer

The dependence of R_H on magnetic field obtained in our experiment resembles the situation when a conductivity is given by two types of carriers, in our case they would be holes, with different mobilities. Searching for an explanation of this behavior it should be noted that PbSe and $Pb_{1-x}Eu_xSe$ are many valley semiconductors possessing four equivalent extrema at the L points of the Brillouin zone. The differences in lattice parameters of BaF_2 and $Pb_{1-x}Eu_xSe$ make a $Pb_{1-x}Eu_xSe$ layer elastically strained. The strain lifts the degeneracy and slightly shifts the valley with the main axis parallel to [111] in respect to the three obliquely oriented valleys (cf. [5,6]). As a result, the carrier transfer between the different valleys takes place exerting an influence on the entire current carriers transport.

The last two samples, enlisted in Table 1, are heavily doped and present different electrical parameters. The carrier concentration, derived from the Hall data, in the n-type sample stays practically constant at about $1.3 \cdot 10^{19}$ cm^{-3}, while in the p-type sample starting from $1.2 \cdot 10^{19}$ cm^{-3} decreases fivefold with the temperature rising from 4 to 300 K. The p-type sample exhibits pronounced Shubnikov de - Haas oscillations (see Fig. 3). At the magnetic field about 10 T the spin split oscillation is seen. A lack of any visible dependence of the position of the spin split oscillation on temperature indicates that the hole - Eu ion exchange interaction in $Pb_{1-x}Eu_xSe$ mixed crystals is very weak. This result is in agreement with the magnetooptical data[2].

Table 1: Carrier concentration n_c and Hall mobility μ in $Pb_{1-x}Eu_xSe$. Carrier concentration is given in cm^{-3}, mobility in $cm^2V^{-1}s^{-1}$. The Hall coefficient, R_H was determined at magnetic field $B = 1$ T.

Type	x	4.2 K		77.4 K		300 K	
		n_c	μ	n_c	μ	n_c	μ
p	0.000[1]	5.5×10^{16}	59300	5.2×10^{16}	11200	2.1×10^{17}	335
p	0.017[1]	5.2×10^{16}	19000	6.6×10^{16}	5900	2.3×10^{17}	290
p	0.024[1]	5.8×10^{16}	7700	8.1×10^{16}	2800	2.9×10^{17}	210
p	0.050[1]	5.6×10^{16}	2100	5.0×10^{16}	1280	8.3×10^{16}	130
n	0.060[2]	1.3×10^{19}	1100	1.2×10^{19}	830	1.1×10^{19}	300
p	0.054[3]	1.2×10^{19}	690	8.9×10^{18}	580	2.3×10^{18}	240

[1] Se enriched
[2] additionally Bi doped
[3] additionally Ag doped

Acknowledgments

This work was supported in part by the State Committee for Scientific Research (Republic of Poland) under grant No. 2P03B10308

References

1. D. L. Partin and J. Heremans. Growth of narrow bandgap semiconductors. In T. S. Moss and S. Mahajan, editors, *Handbook on Semiconductors*, pages 369–450. Elsevier Science B.V., Amsterdam, 1994.
2. G. Bauer, H. Pascher, and W. Zawadzki. *Solid State Comm.*, 7:703–723, 1992.
3. K. H. Herrmann, K. P. Möllmann, J. W. Tomm, H. Böttner, A. Lambrecht, and M. Tacke. *J. Appl. Phys.*, 72:1399–1404, 1992.
4. J. Oswald, B.B. Goldberg, G. Bauer, and P. J. Stiles. *PRB*, 40:3032–3039, 1989.
5. G.L. Bir and G.E. Pikus. *Symmetry and Deformation Effects in Semiconductors.* Nauka, Moscow, 1972.
6. J. Singleton, E. Kress-Rogers, A. V. Lewis, R. J. Nicholas, E. J. Fantner, G. Bauer, and A. Otero. *J. Phys. C: Sol. St. Phys.*, 19:77–92, 1986.

UNIAXIAL STRESS STUDIES OF MOLECULAR CONDUCTORS: ANISOTROPIC MODIFICATION OF FERMI SURFACES AND STRAIN INDUCED SUPERCONDUCTIVITY

J. S. BROOKS, K. HARRIS, S. UJI[a] , J. QUALLS, J. SARRAO , Z. FISK

Physics Department and NHMFL, Florida State University, Tallahassee FL 32306 USA

C.E. CAMPOS[b], P. S. SANDHU, S. VALFELLS

Physics Department, Boston University, Boston MA 02215, USA

P.J.M. van BENTUM , J.A.A.J. PERENBOOM

Research Institute for Materials & High Field Magnet Laboratory, University of Nijmegen, Toernooiveld 1, 6525 ED Nijmegen, the Netherlands

M. TOKUMOTO, N. KINOSHITA, T. KINOSHITA, Y. TANAKA

Electrotechnical Laboratory, Tsukuba, Ibaraki 305, Japan

H. ANZAI, S. TAKASAKI, J. YAMADA

Himeji Institute of Technology, Akahogun, Hyogo 678-12, Japan

We report on recent studies of uniaxial stress/strain in the molecular conductors $\kappa-(BEDT-TTF)_2Cu(NCS)_2$,$\alpha-(BEDT-TTF)_2MHg(SCN)_4$[M = K, NH4], $\eta-Mo_4O_{11}$, and $(TMTSF)_2PF_6$ to values of up to 8 kbar at low temperatures and high magnetic fields. The stress is applied in all cases along the least conducting direction of these molecular layered materials. In the kappa phase superconductor, Tc decreases with increasing strain. In contrast, for the alpha phase materials Tc is either induced, or enhanced. In the latter two materials $\eta - Mo_4O_{11}$, and $(TMTSF)_2PF_6$ we find that although no superconductivity is induced, there are significant changes in the Fermi surface character and in the resistivity and magnetoresistance. The observation of Shubnikov de Haas oscillations allow an in situ determination of how the closed orbit Fermi surface character changes with stress. We discuss the significant difference of strain vs. hydrostatic pressure in such studies, and we compare our results with recent strain and pressure dependent band structure calculations.

[a]Permanent address: Tsukuba Magnet Laboratory, NRIM, Tsukuba, Ibaraki 305, JAPAN.

[b]Present address: Department of Electrical Engineering, Princeton University, Princeton NJ 08544, USA

1 Introduction

We use the term molecular conductors in this paper to describe a class of materials in which molecules make up the basic crystalline order, and in which charge transfer occurs between different molecular units to create mobile carriers in the materials. Both organic[1 2] and inorganic[3] materials can exhibit these features. In the materials we consider therefore, there are conducting molecular layers separated by non-conducting molecular layers. Because of the high quality which results from the synthesis of these materials, the carriers are highly mobile, and the materials represent high quality and also very anisotropic metals. In terms of the ultimate ground state properties of the different materials studied, high anisotropy and unit cell volume play seemingly important roles in the outcome: density wave (insulating or semi-insulating) behavior, superconductivity, or normal state metallic character. Hydrostatic pressure has traditionally played the role of stabilizing metallic and /or superconducting phases. Due to the highly anisotropic nature, i.e. low dimensionality , of many of the materials, they are unstable metals which undergo transitions to charge or spin density wave phases at low temperatures. Hydrostatic pressure can sometimes remove these instabilities, but in the case where a superconducting state is stabilized, further increases in pressure generally cause the transition temperature and effective mass to decrease.[4] Furthermore, hydrostatic pressure generally causes the Brillouin zone to uniformly increase[5] due to the corresponding decrease in the lattice parameters.

Given the nature of hydrostatic pressure, we were motivated to pose the question: which lattice parameters are most important in determining the nature of the ground state in any particular case? Hence the purpose of research described in this paper has been to alter the anisotropy of materials by the application of stress along some principal axis, and to gauge the effects by measuring the behaviour of the resistivity and the Fermi surface character by application of high magnetic fields to measure the Shubnikov de Haas oscillations. We have selected several well studied materials, the stacking patterns for which are shown in Fig. 1, for this purpose.

A)$\kappa-(BEDT-TTF)_2Cu(NCS)_2$,. This material is a superconductor below 10 K and has an alternating stacking of dimers of (BEDT-TTF) molecules in plane. The effective mass as determined from Shubnikov de Haas studies is about 3.5 m_0 B) $\alpha-(BEDT-TTF)_2MHg(SCN)_4$. Here, for M = NH_4, the material is a 1 K superconductor with a mass of about 2.1 m_0. For M = K, the material has a low temperature density wave state below 8 K and the effective mass is about 1.5 m_0. In this class of materials the stacking of the dimers is "chevron-like". C) $\eta-Mo_4O_{11}$. This material is comprised of Mo atoms

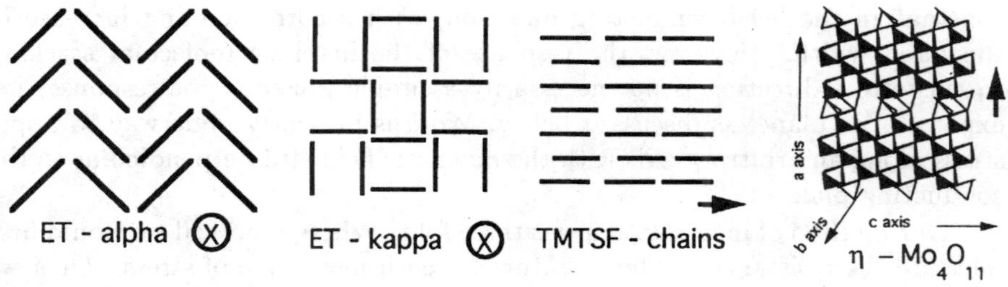

Figure 1: Schematic representation of stacking patterns for the materials investigated. For the organic materials, the thick lines represent the positions of the planar cation molecules BEDT-TTF ("ET") and TMTSF. Anion structures are not shown. Arrow tails and heads indicate the least conducting direction in the crystals, as well as the direction of applied magnetic field and stress.

enclosed in either tetra or octrahedral cages of oxygen atoms. The octrahedral molecular units constitute the conducting layers, which are separated by the tetrahedral units. It undergoes two charge density wave (CDW) transitions at 109 K and 30 K with decreasing temperature, Nearly all of the Fermi surface is removed at low temperatures, leaving only small closed orbit pockets. D) $(TMTSF)_2PF_6$. This material has a spin density wave (SDW) transition at 12 K, below which the resistance increases in an insulating fashion with lower temperature. Hydrostatic pressure in excess of 6 kbar restores the metallic state, which becomes superconducting below 1 K. In the high pressure phase, $(TMTSF)_2PF_6$ exhibits a series of magnetic field induced SDW subphases above about 4 tesla at low temperatures.

2 Experimental methods

The nature of the materials of interest is one of extreme fragility and irregularity. Hence we modified the standard uniaxial arrangement to accomodate these materials[6]. A distinct advantage of our uniaxial stress methodis that in a single low temperature run we can systematically increase the stress (as high as 8 kbar) by use of a gas piston at the top of the stress probe.

So far, the sample configuration is one where the magnetic field and the stress are perpendicular to the conducting planes of the crystal. (This is due to the purely practical reason that presently very high field [30 tesla] resistive solenoids restrict the piston action to the magnet axis direction, and the necessity to study the SdH oscillations with the field perpendicular to the conducting layers.) Hence in all cases discussed the stress reduces the lattice constant in the least conducting direction, with a corresponding increase in the band width. However, the response of the in plane molecular stacking to this stress direction is not necessarily a simple Poisson-type response, i.e. expansion in plane, as discussed below. Work is presently underway to apply stress along an arbitrary axis with the magnetic field still perpendicular to the conducting planse

Our method of investigation was three fold. Where practical we would first measure the resistance vs. temperature for each new value of stress. Then we would determine the magnetoresistance and quantum oscillation (QO) behavior. Finally, we would study the temperature dependence of the oscillations to determine the effective mass parameter[7]. High magnetic fields therefore play a crucial role in that the stress dependence of the QO frequency is a measure of how stress modifies the electronic and physical structure of the materials.

3 Previous results on $\kappa-(BEDT-TTF)_2Cu(NCS)_2$, and $\alpha-(BEDT-TTF)_2MHg(SCN)_4$

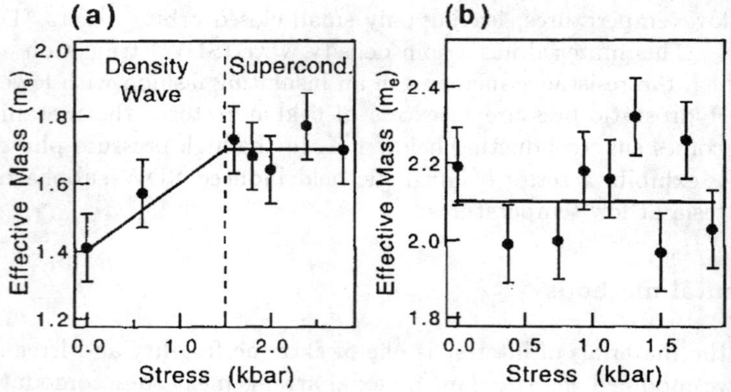

Figure 2: Effects of stress on the effective mass and transition temperature for a) $\alpha-(BEDT-TTF)_2KHg(SCN)_4$ and b) $\alpha-(BEDT-TTF)_2NH_4Hg(SCN)_4$.

In this section we briefly review our previous work[8] for completeness. In the case of $\alpha-(BEDT-TTF)_2MHg(SCN)_4$, we found that the effect of stress perpendicular to the conducting planes was to increase the size of the in-plane unit cell (the Poisson effect). This was manifested in the decrease in the QO frequency with increasing stress[9,10]. In the case of $\alpha-(BEDT-$

$TTF)_2NH_4Hg(SCN)_4$, which is a superconductor under ambient conditions, we observed a three fold increase in T_c with stress. In the case of $\alpha-(BEDT-TTF)_2KHg(SCN)_4$ which is in a density wave state at ambient pressure, we observed the onset of superconductivity with stress, accompanied by an increase in effective mass (see Fig. 2).

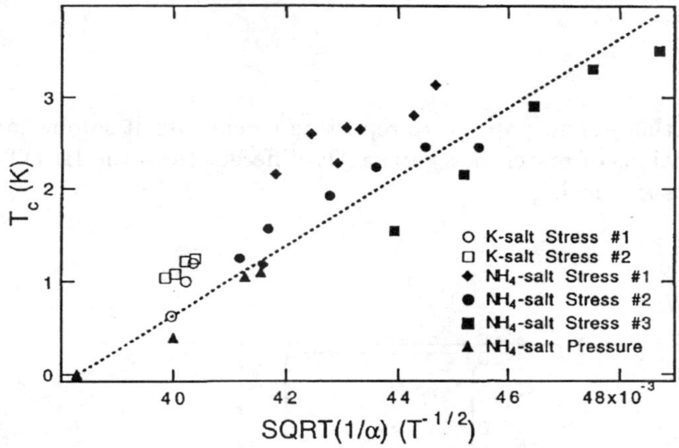

Figure 3: Relation of superconducting transition temperature to the closed orbit Fermi surface parameter for a number of materials.

In contrast, $\kappa-(BEDT-TTF)_2Cu(NCS)_2$, exhibited an increase in QO frequency with stress accompanied by a decrease in the superconducting transition temperature[11]. We also find a striking relationship between the superconducting transition temperature and the Fermi surface parameter (α is the fundamental SdH frequency), which is proportional to the inverse of the Fermi momentum k_F (see Fig. 3).

In a subsequent study of the pressure and stress dependent extended Hückel - tight binding bandstructure of $\kappa-(BEDT-TTF)_2Cu(NCS)_2$, and $\alpha-(BEDT-TTF)_2NH_4Hg(SCN)_4$, the stress (and also pressure) dependence of the QO frequencies could be well described[12]. Here we found that a simple expansion of the in-plane lattice parameters (Poisson effect) accounted for the behavior of $\alpha-(BEDT-TTF)_2NH_4Hg(SCN)_4$, but that for $\kappa-(BEDT-TTF)_2Cu(NCS)_2$, it was necessary to model the effects of stress as causing a sliding of the dimers with respect to each other to yield the observed dependence (slight increase) in the QO frequency.

The onset of superconductivity in the $\alpha-(BEDT-TTF)_2KHg(SCN)_4$

material is still not fully understood. Clearly this material is near an instability due to the nesting condition of the open orbit sheets which cause the density wave transition, and the ambient condition superconductivity which is observed in the isostructural $\alpha-(BEDT-TTF)_2NH_4Hg(SCN)_4$ material. The increase in effective mass in $\alpha-(BEDT-TTF)_2KHg(SCN)_4$ cannot be explained by changes in the band structure alone, hence many body effects must play an important role.

4 New results

The main purpose of the present paper is to report on recent, albeit somewhat preliminary, investigations of materials significantly different from the BEDT-TTF family with α or κ stackings.

4.1 $(TMTSF)_2PF_6$

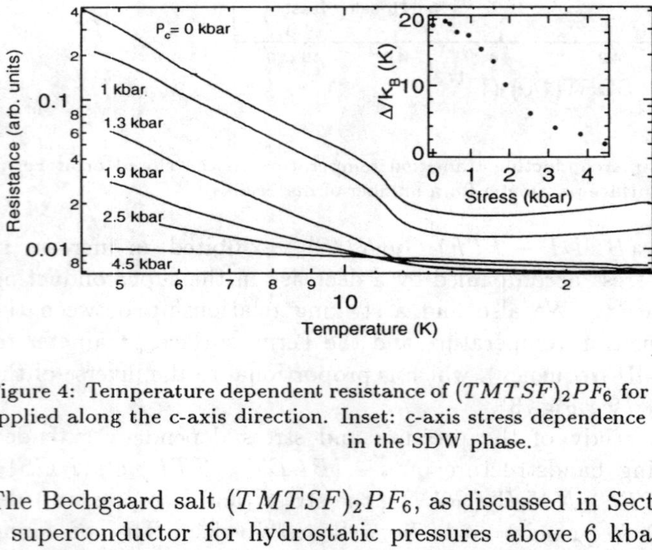

Figure 4: Temperature dependent resistance of $(TMTSF)_2PF_6$ for different vaules of stress applied along the c-axis direction. Inset: c-axis stress dependence of the activation energy in the SDW phase.

The Bechgaard salt $(TMTSF)_2PF_6$, as discussed in Section 1 is a metal and a superconductor for hydrostatic pressures above 6 kbar below 1 K. Under these conditions magnetic fields induce a series of SDW subphases[13]. The mechanisms at work are as follows. First, at ambient pressure the quasi-one dimensional Fermi surface nests nearly perfectly at 12 K to produce a SDW phase transition. The band widths for the a, b , and c crystal directions are $t_a = 250, t_b = 25$, and $t_c = 1.5$ meV respectively. Hence the c-axis transfer integral is nearly negligible, and it is the b-axis bandwidth which determines the sensitivity to the nesting condition. (The a-axis is the most conducting direction.) Hydrostatic pressure most likely increases t_b, thereby destroying the nesting. High magnetic fields on the other hand re-introduce the quasi-one dimensional nature of the electron orbit along the a-axis, and the SDW instability is again realized.

In our initial investigations we studied only the temperature dependence of the resistance of $(TMTSF)_2PF_6$ for different stress values, as is shown in Figure 4. For zero stress, we find the usual behavior of the resistance, namely that at T_{SDW} the resistance increases abruptly, and then continues to increase by several orders of magnitude at low temperatures. We see in Fig. 4 that as stress is increased, the insulating behavior at low temperatures was almost completely removed. In the sample studied however, the resistance still increased slightly with decreasing temperature even at the highest stress values. Hence no sign of truly metallic or superconducting behavior was observed in the temperature range studied.

In the second investigation of $(TMTSF)_2PF_6$ we studied the stress dependence of the magnetorsistance at 1.5 K on a second sample. Our results are shown in Fig. 5. Here we first note that at zero field there is a decrease in the resistance with increasing stress. As one might expect if the Fermi surface is being restored, i.e. the system is being driven into the metallic state,find that the magnetoresistance increases with stress . However, unlike $(TMTSF)_2PF_6$ under hydrostatic pressures above 6 kbar, we found no evidence for the onset of the field induced SDW phase transitions, at least in the range of magnetic field and temperature studied.

A summary of the stress dependence of the magnetoresistance and resistance at zero field and 8 tesla is also shown in Fig. 5. As in the sample studied in Fig. 4, we find that the zero field resistance is reduced nearly two orders of magnitude by stress. The magnetoresistance at 8 tesla also increases with stess, but only by a factor of 2. We will discuss the results of this study in Sect. 5.1

4.2 $\eta - Mo_4O_{11}$

A second investigation has been carried out on the inorganic material $\eta - Mo_4O_{11}$. We were originally attracted to this material due to the very small Fermi surface pockets (the largest yielding an SdH frequency of only about 20 tesla) left at low temperatures after to successive CDW_1 and CDW_2 transitions at 109 K and 30 K respectively.

Hence in a 30 tesla resistive magnet one could go into the quantum limit in a system with a quasi-two dimensional metal. We have recently carried out

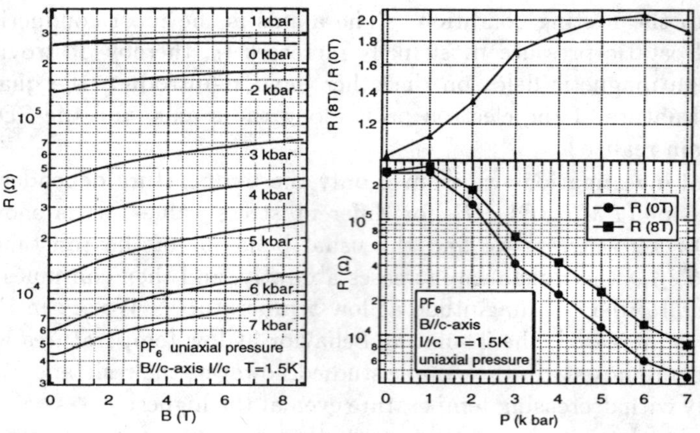

Figure 5: The longitudinal magnetoresistance (left) of $(TMTSF)_2PF_6$ at 1.5 K for increasing stress applied along the c-axis.Stress dependence of zero field resistance (upper right) and 8 telsa magnetoresistance (lower right) of $(TMTSF)_2PF_6$ at 1.5 K.

a systematic study of this material involving magnetoresistance, Hall, hydrostatic pressure, and angular dependent SdH measurements[14]. Many aspects of the low temperature magnetoresistance of this materal can be observed in Fig. 6. The complex oscillatory behavior below 20 tesla is due to a series of orbits which originate from the remaining tiny electron and hole pockects in the CDW_2 phase. Above 20 telsa all of these orbits are in the quantum limit, and the rise in resistance, which is activated, we believe arises from the resulting Landau gap at high fields. We observe that the "peak" associated with the oscillation near 8 tesla moves to higher fields as stress is increased, as does the onset of the M-I transition.

Also in Fig. 6 the results of the Fourier transform of the SdH frequencies is shown. In general, they appear to increase with stress. However their amplitudes decrease rapidly. This is probably due to inhomogeneous strain, although an increase in the effective mass (which has not yet been studied) would yield a similar effect. The general trend is for the frequencies, and therefore the field at which the quantum limit is achieved, to increase with stress. In our model it is the final Landau level gap which causes the M-I transition, hence the position in field of the M-I transition should also increase, as the data show.

We further note that in the $\eta - Mo_4O_{11}$ sample shown in Fig. 6, and also in another sample studied previously, the resistance of the sample seems to be non-monotonic with stress. In general the resistance increases with stress from 0 to about 4 kbar, and then decreases at higher stress values. This is evident in Fig. 6 at high fields. At the highest values of stress the resistance at zero field became less than our ability to measure it, and the signal only appeared above 10 tesla. Although the behavior was somewhat reminiscent of superconductivity, more work is needed to determine the origin of this anomalous behavior.

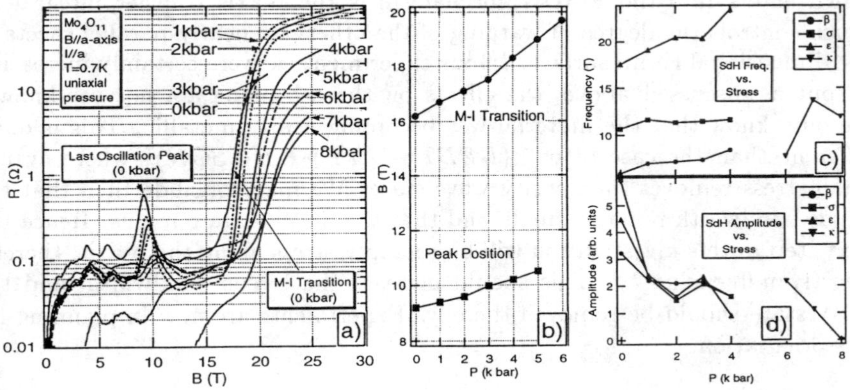

Figure 6: a)Magnetoresistance of $\eta - Mo_4O_{11}$ at low temperatures for increasing values of uniaxial stress. b) Stress dependence of last oscillation peak and metal to insulator transition.c)Stress dependence of SdH frequencies by Fourier transform. d) Stress dependence of SdH oscillation amplitudes.

5 Discussion and Conclusions

We have presented new results on two quite dissimlar materials where uniaxial stress has been used to alter the electronic structure of highly anisotropic materials. We may make a general observation before discussing the materials in detail: it appears that the Poisson-like effect, at least in the cases of $\kappa - (BEDT - TTF)_2Cu(NCS)_2$, and $\eta - Mo_4O_{11}$ does not apply. Here the quantum oscillation frequencies increase, indicating that the in-plane unit cell does not expand uniformly. We know from our stress dependent band computations[12] that it is a sliding of the dimers that explains the behavior of $\kappa - (BEDT - TTF)_2Cu(NCS)_2$,, but no calculations have yet been carried out for $\eta - Mo_4O_{11}$ (or for $(TMTSF)_2PF_6$ for that matter). Hence the response of a molecular crystal to uniaxial stress may be more complex than a simple elastic response.

5.1 $(TMTSF)_2PF_6$

The main surprise (perhaps) is that stress along the c-axis of $(TMTSF)_2PF_6$ does not stabilize the metallic state which exhibits the magnetic field induced SDW transtions (FISDW), at least in the two samples we studied over a somewhat limited range of field and temperature. However, the main parameter which enters into the FISDW mechanism is the b-axis transfer integral t_b, which controls the degree of warping of the otherwise nearly parallel sheets of one dimensional Fermi surface. In our experiment we were certainly increasing t_c, but as discussed above, the effects on the a-b plane are not yet known. We only know that the material was becoming nearly metallic. This is quite different than the case of $\alpha - (BEDT - TTF)_2KHg(SCN)_4$ where even a small stress removes the density wave state. We may conclude then that the b-axis bandwidth remains small, and that the SDW state remains. Hence the next step in this investigation will be to apply stress along the b-axis, thereby directly influencing t_b. This should increase the b-axis bandwidth, and the SDW state should be removed thereby. Experiments are now in planning for this orientation.

5.2 $\eta - Mo_4O_{11}$

One important result of the $\eta - Mo_4O_{11}$ study are that stress along the least conducting axis (the a-axis in this case) apparently increases the size of some of the small pockets on the Fermi surface. This increase is also observed in hydrostatic pressure studies[15]. (More work is needed to make a full comparison of the difference between stress effects and hydrostatic pressure effects as reported in Ref. 15.) The increase in frequency then increases the field needed to reach the quantum limit, and the similar increase with stress of these frequencies and the M-I threshold field support the model that the opening of the final Landau gap is responsible for the rapid increase in resistance above 15 tesla. A second observation is that the increases in QO frequencies indicate a non-trivial response of the crystal structure to the stress tensor, and band calculations will be needed to sort out the mechanisms. A third observation is that the resistance is also a non-trival (non-monotonic) function of stress along the a-axis. In particular it seems that for relatively small values of stress there is a rapid increase in resistance, followed by a less rapid decrease at higher stress values. Finally, the negligible values of resistance observed at high stress values in Fig. 6 at low fields are intriguing , and further work is necessary to determine the origin of this effect.

6 Acknowledgments

We are grateful to J. Rooks who constructed the uniaxial probe, and to the staffs of the Nigmegen High Field Laboratory, and the Tallahassee National High Magnetic Field Laboratory were these studies were carried out. Work at Florida State is supported by NSF DMR-9510427. The work at Nijmegen is part of the research program of the "Stichting voor Fundamenteel Onderzoek der Materie (FOM)", which is financially supported by the "Nederlandse Organisatie voor Wetenschappelijk Onderzoek (NWO)" and has been supported by the European Commission under the contract no. CHGT-CT93-0051.

7 References

1. T. Ishiguro and K. Yamaji, *Organic Superconductors*, Springer Series in Solid State Sciences, Vol. 88, P. Fulde, ed., (Springer–Verlag, Berlin, 1990).
2. J. Williams et al., "Organic Superconductors (Including Fullerenes)", Prentice Hall, Englewood Cliffs, NJ, 1992.
3. C. Shlenker, "Low dimensional electronic properties of molybdenum bronzes and oxides", Kulẁer Academic Pub., 1992.
4. J. Caulfield et al., J. Phys. Condens. Matter bf 6. 2911 (1994).
5. J.S. Brooks *et al.,Phys. Rev.* B **52**, 14457 (1995)
6. C.E. Campos *et al.,Rev. Sci. Instrum.* **66**, 1061 (1995).
7. J. Wosnitza, "Fermi Surfaces of Low-Dimensional organic Metals and Superconductors", Springer, Berlin, 1996.
8. C.E. Campos, Ph.D. Thesis, Boston University, 1995.
9. C.E. Campos *et al.,Synth. Met.* **70**, 803 (1995).
10. C.E. Campos *et al.,Phys. Rev.* B **52**, R7014 (1995).
11. C.E. Campos *et al.,Physica* B **221**, 293 (1995).
12. C.E. Campos *et al.,Phys. Rev.* B **53**, 12725 (1996).
13. W. Kang, S.T. Hannahs, and P.M. Chaikin, Phys. Rev. Lett., **70**, 3091 (1993).
14. S. Hill *et al.*, to be published.
15. S. Ohara *et al.*, Phys. Stat. Sol. (b) **164**,243(1991).

MAGNETS, SEMICONDUCTORS AND ORGANIC CONDUCTORS AT THE LEUVEN PULSED FIELD LABORATORY

FRITZ HERLACH, R. BOGAERTS, W. BOON, I. DECKERS, A. DE KEYSER, N. HARRISON, L. LI, P.H.P. REINDERS, K. ROSSEEL, L. TRAPPENIERS, J. VANACKEN, L. VAN BOCKSTAL, A. VAN ESCH
Laboratorium voor Vaste-Stoffysica en Magnetisme, Katholieke Universiteit Leuven, B-3001 Leuven, Belgium

A. HOUSE, J. SINGLETON
Clarendon Laboratory, Oxford University, Oxford OX1 3PU, England

M. KARTSOVNIK
Institute of Solid State Physics, Chernogolovka, MD 142432, Russian Federation

J. DE BOECK, G. BORGHS
Interuniversitair Micro-Elektronica Centrum, B-3001 Leuven, Belgium

The experimental facility has been updated to allow simultaneous experimentation in six experimental stations, including a 50 mK dilution refrigerator. The continuing development of coils has resulted in 60 T user coils (17 mm bore, 15 ms pulse duration, 400 kJ) and 70 T coils with a 12 mm bore; cryostats are being modified to be accommodated in these. For the FELIX free electron laser, a two-stage 60 T coil with 10 μs rise time has been designed. Semiconductor research is focused on magnetotransport in δ-doped structures including multiple quantum wells, and on diluted semimagnetic materials. Research on the organic charge-transfer salts is mainly concerned with salts based on the (BEDT-TTF) donor molecule. A number of salts have been studied by means of both magnetotransport and magnetisation techniques in temperatures down to 350 mK, in particular α-(BEDT-TTF)$_2$MHg(SCN)$_4$ (where M = K, Tl and NH_4).

1 Update on the Pulsed Field Laboratory

The new generation of high performance coils requires increasingly higher currents in the 25-50 kA range. Together with the intensive use for experimentation, this caused a gradual deformation of the heavy copper bars in the wiring of the capacitor bank, eventually resulting in a spectacular failure. The wiring has now been rebuilt with heavy duty coaxial cabling, eliminating all weak spots that had shown up.

This was taken as a welcome occasion for a general modernisation of the entire installation. The room was divided by fragment-proof walls (high density fibre board with steel plating) into five separate compartments (Fig. 1), each housing one or more experimental stations to allow the running of several experiments in parallel. This is achieved by a decentralised data acquisition system where each experiment is equipped with a shielded box (on wheels) containing all electronic instrumentation and a digital recorder linked by optical fibre to the central computer. All experiments are controlled by one PC with a Windows multi-tasking operating system which also performs all functions needed to control the capacitor bank: user log-in, checking interlocks, selecting capacitance, polarity and charging to a preset voltage, firing, measurement of inductance and resistance, collection and immediate display of the data which are then transferred to a computer network for further analysis. The computer keeps track of inductance and resistance for each individual coil and prevents charging the bank if these values for the cold coil are out of range.

Each of the six experimental stations has its dedicated cryostat, ranging from a disposable nitrogen container in the coil test box to a 50 mK dilution refrigerator (ordered from TBT and to be commissioned in 1996). The ^{3}He cryostat, designed by C. Agosta[1], is a simple plastic insert for an existing He bath cryostat. The temperature range is 350 mK $< T <$ 4.2 K for the field range -60 T $< B <$ +60 T. This has been fitted with efficient vibration damping featuring independent suspension of the pulsed field coil. This cryostat is mainly used for magnetisation and resistivity measurements on organic compounds. The ^{4}He bath cryostat is operated in the range 1.4 K $< T <$ 4.2 K. It is mainly used for magnetotransport measurements on semiconductors; in addition it can be equipped with optical (Far InfraRed) probes. The two flow cryostats with a temperature range 4.2 K $< T <$ 300 K are now mainly used for experiments on high temperature superconductors. One of these was originally designed for FIR transmission experiments; it will now be equipped with new infrared detectors. For optics in the visible range, an Optical Multichannel Analyzer with CCD sensor will be installed. This permits time-resolved as well as integrated measurements.

2 Coils

2.1 Wire-wound coils

At present, user coils with bores ranging from 15 to 21 mm operate at fields up to 50–60 tesla. The pulse duration (half sine wave) is of the order 20 ms. These coils are subjected to severe mechanical loads and therefore require the use of advanced high strength materials. Our standard design is a wire wound coil made from soft copper with optimised layer by layer reinforcement made from

S-glass fibre composite (vacuum impregnated with epoxy)[2]. The magnets are designed using one-dimensional computer codes calculating the thermal and mechanical load in the mid-plane during the pulse, including the effects of pre-stress, plastic deformation and cycling, magnetoresistance, skin effect and heat conduction[3,4].

New design concepts and materials are needed for coils that can operate reliably in the field range above 60 T. In the first place, this is a combination of stronger conductors with carbon fibre epoxy composite or pearlitic steel wires [5]. The higher stiffness of these materials improves fatigue behaviour and thus the service life of the coils, aimed at up to 1000 pulses. As it is mostly the inner part of the coil which is critical, user coils of the next generation will be built with a separate inner section made from stronger materials.

In the context of the European 100 tesla project, the design and testing of advanced coils is pursued in co-operation with Amsterdam (pearlitic steel), Oxford (wire with stainless steel mantle) and Toulouse (insert for dual coil system "coilin-coilex", analytical calculations for scaling[6]). This involves activation of the 10 kV, 500 kJ capacitor bank.

2.2 Coil with ultra short pulse duration for FELIX

A 60 T non-destructive pulsed magnet for the "Free Electron Laser for Infrared eXperiments" (FELIX) at Nieuwegein, The Netherlands, is under development. The resulting combination of high fields with continuously tuneable high intensity infrared radiation will create a unique facility for a variety of new experiments e.g. on semiconductors, organic molecular metals and heavy fermion compounds. This project is carried out in co-operation with the Clarendon Laboratory (Oxford) providing the cryogenics and the Humboldt University (Berlin) providing the detectors; the capacitor banks are built at FELIX.

The pulse duration of the FELIX far infrared macropulses is about 10 μs; this calls for a commensurable rise time of the field in order to sweep through resonances during a single laser pulse. A 60 T coil with 10 μs rise time would require a capacitor bank with extreme performance (high voltage and power), therefore the magnet will be composed of two parts: a conventional outer coil which provides a background field of 30 T with a half period of 1 ms and a fast 30 T insert coil with a rise time to peak of 10 μs. The coil system will be mounted on a helium gas flow cryostat tailored to infrared transmission experiments. The infrared beam will be guided through oversized metal and dielectric waveguides and detected by a fast infrared detector.

3 Semiconductors

3.1 Structures with δ-doped layers

As part of a research programme on III-V materials with single and multiple δ-doped quantum wells, three Si-δ-doped InSb samples were grown heteroepitaxially on GaAs at IMEC (Leuven). These structures contain a single δ-layer at 40 nm, 120 nm and 800 nm from the sample surface. When the δ-layer is located near the sample surface, the transport coefficients clearly exhibit 2D Shubnikov-de Haas oscillations at low magnetic fields and quantum Hall behaviour above 30 T. When the δ-layer is positioned far from the surface, a significant change in the transport properties is observed. At high magnetic fields, the common quantum Hall effect phenomena are no longer observed. The Hall effect deviates strongly from linear behaviour and displays pronounced oscillations and a change in slope, suggesting the presence of holes in the system. The results can be understood in terms of the deformation of the confining potential well due to Fermi level pinning at the sample surface[7].

In close co-operation with the Technical University of Eindhoven, depopulation experiments in parallel magnetic fields have been carried out on Si-δ-doped GaAs structures. Of particular interest is the apparent shift in depopulation field when the magnetic field is rotated from the $B \parallel I$ to the $B \perp I$ configuration. These results have been interpreted by means of a self-consistent calculation of the Poisson and Schroedinger equations in which the effect of the parallel magnetic field has been included[8].

3.2 (Ga,Mn)As

Until a few years ago, no III-V diluted magnetic semiconductor (DMS) had been successfully synthesised because the Mn did not seem to find its place in the III-V lattice. The highest concentrations obtained never exceeded typical doping concentration. A few years ago, Ohno and Munekata reported the successful growth of InMnAs by MBE[9] which motivated us to start a joint project with IMEC on the growth of III-V DMS as these materials have attractive capabilities for applications, and high field studies are required for understanding the interaction between the diluted magnetic ions and the charge carriers. The crystal quality of the first batches of InMnAs however was not adequate; the growth procedure is tedious and concentrations in excess of 3 % resulted in segregation of the constituents. Therefore, the growth of (Ga,Mn)As was started as a new III-V DMS. Mn is known to be metastable in the GaAs host lattice, and therefore the growth parameters are based on the growth of LT(low temperature)-GaAs. Mn/Ga ratios of up to 15 % were obtained with near-

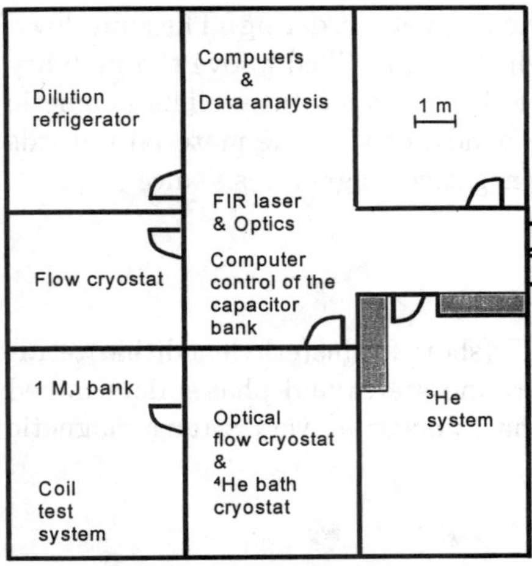

Figure 1: Floor plan of the Pulsed Field Laboratory.

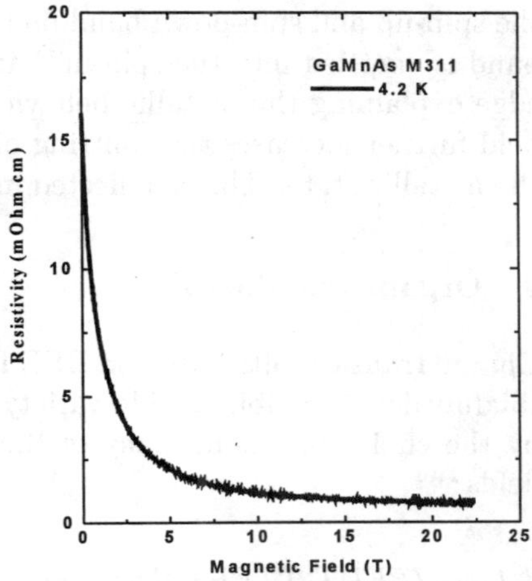

Figure 2: Magnetic field dependence of the resistivity in (Ga,Mn)As showing strong negative magnetoresistance.

perfect crystal quality[10].

The important difference between II-VI and III-V DMS containing Mn is that in the latter case Mn not only provides a local spin, but also donates a hole. This is because it prefers to donate its two 4*s*-electrons and keep its *d*-shell half filled. It thus falls short of one electron to fit into the III-V bonding. This has profound consequences for the magnetic and electrical behaviour of (Ga,Mn)As.

The (Ga,Mn)As samples grown by MBE exhibit a phase transition at temperatures around 50 K. The low temperature phase is characterised by a spontaneous magnetisation with sharp hysteresis[11]. The magnetic phase is attributed to the holes bound to the ionised Mn^{2+} acceptor. For a typical concentration of 8 % of Mn, the hole wave functions overlap and cover approximately 50 Mn ions. By annealing the sample, various mechanisms result in the passivation of Mn which explains the reduction of the ordering temperature.

The resistance in the paramagnetic phase shows a temperature dependence typical for a material near the metal-insulator transition: the resistivity decreases with temperature, but the functional dependence is hardly that of an activated behaviour. Instead, the resistivity follows a $1/T$ dependence as is expected when the Fermi level coincides with the mobility edge[12].

Below the magnetic ordering temperature, the resistivity drops and becomes nearly temperature independent. This is consistent with the splitting of the spin-up and spin-down band due to the magnetic ordering. The spin-down band is emptied into the spin-up band which is now filled above the mobility edge explaining the metallic behaviour at these temperatures. The magnetic field further increases the splitting of the bands, driving the material towards the metallic state. This is reflected in the negative magnetoresistance (Fig. 2).

4 Organic conductors

Charge transfer salts based on BEDT-TTF (short for bis(ethylenedithio)tetrathiafulvalene) exhibit a wide variety of ground states and phases determined by the choice of anion; many of these phases occur at very strong magnetic fields [1,13].

4.1 α-(BEDT-TTF)$_2$KHg(SCN)$_4$

Amongst BEDT-TTF-salts, the family α-(BEDT-TTF)$_2$MHg(SCN)$_4$, where M=NH_4, K, Rb or Tl, has yielded an abundance of physical phenomena. Whilst all possess very similar calculated Fermi surfaces consisting of an open quasi-one-dimensional (Q1D) electron section and a closed Q2D hole pocket, the M=NH_4-salt is a superconductor below 1 K, while those with M=K, Tl or Rb exhibit spin density wave (SDW) ground states at low temperatures. One of the most notable features of the complex magnetic phase diagrams of the M=K, Rb and Tl salts is observed as a sharp kink in the magnetoresistance which has become known as the "kink transition". The kink occurs at ~23 T in the M=K salt, and marks the re-entrance of a metallic phase. By employing pulsed magnetic fields of ~50 T we have measured the dHvA effect in α-(BEDT-TTF)$_2$KHg(SCN)$_4$, both in the SDW regime and much deeper into the high field state [14]. These measurements complement previous pulsed field SdH studies above the kink transition. The dHvA effect has the distinct advantage that it can be related in a quantitative manner to the thermodynamic function of state.

Analysis of the amplitude of the quantum oscillations indicates that α-(BEDT-TTF)$_2$KHg(SCN)$_4$ is only fully transformed into the high field state above ~27 T. Throughout the extended transition, the effective mass of the carriers increases by approximately a factor of two, signalling an enhancement of the electron-electron or electron-phonon interactions. When fitted to the Lifshitz-Kosevich (LK) expression for the dHvA effect, the effective masses of the higher harmonics are found to be anomalously low, indicating that the LK-formalism is no longer appropriate for these materials at very high magnetic fields.

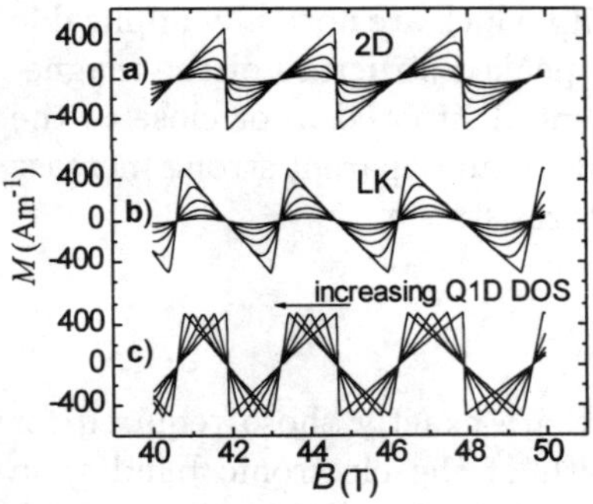

Figure 3: The numerically calculated magnetisation for (a) the 2D-limit, (b) the LK-limit, and (c) taking into account the 1D-contribution to the DOS.

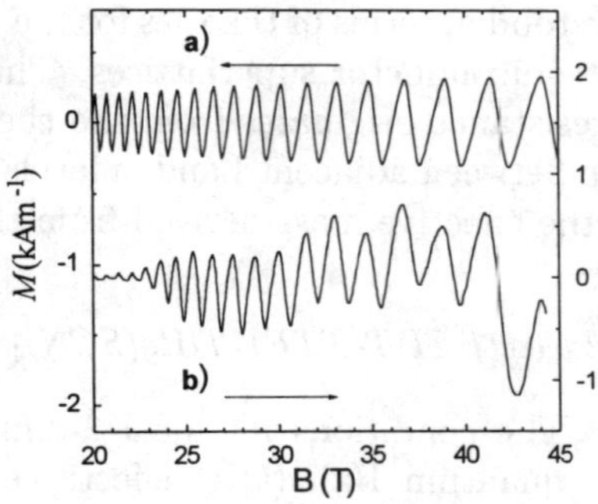

Figure 4: (a) The numerically calculated magnetisation using the parameters m*, τ^{-1} and g_{1D} determined experimentally. (b) The corresponding measured magnetisation. The arrows indicate the appropriate axes.

Using a combination of numerical methods and simple physical models, we have been able to account for many of the departures from LK behaviour [15]. The calculations are based on the thermodynamics of an ideal 2D electron gas, which appears to be an accurate representation of these materials at high magnetic fields (B $>$ 20 T). The pinning of the chemical potential to the Landau levels has a dramatic effect on both the sign and temperature dependence of the harmonics of the magnetisation oscillations (Fig. 3 a and b), and provides an explanation for the apparently low effective masses which have been observed for the higher harmonics in the high field state. The presence of additional states in the Q1D portion of Fermi surface has a pronounced influence on the waveform of the oscillations (Fig. 3 c). A comparison of the numerical calculations with the measurements provides an opportunity by which the DOS associated with the Q1D portion of Fermi surface can be estimated (Fig. 4). Whilst the waveform of the oscillations is significantly perturbed by the effects of an oscillatory chemical potential, the peak to peak amplitude of the free energy is independent of this, and is therefore independent of the size of the DOS of the Q1D portion of the Fermi surface. The analysis of the waveform thus provides a procedure by which the true effective mass and quasiparticle scattering rates can be extracted from experimental data, even in extreme 2D systems in high magnetic fields.

Meanwhile the main features of the longitudinal magnetoresistance in quasi-two-dimensional organic conductors at high magnetic fields can be understood in terms of theories for the conductivity which are normally applicable to semiconductor superlattices. The strongly peaked structure of the magnetoresistance originates from the chemical potential situated in or close to the gap between adjacent Landau levels. This leads to an apparent strong increase of the effective mass derived from the magnetoresistance.

4.2 α*-(BEDT-TTF)*$_2$*TlHg(SCN)*$_4$

The conditions for ideal 2-dimensionality are exactly those required for the quantum Hall (QH) effect; i.e. the width of the electronic band(s) in the longitudinal direction and the broadening of the Landau levels are both significantly less than the separation between the Landau levels. For practical reasons however, magnetoresistance measurements in pulsed magnetic fields have been limited to the longitudinal direction, so that no direct verification of the existence of the QH effect has been possible.

The magnetisation provides an alternative means by which currents within the conducting planes can be investigated. Deep minima in ρ_{xx} (i.e. ideal conducting phases) due to the QH effect cause the induction of persistent eddy currents which contribute to the magnetisation [16]. Recent experiments in this material have revealed the superposition of eddy current "resonances" on top of the conventional dHvA magnetisation waveform. Figure 5 shows a measurement of the derivative dM/dB. The high degree of asymmetry cannot be explained by the dHvA effect but only in terms of eddy currents. The independence of their size on dB/dt indicates a critical current which agrees quantitatively with that expected in the quantum Hall regime. Numerical simulations based on a simple model for the eddy current effect are able to reproduce the experimental data .

4.3 β*"-(BEDT-TTF)*$_2$*AuBr*$_2$.

Magnetotransport measurements extended to 60 Tesla on the β"-(BEDT-TTF)$_2$AuBr$_2$-salt [17] give conclusive evidence that the Fermi surface undergoes a field-induced change in form at around 10 T (Fig. 6). At fields below 10 T, the Fermi surface appears to be characterised by three small Q2D pockets. At fields above 10 T, only one of these frequencies continues to persist while the others vanish to be replaced by two new frequencies. Using angular magnetoresistance measurements, it is found that the frequency visible in the complete field range can be associated with the pocket predicted by band structure calculations centred on the X point of the Brillouin zone. Frequency mixing effects are present in the high field state, and are thought to originate

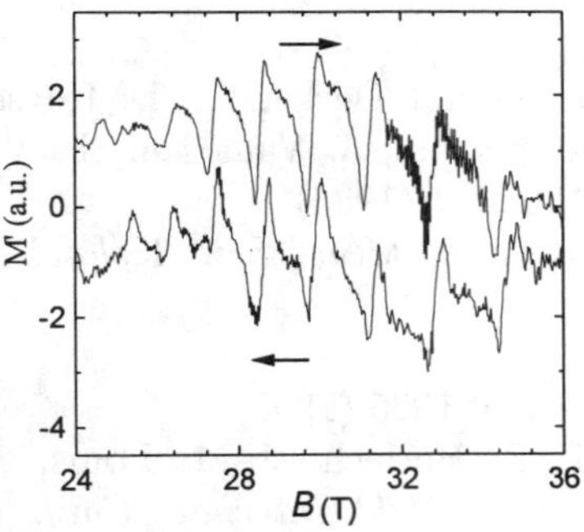

Figure 5: The derivative of the magnetisation during the rising and falling of the magnetic field (the arrows indicate the direction). Note the highly asymmetric waveform.

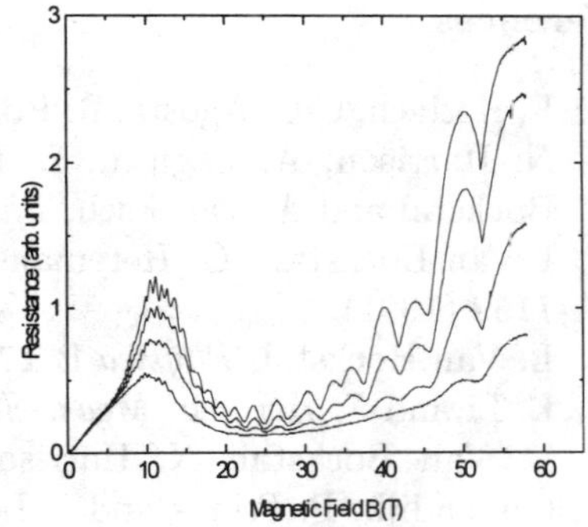

Figure 6: The resistivity of a $AuBr_2$ sample as a function of field for different temperatures.

from oscillations in the chemical potential rather than from the Shoenberg magnetic interaction.

4.4 κ*-(BEDT-TTF)$_2$Cu(NCS)$_2$*

Band structure calculations for this material predict a Fermi surface consisting of a Q2D lens pocket and an open Q1D sheet separated by a small gap enabling magnetic breakdown (MB). Our numerical calculations indicate that in fact the pinning of the chemical potential to sharply defined Landau levels is responsible for the observation of a "forbidden" $\beta - \alpha$ frequency in the magnetisation measurements. Our longitudinal magnetoresistance measurements can qualitatively be explained if the Stark Quantum interference effects are included in the calculations [18].

Acknowledgments

This work was supported by the Belgian Nationaal Fonds voor Wetenschappelijk Onderzoek and by the Commission of the European Union under the Science and TMR programmes. RB and LVB are research fellows of the NFWO; NH, PHPR and JV are research fellows of the Onderzoeksraad at K.U.Leuven.

References

1. F. Herlach, C.C. Agosta, R. Bogaerts, W. Boon, I. Deckers, A. De Keyser, N. Harrison, A. Lagutin, L. Li, L. Trappeniers, J. Vanacken, L. Van Bockstal and A. Van Esch, *Physica* B **216**, 161 (1996).
2. L.Van Bockstal , G. Heremans and F. Herlach, *Meas. Sci. Technol.* **2**, 1159 (1991).
3. L. Van Bockstal, *Physica* B **177**, 44 (1992).
4. L. Li and F. Herlach, *Meas. Sci. Technol.* **6**, 1035 (1995)
5. L. Van Bockstal, N. Harrison, L. Li, F. Herlach, J.-M. Louis, M. Oudendijk, P. Frings and I. Lefever, in *Proc. 14th Internat. Conf. on Magnet Technology, IEEE Trans. Magnetics* (accepted for publication).
6. S. Askenazy, Luc Van Bockstal, F. Dupouy and Fritz Herlach, *Meas. Sci. Technol.* **7**, 680 (1996).
7. A. De Keyser, R. Bogaerts, F. Herlach, V.C. Karavolas, F.M. Peeters, W. van de Graaf and G. Borghs *in these proceedings.*
8. A.F.W. van de Stadt, P.M. Koenraad, J.H. Wolter, R. Bogaerts, A. De Keyser and F. Herlach, *in these proceedings.*
9. H. Ohno, *J. Appl. Phys.* **69**, 6103 (1991).
10. A. Van Esch, W. Van Roy, J. De Boeck, I. Francois, H. Bender, G. Borghs, L. Van Bockstal, R. Bogaerts, F. Herlach. *Materials Science Forum* **182-184**, 623 (1995).
11. L. Van Bockstal, A. Van Esch, J. De Boeck, G. Borghs, R. Bogaerts and F. Herlach, Proc. ICPS 23, Berlin, July 21-26, 1996 (World Scientific)
12. M.A. Dubson and D.F. Holcomb. *Phys. Rev.* B **32**, 1955 (1985).
13. J.S. Brooks, R.G. Clark, R.P. Starrett, R. Newbury, R.H. McKenzie, A.V. Skougarevsky, M. Tokumoto, N. Kinoshita, Y. Tanaka, H. Anzai, S. Takasaki, J. Yamada, M.V. Kartsovnik, A.I. Schegolev, G.J. Athas, P. Sandhu, *Physica* B **216**, 380 (1996).
14. N. Harrison, A. House, I. Deckers, J. Caulfield, J. Singleton, F. Herlach, W. Hayes, M. Kurmoo, P. Day, *Phys. Rev.* B **52**, 5584 (1995).
15. N. Harrison, R. Bogaerts, P.H.P. Reinders, J. Singleton, S.J. Blundell, F. Herlach, *Phys. Rev. B* (accepted for publication).
16. N. Harrison, A. House, M.V. Kartsovnik, A.V. Polisski, J. Singleton, F. Herlach, W. Hayes, N.D. Kushch, submitted to *Phys. Rev. Lett.*.
17. A. House, N. Harrison, S.J. Blundell, I. Deckers, J. Singleton, F. Herlach, W. Hayes, J.A.A.J. Perenboom, M. Kurmoo, P. Day, *Phys. Rev.* B **53(13)**, 9127 (1996).
18. N. Harrison, J. Caulfield, J. Singleton, P.H.P. Reinders, F. Herlach, W. Hayes, M. Kurmoo, P. Day, *J. Phys. C* (accepted for publication).

CARBON NANOTUBES IN MAGNETIC FIELDS

Tsuneya ANDO and Hiroshi AJIKI*

Institute for Solid State Physics, University of Tokyo
7–22–1 Roppongi, Minato-ku, Tokyo 106, Japan

A brief review is given of electronic properties of carbon nanotubes, in particular those in magnetic fields, mainly from a theoretical point of view. The topics include a giant Aharonov-Bohm effect on the band gap and optical absorption spectra, a magnetic-field induced lattice distortion, and a magnetization and susceptibility of ensembles, calculated based on a $\mathbf{k}\cdot\mathbf{p}$ scheme.

1 Introduction

Graphite needles called carbon nanotubes (CN's) were discovered recently[1,2] and have been a subject of an extensive study. A CN is a few concentric tubes consisting of carbon-atom hexagons arranged in a helical fashion about the axis as shown in Fig. 1. The diameter of CN's is usually between 20 and 300 Å and their length can exceed 1 μm. The distance of adjacent tubes is larger than the distance between nearest neighbor atoms in a graphite sheet and therefore electronic properties of CN's are dominated by those of a single layer CN. More recently a monolayer nanotube was produced also.[3,4]

Carbon nanotubes can be either a metal or semiconductor, depending on their diameters and helical arrangement. The condition whether a CN is metallic or semiconducting can be obtained based on the band structure of a two-dimensional graphite sheet and periodic boundary conditions along the circumference direction. This result was first predicted by means of a tight-binding model ignoring the effect of the tube curvature.[5–14] These properties can be well reproduced in a $\mathbf{k}\cdot\mathbf{p}$ method.[15] In this paper a brief review is given on electronic properties predicted based on the $\mathbf{k}\cdot\mathbf{p}$ method.

In Sec. 2 the effective mass equation is introduced and the band structure is discussed with a special emphasis on an Aharonov-Bohm effect. Optical absorption spectra are discussed in Sec. 3. A lattice instability, in particular that induced by a magnetic field perpendicular to the tube axis, is discussed in Sec. 4 and magnetic properties of ensembles of CN's are discussed in Sec. 5.

2 Aharonov-Bohm Effect on Energy Bands

A nanotube is specified by a chiral vector $\mathbf{L}=n_a\mathbf{a}+n_b\mathbf{b}$ with integer n_a and n_b and basis vectors $\mathbf{a}$ and $\mathbf{b}$ ($|\mathbf{a}|=|\mathbf{b}|=a=2.46$ Å) as is shown in Fig. 2. In the

* Present address: Department of Material Physics, Faculty of Engineering Science, Osaka University, Machikaneyama, Toyonaka 560, Japan

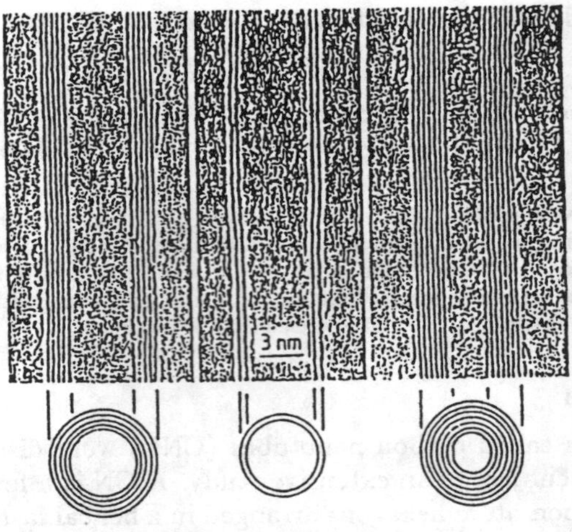

Fig. 1 Some examples of transmission micrograph images of carbon nanotubes.[1] The diameter is 67, 55, and 65 Å from left.

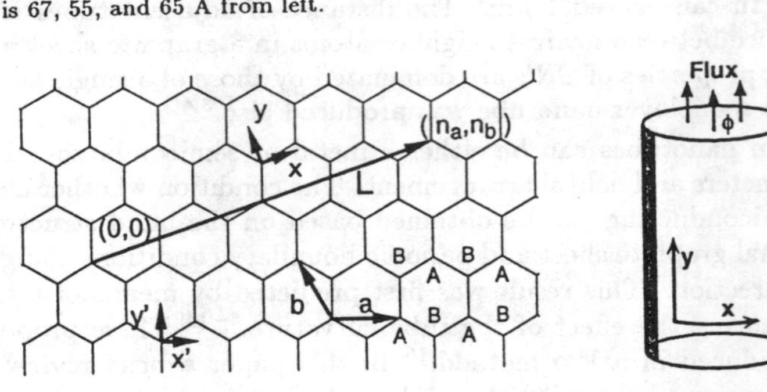

Fig. 2 A single layer of graphite and a carbon nanotube in a magnetic flux passing through its cross section.

(x', y') system of coordinates fixed onto a graphite sheet, we have $\mathbf{a}=(a,0)$ and $\mathbf{b}=(-a/2, \sqrt{3}a/2)$. For convenience we introduce another coordinate system where the x direction is along the circumference and the y direction is along the axis of CN.

A graphite sheet is a zero-gap semiconductor in the sense that the conduction and valence bands consisting of π-states touch at K and K' points of the Brillouin zone, whose wave vectors are given by $\mathbf{K} = (2\pi/a)(1/3, 1/\sqrt{3})$ and $\mathbf{K}'=(2\pi/a)(2/3,0)$. The unit cell has an area $\sqrt{3}a^2/2$ and contains two carbon atoms (denoted as A and B) as shown in Fig. 2.

Electronic states (with an appropriate choice of phases of the Bloch func-

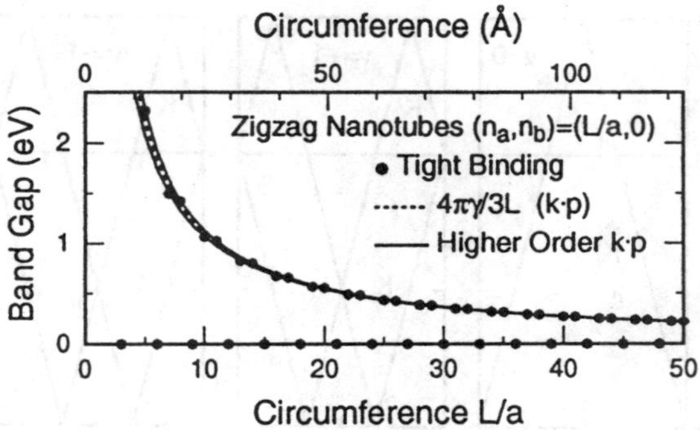

Fig. 3 Energy gap of monolayer CN's specified by $(n_a, n_b) = (m, 0)$. The dots are calculated in a tight-binding model and the dotted line represents $4\pi\gamma/3L$ given in the $\mathbf{k}\cdot\mathbf{p}$ scheme. The solid lines are the results of higher order $\mathbf{k}\cdot\mathbf{p}$ scheme. Use has been made of $\gamma = 6.46$ eV·Å. After Ref. 16.

tions) near K and K' point of 2D graphite are described by the $\mathbf{k}\cdot\mathbf{p}$ equation:

$$\begin{aligned}\gamma(\sigma_x k_x + \sigma_y k_y)F_K(\mathbf{r}) &= \varepsilon(\mathbf{k})F_K(\mathbf{r}),\\ \gamma(\sigma_x k_x - \sigma_y k_y)F_{K'}(\mathbf{r}) &= \varepsilon(\mathbf{k})F_{K'}(\mathbf{r}),\end{aligned} \tag{2.1}$$

where γ is the band parameter, $\mathbf{k} = (k_x, k_y)$ is a wave-vector operator, and σ_x, σ_y, and σ_z are the Pauli spin matrices. In the above equations, the envelope functions are written as

$$F_K(\mathbf{r}) = \begin{pmatrix} F_K^A(\mathbf{r}) \\ F_K^B(\mathbf{r}) \end{pmatrix} \quad \text{and} \quad F_{K'}(\mathbf{r}) = \begin{pmatrix} F_{K'}^A(\mathbf{r}) \\ F_{K'}^B(\mathbf{r}) \end{pmatrix}. \tag{2.2}$$

Equation (2.1) has the form of Weyl's equation for neutrinos.

For nanotubes with sufficiently large diameter, their electronic states near the Fermi level can be obtained by imposing the generalized periodic boundary condition in the circumference direction, $\Psi(\mathbf{r}+\mathbf{L}) = \Psi(\mathbf{r})\exp(2\pi i\varphi)$, with $\varphi = \phi/\phi_0$, where ϕ is an Aharonov-Bohm (AB) magnetic flux passing through the cross section of CN and $\phi_0 = ch/e$ is the magnetic flux quantum. The Bloch functions at K and K' points change their phase by $\exp(2\pi i\nu/3)$ and $\exp(-2\pi i\nu/3)$, respectively, where ν is an integer defined by $n_a + n_b = 3M + \nu$ with integer M and can take 0 and ± 1. This phase change should be canceled by that of the envelope functions and the boundary condition for the envelope function associated with the K point is given by

$$F_K(\mathbf{r}+\mathbf{L}) = F_K(\mathbf{r})\exp\left[2\pi i\left(\varphi - \frac{\nu}{3}\right)\right], \tag{2.3}$$

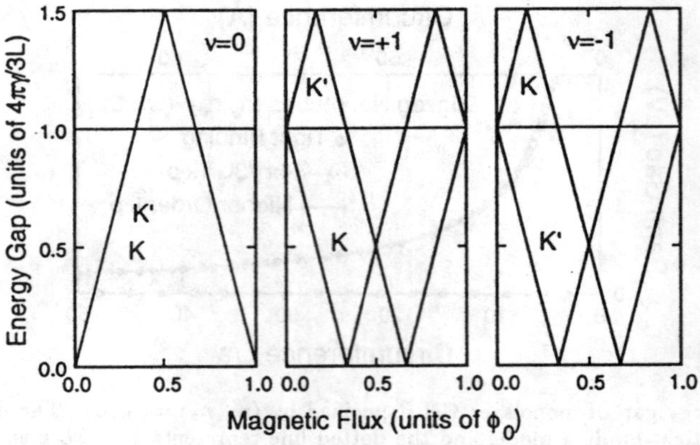

Fig. 4 Energy gap versus magnetic flux passing through the tube cross section for metallic ($\nu=0$) and semiconducting ($\nu=\pm1$) CN.

At the K' point, ν should be replaced by $-\nu$ in this equation.

Energy levels in CN are obtained by putting $k_x=\kappa_{\nu\varphi}(n)$ and $k_y=k$ in the above $\mathbf{k}\cdot\mathbf{p}$ equation as[15]

$$\varepsilon_{\nu\varphi}^{(\pm)}(n,k)=\pm\gamma\sqrt{\kappa_{\nu\varphi}(n)^2+k^2}\quad\text{with}\quad \kappa_{\nu\varphi}(n)=\frac{2\pi}{L}\Big(n+\varphi-\frac{\nu}{3}\Big),\qquad(2.4)$$

where n is an integer and the upper (+) and lower (−) signs represent the conduction and valence bands, respectively. In the absence of a magnetic flux $\varphi=0$, CN becomes metallic for $\nu=0$ and semiconducting with gap $E_g=4\pi\gamma/3L$ for $\nu=\pm1$. Figure 3 compares this gap to that obtained in a tight-binding model.

In the presence of a magnetic flux, the boundary condition is changed by the Aharonov-Bohm effect and the band gap exhibits an oscillation between 0 and $2\pi\gamma/L$ with period ϕ_0 as is shown in Fig. 4. The magnetic field corresponding to ϕ_0 is given by 1400~8 T for circumference 60~800 Å.

In the presence of a magnetic field H perpendicular to the tube axis, the effective field for electrons in a CN is given by the component perpendicular to the surface, i.e., $H(x)=H\cos(2\pi x/L)$ which is periodic in the circumference direction. The parameter characterizing its strength is given by $L/2\pi l$, where $l=\sqrt{c\hbar/eH}$ is the magnetic length or the radius of the smallest cyclotron orbit. In the case $(L/2\pi l)^2\ll1$, the field can be regarded as a small perturbation, while in the case $(L/2\pi l)^2\gg1$, Landau levels are formed on the cylinder surface.

An interesting feature of Weyl's equation lies in the fact that Landau levels are formed at energy $\varepsilon=0$. This has long been known as the origin of a large diamagnetism of graphite. Figure 5 gives some examples of energy bands of a

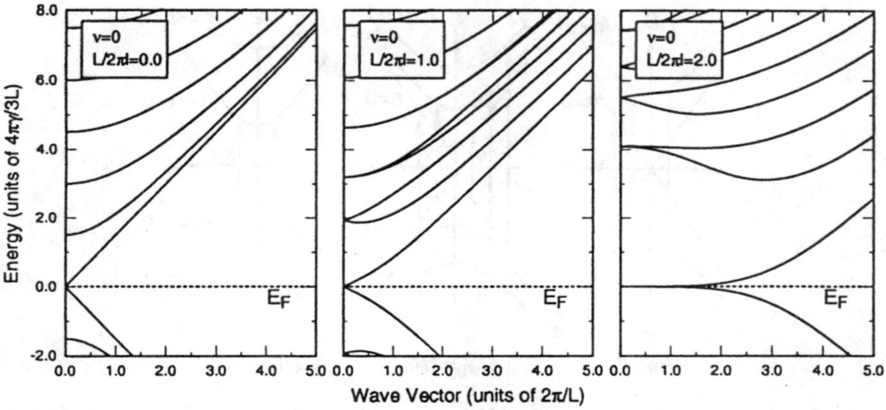

Fig. 5 Some examples of calculated energy bands of a metallic CN in magnetic fields perpendicular to the axis.

metallic CN in perpendicular magnetic fields, which clearly shows the formation of flat Landau levels at the Fermi level in high fields. Note that there is no difference in the spectra between metallic and semiconducting CN's and in the presence and absence of an AB flux for $L/2\pi l \gg 1$, because the wave function is localized in the circumference direction and the boundary condition becomes irrelevant.

3 Optical Absorption

The AB effect on the band gap can be directly observed in optical absorption spectra. There are two different configurations, the x and y polarizations. In the latter case the electric field is along the CN axis and in the former case the field is perpendicular to the axis. Figure 6 shows the energy band of a metallic CN near the K and K' point for flux $\phi/\phi_0=0$, $1/4$, and $1/2$ and the process of optical transitions for the x and y polarizations. Figure 7 gives an example of calculated absorption spectra for metallic CN's in a magnetic flux.

The spectrum changes dramatically as a function of the magnetic flux for the y polarization. In the case of the x polarization, on the other hand, the electric field associated with the incident light induces a large polarization of CN and the resulting depolarization effect strongly modifies the spectra. In fact, the absorption is almost completely suppressed by the depolarization effect.

4 Lattice Distortion

Usually, a metallic one-dimensional (1D) system is unstable and becomes insulating due to various kinds of lattice distortions. A similar instability can occur

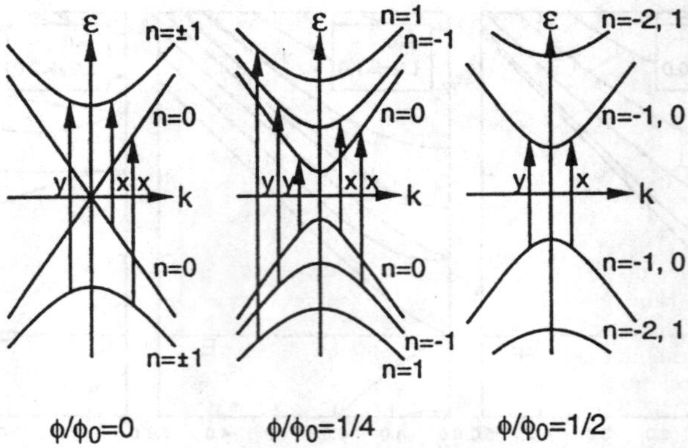

Fig. 6 Energy bands of a metallic CN in the presence of a magnetic flux and allowed optical transitions for the x and y polarizations. The integer n represents the band index.

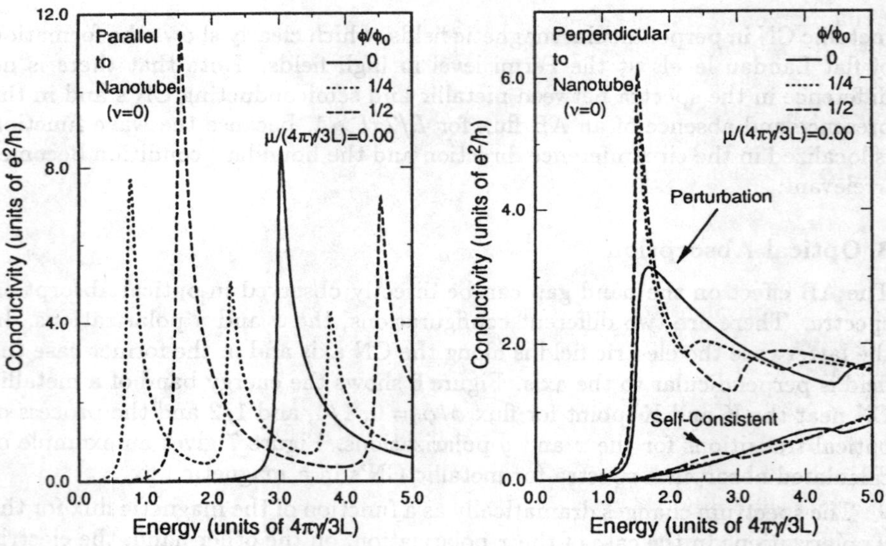

Fig. 7 Calculated optical absorption spectra of a metallic CN in a magnetic flux. In the case that the electric field is parallel to the axis (left), the absorption exhibits a distinct AB effect. In the case of the perpendicular polarization (right) the depolarization effect suppresses the absorption almost completely. After Ref. 17.

in CN's, although the situation is different. There are two kinds of distortions, in-plane and out-of-plane distortions as shown in Fig. 8, leading to an opening up of a gap at the Fermi level in metallic CN's. In the case of the in-plane

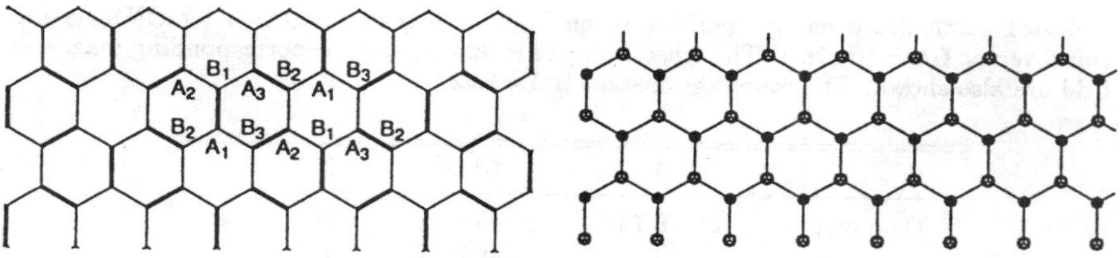

Fig. 8 Lattice distortions in a graphite sheet. For an in-plane distortion (left), the bond denoted by a thin line becomes shorter by u_1 and that denoted by a thick line becomes longer by $2u_1$, leading to a unit cell three times as large as the original. For an out-of-plane distortion (right), an atom denoted by a black dot is shifted down $(-u_2)$ and that denoted by a white circle moves up $(+u_2)$.

distortion called the Kekule distortion, the hexagon network has alternating short and long bonds ($-u_1$ and $2u_1$, respectively) like in the classical benzene molecule.[6,18,19] In its presence the unit cell becomes three times as large as the original one and both K and K' points are folded onto the Γ point in a new Brillouin zone. For an out-of-plane distortion the sites A and B move up and down $(\pm u_2)$ with respect to the cylindrical surface.[7] Consequently, the energy of sites A and B shift in the opposite direction because of a finite curvature of a CN which breaks the mirror symmetry about its surface.

In the presence of lattice distortions, the **k**·**p** equation is given by the 4×4 matrix equation given by

$$\begin{pmatrix} \gamma(\sigma_x k_x+\sigma_y k_y)+\sigma_z\Delta_2 & -i\sigma_y\Delta_1 \\ i\sigma_y\Delta_1 & \gamma(\sigma_x k_x-\sigma_y k_y)+\sigma_z\Delta_2 \end{pmatrix} \begin{pmatrix} F_K(\mathbf{r}) \\ F_{K'}(\mathbf{r}) \end{pmatrix} = \varepsilon \begin{pmatrix} F_K(\mathbf{r}) \\ F_{K'}(\mathbf{r}) \end{pmatrix}, \quad (4.1)$$

which is equivalent to the relativistic Dirac equation.[20] The energy bands are given by

$$\varepsilon_\varphi^{(\pm)}(n,k,\Delta_1,\Delta_2) = \pm\sqrt{\gamma^2\kappa_\varphi(n)^2+\gamma^2k^2+{\Delta_1}^2+{\Delta_2}^2}. \quad (4.2)$$

This shows that the gap given by $E_g=2\sqrt{{\Delta_1}^2+{\Delta_2}^2}$ opens up in the absence of a magnetic flux.

The gap parameters Δ_1 and Δ_2 are determined by the minimization of the total energy given by

$$E = 4\sum_{n=-\infty}^{\infty}\sum_k \varepsilon_\varphi^{(-)}\, g_0\big(|\varepsilon_\varphi^{(-)}|\big) + N\frac{K_1\Delta_1^2}{2f_1^2} + N\frac{K_2\Delta_2^2}{2f_2^2}, \quad (4.3)$$

where $\varepsilon_\varphi^{(-)}$ is the valence-band energy, K_1 and K_2 are the force constants for the in- and out-of-plane distortions, respectively, and N is the total number

Table 1 Calculated energy gap due to an in-plane Kekule distortion for CN's having chiral vector $\mathbf{L}/a=(m,2m)$. The critical magnetic flux φ_c and the corresponding magnetic field are also shown. The coupling constant is $\lambda\approx 1.62$.

m		5	10	20
Diameter	(Å)	6.78	13.6	27.1
Circumference	(Å)	21.3	42.6	85.2
E_g	(meV)	2.38	5.68×10^{-3}	6.46×10^{-8}
u	(Å)	6.29×10^{-5}	1.50×10^{-7}	1.71×10^{-12}
φ_c		7.58×10^{-4}	3.62×10^{-6}	8.23×10^{-11}
H_c	(T)	8.67	1.04×10^{-2}	5.89×10^{-8}

of carbon atoms. Further, f_1 and f_2 are defined by $\Delta_1=f_1u_1$ and $\Delta_2=f_2u_2$. We have introduced a cutoff function $g_0(\varepsilon)$ in order to extract the contribution from the states in the vicinity of the top of the valence band. The results are independent of the choice of cutoff functions as long as the function decays smoothly with energy and the cutoff energy is sufficiently large.

The minimization of the total energy leads to the conclusion that two kinds of distortions cannot coexist and that a distortion having a larger effective coupling constant $\lambda=\sqrt{3}af^2/\pi K\gamma$ occurs, where $(K,f)=(K_i,f_i)$ with $i=1$ or 2. The gap parameter is obtained as

$$E_g(\varphi)=E_g\sqrt{1-\left(\frac{\varphi}{\varphi_c}\right)^2},\quad E_g=\frac{2\pi\gamma}{L}\exp\left(-\frac{L}{a\lambda}-1-C\right),$$
$$\varphi_c=\frac{\phi_c}{\phi_0}=\frac{LE_g}{4\pi\gamma},\qquad C\approx 0.1445972. \tag{4.4}$$

The energy gap vanishes in the critical magnetic flux ϕ_c, where the gap due to the AB effect becomes comparable to that due to the lattice distortion. For $\phi=0$, the gap decreases exponentially as a function of the circumference L/a. Some examples for an in-plane Kekule distortion are given in Table 1.

In the case of out-of-plane distortion, the coefficient $f_2=\Delta_2/u_2$ becomes nonzero because of the finite curvature and is therefore expected to be proportional to a/L. Thus, the coupling constant is proportional to $(a/L)^2$ and the gap decreases very rapidly with the diameter as $\exp[-(L/a)^3]$. We can conclude that unlike ordinary 1D systems, metallic carbon nanotubes are quite stable against lattice distortions and we can expect their metallic properties even at low temperatures except in extremely narrow CN's.

The situation changes drastically in the presence of a high magnetic field perpendicular to the axis. As has been discussed in Sec. 2, Landau levels without dispersion start to appear at the Fermi level corresponding to a cyclotron

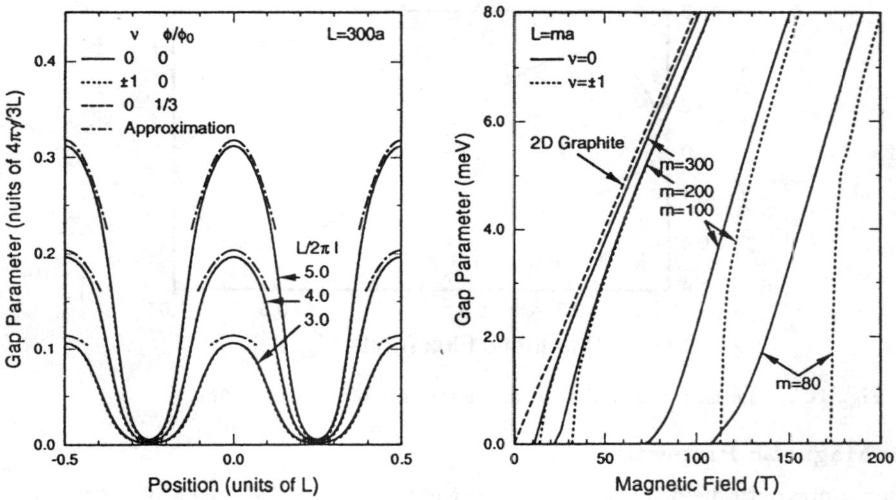

Fig. 9 An example of calculated in-plane lattice distortions induced by a high magnetic field (left) and the dependence of the maximum gap on a magnetic field (right).

motion in the top and bottom part of the cylindrical surface where the effective magnetic field perpendicular to the tube surface is the largest. This enhances the density of states at the Fermi level considerably, leading to a magnetic-field induced distortion.[21]

As has been discussed in Sec. 2, the electron wave function becomes localized spatially and the boundary condition along the circumference direction becomes less important in high magnetic fields. Consequently the distinction between metallic and semiconducting CN's disappears and lattice distortions are possible for semiconducting CN's. Further, the spatial variation of the distortion should also be considered.

The generalized $\mathbf{k}\cdot\mathbf{p}$ equation is the same as (4.1) except that the gap parameters are dependent on the position and should satisfy the boundary conditions:

$$\Delta_1(\mathbf{r}+\mathbf{L}) = \Delta_1(\mathbf{r}) \exp\left(i\frac{2\pi\nu}{3}\right) \quad \text{and} \quad \Delta_2(\mathbf{r}+\mathbf{L}) = \Delta_2(\mathbf{r}). \qquad (4.5)$$

The extra phase factor appearing in the boundary condition for $\Delta_1(\mathbf{r})$ guarantees the fact that the equations remain the same under translation $\mathbf{r}\rightarrow\mathbf{r}+\mathbf{L}$ even for $\nu=\pm1$. Some examples of explicit numerical results for the in-plane distortion are given in Fig. 9. The figure clearly shows that the lattice distortion is induced by a magnetic field in particular for CN's with large diameter for both metallic and semiconducting CN's and that the maximum gap approaches that of a graphite sheet.

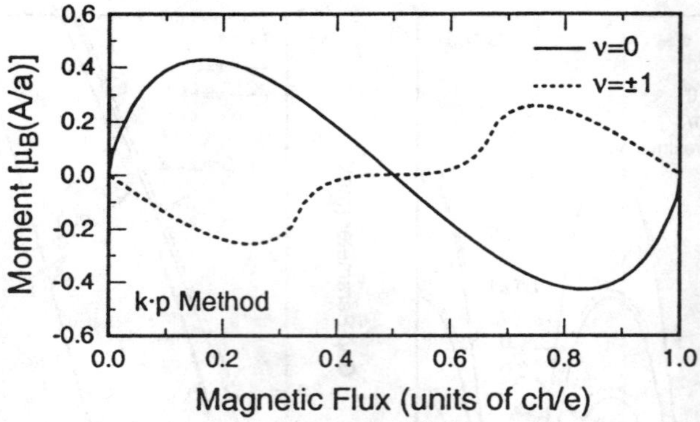

Fig. 10 Magnetic moment of a CN in the presence of a magnetic flux.

5 Magnetic Properties

For a magnetic field perpendicular to the tube axis, CN's usually exhibit diamagnetism similar to that of graphite.[22,23] For a magnetic field parallel to the axis, on the other hand, the magnetic response of CN's becomes completely different. Figure 10 shows an example of calculated magnetization as a function of a magnetic flux passing through the cross section. The most prominent feature is the large paramagnetism of a metallic CN in weak magnetic fields. In fact, the susceptibility diverges logarithmically corresponding to a sudden opening up of a gap in the presence of a flux due to the AB effect.

Realistic samples contain CN's with different layer numbers, circumferences, and orientations. Because interactions between adjacent layers of a multi-shell CN are weak, its magnetic properties are expected to be given by those of an ensemble of single-shell CN's with different circumferences. The distribution function for the circumference, $\rho(L)$, is not known and therefore we shall consider following two different kinds. The first is the rectangular distribution, $\rho(L)=1/(L_{\rm mx}-L_{\rm mn})$ for $L_{\rm mn}<L<L_{\rm mx}$, which roughly corresponds to the situation that CN's with different circumferences are distributed equally and the average layer number of multi-shell CN's is independent of the circumference. When the average layer number of CN's increases with the circumference, on the other hand, the number of CN's with small circumferences is effectively enhanced and the distribution becomes asymmetric. The most extreme case can be realized if we assume that each CN has layers such that the circumference of the inner-most shell is $L_{\rm mn}$. In this case the distribution is given by a triangular form, $\rho(L)=2(L_{\rm mx}-L)/(L_{\rm mx}-L_{\rm mn})^2$ for $L_{\rm mn}<L<L_{\rm mx}$.

Figure 11 shows magnetization and differential susceptibility calculated for the rectangular and triangular distribution with $L_{\rm mn}=22$ Å corresponding to the finest CN so far observed and $L_{\rm mx}=942.5$ Å corresponding to the thickest

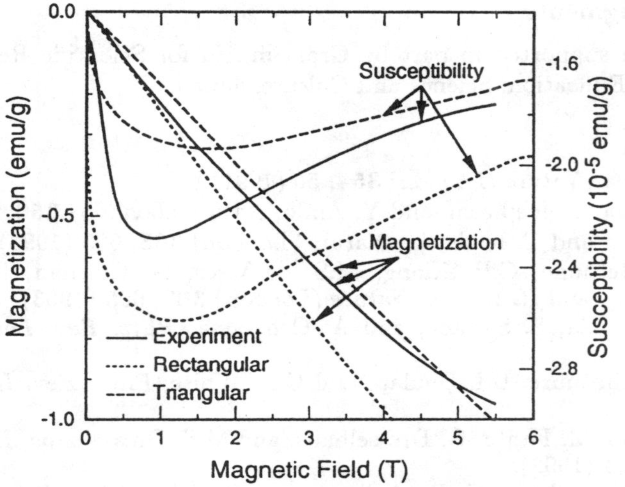

Fig. 11 Calculated ensemble average of magnetic moment and differential susceptibility for CN's with rectangular (dotted lines) and triangular (dashed lines) circumference distribution having $L_{\rm mn} = 22$ Å and $L_{\rm mx} = 942.5$ Å. The solid lines denote experimental results.[24]

CN. The experimental result of Ref. 24 is also included. The calculation can explain the experiments qualitatively, but more detailed information on the distribution of CN's is required for more quantitative comparison.

The magnetic moment is negative (diamagnetic) and its absolute value increases as a function of the magnetic field. This overall dependence is governed by that of the magnetic moment for perpendicular magnetic field and the parallel contribution or the AB effect appears as a slight deviation. This deviation becomes clearer in the differential susceptibility. In fact, the differential susceptibility increases with the decrease of the magnetic field sharply in weak magnetic fields $(L/2\pi L)^2 \lesssim 0.2$. This is a result of the divergent paramagnetic susceptibility of metallic CN's in the parallel field, i.e., the AB effect.

6 Summary and Conclusion

Electronic properties of carbon nanotubes, in particular, electronic states, optical spectra, lattice instabilities, and magnetic properties, have been discussed theoretically based on a $\mathbf{k}\cdot\mathbf{p}$ scheme. The motion of electrons in carbon nanotubes is described by Weyl's equation for a massless neutrino, which turns into the Dirac equation for a massive electron in the presence of lattice distortions. This leads to interesting properties of nanotubes in the presence of a magnetic field including various kinds of Aharonov-Bohm effects.

Acknowledgments

The work is supported in part by Grant-in-Aid for Scientific Research from Ministry of Education, Science and Culture, Japan.

References

1. S. Iijima, *Nature (London)* **354**, 56 (1991).
2. S. Iijima, T. Ichihashi, and Y. Ando, *Nature (London)* **356**, 776 (1992).
3. S. Iijima and T. Ichihashi, *Nature (London)* **363**, 603 (1993).
4. D.S. Bethune, C.H. Kiang, M.S. de Vries, G. Gorman, R. Savoy, J. Vazquez, and R. Beyers, *Nature (London)* **363**, 605 (1993).
5. N. Hamada, S. Sawada, and A. Oshiyama, *Phys. Rev. Lett.* **68**, 1579 (1992).
6. J.W. Mintmire, B.I. Dunlap, and C.T. White, *Phys. Rev. Lett.* **68**, 631 (1992).
7. R. Saito, M. Fujita, G. Dresselhaus, and M.S. Dresselhaus, *Phys. Rev. B* **46**, 1804 (1992).
8. M.S. Dresselhaus, G. Dresselhaus, and R. Saito, *Phys. Rev. B* **45**, 6234 (1992).
9. M.S. Dresselhaus, G. Dresselhaus, R. Saito, and P.C. Eklund, in *Elementary Excitations in Solids*, ed. J.L. Birman, C. Sebenne, and R.F. Wallis (Elsevier Science Publishers B.V., Amsterdam, 1992) p. 387.
10. R.A. Jishi, M.S. Dresselhaus, and G. Dresselhaus, *Phys. Rev. B* **47**, 16671 (1993).
11. K. Tanaka, K. Okahara, M. Okada, and T. Yamabe, *Chem. Phys. Lett.* **191**, 469 (1992).
12. Y.D. Gao and W.C. Herndon, *Mol. Phys.* **77**, 585 (1992).
13. D.H. Robertson, D.W. Berenner, and J.W. Mintmire, *Phys. Rev. B* **45**, 12592 (1992).
14. C.T. White, D.C. Robertson, and J.W. Mintmire, *Phys. Rev. B* **47**, 5485 (1993).
15. H. Ajiki and T. Ando, *J. Phys. Soc. Jpn.* **62**, 1255 (1993).
16. H. Ajiki and T. Ando, *J. Phys. Soc. Jpn.* **65**, 505 (1996).
17. H. Ajiki and T. Ando, *Physica B* **201**, 349 (1994); *Jpn. J. Appl. Phys. Suppl.* **34**, 107 (1995).
18. K. Harigaya, *Phys. Rev. B* **45**, 12071 (1992).
19. K. Harigaya and M. Fujita, *Phys. Rev. B* **68**, 16563 (1993).
20. N.A. Viet, H. Ajiki and T. Ando, *J. Phys. Soc. Jpn.* **63**, 3036 (1994).
21. H. Ajiki and T. Ando, *J. Phys. Soc. Jpn.* **64**, 260 (1995).
22. H. Ajiki and T. Ando, *J. Phys. Soc. Jpn.* **62**, 2470 (1993). [Errata, *J. Phys. Soc. Jpn.* **63**, 4267 (1994)].
23. J.P. Lu, *Phys. Rev. Lett.* **74**, 1123 (1995).
24. J. Heremans, C.H. Olk, and D.T. Morelli, *Phys. Rev. B* **49**, 15122 (1994).

ELECTRONIC STATES OF SEMIMETALLIC GRAPHITE AND BISMUTH UNDER THE EXTREME QUANTUM LIMIT IN ULTRAHIGH MAGNETIC FIELDS

N. MIURA, Y. SHIMAMOTO [a], H. NAKAGAWA,
T. TAKAMASU [b] and H. NOJIRI [c]
Institute for Solid State Physics, University of Tokyo
Roppongi, Minato-ku, Tokyo 106, Japan

We present the recent experimental results of the magneto-spectroscopy in graphite and Bi under high magnetic fields up to several megagauss, generated by 3 different methods: electromagnetic flux compression up to 500T, the single turn coil technique up to 200T, and wire-wound long pulse magnets up to 50T. In the strip-line transmission spectra at a mm wave range with $B \parallel E \parallel$ c-axis, we observed a structure which is ascribed to the phase transition at around 38T in low temperatures. The phase transition was also recognized by the shift of the magneto-plasma peak near 40μm wavelength depending on the magnetic field and temperature. These observations are consistent with the existing CDW model. In the infrared magneto-spectra up to 500T, on the other hand, it was found that one of the lowest subband (0+) crosses the Fermi level in a field below 62T, whilst the subband (0-) remains below the Fermi level up to 208T. This is a puzzling result on the basis of the existing CDW model, which assumes a nesting in the (0+) band. In the DC magneto-resistance, a large monotonous increase was observed from about 60T to 100T. In Bi, an anomalous structure was observed in the mm wave transmission, and in the strip-line transmission at about 90T, corresponding to the semimetal to semiconductor transition.

1 Introduction

Bismuth and graphite are typical semimetals which exhibit various quantum phenomena in high magnetic fields. The both materials are very anisotropic, and some components of the effective mass parameters are very small, so that the extreme quantum limit is realized in high magnetic fields. We have measured infrared magneto-optical spectra, the mm wave transmission through a strip-line system and the magneto-resistance in graphite and Bi under the extreme quantum limit.

Graphite exhibits an electronic phase transition accompanying the dramatic increase of resistivity at low temperatures in high magnetic fields ap-

[a]Present address: Hitachi Ltd., Central Research Laboratory, Higashi-Koigakubo, Kokubunji, Tokyo 185, Japan
[b]Present address: Tsukuba Magnet Laboratories, National Research Institute for Metals, Sengen, Tsukuba 305, Japan
[c]Present address: Institute for Material Research, Tohoku University, Katahira, Aobaku, Sendai 980-77, Japan

plied parallel to the c-axis (perpendicular to the cleaved surface) [1–3]. The mechanism is attributed by Yoshioka and Fukuyama to the formation of the charge density wave (CDW) which is associated with the nesting in one of the lowest Landau subbands (0+) in the quantum limit, among the only 4 subbands (0 ±) and (-1 ±) [4]. According to the Yoshioka-Fukuyama (YK) theory, the phase transition should be reentrant type, and in higher fields, the normal state would reappear. Ochimizu *et al.* observed a large decrease of the resistivity in the insulating phase at a field much higher than the critical field of the phase transition, which seems to be a symptom of the reentrance to the normal state [5]. However, up to the highest field of 52T used for the transport measurement, the second transition has not been observed.

Moreover, there should be more phase transitions associated with the nesting between the subbands other than the (0+) band. Yaguchi *et al.* actually found several temperature-dependent structures in the magneto-resistance below 1K, and attributed some of them to the other phase transitions [6]. It has been a puzzling problem that while the CDW should occur in the direction of the c-axis, the resistance anomaly and its breakdown phenomena were observed in the direction perpendicular to the c-axis [3]. Recently, Yaguchi *et al.* also found that the conductivity parallel to the c-axis shows a much larger increase at the transition than in the perpendicular direction, and the transition field is close to the one which was found at lower temperatures [7]. In order to clarify the problems mentioned above, transport measurements at higher field are needed.

Bi exhibits a semimetal to semiconductor (SM-SC) transition at about 90T, when the magnetic field is applied parallel to the binary axis [8,9]. Around this transition field, the zero-gap state is realized, and there is a possibility that the excitonic insulator phase is realized [10]. Though such a SM-SC transition was actually observed in the far-infrared transmission, it has not become clear whether the excitonic phase really exists or not in Bi. Besides the SM-SC transition, it has been observed that the magneto-resistance shows a remarkable decrease above about 20T, whose origin has not been known [11].

In this paper, we study the electronic states in the both materials by means of infrared and mm wave spectroscopy and the DC magneto-resistance measurements.

2 Experimental Techniques

The high magnetic fields were produced by three different methods. The first technique is with a use of wire-wound long pulse magnets. By reinforcing a coil using glass wool and impregnation with ice, and by discharging a capacitor bank of 300kJ (5-10kV), long pulse fields up to about 50T are generated with a duration time of about 10ms [12]. Higher fields in the megagauss range are generated by the single-turn coil technique[13]. By the direct discharge of a very fast pulsed current (duration time is about 5μs) of the order of 2MA from a fast capacitor bank of 100kJ (40kV), high magnetic fields are produced before the coil destruction occurs. Although the coils are destroyed every time, samples and cryostats are not damaged, and we can repeat the experiments on the same sample. The maximum field of 150T is obtained with a bore of 10mm and 200T with a bore of 6mm. The technique to produce the highest fields is electromagnetic flux compression [13]. Using a one-turn primary coil and a liner set coaxially in the primary coil, and compressing a seed field with the liner which is squeezed by the electromagnetic force between the primary coil and the liner, we can produce very high fields up to 550T [13].

Although the duration time of the megagauss fields produced by the latter two methods is only of the order of several microseconds, fairly accurate measurements can be performed by optical means such as infrared cyclotron resonance, magneto-optical spectroscopy, or the Faraday rotation. In this work, the transmission measurements were performed for thin single crystals of graphite and Bi, in the infrared, far-infrared and mm wave range using molecular gas lasers (3 - 119μm), a carcinotron (2mm) and Gunn diodes (2 and 3mm) as radiation sources. Different detectors with very fast response time were employed for the detection.

Besides the direct transmission, we also measured the transmission through a strip-line transmission system consisting of a sample in one of the walls of the wave guide. We developed a new type of strip-line system usable in pulsed fields, which does not allow a use of any bulk metal [14]. As shown in Fig. 1, the strip-line system is made from Teflon rods with a diameter of 5mm. The central part is shaped to a thin flat plate, both sides of which are tapered continuously. The thickness of the plate D is made sufficiently smaller than the wavelength of the radiation λ. For the λ=2-3mm, D was chosen to be 0.2-0.3mm. On one of the walls of the plate, the sample is attached. The other part of the plate and the tapered part are covered with thin aluminium foils to ensure a good reflection. The radiation transmitting through the plate is reflected many times by the sample, so that the change in the conductivity of the sample is sensitively detected in the transmission signal. It was found that the use of the thin foils did not cause any problem arising from the possible Joule heating and the diamagnetic force exerted by the pulsed field. In such an arrangement as shown in Fig. 1, the attenuation coefficient is proportional to the real part of $1/\sqrt{\epsilon_{zz}}$, where ϵ_{zz} is the component of the dielectric tensor in the direction of the wave propagation (z-direction) [15]. Therefore, we can obtain information

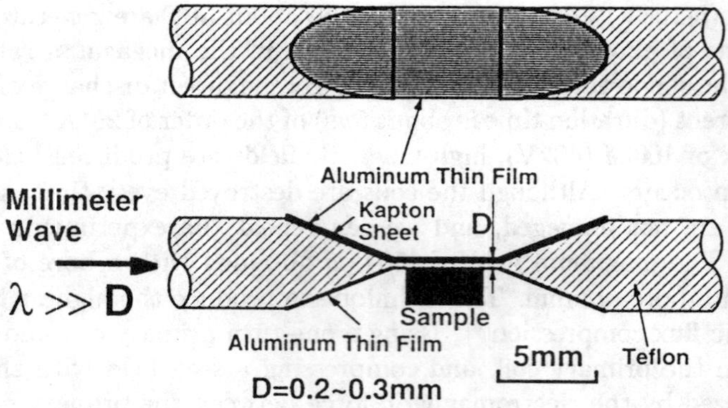

Fig.1: Design of a strip-line system usable in pulsed magnetic fields

on the conductivity parallel to the plate wall. For the direct transmission experiment on graphite, we used samples of Kish (single crystal) graphite, while we used 'super-graphite' for the strip-line experiments. The super-graphite has a considerable thickness along the c-axis, and its properties are much closer to single crystals than HOPG, so that we can mount the sample in such a way that the c-axis is oriented along the z-direction. Thus we can measure the change in conductivity along the c-axis, applying magnetic field to the same direction. This is a convenient feature for studying the phase transition. As the mm wave transmission through the strip-line system reflects the DC conductivity to some extent, we can study the non-resonant conductivity at low frequencies. In the case of Bi, we mounted the sample, so as to align the binary axis parallel to the magnetic field and the wave propagation.

In order to measure the conductivity in megagauss fields, we also developed techniques for the direct transport measurements with contacts. Generally, transport measurements are very difficult in megagauss fields. If there is a loop in the measuring circuit inside of the magnetic field, a large voltage is induced by the fast rise of the high field; for a loop of 1mm radius, the voltage is 250V at a 100T pulse, which is much higher than the voltage applied to the sample for the measurement. An attempt has been made to measure the DC magneto-resistance directly in the megagauss fields generated by the single turn coil technique [16,17]. We reduced the size of the loop as much as possible by using small samples (~1mm) and making all the lead wires exactly parallel to the field. Owing to the fact that the magnetic fields are applied to the sample many times under the same condition, we can eliminate the effect of the induced voltage to some extent, by taking the difference in the signal between the two measurements, one with and the other without the bias current. To each of the lead wires for the current and the voltage contacts, a preamplifier was connected to reduce the effect of the stray capacitance, and differential amplifiers were connected to them to measure the current and the voltage. Both the DC current and the AC current with frequencies of 10-20MHz were employed for the measurements.

3 Results and Discussion

3.1 Graphite

First, we discuss the strip-line transmission in graphite. Figure 2 shows the traces of the transmitted signal measured for super graphite in long pulse fields up to 45T in magnetic field applied parallel to the c-axis, at a frequency of 146GHz (λ=2.05mm). The signal is considered to correspond to the conductivity in the c-axis. The transmission first decreases abruptly with increasing field corresponding to the increase of the magneto-resistance up to about 10T. At a temperature of 4.2K, the transmission shows a drastic decrease at 38T. This is considered to be due to the phase transition. Figure 3 shows the transmission spectra at 100GHz (λ=3.0mm) up to 100T generated by the single turn coil technique. The spectra up to 40T are consistent with those shown in Fig. 2. At 5.1T and 9.8T, sharp structures are observed at 42T and 50T, respectively. These correspond to the phase transition. In the fields higher than these transition fields, we can see no significant structure, which is expected to appear based on the reentrant phase model. The reason of the absence of the reentrant transition may be due the temperature rise of the sample caused by the Joule heating. The temperature rise in the pulsed fields is estimated to be about 5K. At higher temperatures a broader structure is observed at around 55T. This is probably due to the Shubnikov-de Haas peak at a field where the (0+) level crosses the Fermi level.

Next, we present the result of the infrared absorption due to the magneto-plasma effect in single crystal graphite. The reflection and absorption shows a large structure when the dielectric constant crosses zero due to the magneto-plasma effect [18]. The reflection takes the minimum at the crossing point while the transmission takes the maximum. The electronic transition between the (0$\pm$) and (-1$\pm$) subbands gives the dominant contribution to the structure. As the photon energy increases, the extremum shifts to the higher field. We measured the magneto-transmission spectra at photon energies for which the plasma peak appears near the phase transition. Figure 4 shows the transmis-

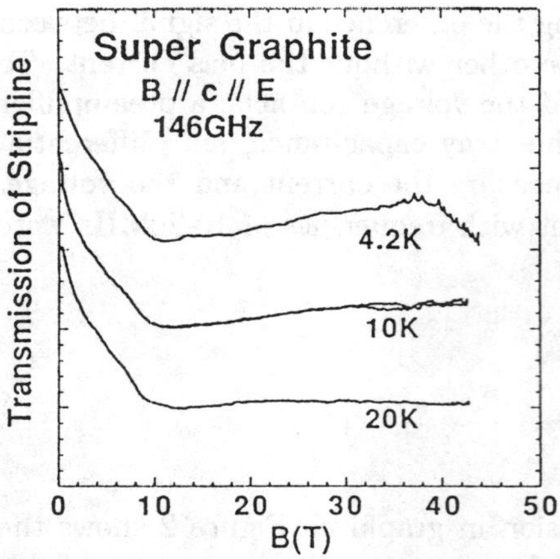

Fig.2: Strip-line transmission signal for super graphite in long pulse fields up to 45T. The frequency of the radiation is 146GHz. $B \parallel$ c. $E \parallel$ c.

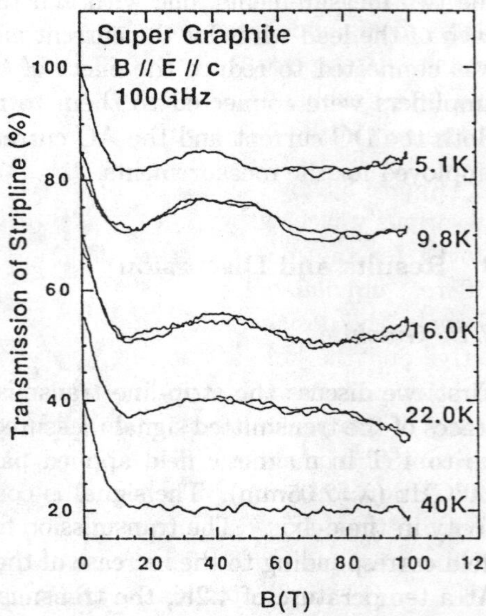

Fig.3: Strip-line transmission signal for super graphite up to 100T generated by the single turn coil technique. The frequency of the radiation is 100 GHz. $B \parallel$ c. $E \parallel$ c.

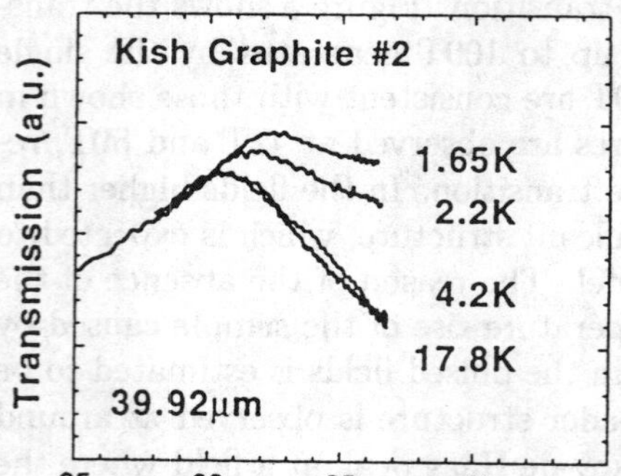

Fig.4: Transmission spectra at a wavelength of 39.92μm showing the magnetoplasma peak at varous temperatures. $B \parallel$ c. $E \perp$ c.

sion spectra at a wavelength of 39.92μm. A large peak at 30-35T corresponds to the zero cross of the dielectric constant. The peak clearly shows a shift to higher fields and broadening as the temperature is lowered across the phase transition temperature. This indicates that the transition between the (0$\pm$) and (-1$\pm$) subbands is suppressed in the CDW state. suggesting the opening of the gap in the (-1$\pm$) subband.

Figure 5 shows the magneto-absorption spectra in single crystal graphite at various wavelengths. The electromagnetic flux compression system was employed for the measurements at 5.527μm and 3.39μm [19]. Two traces with different circular polarizations are shown on the both sides of the B=0 point. Namely, the positive and negative magnetic fields correspond to the electron-active and the hole-active polarizations, respectively. Several prominent absorption peaks are observed as well as many small peaks in the low field region. Many of the peaks are similar in shape just shifting to higher fields with increasing photon energy. We compared the experimental data with a theoretical calculation following Nakao's treatment [20] based on the Slonczewski-Weiss-McClure model [21,22]. Considering the possible electronic transition for each polarization, we can assign the peaks at high fields as listed in Table 1, where (0$\pm$) $\rightarrow$ (1$\pm$) indicates an electronic transition from the transition from the (0$\pm$) subband to (1$\pm$) subband, *etc.*, as shown in Fig. 6 for a field of 50T. The resonant fields and the line shape were compared with the calculation. It was found that they are reasonably in good agreement with the calculation. As for the absorption intensity. However, it was necessary to assume that the hole concentration is increased by about 10% from the neutral condition, as was first pointed out by Nakamura *et al.* [23]. Figure 7 shows the experimental points and the theoretical lines of the resonant photon energy as a function of magnetic field. However, there are some disagreement between theory and experiment. The discrepancy is particularly large at high fields. Furthermore, the strength of the transition indicates several noteworthy points. The fact that the (0$\pm$, 1$\pm$) transition is observed at 62T shows that the (0+) subband at the K-point crosses the Fermi level E_F in a field below 62T. On the assumption that the hole concentration is 10% larger than the electron concentration, the crossing field is estimated to be 54T, being consistent with the above observation. However, this is in contradiction with previous theoretical prediction that the (0+) level should cross E_F at $\sim$65T and with the YF theory which presumes the (0+) subband containing carriers up to 60T. It is therefore likely that the nesting in the (-1-) band may be the origin of the CDW state [24]. On the other hand, the observation of the (0-, 1-) transition at 208T indicates that the (0-) subband remains still below E_F. Considering $g \sim 2$ in graphite, the spin Zeeman energy $g\mu_B H$ is not so large that the difference in the Fermi level

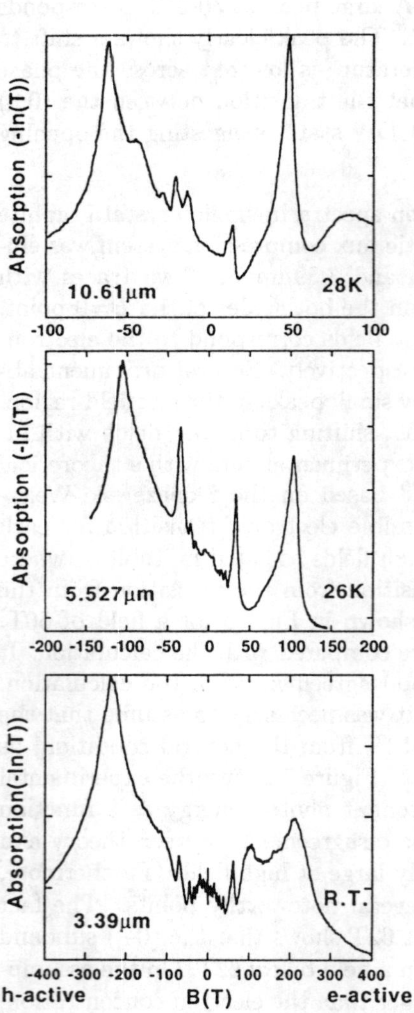

Fig.5: Magneto-absorption spectra in single crystal graphite at three different wavelengths. The electromagnetic flux compression system was employed for the measurements at 5.527μm and 3.39 μm.

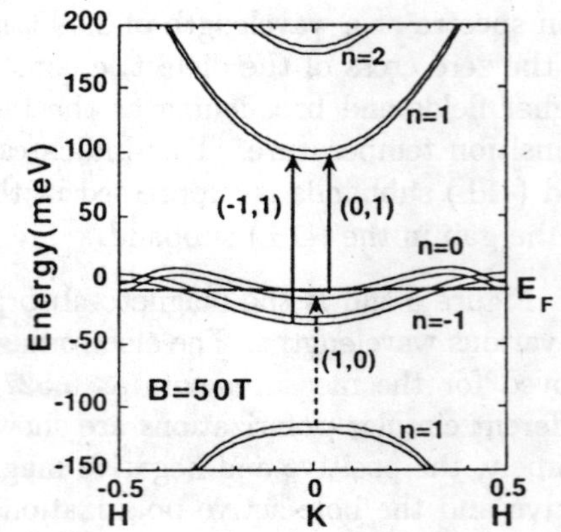

Fig.6: Landau subbands and allowed transition in graphite when the magnetic field of 50T is applied to the c-axis.

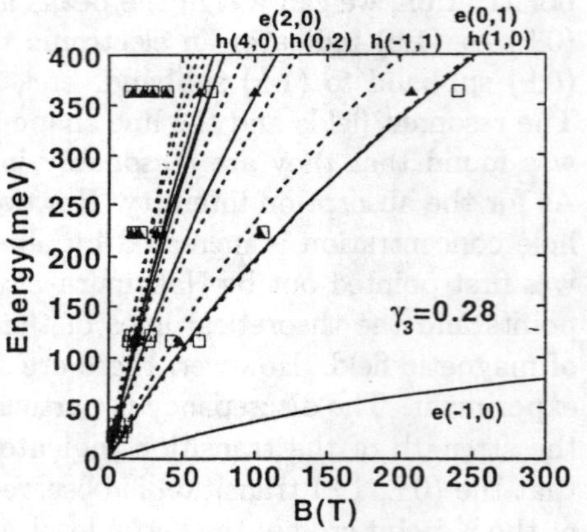

Fig.7: Resonant photon energy as a function of magnetic field for graphite. The solid and broken lines denote the calculated lines for the electron-active and hole-active polarizations, respectively. The corresponding experimental data for each polarization are represented by closed triangular points and open square points.

Table 1: Resonance fields and their assignment for the principal resonance absorption lines in graphite at various wavelengths.

Polarization	Wavelength μm	Resonant field T	Transition
e-active	10.61	48	$(0\pm) \rightarrow (1\pm)$
h-active	10.61	43	$(-1\pm) \rightarrow (1\pm)$
h-active	10.61	62	$(1+) \rightarrow (0+)$
e-active	5.527	103.5	$(0-) \rightarrow (1-)$
h-active	5.527	105.8	$(1+) \rightarrow (0+)$
e-active	3.39	208	$(0-) \rightarrow (1-)$
h-active	3.39	240	$(1+) \rightarrow (0+)$

crossing fields between the two spin split subbands is too large. Therefore, the reason why the (0-) level is pinned below E_F is still an open question.

Figure 8 shows the DC magneto-resistance measured up to 100T using the single turn coil technique. Although one shot signal involves a lot of noise due to the large dB/dt and the triggering pulse, we can eliminate most of the noise by repeating a blank shot under the same condition except the bias current through the sample set to be zero, and subtracting it from the signal with the bias current [16]. In the low field region, there is a hysteresis between the traces during the up and down sweeps of the field. This is because of the fast rise of the field and a temperature rise of the sample. Nevertheless, we can obtain a good agreement between the two traces, which indicates the reliability of the measurement. The decrease of the magneto-resistance from about 40T, which we observed in the long pulse field experiment [5], was not observed. On the contrary, it was found that the magneto-resistance increases gradually from about 60T. The difference between the long pulse field data and the present one at 40 - 60T is probably due to the temperature rise in the present experiment. The temperature rise when the field is swept from zero to 100T (in 2μs) is estimated to be about 20K. If the (0+) level crosses E_F at about 54T, as mentioned above, the increase from 60T is interpreted as the Shubnikov-de Haas effect at the crossing field. The temperature rise may also explain the reason why we have not observed any further phase transitions in the high field region. Apart from the temperature rise, this experiment demonstrates that the DC transport experiment can be made with a reasonable precision up to about 100T using the single turn coil technique.

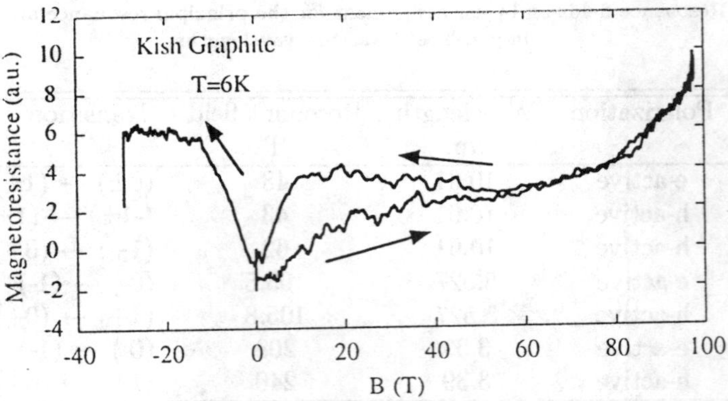

Fig.8: DC magneto-resistance measured for single crystal graphite up to 100T using the single-turn coil technique. $B \parallel$ c.

3.2 Bi

The non-resonant high frequency conductivity was measured in Bi by means of the strip-line transmission. Figure 9 shows the transmission as a function of magnetic field up to 100T at 13K and 9.1K. The magnetic field was applied parallel to the binary axis. The transmission decreases with increasing field below about 20T. This corresponds to the increase of the magneto-resistance. A change corresponding to the decrease of the magneto-resistance from about 30T is only vaguely seen in the figure. A small but clear kink is observed at around 85T. This kink corresponds to the semimetal to semiconductor transition observed in the previous far-infrared transmission experiments [8]. The anomaly at around 85T was also observed in the direct transmission of mm waves at wavelengths of 2 and 3mm. The transmission of the far-infrared and mm waves shows different spectra among different samples, depending on the thickness, temperature and sample mounting. This is due to the interference effect. An oscillatory change due to the Alphen wave interference was observed up to high magnetic field at a wavelength of 119μm. The oscillation was found to disappear at around 80T [14], which is another indirect evidence of the semimetal to semiconductor transition. The excitonic insulator phase has not been observed yet. However, the high field state in Bi would still be an interesting subject in ultra-high magnetic fields.

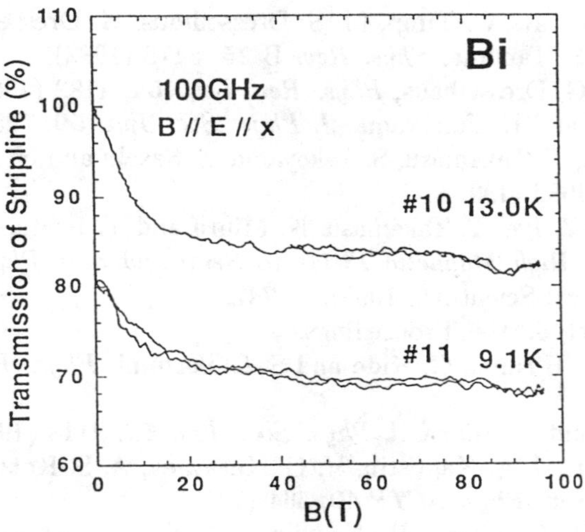

Fig.9: Strip-line transmission signal for two samples of Bi up to 100T. The frequency of the radiation is 100GHz. $B \parallel$ the binary axis.

4 Conclusion

We have measured magneto-absorption and strip-line transmission of infrared, far-infrared and mm waves in graphite and Bi under high magnetic fields in the megagauss range. We found that the plasma peak is modified by the phase transition. In the cyclotron resonance in very high fields, it was found that the (0+) level crosses the Fermi level below 64T, while the (0-) level remains below the Fermi level even at 208T. The present results suggest that the (0-) level rather than the (0+) level is responsible for the phase transition. The DC magneto-resistance was directly measured up to 100T, and the Shubnikov-de Haas peak was observed at around 60T. In Bi, the field induced semimetal to semiconductor transition was observed at about 85T both in the magneto-absorption and the strip-line transmission.

References

1. S. Tanuma, R. Inada, A. Furukawa, O. Takahashi and Y. Iye, in *Physics in High Magnetic Fields*, eds. S. Chikazumi and N. Miura (Springer Verlag, Berlin, 1981) p. 316.

2. Y. Iye, P. Tedraw, G. Timp, M. S. Dresselhaus, G. Dresselhaus, A. Furukawa and S. Tanuma, *Phys. Rev.* B **25**, 5478 (1982).
3. Y. Iye, and G. Dresselhaus, *Phys. Rev. Lett.* **54**, 1182 (1985).
4. D. Yoshioka and H. Fukuyama, *J. Phys. Soc. Jpn.* **50**, 725 (1981).
5. H. Ochimizu, T. Takamasu, S. Takeyama, S. Sasaki and N. Miura , *Phys. Rev.* B **46**, 1986 (1992).
6. H. Yaguchi, Y. Iye, T. Takamasu, N. Miura and T. Iwata, in *Proc. 11th Int. Conf. High Magnetic Fields in Semiconductor Physics*, ed. D. Heiman (World Scientific., 1995), p. 742.
7. H. Yaguchi et al., this Proceedings.
8. N. Miura, K. Hiruma, G. Kido and S. Chikazumi, *Phys. Rev. Lett.* **49**, 1339 (1982).
9. K. Hiruma and N. Miura, *J. Phys. Soc. Jpn.* **52**, 2118 (1983).
10. N. B. Brandt, G. A. Kapustin, V. G. Karavaev, A. S. Kotosonov and E. A. Svistva, *Sov. Phys. JETP* **40**, 564 (1975).
11. N. Miura, R. G. Clark, R. Newbury, R. P. Starrett and A. V. Skougarevsky, *Physica* B **194-196**, 1191 (1994).
12. S. Takeyama, H. Ochimizu, S. Sasaki and N. Miura, *Meas. Sci. Technol.* **3**, 662 (1992).
13. N. Miura, *Physica* B **201**, 40 (1994).
14. Y. Shimamoto, Ph. D Thesis (University of Tokyo, 1996).
15. M. von Ortenberg, in *Infrared and Millimeter Waves*, ed. K. J. Button (Academic Press, New York, 1980), vol. 3, p. 275.
16. T. Takamasu, H. Nakagawa and N. Miura , *Physica* B **216**, 362 (1996).
17. H. Nakagawa, Master Thesis (University of Tokyo, 1996)
18. K. Nakamura, G. Kido and N. Miura, *Solid St. Commun.* **47**, 349 (1983).
19. Y. Shimamoto, H. Nojiri and N. Miura, in *Proc. 11th Int. Conf. High Magnetic Fields in Semiconductor Physics*, ed. D. Heiman (World Scientific., 1995), p. 746.
20. K. Nakao, *J. Phys. Soc. Jpn.* **40**, 761 (1976).
21. J. C. Slonczewski and P. R. Weiss, *Phys. Rev.* **109**, 272 (1958).
22. J. W. McClure, *Phys. Rev.* **108**, 612 (1957).
23. K. Nakamura, T. Osada, G. Kido, N. Miura and S. Tanuma, *J. Phys. Soc. Jpn.* **52**, 2875 (1983).
24. K. Sugihara, *Phys. Rev.* B **29**, 6722 (1984).

NON-OHMIC TRANSPORT BEHAVIOR IN A NEW MAGNETIC FIELD INDUCED INSULATING PHASE OF GRAPHITE

H. YAGUCHI, T. TAKAMASU, Y. IYE, N. MIURA
Institute for Solid State Physics, University of Tokyo, Roppongi, Minato-ku, Tokyo 106 Japan

We have measured the longitudinal magnetoresistance ρ_{zz} of graphite in high magnetic fields up to ~37 T at low temperatures down to ~0.6 K. A drastic increase in ρ_{zz} was newly found in the high field region. We also found prominent non-ohmicity in the current-voltage characteristic in the relevant magnetic field range. This non-ohmic behavior which is reminiscent of sliding motion of a pinned charge density wave or spin density wave suggests that the direction of the density wave is mainly parallel to the *c*-axis. The newly found anomaly in ρ_{zz} could be understood as a magnetic field induced $2k_F$-type density wave along the *c*-axis.

1 Introduction

The electron system in graphite is known to undergo an electronic phase transition.[1] It manifests itself as an abrupt increase in the transverse magnetoresistance ρ_{xx} in a high magnetic field applied parallel to the *c*-axis. The onset field is strongly dependent on temperature (*e.g.* the onset field is ~27 T and ~34 T at 1.5 K and 4.2 K, respectively.), which suggests that the transition is due to some kind of many body effect.[2, 3] Yoshioka and Fukuyama (YF)[4] ascribed this transition to a $2k_F$-type charge density wave (CDW) or spin density wave (SDW) instability of a one dimensional electron system created by a strong magnetic field. Most of the experimental studies made so far qualitatively support the theoretical interpretation proposed by YF. However, Iye and Dresselhaus[5] found non-linear transport behavior in ρ_{xx}. Although this result is reminiscent of sliding motion of a pinned CDW or SDW,[6] it suggests that the direction of density wave is mainly perpendicular to the *c*-axis. In spite of the theoretical interpretation of a CDW or SDW state along the *c*-axis, no attempts to investigate the longitudinal magnetoresistance ρ_{zz} have been made so far. In the present work, we have measured ρ_{zz} in high magnetic fields up to ~37 T at low temperatures down to ~0.6 K. At higher fields, a new striking anomaly was found. We also investigated the current-voltage characteristic in the relevant magnetic field range and found prominent non-ohmic behavior.

2 Experimental

The samples used in the present study were single crystal graphite flakes (Kish

Graphite). The dimensions of samples were approximately 0.5 mm x 1.0 mm x 0.2 mm. The electrical contacts were made with silver paint. High magnetic fields were generated using a non-destructive pulsed magnet. The maximum field was ~ 37 T and the pulse duration was ~ 10 ms. Low temperatures were achieved using ^{4}He and ^{3}He inserts. The resistance as a function of magnetic field was measured by a standard four probe method, with an alternating current of 0.5-1.0 mA at 200 kHz. The current-voltage characteristic was investigated applying a triangular shaped current at 2-10 kHz at the peak of the nearly sinusoidal magnetic field pulse. The variation of the magnetic field was no more than 1% during the current sweep.

3 Results and Discussion

Figure 1 shows the longitudinal magnetoresistance ρ_{zz} at various temperatures. A sudden increase, indicated as α, is seen the field corresponding to the onset of the transition observed in the transverse magnetoresistance ρ_{xx}. It has been considered to be a field induced CDW or SDW state along the *c*-axis due to the mechanism mentioned in the introduction.[4] However this interpretation does not fit to the result of non-ohmic transport behavior in the plane-conductivity which suggests that the direction of the density wave is perpendicular to the *c*-axis.[4] (We also reproduced this non-ohmic behavior.) In higher fields below ~1.4 K, there appears another novel feature: ρ_{zz} starts to increase gradually at a field slightly higher than the onset of α, indicated by an arrow in Fig. 1 and 2, and then shows a steep increase. This is probably also due to some kind of many body effect because of its strong temperature dependence. Figure 2 shows the traces of the longitudinal magnetoresistance ρ_{zz} in an expanded scale. At 0.6 K, ρ_{zz} begins to increase at ~26 T and ρ_{zz} at 36 T is about ten times as large as that at 26 T. This feature may correspond to β (inset of Fig. 2) seen in the transverse magnetoresistance.[3] We measured longitudinal magnetoresistance on two samples. The result of the other sample is basically identical to the first.

Figure 3 shows the current-voltage characteristic for the longitudinal magnetoresistance ρ_{zz} at 1.0 K. Prominent non-ohmicity is seen above ~30 T which corresponds to the onset field of the new anomaly in ρ_{zz}. Figure 4 shows the magnetic field dependence of the threshold field for non-ohmicity. At temperatures of 0.7 K and 1.4 K, similar non-ohmicity in current-voltage characteristic is seen above the onset field of the new anomaly. The threshold fields are distributed in the range of 0.2-2 V/cm. These facts suggest that the non-ohmicity is associated with the anomaly in ρ_{zz}.

The possibility of Joule heating is ruled out: The current-voltage characteristic does not depend on the sweep frequency of the triangular shaped current or the peak current applied. That of the Zener tunneling is also excluded as a possible

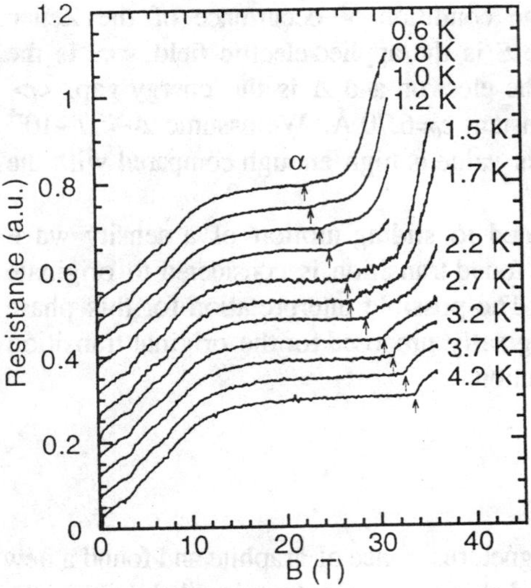

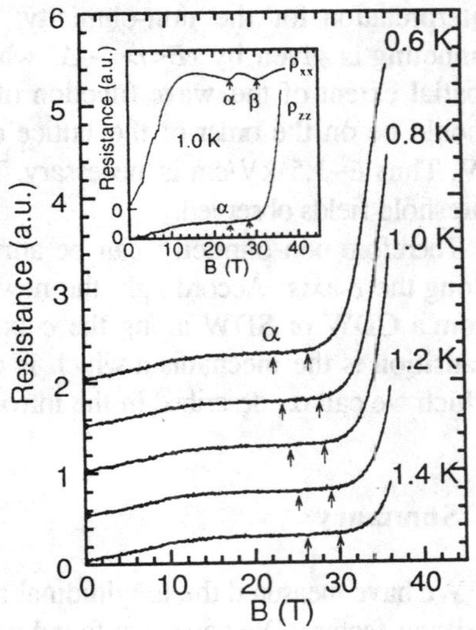

Fig. 1. (left) The longitudinal magnetoresistance as a function of magnetic field at various temperatures. Each trace is shifted vertically for visual clarity.

Fig. 2. (right) The longitudinal magnetoresistance as a function of magnetic field at ^{3}He temperatures. Each trace is shifted vertically for visual clarity. Inset : The comparison of ρ_{xx} and ρ_{zz} at 1.0 K. Note that the sample used for each measurement is different.

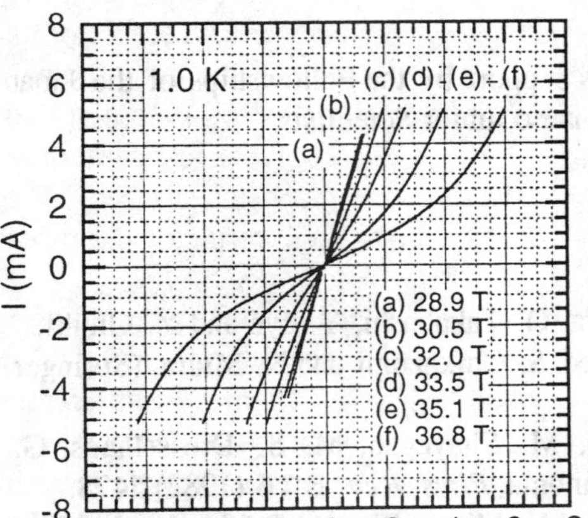

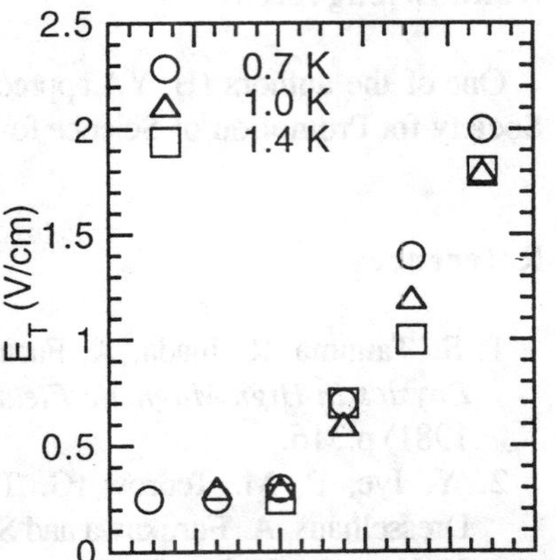

Fig. 3. (left) The current-voltage characteristic in ρ_{zz} at 1.0 K.

Fig. 4. (right) Threshold field as a function of magnetic field at 0.7 K, 1.0 K and 1.4 K.

interpretation for the non-ohmicity: The condition of occurrence of the Zener tunneling is given by $eE<r> \sim \Delta$, where E is the applied electric field, $<r>$ is the spatial extent of the wave function of the electron and Δ is the energy gap. $<r>$ should be on the order of the lattice constant c_0=6.70 Å. We assume $\Delta \sim k_B T \sim 10^{-4}$ eV. Thus $E \sim 1.5$ kV/cm is necessary. This value is high enough compared with the threshold fields observed.

Therefore non-ohmicity can be attributed to sliding motion of a density wave along the c-axis. Accordingly the newly found transition is considered to originate from a CDW or SDW along the c-axis. The possible interpretation for this phase transition is the mechanism which is originally proposed for the original transition which we call α, described in the introduction.

4 Summary

We have measured the longitudinal magnetoresistance of graphite and found a new striking feature. We have also found remarkable non-ohmic transport behavior in the current-voltage characteristic, which is reminiscent of depinning of a CDW or SDW. These experimental results can be explained in the framework of YF theory which was proposed for the original transition α.

Acknowledgement

One of the authors (H. Y.) appreciates support by the Fellowships of the Japan Society for Promotion of Science for Japanese Junior Scientists.

References

1. S. Tanuma, R. Inada, A. Furukawa, O. Takahashi, Y. Iye and Y. Onuki, "*Physics in High Magnetic Fields*", ed. S. Chikazumi and N. Miura, (Springer, 1981) p.316.
2. Y. Iye, P. M. Tedrow, G. Timp, M. Shayegan, M. S. Dresselhaus, G. Dresselhaus, A. Furukawa and S. Tanuma, *Phys. Rev.* B **25** (1982) 5478.
3. H. Yaguchi, Y. Iye, T. Takamasu and N. Miura, *Physica* B **184** (1993) 332.
4. D. Yoshioka and H. Fukuyama, *J. Phys. Soc. Jpn.* **50** (1981) 725.
5. Y. Iye and G. Dresselhaus, *Phys. Rev. Lett.* **54** (1985) 1182.
6. G. Grüner, "*Density Waves in Solids*" (Addison-Wesley, 1994).

HIGH MAGNETIC FIELD INDUCED CDW-TO-NORMAL PHASE TRANSITION IN η-Mo_4O_{11}

M. INOUE, M. SASAKI, Y. HARA, H. NEGISHI
Department of Materials Science, Faculty of Science, Hiroshima University, Higashi-Hiroshima 739, Japan

T. TAKAMASU, N. MIURA
Institute for Solid State Physics, University of Tokyo, Tokyo 106, Japan

G. MACHEL, M. VON ORTENBERG
Humboldt-Universität zu Berlin, Invalidenstraße 110, D-10115 Berlin, Germany

Real and imaginary components of ac resistivity and Hall effect of CDW material η-Mo_4O_{11} have been measured over the frequency range 50-500 kHz in pulsed magnetic fields up to 38 T. The observed frequency dependent magnetotransport properties are discussed by considering the characteristic nature of pinned and depinned CDW condensates. Numerical simulations are made to give a satisfactory agreement with the observations.

1 Introduction

Quasi-two-dimensional conductor η-Mo_4O_{11} undergoes charge-density-wave (CDW) transitions at T_{c1}=105 K and T_{c2}=35 K due to partial nesting of the anisotropic Fermi surfaces of electrons and holes. In the CDW states, some of the electrons and holes are condensed into a collective mode as "CDW condensates" and others remain as "normal carriers" contributing to the dc transport.[1] Magnetotransport properties under high magnetic fields have anomalous behaviors, in which the magnetoresistance and Hall resistivity show a characteristic hysteresis when the field sweep exceeds a threshold value H_c ($\approx$ 10 T).[2] Such behaviors are expected to be CDW-related phenomena, but their origin is still uncertain.

For further understanding of the unusual magnetotransport, we have carried out similar dc and ac resistivity along the b-axis and Hall effect measurements for chemical-vapor-grown single crystals of η-Mo_4O_{11} under pulsed magnetic fields applied along the a-axis up to 38 T at 4.2 K; in the ac method a phase detection technique was used over the frequency range 50-500 kHz.

2 Results and Discussions

Figure 1 shows the magnetic field dependence of (a) the real part Re$\Delta\rho$ and (b) imaginary one Im$\Delta\rho$ of ac magnetoresistance change at different frequencies. The real components are positive and increase reversibly as the magnetic field is varied

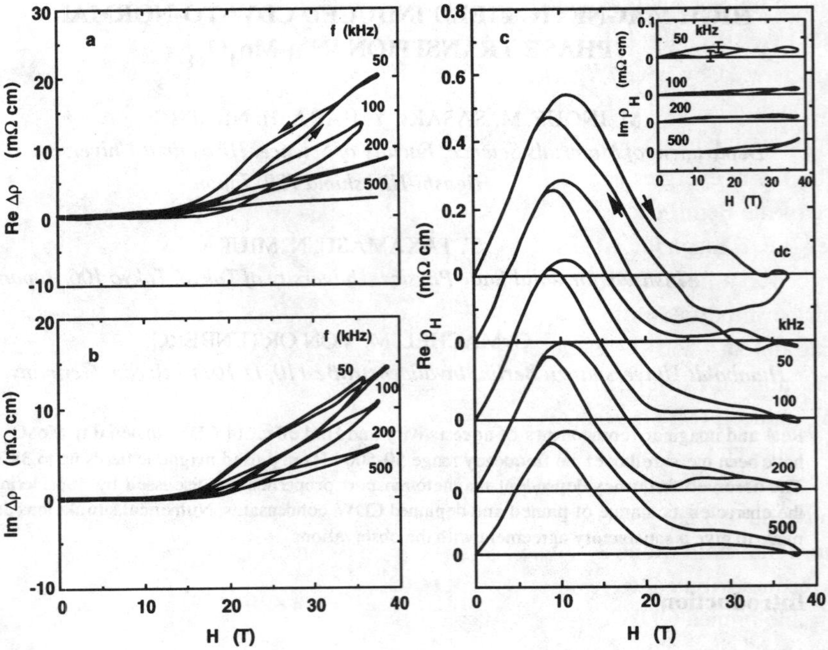

Fig. 1. (a) Real part and (b) imaginary one of the transverse magnetoresistance $\Delta\rho$ and (c) real component of the Hall resistivity ρ_H at various frequencies; inset shows the imaginary part of ρ_H.

up to a characteristic field H_c ($\approx$ 10 T), while the imaginary ones are zero. Once the sweep range exceeds H_c, a hysteresis appears, where the value for up-sweep is always lower than that for down-sweep, which is similar to the dc data reported.[2] The magnitudes of Re$\Delta\rho$ and Im$\Delta\rho$ decrease appreciably with increasing frequency, and the characteristic hysteresis in Re$\Delta\rho$ disappears at high frequency (500 kHz); the Cole-Cole plot of Im$\Delta\rho$ vs. Re$\Delta\rho$ shows a monodisperse curve with a characteristic frequency of ~120 kHz. In Fig. 1(c) are shown the results of the dc and ac Hall resistivity ρ_H; the imaginary part of ρ_H is shown in the inset. As in the magnetoresistance, the real component of the Hall resistivities also exhibits the hysteresis for field sweeps above H_c and tends to level off at higher fields, while the imaginary parts (inset) are very small and less frequency dependent, which contrast to the ac magnetoresistance.

Based on a theoretical band model[3] and transport data of η-Mo_4O_{11}, in earlier work we have proposed the possible nesting and band scheme,[1] the basic idea of which is as follows; the carrier transport in the CDW states are primarily due to two types of hole bands (denoted by h_3- and h_4-bands) and of electron bands (e_1- and e_5-bands) (see Ref. 1). In addition, the carriers that are condensed into the CDW are characterized by a CDW-band (e_6- band) with an energy gap $\Delta_2(T)$ at the Fermi level and only thermally activated carriers contribute to the conductivity. At an appropriate high magnetic field the Zeeman-splitting of the above bands is comparable to the Fermi energies and $\Delta_2(T)$, which leads to a significant magnetic field dependence of the Fermi energies of the up- and down-spin bands, and further the gap vanishes (i.e. a magnetic field driven destruction of the CDW state); the hysteresis is ascribed to a field dependent pinning and depinning process. Here the above model is further modified as follows; provided that the band edges of the 2D nested or vanishing hole h_5- and the electron e_6-bands have a large g-value, these carriers will become normal carriers in high magnetic fields [which we refer to as the magnetic field induced (MI) h_5- and e_6-bands]. This effect may be referred to as "magnetic field induced CDW-to-normal phase transition".

Furthermore, we should note that in the CDW state below T_{c2}, a linear response due to normal carriers is observed when an external electric field E is less than a threshold value E_T, while a nonlinear response appears when $E > E_T$,[4] which has been attributed to two delayed processes, a deformation process of highly polarizable pinned CDW condensates that are screened by surrounding normal carriers and a collective sliding motion of two kinds of CDW condensates formed at T_{c1} and T_{c2}. Experimentally, we have found the appreciable frequency-dependent magnetotransport for a higher magnetic field above H_c (Fig. 1). According to Grüner,[5] the threshold field E_T is defined by $eE_T\lambda = \phi_p$, where λ is a one CDW period and ϕ_p the pinning potential. In an ac electric field with angular frequency ω, the pinning potential is effectively reduced to $\phi(\omega) = \phi_p - (1/2)K\lambda^2 = \phi_p - (1/2)m^*\omega^2\lambda^2$, where K is a restoring force constant and m^* an effective mass of the CDW condensates; the pinning frequency ω_0 is defined by $\phi(\omega_0)=0$ or $\phi_p = (1/2)m^*\omega_0{}^2\lambda^2$. With this picture and the observed frequency-dependent magnetotransport, we may consider that in the presence of a high magnetic field the effective pinning potential is reduced by the Zeeman energy, or $\phi(\omega, H) = \phi_p - (1/2)m^*\omega^2\lambda^2 - g^*\mu_B H$, where g^* is an effective g-value of the CDW and μ_B the Bohr magneton; the critical magnetic field H_c is defined by $\phi(\omega, H_c) = 0$. Thus it is inferred that when the applied magnetic field exceeds H_c, the CDW sliding motion occurs; in other words, when a thermal energy of a CDW phason is larger than the effective potential energy, or $k_BT > \phi(\omega, H > H_c)$, then the CDW condensates undergo translational motion, contributing to the carrier current.

Using the above model, we have computed the dc and ac resistivity and Hall resistivity (real and imaginary components) as a function of magnetic field H and ac frequency ω at 4.2 K. Typical results of numerical computations are shown in Fig. 2, in satisfactory agreement with the observations. Here the oscillatory behaviors (Schubnikov de Haas oscillations)[2] oberved in the magnetoresistance and Hall resistivity have been neglected.

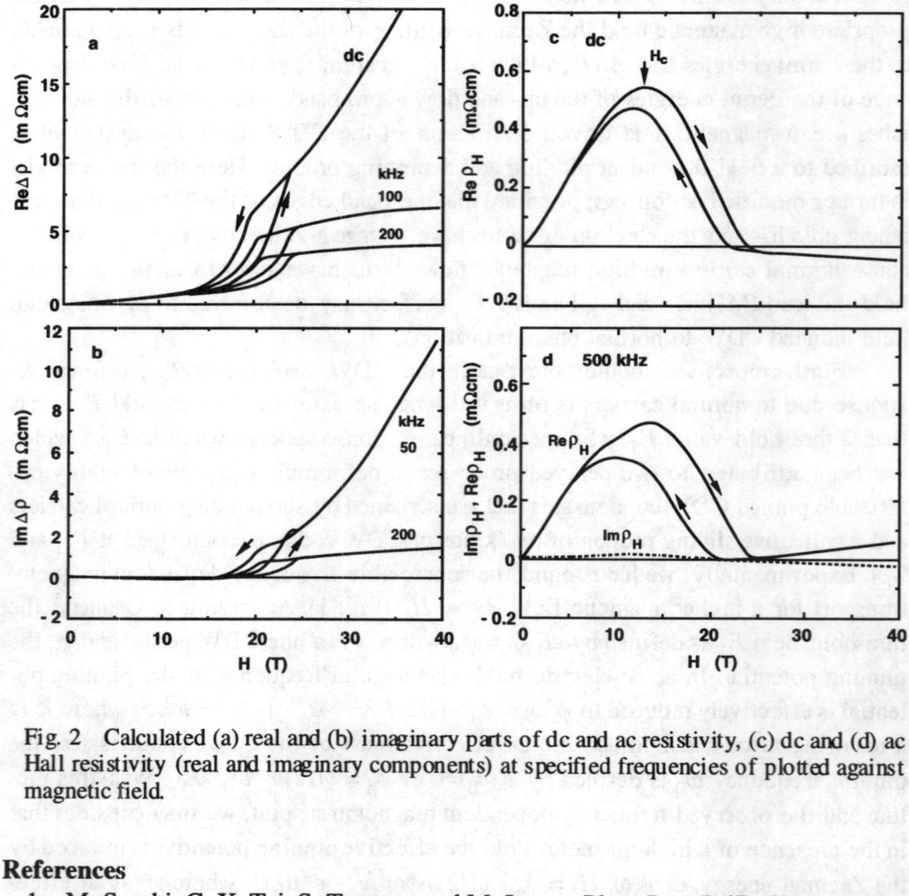

Fig. 2 Calculated (a) real and (b) imaginary parts of dc and ac resistivity, (c) dc and (d) ac Hall resistivity (real and imaginary components) at specified frequencies of plotted against magnetic field.

References

1. M. Sasaki, G. X. Tai, S. Tamura, and M. Inoue, *Phys. Rev. B* **47**, 6216 (1993).
2. M. Inoue, G. Machel, I. Laue, M. von Ortenberg, and M. Sasaki, *phys. stat. sol.* (b). **172**, 431 (1992).
3. E. Canadell, M.-H. Whangbo, C. Schlenker, and C. Escribe-Filippini, *J. Inorg. Chem.* **47**, 6216 (1989); E. Canadell and M.-H. Whangbo, *Chem. Rev.* **91**, 965 (1991).
4. M. Sasaki, G. Houzaki, M. Koyano, and M. Inoue, *J. Phys. Chem. Solids* **57**, 281 (1996).
5. G. Grüner, *Rev. Mod. Phys.* **60**, 1129 (1988).

ELECTRON TRANSPORT IN STRAINED Si/SiGe AT HIGH MAGNETIC FIELDS

KHALID ISMAIL[a,b]

[a] *IBM T.J. Watson Research Center, Yorktown Heights, NY 10598, USA*

[b] *Faculty of Engineering, Cairo University, Giza, Egypt*

Modulation-doped Si/SiGe heterostructures, where electron transport is confined to a strained Si quantum well is used as test vehicle to examine transport properties in a valley-degenerate system at high magnetic fields. In samples with mobility in excess of 3 x 10^5 cm^2/Vs, first the g-factor and the valley splitting are determined using tilted field experiments. The values found differ from those in bulk Si inversion layers. The integer and fractional quantum Hall effects are observed and show some similarities and some striking differences when compared to the more studied GaAs/AlGaAs heterostructure system. In particular, the disappearance of the $\nu = 5/3$ filling factor, the observation of $\nu = 1/2$ filling factor in a single quantum well, and the observation of a non-reentrant insulator transition in a specific density range and at high magnetic fields, are all exciting observations which may shed some light on the quantum transport theory in valley-degenerate materials.

Recent advances in the epitaxial growth of high quality Si/SiGe heterostructures have attracted significant attention in the past few years. This has allowed the study of fundamental transport properties in such heterostructures and the exploration of their use in device applications.[1] This material system has been exploited in order to achieve very high electron mobility at low temperature[2], which is more than an order of magnitude higher than in Si MOSFET structures.

Despite the fact that our highest mobility in strained Si (5.8 x 10^5 cm^2/Vs) is still much lower than that achieved in GaAs/AlGaAs modulation-doped samples, Si remains of great interest for studying quantum transport. It differs significantly from GaAs in terms of various aspects: first, the in-plane electron effective mass is about three times higher than in GaAs. Second, its perpendicular effective mass (split by the strain and quantization) is 14 times larger than in GaAs. This has the direct consequence of much stronger confinement of the wavefunction. Third, the ground state in Si consists of a two-fold valley degenerate band, while in GaAs there is only one band. Fourth, the Zeeman energy splitting in Si is much larger than in GaAs.

Various groups have reported interesting transport measurements in Si MOSFET structure in order to investigate the above effects.[3-5] Unfortunately, that material system is limited in terms of the electron mobility (2 - 4 x 10^4 cm^2/Vs) due to the strong scattering at the Si/SiO_2 rough interface. Such scattering results in disorder broadening of energy levels and obscures various interesting phenomena requiring high purity material.

We here report on electron transport experiments using samples grown by ultra-high vacuum chemical vapor deposition (UHV-CVD) with mobility in the range of 3.2 - 5.2 x 10^5 cm^2/Vs. The spin and valley splitting are estimated from tilted field experiments, and the integer and fractional quantum Hall effects are investigated in the temperature range of 0.4 - 4.2 K under a magnetic field of up to 35 T.

Material Characterization

The strained Si/SiGe material is processed in a similar way to Di devices. The ohmic contacts are formed by ion implantation of Phosphorous ions, amorphization, and solid phase regrowth at 600 °C. Hall patterns are defined using shallow trench reactive ion etching. Schottky metal is evaporated on the surface to form Schottky contacts. On the back of the wafer, Al is evaporated to form an ohmic contact to the p-substrate. Both front an backgate are used to tune the electron density in our samples.[6] The transport measurements are performed in a He3 cryostat in the temperature range of 0.4 - 4.2 K. AC lockin techniques are used in the measurements except in few cases where a dc measurement was performed. Magnetoresistance measurements reveal Shubnikov deHaas oscillations starting at 0.2 T in samples with ungated mobility in the range of 3 - 4 x 10^5 cm^2/Vs. Spin and valley splitting are resolved above 1 and 2 T, respectively.

Spin and Valley Splitting

Before describing the integer and fractional QHE measurements, it is mandatory to have an estimate of the spin and valley splitting in this material. The assumption that the values measured in Si inversion layers are valid for our system is unjustified because of various essential differences between the two systems: first, the confining potential in Si/SiGe is much smaller than in Si inversion layers. This fact means that the Si/SiGe interfaces represent a smaller change to the periodic potential compared to the Si/SiO_2, which could lead to a reduction in valley splitting. Second, the confining potential that forms the two-dimensional electron gas (a rectangular quantum well) is more symmetric in Si/SiGe compared to Si-MOS (a triangular well), particularly at the relatively low electron concentrations achieved in our system. Since it is the destruction of the inversion symmetry that lifts the valley degeneracy, a more symmetric potential should lead to smaller splitting. Third, the spin orbit interaction is known to change with strain, which could affect the Zeeman splitting in the strained Si quantum well, compared to the relaxed bulk Si.

We have performed tilted field measurements in order to estimate both the g-factor and the valley splitting. Figure 1 shows tilted field measurements, which uses the angle of coincidence, at which the Landau levels and the spin-split levels cross in order to determine both the g-factor and the valley splitting. From this measurement we get a g-factor of 3.4 and a valley splitting of 50 μeV. The g-factor is measured to increase as a function of carrier density and is in the range of 2.8 - 3.8 for a carrier density in the range of 5 x 10^{11} cm^{-2} to 3 x 10^{11} cm^{-2}.[7]

The Integer and Fractional QHE

The integer quantum Hall effect (IQHE) is understandable in the framework of a non-interacting electron system, whereas the fractional quantum Hall effect (FQHE) is attributed to a Coulomb electron-electron interaction and electron condensation in a state of an incompressible fluid.[8] Jain[9] has provided a new picture that is capable of explaining the IQHE and the FQHE within the same framework. this picture, which is based on attaching 2 flux quanta to each electron, forming together a 'composite fermion', has the power of predicting the observed fractions $\left(\nu = \frac{n}{2nm \pm 1}\right)$, and the order of their stability.

Si differs from GaAs in that the condensed ground state and its quasiparticle excitation depend not only on the spin degree of freedom, but also on the valley degeneracy. The effect of the latter has been discussed by various groups.[10-12] The valley degeneracy introduces an SU(2) symmetry and, thus, behaves like an isospin[12] and automatically gives the limiting case equivalent to the g-factor → 0.[10]

The important questions that have been recently addressed in GaAs/AlGaAs are related to the spin split state of the various fractional filling factors, and to the observation of even-denominator fractions. It has been established both theoretically[13] and experimentally[14-16] that some fractions (e.g. 4/3, 2/3, 8/5, ...) are spin unpolarized, while others (e.g. 5/3, 1/3, ...) are spin polarized.

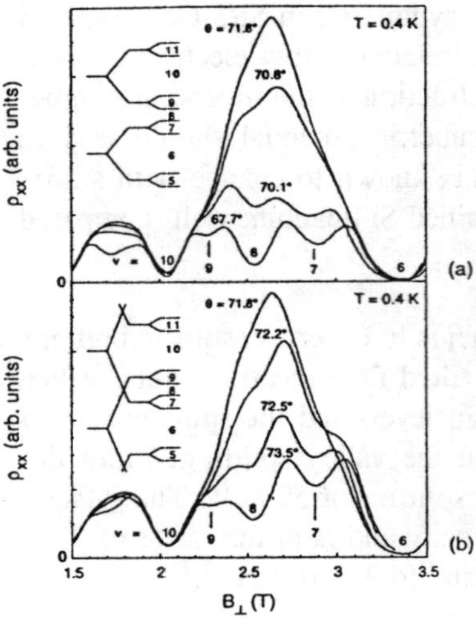

Fig. 1 Longitudinal magnetoresistance of a strained Si/SiGe sample, as a function of perpendicular magnetic field, for various tilt angles. The two angles of coincidence between $\nu = 7$ and $\nu = 8$ are 69.7 and 72.3 degrees.

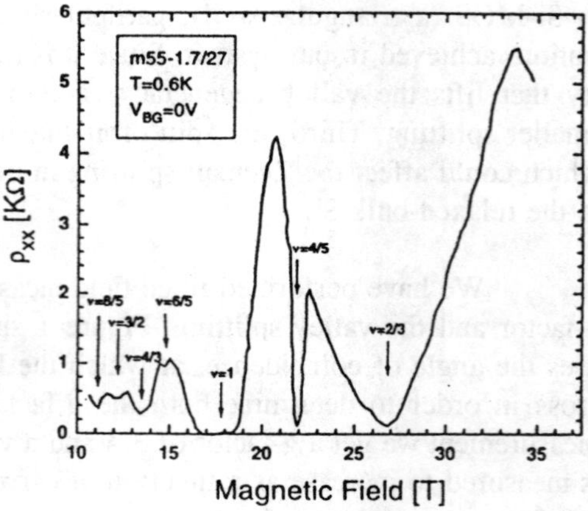

Fig. 2 Longitudinal magnetoresistance of a strained Si/SiGe sample, showing the fractional filling factors.

This has been confirmed by various ways of controlling the Zeeman energy (tilted magnetic field[14], hydrostatic pressure[15], and varrying the quantum well thickness).[16] The fraction $\nu = 1/2$ is observed in double layer systems[17] and is explained in terms of the Coulomb interaction between the electrons in the two quantum wells. Belkhir and Jain[18] have extented the composite fermion picture to explain the observation $\nu = 1/2$ in a single layer system in the limit of no Zeeman splitting.

Our high mobility samples allow us to examine all the above effects in Si. Fig. 2 shows ρ_{xx} of a 3.8 x 10^5 cm^{-2}/Vs mobility sample as a function of magnetic field. The strongest observed dips in ρ_{xx} are at filling factors 2, (not shown in the figure) and 8/5, 3/2, 4/3, 6/5, 1, 4/5 and 2/3. The gap at $\nu = 4/3$ and $2/3$ has been previously deduced to be 2.9 K and 4.5 K, respectively. In previous work[19], the 8/5 was not well resolved and was mixed up with 11/7, due to the lower quality of the sample and to the presence of a parallel conduction. After correcting for the latter, an unambiguous strong drop in ρ_{xx} at $\nu = 2/5$ was observed. Smaller dips at $\nu =$

3/2 and *6/5* can also be seen in Fig. 2. A corresponding bending in ρ_{xy} (but not an obvious plateau) are observed in both cases.

By applying a negative voltage to the back side, the charge density is reduced. Fig. 3 shows ρ_{xx} and ρ_{xy} for that case. It is striking that a $\nu = 1/2$ is so well resolved at this temperature in our single quantum well system. A clear dip in ρ_{xx} and a quantized plateau at $\frac{2h}{e^2}$ can be observed. When the gate is positively biased, the charge density is increased (and so does the mobility[7]) and the following additional fractionals are observed: $\nu = 3/2$, and a weaker $\nu = 6/5$. The striking difference between the above results and those obtained in GaAs is the strong $\nu = 1/2$, the presence of $\nu = 3/2$ (although weak), and the disappearance of prominent fractions such as 5/3 and 3/5. On a first glance, the much stronger Zeeman energy splitting in Si should have resulted in an increase in the gap of the 5/3 fractional, which is known to be spin polarized. Since that is not the case, we believe that the main difference is due to the valley degeneracy.

The observed fractions can all be described by $2\left(\frac{n}{mn \pm 1}\right)$ and its electron-hole symmetrie state $2\left(1 - \frac{n}{2nm \pm 1}\right)$, where the factor 2 in the numerator accounts for the valley degeneracy. Despite the extreme simplicity of this picture, it describes very well the expected observations. The disappearing 5/3 state, which has been previously observed as a function of reducing the Zeeman energy by reducing the thickness of the quantum well[16], is also expected in the case of the two-fold valley degeneracy, which yields the limit of zero Zeeman energy. In this limit, the quasi-particles become spatially unbounded objects with infinite spin, termed 'skyrmions".[10]

Upon reducing the electron density further down (Fig. 4), we have observed a clear transition in ρ_{xx} to an insulating phase that is non-reentrant (within our magnetic field range). Metal-insulator transitions have been previously discussed in Si MOSFETs, an have been attributed (without solid proof) to the creation of a Wigner crystal.[6] In our case, ρ_{xy} behaves normally, and it is only ρ_{xx} that undergoes a dramatic increase (see the inset for a logarithmic dependence on B). The position at which ρ_{xx} takes off is a function of electron density and moves from 34 T to 28 T as the electron density is changed between 2.6 - 2.1 x 10^{11} cm^{-2}. Such an insulating phase has been predicted to occur for $\nu < 1/5$, but in our case it almost coincides with the position of the missing $\nu = 1/3$. Further study (in particular temperature dependence) is needed to understand this effect more.

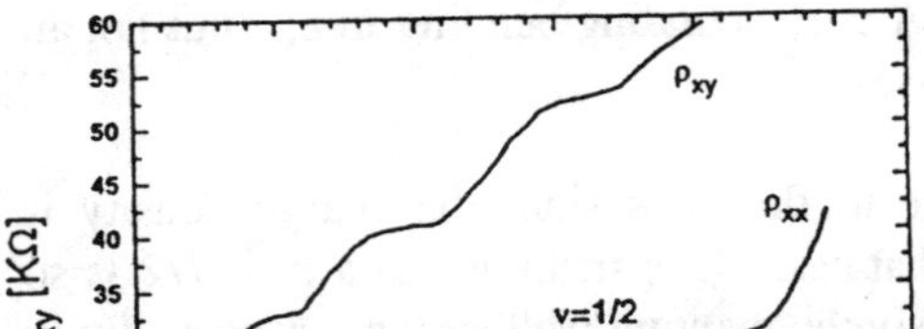

Fig. 3 Longitudinal and transverse magnetoresistance for the same sample in Fig. 1 after applying negative gate bias, which reduces the electron density. The fraction $v = 1/2$ is clearly observed.

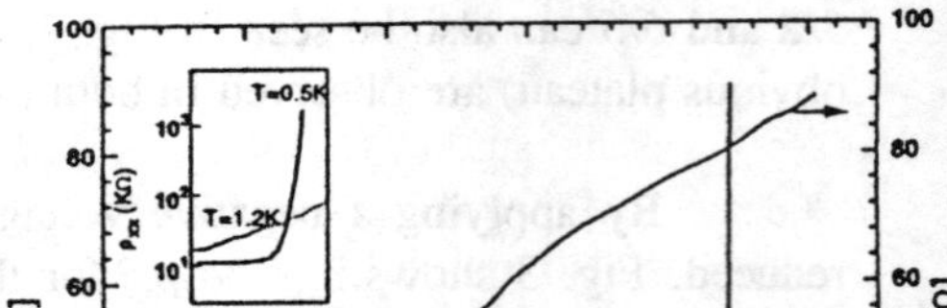

Fig. 4 Longitudinal and transverse magnetoresistance at low electron density (2.45 x 10^{11} cm^{-2}) showing an insulating phase for B > 31 T at T = 0.5 K. The insert shows the logarithmic dependence on the magnetic field at tow temperatures.

Summary

In summary, strained Si quantum wells with very high electron mobilities have been used as a test vehicle to examine electron transport properties in the quantum limit. The g-factor in this system is determined to range from 2.8 to 3.8 as the carrier density is reduced from 53 x 10^{11} cm^{-2}. The valley splitting at about 3 T is determined to be about 50 μeV, about an order of magnitude lower than in Si inversion layers. In the fractional quantum Hall regime, the observed fractions follow a different hierarchy from that of GaAs, which we attribute to the valley degeneracy. A metal-to-insulator transition is observed at a field corresponding to where filling factor 1/3 would have been expected.

Acknowledgement

The author would like to thank J.O. Chu for material growth, and S. Koester, M. Arafa and S.F. Nelson for their help with the measurement. The author is also grateful to the staff at the Francis Bitter National Magnet Lab, where the high magnetic field data were taken, and to J.P. Cheng for his help during these measurements. The explanation of the observed fractional filling factors is due to J. Jain, to whom the author is very grateful.

References

1. K. Ismail, S. Rishton, J.O. Chu, K. Chan, S.F. Nelson, and B.S. Meyerson, Elect. Dev. Lett. **14**, 348 (1993).
2. K. Ismail, M. Arafa, K.L. Saenger, J.O. Chu, and B.S. Meyerson, Appl. Phys. Lett. **66**, 1077 (1995).
3. M. D'Iorio, J. Campell, V. Pudalov, and S. Semenchinsky, Surf. Science **263**, 49 (1992).
4. S.V. Kravchenko, J. Perenboom, and V. Pudalov, Phys. Rev. B **44**, 13513 (1991).
5. V.T. Dolgopolov, G. Kravchenko, S. Kravchenko, and A. Shashkin, Surface Sciene **305**, 96 (1994).
6. K. Ismail, M. Arafa, F. Stern, J. Chu, and B. Meyerson, Appl. Phys. Lett. **66**, 842 (1995).
7. S. Koester, K. Ismail, and J.O. Chu, to be published in Phys. Rev. B.
8. R.B. Laughlin, Phys. Rev. B **23**, 5632 (1981); B.I. Halperin, Phys. Rev. B **25**, 2185 (1982).
9. J.K. Jain, Phys. Rev. Lett. **63**, 199 (1989).
10. S.L. Sondhi, A. Karlhede, and S. Kivelson, Phys. Rev. B **47**, 16419 (1993).
11. M. Rasolt, B. Halperin, and D. Vanderbilt, Phys. Rev. Lett. **57**, 126 (1986).
12. M. Rasolt, Solid State Physics **43**, 93 (1990).
13. P.A. Maksym, J. Phys.: Condens. Matter **1**, 6299 (1989).
14. J.P. Eisenstein, H. Stormer, L. Pfeiffer, and K. West, Surface Science **229**, 21 (1990); R.G. Clark, S. Haynes, A. Suckling, J. Mallet, and P. Wright, Phys. Rev. Lett. **62**, 1536 (1989).
15. N.G. Morawicz, K. Barnham, A. Briggs, C. Foxon, J. Harris, S. Najda, J. Portal, and M. Williams, Semicond. Sci. Technol. **8**, 333 (1993).

16. C.V. Brown, C.J.G.M. Langerak, P.C. Main, L. Eaves, T.J. Foster, and M. Henini, Physica B **184**, 81 (1993).
17. G.S. Boebinger, S. Murphy, J. Eisenstein, L. Pfeiffer, K. West, and S. He, Surface Science **305**, 8 (1994).
18. L. Belkhir and J.K. Jain, Phys. Rev. Lett. **70**, 643 (1993).
19. R.B. Dunford, R. Newbury, F. Fang, R. Clark, R. Starrett, J. Chu, K. Ismail, B. Meyerson, Solid State Commun. **96**, 57 (1995).

MAGNETIC FIELD INDUCED METAL TO INSULATOR TRANSITION IN Si/SiGe SHORT PERIOD SUPERLATTICES

G. BRUNTHALER[a], T. DIETL[a,b], A. PRINZ[a], G. STÖGER[a], M. SAWICKI[b], J. JAROSZYNSKI[b], F. SCHÄFFLER[a] and G. BAUER[a]

[a]*Institut für Halbleiterphysik, Johannes Kepler Universität, A-4040 Linz, Austria*
[b]*Institute of Physics, Polish Academy of Sciences, al. Lotników 32/46, PL-02668 Warszawa, Poland*

We report on magnetoconductance studies of strain symmetrized short period Sb-doped $Si/Si_{0.55}Ge_{0.45}$ superlattices (SL´s). Several samples with doping concentrations ranging from $2x10^{18}$ to $6x10^{18}$ cm^{-3} were investigated, covering the value of $3.7x10^{18}$ cm^{-3} where in bulk Si:Sb a metal to insulator transition occurs. The two lower doped superlattice samples (2 and $2.5x10^{18}$ cm^{-3}) exhibita variable range hopping conduction at low temperatures, i.e. insulating behaviour, whereas the highest doped on ($6x10^{18}$ cm^{-3}) stays metallic in millikelvin temperatures up to B = 8 T. In the medium doped one ($4.5x10^{18}$ cm^{-3}) a magnetic field induced metal-to insulator transition (MIT) is observed, which shows that despite the absence of coherent vertical transport (the miniband width being a factor of 10 smaller than the momentum relaxation rate) the system behaves three-dimensional with respect to disorder. The low temperature magnetoconductance is governed by single electron back scattering effects and disorder induced electron-electron interaction. The magnetic field induced MIT in the Si/SiGe SL's occurs for magnetic fields oriented parallel to the interface planes and is absent for the perpendicular configuration, which implies that it is driven by the influence of spin-splitting on the electron-electron interaction and not caused by an orbital effect.

1 Introduction

The metal-to-insulator transition has been the subject of extensive investigations, both experimentally as well as theoretically.[1-5] The transformation from a metal to an insulator involves the nature of electronic states which contribute to charge transport: these are extended in the metallic state but localized in the insulating one. The localisation may be driven by static disorder (Anderson localization), by strong electron-electron correlations (Mott-Hubbard transition) or by strong electron lattice coupling.[1]

In particular, doped semiconductors have attracted a lot of interest, because the average spacing between the impurity atoms can be varied continuously. Indeed in a large number of systems, activated hopping conductivity was found at low doping concentrations (due to the transport taking place between isolated donor or acceptor levels). With increasing doping concentration, the impurity wave functions overlap, and finally the impurity band merges with the conduction or valence band. Despite the problems associated with the randomness of the impurities, the data collected by Edwards and Sienko[2] have made it clear that the original criterion by Mott on the crossover between the insulating and metallic behaviour $N_c^{1/3} \approx 0.25\ (a_B^*)^{-1}$ gives a remarkably good description for the impurity concentration N_c at which the transition occurs for a wide variety of systems (a_B^* denotes the effective Bohr radius).

On the other hand, detailed experimental and theoretical studies, especially on doped bulk Si:P, Si:Sb and on Si:B, have demonstrated that disordered Fermi liquid theory and scaling concepts account for many of the observed transport phenomena.[6-8] In particular, these studies made it clear that quantum corrections like backscattering and disorder modified electron-electron interactions[4-8] play a crucial role for the transport properties near the metal-insulator transition (MIT) in these semiconductors.

A lot of these studies was performed in high magnetic fields where it was commonly thought that the metal-to-insulator transition is simply induced by impurity wave function shrinkage.[9] However, whereas this is true for systems like GaAs with a comparably small effective mass (0.067 m_0) *and* a small Landé g-factor (0.4), in doped Si, the situation is quite different. In n-Si, the ratio of cyclotron resonance to spin splitting m_0/m^*g^*, for $m^* = 0.19m_0$ and $g^* = 2.0$, is by a factor 15 smaller as compared to GaAs.

Apart from studies of the MIT in bulk semiconductors, a large amount of work has been devoted to low-dimensional systems. Much evidence was found that all states are localized at low T in disordered two dimensional systems.[1] For the two-dimensional (2D) electron gas in Si inversion layers the interplay between backscattering and electron-electron interaction was confirmed.[10] There have been numerous investigations on the relation between the quantum Hall effect and localisation-delocalisation transition near the centres of the Landau levels.[11,12] More recently, the presence of the MIT in high-mobility Si MOS-FETs has been suggested to occur even in the absence of quantising magnetic fields.[13]

In the above contexts, studies on semiconductor superlattices are of considerable interest: these systems have a highly anisotropic superlattice bandstructure, which depending on the parameters represent examples of an effective dimensionality between two and three. Furthermore, fluctuations of the barrier potential heights (due to composition fluctuations) as well as of the well and/or barrier width may provide additional source for disorder and thus for Anderson localisation.[14]

So far mainly studies on the modification of weak localization phenomena in superlattices were performed, mainly in GaAs/GaAlAs. In a series of pioneering studies Szott et al.[15] have established the influence of the superlattice structure on the parallel charge conduction, and have found that a weak localization theory, which explicitly accounts for the superlattice periodic structure, explains the experimental results. Recently, Hilber et al.[15] have studied the magnetic field induced MIT in GaAs/GaAlAs superlattice and its orientational dependence on the direction of the applied magnetic field with respect to the growth direction.

We report here on the observation of a magnetic field induced metal-insulator transition in uniformly Sb-doped short period Si/SiGe superlattices (SL). Our system lies in between the n-type bulk Si systems and the modulation doped or δ-doped Si, and has a highly anisotropic superlattice bandstructure. Apart from the influence of anisotropy, our system is suitable to examine role of spin effects, of the valley degeneracy, and of an open Fermi-surface on the electron transport phenomena in the vicinity of the MIT.

Our SL's are in the limit of strong disorder (the miniband width is a factor of 10 smaller than the in plane momentum relaxation rate $\hbar/\tau$, i.e. with respect to coherent vertical transport the SL's are *2D-like*). The temperature dependence of the high doped and low doped samples indicate metallic and variable range hopping (insulating) behaviour, respectively, whereas the medium doped one (4.5×10^{18} cm^{-3}) shows a magnetic field induced MIT, but only for fields *parallel* to the interfaces.

2 Experimental

The samples investigated were grown by molecular beam epitaxy (MBE) on (001) Si substrates on which a SiGe buffer with graded Ge composition from x= 0.05 to x= 0.22, with a total thickness of 1 μm, followed by one with constant Ge content (x=0.17, d= 0.5 μm) was grown. The strain-symmetrized structures consist of 150 repetitions of Si wells (25 Å) and $Si_{0.55}Ge_{0.45}$ barriers (14 Å) on top of a SiGe graded buffer layer. The nominal Sb-doping concentration in three different samples was N_D = 2.0, 2.5, 4.5 and 6.0 x 10^{18} cm^{-3}, which was checked by secondary ion mass spectroscopy. The layer thicknesses were controlled by a dynamical simulation of high resolution x-ray diffraction data. The Sb doping levels in the superlattice samples cover the range around the critical concentration (3.7×10^{18} cm^{-3}) for the MIT in bulk Si:Sb.[17,18]

3 Angular dependence of the magnetoconductance of the Si/SiGe SL's

Temperature dependent resistivity and Hall effect measurements at 0.3 T from room temperature down to 10 K show a slight variation of the carrier concentration which agrees within 10% with the Sb-doping concentration from the SIMS analysis. The mobility in these samples reaches values between 100 and 200 cm^2/Vs at 10 K.

In the strain symmetrized Si/SiGe SL's the biaxially strained Si layers form the wells, with two valleys being shifted downwards in energy with respect to the remaining four ones. The relevant conduction band in plane effective masses are $m_{||} = 0.19\ m_0$, which is nearly a factor of 5 smaller than the longitudinal mass $m_{\perp} = 0.92\ m_0$ along the [001] growth direction. According to selfconsistent calculations for all our samples only the lowest electric subband is occupied in the absence of disorder, and is located 35 meV above the Si conduction band edge. The miniband width *w* of about 2.5 meV of the lowest subband is substantially smaller than the Fermi energy of 11 meV for the sample with $N_D = 4.5\times10^{18}$ cm^{-3}. Hence we deal with an open Fermi surface. The second subband lies about 100 meV above the first one, as calculated by envelope function calculations and verified experimentally by the observation of intersubband transitions using infrared-spectroscopy. Thus all transport phenomena involve only the lowest electric subband.

Not only due to the large longitudinal effective mass but also due to the large $Si_{0.55}Ge_{0.45}$ barrier height of 300 meV, a weak coupling between carriers in the Si wells results, and the vertical transport mobility is substantially reduced. This results in a rather large anisotropy of the electron transport.

The in-plane electron mobility of about $\mu = 150$ cm^2/Vs at 10 K corresponds to a momentum relaxation time, $\tau = m_{||}\mu/e$ of about 20 fs, a value which is nearly identical to the "vertical" electron flight time, i.e. the electron flight time τ_f between the interfaces of a single QW ($\tau_f = d_w(m_\perp/2E_0)^{1/2}$) which is about 22 fs. Thus, in contrast to more standard anisotropic systems, the large anisotropy under consideration becomes only relevant over a length scale comparable to the mean free path. Furthermore, our SLs are in the limit of strong disorder as the transport time τ corresponds to a scattering broadening of the density of states of $\hbar/\tau = 33$ meV, which is more than 10 times larger than the miniband width of w = 2.5 meV. Correspondingly, coherent vertical transport ($w>\hbar/\tau$) will be hindered.[19]

For the following theoretical description of the magnetoconductance we emphasise that the product of Fermi-wave vector and mean free path, $k_F\ell$, characterises disorder in the sense of diagrammatic perturbation theory.[4,7,8] For the SL sample with a doping concentration of 4.5 x 10^{18} cm^{-3}, we obtain $k_F\ell = 0.66$ at T = 10 K. As this value of $k_F\ell$ is smaller than one, significant deviations from the Fermi liquid behaviour can be expected.

The low temperature magnetoconductance of the SL samples exhibits an interesting angular dependence as shown in Fig.1, indicative of a positive and a negative component. In Fig.1 data of the magnetoconductivity for SLs with n = 4.5 and 6.0 x 10^{18} cm^{-3} are shown for T = 1.4 K. The orientation of the magnetic field B is described by the angle θ to the growth direction. The measurement current was perpendicular to the magnetic field for all angles. For B perpendicular to the layers, a positive magnetoconductivity at lower magnetic fields and a negative one at higher B is observed.

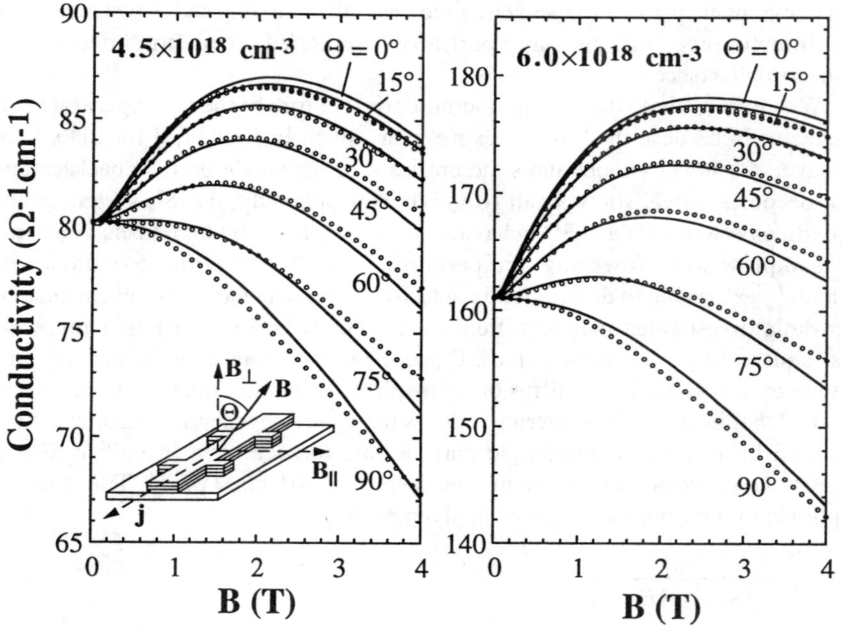

Figure 1. Angular dependence of the magnetoconductivity at 1.4 K for Si/SiGe superlattices with a homogeneous Sb doping concentrations of 4.5 and 6.0 x 10^{18} cm^{-3}. As indicated in the inset, 0° denotes the orientation of magnetic field perpendicular to the layers. The circles represent the experimental data, the solid lines a least square fit according to the anisotropic 3D description.

For the parallel field configuration, in both samples, only the negative component remains. In the case of bulk and δ-doped Si, the positive part has been assigned to single particle quantum interference from back scattering (i.e. an orbital effect) whereas the negative one is due to the Cooperon channel of the electron-electron interaction, and the destructive influence of the spin splitting of the triplet state in the diffusion channel of the electron-electron interaction.[4,7,8]

The influence of the spin splitting on the magnetoresistance is not obvious but can be explained by the following arguments. The particle and hole in the diffusion channel can have either total spin zero or one, the latter with components -1, 0 or +1 parallel to the magnetic field. Without magnetic field, the particles and holes at the Fermi energy (only those contribute) have the same momentum $\hbar k_F$. Thus they have the same wavelength, and consequently contribute to interference effects of their interaction amplitudes, which create a contribution to the conductivity. For a magnetic field $g\mu_B B > k_B T$,, the particle and hole at the Fermi energy with total spin components parallel to magnetic field have different $\hbar k_F$ and wavelengths. Thus the interference effects are partially quenched, causing the negative magnetoconductance.

We found that the magnetoconductivity for higher temperatures can satisfactorily be described by the expressions given in Refs.[7,8] for 2D systems. However, for lower temperatures the prefactor of the single particle backscattering term becomes unrealisticly small ($\alpha n_v = 0.3$). Apparently, our SL system is at the borderline between 2D and 3D behaviour, changing from 2D to 3D behaviour in our SL samples with lowering temperature. The 3D version for anisotropic systems[7,20,21] is able to describe satisfactorily the field and angular dependence over the whole investigated magnetic field range. The solid lines in Fig. 1 represent a least square fit of the anisotropic 3D magnetoconductance including the single particle back scattering, the diffusion spin-splitting and the Cooperon term.[7,8] It was assumed that inter-valley scattering reduces the valley degeneracy factor αn_v to 1 as observed in analysis of the single particle interference term in bulk n-Si[22] and Si/SiGe QWs with mobilities in the range of 10^4 cm^2/Vs.[23] The Coulomb amplitude in the Cooperon channel is given by:[8,24]

$$g_4 = \frac{g_3}{1 + g_3 \ln\left(E_F/\varepsilon\right)} \tag{1}$$

with $\varepsilon = \max\{k_B T,\ 2\hbar D_c/\pi \ell_B^2,\ g^* \mu_B B/\pi\}$, parallel diffusion constant $D_{\|} = E_F \mu/e$, the diffusion constant in the plane perpendicular to the magnetic field $D_c = (D_{\|}(D_{\perp} \sin^2\Theta + D_{\|} \cos^2\Theta))^{1/2}$, the averaged diffusion constant $D_a = (D_{\|}^2 D_{\perp})^{1/3}$, and $\ell_B = (\hbar/eB)^{1/2}$ the magnetic length. In D_c and D_a enters also the perpendicular diffusion constant $D_{\perp}$, which in our case describes the motion of electrons between neighbouring QWs.

The electron Landé factor g^* is expected to be isotropic and should not be rescaled due to correlation effects in confined electron systems.[7] This is in contrast to the spin splitting which enters into the magnetic susceptibility, where a renormalized value of g^* should be used.[25-27] For our analysis, we take the bulk value of n-Si, $g^* = 2.0$ to describe the spin-splitting of the diffusion channel.

The fit of the magnetic field and angular dependence, as indicated by the solid lines in Fig. 1, gives for $n = 4.5 \times 10^{18}$ cm^{-3} (and 6.0×10^{18} cm^{-3}, respectively) the anisotropy factor $r_D = D_{\|}/D_{\perp} = 40$ (110), phase coherence time $\tau_\varphi = 14$ ps (15 ps) and the Coulomb amplitude in the diffusion channel $g_3 = 1.1$ (1.3). All other parameters follow directly from these fitted values and equation (1) and from electron concentration n and temperature T.

As can be seen from the good quality of the fit in Fig. 1, the standard 3D theory of weak localization is able to describe the behavior of our samples with higher doping concentrations. The magnitude of the magnetoconductance is proportional to

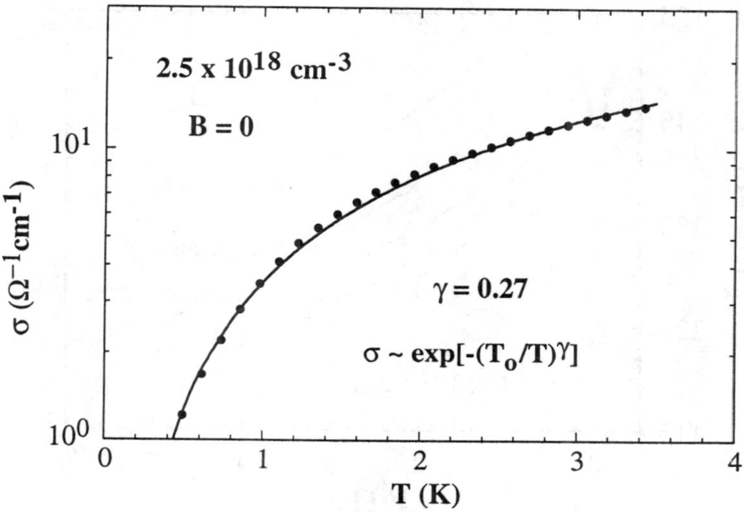

Figure 2. The SL structure with the lower doping concentration of 2.0 x 10^{18} cm^{-3} shows a strong decrease of the conductivity towards lower temperatures, which can be described by $\sigma = \sigma_0 \exp(-(T_0/T)^{\alpha})$ typical for variable range hopping in the insulating phase.

the square root of the anisotropy factor r_D and thus larger than in bulk 3D systems. This interesting property has already been noted in the context of lateral transport in GaAs/AlGaAs SLs.[15]

From our data, the vertical diffusion constant $D_\perp = D_\parallel/r_D$ can be deduced, which is difficult to be measured directly in Si/SiGe SLs due to problems with doping of the graded buffer layer. From $D_\perp$ the escape time from one well into another can be estimated. For $\hbar/\tau > w$, the subsequent transitions between wells are uncorrelated, so that $\tau_{esc} \approx (d_w + d_b)^2/D_\perp$. This gives τ_{esc} = 5.1 ps and 6.9 ps for samples with 4.5 and 6.0 x 10^{18} cm^{-3}, respectively, a value much longer than the escape time for coherent coupling between the wells, $2\hbar/w$ = 0.52 ps. Due to the strong disorder $\hbar/\tau > w$, the coherent transport is hindered and the transition rate should be described by a tunnelling probability through a barrier with width d_w. This gives an estimate of the escape time of 9 ps, assuming that only the lowest subband energy E_0 determines the tunnelling energy and that the parallel momentum is conserved. The escape time evaluated from $D_\perp$ is lower, which may come from additional momentum non-conserving tunnelling processes.[26]

Only the SL systems with higher electron concentration (4.5 and 6.0 x 10^{18} cm^{-3}) could be well described by the magnetoconductivity contributions described

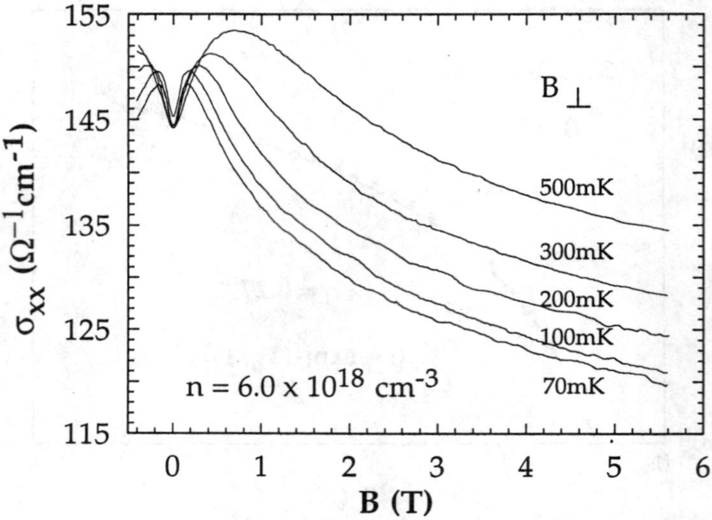

Figure 3. Magnetoconductivity of Si/SiGe superlattice with Sb doping concentration of 6.0 x 10^{18} cm^{-3}. at temperatures between 70 mK and 500 mK.

above. For the lower doped samples (2.0 and 2.5 x 10^{18} cm^{-3}) the quantum corrections become very large and the assumptions of small perturbation and $k_F\ell \geq 1$ are not valid any more. This is demonstrated in Fig. 2 for n = 2.0 x 10^{18} cm^{-3}. The conductivity at zero magnetic field decreases dramatically to lower temperatures and can be described by $\sigma = \sigma_0 \exp(-(T_0/T)^{\alpha})$ with $\alpha = 0.27$. This behavior is typical for a variable range hopping mechanism and demonstrates that the two lower doped samples are in the insulating phase.

In contrast, the sample with n = 6.0 x 10^{18} cm^{-3} exhibits metallic conductivity as shown in Fig.3. The conductivity at zero magnetic field does not change noticeable down to 70 mK, and the magnetoconductivity up to 6 T decreases only moderately, giving no indication for a magnetic field induced MIT either.

The data shown above seem to imply that the 3D theory of weak localisation modified through an anisotropic diffusion tensor does indeed describe the magentoconductivity of the samples under investigation, despite the fact that they are in the strong disorder limit $\hbar/\tau > w$. However, also the 2D expression for the weak localisation theory yield a reasonable fit to the data as shown above. This apparent ambiguity finds its solution in the following observations: for an escape time of τ_{esc} = 5.1 ps as given above, the critical temperature for 3D behaviour following from its comparison with the thermal coherence time $\hbar/k_BT$ would be 1.9

K, i.e. it is just fulfilled for the data measured at T= 1.4 K. Thus for temperatures higher than 1.9 K, τ_{esc} becomes longer than the thermal coherence time and thus the system should behave as two-dimensional. Furthermore, for the huge anisotropy of the diffusion tensor components, the weak localisation expressions for 2D and 3D become quite similar, which explains the successful description of the magnetoconductance data with a 2D theory.

4 Magnetic field induced metal-to insulator transition

The millikelvin magnetoconductance data on the SL samples with the lowest doping concentrations, which exhibit variable range hopping behaviour (Fig.2), and the metallic behaviour (Fig.3) observed for the highest doped SL clearly demonstrate that a metal-to insulator-transition indeed occurs in the short period Si/SiGe SL's. Consequently, the SL sample with a Sb doping concentration of 4.5 x 10^{18} cm^{-3}, a value which is slightly above the one of 3.7x10^{18} cm^{-3} for which in bulk Si a MIT was observed, is the most interesting one, from the point of view of a magnetic field induced MIT.

This is evident from Fig.4 where the magnetoconductivity in the millikelvin-temperature region are compared for three SL samples (with N_D= 2, 4.5 and 6x10^{18} cm^{-3}) as a function of $\sqrt{T}$. The data are shown for two magnetic field configurations, perpendicular as well as parallel to the SL layers. For the perpendicular orientation of the 4.5x10^{18} cm^{-3} doped SL (Fig. 4a), the conductivity extrapolated to T = 0 K is clearly above zero for all investigated fields and even seems to saturate. On the other hand, for the in-plane configuration (Fig. 4b) the conductivity (inverse inplane resistivity) extrapolated to zero temperature vanishes. These extrapolated magnetoconductivites indeed indicate a magnetic field induced MIT transition which occurs for fields between 1 and 2 T. At the same time, over a wide temperature range a dependence of $\sigma \propto T^{1/2}$ is observed, characteristic for 3D systems.[6,17,27,28] Because of the huge diffusion anisotropy, the magnitude of $d\sigma/dT^{1/2}$ is in the Si/SiGe SL's at criticality by about a factor of 8 larger than in bulk Si.[12,16,17] This strong temperature dependence makes extrapolation to T = 0 inaccurate, and precludes a precise determination of critical field and exponents.

The conductivity data in Fig.4 show a deviation from the general dependence of σ vs T at the lowest temperatures. Thermal decoupling of the electron gas from the cold source seems to be involved at temperatures of about 100 mK, presumably caused by high frequency electromagnetic noise, which has been noted before by several investigators. In particular, similar effects were observed in recent investigations of bulk Si:P[27] as well as in bulk Si:Sb.[17] In the latter case it has been assigned to a possible electronic phase transition or a spontaneous lifting of the valley degeneracy from 6 to 2.[17] This explanation can be excluded by our experiment

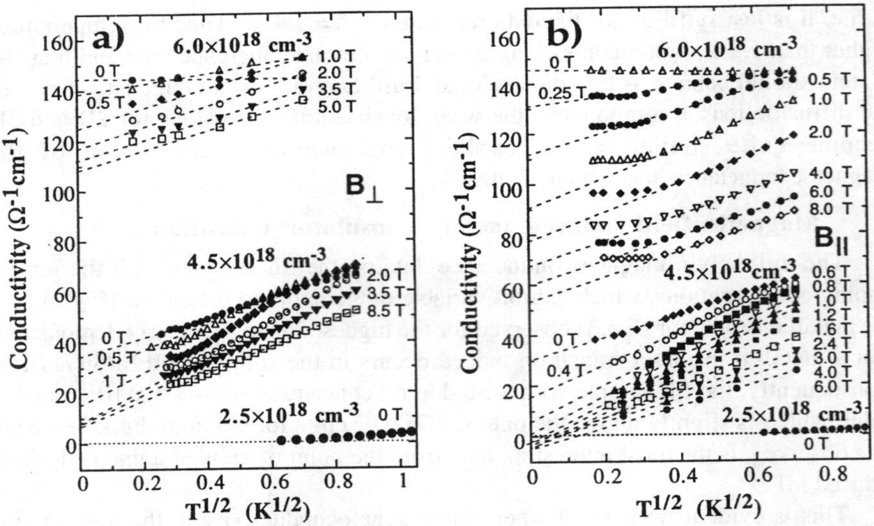

Figure 4. Temperature dependence of the conductivity for sample with 2.0, 4.5 and 6.0 x 10^{18} cm^{-3} at different magnetic fields on a square root of T scale for the (a) perpendicular and (b) parallel orientation of the magnetic field.

as the lifting of the valley degeneracy occurs already due to the biaxial strain in the Si/SiGe SLs.

The observation of the field-induced MIT in the parallel magnetic field configuration and its absence in the perpendicular one proves rather directly that in the Si/SiGe SL-structures the MIT is driven by a spin mechanism and not by orbital effects. As shown above, the decrease in magnetoconductivity (Fig.1) in parallel magnetic fields is indeed due to the influence of the spin splitting on the diffusion term, whereas for the perpendicular field orientation this phenomenon is partly counterbalanced by an increase caused by the destruction of the single electron backscattering effect with increasing magnetic field. This behaviour is in marked contrast to n-GaAs/GaAlAs SL's with a large ratio of cyclotron to spin-Zeeman energy $m_e/m^*g^* = 37$ (instead of 2.6 for Si). Indeed, in n-GaAs/GaAlAs SLs the MIT occurs at lower magnetic field when it is oriented perpendicular to the layers[16] indicating the dominance of orbital effects (wave function shrinkage).

In order to find out whether the behavior of the conductance in the vicinity of the MIT is consistent with theoretical predictions for 3D systems, we replotted the experimental results taken between 90 and 500 mK in the coordinates suggested by Finkelstein´s renormalization group approach:[5,29] The experimental results agree well with the scaling law, but it turns out that due to the strong dependence of σ

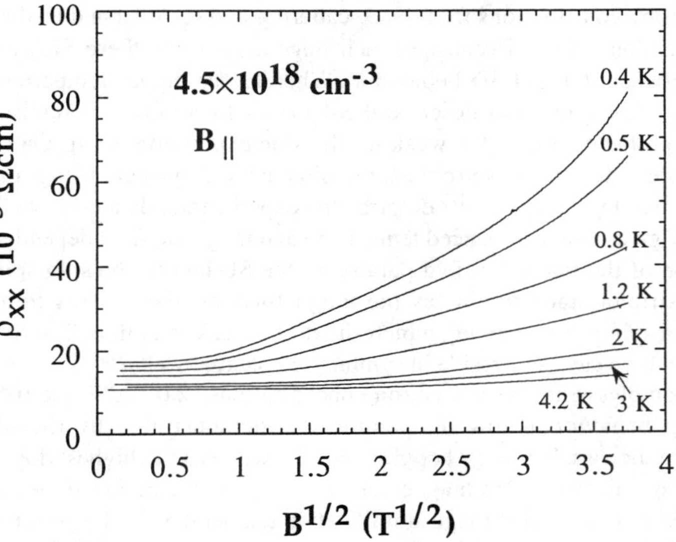

Figure 5. In plane magnetoresistance ρ_{xx} versus magnetic field for different temperatures.

versus T, our data are not sensitive to the choice of the critical magnetic field $B_c \approx 1.6$ T and $\nu \approx 1$ but provide an accurate value of $z = 1.6 \pm 0.2$. A similar analysis for bulk n-Si at B = 0 yielded z = 3.1,[30] while that for the MIT of spin-polarised electrons in diluted magnetic semiconductors (DMS) results in $z = 1.7 \pm 0.4$.[31] This is consistent, as DMS and n-Si in the magnetic field belong to the same universality class, whereas n-Si without magnetic field belongs to another one.[5,29,31]

As an experimental fact, we plot in Fig. 5 the magnetoresitance in paralell magnetic fields up to 14 T with the temperature as a parameter. The strong increase of ρ_{xx} at low temperatures points to the presence of the MIT.

There is no doubt that in n-Si, the observed effects are driven by the influence of spin-splitting on the disorder modified electron-electron interactions. Furthermore, near the MIT, there is an additional spin mechanism responsible for a negative magnetoresistance related to the spin-splitting ($\hbar\omega_s = g\mu_B B$) induced redistribution of the carriers between the two spin subbands:[30,33,34] it affects the conductivity by its influence on the mean free path of the carriers, which appears for $\hbar\omega_s > E_F$, and by changing the distance of E_F from the mobility edge E_m, i.e. for $\hbar\omega_s > E_F - E_m$.[35]

5 Conclusions

The magnetoconductivity of $Si/Si_{0.55}Ge_{0.45}$ SL structures with electron concentrations of 4.5 and 6.0 x 10^{18} cm^{-3} and its orientational dependence can be well described by contributions from backscattering and disorder-induced electron-electron interaction effects. Because of their huge anisotropy, these SL´s are at the borderline between 2D and 3D behaviour. Whereas for higher temperatures, the corresponding 2D expressions describe the data, for lower ones, unrealistic small values of the prefactor αn_v in the weak localisation expression occur. On the other hand, the expressions for anisotropic three dimensional systems using a rescaled diffusion constant $D_a = (D_{\|}^2 D_{\perp})^{1/3}$ describe the experimental data quite well, where the anisotropy gives rise to enhanced temperture and magnetic field dependence. For the orientation of the magnetic field parallel to the SL layers, the spin splitting of the diffusion-triplet state dominates the magnetoconductivity. This term is not weakened by a spin-orbit coupling, which should be weak in both bulk n-Si and our superlattices, where electrons reside in symmetric potential wells.

For the samples with lower electron concentrations (2.0 and 2.5 x 10^{18} cm^{-3}) the insulating behavior at low temperatures is demonstrated by the observed conductivity in the variable range hopping regime, whereas the highest doped (6.0 x 10^{18} cm^{-3}) stays metallic. The magnetoconductivity of the higher doped samples follows a $T^{1/2}$ law over a wide range at millikelvin temperatures. The presence of the MIT and thus of extended states implies that the studied Si/SiGe SL's behave three dimensional with respect to localisation. This indicates that the zero temperature localisation in the SL's is governed by a rescaled diffusion constant $D=(D_{\|}^2 D_z)^{1/3}$.

In addition we note that for the given anisotropy of the superlattice band structure, any orbital effect would shift the MIT to higher magnetic fields, when the field is tilted away from the growth direction, which is just opposite to the observed behaviour in the Si/SiGe SL's. This demonstrates the importance of spin splitting phenomena for the observed MIT in parallel fields. In particular, our measurements for various angles of the magnetic field with respect to the interface planes reemphasise the quantitative importance of the anti-localising triplet terms of the electron-electron interaction at the localisation boundary of n-Si.

We suggest that at non zero temperatures a SL remains *three-dimensional* as long as an electron can move from one well to another before the phase coherence is destroyed, i.e. the escape time τ_{esc} is shorter than both the phase coherence time τ_φ and the thermal coherence time $\hbar/k_B T$. In our SLs, this condition is fulfilled in the low temperature range.

Acknowledgments: We would like to thank B.L. Altshuler, V.Pudalov, M. Helm and P. Wölfle for helpful discussions. This work was supported by FWF Vienna, Austria, Polish KBN Grant No 2P03B 06411, and Austrian Ost-West Programm.

References:

1. T.V.Ramakrishnan, J.Solid State Chem. **111**, 4 (1995).
2. P.P.Edwards and M.J.Sienko, Int.Rev.Phys.Chem. **3**, 83 (1983).
3. M.P.Sarachik, in *Metal Insulator Transitions revisited*, eds. P.P.Edwards and C.N.R.Rao, (Taylor and Francis, London 1995), p.79.
4. P.A. Lee and T.V. Ramakrishnan, Rev. Mod. Phys. **57**, 287 (1985).
5. D. Belitz and T.R. Kirkpatrick, Rev. Mod. Phys. **66**, 261 (1994).
6. T.F. Rosenbaum, R.F. Milligan, M.A. Paalanen, G.A. Thomas, R.N. Bhatt and W. Lin, Phys. Rev. B **27**, 7509 (1983).
7. B.L. Altshuler and A.G. Aronov, in *Electron-Electron Interactions in Disordered Systems*, ed. A. L. Efros, M. Pollak (North-Holland, Amsterdam, 1985), p. 1;
8. H. Fukuyama, *ibid*, p. 155;
9. M.C. Maliepaard et al., Phys. Rev. Lett. **61**,369 (1988).
10. M.J. Uren, R.A. Davies, and M. Pepper, J. Phys. C **13**, L985 (1980); A.Long, and M.Pepper, J.Phys.C, Solid State Phys. 17, L425 (1984).
11. M. Lee, S.A. Solin, and D.R. Hines, Phys. Rev. B **48**, 11921 (1993).
12. H.W.Jiang et al. Phys.Rev.Lett. 71, 1439 (1993); T.Wang et al. ibid. 72, 709 (1994).
13. S.V. Kravchenko et al., Phys. Rev. B **51**, 7038 (1995).
14. T.Ando and H.Fukuyama, Anderson Localisation, in Springer Proceedings in Physics Vol 28 (Springer Berlin, 1988).
15. W. Szott, C. Jedrzejek, and W.P. Kirk, Phys. Rev. B **48**, 8963 (1993), and references therein.
16. W. Hilber, M. Helm, F.M. Peeters, K. Alavi and R.N. Pathak, Phys. Rev. B **53**, 6919 (1996).
17. A.P. Long and M. Pepper, J. Phys. C: Solid State Phys. **17**, L425 (1984); D.M. Finlayson, P.J. Mason, and D.P. Tunstall, J. Phys.: Condens. Matter **2**, 6735 (1990).
18. U.Thomaschefsky, Ph.D.thesis, Cornell University, 1990. U.Thomaschefsky and D.F.Holcomb, Phys.Rev.B45, 13356 (1992).
19. e.g., S.-R.E. Yang and S. Das Sarma, Phys. Rev. B **37**, 10090 (1988);
20. B.L. Altshuler, A.G. Aronov, A.I. Larkin, and D.E. Khmelnitskii, Zh. Eksp. Teor. Fiz. **81**, 768 (1981) [Sov. Phys. JETP **54**, 411 (1981)]; see also, Y. Ootuka and A. Kawabata, Prog. Theor. Phys. (Kyoto), Suppl. **84**, 249 (1985).
21. R.N. Bhatt, P. Wölfle, and T.V. Ramakrishnan, Phys. Rev. B **32**, 569 (1985).
22. Y. Ootuka and A. Kawabata, Prog. Theor. Phys. (Kyoto), Suppl. 84, 249 (1985).
23. G. Brunthaler, to be published
24. M. Sawicki et al., Phys. Rev. Lett. **56**, 508 (1986).
25. A. Isihara, Solid State Physics **42**, 271 (1989).

26. B. Movaghar, J. Leo, A. MacKinnon, Semicond. Sci. Technol. 3 (1988) 397.
27. H. Stupp et al., Phys. Rev. Lett. **71**, 2634 (1993); M. Hornung et al., Europhys. Lett. **28**, 43 (1994).
28. P. Dai, Y. Zhang, S. Bogdanovich, and M. Sarachik, Phys. Rev. B **48**, 4941 (1993), Phys.Rev.B 49, 14039 (1994)..
29. A.M. Finkelstein, Zh. Eksp. Teor. Fiz. **86**, 367 (1984) [Sov. Phys. JETP **59**, 212 (1984)]; C. Castellani, C. Di Castro, P.A. Lee, and M. Ma, Phys. Rev. B **30**, 527 (1984).
30. H. Stupp et al., Phys. Rev. Lett. **72**, 2122 (1994).
31. T. Wojtowicz et al., Phys. Rev. Lett. **56**, 2419 (1986); the critical exponent z is related to G^* determined in that paper by $z = 3 - 2/G^*$.
32. J. Jaroszynski et al. Phys.Rev.Lett. **75**, 3170 (1995).
33. Y.Shapira and R.L.Kautz, Phys.Rev.B **10**, 4781 (1974).
34. H.Fukuyama and Y.Yosida, Physica **105 B+C**, 132 (1981).
35. B.A.Aronzon, N.K.Chumakov, T.Dietl and J.Wrobel, Sov.Phys.JETP **78**, 216 (1994).

MAGNETIC FIELD DRIVEN INSULATOR TO QUANTUM HALL TRANSITION OF 2D HOLES IN A STRAINED SIGE/GE QUANTUM WELL

S.-H. SONG, D. SHAHAR, and D.C. TSUI
Department of Electrical Engineering, Princeton University
Princeton, New Jersey 08544, U.S.A.

Y.H. XIE and DON MONROE
Lucent Technologies Bell Labs Innovations, Murray Hill, New Jersey 07974, U.S.A.

We study the magneto-transport property of highly disordered two dimensional hole system in a strained Ge quantum well. We observe a clear transition from a low magnetic field insulator to the $\nu = 3$ quantum Hall state as well as to the $\nu = 1$ state. Based on our results, we construct an experimental phase diagram in the disorder versus magnetic field plane.

1 Introduction

In a recent theoretical paper, Kivelson, Lee and Zhang (KLZ)[1] considered the global properties of the spinless two dimensional electron system in a magnetic field at low temperature. Based on a law of corresponding states that relates the properties of the system at different filling factors, and a selection rule that concerns the possible continuous phase transitions in the integer quantum Hall effect (IQHE), they were able to construct a global phase diagram in the disorder versus magnetic field (B) plane. An important consequence is that only $\nu = 1$ to insulator transition is allowed.

In experiments, qualitatively different phase diagrams were obtained for different materials or even for similar materials by various groups[2,3,4]. While theoretical attempts[5] were made to resolve the difference, this issue is still controversial. Evidently, studying more samples of various materials is essential to our understanding of the phase diagram.

2 Sample and Experiment

We conducted a magneto-transport study of a new type of 2D system. Our system is composed of holes in a strained Ge quantum well sandwiched between $Si_{0.4}Ge_{0.6}$ layers above and below. A built-in coherent strain due to the lattice mismatch between Si and Ge removes the heavy and light hole degeneracy, normally present in the unstrained material. As a result, the heavy hole ($M_J = \pm\frac{3}{2}$) band becomes the lowest doubly degenerate states for holes.

The strain split light hole ($M_J = \mp \pm\frac{1}{2}$) band is located 112mev higher in energy as estimated from a simple $\boldsymbol{k} \cdot \boldsymbol{p}$ Luttinger model calculation[6]. Using the effective mass value of $0.1m_e$ measured from cyclotron resonance, the Fermi energy is estimated to be approximately 4meV for the highest hole density ($p = 1.7 \times 10^{11}$ cm^{-2}) presented here. Therefore, the band nonparabolicity effect is believed to be negligible in our study.

Using silicon oxynitride grown by PECVD as an oxide layer, we prepared a Hall bar patterned MOSFET from the wafer grown by MBE technique[7]. We utilize the metallic front gate to vary the hole density, or equivalently the effective disorder in our sample. A standard four-terminal AC technique with the excitation current of 0.1nA was used to measure the diagonal, ρ_{xx}, and Hall, ρ_{xy}, resistivities in a dilution refrigerator with a 15T magnet at temperatures from 30 to 350mK.

3 Results and Discussion

Throughout our range of study, we identified a clear transition magnetic field, B_c that separates the low B insulator and the IQHE phase. We used the value of B_c together with the position of the ρ_{xx} peaks between the IQHE minima at different gate voltages, to construct a phase diagram for our sample. This phase diagram has the features: a) at $p \geq 1.2 \times 10^{11}$ cm^{-2}, a direct insulator to $\nu = 1$ IQHE is observed, b) for $p < 1.28 \times 10^{11}$ cm^{-2}, a direct insulator to $\nu = 3$ IQHE is observed.

The observation of the low B insulator to the $\nu = 3$ IQHE transition is a clear violation of the KLZ selection rule, which stems from the floating picture suggested by Khmel'nitzkii and Laughlin[8]. Further theoretical efforts are needed to understand the nature of this transition.

4 Summary

We construct an experimental phase diagram for 2D holes in a strained Ge quantum well. Our diagram is in qualitative disagreement with the theoretical one based on the floating scenario.

Acknowledgments

We would like to thank D.S. Kim for technical assistance with the MOSFET devices and Kun Yang for useful discussions. This work was supported by the NSF.

References

1. S. Kivelson, D.H. Lee, and S.C. Zhang, Phys. Rev. B **46**, 2223 (1992)
2. S.V. Kravchenko, Whitney Mason, J.E. Furneaux, and V.M. Pudalov, Phys. Rev. Lett. **75**, 910 (1995)
3. I. Glozman, L.W. Wong, C.E. Johnson, and H.W. Jiang, Proc. 6th Int. Conf. on Hopping and Related Phenomena, 327 (Jerusalem, 1995)
4. D. Shahar, D.C. Tsui, and J.E. Cunningham, Phys. Rev. B **52** R14, 372 (1995)
5. M.M. Fogler and B.I. Shklovskii, Phys. Rev. B **52**, 17366 (1995); A.M. Tikofsky and S.A. Kivelson, Phys. Rev B **53**, 13275 (1996)
6. See, for example, T. Manku and A. Nathan, Phys. Rev. B **43**, 12634 (1991)
7. Y.H. Xie, Don Monroe, E.A. Fitzgerald, P.J. Silverman, F.A. Thiel, and G.P. Watson, Appl. Phys. Lett. **63**, 2263 (1993)
8. D.E. Khmel'nitzkii, JETP Lett. **38**, 552 (1983); R.B. Laughlin, Phys. Rev. Lett. **52**, 2304 (1984)

INTERBAND TRANSITIONS OF $Si_{1-x}Ge_x$ FILMS AND $(Si_mGe_n)_p$-SUPERLATTICES CHARACTERIZED BY OPTICAL AND MAGNETO-OPTICAL REFLECTANCE SPECTROSCOPY

M. Vergöhl, K. Dettmer, J. Schoenes, and F.R. Keßler
Institut für Halbleiterphysik und Optik, and Hochmagnetfeldanlage, Mendelssohnstraße 3, 38106 Braunschweig, Germany

We have investigated optically (wavelength modulated reflectivity) and magneto-optically (mo Kerr-effect) in the spectral range of the E_1, and E_0'-interband transitions different $Si_{1-x}Ge_x$ thin films and $(Si_mGe_n)_p$-superlattices on (001)-Si substrate. The off-diagonal element of the dielectric tensor $\tilde{\varepsilon}_{xy}$, at B=10 T was determined by inverting numerically the matrix equation for the mo response of the multilayer. An effective medium approximation was used for the superlattices. Here a new method to fit the spectra was applied. Both the differential dielectric function and the off-diagonal element were fitted with the same set of parameters. Thus, the E_1, $E_1 + \Delta_1$ and E_0'-triplet in Si could be resolved even at 300 K. We fit the sum-g-values individual of the three interband transitions. From these results, we were able to determine the E_1-interband matrix element as $\Pi(x) = 0.06 + 0.3 \times x^2 [a.u.]$. In the superlattice the Ge and Si E_1-transitions nearly vanish, and another structure appears at 2.9 eV. For this structure, two critical points are resolved with gap energies and sum-g-values comparable to the values of the E_1-transitions of $Si_{1-x}Ge_x$. This indicates that the structure is due to the E_1-interband transitions of $Si_{0.6}Ge_{0.4}$ at the Si/Ge-interfaces.

1 Theoretical considerations

The complex mo Kerr effect of a multilayer system containing N plane-parallel homogeneous layers is given by the matrix expression [1] (indices $\pm$ correspond to the $\sigma^{\pm}$-Eigenmodes of the Faraday configuration):

$$\tilde{\Theta}_K = \Theta_K + i\,\mathrm{artanh}(\Psi_K) = \frac{1}{2i} \cdot \ln\left(\frac{\tilde{m}^{+}_{res,11} \cdot \tilde{m}^{-}_{res,21}}{\tilde{m}^{-}_{res,11} \cdot \tilde{m}^{+}_{res,21}}\right), \quad \text{with} \qquad (1)$$

$$\begin{pmatrix} \tilde{m}^{\pm}_{res,11} & \tilde{m}^{\pm}_{res,12} \\ \tilde{m}^{\pm}_{res,21} & \tilde{m}^{\pm}_{res,22} \end{pmatrix} = \prod_{i=0}^{N} \begin{pmatrix} e^{+i\Phi_i^{\pm}} & 0 \\ 0 & e^{-i\Phi_i^{\pm}} \end{pmatrix} \times \frac{1}{\tilde{t}^{\pm}_{i,i+1}} \begin{pmatrix} 1 & \tilde{r}^{\pm}_{i,i+1} \\ \tilde{r}^{\pm}_{i,i+1} & 1 \end{pmatrix},$$

$$\text{and} \quad \Phi_i^{\pm} = \frac{\omega}{c}\tilde{n}_i^{\pm}, \; \tilde{r}^{\pm}_{i,i+1} = \frac{\tilde{n}_i^{\pm} - \tilde{n}_{i-1}^{\pm}}{\tilde{n}_i^{\pm} + \tilde{n}_{i+1}^{\pm}} \quad , \quad \tilde{t}^{\pm}_{i,i+1} = \frac{2\tilde{n}_i^{\pm}}{\tilde{n}_i^{\pm} + \tilde{n}_{i+1}^{\pm}}.$$

From Eq. (1), the values $\tilde{n}^{\pm}$ of the homogeneous films were calculated and, with Maxwell's relation for non-magnetic material, $\tilde{\varepsilon}^{\pm} = (\tilde{n}^{\pm})^2$, the off-diagonal element of the dielectric tensor, $\tilde{\varepsilon}_{xy} = (\tilde{\varepsilon}^{+} - \tilde{\varepsilon}^{-})/2i$, has been determined. For this purpose, the film thickness, the optical constants at $B = 0$ of both the film and the substrate, and the mo off-diagonal element of the dielectric tensor of the substrate had to be known. The latter was taken from measurements of the substrate alone, the other values were taken from optical reflectivity measurements. The optical constants and the film thickness of the superlattice SL64 = $(Si_{16}Ge_{10})_{33}$ have been determined from wavelength modulated (wm) reflectivity data, using literature data for $\tilde{\varepsilon}(E)$ [2] and an effective medium consisting of Si, Ge and $Si_{0.60}Ge_{0.40}$. Self-consistency of the mo spectra was confirmed by applying Kramers-Kronig transformations to the real and imaginary part of $\tilde{\varepsilon}_{xy}$. In the polarization ($\sigma^{\pm}$-)modulation, the mo dielectric function $\tilde{\varepsilon}^{\pm}(E,B) = \tilde{\varepsilon}(E) \pm \Delta\tilde{\varepsilon}(E,B)$ is given by ($\Delta\tilde{\varepsilon} \ll \tilde{\varepsilon}$):

$$\tilde{\varepsilon}^{\pm}(E,B) = \sum_{j=1}^{M} \left[\tilde{\varepsilon}_j(E) \pm \frac{1}{2}\bar{g}_j \mu_B B_0 \cdot \left(\frac{\partial \tilde{\varepsilon}_j(E)}{\partial E} + 2i\,\frac{\mathrm{Im}(\tilde{\varepsilon}_j(E))}{E}\right)\right] \qquad (2)$$

with j running the M Van-Hove singularities involved. For $\tilde{\varepsilon}_j(E)$, we used analytical line shapes for the dispersion of the j-th critical point (CP). In the wm, the differential dielectric function is proportional to

$$\Delta\tilde{\varepsilon}_{j,zz} \sim F_j \cdot \tilde{f}(E, E_{G,j}, \Gamma_j) \cdot \Delta\omega_{wm}\,, \quad \text{with} \quad \Delta\omega_{wm} = const \qquad (3)$$

and the splitting of the dielectric function in a magnetic field ($\Delta\tilde{\varepsilon} = 2i\tilde{\varepsilon}_{xy}$):

$$\Delta\tilde{\varepsilon}_j \sim F_j \cdot \tilde{f}(E, E_{G,j}, \Gamma_j) \cdot \Delta\omega_{S,j}\,, \quad \text{with} \quad \hbar\Delta\omega_{S,j} = \bar{g}_j \mu_B B. \qquad (4)$$

For each CP, three parameters determine the complex line shapes of $\Delta\tilde{\varepsilon}_{zz}$ and $2i\tilde{\varepsilon}_{xy}$: gap energy E_G, phase factor φ, and broadening parameter Γ. On assumption that these parameters are not influenced by the magnetic field (E_G has to be corrected by the Landau-shift), a significant improvement of the resolution of a multiplet of CP's is obtained if these three parameters are fitted simultaneously to both $\Delta\tilde{\varepsilon}_{zz}$ and to $\Delta\tilde{\varepsilon} = 2i\tilde{\varepsilon}_{xy}$.

2 Results and Discussion

In Ref.[3], the second derivative of the dielectric function of pure Si was fitted with two CP's in the E_1-interband region. One CP was used for the $E_1/(E_1 + \Delta_1)$-doublet, the other for the E_0'-transition. According to this model, we tried to fit the differential (B=0) dielectric function $\Delta\tilde{\varepsilon}_{zz}$ together with the off-diagonal element $\tilde{\varepsilon}_{xy}$ using two CP's. Indeed, $\Delta\tilde{\varepsilon}_{zz}$ can be fitted well, but the fit of $\tilde{\varepsilon}_{xy}$ is very poor. The latter shows an energy upshift of ≈ 200 meV compared to the field free spectra, as well as a different shape of the spectra. In

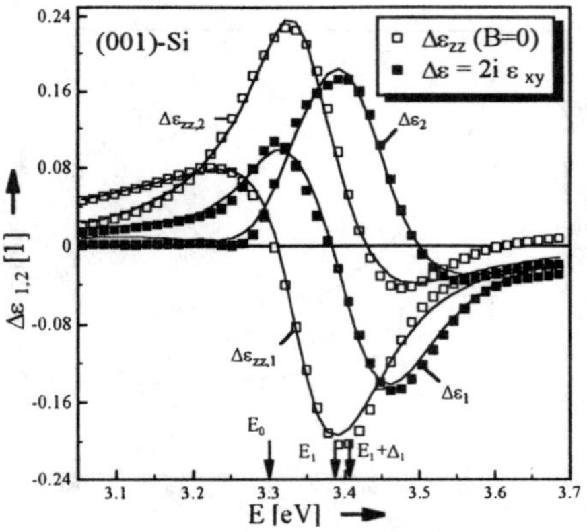

Figure 1: *Line shape fit to* $\Delta\tilde{\varepsilon}_{zz}$ *and* $2i\tilde{\varepsilon}_{xy}$ *of (001)-Si. For* E_1 *and* $E_1 + \Delta_1$*, excitonic line shapes were used, for* E_0'*, a 2D-CP was used. Index 1: real part, Index 2: imaginary part.*

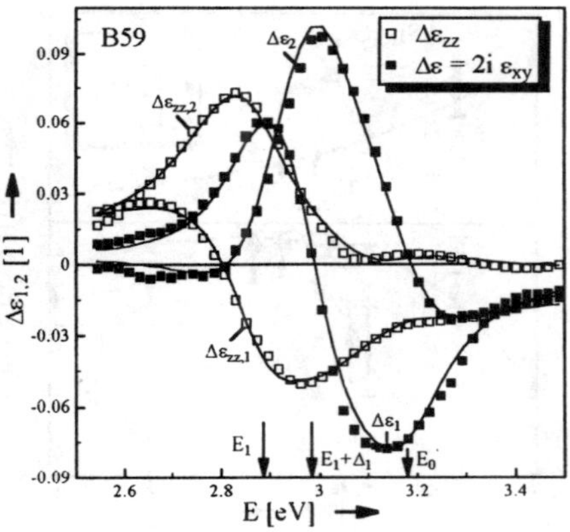

Figure 2: *Line shape fit to the modulated dielectric function and to the off-diagonal element* $\tilde{\varepsilon}_{xy}$ *of sample B59 with* x*=0.38. Three 2D-CP's were used. Index 1: real part, Index 2: imaginary part.*

Ref.[4,5], the discrepancy was attributed to a large difference of $\bar{g}$-values between E_1 and $(E_1 + \Delta_1)$. Therefore, we separated the E_1-doublet of Ref.[3] into two CP's, with a positive and a negative $\bar{g}$-factor, respectively. Fig. 1 shows that the fit is acceptable now.

For all samples with a higher Si content, by combining the $E_1/(E_1 + \Delta_1)$-doublet, the optical spectra at B=0 could be fitted well with only two CP's. However, three CP's (E_1, $E_1 + \Delta_1$,E_0') are required for all composition parameters x in order to fit both the optical and the mo spectra. Fig. 2 shows the result for a $Si_{0.38}Ge_{0.62}$ layer. For the E_1 doublet with an energy separation closer than the lifetime broadening, the value for the spin-orbit splitting Δ_1 was kept fixed. This is the case for the E_1-CP's with $x \leq 0.3$. In Fig. 3, the $\bar{g}$-values of the E_1, $(E_1+\Delta_1)$ and E_0' CP's are plotted as a function of x. In the superlattice SL64, only the strong structure, which appears in the mo spectra at 2.9 eV, has been fitted. The values of the gap energies and of the $\bar{g}$-factors are close to the values of the E_1-CP's of a $Si_{0.42}Ge_{0.58}$-film. This indicates that the mo spectra in this reduced spectral region are mainly determined by the dispersion of the E_1-like mixed-crystal transitions, which occur at the Si/Ge interfaces. From an extended $\underline{k} \cdot p$ theory [6], the E_1-interband matrix element Π of (001)-$Si_{1-x}Ge_x$ was calculated from the difference of the $\bar{g}$-factor between E_1- and $(E_1 + \Delta_1)$, using the combined effective mass of conduction band and valence band:

$$\Delta\bar{g} = \frac{m_0}{m^*_{CV}(E_1 + \Delta_1)} - \frac{m_0}{m^*_{CV}(E_1)} = \frac{-E_p \cdot \Delta_1}{2E_1 \cdot (E_1 + \Delta_1)} + \frac{4\Pi^2}{\Delta_1} \qquad (5)$$

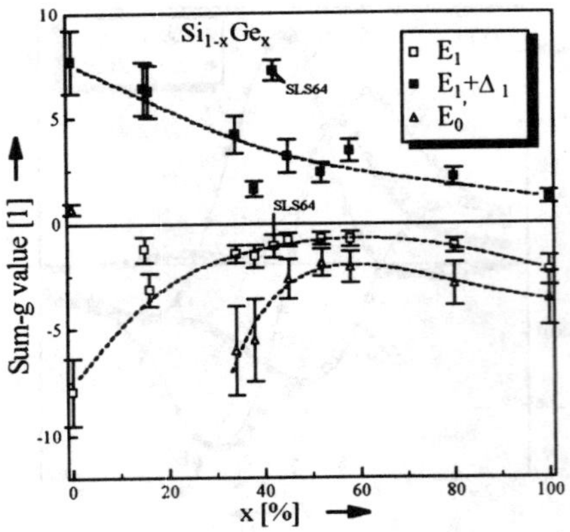

Figure 3: *$\bar{g}$-values of the transitions E_1, $E_1 + \Delta_1$ and E_0' of $Si_{1-x}Ge_x$ mixed crystals. Also marked are the values for the E_1-like transitions for the superlattice SLS64.*

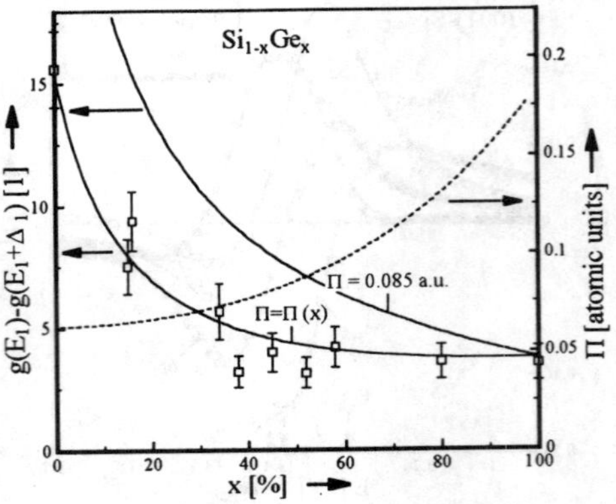

Figure 4: *Fitted values of the difference of the $\bar{g}$-factor of the E_1-transition to that of the $E_1 + \Delta_1$-transition. Solid line: Calculated with Eq. (5). Dotted: transition matrix element $\Pi(x)$ as used for calculation.*

We used literature data for the interaction energy E_p and the spin-orbit splitting Δ_1 of Si and Ge. The values for $Si_{1-x}Ge_x$ have been linearly interpolated. Fig. 4 shows the result of $\Delta\bar{g}(x) = \bar{g}(E_1) - \bar{g}(E_1 + \Delta_1)$. Using the matrix element $\Pi(x) = 0.06 + 0.3 \times x^2$ (dotted line), the curve of $\Delta\bar{g}$ can be fitted quite well. The top line is calculated using the interband matrix element of (111)-Si. For the superlattice, a value similar to that of $\Pi(x = 0.42)$ was found.

Acknowledgments

This work was supported by the Volkswagen Stiftung.

References

1. R. Nies, F.R. Keßler, *phys. stat. sol.* (a),**111**, 639, (1989)
2. J. Humlićek, M. Garriga, M. I. Alonso, M. Cardona, *J. Appl. Phys.* **65** (7), 2827, (1989)
3. P. Lautenschlager, M. Garriga, L. Viña, M. Cardona, *Phys. Rev. B* **36** (9), 4821, (1987)
4. F.R. Keßler, J. Metzdorf, Landau Level Spectroscopy: Interband Effects and Faraday Rotation, Elsevier Verlag, Amsterdam, (1991)
5. J.Metzdorf, Wiss. Berichte HMFA **8 A**, (1983)
6. A. Fujimori, H. Fukutani, G. Kuwabara, *J. Phys. Soc. Jpn.* **45** (4), 910, (1978)

HIGH-MAGNETIC FIELD RESEARCH IN SEMICONDUCTING MATERIALS AT NRIM

G. KIDO, K. TAKEHANA, M. OSHIKIRI,
J. VANACKEN[1], T. TERASHIMA, D. SYMONS[†],
T. ASANO, T. TAKAMASU, H. AOKI
National Research Institute for Metals, Sakura 3-13, Tsukuba, Ibaraki 305, Japan

R. NICHOLAS
Physics Department, Oxford University, Clarendon laboratory, Park Rd. Oxford, UK

A 36.5 T field became possible in steady state using a hybrid magnet system at National Research Institute for Metals. Magneto-optical measurement in far-infrared region has been performed with a Fourier transform spectrometer. Magnetotransport measurements are proceeded by the use of 16 T and 20 T superconducting magnets with dilution refrigerator. The cyclotron resonance are investigated with conventional pulsed magnet and far-infrared laser and Gunn diode.

Introduction

High magnetic fields are quite useful to discover novel phenomena in condensed matters as well as to test superconducting wires. Recently, various kinds of high-field magnets, such as a 36 T hybrid magnet [1, 2], a 73 T pulsed magnet [3] and a 21 T superconducting magnet [4], were constructed at National Research Institute for Metals (NRIM) and these magnets began to be used in measurements of magneto-transport effect [5], magnetooptics [6], cyclotron resonance [7], magnetization processes and the critical current. Besides these magnets, commercially available superconducting magnets combined with dilution refrigerator have been mainly used in the fermiological studies of organic conductors [8], heavy fermion compounds [9] and superconducting materials [10] as well as in the semiconductor studies. Other researches such as the application of high magnetic field on chemical reaction, material strength, magnetic separation etc. will be studied using a high-repetition 40 T pulsed magnet and a liquid-helium-free 10 T superconducting magnet with large room temperature bore.

The present paper will introduce the magnet development and the high-field

Present address *Laboratorium voor Vaste-Stoffysica en Manetism, Kathoric University of Lueven
[†]High Magnetic Field Laboratory, CNRS, Grenoble

experiments in semiconductors at NRIM.

2. Magnet development

2.1 Hybrid magnet

Figure 1 schematically shows the NRIM hybrid magnet system which is designed to generate 40 T in a 30 mm or 35 T in a 50 mm room temperature bore. A high power resistive coil is placed concentrically in a superconducting magnet with a 400 mm room temperature bore. The superconducting coil, which is graded in four and the inner two parts are made of $(Nb\text{-}Ti)_3Sn$ wire, is immersed in the 4.2 K liquid helium bath. The superconducting magnet is excited by a 1500 A current source (60 V) with a sweep rate of 0.5 T/min. The superconducting magnet has been excited up to 14.2 T more than 30 times without quenching. The resistive magnet, which are a polyhelix type composed of 18 and 15 helices, are designed to produce 25 T (30 mm) and 20 T (50 mm) and with a 15 MW power source [1, 2]. The resistive coil is cooled by circulating de-ionized water between adjacent helices along the axial direction through four pumps with a rate of approximately 700 m^3. The de-ionized water is directly cooled by three turbo-refrigerators and the second water of which is cooled by a cooling tower. The hybrid magnet system has been tested step by step, and the highest field of 36.3 T was recorded in September, 1996.

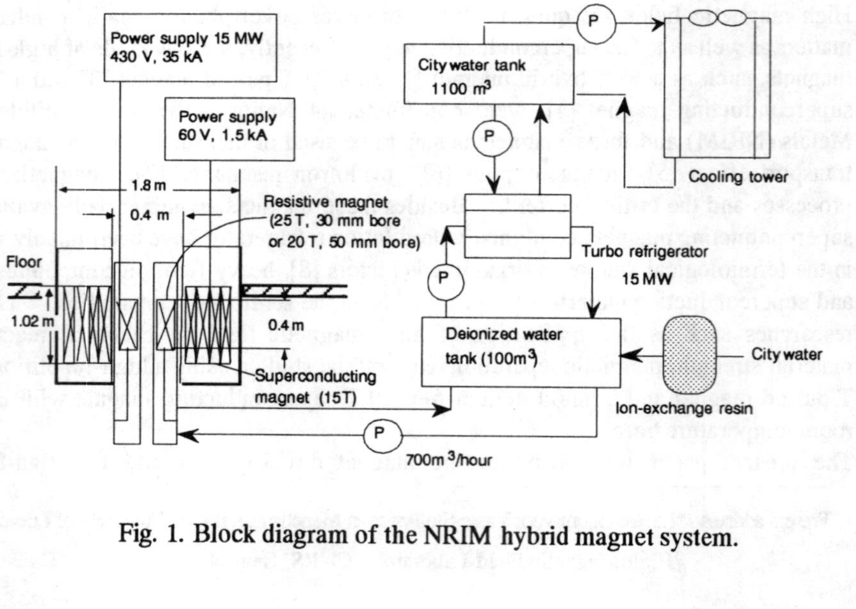

Fig. 1. Block diagram of the NRIM hybrid magnet system.

2.2 Pulse magnet

Higher fields than 30 T can be generated by the use of pulsed magnet. Several scales of capacitor banks are installed for the energy source of pulsed magnets: 1.6 MJ (5 kV), 200 kJ (10 kV), 125 kJ (5 kV) and 50 kJ (2 kV). Figure 2 shows the cross-section of the coil for 1.6 MJ bank. The wire is made of hard drawn Cu-Ag alloy with tensile strength of 100 kg/mm^2 and is reinforced by fiber-glass [11]. The magnet is operated at the nitrogen temperature to reduce the resistance of the coil. The repetition time of the field, however, takes more than 3 hours after the generation of 65 T with 100 ms duration [1]. The cyclotron resonance was carried out with small solenoid coil excited by a small capacitor bank because a ten milli-second pulse is enough to measure the transmission intensity of the far-infrared radiation from laser and Gunn diode. Recently, a water-cooled magnet was fabricated with Cu-Ag Bitter plates. Higher field than 40 T is generated in every 5 minute for 23 mm clear bore.

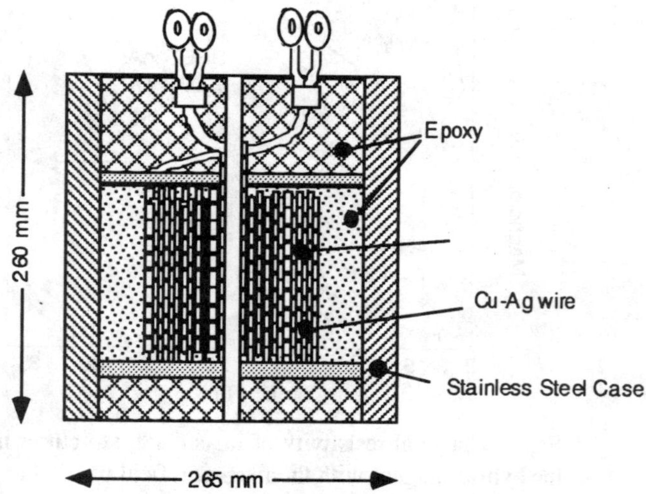

Fig.2. Cross-section of thepulsed magnet for 1.6 MJ capacitor bank.

2.3 Superconducting magnet

A 21 T superconducting magnet with 50 mm cold bore was constructed for the evaluation of advanced superconductors and the test of small size high-field superconducting magnet [4]. Since the discovery of Bi-oxide superconducting material at NRIM, the superconductor tape has been developed for the 1 GHz NMR magnet. Double pancakes which are made of silver sheathed Bi/2212 multifilamentary tapes are stacked to constitute a superconducting magnet. Up to now, the Bi-oxide magnet generated 1.8 T in 21 T resulting the world highest record of 22.8 T only be the superconducting magnet [12].

3. Experiments in semiconductors

3.1 Magnetotransport

The electronic states of a sheet of carriers at the heterojunction form two dimensional subbands in a conflict potential. Attracting phenomena are caused by the application of high magnetic field to the two dimensional subbands. Figure 3 shows the magnetoresistivity of the type II InAs/GaSb system with different confined geometry

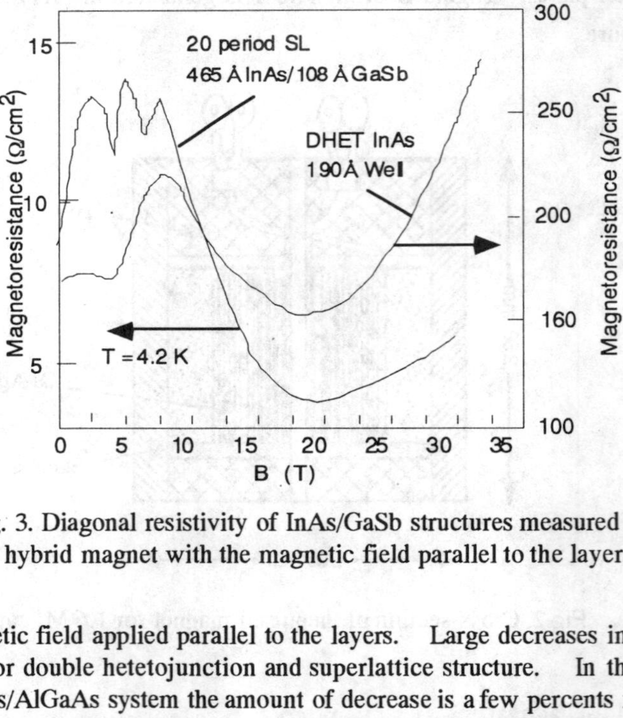

Fig. 3. Diagonal resistivity of InAs/GaSb structures measured in the hybrid magnet with the magnetic field parallel to the layers.

with the magnetic field applied parallel to the layers. Large decreases in resistance are observed for double hetetojunction and superlattice structure. In the previous report for GaAs/AlGaAs system the amount of decrease is a few percents for parallel field. The decreased may be due to the enhancement of mixing of the electron and hole dispersion relation [13] and mixing of the electronic and magnetic quantizations by the parallel filed. More measurements and theoretical calculations will be necessary to understand the large decrease in resistance.

At present, very low temperature experiments are performed using 16 T and 20 T superconducting magnets each of which a dilution refrigerator is incorporated. We have studied the current-induced breakdown phenomena of integer and fractional quantum Hall effects [14]. Figure 4 shows diagonal resistivity ρ_{xx} of a high mobility sample as a function of magnetic field, in which a clear Shubnikov-de Hass oscillation is seen and ρ_{xx} becomes zero at several integer and fractional filling factors. When the magnetic fields is fixed at one of these regions and the current is increased, ρ_{xx} abruptly increases at a critical current to break the non-dissipative current state of the quantum Hall effect. Measurements of the critical current with many samples at different filling factors ν including several fractional fillings revealed that the magnetic field dependence of the critical current is not simply determined by the applied magnetic field, though the critical current has been believed to change its value as a function of magnetic field. Especially at odd integer filling in tilted magnetic field, the critical current is considered to be determined by macroscopic spin structures, the skyrmions [15].

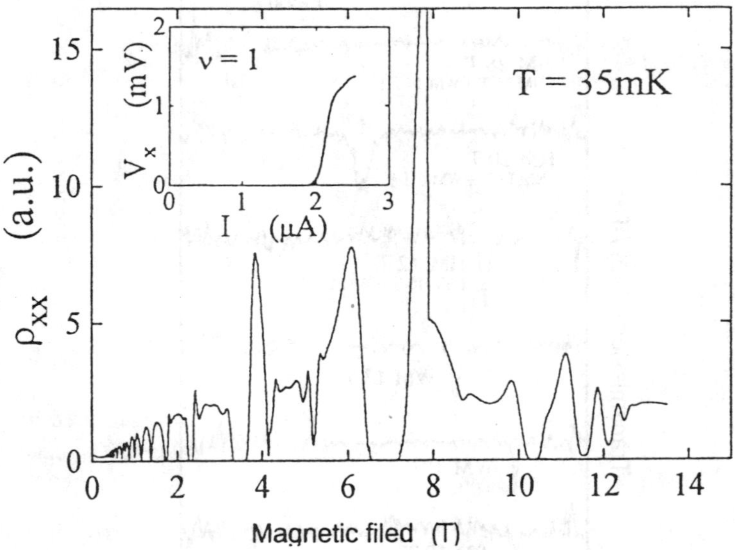

Fig 4. Diagonal resistivity ρ_{xx} vs magnetic field in a high mobility sample 35 mK. When the current is increased at the fxed magnetic field where ρ_{xx} takes minima, the resistivity increases abruptly above the critical value (inset).

3.2 Magnetooptics

An optical spectroscopy in the far-infrared region is quite informative to investigate semiconductors in high magnetic fields, because energy range of the Zeeman splitting and cyclotron resonance corresponds to 0 - 40 cm-1 against 0 - 30 T for $g^* = 2$ and $m_c^* = m_e$. We have developed a Fourier-transform spectrometer for milli- to submillimeter range using a lamellar grating [6]. The spectrometer was employed to calibrate the field strength of the hybrid magnet by measuring the ESR spectrum for $Cd_{1-x}Mn_xSe$ in a continuous field as shown in Fig. 5. The coil constant of superconducting magnet part (SM) and that of resistive magnet (WM) were determined by the independent excitation. Afterward the field strength by the calculation in the hybrid mode (HM) was compared with that by the ESR signal directly.

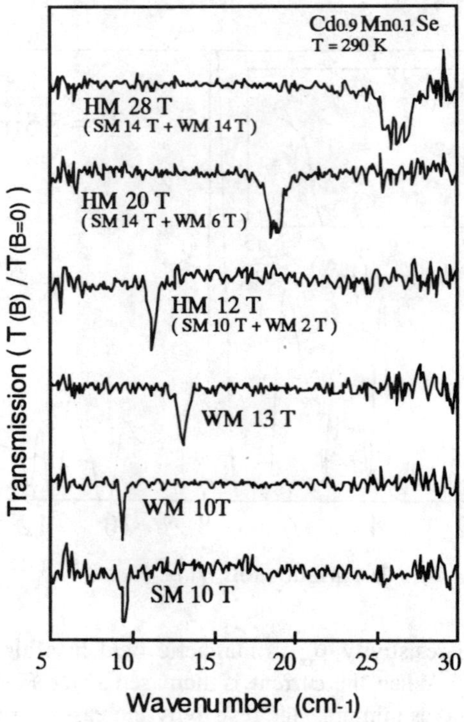

Fig. 5. ESR lines in $Cd_{0.9}Mn_{0.1}Se$ at room temperature. WM, SM and HM represent the excitation mode of the hybrid magnet.

3.3 Cyclotron resonance

GaP is known to have a "camel's back" structure at the bottom of the conduction band on the Δ-axis near the zone boundary X point. So that the Landau energy level, which is doubly degenerated at low magnetic fields, splits as the field increases as shown in Fig. 6. We measured the transmission spectra for S-doped epitaxial layer of GaP with B//<100>. over a wide range of energy (119 - 1500 μm) in pulsed fields. Transmission minima in Fig. 7 correspond to the cyclotron resonance fields. The absorption peaks PT-1~PT-5 originate from the cyclotron resonance in the Landau energy structure due to the array-like energy surface whose major axis is parallel to the magnetic field. The value of m_t unchanged by the magnetic field and was determined to be (0.251 ± 0.005) m_0, which agrees with the previous result [16]. On the other hand, weak absorption peaks labeled as PL-1~PL-4 are from the array-like energy surface whose major axis is perpendicular to the applied field. The effective masses m_t, m_l and the camel's back height ΔE were determined to be m_l =(0.90 ± 0.04) m_0 and ΔE = (2.7 ± 0.1) meV assuming that the absorptions PL-1, PL-2, PL-3 and PL-4 corespond to the transitions $0^+ \rightarrow 2^-, 0^- \rightarrow 2^+, 0^- \rightarrow 1^+$ and $1^- \rightarrow 2^+$. The field dependence of the Landau energies illustrated in Fig. 6 were derived using the present values [7].

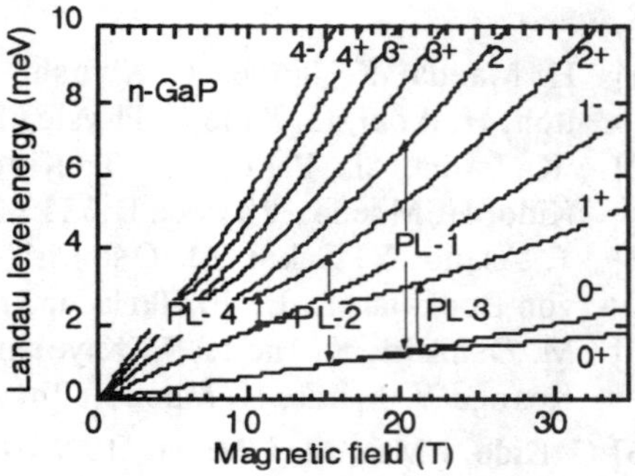

Fig.6. Calcurated Landau levels.

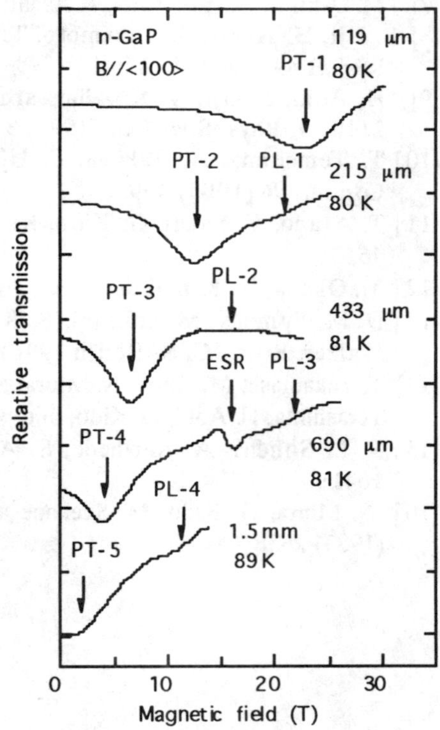

Fig. 7. Transmission change due to the cyclotron resonance in n-GaP.

References

[1] H. Maeda, K. Inoue, T. Kiyoshi, T, Asano, Y. Sakai, T. Takeuchi, K. Itoh, H. Aoki, G. Kido, , Physica B **216** (1996) 141.
[2] K. Inoue, T. Takeuchi, T. Kiyoshi, T, Asano, Y. Sakai, K. Itoh, G. Kido, H. Maeda, Physica B **211** (1995) 17.
[3] T. Asano, Y. Sakai, M. Oshikiri, K. Inoue, H. Maeda, G. Heremans, L. von Bockstal, L. Li, F. Herlach, Physica B **201** (1994) 556.
[4] M. Oshikiri, K. Inoue, T. Kiyoshi, T. Takeuchi, K. Itoh, K. Tanaka, M. Kosuge Y. Iijima, H. Maeda, Physica B **201** (1994) 551.
[5] G. Kido, J Mag. Mag. Mater. (1996) (EMMA-95, Wienna 1995) .
[6] K. Takehana, M. Oshikiri, G. Kido, A Takazawa, M. Sato, K. Nagasaka, M. Hase and K. Uchinokura, Physica B **216** (1996) 354.
[7] M. Oshikiri, K. Takehana, T. Asano and G. Kido, J. Phys. Soc. Jpn. **65** (1996).
[8] S. Uji, H. Aoki, M. Tokumoto, T. Kinoshita, Y. Tanaka, H. Anzai, Phys. Rev. B 49 (1994) 732
[9] H. Aoki, S. Uji,, A.K. Albessard, Y. Onuki, Phys. Rev. Lett. 71 (1993) 2110, J. Phys. Soc. Jpn. (1993) 3157
[10] T. Terashima, H. Takeya, S. Uji, K. Kadowaki, H. Aoki, Solid Sate Comm., **96** (1995) 459.
[11] T. Asano, Y. Sakai, G. Kido, K. Inoue H. Maeda,, Physica B 211 (1995) 46.
[12] M. Okada, H. Kumakura *et al*., Appl. Phys. Lett. submitted.
[13] D. M. Symons, M. Lakrimi, S. Khym,. R. J. Nicholas, N. J. Mason, P. J. Walker, Proc. ICPS (Berlin 1996) submitted.
[14] T. Takamasu, M. Ohno, N. Miura, M. Kato, A. Endo, S. Katsumoto, Y. Iye, T. Terashima, H. Aoki, G. Kido, ibid.
[15] S. L. Sondhi, A. Karlhede, S. A. Kivelson, Phys. Rev. B **47** (1993-II) 16419
[16] N. Miura, G. Kido, M. Suekane and S. Chikazumi: J. Phys. Soc. Japan **52** (1983) 2838

RECENT MAGNETO-OPTICAL INVESTIGATIONS AT THE GRENOBLE HIGH MAGNETIC FIELD LABORATORY

G. MARTINEZ
Grenoble High Magnetic Field Laboratory, MPI-FKF /CNRS, 25, Avenue des Martyrs, BP 166, 38042 Grenoble Cedex 9, France

New results recently obtained in our laboratory are reviewed. They concern the optical properties of the heterostructures made with partially ordered $GaInP_2$ and GaAs which reveal very unusual properties of ligth up-conversion and those of the two dimensional electron-bound hole luminescence system which is known to be sensitive to correlations and which is used to probe the existence of Skyrmions. In both cases the study of the effect of the magnetic field appears to be essential to understand the physics relevant to all these effects.

1 Introduction

It is always difficult to review, within a restricted space, all the works done in a large laboratory like ours. Rather than making an exhaustive list of subjects, I choose to focus this report to two specific subjects treated in many more details. Fortunately, there are other colleagues who will also report in this conference and this will attenuate the selective character of my presentation.

Since the discovery of the spontaneous ordering[1] in ternary compounds such as $GaInP_2$ when grown under special conditions there have been a lot of interest to understand and explain the different physical properties related to the degree of ordering. This interest was recently more stimulated by the report[2] that the heterostructure made between GaAs and such an alloy demonstrated a visible ligth up-conversion efficiency quite important. Besides the obvious important applications it remains to answer questions like what is the exact mechanism involved in this process, to what extent the degree of ordering is a fundamental parameter governing the efficiency and /or is the relative band ordering of the conduction bands of GaAs and $GaInP_2$ playing a crucial role. I will review our most recent experimental results which bring interesting answers to those questions.

The two-dimensional electron gas (2DEG) is a well known system showing, under magnetic fields, spectacular phenomena related to correlations between electrons such as the fractional quantum Hall effect (FQHE)[3]. The elegant model proposed[4] to explain it was not indeed able to explain other correlation effects observed around odd integer values for the filling factor ν ($\nu = n/n_B$ with n equal to the number of electrons per unit area and $n_B = eB/hc$ the number of states per unit area in each Landau level). Instead a spin polarized state concept was early proposed[5] to describe the fundamental state of the 2DEG and to explain the g-factor enhancement between the two lover spin states at $\nu = 1$ for instance. This picture was questioned and a new concept[6-7] involving finite-size Skyrmions with large effective spin was proposed. In the same time Quantum Monte Carlo simulations including corrections with Landau level mixing and based on the first concept reproduce[8] quite well the experimental results. There have been recently different reports from the experimental side claiming to show "evidence" for the formation of Skyrmions. This was resulting from NMR[9] investigations, transport[10] in tilted fields or interband optical transmission[11] measurements. Since there is, obviously very often, not a unique way to interpret data, it is worthwhile to look at different experimental tests. In that respect the 2DEG-bound hole luminescence observed in appropriate samples is an interesting system because it has demonstrated its ability to reflect[12-13] a large sensitivity to correlations effects and turns out to be even more suitable to bring new light into that problem. I will review in the second part recent results obtained in this system which do not support the Skyrmion picture.

2 Up-conversion in GaAs/ $GaInP_2$ heterostructures

The experiments have been carried out on a series of $GaInP_2$ epilayers of thickness ≅ 1 μm grown by MOCVD on a buffer layer of GaAs deposited on [100] oriented semi-insulating GaAs at temperatures of 650 (sample T650), 700 (sample T700) and 750 °C (sample 750) respectively. Other samples grown in different conditions and various angles of orientation with respect to the [100] axis (series S) show a degree of ordering decreasing with increasing angle. The luminescence was excited with the 514.5 nm of an Ar^+ laser (energy higher than the band gap of $GaInP_2$) or with laser line energies below that gap but higher than the GaAs gap and provided by a Ti:sapphire laser. The normal luminescence (Fig. 1) is composed of a peak at high energy corresponding to the gap of $GaInP_2$ and a group of luminescence lines at an energy close to the gap of GaAs. Whereas the energies of these lines do not depend on the sample, the high energy line has a position decreasing in energy with the degree of ordering and somewhat dependent on the excitation intensity[14]. For instance a difference of as much as 100 meV is observed[15] between the disordered and ordered compounds. When ordering the alloy goes from the zinc-blende structure to a phase of a tetragonal symmetry with layers of GaP and InP growing sequentially along the ($1\bar{1}\bar{1}$) direction. This doubles the lattice parameter along that direction and theories have suggested[16] that the related folding of the Brillouin zone should lower the gap. With such a criteria the samples of the T-series should be as ordered as the sample S1 which was grown in the S-series without misorientation. However under excitation conditions as defined by Driessen[2], only samples from the T-series show up-conversion and not S1 (Fig. 1 and Ref. 15). When observable, the up-converted signal is made from different components and one of these is quite similar to the normal luminescence line. This leads to think that the former spectrum probes the degree of ordering at the interface whereas the latter probes it at the surface of the sample. The main features displayed in Fig. 1 have been investigated as a function of magnetic field B up to 23 T for all samples[15].

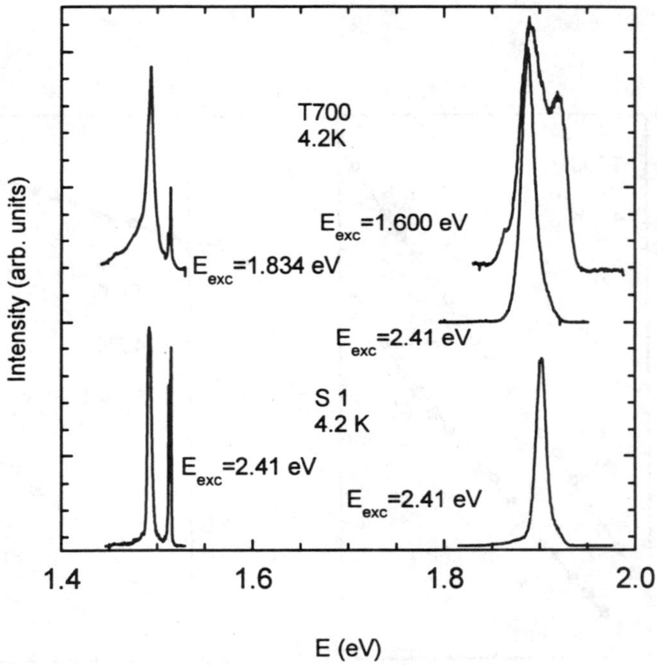

Figure 1 : Luminescence spectra of T700 (upper curves) and S1(lower curves) samples with different excitation energies.

The variation of the normal luminescence at 1.9 eV as a function of B shows a quadratic dependence characteristic of the excitonic nature of this transition. This variation is similar for all investigated samples[15]. However the dependence with B of the luminescence around 1.49 eV or that of the components of the up-converted spectra for the samples of the T-series shows a linear variation (Fig. 2) involving transitions between free and bound particles. For instance, in Fig. 2a, the linear fit gives slopes of 0.83 ± 0.02 meV/T and 0.7 ± 0.02 meV/T for the 1.492 eV and the 1.489 eV lines respectively. In Fig. 2b the slopes are 0. 25 ± 0.02 meV/T and 0.67 ± 0.02 meV/T for the 1.92 eV and the 1.892 eV lines respectively. Assuming that the energy of the free particle, with effective mass m^*, varies like $\hbar e B/(2m^*)$ one can assign these transitions and identify their origin.

The 1.492 eV line corresponds to a transition in GaAs between a free electron and an acceptor like carbon. The 1.489 eV line corresponds to a recombination between very likely the same type of acceptor and an electron with an effective mass of 0.08 ± 0.003. This is in reasonable agreement with the reported[17] value of about 0.09 for electrons in ordered $GaInP_2$. This transition should correspond to a transition across the interface and is only seen in samples showing up-converted luminescence.

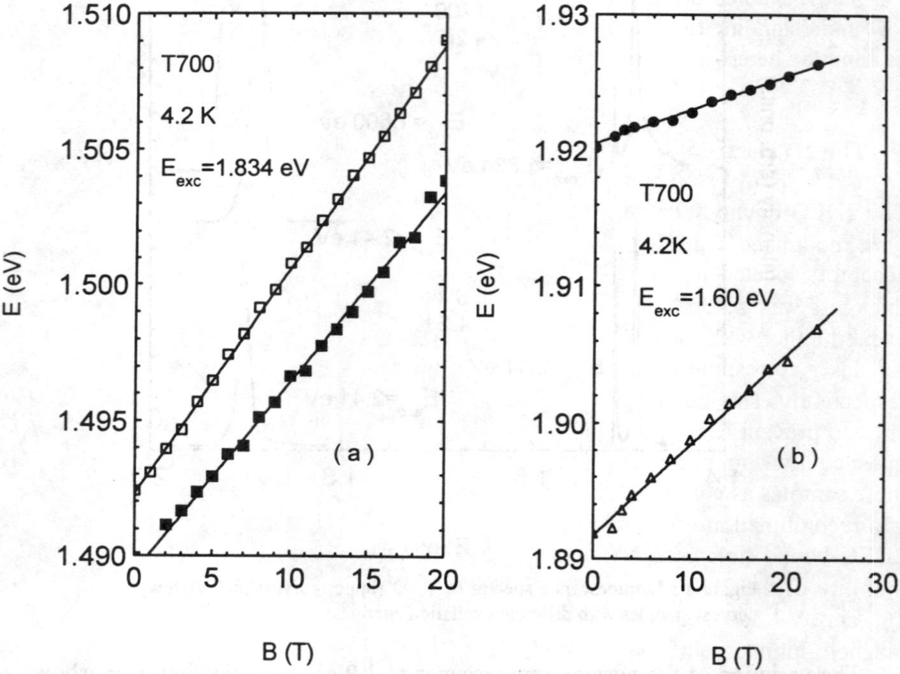

Figure 2 : Luminescence of the T700 sample in up-conversion conditions as a function of the magnetic field : (a) luminescence observed in the GaAs energy range for the two lower energy lines and (b) luminescence of the different components of the up-converted signal in the $GaInP_2$ energy range.

The analysis of the up-converted luminescence spectra is also quite interesting. The transitions are clearly originating from the $GaInP_2$ side and the similarity between the slopes of the 1.892 eV and 1.489 eV lines leads to conclude that the former one implies the recombination of a free electron in $GaInP_2$ with a bound hole. On the other hand the 1.92 eV line corresponds to the recombination of a free hole in $GaInP_2$ with a bound electron. These localized states in $GaInP_2$ are consistent with defects due to a partial ordering of the samples[15] and their energy depths are sample dependent. One observes systematically that, in all samples where the up-conversion is detected, the energy of the transition across the interface is lower than that occurring in GaAs showing that the system has a type II alignment, the conduction band of $GaInP_2$ being lower than that of GaAs by few meV ($\cong$ 3 meV). This may not be the only condition to obtain the up-conversion but if it is always true, this would favor the model of two step process[18] against the Auger process[2] to explain the up-conversion. On the other hand the disorder in $GaInP_2$ seems essential to see this phenomenon.

These studies demonstrate the importance of the magnetic field investigations for understanding and assigning the luminescence structures appearing in this type of complex heterostructures.

3 The 2D electron - bound hole luminescence

The 2DEG-bound hole magneto-luminescence has been investigated in a series of n-type modulated GaAlAs/ GaAs heterojunctions with a δ doped layer of carbon acceptors located in GaAs 35 nm away from the heterojunction. The buffer layer thickness was $\cong$ 30nm and the GaAs well width 500 nm. The luminescence was excited either with the 514.5 nm line of an Ar^+ laser or with the 647.1 nm line of a Kr^+ laser corresponding to energies higher or lower than the band gap of GaAlAs respectively. The incoming power was kept constant for each run below 5×10^{-2} W cm^{-2} to prevent heating of the sample but was varied from run to run to monitor the quasi equilibrium population of the 2DEG carrier density[12]. The large GaAs width of these samples as compared to that used in former experiments[12] insures a direct band gap recombination efficiency at least one order of magnitude lower than that of the 2DEG-bound hole recombination, a condition fulfilled for all values of B[19]. The mobility of the 2DEG measured in dark conditions reaches typically values of 4×10^5 cm^2 V^{-1} sec^{-1}. Due to the very simple structure of the acceptor states, this magneto-luminescence is strongly spin polarized[20] and as a result is a natural tool to investigate the spin polarization of the 2DEG provided the corresponding luminescence constitutes the main radiative channel in the sample and the intensity is recorded in an absolute way. In order to fulfill this condition a special optical design has been developed to perform circular polarized magneto-optical measurements under high magnetic fields and at low temperatures down to 50 mK.

The results obtained can be visualized with the analysis of moments[13] and are displayed in Fig.3 for a particular sample. The σ^- polarized spectra have been registered systematically though for very low temperatures the nature of the acceptor hole level does not require it. As this is clearly apparent in Fig. 3b the intensity has important singularities for $\nu > 1$ around integer filling factors which are also accompanied by abrupt upward energy discontinuities of the energies (Fig. 3a). At the same time the weak luminescence originating from the second electric subband shows strong oscillations in intensity (see below) and a complete analysis of all these concomitant singularities shows that they reflect filling factor dependent changes of the electric field of the heterojunction[19]. However, for $\nu < 1$, the actual system evolves in a smooth way and we propose to analyse this evolution.

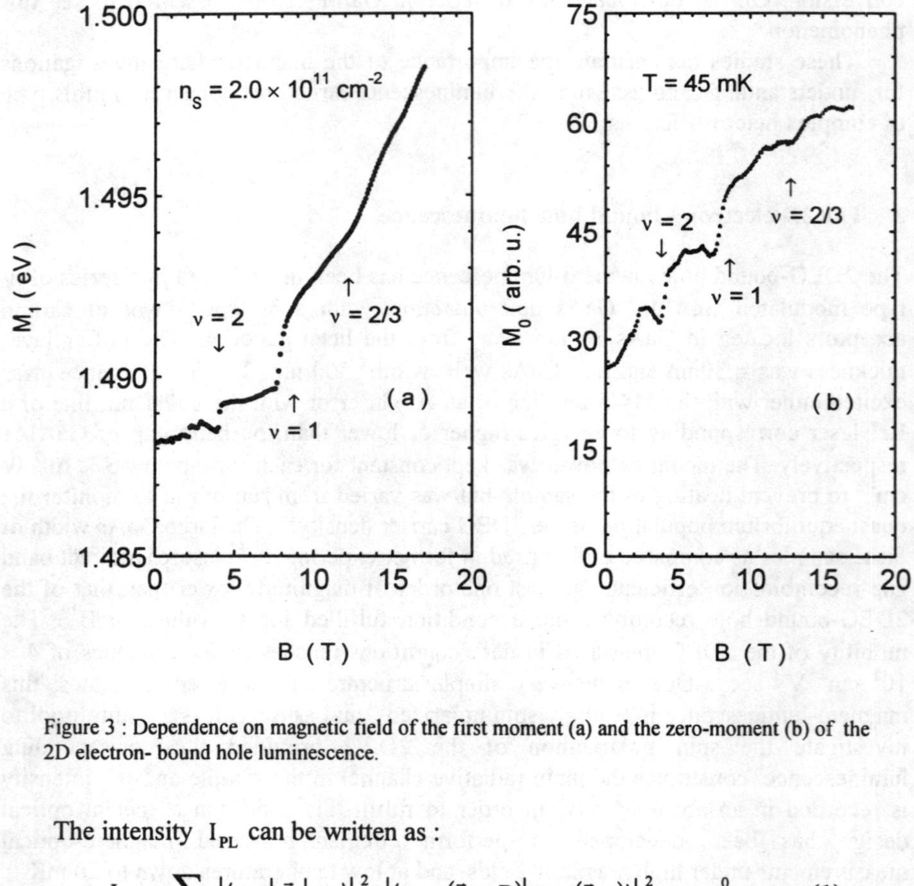

Figure 3 : Dependence on magnetic field of the first moment (a) and the zero-moment (b) of the 2D electron- bound hole luminescence.

The intensity I_{PL} can be written as :

$$I_{PL} \propto \sum_{N,m} \left|\langle u_c \mid \vec{p} \mid u_v \rangle\right|^2 \left|\langle \psi_{el}(\vec{r},z,B) \middle| \psi_h(\vec{r},z)\rangle\right|^2 n_{1N}\, n_A^0 \qquad (1)$$

where n_{1N} is the number of electrons in the Landau level N (degeneracy m) of the first electric subband and n_A^0 the number of neutral acceptors (or bound holes). In Eq. 1, u_c and u_v are the lattice periodic part of the conduction and valence band Bloch functions, $\vec{p}$ the momentum operator, and z and the 2D vector $\vec{r}$ span the real space. The hole wave function $\psi_h(\vec{r}, z)$ is taken as the ground state hydrogenic wave function with a Bohr radius of about 3 nm[21]. The evaluation of the overlap matrix element is easy when the radial part of the electron wave function $\psi_{el}(\vec{r}, z, B) = \phi_{N,m}(\vec{r}, B)\,\chi_1(z)$ is expressed in the symmetric gauge and $\chi_1(z)$ is taken as the Fang and Howard wave function[22] which only depends on a single parameter b describing its extension along the z direction.

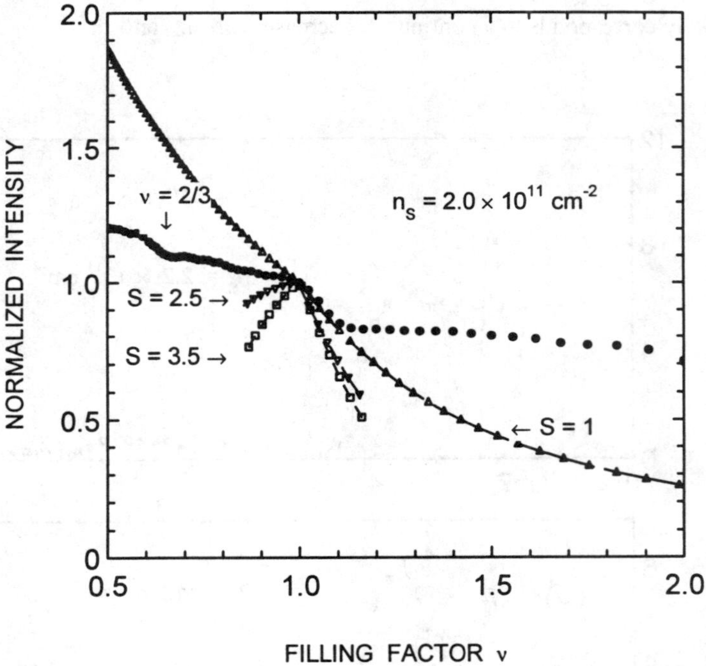

Figure 4. Normalized (at ν = 1) intensity of the 2D electron-bound hole luminescence. Comparison of the experimental data (full dots) with different models corresponding to an effective spin S = 1, 2.5 and 3.5 respectively.

Assuming that the total number of 2D electrons $n_s = \sum_N n_{1N}$, n_A^0 and b are constant, the different contributions from each Landau level are easily calculated. The description supposes the spin polarization of each level given by the Ando and Uemura model[5] (effective spin S=1) with the appropriate magnetic field scaled exchange energy. This is represented in Fig. 4. There are clearly some discrepancies between theoretical and experimental curves for ν >1 but they are expected since, as already mentioned, the parameter b should vary in a discontinuous way in that region. However for $\nu < 1$, the model predicts a continuous increase of the intensity as observed experimentally. If, on the other hand, one assumes a spin polarization of the 2DEG determined by the Skyrmion picture with an effective spin S equal to 2.5 or 3.5 as proposed by different authors[9, 11], I_{PL} decreases for $\nu < 1$ (Fig. 4). The experimental increase is observed in all conditions on different samples[19] which seems to be not consistent with the Skyrmion picture though other correlated effects are observed in the experiment (see the structure at $\nu=2/3$ in Fig. 3). Of course the observed increase is less pronounced than the one predicted by the model with S=1, but this likely corresponds to a continuous decrease[19] of n_s and n_A^0.

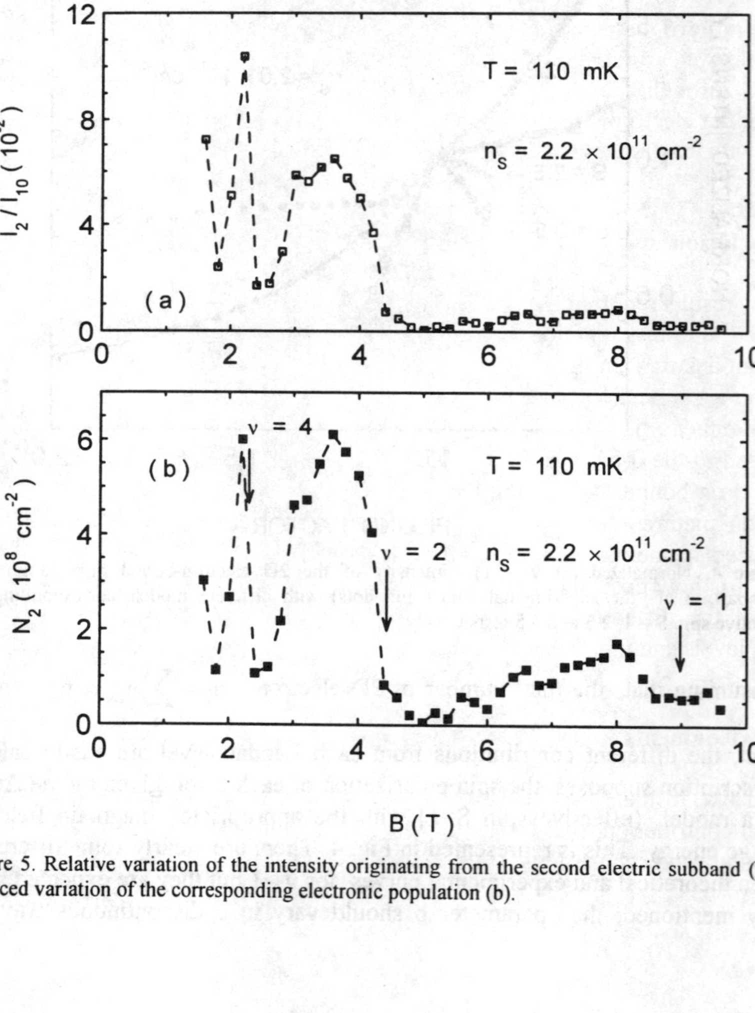

Figure 5. Relative variation of the intensity originating from the second electric subband (a) and deduced variation of the corresponding electronic population (b).

The variation of the intensity of the second electric subband shows strong oscillations with B (Fig. 5). They correspond to the discontinuities observed in M_1 and M_0 in Fig. 3. One can deduce from the model proposed (Eq. 1) the quasi-equilibrium population N_2 of this second subband. Indeed the ratio of the intensity I_2 of this subband with respect to that of the first Landau level of the first electric subband I_{10} represented in Fig. 5a is given by :

$$\frac{I_2}{I_{10}} = \frac{|M_2|^2}{|M_{10}|^2} \times \frac{N_2}{N} \qquad (2)$$

which simply involves the ratio of matrix elements of the optical transitions of the two corresponding electronic Landau levels and the ratio of their respective population. N is equal to n_B for $\nu > 1$ and n_s for $\nu < 1$. It can be demonstrated[19] that the ratio of the matrix elements is practically not dependent of b which provide an absolute measurement of N_2. This is represented in Fig. 5b. From there one can deduced through the rate equations the values of the different relaxation times between the electronic levels, their variation with the filling factor and compare the results with existing models. This analysis is now in progress.

4 . Conclusion

The new results reported here concern different systems and different basic problems to which the investigation under high magnetic fields can bring original and important answers. In the case of $GaInP_2$ these studies have demonstrated that there could be a conduction band alignment of a few meV between the compound and GaAs of either type I or II and that the observation of the up-conversion process was associated to the trapping of holes or electrons in the compound. In the case of the 2D electron-bound hole luminescence, the new results are not supporting the Skyrmion picture showing that the fundamental state of the 2DEG is reasonably reproduced if one assumes a conventional model of a spin polarized ground state. Finally it is now possible to measure directly the population of excited levels and to study all the different mechanisms of relaxation between the various electronic Landau levels and their variation as a function of the magnetic field.

Acknowledgments

I would like to thank my main collaborators, C. Hartmann and J. Zeman for their essential contribution to the work I am reporting on and Professor P. Y. Yu for very stimulating and constructive discussions. The Grenoble high Magnetic Field

Laboratory is Laboratoire conventionné avec l' Université Joseph Fourier de Grenoble.

References

1. A. Gomyo, K. Kobayachi, S. Kawata, I. Hino, T. Susuki and T.Yuasa, *J. Cryst. Growth* **77**, 367 (1986).
2. F. A. J. M. Driessen, *Appl. Phys. Lett.* **67**, 2813 (1995)
3. D. C. Tsui, H. L. Störmer and A. C. Gossard, *Phys. Rev. Lett.* **48**, 1559 (1982).
4. R. B. Laughlin, *Phys. Rev. Lett.* **50**, 1395 (1983)
5. T. Ando and Y. Uemera, *J. Phys. Soc. Japan* **37**, 1044 (1974)
6. S. L. Sondhi, A. Karlhede, S. A. Kivelson and E. H. Rezayi, *Phys. Rev.* B **47**, 16419 (1993)
7. H. A. Fertig, L. Brey, R. Côté and A. H. MacDonald, *Phys. Rev.* B **50**, 11018 (1994)
8. B. Kralik, A. M. Rappe and S. G. Louie, *Phys. Rev.* B **52**, R11626 (1995)
9. S. E. Barrett, G. Dabbagh, L. N. Pfeiffer, K. W. West and R. Tycko, *Phys. Rev. Lett.* **74**, 5112 (1995)
10. A. Schmeller, J. P. Eisenstein, L.N. Pfeiffer and K. W. West, *Phys. Rev. Lett.* **75**, 4290 (1995)
11. E. H. Aifer and B. B. Goldberg, D. A. Broido, *Phys. Rev. Lett.* **76**, 680 (1996)
12. H. Buhmann, W. Joss, K. von Klitzing, I. V. Kukushkin, G. Martinez, A. S. Plaut, K. Ploog and V. B. Timofeev, *Phys. Rev. Lett.* **65**, 1056 (1990)
13. I. V. Kukushkin, N. J. Pulsford, K. von Klitzing, R. J. Haug, K. Ploog, H. Buhmann, M. Potemski, G. Martinez and V. B. Timofeev, *Europhys. Lett.* **22**, 287 (1993)
14. J. E. Fouquet, M. S. Minsky, and S. J. Rosner, *Appl. Phys. Lett.* **63**, 3212 (1993)
15. J. Zeman , G. Martinez, P. Y. Yu, and K. Uchida, to be published.
16. S. H. Wei, and A. Zunger, *Appl. Phys. Lett.*, **56**
17. P. Emanuelsson, M. Drechsler, D. M. Hofmann, B. K. Meyer, M. Moser, and F. Sholz, *Appl. Phus. Lett.* **64**, 2849 (1994)
18. Z. P. Su, K. L. Teo,P.Y.Yu, K. Uchida and D. Dong, to be published.
19. C . Hartmann, G. Martinez, A. Fischer, W. Braun, K. Ploog, to be published.
20. G. Martinez, Proceedings of the Yamada Conference, Tiomura 1992, Ed. N. Miura, *Physica* B **184**, 51 (1995).
21. W. O. G. Schmitt, E. Bangert and G. Landwehr, *J. Phys. :Condens. Matter* **3**, 6789 (1991).
22. F. F. Fang and W. E. Howard, *Phys. Rev. Lett.* **16**, 797 (1966).

HOW TO DEFINE *STRONG* MAGNETIC FIELDS IN PHYSICS?

MICHAEL von ORTENBERG
Chair for Magnetotransport in Solids
HUMBOLDT-University at Berlin
Invalidenstrasse 110
D-10115 Berlin, Germany

Strong magnetic fields provide an excellent tool for the investigation of electronic states in semiconductors. Them qualification *strong*, however, is defined only relation to the sample properties. Examples of experiments in *strong* magnetic fields are presented for spectroscopy in DC and pulsed fields up into the megagauss range using non destructive solenoids and the semidestructive single-turn coil technique.

1 Introduction

1.1 High Magnetic Fields as Experimental Tool

During recent years an ever increasing activity in the application of high magnetic fields in physical investigation has been observed leading world-wide to an increased number of initiatives for high magnetic field applications [1]. Despite the extension to new application areas as biology and medicine semiconductor physics is still the dominant part of high magnetic field investigations [2].

In contrast to electric fields the effect of constant magnetic fields is quasi isoenergetic. The electronic energy levels of a semiconductor system exposed to magnetic fields may, however, exhibit due to these fields a pronounced quantum structure competing with the quantum effects of different origin. In *strong* magnetic fields, however, these field effects are clearly manifested, and in *extremely strong* magnetic fields they become so dominant, that the structural field dependence of these effects is considered as a new tool of investigation. In this way the sample system defines the qualification *strong* and not the absolute value of the magnetic field. It should be noted that in the progress of physics many relevant discoveries and effects are related to the application of high magnetic fields. The progress in generation of high magnetic fields aims now to extend the magnetic peak fields to higher and higher values into the megagauss regime to qualify more and more sample systems to be under the *strong* field limit for improved conditions for investigations.

To quantify the strong magnetic field condition the magnetic field provides in the magnetic length $l = (\hbar/eB)^{1/2}$ a unique quantity depending only on natural constants and the strength of the magnetic field. For B = 100 T we obtain *l* = 2.56 nm, so that even for conventional peak fields *l* ranges in the nanometer scale. In this way we have a natural length measure to probe nanostructured potential fluctuations. In the classical picture this means that we fit the smallest cyclotron orbit of an electron into the two dimensional potential to be probed. The highest fields generated nowadays for experimental application are of the order 600 to 800 T, so that for these fields *l* gets close or even smaller than the lattice constant for some crystalline materials. Under these conditions we expect a completely different behaviour of the material. The down scaling in field of these effects is tested by the use of "man made" periodic structures in form of superlattices in one dimension. The two and three-dimensional realisation, however, is still a promising target for the future.

In the same way as *l* probes ordinary space, the inverse quantity 1/*l* manifests in a suitable measure for the reciproke *k*-space. In the same way as *l* approaches the lattice constant 1/*l* covers the Brillouin zone, so that a more extent mapping of the energy band structure within that zone is possible.

1.2 The High Magnetic Field Facility at the Humboldt University at Berlin

The *HUMBOLDT High Magnetic Field Center* combines in a complementary way three of the most successfully applied and most sophisticated magnetic field generators: DC superconducting solenoid up to 20 T, pulsed fields using non destructive coils up to 50 T, and pulsed fields using the semidestructive technique of single turn-coils up to 240 T in 10 mm diameter and 301 T in 5 mm diameter. The 20 T-magnet has about 859 kJ energy stored in the coil at peak field, whereas the capacitors banks for the non destructive and destructive fields have only 400 kJ and 200 kJ respectively [3]. Due to the different time periods of the energy release the ordinary operation of the different generators is more or less spectacular. In failure, however, all of these field generators are with respect to energy wild beasts and if released in an uncontrolled way the result is a complete experimental disaster. Already in standard operation at 60 kV the megagauss equipment produces a tremendous shock wave resulting in considerable destruction of the external experimental environment and plasma pinch effects in the inner part of the coil, so that completely new measuring techniques have to be applied.

For demonstration of the efficiency we present in the following four different scientific problems.

2 High Magnetic Field Experiments

2.1 Fermi Surface of β-(ET)$_2$I$_3$ and κ-(ET)$_2$I$_3$

Both β-(ET)$_2$I$_3$ and κ-(ET)$_2$I$_3$ belong to the group of superconducting transfer salts of the *Bechgaard* family [4]. Both materials grow in layered structure, so that the electronic system is essentially reproduced within the model of Q2D. Whereas the κ-material has been rather intensively investigated leading to a consistent model of the Fermi surface [5], for β-(ET)$_2$I$_3$ we present here first results of the corresponding investigations. Due to the quasi metallic properties of the organic compound no optical transmission experiments are possible, so that the complementary physical quantity of the reflectivity has to be measured. As a matter of fact the more elaborated experimental set-up used the strip-line transmission of a microscopic wave guide made from sample material [6]. The crystals were oriented with both the surface and layer structure perpendicular to the external magnetic field to ensure the cyclotron motion within the highly conductive layers. Tests on κ-(ET)$_2$I$_3$ corroborated the efficiency of the experimental method also for the *Bechgaard* salts and were consistent with the generally accepted model for the Fermi surface [7], so that the new results for β-(ET)$_2$I$_3$ are expected to be most realistic. In fig. 1 we have reproduced the strip line data for 81 GHz radiation as a function of the magnetic field [8]. The spectrum exhibits clearly three different resonances. This means that in this experiment the magnetic field intensity of less than 10 T has to be considered for this sample as "strong" magnetic field. As shown by the insert all of them "scale" to zero energy at zero magnetic field indicating the

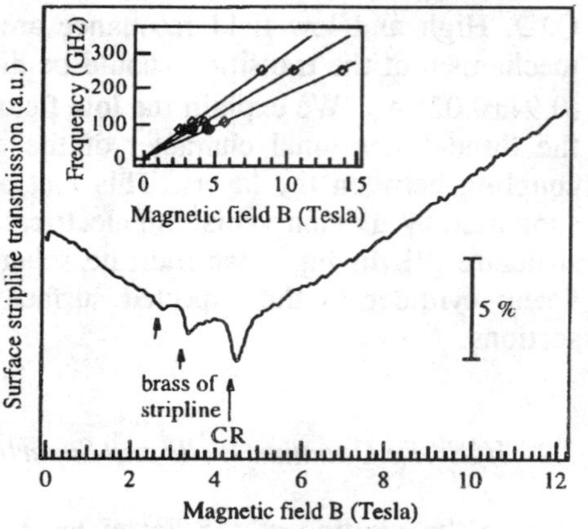

Figure 1: The strip-line transmission of κ-(ET)$_2$I$_3$ showing two sample resonances and one of the sampleholder as a function of the magnetic field. The insert demonstrates the scaling of all resonances.

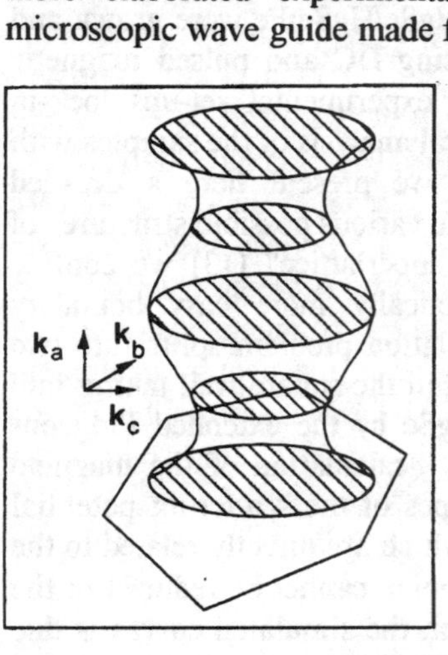

Figure 2: The proposed Fermi surface of κ-(ET)$_2$I$_3$.

resonances as due to spin and/or orbital transitions of quasi free electrons. Whereas the center resonance could be ruled out as due to the reference mirror of the strip line, the two other resonances reflect definitely the sample system. The high field resonance corresponds to a cyclotron mass of m_c = (1.53±0.02)m_o and is considered to be the fundamental resonance of the Fermi cylinder representing the Q2D. High and low field resonance are not related by a harmonic, so that the mechanism of the transition should be different. The corresponding mass is m_c = (0.94±0.02) m_o. We explain the low field resonance tentatively as reminiscence of the three-dimensional character of the electron system due to a non vanishing coupling between the layers. This fact of remaining three-dimensionality is also supported by the non vanishing electrical conductivity perpendicular to the layered structure [9]. In fig. 2 we indicate schematically the possible deformation of the Fermi cylinder to the expected surface with indication of the extremal cross-sections.

2.2 *HgSe:Fe Quantum Wells and Superlattices*

Due to the pinning of the Fermi level to the Fe^{++}-donor in the mixed valence regime about 210 meV above the band edge of the zero-gap semiconductor HgSe the ionized Fe^{3+} arrange in such a way that charged clusters are avoided. This fact results in a dramatic increase in the carrier mobility, so that a high carrier system with $n = 5*10^{18}$ cm^{-3} with extremely narrow energy levels is available [10]. Using MBE technique quantum wells and superlattices of HgSe/HgSe:Fe were grown and investigated by DC- and FIR-magneto transport using DC and pulsed magnetic fields up to 15 T and 50 T respectively. In both experimental set-ups lock-in technique was applied [11]. After a thorough structural analysis of the samples with preliminary magneto transport investigation [12] we present here a detailed analysis of the magneto transport properties. From the various possible structures of "T" and "I" quantum wells as well as the "comb superlattice" [13] we confine ourselves to the "I"-type, because of the theoretically more easy boundary conditions of the corresponding problem. The simulation problem splits into two parts. First, the computation of the energy levels within the square well taking into account of the detailed energy band structure of HgSe by the extended Pidgeon-Brown model in a 8x8 ***k•p***-matrix [14], second, by calculation of the magneto conductivity using Ando's theory for different shapes of the scattering potential [15]. It should be noted that for finite values for k_z, which are directly related to the width of the potential well, the matrix of the hamiltonian cannot be reduced in the dimensionality. The modulation of the Hall voltage in the simulated curves is due to the carrier fluctuation between quasi-free and bound states in the mixed valence regime [16]. In fig. 3 we have plotted both the experimental recordings and the simulated data of the magneto resistance for a sample with a well width of d = 12 nm, where only two subbands are occupied. The comparison of simulation and data favours definitely "constant scattering" to "short range scattering" [13]. In fig. 4 we reproduce the correspondent results for samples with larger width of the square well, namely d = 25 nm and d = 50 nm in the left and right part of the figure, respectively. The agreement of simulation and data is excellent and demonstrates the reality of the applied model even in the case of complex multi-subband occupation. The surprising result of the simulation is, that the data could only be fitted assuming an increasing Fermi energy and hence donor energy of the Fe^{2+}/Fe^{3+}-states with decreasing width d of the square well potential. This means that the correlation energy of the Fe^{3+}-donors depends definitely on the degree of dimensionality. This is from experimental point of view a completely novel result and is summarised in Table 1.

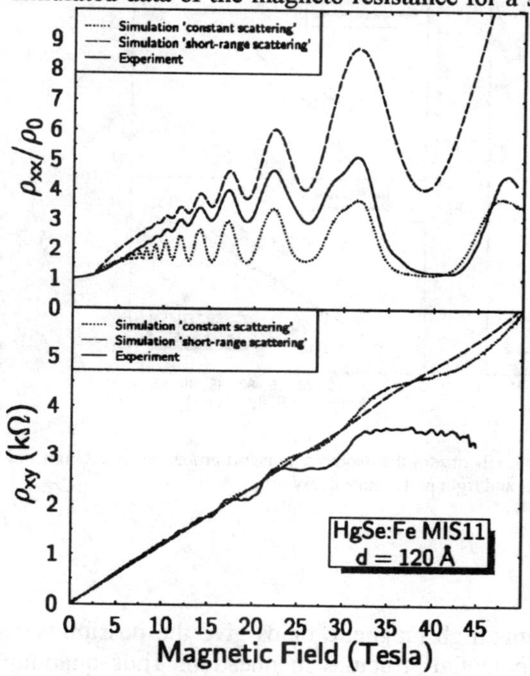

Figure 3: Simulation and experimental magneto transport data for a 12 nm quantum well of HgSe:Fe.

Table 1: Relative Fermi Energy for different Width d of the Quantum Well.

Width d (nm)	Fermi Energy (meV)
12	234
25	228
50	212

The analysis of "T"-wells and "comb superlattices" is in progress, but the simulation is much more complicated since 2D and 3D effects are simultaneously present in one and the same sample.

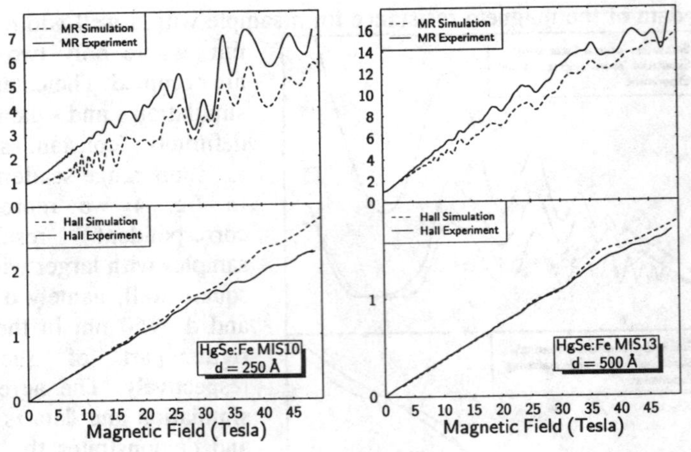

Figure 3: Experimental results in comparison with theory for the magneto transport properties of a 25 nm and a 50 nm thick HgSe:Fe quantum well in the left and right part, respectively.

2.3 ZnSe/ZnMnSe Quantum Wires

Especially for nanostructured systems high magnetic fields give the possibility to probe the geometrical and hence potential structure in question. Thus quantum wires characterised perpendicular to the magnetic field by a geometrical parameter D should definitely change in physical properties if l becomes comparable to D. We have fabricated ZnSe/ZnMnSe quantum wires applying selforganized MBE-growth on patterned substrates as shown in fig. 5. Using both photoluminescence and cathodoluminescence measurements we were able to identify unambiguously the contribution of the quantum wire [17]. For very high magnetic fields when $l \ll$ D

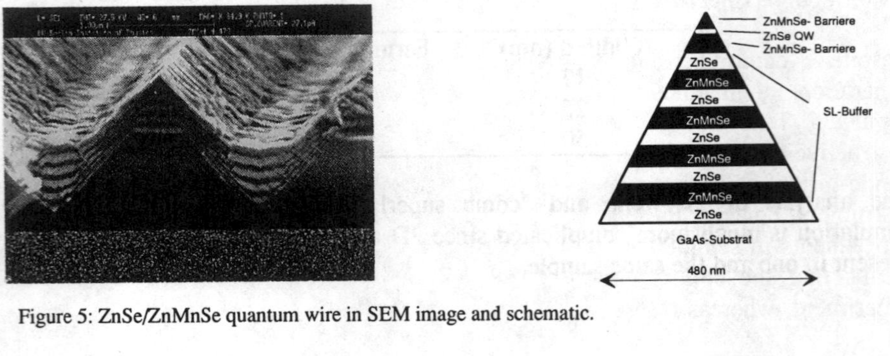

Figure 5: ZnSe/ZnMnSe quantum wire in SEM image and schematic.

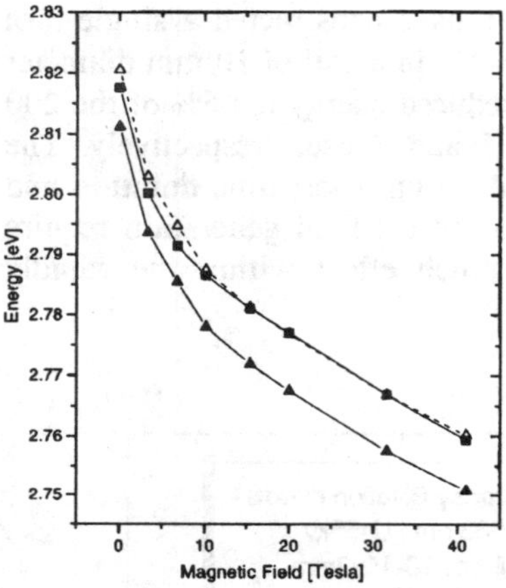

Figure 6: The magneto luminescence data of a ZnSe quantum wire in comparison with the corresponding data of an unpatterned reference sample (▪). The broken curve corresponds to the renormalized data of the quantum wire.

the quantum wire becomes indistinguishable from a corresponding quantum well on a reference sample grown simultaneously on an unpatterned substrate. To test this prediction we have performed magneto-luminescence measurements in pulsed fields up to 40 T and reproduce the results in fig. 6 [18]. The quantum wire has a width of D = 20 nm so that for B = 40 T the cyclotron orbit of the lowest Landau-level has a diameter of $2l$ = 8.2 nm and is hence considerably smaller than D. The magneto luminescence data for a quantum wire (full triangels) and the unpatterned quantum well (full squares) of the thickness of 4 nm do definitely not converge at high fields. This at first surprising fact is, however, easily understood by the different strain conditions for the quantum wire and the quantum well. Evidently the high field offset is only due to strain, so that for direct comparison of quantum wire and quantum well this offset should be compensated by renormalization as indicated by the open triangles. In this way we estimate the effect of the additional quantization in the quantum wire at zero magnetic field as ΔE_{QW} = 2.5 meV at zero magnetic field. This value is in fair agreement with the rough estimate of the confinement energy of $\Delta E_C = h^2/(8m_{exc}D^2)$ = 5 meV. The production of quantum wires having smaller D is in progress.

2.4 Megagausspectroscopy

Despite comprehensive initiatives the generation of magnetic fields higher than about 75 T is not yet possible using non destructive coils. The introduction of the single-turn coil as scientific research instrument, however has opened a wide area of experiments in semidestructive technique: this means that the coil is destroyed during the experiment, whereas the set-up inside the coil

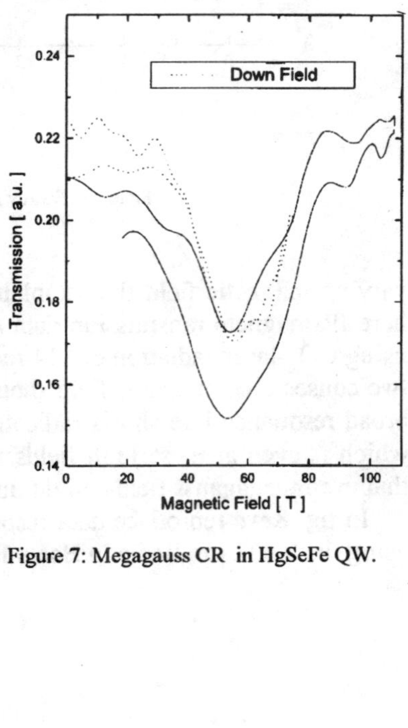

Figure 7: Megagauss CR in HgSeFe QW.

in the generated magnetic field is saved [19]. We have constructed a single-turn coil equipment generating so far a peak field of 240 T in a coil of 10 mm diameter [20] and 300 T in a coil of 5 mm diameter with reduced energy to 66% of the 200 kJ-bank [21] and pulse lengths in the order of 5 and 3 μsec, respectively. The conditions of very tiny space for the experimental set-up, short time duration and the presence of a strong electromagnetic pulse during the field generation require sophisticated experimental techniques. Due to pinch effect within the rapidly

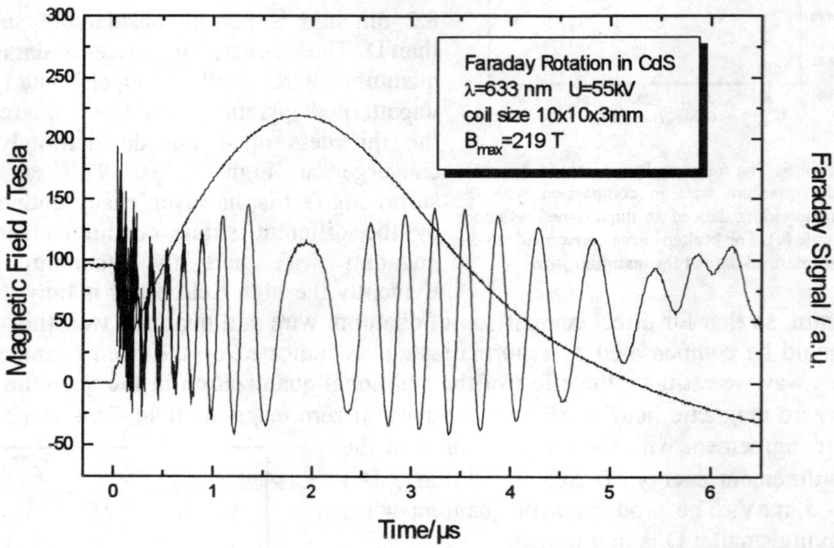

Figure 8: Farady rotation in CdS in Megauss fields.

varying magnetic field the sampleholder and sample may be affected. We present here IR-magnetotransmission data at He-temperatures of a HgSe:Fe quantum well using CO_2-laser radiation of 114 meV energy. In fig. 7 we have plotted the data of two consecutive magnetic field shots at a temperature of about 30 K. The relatively broad resonance line shows indications of fine structure due to different subbands, which is even at these high fields not yet completely resolved. This demonstrates that even megagauss fields might qualify as not sufficiently *strong*.

In fig. 8 we reproduce data recorded up to 219 T for the Faraday effect in CdS using the 0.633 μm line of a HeNe-laser to demonstrate the quality of data recorded in the multi megagauss range. As a matter of fact the CdS-sample exhibits completely classical behaviour despite the magnetic field being more than 200 T.

3 Summary

High magnetic fields are an efficient tool for the investigation of electronic energy levels in semiconductors. This application has stimulated the creation of field generators for higher and higher fields now covering also in a semidestructive way the multi-megagauss range. The qualification *strong* for the magnetic field, however, is not absolute, but can be defined only in connection with the properties of the sample system under investigation.

Acknowledgement

The author likes to thank all members of his group and the international network of cooperations.

References

1. F. Herlach and J.A.A.J. Perenboom, *Physica* B **211**, 1 (1995)
2. Report of the *European Workshop: Science in 100 Tesla*, Leuven, May 15-17 (1992), *ed:* Katholieke Universiteit Leuven Belgium
   Report of the *Second European Workshop on Science in !00 T (Leuven II)*, Leuven, 30 September-1 October (1994), *ed:* L. Challis, J. Franse, F. Herlach, P. Wyder
3. M. von Ortenberg, O. Portugall, H.-U. Mueller, N. Puhlmann, M. Thiede, G. Machel, M. Barczewski, J. Breitlow-Hertzfeld, *Physica* B **201**, 572 (1994)
4. K. Bechgaard, C.S. Jacobsen, K. Mortensen, J.H. Pedersonn, N. Thorup, *Solid State Commun.* **33**, 1119 (1980)
5. K. Oshima, T. Mori, H. Inokuchi, H. Urayama, H. Yamochi, G. Saito, *Phys. Rev.* B **38**, 938 (1988)
6. M. von Ortenberg *in: Infrared and Milimeter Waves*, Vol **3**, *ed:* K. J. Button, Academic Press, New York (1980), 275
7. T. Ishiguro, K. Yamaji, *in: Organic Superconductors*, Springer, Berlin (1990)
8. O. Haubrich, S. Luther, M. von Ortenberg, D. Schweitzer, *Digest of the 20th Intnl. Conf. on Infrared and Millim. Waves*, Orlando 1995, *ed:* R.J. Temkin, 241 (1995)

9. M. Dressel, D. Schweitzer, H.J. Keller, *Phys. Bl.* **50**, 1145 (1994)
10. M. von Ortenberg, *Semicond. Sci. Technol.*, **8**, S16 (1993)
11. G. Machel and M. von Ortenberg, *Physica* B **211**, 355 (1995)
12. D. Schikora, Th. Widmer, K. Lischka, P. Schaefer, G. Machel, S. Luther, M. von Ortenberg, *Phys. Rev.* B **52**, 12072 (1995)
13. G. Machel, S. Luther, M. Thiede, M. von Ortenberg, D. Schikora, T. Widmer, K. Lischka, *Proc. 23th Int. Conf. on the Physics of Semiconductors,* Berlin (1996) in press
14. J. Kossut, *Semiconductors and Semimetals*, **25**, 183 (1988)
15. T. Ando, A.B. Fowler, F. Stern, *Rev. Mod. Phys.* **54**, 437 (1982)
16. M. von Ortenberg, O. Portugall, W. Dobrowolski, A. Mycielski, R.R. Galazka, *J. Physics C: Solid State Physics* **21**, 5393 (1988)
17. L. Parthier, H. Wißmann, S. Luther, G. Machel, M. von Ortenberg, *Proc. 23th Int. Conf. on the Physics of Semiconductors*, Berlin (1996) in print
18. L. Parthier, S. Luther, M. von Ortenberg, K. Uchida, H. Kunimatsu, N. Miura, publication in preparation
19. F. Herlach and R. Mc. Broom, *J. Phys. E: Sci. Instrum.* **6**, 652 (1973)
20. M. von Ortenberg, O. Portugall, N. Puhlmann, H.-U. Mueller, M. Barczewski, I. Stolpe, M. Thiede, G.Machel, F.Herlach, *Proc of the 7th Intnl. Conf. on Megagauss Magnetic Field Generation and Related Topics*, Sarov, August 5-10 (1996), Invited Paper in print
21. N. Puhlmann, O. Portugall, H.-U. Mueller, M. Thiede, I. Stolpe, M. Barczewski, M. von Ortenberg, publication in preparation

PRECISION MEASUREMENTS OF THE QUANTUM HALL EFFECT

E. BRAUN, B. SCHUMACHER, P. WARNECKE
Physikalisch-Technische Bundesanstalt, Bundesallee 100,
38116 Braunschweig, Germany

When Klaus von Klitzing discovered the quantum Hall effect in 1980 he pointed out its great importance for metrology, both as a new experiment for the determination of the fine structure constant α and as a resistance standard. It was a challenge for metrologists all over the world to work out methods to measure this effect more and more precisely. The tremendous development of resistance measurement techniques in the past 16 years, which was stimulated by the discovery of the quantum Hall effect is described. The reproducibility of measurements on the quantized Hall effect has reached a level of uncertainty of a few parts in 10^{10}. Experiments on the dependence of the quantized Hall resistance in various samples of different materials, on the width of the sample and on the step number are summarized. Experimentally determined values of the von Klitzing constant are compared with those which can be deduced from recent determinations of the fine structure constant. The most fascinating thing about the quantum Hall effect is that although no quantitative theory exists its high reproducibility has led to the worldwide agreement on this effect to be the basis for the maintenance of the unit of resistance.

1 Introduction

When von Klitzing intended to publish his discovery, his preliminary title was "Realization of a resistance standard based on natural constants" [1]. However, the final version of his first paper was entitled "New Method for High-Accuracy Determination of the Fine-Structure Constant based on Quantized Hall Resistance" [2]. Both aspects are at present - 16 years after the discovery - the main points of interest.

The fact that a new macroscopic quantum effect, the quantum Hall effect, was found in which a physical quantity seemed to be determined by a fundamental constant stimulated the specialists in precision measurement. The measurement of fundamental constants which are independent of surrounding parameters, allows methods of high precision to be developed. In addition, the experimental set-up can be checked for systematic deviations when the fundamental constant is determined at different places. Of course not only the experimental set-up is the challenge. The values of fundamental constants determined with the smallest possible uncertainties confirm theories in various fields of physics, especially if the same constant is determined. We will see that the discovery of the quantum Hall effect has made it possible to compare solid-state experiments conducted in small areas fractions of mm^2 in size with experiments at a nuclear reactor where neutrons are travelling through paths of tens of meters. If, in addition, the fundamental constant involved has the dimension of an important physical and technical quantity - e.g. resistance -

then a possible practical, e.g. metrological, application is another aspect speaking for a high-precision determination.

In the case of the quantized Hall resistance, all aspects were given and a race for the "right number" for this fundamental constant and the smallest uncertainty attainable started in the first decade after its discovery.

2 Realization of the Ohm

Resistance measurement is the comparison of an unknown resistance with a well-known one. The best known resistance is the as-maintained unit, the ohm. The discovery of the quantized Hall resistance therefore was also a strong stimulus to improve realizations of the ohm. In fact, it turns out that the knowledge of the ohm is the limiting part in the determination of the von Klitzing constant.

The ohm is a derived unit in the SI system. The electrical base unit is the ampere. However, the ampere is not an independent unit. The ampere is determined by mechanical units and by the definition of a fundamental constant, the permeability of vacuum μ_o. A conductor has by definition the resistance 1 ohm if a voltage drop of 1 volt occurs when a current of 1 ampere passes the conductor. The wording of this definition does not give the possibility of realizing the ohm with low uncertainty. However the Australian scientists Thompson and Lampard discovered a theorem which allows the realization in a different way [3]. They showed that, in a first step, the realization of a capacitance is possible via a cross-capacitor. If only a length is precisely measured, the capacitance can be calculated using the definition of the permittivity of vacuum $\varepsilon_o = 1/\mu_o c^2$. (The speed of light in vacuum c is also a defined fundamental constant.) This capacitance is in the order of a few pF. In a second step, this capacitance is linked at a certain angular frequency ω to an impedance $R = 1/\omega C$. In a third step, this impedance is linked to a dc resistance which after a step down leads to 1 Ω. This realization is done at various National Metrological Institutes (NMIs) with an uncertainty of a few parts in 10^8.

3 Resistance Measurement Methods

As already mentioned, a resistance measurement is the comparison of an unknown resistance with a known one. Resistance measurements range from 10^{-6} to 10^{14} Ω. Methods have been developed to scale within this range on a decimal ladder. Since the quantized Hall resistances occur at non-decimal values, they introduce an additional problem into the precision measurement, the comparison of a non-decimal value with a decimal one.

The main principles developed for the comparison of resistances are described in the following.

3.1 Potentiometric method

The straightforward method of comparing two resistances R_X and R_N is to connect them in series, the same current passing through both of them. Then the ratio R_X/R_N results from the measurements of the two voltage drops U_X and U_N at R_X and R_N. The voltage measurements have to be done so that no current flows through the voltmeter, i.e. the voltages are compensated by an adjustable external voltage source using a null indicator, or U_X and U_N are directly measured by means of a high-impedance voltmeter. In order to eliminate the thermal EMFs, the voltage drops have to be measured for both current directions. One of the disadvantages of the potentiometric method is that, in principle, the current has to be stable. However, a linear drift can be eliminated by combining the results of a sequence of measurements.

The simplest way of performing the described potentiometric ratio measurements is to use a digital voltmeter (DVM). The first determinations of the von Klitzing constant in 1980 using the potentiometric method and a digital voltmeter led to results in the order of 10^{-6} [2,4].

The potentiometric measurements can easily be automated by using computerized switches for scanning the voltage drops and for current reversal [5,6]. There are modern 8 1/2-digit voltmeters which allow resistance measurements with uncertainties well below 10^{-7} [7].

3.11 The Josephson Potentiometer

Potentiometric resistance measurements can be improved considerably and the smallest possible uncertainty achieved when advantage is taken of the ultimate linearity of a Josephson potentiometer [8, 9, 10, 11], i.e. a Josephson voltage U_J is used as compensating voltage in combination with a nanovoltmeter as a null detector. The Josephson voltage

$$U_J = n \cdot (h/2e) \cdot f \qquad (1)$$

(n step number, f microwave frequency) is produced by integrated arrays of several thousands of series-connected Josephson junctions. When appropriate values of n and f are chosen, the voltage U_J can be adjusted between several mV and 10 V. The uncertainty of U_J is given only by the uncertainty of the frequency f which is far beyond any discussion. With the Josephson potentiometer the voltage drops at the known and the unknown resistance are subsequently compensated in the same way as with a usual potentiometer. A prerequisite - also for the Josephson potentiometer - is the stability of the current through both resistors. Using a Josephson potentiometer for ratio measurements far from the 1:1 ratio makes no sense because of the decreasing signal-to-noise ratio.

3.2 *Hamon transfer standards (resistance networks)*

Since the potentiometers allow only a one-to-one comparison to be made with the highest precision, there is a need for having a scaling method to cover a broader range of resistance values. The resistance of N resistors connected in series, each of nominal value R_o is $R_s = N{\cdot}R_o$, the resistance in parallel connection is $R_p = 1/N{\cdot}R_o$, therefore the ratio is given by $R_s/R_p = N^2$. Hamon [12] showed that if the relative resistance differences between these resistors are smaller than δ, the ratio R_s/R_p is accurate within a relative uncertainty smaller than δ^2. In practice the Hamon networks are more complicated than described above, because of additional potential compensating resistors. The series-parallel switching is often made using mercury-wetted contacts which allow switching without mechanical shocks.

If one wants to compare, for example, $R_H(2) \approx 6453\ \Omega$ with 1 Ω, a Hamon network with 80 resistors, each of about 80 Ω, is the starting point. In fact, similar but more sophisticated networks were used by several national metrological laboratories to scale between quantized Hall resistance values and decadic ones [13, 14, 15, 16].

3.3 *Current comparator methods*

3.31 *Direct current comparator*

In contrast to the potentiometric method where the resistors to be compared are connected in series and supplied by the same current, the current comparator method [17] for the comparison of two resistances R_P and R_S uses two different currents, I_P and I_S. I_P and I_S are established in such a way that $I_P/I_S = R_S/R_P$. In this case the voltage drops U_P and U_S at the resistors R_P and R_S are equal. The ratio of the currents is controlled by means of a direct current comparator. Its principle is that the two currents are fed into two coils with different windings. The turn number ratio N_P / N_S of these two windings equals the ratio R_P / R_S. Together with a third flux-sensing winding both windings are wound onto a toroidal magnetic core. The current-carrying windings are connected in opposite direction, therefore the resulting net flux is zero. Even if the primary current drifts, a stable current ratio $I_P / I_S = N_S / N_P$ is generated because the secondary current source is controlled by the flux detection system via a feed-back system. An advantage of the current comparator (CC) is that its linearity and stability are given by winding ratios which are more stable than the ratios of resistive dividers. CCs are commercially available and used in many standards laboratories for high-precision resistance measurements.

The uncertainty which can be reached with a direct current comparator working at room temperature is typically several parts in 10^7. This uncertainty results from stray fluxes which cause errors of the winding ratios, and from the limited sensitivity of the flux detector.

3.32 Cryogenic current comparator

A substantial improvement can be achieved by using cryogenic measuring techniques, i.e. cryogenic current comparators (CCCs) [18] which take advantage of the superior properties of superconducting current transformers and flux sensors. The working principle of a CCC is described in [19] as follows "Consider a superconducting tube through which a wire passes, carrying a direct current. This current generates a magnetic field, which, in order to screen the field from the interior of the walls of the tube, in turn induces a supercurrent to flow up the inside surface of the tube and back down the outer surface. Complete screening requires that the walls are at least several penetration depths thick, which is not difficult to achieve. The key point for the usefulness of this device is that the distribution of screening current on the outer surface of the tube is independent of the position of the current-carrying wire within the cross section of the tube, at least for all points on the outer surface which are more than a few tube diameters from either of its ends. Suppose now that another wire is included in the tube which carries a variable current. The magnetic field produced by the supercurrent on the outer surface of the tube resulting from both wire currents can be measured with a SQUID (Superconducting **Qu**antum **I**nterference **D**evice), and zero field will be found when the two currents are exactly equal but opposite in direction. Because the external current distribution is extremely independent of the wire positions, and because SQUIDs are such sensitive devices for magnetic field detection, such a current comparator offers extremely high precision for establishing a 1:1 current ratio." This is also valid for other ratios.

The geometry of most CCCs is very similar to the so-called "Delahaye-type" [20]. The windings form a rigid body of toroidal shape with rectangular cross section. The toroid formed by the windings is covered with a superconducting shield overlapping three turns, which is insulated at the overlap. So the superconducting shield looks like "a snake swallowing its tail". The SQUID magnetometer is coupled to the current comparator via a superconducting flux transformer.

For the application of CCCs in high-precision electrical measurements it is very important to check experimentally the accuracy of the winding ratios, since due to the manufacturing procedure imperfect superconducting shielding and insulation may result. Binary CCCs with 1, 1, 2, 4, 8, ... windings have the advantage that every winding can easily be self-calibrated. To self-calibrate a binary CCC, the sum of the windings $n_1+n_2+n_3+\cdots+n_i$ is connected in series to the winding n_{i+1}, however, in the opposite sense. An increasing, rather large current is fed into the connected windings and the SQUID output observed. Ratio errors as small as several parts in 10^{11} have been detected.

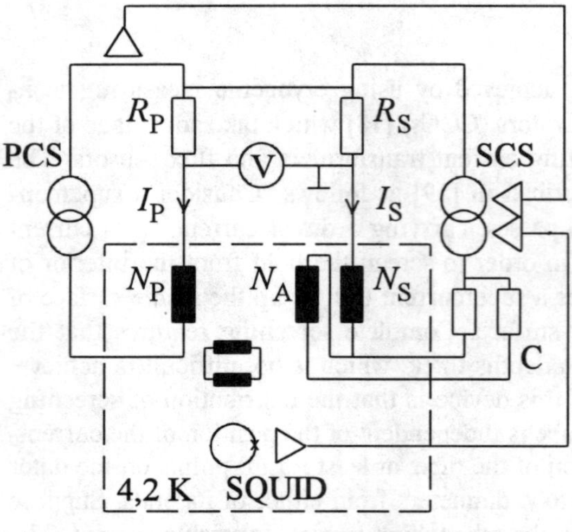

Fig. 1: Schematic diagram of a cryogenic current comparator

Resistance bridges as shown in Fig. 1 based on the CCC achieve uncertainties of less than 10^{-9}.

4 Measurement Results

The first measurements of quantized Hall resistances in 1980 had been made using high-resolution DVMs. As already mentioned, the limiting uncertainty was about 10^{-6} [2,4]. Potentiometric 1:1 methods soon showed the high reproducibility of the quantized values in the order of parts in 10^8 [21]. During the following decade the cryogenic current comparator turned out to be the instrument for the lowest uncertainty and for the most reliable results [20, 22].

Since no quantitative theory existed, from experimental experience "Technical Guidelines for Reliable Measurements of Quantized Hall Resistances" have been formulated [23]. All measurements described in the following are in accordance with these guidelines.

4.1 Dependence on sample width, material and step number

Jeanneret et al.[24] made precision measurements of the quantized Hall resistances $R_H(2)$ and $R_H(4)$ as a function of the width of the sample, which varied between 1 mm and 10 μm. The experimental results show that a possible dependence would be smaller than $3 \cdot 10^{-11}$ at $R_H(2)$ and smaller than $6 \cdot 10^{-11}$ at $R_H(4)$, which is in both cases well below the experimental uncertainty which was in the order of 10^{-9}.

Jeckelmann et al. [25] measured the Hall resistance at step i=4 of a silicon MOSFET against that of a GaAs heterostructure. The result is

$$R_H(4, \text{Si}) \,/\, R_H(4, \text{GaAs}) = 1 - 1{,}2 \cdot 10^{-10} \pm 3{,}6 \cdot 10^{-10}.$$

Jeckelmann et al. [25] also measured the Hall resistance of various silicon and GaAs samples at the steps i = 1, 2, 3, 4, 6 and 8. They compared the mean value $< iR_H(i) >$ of the products $iR_H(i)$ with $2R_H(2)$. The result is

$$iR_H(i) \,/\, 2R_H(2) = 1 - 1{,}1 \cdot 10^{-10} \pm 2{,}9 \cdot 10^{-10}.$$

Hartland et al. [26] also reported on a direct comparison of the quantized Hall resistance $R_H(4)$ in a silicon MOSFET with $R_H(2)$ in a gallium arsenide heterostructure. His result is

$$2R_H(2, \text{GaAs}) \,/\, 4R_H(4, \text{Si}) = 1 - 0{,}22 \cdot 10^{-10} \pm 3{,}5 \cdot 10^{-10}.$$

All results reported here have been obtained with uncertainties below $4 \cdot 10^{-10}$. No deviation of the quantized values within these uncertainties depending on sample width, material or step number could be observed.

4.2 Determination of the von Klitzing constant

Since there is so far no theory which describes the effect quantitatively, the fundamental constant determining the quantized Hall resistance is called *von Klitzing constant*. Very often the assumption is made that the von Klitzing constant is identical with h/e^2. From this quotient the fine structure constant α can be derived by simple calculation

$$\alpha^{-1} = (2/\mu_o c)\, h/e^2 \qquad (2)$$

Up to now only three independent determinations of the von Klitzing constant with uncertainties lower than 10^{-7} have been made at the National Physical Laboratory (NPL), Teddington, UK [27], at the National Measurement Laboratory (NML), Lindfield, Australia [28], and at the National Institute of Standards and Technology (NIST), Gaithersburg, USA [29]. The results are shown in Fig.2. As can be seen, the results are in excellent agreement, the uncertainties range between 2 and 5 parts in 10^8.

Using the assumption mentioned above, these values for R_K can be compared with the most recent determination for the fine-structure constant derived from the measurement of the anomalous magnetic moment a_e of the electron [30]. This value for α has an extremely low uncertainty of $3{,}8 \cdot 10^{-9}$ and is in agreement with the R_K determinations.

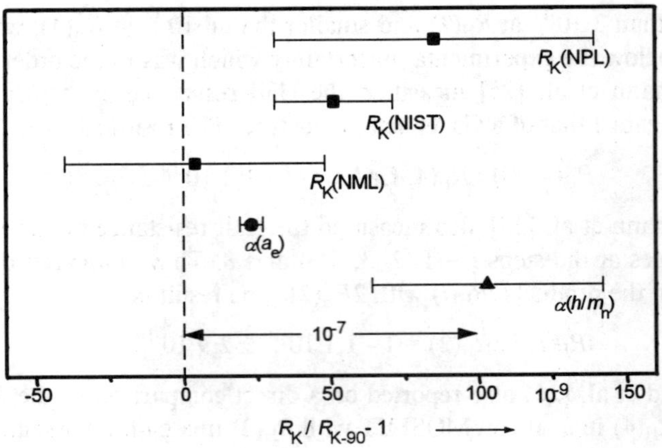

Fig. 2: Deviations of determinations of R_K from the recommended value R_{K-90} assuming $\alpha^{-1} = (2/\mu_0 c)\, R_K$.

Another comparison can be made with an α value determined from the ratio of Planck's constant h and the neutron mass m_n [31]. This ratio is experimentally determined by measuring the velocity and the wavelength of neutrons and using the de Broglie relation $h/m_n = v_n \lambda_n$. This value overlaps with two of the R_K determinations, but is by two standard deviations higher than the α which has been deduced from the anomalous magnetic moment of the electron.

On the basis of the present state of experiments, that means on a level of uncertainty of a few parts in 10^8, the answer to the question of whether the von Klitzing constant and h/e^2 are identical is that there are strong indications that the assumption is correct. In particular, there is no experiment which indicates that the assumption is wrong.

5 Metrological Application

The fact that the quantized Hall resistance values can be reproduced with uncertainties much lower than those of the corresponding units gave rise to an intense discussion on whether or not changes should be made in the base units of the SI System. Since a similar quantum effect, the Josephson effect, provides quantized voltage values with high precision, the temptation was great to use these macroscopic quantum phenomena for a new set of base units.

However, the organs of the Metre Convention decided not to change the SI-System, but to recommend exact conventional values for the characteristic fundamental constants appearing in the equations for the Josephson voltage and the quantized Hall resistance [32]. The following value has been chosen for the maintenance of the ohm:

$$R_{K\text{-}90} = 25\,812{,}807\ \Omega.$$

The index of $R_{K\text{-}90}$ indicates that this value has been in use at the NMIs since 01.01.1990. On the basis of the experimental results available at the time when the recommended value was chosen, it was assumed that within an uncertainty of $2\cdot10^{-7}$ the recommended value agrees with the von Klitzing constant. On the basis of the experiments mentioned in this paper, one comes to the conclusion that R_K is 2 to 5 parts in 10^8 higher than $R_{K\text{-}90}$ with an uncertainty which is roughly 10 times smaller than that estimated for $R_{K\text{-}90}$.

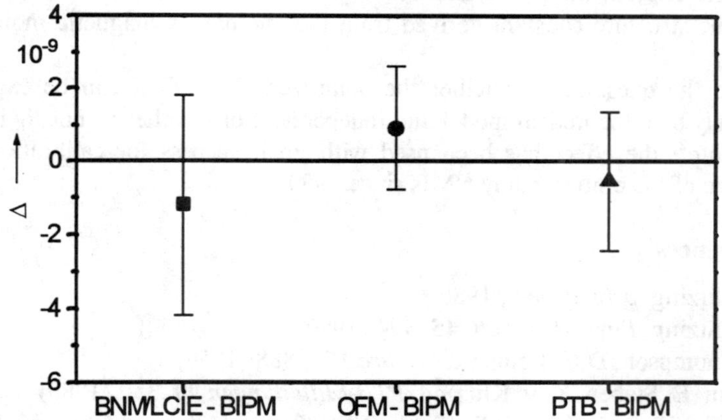

Fig. 3: Relative difference Δ between two calibrations of the same 100 Ω resistance standard with two different quantum Hall apparatus.

The majority of the larger National Metrological Institutes have installed a quantum Hall effect system as the basis for the maintenance of the ohm. In order to assure global traceability of resistance measurement, these quantum Hall systems are compared in a bilateral way through the International Bureau of Weights and Measures (BIPM) in Sèvres near Paris. The BIPM has a travelling quantum Hall effect apparatus, and three of these comparisons were made up to now: at the Laboratoire Central des Industries Électriques (LCIE), Paris, France [33], at the Office Fédéral de Métrologie (OFM), Bern, Switzerland [34], and at the Physikalisch-Technische Bundesanstalt (PTB), Braunschweig, Germany [35]. The same 100 Ω resistance standard was calibrated with the two different quantum Hall effect apparatus. The results are shown in Fig.3. All realizations of the unit of resistance agreed within $1 \cdot 10^{-9}$.

6 Conclusions

The status of precision measurement of the quantum Hall effect 16 years after its discovery is sketched out.

The cryogenic current comparator has become the most effective experimental set-up. Relative uncertainties in the reproducibility of a resistance measurement with this instrument are in the order of a few parts in 10^{10}.

Following technical guidelines for reliable measurements no dependence of the quantized values on sample width, material or step number has been found. The values of the experimentally determined von Klitzing constant are in agreement with the fine structure constant derived from the anomalous magnetic moment of the electron

Despite the question of whether the quantized Hall effect can be explained quantitatively by a theoretical model and independent of whether or not R_K is identical with h/e^2, the effect has been used with great success for calibrations and maintenance of the ohm in many NMIs since 1990.

7 References

1. K. v. Klitzing, *priv. comm.*, 1980
2. K. v. Klitzing, *Phys. Rev. Lett.* **45**, 494 (1980)
3. A.M. Thompson, D.G. Lampard, *Nature* **177**, 888 (1956)
4. E. Braun, E. Staben, K. v. Klitzing, *PTB-Mitteilungen* **90**, 350 (1980)
5. M.E. Cage, R.F. Dziuba, B.F. Field, *IEEE Trans. Instrum. Meas.* **IM-34**, 301 (1985)
6. J. Kinoshita, H. Nishinaka, K. Segawa, C.T. van Degrift, T. Endo, *IEEE Trans. Instrum. Meas.* **IM-40**, 249 (1991)
7. K.C. Lee, M.E. Cage, P.S. Rowe, *Journal of the National Institute of Standards and Technology* **99**, 227 (1994)
8. K. Inagaki, Y. Sakamoto, T. Endo, *IEEE Trans. Instrum. Meas.* **IM-38**, 276 (1989)
9. P. Warnecke, J. Niemeyer, F.W. Dünschede, L. Grimm, G. Weimann, W. Schlapp, *IEEE Trans. Instrum. Meas.* **IM-36**, 249 (1987)
10. J. Kinoshita, K. Inagaki, Y. Murayama, T. Endo, C. Yamanouchi, K. Yoshihiro, J. Wakabayashi, S. Kawaji, *IEEE Trans. Instrum. Meas.* **IM-36**, 230 (1987)
11. T. Endo, Y. Murayama, M. Koyanagi, J. Kinoshita, K. Inagaki, C. Yamanouchi, K. Yoshihiro, *IEEE Trans. Instrum. Meas.* **IM-34**, 323 (1985)

12. B.V. Hamon, *J. Sci. Instrum.* **31**, 450 (1954)
13. G. Boella, G. Marullo Reedtz, *IEEE Trans. Instrum. Meas.* **IM-40**, 245 (1991)
14. G.W. Small, *IEEE Trans. Instrum. Meas.* **IM-32**, 446 (1983)
15. G.W. Small, B.W. Ricketts, P.C. Coogan, *IEEE Trans. Instrum. Meas.* **IM-38**, 245 (1989)
16. R.F. Dziuba, R.E. Elmquist, *IEEE Trans. Instrum. Meas.* **IM-42**, 126 (1993)
17. N. L. Kusters, M.P. MacMartin, *IEEE Trans. Instrum. Meas.* **IM-19**, 291 (1970)
18. I. Harvey, *Rev. Sci. Instrum.* **43**, 1626 (1972)
19. J.C. Gallop, *SQUIDS, The Josephson Effects and Superconducting Electronics*, Adam Hilger, Bristol, Philadelphia and New York, 1990
20. F. Delahaye, *Metrologia* **29**, 81 (1992)
21. L. Bliek, E. Braun, H.-J. Engelmann, H. Leontiew, F. Melchert, W. Schlapp, B. Stahl, P. Warnecke, G. Weimann, *PTB-Mitteilungen* **93**, 21 (1983)
22. A. Hartland, *Metrologia* **29**, 175 (1992)
23. F. Delahaye, *Metrologia* **26**, 63 (1989)
24. B. Jeanneret, B. Jeckelmann, H.-J. Bühlmann, R. Houdré, M. Ilegems, *IEEE Trans. Instrum. Meas.* **IM-44**, 254 (1995)
25. B. Jeckelmann, A.D. Inglis, B. Jeanneret, *IEEE Trans. Instrum. Meas.* **IM-44**, 269 (1995)
26. A. Hartland, K. Jones, J.M. Williams, B.L. Gallagher, T. Galloway, *Phys. Rev. Lett.* **66**, 969 (1991)
27. A. Hartland, R.G. Jones, D.J. Legg, *CCE-Document 9* (1988)
28. G.W. Small, B.W. Ricketts, P.C. Coogan, B.J. Pritchard, M.M.R. Sovierzoski, communicated at CPEM '96, Braunschweig, and submitted for publication in *IEEE Trans. Instrum. Meas.*
29. A. Jeffery, R.E. Elmquist, L.H. Lee, J.Q. Shields, R.F. Dziuba, communicated at CPEM '96, Braunschweig, and submitted for publication in *IEEE Trans. Instrum. Meas.*
30. T. Kinoshita, communicated at CPEM '96, Braunschweig, and submitted for publication in *IEEE Trans. Instrum. Meas.*
31. E. Krüger, W. Nistler, W. Weirauch, *Metrologia* **32**, 117 (1995)
32. T.J. Quinn, *Metrologia* **26**, 69 (1989)
33. F. Delahaye, T.J. Witt, F. Piquemal, G. Geneves *IEEE Trans. Instrum. Meas.* **IM-44**, 258 (1995)
34. F. Delahaye, T.J. Witt, B. Jeckelmann, B. Jeanneret, to be published in *Metrologia*
35. F. Delahaye, T.J. Witt, E. Pesel, B. Schumacher, P. Warnecke, to be published in *Metrologia*

FAR-INFRARED FACILITY AT THE AUSTRALIAN NATIONAL PULSED MAGNET LABORATORY

R. A. LEWIS, R. J. HERON
Department of Physics, University of Wollongong,
Wollongong NSW 2522, Australia

R. G. CLARK, R. P. STARRETT and A. V. SKOUGAREVSKY
School of Physics, University of New South Wales,
Kensington NSW 2033, Australia

A far-infrared ($40 < \lambda < 1000$ μm) facility which permits measurements to be made at simultaneous high field (>50 T) and low temperature (<400 mK) is described. Representative data from experiments on n-GaAs and organic conductors are given.

1 Introduction

Far-infrared radiation is taken to have wavelength, λ, in the range 40–1000 μm. This corresponds in spatial frequency, $\bar{\nu} \equiv 1/\lambda$, to 10–250 cm^{-1}; in temporal frequency, $\nu = c/\lambda$, to ≈0.3–7.5 THz; and in energy, $E = h\nu$, to ≈1–31 meV.

Many physical phenomena have energies in this range making far-infrared radiation a powerful tool in condensed-matter research. For example, the Boltzmann energy, $k_B T$, is in the far infrared for ≈14–360 K. Turning to semiconductors, energy levels of hydrogenic impurities, which are of the order of the effective Rydberg, $Ry^* = (m^*/m_e)(1/\epsilon^2) \times 13.6$ eV, fall in the far infrared (e.g. for n-GaAs, $m^* = 0.0665 m_e$, $\epsilon = 12.56$ and $Ry^* = 5.72$ meV). Confined levels in semiconductor heterostructures have energies of the same order. In magnetic fields, semiconductors show additional far-infrared phenomena. Bound levels undergo Zeeman splitting, $\Delta E = \mu_B g B$ (for n-GaAs, $\mu_B{}^* \approx 0.87$ meV/T). Free carriers undergo cyclotron resonance, characterised by magnetic length $l = \sqrt{\hbar/eB} \approx 257$ Å $\mathrm{T}^{1/2}$ $/\sqrt{\mathrm{B}}$, comparable to the effective Bohr radius $a_B^* = \epsilon a_B m_e/m^*$ (≈100 Å in n-GaAs). The cyclotron energy $\hbar\omega_c = \hbar e B/m^*$ is typically in the far infrared (for n-GaAs, $\hbar\omega_c \approx 1.74$ meV/T.) Cyclotron resonance studies yield effective mass, carrier density and optical mobility. Phonon energies in semiconductors also occur in or near the far infrared (for GaAs the TO phonon energy is 33.3 meV; LO phonon, 36.2 meV).

To date, most far-infrared semiconductor magnetospectroscopy has been limited to the magnetic fields available in superconducting magnets (<20 T). In strong pulsed magnetic fields[1] the work has been limited to liquid helium or nitrogen temperatures (>4 K). This paper describes an experimental arrangement which allows higher-field, lower-temperature, far-infrared measurements.

2 Experimental Apparatus

The radiation source [a] consists of a mid-infrared CO_2 laser which optically pumps a far-infrared cavity. The two laser cavities are incorporated into a common frame of five invar rods to increase thermal and mechanical stability. Over 80 lines are available in the 9–11 μm range: on the strongest, the power >50 W. The peak power in the far infrared >150 mW. The power output is verified using calibrated mid-infrared [b] and far-infrared [c] power meters.

The laser light passes through a black polyethylene vacuum window into a brass light pipe which leads to a switching mirror. This mirror allows the radiation to be monitored by a power meter or directed onto the sample. Right-angle corners along the light path are furnished from polished brass flats. The light pipe penetrates a blast-proof wall into the magnet room.

The pulsed magnet laboratory has been described in detail elsewhere.[3] It produces non-destructive pulses in excess of 50 T using a selectable (1–32 mF) capacitor bank which may be charged to 7 kV.

A top-loading probe [d] conducts the light to the sample in the bore of the magnet. The probe loads into a He^3 cryostat. A room-temperature quartz window separates the light-pipe vacuum from the He^3 space. Cold quartz filters fitted at 4 K and 1.2 K absorb room-temperature radiation. The final section of the probe is shown in Fig. 1. A cone fits into the end of the thin-walled brass light pipe. The final stainless-steel section is split lengthwise to suppress eddy-current heating. The non-magnetic sample mount incorporates a thermal sensor and magnetic field pulse pick-up coil.

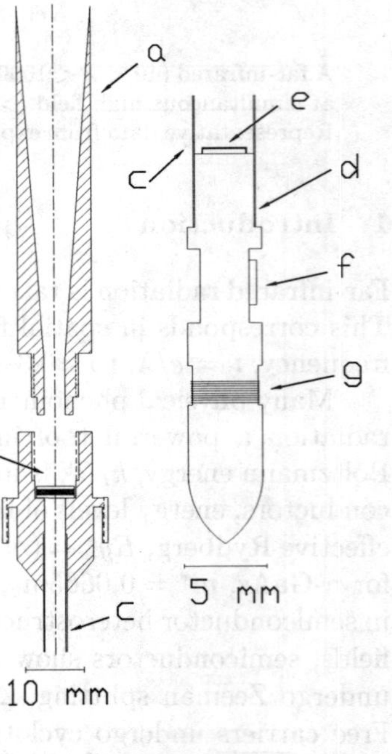

Figure 1: Final section of light pipe. Left, optical coning (1 K); right, sample holder (350 mK). The two parts are joined by a 35-cm split stainless-steel tube. *a*. Cone inside brass light pipe. *b*. Quartz cold filter. *c*. Split tube. *d*. Plastic sample mount. *e*. Sample. *f*. Temperature sensor. *g*. Pick-up coil.

---

[a] Model FIRL 100; Edinburgh Instruments, Riccarton, Currie, Edinburgh EH14 4AP, UK.
[b] Model 200/100 Power Meter; Coherent, Auburn, California 95603, USA.
[c] Calorimeter Model Number MC2500; Scientech, Boulder, Colorado 80303-1399, USA.
[d] Oxford Instruments, Old Station Way, Eynsham, Witney, Oxon OX8 1TL, UK.

3 Applications

3.1 n-GaAs

In proving this new far-infrared facility, bulk *n*-GaAs has been examined as a well-characterised reference material. Some results are presented in Fig. 2.

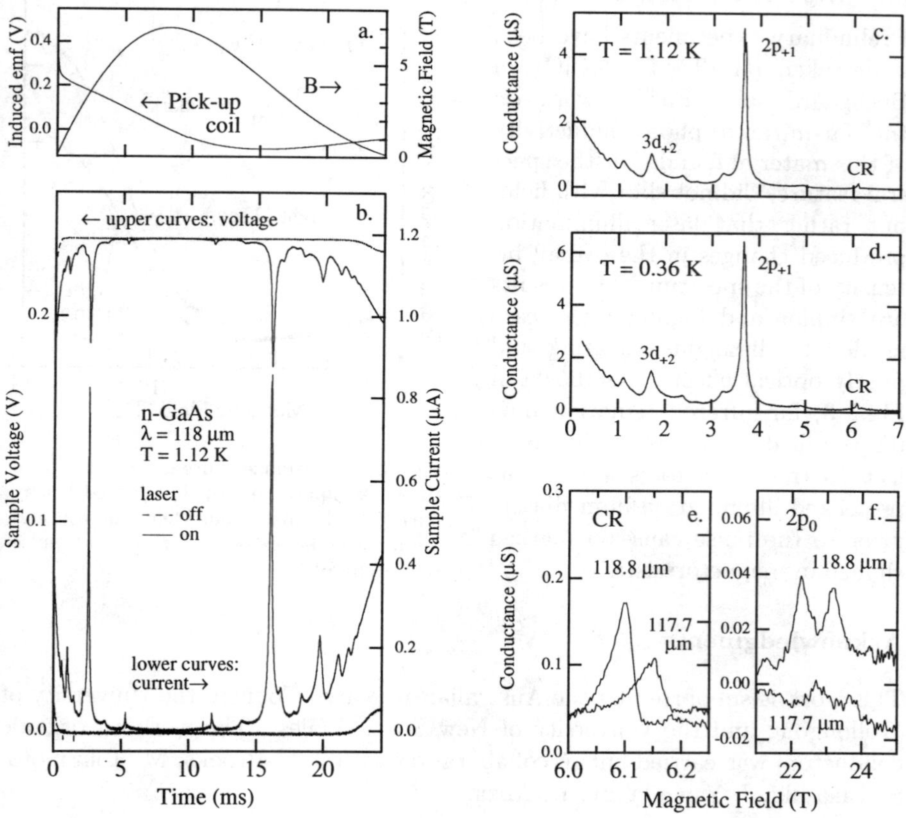

Figure 2: *a.* Pick-up coil signal ($\propto dB/dt$) is integrated to compute magnetic field. *b.* Voltage (upper curves) and current (lower curves) in *n*-GaAs sample during field pulse, with laser on (inner curves) and off (outer curves). Conductance may be measured in constant-voltage (current source impedance $\ll$ sample impedance) or constant-current (vice versa) mode; for optimal signal recovery, impedances are matched. *c.* Sample conductance against field. Only data for decreasing field is shown; up-sweep data is identical. *d.* Similar data at lower temperature. *e.* For two close laser lines cyclotron resonances are separated by 0.05 T; magnetic field resolution exceeds this. *f.* Chemically-split $1s \rightarrow 2p_0$ transition at $\approx$23 T.

In Fig. 2c data is given to 7 T to allow direct comparison with previous work in slowly-swept fields (e.g. Fig. 3 of Ref. 3). Excellent agreement is found. Millikelvin data (Fig. 2d) differs little from 1.12 K (Fig. 2c) or 2 K data (Ref. 3). Chemical splitting is evident for both the $1s \rightarrow 2p_+$ and $1s \rightarrow 2p_0$ transitions.

3.2 Organic Conductors

Preliminary experiments have been undertaken on $(TMTSF)_2ClO_4$, a Bechgaard salt.[4] Earlier work on the far-infrared photoconductivity of this material found that the spectral features did not shift with field, but rather that laser illumination produced changes in the overall intensity of the spectrum.[5] Our results are similar and highlight the care needed to distinguish heating and purely optical effects. As shown in Fig. 3, far-infrared radiation during the field pulse has negligible effect on the magnetoresistance, but prolonged illumination immediately prior to the pulse causes a marked change in magnetoresistance.

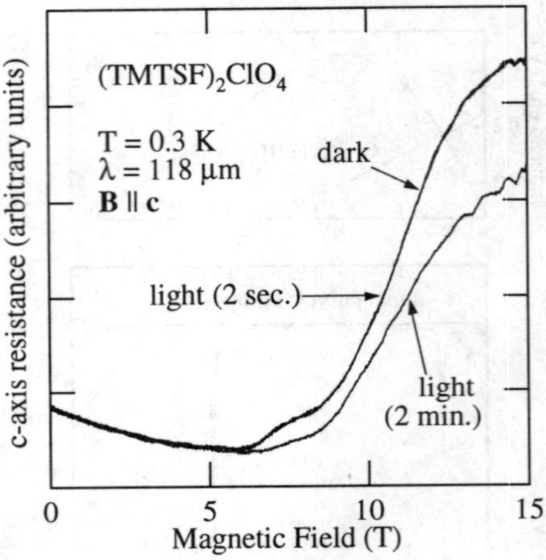

Figure 3: Resistance of organic conductor in the dark (upper curve); illuminated briefly during field pulse (coincident curve); and illuminated before and during field pulse (lower curve).

Acknowledgments

This work is supported by the Australian Research Council, the University of Wollongong and the University of New South Wales. The work on organic conductors was carried out in collaboration with J. S. Brooks, M. Tokumoto, S. Takasaki, J. Yamada and H. Anzai.

References

1. Fritz Herlach and Luc Van Bockstal, *Infrared Phys.* **32**, 145 (1991).
2. R. G. Clark *et al.*, *Physica* B **201**, 565 (1994).
3. C. J. Armistead *et al.*, *Semicond. Sci. Technol.* **4**, 557 (1989).
4. J. S. Brooks, MRS Bulletin **18**, 29 (1993); *Phys. Rev.* **B** in press (1996).
5. J. Lefebvre *et al.*, *Phys. Rev. Lett.* **72**, 3417 (1994).

NOVEL CONCEPTS IN HALL SENSORS

R. G. MANI,[+] K. VON KLITZING,[+] F. JOST,[#] K. MARX,[#]
S. LINDENKREUZ,[§] AND H. P. TRAH[§]
[+]Max-Planck-Institut FKF, Heisenbergstrasse 1, D-70569 Stuttgart
[#]Robert Bosch GmbH, ZWT, P.O. Box 10 60 50, D-70049 Stuttgart
[§]Robert Bosch GmbH, Tübingerstrasse 123, D-72762 Reutlingen

A doubly connected 'anti Hall bar within a Hall bar' type Hall effect device that self compensates for certain spurious offset voltages is described. It is shown that multiaxial, multiport current injection from separated current sources is useful not only for reducing the misalignment offset but also the shear stress induced offset observed in Silicon devices. Measurements suggest field-equivalent offsets below 1 milliTesla in such (001) surface n-Silicon Hall effect devices with current injection in the [110] direction.

Hall effect and certain piezoresistive sensors utilize "off-diagonal" voltages that appear perpendicular to the current in a four terminal Hall plate configuration, upon the application of a magnetic field (B) and stress (S), respectively.[1-10] Such devices typically exhibit an offset voltage across the voltage contacts (at B = 0 in the Hall sensor and at vanishing shear stress in the piezoresistive element) due to effective voltage contact misalignments. This misalignment offset is important because "signal to offset" limits sensor performance in the low magnetic field / weak stress limit, especially when the offset changes with the environment.[7-10] Further complications occur when the B - sensing element exhibits an S-dependent offset (as in silicon based devices) while, analogously, the S - sensor displays a B-dependent offset. Under these conditions, it becomes useful to identify a technique that can reduce the misalignment offset, and separate the Hall voltage from the (symmetric off-diagonal) piezoresistive voltage, or help cancel one of the components across the relevant voltage contacts such that the sensor signal reflects the desired effects.

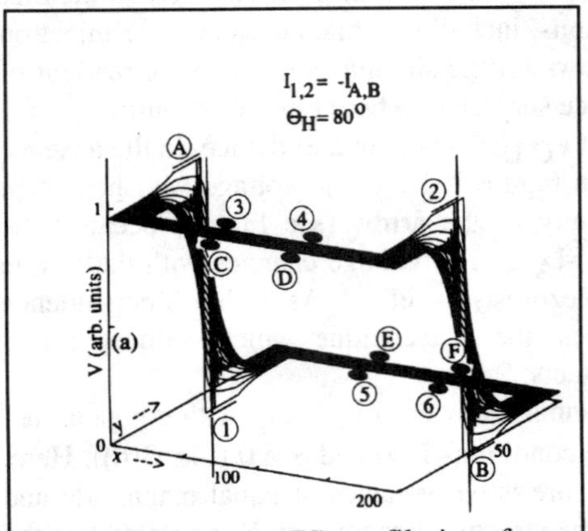

Fig. 1. The voltage (V) profile in a frame shaped device with $I_{1,2} = -I_{A,B}$. Contacts are labelled as in refs. 7,10.

Silicon (Si) devices that respond to external stimuli are especially attractive because of the possibility of integrating advanced functions onto the basic sensing element and manufacturing large quantities at low cost per unit in the standard Si fabrication facilty. They also warrant special attention because Si magnetic sensors have already appeared as position detectors and contactless switches in applications ranging from electric motors to soft-drink machines and automobiles[1,3] - applications where better performance might prove valuable in

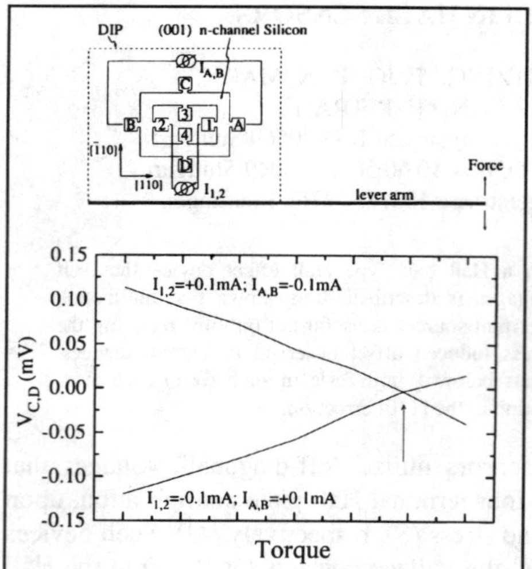

Fig. 2. The torque dependent voltage $V_{C,D}$ in a Si 'anti Hall bar within a Hall bar'. When the specimen is stress free (marked by arrow), the offset due to the misalignment- and stress- contributions vanishes.

reducing manufacturing tolerances elsewhere.

The function of a basic, current compensated 'anti Hall bar within a Hall bar' configuration in reducing the misalignment offset is illustrated in Fig. 1. Here, the numerical simulations correspond to dual current injection at the ends of a frame shaped specimen with $I_{1,2} = -I_{A,B}$, at a Hall angle $\Theta_H = 80^o$. The voltage profile (Fig. 1) within the device indicates that each branch of the device becomes an equipotential as the current is confined to the ends of the device. Under such conditions, one sees that the introduction of a modest misalignment in the Hall voltage contacts (e.g., contacts D and F) need not disturb the measured signal.

Application of this approach to the piezoresistive semiconductor Si showed that although the misalignment offset was reduced as expected, the contribution of another effect - the off-diagonal piezoresistive effect - became significant at low B for certain device alignments within the crystal.[4-6] The study of the B- and S- effects helped identify some new B- and S- sensor configurations, including a biaxial, quadruple injection configuration examined here, which allows for the simultaneous, separate readout of the B- and S- induced voltages in a device subjected to both types of stimuli.

The torque-dependent exterior ($V_{C,D}$) voltage in a Si device, in the absence of a magnetic field (B = 0), is shown in Figure 2. Here, the voltage vanishes when the specimen is stress-free in the vicinity of the arrow (see Fig. 2) because the misalignment offset vanishes for $I_{1,2} = -I_{A,B}$. The voltage changes with the torque on the sample mount due to the shear piezoresistive effect. As such a S-dependence is undesirable in a B-sensor, we describe a technique which reduces the S contribution across the Hall voltage contacts.

Consider measurements in an annular device (Fig. 3(a)), with compensated uniaxial dual current injection under the conditions B=0 and S ≠ 0 (Fig. 3(b)). Here, the voltmeters V_L and V_R indicate piezoresistive voltages of equal magnitude and opposite signs because each branch of the structure becomes an equipotential as the branch current density vanishes. Then, the application of a ramped surface-normal magnetic field tends to rotate V_L and V_R in opposite directions (Fig. 3(e)). In the other case, where compensated double injection occurs perpendicular to the first

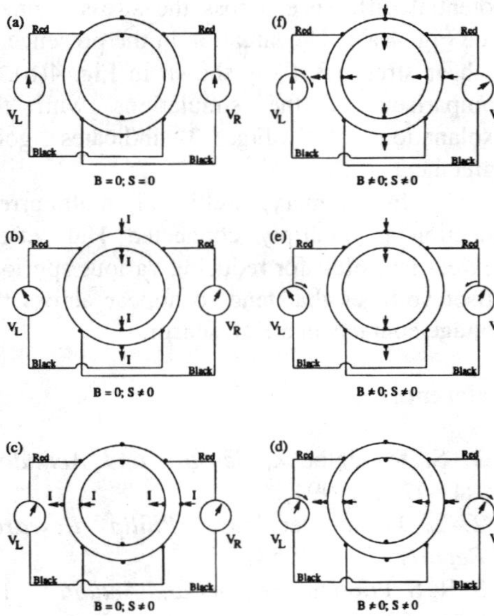

Fig. 3. a) An annulus with contacts. Voltmeters V_L and V_R display the voltage difference. b) An "off-diagonal" piezoresistive voltage is indicated in the absence of a magnetic field (B=0) under dual injection. c) Current injection perpendicular to that in (b). Here, at B = 0, the voltmeters display the same polarity, and a ramped magnetic field rotates the two displays in the same sense (see (d)). e) In configuration (b), a ramped magnetic field turns the clocks in opposite directions. (f) Under simultaneous dual axis injection, V_L displays vanishing piezoresistive voltage because the contributions from configurations (d) and (e) cancel. However, the Hall voltages add up here to produce twice the Hall effect. V_R displays the sum of the piezoresistive voltages, while the Hall voltage from the two configurations cancel.

direction (Fig. 3(c)), V_L and V_R indicate the same sign, and a ramped magnetic field (see Fig. 3(d)) tends to turn both voltmeters in the same direction (clockwise) (as V_L in Fig. 3(e)).

By utilizing electrically separate current sources, one is able to carry out the two (dual injection) experiments simultaneously.[7-10] Thus, in Fig. 3(f), the V_R display does not turn with ramped B because the opposing Hall effects (see Fig. 3(d) and Fig. 3(e)) cancel each other. (V_R shows, however, the sum of the piezoresistive offsets observed in the two configurations). On the other hand, V_L, which exhibits cancellation of the piezoresistive voltage, shows that the magnetic response, or the Hall voltage, is the sum of the effects observed in the two (Fig. 3(d) and 3(e)) configurations. Thus, one obtains vanishing piezoresistive voltage across the Hall voltage contacts, and vanishing Hall voltage across the piezoresistive contacts. For B - sensing applications, this is a desirable characteristic because the sensor output becomes less sensitive to environmental stresses.

Finally, the potential distribution in a frame shaped device under quadruple current injection has been numerically calculated in the presence of a magnetic field as well as shear stress; the equipotential contours are shown in Figure 4. In the presence of a magnetic field (and

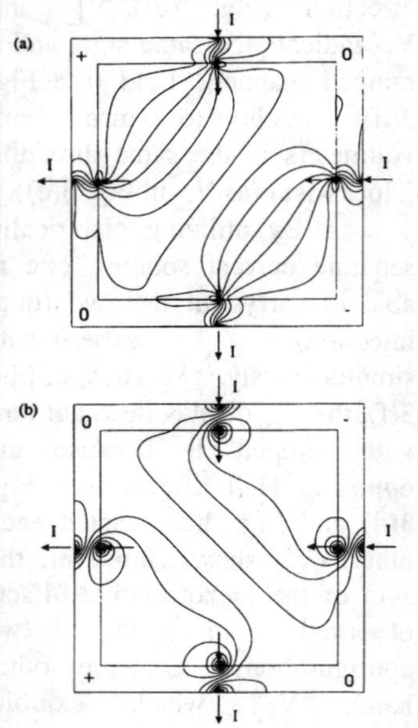

Fig. 4. (a) A contour plot of the numerically calculated potential distribution in the presence of a magnetic field and in the absence of stress. A Hall voltage appears across the upper left and lower right contacts. (b) Applying shear stress to the device (with no magnetic field) yields a potential difference across the stress contacts (lower left and upper right contacts).

S=0), the simulation shows a Hall voltage across the diagonal Hall contacts with no potential difference across the stress contacts (see Fig. 4(a)). The situation in the presence of a shear stress (B=0) is shown in Fig. 4(b). A comparison of the simulations with the explanation of Fig. 3 indicates good agreement.

In summary, multiaxial, multicurrent injection in multiply connected Hall effect devices is useful for reducing various spurious offset voltages that tend to appear across the voltage contacts in a Hall sensor.

References

1. S. Middelhoek, *Sensors and Actuators* A41-42, 1, (1994).
2. L. J. van der Pauw, *Philips Research Reports* 13, 1, (1958).
3. R. S. Popovic, *Sensors and Actuators* 17, 39 (1989).
4. C. S. Smith, *Phys. Rev.* B94, 42 (1954).
5. R. W. Keyes, *Sol. St. Phys.* 11, 149 (1960).
6. Y. Kanda and K. Suzuki, *Phys. Rev.* B43, 6754 (1991).
7. R. G. Mani and K. von Klitzing, *Z. Phys.* B92, 335 (1993); *Appl. Phys. Lett.* 64, 1262 (1994); *Appl. Phys. Lett.* 64, 3121 (1994).
8. R. G. Mani, K. von Klitzing, and K. Ploog in *11th Intl. Conf. High Magnetic Fields in the Physics of Semiconductors*, ed. D. Heiman (World Scientific, Singapore, 1995), p. 146.
9. R. G. Mani, K. von Klitzing, F. Jost, K. Marx, S. Lindenkreuz, and H. P. Trah, *Appl. Phys. Lett.* 67, 2223 (1995).
10. R. G. Mani, *Europhys. Lett.* 34, 139 (1996); *J. Phys. Soc. Jpn.* 65, 2301 (1996); R. G. Mani and K. von Klitzing, *Z. Phys. B*100 (4), xxx (1996).

FERMI EDGE EFFECTS AT EVEN-DENOMINATOR FILLING FACTORS

V. NIKOS NICOPOULOS, DARREN J. T. LEONARD, N. F. JOHNSON
Department of Physics, Clarendon Laboratory, Oxford University, Oxford OX1 3PU, England

F. J. RODRIGUEZ
Departamento de Fisica, Universidad de Los Andes, Bogota A.A. 4976, Colombia

The optical properties of an electron gas in a magnetic field at filling fractions $\nu = \frac{1}{2m}$ $(m = 1, 2, 3 \ldots)$ are investigated using the composite fermion picture. The response of the system to the presence of valence-band holes is calculated. The shapes of the emission spectra are found to differ qualitatively from the well-known electron-hole results at zero magnetic field. In particular, the asymmetry of the emission lineshape is found to be sensitive to the hole-composite fermion plane separation.

1 Introduction

The composite fermion (CF) approach to electronic correlations in the lowest Landau level implies that, at half-filling, the electron gas has properties similar to a Fermi liquid in zero magnetic field [1]. The residual interactions between composite fermions are thought to be weak, but have nevertheless features that characterize the low lying excitations of the system.

The existence of a Fermi surface for composite fermions at half-filling raises then the interesting possibility of optical properties that are closely connected with the existence of a Fermi edge, in particular, optical singularities at the absorption and emission thresholds.

Classic work on the subject has been carried for metals, in which a deep, localized hole is excited or annihilated. This is the well-known X-ray singularity problem. There are two competing features in this problem:

(i) First, the tendency of the electrons in the Fermi sea to form states with the hole, influenced by the attractive Coulomb interaction. This leads to an enhancement (possibly singular) of the absorption and emission spectra at the edge.

(ii) The other effect is the collective response of the Fermi sea to the transient perturbation introduced by the hole. For a localised infinitely heavy hole, this response is one of multiple excitations of electron-hole pairs of vanishing energy around the Fermi sea. In fact, these states dominate the response of the system, because the unperturbed Fermi sea is orthogonal to the low lying states of the system with the hole present (this is the orthogonality catastrophe). These processes in turn affect the nature of the hole state, and induce an asymmetric broadening of the hole spectral function as a function of energy.

The asymmetry reflects the fact that the hole is necessarily accompanied by the small energy electron-hole excitations around the Fermi sea (whose number diverges with vanishing energy), that give the state a larger energy, but no states with energies lower than the energy of the non–interacting hole exists.

A self-consistent treatment then of the excitonic processes, with the new hole spectral function tends to reduce the severity of the singularity, or even cause the absorption (emission) to go smoothly to zero at the edges.

2 The Fermi Edge of Composite Fermions

We would like to understand the corresponding effects when the Fermi sea is that of CF's in a half-filled Landau level in a heterojunction, and the hole is a photo-excited hole near the 2d electron gas. Experiments that probe this regime have been becoming more numerous in recent years. The experiments are thought to create a hole that is not particularly localized in the longitudinal direction with respect to the magnetic field (which is perpendicular to the plane of the electrons) [2]. In general it may be a good approximation to consider the hole displaced by a distance d from the plane of the electron gas. Then we deal with a situation in which a hole is excited out of a valence sub–band Landau level into the Fermi sea of CF's.

Because the hole is not localized, we expect the effects of the orthogonality catastrophe (those in ii) to be mitigated, and crudely model them by a broadening of the hole spectral function. Such symmetric broadening is also due to other processes that give the hole a finite lifetime. This is only a starting point, and we are addressing the subtler effects of the effect of low energy CF excitation on the spectral function of the hole in our work in progress. In the present work, we are after a simple qualitative understanding of the problem.

The features peculiar to the CF problem we are addressing are:

(i) The fact that the hole is subject to the magnetic field, whereas, in the mean field approximation, the CF are in zero field states, and they experience interactions amongst themselves arising from fluctuations away from the mean field.

(ii) The dependence of the optical properties on the strength of the valence hole–CF interaction.

The first point gives a particular form to the dipole moment matrix element, and to the form of the interaction of the incoming photon with the electrons and the valence hole:

(a) Dipole moment matrix element:

$$r_{cv}(\mathbf{K},k) = r_{cv}\delta_{-k,K_y}\sqrt{\frac{2l_B\sqrt{\pi}}{L}}S(\mathbf{K})$$

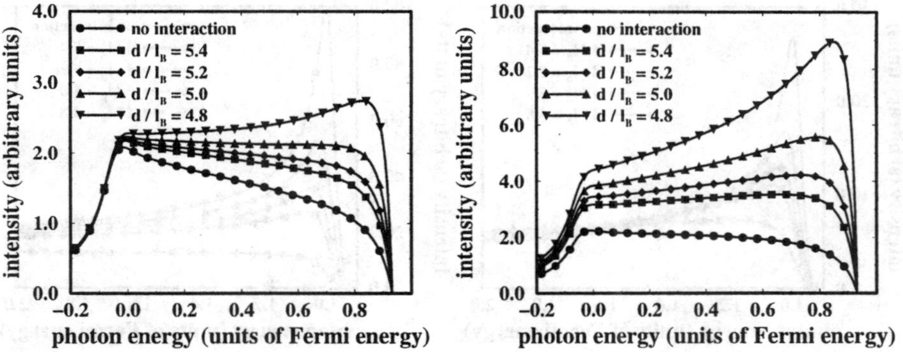

Figure 1: (a) Emission in a magnetic field. The length d parametrizes the strength of the interaction. Note the enhancement at low energies, due to the specific form of the dipole matrix element in a magnetic field. (b) Emission in zero field for the same parameters as in 1(a).

with

$$S(\mathbf{K}) \equiv \exp\left(iK_xK_yl_B^2 - \frac{1}{2}(K_xl_B)^2\right) \tag{1}$$

where $\mathbf{K}$ is the momentum of the CF, and kl_B^2 is the guiding center x–coordinate of the state of the valence hole in the magnetic field (in the Landau gauge). Because of the magnetic field the matrix element is exponentially damped as a function of K_x, and this leads to damping of the optical emission spectrum.

(b) Coulomb matrix element between the valence hole and a CF:

$$V(\mathbf{K},k,k',\mathbf{K}') = -\frac{e^2}{\epsilon_0\epsilon_r}\frac{e^{-|\mathbf{K}-\mathbf{K}'|d}}{|\mathbf{K}-\mathbf{K}'|}e^{-\frac{1}{4}l_B^2|\mathbf{K}-\mathbf{K}'|^2-\frac{i}{2}l_B^2(K_x-K'_x)(k+k')}$$

To obtain simple results we have to simplify this expression while retaining one important feature, which is the dependence on the separation d between the electron plane and the hole.

The optical response of the system is determined by the dielectric constant, which is in turn related to the photon self-energy. For a simple discussion we calculate the latter in the ladder approximation. This tends to underestimate the singularity at the edge, but the singularity in any case is mitigated by the hole broadening we assumed, so the ladder approximation is sufficient for a semi-quantitative discussion.

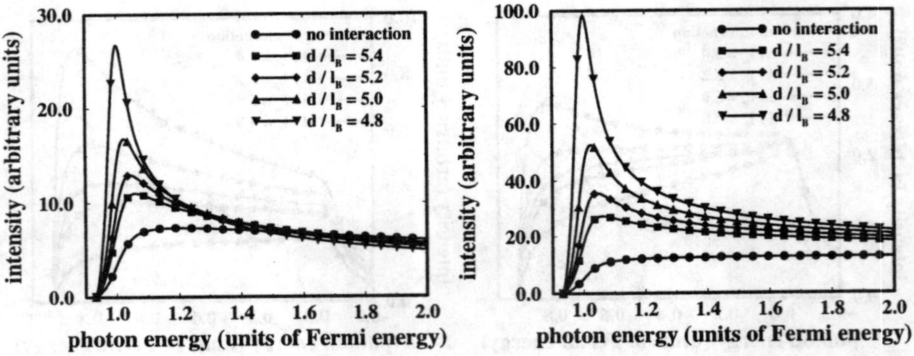

Figure 2: (a) Absorption in a magnetic field. The length d parametrizes the strength of the interaction. (b) Absorption in zero field for the same parameters as in 2(a).

The results are shown in figures 1 and 2.

Two features are of note:

(1) The Fermi edge enhancement, which becomes clear as the separation between the electron plane and the hole is decreased. (2) In addition the particular form of the matrix element for strong magnetic field, gives an enhancement to the low energy part of the emmision spectrum (compare fig. 1(a) to fig. 1(b); the latter is for a flat matrix element, appropriate to the zero B-field case, while the former has the form of the matrix element appropriate for strong B-field).

3 Acknowledgements

We acknowledge the financial support of EPSRC through a Studentship (D.L), EPSRC Grant No. GR/K 15619 (N.F.J. and V.N.N.) and COLCIENCIAS (F.J.R.). We would like to thank Andrew Turberfield for very useful discussions. We are also grateful to Robin Nicholas, Henry Cheng, Taco Portengen and Luis Quiroga. We thank Henry Cheng for showing us his experimental data prior to publication.

References

1. B.I. Halperin, P.A. Lee and N. Read, *Phys. Rev. B* **47**, *7312 (1993)*.
2. I. N. Harris, H. D. M. Davies, R. A. Ford, J. F. Ryan, A. J. Turberfield, C. T. Foxon and J. J. Harris, *Surf. Sci.* **305**, 61 (1994); I.N. Harris, Oxford University, DPhil. Thesis, 1994; A.J. Turberfield (unpublished).

Author Index

A. USHER 017

12th International Conference

High Magnetic Fields in the Physics of Semiconductors II

Volume 1

12th International Conference

High Magnetic Fields in the Physics of Semiconductors II

Volume 1

Würzburg, Germany
29 July–2 August 1996

Editors

G. Landwehr
W. Ossau
Physikalisches Institut der Universität Würzburg, Germany

World Scientific
Singapore • New Jersey • London • Hong Kong

Published by

World Scientific Publishing Co. Pte. Ltd.

P O Box 128, Farrer Road, Singapore 912805

USA office: Suite 1B, 1060 Main Street, River Edge, NJ 07661

UK office: 57 Shelton Street, Covent Garden, London WC2H 9HE

HIGH MAGNETIC FIELDS IN THE PHYSICS OF SEMICONDUCTORS II

ISBN 981-02-3076-1 (Set)
ISBN 981-02-3989-0 (Vol. 1)
ISBN 981-02-3990-4 (Vol. 2)

Printed in Singapore.

PREFACE

This volume contains contributions presented at the XIIth International Conference on High Magnetic Fields in Semiconductor Physics which was held at the University of Würzburg from July 29 to August 2, 1996. In the tradition of previous Würzburg conferences on the same subject - the first conference was held in 1972 -, only the invited papers were presented orally. In order to give an overview, all 37 lecturers were asked not only to review the latest results in their field, but to give in addition a general introduction. Consequently, this volume differs from the usual conference proceedings.

The rapid development in semiconductor physics and technology during the last years resulted in an extensive application of high magnetic fields both in fundamental and applied research. As a consequence, more than 160 contributed papers were presented as posters at the Würzburg meeting.

16 years after its discovery, the Quantum Hall Effect (QHE) is still a subject of high actuality. Many new results on the fractional QHE were presented, especially on breakdown phenomena and on edge effects. In addition to 6 invited papers, 43 contributions were presented. Especially the composite fermion and skyrmion concepts were thoroughly discussed. Another field of high actuality was magnetooptics. Altogether 49 posters were presented, mainly in connection with the investigation of systems of reduced dimensionality. Special emphasis was on quantum dots and microcavities as well as on coupled systems. Magnetotransport also turned out to be of high interest. It is rather obvious that magnetic semiconductors played a prominent role at the conference, too.

Without doubt, the availability of superconducting magnets in most laboratories contributed to the growth of semiconductor physics in high magnetic fields. Because not all experiments can be performed in fields up to 10 or 15 Tesla, high magnetic field laboratories offering larger fields are indispensable. There were reports from 4 laboratories on present work going on at these installations. At present experiments in pulsed fields up to 200 Tesla can be performed with high reliability nowadays, without destroying the samples.

The organizing committee consisted of G. Landwehr (Chairman), W. Ossau (Secretary), K. von Klitzing, J. Hajdu and M. von Ortenberg. The program of the conference was based on suggestions by many colleagues who were consulted. We would like to thank for their active participation.

Financial support of the conference by the following sponsors is gratefully acknowledged:

(institutional)	Deutsche Forschungsgemeinschaft, Bonn Bayerisches Staatsministerium für Unterricht, Kultus, Wissenschaft und Kunst, München
(industrial)	Siemens AG, München Cryophysics, Darmstadt Oswald Magnetfeldtechnik, Miltenberg Oxford Instruments, Wiesbaden

Würzburg, 01 October 1996

G. Landwehr, W. Ossau

CONTENTS

VOLUME 1

I. Quantum Hall Effects

A. Integer Quantum Hall Effect

C. Breakdown Phenomena

II. New Quasi-Particles

A. Composite Fermions

B. Skyrmions

III. Magnetotransport

IV. Tunneling

V. Small Quantum Systems

VOLUME 2

VI. Magneto-Optics

A. Quantum Well Structures

B. Heterostructures and 2D Systems

C. Dots and Microcavities

D. Coupled Systems

E. Excitons

F. Far Infrared Region

VII. Magnetic Semiconductors

VIII. Organic Conductors

IX. Si-Ge

X. Reports from Labs, Experimental Methods

XI. Post-Deadline Papers

QUANTUM HALL EFFECT AT FINITE TEMPERATURES

D.G. POLYAKOV
Institut für Theoretische Physik der Universität zu Köln,
Zülpicher Str. 77, 50937 Köln, Germany

Recent work on the temperature-driven delocalization in the quantum Hall regime is reviewed, with emphasis on the correlation properties of disorder and the role of electron-electron interactions.

1 Introduction

The underlying physics of the quantum Hall effect (QHE) is understood in terms of the Anderson localization in two dimensions (2D): a weakly disordered 2D electron gas experiences a series of metal-insulator transitions with increasing magnetic field B.[1] The field-induced transitions are degenerate in the sense that the metallic phase occurs only at some particular values of B. At zero temperature, the dissipative conductivity $\sigma_{xx}(B)$ vanishes everywhere except at a discrete set of the critical points, where it exhibits peaks of zero width. My purpose is to summarize recent work showing how this "ideal" picture of QHE evolves with lowering T. Let us first go over some key features of the localization in QHE.

At $T = 0$, the crossover to the QHE regime should occur, for weakly interacting electrons, at $\lambda/l \sim 1$, where λ is the magnetic length, l the mean free path. Yet, making contact with most of experiments, we are accustomed to treating QHE in the extreme of high B, where disorder-broadened Landau levels are well separated from each other. The number of conducting phases, which arise in succession as the Fermi level sweeps through a single Landau level, is determined by competition between electron-electron interactions and disorder. Since, to date, there exists no reliable theory of localization in the fractional QHE, I restrict the discussion to the integer QHE with weak enough interactions. Then there is only one extended state per Landau level, and its energy E_c coincides with the center of the level. Within this framework, the localization length $\xi(E)$ diverges as $|E - E_c|^{-\gamma}$, where $\gamma \simeq 2.3$.[2]

Naturally, in a close vicinity of E_c, the critical exponent γ should not depend on the correlation radius of disorder, d; however, the range of the universal scaling behavior shrinks if $\lambda \ll d$. For the most interesting case of the lowest Landau level, $\xi(E)$ then scales as $\xi(\Delta_t)(\Delta_t/|E - E_c|)^\gamma$, only in a narrow band of the width $\Delta_t = \Gamma(\lambda/d)^2$ around E_c, Γ being the width of the Landau level. I refer to this range of E as the "tunneling band", for within it the percolating classical trajectories, which are closed equipotentials in the high-B limit, are strongly coupled via tunneling through saddle points of the random potential. Outside this band, the localization remains classical in the sense that one may neglect the coupling between the critical percolating trajectories, the characteristic radius of which behaves as $R(E) \sim d(\Gamma/|E - E_c|)^{4/3}$.[3] These trajectories almost touch each other at the critical saddle points so as to form the percolation network. However, provided $|E - E_c| \gg \Delta_t$, the gaps that separate them are typically much wider than λ (to generate the power-law scaling, the critical saddle points at given E should have energies of the order of $|E - E_c|$), and so the tunneling plays no role. At the crossover point between the regimes of the quantum and classical localization, $\xi(\Delta_t) \sim d(d/\lambda)^{8/3}$. Clearly, this picture rules out any relation between γ (which is close to 7/3) and the critical exponent of $R(E)$ (exactly equal to 4/3). In the tunneling-dominated regime, the conducting network is essentially represented by that of the Chalker-Coddington model[4] with the characteristic size of the elementary cell $\xi(\Delta_t)$, independent of E. A weak tunneling through chains of non-critical saddle points does not affect the topology of the network.

2 Temperature Scaling: Interactions vs Disorder

In this section, I present an outline of the problem of how σ_{xx} peaks broaden with temperature in the limit of short-range disorder. Since the localization at $E \to E_c$ originates from the interference of multiple scattered waves on the large scale of $\xi(E)$, the problem may be analyzed in terms of the phase breaking. The width of the energy band within which the localization fails to develop, $\Delta_c(T)$, then obeys the simple scaling relation $\xi(\Delta_c)/L_\phi \sim 1$, where L_ϕ is the dephasing length at the critical energy. The notion of the critical broadening of the σ_{xx} peaks implies that the inelastic scattering is strong enough, namely $\Delta_c(T)/T \gg 1$.

In trying to apply, after simple modifications, the standard theory[5] of the phase breaking in dirty metals to the metallic phase in the QHE limit, one can immediately see that the Fermi liquid picture may only be marginally valid. Specifically, it appears that the crucial parameter $\hbar/T\tau_\phi \sim 1$, τ_ϕ being the phase-breaking time associated with electron-electron interactions. In this circumstance, we are only able to say what is the scaling behavior of L_ϕ:

$$L_\phi(T) \sim \lambda(\Gamma/T)^{1/2} \ . \tag{1}$$

Doing so, however, we run into a difficulty: for the critical exponent of $\Delta_c \propto T^\kappa$, Eq. (1) gives $\kappa = 1/2\gamma$, which is approximately half of the thoroughly measured, in samples with short-range disorder, value $\kappa \simeq 0.4$.[6] To fix this flaw, we argue that there exists a crossover temperature $T_c \sim U^2/\Gamma$, below which the Coulomb interaction between electrons cannot be treated perturbatively, however weak it is in comparison with disorder. Here U characterizes the strength of the interaction and scales as $e^2/\varepsilon\lambda$, ε being the dielectric constant. We have to assume the condition $U/\Gamma \ll 1$, in order to ensure that the interaction does not break down the integer QHE.

We first make the elementary observation that the Maxwell relaxation of a charged wave packet, built up from eigenstates with energies close to E_c, is characterized, in the quasi-2D geometry, by the constant velocity of the charge spreading $v_s = 2\pi\sigma_{xx}^p/\varepsilon$.[7] Here $\sigma_{xx}^p \sim e^2/h$ stands for the peak value of the conductivity in the metallic phase. It follows that the dynamics of charged excitations on scales larger than $L_c \sim D_0/v_s \sim \lambda\Gamma/U$ is no longer diffusive ($D_0 \sim \lambda^2\Gamma/\hbar$ is the bare diffusion coefficent, L_c has the meaning of a screening radius); instead, it is governed by the interactions. In particular, the spectral function for the screened density-density correlator in the limit of small q takes the form

$$\mathcal{S}(\omega, \mathbf{q}) = (\varepsilon q/2\pi e^2) \times [v_s q/\pi(\omega^2 + v_s^2 q^2)] \tag{2}$$

[at criticality, the condition of small q depends on ω: q should be small as compared to both L_c^{-1} and ω/v_s; if $qv_s/\omega \gg 1$, the effective velocity v_s in Eq. (2) becomes a function of the ratio q/ω (a similar renormalization of the effective diffusion constant at $U = 0$ was studied in Ref. 8); note also that we deal with spatial scales smaller than $\xi(E)$ at all E involved (cf. Ref. 9)]. This contrasts with the dynamical properties of non-interacting electrons, which are controlled, at the critical point, by the dynamical susceptibility depending on the single parameter q^2D_0/ω.[8] Eq. (2) describes the Coulomb correlations between charged density fluctuations (originating from the dynamical screening[5,10,11]). The relation $\hbar/T\tau_\phi \sim 1$ still holds; however, the Coulomb correlations should affect the dynamical scaling at criticality, unless $L_\phi/L_c \ll 1$, i.e. $T/T_c \gg 1$. Given that in the interaction-dominated regime the dynamical scaling can be explained in terms of the charge-spreading velocity, L_ϕ at lower T obeys the relation

$$L_\phi(T) \sim v_s\tau_\phi \sim e^2/\varepsilon T \tag{3}$$

(borrowing the terminology of thermally driven phase-transitions, one can say that the dynamical critical exponent $z = 1$). This last expression looks universal since σ_{xx}^p does not depend on the strength of disorder. It is worth noting that the assumption of the long-range interaction is absolutely crucial in the picture under discussion. Hence we find that at $T/T_c \ll 1$

$$\Delta_c \sim \Gamma(T/U)^{1/\gamma}\,, \tag{4}$$

which is in decent agreement with the low-temperature data on samples with short-range disorder.[12] In the high-T limit, one should expect a crossover to the regime of broadening governed by electron-phonon interactions.[13]

Now for the localization-length exponent γ, which we keep the same as for non-interacting electrons. To see how it comes about that the interaction strongly influences the dynamical behavior of electrons at criticality, and yet does not change γ, we recall that, at $T = 0$, even a weak Coulomb interaction between localized electrons makes the one-particle density of states $g(E)$ vanish at the Fermi level E_F.[14] In two dimensions, $g(E) = c|E - E_F|\varepsilon^2/e^4$, where $c \sim 1$ is a universal constant. At this point we can already identify T_c with the width of the gap in $g(E)$ deep in the insulating phase. The concept of the Coulomb gap was developed for classical electrons, i.e. point charges. It is evident, however, that the same line of argument applies near the metal-insulator transition as well, only in the range $|E - E_F| \ll e^2/\varepsilon\xi(E)$. Hence, rather remarkably, $g(E_F)$ vanishes however small $|E_F - E_c|$ is. As a consequence, whatever E_F, the Coulomb energy on the scale of the one-particle localization length $\xi(E_F)$ is of the order of the characteristic energy spacing $\delta_c \sim 1/g(E_F + \delta_c)\xi^2(E_F)$ on the same scale. Within the Coulomb glass approach, this naturally implies that the long-range interaction merely leads to the repulsion of the levels of the one-particle states, but it cannot affect the critical behavior of $\xi(E_F)$.[15] This conclusion can be reached on more phenomenological grounds,[16] though essentially by a similar dimension counting. The scaling behavior of $\xi(E_F)$ has been observed by numerical simulation within the Hartree-Fock scheme.[17]. It is worth emphasizing that the range of energies

$$|E - E_F| \ll e^2/\varepsilon\xi(E_F) \,, \tag{5}$$

where the above arguments about the stability of $\xi(E)$ are true, shrinks as $E_F \to E_c$. Thus the question as to the localization properties of the excitations at $E_F = E_c$ remains open. Particularly, it is not clear what kind of the gap in $g(E)$ is observed in the numerical simulations, when E_F is tuned to be precisely E_c.[17] It is possible that the gap at $E_F = E_c$ is a reminiscence of the sharper "polaronic" gap, which pertains to clean systems.[18] Yet, there is no doubt that the true Coulomb gap survives in the range given by Eq. (5), if $E_F \neq E_c$. We conclude, therefore, that it is legitimate to exploit the idea of the one-particle localization at $|E_F - E_c| \sim \Delta_c(T)$, since $|E - E_F|$ does not exceed $e^2/\varepsilon\xi(\Delta_c)$ within the range of the temperature smearing of the Fermi distribution.

Thus, near the metal-insulator transition, the notion of the Coulomb gap matches that of the charge spreading. More specifically, one can introduce the scale-dependent diffusion coefficient $D(L) \sim (e^2/h\varepsilon)L$ [see Eq. (2)], such that the one-particle density of states on the scale of $\xi(E_F)$ obeys the relation $\hbar g D(\xi) \sim 1$. We recognize the latter as a familiar localization criterion (cf. Ref. 17). It is worthwhile to notice that, in this formula, both the diffusion coefficient and the one-particle density of states are renormalized by the interactions, whereas the Einstein relation links σ^p_{xx} and the bare coefficient D_0 via the thermodynamic density of states. At $T \gg T_c$, the Coulomb gap is full to the brim and the Coulomb correlations do not control $L_\phi(T)$ any more (one can then say $z = 2$).

The same behavior of $\Delta_c(T)$ is recovered if we use $L_h(T)/\xi(E)$ as a scaling variable instead of $L_\phi(T)/\xi(E)$, $L_h \sim [\xi(E)L_\phi]^{1/2}$ [with L_ϕ given by Eq. (3)] being the typical hopping length in the insulating phase.[12,19] According to this approach, the conductivity may be represented as $\sigma_{xx}(x) = (e^2/h)F(x)\exp(-x)$, where $F(x)$ is a dimensionless power-law function of the single parameter $x = L_h(T)/\xi(E_F)$. The advantage here is that the dependence $\ln\sigma_{xx} \sim -[e^2/\varepsilon\xi(E_F)T]^{1/2}$ can be microscopically grounded in very general terms for the variable-range hopping regime. It is clear, however, that this scaling form of σ_{xx} explicitly implies the strong coupling limit, in the sense that it is only valid for $\hbar/T\tau_\phi \sim 1$. Otherwise, say if the delocalization would come from weak interactions with phonons, $\sigma_{xx} \ll \sigma^p_{xx}$ at $x \sim 1$, which means that there exists an intermidiate, power-law hopping regime between the metallic phase and that of the variable-range hopping.

In the critical broadening regime, $\sigma^p_{xx}(T)$ displays only a weak temperature dependence; however, it might be interesting in its own right. We first notice that σ^p_{xx} is a poorly defined quantity in coherent samples at zero T. Indeed, the conductance should show strong, sample specific fluctuations of the order of e^2/h (p. 252 in Ref. 1). Thus the widely accepted notion that the conductivity σ^p_{xx} at $T \to 0$ has a universal value may only be relevant if L_ϕ is much shorter than the sample size. We take this limit so as to deal with the self-averaging σ^p_{xx}. Then one can identify two contributions to the temperature deviation $\delta\sigma^p_{xx}(T)$. One is related to the temperature smearing of the Fermi distribution and is apparently $\sim -(e^2/h)[T/\Delta_c(T)]^2$. It is negative and scales as $T^{2(1-\kappa)}$. The other is similar to the weak-localization correction, with the difference that the expansion in terms of the small parameter λ/L_ϕ should be done around the critical point. We obtain

$$\delta\sigma^p_{xx} \sim (e^2/h)(\lambda/L_\phi)^{D_2} \propto T^{\kappa\gamma D_2} \,, \tag{6}$$

where D_2 may be defined as a generalized dimension of the critical eigenstates.[20] For non-interacting electrons $D_2 = 2-\eta$, where $\eta \simeq 0.4$ is the critical exponent of eigenfunction correlations.[8] In the interaction-dominated regime the problem is more delicate, though it is likely that the fractal dimension D_2 in Eq. (6) is just half that for non-interacting electrons. The scaling arguments imply that the "weak delocaliation" correction dominates at $T \to 0$. Strictly speaking, near the critical point, the sign of the correction cannot be obtained within the power-counting analysis. According to numerical simulations,[20] however, the correction in the non-interacting case should be negative for the lowest Landau level $N = 0$. Yet, $\delta\sigma^p_{xx}$ is likely to be positive for $N \geq 1$ (since σ^p_{xx} exceeds the SCBA value only at $N = 0$). Note that $\delta\sigma^p_{xx}$ for higher Landau levels may be much larger than e^2/h, so that at $\delta\sigma^p_{xx} \sim e^2/h$ a crossover to the logarithmic dependence[21] on T should occur (in the high-T regime, the "conventional" weak-localization correction is $h\sigma^p_{xx}/e^2$ times smaller than that originating from the electron-electron interaction).

We might be tempted to conclude that the above concept of the phase breaking due to the electron-electron interaction completely accounts for the broadening of σ_{xx} peaks at low T. In fact, it is by no means obvious that at finite T the interaction alone is able to delocalize electrons in the QHE regime. The subtlety is that we treat the problem of the phase breaking self-consistently, on the assumption that the inelastic scattering occurs due to electromagnetic fluctuations produced by delocalized excitations (Nyquist noise). This approach works perfectly in the weak-localization theory, but in the QHE it still constitutes a challenging problem because of its non-perturbative nature.

3 Long-Range Disorder: Brownian Motion in a Stream

So far we have dealt with the phase-breaking effects in a short-range random potential, in which case $\Delta_c(T)/T \gg 1$. As argued in the introduction, increasing the correlation radius of disorder d brings the new energy scale $\Delta_t = \Gamma(\lambda/d)^2$ into play. The ratio $\Delta_c(T)/\Delta_t$ then becomes relevant; specifically, $\Delta_c(T)$ exhibits the universal scaling behavior in terms of λ/L_ϕ only as long as $\Delta_c(T)/\Delta_t \ll 1$. At $T \ll \Delta_t$, $\Delta_c(T)$ either obeys the relation $\xi(\Delta_c)/L_\phi \sim 1$ or saturates at the level of Δ_t. The crucial observation is that the range of E within which the inelastic scattering prevents the localization, $|E-E_c| \sim \Delta_c(T)$, cannot exceed T if $\Delta_c(T)$ becomes wider than the tunneling band. For this reason, the width of the σ_{xx} peak $\Delta\nu(T) \sim \max\{\Delta_c(T), T\}/\Gamma$, ν being the filling factor, grows linearly with T at $T \gg \Delta_t$, irrespective of any particular mechanism of the inelastic scattering. A sample dependent behavior of $\Delta\nu(T)$, with a crossover to higher κ's as T is increased, indeed was observed in samples with long-range potential fluctuations.[22]

In the classical regime, the following Fokker-Planck equation typifies the entire problem:

$$\frac{\partial f}{\partial t} + \frac{\partial}{\partial \vec{\rho}}\left[\mathbf{v}f - D_i\left(\frac{\partial f}{\partial \vec{\rho}} + \frac{\hbar(\mathbf{v}\times\mathbf{b})}{\lambda^2}\frac{\partial f}{\partial E}\right)\right] = 0\,, \tag{7}$$

where $\mathbf{v}(\vec{\rho})$ is the drift-velocity field, $\partial\mathbf{v}/\partial\vec{\rho} = 0$, $\mathbf{b}$ is the unit vector along the magnetic field. Despite having a simple form, the equation does not allow for any perturbative treatment of the inelastic scattering diffusion, characterized by the temperature-dependent coefficient $D_i(T)$. The point is that, right at the percolation transition, the critical trajectories are strongly coupled by arbitrarily small D_i. The width of the conducting band then satisfies the self-consistent equation $\Delta_c \sim \Gamma[D_i/\Omega(\Delta_c)d^2]^{1/2}$, $\Omega(E)$ being the inverse period of the critical trajectories with the energy E. The fractal dimensionality of the percolating trajectories is known to be 7/4,[3] which gives $\Delta_c \sim \Gamma(D_i/D_0)^p$

with $p = 3/13$ (in the case of smooth disorder, D_0 can be re-expressed as $D_0 \sim < \mathbf{v}^2 >^{1/2} d$). Accordingly, σ_{xx}^p falls off with increasing T as[23,3]

$$\sigma_{xx}^p \sim (e^2/h)(\Gamma/T)[D_i(T)/D_0]^p \ . \tag{8}$$

The self-consistent treatment is necessary only as long as $\Delta_c(T)/T \ll 1$. In the opposite limit, $1 \gg D_i/D_0 \gg (T/\Gamma)^{1/p}$, the height of the peak is determined by the conductivity of the percolation network built up from trajectories with $|E - E_c| \sim T$; apparently, this yields $\sigma_{xx}^p \sim e^2/h$. Interestingly, D_i should be large enough for this regime to occur, yet neither σ_{xx}^p nor $\Delta\nu$ depends on D_i in this case. Whatever D_i/D_0 is, in the classical limit $\Delta\nu \sim T/\Gamma$. A noteworthy feature of the classical delocalization is also the strong enhancement of σ_{xx}^p as compared to the bare value $e^2 D_i/\lambda^2 T$. Notice that the problem can be equivalently formulated in terms of inhomogeneous local conductivities.[24] The above picture of percolation implies a sharp Fermi distribution for percolating particles; however, there remains a challenging question: To what extent is the self-consistent approach adequate in the strongly correlated electron system with long-range disorder?

4 Plateau Regime

In the case of short-range disorder and weak electron-phonon interactions, one can hardly expect any fascinating features of $\sigma_{xx}(T)$ deep in the insulating phase at $|E_F - E_c| \gg \Gamma \gg T$. Indeed, σ_{xx} then behaves[25] as $\sigma_0 \exp(-|E_F - E_t(T)|/T)$ due to activation to the energy level $E_t(T)$, such that $|E_t(T) - E_c| \sim \Gamma \ln^{1/2}(\Gamma/T)$. The pre-exponential factor σ_0 is small in comparison with e^2/h and depends on the electron-phonon coupling constant, though weakly.[25] That is why a lively debate arose concerning the universal activated behavior $\sigma_{xx} = (e^2/h)\exp(-\Delta/T)$, which was reported in a number of experiments.[26] The universality of σ_0 seemed to be intriguing since the activated conductivity occurs solely due to supposedly weak interactions with phonons. A theory that explained this feature was forthcoming in terms of the classical dynamics in a long-range random potential.[27] If $d \gg \lambda$, σ_{xx} is governed by the activation already at $|E_F - E_c| \gg T \gg \Delta_t$, once the Fermi level leaves the tunneling band. Notice, since both the activation exponent and the tunneling one are linear functions of E, the necessary condition of that the activation dominates is simply $T \gg \Delta_t$. Again, we identify the crucial parameter $\Omega(T)\tau_T$, where τ_T is the time it takes for an electron to change its energy by T. The parameter is small provided $D_i/D_0 \gg (T/\Gamma)^{1/p}$. In this limit Δ_c scales as T. The essential idea of this approach is to represent the electron system as a random network of thermal reservoirs connected via ballistic contacts (critical saddle points) with energies V_i scattered around E_c in the tail of the Fermi distribution. This parallels the Landauer-Büttiker formalism, with the difference that the latter was developed for the edge states. The ballistic conductances of the contacts, $G_i = (e^2/h)\exp(-|E_c - E_F + V_i|/T)$, exhibit strong asymmetric fluctuations. However, since $\ln G_i$ are distributed randomly, the conductivity of the network satisfies the exact relation $\sigma_{xx} = \exp\langle\ln G_i\rangle$ (valid in 2D only), which immediately gives $\sigma_0 = e^2/h$.[27] If $\Omega(T)\tau_T \gg 1$, the conducting band $\Delta_c(T)$ gets narrower than T. To put it simply, electrons with $\Delta_c \ll E - E_c$ are now out of play as they have no time to tune their electrochemical potential so that it equals the potential of the adjacent thermal reservoir. Apparently, $\sigma_0 = \sigma^p_{xx}$ in the extreme of high T [see Eq. (8)].

5 Overlapped Levels

According to the Drude formula, σ_{xx} grows with increasing overlap of Landau levels. For this reason, $\xi(E)$ in the middle between adjacent σ_{xx} peaks acquires an exponentially large factor, such that $\ln\xi$ scales as $(h\sigma_{xx}/e^2)^2$. This quickly breaks down QHE at any reasonable T (unless the σ_{xx} peaks are due to the classical percolation, which is probably the case in high-mobility samples). What happens if the number of the coupled levels is limited to two? Let two Zeeman levels be strongly overlapped in the sense that $\Gamma \gg \Delta_s$, where $2\Delta_s$ is the difference of the critical energies $E_c^{\pm}$ corresponding to two spin projections. Then turning on a spin-orbit (SO) interaction drives two systems of electrons with opposite spin into a new quantum Hall phase with an internal degree of freedom. We present arguments that the SO coupling is able to greatly facilitate the inelastic-scattering-induced delocalization.[28]

In the case of short-range disorder, Δ_s obeys the relation[29] $\xi(E_0)/R_{so} \sim 1$, where $E_0 = \frac{1}{2}(E_c^+ + E_c^-)$, R_{so} is the spin-flip scattering length. It follows that, for the weak SO interaction, Δ_s is far larger than the local SO splitting. Numerical simulations carried out in this limit support the conclusion that the critical behavior of $\xi(E)$ at $E \to E_c^{\pm}$ remains the same as for spinless electrons.[30] Hence, whatever the ratio $\Delta_c(T)/\Delta_s$ is, if disorder is short ranged, the SO coupling does not lead to any principal change in the critical broadening of the σ_{xx} peaks. It merely splits the critical point [though from an experimental point of view, it might be of great importance that the SO coupling does yield a sharp numerical growth of $\xi(E)$]. At $d \gg \lambda$, however, the "coherent" contribution to Δ_s cannot exceed Δ_t (the limit $\Delta_s/\Delta_t \sim 1$ then corresponds to the generalized Chalker-Coddington model with strong spin-flip scattering). In the extreme of smooth disorder, Δ_s may be described in completely classical terms as a result of splitting of percolating classical trajectories.[31] Let us formulate an auxiliary percolation problem for particles with spin: Given two sets of closed equipotentials $V(\vec{\rho}) \pm \Delta_s = E$, the particles are allowed to change the classical trajectories if two equipotentials come within a distance δ of each other. The system undergoes a percolation transition at $\delta = \delta_c(E)$, such that $\delta_c(\pm\Delta_s)$ vanishes. The crucial observation is that $\delta_c(E) \sim d\Delta_s/\Gamma$ for $|E| < \Delta_s$, but rapidly grows as $\delta_s \sim d[(|E| - \Delta_s)/\Gamma]^{1/2}$ outside this band, since at $|E| > \Delta_s$ the percolation occurs only due to the coupling across the critical saddle points. This simple game leads us to an interesting result: the strong spin-flip scattering destroys the classical localization within the energy band $|E| < \Delta_s$. Consequently, in the limit of long-range disorder ($\Delta_t/\Delta_s \to 0$), when the tunneling through the saddle points may be neglected, the SO coupling makes $\sigma_{xx}(E_F)$ exhibit a boxlike behavior, namely $\sigma_{xx} \sim e^2/h$ at $|E_F| \leq \Delta_s$, otherwise $\sigma_{xx} \to 0$. Thus, provided $T \ll \Delta_s$ and $R(\Delta_s)$ exceeds both L_ϕ and R_{so} (the latter was evaluated in Ref. 31), two σ_{xx} peaks merge and form the boxlike one. At lower T, when L_ϕ is still large in comparison with $R(\Delta_s)$, there are two peaks, but they are strongly asymmetric. Specifically, σ_{xx} between the peaks falls off with decreasing T only in a power-law manner: $\sigma_{xx} \sim (e^2/h)R(\Delta_s)/L_\phi(T)$. This power-law hopping occurs in a wide range of T and goes over into the variable-range hopping only at very low T.[28] The crucial point is that the strong spin-flip scattering changes the character of localization in the energy band limited by E_c^+ and E_c^-: it gives rise to the Anderson localization (instead of the classical one) with $\xi(E_0)/R(\Delta_s) \sim 1$.[28] We conclude that, in the case of smooth disorder, the SO coupling is capable of strongly changing the conventional picture of QHE.

6 Conclusion

We have discussed inelastic broadening of σ_{xx} peaks in the low-T limit, stressing (i) the crucial role of electron-electron interactions in the integer QHE; (ii) the classical aspects of electron dynamics in samples with long-range disorder.

Acknowledgments

I am grateful to A. L. Efros, J. Hajdu, and K. V. Samokhin for interesting discussions. The work was supported by the Deutsche Forschungsgemeinschaft.

References

1. M. Janßen, O. Viehweger, U. Fastenrath, and J. Hajdu, *Introduction to the Theory of the Integer Quantum Hall Effect*, ed. J. Hajdu (VCH, Weinheim, 1994).
2. B. Huckestein, *Rev. Mod. Phys.* **67**, 357 (1995).
3. M. B. Isichenko, *Rev. Mod. Phys.* **64**, 961 (1992).
4. J. T. Chalker and P. D. Coddington, *J. Phys.* C **21**, 2665 (1988).
5. B. L. Altshuler and A. G. Aronov, in *Electron-Electron Interactions in*

Disordered Systems, ed. A. L. Efros and M. Pollak (North-Holland, Amsterdam, 1985).
6. H. P. Wei, D. C. Tsui, M. A. Paalanen, and A. M. M. Pruisken, *Phys. Rev. Lett.* **61**, 1294 (1988).
7. M. I. D'yakonov and A. S. Furman, *Zh. Eksp. Teor. Fiz.* **92**, 1012 (1987) [JETP **65**, 574 (1987)].
8. J. T. Chalker and G. J. Daniell, *Phys. Rev. Lett.* **61**, 593 (1988).
9. V. L. Berezinskii and L. P. Gor'kov, *Zh. Eksp. Teor. Fiz.* **77**, 2498 (1979) [JETP **50**, 1209 (1979)].
10. S. C. Siak, *J. Phys.: Cond. Matter* **3**, 4213 (1991).
11. L. S. Levitov and A. V. Shytov, cond-mat/9501130.
12. D. G. Polyakov and B. I. Shklovskii, *Phys. Rev. Lett.* **70**, 3796 (1993).
13. T. Brandes, L. Schweitzer, and B. Kramer, *Phys. Rev. Lett.* **72**, 3582 (1994).
14. A. L. Efros and B. I. Shklovskii, in Ref. 5.
15. I. L. Aleiner, D. G. Polyakov, and B. I. Shklovskii, unpublished (1993).
16. D.-H. Lee and Z. Wang, *Phys. Rev. Lett.* **76**, 4014 (1996).
17. S.-R. E. Yang and A. H. MacDonald, *Phys. Rev. Lett.* **70**, 4110 (1993); S.-R. E. Yang, A. H. MacDonald, and B. Huckestein, *ibid.* **74**, 3229 (1995).
18. S. He, P. M. Platzman, and B. I. Halperin, *Phys. Rev. Lett.* **71**, 777 (1993); F. G. Pikus and A. L. Efros, *ibid.* **73**, 3014 (1994).
19. M. P. A. Fisher, *Phys. Rev. Lett.* **65**, 923 (1990).
20. B. Huckestein and L. Schweitzer, *Phys. Rev. Lett.* **72**, 713 (1994); B. M. Gammel and W. Brenig, *ibid.* **73**, 3286 (1994).
21. A. Houghton, J.R. Senna, and S. C. Ying, *Phys. Rev.* B **25**, 6468 (1982); S. M. Girvin, M. Jonson, and P. A. Lee, *ibid.* **26**, 1651 (1982).
22. S. Koch, R. J. Haug, K. von Klitzing, and K. Ploog, *Phys. Rev. Lett.* **67**, 883 (1991); H. P. Wei, S. Y. Lin, D. C. Tsui, and A. M. M. Pruisken, *Phys. Rev.* B **45**, 3926 (1992).
23. D. G. Polyakov, unpublished (1986).
24. S. T. Simon and B. I. Halperin, *Phys. Rev. Lett.* **73**, 3278 (1994).
25. D. G. Polyakov and B. I. Shklovskii, *Phys. Rev. Lett.* **73**, 1150 (1994).
26. R. G. Clark, J. R. Mallet, S. R. Haynes, J. J. Harris, and C. T. Foxon, *Phys. Rev. Lett.* **60**, 1747 (1988); Y. Katayama, D. C. Tsui, and M. Shayegan, *Phys. Rev.* B **49**, 7400 (1994).
27. D. G. Polyakov and B. I. Shklovskii, *Phys. Rev. Lett.* **74**, 150 (1995).
28. D. G. Polyakov and M. E. Raikh, *Phys. Rev. Lett.* **75**, 1368 (1995).
29. D. E. Khmelnitskii, *Helv. Phys. Acta* **65**, 164 (1992).
30. D. K. K. Lee and J. T. Chalker, *Phys. Rev. Lett.* **72**, 1510 (1994); Z. Wang, D.-H. Lee, and X.-G. Wen, *ibid.* **72**, 2454 (1994); C. B. Hanna, D. P. Arovas, K. Mullen, and S. M. Girvin, *Phys. Rev.* B **52**, 5221 (1995).
31. D. G. Polyakov, *Phys. Rev.* B **53**, 15777 (1996).

σ_{xx} and σ_{xy} Experiments Regarding High Frequency Scaling in the Integral Quantum Hall Effect

W.BELITSCH, R.MEISELS, F.KUCHAR
Department of Physics, University of Leoben, Franz Josef Str. 18, A-8700 Austria

G.WEIMANN
Fraunhofer Institut für Angewandte Festkörperphysik, D-79108 Freiburg, Germany

W.SCHLAPP
Deutsche Telekom Forschungsinstitut, D-64276 Darmstadt, Germany

V.UMANSKY
Weizmann Institute of Science, Rehovot, Israel

G.HEIN, K.PIERZ
PTB Braunschweig, D-38023 Braunschweig, Germany

The temperature dependence of σ_{xx} and σ_{xy} of AlGaAs/GaAs is studied in high magnetic fields at microwave frequencies of about 30 GHz. Samples with different mobilities are measured. In contrast to the dc measurements, the plateau width, ΔB, of σ_{xy} is observed to be temperature independent below 2.4 K, where $hf > kT$. In the σ_{xx} measurements the shape of the SdH oscillations does not change with frequency in the temperature range from 2.2 K to 4.2 K.

1 Introduction

The explanation of the quantum Hall effect is closely related to the localization of the states of a two-dimensional electron gas (2DEG) at high magnetic fields in the tails of the Landau levels. The characteristic R_{xy} Hall plateaus and the vanishing longitudinal resistance R_{xx} appear when the Fermi level lies within localized states of the disorder-broadened Landau levels. The extension of the Landau states, their localization length ξ, diverges at the center of a Landau level E_N^* as:[1]

$$\xi = \xi_0 |E - E_N^*|^{-\mu} \tag{1}$$

$\mu \approx 2.34$ is the scaling exponent. At this E_N^*, the states are extended and therefore delocalized. For a sample with a finite size L it is sufficient for a state to obey $\xi(E) > L$ to be delocalized. At non-zero temperatures L is replaced by the inelastic scattering length L_{in} if the latter is smaller. It follows a power law temperature dependence, $L_{in} \sim T^{-p/2}$.

Conduction is determined by the position of the Fermi energy $E_F(B)$ relative to the Landau levels. Therefore there exist magnetic fields B_N^* corresponding to $E_F(B_N^*) = E_N^*$, and $E - E_N^*$ in Eq.1 is replaced by $B - B_N^*$. The half-width $\Delta B'$ of the R_{xx} peaks (between the regions where R_{xx} vanishes) is determined by the transition between localization and delocalization at $\xi(E) = L_{in}$. Therefore:

$$|B - B_N^*| = \Delta B' \sim T^{\kappa} \tag{2}$$

with $\kappa = p/2\mu$. This temperature dependence was first observed by Wei et al.[2]. Since p depends on the details of the inelastic scattering, the value of κ is not universal.[3]

High measuring frequencies are an alternative source of (effective) delocalization. Here, in analogy to the correlation length a correlation time τ_c can be defined: $\tau_c \sim |B - B^*|^{-y}$. As ξ is set in relation to an effective system size, τ_c is set in relation to an effective measuring time $\tau_m \approx 1/2\pi f$. This yields

$$\Delta B' \sim \tau_c^{-1/y} \sim f^{1/y} \tag{3}$$

y is proportional to μ. Experimentally $y = \mu$ was found[4] at low microwave frequencies in a sample with very high disorder. Thus, a measurement of $\Delta B'$ at high frequencies allows a direct determination of the critical exponent μ. hf/kT determines which of the two possible effects dominates.[4] If it is much larger than 1, f should dominate (temperature independent). If it is much smaller, $\Delta B'$ should increase with temperature and the corresponding Hall plateau width ΔB should decrease.

The purpose of this work is to investigate the high-frequency delocalization in a wide range of disorder and at higher frequencies than used in Ref.4.

2 Experiments

To measure σ_{xy} of the 2DEG, a transmission arrangement with crossed rectangular wave guides is used (Fig.1a).[5] The σ_{xx} experiments are performed in a reflection microwave bridge arrangement (Fig1b). In a matched T the microwave is split up into a reference branch and a sample branch. The partial wave, which is reflected from the sample, interferes with the reference wave inside the matched T. The interference signal is detected by a crystal detector. The bridge is balanced by matching the reference branch to the sample branch to get a zero signal at zero magnetic field. AlGaAs/GaAs samples A,B,C and D had carrier concentrations of 2.9, 5.2, 2.5 and 1.4 $\times 10^{11}$cm^{-2} and mobilities of 1.2, 1.4, 1.7 and 6.5 $\times 10^5$cm^2/Vs, respectively. Using samples with gate electrodes, a double modulation method is used which results in a strongly improved signal to noise ratio. There, in addition to the microwave modulation

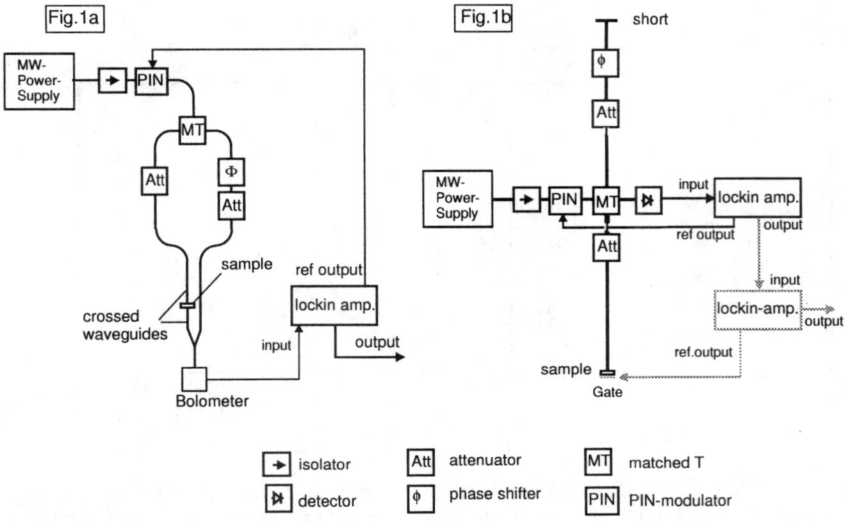

Figure 1: a) σ_{xy} measurements by transmission, b)σ_{xx} measurements by reflection. Dashed part is used for gate modulation.

also the carrier concentration of the 2DEG is (fully) modulated by the gate voltage. This way only the reflected intensity due to the 2DEG is measured.

3 Results and Discussion

Fig. 2 shows the temperature dependence of the Hall plateau width ΔB of σ_{xy} ($i = 2$: sample A,C,D, $i = 4$ for sample B) for a microwave frequency of 35 GHz. In addition, it contains the dc data for sample B. In contrast to the dc data, the microwave plateau width is temperature independent below 2.4 K. There, the delocalization is dominated by the effect of the microwaves. Comparing the results from different samples, it is seen that the transition from temperature independence to a decrease with temperature depends neither on mobility nor on plateau index. Our σ_{xy} data qualitatively follow the scaling function introduced by Engel et al.,[4] i.e. the dependence on kT/hf, although the disorder is much lower in our samples. The value of kT/hf where the transition from frequency to temperature dominated delocalization occurs is 1.5, somewhat smaller than in Ref.4.

From the σ_{xy} data, flat σ_{xx} minima corresponding to the σ_{xy} plateaus are expected. Clear SdH oscillations are observed (Fig.3) in the temperature range of 2.2 to 4.2K for microwave frequencies of 28 to 36GHz, however, withoutout

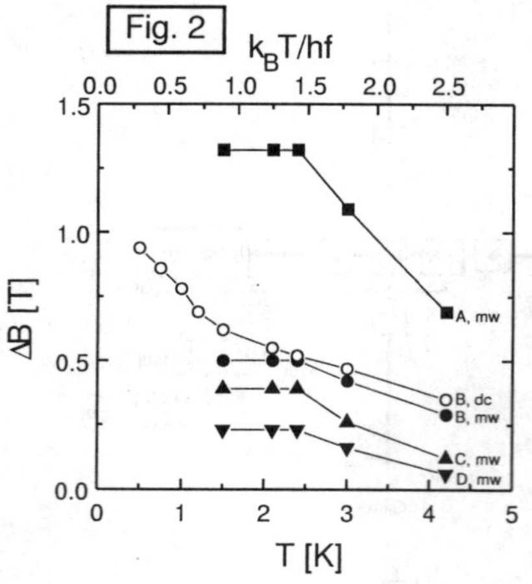

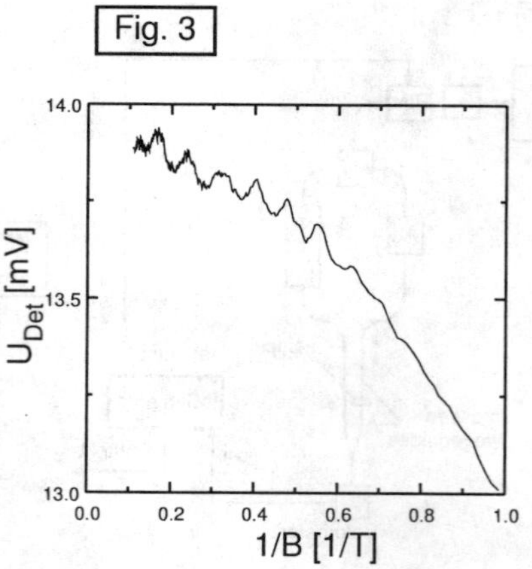

Figure 2: Experimental results: Plateau width ΔB of σ_{xy} vs. temperature for samples with different mobilities and filling factors (see text) for microwaves (mw, $f = 35$GHz) and for dc (open circles, sample B). On top the scaled temperature $k_B T/hf$ is given.

Figure 3: σ_{xx} related signal vs. $1/B$ for sample B, $f = 35$GHz, $T = 2.2$K

the flat minima of dc ρ_{xx} data. Furthermore, no change was observed in the shape of the SdH oscillations in these temperature and frequency ranges. The reason for this increased response to frequency-induced delocalization compared to the σ_{xy} measurements is not clear at this point. For a sample with a mobility of $2.5\times 10^6 \text{cm}^2/Vs$ no oscillations could be resolved. This is possibly due to the stronger effect of the high-frequency delocalization in a low-disorder sample with sharp Landau levels.

This work was supported by the "Fonds zur Förderung der wissenschaftlichen Forschung", Austria.

References

1. B. Kramer and A. MacKinnon, *Rep. Prog. Phys.* **56**, 1496 (1993).
2. H.P. Wei *et al.*, *Phys. Rev. Lett.* **61**, 1294 (1988).
3. S. Koch *et al.*, *Phys. Rev. Lett.* **67**, 883 (1991).
4. L.W. Engel *et al.*, *Phys. Rev. Lett.* **71**, 2638 (1993).
5. F. Kuchar *et al.*, Phys. Scripta **T19**, 79 (1987); F. Kuchar, Adv. Sol. State Phys. **28**, 45 (1988).

A SYSTEMATIC STUDY OF THE TEMPERATURE DEPENDENCE OF QUANTUM HALL PLATEAUX WIDTHS IN HIGH MOBILITY BACK-GATED GaAs-(Ga,Al)As 2DEGs

[1]P. J. GEE, [1]J. SINGLETON, [2]M. P. GRIMSHAW, [2]D. A. RITCHIE,
[2]G. A. C. JONES and [3]J. H. BURROUGHES

[1]*University of Oxford, Clarendon Laboratory, Parks Road, Oxford, OX1 3PU, UK*
[2]*University of Cambridge, Cavendish Laboratory, Madingley Road, Cambridge, CB3 OHE, UK*
[3]*Toshiba Cambridge Research Centre Ltd., 260 Cambridge Science Park, Milton Road, Cambridge, CB4 4WE, UK*

The temperature dependence of peaks in ρ_{xx} has been measured in two high mobility back-gated GaAs-(Ga,Al)As heterojunctions. The back gates allow the mobility and carrier density to be varied in a systematic manner, thus improving experiments[1] carried out using several devices whose carrier density and mobility could only be tuned via illumination. The central objective of these measurements is the search for possible universal behaviour in the temperature dependence of ρ_{xx} in the quantum Hall regime.

The Integer and Fractional quantum Hall effects (IQHE and FQHE)[2,3] observed in the resistivity of a two-dimensional electron system (2DES) at high magnetic fields and low temperatures continue to generate much interest. One unresolved problem concerns the behaviour of the peaks in ρ_{xx} and the corresponding risers in ρ_{xy} between QH plateaux as the temperature T tends to zero. Early work[4-6] using low mobility heterostructures implied that the width of the ρ_{xx} peaks tended to zero at T=0, corresponding to there being only one populated extended state at the centre of each Landau level. However, more recent theory[7] incorporating many body effects indicated that the T dependence of the width of risers in ρ_{xy} and peaks in ρ_{xx} would be qualitatively different in higher mobility samples. To test this prediction, our previous measurements[1] involved many high mobility GaAs-(Ga,Al)As heterojunctions spanning the mobility range 3 to 1200 m^2/Vs; it was found, as suggested, that the widths of the features did not extrapolate to zero at T=0. However, a search for "universal" behaviour of the resistivity features was not straightforward, as many different samples of variable geometry and preparation were measured.

In order to eliminate artefacts arising from differing sample geometry and yet still vary parameters such as mobility, we have extended the resistivity studies to back-gated GaAs-(Ga,Al)As heterojunctions. The heterojunctions are grown on a facet[8] of a GaAs structure containing a layer heavily doped with Si which acts as the gate. The gate voltage V_g allows the carrier concentration (n_e) and mobility (μ_e) to be controlled over a wide range, enabling systematic studies of the T-dependence of ρ_{xx} as a function of n_e and μ_e, whilst retaining consistent contact geometry, Hall bar width, spacer layer width and current; in addition only a single cooldown from room temperature is necessary.

Two 40 μm Hall bars with identical contact geometries were used. Device 1 covered the range 2.6×10^{15} $m^{-2}<n_e<4.8\times10^{15}$ m^{-2}, corresponding to 50 $m^2/Vs<\mu_e<122$ m^2/Vs; the parameter range was extended by device 2, which had $0.50x10^{15}$ m^{-2}<n_e<$3.3x10^{15}$ m^{-2} and 10 m^2/Vs<μ_e<80 m^2/Vs. The Hall bars were designed so that ρ_{xx} data could be recorded using voltage contacts 20 μm apart and 1020 μm apart simultaneously, so that the role of various length scales could be examined. The devices were cooled in the dark and illuminated with a red LED at magnetic field B=0 at T=50 mK. The T dependence (30 mK<T<1.2 K) of ρ_{xx} and ρ_{xy} was examined using a low frequency A.C. current of 20 nA and 0<B<15 T at V_g=-1.25 (52), -0.90 (67), -0.40 (90), 0.00 (103) and +0.50 V (122) in device 1 and at V_g=-0.60 (15), 0.00 dark (23), -0.20 (32), +0.10 (45), +0.40 (60) and +0.70 V (77) in device 2, where the figures in brackets are the corresponding μ_e. Thus the difference in μ_e between each set of data was ≈10 m^2/Vs. It should be noted that in device 2, the measurements cooled in the dark at V_g=0.00 V were virtually identical to those after illumination at V_g=-0.40 V.

The data was analysed by determining the width in B of the ρ_{xx} peaks at each V_g at a fixed ρ_{xx} level[1]. Such an approach was used in preference to that of References 4-6 after a detailed comparison of the two methods; the latter technique measures the distance between the extrema in $d\rho_{xx}/dB$ and is more suitable for very low μ_e samples in which very sharp peaks in ρ_{xx} with well defined points of inflexion occur. The method of References 4-6 is limited by the poor resolution of the extrema in $d\rho_{xx}/dB$ if the ρ_{xx} peak is at a low magnetic field and is probably less systematic in the cases where a change in T also shifts the ρ_{xx} peak position (see data below).

An example of the evolution of the ρ_{xx} peak between ν=3 and ν=4 with increasing T (30mK<T<1 K) is shown in Figures 1a-1c for three different V_g; these data were recorded between voltage probes separated by 1020 μm. Figure 1d shows simultaneous recordings of ρ_{xx} in device 2 (V_g=-0.60 V) for voltage contacts separated by 20 μm and 1020 μm. Note that whilst all of the features occur at the same B, indicating that n_e is constant throughout the device, the relative heights of the peaks between ν=2 and 3 and ν=3 and 4 is different for the two regions, indicating that some of the effects leading to the form of ρ_{xx} have a length scale >20 μm. The difference in ρ_{xx} between the two regions at B=0 is likely to be due to the uncertainty in measuring the device.

Figures 2a and 2b show the width Δ of the ρ_{xx} peak (in Tesla) between ν=3 and 4 (c.f. Figures 1a-1c) at a constant height ρ_{xx}=100 Ω/□ for various values of V_g; data from the voltage contacts 1020 μm apart are in Figure 2a and those from contacts 20 μm apart are in Figure 2b. These data show that the extrapolated ρ_{xx} peak width at absolute zero is not only finite but also different for each V_g, i.e., at each n_e and μ_e. In agreement with our previous findings[1], the variation of the ρ_{xx} peak width is fairly linear for T>200 mK, but the rate at which the width varies with T levels off somewhat below this temperature. Moreover, the general trend is that the ρ_{xx} peak width decreases as μ_e becomes lower; such behaviour is intuitively expected, since in low μ_e devices which exhibit the IQHE, ρ_{xx} minima are wide and peaks are narrow, whereas in higher μ_e devices which exhibit the FQHE, the ρ_{xx} minima are narrow and ρ_{xx} peaks are wide, the latter displaying a rich spectrum of oscillations. Note that Δ varies more slowly with T in the 20 μm long part of

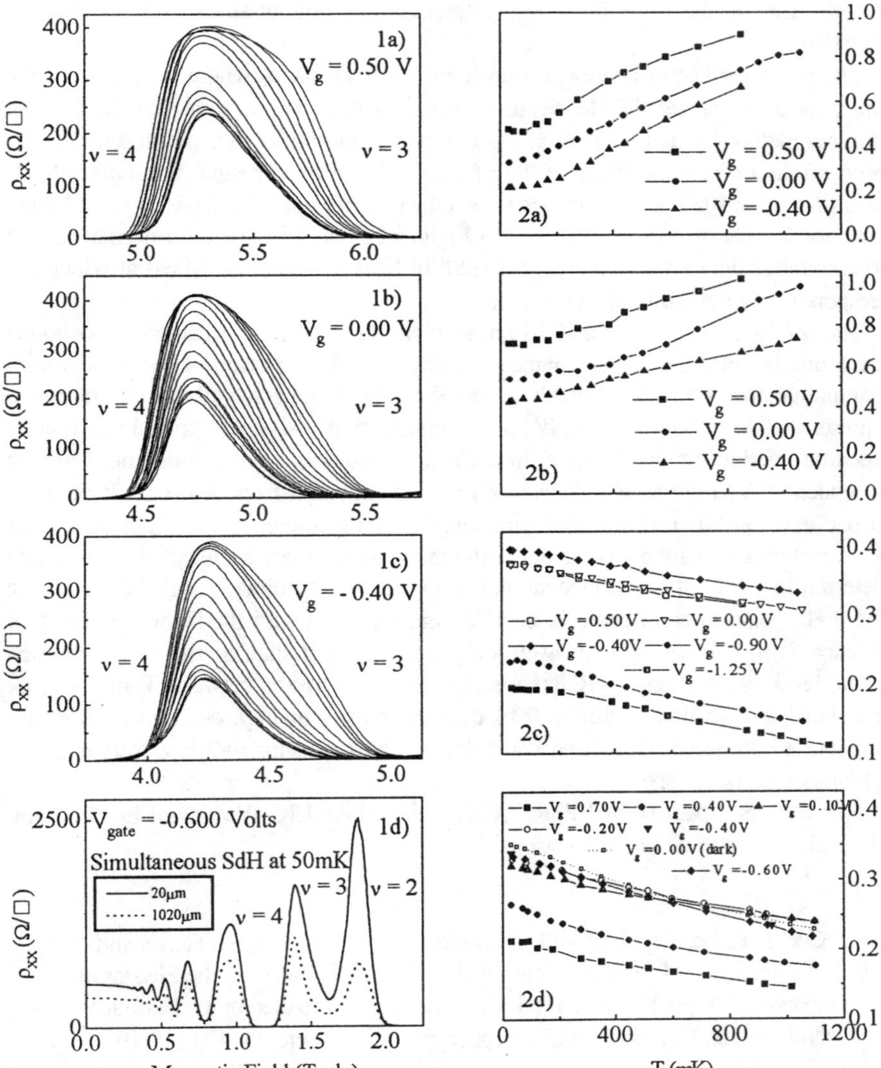

Figure 1-2: Figure 1a-1c shows the temperature dependence (30 mK<T<1.2 K) of ρ_{xx} between ν=3 and 4 from the 1020 μm region of device 1 for various V_g; Figure 1d shows simultaneous recordings of ρ_{xx} in the 20 μm and 1020 μm region at T=50 mK and at V_g=-0.60 V in device 2; Figure 2a-2b show the width Δ in Tesla between ν=3 and 4 at a height ρ_{xx}=100Ω/□ for various V_g in device 1. Figure 2a is the 1020 μm region and Figure 2b is the 20 μm region; Figure 2c-2d show the normalised width variation of the ν=2 minimum from measurements in the 1020 μm region for various V_g in both devices. Figure 2c is device 1 and Figure 2d is device 2.

the device than in the 1020 μm section emphasising the importance of length scales in the system.

The above method of analysis is only appropriate at low μ_e where the ρ_{xx} peaks are simple; as μ_e increases, FQHE features appear within the ρ_{xx} maxima, and so an additional method of analysis is used. As in our previous study[1], we plot the width of a chosen minimum in ρ_{xx} at integer filling factor and at a fixed height. Examples of this analysis from the 1020 μm regions are shown for ν=2 in Figure 2c (device 1) and Figure 2d (device 2). The widths *W* displayed in Figures 2c and 2d have been normalised[1] to take account of the variation in n_e; *W*=(width in field of minimum)/(field at which the corresponding integer filling factor ν occurs).

The results in Figures 2c and 2d are very similar to data in Reference 1 obtained using a number of different heterojunctions; however, the variation of parameters such as sample geometry which inevitably occurred in Reference 1 have been eliminated in the present work. In both devices, *W* for the minimum at ν=2 varies fairly linearly with temperature within the given range. In each sample, the gradient of this line is almost independent of V_g and hence μ_e. However, the gradient varies from device to device (and also between the 20 μm and 1020 μm lengths of the same device), indicating that geometrical effects or inhomogeneities with macroscopic length scales play a major role in determining *W*(*T*). It is also evident on extrapolation to absolute zero that the width of the ν=2 Hall plateau is different for each V_g, in agreement with the theoretical work of Reference 7. In contrast, if there were only one extended state at the centre of each Landau level as implied by References 4-6 (i.e. the risers between Hall plateaux extrapolated to zero field width at *T*=0) then the curves in Figures 2c and 2d would extrapolate to the same normalised width *W*(0)=0.53 (valid for ν=2; see Reference 1); this plainly does not happen.

This work is supported by EPSRC (UK) and Toshiba UK. We should like to thank Dr J. Chalker for stimulating discussions.

1. P.J. Gee, D.R. Leadley, S.O. Hill, J. Singleton, R.J. Nicholas, J.J. Harris and C.T. Foxon, Proc. of 11th Int. Conf. of High Magnetic Fields in the Physics of Semiconductors, pp 252-255, 1994, Cambridge, USA (edited by D. Heiman).
2. K. von Klitzing, G. Dorda and M. Pepper, Phys. Rev. Lett. **45**, 494 (1980).
3. D.C. Tsui, H.L. Störmer and A.C. Gossard, Phys. Rev. Lett. **48**, 1562 (1982).
4. H.P. Wei, D.C. Tsui, M.A. Paalanen and A.M.M. Pruisken, Phys. Rev. Lett. **61**, 1294 (1988).
5. S. Koch, R.J. Haug, K. von Klitzing and K. Ploog, Phys. Rev. Lett. **67**, 883 (1991).
6. H.P.Wei, S.Y.Lin, D.C.Tsui and A.M.M. Pruisken, Phys.Rev.Lett. **45**, 3926 (1992).
7. N. R. Cooper and J. T. Chalker, Phys. Rev. B **48**, 4530 (1993).
8. M. P. Grimshaw, D. A. Ritchie, J. H. Burroughs and G. A. C. Jones, J. Vac. Sci. Technol. B **12**, 1290 (1994).

HIGH–FIELD SPIN–POLARISATION TRANSITION IN THE LANDAU BANDS OF QUASI 2D ELECTRON GASES WITH STRONG LATERAL MODULATION

F. PETIT, M. HAYNE and B. ETIENNE
Laboratoire de Microstructures et de Microélectronique (L2M), C.N.R.S., B.P. 107, 92225 Bagneux Cedex, France

Lateral superlattices (LSLs) of short period are shown to exhibit a novel high–field effect due to the dispersion of the Landau levels (LL) into 1D bands. Whereas Shubnikov–de Haas (SdH) oscillations start at 0.3T, the spin–splitting of the resistivity peaks is inhibited below a threshold field (~6T) around which it appears unusually abruptly. This is supported by a Hartree–Fock calculation of the electronic exchange energy, but a more comprehensive theory which includes the effects of localisation is required to make direct comparison with experiment.

1 Introduction

Two–dimensional electron gases (2DEGs) submitted to a one–dimensional (1D) periodic modulation have been characterized up to now by their low–field properties, such as the Weiss oscillations in modulated III–V heterojunctions[1] or the open–orbit magnetoresistance in Si and InSb MOS structures fabricated on substrates with a large misorientation angle.[2] Here we study the high–field behaviour of 2DEGs strongly modulated at a short length scale, a regime not accessible by standard lithographic techniques at the present time. A. Manolescu and R.R. Gerhardts[3] have proposed that a 1D modulation broadens the LL into finite–width bands (LB) and so should inhibit the exchange enhancement of their Zeeman splitting. We report the observation of this effect and an estimate of the modulation amplitude through a Hartree–Fock calculation of the band structure of a 2DEG submitted to a 1D modulation of arbitrary strength under high magnetic field. However, our experiments show that the absence of disorder in this model is an important limitation which is yet to be addressed.

2 The Samples

The 1D periodic potential, which is obtained in a single technological step by self–organized epitaxy on vicinal substrates,[4] is due to the lateral composition modulation of an AlGaAs alloy embedded in a modulation–doped GaAs/AlGaAs heterojunction. The resulting LSL period is nominally 32 nm, of the order of the electronic zero–field Fermi wavelength, and from previous studies we estimate the unscreened peak–to–peak modulation amplitude to be around 15 meV, comparable with the Fermi energy.[5] On some regions of the wafer the LSL is tilted with respect to the growth direction resulting in a drastically reduced modulation potential but with the same disorder.[6]

Measurements are performed on two mesas A (untilted LSL) and B (tilted LSL) processed on the same wafer. The 2DEG density, increased by brief illumination with an infra–red LED, is 4.5 $\times 10^{11}$ cm^{-2} in the dark and mobility 200000 $cm^2/V/s$ in sample A and 150000 $cm^2/V/s$ in sample B. The quantum lifetime is 0.6 ps in sample A and 0.45 ps in sample B. We insist that this difference is not systematic but is a matter of sample–dependent disorder.

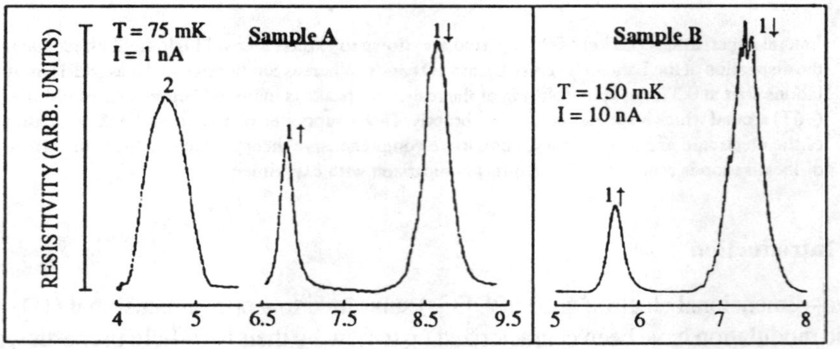

Fig. 1: Shubnikov–de Haas measurements for both samples. The magnetic field (in abcissa) is in Tesla. The current I flows along the modulation direction. LB numbers are indicated (the arrows stand for the spin direction).

3 Results and Model

Figure 1 shows for sample A the resistivity peaks corresponding to the third (N=2) and the second (N=1) LB at 75 mK (after some illumination). The spin–polarization transition is unusually sudden and takes place at about 20 times the SdH onset field, which is inconsistent with the disorder–induced transition picture. On the right part of fig. 1 the N=1 peak for sample B at the same field is quite similar but at much higher temperature (150 mK) and at lower field: the splitting is stronger in sample B despite its higher scattering rate and so more disorder–broadened LB. The N=2 peak remains unsplit for sample B.

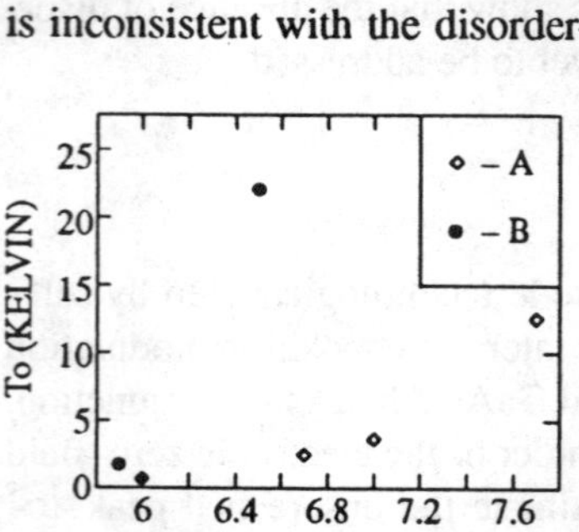

Fig. 2: Characteristic temperature of the hopping conductivity at ν=3 for sample A and B.

By illuminating the samples we have investigated the temperature dependence of the N=1 peak in the range 60 to 600 mK for different values of the field. The dissipative resistance at ν=3 (ν being the LB filling factor) is analysed at the lowest temperatures in terms of variable–range hopping conductivity:[7]

$$\sigma \propto \frac{1}{T}.\exp(-\sqrt{\frac{T_0}{T}}).$$

At the highest considered temperatures σ increases faster than the above law, which may arise from some part of activation in the conduction. Fig. 2 shows the characteristic temperature T_0 for the two samples at different values of the field. The much higher values of T_0 in sample B (an order of magnitude bigger than in A at 6.5 T) demonstrate the effect of a decrease of the modulation. This will be discussed further below.

We have developed a 1D calculation of the LB in a modulated 2DEG taking into account the inter–LB mixing, which cannot be ignored when strong potential gradients are involved. The bands are first calculated with a transfer–matrix method (screening effects coming from Hartree terms are accounted for approximately by a reduction of the modulation amplitude calculated at zero field). The resulting eigenenergies and eigenfunctions are injected in a Fock calculation of the exchange energy, in which the energies and ocupation numbers are computed self–consistently while for simplicity the eigenfunctions are not modified. This model predicts a first order transition between a fully spin–polarised and an unpolarised N=1 LB at 6 T for a 10 meV screened peak–to–peak modulation amplitude in agreement with previous studies.[5] The transition is extremely sensitive to the modulation amplitude, the magnetic field and the temperature. On the contrary disorder broadening of the Landau levels would result in a smooth second order transition and the quantum lifetime introduced in the Gaussian LL broadening would have to be an order of magnitude shorter than the real one for the transition to be well placed (fig. 3).

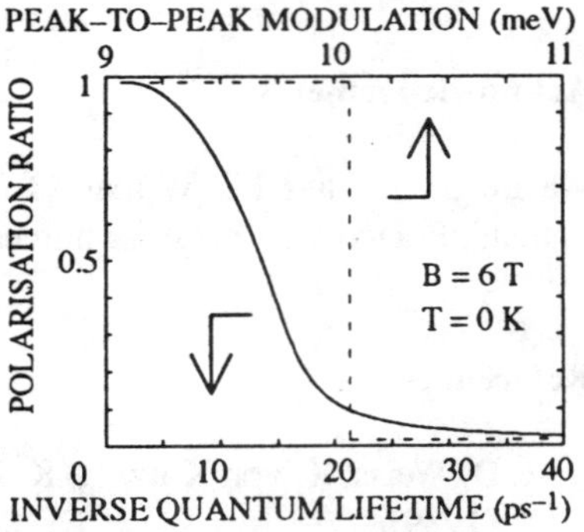

Fig. 3: Variations of the calculated ν=3 polarisation ratio $(n_{1\uparrow}-n_{1\downarrow})/(n_{1\uparrow}+n_{1\downarrow})$ as a function of the modulation amplitude (dotted line, top scale) or the collision rate (full line, bottom scale).

4 Discussion

The sharpness and the position of the peak spin–splitting transition observed contrasts with all the studies performed on GaAs 2DEGs we are aware of. For instance 2DEGs with comparable (sometimes quasi equal) densities and mobilities studied by S. Koch *et al.*[8] showed completely different behaviour. In particular, for a sample with a mobility half ours the N=2 peak was completely split at 2 T (at 40 mK) while our corresponding peak does not show any sign of splitting below 6 T.

The effect of the modulation is also experimentally evidenced in the difference between the two samples. Namely the LB are less disorder–broadened in sample A, so our results imply that they are more dispersed by the modulation. Unfortunately, since the conductivity is dominated by hopping we cannot extract a spin gap or a Landau bandwidth to compare with the calculation. However the latter plainly accounts for the striking behaviour observed, as well as for the position of the transition.

The hopping regime indicates that the tails of localised states attached to the spin-split LB overlap significantly, which in turn means that the spin–gap and disorder broadening are of comparable magnitude. The calculated Landau bandwidth may be larger than these quantities, but no account is taken of the role of disorder and in particular the effect of localisation on dispersed Landau bands is far from clear. Indeed, in unmodulated 2DEGs it is generally considered than only one energy level (per LL) give rise to an extended state whereas the neighboring levels become exponentially localised.[9] Exploring the evolution of this picture when the LL become dispersed LB would participate to a better understanding of the fundamental problem of localisation in the quantum Hall effect. This requires further experimental and theoretical work.

Acknowledgements

We are grateful to F.I.B. Williams for the use of his dilution refrigerator. It's a pleasure to thank François Laruelle for numerous stimulating discussions.

References

1. D. Weiss, K. von Klitzing, K. Ploog and G. Weimann, *Europhys. Lett.* **8**, 179 (1989).
2. T. Cole *et al.*, *Phys. Rev. Lett.* **38**, 722 (1977); T. Evelbauer *et al.*, *Z. Phys. B* **64**, 69 (1986).
3. A. Manolescu and R.R. Gerhardts, *Phys. Rev. B* **51**, 1703 (1995).
4. B. Etienne, F. Laruelle, J. Bloch, L. Sfaxi and F. Lelarge, *J. Crystal Growth* **150** 336–340 (1995).
5. L. Sfaxi, F. Petit, F. Lelarge, A. Cavanna and B.Etienne, *Surface Science* (in press).
6. M. Hayne, F. Petit and B. Etienne, *"The Application of High Magnetic Fields in Semiconductor Physics"*, Würzburg 1996.
7. Y. Ono, *J. Phys. Soc. Japan* **51**, 237 (1982).
8. S. Koch, R.J. Haug, K.v. Klitzing and K. Ploog, *Phys. Rev. B* **43**, 6828 (1991); *Phys. Rev. Lett.* **67**, 883 (1991); *Phys. Rev. B* **46**, 1596 (1992).
9. A.M.M. Pruisken, *Phys. Rev. Lett.* **61**, 1297 (1988).

SPECTROSCOPY OF TWO-DIMENSIONAL ELECTRONIC SYSTEMS IN STRONG MAGNETIC FIELDS WITH MONOCHROMATIC PHONONS

U. WIESNER, S. ROSHKO, W. DIETSCHE, K. v.KLITZING AND K. EBERL
Max-Planck-Institut für Festkörperforschung, Stuttgart, Germany

Tunnel junctions from Al-films have been used in the normal state to generate tunable quasimonochromatic phonons in strong magnetic fields. With this technique, the absorption of phonons by two-dimensional electron gases was investigated.

1 Introduction

The excitation spectra of two-dimensional electron gases (2DEG) in strong magnetic field have been of continuous interest since many years because of their relevance to the electronic transport properties, particularly to the quantum Hall-effects. An alternative technique to the more commonly used optical methods to study these excitations is the spectroscopy with acoustic phonons. Phonons can be generated and detected by suitable techniques in the meV range which is very suitable for the investigation of 2DEG systems. The main differences from optical spectroscopy are the selection rules for the interaction and the fact that acoustic phonons possess momentum.

The spectroscopy with acoustic phonons was successfully done in many solid state systems including 2DEG's using superconducting tunnel junctions as generators and detectors of quasimonochromatic and frequency-tunable phonons.[1] Their use is, however, restricted to nearly zero magnetic field because of the destruction of the superconducting state. In high magnetic fields the phonon spectroscopy was limited to the use of heat pulses, where the phonon spectra consist of broad Planckian distribution with an average phonon energy corresponding to the heater temperature. Using heat pulses, energy gaps of the expected size were indeed found both at integer and at fractional filling factors.[2]

2 Experimental details

Tunnel junctions emit acoustic phonons by the energy relaxation of the electrons after tunneling. The phonon spectrum extends from zero energy up to the energy corresponding to the applied voltage. In the case of superconducting junctions the high energy cut-off is very sharp due to the BCS-singularities in the density of states. Modulation of this cut-off leads to very narrow lines in the differential spectrum. Surprisingly, the phonon emission of tunnel junctions in the normal state was tested only very recently by measuring the binding energies of positively charged acceptors in Si as function of magnetic field.[3] It was found that the emission spectra of the normal state junction were only slightly inferior to the one in the superconducting state. Thus, phonon spectroscopy in strong magnetic fields is possible using the normal state tunnel junctions.

For measuring the phonon absorption of a 2DEG system a tunnel junction is placed on one side of the substrate crystal. It consists of two Al films of 50 nm thickness which are separated by a thin layer of native oxide which forms the tunneling barrier. The junction area was 0.5 x 0.5 mm^2. Phonons are emitted by passing a tunnel current through this barrier, modulation of this current yields the quasimonochromatic phonons. The phonons travel through the substrate (thickness 0.5 mm) and are absorbed by the 2DEG causing an increase of the electron temperature. The 2DEG structures had the shape of a "Hall-bar" of 0.8 mm width and 1.4 mm length. The phonon absorption is monitored by measuring the resistivity changes (R_{xx}). The experiments were done in a ^{3}He cryostat at about 0.4 K.

In this paper we report on results obtained in two types of samples. The first one was a 2DEG in a GaAs heterostructure with a carrier concentration of 2.5 x 10^{11} cm^{-2} and a mobility of 650,000 cm^2/Vsec (dark). Measurements were done at magnetic fields corresponding to the complete filling of 3, 5, 7 and 9 Landau levels. In Fig. 1, we show phonon absorption spectra as function of the bias voltage at the tunnel junction which corresponds to the energy of the phonons. All

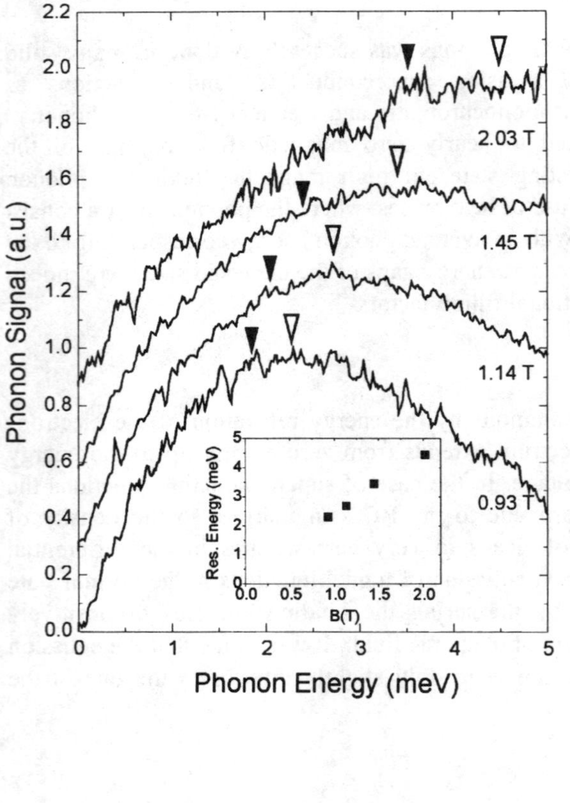

Figure 1: Phonon absorption spectra of a 2DEG in a GaAs heterodstructure. The magnetic field values correspond to filling factors 3, 5, 7 and 9, respectively. The curves are vertically displaced for clarity. The positions where cyclotron resonance is expected are marked by filled arrows. The insert shows the observed maximum position as function of magnetic field (open arrows).

four traces show rather broad maxima. The position of the respective maxima (open arrows) shift with increasing magnetic fields to higher phonon energies as expected for a phonon-cyclotron resonance; see insert. The maxima positions are, however, always slighly higher than the energies of the Landau level splitting ($\hbar eB/m^*$) which are indicated by full arrows in the main figure.

Obviously no sharp cyclotron resonance can be deduced from our results. However, sharp structures are not necessarily expected in a phonon absorption experiment. The phonons will excite magnetoplasmons which have increasing energies at larger wavevectors.[4] The strongest absorption will occur at energies near a roton-like minimum in the dispersion curve which is slightly higher than the cyclotron energy. We consider it to be possible that our broad absorption spectra are caused by the dispersion of the magnetoplasmon curves.

Measurements at filling factor 2/3 were performed with the same sample. There were broad maxima which reach their highest energy at the filling of 2/3. The energy position of the maxima is much higher (3 meV) than expected for the magnetoroton-gap ($\approx$ 0.5 meV),[5] however. Therefore, the bulk of the electron-phonon interaction leading to the broad maximum must be due to other excitations. Higher energy excitations exist at larger wavevectors according to theory[5], but it is not clear if they are responsible for our results. No clear structure of excitations across a gap can be found in the 0.5 meV range. Therefore, we can not at this time verify the magneto-roton excitations observed with heat pulses by Mellor et al.[2]

Another 2DEG system studied by us were Si-MOSFET structures which had a peak mobility of about 10000 $cm^2/Vsec$.[6] Due to the larger effective mass of the electrons one would expect smaller cyclotron energies and larger electron-phonon interaction. It turned out that in this case the behavior of the phonon absorption spectra was significantly different from the one in GaAs. In Fig. 2, a set of phonon absorption curves taken at arbitrary magnetic fields with a carrier concentration of 1×10^{12} cm^{-2} is shown. Very strikingly, the center of the phonon absorption shifts to lower energies with increasing magnetic field. The position of the expected cyclotron resonances are again marked by arrows. No structures could be found near those energies. Obviously, the data cannot be explained by the excitation of magnetoplasmons as in the case of the 2DEG in GaAs but there must be excitations at much lower energies. Low-energy modes have recently been predicted for optical excitation.[7] These modes should be caused by the oscillation of the electrons in the potential of impurities.and would decrease in energy with increasing magnetic fields. An (realistic) average impurity potential parameter of 3 meV is required to find an energy of the low-energy mode of about 2 meV. It is not clear, at present, if such low-energy modes exist for the excitation with phonons and what their energy spectrum would be, but they would offer an attractive explanation of our data.

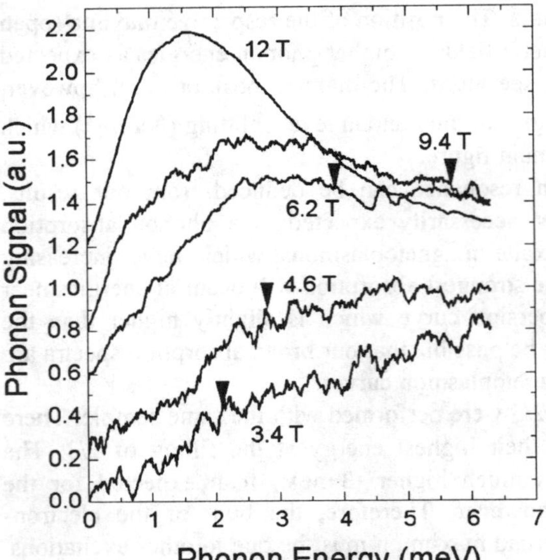

Figure 2: Phonon absorption spectra of a 2DEG in a Si-MOSFET. The absorption maxima shift to lower energies with increasing magnetic field. The arrows mark the expected cyclotron energies E_C at the corresponding magnetic fields [$E_C(12\,T) = 7.3$ meV]

3 Conclusions

We have used the newly developed phonon spectroscopy based on normal state tunnel junction to study the phonon absorption of 2DEG systems. No sharp structures corresponding to a phonon-cyclotron resonance were found. The broad structures in the high mobility GaAs are possibly due to the dispersion of the magnetoplasmon excitations, while in low-mobility Si-MOSFET's low energy excitations must exist.

References

1. W. Dietsche in *Condensed Systems of Low Dimensionality*, edited by J.L. Beeby (Plenum Press, New York, 1991) p.327.
2. C. Mellor, R.H. Eyles, J.E. Digby, A.J. Kent, K.A. Benedict, L.J. Challis, M. Henini, C.T. Foxon and J.J. Harris, Phys.Rev.Lett. **74**, 2339 (1995)
3. S. Roshko and W. Dietsche, Sol.State Comm. **98**, 453 (1996)
4. C. Kallin and B.I. Halperin, Phys.Rev. **B30**, 5666 (1984)
5. S.M. Girvin, A.H. MacDonald and P.M. Platzman, Phys. Rev. **B33**, 2481 (1986)
6. Provided by L.J. Challis, Nottingham
7. U. Merkt, Phys.Rev.Lett. **76**, 1134 (1996)

SURFACE ACOUSTIC-WAVE ATTENUATION BY A TWO-DIMENSIONAL ELECTRON GAS IN A STRONG MAGNETIC FIELD

ANDREAS KNÄBCHEN, YEHOSHUA LEVINSON
Weizmann Institute of Science, Department of Condensed Matter Physics, 76100 Rehovot, Israel

ORA ENTIN-WOHLMAN
School of Physics and Astronomy, Raymond and Beverly Sackler Faculty of Exact Sciences, Tel Aviv University, 69978 Tel Aviv, Israel

The absorption of a surface acoustic-wave (SAW) by a two-dimensional electron gas (2DEG) in a strong magnetic field and a smooth random potential (with correlation length Λ and amplitude Δ) is studied in the case where the Fermi energy ϵ_F is close to the center of a spin-split Landau level (LL). It is shown that for the range of SAW frequencies ω used in experiments the absorption is due to extended electron states. These states have a fractal structure and can be described by classical percolation theory. Both piezoelectric and deformation potential interactions of surface acoustic phonons with electrons are considered and the corresponding interaction vertices are derived. The screening of the electron-phonon coupling due to the 2DEG is accounted for by an appropriate dielectric function $\varepsilon(\omega, q)$ whose behavior is also determined by extended states. In the low temperature regime, both the SAW attenuation coefficient Γ and ε are independent of temperature. As a function of the Fermi energy (or the filling factor $\bar{\nu}$), Γ exhibits a double peak centered at the LL. The minimum of the absorption at the center of the LL is due to enhanced screening. The width of the double peak is found to be given by the scaling law $|\Delta\bar{\nu}| \approx (\omega\Lambda/v_D)^{\alpha/2\nu}$, where $v_D = \Delta l_B^2/\Lambda\hbar$ is the average electron drift velocity in the random potential and l_B is the magnetic length. The exponent $\alpha/2\nu = 3/7$ is determined by scaling exponents of two-dimensional percolation theory. For piezoelectric interaction, the attenuation coefficient increases almost linearly with ω above the threshold frequency $\omega_F = (v_D/\Lambda)|\epsilon_F/\Delta|^{2\nu/\alpha}$, while it is exponentially small below ω_F.

1 Interaction Hamiltonians

The electron-phonon interaction Hamiltonian can be written in the general form

$$H = \frac{1}{L}\sum\nolimits_{\boldsymbol{q}} \gamma_{\boldsymbol{q}} e^{i\boldsymbol{q}\boldsymbol{R}} b_{\boldsymbol{q}} + h.c.\,, \tag{1}$$

where L^2 is the normalization area in the plane of the 2DEG and $b_{\boldsymbol{q}}$ ($b^{\dagger}_{\boldsymbol{q}}$) annihilates (creates) a surface phonon with energy $\hbar\omega_q = v_s q$, v_s being the velocity of the SAW. Assuming that the elastic properties of the sample are described by a continuous, isotropic medium, our results for the interaction vertices for the deformation potential and the piezoelectric coupling are

$$\gamma_{\boldsymbol{q}}^{DA} = (\hbar/\rho v_s a_{DA})^{-1/2}\,\Xi q \qquad \text{and} \qquad \gamma_{\boldsymbol{q}}^{PA} = (\hbar/\rho v_s a_{PA})^{-1/2}\,\beta e\,, \qquad (2)$$

respectively, where Ξ is the deformation potential constant, e is electron charge, and ρ is the mass density of the lattice. β is the component of the piezoelectric modulus which corresponds to the given crystal cut and the propagation direction of the SAW. The quantities a_{DA} and a_{PA} [of order unity] can be expressed in terms of the elastic constants of the lattice [1]. The interaction vertices (2) for SAW's exhibit an additional factor $\sqrt{q}$ compared to the corresponding bulk phonon vertices. Additionally, for surface phonons, the deformation potential and the piezoelectric interactions are *in* phase, whereas they are *out* of phase for bulk phonons. For typical SAW frequencies and GaAs-heterostructures, $\gamma_{\boldsymbol{q}}^{PA} \gg \gamma_{\boldsymbol{q}}^{DA}$.

2 Attenuation coefficient and dielectric function

The attenuation coefficient Γ can be written in the form [1]

$$\Gamma = \frac{2}{\hbar v_s}\frac{|\gamma_{\boldsymbol{q}}|^2}{|\varepsilon(\omega_q, q)|^2}\mathrm{Im}\Pi(\omega_q, q)_{\omega_q>0}\,. \qquad (3)$$

The screening of the electron-phonon coupling due to the 2DEG is accounted for by the dielectric function $\varepsilon(\omega_q, q)$. For low SAW intensities, ε can be calculated assuming linear screening. Then, ε is essentially given by the density correlator Π. The density correlator comprises the matrix elements $M_{if}^{\boldsymbol{q}} = \langle f|\exp(i\boldsymbol{qR})|i\rangle$ which give the transition amplitude between two electron states under the influence of the interaction Hamiltonian, Eq. (1).

The electron states of a 2DEG, subject to a strong magnetic field, B, and a smooth random potential $V(\boldsymbol{R})$ with correlation length $\Lambda \gg l_B = \sqrt{c\hbar/eB}$ and magnitude Δ, have the following properties. A state with energy ϵ is localized in the vicinity of an equipotential line [2] (EL) of the potential $V(\boldsymbol{R}) = \epsilon$. The motion of the guiding center of the electron around an EL is determined by the drift velocity $v_D = c|\nabla V|/eB$. Using $|\nabla V| \simeq \Delta/\Lambda$, one obtains $v_D \simeq l_B^2\Delta/\hbar\Lambda$. The period $\mathcal{T}$ associated with one revolution around the EL determines the energy level spacing $\hbar\omega_{\mathcal{T}} \equiv 2\pi\hbar/\mathcal{T}$ for these quasi-classical states. With respect to the length $\mathcal{L}$ of the EL's, one can distinguish between standard ($\mathcal{L} \simeq \Lambda$) and extended ($\mathcal{L} \gg \Lambda$) EL's. For energies $|\epsilon| \lesssim \Delta$, the majority of EL's is standard ones with $\omega_{\mathcal{T}} \simeq \Omega \equiv 2\pi v_D/\Lambda$. Extended EL's appear for $|\epsilon| \ll \Delta$ and have a fractal structure which is described by percolation theory. EL's with energy ϵ and length $\mathcal{L} \gtrsim \mathcal{L}_c(\epsilon) \simeq \Lambda(\Delta/|\epsilon|)^{2\nu/\alpha}$, where $2\nu/\alpha = 7/3$, are exponentially rare.

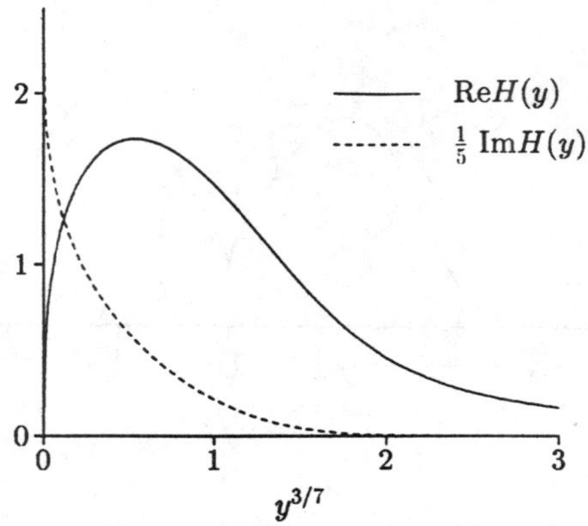

Figure 1: The real and imaginary parts of the function $H(y)$. The scaling of the y-axis corresponds to the dependence of H on ϵ_F.

For typical experimental parameters, $\omega_q \ll \Omega$. Therefore, *real* electronic transitions [contributing to ImΠ, see Eq. (3)] are restricted to extended EL's for which $\omega_T \leq \omega_q$. It can be shown that (virtual) transitions between standard EL's, which might contribute substantially to ReΠ and hence to ε, are not important [1]. Substituting the shortest period for a given energy, $\mathcal{L}_c(\epsilon)/v_D$, into the relation $\omega_T \leq \omega_q$ yields

$$|\epsilon_F| \lesssim \epsilon_\omega \equiv \Delta(\omega_q/\Omega)^{\alpha/2\nu}\,, \qquad |\Delta\bar{\nu}| \simeq \left(\frac{\omega_q}{\Omega}\right)^{\alpha/2\nu}\,, \tag{4}$$

where ϵ has been replaced by the Fermi energy ϵ_F, the relevant electron energy at zero temperature. That is, the absorption of SAW's is restricted to Fermi energies of the order of or smaller than ϵ_ω; ϵ_F is measured from the center of the Landau level (LL). In terms of the filling factor $\bar{\nu}$, one obtains for the width $|\Delta\bar{\nu}|$ of the absorption region the second of the relations (4). The value of the exponent $\alpha/2\nu = 3/7$ agrees with the exponent κ which determines the shrinking of the peaks of the longitudinal conductivity[3] σ_{xx} as the temperature goes to zero, $|\Delta\bar{\nu}| \sim T^\kappa$. In our case the width of the absorption region arises from the frequency ω_q. In this sense, $\hbar\omega_q$ may be considered as an effective temperature which replaces the real temperature T. Frequency scaling in the integer quantum Hall regime has been observed in microwave experiments [4]. For spin-split LL's, the width of the peaks in Reσ_{xx} corresponding to different LL's was found to scale as $|\Delta\bar{\nu}| \sim \omega^\kappa$, with $\kappa \approx 0.41 \approx 3/7$. Keeping ϵ_F (i. e. B) fixed, the relations (4) give $\omega_q \gtrsim \omega_F \equiv \Omega(\epsilon_F/\Delta)^{2\nu/\alpha}$. That is, $\omega_q \simeq \omega_F$

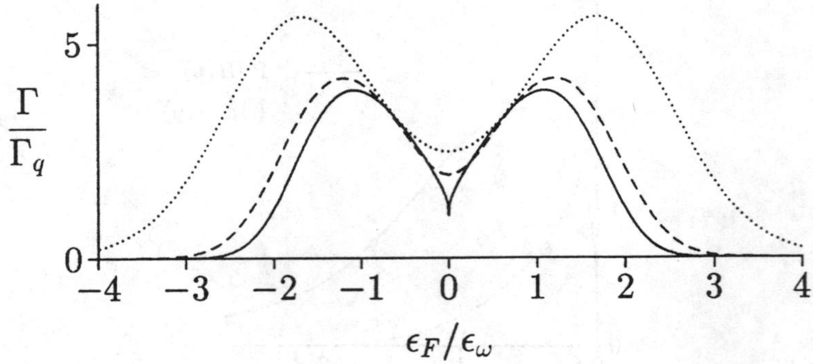

Figure 2: The attenuation coefficient Γ as a function of the Fermi energy near the center of a Landau level $\epsilon = 0$. The three curves correspond to the following temperatures: $T = 0$ (solid line), $T = 0.15 \times \epsilon_\omega$ (broken line)$T = 0$ (solid line), $T = 0.15 \times \epsilon_\omega$ (broken line), and $T = 0.3 \times \epsilon_\omega$ (dotted line).

marks the onset of strong SAW attenuation.

For $T \ll \epsilon_\omega$, the dielectric function and the attenuation coefficient Γ can be expressed in the form

$$\varepsilon(\omega_q, q) = 1 + \frac{e^2}{\varepsilon_o \hbar v_s} 4\pi c H(y_F), \qquad \Gamma = \Gamma_q \mathrm{Im} H(0) \frac{\mathrm{Im} H(y_F)}{|H(y_F)|^2}, \tag{5}$$

where $\Gamma_q = |\gamma_q|^2 \varepsilon_o^2 q (4\pi^2 c e^4 \mathrm{Im} H(0))^{-1}$, $y_F = (\epsilon_F/\epsilon_\omega)^{2\nu/\alpha}$, ε_o is the dielectric constant of the lattice, and $c \simeq 1$. The dimensionless function H is shown in Fig. 1. ImH is exponentially small for $y \gg 1$. The magnitude of ε is determined by the large parameter $e^2/\varepsilon_o v_s \hbar$ (≈ 55 for GaAs). The attenuation coefficient Γ is shown in Fig. 2. In agreement with relations (4), Γ increases strongly once $|\epsilon_F| \lesssim \epsilon_\omega$. The enhanced screening $\sim (\mathrm{Im} H)^2$ at $|\epsilon_F| \ll \epsilon_\omega$ prevents a further rise of Γ and leads eventually to its decrease as ϵ_F goes to zero. For higher temperatures $T \gtrsim \epsilon_\omega$, the screening is weaker. This results in a broadening and an increase of Γ. Eq. (5) shows that $\Gamma \sim \omega_q$ for $\omega_q \gg \omega_F$ [$H \to$ *const.*].

References

1. A. Knäbchen *et al.*, submitted to *Phys. Rev.* B.
2. S. A. Trugman, *Phys. Rev.* B **27**, 7539 (1983).
3. H. P. Wei *et al.*, *Phys. Rev.* B **45**, 3926 (1992).
4. L. W. Engel *et al.*, *Phys. Rev. Lett.* **71**, 2638 (1993).

INTEGER QUANTUM HALL EFFECT IN InGaAs/InP HETEROSTRUCTURES AT MICROWAVE FREQUENCIES

R. GREISER, A. BRENSING, J. PILLATH, W. BAUHOFER
AB Materialien der Mikroelektronik, TU Hamburg-Harburg, Germany

V. HÄRLE, F. SCHOLZ
4. Physikalisches Institut, Universität Stuttgart, Germany

H. HARDTDEGEN
Institut für Schicht- und Ionentechnik, Forschungszentrum Jülich, Germany

Using the crossed waveguide method we have measured the high-frequency Hall conductivity σ_{xy} in two-dimensional electron gases of different InGaAs/InP heterostructures grown by metalorganic vapor phase epitaxy. In agreement with analogous investigations on the AlGaAs/GaAs system we find quantized values of σ_{xy} confirming the existence of quantum Hall plateaus in InGaAs/InP heterostructures up to 35 GHz. Spin-split plateaus are observed for a low electron density sample. All plateau widths are consistently narrower at high frequencies and for high-mobility samples. The temperature dependence of the transition width ΔB between successive spin-degenerate plateaus can be approximated in the temperature interval 1.6 K to 4.2 K by a power-law $\Delta B \propto T^{\kappa}$ with κ in the range between 0.75 and 1.

Introduction

The high-frequency Hall conductivity σ_{xy} of the two-dimensional electron gas in InGaAs/InP heterostructures is of considerable interest regarding the role of localization and the occurence of the quantum Hall effect. There are several experiments[1,2,3] performed for exploring the quantum Hall effect at high frequencies in the AlGaAs/GaAs heterostructure system. For the temperature dependence of the transition width ΔB Wei et *al.*[4] found that due to the long-range potential fluctuation in AlGaAs/GaAs the scaling behaviour with $\Delta B \propto T^{\kappa}$ appears for $T < 200$ mK and has a constant value of κ of 0.42 for spin-split plateaus and 0.21 for spin-degenerate plateaus. In addition, Koch et *al.*[5] reported that κ increases with decreasing mobility. Because of the short-range random alloy potential in the InGaAs/InP system the scaling behaviour should persist up to 4.2 K[4].

Experimental Method

To investigate the high-frequency Hall conductivity σ_{xy} in InGaAs/InP heterostructures we used a crossed waveguide method introduced by Kuchar et *al.*[1]. The frequencies are restricted by the waveguide dimensions to the K_a-band (26.5 GHz to 40 GHz). The waveguide system is placed in a cryostat where temperatures down to 1.5 K and magnetic fields up to 10.5 T can be achieved. Because of the crossed waveguides only the part of the electric field twisted due to a magnetic field perpendicular to the two-dimensional electron sheet is transmitted. Meisels et *al.*[2] showed that the transmitted microwave power P_{trans} which is detected by a bolometer is proportional to ${\sigma_{xy}}^2$.

Results

Using this high-frequency method we found quantized values of σ_{xy} in InGaAs/InP heterostructures. Figure 1 shows the bolometer signal of a measurement of sample HB990[6] (modulation-doped double heterostructure with an InGaAs-channel width of 10 nm) at 1.6 K and 27.9 GHz with plateaus at magnetic fields between 2 and 7 T. The general form of the curve is similar to measurements carried out on AlGaAs/GaAs heterostructures [1]. At samples with high mobility electrons (more than $2 \times 10^5 cm^2/Vs$ at 4 K) also spin-split plateaus with odd filling factors are resolved (Fig.2).

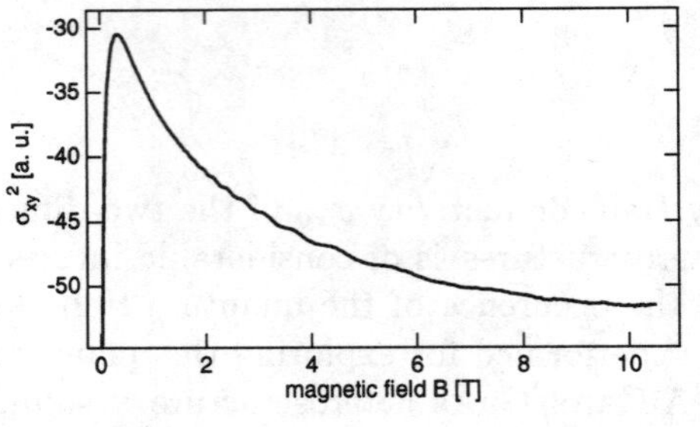

Figure 1: Bolometer signal of sample HB990 at 1.6 K and 27.9 GHz with plateaus between 2 and 7 T. The electron concentration and mobility at 4 K is $n_s = 8 \cdot 10^{11} cm^{-2}$ respectively $\mu = 1.625 \cdot 10^5 cm^2/Vs$.

Corresponding to Meisels et al.[2] the width of the high frequency plateaus are about 30 % narrower than the dc plateau width. They explain this effect with inhomogeneities of the electron sheet concentration n_s. When comparing the hf plateau width with same filling factor of different samples it is perceivable that the plateau width depends on the mobility of electrons respectively the number of traps (Fig.3). This agrees with quantum Hall effect theory which says that the plateau width ΔB is correlated with the number of localized states.

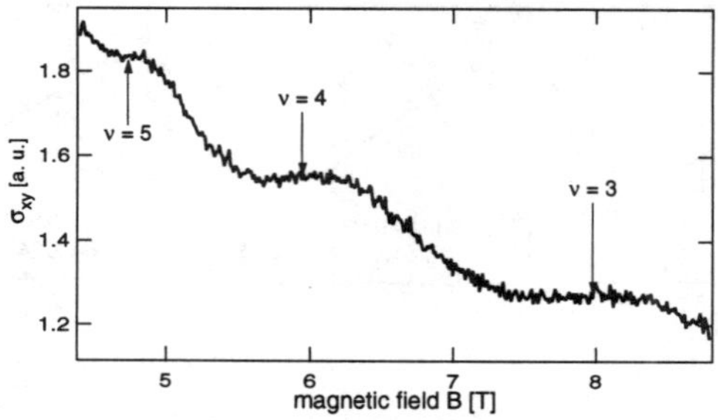

Figure 2: Spin-split plateaus ($\nu = 3, 5$) at a modulation-doped single heterostructure sample $M1628^7$ with $n_s = 5.7 \cdot 10^{11} cm^{-2}$, $\mu \geq 3 \cdot 10^5 cm^2/Vs$.

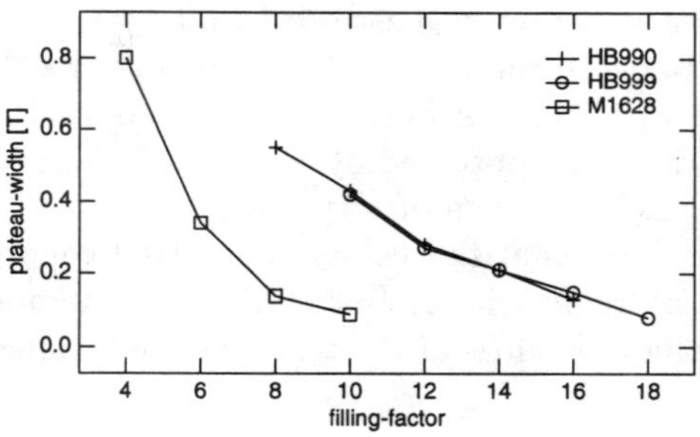

Figure 3: Comparison of the plateau width versus filling factor of samples with different electron mobilities. The samples HB990 and HB999 have nearly the same mobility, the one of M1628 is about three times larger.

The plateaus in σ_{xy} appeared at temperatures between 1.5 to 5 K. Due to this it was possible to investigate the expected power-law behaviour $\Delta B \propto T^{\kappa}$ of the transmission width ΔB between the the plateaus. Wei et *al.*[8] found that the point of maximum gradient of the Hall resistance ρ_{xy} has the same exponent κ in temperature dependence:

$$\left(\frac{d\rho_{xy}}{dB}\right)_{max} \propto T^{-\kappa} \tag{1}$$

In contrast to dc measurements the Hall conductivity σ_{xy} is detected by the microwave method. With

$$\sigma_{xy} = \frac{\rho_{xy}}{\rho_{xx}^2 + \rho_{xy}^2}, \qquad {\rho_{xx}}^2 << {\rho_{xy}}^2 \tag{2}$$

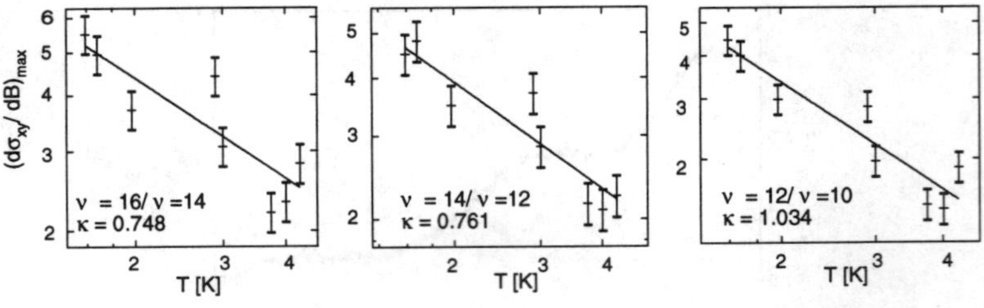

Figure 4: Temperature dependence of the maximum gradient of the transition between different quantum Hall plateaus of sample HB990 with fitted values of κ.

it can be shown that $\frac{d\rho_{xy}}{dB}$ and $\frac{d\sigma_{xy}}{dB}$ are equivalent concerning the temperature dependence. In this connection it is assumed that $\frac{d\rho_{xx}}{dB}$ vanish at the point of maximum gradient between the plateaus. Figure 4 shows the results at three different transitions with fitted κ between 0.748 and 1.034. Our values of κ are higher than the dc values obtained by Wei et *al.*[4] (0.21 for spin-degenerate plateaus) on analogous heterostructure systems.

In summary, we find quantized values of the Hall conductivity σ_{xy} in the InGaAs/InP system at frequencies up to 35 GHz. From temperature dependent measurements we achieved values of the temperature exponent κ between 0.75 and 1.

References

1. F. Kuchar *et al.*, *Phys. Rev.* B **33**, 2965 (1986).
2. R. Meisels *et al.* in *High Magnetic Fields in Semiconductor Physiks III*, ed. G. Landwehr (Springer, Berlin; 1990)
3. L.W. Engel *et al.*, *Phys. Rev. Lett.* **71**, 2638 (1993).
4. H.P. Wei *et al.*, *Phys. Rev.* B **45**, 3926 (1992).
5. S. Koch *et al.*, *Phys. Rev.* B **43**, 6828 (1991).
6. V. Härle in *Untersuchung zur Bandstruktur verspannter GaInAs/InP Quantenfilme*, Dissertation (ISBN 3-8265-0714-2) (Shaker, Aachen; 1995)
7. H. Hardtdegen *et al.*, *J. of Appl. Phys.* **73**, 4489 (1993).
8. H.P. Wei *et al.*, *Phys. Rev. Lett.* **61**, 1294 (1988).

ELECTRON-ELECTRON INTERACTION AND PERCOLATION METAL - INSULATOR TRANSITION IN THE QUANTUM HALL REGIME

HANS NIELSEN

Niels Bohr Institute, Ørsted Laboratory Universitetsparken 5, DK - 2100 Copenhagen Ø, Denmark

PREBEN ALSTRØM, PER JØRGENSEN

Niels Bohr Institute, Blegdamsveij 17, DK - 2100 Copenhagen Ø, Denmark

Under rather general assumptions it is shown that the critical filling factor $\nu = \nu_c$ for a Metal-Insulator transition in the Quantum Hall Regime is the filling factor where the inverse thermodynamic density of states $D_T^{-1} = 0$. This determines the activation energy as function of filling factor. The results are compared with different experiments for high mobility samples.

1 Introduction

For high mobility samples in the Quantum Hall Regime the potential V for the electrons varies very little over a magnetic length l_B since the scale for variations of V is the spacer distance s and in these samples $s > l_B$ is always fullfilled. Since the kinetic energy is quenched by the magnetic field, the energy of an electron is the potential at the position of its guiding center. This means that percolation theory[1] may be applied. According to this theory there is just one energy E_c where extented states exist in an infinite sample. Since a metal is a system where both occupied and unoccupied states exist at the fermi level E_F, it follows that $E_F = E_C$ is necessary in order to have a metal in the zero temparature limit. But especially in the region where linear screening may occur and $E_F > E_C$ the difference $E_F - E_C$ is so small that the system acts as a metal even at low temperatures. In high mobility samples electron - electron interaction has to be taken into account. We will argue that this means that a metal - insulator transition (M/I transition) occurs in the non linear screening regime where the local density is strongly inhomogeneous[2], and the system is divided in regions where the potential satisfies $V \leq E_F$ and regions with $V > E_F$. Experimental results[3,4] seems to show that indeed M/I transitions may occur both just below and just above integer filling where the electron density is supposed to be very inhomogeneous.

2 The Metal Insulator transition

According to percolation theory[1] the energy level E_C of the extented states is determined by $< E - E_C > = 0$ where < > means average over the sample. The probability of finding an energy E is given by the density of states (DOS) which comes from averaging over "metallic regions" with $E \leq E_F$ and "insulator regions" with $E > E_F$. We now do the fundamental assumption for the DOS that only the states below and around E_F is affected by the screening whereas states at a certain distance above E_F is not. This means that we can choose a value $E_0 > E_F$ so that the DOS for $E \geq E_0$ is independent of the filling factor ν. We then write D(E) (the DOS)

$$D(E) = \frac{A_1}{A} D_\upsilon(E) + \frac{A_2}{A} D_0(E) \tag{1}$$

where A_1 and A_2 refers to regions with $E<E_0$ and $E\geq E_0$ respectively. $A = A_1 + A_2$ is the total area of the sample. $D_0(E)$ which represents the DOS for the "insulator region" is independent of ν. The ratio $\frac{A_1}{A_2}$ only depends on the chosen value E_0.

The condition for a M/I transition $< E - E_C > = 0$ with $E_C = E_F$ can now be written

$$\int_{-\infty}^{\infty} (E - E_F)\, D(E)\, d E = 0 \tag{2}$$

by inserting (1.) in (2.) we get

$$E_F = \int_{-\infty}^{\infty} E\, (A_1\, D_\upsilon(E) + A_2\, D_0\,(E)\,)\, d E$$

which may be written

$$\frac{1}{A_1} E_F = \int_{-\infty}^{E_0} E\, D_\upsilon(E)\, d E + \frac{A_2}{A_1} \int_{E_0}^{\infty} E\, D_0(E)\, d E. \tag{3}$$

By taking the derivative of (3.) with respect to ν we get

$$\frac{1}{A_1} \frac{\partial E_F}{\partial \upsilon} \int_{-\infty}^{E_0} E\, \frac{\partial D_\upsilon}{\partial \upsilon}\, d E. \tag{4}$$

Since we have assumed $D_\nu\,(E) = 0$ for $E \geq E_0$ it follows that only $\frac{1}{A_1}$ depends on E_0. By differentiating (4) with E_0 we then get at the M/I transition

$$\frac{\partial E_F}{\partial \upsilon} = 0 \tag{5}$$

$D_T^{-1} = \dfrac{\partial\mu}{\partial\upsilon} = \dfrac{\partial E_F}{\partial \upsilon}$ (μ = chemical potential) is the inverse thermodynamic density of states. It is closely related to the compressibility of the system and has been measured in magnetocapacitance experiments.[5] It is mainly negative but close to integer filling it becomes positive. The value $\nu = \nu_c$ where the sign shift occurs is very well defined since the slope of $D_T^{-1}(\nu)$ is large where $D_T^{-1} = 0$. According to[2] the negative part of D_T^{-1} comes from the exchange term. The electron - impurity interaction gives a positive contribution which depends on the spacer distance s. The value $\nu = \nu_C$ where the two terms cancel is then a function of s. Therefore measurements of the M/I transitions in samples with different spacer distance could be a further test of the theory.

The activation energy $E_A = E_F(\nu_c) - E_F(\nu)$ may be expanded in the vicinity of ν_c with the result

$$E_A = \frac{1}{2}\frac{\partial^2\mu}{\partial\upsilon^2}(\upsilon - \upsilon_c)^2. \tag{6}$$

Since $\dfrac{\partial^2\mu}{\partial\upsilon^2}$ is large at $\nu = \nu_c$ this means that the diagonal part σ_{xx} of the conductivity tensor has a sharp maximum at $\nu = \nu_c$. This seems to be in agreement with fig. 2 in the paper of H.L. Stormer et al.[4] where sharp maxima of $\rho_{xx} \propto \sigma_{xx}$ is observed close to integer filling. As mentioned before E_A is expected to have a small but finite value in regions where linear screening may occur. This means that percolation theory can be applied in regions with linear screening at not too low temperatures as has been done recently by S.H. Simon and B.I. Halperin.[6] The rapid change in E_A close to integer filling has also been observed by measuring ρ_{xx} (T) as function of the magnetic field B and fitting the results to an Arrhenius factor. For $\nu_c < \nu < 1$ we have $E_A = \displaystyle\int_{\upsilon_c}^{\upsilon} \frac{\partial\mu}{\partial\upsilon}\,d\upsilon$ and since $\dfrac{\partial\mu}{\partial\upsilon}$ seems to increase exponentially for $\nu \rightarrow 1$, a strong increase of E_A is expected. Unfortunately no analytical expression exists for $\dfrac{\partial\mu}{\partial\upsilon}$. In the paper by E. Stahl et al.[7] it was found that E_A changes from its maximum value at $\nu_0 = 2$ to $E_A \sim 0$ for $\dfrac{\Delta B}{B} = \dfrac{\Delta\upsilon}{\upsilon} \simeq 0{,}09$. This agrees well with the magneto capacitance measurement which shows

$\Delta\nu = \nu_0 - \nu_c \simeq 0{,}08.$

Finally we consider the DOS for the extended states in a finite sample. If W is a typical length scale then all states with a localization length $\xi \geq W$ will act as extended states. We may consider the sample as embedded in a very large sample. In such a sample[1] the fraction of states with size larger than W is $\sim W^{-p}$ where $p = \frac{1+\beta}{\upsilon}$ is expressed by the universal exponents $\beta = \frac{5}{36}$ and ν which characterize the location length $\xi \sim (E - E_C)^{-\nu}$. (7)

Since the total number of states pr. area is $\frac{eB}{h}$ (neglecting spin) we have $\Delta N \sim \frac{eB}{h} W^{-p}$. From (7) with $\xi \sim W$ follows $\Delta E \sim W^{-\frac{1}{\upsilon}}$ with the result

$$\frac{dN}{dE} \sim \frac{e\,B}{h} W^{-p+\frac{1}{\upsilon}} = e\frac{B}{h} W^{-\frac{\beta}{\upsilon}}. \qquad (8)$$

This formula shows that DOS at the extended states has an universal power law dependence of the sample size and then the conductance at $E_F = E_C$ should also be size dependent.

References

1. S.A. Trugman, Phys. Rev. B **27**, 7539 (1983)
   S.A. Trugman, A. Weinrib, Phys. Rev. B **31**, 2974 (1985)
2. A.L. Efros, Phys. Rev. B **45**, 11354 (1992)
   A.L. Efros et al., Sol. St. Comm. **84**, 91 (1992)
3. A.A. Shashkin et al., Phys. Rev. Lett. **73**, 3141 (1994)
4. H.L. Stormer et al., Sol. St. Comm. **84**, 95 (1992)
5. S.V. Kravchenko et al., Phys. Rev. B **47**, 12961 (1993) and references therein
6. S.H. Simon, B.I. Halperin, Phys. Rev. Lett. **73**, 3278 (1994)
7. E. Stahl et al., J. Phys. C. Solid State **18** L, 783 (1985)

MULTIFRACTALITY OF THE QUANTUM HALL WAVEFUNCTIONS IN HIGHER LANDAU LEVELS

TAKAMICHI TERAO, TSUNEYOSHI NAKAYAMA and HIDEO AOKI*
Department of Applied Physics, Hokkaido University, Sapporo 060, Japan
**Department of Physics, University of Tokyo, Hongo, Tokyo 113, Japan*

To probe the universality class of the quantum Hall system at the metal-insulator transition point, the multifractality of the critical wavefunction ψ is studied for Landau levels $N = 0, 1, 2$. We have found that, while the multifractal spectrum $f(\alpha)$ (and consequently the fractal dimension) does vary with N, the parabolic form for $f(\alpha)$ indicative of a log-normal distribution of $|\psi|$ persists. If we relate the multifractality with the scaling of localization via the conformal theory, the single-parameter scaling is seen to be asymptotically recovered as the range of the random potential is increased, in agreement with Huckestein's irrelevant scaling field argument.

It has been suggested the localization-delocalization transition may be regarded as a phase transition, where the critical wavefunction at the transition point should be self-similar (fractal) on all length scales just as in a phase transition.[1] The self-similarity in critical wavefunctions extends beyond a single scale transformation, so the idea has been subsequently developed into the multifractal analysis.[2,3,4] The critical wavefunction is particularly interesting for the quantum Hall (QH) system,[1] where the delocalized states coalesce into a single point (the centre of each Landau level). The multifractal analysis has also been extended to this system.[4]

On the other hand, the scaling of localization in QH systems is still some way from a full understanding: a finite-size scaling study [5] indicates that the single-parameter scaling holds for the lowest Landau level ($N = 0$), while a two-parameter scaling seems to be required for higher ($N \geq 1$) ones. In an attempt to reconcile the apparent lack of universality, Huckestein has introduced an irrelevant scaling length $\xi_{\rm irr}$.[6,4]

Here we study (i) how the multifractality depends on the Landau index, and (ii) how we can relate this to the scaling, for various values of the correlation length of the random potential σ, which controlls the localization.

We take system sizes as large as $L = (2\pi n_L)^{1/2}\ell = 300\ell$ with ℓ being the magnetic length, where each Landau level is treated separately assuming a strong magnetic field. The Hamiltonian is diagonalized by the Lanczos algorithm with no re-orthogonalization to obtain the mid-band states.[7]

Typical spatial behaviour of $|\psi(\mathbf{r})|^2$ for $N = 0, 1$ Landau levels (Fig.1, left panels) has sparsely distributed spikes particularly for $N = 0$.

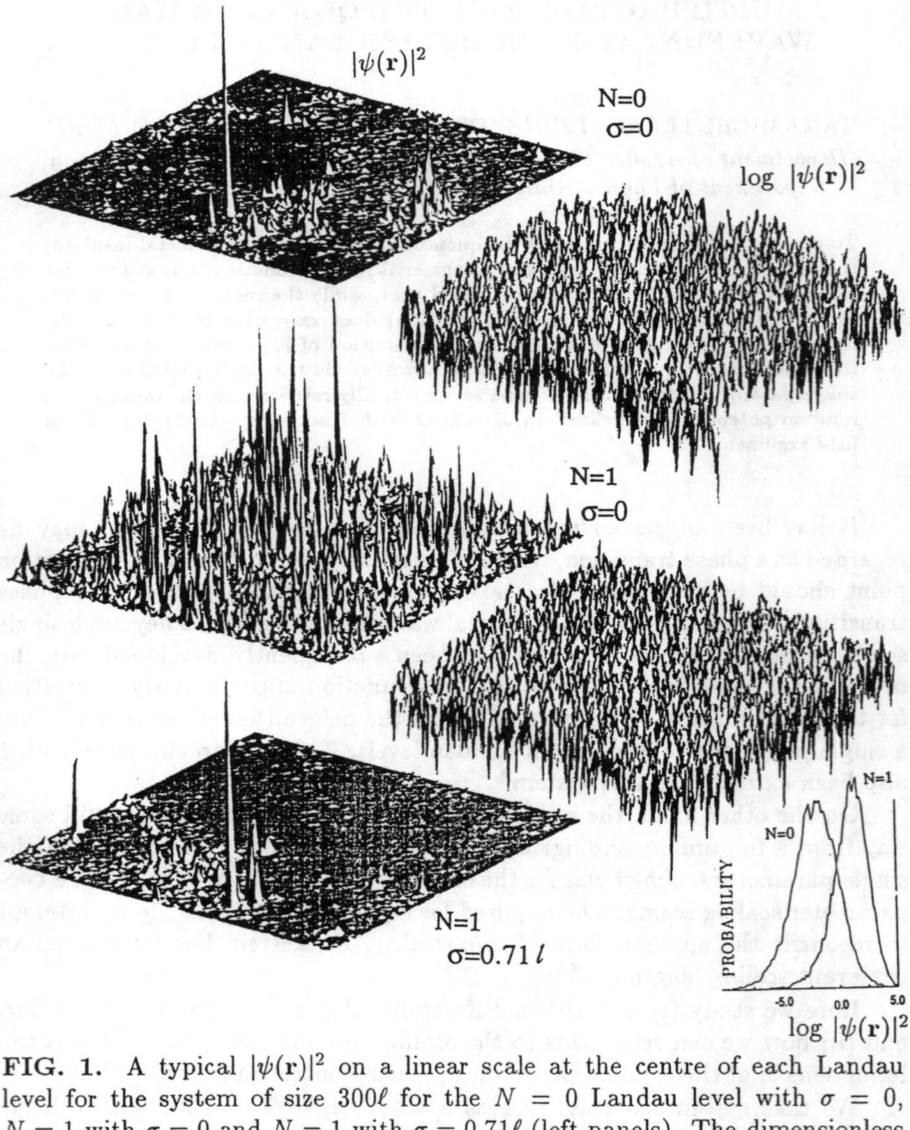

FIG. 1. A typical $|\psi(\mathbf{r})|^2$ on a linear scale at the centre of each Landau level for the system of size 300ℓ for the $N = 0$ Landau level with $\sigma = 0$, $N = 1$ with $\sigma = 0$ and $N = 1$ with $\sigma = 0.71\ell$ (left panels). The dimensionless concentration of scatterers, $c \equiv 2\pi\ell^2 n_i = 10(3)$ for $\sigma = 0(0.71)$. Right panels depict logarithm of $|\psi(\mathbf{r})|^2$ for $N = 0, 1$ with $\sigma = 0$ with a rescaling of the wavefunction to make their log-normal distributions (inset) coincide.

The multifractal spectrum $f(\alpha)$ is obtained numerically by the box - counting procedure [8,9] along with the generalized dimension $D(q) = [f(\alpha(q)) - q\alpha(q)]/(1-q)$. We have checked that the system size is large enough by confirming that $f(\alpha)$'s for two different $L = 200\ell, 300\ell$ agree within statistical errors so that we are indeed looking at the critical wavefunction.[10] We have also performed a sample average. A salient feature in $f(\alpha)$ spectrum (Fig.2) is that, while $f(\alpha)$ does vary systematically with N, $f(\alpha)$ for all of $N = 0, 1, 2$ fits very well to a parabolic form,[8] $f(\alpha) = d - (\alpha - \alpha_0)^2/4(\alpha_0 - d)$. Thus $f(\alpha)$ is characterized by a *single* exponent $\alpha_0(N)$, which decreases from $\alpha_0(0) = 2.31 \pm 0.02$, $\alpha_0(1) = 2.15 \pm 0.02$ down to $\alpha_0(2) = 2.09 \pm 0.01$. A parabolic $f(\alpha)$ indicates a log-normal distribution, $\exp[(\ln P - \alpha_0 \ln\lambda)^2/4(\alpha_0 - d)\ln\lambda]$, for the box probability P for the box size λ. We can visualize this in a typical profile of the $N = 1$

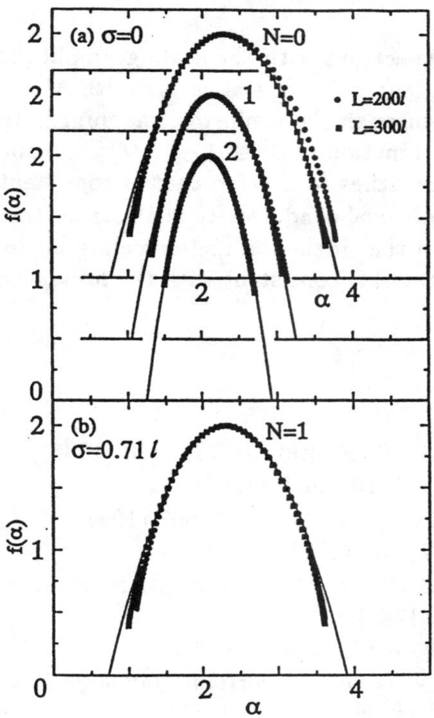

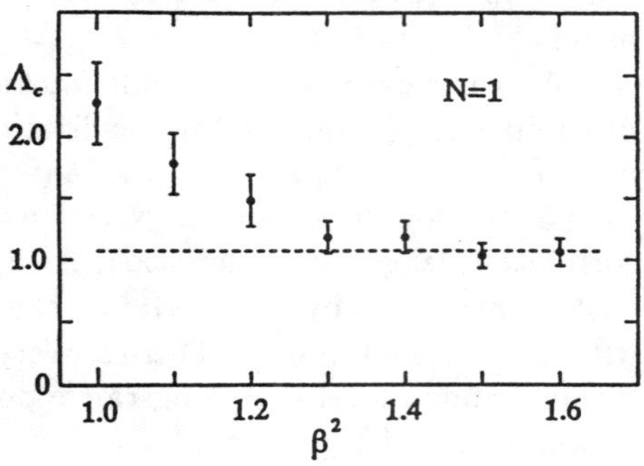

FIG. 3. The scaling amplitude $\Lambda_c = [\pi(\alpha_0 - d)]^{-1}$ for $N = 1$ against $\beta^2 \equiv 1 + (\sigma/\ell)^2$. The horizontal dashed line indicates $\Lambda_c = 1/\ln(1+\sqrt{2})$.

FIG. 2. (a) The $f(\alpha)$ spectra of the wavefunctions at the centre of the $N = 0, 1, 2$ Landau levels for short-range ($\sigma = 0$) random potentials for the system size of 200ℓ (solid circles) or 300ℓ (squares). (b) The same for a finite-ranged random potential ($\sigma = 0.71\ell, \beta^2 = 1.5$) for $N = 1$.

wavefunction on a logarithmic scale in Fig.1(right panels), which shows that the fluctuation in $|\psi(\mathbf{r})|$ indeed becomes similar when a linear transformation on logarithmic scale is performed to make the two distributions coincide.

Concomitantly, the fractal dimension $D(2)$ systematically approaches to $d = 2$ as we go from $N = 0(D(2) = 1.50 \pm 0.06)$ to $N = 1(D(2) = 1.73 \pm 0.05)$ and $N = 2(D(2) = 1.82 \pm 0.05)$. Consequently, the density autocorrelation, $\langle|\psi(\mathbf{r})\psi(\mathbf{r}+\mathbf{R})|^2\rangle \sim R^{D(2)-d}$, becomes longer tailed. The N dependence of the fractality will also have an implication on the anomalous diffusion coefficient $D(q,\omega) \sim (q/\sqrt{\omega})^{D(2)-d}$.

If we turn to finite-ranged potentials in Fig.2(b), the $f(\alpha)$ for $N = 1$ with $\sigma = 0.71\ell$ continues to be parabolic, but the value of α_0 significantly increases to $\alpha_0 = 2.32 \pm 0.03$ with decreased $D(2) = 1.51 \pm 0.07$. The respective values of α_0 and $D(2)$ thus go back to those for $N = 0$, as also seen in Fig.1, bottom panel.

We can now relate the multifractal spectrum with the scaling amplitude $\Lambda_c = \lim_{M\to\infty}\xi_M/M$ for the localization length ξ_M of a strip of width M via $\Lambda_c = [\pi(\alpha_0 - d)]^{-1}$, as given by Janßen through the conformal mapping.[11] In Fig. 3 we plot the result for $N = 1$ as a function of $\beta^2 \equiv 1 + (\sigma/\ell)^2$. As the correlation length σ is increased, Λ_c approaches to a value that is consistent with an argument by Lee *et al*[12] on the universal conductivity at the transition with $\Lambda_c = 1/\ln(1+\sqrt{2})$. This recovery of the single-parameter scaling in the higher Landau level with long-range potential is consistent with the finite-size scaling result of Huckestein.[4]

References

1. H. Aoki, J. Phys. C **16**, L205 (1983); Phys. Rev. B **33**, 7310 (1986).
2. C. Castellani and L. Peliti, J. Phys. A **19**, L429 (1986).
3. M. Schreiber and H. Grussbach, Phys. Rev. Lett. **67**, 607 (1991).
4. B. Huckestein, Rev. Mod. Phys. **67**, 357 (1995).
5. H. Aoki and T. Ando, Phys. Rev. Lett. **54**, 831 (1985); T. Ando and H. Aoki, J. Phys. Soc. Jpn. **54**, 2238 (1985).
6. B. Huckestein, Phys. Rev. Lett. **72**, 1080 (1994).
7. J. K. Cullum and R. A. Willoughby, *Lanczos algorithms for large symmetric eigenvalue computations* (Birkhäuser, Basel, 1985).
8. W. Pook and M. Janßen, Z. Phys. B **82**, 295 (1991).
9. A. Chhabra and R. V. Jensen, Phys. Rev. Lett. **62**, 1327 (1989).
10. Y. Avishai et al, Phys. Rev. B **51**, 13419 (1995).
11. M. Janßen, Int. J. Mod. Phys. B **8**, 943 (1994).
12. D.H. Lee et al, Phys. Rev. Lett. **70**, 4130 (1993).

UNEXPECTED STABILITY OF THE INTEGER QUANTUM HALL EFFECT AGAINST DISSIPATION

FLORIAN GAGEL,[1,2] KLAUS MASCHKE[1]
[1]*Institut de Physique Appliquée, Ecole Polytechnique Fédérale, CH 1015 Lausanne, Switzerland*
[2]*Institut für Physik, Technische Universität Chemnitz-Zwickau, D-09107 Chemnitz, Germany*

The influence of dissipative scattering on the integer quantum Hall effect is investigated numerically. Dissipation is implemented by connecting the sample to auxiliary phase destroying electron reservoirs. We find a surprising stability of the Hall plateaus against dissipation, even in the regime of strong backscattering.

The most intriguing experimental aspect of the quantum Hall effect is its extraordinary insensitivity against elastic as well as inelastic scattering processes. In the present contribution we investigate the dc transport in a quantum Hall device in presence of dissipation. Our study is based on the Landauer-Büttiker approach which provides a consistent description of the transport situation in the quantum Hall regime.[1] Dissipation within the sample is simulated following an idea of Büttiker who proposed to simulate dissipation by introducing fictitious virtual voltage probes. [2,3,4]

We consider the four-terminal Hall device depicted in Fig. 1. It consists of a sample region which is linked at its four sides to waveguides and which establish the connections with the contacts. The Schrödinger equation of the system is solved in a finite-difference scheme with grid spacing a. With the energy scale $\epsilon = 2\, m_0\, E\, a^2/\hbar^2$, the Anderson Hamiltonian reads

$$\hat{H} = \sum_{n,m;k,l} |n,m\rangle h_{n,m;k,l} \langle k,l|, \tag{1}$$

$$h_{n,m;k,l} = \begin{cases} 4 + v_{nm}, & \text{if } (n,m) = (k,l), \\ -\exp(2\pi i A_{n,m;k,l}), & \text{if } (n,m) \text{ is nn of } (k,l), \\ 0 & \text{else.} \end{cases}$$

The magnetic field parameter measured by $\alpha = (e/h)Ba^2$ is represented by the Peierls phases $A_{n,m;k,l}$. The best way to account for the translational invariance along the direction of propagation x in the waveguides is to use the Landau gauge $A_y = 0$, $A_x = -By$ in all four waveguides. This can be achieved by extending the crossing area over one more layer in the upper and the lower waveguide, respectively; a possible resulting gauge in this buffer area is indicated by the arrows in Fig. 1.

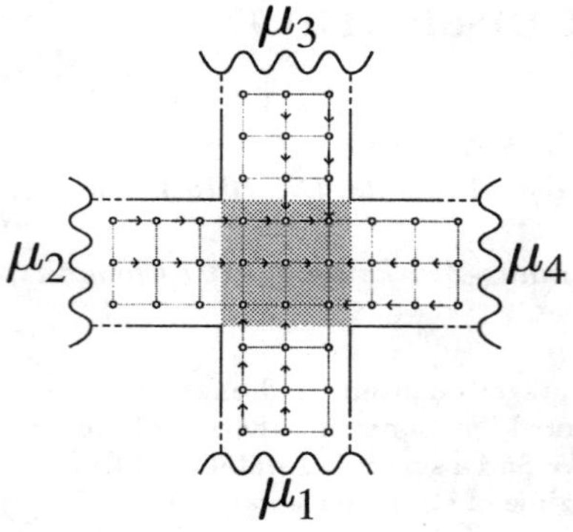

Figure 1: Four-terminal Hall device. The arrows represent the vector potential **A** on the grid. Note that the number of grid points is much larger in the actual calculations.

For small driving voltages $U_{ij} \equiv (\mu_i - \mu_j)/(-e)$, μ_i being the chemical potential of the electron reservoir connected to waveguide i, we have for the corresponding currents

$$I_i = \frac{2e}{h} \sum_j \hat{T}_{ij}\, \mu_j . \tag{2}$$

In absence of dissipation inside the device, the terminal conductance matrix $\hat{T} \equiv (\hat{T}_{ij})$ is related to the scattering matrix S by [1,5]

$$\hat{T}_{ij} = \begin{cases} \sum_{m_i,n_j} |S_{m_i,n_j}|^2 & \text{if } i \neq j, \\ -\sum_{k\neq j} \hat{T}_{kj} & \text{if } i = j, \end{cases} \tag{3}$$

where S_{m_i,n_j} denotes the outgoing scattering amplitude in channel m of terminal i for unit incoming amplitude in channel n of terminal j. Here, the channels are defined by the propagating modes in the wave guides. An efficient method to calculate the propagating modes in presence of a magnetic field has been presented in Ref. [5].

Dissipative effects in the sample region can be implemented by adding side channels which link the sample region to supplementary electron reservoirs. [2] These channels can be described by linear Anderson chains.[6] This approach is equivalent to adding imaginary self-energies to the on-site potentials in the Hamiltonian, which is numerically advantageous. Practically, we have to solve the system of inhomogeneous equations

$$(v_j - \epsilon - i\frac{\delta}{2})\, \psi_j + \sum_{k\neq j} h_{jk}\, \psi_k \;=\; i\, C_{j,+}\delta, \tag{4}$$

where δ is the group velocity in the side channels, $C_{j,+}$ is the amplitude of the incoming wave in the side channel j connected to site j in the sample, and ψ_k is the amplitude of the wave function at site k. The matrix elements of the scattering matrix involving the side channels can be calculated from Eq. 4 for the different scattering boundary conditions.

Defining the matrix $T_{ij} = |S_{ij}|^2$ which equals the matrix $\hat{T}$ defined in Eq. 3 with the exception of the diagonal elements, the effective transmission rate $\tilde{T}$ between two *external* reservoirs l an k can then be obtained by summing over

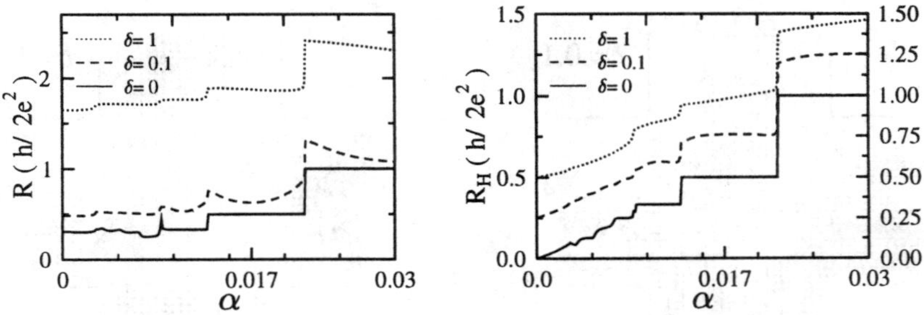

Figure 2: Longitudinal resistance (left) and Hall resistance (right) for different dissipation parameters δ. The curves for R_H belonging to $\delta = 0.1$ and $\delta = 1$ have been shifted by 0.25 and 0.5 units of $h/2e^2$, respectively.

all possible intermediate trajectories. We thus obtain

$$\tilde{T}_{kl} = T_{kl} + \sum_i T_{ki}T_{il} + \sum_{i,j} T_{ki}T_{ij}T_{jl} + \ldots . \tag{5}$$

The sums run over all reservoirs with the exception of l, k. The above expression can be summed as a geometric series. The transport in the dissipative Hall device is then described as in Eq. 2, but with modified rates $\tilde{T}_{ij}$ instead of $\hat{T}_{ij}$.

Our numerical calculations have been performed for the device shown in Fig. 1. Sites inside the shaded crossing area are connected to independent dephasing reservoirs via side chains. The dephasing rates δ are assumed to be the same at all sites. Imposing current conservation also at the upper and lower voltage terminals 1 and 3 ($I_1 = I_3 \equiv 0$), the solution of Eq. 2 yields the two-terminal resistance $R = U_{24}/I_2$ and the Hall resistance $R_H = U_{13}/I_2$.

Figs. 2 and 3 show our results for the Fermi energy $\epsilon_F = 0.4$ and a crossing area of 24×24 sites. The magnetic-field dependence of R and R_H is presented in Fig. 2 for different dissipation parameters δ. The two-terminal resistance R is much more influenced by the dissipation than the Hall resistance R_H. Quite surprisingly, the low-order Hall plateaus are still well resolved for rather large dissipation parameters (see e.g. the curve for $\delta = 1$ in Fig. 3), even if the Hall resistance is somewhat lowered with respect to the ideal quantized value $R_H = h/(2e^2 n)$, $n = 1, 2, \ldots$.

The effect of dissipative backscattering can clearly be seen in the current distributions, which are shown in Fig. 3 for a magnetic field parameter corresponding to the $n = 2$ Hall plateau. The current distribution is completely

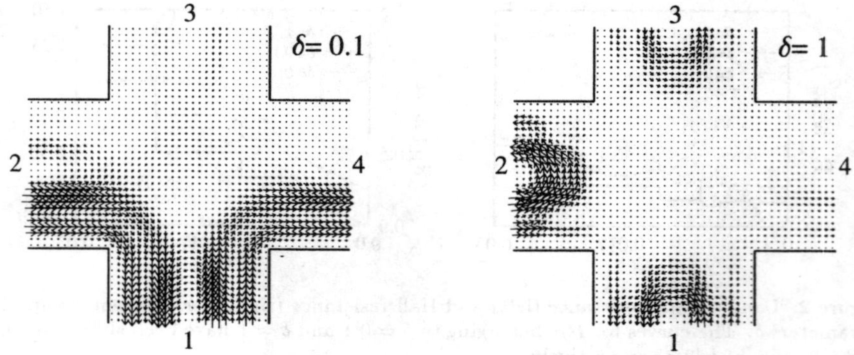

Figure 3: Current distributions for $\alpha = 0.017$ and different dissipation parameters δ.

changed for $\delta = 1$ where backscattering becomes very strong. While it is evident from Fig. 3 that the strong increase of the two-terminal resistance is resulting from the dissipative backscattering at terminal 2, the relative stability of the quantum Hall plateaus shown in Fig. 2 becomes the more surprising, and must be attributed to an intricate compensation of the specific magnetic-field dependences of the rates $\tilde{T}_{ij}$ in Eq. 2.

The presented results show that the Hall plateaus survive even for rather strong dissipation, where backscattering leads already to a significant increase of the two-terminal resistance. While the usual explanation of the stability of the quantum Hall effect is based on the suppression of backscattering by the magnetic field in the plateau regime, and thus relies on the conservation of *integer* elements of the terminal transmission matrix $\tilde{T}$, we find that the plateaus still exist, even though $\tilde{T}$ now contains α dependent *non-integer* elements.

This work was supported by the Swiss National Science Foundation under Grant Nos. 21-39114.93 and 21-45253.95.

References

1. M. Büttiker, Phys. Rev. Lett. **57**, 1761 (1986); Phys. Rev. B **38**, 9375 (1988); Phys. Rev. Lett. **62**, 229 (1989)
2. M. Büttiker, Phys. B **32**, 1846 (1985); Phys. B **33**, 3020 (1986);
3. K. Maschke, M. Schreiber, Phys. Rev. B **44**, 3835 (1991)
4. K. Maschke, M. Schreiber, Phys. Rev. B **49**, 2295 (1994)
5. F. Gagel, K. Maschke, Phys. Rev. B **52**, 2013 (1995)
6. J. L. D'Amato and H. M. Pastawski, Phys. Rev. B **41**, 7411 (1990)

LEVEL STATISTICS IN THE QUANTUM HALL REGIME

M. BATSCH[1,2] and L. SCHWEITZER[1]
[1] *Physikalisch-Technische Bundesanstalt, Bundesallee 100, D-38116 Braunschweig, Germany*
[2] *I. Institut für Theoretische Physik, Jungiusstraße 9, D-20355 Hamburg, Germany*

The statistical properties of energy eigenvalues in the critical regime of the lowest Landau band are investigated. The nearest neighbour level spacing distribution $P(s)$ and the Mehta quantity I_0, from which the scaling exponent ν of the correlation length can be derived, are calculated. While $P(s)$ disagrees with recent analytical predictions, ν is found to be in agreement with theoretical and experimental results.

1 Introduction

In recent years the investigation of energy level statistics in systems showing a localisation-delocalisation transition has made remarkable progress. For the metallic side of the transition the spectral fluctuations only depend on the symmetry of the system (orthogonal, unitary or symplectic) and are well described by Random Matrix Theory (RMT) [1]. On the insulating side, the energy levels are uncorrelated and exhibit Poissonian behaviour. At the transition point, new critical ensembles were found for the 3d orthogonal [2,3,4] the 2d symplectic [5,6], and the 3d unitary case [7]. In the quantum Hall system no complete metal-insulator transition is observed, and the situation is unclear [8,9,10,11].

This is due to the fact that the localisation length of the electronic wavefunctions diverges at the centre of the disorder broadened Landau bands. Hence, in the thermodynamical limit there is no range of extended states but only one energy where the states are multifractal [8,12,13]. Therefore, the question is whether there exists a scale-invariant level statistics in *finite systems* in the absence of true extended states.

As a result of analytical theories [14,15] the asymptotic behaviour of the nearest neighbour level spacing distribution $P(s)$, where the energy difference s is measured in units of the mean level spacing Δ, was proposed to obey $P(s) \propto \exp(-cs^{2-\gamma})$ with $\gamma = 1 - \frac{1}{\nu d}$, where ν is the critical exponent and d the Euclidean dimension. Fits to the numerical data in 3d orthogonal [4], 3d unitary [7] and 2d symplectic [5,6] systems did not result in a sufficient agreement. Investigations that seem to show a reasonable fit [11,16] require, however, values for ν which are different from those commonly accepted. Previously, a simple exponential decay of $P(s)$ was proposed [2], $P(s) \propto \exp(-\kappa s)$, where $\kappa > 1$ is a numerical constant. From the calculations in 3d orthogonal [4] and 3d unitary

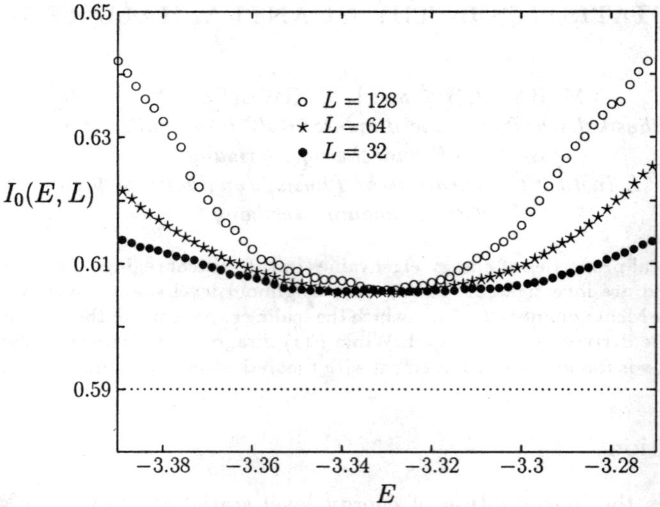

Figure 1: Energy and size dependence of I_0 at the critical energy in the lowest Landau band

systems [7], $\kappa \approx 1.9$, and for the 2d symplectic, $\kappa \approx 4.0$ was obtained [5]. In our opinion, κ should reflect the fractal structure of the critical wavefunctions. Therefore, it is very interesting to study the asymptotics of $P(s)$ in the 2d QHE situation where no complete localisation-delocalisation transition occurs.

2 Model and method

We investigate a 2d Anderson model with the magnetic field included via a phase-factor in the transfer matrix elements V (see Ref.[8] for details). The diagonal disorder potentials are taken at random with constant probability from $[-W/2, W/2]$ and $W/V = 2.5$. The strength of the magnetic field equals 1/8 flux quanta per unit cell. We diagonalise the corresponding Hamiltonian matrix for finite squares of size $L^2 = 32^2$, 64^2, 128^2, and 160^2 using a Lanczos procedure. To obtain a reliable statistic, up to 3^6 realizations were calculated.

3 Results and discussion

To find the energy range of the eigenstates with localisation length larger than the system size and to extract the critical exponent ν, we employ the quantity I_0 known from RMT [1]. The calculation does neither require a fitting procedure

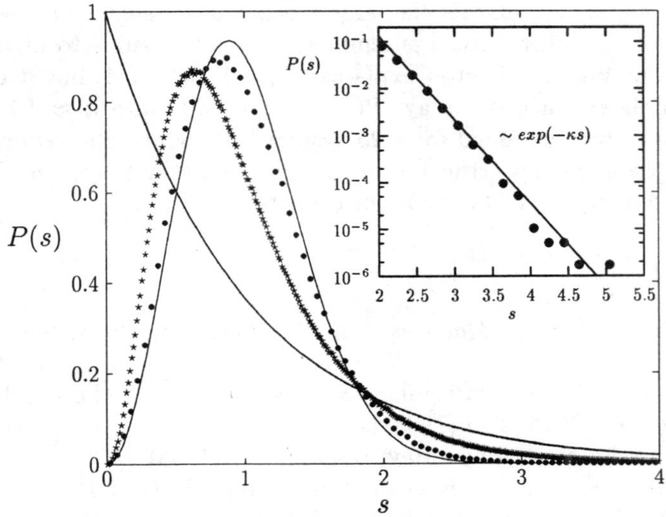

Figure 2: The critical QHE-$P(s)$ ($\bullet$) in comparison with the critical 3d unitary ($\star$), the Poisson and the GUE result. The inset shows a $\exp(-\kappa s)$ decay with $\kappa \approx 4.1$

as the Brody level repulsion parameter [9], nor some cutoff value, as for example the integrated Δ_3 statistics [17]. I_0 is obtained by integrating over $P(s)$

$$I_0 = \int_0^\infty \int_s^\infty \int_{s'}^\infty P(s'')ds''ds'ds. \tag{1}$$

The value of I_0 for the Gaussian unitary ensemble is $I_0 = 0.590$ and for uncorrelated (localised) levels $I_0 = 1$. Multifractal states are expected to have a value larger than 0.59. From our calculations we find $I_0 = 0.605 \pm 0.002$ near the critical energy independent of the considered system sizes. Fig. 1 displays the energy dependence of I_0 in the energy range about the critical point. We have carefully checked that I_0 does not dependent on the width of the energy interval from which the eigenvalues are taken. For all system sizes one observes a minimum $I_0 \approx 0.605$ at the critical energy $E_c/V = -3.33$. States away from E_c become more localised when the system size is enlarged. From the scaling behaviour of $I_0(E, L)$ we extract the critical exponent $\nu = 2.4$ in good agreement with both theoretical [13] and experimental [18] results.

The nearest neighbour level spacing distribution $P(s)$ is shown in Fig. 2. It clearly deviates from the Wigner surmise although it shows a quadratic small-s behaviour as expected for a unitary symmetry. Our result for the lattice model resembles that obtained for a continuum model [9,11]. Due to the large number of realizations we are able to extract the large-s behaviour displayed in the inset of Fig. 2. As for the other critical ensembles, it is not possible to fit the tail of $P(s)$ by the proposal [15] $P(s) \sim \exp(-cs^{2-\gamma})$ with $\nu = 2.4$, but it can be fitted by a simple exponential decay, $P(s) \propto \exp(-\kappa s)$, with $\kappa \approx 4.1$. This value is similar to the one found for a 2d system with symplectic symmetry [5]. In conclusion, there exists a critical level statistics in 2d QHE systems despite the absence of a complete metal-insulator transition.

References

1. M. L. Metha, *Random Matrices*, 2nd ed. (Academic Press, San Diego, 1991).
2. B. L. Altshuler, I. Kh. Zharekeshev, S. A. Kotochigova, and B. Shklovskii, Sov. Phys. JETP **67**, 625 (1988).
3. B. I. Shklovskii *et al.*, Phys. Rev. B **47**, 11487 (1993).
4. I. Kh. Zharekeshev and B. Kramer, Jap. J. Appl. Phys. **34**, 4361 (1995).
5. L. Schweitzer and I. Kh. Zharekeshev, J. Phys.: Condens. Matter **7**, L377 (1995).
6. S. N. Evangelou, Phys. Rev. Lett. **75**, 2550 (1995).
7. M. Batsch, L. Schweitzer, I. Kh. Zharekeshev, and B. Kramer, Phys. Rev. Lett. (1996).
8. B. Huckestein and L. Schweitzer, in *High Magnetic Fields in Semiconductor Physics III: Würzburg 1990*, edited by G. Landwehr (Springer Series in Solid-State Sciences 101, Springer, Berlin, 1992), p. 84.
9. Y. Ono, in *High Magnetic Fields in the Physics of Semiconductors*, edited by D. Heiman (World Scientific Publ., Singapore, 1995), p. 260.
10. M. Feingold, Y. Avishai, and R. Berkovits, Phys. Rev. B **52**, 8400 (1995).
11. T. Ohtsuki and Y. Ono, J. Phys. Soc. Jpn. **64**, 4088 (1995).
12. W. Pook and M. Janßen, Z. Phys. B **82**, 295 (1991).
13. B. Huckestein, B. Kramer, and L. Schweitzer, Surf. Science **263**, 125 (1992).
14. V. E. Kravtsov, I. V. Lerner, B. L. Altshuler, and A. G. Aronov, Phys. Rev. Lett. **72**, 888 (1994).
15. A. G. Aronov, V. E. Kravtsov, and I. V. Lerner, JETP Lett. **59**, 39 (1994).
16. S. N. Evangelou, Phys. Rev. B **49**, 16805 (1994).
17. B. Huckestein and L. Schweitzer, Phys. Rev. Lett. **72**, 713 (1994).
18. S. Koch, R. J. Haug, K. v. Klitzing, and K. Ploog, Phys. Rev. Lett. **67**, 883 (1991).

COLLAPSE OF SPIN SPLITTING IN A TWO-DIMENSIONAL ELECTRON GAS NEAR SILICON VICINAL SURFACE IN QUANTUM HALL EFFECT REGIME

G.M.GUSEV, Z.D.KVON

Institute of Semiconductor Physics, 630090, Novosibirsk, Russia

A.A.SHASHKIN, V.T.DOLGOPOLOV

Institute of Solid State Physics, Chernogolovka, 142432, Russia

In the present work the collapse of spin-splitting in a two-dimentional electron gas near a vicinal silicon surface in quantum Hall effect regime was observed and studied. It was shown that this collapse arises under the influence of the superlattice potential when Fermi surface intersects boundary of the Brillouin zone.

The behaviour of spin splitting in a 2D-electron gas in the QHE regime has attracted new attention due to the prediction and observation of unusual spin phenomena. Recently [1] it has been shown that under certain conditions of the disorder, the g-factor and the exchange interaction, the collapse of spin splitting can arise in a 2D-electron gas in quantizing magnetic fields.

In the present work the collapse of a spin splitting in a 2D-electron gas near high index silicon surface under the influence of surface superlattice has been observed and studied. The experimental samples were Si-MOSFET Hall bars fabricated on a vicinal silicon surface, tilted at an angle of 9.5° from the exact (100) surface around the [011] direction. The thikness of SiO_2 was 120nm. The gate voltage and magnetic field dependence of the diagonal ρ_{xx} and Hall ρ_{xy} components of the resistivity has been measured at temperatures 20mK–1.5K.

Fig.1 shows the mobility of one of these samples as a function of the gate voltage $\mu(V_g)$. It can be seen that this dependence has a maximum at V_g=2.5 and then the mobility decreases as V_g is further increased. This is a conventional mobility behavior for Si-MOSFET at T<10K. However, at V_g=14V a feature appears on the $\mu(V_g)$ dependence. Then, at V_g=18.5 the dependence becomes again an ordinary one. The behaviour described is due to the superlattice potential on the vicinal silicon surface. The appearence of the feature coincides with the start of a minigap, while its vanishing corresponds to the start of a second miniband. The dependences of ρ_{xx} vs V_g in a high magnetic field (B=10T) are shown in Fig.2. One can see that for $V_g < V_g'$ (V_g' is the gate voltage at which the minigap begins) ρ_{xx} has a deep minimum corresponding to spin splitting. However for $V_g > V_g'$ the minimum vanishes completely and instead of it a maximum appears. Thus we observe a spin splitting collapse arising at $V_g > V_g'$. A more detailed picture of the spin splitting

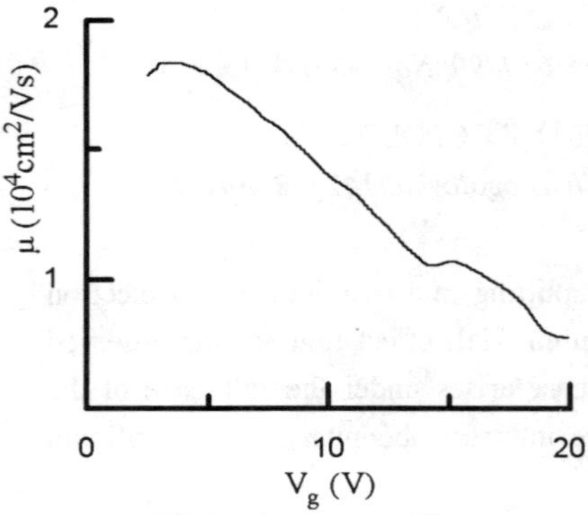

Figure 1: μ versus V_g

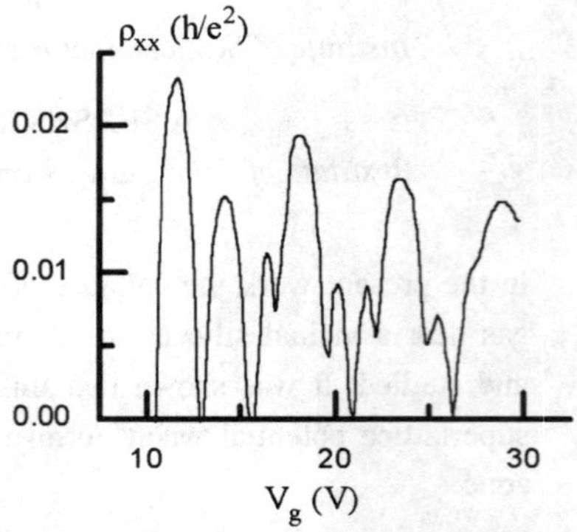

Figure 2: Dependence of ρ_{xx} on V_g.

transformation is shown in Fig.3a–3d that gives $\rho_{xx}(B)$ dependencies for different gate voltages corresponding to different configurations of the 2D-Fermi surface. The first curve in Fig.3a (V_g=13V) corresponds to a situation when the Fermi energy is located in the first miniband. The curve in Fig.3b (V_g=14V) corresponds to the intersection of the Fermi surface by the boundary of the Brillouin zone. The dependence in Fig.3c (V_g=16V) reflects the situation when the Fermi level lies in the minigap. And, finally, the dependence in Fig.3d is for the case when the Fermi level is located in the second miniband. Analyzing the curves in Fig.3 we can come to two conclusions:

1. The collapse of spin splitting begins as the Fermi surface is just beginning to intersect the boundary of the Brillouin zone.

2. This collapse occurs in a very narrow interval of gate voltages near V_g'.

In accordance to [1], what we observe is probably a phase transition which is due to a sharp destruction of the exchange enhancement of g-factor in a 2D-electron gas near silicon suface under the influence of the superlattice potential resulting from the new translation symmetry of a high index surface. It is also necessary to note that the collapse of spin splitting results in an anomalous behaviour of Hall resistance rxy (see Fig.4). As one can see from this figure in the region of the collapsing state the ρ_{xy}(B) is not described by the conventional

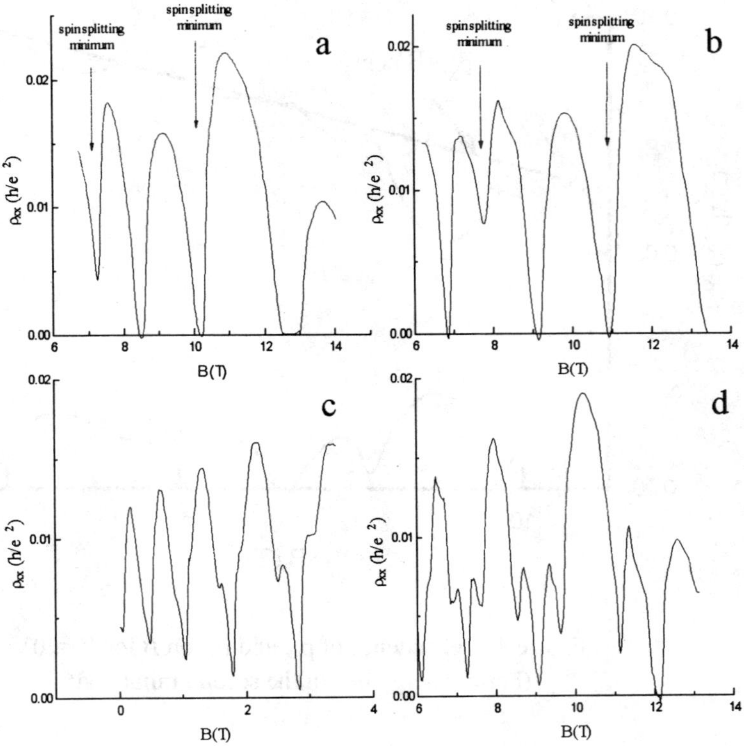

Figure 3: Dependences $\rho_{xx}(B)$ for different gate voltages:
a – V_g=13V, b – V_g=14V, c – V_g=16V, d – V_g=20V

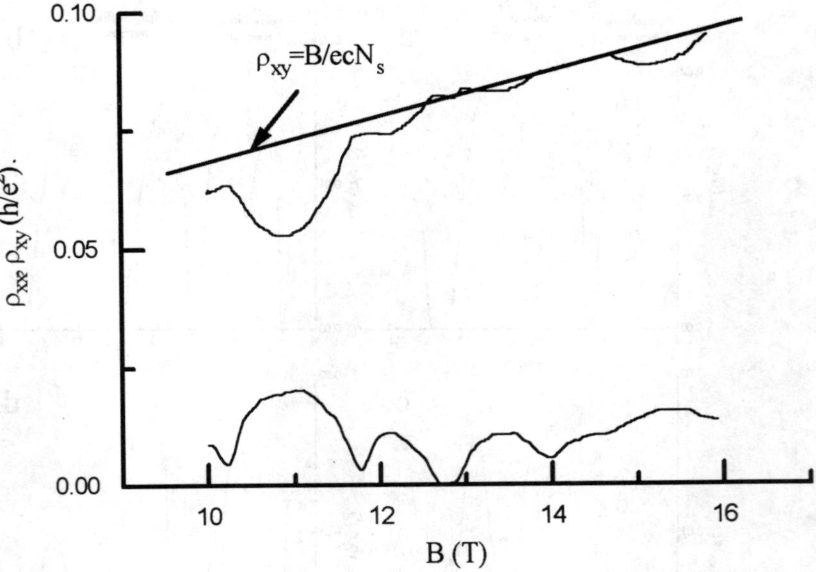

Figure 4: Dependence of ρ_{xx} and ρ_{xy} on B for V_g=20V
(Fermi energy lies in the second miniband)

equation $\rho_{xy} = B/ecN_s$ (N_s is the electron concentration) but has a deep and wide minimum. In this minimum the experimental value of ρ_{xy} is 40% less than the value obtained from the equation given above.

Reference

1. M.M.Fogler, B.I.Shklovskii, Phys.Rev. B52, 17366 (1995)

QUANTUM HALL EFFECT AND VALLEY DEGENERACY IN (001) SI-MOSFETS UNDER UNIAXIAL STRESS

J. LUTZ, F. KUCHAR
Institute of Physics, University of Leoben
Franz Josef Str. 18, 8700 Leoben, Austria

The influence of the valley degeneracy on the Shubnikov-de Haas and quantum Hall effect has been studied in Si MOSFETs under uniaxial stress. The stress allows to vary the relative positions of the conduction band valleys and therefore the valley degeneracy of the two-dimensional electron gas. Due to the electron-electron interaction the influence of the stress on the energetic position of the valleys is non-linear. This results in a complex filling mechanism for the different valleys with increasing carrier density. From the evaluation of the Shubnikov-de Haas data we clearly obtain a four-fold valley degeneracy in measurements with stress in [100] direction. Due to the magnetic field two additional effects occur when in all occupied valleys the Fermi energy is positioned in the localized tails of the Landau levels: (i) a broadening of the Hall plateaus up to a factor of three and (ii) the number of electrons in the occupied valleys does not change with increasing stress. The broadening of the Hall plateaus up to a factor of three for stress in [110] direction is an indication for a valley degeneracy of six.

1 Introduction

The conduction band of the two dimensional electron system (2DES) in an n-channel (100) silicon MOSFET has (like in three dimensions) six lowest minima (valleys). Due to the quantisation in one dimension two of them are centred in the k space and their energy contour is a circle. The other valleys are positioned near the boundary of the Brillouin zone and show elliptical energy contours. Because the quantisation energy of the valleys is inversely proportional to the effective mass of the electrons perpendicular to the plane the energy of the centred valleys is lower than the energy of the four elliptical valleys. The relative energetic position of the valleys can be varied by the application of the external parameter uniaxial stress. Therefore the Si MOSFET is a powerful tool to study the electron-electron interaction in different valleys[1].

But the understanding of the interacting 2D electron system in silicon, the prototype many valley semiconductor, was far away from being satisfactory because of contradictory results from cyclotron resonance[2] (CR) and magnetotransport experiments[3]. CR measurements showed two different masses and agreed with the theoretical model of Takada and Ando[4]. In the transport experiments no evidence was found for a valley degeneracy $g_v > 2$. This was explained by a theory of Kelly and Falicov[5] where the phonon-mediated intervalley electron exchange interaction results in charge density waves (CDW). The problem of this theory was the assumption of a unrealistically strong electron-phonon interaction.

In the present work Si MOSFETs have been studied under uniaxial stress using high-magnetic field experiments, i.e., evaluating the Shubnikov-de Haas (SdH) effect and quantum Hall effect (QHE).

2 Samples and experimental parameters

Silicon (001) MOSFETs were used for the experiments with standard Hall bar geometry. Length and width were 800 and 80 µm, resp. Two potential probes on each side were spaced by 160 µm. The thickness of the gate oxide was 680 nm. The maximum carrier density was $6 \times 10^{12} cm^{-2}$ at a gate voltage of 200 V and the maximum mobility was $2 \times 10^4 cm^2/Vs$ around a gate voltage of 10 V.

All experiments were performed at T = 2.1 K and the magnetic field of B = 10.7 T was perpendicular to the 2DES plane. Stress was applied either in [100] direction up to a value of 2.5 kbar or in [110] direction up to 6 kbar[6].

3 Experimental results

Our SdH data show well defined minima for the Landau splitting up to the highest electron density[7].

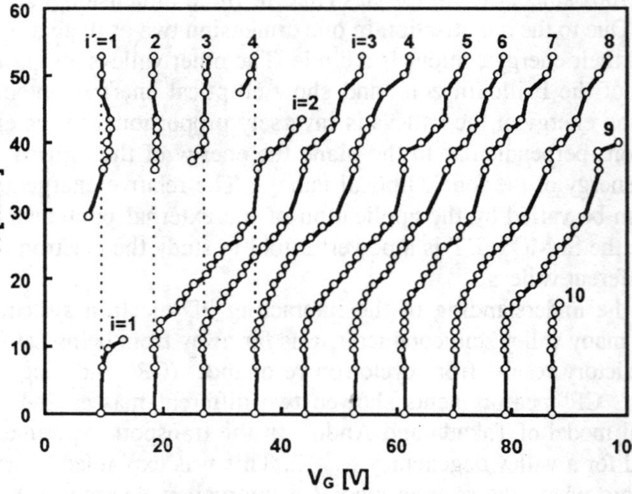

Figure 1: Positions of the minima of the SdH oscillations as a function of applied stress X in [110] direction. The highest stress value is 6 kbar (X [meV] = 8.82 X [kbar][7]).

With increasing stress all minima are shifted to higher gate voltages but at low gate voltages the structure of the SdH data is complex. As a prominent feature the minimum for the spin splitting in the first Landau level is positioned at $V_G = 18$ V for zero stress. This minimum is shifted up to a gate voltage of 50 V for the highest stress. At medium stress and gate voltages below 25 V the longitudinal resistance of the Hall bar is larger than 10 MΩ. But with increasing stress up to the highest values a new minimum appears again at the same gate voltage of 18 V. Up to four new minima can be observed for the highest stress values. As an example these results are summarised in Fig. 1 for stress parallel [110]. In Fig. 1 the positions of the SdH minima are plotted in a diagram of stress X vs gate voltage. For gate voltages above 50 V the period of the SdH oscillations changes maximally to a value 10% smaller than without stress. This results in constant slopes for the shift of the minima and shows that $g_v > 2$ is possible without a significant change in the SdH period. The reason for this behaviour is the non-linear effect of the electron-electron interaction between two valleys on their energetic position.

3 Interpretation and Summary

Fig. 2 shows the relative electron density of the central valley as a function of the total electron density n_S for two constant stress values. At high stress and low electron density all electrons are in the stress lowered valleys. With increasing electron density the Fermi energy increases and the ground energy of the stress lowered valleys increases faster than the ground energy of the central [001] valleys. Therefore, if for a constant stress value the electron density is high enough to start the population of the central valleys, an increase of the electron density causes a transfer of electrons from the stress lowered valleys into the central valleys. Only a small increase of the electron density provokes that 80 - 90 % of the electrons are transferred into the central valleys.

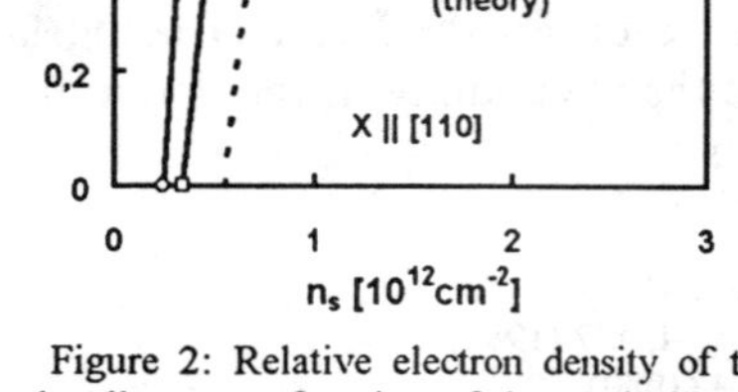

Figure 2: Relative electron density of the central valleys as a function of the total electron density of the 2DES. The broken line shows a theoretical result from Takada and Ando[8].

A further increase of the electron density raises the number of electrons in the central valley and the transfer rate of electrons from the stress lowered into the central valleys is small. As a consequence the period of the SdH oscillation is nearly the same than without stress. This is the reason for the constant period and phase shift of the SdH oscillations with increasing stress observed in previous experiments[9,10]. Nevertheless the valley degeneracy is four for stress in [100] direction. For stress in [110] direction it is not possible to decide between $g_v = 4$ and 6 using only SdH oscillations because the number of SdH minima where all electrons stay in the stress lowered valleys is to small even at the highest stress.

Due to the magnetic field two additional effects have to be considered. Both effects can only be observed at conditions where at the Fermi energy the electrons in all occupied valleys are localized. Firstly, the increased number of localized states causes a broadening of the width of the Hall plateaus (Fig. 3). Secondly, at these conditions the slope of the shift of the SdH minima with increasing stress changes (Fig. 1). For our samples the conditions for these effects are best fulfilled when the Fermi energy is located at least in one set of the different valleys either in the localized states of the spin or Landau splitting or in the lower tail of the lowest sublevel of the first Landau level. Only there both effects are well pronounced.

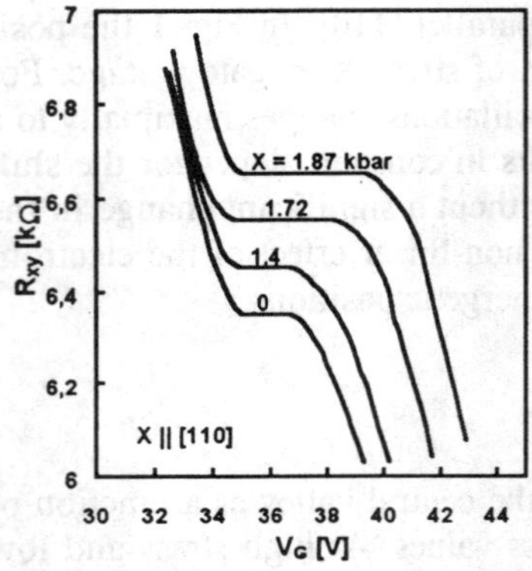

Figure 3: Hall resistance as a function of V_G for four different stress values. The curves for $X > 1.4$ are shifted upwards and for $X = 0$ downwards.

Because the electron density is proportional to the gate voltage the width of the Hall plateaus is proportional to the number of localized states. From our QHE data we deduce a broadening of the plateaus up to a factor of 2.5 for stress in [100] direction and around three for stress in [110] direction (Fig. 3). Assuming that the number of localized states is equal in all the valleys the factor of three means that two more valleys contribute to the QHE at the Fermi energy for stress in [110] direction. Consequently the valley degeneracy is six. A reason for the factor of being larger than two for stress in [100] direction could be the uncertainty in the determination of the width of a Hall plateau or a SdH minimum.

References

[1] Review by T. Ando, A.B. Fowler, F. Stern, *Rev. Mod. Phys.* **54**, 437 (1982)
[2] G. Abstreiter, P. Stallhofer, J.P. Kotthaus, *Surf. Sci.* **98**, 413 (1980)
[3] Th. Englert, G. Landwehr, K. von Klitzing, G. Dorda, H. Gesch, *Phys. Rev.* **B18**, 794 (1978)
[4] Y. Takada, T. Ando, *J. Phys. Soc. Japan* **44**, 905 (1978)
[5] M.J. Kelly, L.M. Falicov, *Suf. Sci.* 73, 303 (1978)
[6] J. Lutz, F. Kuchar, G. Dorda, *Springer Series in Solid State Sciencies* **87**, 41 (1989)
[7] J. Lutz, F. Kuchar, *Phys. Rev. B submitted* (1996)
[8] Y. Takada, T. Ando, *priv. commun.*
[9] G. Dorda, H. Gesch, I. Eisele, *Solid state Commun.* **20**, 429 (1976)
[10] H. Gesch, G. Dorda, P. Stallhofer, J.P. Kotthaus, *Solid State Commun.* **32**, 543 (1979)

COULOMB PROXIMITY PHENOMENA IN TWO DIMENSIONAL CORBINO DISK

V.B.SHIKIN
Institute of Solid State Physics, RAS
142432, Chernogolovka, Moscow District, Russia.

It is shown, that in presence of the contacts between 2D electron system and metallic probes the distribution of 2D electron density becomes non uniform (Coulomb proximity effect). The classical description of this effect for the Corbino sample is developed. The results of these calculations are used for an estimation of the contact potential's difference in experiments[1].

One of the interesting new results, which has been obtained using a linear electro-optical effect[1], is the non-uniform distribution of the electro-potential and electron density along 2D system (2DEG) in Corbino disk. The scale of this non-uniformity is comparable with the size of 2DEG. The effects take place both in Quantum Hall Effect (QEH) regime and also without any magnetic field. All measurments have been carred out in AC regime. Under QEH conditions the results of AC measurments are not sensitive to frequency ω, if $\omega < 5kHz$..

In a given note we discuss one of the possible reasons for such phenomena, so called Coulomb proximity effect. As it will be shown below the contacts of 2DEG with metallic probes lead to electro-statical long-distance perturbation of the electron density along 2DEG. This perturbation (coulomb proximity effect) has the same nature, as the contact phenomena in three dimensional electrostatics[2] and it could be responsible for the non-uniform behaviour of 2DEG[1].

1. Realisation of this program will be developed for the special hetero-structure with a step-like distribution of the donor's density $n_d(x)$. These donors are distributed in the plane z = O and follow to low

$$n_d(x) = N_d \qquad /x/ > w \quad ; \qquad n_d(x) = n_d \qquad /x/ < w \tag{1}$$

Here $2w$ is the width of the step in the donor's distribution. This step plays the role of the central part for the Corbino in one dimensional approximation.

In degenerated case the equilibrium condition for electron system looks now as

$$e\phi(x) + \frac{\pi\hbar^2}{2m_*}n(x) = const \tag{2}$$

$$e\phi(x) = \frac{2e^2}{\kappa} \int_{-\infty}^{+\infty} \delta n(s) \ln(x-s) ds \tag{3}$$

$$\delta n(x) = n(x) - n_d(x) \tag{4}$$

Here κ is the dielectric const, $n_d(x)$ is from (1), a thermodynamical part of electro-chemical potential is written in Thomas-Fermi approximation.

Using Fourier transform of the eq.(2)-(4) it is possible to solve the equation (2) relatively $\delta n(x)$. As a result in limit

$$a_b^* \ll w, \qquad a_b^* = \frac{\kappa \hbar^2}{m_* e^2} \tag{5}$$

this solution looks like

$$\delta n_0(x) = \frac{a_b^* w}{2\pi} \frac{(N_d - n_d)}{(w^2 - x^2)} \equiv \frac{\kappa w \phi_{ab}}{\pi^2 e(w^2 - x^2)}, \qquad \frac{\pi \hbar^2}{2m_*}(N_d - n_d) = e\phi_{ab} \tag{6}$$

The approximation (6) "works" far enough from the points $x = \pm w$. The indicated in (6) contact potential difference ϕ_{ab} will be used below as a fitting parameter.

2.Let us consider now the experiments[1]. These data show in particular, that the Corbino sample with the nominal size of 2DEG $2w = 700\mu$m, is electrically active under the QEH conditions near the edges of 2DEG only. The width of these "active" stripes is about $100\mu m$. As to the central part of Corbino, which could be identifed with the non-compressible channel $2a = 500\mu$m, having an integer filling factor $\nu = 2$, this channel is electrically passive in experiments[1]. Our interpretation of the data[1] leads to an estimation (using an information about w and a) the value of the coefficient ϕ_{ab} from (6). Such a possibility comes, if take into account the results of the paper[3]. Using the general determinations from this paper and information about $\delta n(x)$ (6), we have the following relation between a/w and ϕ_{ab}

$$(a/w)^3 = \frac{3\pi \hbar \omega_c}{2e\phi_{ab}} \tag{7}$$

here ω_c is the cyclotron frequency. Therefore the value a is less w, if $\hbar\omega_c < e\phi_{ab}$.

For the experimental value[1] $w/a = 7/5$, we have from (7)

$$e\phi_{ab} = 12.9\hbar\omega_c \tag{8}$$

If the magnetic field $H = 8.5$ T, which corresponds to filling factor[1] $\nu = 2$, the value $e\phi_{ab}$ from (8) has a scale about 1800° K.

3.Some AC details of experiments[1] need in comments. First of all, a symmetrical AC voltage between metalic probes and additional gate has an amplitude $V_{ac} \leq 0.5$ V. This amplitude is big enough respectively to ϕ_{ab}, and the question comes, how is it possible to use AC voltage as a perturbation in our estimations of ϕ_{ab} from the data[1]?

The answer is the following. An existance of low frequency V_{ac} between metallic probes and additional gate along Corbino disk leads mainly to periodical perturbation of the uniform electron density

$$\delta n(t) = \kappa V_{ac}/(4\pi ed) \tag{9}$$

where d is the distance between 2DEG and additional gate. In experiments[1] $\kappa \sim 5$, $d \sim 500\mu m$, $V_{ac} \sim 0.5V$. Corresponding $\delta n(t)$ has a scale $\delta n_{max} \sim 10^8 cm^{-2}$. This perturbation has to be compared with the equilibrium uniform electron density n_s, which is about $10^{11} cm^{-2}$. We can see, that $\delta n(t) \ll n_s$ and , therefore V_{ac} can be regarded as a small perturbation in our calculations.

The second question is more interesting. Experiment[1] shows frequently independent potential map for $\omega < \tau_*^{-1}$, where adiabatic frequency τ_*^{-1} ia about 5 KHz. What is the origin of this time?

It cannot be the time τ_{fast}

$$\tau_{fast} \sim w/\sigma_{xx}^{peak}, \quad \sigma_{xx}^{peak} = e^2/(\pi^2\hbar) \tag{10}$$

For $w \geq 10^{-2}cm$, $\tau_{fast} \sim 10^{-8}sec.$, $\tau_{fast} \ll \tau_*$.

It cannot be also τ_{slow}, which is traditional for 2DEG with the integer filling factor.

$$\tau_{slow} \sim \tau_{fast} \exp(+\hbar\omega_c/2T) \tag{11}$$

If $\hbar\omega_c/2T \leq 100$, $\tau_{slow} \gg \tau_*$.

The time τ_{slow} (11) has been estimated, using the definition $\tau \sim \sigma_{xx}/w$ and $\sigma_{xx} \simeq \sigma_{xx}^{peak} \exp(\hbar\omega_c/2T)$. But this way is not self-consistent. To be more correct , it is necessary to start from the determinations

$$e\partial n/\partial t + div\,\overrightarrow{j} = 0 \quad ej_i = \sigma_{ik}\partial\mu/\partial x_k, \tag{12}$$

where μ is electro.chemical potential for 2DEG in magnetic field

$$\mu(x) = e\phi(x) - T ln S(H,T,\nu(x)), \tag{13}$$

here

$$2S(H,T,\nu) = (\frac{1}{\nu}-1)+[(\frac{1}{\nu}-1)^2+4\epsilon(\frac{2}{\nu}-1)]^{1/2}, \quad \epsilon = \exp(-\hbar\omega_c/T) \ll 1, \tag{14}$$

$$\nu < 2, \quad \nu(x) = \pi l_h^2 n(x), \quad l_h^2 = c\hbar/eH$$

Near filling factor $\nu = 1$ the main contribution in $\nabla\mu$ comes from the term $T \ln S(\nu)$. Under these conditions the system (12), (13) is redused to

$$\partial n/\partial t = D_{plato}\partial^2 n/\partial x^2, \quad D_{plato} = \pi l_h^2 T \sigma_{fast}/e^2 \tag{15}$$

Now, if the width of 2D stripe with an integer filling factor in the center of Corbino is equal to "2a", then the typical diffusion time τ_{plato} has an estimation

$$\tau_{plato} \sim a^2/D_{plato} = \tau_{fast}\frac{a}{l_h}\frac{e^2}{l_h T} \tag{16}$$

Under the conditions: $a/l_h \gg 1$ and $e^2/l_h T \gg 1$

$$\tau_{fast} \ll \tau_{plato} \ll \tau_{slow} \tag{17}$$

For the parameters[1] $\tau_{plato} \leq 10^{-5} sec.$

Thus in a given note an existance in 2DEG of the longdistance coulomb proximity phenomena is indicated. The basical reasons for these phenomena are the same as in 3D metallic systems. But due to special 2D electrostatics the density's perturbation embrases a long part of 2DEG far from the probes contacts. This 2D-pequliarity allows us to say about a special kind of the proximity phenomena interesting both theoretically and experimentally. The results above can be used for an interpretation of the data[1]. In particular it becomes possible to estimate the contact potential's difference $e\phi_{ab}$ (about 1800 K) for the non-gated Corbino with the metallic probes.

Acknowledgments

The author thanks Prof. K. von Klitzing and Prof. W.Dietsche for stimulating discussions. This activity is supported by RFBI 95 02 06108 and RFBI 96 02 19568 Grants.

References

1. W.Dietsche, K. v. Klitzing and K.Ploog, Workbook Program of the Conference "Electronic Properties of Two Dimensional Systems" Nottingham U.K. 1995 p.311.
2. L.D.Landau, E.M.Lifshitz Electrodynamics of Continuous Media, GosTechIzdat, Moscow, 1957
3. D.B.Chklovskii, K.A.Matveev,B.I.Shklovskii, Phys.Rev.B47, (1993) 12 605

INTERNAL DEGREES OF FREEDOM IN 0D, 1D AND 2D ELECTRON GASES IN STRONG MAGNETIC FIELDS

HIDEO AOKI
Department of Physics, University of Tokyo,
Hongo, Tokyo 113, Japan

We give an overview of theoretical studies on electron gases in 0D (quantum dots), 1D (quantum wires) and 2D (fractional quantum Hall system) in strong magnetic fields with a special emphasis on the spin degrees of freedom and double-layer systems.

1 Introduction

Strong magnetic fields have an ability of totally changing the electronic structures, where the fractional quantum Hall(FQH) effect [1] is a most spectacular example. If we introduce internal degrees of freedom such as the spin degrees of freedom or double-layer systems, the physics of correlated electrons becomes even more fascinating. Here we give an overview on two-dimensional (2D), 1D (quantum wires) down to 0D (quantum dots) systems.

2 2D — single- and double-FQH systems

Spins: Laughlin's quantum liquid is spin fully polarized due to the electron correlation rather than the Zeeman energy which is only a few percent of the typical Coulomb energy. The spin ferromagnetism readily realised here strongly contrasts with the difficulties encountered in the conventional 'strongly correlated electrons' as exemplified the Hubbard model. There, the spin ferromagnetism is realised around a singular Nagaoka's limit (with an infinite repulsion + infinitesimal doping), and a 'generalized Hund's coupling' [2] dictates that only when the system has flat band(s) does the magnetism become stable.[4] In this sense, the FQH ferromagnetism may be thought of as coming from an exotic 'flat band', i.e., the Landau level.

To restore the broken SU(2) symmetry, the spin-polarized liquid has to have Nambu-Goldstone bosons, i.e., magnons. The low-lying states are found to conform well-defined series of multiplets of attractively interacting magnons (Fig.1a) from a numerical result.[5]

Specifically, the ferromagnetic spin wave poses intriguing questions: (i) how the composite-fermion picture, which has so successfully been applied to the FQH liquid, can describe the low lying excitation. (ii) how the spin

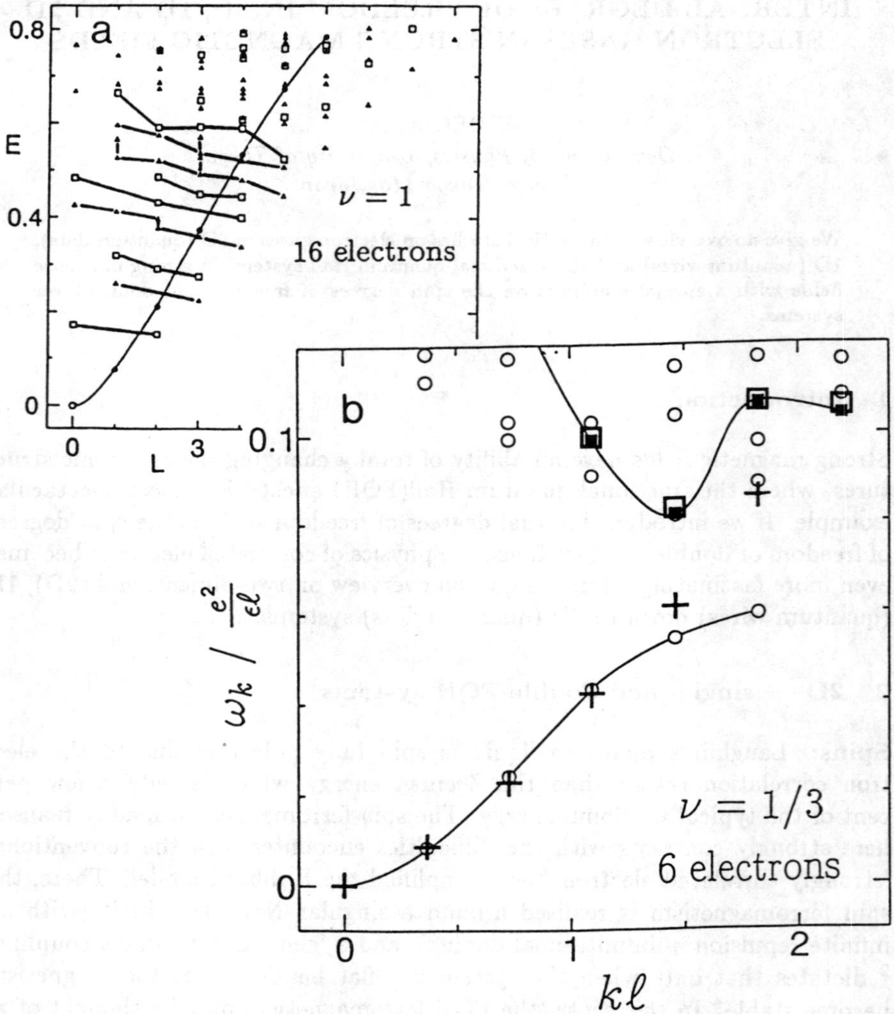

Fig.1. (a) Magnon multiplets (each connected by thick lines) for one-magnon states (•), two-magnon states (□) and three-magnon states (△) in a $\nu = 1$ system. (b) The excitation spectrum for a FQH system, which comprises one-spin-flip ($\Delta S_{\rm tot} = -1$) excitations (∘, connected by a curve as a guide to the eye) and charge ($\Delta S_{\rm tot} = 0$) excitations (■) for a $\nu = 1/3$ system with 6 electrons. The composite-fermion result for the spin wave is shown by crosses, while the composite-fermion result by Kamilla et al[6] for the magnetoroton by open squares.

stiffness arises from 'exchange interactions' among the cyclotron orbits. The spin stiffness in the flat-band Hubbard model is related to peculiar exchange interactions [4], so that the analogy persists here.

Nakajima and Aoki have looked into this.[3] We can map the $\nu = 1$ liquid to $\nu = 1/3, 1/5, ...$ liquid by attaching two, four, ..., flux quanta to each electron to convert it into a composite fermion. In the FQH system, the electron-electron interaction is decomposed into Haldane's pseudopotentials labeled by the relative angular momentum. Attaching the fluxes shifts the relative angular momentum upwards, so that composite particles feel smaller interactions. The single-mode approximation for the spin wave for the composite fermions reproduces quite accurately the numerical result for finite systems for the wavenumber k up to the inverse magnetic length ℓ (Fig.1b). The softening of the spin stiffness as we go from $\nu = 1$ to $\nu = 1/3, 1/5, ...$ may be used as an input for an effective nonlinear σ Lagrangian (a field theoretic model for continuum magnets). In the figure we display all the other low-lying (spin and charge) excitations, and the result of a composite-fermion theory for density (magnetoroton) excitation by Kamilla et al [6] is also shown. We can now see that both the spin wave and magnetoroton are reproduced by composite-fermion descriptions. Recently, these two modes are also experimentally observed.[7] For $k \sim 1/\ell$ the spin wave and the individual (or 'Stoner') excitation continuum seem to hybridize.

Recently, spin excitations have received a more general attention in terms of 'skyrmions', which is a standard spin-texture excitations in the σ model. The hedgehog-like excitation has been applied to the excitations around $\nu = 1$, and, more recently, around fractional fillings.[8] Microscopic analysis of skyrmions has also been attempted.[9] Effect of g factor for $\nu > 1$ has also been studied.[10]

Double layers: Another way to introduce multicomponents is the double-layer FQH systems,[11] which are realised in double quantum wells or wide single wells. The pseudospin (=layer) degrees of freedom might seem similar to the real spin, but the interlayer tunnelling that splits the states into symmetric/antisymmetric combinations with a gap Δ_{SAS} makes the physics even richer. When the interlayer tunnelling is absent, a finite layer separation d degrades the symmetry of the system from SU(2) to U(1). The correlated electron state with a broken U(1) symmetry has been discussed in analogy with a BCS state or an excitonic insulator.[12]

When we switch on the tunnelling, which acts as a transverse magnetic field for pseudospins, the system may be regarded as an easy-plane magnet. We can identify the presence of the charge-excitation (pseudospin wave) gap as the QHE region on the phase diagram against d and Δ_{SAS}. Experimentally, d can be reduced with Δ_{SAS} kept small if we go to double-layer hole gases.[13]

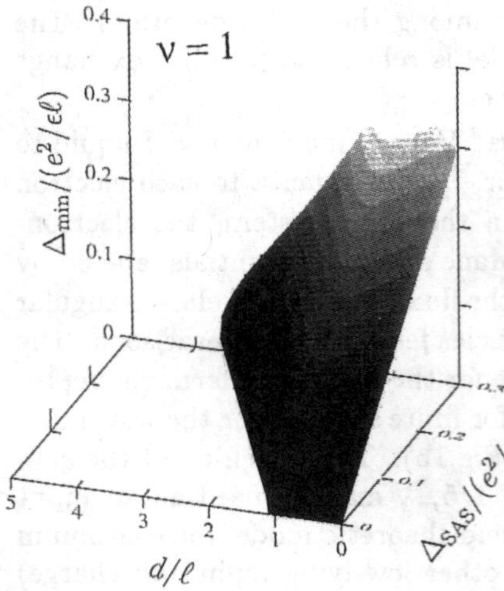

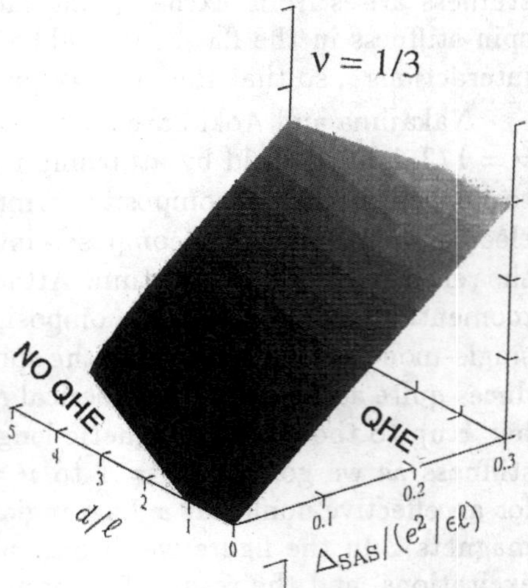

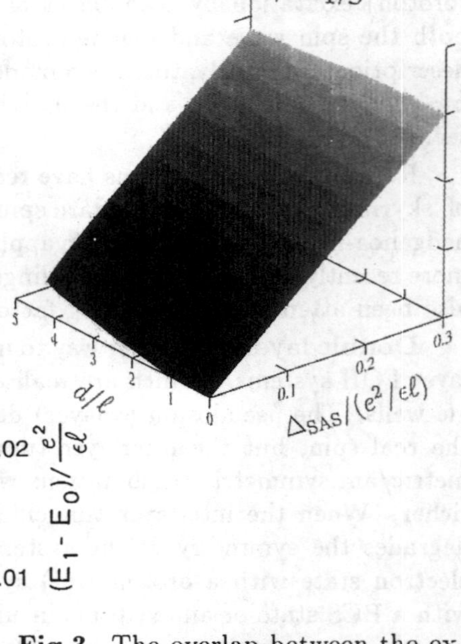

Fig.2. The composite-fermion result for the energy gap (minimum excitation energy) for $\nu = 1/3, 1/5$ derived from the $\nu = 1$ result (left), where the finite gap region corresponds to the pseudospin-polarised quantum-Hall regime.

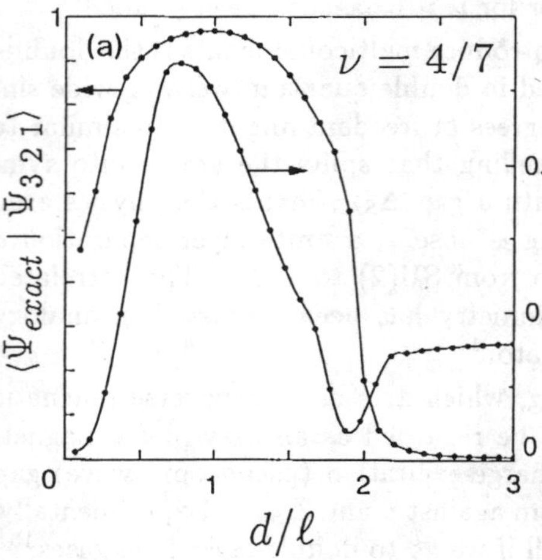

Fig.3. The overlap between the exact ground state for an 8 electron system and ψ_{321} is shown along with the energy, E_1 of the lowest spin-polarized state.

The composite fermions, allowed to tunnel across the two layers, can again describe the pseudospin wave in tunnelling-coupled bilayers accurately,[14] and predicts (Fig.2) that the QHE region should widen as we go from $\nu_{\text{total}} = 1$ to $1/3, 1/5, ...$, since the inter-composite particle interactions are weakened as more fluxes are attached, thereby making the effect of tunnelling more important. The wavenumber at which the pseudospin wave attains its minimum also strongly depends on Δ_{SAS} and d as well as on ν.

Another manifestation of the U(1) symmetry degraded from SU(2) is the 'meron',[15] a counterpart to the skyrmion for a finite d. Ho[16] has described the U(1) symmetry as a p-wave pairing of composite fermions in analogy with the superfluid ^{3}He. This reminds us of the Hubbard model, in which the second SU(2) symmetry (related to the particle-hole symmetry at the half filling[17]) is degraded to U(1) when doped, where a d-wave 'correlated pairing' superconductivity is recently detected from a quantum Monte Carlo study.[18]

Spins in double layers: If we further move on to spinful electrons in double-layer systems to combine the two kinds of degrees of freedom, we can show [19] that there is a spin-*un*polarized quantum liquid specific to such cases, where the simplest realisation, ψ_{lmn} with $\nu = 4/(l + m + 2n) = 4/7$ (with the exponents for intralayer like spins($l = 3$), intralayer unlike spins($m = 2$) and interlayer particles($n = 1$) all differ), is indeed stabilised (Fig.3) for $d \sim$ magnetic length. This indicates that the concept of Laughlin's liquid surprisingly versatile as to accommodate simultaneously two kinds of degrees of freedom.

3 1D — quantum wires in strong magnetic fields

A striking feature brought to light in mesoscopic systems [20] is peculiar quantum transports in quantum wires. In 1D the interactions between the electrons is so crucial due to the strong constraint in the phase space that the system becomes universally the Tomonaga-Luttinger (TL) liquid as far as low-lying excitations are concerned.[21] This appears in the low-temperature conductivity of dirty TL liquids [22,23], $\sigma(T) \sim T^{1-K_\rho}$, where the power law is due to the degraded Fermi singularity in the TL liquid with K_ρ being the charge critical exponent. A recent experiment [24] for high-quality quantum wires seems to support the result.

Now we can ask what happens if we degrade the spin-rotational symmetry. An application of a magnetic field indeed realises this, where the Fermi velocities (v_F's) become spin dependent due to the Zeeman splitting. The system is then described as a spin-split TL model. This destroys the 'spin-charge separation' usually evoked in 1D, and even the quantities such as the conduc-

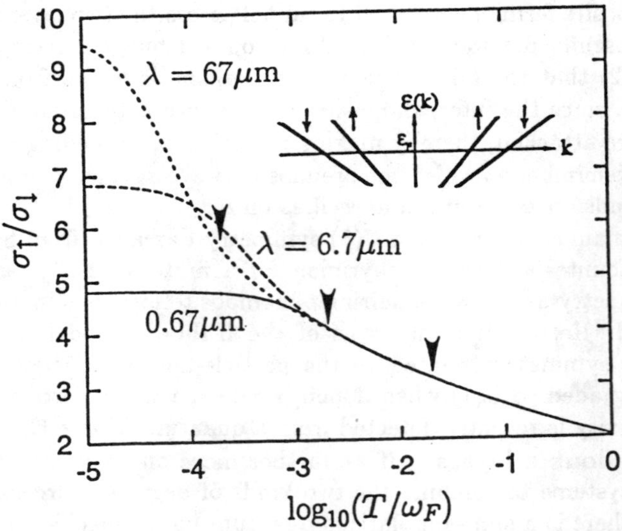

Fig.4 Temperature dependence of the ratio $\sigma_\uparrow/\sigma_\downarrow$ for various values of the mean free path $0.67\mu m \le \lambda \le 67\mu m$ (where $\lambda = 67\mu m$ is taken from Ref.[20]) for $v_{F\downarrow}/v_{F\uparrow} = 0.7, \omega_F = 100K$, and the electron-electron interaction $g = 0.5$. The arrows indicate the position of $T = \hbar v_F/\lambda$ for respective value of λ. The inset schematically depicts the Zeeman-split Tomonaga-Luttinger liquid.

tivity begin to involve the spins. Specifically, we can shown that the ratio of the up-spin and down-spin conductivities in a dirty system increases at low temperatures like $\sigma_\uparrow/\sigma_\downarrow \propto T^{-(K_\uparrow - K_\downarrow)}$, resulting in a spin-polarized current.[25]

At still lower temperatures, a renormalization-group treatment is required. We can than show[26] that the ratio $\sigma_\uparrow/\sigma_\downarrow$ begins to saturate around $T \sim \hbar v_F/\lambda$ (with λ being the mean free path), but we can still have a large ratio (Fig.4).

The generation of spin-polarized currents has been of a long-standing interest for academic[27] as well as practical points of view such as the spin-polarized STM[28] or the Mott-detector[29]. Fasol and Sakaki[30] have proposed another mechanism that involves the curvature in the band dispersion, while the mechanism proposed here is purely an electron-correlation effect.

For the present mechanism usual GaAs quantum wires have too small a g-factor (~ -0.4), but InSb quantum wires (whose g-factor is as large as ~ -50 in the bulk) will produce a significant Zeeman splitting (~ 3.0meV for $B = 1$T).

4 0D — quantum dots in the FQH regime

When a few electrons are confined on a scale of the order of the magnetic length, the ground-state energy as a function of the total angular momentum, L, exhibits downward cusps at specific L values which are known as the 'magic numbers' [31,32,33]. The magic numbers are caused by the fact that the electron correlation in a confined space is severely constrained due to the Landau quantization, which dictates that the favourable electron configuration (triangular for three electrons) requires the preferential values of L. We can even regard such configurations as 'electron molecules'.[34] Unfortunately, the magic numbers cannot be observed by the optical absorption due to Kohn's theorem dictating that the absorption, which picks up the centre-of-mass motion, cannot be affected by electron-electron interactions [32,35].

It is then intriguing to ask what happens if we laterally confine a double-layer FQH system to form 'vertically-coupled quantum dots'. Palacios and Hawrylak [36] have evoked the pseudospin when there is no inter-dot tunnelling. We have looked into the tunnelling-coupled double dots from the exact diagonalization method for finite systems (with three or four electrons) to show [37] the following:

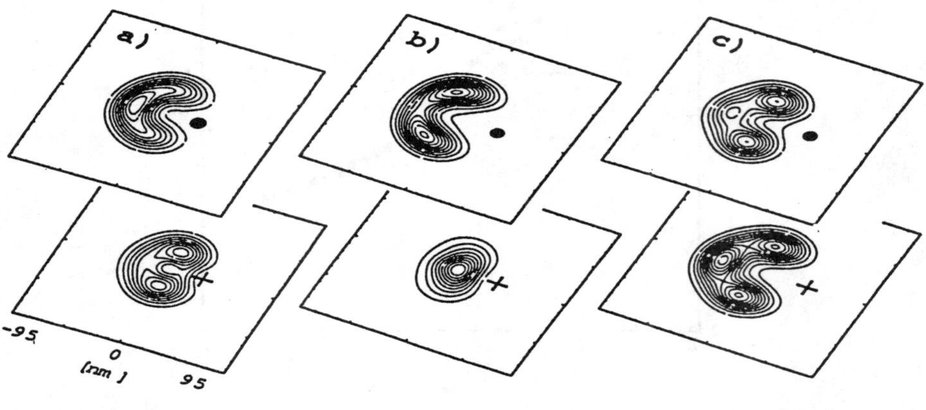

Fig.5 Total angular momentum of the ground state against magnetic field B for a four-electron double dot. Contours depict the pair correlation function for $L = 10$ (a), $L = 12$ (b) or $L = 14$ (c), where solid circles denote the position of a fixed electron in the upper layer at a radius at which the charge density is peaked, while '+' represents their projection on the the lower layer. We assume $\Delta_{SAS} = 0.5$meV, $\hbar\omega_{+(-)} = 2.0(2.2)$meV , and $d = 20$nm.

(i) A new series of magic numbers emerge as a result of the competition of intra-/inter-layer correlations, in a manner dependent on the strength of the inter-layer tunnelling (Fig.5).
(ii) Kohn's theorem become degraded in a double dot having different confining potentials, so that the magic numbers do become observable as jumps in the position/intensity in the absorption[38] (Fig.6). The jumps are significant for a realistic asymmetry in the confining potential between the two layers (5-10% for the layer separation of a few hundred nanometres).

From the correlation function (contours in Fig.5) we see that the new series of magic numbers appear when the tunnelling and layer separation favour the cyclotron orbits to develop a staggered correlation or an 'electron molecule'

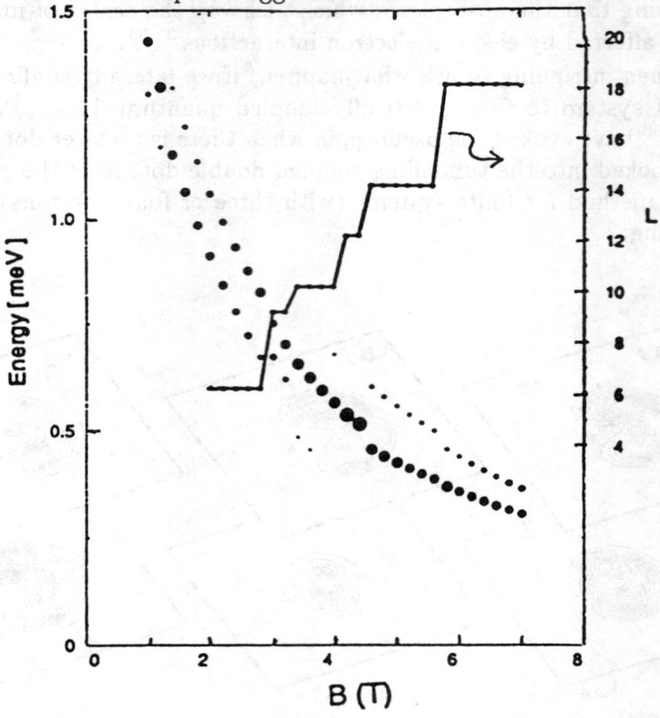

Fig.6 FIR absorption spectrum (left scale) and the total angular momentum (right scale) of vertically coupled dots with four electrons for $\hbar\omega = 2.0$ (2.2) meV for the upper (lower) layer, $d = 20$ nm and $\Delta_{SAS} = 0.5$ meV. Size of the circle represents the absorption intensity.

formed *across* the layers. For smaller magic number L($\sim$ squared lateral spatial extent of the wave function) the strong intralayer interaction has to be avoided by developing an inter-layer correlation at the cost of the mixing of states across the Δ_{SAS} gap. These phenomena persist when the number of electrons is varied, although the magic numbers change accordingly.[40]

Thus a competition of the intra-/inter-layer electron correlation and the interlayer tunnelling is at work, and it would be interesting to relate the double dots to the double-layer FQH liquids and crystals. Hu et al [39] have studied the Hartree-Fock approximation. A composite-fermion analysis, attempted for single dots,[41] will be also interesting. Technically, we believe that the structures considered here are within the scope of current fabrication technology [42].

In summary, the application of strong magnetic fields in 0D, 1D and 2D with internal (spin or layer) degrees of freedom provides fascinating opportunities for what might be called the 'electron correlation engineering'. The author wishes to acknowledge collaborations with Tatsuya Nakajima on FQH systems, Takashi Kimura and Kazuhiko Kuroki on the quantum wire, and Peter Maksym and Hiroshi Imamura on double quantum dots. This work was in part supported by a Grant-in-Aid from the Ministry of Education, Science and Culture, Japan.

References

1. (a) R.E. Prange and S. Girvin (eds.): *The Quantum Hall Effect*, 2nd edn (Springer-Verlag, New York, 1990); (b) S. Das Sarma and A. Pinczuk (eds.): *New Perspectives in Quantum Hall Effects* (Wiley, 1996).
2. K. Kusakabe and H. Aoki, J. Phys. Soc. Japan **61**, 1165 (1992).
3. T. Nakajima and H. Aoki, Phys. Rev. Lett. **73**, 3568 (1994).
4. K. Kusakabe and H. Aoki, Phys. Rev. Lett. **72**, 144 (1994).
5. T. Nakajima and H. Aoki, Physica B **201**, 327 (1994).
6. R.K. Kamilla, X.G. Wu and J.K. Jain, Phys. Rev. Lett. **76**, 1332 (1996); to be published in Phys. Rev. B.
7. A.J. Turberfield et al, preprint.
8. R.K. Kamilla, X.G. Wu and J.K. Jain, Solid State Commun., to appear.
9. A.H. MacDonald et al, Phys. Rev. Lett. **76**, 2153 (1996).
10. R.R. Du et al, Phys. Rev. Lett. **75**, 3926 (1995).
11. K. Moon et al., Phys. Rev. B **51**, 5138 (1995).
12. K. Yang et al, Phys. Rev. Lett. **72**, 732 (1994); preprint (cond-mat/9605153); J. Hu, E. Dagotto and A.H. MacDonald, preprint; Z.F. Ezawa, Tohoku University preprint TU-501.

13. A.R. Hamilton et al, preprint (cond-mat/9606122).
14. T. Nakajima and H. Aoki, Phys. Rev. B **52**, 13780 (1995).
15. S.M. Girvin and A.H. MacDonald, in Ref.[1b].
16. T.-L. Ho, Phys. Rev. Lett. **75**, 1186 (1995).
17. C.N. Yang, Phys. Rev. Lett. **63**, 2144 (1989).
18. K. Kuroki et al, preprint (cond-mat/9606143).
19. T. Nakajima and H. Aoki, Phys. Rev. B **51**, 7874 (1995).
20. See, e.g., B.L. Altshuler et al (eds.) *Mesoscopic Phenomena in Solid* (North-Holland, 1991).
21. See, e.g., V.J. Emery in *Highly Conducting One-Dimensional Solids*, ed. by J.T. Devreese *et al.* (Plenum, New York, 1979).
22. M. Ogata and H. Fukuyama, Phys. Rev. Lett. **73**, 468 (1994); A. Kawabata, J. Phys. Soc. Jpn.**63**, 2047 (1994).
23. A. Luther and I. Peschel, Phys. Rev. Lett. **32**, 992 (1974).
24. T. Tarucha et al, Solid. State. Commun. **94** 233 (1995).
25. T. Kimura, K. Kuroki and H. Aoki, Phys. Rev. B **53**, 9572 (1996).
26. T. Kimura, K. Kuroki and H. Aoki, to be published in Proc. 23rd Int. Conf. on Physics of Semiconductors, Berlin, 1996.
27. V.M. Edelstein, Solid. State. Commun. **73** 233 (1990).
28. R. Wiesendanger et al., Phys. Rev. Lett. **65** 247 (1990).
29. K. Koike et al., Jpn. J. Appl. Phys. **22**, 1332 (1983).
30. G. Fasol and H. Sakaki, Phys. Rev. Lett. **70**, 3643 (1993).
31. S.M. Girvin and T. Jach, Phys. Rev. B **28**, 4506 (1983).
32. P.A. Maksym and T. Chakraborty, Phys. Rev. Lett. **65**, 108 (1990).
33. P.A. Maksym, Physica B **184**, 385 (1993).
34. P.A. Maksym, Phys. Rev. B **53**, 10871 (1996).
35. L. Brey et al, Phys. Rev. B **46**, 10647 (1989).
36. J.J. Palacios and P. Hawrylak, Phys. Rev. B **51**, 1769 (1995).
37. H. Imamura, P.A. Maksym and H. Aoki, Phys. Rev. B **53**, 12613 (1996).
38. In mesoscopic systems a care must be taken about depolarization effects (see, e.g., J.A.A.J. Pernboom et al, Phys. Rep. **78**, 173 (1981); O. Keller and T. Garm, Phys. Rev. B **52**, 4670 (1995). Here we assume a uniform internal field.
39. J. Hu, E. Dagotto and A.H. MacDonald, preprint.
40. P.A. Maksym, H. Imamura and H. Aoki, to be published in Proc. 23rd Int. Conf. on Physics of Semiconductors, Berlin, 1996.
41. J.K. Jain and T. Kawamura, Europhys. Lett. **29**, 321 (1995); R.K. Kamilla and J.K. Jain, Phys. Rev. B **52**, 2798 (1995).
42. K. Bollweg et al., Surface Science, in the press.

GROUND STATE OF A TWO-DIMENSIONAL ELECTRON LIQUID AT HIGH LANDAU LEVELS

M. M. FOGLER, A. A. KOULAKOV, and B. I. SHKLOVSKII
Theoretical Physics Institute, University of Minnesota, 116 Church St. Southeast, Minneapolis, Minnesota 55455, USA

Abstract

We consider a clean two-dimensional electron liquid in a weak magnetic field where N lower Landau levels are completely filled, while the upper level is only partially filled. The system of electrons at the upper Landau level beares a close analogy with a gas of narrow interacting rings. In such a system the uniform distribution of the guiding centers is unstable, and domains with filling factor equal to one and zero are formed. The shape of domains depends on the filling factor: for filling factors of the upper Landau level close to one half, the domains have the form of parallel stripes; away from a small interval around the half-filling, a "bubble" phase is more favorable. The optimal distance between nearest bubbles is approximately three cyclotron radii. We compare the energies of the Laughlin liquid and the "bubble" phase for the upper Landau level filling factors $\nu_N = \frac{1}{3}$ and $\frac{1}{5}$. We conclude that the charge density wave is lower in energy for the Landau level indices $N \geq 2$ at $\nu_N = \frac{1}{3}$ and for $N \geq 3$ at $\nu_N = \frac{1}{5}$. This implies that the $\frac{1}{3}$ quantum Hall plateaus cannot be observed for $N \geq 2$, in agreement with the experiment.

1 Introduction

The nature of the ground state of an interacting two-dimensional (2D) electron gas in magnetic field has attracted much attention. The studies have been focused mostly on the case of very strong magnetic fields where only the lowest Landau level (LL) is occupied, so that the filling factor $\nu = k_F^2 l^2$ does not exceed unity (here k_F is the Fermi wave-vector of the 2D gas in zero magnetic field and l is the magnetic length, $l^2 = \hbar/m\omega_c$). The physics at the lowest LL turned out to be so rich that, perhaps, only at $\nu = 1$ the ground state has a simple structure. Namely, it corresponds to one fully occupied spin subband of the lowest LL. The charge density in such a state is uniform. The case of a partial filling, $\nu < 1$, is much more interesting. Using the Hartree-Fock (HF) approximation, Fukuyama *et al.* [2] found that a uniform uncorrelated spin-polarized electron liquid is unstable against the formation of a charge density wave (CDW). The optimal CDW period was later found to coincide with that of the triangular Wigner crystal[3]. Subsequently, however, it turned out that non-HF trial states suggested by Laughlin[4] for $\nu = 1/3, 1/5$ to explain the fractional quantum Hall effect are by a few percent lower in energy.

In this paper we consider the case of weak magnetic fields or high LL indices N and propose a new candidate for the ground state of a partially filled upper LL. This is a charge density wave with the spatial period of the order of the cyclotron radius $R_c = \sqrt{2N+1}\,l$. By explicit comparison of the interaction energies of our state and of the conventional trial states for the Wigner crystal and $\frac{1}{3}$, $\frac{1}{5}$ Laughlin liquids, we demonstrate that the proposed state has the lowest energy at $N \geq 2$.

We restrict our analysis to spin-polarized states at the upper LL. A justification for this comes from the fact that at least one kind of partially polarized states, so called Skyrmions[5] or, more generally, spin textures[6] are known to be confined to the lowest LL[7]. Note that for spin polarized states, it is sufficient to consider only the upper LL filling factors ν_N in the range $0 < \nu_N < \frac{1}{2}$.

It is convenient to describe our CDW state in terms of the local filling factor, proportional to the guiding center density at a given point $\mathbf{r}$:

$$\hat{\nu}(\mathbf{r}) = 2\pi l^2 \sum_i \delta\left(\mathbf{r} - \hat{\mathbf{R}}_i\right) \tag{1}$$

The summation here is carried over the electrons at the upper LL and $\hat{\mathbf{R}}_i = \mathbf{r}_i + \frac{l^2}{\hbar}\mathbf{z} \times \hat{\mathbf{P}}_i$ is the guiding center operator, with $\hat{\mathbf{P}}_i$ being the canonical momentum of i-th particle.

In the proposed CDW state, the electron liquid consists of periodically positioned large domains with the local filling factor equal to one and zero. The shape of domains depends on the average filling factor ν_N. In a narrow interval around the half-filling, the stripe pattern forms (Fig. 1b), at other ν_N's, one has the "bubble" phase (Fig. 1a). In fact, the stripes appear only at $N > 5$ where in experimental practice the magnetic field is rather low and impurity effects may inhibit the CDW formation. For this reason, hereon we will consider the bubble phase only. We found, for example, that the optimal distance between neighboring bubbles (for not too small ν_N) is approximately $3.3R_c$, which corresponds to the optimal number of

$$M_{\text{opt}} = 3\nu_N N \tag{2}$$

electrons per bubble.

It is important to keep in mind that the actual charge density variations in our CDW state are not too large (of the order of 20%). This is because the characteristic spread of electronic wavefunction, R_c, is of the same order of

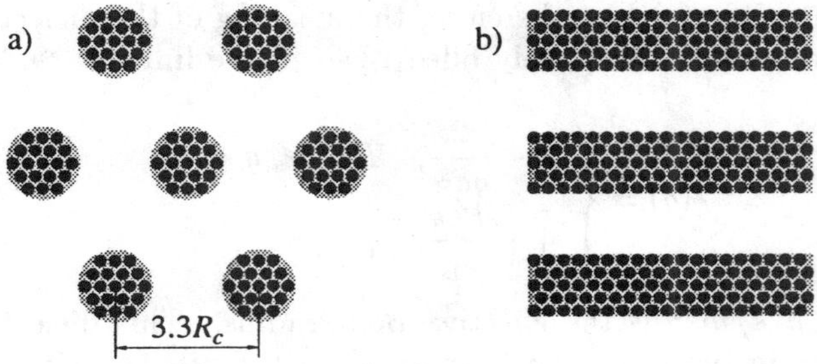

Figure 1: (a) "Bubbles". (b) "Stripes". The black dots represent the guiding centers of cyclotron orbits.

magnitude as the CDW period. Thus, our state is not a conventional charge density but rather a guiding center density wave.

The paper is organized as follows. In Sec. 2 we discuss the interaction between the electrons at the upper LL and demonstrate that the system possesses the CDW instability. In Sec. 3 we formulate the trial wavefunction of the CDW state and compare its energy with energies of other candidates for the ground state. In Sec. 4 we discuss the consequences of the proposed ground state structure on experimentally measurable quantities.

2 Electron-electron interaction at high Landau levels

In a strong magnetic field where the cyclotron gap $\hbar\omega_c$ is large, the LL mixing can be ignored and the system of electrons at the upper LL is described by the following Hamiltonian

$$\hat{H} = \frac{1}{2L_x L_y} \sum_{\mathbf{q}} \rho_N(\mathbf{q})\tilde{v}(\mathbf{q})\rho_N(-\mathbf{q}), \tag{3}$$

where $\rho_N(\mathbf{q})$ is the projection of the density operator onto the N-th LL, and $\tilde{v}(q)$ is the Coulomb potential $\tilde{v}(q) = 2\pi e^2/\kappa q$, κ being the background dielectric constant. Here and below we denote Fourier transforms by tilde. In weak magnetic fields, however, the LL mixing can not be ignored. Nevertheless, as it was shown by Aleiner and Glazman [8], the fast degrees of freedom associated with the inter-LL electron-hole excitations can be integrated out, and it is possible to introduce the effective Hamiltonian of the form (3) with renormalized potential $\tilde{v}(q) \to \tilde{u}(q) = \tilde{v}(q)/\epsilon(q)$. Physically, the bare Coulomb interaction amongst the electrons at the upper partially filled LL gets renormalized because electric fields become screened by the lower completely filled

LLs. The quantity $\epsilon(q)$ has, therefore, the meaning of the dielectric constant of the system with N completely filled LLs. In the limit $N \gg 1$, it is given by[8,9]

$$\epsilon(q) = \begin{cases} 1 + \dfrac{2}{qa_{\rm B}}, & R_c^{-1} \ll q \ll k_{\rm F} \\ 1 + \dfrac{R_c^2 q}{a_{\rm B}}, & q \ll R_c^{-1}, \end{cases} \tag{4}$$

where $a_{\rm B} = \hbar^2\kappa/me^2$ is the effective Bohr radius. This dielectric function tends to unity in the short and long wavelength limits ($q \to \infty$ and $q \to 0$), where the screening is thus not important. However, provided N is much larger than r_s^{-1}, where $r_s = \sqrt{2}/k_{\rm F}a_{\rm B}$ is the gas parameter, a wide range of intermediate length scales $a_{\rm B} < q^{-1} < R_c^2/a_{\rm B}$ exists, where $\epsilon(q)$ is large. Thus, the interaction potential $u(r)$ deviates from the usual Coulomb law at distances $a_{\rm B} < r < R_c^2/a_{\rm B}$. We will be interested in properties of not too dilute systems ($\nu_N > 1/N$) where the average interparticle distance is smaller than, say, several cyclotron radii. At such r, the interaction potential $u(r)$ can be approximated by[8]

$$u(\mathbf{r}) \simeq \pi e^2 a_{\rm B}\delta(\mathbf{r}) + E_{\rm h}, \tag{5}$$

where $E_{\rm h} = (e^2 a_{\rm B}/\kappa R_c^2)\ln(R_c/a_{\rm B})$, i.e., as the sum of a short-range and a constant terms. The constant term ($E_{\rm h}$) is, of course, irrelevant for the structure of the ground state, which means that the bare interaction potential is effectively short-range.

The knowledge of the bare potential $u(r)$ is not yet sufficient for visualizing the properties of the electron-electron interaction at the upper LL. Indeed, due to a very fast rotation in their cyclotron orbits, the electrons appear not as essentially point-like charges but rather as narrow rings of radius R_c. The dynamics of a single LL, therefore, reduces to the dynamics of the guiding centers of the cyclotron motion. Another way to see that is to notice that

$$\rho_N(\mathbf{q}) = \frac{\nu(\mathbf{q})}{2\pi l^2}\left[L_N\left(\frac{q^2l^2}{2}\right)\right]^2 e^{-q^2l^2/2}, \tag{6}$$

where $L_N(x)$ is the Laguerre polynomial, and so[13]

$$\hat{H} = \frac{1}{2}\sum_{i\neq j} u_{\rm H}(\hat{\mathbf{R}}_i - \hat{\mathbf{R}}_j), \tag{7}$$

where $u_{\rm H}(\mathbf{r}) = \int \frac{d^2\mathbf{q}}{(2\pi)^2} e^{i\mathbf{qr}}\tilde{u}(\mathbf{q})\left[L_N(q^2l^2/2)\right]^2 e^{-q^2l^2/2}$ can be identified as the direct (Hartree) interaction potential of two "rings" whose centers are separated by the distance $\mathbf{r}$. Since the bare potential is short-range, $u_{\rm H}(r)$ is

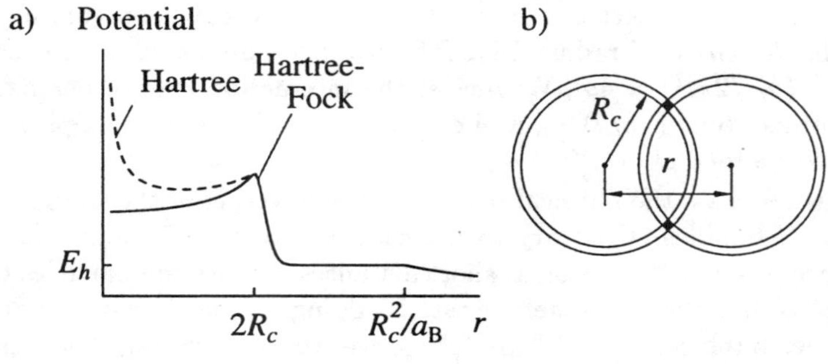

Figure 2: (a) Schematic plot of Hartree and Hartree-Fock interaction potentials of two electron orbitals ("rings") (b) Two rings overlap only if $\mathbf{r} < 2R_c$.

appreciable only when the rings overlap (Fig. 2a), i.e., when $r < 2R_c$ (Fig. 2b) and sharply drops near $r = 2R_c$ where the two rings cease to overlap.

One can also calculate the Hartree-Fock interaction potential defined by $u_{\rm HF}(r) = u_{\rm H}(r) - u_{\rm ex}(r)$, where $u_{\rm ex}(r) = \tilde{u}_{\rm H}(r/l^2)/2\pi l^2$ (see Ref. 10). This potential is also shown in Fig. 2a. As one can see, the exchange interaction preserves the sharp drop near $r = 2R_c$ but removes a strong repulsion near $r = 0$. For $r_s \sim 1$, $u_{\rm HF}(r)$ can be approximated by [10]

$$u_{\rm HF}(\mathbf{r}) \approx \frac{e^2 a_{\rm B}}{\pi^2 \kappa R_c} \frac{\Theta(2R_c - r)}{\sqrt{4R_c^2 - r^2}} + E_{\rm h}, \quad a_{\rm B} \ll r \ll \frac{R_c^2}{a_{\rm B}}, \tag{8}$$

Clearly, such an interaction potential is very different from the Coulomb law, and one may suspect that this difference may have dramatic consequences for the ground state structure of the system. Indeed, such an unusual potential leads to the CDW instability.

Let us illustrate this on the example of a box-like interaction potential

$$u(r) = u_0 \Theta(2R_c - r), \tag{9}$$

which has essentially the same properties as the actual one: the absence of a strong repulsion at short distances and the sharp drop at $r = 2R_c$. We will show that the energy of the bubble phase is lower than the energy of a uniform uncorrelated state. Indeed, since even the nearest bubbles in the bubble phase (Fig. 1a) are separated by the distance of $\simeq 3.3R_c$, electrons of different bubbles barely interact with each other. One can say, with a minor oversimplification, that a given electron interacts only with the electrons of the same bubble. The number of the latter electrons is $3\nu_N N$ [Eq. (2)], which means that the interaction energy per particle is $(u_0/2)(3\nu_N N)$. On the other hand, in the uniform liquid each electrons interact with all those whose guiding centers are within the circle of radius $2R_c$. The average number of such electrons is $\pi(2R_c)^2 \times (\nu_N/2\pi l^2) = 4\nu_N N$. Hence, the interaction energy per particle in the uniform state is $(u_0/2)4\nu_N N$, i.e., larger than in the bubble phase, which is what we claimed above.

Let us reiterate the physical reason for the lowering of the energy with the formation of bubbles. Contrary to the case of, say, the Coulomb interaction, the absence of strong repulsion at short distances enables the particles to group in compacted bubbles at no energy cost. By doing so, they avoid the particles in all the other bubbles, which ultimately proves to be advantageous energywise.

Concluding this Section, we would like to emphasize two important points. First, the screening by the lower LLs does not by itself lead to the CDW instability. The latter is due to the ring-like shape of wavefunction. The screening by the lower LLs merely gives rise to the peculiar form of the interaction potential [Eq. (8)], which makes the presence of the instability very obvious. Second, and more important point is the relation between the interaction energy per particle and $\hbar\omega_c$. In the considered case of $r_s \sim 1$, the interaction energy per particle for $\nu_N \lesssim \frac{1}{2}$ is of the order of $(e^2 a_{\rm B}/\kappa R_c^2)N \sim e^2/\kappa R_c$ [we used Eq. (8) and $N a_{\rm B} \sim R_c$ valid for $r_s \sim 1$]. This estimate remains essentially correct (up to a logarithmic factor) at smaller r_s as well [10]. It is easy to see that $e^2/\kappa R_c = (r_s/\sqrt{2})\hbar\omega_c$ and thus, the interaction energy per particle is smaller than the cyclotron gap provided $r_s \lesssim 1$. This means that the interaction does not destroy the LLs and our discussion of the electrons' distribution at the upper partially filled Landau level makes perfect sense.

3 Numerical study

The quasiclassical image of the CDW state (Fig. 1) is too crude for the purpose of accurate energy comparison of different trial states, and we have to refine our definition of the bubble state. To this end we introduce a trial wave-function of the bubble $\Psi^N\{\mathbf{r}_k\}$. We do so in two steps. First, we construct the wave function $\Psi^0\{\mathbf{r}_k\}$ of one bubble consisting of M electrons at the *lowest* LL (we adopt the cylindrical gauge):

$$\Psi^0\{\mathbf{r}_k\} = \prod_{1\le i<j\le M} (z_i - z_j) \times \exp\left(-\sum_{i=1}^{M} \frac{|z_i|^2}{4l^2}\right), \qquad (10)$$

where $z_j = x_j + iy_j$ is the complex coordinate of the j-th electron. Second, we define the wave function of a bubble at the N-th LL by raising every electron

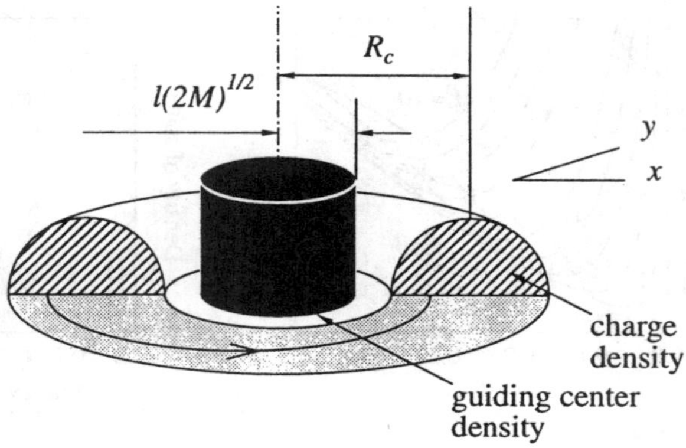

Figure 3: Distributions of the charge and guiding center densities in a single bubble. The charge density (the toroidal figure, half of which is removed for clarity) is formed by electrons moving in the cyclotron orbits centered inside of the bubble. One of such cyclotron orbits is shown by the arc.

onto this LL:

$$\Psi^N\{\mathbf{r}_k\} = \prod_{i=1}^{M} \left(\frac{\partial}{\partial z_i} - \frac{z_i^*}{4} \right)^N \Psi^0\{\mathbf{r}_k\}, \tag{11}$$

The wavefunction (11) represents the most compact arrangement of M electrons at the given LL. The distributions of the charge and the guiding center densities in the bubble are shown in Fig. 3.

Finally, the wavefunction of the bubble phase is given by the antisymmetric combination of the wavefunctions of the individual bubbles centered at the triangular lattice sites.

The cohesive energy of the bubble phase, i.e., the energy per particle with respect to that in the uniform uncorrelated liquid, for any given ν_N can be calculated numerically (see Ref. 11 for details). The result is, of course, different for different M (see Fig. 4). Therefore, one has to find $M_{\rm opt}$ corresponding to the lowest energy. The results for two particular filling factors $\nu_N = \frac{1}{3}, \frac{1}{5}$ are given in Table 1. The calculations are done for the case $r_s = \sqrt{2}$, which corresponds to the practical range of electron densities in GaAs heterostructures. One can see that the optimum number of electrons per bubble is in perfect agreement with Eq. (2).

Note that the Wigner crystal is a particular case of the bubble phase with $M = 1$. The fact that $M_{\rm opt}$ is larger than one at $\nu_N \gg 1/N$ means that the Wigner crystal is not the ground state.

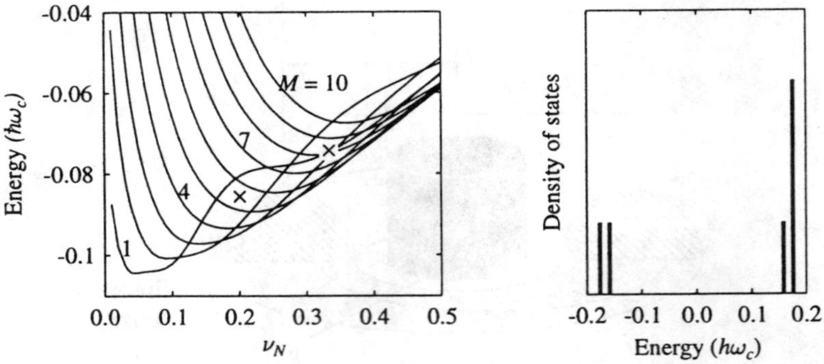

Figure 4: (Left) The cohesive energy of the CDW as a function of ν_N for different numbers of electrons in a bubble M. The calculations are made for $N = 5$ and $r_s = \sqrt{2}$. The crosses mark the Laughlin liquid energies. (Right) The tunneling density of states of the bubble phase with $M = 2$ electrons per bubble and other parameters as follows: $N = 2$, $\nu_N = \frac{1}{3}$, and $r_s = \sqrt{2}$. The zero energy corresponds to the Fermi level.

Let us now compare our state with the Laughlin liquid at $\frac{1}{3}$ and $\frac{1}{5}$ fillings, whose cohesive energy is also listed in Table 1. For this calculations we used the trial wavefunction of Laughlin liquid proposed by MacDonald and Girvin (see Ref. 11 for details). As one can see, at $N = 0$ and $N = 1$ the Laughlin liquid is more energetically preferable. At large N however the CDW wins. The transition to the bubble state occurs at $N = 2$ for $\nu_N = \frac{1}{3}$ and at $N = 3$ for $\nu_N = \frac{1}{5}$. The difference in the energies of these two states at $N = 2$ and $\nu_N = \frac{1}{5}$ is very small. It should be noted in this connection that the CDW trial state can be further improved by introducing the magnetophonon correlations[15]. Hence, the question of what exactly phase dominates at $\nu = 4\frac{1}{5}$ remains open. Evidence that the Laughlin liquid fails to be the ground state at $N > 1$ has been presented before[12,13,14]. However, no alternative candidate for the ground state has been proposed.

In Sec. 2 we argued that interaction does not destroy the LLs if r_s is smaller than unity. According to Table 1, the cohesive energy per electron at $r_s = \sqrt{2}$ is of the order of $0.1\hbar\omega_c$, i.e., significantly smaller than $\hbar\omega_c$. This suggests that the present theory applies not only to a theoretical case of $r_s \ll 1$ but also to the experimental practice where $r_s \sim 1$.

$\nu_N = 1/3$

N	M_{opt}	E^L	E^{CDW}	$\delta E/E^{CDW}$
0	1	-0.1158	-0.1037	-11.7%
1	1	-0.1520	-0.1424	-6.7%
2	2	-0.1141	-0.1188	4.0%
3	3	-0.0948	-0.1018	6.9%
4	4	-0.0822	-0.0896	8.2%
5	5	-0.0743	-0.0805	7.7%

$\nu_N = 1/5$

N	M_{opt}	E^L	E^{CDW}	$\delta E/E^{CDW}$
0	1	-0.0903	-0.0880	-2.7%
1	1	-0.1731	-0.1692	-2.3%
2	1	-0.1423	-0.1396	-1.9%
3	2	-0.1143	-0.1202	4.9%
4	2	-0.0979	-0.1050	6.8%
5	3	-0.0856	-0.0946	9.7%

Table 1: The cohesive energies of the Laughlin liquid E^L and the CDW E^{CDW} for different N. The right column shows their relative difference: $(E^L - E^{CDW})/E^{CDW}$. All the energies are given in the units of $\hbar\omega_c$. M_{opt} is the optimum number of electrons per bubble. The gas parameter r_s is equal to $\sqrt{2}$, which corresponds to the electron density of $1.6 \cdot 10^{11}\text{cm}^{-2}$ in GaAs-GaAlAs heterostructures.

4 Experimental consequences

One of the important consequences of the proposed ground state is the absence of the $\frac{1}{3}$ fractional quantum Hall plateaus at $\nu > 4$, which is in agreement with the existing experiment.

Another consequence of the CDW state is a peculiar shape of the one-particle density of states (see Fig. 4 right). It contains several peaks corresponding to different energy "shells" inside the bubble and also a pronounced gap centered at the Fermi energy. The existence of the latter was predicted earlier in Ref. 8. Our calculations for the gap width seems to be in a good agreement with the width of the pseudogap observed in the tunneling conductance of the double-well system [16].

The CDW at the upper LL strongly affects the low-temperature transport properties of the 2D gas. Due to the pinning of the CDW by disorder, the dissipative conductivity σ_{xx} has narrow peaks at half-integer fillings even in high-mobility heterostructures [10]. At higher temperatures the depinning of the CDW becomes possible. The effect of this phenomenon on the transport properties remains to be studied. At the moment, we can only estimate [10] the temperature T_c at which the CDW melts into a perfectly uniform electron liquid, $k_B T_c \approx 0.03\hbar\omega_c$ (for $\nu_N = \frac{1}{2}$, $r_s \sim 1$, and $N > 2$). Above T_c, the pinning effects disappear completely; the peaks in σ_{xx} become wide; the plateaus in σ_{xy} become narrow.

We appreciate useful discussions with I. L. Aleiner and L. I. Glazman. This work is supported by NSF under Grant DMR-9321417.

References

1. *The Quantum Hall Effect,* edited by R. E. Prange and S. M. Girvin (Springer-Verlag, New York, 1990).
2. H. Fukuyama, P. M. Platzman, P. W. Anderson, *Phys. Rev.* B **19**, 5211 (1979).
3. D. Yoshioka and H. Fukuyama, J. Phys. Soc. Jpn. **47**, 394 (1979); D. Yoshioka and P. A. Lee, *Phys. Rev.* B **27**, 4986 (1983).
4. R. B. Laughlin, *Phys. Rev. Lett.* **50**, 1395 (1983).
5. S. L. Sondhi, A. Karlhede, S. A. Kivelson, and E. H. Rezayi, *Phys. Rev.* B **47**, 16 419 (1993).
6. H. A. Fertig, L. Brey, R. Coté, and A. H. MacDonald, *Phys. Rev.* B **47**, 16 419 (1993).
7. X.-G. Wu and S. L. Sondhi, *Phys. Rev.* B **51**, 14 725 (1995); A. Schmeller, J. P. Eisenstein, L. N. Pfeiffer, and K. W. West, *Phys. Rev. Lett.* **75**, 4290 (1995).
8. I. L. Aleiner and L. I. Glazman, *Phys. Rev.* B **52**, 11 296 (1995).
9. I. V. Kukushkin, S. V. Meshkov, and V. B. Timofeev, Usp. Fiz. Nauk **155**, 219 (1988) [Sov. Phys. Usp. **31**, 511 (1988)].
10. A. A. Koulakov, M. M. Fogler, and B. I. Shklovskii, *Phys. Rev. Lett.* **76**, 499 (1996); M. M. Fogler, A. A. Koulakov, and B. I. Shklovskii, *Phys. Rev.* B **54** (1996).
11. M. M. Fogler and A. A. Koulakov, Report No. cond-mat/9602128.
12. A. H. MacDonald, *Phys. Rev.* B **30**, 3550 (1984); A. H. MacDonald and S. M. Girvin, *Phys. Rev.* B **33**, 4009 (1986).
13. F. D. M. Haldane, in Ref. 1.
14. N. d'Ambrumenil and A. M. Reynolds, J. Phys. C **21**, 119 (1988); L. Belkhir and J. K. Jain, Solid State Commun. **94**, 107 (1995).
15. P. K. Lam and S. M. Girvin, *Phys. Rev.* B **30**, 473 (1984).
16. N. Turner, J. T. Nicholls, K. M. Brown, E. H. Linfield, M. Pepper, D. A. Ritchie, and G. A. C. Jones, Report No. cond-mat/9503040.

LIGHT SCATTERING BY LOW-ENERGY COLLECTIVE EXCITATIONS OF QUANTUM HALL STATES

A. PINCZUK, B.S. DENNIS, A.S. PLAUT,* L.N. PFEIFFER, and K.W. WEST
Bell Laboratories, Lucent Technologies, Murray Hill, New Jersey 07974-0636, USA

Resonant inelastic light scattering is emerging as a major experimental method to study low-energy collective excitations of quantum Hall states. We present a short review of light scattering experiments that study gap excitations of the FQHE state at ν=1/3; and of soft spin-density excitations of electrons in double quantum wells at even values of ν.

1 Introduction

Intensive experimental and theoretical research of low-dimensional electron systems in large magnetic fields continues to uncover intriguing behaviors arising from strong correlations.[1-3] The fractional quantum Hall effect(FQHE) is an archetype of the quantum liquids that occur in low-dimensional electron systems.[4-5] In the FQHE the many-body electron states are incompressible and an energy gap separates the ground state from its excited states. The gaps of the FQHE are manifestations of the strong interactions that lead to condensation of the 2D system into quantum fluids.[5-8]

Gaps of quantum Hall states are represented by dispersive collective modes constructed with neutral pair-states in which a particle (or quasiparticle) is promoted to an excited state leaving behind a hole (or quasihole). For excitations with wave vector $\vec{q}$ there is a displacement between the centers of the cyclotron orbits of the two 'single-particle' states. This dipole length is $x_o=q\ell_o^2$, where $\ell_o=(\hbar c/eB)^{1/2}$ is the magnetic length.[6] The characteristic energies and wave vector dispersions of collective gap modes reveal the remarkable physics that arises from electron-electron interactions in reduced dimensions. Measurements of thermal activation of magnetotransport yield gap energies of the FQHE.[9] These gaps, as well as those determined from acceptor luminescence,[10] are interpreted as energies required to create widely separated quasiparticle-quasihole pairs that have $q \rightarrow \infty$.

Collective gap excitations of the FQHE at ν=1/3 were observed by resonant inelastic light scattering.[11] The success in the fractional quantum Hall regime suggests that these methods, introduced in 1978-79 for studies of 2D electron systems in semiconductors,[12] will continue to play important roles in the research of electron systems of artificial semiconductor quantum structures.

(*) Permanent address: Department of Physics, University of Exeter, Exeter EX4 4QL, UK.

In this short review we highlight recent resonant inelastic light scattering experiments that enable measurements of low-energy collective excitations in the quantum Hall regimes.[13-15]

The light scattering experiments at ν=1/3 offer significant evidence that the state supports the dispersive collective gap excitations predicted by theories of the quantum fluid of the FQHE. In these studies we have observed q=0 gap excitations, and found evidence of the magnetoroton minimum in the mode dispersion. We also consider the recent discovery of *soft*, low-energy, spin-density tunneling excitations in coupled double quantum wells. This light scattering work indicates that at even integer values of ν the electron bilayers have unstable long wavelength spin-density collective modes.[14,15]

In the experimental work we employ a dilution $^3He/^4He$ cryostat that is inserted in the cold bore of a superconductive magnet. *The system has silica windows for optical access.* A conventional backscattering geometry allows only small in-plane components of the scattering wave vector. Incident powers are kept at levels of 10^{-4} to $10^{-3}Wcm^{-2}$. *This experimental configuration enables low-temperature measurements of inelastic light scattering at energy shifts that are can be smaller than 0.1 meV.*

In the Section that follows we review light scattering measurements of gap excitations in the FQHE state with ν=1/3.[11,13] In Sec. 3 we briefly consider *soft* spin-density excitations of coupled electron double-layers at even values of ν.[14,15]

2 Gap Excitations of the FQHE State at ν=1/3

The light scattering observations of collective gap excitations of the FQHE state at ν=1/3 are surprising because for these *intra*-Landau-level excitations the dynamical structure factors, the functions that describe conventional inelastic scattering cross-sections of the electron gas, vanish as q^4 for the small wave vectors in optical experiments.[5,8] The results are also intriguing because the character of q=0 gap excitations of the FQHE is still a subject of conjecture.[8,16-19] In fact, the initial light scattering experiments in the regime of the FQHE were stimulated by ideas introduced to understand the q=0 collective gap mode.[11]

In seminal work,[8] Girvin et al. speculated that two gap modes, each near the magnetoroton minimum at wave vectors close to $1/\ell_o$, may pair to produce a two-roton state with q=0. The long wavelength excitation was also considered within the framework of Landau-Guinzburg theory.[16,17] Working within this framework, Lee and Zhang proposed that the q=0 collective gap excitations consists of two electric dipoles, each a gap mode at wave vector near the magnetoroton minimum, arranged in a configuration that has a quadrupole moment but no net dipole moment.[18]

These conjectures suggested that inelastic light scattering, which as a two-photon process is sensitive to collective excitations that lack an electric dipole moment, could be the method to observe long wavelength gap excitations of the FQHE.[11]

Light scattering experiments that probe the states of the dilute 2D electron gas in large magnetic fields rely on the enhanced sensitivity that is achieved with resonant excitation of the spectra.[20] The intermediate optical transitions that offer resonant enhancements of matrix elements and cross-sections are excitonic states constructed from valence and conduction Landau levels.[21] Resonant inelastic light scattering is thus described within a time-dependent perturbation theory that invokes *virtual* excitonic transitions in the intermediate steps. Here, light scattering takes place because electron-electron interactions introduce processes in which the virtual excitons change state with emission of *real* collective excitations of the 2D electron system.[22]

As an example, consider the scenario in which the long wavelength gap mode of the FQHE is a two-roton state. Within perturbation theory, resonant inelastic light scattering by two-roton states can be thought of as *four-step*(fourth-order) processes in which the exciton-roton interaction acts twice. In such processes exciton-roton interactions could arise from the substantial electric dipole associated with excitations of wave vectors $\vec{k}$ and $-\vec{k}$ near the magnetoroton minimum.

The light scattering matrix elements constructed with virtual intermediate exciton states have characteristic *incoming* and *outgoing* resonance channels.[22,23] The incoming resonance occurs when the energy of the incident(laser) photon energy ω_L overlaps the energy of the intermediate exciton state. In the outgoing resonance it is the scattered photon energy ω_S that is equal to the intermediate exciton energy.

Following the initial observations,[11] resonant inelastic light scattering by q=0 gap excitations of the FQHE state with $\nu=1/3$ was measured in several 250Å GaAs quantum well samples.[13] These structures have very high electron mobilities, in excess of $4\text{x}10^6\text{cm}^2/\text{Vsec}$, and carrier densities in the range $0.7 \leq n \leq 1.3\text{x}10^{11}\text{cm}^{-2}$. Figures 1(a) and 1(c) show results measured at a field of 15.5Tesla. The spectra show a sharp peak(FWHM)≃0.07meV) at 1.38mev that overlaps a broader structure. In Fig. 1(b) we show results typical of filling factors which are not those of the FQHE state. These spectra display a broad structure that acts like luminescence when the incident photon energy is changed.

The results in Fig. 1(c) show that the sharp peak present only at filling factors very close to 1/3 has the marked temperature dependence that is characterisitic of the FQHE. We interpret it as the q=0 gap excitation of the FQHE state. Its energy, 1.38meV, is in excellent agreement with recent numerical evaluations of systems of small numbers of electrons that incorporate the effects of the finite width of the 2D layer.[19,24]

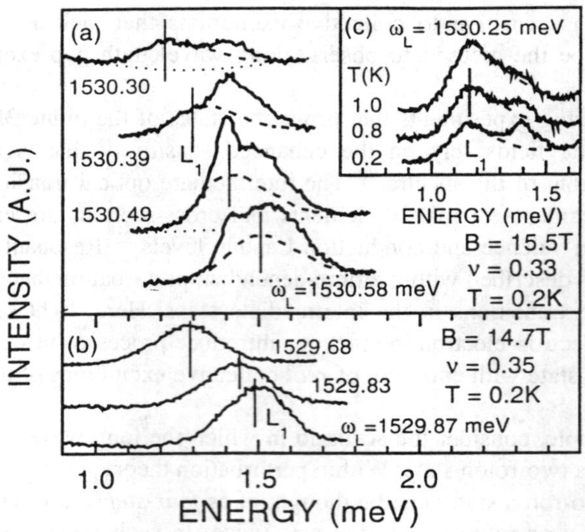

Figure 1: resonant inelastic light scattering spectra in the regime of the FQHE at ν=1/3. The sample is a 250Å GaAs quantum well of high electron mobility. Incident photon energies ω_L are given in meV. The dotted lines are estimated backgrounds. The dashed lines are estimated non-equilibrium luminescence L_1. The vertical lines indicate the peak positions of L_1. (a) spectra measured at the field of ν=0.33, showing the sharp peak at 1.38meV that is interpreted as the q=0 collective gap excitation of the FQHE state. (b) spectra measured at ν=0.35. (c) temperature dependence of spectra at ν=0.33.

The experiments take advantage of the sharp and strong excitonic resonant enhancements that are revealed in the results seen in Fig. 1. The figure also displays the impact of luminescence on these experiments. The broader structure labeled L_1 is at least in part due to excitonic recombination at 0.7meV above the fundamental emission of the 250Å GaAs quantum well. *The large enhancement of the light scattering signal at 1.38meV when it overlaps the center of L_1 is the signature of an outgoing excitonic resonance in the scattering cross-section.*

In Fig. 1(a) the intensities of non-equilibrium luminescence L_1 are comparable to those of the resonant inelastic light scattering. In fact, close examination of the spectra indicates that the broader structure may also incorporate a light scattering contribution. This is apparent in the spectra measured with ω_L close to the outgoing resonance(1530.49 and 1530.39meV). These spectra suggest that light scattering spectra of q=0 gap excitations of the FQHE state may consist of a sharp peak followed by a continuum at higher energy.

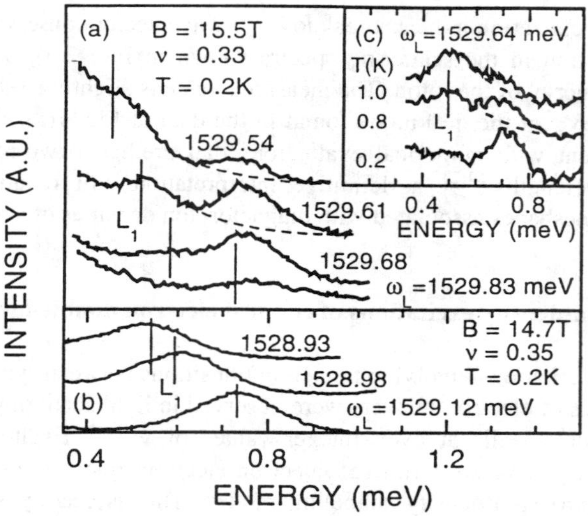

Figure 2: resonant inelastic light scatteering spectra in the regime of the FQHE at ν=1/3. (a) spectra at the field of ν=0.33 showing a narrow band at 0.74meV that is interpreted as the peak in the density of states at the magnetoroton minimum in the dispersion of gap excitations of the FQHE state. (b) spectra measured at ν=0.35. (c) temperature dependence of spectra at ν=0.33.

The existence of such continua of excitations at higher energies is indeed predicted by the numerical evaluations of systems with small numbers of electrons.[7] In recent work Song He and Platzman re-examined the structure of collective gap excitations of the state with ν=1/3.[19,24] This work, based on numerical evaluations of systems of up to ten electrons, indicates a large two-roton component in the long wavelength gap excitations. These results also uncover numerical evidence that the lowest q=0 mode consists of a weakly-bound two-roton state at energy below that of the two-roton continuum.

Figure 2 shows the results obtained in the energy range of the magnetoroton minimum of dispersive gap excitations. The data at ν=0.33 do show a narrow(FWHM≃0.15meV) inelastic scattering band at 0.74meV. This feature, like the one in Fig. 1, has the magnetic field and temperature dependence of the FQHE. We interpret this band as the peak in the density of states that occurs at the magnetoroton minimum. The wave vector of this critical point, $q\sim 1/\ell_o$, is much larger than the wave vectors of light. Inelastic light scattering at the roton is activated by residual-disorder in quantum Hall states. Here, the loss of translational invariance due to localization induces breakdown of wave vector conservation.[25,26]

The peak position in the scattering spectra shown in Fig. 2(a), 0.74meV, is interpreted as the energy of the roton. The measured value is slightly larger than half the energy, 1.38meV, of the q=0 mode found in the data of Fig. 1(a). While these results are consistent with numerical evaluations that predict a two-roton bound state at long wavelengths,[19,24] a definitive interpretation will require a better understanding of lineshapes associated with magnetoroton densities of states.

3 Low-energy spin-density excitations of coupled electron double-layers

Soft, low-energy, excitations involving tunneling transitions between symmetric and antisymmetric states of electron bilayers were observed in light scattering spectra of GaAs coupled double wells at even integer values of ν.[14,15] Excitonic vertex-corrections due to exchange terms of electron-electron interactions drive the softening of the q=0 spin-density tunneling modes. This is readily seen in the expression for the mode energy[27]

$$\omega_{SD}^2 = \Delta_{SAS}^2 - 2(n_s - n_{AS})\beta\Delta_{SAS} \tag{1}$$

where n_s and n_{AS} are the populations of the symmetric and antisymmetric states of the double-layers, Δ_{SAS} is the symmetric-antisymmetric tunneling gap and the factor β parametrizes the strength of vertex-corrections. The second term on the right-side of Eq. 1 represents the vertex-corrections. When this term is equal or larger than Δ_{SAS}^2 the effective tunneling gap collapses. Such unstable collective mode may trigger a quantum phase transition to a highly correlated state.[28] While the quantum phase transition proposed to exist in the double-layers at B=0 is not observed,[29,30] our recent work in the quantum Hall regime uncovered a tunneling spin-density mode which has a remarkably low energy of 0.05meV at filling ν=2.[14,15] Here, a major impact of the perpendicular magnetic field on the strengths of vertex-corrections is the enhancement of the phase space, (n_S - n_{AS}), available for tunneling transitions when symmetric and antisymmetric states are occupied. These experiments stimulated theoretical research that actually interprets the presence of the very low-energy spin-excitation as evidence of a phase transition to a new quantum Hall state with antiferromagnetic spin correlations.[31]

4 Conclusion

The resonant inelastic light scattering work reviewed here shows that the method offers unique insights into low-energy excitations in the quantum Hall regimes. These experiments are difficult and require new approaches that enhance the resonant light scattering signal. The method will continue to give access to elementary excitations of electron states of semiconductor quantum structures created by advanced fabrication technologies. Such light scattering experiments could uncover remarkable manifestations of electron interactions in low-dimensional systems.

Acknowledgements

We are grateful to E. Burstein, J.P. Eisenstein, Song He and P.M. Platzman for helpful discussions. We are also grateful to the authors of Ref. 31 for letting us know their results prior to publication. ASP gratefully acknowledges financial support from the Leverhulme Trust, UK.

References

1. *The Quantum Hall Effect*, 2nd Ed., eds. R.E. Prange and S.M. Girvin, (Springer, New York, 1990).
2. T. Chakraborty and P. Pietilainen, *The Fractional Quantum Hall Effect: Properties of an incompressible quantum fluid*, (Springer-Verlag, Berlin, 1988).
3. *Perspectives in Quantum Hall Effects: Novel Quantum Liquids in Low Dimensional Semiconductor Structures*, eds. S. Das Sarma and A. Pinczuk, (J. Wiley & Sons, New York, 1996).
4. D.C. Tsui, H.L. Stormer and A.C. Gossard, *Phys. Rev. Lett.* **48**, 1559(1982).
5. R.B. Laughlin, in Ref. 1, p. 233.
6. C. Kallin and B.I. Halperin, *Phys. Rev.*B **30**, 5655(1984).
7. F.D.M. Haldane and E.H. Rezayi, *Phys. Rev. Lett.* **54**, 237(1985).
8. S.M. Girvin, A.H. MacDonald and P.M. Platzman, *Phys. Rev. Lett.* **54**, 581(1985).
9. H.L. Stormer and D.C. Tsui, in Ref. 3.
10. I.V. Kukushkin, N.J. Pulsford, K. von Klitzing, K. Ploog, R.N. Haug, S. Koch and V.B. Timofeev, *Europhys. Lett.* **18**, 63(1992).
11. A. Pinczuk, B.S. Dennis, L.N. Pfeiffer and K.W. West, *Phys. Rev. Lett.* **70**, 3983(1993).
12. E. Burstein, A. Pinczuk, and S. Buchner, *Physics of Semiconductors 1978*, ed. B.L.H. Wilson, (The Institute of Physics, London, 1979), p. 1231; G

Abstreiter and K. Ploog, *Phys. Rev. Lett.* **42**, 1308(1979); A. Pinczuk, H.L. Stormer, R. Dingle, J. M. Worlock, W. Wiegmann, A.C. Gossard, *Solid State Commun.* **32**, 1001(1979).

13. A. Pinczuk, B.S. Dennis, L.N. Pfeiffer and K.W. West, unpublished.
14. A. Pinczuk, B.S. Dennis, A.S. Plaut, J.P. Eisenstein, L.N. Pfeiffer and K.W. West, *Bull Am. Phys. Soc.* **41**, 482(1996).
15. A. Pinczuk, B.S. Dennis, A.S. Plaut, L.N. Pfeiffer and K.W. West, work in progress.
16. S.C. Zhang, T.H. Hansson and S. Kivelson, *Phys. Rev. Lett.* **62**, 82(1989).
17. N. Read, *Phys. Rev. Lett.* **62**, 86(1989).
18. D.-H. Lee and S.-C. Zhang, *Phys. Rev. Lett.* **66**, 1220(1991).
19. Song He and P.M. Platzman, Procs. of the 11th Int. Conf. on Electr. Props. of 2D Systems, Surface Science, 1996, in press.
20. A. Pinczuk, B.S. Dennis, L.N. Pfeiffer and K.W. West, *Semicond. Sci. Technol.* **9**, 1865(1994).
21. B.S. Wang, J.L. Birman and Z.B. Su, *Phys. Rev. Lett.* **68** 1605(1992); A.H. MacDonald, E.H. Rezayi and D. Keller, *ibid fB 68*, 1939(1992); X.M. Chen and J.J. Quinn, *ibid* **70**, 2129(1993); E.I. Rashba and M.E. Portnoi, *ibid* **70**, 3315(1993).
22. G. Danan, A. Pinczuk, J.P. Valladares, L.N. Pfeiffer and K.W. West, *Phys. Rev.* B**39**, 5512(1989).
23. A.K. Ganguly and J.L. Birman, *Phys. Rev.* **162**, 806(1967).
24. Song He, private communication.
25. A. Pinczuk, J.P. Valladares, D. Heiman, A.C. Gossard, J.H. English, C.W. Tu, L. Pfeiffer and K. West, *Phys. Rev. Lett.* **61**, 2701(1988); A. Pinczuk, B.S. Dennis, D. Heiman, C. Kallin, L. Brey, C. Tejedor, S. Schmitt-Rink, L.N. Pfeiffer and K.W. West, *ibid* **68**, 3623(1992).
26. I.K. Marmorkos and S. Das Sarma, *Phys. Rev.* B**45**, 13396(1992).
27. R. Decca, A. Pinczuk, S. Das Sarma, B.S. Dennis, L.N. Pfeiffer and K.W. West, *Phys. Rev. Lett.* **72**, 1506(1994).
28. A.H. MacDonald, P.M. Platzman and G.S. Boebinger, *Phys. Rev. Lett.* **65**, 775(1990); L. Brey. *ibid* **65**, 903(1990); H.A. Fertig, *Phys. Rev.* B**40**, 1087(1989); X.M. Chen and J.J. Quinn, *ibid* **45**, 11054(1992).
29. S. Das Sarma and P.I. Tamborenea, *Phys. Rev. Lett.* **73**, 1971(1994).
30. A.S. Plaut, A. Pinczuk, P.I. Tamborenea, B.S. Dennis, L.N. Pfeifer and K.W. West, submitted for publication.
31. Lian Zheng, R.J. Radtke and S. Das Sarma, submitted for publication.

AN INVESTIGATION OF THE COLLECTIVE EXCITATIONS OF THE FQHE USING A NEW BOSON FORMALISM

M. BROWNLIE, K.A. BENEDICT
Department of Physics, University of Nottingham,
University Park, NG7 2RD, United Kingdom

We describe a new formalism, which we believe will be useful for studying the collective excitations of the FQHE system, in which the space of collective excitations is transformed into a subspace of a boson Fock space. We show that the new formalism gives an integral transform method for calculating the three point correlation function (as well as the other higher order correlation functions) for the FQHE ground state. We investigate an approximation scheme, designed to make the transform computationally tractable, demonstrating its use in performing calculations.

1 Introduction:An $a^\dagger$,a Representation for Magnetorotons

In the Magnetoroton theory of Girvin *et al* the first excited states of the FQH system are given by $\bar{\rho}_k^\dagger \psi$ (where ψ is the ground state). The $\bar{\rho}_k$'s, unlike the normal density operators, do not commute with each other but rather they obey the commutation relations

$$[\bar{\rho}_k^\dagger, \bar{\rho}_q^\dagger] = \phi(k,q).\bar{\rho}_{k+q}^\dagger \tag{1}$$

$$\phi(k,q) = 2i.e^{\frac{1}{2}k.q}.\sin(\frac{k \times q}{2}) = e^{\frac{k^* q}{2}} - e^{\frac{kq^*}{2}} \tag{2}$$

and $k \times q = (k_x q_y - k_y q_x)$ and it is obvious from the context whether the two dimensional vector or complex number representations are being used.

The average energy of the magnetoroton state, $\Delta(k)$, can be readily shown, by elementary manipulations of these commutation relations, to be a functional of the static structure factor of the ground state of the system:

$$\Delta(k) = \frac{1}{4.\bar{S}(k)} \sum_q v_q.(e^{\frac{q^* k}{2}} - e^{\frac{k^* q}{2}}).f(k,q) \tag{3}$$

$$f(k,q) = \bar{S}(q).e^{-\frac{|k|^2}{2}}.(e^{\frac{-k^* q}{2}} - e^{\frac{-q^* k}{2}}) + \bar{S}(k+q).(e^{\frac{k^* q}{2}} - e^{\frac{q^* k}{2}}) \tag{4}$$

We begin by trying to build a creation/annhilation operator representation for magnetorotons. We investigate such a picture by expanding the projected density operator as a series in the magnetoroton raising and lowering operators

$$\bar{\rho}_k^\dagger = A(k)(a_k^\dagger + a_{-k}) + \sum_h B(k,h).a_{h+k}^\dagger a_h + \sum_h C(k,h)(a_{k-h}^\dagger a_h^\dagger + a_{h-k}a_{-h}) + ... \tag{5}$$

where the coefficients in the expansion have to fit the projected density commutation relations and the adjoint condition $\bar{\rho}_k^\dagger = \bar{\rho}_{-k}$. We take the vacuum state, $|0\rangle$, as representing the ground state of the system.

Now, suppose that the only pure creation terms in the expansion are the single creation terms (*i.e.* the $C(k,h)$ coefficients are zero), which we might expect to be approximately the case in the region where the magnetoroton is a good approximation to the first excited state. Then we would have that in this picture the magnetoroton state would be given by $\bar{\rho}_k^\dagger \psi = A(k).a_k^\dagger|0\rangle$. Now, knowing the Hamiltonian of the system to be

$$\bar{H} = \frac{1}{2\Omega}\sum_q v_q.\bar{\rho}_q^\dagger \bar{\rho}_q \tag{6}$$

leads to an immediate inconsistency. If we calculate the average energy of the magnetoroton state in the $a^\dagger$,a picture we find that

$$\Delta_a(k) = n_e.v_k.\bar{S}(k) + \ self \ \ interaction \ \ terms \tag{7}$$

where n_e = electron density.

This is not a surprising result; the energy of an excitation of wavevector k depending on the Fourier component of the interaction potential at k and the projected static structure factor at k. The problem is, though, that from equation 3 we can see that the actual energy of the magnetoroton state, $\Delta(k)$, does not contain a term in v_k or $\bar{S}(k)$ although it does contain evaluations of these functions at almost all other values of wavevector - it is as if $\Delta(k)$ contains only interaction terms. The reason why this $a^\dagger$,a picture gives such an inconsistent result is very simple; the vacuum state is not an eigenstate of the Hamiltonian. Indeed we see, even from inspection of the first order terms,

$$\bar{\rho}_k^\dagger \bar{\rho}_k = |A(k)|^2.(a_k^\dagger a_k + a_k^\dagger a_{-k}^\dagger + a_{-k}a_{-k}^\dagger + a_{-k}a_k) + ... \tag{8}$$

that the vacuum state cannot even be used as a reasonable approximation to an eigenstate of this Hamiltonian. Now, we see that to interpret the ground state as a vacuum state is inconsistent with interpreting the action of the projected density operator as behaving like a creation operator. Also, we must ask ourselves why it is that the energy of an excitation at wavevector k does not contain any term in $v(k)$ or $\bar{S}(k)$ but does contain an infinity of terms in other values of wavevector.

2 An Exact Solution

This time we will look for a representation in the boson Fock space which obeys the conditions (1) It is a natural transformation from the wavefunction space into the Fock space - *i.e.* $T(\bar{\rho}_k^\dagger)$ has a sensibly behaved expansion in terms of the $a^\dagger$,a operators. (2) The vacuum state in the Fock space is an eigenstate of the transformed Hamiltonian. Under these conditions and using the fact that the transformed operators must obey the commutation relations (equation 1) we find the exact solution:

$$\bar{\rho}_k^\dagger = \sum_h B(k,h).a_{h+k}^\dagger a_h \tag{9}$$

$$B(k,h) = 2i.e^{-\frac{1}{4}|k|^2}.\sin(\frac{k \times h}{2}) \tag{10}$$

The exact physical nature of these $a^\dagger, a$'s is still under study. We will refer to these generally underdefined magnetic boson operators as gumbons. We can use some simplifying approximations in order to make calculations feasible using this representation. Remember that the Hamiltonian is given by

$$\bar{H} = \frac{1}{2\Omega}\sum_k v_k.\bar{\rho}_k^\dagger \bar{\rho}_k \tag{11}$$

$$\bar{\rho}_k^\dagger \bar{\rho}_k = \sum_h |B(k,h)|^2.(1+n_{h+k}).n_h + \sum_{\substack{h,s \\ h\neq s}} B(k,h).B^*(k,s).a_{h+k}^\dagger a_s^\dagger a_h a_{s+k} \tag{12}$$

We see that the Hamiltonian contains a number operator term (a highly non-linear one) which is always positive and another term which oscillates between positive and negative. This suggests that a zeroth order approximation for the ground state may be found via a random phase argument where we suppose that for the projected static structure factor the oscillations of the second term approximately cancel out to zero leaving the only contribution to come from the always positive number term. This is like considering the oscillating term as a perturbation to the (dominant) number term and it is easy to see that the first order energy correction to a number state caused by the oscillating term is zero, suggesting that the RPA might be quite useful. Using the R.P.A. in this way we can obtain an integral equation relating the projected static structure factor to the average gumbon distribution:

$$2\pi\nu\bar{S}(k) = \int d^2s.|B(k,s)|^2.n(s)[1+n(s+k)] \tag{13}$$

We also obtain the result (independent of the R.P.A.) that the total gumbon number n_g is given by (where ν is the filling factor)

$$\int d^2s.n(s) = n_g = \pi\nu(1-\nu) \tag{14}$$

and so obeys particle hole symmetry. In this picture (except for $\nu = 0$ and $\nu = 1$) the ground state is not a vacuum state but rather it is an interacting sea of bosons and so the $\bar{\rho}_k$ acts to redistribute a collection of bosons that are already there; this explains the absence of the v_k term in the expression for the magnetoroton energy (equation 3).

If we have a known expression for $\bar{S}(k)$ (*i.e.* the calculation for the Laughlin states performed in [1]) then we can find $n(s)$ via a readily derived iterative formula for inverting equation 13. Once $n(s)$ is known it is quite simple to calculate any of the correlation functions (within the R.P.A.) - they will just be integrals involving $n(s)$. This method automatically respects particle hole symmetry, the commutation relations, and certain sum rules. We would not expect the R.P.A. to respect phase properties but we might expect that it may be a reasonable approximation for energy type calculations - *i.e.* for calculating the Hermitian correlation functions or the imaginary parts of the odd correlation functions (such as the three point one).

Conclusion

We have described a new representation for the projected density operators in the lowest Landau level. In this new picture the ground state is represented as a sea of interacting bosons (the number of them present being determined by ν) and the effect of the projected density operator, rather than creating any particles, is to redistribute particles that are already there. We feel that this raises questions as to the interpretation of multiple magnetoroton states - *i.e.* the interpretation of states like $\bar{\rho}_k^\dagger \bar{\rho}_q^\dagger \psi$ as being two magnetoroton particles. We also suggested a simple approximation scheme which allows some elementary calculations of interesting quantities to be made.

References

1. S.M. Girvin *et al.*, *Phys. Rev.* B **33**, 2481 (1986).

A Theory of Phonon Spectroscopy of the Fractional Quantum Hall State

K.A. Benedict
Department of Physics, University of Nottingham, Nottingham, NG7 2RD U.K.

A simplified theoretical model is discussed for the processes occuring in phonon spectroscopy experiments in the fractional quantum Hall regime. There is found to be a regime of pulse duration in which the response of the system is determined by the magnetoroton gap.

1 Introduction

The fractional quantum Hall effect is the most obvious experimental consequence of the incompressible quantum liquid state that occurs in two-dimensional electron systems close to certain commensurate magnetic field strengths. There is currently considerable interest in experimental techniques which might shed light on the collective excitations of this novel phase of 2d matter. One method is to use the bulk acoustic phonons of the semiconductor crystal in which the 2d electron system is embedded. These should couple directly to the collective modes at a well defined, finite, q.

2 Collective Modes of the FQHE

The first theoretical work on the collective modes of the fqh state was that of Girvin, MacDonald and Platzman [1] who used a single mode approximation in the spirit of Feynman's calculation of the collective modes of superfluid helium. By projecting all quantities onto the lowest Landau level they found a low lying branch of collective modes which they dubbed *magnetorotons.* The typical energy of these modes is a small fraction of the Coulomb energy, $e^2/4\pi\varepsilon l_c$ ($l_c = \sqrt{\hbar/eB}$ is the cyclotron length). The magnetoroton energy, $\Delta(q)$, never goes to zero (a consequence of incompressibility) but has a minimum value, Δ^*, at $q^* \approx 1.3/l_c$ which is roughly half the value at $q = 0$. At large q the magnetoroton acquires a single particle character and can be thought of as a bound state (quasi-exciton) composed of oppositely charged Laughlin quasiparticles [2]. Numerical studies on finite systems have confirmed the existence of a well defined collective mode above which lies a multi-particle continuum.

In the single mode approximation a wavefunction for a single quantum excitation of the low lying collective mode with wave vector $\mathbf{q}$ is written as

$$|\mathbf{q}\rangle = \frac{1}{\sqrt{\overline{s}(q)}} \mathcal{P} \widehat{\rho}_{\mathbf{q}} |\psi\rangle$$

where $|\psi\rangle$ is the ground state wave function (well approximated by Laughlin's wave function [2]), $\widehat{\rho}_{\mathbf{q}}$ is the electron density operator, $\mathcal{P}$ is a projection operator onto the lowest Landau level and $\overline{s(q)}$ is the lowest Landau level projected static structure factor. The projected structure factor is large for $q \approx q^*$ but small elsewhere. The gap measured in activated transport experiments is the energy required to produce unbound (charged) Laughlin quasiparticles, i.e.

$$\Delta_{tr} = \lim_{q \to \infty} \Delta(q) > \Delta^*$$

3 Electron Phonon Interactions

The coupling of the bulk acoustic phonons to the electronic system is via the electronic density. Hence the obvious channel by which energy will be transferred from the bulk phonons to the electronic system is via the creation of magnetorotons with wave vectors close to q^* where the coupling is strongest and the the gap smallest. An additional channel opens when there are already some magnetorotons present, either because of a non-zero initial electron temperature or because they have been created by the previous process. The magnetorotons can, themselves, absorb and emit phonons provided momentum and energy can be conserved with a suitable perpendicular component of the phonon momentum. The matrix elements for this type of process are not known in detail at present but some of their general characteristics have been determined [5], in particular it can be seen that the emission of large q phonons is exponentially suppressed.

4 Outline of the Experiments

In the phonon experiments carried out by Mellor et al [4], a pulse non-equilibrium acoustic phonons of duration, t_p, characterized by an effective temperature, T_h, is injected into the semiconductor substrate. A constant bias current is driven through the 2des and the two-terminal resistance recorded as a function of time. When the "cloud" of phonons reaches the 2d layer we can expect some absorption to occur leaving the 2d system in an excited state which then decays after the phonon "cloud" has passed.

The characteristic temperature of the phonon cloud is chosen to be such that

$$k_B T_h \ll \Delta^*$$

so that only very few of the phonons have sufficient energy to create a magnetoroton.

5 Theoretical Model

We can identify two extreme (and uninteresting) regimes:

1. if the pulse is of very short duration then the only effect of the phonon pulse will be to generate a small number of magnetorotons with $q \approx q^*$ which, being electrically neutral, do not result in any change in the two-terminal resistance;

2. if the pulse if of very long duration then the electronic system will come to equilibrium with the phonon cloud and the resistance will simply be given by

$$R \sim e^{-\Delta_{tr}/k_B T_h}$$

i.e. the 2d system will be heated to T_h and the experiment will simply record the activated resistance at that temperature.

The interesting case is when the pulse is of intermediate duration. If we suppose that the interaction is sufficiently weak that only a small number of excitations are created then the dominant mechanism by which a change in resistance will occur will be the absorption of low energy phonons by a magnetoroton previously created at q^*. In order to understand what kind of behavior should be expected a simple model has been examined. Detailed numerical work on more realistic models is in progress.

Each collective mode is assigned an occupation number $\nu(q,t)$ and a rate equation is written down for this using some simple model forms for the transition rates for various processes. The results of this model are as follows (details will be presented elsewhere). While the phonon cloud is passing through the 2d layer the evolution of the occupation number for modes close to q^* is dominated by the direct excitation processes across the gap by the (rare) high energy phonons. The occupation of these modes will grow according to

$$\nu(\mathbf{q},\mathbf{t}) \sim e^{-\Delta(q)/k_B T_h}\left(1 - e^{-t/\tau_{\mathbf{q}}}\right) \qquad q \approx q^*$$

where τ_q is a time scale set by $\overline{s}(q)$. The evolution of the occupation numbers of the larger q (quasi-exciton) modes will indeed be dominated by the absorption of low energy phonons by previously created magnetorotons, this leads to a diffusion of excitations uphill in energy (due to an incomplete balance of phonon absorption and emission) giving

$$\nu(q,t) \sim \nu(q^*,t)\sqrt{\frac{1}{4D_q t}}e^{-\Delta(q)^2/4D_q t} \qquad q \gg q^*$$

where the diffusion constant D_q depends on T_h and the phonon-magnetoroton matrix elements. In other words the total number of collective excitation quanta is determined by the creation process at q^*, the fraction of those which have diffused up to higher energy is determined by D. The dominant dependence of the change in two terminal resistance on the pulse temperature should then be $e^{-\Delta^*/k_B T_h}$ allowing the gap at the magnetoroton minimum to be extracted.

6 Conclusions

A simple model of the effects of the passage of a cloud of non-equilibrium acoustic phonons through a 2d electron system in the fractional quantum Hall regime has been examined and a useful intermediate regime of pulse duration identified from which the magnetoroton gap might be extracted experimentally.

Acknowledgments

This work has been supported by the Engineering and Physical Sciences Research Council of the United Kingdom.

References

1. SM Girvin, AH MacDonald and PM Platzman, Phys Rev **B33**, 2481 (1986)
2. RB Laughlin, Phys Rev Lett, **50**, 1395 (1983)
3. RK Kamilla, XG Wu, JK Jain, *preprint* "Excitons of Composite Fermions", Cond-Mat 9604023
4. CJ Mellor, RH Eyles, JE Digby, AJ Kent, KA Benedict, LJ Challis, M Henini and CT Foxon, Phys Rev Lett, **74**, 2339 (1995)
5. M Brownlie and KA Benedict, *submitted to Phys Rev B*

RAMAN SCATTERING FROM A LAUGHLIN LIQUID AT $\nu = 1/3$

A. J. TURBERFIELD, H. D. M. DAVIES, J. C. HARRIS, R. L. BROCKBANK, M. J. SNELLING and J. F. RYAN

Department of Physics, University of Oxford, Clarendon Laboratory, Parks Road, Oxford OX1 3PU, United Kingdom

We report measurements of inelastic light scattering from 2D electrons below 100 mK at Landau level filling factor $\nu = 1/3$ which reveal collective charge and spin density excitations of the incompressible electron liquid. We find that the temperature scale for the collapse of electron correlation in the spin-polarized electron liquid is determined by thermal excitation of long-wavelength spin waves.

Collective intra-Landau level excitations are of central importance to all theories of the fractional quantum Hall effect (FQHE).[1] The energy of large wavevector $\mathbf{q}_\infty$ excitations may be estimated from measurements of thermally activated resistivity; Pinczuk et al. have observed light scattering from $\mathbf{q} \approx 0$ charge density (CD) and spin density (SD) excitations;[2] phonon absorption measurements have been used to investigate intra-Landau level excitations.[3] There is still almost no experimental information on the dispersion of excitations, however: the existence of a magnetoroton minimum[4] in the density mode dispersion remains unconfirmed. The evolution of the excitation spectrum as the electron temperature is raised is also unknown.

Light scattering from two samples of a high-mobility two-dimensional electron system (2DES) ($\mu \approx 10^7 \mathrm{cm}^2\,\mathrm{V}^{-1}\,\mathrm{s}^{-1}$) at a GaAs/$\mathrm{Ga}_{0.67}\mathrm{Al}_{0.33}$As heterojunction with densities $n = 9.7 \times 10^{10}\mathrm{cm}^{-2}$ and $9.1 \times 10^{10}\mathrm{cm}^{-2}$ was studied at temperatures 50 mK – 2 K. An etalon-tuned Ti:sapphire laser ($< 1\mu$W, linewidth 2 μeV) was used, together with a Dilor quadruple spectrograph with a nitrogen-cooled CCD detector; Raman scattering could be measured down to a Stokes shift of 120 μeV with spectral resolution 20 μeV.

Figure 1 shows Raman spectra measured at 50 mK: each corresponds to a different laser energy within the band of photoluminescence from the 2DES.[5] Highlighted peaks correspond to resonant light scattering from collective excitations of the 2DES; a detailed discussion of their assignment will be published elsewhere.[6] C is observed at different magnetic fields in the two samples but in the same narrow range of filling factors $\nu = 1/3 \pm 1\%$; it is extremely sensitive to temperature and vanishes at $\sim$1 K. M is observed over a wider range of filling factors. C is identified as the $\mathbf{q} \approx 0$ CD mode in agreement with Pinczuk et al.[2] M has the relatively robust behaviour expected for magnetoroton modes; it

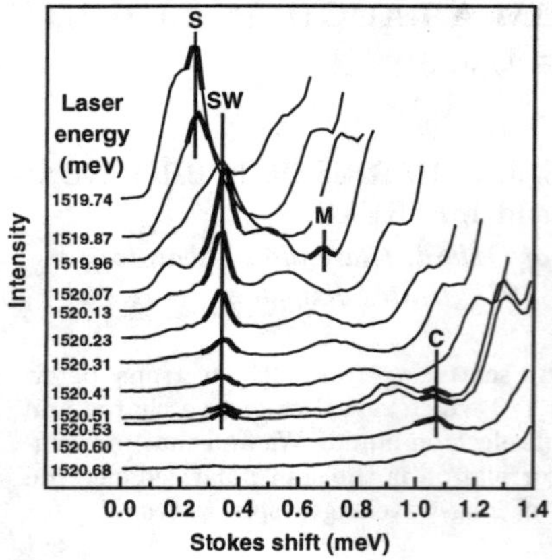

Figure 1: Light scattering from a two-dimensional electron system at 50 mK showing sharply resonant light scattering from spin- and charge-density excitations of the incompressible electron liquid at $\nu = 1/3$.

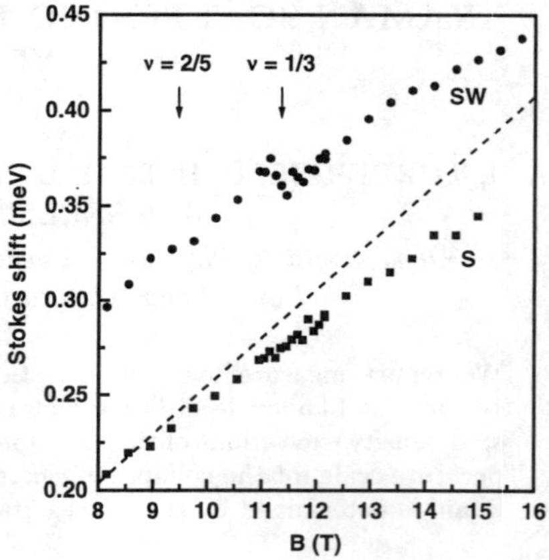

Figure 2: Energies of spin excitations S and SW as a function of magnetic field. The dashed line corresponds to $g^*_{\mathrm{GaAs}} = -0.44$.

is assigned to disorder-activated[7] magnetoroton scattering. Excitation energies $\epsilon_C = 1.07$ meV, $\epsilon_M = 0.68$ meV are consistent with theoretical estimates of 1.1 meV for $\mathbf{q} \approx 0$ CD modes and 0.5 meV for the magnetoroton mode[4,8,9] and are compatible with the quasiparticle gap of 0.50 meV deduced from activated transport measurements.[10]

Excitations S ($\epsilon_S = 0.27$ meV) and SW ($\epsilon_{SW} = 0.35$ meV) are spin density excitations; both have the extreme temperature sensitivity expected for collective modes and vanish at ~1 K. Fig. 2 shows their magnetic field dependence. S is close to the single-particle Zeeman energy as expected for wavevector-allowed scattering from $\mathbf{q} \approx 0$ SD excitations.[2,11] SW increases in energy with field with approximately the same slope as S although at $\nu = 1/3$ there is a slight but distinct dip in its energy; it is interpreted as a $\mathbf{q} \neq 0$ spin wave excitation whose energy is increased above that of S by exchange interaction. Again, this nominally forbidden light scattering is disorder-activated at resonance. The narrow linewidth of SW suggests that it is due to a peak in the spin wave density of states; we suggest that the spin-orbit interaction[12] mixes the SD and CD excitations leading to an avoided crossing near the magnetoroton wavevector $\mathbf{q}_{\mathrm{MR}}$: SW corresponds to scattering from the lower-branch mode, while M is the complementary upper-branch mode. The persistence of

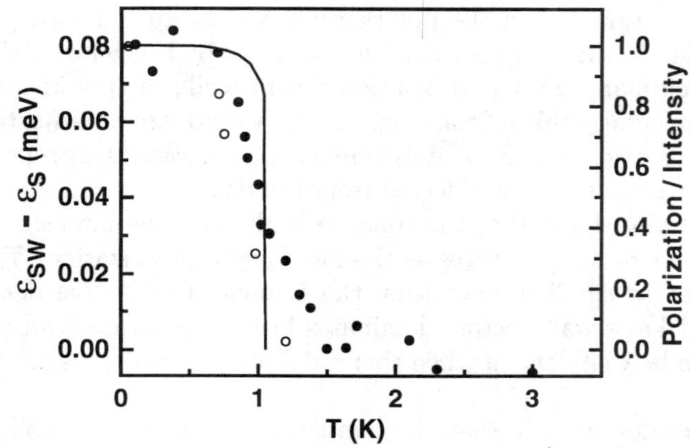

Figure 3: • Exchange enhancement of $\mathbf{q} \neq 0$ spin excitation SW over a range of temperatures; ∘ intensity of light scattering from $\mathbf{q} \approx 0$ charge density excitation C. The solid line shows the calculated spin polarization of the electron system.

SW (and M) across a wide range of filling factors is consistent with our expectation that the density-wave dispersion near $\mathbf{q}_{\mathrm{MR}}$ is relatively insensitive to changes in filling factor. The dip in the energy of SW at $\nu = 1/3$ is consistent with a lowering of the magnetoroton energy due to particularly strong short-range order there.

Figure 3 shows the temperature dependence of SW at fixed filling factor $\nu = 1/3$. As temperature increases the energy ϵ_{SW} decreases continuously to the energy of the long-wavelength spin wave ϵ_S. We interpret this as evidence for a decrease in exchange energy due to a rapid loss of spin polarization of the electron liquid at ~1 K and the collapse of the $\nu = 1/3$ correlated state. In the Hartree-Fock approximation the exchange contribution to the spin wave energy is proportional to the spin polarization.[13] We have modelled our results by comparing the exchange enhancement of the energy of SW to the polarization of the 2DES calculated by assuming a spin wave dispersion

$$\epsilon_{SW}(q) = |g^*|\,\mu_B B + (\nu_\uparrow - \nu_\downarrow)\Delta\left(1 - \exp\left\{-\alpha/\Delta(ql_0)^2\right\}\right)$$

where $\nu_\uparrow$, $\nu_\downarrow$ are the partial filling factors of the two N=0 Landau levels and α and Δ are chosen to match (at zero temperature) the short- and long-wavelength behaviour of the spin wave dispersion calculated numerically by Rezayi[11] corrected for the finite thickness of the electron system.[4,8] The density of reversed spins is calculated self-consistently by integrating the Bose-Einstein distribution function over phase space;[14] this approach is reasonable at low temperatures when the polarization is close to unity (it gives a good approximation to the spin polarization calculated by Kasner and MacDonald[15] at $\nu = 1$ using many-body perturbation theory while the polarization is above 70%). The results of this calculation, in which there are no adjustable parameters, are displayed in Fig. 3; it is remarkably successful in reproducing the sharp depolarization at 1 K inferred from the data.

Figure 3 also shows that the collapse in the C mode intensity occurs over the same range of temperatures as the loss of spin polarization. This suggests that, at least at small wavevectors, the branch of collective density excitations that at large wavevectors dominates low-temperature transport properties ceases to be well defined when thermal fluctuations depolarize the electron spins.

These results indicate that the temperature scale for the collapse of electron correlation in the spin-polarized electron liquid at $\nu = 1/3$ is determined by thermal excitation of long-wavelength spin waves.

References

1. D. C. Tsui *et al.*, *Phys. Rev. Lett.* **48**, 1559 (1982)
2. A. Pinczuk *et al.*, *Phys. Rev. Lett.* **70**, 3983 (1993)
3. C. J. Mellor *et al.*, *Phys. Rev. Lett.* **74**, 2339 (1995)
4. S. M. Girvin *et al.*, *Phys. Rev. Lett.* **54**, 581 (1985)
5. A. J. Turberfield *et al.*, *Phys. Rev. Lett.* **65**, 637 (1990)
6. H. D. M. Davies *et al.*, submitted for publication
7. A. Pinczuk *et al.*, *Phys. Rev. Lett.* **61**, 2701 (1988)
8. A. H. MacDonald and G. C. Aers, *Phys. Rev.* B **29**, 5976 (1984)
9. R. K. Kamilla *et al.*, *Phys. Rev. Lett.* **76**, 1332 (1996)
10. D. R. Leadley *et al.*, Phys. Rev. B **53**, 2057 (1996)
11. E. H. Rezayi, *Phys. Rev.* B **36**, 5454 (1987)
12. E. I. Rashba, *Fiz. Tverd. Tela* **2**, 1224 (1960) [*Sov. Phys. Solid State* **2**, 1109 (1960)]
13. T. Ando and Y. Uemura, *J. Phys. Soc. Jpn.* **37**, 1044 (1974)
14. The integral is truncated at a wavevector $100 l_0^{-1}$ to avoid a logarithmic divergence in the spin wave density.[15]
15. M. Kasner and A. H. MacDonald, *Phys. Rev. Lett.* **76**, 3204 (1996)

Time-resolved phonon absorption of a two-dimensional electron gas in the fractional quantum Hall regime

U. Zeitler, J.E. Digby, C.J. Mellor, A.J. Kent, K.A. Benedict, C.T. Foxon
Department of Physics, University of Nottingham, Nottingham, NG7 2RD, UK

J.J. Harris
Semiconductor Materials I.R.C., University College, London, UK

We have measured the reaction of a two-dimensional electron system (2DES) in the fractional quantum Hall regime to short pulses of ballistic phonons. Using a time-resolved technique we can distinguish between a ballistic response occuring at short time scales and the effect of heating the GaAs substrate at longer time scales. Heating is dominant between fractional filling factors where ρ_{xx} is strongly temperature dependent. At magnetic fields with a weak temperature dependence of ρ_{xx} (fractional filling factors and also around $\nu = 1/2$) we observe a response consistent with the absorption of ballistic phonons by the contacts but no direct ballistic response of the 2DES.

1 Introduction

Ballistic acoustic phonons are a well-suited tool for the study of fundamental electronic properties of low dimensional electron systems.[1] Earlier time-averaged experiments on phonon absorption of two dimensional electron systems (2DES) in the fractional quantum Hall (FQH) regime[2] assigned the response to be a signature of the theoretically predicted magneto-roton minimum in the collective excitations of the 2DES.[3] In this paper we report time-resolved experiments that seek to probe phonon absorption more directly.

2 Experimental Principles

The experiment was performed on a very high mobility 2DES ($n = 1.57 \times 10^{15}\text{m}^{-2}$, $\mu \approx 700\ \text{m}^2/\text{Vs}$) on a 0.4-mm GaAs substrate. On the rear face, a constantan film heater ($600 \times 60\ \mu\text{m}^2$) was evaporated facing the bulk of the 2DES. Applying a short voltage pulse (≈ 100 ns) to the heater creates ballistic phonons with a non-equilibrium phonon temperature $T_p \approx 1...8$ K. They hit the 2DES after the phonon time of flight through the substrate. Subsequently they thermalize inside the substrate within a few μs and finally the thermalized phonons leave the substrate within a few hundred μs via thermal conduction to the mixing chamber. The time-dependent reaction of the 2DES to a phonon pulse was investigated by measuring the voltages $U_{4pt}(t) = R_{xx}(t)I(t)$ and

$U_{2pt}(t) = R_{2t}(t)I(t)$ in a four terminal and two terminal setup with a constant current ($I_0 = 100$ nA) applied in equilibrium. Typically $10^3 - 10^6$ traces were averaged with a repetition rate of the order of kHz.

The intrinsic electronic time response τ_0 of the system is given by the cable capacitance connected to the sample multiplied by the two point resistance R_{2pt} between two contacts. In the FQH regime where $R_{2pt} \approx 30...100$ kΩ this results into $\tau_0 \approx 6...20$ μs. Accepting a loss in sensitivity by R_A/R_{2pt} the electric response can be accelerated by the same factor by connecting a resistor R_A in parallel to the voltage contacts allowing us to distinguish clearly between a ballistic response in the sub-μs range and heating effects on time scales of tens of μs. Here it is important to remark that a fast decreasing (increasing) R_{2pt} of the system will cause a fast discharging (charging) of the capacitance between the two current contacts which in turn leads to an increase (decrease) of the actual current through the 2DEG. The current relaxes back to equilibrium after τ_0. Therefore, in a four terminal measurement, only signals changing slower than τ_0 can be related directly to changes in R_{xx}.

3 Results

In Fig. 1 the reaction of the 2DEG at $\nu = 0.9$ ($B = 7.25$ T) and $T = 250$ mK to a 100-ns pulse of 3.7-K ballistic phonons is plotted in the form of $U_{4pt}(t)/I_0$. The signal is due to a time-dependent heating of the substrate transmitted to the 2DEG. The actual temperature $T_{2DEG}(t)$ can be deduced from the temperature dependence of the four terminal resistance R_{xx} shown in the inset. T_{2DEG} reaches a maximum at about 15μs and then re-cools slowly. Both the maximum temperature and the time scale of the cooling are consistent with a fast heating of the GaAs substrate via thermalization of the hot ballistic phonons within the substrate and a slow cooling via thermal conduction to the heatsink. The effect of heating only depends on the total amount of heat dissipated in the heater, the above statements remaining consistent for different heater powers.

The situation is different at magnetic fields where R_{xx} is only weakly temperature dependent. In Fig. 2 a typical example ($B = 13$ T, $\nu = 1/2, T = 250$ mK) is shown for a 100-ns ballistic phonon pulse with $T_p = 3.7$ K. As expected, no significant signal is observed in $U_{4pt}(t)$ for $t \gg \tau_0$ (Fig. 2a) where the current is constant. In contrast, due to the strong effect of phonons on the contacts, $U_{2pt}(t)$ shows a strong and slowly decaying heating contribution. Additionally, a fast change of R_{2pt} occurs already for $t < 1\mu$s demonstrating the strong sensitivity of the contact resistances to ballistic phonons particularly pronounced in high magnetic fields. This fast decrease of R_{2pt} also yields a strongly increased current through the 2DES observed as a strong positive four terminal signal which disappears for $t \gg \tau_0$.

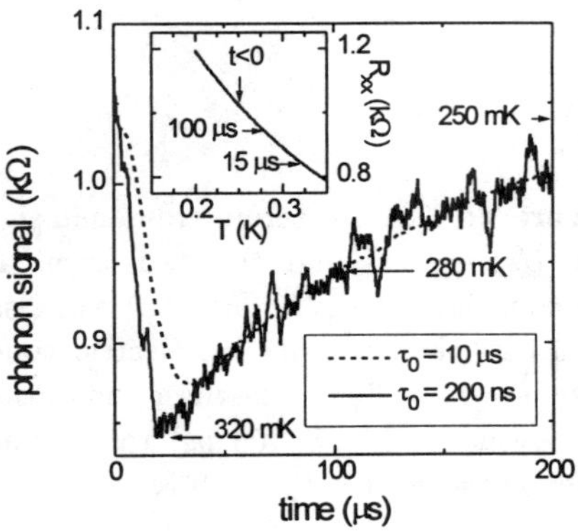

Figure 1: Time-dependent four terminal phonon signal $U_{4pt}(t)/I_0$ at $\nu = 0.9$ for a fast and a slow electric response time. Ballistic phonons with $T_p = 3.7$ K are emitted during 100 ns at $t = 0$. The inset shows the temperature dependence of the four terminal resistance R_{xx}

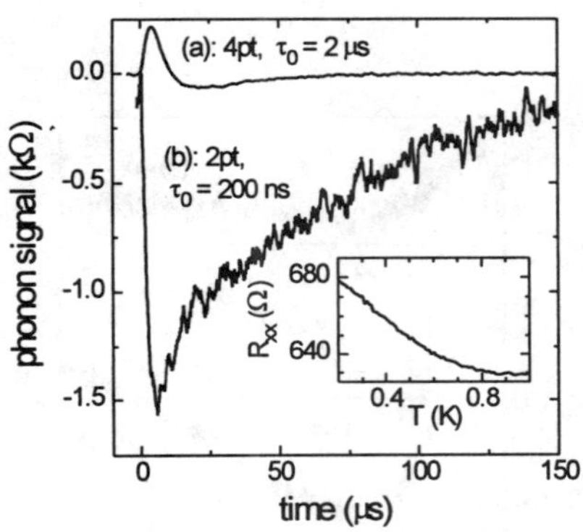

Figure 2: Reaction of the 2DEG at $\nu = 1/2$ to 3.7 K ballistic phonons emitted during 100 ns shown as the four terminal and the two terminal phonon signal. The inset shows the dependence of the four terminal resistance R_{xx} on temperature

At a filling factor of 2/3 (and other fractional filling factors), again no significant heating is observed in $U_{4pt}(t)$ for $t > \tau_0$ (Fig. 3), a typical change in temperature from 250 mK to 320 mK corresponding to a change in R_{xx} of only 10 Ω. However, at short time scales ($t < \tau_0$) we observe an increase in $U_{2pt}(t)$ as well as in $U_{4pt}(t)$. Since a fast increasing $U_{2pt}(t)$ yields a smaller current this means that $R_{xx}(t) = U_{4pt}(t)/I(t)$ is definitely increasing. The signal might then be interpreted as a rapid heating of the 2DES to a non-equilibrium temperature followed by a relaxation to its equilibrium temperature. However, we can not assign this effect with ceratainty to a direct absorption of ballistic phonons by the 2DES. Although ballistic phonons are expected to hit the 2DES after 80 ns (LA) and 130 ns (TA) no change in the two terminal signal is resolved experimentally at these times using short pulses (30 ns) and hot ballistic phonons (5.8 K). Drastic changes only occur after 180 ns, the time of flight for the LA phonons from the heater to the closest contact, see inset in Fig. 3. Moreover, all the signals we observe, are only dependent on the total amount of heat dissipated in the heater strongly proposing a bolometric response of either the contacts or the edges of the 2DES rather than a direct absorption by the bulk of the 2DES containing spectroscopic information.

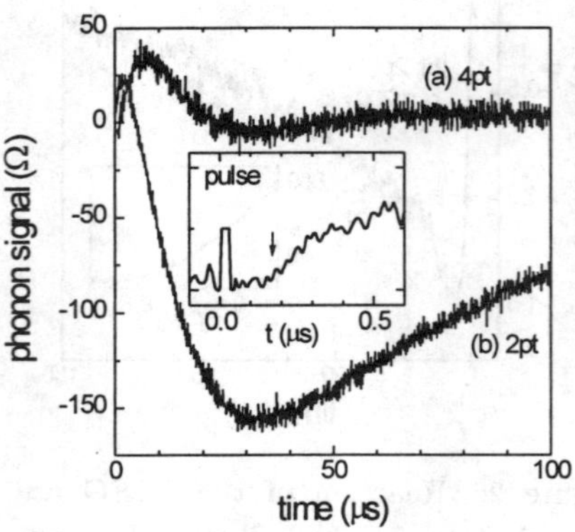

Figure 3.: The four terminal phonon signal U_{4pt}/I_0 (a) and the two terminal phonon signal U_{2pt}/I_0 for a 100 ns pulse of 3.7-K phonons with an electric time response $\tau_0 = 20\mu s$. The inset shows the fast response ($\tau_0 = 100$ ns) to a 30-ns phonon pulse with $T_p = 5.8$ K.

4 Conclusions and Perspectives

The time-resolved response of a 2DES in the FQH regime to a pulse of ballistic phonons has been explained in terms of a heating of contacts by ballistic phonons on a sub-μs time scale and a heating of the GaAs substrate at longer times. At present we have not observed clear experimental evidence for a direct absorption of ballistic phonons by the 2DES. Future work will concentrate on improving the time response and the sensitivity of the experiment and to eliminate contact effects in order to be able to resolve signals which are unambiguously caused by a direct ballistic response of the 2DES.

References

1. see, e.g. L.J. Challis and A.J. Kent, *in: Physics of Phonons*, eds. T. Paszkiewicz and K. Rapcewicz (Plenum, New York, 1984) p159.
2. C.J. Mellor et al., Phys. Rev. Lett. **74**, 2339 (1995).
3. S.M. Girvin, A.H. MacDonald, and P.M. Platzmann, Phys. Rev. B **33**, 2481 (1986).
4. F. Rosch and O. Weiss, Z. Phys. B **27**, 33 (1977).

Experiment-based Global Phase Diagram for the Quantum Hall Effect

S. V. Kravchenko
Physics Department, City College of the City University of New York, New York, NY 10031, USA

Whitney Mason and J. E. Furneaux
Laboratory for Electronic Properties of Materials and Department of Physics and Astronomy, University of Oklahoma, Norman, OK 73019, USA

V. M. Pudalov
Institute for High Pressure Physics, Troitsk, 142092 Moscow District, Russia

Using experimental data obtained from a two-dimensional electron system in silicon, we construct an experiment-based "disorder *vs* filling factor" phase diagram for the integer quantum Hall effect (QHE). Generalizing this diagram to the case of the fractional QHE, we show that it is consistent with the recently observed direct transitions between insulator and fractional QHE at $\nu = 2/5$, 2/7, and 2/9.

In a two-dimensional electron system (2DES), placed in a strong perpendicular magnetic field, B, there exist extended states at the center of each Landau level.[1] We have performed a systematic study of the positions of these extended states in very high-quality silicon MOSFET's with maximum mobility around 4×10^4 cm^2/Vs. Using the obtained map of the extended states, we obtain an experiment-based "disorder *vs* filling factor" phase diagram for the integer QHE. Applying correspondence principles articulated by Kivelson *et al*,[2] we construct a global phase diagram which appears to be consistent with direct transitions between an insulating state and the fractional QHE (FQHE) at $\nu = 2/5$, 2/7, and 2/9, recently observed in high-quality GaAs/AlGaAs heterostructures.[3,11,5,6]

To map the positions of extended states, we used a method suggested by Glozman *et al.*:[7] every time the Fermi energy, E_F, crosses an extended state, the diagonal conductivity, σ_{xx}, exhibits a maximum. σ_{xx} was recalculated from the data for diagonal and Hall resistivities, ρ_{xx} and ρ_{xy}.

Figure 1 shows the resulting positions of the extended states on the (B, n_s) plane (here n_s is electron density). Each set of symbols reflects the behavior of one σ_{xx} peak as a function of n_s and B, and the dotted lines represent the expected positions of the extended states at half-filled Landau levels, $\nu = i + \frac{1}{2}$. The regions between extended states correspond to the QHE with $\sigma_{xx} = 0$ and $s_{xy} \equiv \sigma_{xy} \cdot h/e^2$ shown by the numbers obtained

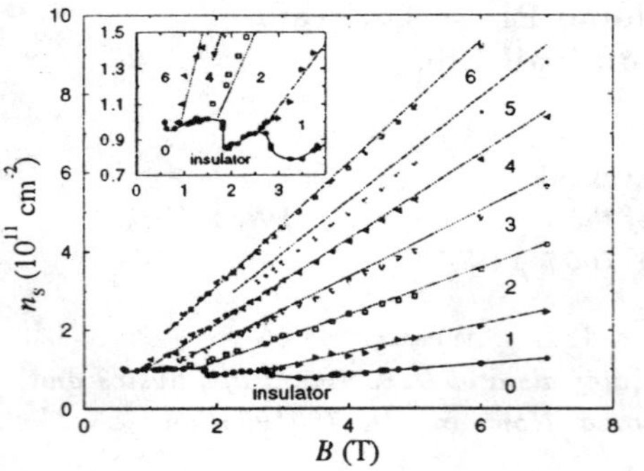

FIG. 1. The map of extended states. Numbers show σ_{xy} in units of e^2/h. The inset shows details at low B and n_s.

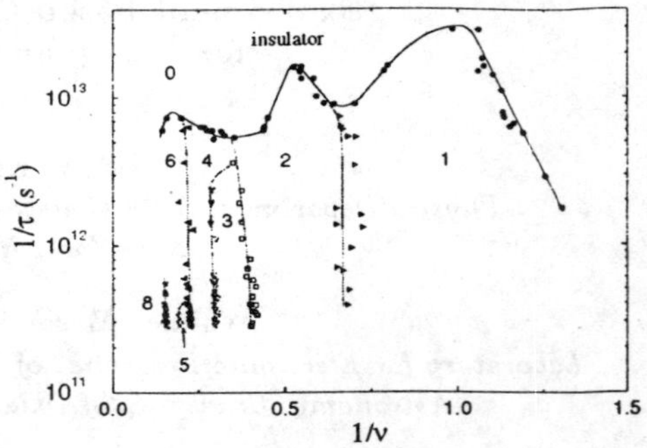

FIG. 2. Disorder vs. filling factor phase diagram recalculated from the data shown in Fig. 1.

from the data for σ_{xy}. The region with $s_{xy} = 0$, separated from regions with $s_{xy} \neq 0$ by the lowest extended state, is insulating. We will call this boundary, shown by the solid line, the "QHE/insulator (QHE/I) boundary".

The two most important features of Fig. 1 are: (i) as the magnetic field is lowered, the higher n_s extended states, separating QHE with different values of s_{xy}, *merge* with the QHE/I boundary, and (ii) the QHE/I boundary oscillates as a function of magnetic field for $B < 3.5$ T reaching maxima where the higher n_s extended states join. Both features are in agreement with the phase diagram obtained by Shashkin *et al.*[8] They are not observed by Glozman *et al.*[7]

Using these experimentally obtained positions of the extended states, we reconstruct a "disorder *vs* filling factor" phase diagram for the QHE to compare with the theoretical results of Kivelson *et al.*[2] To characterize the disorder quantitatively, we used the $B = 0$ electron scattering time extracted from $\mu(n_s)$ at $T = 1.8$ K. The resulting phase diagram is shown in Fig. 2. The level structure in silicon is complex, with both spin and valley splittings associated with each Landau level. Thus, QHE states with $s_{xy} = 1$ to 4 all belong to Landau level zero, and the merging of different extended states belonging to this level is trivial. However, the QHE state with $s_{xy} = 6$ belongs to Landau level one. Therefore, the experimentally obtained phase diagram principally differs from the theoretical one because *it allows a direct transition from insulator to QHE with $s_{xy} = 6$, associated with the next Landau level.* The QHE/I boundary (the upper solid line) oscillates reaching maxima near $\nu = 1$, 2, and 6, and minima at the points where it merges with extended states separating neighboring QHE regions. A possible physical nature of these oscillations was discussed in Ref. [9]

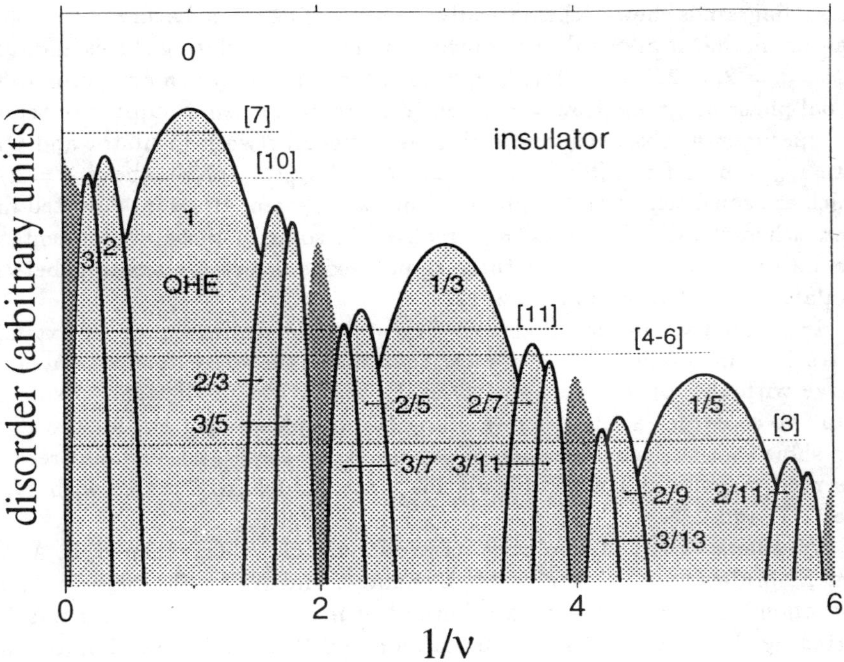

FIG. 3. A suggested phase diagram for the quantum Hall effect. Shaded regions correspond to the QHE with $\sigma_{xx} = 0$ and values of s_{xy} given by numbers. Thick solid lines separating regions of QHE with different s_{xy} designate the extended states. The white region corresponds to the insulator with $s_{xy} = \sigma_{xx} = 0$, and the darker shaded regions correspond to the "metallic" states near $B = 0$ and half-integer filling factors. The horizontal lines trace out transitions between QHE and insulating phases consistent with the experimental references with which they are labeled.

According to the global phase diagram of Kivelson *et al.*, there should be a similarity between the integer QHE (IQHE) at $\nu = 1, 2, 3$ ... and FQHE with ν equal, *e.g.*, to 1/3, 2/5, 3/7 ... This similarity prompts us to generalize the experimentally obtained IQHE phase diagram to FQHE using correspondence principles presented by Kivelson *et al.* For simplicity, we ignored spin and valley splittings. In this case, the regions with $s_{xy} = 1...4$ collapse into a single region with $s_{xy} = 1$, the regions with $s_{xy} = 5...8$ — into a single region with $s_{xy} = 2$ *etc.* We used two main features of the experimental phase diagram shown in Fig. 2: (i) merging of the upper extended states with QHE/I boundary, and (ii) the oscillatory QHE/I boundary. The suggested resulting phase diagram is shown schematically in Fig. 3. The main feature of this phase diagram is that it allows direct transitions between insulating states and QHE with $s_{xy} = 2, 3, 2/3, 2/5, 2/7, 2/9, 2/11$, which are forbidden according to the global phase diagram. However, recently there have been a number of reports on experimental observation of direct transitions between insulator and QHE with $s_{xy} = 6$ and possibly 10,[10] $s_{xy} = 2/5$,[11,6] $s_{xy} = 2/7$[5,6], and $s_{xy} = 2/9$,[3] which are consistent with the proposed phase diagram. Horizontal dotted lines show schematically the transitions realized in some of these experiments. It can be seen that in principle there should exist a direct transition between insulator and QHE with $s_{xy} = 2/11$.

In summary, we report an experimentally determined map of extended states in a magnetic field. We show that the higher-n_s extended states can merge with the lowest one, in contrast to theoretical predictions. Using our data, we construct a "disorder vs filling factor" phase diagram for the IQHE and show that the generalization of our results to the FQHE explains recently observed transitions between insulating states and FQHE states with $s_{xy} = 2/5$, $2/7$, and $2/9$.

We thank V. T. Dolgopolov, A. G. Golitsyna, G. V. Kravchenko, B. A. Mason, K. Mullen, S. Q. Murphy, M. B. Santos, and A. A. Shashkin for useful discussions and help with preparation of this manuscript. This work was supported by the National Science Foundation and Russian Foundation for Basic Research.

References

1. *The Quantum Hall Effect*, ed by R. E. Prange and S. M. Girvin (Springer, New York, 1987).
2. S. Kivelson, D.-H. Lee, and S. C. Zhang, Phys. Rev. B **46**, 2223 (1992).
3. H. W. Jiang *et al*, Phys. Rev. B **44**, 8107 (1991).
4. M. B. Santos *et al*, Phys. Rev. B **46**, 13639 (1992).
5. P. J. Rodgers *et al*, J. Phys.: Condens. Matter **5**, L449 (1993).
6. H. C. Manoharan and M. Shayegan, Phys. Rev. B **50**, 17662 (1994).
7. I. Glozman, C. E. Johnson, and H. W. Jiang, Phys. Rev. Lett. **74**, 594 (1995).
8. A. A. Shashkin, G. V. Kravchenko, and V. T. Dolgopolov, JETP Lett. **58**, 220 (1993).
9. S. V. Kravchenko, W. Mason, J. E. Furneaux, and V. M. Pudalov, Phys. Rev. Lett. **75**, 910 (1995).
10. V. M. Pudalov, M. D'Iorio, and J.W. Campbell, JETP Lett. **57**, 608 (1993); Surf. Sci. **305**, 107 (1994).
11. M. B. Santos *et al*, Phys. Rev. Lett. **68**, 1188 (1992).

QHE–INSULATOR TRANSITION AT THE LOWEST LANDAU LEVEL IN A TWO-DIMENSIONAL ELECTRON SYSTEM

V.M. PUDALOV
Institute for High Pressure Physics, Troitsk, Moscow district, 142092, Russia

Transition from the QHE to insulator in the system with only one filled Landau level was studied on high mobility Si-MOS structures. The peak in the diagonal conductance, σ_{xx}, indicating the QHE-insulator transition at the lowest Landau level was found to exhibit a substructure. The main peak manifests the transition associated with the fractional filling factor $\nu=2/3$, while the additional weaker feature develops near and slightly above $\nu=1/2$. In a more disordered sample the quasiclassical transition moves to higher fillings and overpowers the feature at $\nu=2/3$.

As the magnetic field increases, when the lowest Landau level is partially filled, the 2D electron system usually exhibits an insulating transition due to the shrinkage of the electron wave functions. This fundamental transition, **1⇔0,** from the quantized Hall effect (QHE) phase with only one filled band of extended states ($s_{xy}=1$ [1]) to the insulator, $s_{xy}=0$, is in the focus of experimental interest. Kawaji et al. [2] studied this transition at Landau level filling $\nu \approx 0.9$ and at temperature 1.5 K. Later on, Shashkin et al. reported [3] that this transition occurs as a percolating transition, and develops along and slightly above the classical percolation boundary, $\nu=1/2$. Kravchenko et al. [4] found that the insulating state in the extreme quantum limit is associated with the Hall resistance remaining at the classical value in contrast to the diverging diagonal resistance.

Studying this transition in the high mobility Si MOS structure, we found that both the diagonal resistivity, ρ_{xx} , and conductivity, σ_{xx}, show a substructure as the magnetic field increases above $n_s hc/e$. At first, σ_{xx} shows a peak near the $\nu=2/3$-trajectory, and then at a slightly higher field, σ_{xx} exhibits a second feature approaching $\nu = 1/2$. The total picture may be understood as a passage of two different extended states through the Fermi level: the first one related to the fractional 2/3 level, and the second one related to the integer $\nu=1$ level. The σ_{xx} peak value, $(0.31\pm0.02)\ e^2/h$, corresponds to the critical ρ_{xx} value, h/e^2.

The studies were performed on the (100) Si-MOS sample with peak mobility of 50,000 cm^2/Vs at $n_s \approx 2\times10^{11}$ cm^{-2}. The sample was patterned as a Hall bar with a source - drain length of 5 mm, a width w=0.8 mm and an intercontact distance l=2.5mm (w/l = 0.32). The carrier density was varied by the top gate from 4×10^{10} (deep in the insulating state) to 10^{12} cm^{-2} (well in the metallic state). Measurements were performed at currents in the sub-pA to nA-range in order to insure the nearly current-independent transport coefficients in the insulating state and to avoid Joule heating. The instrumentation for true four-contact measurements at sample resistance up to 10^{11} Ω was described earlier [5]. We now describe the results in detail.

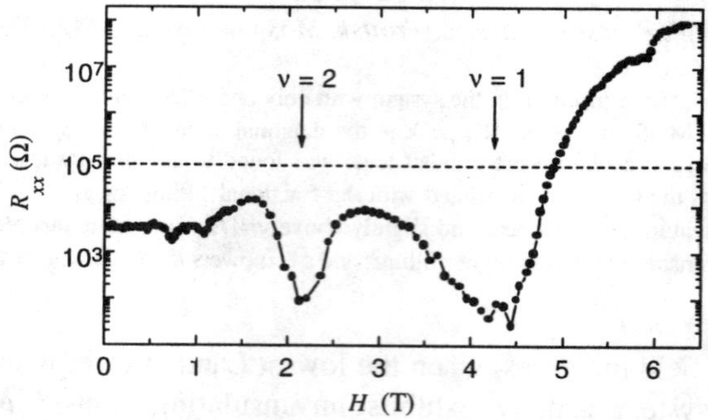

Figure 1: Diagonal resistance vs magnetic field. T=0.035 K, n_s=10.15×10^{10} cm^{-2}. Dashed line shows the level corresponding to $\rho_{xx} = h/e^2$.

Fig. 1 shows the typical Shubnikov-de Haas oscillations in the high mobility Si-MOS sample. In the ultraquantum regime, at $H > 4.4$T, ρ_{xx} diverges with magnetic field, but not simply exponentially. An additional feature in ρ_{xx} is clearly seen at H=5.9 T. In accord with Ref. [4] no features were detected in ρ_{xy}. The latter was found to follow the regular quantum Hall effect behavior in the range of filling factor from 1 to 1/2. The typical conductivity curve, calculated on the basis of ρ_{xx} and ρ_{xy} measurements is shown in Fig. 2.

σ_{xx} vanishes at both sides of the transition: at lower fields due to the QHE state,

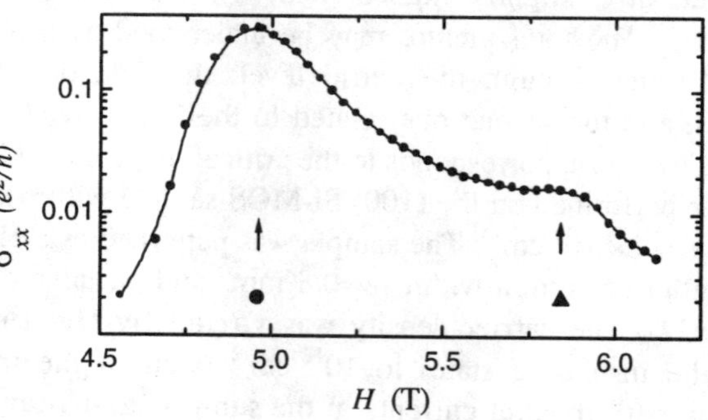

Figure 2: Diagonal conductivity vs magnetic field. T=0.035 K, n_s=10.15×10^{10} cm^{-2}.

and at higher fields due to the insulating state. The peak in σ_{xx} confined between two diminishing tails was found earlier to correspond the QHE-insulator boundary [6-9]. In the framework of the scaling picture [1,10] it indicates the passage of the extended states through the Fermi level. The main peak, at the lower field, reaches 0.31 e^2/h. This value is in agreement with those observed for other QHE - insulator transitions [5-9].

An additional weaker peak is seen on the high-field side of the $\sigma_{xx}(H)$ curve. In Fig. 3 we plotted the positions of the two peaks designated with filled circle and filled triangle in Fig. 2, as a function of electron density. The empty circles depict the position of the ρ_{xx} minima for $\nu=1$, in order to insure the correct calculations of the electron density and filling factor.

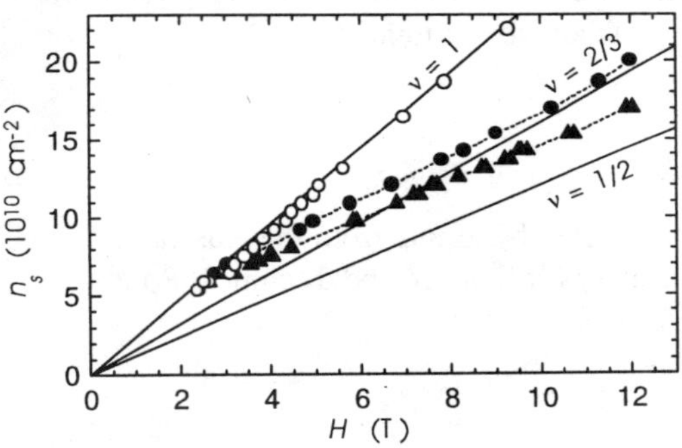

Figure 3: Location of the two peaks in σ_{xx} on the H - n_s diagram. Filled circles and triangles designate the two peaks, as defined in Fig. 2. Continuous lines depict constant filling factor $\nu=1$, 2/3, and 1/2, correspondingly.

Discussion

Recently, an empirical Global Phase Diagram (GPD) was constructed on the basis of a large set of data for the transitions between the integer, fractional QHE and insulating states [6]. The GPD incorporates all of the observed QHE/i transitions at $\nu > 1$, for instance, **0⇔1⇔0⇔2**. From symmetry arguments, a few more transitions may be expected to occur in less disordered systems [6], e.g. **1⇔0⇔2/3⇔0**. The reported features in σ_{xx} may be interpreted in the framework of this diagram. As found earlier [6-9], σ_{xx} exhibits a peak whenever an extended state passes through the Fermi energy. The first peak in σ_{xx} may thus be interpreted as the one associated with the fractional 2/3 level, and the second peak with the integer $\nu=1$ level. Due to strong overlapping and mixing of the $\nu=2/3$ and $\nu=1$ levels, the corresponding trajectories of σ_{xx} peaks deviate from their expected locations depicted by the straight lines in Fig. 3. The two features are worth noting: (i) the lower field peak is much stronger, than the higher field peak. Thus, although the fractional 2/3 state is not well pronounced in transport, it contributes significantly into the transition to insulator. (ii) The peak value for σ_{xx} is equal to $0.31 \pm 0.02\ e^2/h$ for all points in Fig. 3. This is consistent with the critical ρ_{xx} value, which was found to be equal to h/e^2, for the integer and fractional QHE - insulator transitions [6-9].

At lower electron density, the two σ_{xx} peaks shift to lower fields and finally merge with the σ_{xx} minimum at $\nu=1$, as seen in Fig. 3. In the more disordered sample with peak mobility of 25,000 cm^2/Vs, the $\nu=1/2$-peak shifts to lower fields and merges with the 2/3-peak already at field $H \leq 12$T.

Acknowledgments

The work was partially supported by grants from the Russian Foundation for Basic Research and Russian Ministry for Science and Technical Policy .

References

1. S.A. Kivelson, D.H. Lee and S.C. Zhang, *Phys. Rev*. B **46**, 2223 (1992).
2. S. Kawaji, J. Wakabayashi, M. Namiki and K. Kasuda, *Surf. Sci*., **73**, 121
3. A.A. Shashkin, V.T. Dolgopolov and G.V. Kravchenko, *Phys. Rev*. B **49**, (1994).
4. S.V. Kravchenko, J.E. Furneaux and V.M. Pudalov, *Phys. Rev*. B **49**, 2250 (1995).
5. M. D'Iorio, V.M. Pudalov and S.G. Semenchinsky, *Phys. Rev*. B **46**, 15992 (1992).
6. S.V. Kravchenko, W. Mason, J.E. Furneaux and V.M. Pudalov, *Phys. Rev. Lett*., **75**, 910 (1995).
7. J.E. Furneaux, S.V. Kravchenko, W.E. Mason, G.E. Bowker and V.M. Pudalov, *Phys. Rev*. B **51**, 17227 (1995).
8. D. Shahar, D.C. Tsui, M. Shayegan, R.N. Bhatt, and J.E. Cunningham, *Phys. Rev. Lett*., **74**, 4511 (1995).
9. I. Glozman, C.E. Johnson and H.W. Jiang, *Phys. Rev. Lett*. **74**, 594 (1995).
10. D.E. Khmelnitskii, *Phys. Lett*. **106**, 182 (1984)

THE CORRELATED $\nu = 1$ BI-LAYER FRACTIONAL QUANTUM HALL STATE AND FINITE TEMPERATURE PHASE TRANSITIONS

R.J. Hyndman, T. Ihn, B.L. Gallagher, M. Henini, T.J. Foster,
K.A. Benedict, J.S. Chauhan, J.R. Middleton and H. Smith
Department of Physics, University of Nottingham, Nottingham NG7 2RD, UK

We present magnetotransport results for strongly coupled p-type double quantum wells which demonstrate the existence of a correlated $\nu = 1$ quantum Hall state in the limit of weak tunnelling. Our results are consistent with the finite temperature phase transition predicted for this state. We are able to construct an experimental phase diagram for the state which is in very good agreement with theoretical prediction.

1. Introduction

Advances in growth techniques have now made it possible to grow double layer 2D electron and hole gases with very small layer separations and high mobilities. It has long been predicted[1] that a new correlated bi-layer fractional quantum Hall state should occur in such systems in the limit of weak tunnelling at a total Landau level filling factor, $\nu = 1$ (i.e. $\nu = ½$ in each of the two layers). This state involves intra- and inter-layer correlations between the carriers and is predicted to occur when the layer separation, d, is comparable with the magnetic length ℓ_B (which determines the intra-layer Coulombic energy). Recently the detailed theory of this state has been explored in considerable depth[1] by treating the carriers in the two layers as if they carried a fictional spin of $\pm ½$.

The lowest energy charged excitations of the bi-layer correlated $\nu = 1$ states are merons which can be pictured as vortices in the spin orientation and which carry a charge of $\pm$ e/2. At low enough temperatures, these objects are bound together in vortex-antivortex pairs with a total charge of 0 or $\pm$e (a closely bound meron pair carrying charge $\pm$e can be viewed as a skyrmion). At a finite temperature, T_{KT}, a Kosterlitz-Thouless transition is predicted to occur. At T_{KT} the pairs unbind and the energy gap to charge excitations disappears. It is predicted[1] that the bi-layer $\nu = 1$ state should only be stable for $d/\ell_B < 1.6$, since T_{KT} tends to zero at this value.

Evidence for a correlated $\nu = 1$ bi-layer state has been found in n-type double quantum wells[2] which indicates a phase boundary at $d/\ell_B \sim 2$ and in very wide single quantum wells[3]. However because of the low electron mass tunnelling is usually appreciable. In such samples a normal integer $\nu = 1$ state in which the energy gap (Δ_{SAS}) is that between the symmetric and anti-symmetric state of the bi-layer can also occur.

2. Experimental Results

Here we present a systematic study of p-type GaAs/GaAlAs double quantum wells with 100 Å wells separated by a 30 Å AlAs barrier. The samples have mobilities ~500,000 $cm^2\ V^{-1}\ s^{-1}$. The density in the individual wells is controllable by front and back gates. The high hole effective mass leads to a calculated Δ_{SAS} ~ 50 mK which is negligibly small compared to the Coulombic energies.

In Figure 1 we show typical resistivity and Hall resistance results for a sample with equal hole densities in the two wells. The $\nu = 1$ state is clearly not a normal single particle state since all other odd integer states are absent. The state also weakens as the total hole density is increased and the $\nu = 1$ state moves to higher

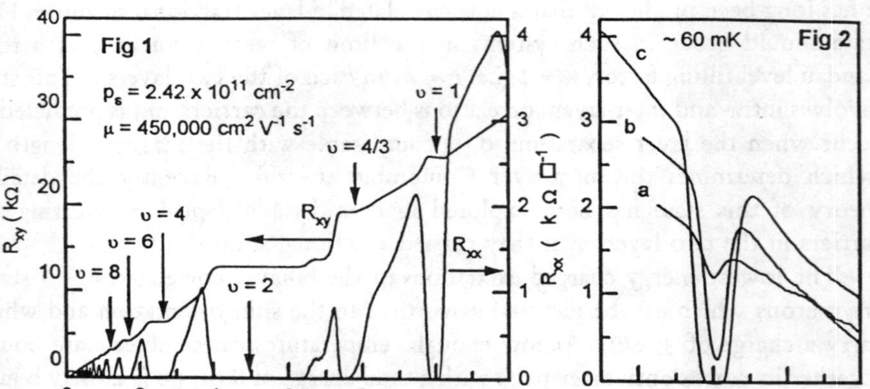

Figure 1: (left) Typical R_{xx} and R_{xy} data at ~60 mK for equal number densities in each well and a total number density of 2.42 × 10^{11} cm^{-2}.

Figure 2: (right) The collapse of the $\nu = 1$ state with increased density for the same sample at densities p_s = (a) 2.99, (b) 2.63, (c) 2.42 × 10^{11} cm^2, corresponding to d/ℓ_B = (a) 1.685, (b)1.67 and (c) 1.60.

field. This is unlike the behaviour of any other (spin-polarised) quantum Hall state, but is what one expects for the correlated bi-layer state. As Figure 2 shows, the $\nu = 1$ state is destroyed as d/ℓ_B (where d is taken as the well centre separation) approaches ~ 1.7, in good agreement with expectation.

Activation plots of $\ell n R_{xx}$ against 1/T are given in Figure 3. This shows very unusual behaviour since at low enough temperature one observes normal activated behaviour indicating the presence of an energy gap, ΔE. But as the temperature is raised the activated behaviour is terminated rather abruptly and no minimum in R_{xx} is observable above a temperature T^* which is much smaller than ΔE. This is in contrast to the behaviour of the adjacent $\nu = 4/3$ state, which shows normal behaviour. This behaviour is just what one would expect for the predicted Kosterlitz-Thouless transition in which one has the disappearance of the energy gap for charged excitations at a finite temperature, T_{KT}. Similar behaviour has rather surprisingly been seen[4] in very wide quantum wells despite the very strong tunnelling.

Fig 3

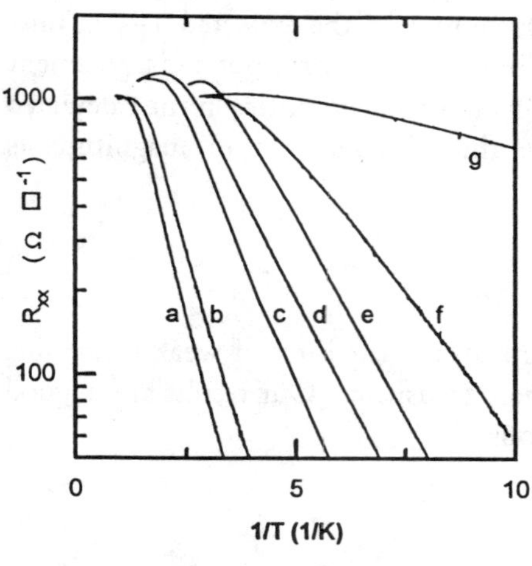

Fig 4

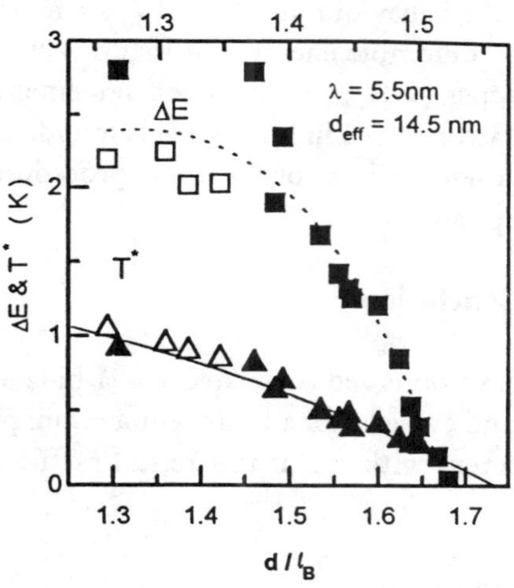

Figure 3: (left) Activation plot of $\nu = 1$ ($\nu = 1/2$ in each layer) for the same sample as Figs. 1 & 2 for a series of total number densities in units of 10^{11} cm^2 with activation energies, ΔE, in Kelvin in brackets: (a) 1.61 (2.9 K), (b) 2.11 (2.8 K), (c) 2.23 (1.7 K), (d) 2.28 (1.4 K), (e) 2.33 (1.3 K), (f) 2.54 (0.54 K), (g) 2.63 (0.2 K).

Figure 4: (right) Dependence of the measured activation energies, ΔE, and temperature T^* on d/ℓ_B for the same sample as the other figures (solid symbols) and a slightly lower mobility sample (open symbols). The solid line is the theoretical prediction for T^* multiplied by a factor of 2.5. The dashed line is a guide to the eye for ΔE. T^* is defined on the point of intersection of the extrapolated activation line and the horizontal at the saturation resistance.

3. Comparison between the Experimental and Predicted Phase Diagram

In Figure 4 we present the dependence of the activation energy and our T^* values as a function of d/ℓ_B. T^* steadily falls towards zero as d/ℓ_B approaches ~ 1.7, while ΔE falls rapidly to zero at this value. Even better agreement with theory is obtained if one replaces the well centre separation with the calculated separation of the maxima of the hole probability densities in each well and ℓ_B by $(\ell_B{}^2 + \lambda^2)^{1/2}$ where λ is the width at half height of the probability densities to define a $(d/\ell_B)_{eff}$. The latter finite thickness correction gives an estimate of the reduction of the intralayer Coulombic energy. The critical $(d/\ell_B)_{eff}$ is then ~ 1.6, in excellent agreement with expectation. The predicted values[1] of ΔE are ~ 10 K but it is to be expected that finite thickness corrections, Landau level mixing and disorder will all tend to reduce this value so our values are not inconsistent with expectation. A more useful comparison can be made between our values of T^* and the predicted behaviour of T_{KT} since this transition temperature should be less influenced by disorder. To do so, we have taken the prediction of reference 1 for the behaviour at a fixed ℓ_B as a function of d and scaled them by the intralayer Coulombic energy. In Figure 4 we have plotted the expected T_{KT} against our upper $(d/\ell_B)_{eff}$ scale. Good agreement in functional form but a disagreement by a factor of ~ 2 in magnitude is found. Given the uncertainties in the calculated magnitude and in our scaling procedure, this disagreement in magnitude is unsurprising.

4. Conclusion

We have observed correlated $\nu = 1$ bi-layer states in the limit of weak tunnelling and find evidence of a finite temperature phase transition. Our results are in good agreement with recent theoretical predictions.

References

1. K. Moon *et al*, *Phys. Rev. B* **51**, 5138 (1995) and references therein.
2. S.Q. Murphy, J.P. Eisenstein, G.S. Boebinger, L.N. Pfeiffer and K.W. West, *Phys. Rev. Lett.* **72**, 728 (1994).
3. Y.W. Suen, H.C. Manoharan, X. Ying, M.B. Santos and M. Shayegan, *Phys. Rev. Lett.* **72**, 3405 (1994).
4. T.S. Lay *et al*, *Phys. Rev. B* **50**, 17725 (1994).

MICROWAVE RESONANCE OF 2D HOLES IN WIGNER SOLID REGIME

C.-C. Li, L. W. Engel†, D. Shahar, D. C. Tsui, and M. Shayegan
Department of Electrical Engineering, Princeton University, Princeton NJ 08544, USA

†*National High Magnetic Field Laboratory, Florida State University, Tallahassee, FL 32306, USA*

We present data on microwave conductivity, $\mathrm{Re}(\sigma_{xx})$, of a high quality 2D hole system for frequencies, f, between 0.2 GHz and 9 GHz. A resonance in $\mathrm{Re}(\sigma_{xx})$ vs f is observed for Landau filling $\nu \leq 0.3$. The resonance, whose linewidth narrows with increasing B and has a FWHM $\sim 1/5$ of the resonance f at $B=$ 14.4 T and $\nu=$ 0.154, is interpreted as a local mode of Wigner solid pinned by residual impurities.

We have measured the microwave conductivity, $\mathrm{Re}(\sigma_{xx})$, for a high-quality 2D hole system (2DHS), for frequency $0.2 \leq f \leq 9$ GHz, and at the temperature of 50 mK. The sample was a heterojunction grown on (311)A GaAs substrate, with holes produced by Si modulation doping. Hole density was 5.37×10^{10} cm^{-2}, and mobility was about 2.5×10^5 cm^2/V·sec at 350 mK. The sample in this work was taken from the same wafer (M259), used by Santos *et al.* [1] in a study of reentrance of insulating behavior around the 1/3 fractional quantum Hall effect (FQHE) in a 2DHS. Effective mass of the 2DHS, $m^*=$ 0.37 m_0, is about five times that of 2D electrons, so that it is of interest to compare the dynamics observed in the two kinds of systems.

This work focuses on the insulator that is observed at high B. This insulator is of special interest since it may be a pinned Wigner crystal, so that the pinning modes may be observed directly in the microwave conductivity. Previous microwave studies by transmission line methods [2,3] (like the one used here), and by surface acoustic wave propagation [4], revealed resonances in the high B insulators of high or intermediate [3] quality 2DES's.

The measurement technique [5] employs cold detectors, ~1 cm from the sample. The attenuation of a coplanar waveguide transmission line[6], patterned onto the surface of a sample, is measured, and the conductivity $\mathrm{Re}(\sigma_{xx})$ is calculated from the measured attenuation, accounting for reflections, for the possiblity that σ_{xx} may have a large imaginary part, and for the distributed capacitive coupling between the 2DHS and the transmission line. The metal pattern of the transmission line is shown in the inset to Fig. 1, and consists of a 45μm wide center strip separated from side planes by gaps of width W=30μm.

We measured dc transport using In/Zn ohmic contacts on the edges of the sample, outside of the transmission line pattern, and the diagonal resistance R_{xx} vs B is shown in Fig. 1, for two temperatures (T) of 30 and 130 mK. The quality of the sample is reflected in the well-developed 1/3 FQHE, with insulating phase reentrant around it, as observed earlier by Santos *et al.* [1].

Fig. 2 shows Re(σ_{xx}) vs B for several f's, for $T = 50$ mK. For the insulator with $\nu > 1/3$ the dependence of Re(σ_{xx}) on f is weak, and no resonance can be detected within the noise of our instruments. The contactless microwave technique allows measurement of linear-regime transport to deeper in the higher B insulator than is possible at dc and low T, since the dc resistance of the sample becomes extremely high. In the insulator for $\nu \leq 0.3$, though, a strong f dependence clearly exists. At 0.20 GHz, Re(σ_{xx}) vs B shows a small peak, about 2 μS high, for B≈7.1 T, just above the 1/3 FQHE. The curve then decreases steadily with B. The 1.07 GHz Re(σ_{xx}) vs B curve, on the other hand, exhibits a maximum around 11 T, before decreasing with B. The higher f curves, which are nearest to the resonance peak frequency f_{pk} at high B, increase strongly with B in the insulator, with the 1.24 GHz curve exhibiting a maximum at 13 T, and the 1.41 GHz curve increasing throughout our experimental B range. A slight dip in the 1.41 GHz curve of the figure appears around $\nu = 1/5$ ($B \approx 11$ T).

Maxima in curves of Re(σ_{xx}) vs f are discernable at the measuring T of 50 mK, for $B \gtrsim 7.4$T (Landau fillings $\nu \lesssim 0.3$). As B is increased from this point the frequency dependence of the resonance becomes sharper and the peak height σ_{pk}, increases. The sharpest and largest resonance occurs at our maximum B of 14.4 T, and Re(σ_{xx}) vs f is shown at this B in Fig. 3. f_{pk} is 1.25 GHz, and the resonance FWHM is only 0.25 GHz, implying an oscillator Q of about 5.

The f_{pk} of 1.25 GHz is within the range of resonance frequencies observed for high B insulators in 2DES's [2,3,4]. In an interpretation of the resonance as a pinning mode[7] of a WC, $f_{pk} = \omega_0^2/(2\pi\omega_c)$, where $\omega_c = eB/m$, and ω_0 is the pinning frequency, a parameter of the restoring potential $M\omega_0^2$, felt by domains of mass M displaced by an electric field. With a given domain size, impurity potential, and carrier density, the effective mass cancels out of the formula for f_{pk}, so the rough agreement of f_{pk} in 2DHS and 2DES is not surprising.

The observed Q of the resonance gives an upper limit of the damping of an individual oscillator, since inhomogenous broadening also contributes to the observed width[8]. High observed Q implies not only that oscillators are slightly damped, but that the oscillators in the sample are quite similar to one another. Well-separated pinning centers may satisfy this requirement, as would many like traps for individual holes. We argue against the second possibility, since in that case ω_0 would be constant, and the observed f_{pk} would decrease with B, contrary to our observation. The well-separated pinning centers would be most likely to produce strong pinning [7], in which the domain size L is fixed by the impurity spacing. Then the observed f_{pk} dependence on B would be explained by an increase in WC stiffness as B is increased[3]. Strong pinning for a 2DES was proposed by Li *et al.* [9] as an explanation of the pinning energies deduced from static dielectric constant measurements.

Finally, the oscillator model predicts a per-carrier oscillator strength[3,4,7], allowing us to assess the fraction of holes participating in the resonance by integrating Re(σ_{xx}) vs f plots. The integral of such a plot is expected to be $ne^2\omega_0^2/2m^*\omega_c^2 = (ne/B)\pi f_{pk} = 24$ μS-GHz, for f_{pk} of 1.25 GHz, $B = 14.4$ T, and all holes participating in the resonance. A numerical integral of the Re(σ_{xx})vs f data, for f between 0.5 and 2.0 GHz, gives 8.4 μS-GHz at $B = 14.4$ T, so according to the oscillator model, about a third of the holes participate in producing the resonance.

Acknowledgments

This work is supported by the National Science Foundation.

References

1. M. B. Santos, J. Jo, Y. W. Suen, L. W. Engel, and M. Shayegan, Phys. Rev. B **46**, 13639 (1992).
2. F. I. B. Williams, E. Y. Andrei, R. G. Clark, G. Deville, B. Etienne, C. T. Foxon, D. C. Glattli, J. J. Harris, E. Paris P. A. Wright in *Localization and Confinement of Electrons in Semiconductors*, Ed. F. Kuchar, H. Heinrich and G. Bauer, (Springer-Verlag, Berlin 1990).
3. L. W. Engel, C.-C. Li, D. Shahar, D. C. Tsui, and M. Shayegan, (unpublished).
4. M. A. Paalanen, R. L. Willett, R. R. Ruel, P. B. Littlewood, K. W. West, L. N. Pfeiffer and D. J. Bishop, Phys. Rev. B **45**, 11342 (1992); *ibid.* B **45**, 13784 (1992).
5. L. W. Engel, D. Shahar, Ç. Kurdak, and D. C. Tsui, Phys. Rev. Lett. **71**, 2638 (1993).
6. C. P. Wen, *IEEE Trans. Microwave Theory Tech.*, **MTT-17**, 1087 (1969).
7. H. Fukuyama and P. A. Lee, Phys. Rev. B **17**, 535 (1978); *ibid.* B **18**, 6245 (1978).
8. A. J. Millis and P. B. Littlewood, Phys. Rev. B **50**, 17632 (1994).
9. Y. P. Li, D. C. Tsui, T. Sajoto, L. W. Engel, M. Santos and M. Shayegan, Solid State Commun. **95**, 619 (1995).

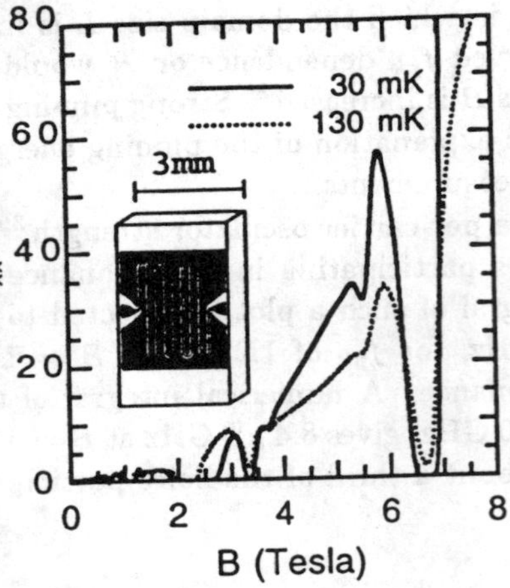

Figure 1: dc magnetoresistance vs magnetic field, at T= 30 mK and 130 mK. inset: the pattern of transmission line on sample surface, with black indicating metal film.

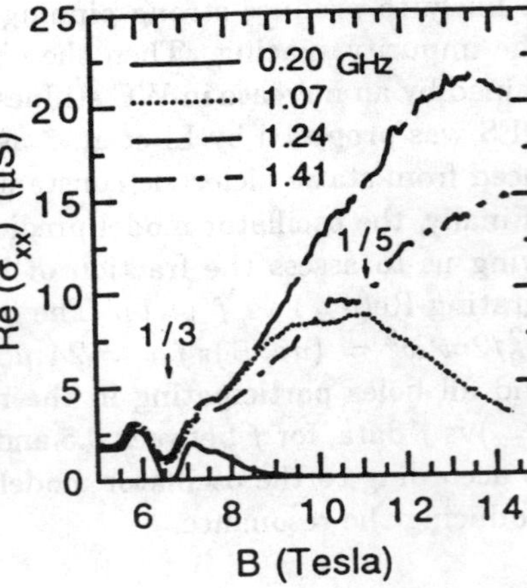

Figure 2: Real part of diagonal conductivity vs magnetic field for several frequencies, T= 50 mK.

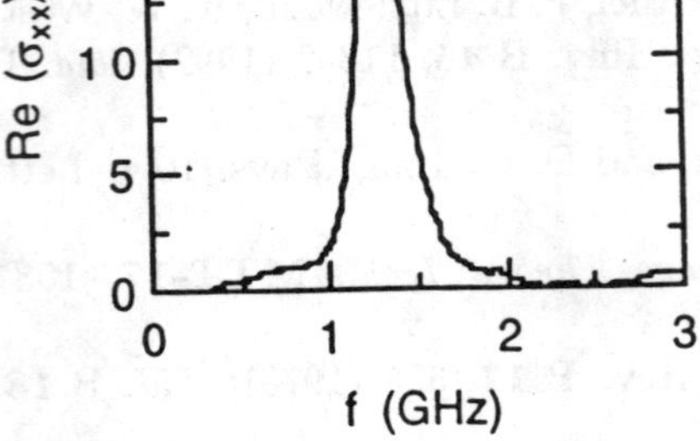

Figure 3: Real part of diagonal conductivity vs frequency for B=14.4 Tesla,T=50 mK.

Rf Absorption Study of the Magnetically-Induced Insulating Phase of a 2DHS

P.F. Hennigan, C.J. Mellor, M. Henini
Department of Physics, University of Nottingham, Nottingham, NG7 2RD, UK

A. Beya, F.I.B. Williams
SPEC, Centre d'Etudes de Saclay, CEA, 91191 Gif sur Yvette Cedex, France

The magnetically-induced insulating phase of a high mobility two dimensional hole system (2DHS) has been investigated using rf absorption. At filling factors, ν, above 1/3 clear signatures of the fractional quantum Hall effect (FQHE) are observed. Below $\nu = 1/3$, the nature of the absorption changes and a narrow absorption feature is observed at a frequency that increases with magnetic field. This absorption is interpreted as absorption by the magnetophonons of the pinned magnetically-induced Wigner solid (MIWS). This is the first observation of the MIWS in a 2DHS using rf absorption. At $\nu = 1/5$ the frequency and amplitude of the absorption decrease, suggesting competition between the FQHE state and the MIWS. No evidence for a re-entrant phase around $\nu = 1/3$ is seen.

1 Introduction

Studies of the magnetically-induced insulating phase of two-dimensional electron (2DES) and hole (2DHS) systems in high quality GaAs/AlGaAs heterostructures have been made using transport [1], optical [2], surface acoustic wave (SAW) [3] and rf absorption techniques [4]. The possibility that the insulating phase is a Wigner solid pinned to impurities in the system has been a point of great debate. In zero magnetic field, the ratio of the correlation energy to the kinetic energy of the carriers can be characterised by the parameter $r_s = a/a_B$ where $a = (\pi n_s)^{-1/2}$ and $a_B = 4\pi\varepsilon_0\varepsilon\hbar^2/(me^2)$, ($n_s$ is the areal density, m is the mass of the carrier). A zero-field Wigner transition is expected when $r_s > 33$ [5]. For electrons in GaAs $r_s \sim 2-3$, whereas for holes $r_s \sim 10-15$. Thus holes are closer to the critical value of r_s and therefore are expected to crystallize at a higher filling factor than electrons; the magnetic field helping to confine the carriers. In low density 2DHS, the insulating phase appears close to a Landau level filling factor $\nu = 1/3$ rather than $\nu = 1/5$ as observed in a 2DES. Therefore the study of the MIWS in a 2DHS can be taken much further into the insulating region than is possible with a 2DES of the same carrier density and range of magnetic fields.

A solid differs from a liquid by its elastic response to shear, leading to transverse phonons and an inability to flow round obstacles. In a strong magnetic field the longitudinal, transverse and pinned modes become coupled due

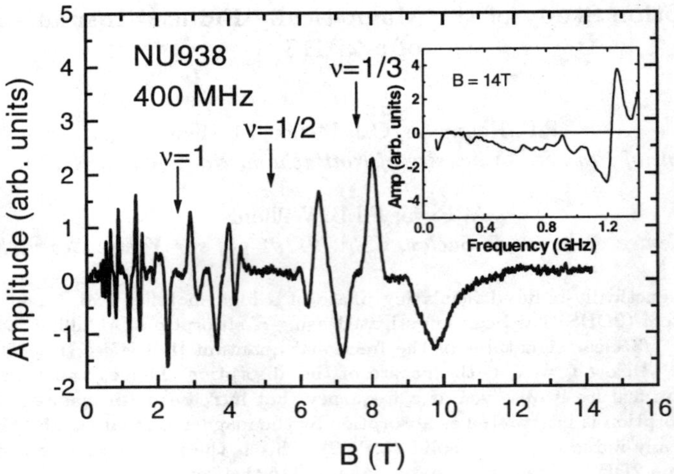

Figure 1: Detected output at 400MHz. Inset shows frequency sweep at 14T.

to the Lorentz force. This leads to a magnetophonon mode of the pinned Wigner solid that has a finite frequency at zero wavevector due to the pinning. At non-zero wavevector the energy of the mode increases with the in-plane wavevector. The frequencies are in the GHz region and can be probed using rf absorption and SAW techniques. SAW absorption occurs when the energy and wavevector of the SAW match those of the magnetophonon excitations. Due to the relatively low SAW velocity only one point on the magnetophonon curve has been seen with this technique. In contrast, rf absorption uses a meandering stripline above the sample to impose a harmonic set of wavevectors on the system. The frequency can be continuously varied, absorptions occurring whenever energy and momentum conservation can be satisfied. This allows the possibility of mapping the magnetophonon dispersion curve. As modes will only exist if the system exhibits a finite shear modulus, evidence for a solid phase can be obtained.

2 Experimental

The GaAs/AlGaAs heterojunction sample was placed in the mixing chamber of a dilution refrigerator below 50mK. The sheet density of the 2DHS

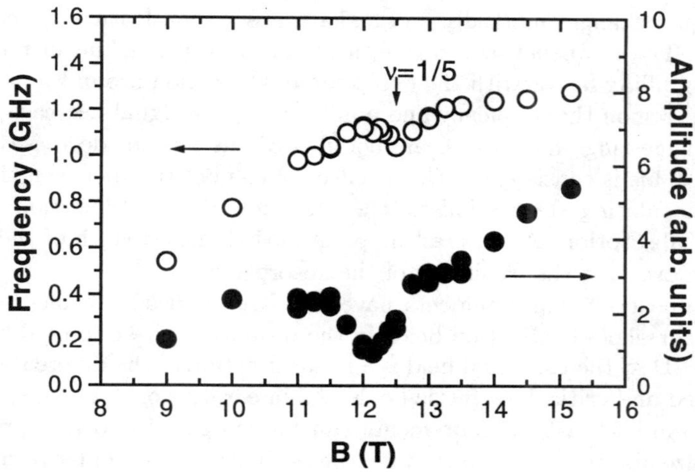

Figure 2: Frequency and amplitude of first absorption line as a function of magnetic field.

was $6 \times 10^{10} cm^{-2}$ and the low temperature mobility was greater than $8 \times 10^5 cm^2V^{-1}s^{-1}$. A meander line of $16\mu m$ period was placed on top of the sample. This rf transmission line imposes a spatially varying rf field on the 2DHS. A low frequency voltage of less than 1V amplitude was applied to the backgate to modulate the rf absorption of the 2DHS. This enables any change in transmission through the meander line due to the 2DHS to be detected using lock-in-amplifier techniques. The technique provides a high level of immunity from the effect of reflections along the coaxial transmission line used to get the rf signal to and from the low temperatures. The homodyne spectrometer[4] used in this study had a frequency range of 0.4-1.4GHz. To prevent carrier heating, signal levels in the meander line were kept to less than -70dBm, approximately 10^{-3} of which is absorbed by the sample.

3 Results and Discussion

At filling factors higher than 1/3 we clearly detect the integer and fractional quantum Hall effects in the rf absorption (Fig. 1). It should be remembered that due to the modulation technique we detect the derivative of the absorption with respect to the backgate voltage. Below $\nu = 1/3$ the frequency dependence of the signal changes markedly and a sharp absorption feature appears (inset Fig. 1). The frequency and strength of the absorption line increases with decreasing filling factor with the exception of the region around $\nu = 1/5$ (Fig. 2). In this region the frequency and amplitude of the signal decrease and then recover suggesting that the shear modulus of the system decreases close to $\nu = 1/5$. This is consistent with the effect of competition between the FQHE and the insulating state. This is the first time such competition has been seen in rf absorption. An interesting point is that the strength of the signal is affected more than the frequency of the absorption.

Non-linear I-V measurements have also been carried out on this sample allowing threshold conduction fields in the insulating phase to be determined. As in the 2DES the threshold field is a linear function of the difference between the applied and critical magnetic fields. As in earlier work[4] on the 2DES, the rf results and the I-V measurements can be compared and an approximate domain size obtained. For our sample the estimate yields similar results to the 2DES with a domain containing of order 100 carriers.

4 Conclusions

Clear observation of rf absorption in the insulating phase of a 2DHS has been found. This is consistent with the existence of a pinned Wigner solid at filling factors less than $\nu = 1/3$. No evidence of a re-entrant state around $\nu = 1/3$ has been seen. Close to $\nu = 1/5$ the absorption frequency and the amplitude of the absorption decrease providing evidence of competition between the FQHE and the MIWS.

Acknowledgments

We gratefully acknowledge the funding from the E.P.S.R.C. and the support of the C.E.A. that has made this collaboration possible.

References

1. M.B. Santos, *et al*, *Phys. Rev. Lett.* **68**, 1188 (1992).
2. I.V. Kukushkin, *et al*, *Phys.Rev. Lett.* **72**, 3594 (1994).
3. M.A. Paalanen, *et al*, *Phys. Rev.* B **45**, 11342 (1992).
4. F.I.B. Williams, *et al*, *Phys. Rev. Lett.* **66**, 3285 (1991).
5. D. Ceperley, *Phys.Rev.* B **18** 3126 (1978); M. Imada. and M. Takahashi *J. Phys. Soc. Jpn.* **53** 3770 (1984).

ELECTRON-HOLE CHAINS IN UNBALANCED DOUBLE LAYERS

H. AKERA
Department of Applied Physics, Hokkaido University, Sapporo 060, Japan

A.H. MACDONALD
Department of Physics, Indiana University, Bloomington, Indiana 47405, USA

D. YOSHIOKA
Department of Basic Science, University of Tokyo, Komaba, Tokyo 153, Japan

We propose novel strongly correlated states for unbalanced double-layer electronic systems in strong magnetic fields. Our proposal is based on simple arguments, corroborated in part by numerical exact-diagonalization calculations. The ground state of an electron-hole pair in quantized cyclotron orbits with different radii possesses an electric dipole moment. The resulting attractive interactions between pairs creates the possibiliy of states with electrons and holes alternating along chains.

Studies of the fractional quantum Hall effect have been extended in recent years to include double-layer electronic systems (DLES's) composed of two nearby two-dimensional electron layers.[1] Depending on density, magnetic field strength, and layer separation, DLES's have already been shown to possess a wide variety of interesting properties. In this paper we propose that in a strong magnetic field the properties of *unbalanced double layers*, in which the electron densities in the two layers are substantially different, will differ qualitatively from those of DLES's studied previously.

Specifically we consider the case where Landau level $n_L > 0$ in one layer is degenerate with the lowest Landau level ($n_S = 0$) in the other layer. For simplicity we assume that the strong field produces Landau level separations and spin splittings large enough to permit us to focus on a single set of orbitally degenerate levels with fixed kinetic energy and spin polarization in each layer. (The full Landau levels of the high-density layer are assummed to be frozen.) The quantum orbitals in higher Landau levels have ring-shaped electron-density profiles; the radii of the rings corresponds to cyclotron orbit radii for classical particles with the same kinetic energy. In the $n = 0$ Landau level the ring structure is completely obscured by quantum smearing. In the quantum limit where the kinetic energy is quantized, the effective Coulomb interaction between electrons in the different Landau levels should have a non-monotonic dependence on the separation between orbit centers with a local *minimum* occurring for coincident centers. This is entirely different from the previously studied case in which both layers have $n = 0$. In our previous studies,[2] it was shown that the corresponding Haldane pseudopotentials[3] indeed have a local *minimum* at zero relative angular momentum when $n_L > 0$ and the layer distance is small. Our numerical exact-diagonalization study for small systems suggested that the attractive interaction at short distances can produce states in which electrons pair.

In this paper we suggest another interesting type of correlated state, which could occur when holes are introduced in the n_Lth Landau level. In the unbalanced case ($n_L > 0$) for sufficiently small layer separations the exciton ground state occurs at a finite distance between orbital centers, corresponding to the maximum appearing in the effective electron interaction, and possesses finite (in-plane) electric dipole. The resulting attractive interactions between excitons open the possibiliy of states with unusual correlations. We will present results of numerical diagonalization calculations which show the presence of strong correlations with electrons and holes alternating along chains.

States of a single exciton are labeled by the center-of-mass wavevector ($\vec{K}$) and in the quantum limit the electron and hole in an exciton have their cyclotron orbit centers displaced by $\vec{z} \times \vec{K}\ell^2$ with $\vec{z} = (0,0,1)$. If we neglect the possibility of tunneling between the layers (we can enforce this condition artificially by assuming that the two Landau levels of interest have opposite spins), the energy $E_{\vec{K}}$ of an exciton with an electron in Landau level n_S and a hole in Landau level n_L is given by[4,5]

$$E_{\vec{K}} = -\int_0^\infty q dq V(q) e^{-q^2\ell^2/2} J_0(Kq\ell^2) L_{n_L}(q^2\ell^2/2) L_{n_S}(q^2\ell^2/2). \quad (1)$$

Here $\ell = (c\hbar/eB)^{1/2}$ and $V(q) = e^2 \exp(-qd)/\varepsilon q$ for thin layers separated by a distance d. $J_0(x)$ is the Bessel function and $L_n(x)$ is the Laguerre polynomial. For nearby unbalanced double layers with $n_S = 0$ and $n_L > 0$, the lowest-energy state should be expected to occur when the orbit center separation is close to the radius of the larger orbit $\sqrt{2n_L+1}\ell$. The energy dispersion is plotted in Fig.1 and has a minimum around $K\ell = \sqrt{2n_L+1}$, as expected, when d is less than $\sim \ell$. The associated in-plane electric dipole gives rise to attractive interactions between excitons, the consequences of which we have investigated by exact-diagonalization calculations.

We assume that the density in the two layers has been adjusted so that the number of holes in the Landau level of the high-density layer equals the number of electrons in the low-density layer. Standard fractional-quantum-Hall-effect exact-diagonalization codes[6] are readily extended to unbalanced double-layer systems by inserting the form factors $L_{n_L}(q^2\ell^2/2)L_{n_S}(q^2\ell^2/2)$.

An example of the ground-state correlations is shown by plotting in Fig.2

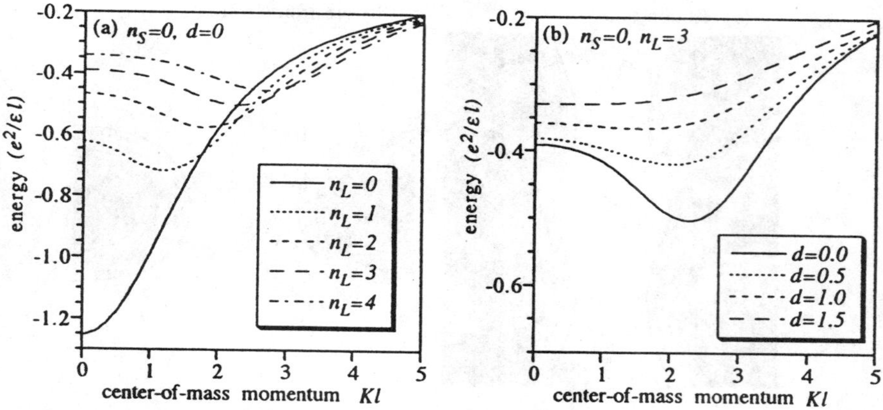

Figure 1: Energy dispersion of a magnetoexciton with an electron in Landau level n_S and a hole in Landau level n_L for unbalanced double layer systems. (a) n_L dependence. (b) layer separation d dependence (in units of ℓ).

the (a) electron-hole and (b) electron-electron correlation functions for $n_L = 3$, $d = \ell$ and for the aspect ratio of the cell at which the energy is minimized ($a_y/a_x = 1.07$). These results are for two electrons and two holes in a cell with 20 single-electron states per Landau level per layer (periodic boundary conditions have been applied.). Electrons and holes alternate along the vertical axis (ring-shaped orbits of holes are evident in the electron-hole correlation function). An effect of quantum fluctuations is seen in the electron-electron correlation function in Fig.2(b): the electron at the center of a side fluctuates perpendicular to the chain relative to the electron at a corner.

The electron-hole chain is found in a wide range of n_L, d, and the filling factor. In particular, it exists even when the exciton dispersion has no pronounced minimum at non-zero momentum for larger layer separations, for example, at $d = 1.5\ell$ ($n_L = 3$). Although in this case the single-exciton state has no dipole, the chain correlation is stabilized by attractive interactions between polarized excitons. When the layer separation becomes even larger or the orbit radius becomes smaller (smaller n_L), the ground state undergoes a transition into a triangular-lattice state, which is essentially the dipole density wave proposed previously[7,8] for $n_L = n_S = 0$.

In conclusion we have suggested that unbalanced double-layer electronic systems have the possibiliy of showing novel strongly-correlated states in strong magnetic fields.

One of the authors (H.A.) would like to thank Y. Takagaki, S. Tarucha,

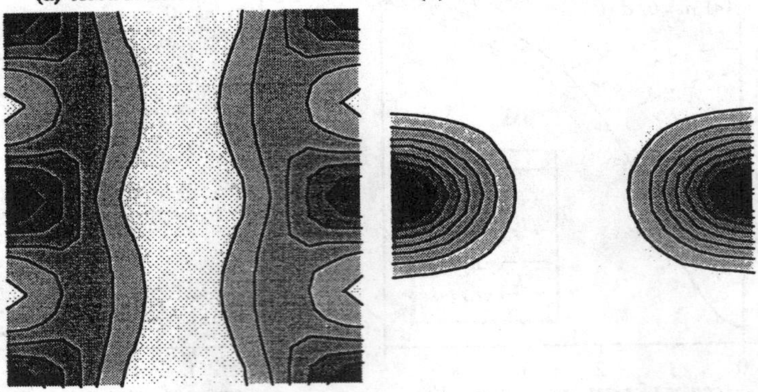

Figure 2: Electron-hole chain with two electrons and two holes in a periodic cell, for an unbalanced double layer with $n_S = 0, n_L = 3$ and layer separation $d = \ell$. The filling factor of Landau level n_S is 1/10. A contour plot of the ground-state correlation function with one electron located at the lower left corner of the cell $g(\vec{r})$ is given for (a) electron-hole and (b) electron-electron pairs. Darker area has larger values.

and N. Tokuda for valuable discussions. This work was supported in part by the Grant-in-Aid for Encouragement of Young Scientists from the Ministry of Education, Science, Sports and Culture, Japan and by the US National Science Foundation under grant DMR-9416906.

References

1. S. M. Girvin and A. H. MacDonald, in *Novel Quantum Liquids in Low-Dimensional Semiconductor Structures*, eds. Sankar Das Sarma and Aron Pinczuk (Wiley, New York, 1996) and work cited therein.
2. H. Akera, A.H. MacDonald, and D. Yoshioka, *Physica* B **201**, 255 (1994); **212**, 273 (1995).
3. F.D.M. Haldane, in *The Quantum Hall Effect*, eds. R.E. Prange and S.M. Girvin (Springer-Verlag, New York, 1987) p.303.
4. I.V. Lerner and Yu.E. Lozovik, *Zh. Eksp. Teor. Fiz.* **78**, 1167 (1978) [*Sov. Phys.-JETP* **51**, 588 (1980)].
5. C. Kallin and B.I. Halperin, *Phys. Rev.* B **30**, 5655 (1984).
6. D. Yoshioka, *Phys. Rev.* B **29**, 6833 (1984).
7. D. Yoshioka and H. Fukuyama, *J. Phys. Soc. Jpn.* **45**, 137 (1978).
8. D. Yoshioka and A.H. MacDonald, *J. Phys. Soc. Jpn.* **59**, 4211 (1990).

DYNAMICS OF COMPRESSIBLE EDGE AND BOSONIZATION

J.H. HAN, D.J. THOULESS
Department of Physics, Box 351560, University of Washington, Seattle, Washington 98195

We work out the dynamics of the compressible edge of the quantum Hall system by introducing a generalized version of Wen's hydrodynamic quantization approach. We rederive Aleiner and Glazman's earlier result based on classical hydrodynamics. We also analyze the dynamics starting with the second-quantized Hamiltonian in the lowest Landau level and work out the time development of various density mode operators. It turns out the density modes are strongly coupled and there is one stable, charged mode to which all other so-called acoustic modes flow.

1 Introduction

The theory of edge excitations in the fractional quantum hall liquid had been first derived by Wen[1] based on the gauge invariance of the Chern-Simons term for a bounded system. He and many of the subsequent authors visualized the edge to be where a sharp change of density takes place. With a sharp edge, there are so few degrees of freedom that one cannot distinguish between collective modes and single-particle excitations. This leads to bosonization of one-dimensional fluids.[1,2]

Experimentally, a sharp edge is an exception rather than a norm. It was demonstrated by Chklovskii *et al.*[3] that with the gate-confined 2DEG, the edge is a rather wide, smoothly varying region of density. The dynamics of the smooth edge had been worked out by Aleiner and Glazman (AG) [4] using classical hydrodynamics of charged fluid.

In this paper, we work out the dynamics of the wide edge in the quantum-mechanical setting and examine the applicability of bosonization scheme to such a system.

2 Phenomenological approach to edge dynamics

The geometry of the sample is that of a semi-infinite Hall bar where the boundary separating the incompressible and compressible (IC) liquid is defined as $y = 0$ and the compressible strip extends in the positive y direction. The equilibrium density, $n_0(y)$, runs from $\nu/2\pi$ to zero.

In the presence of a relatively weak local electric field fluctuation and a strong external magnetic field, one can write $v_x = \delta E_y/B$ and $v_y = -\delta E_x/B$. To completely specify the motion, we need the equation of continuity and

$n(x, y + h(x, y, t), t) = n_0(y)$ for the density $n(x, y)$. This equation parameterizes the dynamics by identifying points for a given density $n_0(y)$ with $(x, y+h(x, y))$. For small fluctuations, one can show the following equation holds $(h(x, y, t) \equiv \partial\eta(x, y, t)/\partial x)$.

$$\frac{\partial^2 \eta}{\partial x \partial t} = v_y - v_x \frac{\partial^2 \eta}{\partial x^2} = \frac{1}{B}\frac{d}{dx} V(x, y + \frac{\partial \eta(x, y, t)}{\partial x}, t), \tag{1}$$

where $V(x, y + h)$ is the local potential (fluctuation) at time t. The corresponding action can be found and the proper set of conjugate fields identified: $[h(x, y), \eta(x', y')] = i\delta(x - x')\delta(n_0(y) - n_0(y'))$. Note that all of our analyses reduce to the sharp-edge results in the limit $\partial n_0(y)/\partial y = -(\nu/2\pi)\delta(y)$.

The amount of energy necessary to cause a density fluctuation and also to move some of the electrons on the incompressible side against the confining field is given by

$$H = \frac{e^2}{2\kappa} \int \frac{\partial n_0}{\partial y}\frac{\partial n_0}{\partial y'}\frac{h(r)h(r')}{|r - r'|} d^2r d^2r' + \frac{\nu v_F}{4\pi} \int h(x, 0)^2 dx. \tag{2}$$

The eigenstates and eigenvalues for the Coulomb part of eq. (2) were worked out by AG, and we will denote them as $(\omega_\alpha(q), h_{q\alpha}(y))$ where q is an integer multiple of $2\pi/L$ and L is the length of the sample. In the basis of eigenfunctions, $h(x, y) = L^{-1/2} \sum_{q\alpha} c_{q\alpha} h_{q\alpha}(y) e^{-iqx}$, and the Hamiltonian becomes

$$H = \sum_{q\alpha} \left(\frac{\omega_\alpha(q)}{2q} + \frac{\nu v_F}{4\pi} h^2_{q\alpha}(0) \right) c^*_{q\alpha} c_{q\alpha} + \sum_{q, \alpha \neq \beta} \frac{\nu v_F}{4\pi} h_{q\alpha}(0) h_{q\beta}(0) c^*_{q\alpha} c_{q\beta}, \tag{3}$$

with the commutator $[c^*_{q\alpha}, c_{q\alpha}] = q$. The boundary term modifies the Hamiltonian but the perturbation is quadratic in the field. Therefore one can rotate to a basis where one recovers a set of independent harmonic oscillators. Each mode corresponds to a coherent density fluctuation over the IC boundary and the compressible strip. For each mode, we can construct a fermion operator according to the usual prescription of bosonization.

3 Microscopic approach to edge dynamics

Our starting point is the Coulomb Hamiltonian,

$$H = \frac{e^2}{\kappa L} \sum_{k_1 k_2 q} e^{-\frac{1}{2}q^2} K_0(|qk_1 - qk_2|) c^\dagger_{k_1+\frac{q}{2}} c_{k_1-\frac{q}{2}} c^\dagger_{k_2-\frac{q}{2}} c_{k_2+\frac{q}{2}} \tag{4}$$

expanded in the lowest Landau level. The kernel, K_0, is the zeroth-order modified Bessel function. The operators are not written in the normal-ordered form, but the difference is a constant which can be ascribed to a positive charge background. Since the eigenfunctions derived in the previous section are a complete set, we can write the kernel in the factorized form, $K_0(|qk-qk'|)=(\kappa/2e^2|q|)\sum_{\alpha=0}^{\infty}\omega_\alpha(q)h_{q\alpha}(k)h_{q\alpha}(k')$. The Hamiltonian factorizes as well,

$$H = 1/2\sum_{q\alpha}\omega_\alpha(q)e^{-\frac{1}{2}q^2}P_\alpha^\dagger(q)P_\alpha(q), \tag{5}$$

where $P_\alpha^\dagger(q)=(1/\sqrt{|q|L})\sum_k h_{q\alpha}(k)c_{k+q/2}^\dagger c_{k-q/2}=P_\alpha(-q)$. We find that the microscopic Hamiltonian is reduced to an extremely simple form which resembles the harmonic oscillators of the previous section. It is tempting to suppose that each $P_\alpha^\dagger(q)$ creates a quantum of fluctuation.

To examine the validity of the hydrodynamic result, one needs to look at the equation of motion of $P_\beta^\dagger(p)$,

$$-i\dot{P}_\beta^\dagger(p)=\frac{1}{2}\sum_{q\alpha}\omega_\alpha(q)\{P_\alpha^\dagger(q)[P_\alpha(q),P_\beta^\dagger(p)]+[P_\alpha(q),P_\beta^\dagger(p)]P_\alpha^\dagger(q)\}. \tag{6}$$

We have absorbed the overlap integral $e^{-q^2/2}$ into $\omega_\alpha(q)$. We will linearize the r.h.s. of eq. (6) by taking the expectation value of the commutators in the ground state. For the ground state, we assume $|g.s.\rangle=\prod_k(\sqrt{1-n_k}+\sqrt{n_k}c_k^\dagger)|vac\rangle$. There is no consensus as to the nature of the ground state of the compressible region at $T=0$, but we believe our wavefunction is a reasonable description at finite T, where possible many-body correlations of the electrons inside the compressible region are lost.

The commutators can now be worked out straightforwardly. For $p=q$, it is equal to $\delta_{\alpha\beta}$ and reproduces the hydrodynamic piece. The $p\neq q$ part leads to the coupling of various density modes. We find it more convenient to carry out the analysis in terms of $P_\alpha^\dagger(p)/\sqrt{p}$, so refer to it as $P_\alpha^\dagger(p)$. With $\nu=1$ and assuming the existence of just two modes, $\alpha=0$, and 1, we have

$$\begin{aligned}
-i\dot{P}_0^\dagger(p) &= \omega_0(p)P_0^\dagger(p)+\frac{4\sqrt{2}}{\pi^2}\omega_1(p)\sum_q P_1^\dagger(q),\\
-i\dot{P}_1^\dagger(p) &= \omega_1(p)P_1^\dagger(p)+\frac{4\sqrt{2}}{\pi^2}\sum_q \omega_0(q)P_0^\dagger(q).
\end{aligned} \tag{7}$$

This is the crucial difference that arises as a result of the microscopic calculation. The modes which were thought to be independent in the hydrodynamic model are in fact strongly coupled. Since a lot of the experiments to detect the edge waves are done at the fundamental frequency, the sum in the eq. (7) can be replaced by a single entry with $q=p=2\pi/L$, and we end up with a simple coupled pendulum problem. The coupling results in two frequencies, one of which is larger than ω_0 and the other lower than ω_1. The low frequency mode is over-damped due to dissipation.[4] Due to the coupling in eq. (7), the acoustic mode propagates even more slowly than was expected in the classical analysis, thus making it harder to detect. Experiments on edge magnetoplasmon waves have not been able to detect the acoustic modes even in very high mobility samples.[5] The single-particle operators associated with acoustic modes will likewise be short-lived. With $\nu = 1/3$, there is also a coupling within the same mode, $P_\alpha(p) \leftrightarrow P_\alpha(p')$, but otherwise, the analysis runs similar to $\nu = 1$.

Tunneling of electrons to the fractional edge is a powerful method to probe the microscopic structure of the edge. Recently, Chang *et al.*[6] carried out such a tunneling experiment for a sharp edge. He obtained good agreement with theory, but some of the exponents he observed were not strictly integral, e.g., 2.7 instead of 3. Residual effects of the acoustic mode may be partially responsible for the deviation.

Acknowledgments

We are grateful to Ping Ao, Boris Spivak, Kiril Tsemekhman, Vadim Tsemekhman, and Carlos Wexler for useful discussions throughout the development of our work. This work was supported in part by the National Science Foundation under grant DMR-9528345.

References

1. X. G. Wen, Phys. Rev. B **41**, 12838 (1990); X. G. Wen, Phys. Rev. B **43**, 11025 (1991)
2. X. G. Wen, Int. J. Mod. Phys. B **6**, 1711 (1992).
3. D. B. Chklovskii, B. I. Shklovskii, and L. I. Glazman, Phys. Rev. B **46**, 4026 (1992).
4. I. L. Aleiner and L. I. Glazman, Phys. Rev. Lett. **72**, 2935 (1994)
5. V. I. Talyanskii *et al.*, Phys. Rev. B **50**, 1582 (1994); V. I. Talyanskii *et al.*, J. Phys.: Condens. Matter **7**, L435-L443 (1995).
6. A. M. Chang, L. N. Pfeiffer, and K. W. West, preprint.

INTERACTING ELECTRONS IN HIGH LANDAU LEVELS

R. MOESSNER, J.T. CHALKER

Theoretical Physics, 1 Keble Road, Oxford OX1 3NP, England

We study a two-dimensional electron system in a magnetic field with a fermion hardcore interaction and without disorder. Projecting the Hamiltonian onto the nth Landau level, we show that Hartree-Fock theory is exact in the limit $n \to \infty$, for the high temperature, uniform density phase of an infinite system; for a finite-size system (in which the Larmor radius is larger than the system size as $n \to \infty$), it is exact at all temperatures. In addition, we show that a charge-density wave arises below a transition temperature T_t. Using Landau theory, we construct a phase diagram and find that it contains both unidirectional and triangular charge-density wave phases.

1 Introduction

For a two-dimensional electron system (2DES) in a high magnetic field B in the absence of interactions and disorder, each Landau level is macroscopically degenerate. Interactions between the electrons lift this degeneracy and, in the lowest Landau level ($n = 0$), give rise to phenomena such as the fractional quantum Hall effect.

Here,[a] we consider what happens if it is a high Landau level (n large) that is partially occupied.[b] In this case a second, much larger lengthscale appears besides the magnetic length $l_M = \sqrt{\hbar/eB}$: the Larmor radius $R_L \simeq l_M\sqrt{2n}$, which is the approximate size of the smallest wavepacket that can be constructed in the nth Landau level.

As the area of a semiclassical cyclotron orbit is proportional to R_L^2, there is massive spatial overlap between the single-particle states. As a result, fluctuations become unimportant in certain regimes and a mean field theory can be used to describe the system exactly. Formally, the parameter n^{-1} can be used as a small parameter in perturbation theory as $n \to \infty$.

[a] A more detailed account of this work will be published elsewhere.[1]

[b] Koulakov, Fogler and Shklovskii[2] have recently studied the related problem of electrons in a weak magnetic field at $T = 0$ within the Hartree-Fock approximation using an effective interaction derived by Aleiner and Glazman.[3]

2 The limit $n \to \infty$

2.1 *The Hamiltonian*

For simplicity, we consider a repulsive fermion contact potential, the Fourier transform of which is $\tilde{V}_{HC} = -vq^2$, where $v \ll \hbar\omega_c$ is the strength of the potential and ω_c is the cyclotron frequency. The Hamiltonian, in the absence of Landau level mixing, reads ($l_M \equiv \hbar \equiv 1$)

$$H = \frac{1}{2}\sum_{k,m,l} M(m,l)\, a^\dagger_{k+l} a^\dagger_{k+m} a_{k+m+l} a_k \tag{1}$$

$$M(m,l) \equiv \frac{2\pi}{L}\int \frac{d^2\mathbf{q}}{(2\pi)^2}\, \tilde{V}(\mathbf{q})\, \alpha_n(\frac{q^2}{2})\, e^{-iq_x m}\delta(q_y - l). \tag{2}$$

Here, $a^\dagger_p$ creates an electron in a Landau orbital centred at $x = p$, and $\alpha_n(q^2/2) = \left[L_n(q^2/2)exp(-q^2/4)\right]^2$, L_n being the Laguerre polynomial. We assume periodic boundary conditions in the y-direction with period L.

For $n \to \infty$, we can replace $\alpha_n(q^2/2)$ by its WKB envelope and find that $M(m,l)$ vanishes due to the rapidly oscillating exponential in its integrand unless $m \ll n^{-1/2}$ or $l \ll n^{-1/2}$. The implications of this for a system on a cylinder of fixed circumference L are as follows. Since m and l are integer multiples of $2\pi/L$, only terms with $l = 0$ or $m = 0$ survive in the limit $n \to \infty$. These are the Hartree and Fock terms, respectively. Therefore, the Hamiltonian becomes a pure Hartree-Fock Hamiltonian in the large-n limit, and its eigenstates are single Slater determinants of Landau orbitals, which have a uniform charge density around the circumference of the cylinder.

2.2 *Diagrammatic analysis*

Turning our attention to the infinite system in the uniform high temperature phase, we find that the diagrammatic calculations for the self-energy and the two-particle vertex simplify greatly and can be resummed *exactly*. For the self-energy, all diagrams with closed fermion loops or intersecting interaction lines are found to vanish, so that only the Fock diagrams have to be summed; the result is simply a shift in the chemical potential.

In the sum for the two-particle vertex, only the ladder diagrams survive the large-n limit. Summing these, we find that the two-particle vertex has a pole at a temperature $T_c = \nu(1-\nu)\tilde{u}_F(p)/2\pi$, where ν is the filling fraction of the partially filled Landau level, and $\tilde{u}_F(p)$ is the Fourier transform of the Fock potential. This indicates that the uniform phase is unstable towards the formation of a charge-density wave (CDW) at the wavevector p_M for which

$\tilde{u}_F(p)$ is maximal. For our potential, this wavector is $p_M = 2.16/R_L$: the periodicity of the CDW is of order of the Larmor radius R_L. In this sense, R_L is the dominant lengthscale for $n \to \infty$.

Yet, we also find that corrections to the Hartree-Fock approximation for a potential of finite range R are small in R/l_M, not R/R_L. This indicates that the magnetic length is still an important lengthscale for $n \to \infty$. For the screened Coulomb interaction potential,[3] this restricts the validity of the Hartree-Fock approximation to weak magnetic fields ($R/l_M \simeq 1/10$ for $B \simeq 70$mT).

3 The phase diagram for the charge-density wave states

Next we examine which CDW state has the lowest free energy.[c] This problem has been studied previously for the lowest Landau level.[4–5] Fukuyama, Platzmann and Anderson have derived an expression for the free energy of a triangular CDW;[4] by the same method, we have derived one for a unidirectional CDW.

Unlike in the $n = 0$ case, we find that in high Landau levels, it is not favourable for the charge-density wave to acquire a wavevector different from p_M. This feature simplifies the calculation significantly as the only quantity that remains to be varied is the order parameter.

Minimizing the free energy for the different CDW phases, we find that the unidirectional CDW is more favourable at $\nu = 1/2$ for T just below $T_t \equiv T_c(\nu = 1/2)$. As the filling fraction is lowered at constant T, there is a first-order phase transition to the triangular CDW at T_{12}. When ν is lowered even further, the uniform density state becomes more favourable. We have also examined competition between the striped phase and a square CDW, using the expression derived by Gerhardts and Kuramoto;[5] we find that at $\nu = 1/2$ for T just below T_t, the square CDW has a higher free energy than the unidirectional one for $n \to \infty$. The resulting phase diagram is displayed in Fig. 1 for $\nu \leq 1/2$. The results for $\nu > 1/2$ follow from particle-hole symmetry.

As the temperature is lowered far below T_t, the order parameter grows and our expansion ceases to be valid. An additional difficulty is that higher harmonics will start to play a role: the density profile must be constructed by occupying states in the nth Landau level, and must in particular not require a filling fraction larger than one at any point. For the hardcore potential, the cost for higher harmonics is small. We therefore expect that these considerations alter only the shape of the CDW without changing its structure or periodicity.

[c] We do not consider here correlated, large-scale fluctuations that might result, for example, in defect-mediated melting of the CDW.

In this way, we expect competing unidirectional and triangular patterns to persist at wavevector p_M down to zero temperature.

Extrapolating the phase-separation line T_{12} beyond the range of validity of our expansion to $T = 0$, we obtain a first-order transition at $\nu = 0.35$ for $T = 0$. This is close to the value $\nu = 0.39$ obtained previously for the screened Coulomb interaction,[2] separating the unidirectional and the triangular patterns.

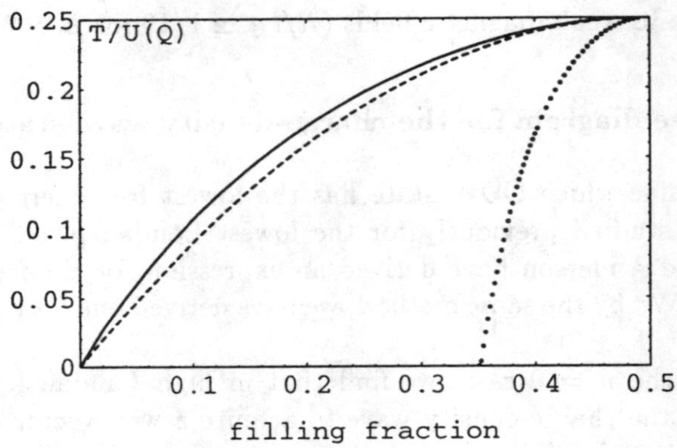

Figure 1: Phase diagram for the charge-density waves. The solid line, T_1, separates the uniform and triangular phases, the dots, T_{12}, the triangular and unidirectional phases. The dashes, T_c, denote the spinodal line. Our expansion is valid near the triple point.

Acknowledgements

One of us (JTC) is grateful to L. Glazman and B. I. Shklovskii for helpful discussions. This work was supported in part by EPSRC grant GR/GO 2727.

References

1. R. Moessner and J. T. Chalker, to appear in *Phys. Rev.* B
2. A. Koulakov, M. M. Fogler and B. I. Shklovskii, *Phys. Rev. Lett.* **76**, 499 (1996).
3. I. L. Aleiner and L. I. Glazman, *Phys. Rev.* B **52**, 11296 (1995).
4. H. Fukuyama, P. M. Platzmann and P. W. Anderson, *Phys. Rev.* B **19**, 5211 (1979).
5. D. Yoshioka and H. Fukuyama, *J. Phys. Soc. Japan* **47**, 394 (1979), D. Yoshioka and P.A. Lee, *Phys. Rev.* B **27**, 4986 (1983) and references therein. R. R. Gerhardts and Y. Kuramoto, *Z. Phys.* B **44**, 301 (1981).

SPIN-REVERSED QUASIPARTICLES IN THE FRACTIONAL-QUANTUM-HALL-EFFECT STATE

S.I.DOROZHKIN and M.O.DOROKHOVA
Institute of Solid State Physics RAS, Chernogolovka, Moscow district, 142432, Russia, E-mail: dorozh@issp.ac.ru

R.J.HAUG
Max-Planck-Institut für Festkörperforschung, Heisenbergstr. 1, D-70569 Stuttgart 80, Germany

K.PLOOG
Paul-Drude Institut für Festkörperelektronik, Hausvogteiplatz 5-7, 10117 Berlin, Germany

Changes of spin polarization of a two-dimensional electron system (2DES) in the GaAs/AlGaAs heterostructures caused by appearance of the 1/3- and 2/3-filling fractional-quantum-Hall-effect (FQHE) states have been investigated using a capacitive spectroscopy method in tilted magnetic fields. Effect of an in-plane magnetic field component connected with FQHE states has been found at filling factors $\nu > 2/3$ only, which is consistent with existence of spin-reversed quasielectrons and spin-polarized quasiholes in the 2/3-filling FQHE state.

1 Introduction

Interplay of the Coulomb and Zeeman energies, when compared one to the other, can lead to nontrivial spin configurations of a 2DES in a FQHE state at a filling factor of magnetic levels $\nu = \nu_f$ (see, for example [1], and references therein). Deviation of ν from ν_f is generally accepted to create quasielectrons ($\nu > \nu_f$) or quasiholes ($\nu < \nu_f$) in the FQHE state with a spin defined by a change of the total spin of 2DES. In particular, reduction of the polarization is considered as the appearance of spin-reversed quasiparticles. In this paper we are concerned with experimental test of the prediction [1] that quasielectrons (quasiholes) in the fully spin-polarized $\nu_f = 2/3$ FQHE state are spin-reversed (spin-polarized). To characterize the spin of 2DES we have employed the method of capacitive spectroscopy.

2 Experiment

The method of capacitive spectroscopy allows one to get information about the lowest-energy state of a 2DES. It is based on the fact that the measured value of the capacitance C between the 2DES and the gate depends on the

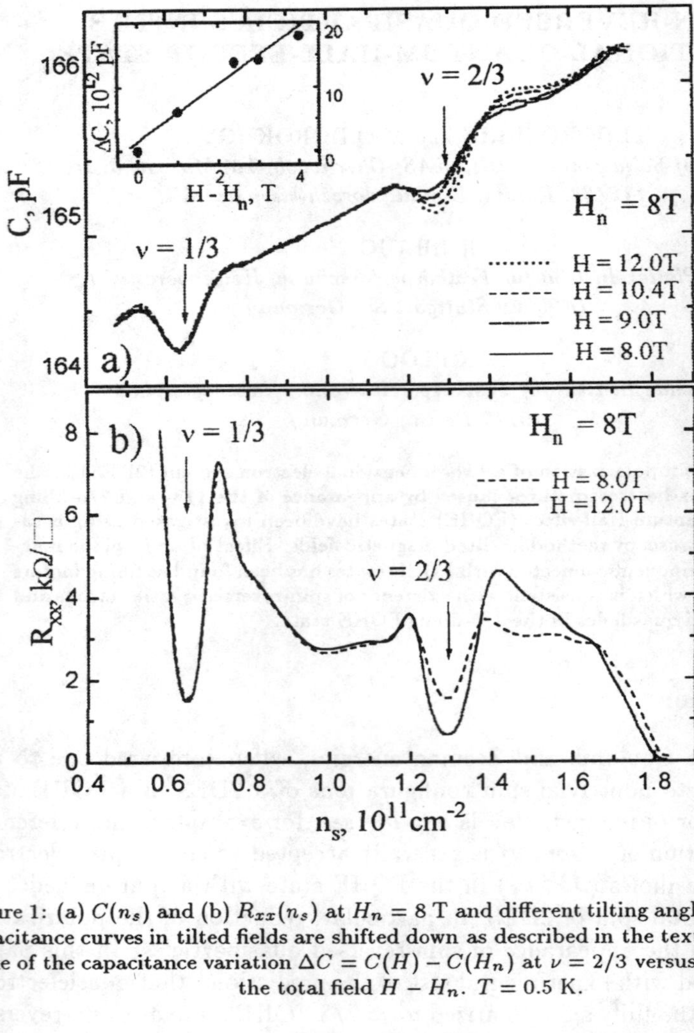

Figure 1: (a) $C(n_s)$ and (b) $R_{xx}(n_s)$ at $H_n = 8$ T and different tilting angles. Experimental capacitance curves in tilted fields are shifted down as described in the text. Inset: absolute value of the capacitance variation $\Delta C = C(H) - C(H_n)$ at $\nu = 2/3$ versus the increment of the total field $H - H_n$. $T = 0.5$ K.

derivative $d\xi/dn_s$ of 2DES chemical potential ξ over an electron areal density n_s: $S/C = S/C_g + (1/e^2)d\xi/dn_s$. Here S is a gated area of the sample and $C_g = \chi S/4\pi d$ is a geometrical capacitance determined by the effective distance d between 2DES and the gate (χ is a dielectric constant). The variation of the Zeeman energy caused by addition of the in-plane magnetic field component leads to changes in the capacitance which provide information on the spin polarization of a 2DES.

Our samples are field-effect-transistors based on a single remotely doped

GaAs/AlGaAs heterojunction with a metal gate. Capacitance between 2DES and the gate and magnetoresistance of 2DES were measured *in situ* as a function of n_s for different angles between the magnetic field and the sample plane keeping the same normal component H_n of the field. Typical experimental results are shown in Fig.1. Dips in the capacitance at $\nu = 1/3$ and $2/3$ arise from the chemical potential discontinuity at the FQHE states. Their width is determined by the scale of inhomogeneity of n_s in the sample [2,3].

In-plane magnetic field H_p shifts the capacitance curves as a whole and strongly modifies both the $C(n_s)$ and $R_{xx}(n_s)$ dependencies at $\nu > 2/3$. The shift was found to be temperature independent and has been determined for each tilting angle from similar measurements at $T = 4.2$ K where the FQHE signatures were absent. The maximum value of the shift for data of Fig.1a is 0.07 pF at $H = 12$ T. By compensation of this shift we exclude all effects of H_p except those related to the FQHE. Both, the capacitance and the magnetoresistance data in Fig.1 show that, in contrast to $\nu_f = 2/3$, the $\nu_f = 1/3$ FQHE state does not depend on H_p. This result implies the absence, in our range of parameters, of the qusiparticles with a large number of reversed spins (skyrmions) recently predicted for the $\nu_f = 1/3$ FQHE state.

3 Evaluation and discussion

Our results are consistent with an assumption that all variations ΔC in the capacitance with H_p which can be seen in Fig.1a originate from the variations of the Zeeman energy only. This, in particular, is supported by the linear dependence of ΔC on $H - H_n$ shown in the inset in Fig.1a for the case of $\nu = 2/3$. Observed variation of the capacitance around $\nu = 2/3$ agrees well with that expected from the results of small-system calculations [1]. Absence of the effect at $\nu < 2/3$ corresponds to spin-polarized quasiholes. The variation of the chemical potential discontinuity with H_p at $\nu = 2/3$ expected from the results[1] is $\alpha g \mu_B (H - H_n)$, where μ_B is the Bohr magneton. Naive extrapolation of the results [1] to the thermodynamical limit gives $\alpha = 2$. In the presence of the carrier density fluctuations with gaussian distribution in a sample, the corresponding change in the capacitance reads as [2,3]:

$$\Delta C = -\frac{C^2 \alpha g \mu_B (H - H_n)}{e^2 S \sqrt{2\pi}\sigma} \exp\left(-\frac{(n_s - n_s(\nu = 2/3))^2}{2\sigma^2}\right). \quad (1)$$

For this sample the value of the carrier density dispersion $\sigma = 4.2 \times 10^9 \text{cm}^{-2}$ has been determined in Ref. [2] from the width of the FQHE signature in capacitance at $\nu = 1/3$. Fig. 3 shows that equation (1) fits well the dip in ΔC at $\nu = 2/3$ if $\alpha = 1$. Such a behavior of the chemical potential discontinu-

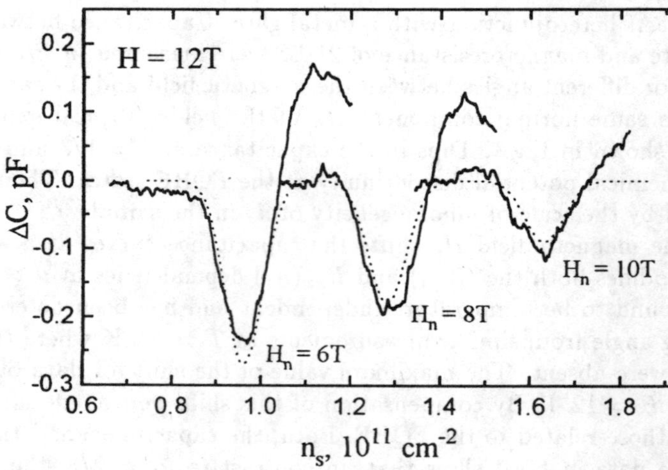

Figure 2: Experimental differential capacitance $\Delta C = C(H) - C(H_n)$ (solid lines) versus areal electron density at different H_n shown at the curves and $H = 12$ T. Results of calculations according to equation (1) with $\alpha = 1$ are shown by dotted lines.

ity is equivalent to the jump of the derivative $dn_{s\downarrow}/dn_s$ by unity at $\nu = 2/3$ and proves a decreasing spin polarization of the 2DES at $\nu > 2/3$. Here $n_{s\downarrow}$ is an areal density of spin-reversed electrons. If we assume (keeping in mind results[1]) that the 2DES is fully spin-polarized at $\nu \leq 2/3$ ($dn_{s\downarrow}/dn_s = 0$) then in the vicinity of $\nu = 2/3$ all electrons added to the system at $\nu > 2/3$ have reversed spins ($dn_{s\downarrow}/dn_s = 1$). The spin polarization at higher carrier density, where $\Delta C > 0$, varies in a more complicated manner which will be discussed elsewhere.

Acknowledgments

This work was supported in part by Russian Foundation for Fundamental Research (grant No 95-02-06107). S.I.D. and M.O.D. gratefully acknowledge a grant from INTAS.

References

1. T. Chakraborty,. *Surface Science* **229**, 16 (1990).
2. S.I. Dorozhkin, R.J. Haug, K. von Klitzing, and K. Ploog, *Phys. Rev.* B **51**, 14729 (1995).
3. F.G. Pikus and A.L. Efros, *Phys. Rev.* B **47**, 16395 (1993).

WIGNER CRYSTALS IN VERY STRONG MAGNETIC FIELDS

P. A. MAKSYM
Department of Physics and Astronomy, University of Leicester,
Leicester LE1 7RH, U.K.

The quantum states of a 2D Wigner crystal in a very strong magnetic field are studied theoretically. These states are usually obtained from a harmonic expansion of the Hamiltonian about the global minimum of the potential energy. However there are local minima which are very close to the global minimum. This leads to the possibility that a mixed symmetry state could lower the total energy. This is examined and it is found that states localised on all the minima are allowed to mix only when the filling factor is an odd denominator fraction. The experimental implications are discussed.

1 Introduction

There is experimental evidence [1,2,3] for magnetic field induced Wigner crystallisation of 2D electron systems at filling factors less than about 0.2, with indications of re-entrant behaviour and liquid-like features appearing at odd denominator filling beyond the threshold field for Wigner crystallisation [1]. Hints of re-entrant behaviour have also been seen in finite size calculations of the pair distribution function [4] but the fundamental reasons for this behaviour are not yet clear. The purpose of the present work is to examine the possibility that the re-entrant behaviour is related to the presence of local potential energy minima that are very close to the global minimum. Evidence is presented to demonstrate that such minima exist and this raises the question of whether the quantum ground state is localised only around the global minimum or whether there is also some mixing with states localised on the local minima. The mixing is found to be governed by selection rules and would be particularly strong at odd denominator filling. This seems to be consistent with the experimentally observed re-entrant behaviour.

2 Theoretical formalism

N electrons are taken to be in a rectangular cell of width a and height b. The wavefunction, Ψ, satisfies generalised periodic boundary conditions so that $T_y\Psi = \Psi$ and $T_x\Psi = \exp(-ia\sum_j y_j/l^2)\Psi$, where T_x and T_y are N-electron translation operators that shift all the electron x and y co-ordinates by a and b respectively and l is the magnetic length. The magnetic field is taken to be so strong that Ψ can be represented accurately in the zeroth Landau level and the system is spin-polarised. Thus Ψ has the form $\hat{A}\sum_{X_1\cdots X_N} c(X_1\cdots X_N)\prod_i \phi_{X_i}(\mathbf{r}_i)$, where $\hat{A}$ is the antisymmetrization operator, $\phi_{X_i}(\mathbf{r}_i)$ are zeroth Landau level, single electron states that satisfy the generalised periodic boundary conditions and X_i is the centre co-ordinate.

The $c(X_1\cdots X_N)$ are determined from the projection of the Hamiltonian onto the zeroth Landau level. In general the projected Hamiltonian [5] has the form $(1/2)\sum_{i\neq j}\sum_{\mathbf{q}}\tilde{V}_{\mathbf{q}}\exp(i\mathbf{q}\cdot(\mathbf{R}_i-\mathbf{R}_j))$, where $\mathbf{R}$ is the centre co-ordinate operator, $\mathbf{R}=(\hat{X},\hat{Y})$, and $\tilde{V}_{\mathbf{q}}=V_{\mathbf{q}}\exp(-q^2l^2/2)$ where $V_{\mathbf{q}}$ is a Fourier component of the interaction potential. In the present work the projected Landau level formalism is used in the form introduced by Tsukada [6] where $\hat{X}=X$, $\hat{Y}=-il^2d/dX$ and the $c(X_1,\cdots X_N)$ obey the eigenvalue equation

$$\frac{1}{2}\sum_{i\neq j}\sum_{\mathbf{q}}\tilde{V}_{\mathbf{q}}\exp[i\mathbf{q}\cdot(\mathbf{R}_i-\mathbf{R}_j)]c(X_1,\cdots X_N)=\lambda c(X_1,\cdots X_N). \qquad (1)$$

In the harmonic approximation this simplifies to the general form

$$\sum_{i\neq j}\sum_{\alpha\beta}A_{ij\alpha\beta}(R_{i\alpha}-R^0_{i\alpha})(R_{j\beta}-R^0_{j\beta})c(X_1,\cdots X_N)=(\lambda-V_0)c(X_1,\cdots X_N),$$

where the $A_{ij\alpha\beta}$ are Taylor expansion coefficients, $R_{i\alpha}$ are the components of $\mathbf{R}_i$, the $\mathbf{R}^0_i$ are the positions that minimise the projected potential and V_0 is its minimum value. The equilibrium positions can be eliminated from this equation with the aid of the substitutions $X_i=X^0_i+X'_i$ and

$$c(X_1,\cdots X_N)=\exp\left(i\sum_j Y^0_jX_j/l^2\right)c'(X_1,\cdots X_N) \qquad (2)$$

which results in the equation

$$\sum_{i\neq j}\sum_{\alpha\beta}A_{ij\alpha\beta}R'_{i\alpha}R'_{j\beta}c'(X'_1,\cdots X'_N)=(\lambda-V_0)c'(X'_1,\cdots X'_N). \qquad (3)$$

The phase that appears in Eq. 2 has particularly interesting consequences which are examined in the next section.

3 Results and discussion

The two issues considered here are the near degeneracy of the local minima of the projected potential and the symmetry of the wave function. In general it is very difficult to locate *all* the local minima of the projected potential analytically. However it is easy to show that the potential is minimised when the electrons form a lattice. Each way of writing N as a product of 2 factors corresponds to a different primitive lattice and in addition centred lattices occur when one of the factors is even. It is reasonable to suppose that there are also other minima which correspond to a lattice plus defects but intuition suggests that the perfect lattice has the lowest energy. To ensure the energy does not diverge the lattice is taken to be in the presence of a neutralising positive background. The total energy of the system is then $NE_0 + (1/2)\sum_{i\neq j}\sum_{\mathbf{q}\neq 0}\tilde{V}_{\mathbf{q}}\exp[i\mathbf{q}\cdot(\mathbf{R}_i^0 - \mathbf{R}_j^0)]$ where E_0 is a constant determined by the interaction of the lattice with the background. This sum is simple to evaluate numerically. Results are shown in Figure 1 for $2^5 3^4 5^2 7^2 = 3,175,200$ electrons at $1/7$ filling. The electron density is $10^{11}\mathrm{cm}^{-2}$ and the aspect ratio of the system is $1/\sqrt{3}$. This ensures that the minimum

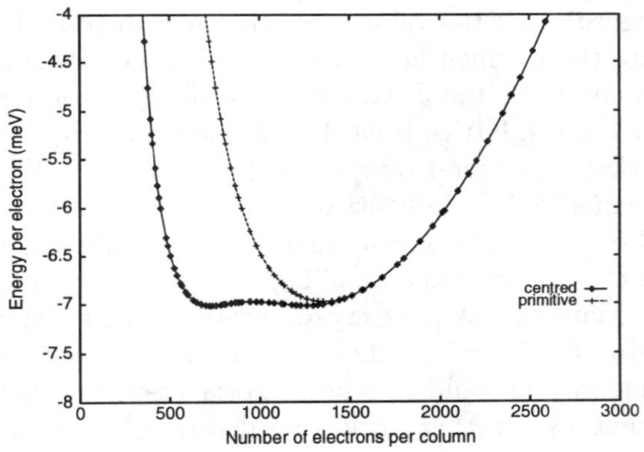

Figure 1: Energy per electron for centred and primitive lattices.

energy lattice is triangular. The figure shows the energy per electron against the number of electrons, S, per column of a lattice. It is clear that there is a range of lattice types which are very close together in energy. The centred lattices exhibit an interesting double minimum with an energy difference approximately 0.0002meV per electron and the region between minima is flat to about 0.05meV. This is smaller than the energy difference of about 0.1meV per electron found in quantum dots in the high field limit where a mixed symmetry state has recently been identified[7]. This gives further support to the idea that similar mixing could occur in the Wigner crystal regime.

To examine the symmetry properties of states localised on the lattices it is necessary to consider the effect of the antisymmetrization operator. An antisymmetric state can be obtained by antisymmetrizing with respect to the $\mathbf{r}_i$ or the X_i and it is most convenient to do the latter. To do the antisymmetrization it is necessary to know how the $c(X_1, \cdots X_N)$ transform under permutations and the most important permutations are those that cycle the electrons within columns. The coefficient c' is determined by a harmonic oscillator equation and should be symmetric in the ground state. However the phase factor in Eq. 2 is not invariant and a permutation that cycles the electrons through one lattice position in the column direction changes the phase of c by a factor $\exp[-ib\sum_j X_j/(l^2S)]$. The sum over j is a conserved quantity because the electron-electron interaction conserves the centre co-ordinate. Therefore the phase factor can be written as $\exp(-2\pi iJ/S)$ where J is a quantum number that characterises the lattice. The same quantum number characterises the Slater determinants that have been used for finite size studies of fractional quantum Hall systems in the toroidal geometry. By considering the effect of the remaining permutations it can be shown that antisymmetric states are allowed to occur regardless of the value of J. However the only lattices that are allowed to mix are those which have the same J value. Because of magnetic translational symmetry [5,8,9] the J values at $1/k$ filling are only unique modulo N. This value for a given lattice is easily obtained from the sum $\sum_j X_j^0$. For odd k it is found that $J \equiv 0 \bmod N$ when N is odd and $J \equiv N/2 \bmod N$ when N is even, independent of S. In contrast, the J value for even k depends on whether S is odd or even. This means that mixing of different lattice types is greatly enhanced at odd denominator filling.

In summary, triangular Wigner crystal states at odd denominator filling could be affected by mixing with other types of lattice. This would result in some loss of symmetry and could be related to the re-entrant behaviour that is observed experimentally. Further work is needed to calculate matrix elements and assess the strength of this effect.

References

1. H. Buhmann et al., Phys. Rev. Lett. **65**, 1056 (1990).
2. A. G. Davies et al., Physica B **184**, 56 (1993).
3. I. N. Harris et al,. Europhys. Lett., **29**, 333 (1995).
4. P. A. Maksym, J. Phys: condensed matter, **4**, L47 (1992); **4**, 6027 (1992).
5. F. D. M. Haldane in "The Quantum Hall Effect", ed. R. E. Prange and S. M. Girvin, Springer, New York (1987).
6. M. Tsukada, J. Phys. Soc. Japan, **41**, 1466 (1976).
7. P.A. Maksym, Phys. Rev. B **53**, 10871 (1996).
8. P. A. Maksym, J. Phys. C, **18**, L433 (1985).
9. T. Chakraborty and P. Pietiläinen, "The Quantum Hall Effects", Springer, Berlin (1995).

BREAKDOWN OF THE QUANTUM HALL EFFECT AT HIGH CURRENTS

S. KAWAJI
Department of Physics, Gakushuin University, Mejiro, Toshima-ku, Tokyo 171, Japan

Properties of breakdown of the integer quantum Hall effect at high currents measured for butterfly-type Hall bars, conventional rectangular Hall bars and butterfly-type Hall bars with a constriction fabricated from GaAs/AlGaAs heterostructure wafers with various electron concentrations and mobilities are presented. Critical breakdown current measured for butterfly-type Hall bars is proportional to sample width. The result shows that the edge channel picture is not a realistic description of the quantum Hall effect. The critical breakdown Hall fields $F_{cr}(i=2$ and $4)$ and $F_{cr}(i=1$ and $3)$ are proportional to $B^{3/2}$ where B is the magnetic field at the center of each quantized Hall plateau. The B-dependence of F_{cr} is described based on a model supported by activation energy measurements where the broadening of extended state band increases with increasing Hall electric field.

1. Introduction

In the integer quantum Hall effect (QHE), the Hall conductivity is quantized as $\sigma_{xy} = -ie^2/h$ for an nteger i with zero diagonal conductivity $\sigma_{xx}=0$ in degenerate two dimensional electron systems (2DES) at low temperatures and in strong magnetic fields.[1] The quantum Hall effect was predicted by Ando *et al*[2] in 1975 by their theoretical study of the Hall effect in Landau levels in 2DES with attractive and repulsive short range scatterers. In 1980 Kawaji and Wakabayashi[3] obtained results by Hall current measurements for Si-MOSFETs which show that the Hall conductivity is quantized when the diagonal conductivity is zero as predicted by Ando *et al.* Their results of temperature dependence of the conductivities have shown that Anderson localization causes the quantum Hall effect as theoretically studied by Ando *et al.* In the same year von Klitzing *et al*[4] made a precision measurement of the Hall resistance $R_H=V_H/I$ and showed that $R_H(4) = h/4e^2$ with an accuracy of 3 ppm to the recomended value. Von Klitzing *et al* proposed that this is a new method for the determination of the fine structure constant $\alpha = \mu_0 ce^2/2h$ where μ_0 is the permittivity of vacuum and c the light velocity. This proposal opened a new era in the electrical resistance standards.

The quantized Hall resistance has been used as an international resistance standard since 1990.[5] In order to carry out a high precision measurement of the quantized Hall resistance(QHR), we must pass a high electric current through a Hall bar to generate a high Hall voltage across the sample. When the current exceeds a critical value, the QHE breaks down.[1] In the early 1980s several experimental and theoretical studies of the breakdown of the QHE were reproted.[6] Since 1992, we have carried out experiments on breakdown of the QHE with a renewed interest. Our aim is to observe intrinsic breakdown of the QHE or to observe the breakdown as a local property of the 2DES in a strong magnetic field.

The QHE state is characterized by non-dissipative electric current whose direction is orthogonal to the direction of the electric field, i.e. a Hall electric field, where the diagonal resistance in the Hall bar is zero, i.e. $R_{xx}=0$. Such non-dissipative electronic state only appears near the central part of the Hall bar sample. Joule heat is always generated at both ends of the sample, i.e. in transition regions between the 2DES and two current electrodes, the source electrode and the drain electrode. The breakdown of the QHE which we want to measure is the appearence of dissipation in a part of a Hall bar other than the parts near the current electrodes. In order to do this, we have to eliminate effects of dissipation at the current electrodes. Based on our knowledge that in high-precision measurements of the QHR in Si-MOSFETs the resistances at the source and the drain electrodes appear in narrow Hall bar samples,[7] we have fabricated butterfly-type Hall bars from $GaAs/Al_{0.3}Ga_{0.7}As$ heterostructure wafers.[8-10] Our results are summarerized as follows. (1) Critical breakdown current I_{cr} is proportional to the width w of samples. In other words, the breakdown at high currents is caused by average current density J_{cr} or average Hall electric field F_{cr}.[8-10] (2) The average critical Hall electric field F_{cr} is proportional to $B^{3/2}$ where B is the magnetic field at the plateau center with given even plateau quantum number i for $R_H(i)=h/ie^2$.[10] (3) The critical electric field does not depend on electron mobility μ in the range $13.5<\mu(m^2/Vs)<27$.[10] (4) The dependence $F_{cr} \propto B^{3/2}$ is in accord with the inter-Landau level transition mechanism theoretically proposed by Eaves and Sheard[11] but their theory gives more than one order of magnitude larger critical field F_{cr}(E-S) than our experimental results F_{cr}(Exp).[9,10,12]

Our experimental critical electric fields are no smaller than those reported by other authors at the same magnetic field[12] except the result by Bliek *et al*[13] who observed the breakdown in Hall bars with a narrow constriction.

Recently Balaban *et al* claimed that the critical current I_{cr} measured in high mobility samples (μ = 70 m^2/Vs and 90 m^2/Vs) depends sublinearly on the sample width w.[14] They also reported that samples made from a low mobility wafer showed a linear dependence of I_{cr} on w but the dependence changed into sublinear and I_{cr} decreases after long illumination.[15]

This paper briefly describes our recent experimental results on breakdown of the QHE obtained by using butterfly-type Hall bars, conventional rectangular Hall bars and butterfly-type Hall bars with a constriction, and results on Hall electric field dependence of activation energy. The results show that (1) the sublinear dependence of I_{cr} on w is caused by a short channel effect, (2) the critical breakdown currents in Hall bars with a constriction are not larger than those observed in Hall bars without a constriction, and (3) the dependence $F_{cr} \propto B^{3/2}$ arises from the broadening of extended states band at the center of a Landau level due to the Hall electric field.

2. Experimental

2. 1 Electrode structures

Electrode structures are shown in Fig.1.[16] A conventional rectangular Hall bar is shown in (a). We fixed the length of 2D channel to L=600 μm. The width w ranges from 15 to 120 μm. Each Hall bar has three pairs of potential probes separated by 150 μm. In butterfly-type Hall bars, (b), the central parallel part has the same dimensions as the rectangular Hall bars. However, the width of the current electrodes is W=400 μm and the length of the 2D channel between the current electrodes is L=2900 μm in every butterfly-type Hall bar. The width of each butterfly-type Hall bar is linearly narrowed from both current electrodes to the ends of the central rectangular part as shown in (b). In a butterfly-type Hall bar with a constriction at the center as shown in (c), the width of the current electrode is W=400 μm, the length of 2D channel is L=2500 μm, the length and the width of the central parallel part are ℓ=1100 μm and w = 120 μm, respectively. At the center of this Hall bars, we make a constriction whose length ℓ' and width $w'$are ℓ' =5 μm and w'=10 or 20 μm, respectively. The Hall bar samples were prepared by photolithography and wet chemical etching from several $GaAs/Al_{0.3}Ga_{0.7}As$ heterostructure wafers which have electron mobilities ranging from 13.5 to 110 m^2/Vs.

2.2 Breakdown measurements

Diagonal resistance R_{xx} in quantized Hall plateau region was observed in a trace of magnetic field dependence at temperatures 0.5 ~ 0.7 K. The critical current I_{cr} was

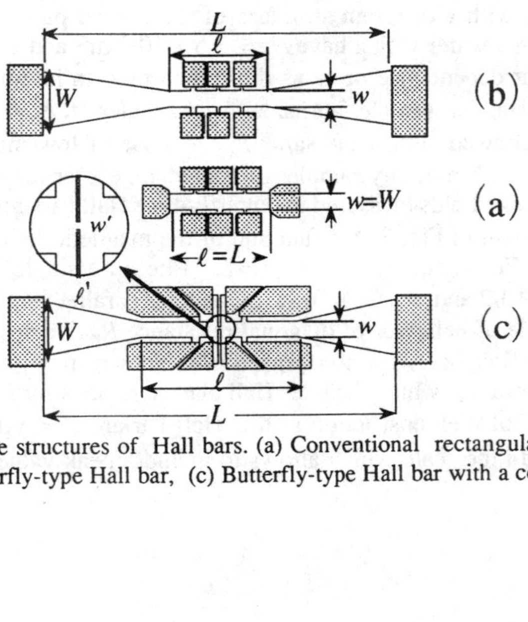

Fig. 1 Electrode structures of Hall bars. (a) Conventional rectangular Hall bar, (b) Butterfly-type Hall bar, (c) Butterfly-type Hall bar with a constriction.[16]

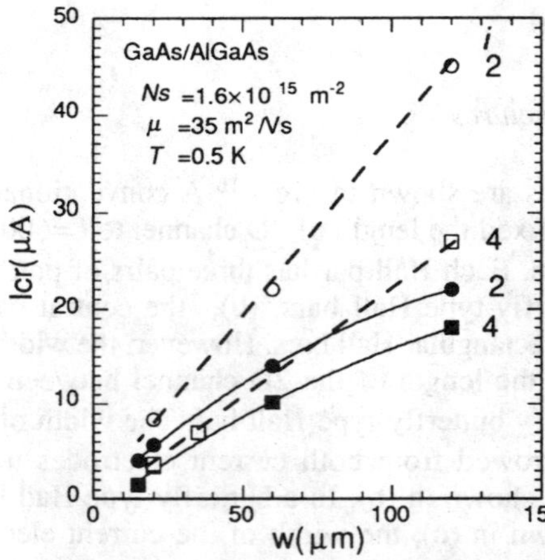

Fig. 2 Critical breakdown current I_{cr} versus sample width w in butterfly-type Hall bars (open symbols) and rectangular Hall bars (full ymbols).[16]

determined from the current at which the plateau disappears, i.e. the diagonal resistance >0.1 Ω as was made in our former works.[8-10] The rise in the diagonal resistance at the breakdown is so abrupt that the critical current can be determined without any uncertainty affecting the results. Sample width dependences of the critical breakdown current measured in conventional rectangular Hall bars and butterfly-type Hall bars made from the same heterostructure wafer are compared in Fig.2. The conventional rectangular samples show a sublinear dependence of I_{cr} on w, while the butterfly-type samples show a linear dependence. We have extended breakdown measurements for higher-mobility butterfly-type samples. Butterfly-type Hall bars with a different structure of the central part as was used in ref. 9 fabricated from a wafer which have N_s = 3.5 x 10^{15} m^{-2} and μ = 75 m^2/Vs showed the same linear dependence on w as those measured in low-mobility samples. A butterfly-type Hall bar sample fabricated from a wafer of N_s = 3.0 x 10^{15} m^{-2} and μ = 110 m^2/Vs showed almost the same F_{cr} as those of low-mobility samples. The critical field in high-mobility samples did not change after long illumination.[17] The critical breakdown fields measured in butterfly-type Hall bars for the plateaus i= 1, 2, 3 and 4 are shown in Fig. 3 as a function of the magnetic field at the center of the Hall plateau. Fig.3 shows that F_{cr} (i=2, 4) lie on a single line in log-log plot with slope of ~3/2 and F_{cr} (i=1, 3) lie on another parallel single line.[17]

An example of behavior of diagonal resistance R_{xx} versus magnetic field of a butterfly-type Hall bar with a constriction is shown in Fig. 4. Figure 4 shows whole behavior of R_{xx} which includes Hall plateau regions for i=4 and 6. The length and the width of the constriction in this Hall bar are ℓ' = 5 μm and w' = 20 μm, respectively. In the scale which allows to include peak values of R_{xx} , this Hall

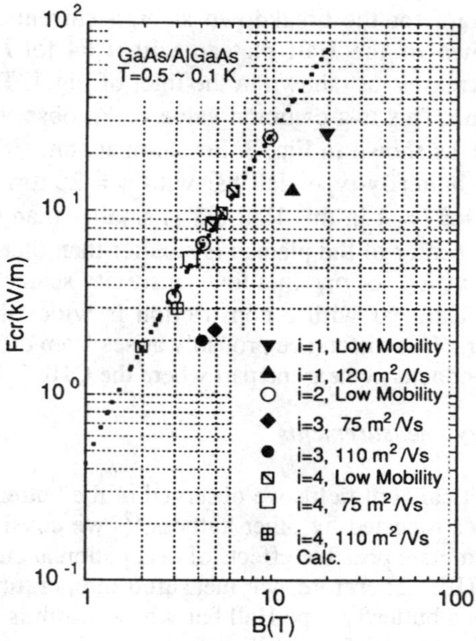

Fig.3 Critical breakdown Hall electric field F_{cr} versus magnetic field B at the plateau center in butterfly-type Hall bars.[17] Open symbols: i = 2 and 4, full symbols: i = 1 and 3.

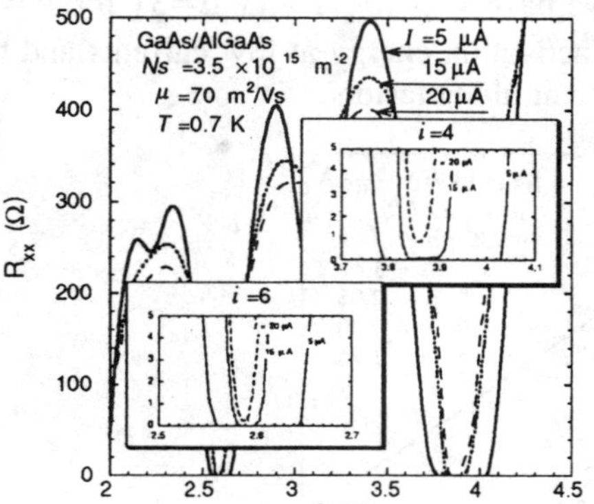

Fig. 4 Diagonal resistance R_{XX} versus magnetic field B in a butterfly-type Hall bar with a constriction of ℓ = 5 μm and w' =20 μm.[16]

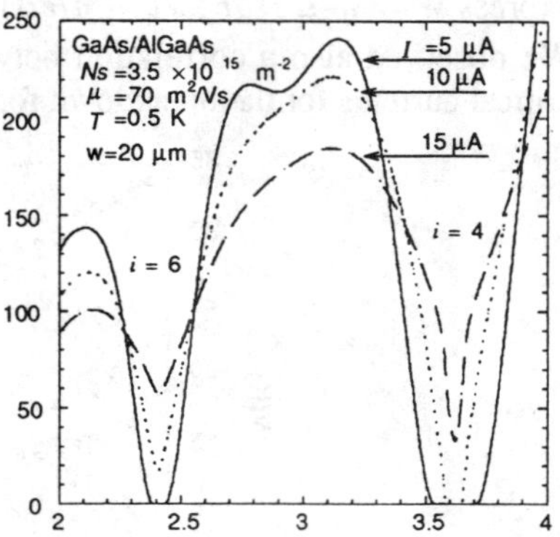

Fig. 5 Diagonal resistivity ρ_{XX} versus magnetic field B in a butterfly-type Hall bar with ℓ=600 μm and w = 20.[16]

bar appears very strong for the breakdown at high currents. However, we can observe the breakdown of the Hall plateau with i =4 for I = 20 μA when the sensitivity of R_{xx} increases as shown in the inlet of Fig.4. Then the critical field for i = 4 is 6.5 kV/m. This is a common value of F_{cr} observed in butterfly-type Hall bars at B=3.85 T as shown in Fig. 3. For comparison, similar behavior of ρ_{xx} observed in a simple butterfly-type Hall bar with w = 20 μm are shown in Fig. 5. The critical current for i = 4 in this Hall bar is smaller than that in Fig. 4 as the magnetic field at the center of the plateau is smaller than that in Fig. 4. When we observe the behavior of R_{xx} or ρ_{xx} in a low sensitivity scale, the plateau width at high currents in the Hall bar with a constriction is wider than that in a simple butterfly-type Hall bar. This difference probably arises from difference between two samples in thermal conductance near the part where the QHE is broken down.

2.3 Activation energy measurements

As the breakdown critical Hall fields we observed in the butterfly-type Hall bars are no smaller than values reported by other authors,[12] we consider that the butterfly-type Hall bars can eliminate possible effects of dissipation at current electrodes in the breakdown of the QHE. Therefore, we measured temperature dependence of the diagonal resistance in a butterfly-type Hall bar whose width is 35 μm. The diagonal resistance was measured for different filling factors around $\nu = 2$ at various currents. From temperature dependences, activation energy $W(\nu, I)$ was determined by $R_{xx}(T)=R_0\exp(-W/T)$. The activation energy linearly decreases with increasing Hall electric field F as shown in Fig. 6 at three filling factors near $\nu = 2$.[18] When we express $W=W_0-\alpha e\ell_B F$, $\ell_B=(\hbar/eB)^{1/2}$, we have $W_0=\hbar\omega_c/2$ and $\alpha=31$ for $\nu = 2$. We observed also a correlation between activation energy at low currents and the critical currents for the breakdown for different filling factors.

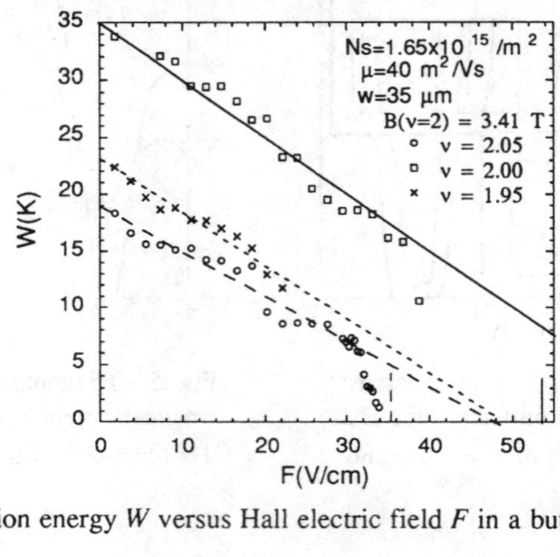

Fig. 6 Activation energy W versus Hall electric field F in a butterfly-type Hall bar.[18]

3. Discussions

As shown in the sample-width dependences of the critical breakdown current in Fig.2, a linear dependence was observed in buttrfly-type Hall bars, while a sublinear dependence was observed in rectangular Hall bars. Our experimental results of F_{cr} in high-mobility butterfly-type samples with μ = 75 m^2/V and 110 m^2/V showed almost the same results as those observed in low-mobility samples as shown in Fig. 3. Long illumination did not decrease critical current I_{cr} in butterfly-type samples. Based on these results we consider that the sublinear dependence of I_{cr} on w, which Balaban *et al* observed,[14,15] probably arises from effects of current electrodes in short samples, though, unfortunately, they did not describe the lengths of their samples.

Our results measured in the butterfly-type Hall bars with a constriction similar to Bliek *et al*'s[13] show that the critical currents determined at low diagonal resistance, as shown in Fig. 4, are almost the same as measured in the butterfly-type Hall bars. Therefore, our results of $F_{cr}(i$=2,4) and $F_{cr}(i$=1,3) obtained by the butterfly-type Hall bars are an intrinsic propertiy or a local property of the breakdown of the QHE.

An important result of our experiments made by using butterfly-type Hall bars is that the onset of the breakdown of the QHE is not caused by the current but the average current density or the average Hall electric field. This characteristic feature does not depend on electron mobility which ranges between 13.5 and 110 m^2/Vs. The result strongly supports that the edge states do not play a principal role in quantization of the Hall resistance. In other words, the edge state picture is not a realistic description of the QHE though the edge channel picture might be able to describe phenomena in Hall bars at small currents I< 0.1 μA.[19] We note here that quantized Hall conductivity can be observed without an edge by measurements using microwave frequency[20] and by measurements for Corbino disk samples.[21]

Magnetic field dependence of the critical breakdown field, $F_{cr} \propto B^{3/2}$, is a key to discuss a possible mechanism of the breakdown. The mechanism of abrupt phonon emission analogous to Cerenkov radiation, proposed by Streda and von Klitzing[22] and Stormer *et al*[23], and an electron heating mechanism such as discussed by Komiyama *et al*[24] lead to a dependence of $F_{cr} \propto B$. Our result of $F_{cr} \propto B^{3/2}$ appears to suggest that a possible mechanism of the breakdown is production of dissipative carriers by inter-Landau level tunneling of electrons due to small scatterers from the highest filled Landau level to the lowest empty Landau level. The spacial extent of the electron wave function of a Landau level along the Hall electric field is proportional to $B^{-1/2}$ and is described by a harmonic oscillator wave function.[1] The energy splitting between adjacent Landau levels is proportional to B irrespective of their origin, i.e. Landau gaps $\hbar\omega_c$ and spin gaps $g_{eff}\mu_B B$. Therefore, the critical electric field at the onset of electron tunneling is expected to be proportional to $B^{3/2}$ irrespective of tunneling processes, i.e. a process due to a single transition or a process due to higher-order multiple transitions. However, as we discussed,[12] Eaves and Sheard's formula leads a result which is more than one order of magnitude larger than our experiments,i.e. $\{F_{cr}(\text{E-S}) / F_{cr}(\text{Exp})\}_{i=4} \approx 20$. According to their expression, F_{cr} $(i = 2)$ should be larger than F_{cr} $(i = 4)$, i.e. $\{F_{cr}(\text{E-S})_{i=2} / F_{cr}(\text{E-S})\}_{i=4} = 1.5$, while they are same in our experimental results. Kawabata[25] calculated the critical field due to higher-order tunneling and obtained a result F_{cr}(Kawabata)/F_{cr}(E-S) ~1/3. The relation $F_{cr} \propto B^{3/2}$ is held in the higher-order tunnelung. The higher-order tunneling may also eliminate the dependence of the Landau level quantum number in the inter-Landau level transition.[25] However, the field for higher order tunneling is still too high to explain our experiments. Trugman *et al*[26] calculated enhancement of the inter-Landau level transition by spacially extended scatterers but their result leads to a weaker magnetic field dependence of $F_{cr} \propto B$. Recently Ishikawa *et al* calculated the inter-Landau level tunneling caused by Landau level broadening due to Hall electric field and obtained that the level width is given by $\Gamma = (2\pi)^{1/2} e \ell_B F$.[27] They consider that the electron tunneling occurs at the field when the Landau level width is equal to the energy splitting. Then the critical field is given by F_{cr}(Ishikawa)= $\hbar\omega_c / (2\pi)^{1/2} e \ell_B$ which is independent of the plateau quantum number i, but it is close to F_{cr}(E-S). Thus we conclude that the inter-Landau level tunneling cannot explain our experiments.

In the QHE state the Fermi level, E_F, resides within the region of localized states between centers of Landau levels and non-dissipative electrons flow through extended states near the centers of filled Landau levels. The activation energy W of R_{xx} is considered to be the difference in energy between an edge of the extended state band (mobility edge) and E_F. As was reported first by Ebert *et al*,[28] we observed a correlation between the activation energy and the breakdown critical current against the change in magnetic field. As a percolation problem, Trugman discussed that Hall electric field F increases extended states in numbers proportional to a power of F.[29] Stormer *et al* discussed a possibility of breakdown caused by the increase of extended states due to the increase in Hall electric field.[23] Recently Riess theoretically discussed the breakdown caused by electric field dependent mobility edge.[30] In their discussions,[29,23,30] however, magnetic field dependence of F_{cr} is not clear.

Results of the activation energy measurements in Fig. 6 show that the activation energy linearly decreases with incresing Hall electric fields as $W(\nu=2) = \hbar\omega_c / 2 - \alpha e \ell_B F$ where α =31. Similar dependences of activation energy on Hall electric field are reported by Komiyama *et al*[24] and Boisen *et al.*[28] Our result of the magnetic field dependence, $F_{cr} \propto B^{3/2}$, can be explained by a model based on the experimental results in Fig. 6 as follows. The half width of the extended state band γ increases as $\gamma = \gamma_0 + a e \ell_B F$ where γ_0 is the half width for $F = 0$. At the integer filling of Landau levels whose density of states is symmetric, the activation energy can be given by $W = \hbar\omega_c / 2 - \gamma$. Overlapping of wave functions between electrons at the Fermi level and the lower mobility edge in the higher Landau level gives rise to the breakdown of the QHE. Then the critical field is expressed by

$$F_{cr} = \frac{\hbar\omega_c / 2 - \gamma_0}{a' e \ell_B} \quad . \tag{1}$$

where $a' > a$ for the breakdown occurs at a finite W. The critical field $F_{cr}(i=2,4)$ in Fig, 3 is reproduced by a dotted line when we take $a'=38$ and $\gamma_0=0.34$ meV. Similar empirical formula can reproduce the critical field $F_{cr}(i=1,3)$ in Fig. 3. It is interesting to note here that a breakdown mechanism based on localization works well independent of the electron mobility of samples in a wide range.

In a previous paper,[12] the effective g-factor was evaluated as $g_{eff} \approx 10$ from the experimental results $F_{cr}(i=2,4)/F_{cr}(i=1,3) \approx 3$ in Fig. 3 assuming $F_{cr}(i=2,4)/F_{cr}(i=1,3) = \hbar\omega_c / g_{eff}\mu_B B$. This result does not change in the present breakdown mechanism as $2\gamma_0$ is much smaller than the Landau gap and the Zeeman gap.

In conclusion we have carried out measurements of the critical current at the onset of the breakdown of the QHE in butterfly-type Hall bars, rectangular Hall bars, and butterfly-type Hall bars with a narrow constriction, made from different GaAs/ $Al_{0.3}Ga_{0.7}As$ heterostructure wafers. The butterfly-type Hall bars show the intrinsic properties of the breakdown which does not depend on electron mobility ranging from 13.5 to 110 m^2/Vs. It is found that the onset of the breakdown is caused not by the current but by the average current density or the average Hall electric field and the critical field is proportional to $B^{3/2}$ where B is the magnetic field at the plateau center. The former shows that the picture of edge state transport is not a realistic description of the QHE. The latter is explained, based on the observed Hall electric field dependence of the activation energy of R_{xx}, by a mecahnism in which the broadening of extended state band in the center of a Landau level due to electric field is a principal cause of the breakdown. However, further experimental and theoretical studies are needed to lay the foundations of the breakdown mechanism based on the electron localization in 2DES in strong magnetic fields.

Acknowledgements

The aouthor would like to thank colleagues in Gakushuin University, in particular A. Kawabata for discussions and T. Fukase and T.Goto in IMR, Tohoku University, for their help in high magnetic field experimentds.This work is supported in part by Grants-in-Aid from the Ministry of Education, Science, Sports and Culture, Japan.

References

1. S. Kawaji, *Proc. Int. Symp. Foundation of Quantum Mechanics, Tokyo, 1983*, ed. S. Kamefuchi *et al* (Physical Society of Japan, Tokyo, 1984) p. 327.
2. T. Ando, Y. Matsumoto and Y. Uemura, *J. Phys. Soc. Jpn.* **39**, 279 (1975)
3. S. Kawaji and J.Wakabayashi, *Proc. Int. Conf.Physics in High Magnetic Fields, Hakone, 1980* eds. S. Chikazumi and N. Miura (Springer, Berlin, 1981) p. 284.

4. K. von Klitzing, G. Dorda and M. Pepper, *Phys. Rev. Lett.* **45**, 449(1980)
5. T. J. Quinn, *Metrologia* **26**, 69 (1989)
6. See references in ref. 9.
7. K. Yoshihiro, J. Kinoshita, K. Inagaki, C. Yamanouchi, T. Endo, Y. Murayama, M. Koyanagi, A. Yagi, J.Wakabayashi and S. Kawaji, *Phys. Rev. B* **33**, 6874 (1986)
8. S. Kawaji, K. Hirakawa and N.Nagata, *Physica B* **184**, 17(1993)
9. S. Kawaji, K. Hirakawa, N.Nagata, T. Okamoto, T. Goto and T.Fukase, *J. Phys. Soc. Jpn.* **63**,2303(1994)
10. T. Okuno, S. Kawaji, T. Ohrui, T. Okamoto, Y. Kurata and J. Sakai, *J. Phys. Soc. Jpn.* **64** ,1881(1995)
11. L. Eaves and F. W. Sheard, *Semicond. Sci. Technol.* **1**, 346(1986)
12. S. Kawaji, *Semicond. Sci. Technol.* **11** (1996) to be published.
13. L. Bliek, E. Braun, G. Hein, V. Lose, J. Niemeyer, G. Weiman and W.Schlapp, *Semicond. Sci. Technol.* **1** ,110 (1986).
14. N.Q. Balaban, U. Meirav, H. Strikman and Y. Levine, *Phys.Rev. Lett.* **71**, 1443 (1993).
15. N.Q. Balaban, U. Meirav, H. Strikman, *Phys. Rev. B* **52**,R5503(1995)
16. A. Fukano, S. Kawaji , T. Okamoto, Y. Kurata and J. Sakai, submitted to *J. Phys. Soc. Jpn.*
17. S. Kawaji, A. Fukano, T. Okamoto, T. Goto and T. Fukase (unpublished).
18. T. Shimada, A. Fukano, T. Okamoto and S. Kawaji (unpublished).
19. K. von Klitzing, *Physica B* **116**,111(1995)
20. F. Kuchar, R. Meisel, G. Weimann and W. Schlapp, *Phys. Rev. B* **33**, 2965 (1986).
21. V. T. Dolgopolov, N. B. Zhitenev, A. A. Shashkin, *Pis'ma Zh. Eksp. Teor. Fiz* **52**, 826 (1990), *JETP Lett.* **52**,196(1990).
22. P. Streda and K. von Klitzing, *J. Phys. C* **17**,L483(1984).
23. H. L. Stormer, A. M. Chan, D. C. Tsui and J. C. M. Hwang, *Proc. 17th Int. Conf. Physics of Semicond., San Francisco, 1984*, ed. D. C. Chadi and W. A . Harrison(Springer, Berlin,1985) p.267
24. S. Komiyama, T. Takamasu, S. Hiyamizu and S. Sasa, *Solid State Commun.* **54**, 479(1985).
25. A. Kawabata (private communication).
26. S. A. Trugman, *Phys. Rev. Lett.* **62** 579 (1989).
27. K. Ishikawa, N. Maeda and K. Tadaki , submitted to *Phys. Rev. B*.
28. G. Ebert, K von Klitzing, K. Ploog and G. Weimann, *J.Phys.* **C16**,5441(1983).
29. S. A. Trugman, *Phys. Rev.* **B 27**, 7539 (1983).
30. J. Riess, *Physica* **B 190**, 366 (1993).
31. A. Boisen, P. Boggild, A. Kristensen and P.E. Lindelof, *Phys. Rev.***B 50**, 1957 (1994).

Theory of the Breakdown of the Quantum Hall Effect

V. Tsemekhman, K. Tsemekhman, C. Wexler, J.H. Han, and D.J. Thouless
Department of Physics, University of Washington, Box 351560, Seattle, WA 98195

Breakdown of the Quantum Hall Effect at high values of injected current is explained as a consequence of an abrupt formation of a metallic "river" percolating from one edge of the sample to the other. Such river is formed when lakes of compressible liquid, where the long-range disorder potential is screened, get connected due to the strong electric field. Our theory predicts critical currents consistent with experiment values and explains various features of the breakdown.

In this paper we discuss the breakdown of the quantum Hall effect (QHE) at high injected currents. First observed [1] as a sudden onset of the dissipation when the injected current exceeds some critical value, the breakdown of the integer QHE has been intensively investigated since then [2]. However, the breakdown of the dissipationless regime has not been given a clear explanation. Existing theories of this effect [3] include production of hot electrons, inter-Landau level transitions in the high local electric field (due to tunnelling or emission of phonons), increase in the number of the delocalized states in the Landau level. These theories produce critical currents orders of magnitude higher than observed experimentally or need some artificial assumptions to get reasonable results. Also, none of them can explain hysteresis and localized nature of the breakdown phenomenon, nor the existence of the transient switching and broadband noise before the breakdown, as well as peculiar steps in the magnetic field dependence in the regime of critical current.

Experiments on the breakdown of the QHE are performed on GaAs heterostructures or on high-quality MOSFET devices both characterized by the presence of long-range fluctuations of the disorder potential. In the GaAs systems the disorder potential due to the remote dopants has predominantly long wave-length fluctuations ($\lambda > d > l_H$, where λ is the wavelength of the fluctuations, d is the spacer thickness, and l_H is the magnetic length). Long-range potential fluctuations are also present in high-mobility MOSFET devices though their nature is not that evident [4]. When the magnetic field and the density of the 2DEG correspond to a filling factor close to an integer number, these fluctuations are not screened by the electrons in most parts of the system [5]. The region with completely filled Landau level percolates through the sample, leading to the QHE. In the percolating incompressible region electric fields are typically about the same as those of the bare disorder potential [6] $E_{inc} \simeq \sqrt{<E_{bare}^2>} = \sqrt{n_0/32\pi} e/\epsilon\epsilon_0 d$. For a typical clean sample this leads to: $E_{inc} \simeq 0.1\hbar\omega_c/el_H$, where ω_c is the cyclotron frequency. There are, however, isolated areas where the long wave-length fluctuations of the disorder potential are screened. Temperatures in experiments on the breakdown are usually high ($T \sim 1$ K) compared to the energy scales of short wave-length fluctuations of the disorder potential (that are left unscreened) and of the residual inter-electron interaction. Under such conditions a compressible liquid fills these isolated areas of screened potential which, therefore, behave like metallic lakes. In each of those lakes, all states of the highest available Landau level are partially occupied, and the screened potential fluctuates around the Fermi level ϵ_F with an amplitude of the order of T which is much smaller than the amplitude of the potential fluctuations in the incompressible region. Below we consider the potential in the compressible lakes to be flat. The main idea of our theory is that at high enough currents the insulating region separating two lakes, suddenly breaks down due to the high electric field; this leads to the connection of the lakes. When such connected lakes form a metallic river percolating from one edge of the sample to the other, abrupt onset of the dissipative regime is observed.

In general, the distance between two adjacent lakes can be large. We will see from the results of the numerical simulations that the lakes closest to each other get connected first. Let us, therefore, consider a simple model in which two metallic lakes are separated by a narrow insulating region with one parabolic potential fluctuation characterized by the root mean square electric field $E = E_{inc} \simeq 0.1\hbar\omega_c/l_H$. In equilibrium, all electrons in the lakes have energies equal to ϵ_F (up to T), and all electrons in the incompressible region separating them have energies below the Fermi energy. Injection of the current into the 2DEG leads to the appearance of the Hall electric field troughout the sample. In the insulating region this external field can be considered to be uniform on the scale of the distance between the lakes. In the lakes the charges are redistributed to screen the external field. Now, however, the two lakes are at different potentials (Fig. 1a). One can easily show that when the external field $E_0 = \sqrt{3}E_{inc}$ the potential monotonically drops from one lake towards the other. At this point, electrons can freely move from the incompressible region into the lake, as their energies are higher than those of the partially filled states in the lake. In this simplified picture, E_0 is a critical field at which the two lakes get connected. Possible existence of fluctuations in the separating region does not change the picture.

The above value of the electric field necessary to connect the lakes does not need to be equal to the critical average electric field $E_c = V_H/W$, where V_H is the measured Hall voltage, and W is the width of the sample. It is easy to see that the electric field in the narrow region between the two lakes can be much larger than the average Hall electric field: if the size of the

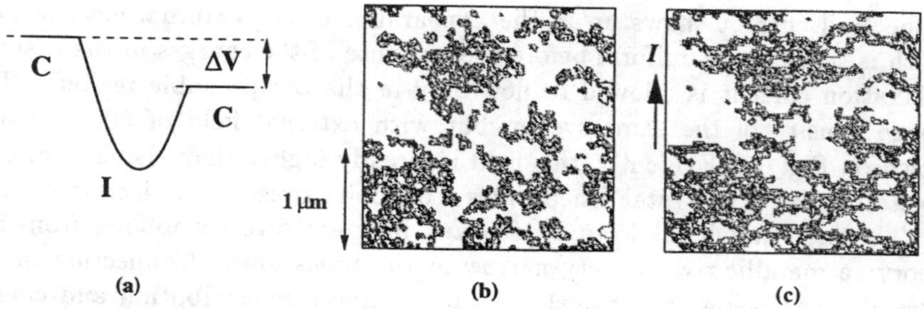

Figure 1: a. Distribution of the potential in the two adjacent islands and between them before breakdown; b. and c. Charge distribution from numerical simulations for the 2.5 μm $\times$2.5 μm sample with d = 25 nm, l_H = 10 nm, ν = 1.93, T = 0.5 K. White areas are the ν = 2 incompressible liquid, grey areas the $\nu < 2$ partially filled compressible liquid. b. at zero electric field, c. at E = 500 V/cm

separating region is smaller than the size of the lakes, the equipotential lines are highly squeezed between the two lakes. Therefore, an average electric field *much smaller* than E_0 is needed to connect the lakes. When the two lakes merge, the electric field is expelled from the former incompressible region now covered with compressible liquid. Therefore, the electric field becomes stronger outside the newly formed large lake, facilitating further connections as the injected current is increased. One would expect that after the metallic region acquires some critical length in the direction of the Hall electric field upon the increase of the injected current, the process of further connection must take an avalanche form. Finally, when the compressible liquid forms a river flowing from one edge to the other, dissipation drastically increases. It is important to notice that just before the formation of the metallic river, the longitudinal conductivity is still determined by the variable-range hopping. This is the reason why the longitudinal conductivity jumps several orders of magnitude at the critical current: the mechanism of the conductivity abruptly switches from hopping to the metallic one.

To see how the breakdown occurs in real systems we performed numerical simulations (the details will be published elsewhere). In equilibrium, the free energy of the interacting 2DEG in the strong magnetic field and in the smooth disorder potential was minimized with respect to the occupation numbers at finite temperature. The disorder potential was created by a randomly chosen distribution of charged impurities in a plane a distance d away from the 2DEG. Coulomb interaction between electrons was taken into account in the mean-field approximation. Fig. 1b shows the charge distribution for a clean sample at ν = 1.93. Lakes of compressible liquid (ν < 2, grey areas) are immersed into the percolating incompressible region (ν = 2, white areas). The effect of the Hall current shows up in the appearance of the external electric field which is taken to be uniform before the response of the charges in the system. Relaxation current is allowed to flow only in the compressible regions. The charge density of the same system but with external field of 500 V/cm is shown on Fig. 1c. Evidently, this field is already higher than the critical one: a metallic river is well established. This is in good agreement with experiments.

The fact that breakdown is localized in space directly follows from our theory: a metallic river is very narrow at the breakdown. Connection of the lakes at fields below the critical one causes charge redistribution and energy relaxation. The amount of dissipated energy varies with the size of the connecting lakes. This energy relaxation shows up as a broadband noise. The transient switching is the result of the formation and immediate disconnection of the river at the fields just below critical. This happens since after the river is formed, relaxation is available all over the sample width. The system can then relax to a new steady state with incompressible region still percolating. Our numerical simulations show such connection/disconnection processes. As in any dielectric breakdown, hysteresis is the result of the irreversible change of the system properties (namely the pattern of rivers and lakes) after the current was allowed to flow through the system. Finally, the peculiar steps observed in the magnetic field dependence of the longitudinal voltage in the critical current regime show the evidence of the opening of new metallic channels when the system is moved away from the center of the plateau. In equilibrium, as the difference of the filling factor from the integer becomes larger, the size of the lakes of compressible liquid increases while their separation becomes smaller. Therefore, in the critical regime the number of the percolating narrow metallic channels also grows, each giving its own discrete contribution in the voltage drop.

In conclusion, we propose a theory of the breakdown of the QHE based on the existence of the compressible regions in the inhomogeneous 2DEG. The predictions of the theory agree with experimental data.

This work was supported by the NSF grant DMR-9528345.

1. G. Ebert *et al*, J. Phys. C **16**, 5441 (1983).
2. M. Cage, J. Res. Natl. Inst. Stand. Technol. **98**, 361 (1993), and references therein.
3. The Quantum Hall Effect, ed. by R. E. Prange and S. M. Girvin, Springer-Verlag, New York (1990), and references therein.
4. K. von Klitzing *et al*, Proc. of the 17th Int. Conf. on the Physics of Semiconductors, ed. D. J. Chadi and W. A. Harrison, 1984.
5. A. Efros, Solid State Commun., **65**, 1281 (1988); A. Efros, Solid State Commun., **67**, 1019 (1988).
6. N. R. Cooper and J. T. Chalker, Phys. Rev. B, **48**, 4530 (1993).

THEORETICAL DETERMINATION OF THE CRITICAL CURRENT FOR THE BREAKDOWN OF THE QUANTUM HALL EFFECT

C. CHAUBET, F. GENIET* AND A. RAYMOND

Groupe d'Etude des Semiconducteurs, Université Montpellier II, place Eugène Bataillon, 34095 Montpellier cedex 5.

**Laboratoire de Physique Mathématique, Université Montpellier II..*

We calculate $V_{xx}(I)$ curves in the plateaus regime of the Quantum Hall Effect (QHE), using an original theoretical model. We place our investigation in the framework of a recent experimental work published by N.Q. Balaban et al.[1], and we reproduce exactly their experimental data. We first calculate the inter Landau level transition probabilities induced by phonon scattering, and we deduce the new statistical equilibrium which results in a higher electronic temperature. The longitudinal conductivity σ_{xx} is therefore calculated in the Kubo formalism and is found to increase drastically with the applied current through the sample.

1 The Quasi elastic inter Landau level scattering

We study a two dimensional electron gas (2DEG) confined near the interface of a GaAlAs-GaAs heterojunction[1] by an electrostatic potential U(z). In the plateaus regime of the QHE[2], the magnetic field is perpendicular to the 2DEG and the Hall electric field F_H lies in the y direction when the current flows along the x direction. No electric field exists parallel to the current. This quantum system is described by the following Hamiltonian

$$H = \frac{\left(\vec{p} + e\vec{A}\right)^2}{2m^*} + e\vec{F}_H.\vec{r} + U(z) \tag{1}$$

We note ν = (n,k_x) the eigenfunctions which are composed of a plane wave part on the x axis, an harmonic oscillator on y and an envelop function on z. The eigen-energies vary linearly versus the momentum k_x with a slope $V = F_H / B$. The electric field bends the Landau levels and increases the overlap between wave functions of different levels. Between two given states ν and μ, the transition probability is :

$$\underset{\upsilon\to\mu}{W} = \frac{2\pi}{\hbar}\left|\langle\upsilon|H_{int}|\mu\rangle\right|^2 \delta\left(\varepsilon_\mu - \varepsilon_\upsilon\right) \tag{2}$$

where H_{int} is the interaction Hamiltonian. For longitudinal acoustic phonon in the deformation-potential approximation, the probability can be expressed as [3] :

$$\underset{\upsilon\to\mu}{W}(q_x) = \frac{2\pi}{\hbar}\sum_{\vec{q}} C(q)^2 \left|f_{\upsilon\mu}(\vec{q})\right|^2 \delta\left(\varepsilon_\mu - \varepsilon_\upsilon \pm \hbar c_s q\right) \tag{3}$$

where C(q) is the strength for a given $\vec{q}$ -phonon scattering with fixed q_x momentum, and $f_{\upsilon\mu}(\vec{q})$ is the form factor. For the two first levels it is written[3] :

$$\left|f_{01}(q)\right|^2 = \frac{1}{\left(1+\frac{q_z^2}{b^2}\right)^3}\left(\frac{q_x^2+q_y^2}{2\beta}\right).exp\left(-\frac{q_x^2+q_y^2}{2\beta}\right) \quad (4)$$

where b is the Fang-Howard parameter of the envelop wave function on the z axis.

Results are presented in figure 1. The parameters used in the calculations are experimental ones described in part 3. It appears clearly, as previously discussed[3], that the dependence on the Hall field is drastic for the absorption process. For the emission process the low field probability is controlled by the emission of phonon along the z-axis. Indeed the form factor is maximal when q_x and q_y are equal to zero, and $q = q_z$ (see eq. (4)). Consequently if the 0-1 transition is undoubtely quasi elastic, the 1-0 transition is purely inelastic. Let us notice that radiative recombinations also occur slightly less efficient than phonon emission, and independant of the electric field[3].

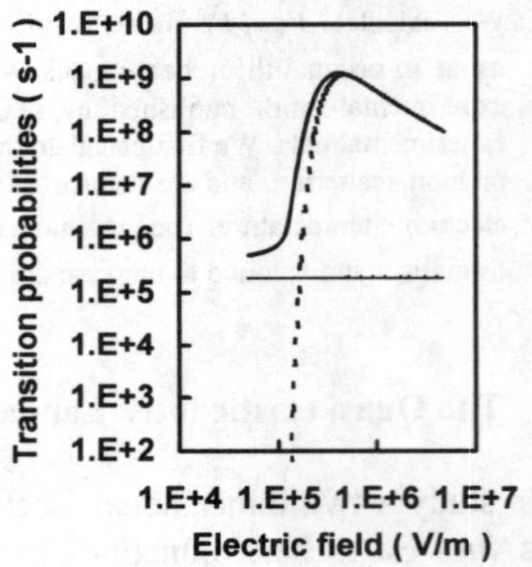

Fig.1: Transition probabilities for phonon emission (full line), phonon absorption (dashed line) and radiative transition (horizontal line).

2 The statistical equilibrium and the electronic temperature

In the experiment we are concerned with[1], the Fermi level ε_F is located in the middle of the energy gap between the two first Landau levels. The temperature is such that electrons do not statistically occupy the upper level. However, the total quantization is altered at sufficiently high electric field when phonons are absorbed. The following master equation governs the statistics of the Landau levels[3] :

$$\frac{\partial \rho_\upsilon}{\partial t} = \sum_{\upsilon'}\left(\rho_{\upsilon'}W_{\upsilon'\upsilon}(1-\rho_\upsilon) - \rho_\upsilon W_{\upsilon\upsilon'}(1-\rho_{\upsilon'})\right) \quad (5)$$

We assume that the mean value ρ_υ of the ν state density operator is independant of k_x . Indeed, since the drift velocity V is larger than the sound velocity in our calculations, intra level scattering via acoustic phonons are numerous[4]. Then all the states of a same level can be described by the same occupation number, which represents now the occupation function of the given level. The steady state solution can therefore be expressed with an electronic temperature (Fig. 2).

3 Transport properties and breakdown of the QHE

We assume that the abrupt jump of the longitudinal resistivity results from the loss of the total quantization. As soon as empty states exist in the lower level, and electrons occupy the upper one, the diffusion creates the dissipation. The diffusion can be accounted for, by a broadening parameter Γ of Landau levels. We have evaluated the width Γ using the classical formula $\Gamma = e\hbar/m^* \mu$ where μ is the mobility. The conductivity σ_{xx} is calculated in the linear formalism[5],

$$\sigma_{xx} = -\frac{e^2}{\pi^2\hbar}\int d\varepsilon \frac{\partial f}{\partial \varepsilon}\sum_n \left(n+\frac{1}{2}\right) exp\left(-\left(\frac{(\varepsilon-\varepsilon_F)}{\Gamma}\right)^2\right) \qquad (6)$$

Let us now analyse the physical situation. The Hall field distribution in the sample is known, as demonstrated by N. Balaban et al.[1]. It obeys the law :

$$F_H(y) = \frac{IR_H}{2}\left(ln\frac{W}{\delta}+\frac{\xi+\delta}{2\delta}\right)^{-1}\frac{W}{(W/2)^2-y^2} \qquad (7)$$

and takes its highest value at the very edge of the sample. Nevertheless the electric field that feels the edge wave function has to be taken at a cutoff distance δ from the edge, to take into account the finite spread of the wave function along y. The cutoff distance must be close to one magnetic radius[1] which is the characteritic length of the wave function in the Hall field direction. This merely signifies that the wave function feels an average electric field. However no precise value can be derived for δ.

As concluded by S. Kawaji et al.[6] we emphasize that the Hall electric field is the good parameter to describe the breakdown of the QHE. Using the field distribution, the current in the sample can be expressed as a function of the electric field. Hence, for a given current value, we calculate the highest value of the Hall field across the sample. We deduce from the QUILLS model, first the electronic temperature and then the conductivity, and finally, we calculate the longitudinal electric field from tensor equation (8). The potential drop is deduced from the sample geometry.

$$j_y = \sigma_{xx}F_h + \sigma_{xy}F_{//} = 0 \qquad (8)$$

Results of calculations shown in Fig. 3 must be compared with the experimental curve. We used experimental parameters[1] : channel width W = 18 μm, μ = 70 m^2/Vs, Ns = 2.6 10^{11} cm^{-2}, B=5.05 T, the filling factor is 2, T = 30 mK. We have drawn two different $V_{xx}(I)$ curves for two different values of the cutoff parameter : δ equal to one magnetic radius l_B , and $\delta = \sqrt{2}\ l_B$. The curve obtained with $\delta = \sqrt{2}\ l_B$, reproduces exactly the experimental results. However, we have to be prudent : one cannot exactly know the value of this parameter. Nevertheless, in so far as the value of the only parameter we used has a good physical meaning and allows to reproduce exactly the experimental results, we can think that the basic physics we developped is relevant. Actually, we obtained further agreement for different breakdown curves using the same $\sqrt{2}\, l_B$ parameter. Let us finally remark that we are aware of the linear model limitations (eq. 6 and 8) which cannot account for step like structure[7,8] and bistability[8,9] observed above the critical current. More, the linear approximation is undoubtly responsible for the too smooth increase of V_{xx} at the breakdown (fig.3).

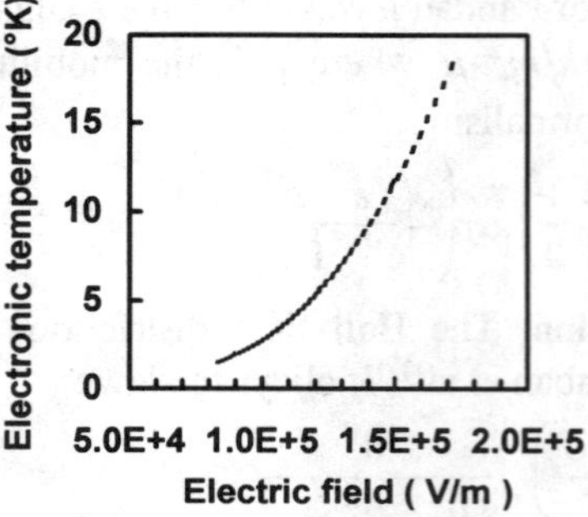

Fig.2 : Calculated electronic temperature

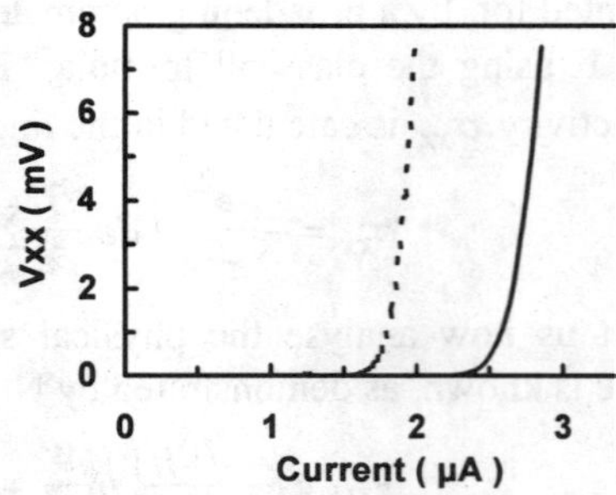

Fig. 3: Calculated breakdown curves. dashed line is obtained for $\delta = l_B$, and full one for $\delta = \sqrt{2}\, l_B$.

In conclusion, we have shown that acoustic phonon scattering allows a precise calculation of the critical current in samples exhibiting a well established Hall field distribution1[10]. This work corroborates recent experiments[11] which established that the breakdown first appears at the edge of the sample.

Acknowledgements

We are very gratefull to N.Q. Balaban for helpfull precisions on their samples, and providing experimental data.

References

1. N.Balaban, U.Meirav, H.Shtrikman; Y.Levinson,Phys.Rev.Let.**71**, 1443 (1993).
2. K. von Klitzing, G. Dorda and M. Pepper, Phys. Rev. Lett. **45**, 449 (1980).
3. C. Chaubet, A. Raymond and D. Dur, Phys. Rev. B **52**, 11178 (1995).
4. P. Streda, K. von Klitzing,, J. Phys. C, Solid State Phys. **17**, L483 (1984).
5. T. Ando and Y. Uemura, J. Phys. Soc. Jap. **36**, 1521, **37**, 622, **37**, 1233 (1974).
6. S. Kawaji et al., Journal of the Phys. Soc. of Japan **63**, 2303 (1994).
7. M.E. Cage, Marullo Reedtz, Yu , Van Degrift, Semic. Sci. Tech. **5**, 351 (1990).
8. G. Boella et al. Phys. Rev. B **50**, 7608 (1994).
9. F.J. Alhers et al. Semicon. Sci. Technol. **8**, 5379 (1994).
10. N.Q. Balaban, U. Meirav, and H. Shtrikman, Phys. Rev. B **50**, 1957 (1994).
11. A.A. Shashkin et al. Phys. Rev. B **49**, 5379 (1994).

EXPLANATION OF RESISTANCE JUMPS NEAR THE BREAKDOWN OF THE INTEGER QUANTUM HALL EFFECT

J. RIESS

Centre de Recherches sur les Très Basses Températures, associé à l'UJF, CNRS, BP 166, 38042 Grenoble-Cedex 9, France

We propose an explanation for the occurrence of resistance jumps (bistabilities) near the breakdown of the integer quantum Hall effect, focussing on recent experiments in narrow constrictions of GaAs/GaAlAs Hall bars. We show that the resistance jumps including their dependence on the filling factor ν can be explained from the typical form through which $\sigma_{xx}(\nu)$ and $\sigma_{xy}(\nu)$ at a given ν depend on the macroscopic electric field E.

1 Introduction

Longitudinal resistance jumps (bistabilities) near the breakdown of the integer quantum Hall effect (IQHE) have been observed by several groups in the past. Recently this phenomenon has been studied in detail in narrow constrictions of GaAs heterojunctions (typically of dimension $L_x = L_y = 50 - 200$ μm). Here jumps slightly below the filling factor $\nu = 2$ have been observed [1,2,3,4]. Experimentally one measures the potential drop $U_x = E_x L_x$ along the constriction. The longitudinal resistance is $R_x = U_x/I = E_x/j_x$ (for $L_x = L_y$), where I is the current and j_x the current density.

Jumps of R_x occurred when I was increased beyond the breakdown current I_b, or when I decreased from a value somewhat above I_b. Quite generally, such R_x-jumps are the result of an N-shaped $j_x - E_x$ relation. In the following we will show that the general behaviour of σ_{xx} and σ_{xy} as a function of $|\mathbf{E}|$ in a quantum Hall system (for a microscopic theory see [5]) can lead to such a $j_x - E_x$ relation, provided the filling factor ν is slightly *below* an integer value n.

2 Equation for the resistance jumps

In the constriction we have

$$j_x = \sigma_{xx} E_x + \sigma_{xy} E_y, \quad j_y = \sigma_{xx} E_y - \sigma_{xy} E_x = 0, \tag{1}$$

whence

$$j_x = [\sigma_{xx} + (\sigma_{xy})^2/\sigma_{xx}] E_x = (1/R_x) E_x. \tag{2}$$

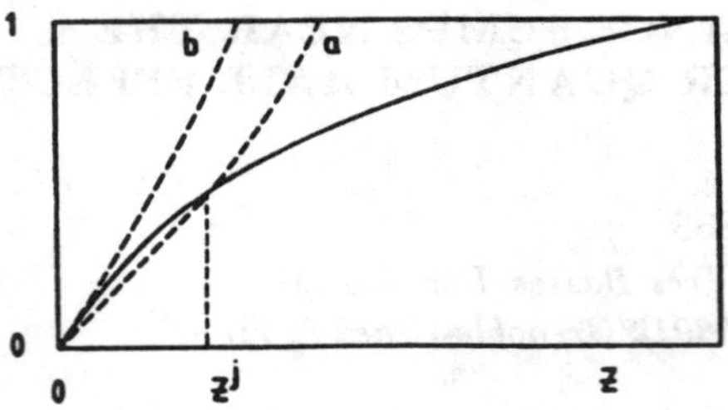

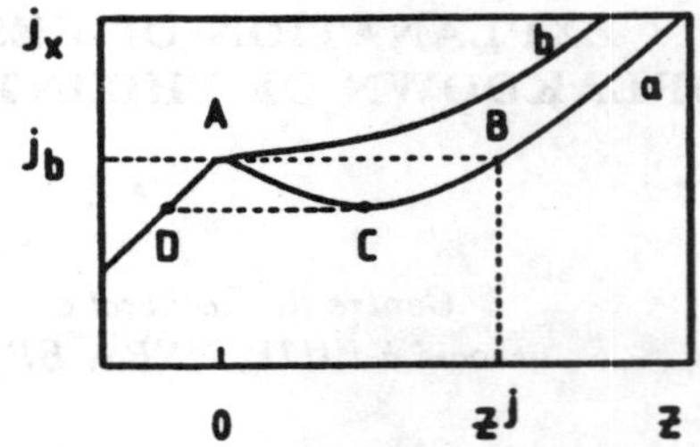

Figure 1: The full line illustrates the qualitative behaviour of σ_{xx}/σ^s as a function of $z = |\mathbf{E}| - E_b$ (σ^s denotes the saturation value of σ_{xx} for $z \gg 0$), and of the function $f(z)$ (see eq. 5). The dashed line shows the corresponding function $f^0(z)$ (eq. 7), for which two situations are possible: a) intersection with $f(z)$; b) no intersection.

Figure 2: Qualitative illustration of the two possible behaviours of the current j_x: a) N-shaped behaviour; b) monotonic increase with z.

The variables R_x and E_x in (2) are directly accessible from the experiment. On the other hand, the microscopic theory [5] gives the qualitative behaviour of σ_{xx} and σ_{xy} as a function of the variable $|\mathbf{E}|$. Therefore we also need the following equations (which follow from (1)):

$$E_x = |\mathbf{E}|\sigma_{xx}/[(\sigma_{xx})^2 + (\sigma_{xy})^2]^{1/2}, \tag{3}$$

$$j_x = |\mathbf{E}|[(\sigma_{xx})^2 + (\sigma_{xy})^2]^{1/2}. \tag{4}$$

In the quantum Hall regime, E_x vanishes hence $j_x = (ne^2/h)E_y$. If $|j_x|$ increases beyond the breakdown value $j_b = (ne^2/h)E_b$, $|\mathbf{E}|$ increases above the breakdown field E_b, and σ_{xx} starts to increase monotonically from zero to higher values (until saturation), hence $|E_x|$ increases as well [see (3)]. Fig. 1 shows the typical behaviour of σ_{xx} as a function of $|\mathbf{E}|$. Its behaviour as a function of E_x is similar. (From (3) one obtains $\sigma_{xx} \approx E_x(ne^2/h)/E_b$ for sufficiently low values of E_x).

The behaviour of σ_{xy} is different. Consider $\nu = n - \delta$, $\delta > 0$. Here, when $|\mathbf{E}|$ increases beyond E_b, $|\sigma_{xy}|$ monotonically *decreases* from the plateau value ne^2/h towards $(e^2/h)(n - \delta)$. [These are just the two limits of $|\sigma_{xy}|$ which are reached for $|\mathbf{E}| \leq E_b$ and $|\mathbf{E}| \gg E_b$, see [5]. In the second limit the Hall velocity of each state has just the unperturbed classical value $c|\mathbf{E}|/B$, which leads to $|\sigma_{xy}| = (e^2/h)\nu$]. As a consequence,

$$|\sigma_{xy}| = (e^2/h)[n - \delta f(|\mathbf{E}|)], \tag{5}$$

where $f(|\mathbf{E}|)$ increases monotonically from zero (at $|\mathbf{E}| = E_b$) to one (at some saturation value higher than E_b), see Fig. 1.

Expressing σ_{xy} in (4) by (5) and writing $|\mathbf{E}| = E_b + z$ we obtain

$$j_x = |E_b + z|\{[\sigma_{xx}(z)]^2 + (e^4/h^2)[n - \delta f(z)]^2\}^{1/2}. \qquad (6)$$

When z increases from zero to higher values, the first term in the curly bracket *increases* whereas the second *decreases*. Therefore it is possible that $j_x(z)$ first decreases below j_b, before it increases again at higher z.

To investigate this in more detail we keep $j_x = j_b$ fixed in (6) and solve the resulting equation for $f(z)$. This defines a function $f^0(z)$, whose meaning is the following: for a given $z > 0$ the value of j_x is smaller (bigger) than j_b when $f(z)$ in (5) is bigger (smaller) than $f^0(z)$. A simple calculation gives

$$f^0(z) \approx zn/(E_b\delta) + [\sigma_{xx}(z)h/e^2]^2/(2n\delta), \quad (0 < z \ll E_b). \qquad (7)$$

Therefore, if $f^0(z)$ increases sufficiently slowly (curve (a) in Fig. 1), such that it intersects with $f(z)$, the current $j_x(z)$ has an N-shaped behaviour (Fig. 2a), hence jumps of R_x occur. In the opposite case (curve (b) in Fig. 1), the current j_x monotonically increases with z (Fig. 2b)) i.e. no resistance jumps can occur.

Since E_x increases when z increases ($z > 0$), j_x has an N-shaped behaviour as a function of E_x, whenever it has an N-shaped behaviour as a function of z. The value z_j of Fig. 1 corresponds to point B in Fig. 2. It is obtained by solving equation 6 for z ($z > 0$), setting $j_x = j_b$ on the left hand side. The corresponding value $E_x(z^j)$ is then obtained from (3). The resistance jump is

$$R^j = E_x(z^j)/j_b = (1/j_b)(E_b + z^j)/\left\{[\sigma_{xx}(z^j)]^2 + [\sigma_{xx}(z^j)]^2\right\}^{1/2}. \qquad (8)$$

3 Discussion

Our analysis shows that the existence of an intersection point z^j in Fig. 1, hence of a resistance jump R^j, depends on the detailed z-dependence of $f(z)$ and of $\sigma_{xx}(z)$. This seems to explain why such jumps have been seen only in particular samples under particular experimental conditions, even if δ was positive. From the n-dependence of $f^0(z)$ [eq. 7] the existence of an intersection point z^j seems favored by small n. This might explain why in [2,3,4] R_x-jumps are observed for $n = 2$ but not for $n = 4$.

The form of $\sigma_{xx}(z)$ and of $f(z)$ imply that a jump of R_x always starts at (or near) $z = 0$, i.e. at (or near) zero resistance R_x, when the current is *increased* (since the system jumps from A to B in this case). When the current is *decreased* from values above j_b to a value below j_b, the system jumps from C

to D in Fig. 2. The corresponding resistance jump $R^{CD} = E_x^C / j_D$ is expected to be smaller than the corresponding resistance jump R^j from A to B (since $E_x = 0$ for $z \leq 0$). Further, according to our analysis only jumps for positive δ are possible, since for negative δ both σ_{xx} and $|\sigma_{xy}|$ monotonically increase with z, hence j_x monotonically increases as well. All these conclusions are consistent with the experiments.

We have shown that resistance jumps are caused by the fact that σ_{xx} and σ_{xy} change with different signs as a function of a parameter. Instead of E_x or $|\mathbf{E}|$ as discussed above, this parameter can also be ν (B). In this case j_x is fixed and j_b changes, due to the change of ν. When ν decreases such that j_b attains the value of j_x, (i.e. such that ϵ_F crosses a mobility edge), resistance jumps can occur according to our analysis, provided this mobility edge is situated at a fraction δ *below* a fully occupied band. This effect has been observed in [1].

A more detailed analysis of the resistance jumps can be found in [6].

As we have shown in previous articles [5], the general form of the $|\mathbf{E}|$-dependence of the conductivities $\sigma_{xx}(\nu)$ and $\sigma_{xy}(\nu)$ on which our analysis is based has a microscopic origin which resides in the general properties of the solutions of the time-dependent Schrödinger equation of the electrons in the presence of disorder and of a macroscopic electric field $\mathbf{E}$. This leads to $|\mathbf{E}|$-dependent mobility edges in a broadened Landau band, implying the breakdown of the IQHE (onset of dissipation), when $|\mathbf{E}|$ exceeds a threshold value (which depends on ν).

Acknowledgments

The author thanks J. Peinke for discussions, F.-J. Ahlers for constructive comments, H. Scherer for sending a copy of his thesis, and L. Bliek for Ref. 4.

References

1. G. Hein, P. Schneider, L. Schweitzer, F.-J. Ahlers, L. Bliek, H. Nickel, R. Lösch, and W. Schlapp, Surf. Sci. **263**, 293 (1992).
2. F.-J. Ahlers, G. Hein, H. Scherer, L. Bliek, H. Nickel, R. Lösch, and W. Schlapp, Semicond. Sci. Technol. **8**, 2062 (1993).
3. H. Scherer, Thesis, Techn. Univ., Braunschweig, Germany, 1995.
4. G. Nachtwei, A. Linke, M. Thiede, H. Scherer, C. Breitlow, F.-J. Ahlers, L. Bliek, H. Künzel, and J. Böttcher, Proc. 7th Int. Conf. on *"Narrow Gap Semiconductors"*, Santa Fé, 1995.
5. J. Riess, Physica B **204**, 124 (1995), and references therein.
6. J. Riess, Z. Naturforsch. **50a**, 1090 (1995).

CURRENT SCALING OF CONDUCTIVITY IN THE QUANTUM HALL EFFECT

V. Kagalovsky
Max-Planck-Institut für Kernphysik, Postfach 10 39 80, D-69029, Heidelberg, Germany

We consider current scaling at low temperatures and show that above a crossover current density j_{cr} an electron can be activated into extended state propagating along constant potential contour. The width of the conductivity peak σ_{xx} between Hall plateaus scales then with current as j^{μ} with $\mu = 3/14$ at $j < j_{cr}$ and crosses over to $\mu = 3/7$ at $j > j_{cr}$. We find that j_{cr} decreases for longer range potentials or for higher magnetic fields. These results are consistent with data showing $\mu \approx 0.3$ and $\mu \approx 0.21$ for GaAs/Al_xGa_{1-x}As and In_xGa_{1-x}As/InP samples respectively.

The behavior of transport coefficients in the Quantum Hall Effect (QHE) as a function of temperature T, current j and frequency is a subject of great interest. In this work we address the question of current induced broadening of the conductivity peaks. In a number of experiments [1,2] it was found that the width in the filling fraction of the peaks of the longitudinal conductivity σ_{xx} scales with current by a power law: $\Delta\nu \sim j^{\mu}$. Experiment[1] on GaAs/Al_xGa_{1-x}As heterostructure shows that $2\mu = 0.59$ while other data[2] obtained on In_xGa_{1-x}As/InP sample yields $\mu = 0.23 \pm 0.02$. The main distinction between the two types of heterostructures, as emphasized in [3], is the longer range of potential fluctuations in GaAs/Al_xGa_{1-x}As system. It is however not obvious why this results in different μ values.

There are two theoretical approaches to this question: the first one, used to treat experimental results [2], combines scaling behavior of an inelastic scattering length [4,5] $L_{in} \sim T^{-p/2}$ with effective temperature T_e of the electron gas. The smearing of the Fermi level (characterizing by T_e) is given by the energy acquired by an electron diffusing a distance L_{in} in the electric field ε_{el} . Therefore, $k_B T_e \sim e\varepsilon_{el} L_{in}$ and using the experimental result $T_e \sim j^{1/2}$ the value $p = 2$ was obtained[2]. This value disagrees with the result $p = 3$ obtained in [6] where L_{in} due to electron-phonon scattering was considered.

A different theoretical approach [7] proposed by Polyakov and Shklovskii (PS) is based on the theory of hopping in a strong electric field[8,9]. This theory claims that in a strong electric field the local Fermi distribution with effective temperature $\sim e\varepsilon_{el}\xi$ is formed. The localization length ξ diverges when the energy E of the electron approaches the energy E_c of a delocalized state in each Landau band (which we will chose equal to zero for convenience). It is well established that $\xi \sim |E|^{-7/3}$, based on experimental data[10], semiclassical derivations [11,12] and numerical simulations [13]. PS use their result obtained in the same paper [7] for the temperature dependence of the width of σ_{xx} (as function of the filling factor) $\Delta\nu \sim T^{3/7}$ and replace the temperature T by $e\varepsilon_{el}\xi/2$. This finally yields

$$\Delta\nu = \left(\frac{\varepsilon_{el}}{\varepsilon_1}\right)^{3/14} , \qquad (1)$$

where $\varepsilon_1 = Ae/\kappa\xi_0^2$, A is a numerical coefficient, κ is a dielectric constant and ξ_0 is a constant derived from the expression for the localization length $\xi = \xi_0|\Delta\nu|^{-7/3}$. It is interesting to mention that this theoretical prediction (which considers the transport from the insulator side) coincides with the experimental result [2] measured on the metal side and disagrees with the value derived from experimental results [1] treated using the PS approach [7].

In order to understand this difference one has to consider the interplay between the divergence of the localization length as energy approaches the percolation threshold and the change of the electric field energy at the opposite sides of the cluster (wave function of an electron is localized along constant potential contours which are the perimeters of the clusters [14]). This energy change is equal to $e\varepsilon_{el}L(\varepsilon_{el})$ ($L(\varepsilon_{el})$ is a typical cluster size in the presence of the electric field. If $E \gg e\varepsilon_{el}L(\varepsilon_{el})$ it means that at the cluster's length, an electron does not feel the electric field and one can apply the PS picture with electron jumping due to the quasi-Fermi-level inclined by the electric field leading to Eq. (1).

Now let us consider the case when $E < e\varepsilon_{el}L(\varepsilon_{el})$. In a strong electric field the typical cluster size in the field direction is $L_{el} = c\varepsilon_{el}^{-4/7}$ (c is a constant) [14]. The natural measure for the effective temperature in this case is $\sim e\varepsilon_{el}L_{el}$. Hence, a strong electric field means now a high effective temperature. It was shown in [15] that at high temperatures conductivity becomes thermally activated $\sigma_{xx} \sim \exp(-E/T)$. We will discuss now how the arguments explaining this activated behavior can be applied also to the present case. The clusters of a large size in the longitudinal direction are strongly suppressed by electric field, and therefore tunneling between different clusters is also negligible. Nevertheless, the the critical cluster (percolating through the whole system) corresponding to the extended state should exist, otherwise the QHE would not persist at high electric fields (the existence of extended state is a necessary condition for the transition between plateaus). The electron, therefore, can be activated into extended state propagating along the perimeter of one particular cluster, i.e. the effective temperature is of the order of $e\varepsilon_{el}L_{el}$. All mentioned above leads to a modified form of conductivity expression

$$\sigma_{xx} \sim \exp[-E/e\varepsilon_{el}L_{el}] = \exp[-(\varepsilon_2/\varepsilon_{el})^{3/7}], \tag{2}$$

where $\varepsilon_2 = (E/ec)^{7/3}$. By equating $\varepsilon_2 \sim \varepsilon_{el}$ we obtain the exponent $\mu = 3/7$, twice larger than low field PS result.

An interesting question is to find the crossover electric field at which this transition occurs. It was shown [12] that there exist a crossover temperature T_1 above which the tunneling probability is modified into activation form. It is natural to assume that analogous crossover occurs for current scaling when the PS effective temperature becomes of the order of T_1. We take the localization length ξ at energy equal to the PS half width $\Delta\nu = (\varepsilon_{el}/\varepsilon_1)^{3/14}$ and immediately obtain a crossover electric field

$$\varepsilon_{cr} = \frac{\kappa T_1^2}{e^3 B}, \tag{3}$$

where B is a numerical constant. It can be easily shown that the same estimate for ε_{cr} could be obtained by equating $E \sim e\varepsilon_{cr}L(\varepsilon_{cr})$. Using the result [12] $T_1 = \ell^2 K/2\pi$ obtained in a strong magnetic field limit ($\ell = \sqrt{\hbar c/eB}$ is the magnetic length, K is an average curvature of the potential, $K \sim 1/\lambda^2$ and λ is a range of potential fluctuations) we obtain $\varepsilon_{cr} \sim 1/(B\lambda^2)^2$.

We can now estimate the crossover current for the experiment of Koch *et al*[1]. The value of the crossover temperature [12] $T_1 = 0.1 - 1$K, $\kappa \approx 10$ and therefore $\varepsilon_{cr} = 0.01 - 1$[V/m]. The Hall resistance is $R_H \approx 2.6 \times 10^4[\Omega]$. The measurements were performed between $n = 3$ and $n = 4$, so $R \approx 10^4[\Omega]$ and the crossover current density is $j_{cr} = \varepsilon_{cr}/R \approx 10^{-6} - 10^{-4}$[A/m], which is in the range of measurements[1].

Concerning the experiment of Koch *et al.* [1] it seems plausible that j_{cr} is so close to the saturation current observed in the experiment that there was no range where the pure PS regime existed and that is why the fit of the data showed $\mu \approx 0.3$ demonstrating transition between low 3/14 and high filed values 3/7. On the other hand for short-range potential in the experiment of Wei *et al*[2], j_{cr} is extremely large due of the large curvature of the short-range potential and is out of range of experimental currents.

Now we address the relation between temperature and electric field. It was shown[7] that the dependence on electric field leads to the same broadening as if there was a temperature $T_{eff} \sim \varepsilon_{el}^{1/2}$ (leading to the following relation between temperature and current critical exponents $\kappa = 2\mu$). This relation is universal which was confirmed by experiments with different materials[1,2,16,17]. In our case comparing high temperature ($\kappa = 1$) [15] and high field dependences we find $T_{eff} \sim (\varepsilon_{el}^{1/2})^{6/7}$ ($\kappa = (7/6)2\mu$) which could be responsible for a small deviation ($2\mu = 0.59$, $\kappa = 0.62$) from a universal behavior[1].

In conclusion, we have shown that the conductivity mechanism in the QHE is sensitive to current density and can modify the apparent exponent of the data. We account existing experimental results for μ. Further experiments can probe full current density dependence of the width of σ_{xx} and clarify the change in the conductivity mechanism at j_{cr}.

References

1. S. Koch, R. J. Haug, K. v. Klitzing and K. Ploog, *Semicond. Sci. Technol.* **10**, 209 (1995);
2. H. P. Wei, L. W. Engel and D. C. Tsui, *Phys. Rev. B* (1994);
3. H. P. Wei, S. Y. Lin, D. C. Tsui and A. M. M. Pruisken, *Phys. Rev. B* **45**, 3926 (1992);
4. H. Aoki and T. Ando, *Phys. Rev. Lett.* **54**, 831 (1985); *Surf. Sci.* **170**, 249 (1986);
5. A. M. M. Pruisken, *Phys. Rev. Lett.* **61**, 1297 (1988);
6. H. L. Zhao and S. Feng, *Phys. Rev. Lett.* **70**, 4134 (1993);
7. D. G. Polyakov and B. I. Shklovskii, *Phys. Rev. Lett.* **70**, 3796 (1993); *Phys. Rev. B* **48**, 11167 (1993);
8. B. I. Shklovskii, *Fiz. Tekh. Poluprovodn.* **6**, 2335 (1972) *Sov. Phys. Semicond.* **6**, 1964 (1973)];
9. B. I. Shklovskii, E. I. Levin, H. Fritzsche and S. D. Baranovskii, in *Transport, Correlation and Structural Defects*, edited by H. Fritzsche (World Scientific, Singapore,1990);
10. S. Koch, R. J. Haug, K. v. Klitzing and K. Ploog, *Phys. Rev. B* **46**, 1596 (1992);
11. G. V. Milnikov and I. M., *JETP Lett.* **48**, 536 (1988);
12. V. Kagalovsky, B. Horovitz and Y. Avishai, *Europhys. Lett.* **31**, 467 (1995) ;
13. J. T. Chalker and P. D. Coddington, *J. Phys. C* **21**, 2665 (1988);
14. S. A. Trugman, *Phys. Rev. B* **27**, 7539 (1983);
15. B. I. Shklovskii (private communication);
16. A. A. Shashkin, V. T. Dolgopolov, G. V. Kravchenko, M. Wendel, R. Schuster, J. P. Kotthaus, R. J. Haug, K. v. Klitzing, K. Ploog, H. Nickel and W. Schlapp, *Phys. Rev. Lett.* **73**, 3141 (1994);
17. E. Chow, H. P. Wei, S. M. Girvin and M. Shayegan, preprint (1995).

NONLOCAL BREAKDOWN OF THE INTEGER QUANTUM HALL EFFECT

Y. KAWAGUCHI [a], S. KOMIYAMA [a], T. OSADA [b], and Y. SHIRAKI [b]

[a] Department of Pure and Applied Sciences, University of Tokyo, Komaba 3-8-1, Meguro-ku, Tokyo, 153, Japan
[b] RCAST, University of Tokyo, Komaba 3-6-1, Meguro-ku, Tokyo 153, Japan

Abstract

We investigate the breakdown of the integer quantum Hall effect, and suggest that a bootstrap-type electron heating model is a probable mechanism of the breakdown.

In the integer quantum Hall (IQHE) regime, sudden onset of dissipation occurs when the current increases beyond a certain threshold, I_c [1,2]. In spite of extensive studies of this breakdown phenomenon [3-8], there is no broad consensus on the mechanism of the breakdown phenomenon.

In a recent set of experiments, we have reported remarkable device-length dependence of the breakdown [9,10] and reported that the device-length dependence originates from the absence of the breakdown over a macroscopic region in the two dimensional electron gas (2DEG) channel on the side of the electron-injecting corner of the Hall-bars [11]. Our experimental results suggest the bootstrap-type electron heating model [3] to be a promising mechanism. In this report, we present new experimental results on the sample dependence and the effect of illumination.

Samples are fabricated from two wafers of GaAs/AlGaAs single heterostructures, in which the critical current of the breakdown, Ic, proved to linearly scale with sample width W [9-11]. The 4.2K-mobility of 2DEG, μ_H, and the 2DEG sheet density, n_s, are μ_H=80m^2/Vs and n_s=2.6X10^{15}m^{-2} in one wafer (Wafer 1) and μ_H=10m^2/Vs and n_s=5.2X10^{15}m^{-2} in the other wafer (Wafer 2).

Figure 2(a) displays the longitudinal voltage V_{65} at 4.2K and at the center of the ν=2 Hall plateau as a function of current in a sample fabricated from Wafer 1. The sample is a Hall-bar shown in Fig.1(a). The voltage V_{65} sharply increases at about 3 μA to exhibit well-defined breakdown characteristics for the "positive polarity" of current in which electrons are injected from the right-hand current contact. However, the voltage V_{65} does not increase appreciably for the opposite polarity of current. The asymmetric breakdown characteristic is independent of the magnetic field polarity as demonstrated by the solid and dotted lines in Fig.2(a). The characteristics are also independent of the side of the 2DEG channel on which the voltage is measured [11].

Similar current-polarity dependent breakdown characteristics are observed also in a lower-mobility sample. Fig.2(b) displays the longitudinal voltages V_{12} and V_{34} in a sample fabricated from Wafer 2 at 1.5K and at the center of ν=4 Hall plateau. The sample is a Hall-bar shown in Fig.1(b).

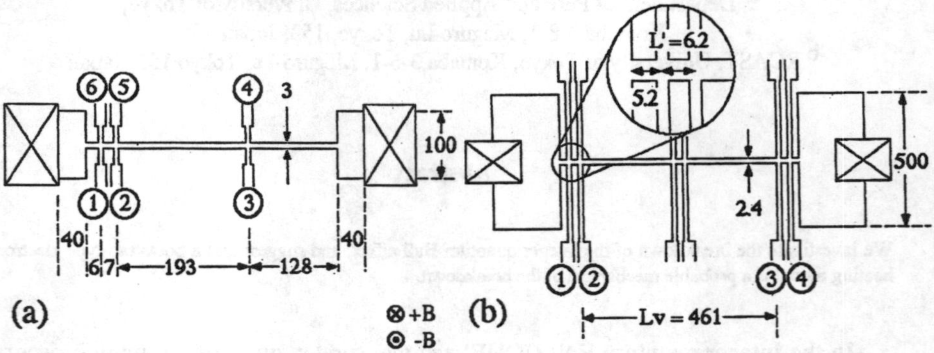

Fig.1 Schematic view of samples. The dimensions are given in unit of μm.

The inset of Fig.2(b) shows the longitudinal voltage V_{23} by the values normalized by the inter-probe distance Lv = 461μm. The longitudinal voltages V_{12} and V_{34} sharply increase at a critical current of about 3 μA for both polarities of current. However, the increase in V_{12} is distinctly steeper than that in V_{34} for the positive polarity of current . The feature is opposite for the negative current polarity.

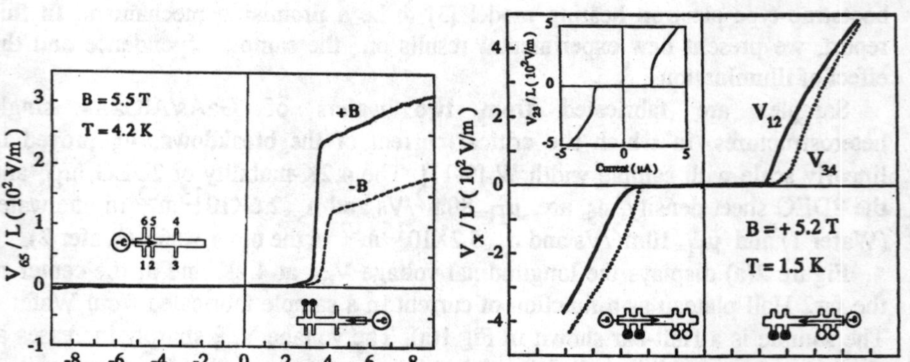

Fig.2(a) Longitudinal voltage V_{65} in the sample (a) shown in Fig.1(a).
(b) Longitudinal voltage V_{12} and V_{34} in the sample shown in Fig.1(b).
The inset shows V_{23}.

Figure 3 shows the effect of illuminating a sample by an infrared photo diode. The sample is the one used for the data of Fig.2(a). After the illumination, in which the carrier concentration increased from $2.6X10^{15}m^{-2}$ to $3.1X10^{15}m^{-2}$, the critical current increases by a factor of two. At the same time, the asymmetry of the breakdown with respect to the current polarity is further strengthened.

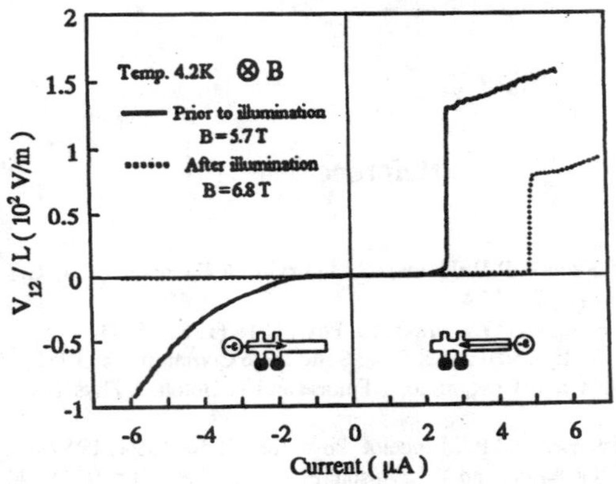

Fig.3 The effect of illumination.

The present experimental results directly indicate that the Hall electric field $E_H(r)$ does not locally affect the increase of the longitudinal conductivity $\sigma_{xx}(r)$ in the QHE breakdown, because $E_H(r)$ in the 2DEG channel is independent of the current polarity. Thus a model like the Zener-type tunneling [12-14] may be difficult to explain the breakdown characteristics observed here. We can interpret these results in the framework of bootstrap-type electron heating model [3]. Based on this model, the electron distribution is characterized by an effective electron temperature $T_e(r)$. Suppose that cold electrons ($T_e \sim 0$) are injected from the wide 2DEG pad region into the narrow 2DEG channel. Even if the energy gain rate and the energy loss rate are both negligibly small, the electron system can be unstable against infinitesimal fluctuation in the electron distribution[3,11]. In such a case, the fluctuation triggers multiplication of excited electrons in the higher Landau level, elevating Te. During the avalanche-type multiplication, electrons have to travel a macroscopic distance along the narrow channel, until the electron system finally reach a new stable (breakdown) state with a higher Te.

In the lower mobility wafer, we suppose that the electron multiplication process is more rapid due to an effect of stronger random potential. The illumination improves the homogeneity of the sample, reducing the extent of random potential

fluctuations. So the illumination may work to slow down the process of avalanche-type breakdown, leading to the enhancement of the presently observed current-polarity dependent breakdown of the QHE.

References

[1]M.E.Cage, R.F.Dziuba, B.F.Field, E.R.Williams, S.M.Girvin, A.C.Gossard, D.C.Tsui and R.J.Wagner: Phys. Rev. Lett. **51** (1983)1374
[2]G.Ebert, K.von Klitzing, K.Ploog and G.Weimann: J. Phys. C **16** (1983) 5441
[3]S.Komiyama, T.Takamasu, S.Hiyamizu and S.Sasa: Solid State Commun. **54** (1985)479
[4]S.Kawaji, K.Hirakawa, M. Nagata, T.Okamoto, T.Fukase and T. Gotoh: J. Phys. Soc. Jpn. **63** (1994)2303
[5]A.Boisen, P.Boggild, A.Kristensen and P.E.Lindelof: Phys. Rev. B **50** (1994) 1957
[6]N.Q.Balaban, U.Meirav, H.Shtrikman and Y.Levinson: Phys. Rev. Lett. **71** (1993) 1443
[7]C.L.Jones, A. Usher, Elliott, W.G.Herrenden-Harker, A. Potts, R.Shepherd, T.S.Cheng, and C.T.Foxon: Solid State Commun. **97** (1996) 763
[8]N.Q.Balaban, U.Meirav, and H.Shtrikman: Phys.Rev.B **52** (1995) R5503
[9] Y.Kawaguchi, F.Hayashi, S.Komiyama, T.Osada, Y.Shiraki and R.Itoh:Jpn.J.Appl.Phys. **34** (1995) 4309
[10] Y.Kawaguchi, S.Komiyama, T.Osada, and Y.Shiraki: Physica B (1996) in press
[11] S.Komiyama, Y.Kawaguchi, T.Osada, and Y.Shiraki: Phys. Rev. Lett. **77** (1996)558
[12]D.C.Tsui, G.J.Dolan and A.C.Gossard: Bull. Am. Phys. Soc. **28** (1983) 365
[13]L.Eaves and F.W.Sheard: Semicond. Sci. Technol. **1** (1986)346
[14]O.Heinonen, P.L.Taylor and S.M.Girvin: Phys. Rev. B **30** (1984) 3016

BREAKDOWN PHENOMENA OF INTEGER QUANTUM HALL EFFECT IN TILTED MAGNETIC FIELDS

M. OHNO, T. TAKAMASU*, N. MIURA, M. KATO, A. ENDO
S. KATSUMOTO, Y. IYE, T. TERASHIMA*, H. AOKI* and G. KIDO*
Institute for Solid State Physics, University of Tokyo
7-22-1 Roppongi, Minato-ku, 106 Tokyo, Japan

**National Research Institute for Metals*
1-2-1, Sengen, Tsukuba-shi, 305 Ibaraki, Japan

Current-induced breakdown of integer quantum Hall effect was studied in tilted magnetic fields. Magnetic field dependence of critical electric field of the breakdown was analized in terms of energy gap enhanced by exchange interaction.

Reccently, the ground and the excited states of two-dimensional electron gas at odd integer fillings have been extensively studied[1-5]. Especially in GaAs systems, the g-factor is so small compared with the Coulomb interaction energy that the energy gap between the ground state and the lowest excited state is not simply determined by $g\mu_B B$. In fact, Usher et al. have reported that the effective g-factor in the system is largely enhanced compared to the bare g-factor in bulk materials. Moreover, it has been predicted that the macroscopic spin structures of the excited states, skyrmion, may appear when the g-factor of the system is negligilly small[1]. Evidences of such a macroscopic spin structure were observed as the change of the energy gap[2,3] or the spin polarization[4,5].

In recent works, we have revealed that the energy gap can be estimated from the critical electric field of the current-induced breakdown of the quantum Hall effect[6]. In this paper, we present results of measurements of the breakdown phenomena in tilted magnetic field to clarify the nature of the spin polarized states in a wide range of magnetic field.

Samples we used are two different GaAs/AlGaAs heterostructures grown by MBE. Mobilities and carrier densitiews of the samples in a dark condition are μ=4.26×10^5 cm^2/Vs and n_s=1.89×10^{11} cm^{-2} in sample #1 and μ=4.13×10^5 cm^2/Vs and n_s=0.57×10^{11} cm^{-2} in sample #2, respectively. Voltage vs. current characteristics of the samples were measured at several odd and even integer fillings using pulsed and superconducting magnets. Critical electric field of the breakdown was determined from the maxima of the derirative of V vs. I curves as a function of magnetic field. To observe the quantum Hall effect with the integer filling factors ν, we varied the total magnetic field

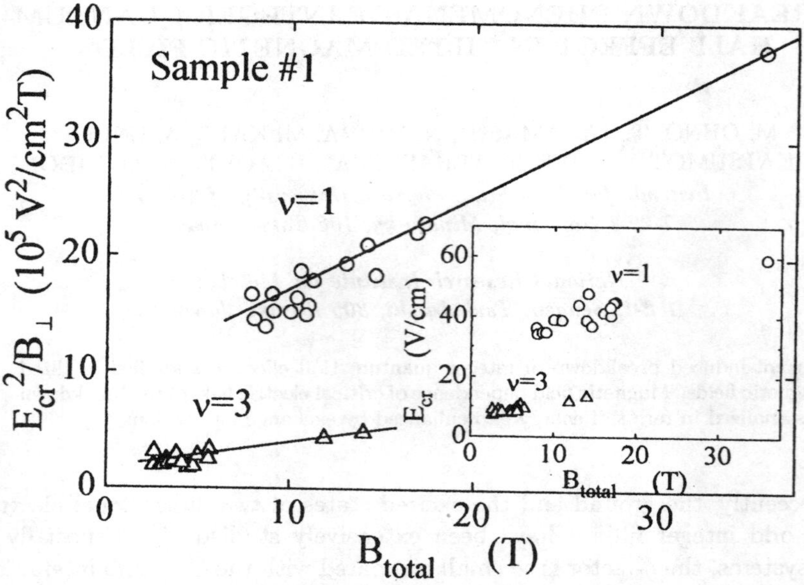

Figure 1: Critical electric field E_{cr} (inset) and $E_{cr}^2/B_\perp$ as a function of the total applied magnetic field measured at ν=1 and 3 for sample #1.

B_{total} and the perpendicular component to the 2D layer $B_\perp$, controlling the carrier densities with LED illumination and changing the tilt angle θ between magnetic field and the normal direction of the two dimensional electron system.

In the inset of Fig. 1, we show the measured E_{cr} as a function of B_{total} for sample #1. Similarly to the other samples reported in ref.6, E_{cr} depends linearly on the normal component of the applied magnetic field $B_\perp$ for even integer fillings. However, at odd integer fillings, E_{cr} is scaled not only by $B_\perp$ but also by B_{total}. In ref.7, we analyzed these data and found that the critical electric field E_{cr} in the heating model of the breakdown [8] can be written as,

$$E_{cr} = (\frac{1}{\tau_\varepsilon} \cdot \frac{1}{2\pi l_c^2} \cdot \frac{\Delta}{\sigma_{xx}^0})^{\frac{1}{2}} \propto (B_\perp \cdot \Delta)^{\frac{1}{2}}$$

Thus, the energy gap Δ is considered to be proportional to $E_{cr}^2/B_\perp$. In fact, we obtained scaled relation between $E_{cr}^2/B_\perp$ and B_{total} as shown in Fig.1. In this sample, the energy gap seems to be almost proportional to B_{total} and slopes at ν=1 and 3 are small compared with that at even integer fillings. Moreover, the curves are different in the absolute value and slopes between

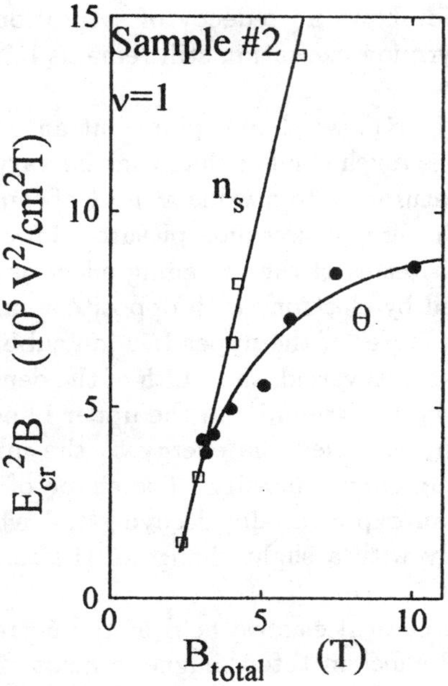

Figure 2: The same plot as Fig. 1 for ν=1 of sample #2.

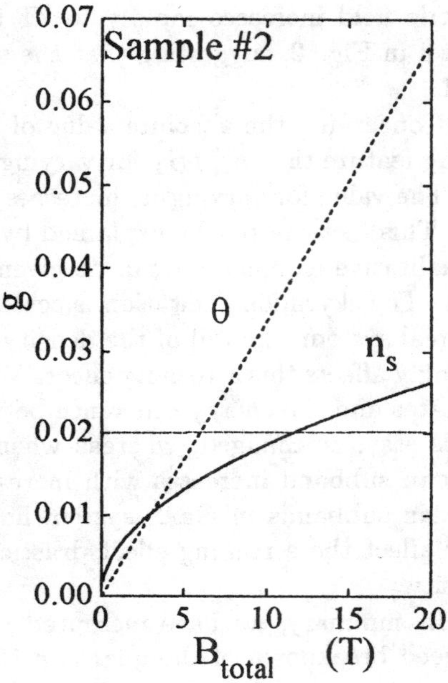

Figure 3: Effective g-factor, $\tilde{g}$, as a function of total magnetic field B_{total} for sample #2 for variable n_s (solid line) and for variable θ(broken line).

ν=1 and 3 . On the other hand, the same plot for another sample #2 looks different (Fig. 2). When θ is varied keeping n_s constant, E_{cr} ($E_{cr}^2/B_\perp$) clearly shows a nonlinearity as a function of B_{total}. We can explain this nonlinear relationship in terms of skyrmion as follows. The number of spins which gather in the skyrmion structure is considered to be a function of the effective g-factor, $\tilde{g} \equiv E_Z/E_c$, where E_Z is the Zeeman energy $g\mu_B B$ and E_c is Coulomb energy[1]. The skyrmion excitation can be observed only under the condition that $\tilde{g} < 0.02$. When θ is varied keeping n_s constant, $\tilde{g}$ increases linearly with the total magnetic field as shown in Fig. 3, and this condition is fulfilled only below 6T for the sample #2. This magnetic field coincides with the saturation field of $E_{cr}^2/B_\perp$ in Fig.2. When the carrier density of the sample is changed keeping θ constant (perpendicular to the 2D layer), $\tilde{g}$ increases with the square root of B_{total} and it exceeds 0.02 at B_{total} ~12T. When n_s is varied, the critical electric field increases rapidly to 7T and shows no tedency of saturation as shown in Fig. 2, suggesting that the skyrmion excitation still remains in this field.

Concerning the absolute value of $E_{cr}^2/B_\perp$, we should point out an interesting feature that $E_{cr}^2/B_\perp$ for varying θ is much smaller than that for varying n_s. The value for varying n_s increases, saturating to a value at ν=1 of sample #1. This fact can not be explained by the simple skyrmion picture. However, a qualitative explanation can be given in terms of the screening effect as follows. The skyrmion excitation is screened by electrons with opposite spin. In the real system, the tail of the density of states of the upper Landau subband strongly affects this screening effect. When θ is varied, the width of the density of states and the energy difference between the ground and the upper Landau levels stays unchanged, whereas when n_s is varied the energy to the upper Landau subband increases with increasing carrier density. The shape of the Landau subbands in GaAs system has an exponentially decaying tail which may affect the screening effect drastically with a slight change of the carrier density.

In summary, we have measured the critical electric field of the current-induced breakdown of the quantum Hall effect in tilted magnetic fields. The carrier density and the angle dependence of the critical field can be explained partially in terms of the exchange enhancement of the energy gap including the existence of the skyrmion state.

References

1. A. Usher, R. J. Nicholas, J. J. Harris and C. T. Foxon, Phys. Rev. **B41**, 1129(1989).
2. S. L. Sondhi, A. Karlhede, S. A. Kivelson and E. H. Rezayi, Phys. Rev. **B47**, 16419(1993).
3. A. Schmeller, J. P. Eisenstein, L. N. Pfeiffer and K. W. West, Physical Rev. Lett. **75**,4290(1995).
4. E. H. Aifer, B. B. Goldberg and D. A. Broido, Phys. Rev. Lett. **76**,680(1996).
5. S. E. Barret G. Dabbagh, L. N. Pfeiffer, K. W. West and R. Tycko, Phys. Rev. Lett. **74**, 5112(1995).
6. T. Takamasu, H. Dodo, M. Ohno and N. Miura, Solid State Commun. **96**,121(1995).
7. T. Takamasu et al., to be submitted to 12th HMSP.
8. S. Komiyama, T. Takamasu, S. Sasa and Hiyamizu, Solid State Commun. **54**, 47(1986).

Magnetic Field Dependence of the Nonlinear $\rho_{xx}(j)$-Behaviour in the Breakdown of the Quantum-Hall-Effect on the $i = 2$-Plateau

H. Scherer, F.-J. Ahlers, L. Bliek

Physikalisch-Technische Bundesanstalt,
Bundesallee 100, 38116 Braunschweig, FRG

We studied the current-voltage characteristics of GaAs/Al$_x$Ga$_{1-x}$As Hall-bars with constricted current paths in the magnetic field range near the filling factor $\nu = 2$. In the breakdown behaviour at $\nu < 2$, we found typical hysteretic $\rho_{xx}(j)$-curves, accompanied by fluctuations of ρ_{xx} corresponding to jumps between the two branches of the hysteresis loops. We investigated the evolution of the hysteresis loops with respect to their filling factor dependence in the range $1.95 < \nu < 2.0$. Using a simple electron-heating model, we calculated both the critical current and the corresponding dissipative resistance value and could explain their filling factor dependence at the low current side of the hysteresis loops.

1 Sample Properties

All samples were Hall-bars, prepared from the same wafer of MBE-grown GaAs/Al$_x$Ga$_{1-x}$As heterostructure material ($x = 0.32$), showing mobilities of 27 m^2/Vs and carrier concentrations of $3.8 \cdot 10^{15}$ m^{-2}. Electrical contacts with good ohmic behaviour were made by alloying tin pills into the material. A wet etching process was used to provide constrictions in the current path in such a way that square shaped current 'bridges' of dimensions $L \times W = 50 \times 50$ μm, 100×100 μm and 300×300 μm were formed between neighboring potential contacts (see inset in Fig. 1).

2 Experiment

The measurements were performed in a dilution refrigerator in the temperature range between 30 mK and 1.5 K. The sample was immersed in the liquid helium so that efficient heat removal from the sample was ensured and consequently lattice heating effects due to Joule heating were not significant. Commercial current sources and voltmeters were used to record $U_x(I)$-curves for an investigation of the current-induced breakdown behaviour in the magnetic field range near the filling factor $\nu = 2$.

3 Results and Discussion

In all samples we found the current-induced breakdown behaviour for $\nu \lesssim 2$ to be accompanied by hysteretic $\rho_{xx}(j)$-curves (Fig. 1). Furthermore, if the current was set to a fixed value inside a hysteresis loop, fluctuations of the longitudinal resistance occured [1] that corresponded to jumps between the two branches of the loop. Similar observations have been reported by other groups [2]. The observation of hysteretic $\rho_{xx}(j)$-curves and instabilities strongly suggests S-shaped $U_x(I)$-characteristics in the breakdown regime of the IQHE [3]. In this paper, we present a closer investigation of the filling factor dependence $(1.95 < \nu < 2.0)$ of the nonlinear behaviour in the breakdown regime.

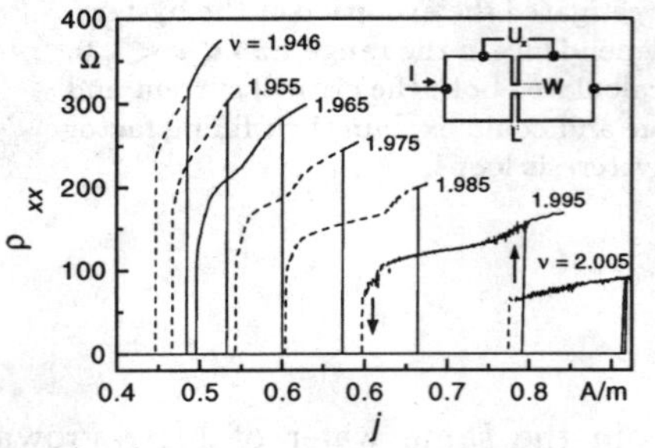

Figure 1: Hysteretic $\rho_{xx}(j)$-curves for $\nu \lesssim 2$ and sample geometry (inset).

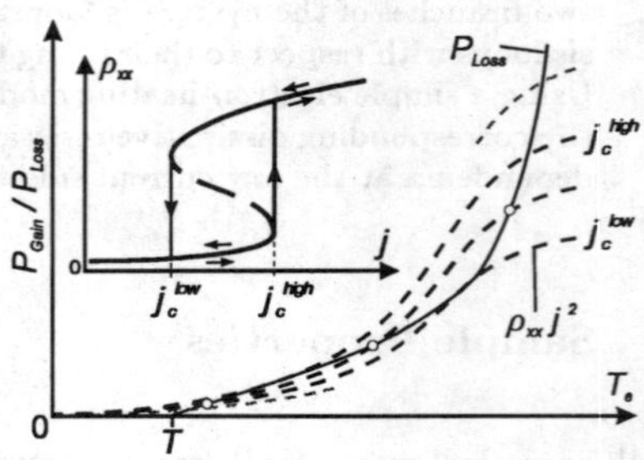

Figure 2: Electron-heating mechanism (after [3]) and resulting $\rho_{xx}(j)$-curve (inset).

Increasing the magnetic field away from the center of the $i = 2$-plateau (Fig. 1), we found the well-known decrease in critical current. Additionally, the width of the hysteresis loop diminished, while its height (i.e. the ρ_{xx}-'jump' height to the dissipative resistance value) grew considerably. It is important to mention that the quantitative evolution of the hysteresis effects with respect to the critical current densities and loop heights was similar for all the samples in spite of their different constriction sizes, so that the quantitative breakdown behaviour can be assigned to intrinsic properties of the heterostructure material rather than to the sample geometry.

In order to explain the occurence of S-shaped $U_x(I)$-traces, we applied an electron-heating model proposed in [3]. In this model, the power balance equation $P_{Gain} = P_{Loss}$ is evaluated, with the Joule heating power $P_{Gain} = \rho_{xx} j^2$ and P_{Loss} representing the energy loss rate of the electrons to the phonon bath of the crystal lattice. This leads to an equation of the form

$$\rho_{xx}(T_e) j^2 = P_{Loss}(T_e) , \tag{1}$$

where T_e stands for the effective temperature of the electron system ($T_e \geq T$).[a] Following the arguments of Komiyama *et al.* [3], ansatz (1) was evaluated by using $\rho_{xx}(T)$ as measured in the ohmic regime, i.e. at low currents where electron heating is insignificant ($T_e \cong T$). Further, P_{Loss} was calculated using the relaxation time approximation $P_{Loss} = [U(T_e) - U(T)]/\tau_E$ with the energy density U of the electron system relaxing at a rate τ_E^{-1}. In a certain range of the current density j, three intersections occur between the curves of $P_{Gain}(T_e)$ and $P_{Loss}(T_e)$, leading to the S-shape of $\rho_{xx}(j)$ (see inset of Fig. 2).

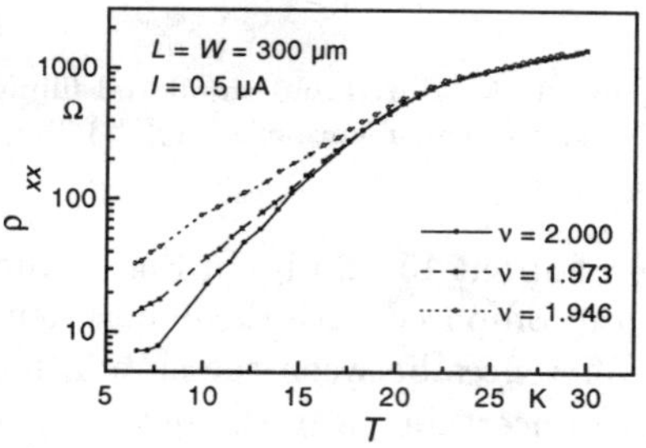

Figure 3: $\rho_{xx}(T)$, measured at different filling factors ν.

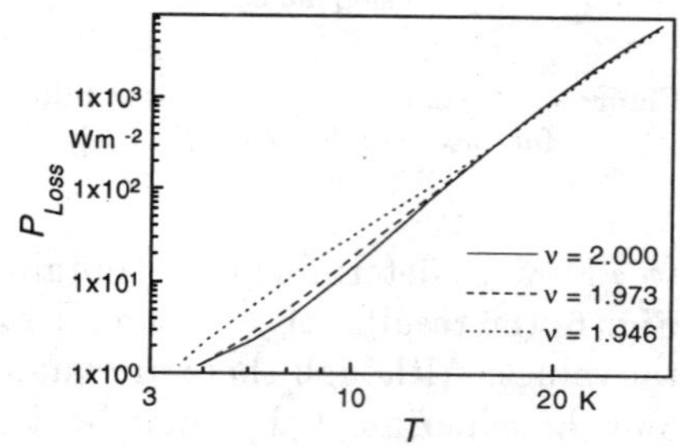

Figure 4: $P_{Loss}(T)$, calculated for different filling factors ν.

We expanded this model by including the filling factor dependence of the parameters $\rho_{xx}(T)$ and $U(T)$ into the evaluation of (1). Whereas $\rho_{xx}(T)|_\nu$ was measured under ohmic conditions (see Fig. 3), $P_{Loss}(T)|_\nu$ was calculated (see Fig. 4) using

$$U(T,\nu) = \int_{-\infty}^{\infty} \epsilon\, D(\epsilon,\nu)\, f(\epsilon, T, \mu(\nu))\, d\epsilon\,. \qquad (2)$$

Here, D is the Gaussian density of states [b], and

$$\tau_E^{-1} = 1.5 \cdot 10^7 \mathrm{K}^{-2}\mathrm{s}^{-1} \cdot T_e^2 \qquad (3)$$

is introduced as a reasonable approximation for the energy relaxation rate.[c] Thus, performing the evaluation described above, we were able to obtain a prediction for the filling factor dependence of the critical current and the corresponding dissipative resistance value at the low current side (j_c^{low}) of the hysteresis loops.[d] Furthermore, effective electron temperatures corresponding

[a] In the following, T means the lattice temperature.

[b] The linewidth $\Gamma = 0.7$ meV was estimated from the carrier mobility.

[c] As in [3], we also used this approximation which provides the best agreement with our experimental results.

[d] Due to uncertainties in the measurements of $\rho_{xx}(T)$ at low T, this was not feasible for the high current side, i.e. at j_c^{high}.

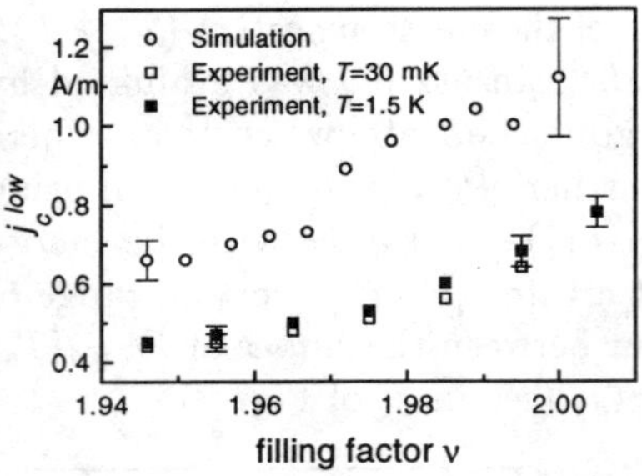

Figure 5: Measured and simulated filling factor dependence of j_c^{low}.

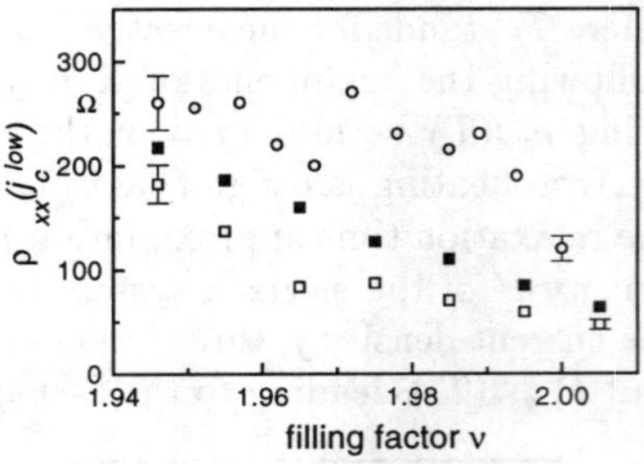

Figure 6: Measured and simulated filling factor dependence of $\rho_{xx}(j_c^{low})$.

to j_c^{low} were determined. We obtained values of about 15 - 20 K. In Fig. 5 and Fig. 6, the results for j_c^{low} and for $\rho_{xx}(j_c^{low})$ are compared with the experimental values. Although there are quantitative differences between the simulation and the experiment that may be attributed to uncertainties in the determination of $\rho_{xx}(T)$ and other simulation parameters, we obtain a good qualitative agreement with respect to the filling factor dependence of both quantities, supporting the applicability of the proposed electron-heating model.

In conclusion, our results confirm that an electron-heating effect might be the mechanism responsible for the occurence of S-shaped $\rho_{xx}(j)$-characteristics and consequent hysteresis effects in the breakdown of the integer quantum-Hall-effect.

Acknowledgments

This work has been supported by the Deutsche Forschungsgemeinschaft.

References

1. F. J. Ahlers, G. Hein. H. Scherer, L. Bliek, H. Nickel, R. Lösch, and W. Schlapp, *Semicond. Sci. Technol.* **8**, 2062 (1993).
2. for example: G. Ebert *et al.*, *J. Phys.* C **16**, 5441 (1983), V.G. Mokerov *et al.*, *JETP Lett.* **47**, 71 (1988), A. Boisen *et al.*, *Phys. Rev.* B **50** (3), 1957 (1994).
3. A.Vl. Gurevich, R.G. Mints, *JETP Lett.* **39**, 381 (1984), S. Komiyama, T. Takamasu, S. Hiyamizu, S. Sasa, *Solid State Comm.* **54**, 479 (1985).

RELATION BETWEEN CRITICAL ELECTRIC FIELD OF THE BREAKDOWN AND MAGNETIC FIELD IN INTEGER AND FRACTIONAL QUANTUM HALL EFFECTS

T. TAKAMASU*, M. OHNO and N. MIURA
Institute for Solid State Physics, University of Tokyo
7-22-1 Roppongi, Minato-ku, 106 Tokyo, Japan

Relation between critical electric field of the breakdown phenomena of the integer and fractional quantum Hall effects and magnetic field was studied at odd integer fillings in tilted magnetic fields. Analysis of the relation in terms of the electron heating model revealed that the critical electric field can be scaled with the energy gap and cyclotron radius, which opens up a new method to estimate the energy gap from the critical electric field of the breakdown.

Current-induced breakdown of the integer quantum Hall effect is one of the most remarkable phenomena, which gives much information of the ground and excited states. In previous works, we have reported that the breakdown phenomena can be observed in the fractional quantum Hall effect and the critical electric field E_{cr} of the breakdown can be scaled with the effective magnetic field B_{CF} for the composite Fermion[1]. From this fact, we can infer that the critical electric field of the breakdown is closely related with the energy gap in the density of states through a mechanism of the breakdown phenomena. On the other hand, recent studies of quantum Hall effect at odd integer fillings revealed that the energy of the lowest excited state of the spin polarized quantum Hall states is not a simple spin-flip state. Especially, in GaAs/AlGaAs heterostructures, the g-factor of electrons is small enough to be enhanced by the exchange interaction, which seems to be easily modified by the change of external conditions such as impurity concentration or the total magnetic field. Utilizing such states, it is easy to distinguish if the linear dependence of the critical electric field of the breakdown E_{cr} arises from the magnetic field itself or from the energy gap. In this paper, we will discuss the relation between E_{cr} and B for the integer and fractional quantum Hall effect in terms of electron heating model.

Samples we used are GaAs/AlGaAs heterostructure devices shaped into a standard Hall-bar geometry. Mobility and carrier density at dark condition is μ=4.26$\times$10^{5}cm^2 and n_s=1.89$\times$10^{11}cm^{-2}, respectively. Critical electric field at several even and odd filling factors ν were measured as a function of magnetic

*present address is National Research Institute for Metals, 3-12 Sakura, Tsukuba, Ibaraki 305, Japan.

field by changing carrier density of the sample with LED illumination and also by changing the tilt angle of the sample to applied magnetic fields.

In Fig. 1(a), we show measured E_{cr} as a function of the normal component of the applied magnetic field to the electron system $B_{\perp}$. It is clear that E_{cr} at even integer fillings ν=2 and 4 depends linearly on $B_{\perp}$. On the other hand, at odd integer fillings, especially at ν=1, E_{cr} is not scaled by $B_{\perp}$; data points are scattered in this plot. Figure 1(b) shows the same data plotted as a function of total applied magnetic field B_{total}. If the critical electric field is scaled by the energy gap, these data at odd integer fillings should be scaled by B_{total}, however, the scatter of these data is not reduced with this plot. From this fact, we can consider two possibilities. One is that the exchange interaction changes the energy gap even in such high magnetic fields and, as a result, the energy gap when θ is varied is different from that when n_s is varied at the same magnetic field. Another possibility is that the critical electric field of the breakdown does not solely depend on the energy gap, other factors changing E_{cr}. To clarify this, firstly, we discuss the mechanism of the breakdown below.

In the heating model of the breakdown[2], energy balance of the electron system is determined by the equation,

$$\sigma_{xx} E_y^2 = \frac{\Xi(T_e) - \Xi(T_L)}{\tau_\varepsilon} \tag{1}$$

where $\Xi(T_e)$ is the total energy of the electron system at an electron temperature T_e. From this equation, a pair of solution (σ_{xx}, T_e) can be obtained for a given Hall electric field E_y. If the mobility of the sample is high and the density of states has a δ-function like shape, temperature dependence of σ_{xx} and Ξ can be written in a simplified form as

$$\sigma_{xx} = \sigma_{xx}^0 \exp(-\frac{\Delta}{k_B T}) \tag{2}$$

$$\Xi = \int_{-\infty}^{\infty} \varepsilon f(\varepsilon) D(\varepsilon) d\varepsilon \simeq \frac{\Delta}{2\pi l_c^2} \exp(-\frac{\Delta}{k_B T}) \tag{3}$$

where Δ is a half of the energy gap between two Landau levels and l_c is the cyclotron radius. In the limit of low lattice temperatures, $\Xi(T_L)$ becomes zero. Then, the energy balance equation (1) has nonzero solution (σ_{xx}, T_e), only when the energy loss and the gain completely coincide with each other, giving the breakdown. Thus, in this simplified m odel, E_{cr} is given as

$$E_{cr} = (\frac{1}{\tau_\varepsilon} \cdot \frac{1}{2\pi l_c^2} \cdot \frac{\Delta}{\sigma_{xx}^0})^{\frac{1}{2}} \propto (B_{\perp} \cdot \Delta)^{\frac{1}{2}}. \tag{4}$$

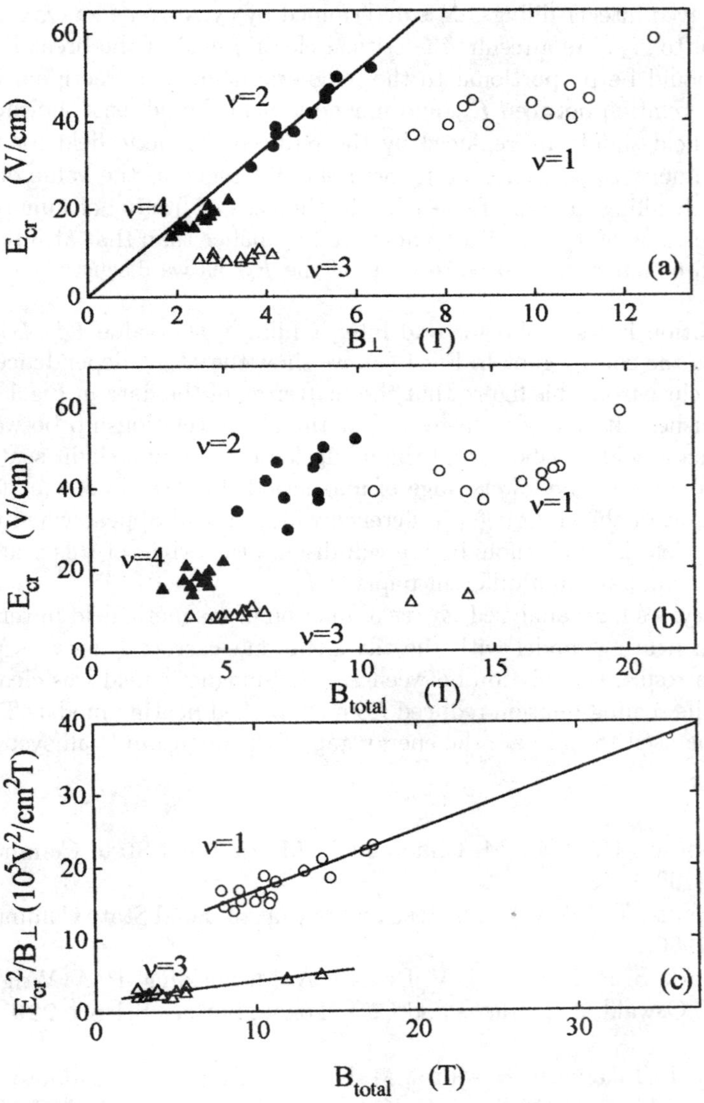

Figure 1: (a) Critical electric field E_{cr} of the quantum Hall effect measured at several integer filling factors in tilted magnetic field as a function of the normal component of the magnetic field $B_{\perp}$. (b) E_{cr} as a function of the total applied magnetic field B_{total}. (c) The value of $E_{cr}^2/B_{\perp}$ as a function of B_{total}.

In the case of even integer fillings, Δ is determined by cyclotron energy, which is proportional to $B_{\perp}$. As a result, the critical electric field of the breakdown phenomena should be proportional to the magnetic field. For fractional fillings, the same relation between E_{cr} and magnetic field should exist, however, the magnetic field should be replaced by the reduced magnetic field for the composite Fermions B_{CF}. Here, we ignored the difference in the value of τ_{ε} and σ_{xx}^0 at each filling factors. Especially, in the case of fractional quantum Hall effect, the value of σ_{xx}^0 has been known to be smaller than that at integer fillings[3], which enhances the absolute value of the E_{cr} as we discussed in ref. 4.

If this relation is also valid for odd integer fillings, the value $E_{cr}^2/B_{\perp}$ is proportional to the energy gap. In Fig.1 (c), we show the B_{total} dependence of $E_{cr}^2/B_{\perp}$. It is clear from this figure that the scattering of the data in Fig.1 (a) and (b) diminishes. It strongly suggests that the above relationship between E_{cr} and magnetic field is valid for all the filling factors. Only a slight scatter of the data rema ins in the lower range of magnetic field. We think that this scatter is a result of the energy gap difference which should appear when the total magnetic field is small enough. We will discuss the origin of this scatter with the additional data in a different paper[5].

In summary, we have analyzed E_{cr} as a function of magnetic field in terms of the electron heating model with the tilt angle and carrier density as parameters. As a result, the relation between E_{cr} and magnetic field was clearly explained by the scaling relation reduced from simplyfied heating model. This relation may be used to estimate the energy gap of the quantum H all system.

References

1. T. Takamasu , H. Dodo, M. Ohno and N. Miura, Solid State Commun. **96**, 121 (1995).
2. S. Komiyama, T. Takamasu, S. Sasa and Hiyamizu, Solid State Commun. **54**, 47 (1986).
3. R. G. Clark, S. R. Haynes, J. V. Branch, A. M. Suckling, P. A. Wright, P. M. W. Oswald, J. J. Harris and C. T. Foxon, Surface Science **229**, 25 (1990).
4. M. Ohno, T. Takamasu, N. Miura, M. Kato, A. Endo, S. Katsumoto, Y. Iye, T. Terashima, H. Aoki and G. Kido, to be submitted to 12th HMSP.
5. T. Takamasu, H. Dodo, M.Ohno and N. Miura, in Proceedings of 11th International Conference on Electronic Properties of Two-dimensional Systems (1995, Nottingham).

TEMPERATURE DEPENDENCE OF THE 'ZERO-RESISTANCE' QUANTUM HALL STATE PROBED BY MILLI-KELVIN TORQUE MAGNETOMETRY

J.P. WATTS, A. USHER
Department of Physics, University of Exeter, Stocker Road, Exeter, EX4 4QL, UK

M. ELLIOTT, W.G. HERRENDEN-HARKER
Department of Physics and Astronomy, University of Wales College of Cardiff, PO Box 913, Cardiff CF1 9HQ, UK

T.S. CHENG, C.T. FOXON
Department of Physics, University of Nottingham, Nottingham, NG7 2RD, UK

Torque magnetometry provides a contactless method of measuring the extremely small diagonal resistivity which accompanies the quantum Hall effect (QHE) at very low temperatures. In our experiments we induce circulating 'eddy' currents in the two-dimensional electron system by sweeping the magnetic field to an integer Landau-level filling factor, and detect these currents by measuring the magnetic moment associated with them. For a short period after the field sweep is stopped, the eddy current decays rapidly, which we believe to be due to high-current breakdown of the QHE, but at longer times the decay fits well to a single slow exponential, indicating that the QHE has recovered. In this paper, we examine the temperature dependence of this exceedingly small resistivity.

1. Introduction

When a two-dimensional electron system (2DES) at low temperature is subjected to a sweeping magnetic field around a Shubnikov - de Haas resistivity minimum, a circulating 'eddy' current is induced whose magnetic moment can be several orders of magnitude greater than the de Haas - van Alphen oscillations in the system. These eddy currents were first observed by J.P. Eisenstein *et al.*,[1] and T. Haavasoja *et al.*[2] used their decay times to estimate the sample resistivity. This work was continued by C.L. Jones *et al.*,[3,4] who reported unprecedentedly small diagonal resistivities, and compared the currents needed to cause breakdown of the quantum Hall effect (QHE) in their samples with those observed using more conventional techniques. If the field sweep is halted, the current decays and may have a time constant of several hours at temperatures below 100 mK. The rate of decay of these currents is inversely proportional to the value of the diagonal resistivity, so decay measurements of this sort provide useful information about the conduction mechanisms in well-resolved Shubnikov - de Haas minima, which are not easily available from conventional techniques. In this paper we report measurements of the temperature dependence of the eddy current decays in the regime in which the conduction is expected to occur through variable-range hopping. We have previously reported decay times of several hours for $\nu = 1$ and 2.[3] In this report we concentrate on the weaker $\nu = 4$ eddy current, in which relatively short decay times are observed, which enable us to examine the temperature dependence of the decay without encountering problems of long-term instrumental drift.

2. Experimental Details

In these experiments, a torque magnetometer has been used of similar design to that reported previously,[3,4] consisting of a moveable rotor suspended by fine wires, upon which the sample was placed. This assembly was placed in a magnetic field so that the normal of the sample makes an angle of ~20° to the field direction, so that any anisotropic magnetisation in the sample caused a torque, and a consequent movement of the rotor. Such movements were detected capacitively, as described in reference 3. A small coil was fitted to the magnetometer to provide a calibration of the system sensitivity, which was sufficient to measure torques as small as 10^{-12} Nm above the system noise.

The magnetometer was placed in the dilute phase of a dilution refrigerator, and temperatures below 50 mK were obtained. The 2DES studied was an uncontacted modulation-doped GaAs-(Al,Ga)As heterojunction, having area of 0.5 cm^2 and 2DES density 3.6×10^{15} m^{-2}. Transport measurements on a contacted sample confirmed the density and gave the sample mobility as 27 m^2 $(V\ s)^{-1}$.

The eddy currents were induced by sweeping the magnetic field toward the resistivity minimum at a constant rate, and their decay was observed once the field sweep was stopped at exactly $\nu = 4$. This procedure was carried out with the field sweeping in both directions, and at temperatures in the range from 40 mK to 700 mK.

3. Results

Figure 1 shows the magnetometer output for a continuous field sweep up to 15 T, prior to the detailed examination of the $\nu = 4$ decay. Clear eddy currents were observed at filling factors 1 to 6.

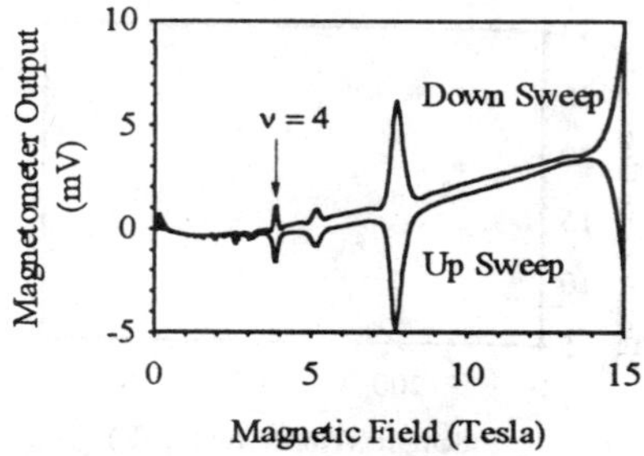

Figure 1: Magnetometer output (proportional to torque) from a magnetic field sweep at 100 mK, showing eddy current peaks in the up and down sweep directions.

When the magnet was swept to the field corresponding to an eddy current peak, the eddy current was observed to decay with time, as shown by figure 2. It proved impossible to fit these data to a single exponential, so the system clearly has at least two separate decay times which we discuss below. A single exponential was fitted to the slow decay, and this gave a decay time which was 21 ± 2 s for the decay in figure 2.

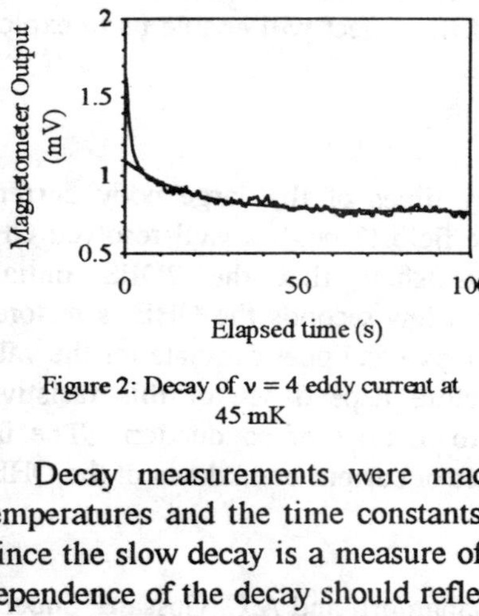

Figure 2: Decay of ν = 4 eddy current at 45 mK

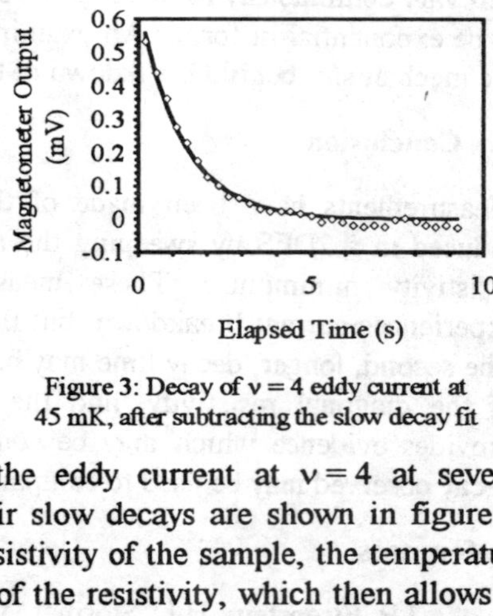

Figure 3: Decay of ν = 4 eddy current at 45 mK, after subtracting the slow decay fit

Decay measurements were made on the eddy current at ν = 4 at several temperatures and the time constants of their slow decays are shown in figure 4. Since the slow decay is a measure of the resistivity of the sample, the temperature dependence of the decay should reflect that of the resistivity, which then allows us to examine the conduction mechanism. The temperature dependence of figure 4 shows a slight increase in time constant (corresponding to a decrease in resistivity) as the temperature is reduced. However the resistivity is clearly not exponentially activated in this temperature range which suggests that the system is in the variable-range hopping regime. The scatter in the data is a consequence of temperature instability. Further refinement of the experimental technique will enable us to examine the temperature dependence in more detail.

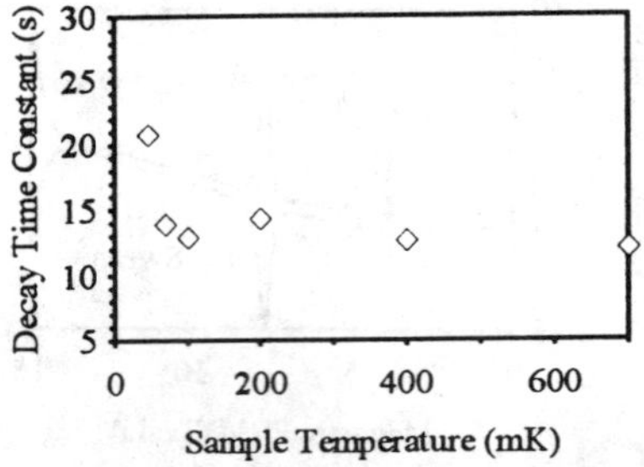

Figure 4: Temperature dependence of the long ν = 4 eddy current decay

When the magnetic field is swept, the current induced is sufficiently large to cause current breakdown of the QHE. When the field sweep is halted, the current quickly drops until it is small enough for the QHE to recover. At this time, the resistivity drops to an extremely small value, and the eddy current decays at the slower rate expected. Figure 3 shows the first few points of the decay presented in figure 2 in more detail: an exponential has been fitted to these data, having a time constant of around 2 seconds, but it seems likely that the resistivity of the sample will vary continuously in this regime, so one would not necessarily expect the decay to be exponential in form. An examination of this effect will enable us to explore the mechanisms behind breakdown of the QHE.

4. Conclusion

Measurements have been made of the decay times of the large eddy currents induced in a 2DES by sweeping the magnetic field through a well-resolved QHE resistivity minimum. These measurements show that the 2DES initially experiences current breakdown, but that after a few seconds the QHE is restored. The second, longer, decay time may be used to give an upper estimate for the value of the diagonal resistivity, and the temperature dependence of this resistivity provides evidence which may be compared to theories of conduction. The fast decay observed may be used to compare theories of current breakdown of the QHE.

References

1. J.P. Eisenstein, H.L. Störmer, V. Narayanamurti and A.C. Gossard, Superl. Microstr. **1**, 11 (1985).
2. T. Haavasoja, H.L. Störmer, V. Narayanamurti, A.C. Gossard and W. Wiegmann, Surface Science **142**, 294 (1984).
3. C.L. Jones, A. Usher, M. Elliott, W.G. Herrenden-Harker, A. Potts, R. Shepherd, T.S. Cheng and C.T. Foxon, Sol.-St. Comm. **95**, 409 (1995).
4. C.L. Jones, A. Usher, M. Elliott, W.G. Herrenden-Harker, A. Potts, R. Shepherd, T.S. Cheng and C.T. Foxon, Sol.-St. Comm. **97**, 763 (1996).

DYNAMIC CONDUCTANCE IN QUANTUM HALL SYSTEMS

M. BÜTTIKER AND T. CHRISTEN
Université de Genève, 24 Quai E. Ansermet, CH-1211 Geneva Switzerland

In the framework of the edge-channel picture and the scattering approach to conduction, we discuss the low frequency admittance of quantized Hall samples up to second order in frequency. The first-order term gives the leading order phase-shift between current and voltage and is associated with the displacement current. It is determined by the emittance which is a capacitance in a capacitive arrangement of edge channels but which is inductive-like if edge channels predominate which transmit charge between *different* reservoirs. The second-order term is associated with the charge relaxation. We apply our results to a Corbino disc and to two- and four-terminal quantum Hall bars, and we discuss the symmetry properties of the current response. In particular, we calculate the longitudinal resistance and the Hall resistance as a function of frequency.

1 Introduction

In this work we discuss the low frequency response of two-dimensional conductors subject to a perpendicular quantizing magnetic field[1] and subject to time-dependent oscillating voltages applied to the contacts of the sample. We use an approach[2–4] closely related to the scattering approach to dc conduction[5] in a mesoscopic phase-coherent multiterminal sample. The approach treats current and voltage contacts on equal footing and views the entire sample and the nearby conductors as one electric entity.

It is well-known from mesoscopic conduction theory that all dc resistances are rational functions of transmission probabilities for carriers to reach a contact if they are injected at another contact.[5] Moreover, only the states at the Fermi surface which connect different contacts count. In a conductor subject to a quantizing magnetic field the only states at a Hall plateau which provide such a connection are the *edge states* which are the quantum mechanical generalization of the semi-classical 'skipping orbits' of the charge carriers along the sample boundary. As a consequence the dc conductance is quantized and determined by the topology of the edge-state arrangement.[5] For a lucid discussion of the current patterns and Hall potentials in the presence of transport, we refer the reader to a recent work by Komiyama and Hirai.[6]

Most of the early work on edge states relies on a picture valid for noninteracting electrons which must be modified in very pure samples with smooth boundaries.[7] Interaction leads to a decomposition of the electron gas into compressible and incompressible regions, where the former ones are associated with the edge states. Interaction must also be considered to describe edge states in the fractional quantum-Hall regime.[8] A detailed description of the contacts for chiral edge-state models has recently been provided by Kane and Fisher.[9] While in these works the main concern is with short-range interactions, we emphasize in our work the long-range part of the Coulomb interaction. For this case, edge states are important since they contribute to screening, in contrast to the bulk states.[6,10] Due to the Coulomb interaction, the ac response is not only determined by the topology of the edge-state arrangement but also by its geometry.

In the following, we concentrate on a two-dimensional electron gas with spin-split Landau levels. Coulomb interaction is taken into account by consideration of the capacitances of all edge channels and all gates. We assume that additional contributions due to parasitic capacitances associated with contacts can be neglected, that the contacts are quantized, and that inter-edge scattering is absent.

2 Low-frequency admittance

2.1 Current conserving and gauge-invariant electric transport

The gates together with the edge channels which are separated by regions of incompressible electron gas can be seen as an arrangement of mesoscopic conductors embedded in a dielectric medium. We describe an N-terminal sample at a Hall plateau by an arrangement of such conductors $j = 1, ..., M$ connected to contacts $\alpha = 1, ..., N$ (see, e.g., Fig. 1). We are interested in the dynamical conductance $G_{\alpha\beta}(\omega)$ which determines the Fourier amplitudes of the current, δI_α, at a contact α in response to an oscillating voltage $\delta V_\beta \exp(-i\omega t)$ at contact β

$$\delta I_\alpha = \sum_\beta G_{\alpha\beta}(\omega)\, \delta V_\beta \ . \tag{1}$$

The voltage variation δV_β is related to the variation of the electro-chemical potential $\delta\mu_\beta$ in reservoir β by $\delta\mu_\beta = e\delta V_\beta$, where e is the electron charge.

Since all the present gates and conductors are included in the description, and due to the long-range nature of the Coulomb interaction the theory for the conductance has to satisfy two basic requirements, namely *current conservation* and *gauge invariance*. Current conservation corresponds to zero total current, $\sum_\alpha \delta I_\alpha = 0$. Hence the total charge is conserved. Gauge invariance means that a global voltage shift, $\delta V_\alpha \mapsto \delta V_\alpha + \delta V$, does not affect the currents. In particular, we fix the voltage scale such that $\delta V_\alpha \equiv 0$ corresponds to the equilibrium state where all electrochemical potentials μ_α are equal. Current conservation and gauge invariance demand the admittance matrix $G_{\alpha\beta}$ to satisfy $\sum_\alpha G_{\alpha\beta} = \sum_\beta G_{\alpha\beta} = 0$. For example, the admittance of a two-terminal device is determined by a single scalar quantity, $G_{\alpha\beta} = (-1)^{\alpha+\beta} G$. Due to microreversibility the linear response coefficients must additionally satisfy the Onsager-Casimir reciprocity relations $G_{\alpha\beta}|_B = G_{\beta\alpha}|_{-B}$ where B is the magnetic field. We are interested in the low-frequency limit and write

$$G_{\alpha\beta}(\omega) = G^{(0)}_{\alpha\beta} - i\omega E_{\alpha\beta} + \omega^2 K_{\alpha\beta} + \mathcal{O}(\omega^3) \ . \tag{2}$$

While the theory for the dc-conductances $G^{(0)}_{\alpha\beta}$ of quantized Hall systems is well established,[5] the *emittance* matrix $E_{\alpha\beta} \equiv i(dG_{\alpha\beta}/d\omega)_{\omega=0}$ has been investigated only recently.[2] Below we recall these results and, in addition, provide new results for the second order term $K_{\alpha\beta} \equiv (1/2)\,(d^2G_{\alpha\beta}/d\omega^2)_{\omega=0}$. To get some feeling for the emittance and the second order term, consider a two-terminal device with a macroscopic capacitor C in series with a resistor R. For this purely capacitive structure with zero dc-conductance one finds that the emittance is the capacitance, $E = C$, and that $K = RC^2$ contains the RC time constant associated with charge relaxation. These simple results must be modified for *mesoscopic* conductors and conductors which connect *different* reservoirs.[3] Firstly, it is not the geometrical capacitance but rather the *electro-chemical capacitance* which relates charges on mesoscopic conductors to voltage variations in the reservoirs. Secondly, conductors which connect different reservoirs allow a transmission of charge which leads to inductance-like contributions to the emittance.

2.2 Transmission properties of edge channels

An important property of a single edge channel is its uni-directional transparency and the absence of backscattering.[5] The scattering matrix $s_{\alpha\beta}$ relates the out-going current amplitude at contact α to the incoming current amplitude at contact β. The part of the scattering matrix associated with edge channel j can be written in the form $s^{(j)}_{\alpha\beta} = \Delta_{\alpha j} t_j \Delta_{j\beta}$ with a scattering amplitude $t_j = \exp(i\phi_j(E))$. The energy dependence of the phase $\phi_j(E)$ determines the density of states of the edge channel, $dN_j/dE = (1/2\pi)(d\phi_j/dE)$ at the Fermi energy. We mention that for noninteracting particles the density of states is related to the electric field at the edge.[5] $\Delta_{j\alpha}$ and $\Delta_{\alpha j}$ are the injection and emission probability, respectively, of channel j at contact α:

$$\Delta_{j\alpha} = \begin{cases} 1 & \text{if contact } \alpha \text{ injects into channel } j \\ 0 & \text{otherwise} \end{cases} \tag{3}$$

$$\Delta_{\alpha j} = \begin{cases} 1 & \text{if channel } j \text{ emits into contact } \alpha \\ 0 & \text{otherwise} \end{cases} \ . \tag{4}$$

Since an edge channel is always connected to one contact at each of its ends, one has $\sum_\alpha \Delta_{k\alpha} = \sum_\alpha \Delta_{\alpha k} = 1$. Furthermore, $\sum_k \Delta_{k\alpha}$ $(= \sum_k \Delta_{\alpha k})$ is the number of edge channels which enter (leave) the sample at contact α. Clearly, the transmission probabilities are functions of the magnetic field B and obey the micro-reversibility conditions $\Delta_{\alpha k}(B) = \Delta_{k\alpha}(-B)$. This follows directly from the fact that an inversion of the magnetic field inverses the arrows attached to the edge states. In terms of the $\Delta_{j\alpha}$ and $\Delta_{\alpha j}$ the dc conductance becomes[2]

$$G^{(0)}_{\alpha\beta} = \frac{e^2}{h} \sum_k \left(\Delta_{k\alpha}\,\Delta_{k\beta} - \Delta_{\alpha k}\Delta_{k\beta} \right) \quad . \tag{5}$$

Hence the topology of the edge-channel arrangement determines the dc conductance. Note that current conservation, gauge invariance, and the Onsager-Casimir relations follow immediately from the properties of $\Delta_{j\alpha}$ and $\Delta_{\alpha j}$.

2.3 Screening properties: a discrete model

While the dc-conductance is completely determined by the overall transmission, the ac-conductance contains information on the nonequilibrium state within the sample. Application of an oscillatory voltage at a sample contact leads to a periodic charging of the sample, and the induced electric fields have to be taken into account self-consistently. In order to treat the dynamic charge and potential distribution in the nonequilibrium state we spatially discretize the two-dimensional electron gas. More concrete, we assign to each edge channel j a single electrochemical potential of the carriers, a single electrical potential, and a total charge. We neglect spatio-temporal effects inside edge channels such as edge plasmons with finite wave numbers.[11,12] However, our discussion takes into account the zero wave-number modes of the edge states. The electrochemical potential shift δV_j in edge channel j can be written in terms of the voltages δV_β in the reservoirs, $\delta V_j = \sum_\beta \Delta_{j\beta}\,\delta V_\beta$. To simplify notation we write quantities associated with edge channels as M-dimensional vectors, e.g. $\delta \mathbf{V} = (\delta V_1, ..., \delta V_M)^t$. The nonequilibrium distribution of charges and electric potentials $\delta \mathbf{q} = (\delta q_1, ..., \delta q_M)^t$ and $\delta \mathbf{U} = (\delta U_1, ..., \delta U_M)^t$, respectively, are related by $\delta \mathbf{q} = \mathbf{C}\,\delta \mathbf{U}$, where $\mathbf{C}$ is the geometrical capacitance matrix for the edge channels and gates. As usual, $\mathbf{C}$ can be obtained from the Poisson equation. A relation between the charges and the voltages of the reservoirs, $\delta \mathbf{q} = \mathbf{C}_\mu\,\delta \mathbf{V}$, defines the electrochemical capacitance matrix $\mathbf{C}_\mu$. Note that while $\mathbf{C}$ is real and frequency independent, $\mathbf{C}_\mu(\omega)$ is a function of frequency. Here, we define $\mathbf{C}$ by assuming that each edge channel is connected to its own reservoir, hence charge conservation and gauge invariance imply for both $\mathbf{C}$ and $\mathbf{C}_\mu(\omega)$ the sum rules $\sum_k C_{kl} = \sum_l C_{kl} = \sum_k C_{\mu,kl} = \sum_l C_{\mu,kl} = 0$.

Both matrices are symmetric and even functions of the magnetic field. Further we need the dynamic densities of states of the edge channels. In the present notation they can be summarized in a diagonal matrix $\mathbf{D}(\omega)$ with elements $D_{jj}(\omega) = e^2(dN_j/dE)(1 + i\tau_j\omega/2) + \mathcal{O}(\omega^2)$.[13] Here, $\tau_j = h(dN_j/dE)$ is the dwell time of carriers in edge channel j. The specific form of the $D_{jj}(\omega)$ can be understood by noticing that the charging of edge channel j is a relaxation process with a time $\tau_j/2$. The nonequilibrium charge is $\delta\mathbf{q} = \mathbf{D}(\delta\mathbf{V} - \delta\mathbf{U}) = \mathbf{C}\delta\mathbf{U} = \mathbf{C}_\mu\delta\mathbf{V}$ from which one obtains the dynamic electrochemical capacitance

$$\mathbf{C}_\mu(\omega) = \mathbf{C}_\mu(0) + i\omega\frac{h}{2e^2}\mathbf{C}_\mu^2(0) + \mathcal{O}(\omega^2) \ , \tag{6}$$

where $\mathbf{C}_\mu(0) = [\mathbf{1} + \mathbf{C}\mathbf{D}^{-1}(0)]^{-1}\mathbf{C}$ is the static part of the electrochemical capacitance matrix. The dependence of the effective capacitance $\mathbf{C}_\mu(0)$ on the densities of states is reasonable: small densities of states make it more difficult to charge the conductor. For large densities of states, on the other hand, one recovers the geometrical capacitance.

2.4 The dynamic response

The low frequency admittance can be calculated with the help of a recent theory which generalizes the scattering approach to ac conduction.[4] The ingredients which we need are the current response to an external voltage shift $\delta\mathbf{V}$ and the current response to an internal voltage shift $\delta\mathbf{U}$ for carriers at the Fermi energy E_F and for carriers with energies $E_F + \hbar\omega$. We do not enter into details here but present only the result. We find

$$G_{\alpha\beta}(\omega) = G^{(0)}_{\alpha\beta} - i\omega\sum_{k,l=1}^{N}\Delta_{\alpha k}\, C_{\mu,kl}(\omega)\,\Delta_{l\beta} \ . \tag{7}$$

Using the micro-reversibility properties of the $\Delta_{\alpha k}$ and the symmetry of the capacitance matrix it is again easy to prove current conservation, gauge invariance, and the Onsager-Casimir reciprocity relations. The results (6) and (7) yield the admittance to order $\mathcal{O}(\omega^2)$. The emittance $E_{\alpha\beta} = \sum_{kl}\Delta_{\alpha k}C_{\mu,kl}(0)\Delta_{l\beta}$ can be understood as the sum of all 'two-body' Coulomb interactions between edge states and/or gates.[2] In a pure capacitive arrangement where $\Delta_{k\alpha} = \Delta_{\alpha k}$ the emittance matrix is a capacitance matrix, i.e. diagonal elements are positive and off-diagonal elements are negative. As we will see below, this can change in the presence of edge channels which connect different reservoirs. The result for the second-order term, $K_{\alpha\beta} = (h/2e^2)\sum_{kjl}\Delta_{\alpha k}C_{\mu,kj}(0)$

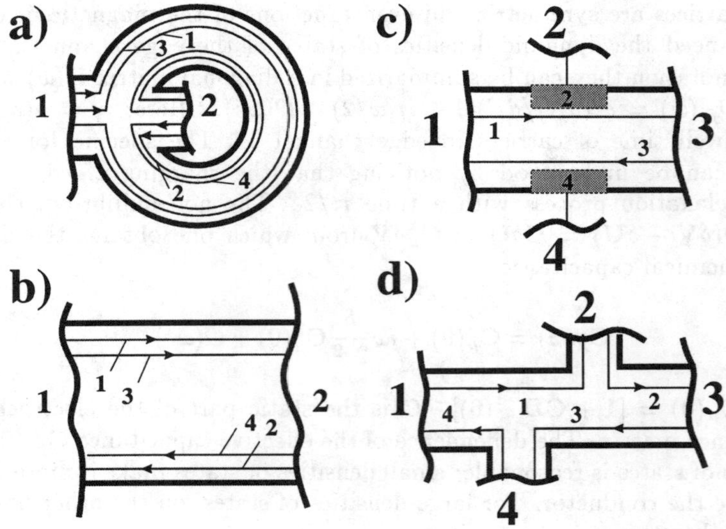

Figure 1: Two-terminal samples: a) Corbino disc and b) quantum Hall bar. Four-terminal samples: c) two-terminal Hall bar with two additional gates; d) four-terminal Hall bar

$C_{\mu,jl}(0)\Delta_{l\beta}$, can thus similarly be interpreted as a 'three-body' interaction with an intermediate step via edge channel (or gate) j. This simple result, of course, relies on the discrete model and is modified as soon as one includes internal spatio-temporal edge excitations.

3 Examples

3.1 Two-terminal devices

Let us discuss the Corbino disc and the quantum Hall bar with m channels along an edge as shown in Fig. 1 for a magnetic field B such that $m = 2$. Channels at the upper and the lower edges are labeled by odd and even numbers, respectively. Equation (6) permits now to treat the case where different edge states are at different potentials. For the sake of clarity, however, we assume that the capacitances between edge channels at the same edge are very large such that they are at equal potentials. We take thus each set together to form a single m-channel conductor associated with only even or only odd labels. The geometric capacitance between the two sets of channels is denoted by C ($= \sum_{kl} C_{2k\,2l}$). While for the disc the reservoirs are capacitively coupled, for the bar all edge channels transmit charge between different reservoirs. Hence, the dc-conductance of the disc and the bar are $G(0) = 0$ and $G(0) = me^2/h$, respectively. The dynamic capacitance between the two sets of edge channels is $C_\mu(\omega) = C_\mu(0) + i\omega RC_\mu^2$ with a static electrochemical capacitance given by $1/C_\mu(0) = 1/C + (\sum_{\text{even}} D_i)^{-1} + (\sum_{\text{odd}} D_i)^{-1}$. The charge relaxation resistance is given by

$$R = \frac{h}{2e^2}\left(\frac{\sum_{i=\text{odd}} D_i^2}{(\sum_{i=\text{odd}} D_i)^2} + \frac{\sum_{i=\text{even}} D_i^2}{(\sum_{i=\text{even}} D_i)^2}\right) . \tag{8}$$

Due to the purely capacitive coupling the low frequency admittance of the Corbino disc is

$$G(\omega) = -i\omega C_\mu(\omega) = -i\omega C_\mu(0) + \omega^2 RC_\mu^2(0) . \tag{9}$$

As one expects, this result is equivalent to the low frequency admittance of a mesoscopic capacitor with m channels on each side.[4] On the other hand, one finds for the bar

$$G(\omega) = m\frac{e^2}{h} + i\omega C_\mu(\omega) = m\frac{e^2}{h} + i\omega C_\mu(0) - \omega^2 RC_\mu^2(0) \tag{10}$$

While the emittance of the disc behaves like a capacitance, the emittance of the bar is inductive-like. Inductive behavior is a typical property of conductors where transmission between different reservoirs dominates.[3] The crossover between the two limits of a purely transmitting and a purely capacitive arrangement can be investigated by considering tunneling between edge channels. In a recent work, [14] we have shown that the emittance of a quantum point contact without magnetic field exhibits a change of the sign as a function of the transparency.

Note that if all D_j are of the same order of magnitude, the charge relaxation resistance R scales like $1/m$. Nevertheless, if C_μ scales with m (as is the case on a Hall plateau and for large C) one has $G(\omega) \propto m$. At a transition between two Hall plateaus, however, $D_j \to \infty$ and one expects the charge relaxation resistance to be of the order of a single resistance quantum.

3.2 Four-terminal devices

Figure 1c. shows a two-terminal bar with two additional symmetric gates each located close to an edge state. Macroscopic gates have huge densities of states which means that $C/D_2 = C/D_4 = 0$, where $C = |C_{12}| = |C_{34}|$ is the geometric capacitance between a gate and the nearest edge channel. Note that the off-diagonal elements of the capacitance matrix are negative. We assume that the other off-diagonal capacitance elements are much smaller, i.e. $|C_{24}| = C_g \ll C$ and $C_{13} = C_{14} = 0$. Now, let us ask for the current through contact 1 in response to a voltage oscillation at gate 2. We find to leading order of C_g for the associated admittance $G_{12}|_B = \omega^2(h/2e^2)C_g C_3 + \mathcal{O}(\omega^3)$ with $C_3 = 1/(C^{-1} + D_3^{-1})$. Clearly, since edge channel 3 emits into reservoir 1, and gate 2 interacts only via gate 4 with this edge channel, the response is proportional to $C_g C_3$ and of second order in frequency. On the other hand, by inverting the magnetic field, edge channel 1 emits into reservoir 1. Now, the relevant capacitance turns out to be $C_1 = 1/(C^{-1} + D_1^{-1})$; to leading order of C_g we find $G_{12}|_{-B} = i\omega C_1 + \omega^2(h/e^2)C_1^2 + \mathcal{O}(\omega^3)$. Thus while $G_{12}|_{-B}$ exhibits a capacitive response given by C_1, the capacitive response in $G_{12}|_B$ is absent. Such a dramatic magnetic field asymmetry has been verified in an experiment by Chen et al. [15] in the absence of gate 4, i.e. for $C_g = 0$.

The result (7) implies that the conductance $G_{\alpha\beta}$ between two contacts which are capacitively coupled to the sample is a symmetric function of the magnetic field. This is in general true up to $\mathcal{O}(\omega)$.[3] To higher order in frequency, however, the magnetic-field symmetry of the current response can change in the presence of voltage probes. As an example, we assume $C_g = 0$ and that contact 3 in Fig. 1c. acts as a voltage probe. Using $\delta I_3 = 0$, we eliminate δV_3 from the Eqs. (1) and obtain the reduced admittance matrix $\tilde{G}_{\alpha\beta}$ associated with the contacts 1, 2, and 4. We find $\tilde{G}_{24}|_B = 0$ and $\tilde{G}_{42}|_B = \tilde{G}_{24}|_{-B} = \omega^2(h/e^2)C^2$. Hence, the impedance of a purely capacitive arrangement is in general not symmetric but only governed by the Onsager-Casimir symmetry relations, $G_{\alpha\beta}|_B = G_{\beta\alpha}|_{-B}$. This conclusion has been drawn by Sommerfeld et al.[12] who found an asymmetric capacitance of a three-terminal Corbino disc. Our theory predicts such an asymmetric behavior in a simple way.

Another interesting question concerns the Hall resistance $R_H = (\delta V_2 - \delta V_4)/\delta I_1$ of the sample shown in Fig. 1c. If C_g vanishes one expects exact quantization because the gates do nothing but measure the edge potentials. An expansion to first order in C_g yields the low frequency resistance

$$\begin{aligned} R_H &= \frac{h}{e^2}\left(1 - C_g(C_1^{-1} + C_3^{-1})\right) \\ &+ \frac{1}{2}\omega^2 \frac{h^3}{e^6} \frac{C_g(C_1^{-1} + C_3^{-1})}{D_1^{-1}D_3^{-1} + C^{-1}(C^{-1} + D_1^{-1} + D_3^{-1})} + \mathcal{O}(\omega^3) \qquad (11) \end{aligned}$$

In the case where the two gates are decoupled, one concludes indeed that the resistance is quantized, $R_H(\omega) = h/e^2$. However, due to the Coulomb coupling between the gates this result is modified even for the dc case.

The integer quantum Hall effect[1] corresponds to the quantization of the

Hall resistance and the vanishing of the longitudinal resistance of the ideal four-probe quantum Hall bar of Fig. 1d. at zero frequency. Two of the contacts serve as voltage probes, whereas the two remaining contacts are used as source and sink for the current. With the help of our theory the dc-results can be extended to the low-frequency case. If the contacts 3 and 4 are the voltage probes, the longitudinal resistance is defined by $R_L = (\delta V_4 - \delta V_3)/\delta I_1$. The Hall resistance is defined by $R_H = (\delta V_2 - \delta V_4)/\delta I_1$, provided the contacts 2 and 4 are voltage probes. Let us assume that the elements $C_{\mu,kl}$ of the electrochemical capacitance matrix are known. For the specific geometry of Fig. 1d. we assume $C_{\mu,12} = C_{\mu,24} = C_{\mu,34} = 0$ and calculate the low frequency admittance. From Eqs. (1), (6), and (7) we find then the longitudinal resistance

$$R_L = i\omega \frac{h^2}{e^4} C_{\mu,13} + \frac{\omega^2}{2} \frac{h^3}{e^6} C_{\mu,13}(C_{\mu,23} - C_{\mu,14}) \ . \tag{12}$$

The leading term of the longitudinal resistance is determined by the Coulomb coupling between the current circuit and the voltage circuit which are represented by edge channels 1 and 3, respectively. This is even true for finite $C_{\mu,24}$.[2] Moreover, for the specific symmetry $C_{\mu,23} = C_{\mu,14}$ the dissipation vanishes also in $\mathcal{O}(\omega^2)$. For the Hall resistance one finds

$$R_H = \frac{h}{e^2} - i\omega \frac{h^2}{e^4} C_{\mu,13} + \frac{\omega^2}{2} \frac{h^3}{e^6} C_{\mu,13}(C_{\mu,23} + C_{\mu,14}) \ . \tag{13}$$

We mention, that if we had chosen the symmetry[2] $C_{\mu,13} = C_{\mu,24}$, the imaginary part of the Hall resistance would have vanished in first order of ω.

4 Conclusion

We presented a theory of ac-conductance for quantized Hall samples which takes the Coulomb interaction into account self-consistently. It is based on a discretized model where each edge channel is described by a homogeneous mesoscopic conductor. The whole sample is viewed as an electric entity and the results satisfy the important requirements of gauge invariance and current conservation. Obviously, the theory provides a manifold of experimental tests. The internal dynamics enters by a single relaxation process associated with the uniform charging or decharging of edge channels with a certain relaxation time. A next step should be to include internal dynamics due to spatio-temporal excitations[11] along the edge channel within a microscopic theory.

Acknowledgments

This work was supported by the Swiss National Science Foundation under grant Nr. 43966.

References

1. K. von Klitzing, G. Dorda, and M. Pepper, Phys. Rev. Lett. **45**, 494 (1980).
2. T. Christen and M. Büttiker, Phys. Rev. B **53** 2064, (1996).
3. M. Büttiker, J. Phys.: Condens. Matter **5**, 9361 (1993).
4. M. Büttiker, H. Thomas, and A. Prêtre, Phys. Lett. A **180**, 364 (1993); Z. Phys. B **94**, 133 (1994).
5. For a review see: M. Büttiker, in 'Semiconductors and Semimetals', Vol. 35, edited by M. Reed, Academic Press, p.191, (1992).
6. S. Komiyama and H. Hirai, Phys. Rev. B **54**, July (1996).
7. D. B. Chklovskii, B. I. Shklovskii, and L. I. Glazman, Phys. Rev. B **46**, 4026 (1992); N. R. Cooper and J. T. Chalker, Phys. Rev. B **48**, 4530 (1993); K. Lier and R. R. Gerhardts, Phys. Rev. B **50**, 7757 (1994).
8. C. W. J. Beenakker, Phys. Rev. Lett. **64**, 216 (1990); A. M. Chang, Solid State Commun. **74**, 871 (1990); A. H. McDonald, Phys. Rev. Lett. **64**, 222 (1990); X. G. Wen, Phys. Rev. B **43**, 11025 (1991).
9. C. L. Kane and M. P. A. Fisher, Phys. Rev. B **52**, 17393 (1995).
10. E. Yahel, D. Orgad, A. Palevski, and H. Shtrikman, Phys. Rev. Lett. **76**, 2149 (1996); A. H. McDonald, T. M. Rice, and W. F. Brinkman, Phys. Rev. B **28**, 3648 (1983); S. Takaoka et al., Phys. Rev. Lett. **72**, 3080 (1994).
11. V. I. Talyanskii et al., J. Phys.: Condens. Matter **7**, L435 (1995).
12. P. K. H. Sommerfeld, R. W. van der Heijden, and F. M. Peeters, Phys. Rev. B **53**, 13250 (1996).
13. A. Prêtre, H. Thomas, and M. Büttiker, Phys. Rev. B (unpublished).
14. T. Christen and M. Büttiker, Phys. Rev. Lett. **76**, July (1996).
15. W. Chen, T. P. Smith III, M. Büttiker, and M. Shayegan, Phys. Rev. Lett. **73**, 146 (1994).

Quantum Hall Effect: Current Distribution and Existence of Extended States

K. Tsemekhman, V. Tsemekhman, C. Wexler, and D.J.Thouless
Department of Physics, University of Washington, Box 351560,Seattle, WA 98195

The spatial distribution of Hall current is discussed. It is demonstrated that when the Hall conductivity is quantized no part of the non-equilibrium Hall current flows in the edge channels. Studying the location of extended states we show that the Hall current is in general distributed throughout the bulk of the QH sample.

One of the still unanswered questions in the field of QHE is the problem of spatial current distribution. Historically the first approach [1,2], assumed that electrons were injected into the delocalized states at the Fermi level. Therefore, the Hall current is carried by states in a narrow region near the sample boundary. In the other approach the Hall current density is proportional to the electric field created by charges concentrated at the edges of the system [3,4]. In this case the Hall current is distributed throughout the sample. That delocalized states exist in the presence of disorder is assumed in both approaches. However, it is important where such states are located and under what conditions they are really present [11,10]. Both approaches find experimental support. Measurements of equilibration rates [5] and of non-local conductivity [6] are explained in terms of edge currents. However, studies of the breakdown of the QHE [7] and of the Hall voltage distribution [8] favor the picture of bulk currents. In this work we demonstrate that when the measured σ_{xy} is quantized, almost the entire Hall current flows in the bulk of the system. Inter-electron interaction is crucial for establishing such current distribution.

Delocalized trajectories in the equilibrium state

We consider a typical QHE system: a two-dimensional electron gas (2DEG) formed at the interface of the GaAs/GaAlAs heterostructure, in a strong perpendicular magnetic field B, with the length of the system L_y being much larger than its width L_x. We study the electric field distribution and location of extended classical trajectories in equilibrium and steady states of the 2DEG. A random disorder potential is created by remote Si donors. The fluctuations of the random potential are not screened in most parts of the system. A percolating region of incompressible liquid with N completely filled Landau levels (LL) is formed. Most of the states get localized by the potential fluctuations. In smooth potentials and strong magnetic fields, electrons move along the equipotentials. In 2D disordered system all these equipotential lines, except for one, are closed. Therefore, in an *infinite* system for each LL there exists a single energy at which a state is extended in both directions. How-

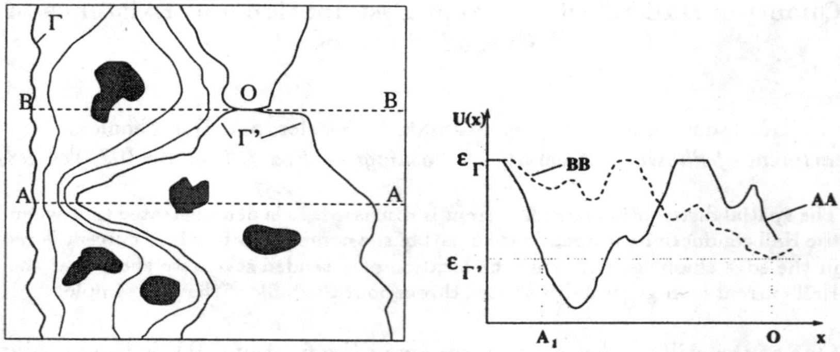

Figure 1: a. Delocalized states between the boundary of edge channels Γ and 'critical trajectory' Γ′; b. Potential along the horizontal cross sections *AA* and *BB*

ever, in finite size samples these states can percolate only in one direction, namely, the shorter one[10]. Therefore, in this case there are no states extended vertically (i.e. in the y-direction), where the Hall current is expected to flow. Such states apppear due to the confining potential created by the positive background, electrons, and gate voltage. With the fall of this potential from, e.g., left edge into the sample, the electron density grows from 0 to the bulk value through alternating vertically percolating compressible and incompressible strips[9]. In the compressible strips the population grows and the potential is flat, while across the incompressible regions the potential falls by $\hbar\omega_c$ and LL are completely filled. After the inner-most vertically percolating compressible region the incompressible bulk is formed. There exists, therefore, an 'infinite in y-direction' geometric boundary line Γ such that in the region to the right of it the potential falls due to unscreened fluctuations of disorder potential (Fig.1b). Γ itself is an equipotential at energy ϵ_Γ. We seek to find the right boundary of this region.

In the absence of boundaries one trajectory is percolating in both directions. We denote its energy by ϵ_0. Applying the confining potential in the x-direction repels this trajectory from the boundary closing it somewhere away from the edge (Fig.1a). Denote this trajectory by Γ′. One can present a mathematical proof based on continuity of the following theorem: For each energy between the energies ϵ_0 and ϵ_Γ there exists one and only one equipotential line open in the y-direction and no trajectory at energies lower than ϵ_0 can percolate vertically. In a first approximation trajectory Γ′ has to percolate in both vertical and horizontal directions. This is a 'critical trajectory' with at least one saddle point O on it. An equilibrium picture of extended trajectories emerges as pictured in Fig.1a. The critical trajectory Γ′ travels throughout the system towards the 'saddle point'. It approaches the boundaries only to get around closed equipotentials with a large localization length. Since such trajectories are very sparse, the critical trajectory and other extended equipotentials with energies close to ϵ_0 spend most of the time inside the system rather than at the boundary.

Absence of non-equilibrium current in the edge channels

We first define 'edge channels' as the region from the left physical boundary to the line Γ (and a corresponding region at the right edge). We will consider the conditions of a measurement such that σ_{xy} is quantized and show first that no non-equilibrium current is flowing in the 'edge channels'.

In equilibrium, although the net current is zero, orbital current flows along the extended trajectories at the edges. Non-trivial contributions come from incompressible edge regions. As was shown by Trugman[11], the current carried by extended trajectories with energies in the interval $\triangle V$ is: $\triangle I = -\frac{e^2}{h}\frac{\triangle V}{e}$. If, in equilibrium N Landau levels are filled the overall fall of the potential from the depletion layer boundary to the line Γ is measured by $(N-1)\hbar\omega_c$. Therefore, the equilibrium diamagnetic current carried by states to the left of Γ at the n–th Landau level is equal to $\frac{e}{h}(N-n)\hbar\omega_c$. It is very important that all LL's are equilibrated, i.e. filled up to the same chemical potential.

In the steady state with the Hall current present in the system, there are two well defined chemical potentials — the ones at the left and right edges, μ_{left} and μ_{right} — such that the Hall voltage is $V_H = \mu_{left} - \mu_{right}$. The condition for Landau levels to still be equilibrated at each edge is the necessary one for the quantization of σ_{xy} [2]. On the other hand, if this condition is satisfied, there is no change in the current carried by the states to the left from the line Γ and to the right of its counterpart at the right edge, compared to the equilibrium state. Namely, these currents are still $\frac{e}{h}(N-n)\hbar\omega_c$ for the n–th LL. Complete equilibration at the edge leads, therefore, to complete absence of *non-equilibrium* current in the 'edge channels'. If equilibration is not perfect, some portion of *non-equilibrium* current flows in the edge channels. However, in this case, the measured Hall conductance is not quantized in the units of e^2/h which reconciles interpretations of different experiments. We conclude that experimental conditions are crucial in determining the Hall current distribution.

Current distribution in the steady state

We demonstrate how, under the conditions of complete equilibration at each edge, the Hall current is distributed in the sample. The same arguments that were applied in equilibrium demonstrate that there is a critical trajectory Γ' with at least one saddle point in the steady state system. If the Hall field is pointing from right to left, $\epsilon_{\Gamma'}$ will satisfy the inequality: $(\epsilon_\Gamma - \epsilon_{\Gamma'})_{non-equil} > (\epsilon_\Gamma - \epsilon_{\Gamma'})_{equil}$. According to the theorem formulated above, due to redistribution of electric charge, some of the equipotentials that were closed in equilibrium, become vertically percolating. At a small Hall voltage charge redistribution should not be significant. Such process costs substantial electrostatic energy which cannot be compensated by a negligible energy release due to hopping. Yet, it results in appearance of vertically percolating equipotentials at all energies in an interval of size V_H. Therefore, a small perturbation of the potential results in a complete change in the structure of equipotentials. Formation of delocalized states may occur either in the spatial region where vertically percolating trajectories already exist in equilibrium or in the place of localized ones. The first mechanism is acting along the horizontal cross section BB (Fig.1a) passing through the saddle point: the slope of potential increases to adopt more equipotential lines; all of these lines will still be open in the vertical direction. The second mechanism of delocalization takes place in regions where critical line Γ' approaches line Γ (cross section AA on Fig.1a). There, it is impossible to squeeze the total external potential drop into the narrow region since this would require another charge redistribution. Delocalization is possible due to the fact that the slope of the potential immediately to the right of line Γ' is determined by the field of the disorder potential rather than by the much weaker external field (Fig.1b). We conclude that the current injected into the system flows in the region between the lines Γ and Γ' as they were defined in equilibrium and in the dominant area of the sample has no preference for the edge regions. In the other words, current is *distributed* throughout the system.

This work was supported by the NSF grant DMR-9528345.

1. B. I. Halperin, Phys. Rev. B **25**, 2185 (1982).
2. M. Büttiker, Phys. Rev. B **38**, 9375 (1988).
3. D. J. Thouless, Phys. Rev. Lett. **71**, 1879 (1993).
4. C. Wexler and D. J. Thouless, Phys. Rev. B **49**, 4815 (1994).
5. B. W. Alphenaar *et al*, Phys. Rev. Lett. **64**, 677 (1990).
6. P. L. McEuen *et al.*, Phys. Rev. Lett. **64**, 2062 (1990)
7. N. Q. Balaban *et al* Phys. Rev. Lett. **71**, 1443 (1993).
8. P. F. Fontein *et al.*, Phys. Rev. B **43**, 12090 (1991), and references therein.
9. D. B. Chklovskii, B. I. Shklovskii, and L. I. Glazman, Phys. Rev. B **46**, 4026 (1992).
10. R. F. Kazarinov and S. Luryi, Phys. Rev. B **25**, 7526 (1982).
11. S. A. Trugman, Phys. Rev. B **27**, 7539 (1983).

INTER-LAYER EDGE TUNNELING AND TRANSPORT PROPERTIES IN DOUBLE-LAYER QUANTUM HALL SYSTEMS

D. YOSHIOKA

Department of Basic Science, University of Tokyo, Komaba, Tokyo 153, Japan

A.H. MACDONALD

Department of Physics, Indiana University, Bloomington, IN 47405, USA

A theory of transport in the quantum Hall regime is developed for separately contacted double-layer electron systems. Inter-layer tunneling provides a channel for equilibration of the distribution functions in the two layers and influences transport properties through the resulting influence on steady-state distribution functions. Resistences for various configurations of the electrodes are calculated as a function of the inter-layer tunneling amplitude. The effect of misalignment of the edges of the two layers and the effect of tilting the magnetic field away from the normal to the layers on the inter-layer tunneling amplitude near the sample edges are investigated. The results obtained in this work is consistent with recent experiments.

1 Introduction

Recently it has become possible to fabricate a multi-layer quantum Hall system, where current and voltage leads are attached separately to each layers.[1] For such a system it is expected that inter-layer interaction affects the resistances. For example, in a recent experiment by Ohno *et al.*, [2] the Hall resistance and the longitudinal resistance on one layer is affected considerably by the presence of the other. Inspired by this experiment, we have investigated how the interlayer tunneling affects the resistance in the quantum Hall regime, where the electron distribution near the system edges determine the electrical conduction.[3] In Sect.2 we explain our theory briefly. We show the results where only one of the layer have leads attached. It is shown that the resistances are non-local, and depends on the strength of the inter-layer tunneling probability. Then in Sect. 3 we give our new result on how misalignment of the sample edges and tilting of the magnetic field affects the hopping probability. Finally brief discussions are given.

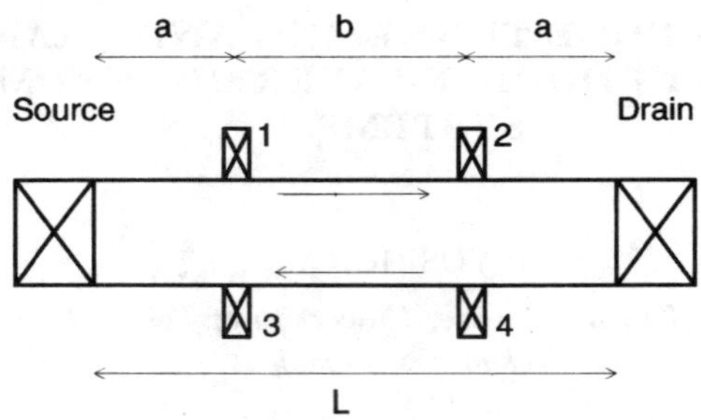

Figure 1: Geometry of the sample. Two identical layers are stacked vertically. The leads (source, drain, 1, 2, 3, and 4) can contact both layers or either layer. The drift directions along the edges are indicated by arrows. The lengths L, a, and b specify the length of the sample and the positions of the leads.

2 Resistance

We consider a Hall bar type sample shown in Fig.1. The energy levels in the bulk are aligned in the two layers, and we consider the system in the lowest quantum Hall regime. For a single layer system the quantum Hall effect can be understood as a consequence of the spatial separation between left-going and right-going states on opposite edges of the sample.[4] At each edge we can define the chemical potential, which stays constant between the leads, and which differs between the opposite edges. However, in the case of a double layered system with finite inter-layer tunneling, the local chemical potential in each layer is generally not constant between the leads. We take x-axis along the edge and introduce $\mu_\sigma(x)$ for the local chemical potential, where $\sigma = \pm$ represents the layer index. Then the following equation governs the development of the chemical potentials between the leads:

$$\frac{d\mu_\sigma(x)}{dx} = -\frac{1}{\xi}[\mu_\sigma(x) - \mu_{-\sigma}(x)], \tag{1}$$

where ξ is a parameter meaning the relaxation length for interedge equilibration. The current along an edge in layer σ at position x is given by

$$I_\sigma(x) = \frac{e}{h}[\mu_\sigma(x) - \epsilon_0], \tag{2}$$

with ϵ_0 being a reference energy. The voltage leads will not affect the chemical potential, since the current will not flow through the lead. On the other hand the current leads give discontinuous change to the chemical potential. Following these rules we can solve for the chemical potentials of both layers, once the external current through each leads are given for any configuration of the leads.

As an example of such a solution, the followings are the obtained resistances for the case where only the minus layer has leads attached. Here current is fed to the minus layer through the source and extracted through the drain, and μ_- is observed at voltage leads 1 to 4 (see Fig.1). The longitudinal resistance $R_{xx} = R_{12} = R_{34}$ and the Hall resistance $R_{xy} = R_{13}$ are given as follows:

$$R_{xx} = \frac{h}{e^2} \frac{\exp(-2a/\xi)[1 - \exp(-2b/\xi)]}{2[1 + \exp(-2L/\xi)]}, \tag{3}$$

$$R_{xy} = \frac{h}{e^2} \frac{1 + \exp(-2a/\xi) + \exp[-2(L-a)/\xi] + \exp(-2L/\xi)}{2[1 + \exp(-2L/\xi)]}. \tag{4}$$

where a and b specify the voltage probe positions as indicated in Fig. 1. Thus only in the strong (weak) coupling limit, where $L/\xi = \infty(0)$, the resistances are quantized: $R_{xx} = 0$ and $R_{xy} = h/2e^2(h/e^2)$.

3 Relaxation length ξ

The results for the resistances depend on the ratio L/ξ. Thus we need to know typical size of this parameter, and how it depends on various factors. The tunneling between the layers occur between state with the same energy. Let's assume such pair of states are given by wave functions $\phi_\sigma(x, y, x)$ in the absence of tunneling. When the two edges of the two layers are aligned perfectly, they satisfy $\phi_+(x, y, z) = \phi_-(x, y, z - d)$, where d is the interlayer separation. In the presence of the tunneling, they couple each other, and symmetric and antisymmetric combination of these states will become the eigen states. From the energy difference of these states, Δ_{SAS}, the correlation length is estimated to be $\xi = \hbar v/\Delta_{SAS}$, where $v \simeq \ell\omega_c$ is the velocity of edge states, with ℓ being the magnetic length and ω_c being the cyclotron frequency. This estimate gives $L/\xi \simeq 50$ for $L = 200\mu$m, and $\Delta_{SAS} = 0.02$meV, which are typical values for an experimental situation. [2] This value of L/ξ is quite large, and the system can be considered to be in the strong coupling regime. However, various factors reduce this value. Among them here we consider misalignment of the edges and in-plane magnetic field. We assume the center coordinate of ϕ_+ projected on to the xy-plane cross many times with that of ϕ_-. Among those crossings we focus on one of them, approximate the trajectory of the center coordinates around the crossing by straight lines, and calculate the contribution to the tunneling amplitude from a section of length $L_s (\gg \ell)$ around the crossing. We take the coordinate such that the x-axis bisects the two edges, so the two edges make angle $\pm\theta/2$ from the x-axis. The in-plane field B_x and B_y also refer to this coordinate. The tunneling amplitude, $I(B_x, B_y, \theta)$, is proportional to the overlap integral of ϕ_+ and ϕ_-. If we approximate ϕ_σ by the lowest Landau level wave function, the relative amplitude at $\theta = 0$ is given as follows:

$$\frac{I(B_x, B_y, 0)}{I(0,0,0)} = \frac{2\ell}{L_s} \frac{\sin(L_s\beta_y/2\ell)}{\beta_y} \exp(-\frac{\beta_x^2}{4}), \tag{5}$$

where $\beta_x = B_x d/B_z\ell$, and $\beta_y = B_y d/B_z\ell$. So in this case, B_y has a drastic effect: the amplitude is deminished by a small B_y, but B_x dependence is small. For $\theta \gg (\ell/L_s)^2$ the amplitude is much reduced and given by

$$\frac{I(B_x, B_y, \theta)}{I(0,0,0)} = \frac{\ell}{L_s}\sqrt{\frac{2\pi}{\sin\theta}} \exp[-\frac{(\beta_x^2 + \beta_y^2)}{4}]. \tag{6}$$

In this case $I(B_x, B_y, \theta) \ll I(0,0,0)$, and the effect of the in-plane magnetic field is quite weak.

4 Discussion

In recent experiments Ohno *et al.*[2] observed $R_{xy} \simeq h/e^2$ in the presence of finite R_{xx} in the situation considered in Sec.2. This result is at a first glance inconsistent with our result, since $L/\xi \simeq 50$, if estimated using $\Delta_{SAS} =$ 0.02meV. However, the inconsistency is resolved, if the edges are not aligned perfectly. In this case the tunneling amplitude is reduced as shown in Eq. 6. In the experiments they observed that the effect of in-plane field is quite small. This insensitivity to the tilting is also in accordance with Eq. 6. Thus our theory is consistent with the experiments. It is general enough to treat various other configurations of leads. We hope such situations are realized experimentally, and results are compared with our theory.

References

1. J.P. Eisenstein, L.N. Pfeiffer, and K.W. West, *Appl. Phys. Lett.* **57**, 2324 (1990).
2. Y. Ohno and H. Sakaki, Surf. Sci. (to be published), and Y. Ohno, M. Foley, and H. Sakaki, unpublished.
3. D. Yoshioka and A.H. MacDonald, *Phys. Rev.* B **53**, 16168 (1996)
4. M. Büttiker, *Phys. Rev.* B **38**, 9375 (1988).

PHASE COHERENCE IN THE QUANTUM HALL EFFECT REGIME

T. MACHIDA, H. HIRAI, and S. KOMIYAMA
Department of Pure and Applied Sciences, University of Tokyo,
Komaba 3-8-1, Meguro-ku, Tokyo 153, Japan

T. OSADA and Y. SHIRAKI
RCAST, University of Tokyo,
Komaba 4-6-1, Meguro-ku, Tokyo 153, Japan

We study the transition between successive integer quantum Hall effect states, using AlGaAs/GaAs heterostructure Hall bars with Schottky cross gates at low temperatures down to 25mK. The transition width decreases with increasing the length as well as the width of the scattering region. Using these results, we show that the electron states in the separated scattering regions strongly interfere with each other mediated by edge states. The experimental results indicate that the phase coherence of edge states extends over a distance far larger than 200μm at 25mK.

The phase coherence length of two dimensional electron gas (2DEG) systems in high magnetic fields is one of the most interesting subjects, which has not yet been satisfactorily studied [1]. In this work, we study gate-bias spectra of Hall-bars with Schottky front gates at low temperatures down to 25mK [2,3].

By negatively biasing the gate while maintaining the gate-free 2DEG regions in a QHE state, the 2DEG region underneath the gate behaves as a well-defined scattering region, while the gate-free 2DEG region serves as ideal leads [4]. Our experiments consist of two parts. First, we study transmission coefficients of edge channels and their dependence on the width and the length of the scattering region, and present a simple picture connecting the transmission probability of edge channels with the localization length [5,6] of bulk Landau states [2]. Second, utilizing the results obtained in the first part, we study the phase coherence of edge states in the QHE state, not in the transition region, and suggest that the phase coherence length is much larger than 200μm at 25mK [3].

Hall bars studied in the first set of experiments are patterned from a AlGaAs/GaAs heterostructure wafer (μ=80m^2/Vs, n_s=2.6×10^{11}cm^{-2}) with a

Schottky cross gate. Figure 1(a) and (b) display the four-terminal longitudinal resistance R_{xx} against the gate-bias voltage at a base temperature of 25mK, and represent the transition from the QHE state of the Landau level filling factor of ν=4 to that of ν=3. The full-width at half maximum of the R_{xx} peak, ΔR_{xx}, significantly decreases with increasing the length L as well as the width W of the gated 2DEG region [7]. This dependence of ΔR_{xx} on W and L enables us to derive the localization length ξ of bulk Landau states as a function of V_G. Similarly, from the dependence of ΔR_{xx} on the temperature T, we derive the phase coherence length L_ϕ as a function of T. We show that L_ϕ increases to reach a value comparable to 300μm with decreasing temperature down to 25mK. The size of L_ϕ as well as the saturationless increase of L_ϕ with decreasing T should be compared with the extensively studied characteristics of L_ϕ in zero or low magnetic fields [8-13].

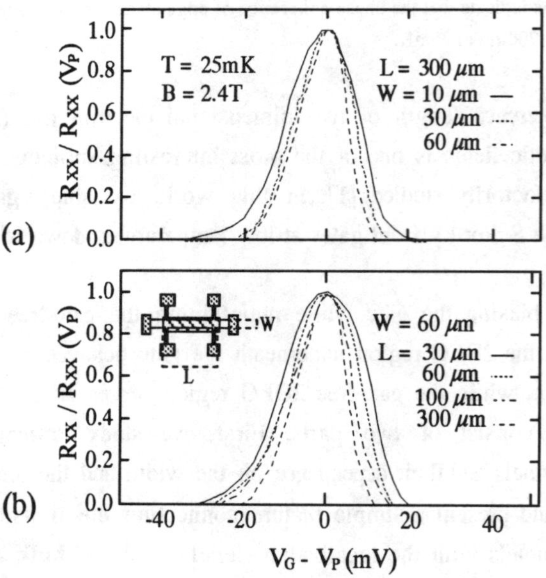

Figure 1: Longitudinal resistance as a function of the gate-bias voltage V_G. The dependence on the channel width (a), and on the gate length (b). The Hall-bar is schematically represented in the inset.

In the second set of experiments, we study 30μm-wide Hall bars, on which a pair of separated Schottky gates, PG1 and PG2, are placed leaving a 100μm-long gate-free 2DEG region, as shown in the inset of Fig. 2. In addition, voltage probes with Ohmic contacts are electrically connected or disconnected to the gate-free region by using additional control gates, CG1 and CG2. Figure 2 shows the differential resistance $dR_{15;24}/dV_{PG}$ obtained by modulating the probe-gate bias voltage V_{PG} with an amplitude of 1mV at T=25mK. The data are displayed as a function of the background dc-voltage V_{PG}. The solid line indicates the data obtained by biasing only PG1. The dashed line represents the data taken in the condition when both PG1 and PG2 are biased, while CG1 and CG2 are open so that the voltage probes 3 and 7 are electrically connected to the gate-free region. These two spectra are substantially equal to each other indicating that the separated scattering regions behave independently to respectively yield a classically additive longitudinal resistance. The dotted line indicates the data obtained when both of PG1 and PG2 are biased, while CG1 and CG2 are closed so that the phase coherence of edge states is undisturbed. In this condition, the spectrum is significantly sharpened. This definitely suggests that the electron

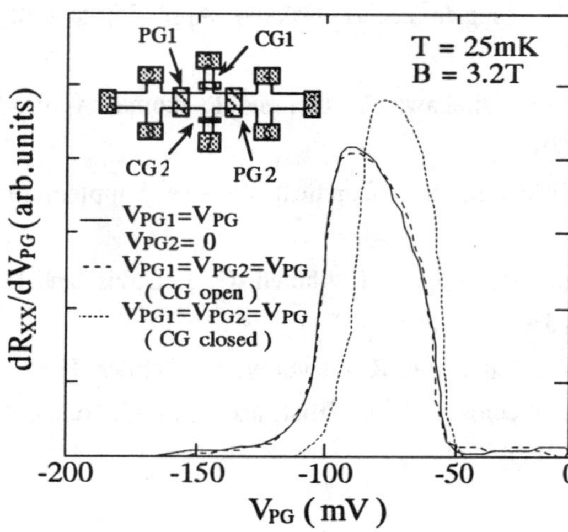

Figure 2: Differential longitudinal resistance in a Hall bar with double-gate geometry, as a function of the probe-gate bias voltage V_{PG}. The inset is the schematical representation of the sample.

states in the two separated gated regions strongly interfere with each other, being mediated by edge states. The edge states are suggested to be coherent over a length much larger than 200μm [3].

References

1. S. Koch, R. J. Haug, K. von Klitzing, and K. Ploog, Phys. Rev. Lett. **67**, 883 (1996).
2. T. Machida, H. Hirai, S. Komiyama, and Y. Shiraki, submitted to Phys. Rev. B.
3. T. Machida, H. Hirai, S. Komiyama, T. Osada, and Y. Shiraki, subitted to Phys. Rev. Lett.
4. M. Büttiker, Phys. Rev. B **38**, 9375 (1988).
5. H. Aoki and T. Ando, Surface Science **170**, 249 (1986).
6. A. M. M. Pruiskin, Phys. Rev. Lett. **61**, 1297 (1988).
7. T. Ando, Phys. Rev. B **49**, 4679 (1994).
8. A. Yacoby, U. Sivan, C. P. Umbach, and J. M. Hong, Phys. Rev. Lett. **66**, 1938 (1991).
9. Y. K. Fukai, S. Yamada, and H. Nakano, Appl. Phys. Lett. **56**, 2123 (1990).
10. T. Hiramoto, K. Hirakawa, Y. Iye, and T. Ikoma, Appl. Phys. Lett. **54**, 2103 (1989).
11. S. Kawaji, Progress of Theoretical Physics Supplement, **84**, 178 (1985).
12. T. J. Thornton, M. Pepper, H. Ahmed, G. J. Davis, and D. Andrews, Phys. REv. B **36**, 4514 (1987).
13. C. J. Ford, T. J. Thornton, R. Newbury, M. Pepper, H. Ahmed, D.C. Peacock, D. A. Piechis, J. E. F. Frost, and G. A. C. Jones, Appl. Phys. Lett. **54**, 21 (1989).

QUANTUM HALL RESISTANCES WITH DIFFERENT RATIONAL FRACTIONS OF h/e^2 REALIZED BY CONTROLLING THE CURRENT DISTRIBUTION VIA SCHOTTKY GATES

M. Blöcker
Physikalisch-Technische Bundesanstalt,
Bundesallee 100, D-38116 Braunschweig, Germany

G. Nachtwei
Max-Planck-Institut für Festkörperforschung,
Heisenbergstr. 1, D-70569 Stuttgart, Germany

F. J. Ahlers and L. Bliek
Physikalisch-Technische Bundesanstalt,
Bundesallee 100, D-38116 Braunschweig, Germany

We have investigated the magnetotransport properties of samples with a multiple connected geometry and Schottky gates in the quantum Hall regime. With these structures which consist of two parallel Hall bars shorted by regions of 2DEG we demonstrate the possibility of realizing Hall resistances with values equal to different rational fractions of h/e^2 at fixed magnetic field. This is achieved by controlling the partition of the fixed total current onto the two sample arms via the voltages applied to the surface gates across each of the Hall bars.

We investigated two samples structured out of different GaAs/$Al_xGa_{1-x}As$ heterostructure wafers by a standard photolithography process including wet chemical etching and lift-off processes. Figure 1 shows the resulting sample geometries with mesa (gray), current and voltage probes (white; Ni/AuGe/Ni) and surface gates (hatched; Cr/Au).

The measurements were carried out at fixed temperature (sample A: $T =$ 1.3 K; sample B: $T = 50$ mK) and fixed current $I_{SD} = 1\ \mu$A. At fixed magnetic field corresponding to a filling factor of $\nu_0 = 4$ in the undisturbed sample regions the gate voltage dependence of the potential difference between two contacts was recorded. The two gate voltages where applied with respect to different current contacts in order to avoid shifting the effective gate voltage by the Hall voltage.

If the gate voltages are set to values corresponding to integer filling factors ν_{Gi} $(i = 1, 2)$ below the gates, Büttiker's [1] edge channel picture of the QHE can be applied to the structures shown in figure 1. This leads to the following expression for the four-terminal Hall resistances (e.g. $R_{56} = U_{56}/I_{SD}$ for

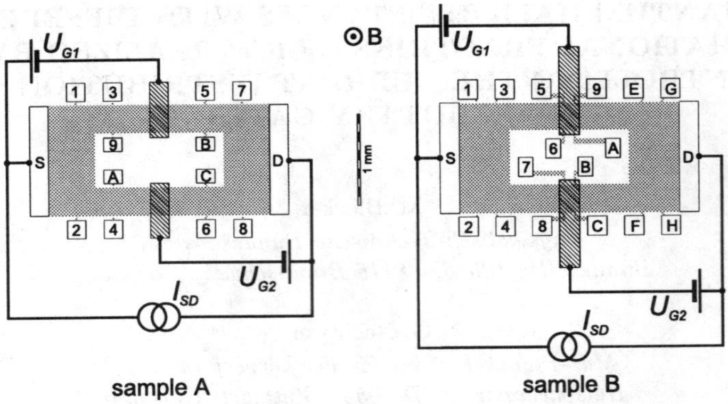

Figure 1: Schematic view (to scale) of the sample geometries, contact denominations and measurement setup of the 'windowed' Hall bars.

sample B) in the sample regions below the gates:

$$R_{Hall,1} = \frac{U_{Hall,1}}{I_{SD}} = \frac{\nu_0 - \nu_{G2}}{\nu_{G1}(\nu_0 - \nu_{G2}) + \nu_{G2}(\nu_0 - \nu_{G1})} \frac{h}{e^2} \tag{1}$$

$$R_{Hall,2} = \frac{U_{Hall,2}}{I_{SD}} = \frac{\nu_0 - \nu_{G1}}{\nu_{G1}(\nu_0 - \nu_{G2}) + \nu_{G2}(\nu_0 - \nu_{G1})} \frac{h}{e^2} \quad . \tag{2}$$

The same equations hold for samples of type A where the Hall voltage in the regions below the gates can be measured between voltage probes located diagonally on different sides of the gate (e.g. contact pair $3 - B$) because of the vanishing longitudinal voltage.

It has to be stressed that equations (1) and (2) can be derived as well by considering quantized Hall voltages in sample regions of integer filling factor and applying Kirchhoff's laws [2]. In accordance with findings by Thouless [3] either applying a local or a non-local transport model leads to the same results for the potential distribution on the sample edges in the quantum Hall regime.

The four-terminal resistances shown in figure 2 show good agreement with the values given by equations (1) and (2). The measurements can be interpreted by considering that the four-terminal resistances are also given by:

$$R_{Hall,i} = \frac{1}{\nu_{Gi}} \frac{h}{e^2} \frac{I_i}{I_{SD}} \qquad ; \; i = 1,2 \quad . \tag{3}$$

The Hall resistances are a measure for the portion of the total current flowing through the corresponding sample arm. The distribution of the current onto the sample arms is controlled via the gate voltages.

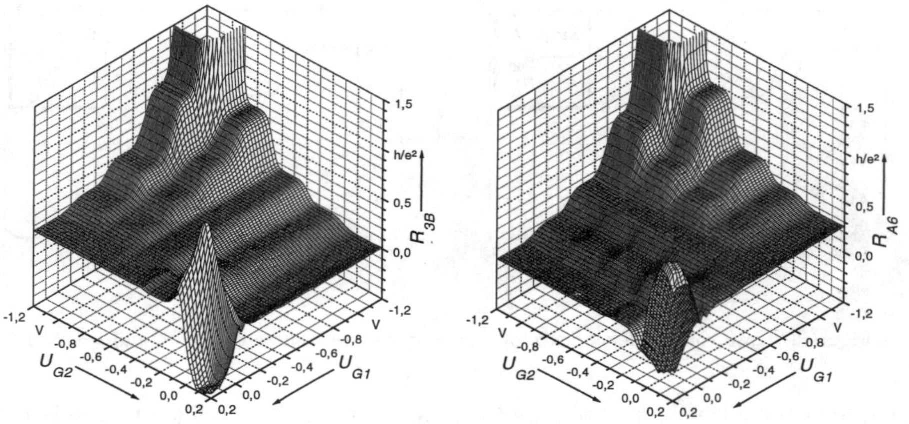

Figure 2: Four-terminal resistances R_{3B} and R_{A6} on sample A at $\nu_0 = 4$.

Two features of the measurement results should be emphasized here:

Firstly, if the gate voltages are set to values corresponding to filling factors $\nu_{G1} = \nu_{G2} = \nu_0$ the equations (1) and (2) make no predictions for the four-terminal resistances. This is in accordance with results on samples with an inner edge but without Schottky gates [4]. There, the inner and outer edges of the sample are decoupled from each other in the quantum Hall regime. Measurement results in that region of filling factors only show the spurious internal charging of the voltmeters.

Secondly, if the filling factors are set to e.g. $\nu_{G1} = \nu_0$ and $\nu_{G2} < \nu_0$ there are still current carrying states below gate 2 and yet the total current flows completely through the other sample arm. This resembles a model of the QHE proposed by Woltjer et al. [5] in which the current flows predominatly through sample regions with integer filling factor.

Assuming a filling factor $\nu_{G1} = \nu_0/2$, equations (1) and (2) reduce to:

$$R_{Hall,1} = \frac{2}{\nu_0}\left(1 - \frac{\nu_{G2}}{\nu_0}\right)\frac{h}{e^2} \tag{4}$$

$$R_{Hall,2} = \frac{1}{\nu_0}\frac{h}{e^2} \quad . \tag{5}$$

The measurements shown in figure 3 show agreement with these equations for integer filling factors. To explain the observed differences between the two samples at non-integer filling factors below gate 2, the longitudinal voltage drop below gate 2 has to be considered. For sample A this can be done within the

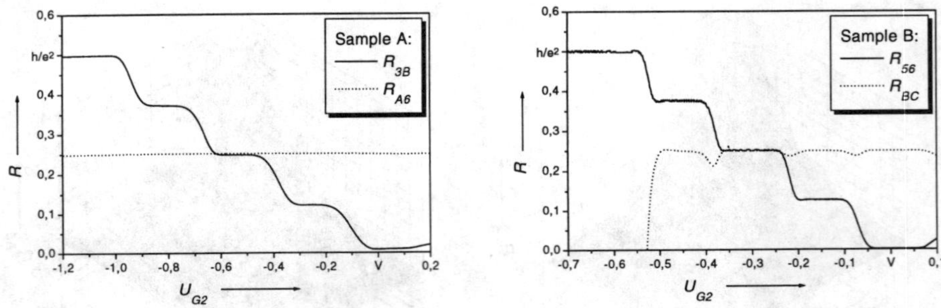

Figure 3: Gate voltage dependence of four-terminal resistances at $\nu_0 = 4$ and $\nu_{G1} = 2$.

scope of the edge channel picture by accounting for the modified transmission of edge states by replacing ν_{G2} in equations (1) through (5) by a monotonous function $T_2(\nu_{G2})$ for which $T_2(\nu_{G2}) = \nu_{G2}$ for integer ν_{G2} holds. In agreement with equation (5) R_{A6} remains constant even at non-integer filling factors.

For sample B problems arise with the edge channel picture at non-integer filling factors because the backscattering can't be accounted for in the same way as in sample A. By the application of Kirchhoff's laws however, a modified expression for the Hall resistance can easily be derived:

$$R_{BC} = \frac{1}{1 + \frac{e^2}{h}\, R_{xx,2}\, T_2}\, \frac{1}{\nu_0}\, \frac{h}{e^2} \quad . \tag{6}$$

Here $T_2(\nu_{G2})$ takes account of the filling factor dependence of the Hall voltage. The rise of the longitudinal resistance $R_{xx,2}$ below gate 2 in the vicinity of half-integer filling factors leads to the observed deviations of R_{BC} from $\frac{1}{4}\, h/e^2$.

Acknowledgments

We would like to thank M. Cukr, H. Nickel, L. Bartholomäus, C. Breitlow and A. Linke for their help in sample preparation and measurements.

References

1. M. Büttiker, *Phys. Rev.* B **38**, 9375 (1988).
2. G. Nachtwei *et al.*, *Semicond. Sci. Tech.* **11**, 89 (1996).
3. D.J. Thouless, *Phys. Rev. Lett.* **71**, 1879 (1993).
4. H. Wolf *et al.*, *Semicond. Sci. Tech.* **5**, 1046 (1990).
5. R. Woltjer *et al.*, *Europhys. Lett.* **2**, 149 (1986).

BULK CONDUCTION IN THE TRANSITION REGIONS BETWEEN HALL PLATEAUS IN A TWO-DIMENSIONAL ELECTRON GAS AT HIGH MAGNETIC FIELDS

QINGHONG DU, HONGQI XU, P. RAMVALL, and P. OMLING
Department of Solid State Physics, Lund University, Box 118,
S-221 00 LUND, Sweden

Using a partially-gated Hall-bar device, We demonstrate the existence of two different regimes of electron transport in the transition regions between Hall plateaus at high magnetic fields. One regime is found in the high-magnetic-field tail of a Shubnikov-de Haas peak, while the other one is in the remaining magnetic field region of the peak. The magnetoresistance measured in the two regimes is shown to arise from different physical origins.

1 Introduction

On the basis of one-particle theory, Büttiker, using the Landauer formalism, showed that the bulk states give no contribution to current transport in a Hall-bar device.[1] Many experiments can be explained by the simple, one-particle Landauer-Büttiker theory. In particular, we showed, using a partially-gated device, that the bulk states can indeed be important only to facilitate backscattering, but not for current transport.[2] On the other hand, models that assume both edge and bulk conduction were also proposed for quantum Hall conductors. Although reasonable agreement with a few experiments was found, the mechanism for the bulk conduction was not specified in the models.

In this contribution, we present an investigation on this controversial problem and demonstrate the existence of two different regimes of electron transport in the transition regions between Hall plateaus at high magnetic fields. In the investigation, we measured the longitudinal magnetoresistance of a Hall-bar device partially covered by a Schottky gate on top. By cutting off the coupling between the bulk states and the edge channels on one side of the Hall bar, with the use of the Schottky gate, while measuring the longitudinal magnetoresistance on the other side, we were able to identify a clear contribution of the bulk states to the current transport.

2 Experiment

The device structure is schematically shown in Fig. 1. The mesa was defined by wet etching a GaAs/$Al_xGa_{1-x}As$ heterostructure. The sheet electron concentration and the mobility of the 2DEG at 4.2 K are $n_{2D} = 3.5 \times 10^{11}/cm^2$

and $\mu = 3.2 \times 10^5$ cm^2/V·s, respectively. The Hall bar is 80 μm wide, while the voltage probes are slightly wider. Ohmic contacts were produced by deposition of AuGe/Ni, using a lift-off technique and annealing at 470 ^oC for 2 minutes. A Ti/Au Schottky gate was then deposited on one side of the Hall bar. The measurements were performed at the temperature $T = 0.3$ K by the standard lock-in technique with a current of 10 nA root-mean-square.

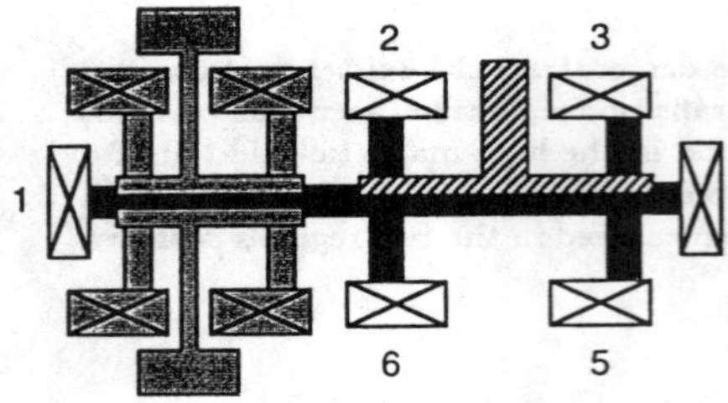

Fig. 1. Top View of the device. The grey parts are not used. Contacts 1 to 6 are ohmic and the hatched region represents the Schottky gate placed on top of the Hall bar. The width of the Hall bar is 80 μm, while the width and length of the connections to contacts 2, 3, 5 and 6 are 95 and 22 μm, respectively. The distance between contacts 2 and 3 and between contacts 5 and 6 is 680 μm.

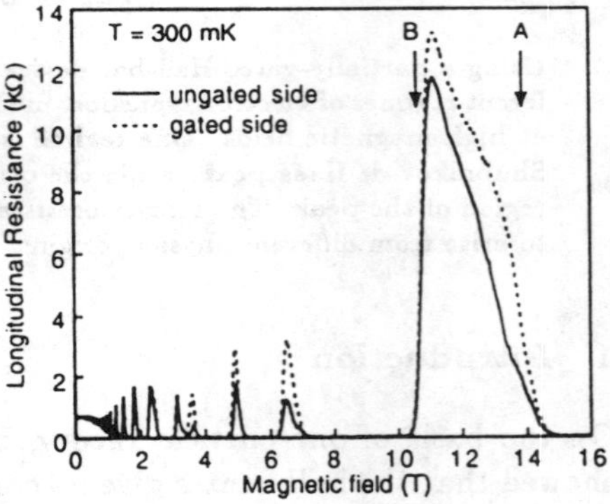

Fig. 2. Longitudinal resistances, $R_{14,23}$ (dashed curve) and $R_{14,65}$ (solid curve), measured at the gate voltage $V_g = 0$ on the two sides of the Hall bar as functions of the magnetic field B. Arrows A and B mark the magnetic field positions of $B = 13.89$ and 10.60 T, respectively.

3 Results and Discussion

Fig. 2 shows the characteristics of the sample, where the Shubnikov-de Haas (SdH) spectra measured on both the gated and ungated sides at zero gate voltage ($V_g = 0$) are plotted.

Fig. 3(a) shows the longitudinal resistances, $R_{14,23}$ (dashed line) and $R_{14,65}$ (solid line), measured on both sides of the Hall bar as a function of gate voltage V_g at magnetic field $B = 13.89$ T (denoted by arrow A in Fig. 2). As V_g becomes more negative, both $R_{14,23}$ and $R_{14,65}$ decrease and reach their minimum value very close to zero at $V_g \approx -0.12$ V and remain at this minimum until $V_g \approx -0.22$ V. $R_{14,23}$ and $R_{14,65}$ are seen to increase when V_g is further lowered. This result can be easily explained in the framework of the one-particle, Landauer-Büttiker theory.[2] The zero value of resistance $R_{14,23}$ is in fact an indication of a complete decoupling between the edge channel on the gated side and the bulk states of the last occupied, spin-split, Landau level in the ungated region, while the fact that both $R_{14,23}$ and $R_{14,65}$ vanish at gate voltages between -0.12 and -0.22 is evidence that the last occupied, spin-split, Landau level at low filling fraction ν of the last occupied, spin-split, Landau level is important only to facilitate backscattering, but not for current transport [2].

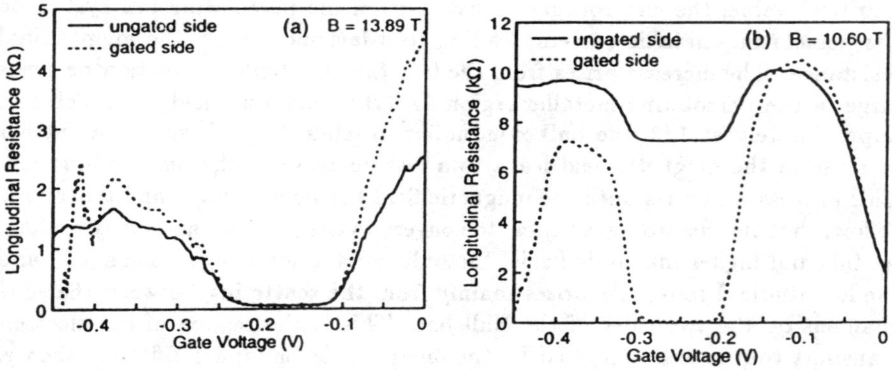

Fig.3. Longitudinal resistances measured on the two sides of the Hall bar, $R_{14,23}$ (dashed curve) and $R_{14,65}$ (solid curve), as functions of the gate voltage V_g at (a) the magnetic field $B = 13.89$ T (denoted by arrow A in Fig. 2) and (b) the magnetic field $B = 10.60$ T (denoted by arrow B in Fig. 2).

In sharp contrast, the longitudinal resistances $R_{14,23}$ and $R_{14,65}$ measured at $B = 10.60$ T (denoted by arrow B in Fig. 2), as displayed in Fig. 3(b), show that the above conclusion does not hold when the filling fraction ν of the last occupied, spin-split, Landau level becomes high. Although the resistance $R_{14,23}$ measured on the gated side shows a behavior similar to that at $B = 13.89$ T, the resistance $R_{14,65}$ measured on the ungated side is different from its corresponding curve at $B = 13.89$ T. As V_g decreases, $R_{14,65}$ shown in Fig. 3(b), does not drop to zero when the coupling between the edge channel on the gated side and the bulk states of the last occupied, spin-split, Landau level is cut off. Instead, it remains at a high, constant value until the lowest, spin-split, Landau level in the gated region is aligned with the last occupied, spin-split, Landau level in the ungated region. This finding is clear evidence of bulk conduction when the filling fraction ν has a high value and can not be explained by the simple, one-particle, Landauer-Büttiker theory.

In fact, our experiment indicates the existence of two different regimes of electron transport in the transition regions at high magnetic fields. We interpret the two regimes as a percolating-empty and a percolating-metallic regime, based on recent theoretical models that take self-consistently calculated electrostatic potential into account. At a magnetic field for which the last occupied, spin-split, Landau level is well below the Fermi energy, the filling fraction ν is close to 1 and the electron gas is a completely-full or percolating-full system. The bulk conduction is negligible and the longitudinal resistance is zero. As the magnetic field increases, the last occupied, spin-split, Landau level is pushed up and ν decreases. When the magnetic field passes through a critical value, the electron gas transfers from the percolating-full system to the percolating-metallic system, leading to a fast increase in the longitudinal resistance. The increase arises from the fact that the bulk conduction becomes large in the percolating-metallic region. At the magnetic field for which ν is approximately at 1/2, the bulk conduction reaches its peak value. A further increase in the magnetic field leads to a fast decrease in the bulk conduction. This process continues until the magnetic field becomes so high, and ν becomes so low, that the electron gas begins to convert to the percolating-empty system. At this and higher magnetic fields, the bulk conduction is again negligible and the longitudinal resistance arises mainly from the scattering between the edge channels on the two sides of the Hall bar. This is the regime of the electron transport that is well described by the one-particle Landauer-Büttiker theory.

4 Conclusion

In conclusion, we have demonstrated the existence of two different regimes of electron transport in transition regions between Hall plateaus at high magnetic fields, namely the percolating-empty and percolating metallic regimes. The two regimes are identified according to whether or not there is bulk conduction in the 2DEG. In the percolating-empty regime, the bulk conduction is negligible and the magnetoresistance arises solely from the scattering between edge channels via the bulk states. This is the regime one should observe in the high-magnetic-field tail of a SdH peak. In other magnetic field regions of the peak, the electron transport in the 2DEG is in the percolating-metallic regime. In this regime, the resistance can, in principle, arise from both the edge channel scattering and the bulk conduction.

References

1. M. Büttiker, *Phys. Rev.* B **38**, 9375 (1988).
2. P. Ramvall, Qinghong Du, Hongqi Xu, L. Samuelson, and P. Omling, *Phys. Rev.* B **53**, 1933 (1996).

COMPARISON OF NONLOCAL RESISTANCE BETWEEN Si-MOS FET AND GaAs/AlGaAs HETEROSTRUCTURE

S.TAKAOKA, K.OTO, T.SHIRAE, H.ISHIDA and K.MURASE,
Department of Physics, Faculty of Science, Osaka University
1-1 Machikaneyama, Toyonaka, 560, Japan

K.MORIMOTO and Y.HIRAI
Matsushita Electric Industrial Co.,Ltd.
3-15 Yagumo-nakamachi, Moriguchi, 570, Japan

We investigate the nonlocal resistance in the two typical two dimensional electron system (2DES) of GaAs/AlGaAs heterostructure and Si-MOSFET. From the dependence of the nonlocal resistance peaks on the separation length between the current path and voltage probes, the energy relaxation length between the edge and bulk currents is estimated as the order of tens mm and hundreds of μm, for GaAs/AlGaAs and Si-MOSFET, respectively. However there is considerable specimens dependence, especially in GaAs/AlGaAs system, which may be due to the inhomogeneity of 2DES.

1 Introduction

The concept of the edge and bulk currents[1,2] can explain various kinds of the curious electrical transport properties of two dimensional electron system (2DES) in the quantum Hall regime[3-6]. The nonlocal resistance[7-11], where the nominal current path is separated from voltage probes, is one of these phenomena. The most of the investigations of the nonlocal resistance have been performed in GaAs/ AlGaAs heterostructure and there have been few studies in Si-MOSFET[12,13], which is another typical 2DES and the quantum Hall effect is clearly observed. However, the electron mobility of Si-MOSFET at low temperatures (μ~10^4cm^2/Vs) is smaller than that of GaAs/AlGaAs heterostructure (μ~10^6cm^2/Vs) by two orders of magnitude. We have investigated and compared the nonlocal resistance between two typical 2DES of GaAs/AlGaAs heterostructure and Si-MOSFET.

2 Experiments

The Si-MOSFETs were made from p-type Si substrate (p~5×10^{14}cm^{-3}).

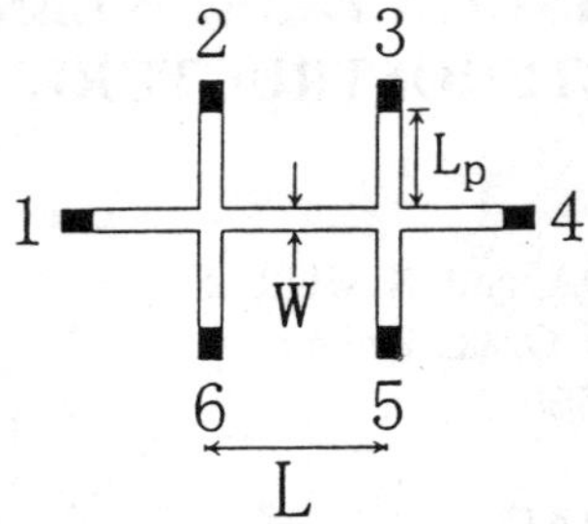

Figure 1: Schematic view of sample and probe numbers.

Table 1. Typical electrical characteristics of GaAs/AlGaAs wafers at 0.5K. Critical current to break quantum Hall effect at ν=2 is measured in the samples with W=50μm.
μ : electron mobility. N_s : electron concentration.

wafer	N_s (cm^{-2})	μ (cm^2/Vs)	Nonlocal R with L_p=0	Critical current
#A	2.9×10^{11}	4.9×10^{5}	disappear	7 μA
#B	3.9×10^{11}	7.5×10^{5}	appear	0.8 μA
#C	2.4×10^{11}	1.2×10^{6}	appear	1.8 μA
#D	1.3×10^{11}	1.7×10^{6}	disappear	3.5 μA

The thickness of gate–oxide is 100nm and the gate electrode is n^+ poly–Si. The samples of GaAs/AlGaAs was fabricated by UV–lithography and wet chemical etching techniques. The typical electrical properties of the wafers used in this study are listed in Table 1. The magnetoresistance up to B=12T was measured at 0.5K by applying AC current at 15Hz. The three kinds of multi–terminal samples have the same sizes in both materials as follows; The length of separation path (L) between the current path and voltage probes, the width of path (W) and the length of voltage and current probe (Lp) as shown in Fig.1 is, respectively, (L=200μm, W=20μm, Lp=90μm), (L=20μm, W=2μm, Lp=9μm) and (L=5μm, W=0.5μm Lp=2.25μm), with keeping the ratio (L/W) and (Lp/W) constant in order to show the same nonlocal R(26,34) and longitudinal resistance R(14,23) calculated from the Landauer–Büttiker (LB) formula with neglecting the energy relaxation[2,7,11]. R(mn,kl) is the four–terminal resistance, where current flows between m–th and n–th probes and voltage is measured between k–th and l–th probes.

3 Results and Discussion

In Fig.2, the magnetic field dependence of nonlocal and longitudinal resistance in Si–MOSFET and GaAs/AlGaAs samples at T=0.5K is shown. The electron mobility (μ) and concentration (N_s)is $1.0\times10^4 cm^2$/V.s and $9\times10^{11} cm^{-2}$, respectively at Vg=4.5V in Si–MOSFET. The peaks of nonlocal resistance decrease and those of longitudinal resistance increase with increasing L in both materials generally, which is due to the energy relaxation between the edge and bulk currents. The peak

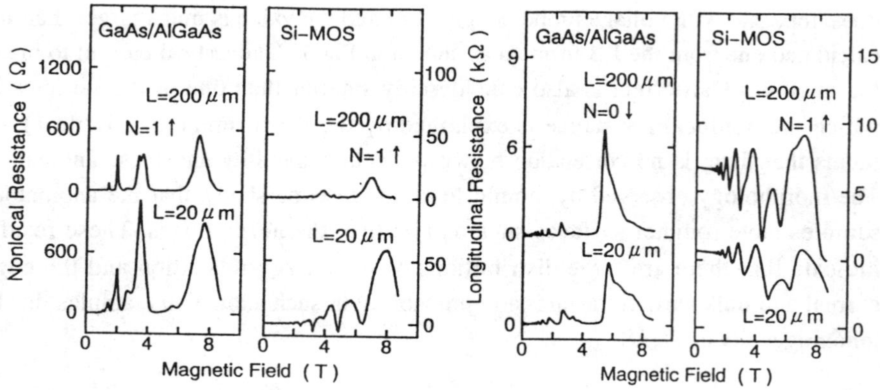

Figure 2: Magnetic field dependence of nonlocal resistance and longitudinal resistance of Si-MOSFET and GaAs/AlGaAs (#C) with L=200μm and 20μm.

in Si-MOSFET is much smaller than one in GaAs/AlGaAs for N=1 and up-spin state (N: Landau quantum number). From the separation length dependence of the nonlocal peaks, the energy relaxation length is estimated as 10~20 mm and ~0.1mm, for GaAs/AlGaAs and Si-MOSFET, respectively.

However there is considerable wafer dependence, especially in GaAs/AlGaAs system[14]. In certain samples made from #B and #C, the nonlocal resistance with zero probe length (Lp=0) does not disappear, nevertheless the nonlocal resistance in such a sample geometry will be expected to be almost zero from the LB formula[11] and the resistance vanishes in Si-MOSFET.

Furthermore, the nonlocal resistance

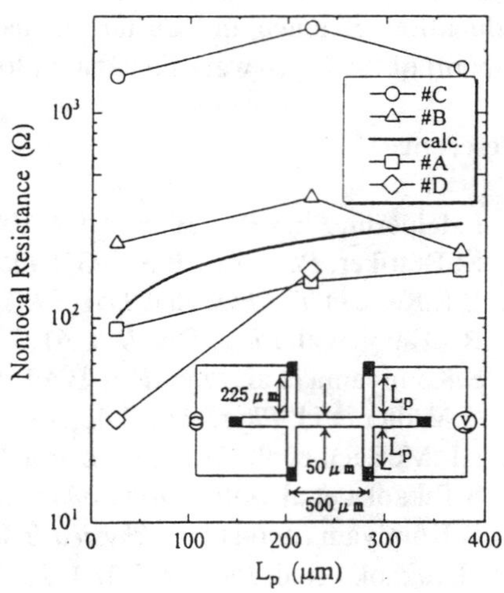

Figure 3: Voltage probe length (Lp) dependence of nonlocal resistance peak at N=1 and up-spin state in various GaAs/AlGaAs samples as shown in the inset. Solid curve is the calculation from the LB formula.

does not scale with voltage probe length (Lp) and the value is much larger than the calculated one from the LB formula as shown in Fig.3. The critical current to break the quantum Hall effect is also considerably smaller than one of the samples in which the nonlocal resistance is explained by the LB formula (see Table 1). It seems that there is no correlation between electron mobility and these anomalies. The morphology observed by atomic force microscope shows that the anomalous samples have rougher surfaces such as pits than the normal ones. These results indicate that there are large distributions of the energy relaxation and the edge current and bulk current transmission probability in such anomalous samples due to inhomogeneity of 2DES.

Acknowledgements

We thank S.Nishi of OKI Electric Industrial Co., Ltd., H.Okada of Sumitomo Electric Industries and H.Yamaguchi of NTT Basic Research Laboratories for providing GaAs/AlGaAs heterostructure wafers. This work is partially supported by a Grand–in–Aid for Scientific Research on Priority Area, from the Ministry of Education, Science and Culture (Japan). We (S.T. and K.O.) acknowledge the support of the Ogasawara Foundation for Promotion of Science and Engineering.

References

1. I.Halperin, *Phys.Rev.* B **25**, 2797 (1982).
2. M.Büttiker, *Phys.Rev.* B **38**, 9375 (1988).
3. B.E.Kane et al. *Phys. Rev. Lett.* **59**, 1353 (1987).
4. R.J.Hang et al. *Phys. Rev. Lett.* **61**, 2797 (1988).
5. S.Komiyama et al. *Phys. Rev.* B **40**, 12566 (1989).
6. G.Müller et al. *Phys. Rev. B* **42**, 7633 (1990).
7. P.L.McEuen et al. *Phys. Rev. Lett.* **64**, 2062 (1990).
8. S.Takaoka et al. *Solid State Commun.* **75**, 293 (1990).
9. K.Komiyama and H.Nii, *Physica B* **184**, 7 (1993).
10. S.Takaoka et al. *Physica B* **184**, 21 (1993).
11. K.Oto et al. *Surf. Sci.* **263**, 303 (1992)
12. P.C. van Son et al. *Surf. Sci.* **263**, 284 (1992)
13. P.C.Main et al. *Phys. Rev.* B **50**, 4450 (1994)
14. K.Tsukagoshi et al. *Solid State Commun.* **80**, 797 (1991)

NON LOCAL TRANSPORT IN PBTE WIDE PARAPOLIC QUANTUM WELLS

J. OSWALD, G. HEIGL, G. SPAN, A. HOMER, P. GANITZER
Institute of Physics, University of Leoben, Franz Josef Str. 18
A-8700 Leoben, Austria

D.K. MAUDE, J.C. PORTAL
High Magnetic Field Laboratory, CNRS, 25 Avenue des Martyrs
BP 166 Grenoble, France

The results of non-local experiments in different contact configurations are discussed in terms of a non-local behaviour of the contact arms. It is shown that the observed reproducible fluctuations can be understood to result from fluctuations of a non-local bulk current in the contact arms. The fluctuations are explained by edge channel backscattering because of potential fluctuations in the bulk region.

1 Basic considerations about non-local transport

Basically the presence of edge channels (EC) makes it possible to have transport of current to a remote region. However, a potential difference in a remote region requires dissipation which leads automatically to a potential difference between opposite EC's of the remote path. As a result, a net edge current to the remote region appears, which finally requires an independent return path in order to maintain the current balance. The modelling of an EC-related non-local behaviour must therefore include also an alternative current path in addition to the EC's. This is schematically shown in the equivalent circuit of Fig.1. In the standard Quantum Hall (QH) systems such an alternative path occurs only in the regime between (QH) plateaus[1]. In this regime the states of the upper Landau level (LL) become extended into the bulk region and are almost de-coupled from the EC's of the lower LL's. In this way we have an independent bulk- and EC-system even in the standard single valley situation. However, well developed DOS-gaps between the LL's are required and therefore the observation of non-local EC-transport in standard single valley systems must be closely related to the appearance of the QHE. The presence of the bulk conduction in addition to edge channels leads to different potentials at the contacts (non-zero R_{xx}) which causes also a non-zero bulk current in the voltage probes according to the right part of Fig.1. Consequently this bulk current in the voltage probes leads to a non-local behaviour in the contact arms.

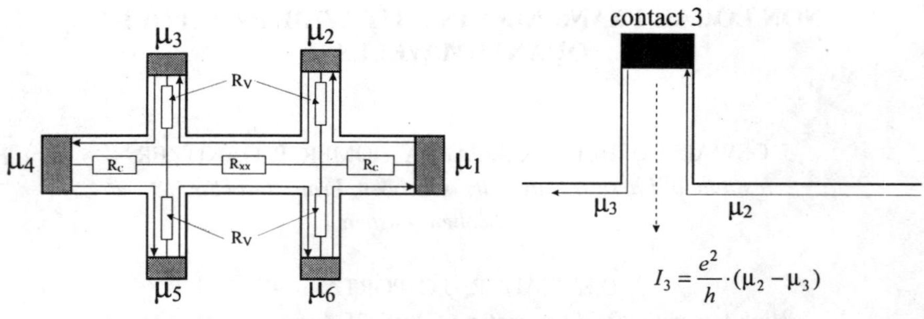

Figure 1 left: Equivalent circuit of a Hall bar sample combining EC- and bulk transport. The EC's are represented by arrows connecting directly the reservoirs of the contacts. The bulk system is represented by the resistor network. **Right:** Situation at contact 3. From the difference of potentials μ_2 and μ_3 a net edge current to the reservoir of contact 3 appears. This causes a bulk current (dashed arrow) leaving contact 3.

An influence of this non-local current on the total sample behaviour can result in two ways. (i) The non-local voltage drop in the contact arms adds directly to the local voltage between the voltage probes. (ii) The current in the contact arms leads to a global current redistribution between the EC-system and the bulk system of the sample which finally causes also a change of the local signals.

2 Wide quantum wells based on PbTe

In quasi-3D wide quantum wells the energy splitting between individual Landau subband levels will be insufficient to create an independent bulk- and EC-conduction within a single electron system. However, the presence of independent electron systems in different valleys of a many valley semiconductor like PbTe can also meet the requirements for non-local transport. One major advantage of PbTe is the possibility to achieve a high electron mobility ($\mu > 10^5$ $cm^2V^{-1}s^{-1}$) even without remote doping. The details about PbTe can be obtained from reference 2. The most important fact to mention here is that we get two sets of subbands and two sets of LL's[3]. For typical sample parameters we get a thickness of the conducting channel of about 330nm and an electron sheet density of about $10^{13}cm^{-2}$. This results in a subband splitting of the order of 1 meV for the small effective mass and about an order of magnitude smaller for the large effective mass. Consequently the small mass can lead already to EC-conduction while the large mass will create mainly a bulk system. In this way we can get a system which also fits to the equivalent circuit in Fig.1. The necessary suppression of intervalley scattering must be considered as an assumption and cannot be discussed within the scope of this paper. It is important to note that a subband splitting of about 1 meV is already of the order of the native potential fluctuations which result to be of the order of 1...2meV[4]. Therefore we can expect that a system of coupled magnetic bound states is created in the bulk region, which can provide an effective mechanism for EC-backscattering in the narrow contact arms of the potential probes.

Conductance fluctuations in PbTe WPQW's in local and non-local contact configuration have been reported for the first time in reference 3 and 5, where also further details about the experiments and the samples can be found. The fluctuation amplitude in the magneto conductance was found to be close to e^2/h although the samples have been of macroscopic size (butterfly Hall bar of width: 100 μm, length between voltage probes: 200 μm, total length between the current contacts: 600μm). In this paper we concentrate on the results of non-local experiments which are shown in Fig.2.

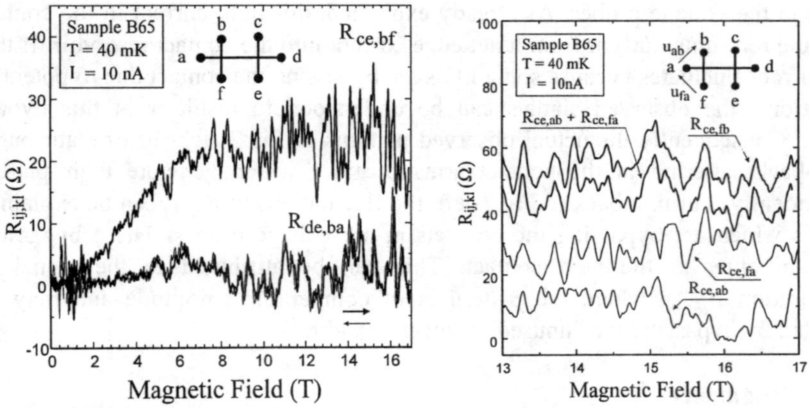

Figure 2 left: non-local magnetotransport data for 2 different contact configurations. The insert shows the contact scheme. **Right:** non-local data from different parts of the sample edge. The traces are shifted vertically for better visibility

Fig.2-left shows the data of two different non-local configurations $R_{ij,kl}$. *ij* denotes the current contacts and *kl* the voltage probes. In configuration $R_{ce,bf}$ the increase of the signal to an average value of about 20 Ω with increasing field indicates that edge state conduction is established at high fields. Because of the bulk conduction between the remote contacts b and f also a considerable amount of dissipation has to occur between these contacts. It is interesting to note that in this configuration there is a direct path along the edge from contact c to b and from e to f. Therefore the contact pair b - f seems to provide the first effective chance for EC's to be scattered to the bulk in order to produce dissipation. In the second configuration $R_{de,ba}$ there is always an "unused" contact between the current and remote voltage contacts along the sample edge. Consequently the EC's have the chance to be scattered to bulk before reaching the non-local voltage probe. Therefore most of the edge current is lost to the bulk before it reaches the voltage probes and therefore there is much less dissipation between the voltage probes in configuration $R_{de,ba}$. The signal below 2 Tesla is non reproducible noise which has no meaning for the interpretation. We have checked configuration $R_{ce,bf}$ again but measuring also the fluctuations between contacts f and a and contacts a and b separately (Fig.2-right). One can clearly see that the sum of $R_{ce,ab}$ and $R_{ce,fa}$ is equal to $R_{ce,fb}$ but the individual fluctuation patterns are significantly different. This strongly indicates that the individual patterns are generated independently. Remembering the general arguments given in section I, it is likely that the non-zero bulk resistance of the contact arms is responsible for the observed signals. This means that the fluctuation pattern can be generated by fluctuations of the bulk current in the voltage probes. As already explained, this bulk current in the contact arm is the returning current from a net edge current into the contact reservoir. If this edge current fluctuates because some EC's are bypassing the contact due to potential fluctuations, the observed signals can be understood to result from this bypass process. Consequently the actual observed patterns of Fig.2-right can be attributed to EC-backscattering in the contact arms b and f which generate their pattern independently. Turning back to Fig.2-left, the fluctuations in $R_{ce,bf}$ can be explained by EC's which are bypassing the contacts at the voltage probes. These bypassing EC's can enter at the next contact. This can be attributed to the signal of configuration $R_{de,ba}$ where the pattern is of comparable amplitude and may be generated by bypassing the "unused" contacts c and f.

Acknowledgments

Financial support by Fonds zur Förderung der wissenschaftlichen Forschung (FWF) Vienna, Project: P10510 NAW.

References

1. P.L. McEuen et al., Phys. Rev. Lett. **64**, 2062 (1990)
2. J.Oswald, P.Pichler, B.B.Goldberg, G.Bauer, Phys. Rev. B **49** 17029 (1994)
3. J. Oswald, et al., Proc. of 11th Int. Conf. on the Electronic Properties of Two Dimensional Systems, Nottingham 1995, to be published in Surf. Sci.
4. Estimate based on: B. I. Shklovskii and A. L. Efros, *Electronic Properties of Doped Semiconductors* (Springer-Verlag, New York, 1984)
5. J. Oswald, et al., Proc. of 3rd Int. Symposium on „New Phenomena in Mesoscopic Structures", Hawaii, Dec. 4-8, 1995, Physica B, in print

EXCITATIONS OF COMPOSITE FERMIONS

J.K. JAIN and R.K. KAMILLA
Department of Physics, State University of New York at Stony Brook, Stony Brook, New York 11794-3800, USA

The incompressible fractional quantum Hall states are understood as filled Landau levels of composite fermions (CFs), and the low-energy excitations as states in which composite fermions are excited to higher CF Landau levels. This understanding provides new information about the dispersion of the low energy neutral excitations, stability of the fractional quantum Hall states and the gap to charged excitations, reviewed in this article.

1 INTRODUCTION

Not long after the discovery of the fractional quantum Hall effect (FQHE) [1], Laughlin [2] proposed the following wave function for the ground state at $f = \frac{1}{(2m+1)}$:

$$\psi_{1/(2m+1)} = \prod_{j<k}(z_j - z_k)^{2m+1} \exp[-\frac{1}{4}\sum_i |z_i|^2] \tag{1}$$

where $z_j = x_j + iy_j$ denotes the coordinates $(x_j\ ,y_j)$ of the jth electron, and the lengths are expressed in unit of the magnetic length $l_0 = \sqrt{\frac{\hbar c}{eB}}$. This wave function was studied for small systems and found to be quite accurate. Laughlin also wrote trial wave functions for the charged excitations (quasielectrons and quasiholes, collectively referred to as quasiparticles), which provided a reasonable estimate of the gap relevant to transport measurements, called the transport gap below. The main source of our knowledge about the transport gaps of other FQHE states was from finite size exact diagonalization studies. These are, however, limited to very few particles, typically less than ten. Only the gap of the 2/5 state was computed with good accuracy [3], but those of other states were not known. Several experimentally observed states, e.g. the ones at 4/9, 5/11 etc. are not accessible at all in the exact diagonalization studies.

Girvin, MacDonald and Platzman [4] (GMP) employed a single mode approximation (SMA) to investigate the neutral excitation of the Laughlin state, called the collective mode or exciton. Comparison with exact results in finite system studies found the SMA to work well in an intermediate range of wave vectors near a minimum in the dispersion, termed the "roton" minimum, which stimulated experimental studies [5,6]. At wave vectors beyond the minimum, however, the SMA was not satisfactory. For large wave vector k, the energy of the true mode should be roughly independent of k (equal to the transport gap), since it contains a far separated quasielectron-quasihole pair; the energy of the SMA wave function, however, continues to grow with k. Much less was known about the collective modes of the other FQHE states. The SMA was tested in finite system studies, but did not give a satisfactory description of the low energy neutral excitations [7,8].

In the last few years, significant progress has been made on the experimental front. The transport gaps have been measured in a number of studies [9]. Further, Raman scattering [5] and phonon absorption [6] experiments have reported observation of the neutral modes in the FQHE regime. Substantial progress has also been made recently in our theoretical understanding of the low energy excitations using the framework of the composite fermion theory. This article will provide a brief review of some recent results.

2 Composite Fermion Theory

The composite fermion (CF) theory [10] describes all FQHE on an equal footing by mapping them on to IQHE of new particles called composite fermions. The basic principle of the CF theory is that, in a range of filling factors, the electrons in the lowest Landau level (LL) find it energetically favorable to capture an even number of vortices of the many particle wave function. The bound state of an electron and the vortices behaves as a particle, called composite fermion. The vortices produce phases as the composite fermions move around, which partly cancel the Aharonov-Bohm phases originating from the external magnetic field, and, as a result, the composite fermions experience an effective magnetic field given by

$$B^* = B - 2m\rho\phi_0 \, , \tag{2}$$

where B is the external field, $2m$ is the number of vortices carried by composite fermions, $\phi_0 = hc/e$ is the flux quantum, and ρ is the electron (or CF) density. The residual interaction between the composite fermions is weak, and the strongly correlated liquid of electrons maps into a weakly interacting gas of composite fermions. An effective single-particle description of the electron state thus becomes possible in terms of composite fermions. The energy levels of composite fermions are analogous to the LLs of non-interacting electrons in this *weaker* magnetic field, and are called CF-LLs. Defining the CF filling factor as $\nu^* = \rho\phi_0/B^*$, in analogy to the electron filling factor $\nu = \rho\phi_0/B$, the above equation can also be expressed as

$$\nu = \frac{\nu^*}{(2m\nu^* + 1)} \ . \tag{3}$$

The IQHE of composite fermions at $\nu^* = n$ manifests as the FQHE of electrons at $\nu = \dfrac{n}{(2mn+1)}$.

The trial wave functions for the CF states are constructed as follows. Let us denote the ground state of non-interacting electrons at $\nu^* = n$ by Φ_n. The wave function of n filled CF-LLs is obtained by attaching $2m$ vortices to each electron in the state Φ_n, which amounts to a multiplication by the Jastrow factor

$$D^m \equiv \prod_{j<k} (z_j - z_k)^{2m} \ , \tag{4}$$

which produces

$$\Phi_n^{UP-CF} = \prod_{j<k} (z_j - z_k)^{2m} \Phi_n \ . \tag{5}$$

This wave function is not strictly in the lowest LL and must be projected on to the lowest LL (LLL) to obtain a wave function appropriate to the limit $B \to \infty$:

$$\Phi_n^{CF} = \mathcal{P} \Phi_n^{UP-CF} \ . \tag{6}$$

Here $\mathcal{P}$ is the projection operator. Φ_n^{CF} describes the electron ground state at $\nu = \dfrac{n}{(2mn+1)}$. It has been tested for several FQHE states in small system exact calculations and found to be extremely accurate [11,12].

The analogy between the IQHE and the FQHE states can also be used to write wave functions for the excitations. For the n filled LL state at $\nu^* = n$, the low energy excitations are obtained by exciting one electron from the topmost filled nth LL to the lowest unoccupied $(n+1)$st LL. In other words, it is an exciton. By analogy, the low energy excitations of the FQHE states, which are filled CF LLs, are excitons of composite fermions. They contain a single excited composite fermion in the $(n+1)$st CF-LL and a hole left behind in the nth CF-LL, as shown schematically in Fig. 1. Their wave functions are obtained from the wave functions of the excitons of the $\nu = n$ IQHE state precisely in the same manner as the FQHE ground state is obtained from the $\nu = n$ IQHE ground state, i.e., by multiplication by the Jastrow factor, followed by LLL projection.

We work in the spherical geometry [13,14] where N electrons move on the surface of a sphere under the influence of a radial magnetic field. The flux

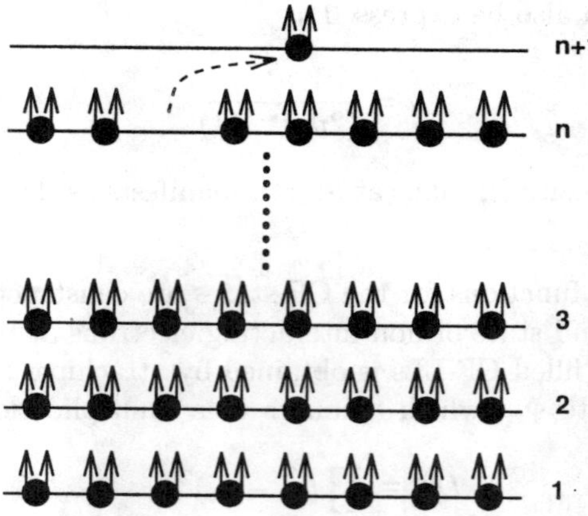

Figure 1: Schematic view of a CF exciton. Each filled circle with two arrows depicts a composite fermion as an electron carrying two flux quanta. All composite fermions on a given line are in the same CF Landau level. A CF exciton of the n filled CF-LL state [which corresponds to $\nu = n/(2n+1)$ FQHE state of electrons] is obtained by promoting a single composite fermion from the topmost occupied [nth] CF Landau level to the lowest unoccupied [$(n+1)$th] CF Landau level.

through the surface is $2q\phi_0$, where $2q$ is an integer. The orbital angular momentum L is a good quantum number, and is related to the wave vector of the planar geometry by $kl_0 = \frac{L}{\sqrt{q}}$ [14]. For each L, the wave function of the electron exciton at $\nu^* = n$ [and consequently also of the CF exciton at $\nu = n/(2mn \pm 1)$] is uniquely determined, i.e., contains no adjustable parameters.

We give below only the results of our study and refer the reader to literature for details of our exact diagonalization and Monte Carlo calculations and also of how the projection is carried out. A perfect two dimensional system is assumed in the calculations, and the magnetic field is taken to be sufficiently large that the Hilbert space can be restricted to the lowest LL and the electrons can be taken to be fully polarized.

3 Comparison with exact results

First we compare the predictions of the CF theory with exact results which can be obtained for small systems by exact numerical diagonalization of the Hamiltonian in the LLL subspace. We have computed the exciton energy using the CF wave function, denoted by ΔV_{CF}, for finite systems [11,12,15]. (All energies will be quoted in units of $e^2/\epsilon l_0$, where ϵ is the background dielectric

Table 1: The exact gap compared to the CF prediction for several systems.

ν	N	CF gap	exact gap
$\frac{2}{5}$	6	0.07615(61)	0.07505
	8	0.07021(87)	0.06809
	10	0.0681(12)	0.0673
$\frac{3}{7}$	9	0.0691(14)	0.0681
	12	0.0518(12)	0.0525

constant.) It is measured relative to the CF ground state energy. Fig. 2 gives the exact Coulomb energies, ΔV_{ex}, along with the energies of the CF wave functions. The range of L over which the exciton extends in the exact diagonalization studies agrees with that predicted by the CF theory. Moreover, ΔV_{CF} is within a few percent of the exact energy ΔV_{ex}.

Table I gives the exact transport gaps as well as those computed from the CF theory [16]. A comparison again shows a similar level of accuracy.

4 Large system studies

We have recently developed projection techniques that allow us to study large systems [15,16]. We apply these to obtain thermodynamic limits for the lowest energy of the exciton and the transport gaps.

First we compute the energy of the CF exciton of the $\frac{1}{(2m+1)}$ state by generalizing a method developed by Bonesteel [17]. This method relies on the special feature of the unprojected CF wave function that it contains no more than one electron in the second LL, and none at all in the higher LLs. The projected CF wave function can then be written as

$$\chi^{CF} \propto (K - \hbar\omega_c)\chi^{UP-CF} \tag{7}$$

where K is the kinetic energy operator and $\hbar\omega_c$ is the cyclotron energy. The energy of this wave function is evaluated by Monte Carlo. The CF-exciton energy is shown as a function of L for a 20 electron system in Fig. 3. There is a deep minimum at the expected position of $kl_0 \approx 1.4$. The CF theory also identifies additional minima in the exciton dispersion at $kl_0 \approx 2.7$ and $kl_0 \approx 3.5$. The thermodynamic limit of the energy at the minimum is estimated to be $0.063(3)e^2/\epsilon l_0$ [Fig. 4].

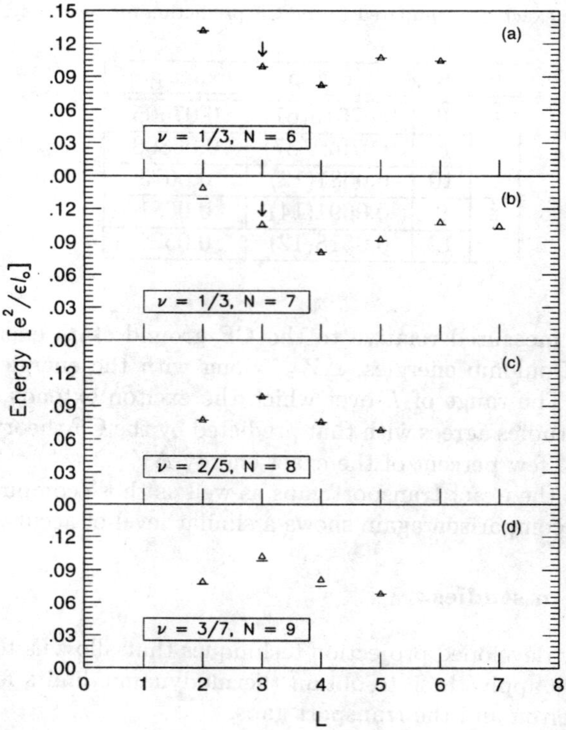

Figure 2: The exact energies ΔV_{ex} (dashes) and projected CF energies ΔV_p (triangles) are shown for the exciton at $\nu = 1/3$, 2/5 and 3/7. The energy of the ground state at $L = 0$ has been set to zero.

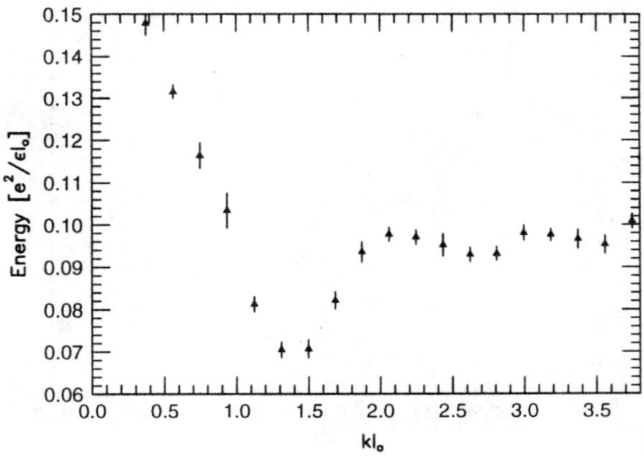

Figure 3: ΔV_{CF} for 1/3 CF state with $N = 20$. The statistical errors of Monte-Carlo estimation of the energies are shown on each point.

We have done a similar study for $\nu = 1/5$, $1/7$, and $1/9$. Interestingly, for $\nu = 1/9$ the energy of the CF exciton is negative, as shown in Fig. 5. This implies that the Laughlin state is unstable to spontaneous creation of CF excitons. As a consequence, FQHE at 1/9 is ruled out. The wave vector of the exciton where the instability takes place is close to the reciprocal lattice vector of the Wigner crystal, suggesting that the instability is toward the formation of a Wigner crystal. At $\nu = 1/5$ and $1/7$, the minimum energy of the CF exciton is positive, although it is rather small at $\nu = 1/7$. Table II lists the minimum energies at all these filling factors.

Earlier studies have compared the Laughlin wave function to a trial wave function for a Wigner crystal and concluded that the latter has lower energy for $\nu < 1/(6.5 \pm 0.5)$ [18]. However, this is the first time that such an instability has been found within the framework of a single theory.

We have also estimated the thermodynamic values of the transport gaps for several FQHE states using a more general projection technuque [16], which we list in Table III. Systems of up to 50 composite fermions were used in this study.

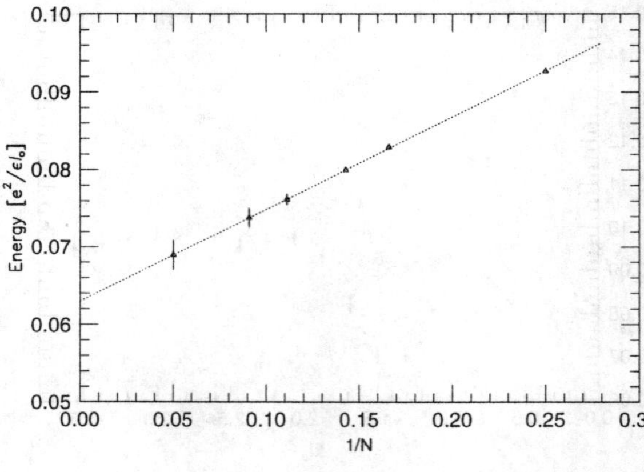

Figure 4: Estimation of ΔV_{CF} at the minimum of the exciton dispersion at $\nu = 1/3$. For each value of N, the minimum was determined by smooth polynomial interpolation in the neighborhood of the minimum to reduce the sensitivity of our thermodynamic estimation to the discreteness of L for finite systems. A linear χ^2-fit gives a thermodynamic estimate of $0.063\ e^2/\epsilon l_0$, with an uncertainty of 2 (3) in the last digit if four (three) largest systems are used.

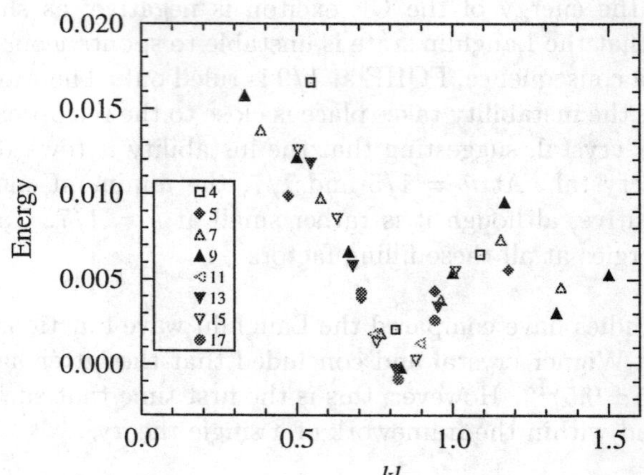

Figure 5: Exciton dispersion for $\nu = 1/9$. The exciton energy is plotted for several different N as a function of the wave vector k.

Table 2: The minimum gap to neutral excitation for several filling factors. The SMA prediction is also given for comparison.

ν	CF gap	SMA gap
$\frac{1}{3}$	0.063(3)	0.078
$\frac{1}{5}$	0.0095(6)	0.017
$\frac{1}{7}$	0.0009(5)	0.0063
$\frac{1}{9}$	< 0	0.0032

Table 3: The transport gaps for several FQHE States.

ν	CF gap
$\frac{1}{3}$	0.105(3)
$\frac{2}{5}$	0.058(5)
$\frac{3}{7}$	0.047(4)
$\frac{4}{9}$	0.035(6)
$\frac{5}{11}$	0.021(9)

5 Conclusion

In short, the following unified picture becomes possible in the CF framework: the incompressible ground states correspond to an integer number of filled CF-LLs and the excited states are obtained by exciting composite fermions to higher CF-LLs. Not surprisingly, the same physics provides the key to understanding both the ground state and its excitations.

These studies establish that the CF theory not only provides a very simple qualitative understanding of the FQHE but also gives a very accurate quantitative description: its predictions are reliable to within a few percent. This is especially impressive considering that the theory does not contain any adjustable parameters.

The dispersion of the neutral mode has also been calculated in a Chern-Simons (CS) field theoretical scheme in several recent papers. Here, the composite fermions are modeled as electrons carrying (gauge) flux quanta [10,19], which simulate vortices. The qualitative physical picture of composite fermions is obtained at a mean-field level, and a perturbation theory is carried out at the level of random-phase-approximation (RPA) [20]. Although the trial-wave-function and the CS schemes are both attempting to describe the same physics from somewhat different starting points, a precise relationship between them is not known at the moment. There is qualitative agreement between the results of our calculations and those of the Chern-Simons theory in that both obtain flat dispersion at large k and finite energy in the limit $k \to 0$.

Several important effects left out in the above study. Modification in the Coulomb interaction due to the finite width of the quantum well[21], LL mixing [22], and disorder are all known to change the numerical values of the excitation energies and must be incorporated before a comparison with experiment may be made. All these effects reduce the excitation energy and our estimates provide the upper limits.

Acknowledgments

This work was supported in part by the National Science Foundation under Grant no. DMR93-18739 and by a fellowship from the John Simon Guggenheim Foundation. We also acknowledge grant of computing time on the SGI Power Challenge cluster at the NCSA, University of Illinois, Urbaba-Champaign.

References

1. D.C. Tsui, H.L. Stormer and A.C. Gossard, Phys. Rev. Lett. **48**, 1559 (1982).
2. R.B. Laughlin, Phys. Rev. Lett. **50**, 1395 (1983).
3. N. d'Ambrumenil and R. Morf, Phys. Rev. B **40**, 6108 (1989).
4. S.M. Girvin, A.H. MacDonald and P.M. Platzman, Phys. Rev. Lett. **54** 581, 1985; Phys. Rev. B **33**, 2481 (1986).
5. A. Pinczuk *et al.*, Phys. Rev. Lett. **70** 3983, 1993; Semiconductor Science and Technology, vol. **9**, 1865 (1994).
6. C.J. Mellor *et al.*, Phys. Rev. Lett. **74**, 2339 (1995).
7. W.P. Su and Y.K. Wu, Phys. Rev. B **36**, 7565 (1987).
8. S. He, S.H. Simon and B.I. Halperin, Phys. Rev. B **50**, 1823 (1994).
9. R.R. Du *et al.*, Phys. Rev. Lett. **70**, 2944 (1993); H.C. Manoharan, M. Shayegan and S.J. Klepper, Phys. Rev. Lett. **73**, 3270 (1994).
10. J.K. Jain, Phys. Rev. Lett. **63**, 199 (1989); Phys. Rev. B **41**, 7653 (1990); Science **266**, 1199 (1994).
11. G. Dev and J.K. Jain, Phys. Rev. Lett. **69**, 2843 (1992).
12. X.G. Wu and J.K. Jain, Phys. Rev. B **51**, 1752 (1995).
13. T.T. Wu and C.N. Yang, Nucl. Phys. B **107**, 365 (1976); T.T. Wu and C.N. Yang, Phys. Rev. D **16**, 1018 (1977).

14. See, F.D.M. Haldane in *The Quantum Hall Effect*, edited by R.E. Prange and S.M. Girvin (Springer-Verlag, NY 1990); G. Fano, F. Ortolani and E. Colombo, Phys. Rev. B **34**, 2670 (1986).
15. R.K. Kamilla, X.G. Wu, and J.K. Jain, Phys. Rev. Lett. **76**, 1332 (1996); Phys. Rev. **B** 54, 4873 (1996).
16. J.K. Jain and R.K. Kamilla, SISSA preprint cond-mat/9607147.
17. N.E. Bonesteel, Phys. Rev. B **51**, 9917 (1995).
18. P.K. Lam and S.M. Girvin, Phys. Rev. B **30**, 473 (1984).
19. A. Lopez and E. Fradkin, Phys. Rev. B **44**, 5246 (1991); B.I. Halperin, P.A. Lee and N. Read, Phys. Rev. **B** 47, 7312 (1993).
20. A. Lopez and E. Fradkin, Phys. Rev. B **47**, 7080 (1993); S.H. Simon and B.I. Halperin, Phys. Rev. B **48**, 17368 (1993); *ibid*, **50**,1807 (1994); S. He, S.H. Simon and B.I. Halperin, Phys. Rev. B **50**, 1823 (1994); X.C. Xie, *ibid.*, **49**, 16833 (1994); L. Zhang, *ibid.*, **51**, 4645 (1995).
21. F.C. Zhang and S. Das Sarma, Phys. Rev. B **33**, 2903 (1986); D. Yoshioka, J. Phys. Soc. Jpn. **55**, 885 (1986).
22. D. Yoshioka, J. Phys. Soc. Jpn. **53**, 3740 (1984); X. Zhu and S.G. Louie, Phys. Rev. Lett. **70**, 339 (1993) V. Melik-Alaverdian and N.E. Bonesteel, Phys. Rev. B **52**, R17032 (1996); Rodney Price and S. Das Sarma, Phys. Rev. B (in press).

COMPOSITE FERMIONS IN EXPERIMENT

H. L. STORMER
Bell Laboratories, Lucent Technologies
Murray Hill, N.J. 07974 U.S.A.

A brief summary of experimental data on composite fermions is presented. Measurements of fermi-wave vector, effective mass and g-factor are summarized in a cursory fashion. While the composite fermion model is very successful in interpreting qualitatively and in some cases quantitatively the experimental results, there remain many mysteries and many opportunities for experiment and theory alike.

1. Introduction

Theoretical work of the past years[1-3] has cast a new light on the fractional quantum Hall effect(FQHE) and offers a fresh view from a different angle on this intriguing electronic state. In this model electron-electron interaction forces an even number of flux quanta onto each electron in a high magnetic field. The resulting new particles, consisting of electron and flux quanta, are appropriately termed composite fermions(CFs). In the presence of a residual magnetic field (deviation from an exactly even number of flux quanta per electron), the energy spectrum of the composites develops "Landau levels". The liquids of the FQHE then coincide with exact integer filling of these "Landau levels" of CFs. Recent theoretical studies by Halperin, Lee and Read[3] associate a definite mass (about ten times the GaAs mass) with the composites and describe the states at even-denominator filling as being a Fermi sea of CFs with a well-defined Fermi wavevector, in many respects equivalent to the state at *zero magnetic field.*

This contribution briefly reviews several of the experimental data which support the existence of such bizarre new particles. The overview is far from comprehensive and the reader is referred to other articles in this volume for further reference.

2. Fermi wave vector of CFs.

The fermi wave vector, k_F, of the CF system has been observed in several experiments.[4-9] They all rely on a resonance between an imposed length L and the classical cyclotron radius r_c of the particle. Since $r_c = m^* v_f/eB = \hbar k_F/eB$ (m^* = effective mass, v_f = fermi velocity), a measurement of r_c provides a measure of k_F.

The characteristic length L is either provided by the wave length of a surface acoustic wave (SAW)[4-6], by the periodicity of an anti-dot superlattice[7], by the distance between quantum point contacts[8], or by the spacing in an array of cavities[9].

Such experiments have been performed on CFs around $\nu = 1/2$ on several samples. For CFs the cyclotron radius is $r_c^{CF} = \hbar k_F^{CF}/eB_{eff}$, where the effective magnetic field $B_{eff} = B - B_{1/2}$, with $B_{1/2}$ being the magnetic field at exactly $\nu = 1/2$. As an example we present the results from anti-dot superlattice resonances.[7]

Anti-dot superlattices consist of periodic arrays of holes etched into the 2DEG. In this case the classical cyclotron orbit of the particle resonates with the period of the superlattice. Depending on the ratio of dot size to dot spacing the electronic transport through the specimen shows resonances when the particle orbits around 1, 4, 9 or 21 of the dots. Such oscillations are clearly observed for electrons around $B = 0$ and well understood in semiclassical terms.[10]

Fig. 1 shows the magneto-resistance R_{xx} of a $d = 600nm$ period anti-dot superlattice in comparison with R_{xx} of the unprocessed bulk part of the sample, devoid of dots. A striking difference between both specimens is apparent in R_{xx} near $B = 0$, $\nu = 3/2$ and $\nu = 1/2$. The bulk sample clearly exhibits local minima at these field positions, whereas the resistance in the anti-dot trace shows overall maxima with peaks of varying strength superimposed at the same fields. The features around $B = 0$ are clearly identifiable as the well-known dimensional resonances of the electrons. The peaks around $\nu = 1/2$ are the sought-after dimensional resonances of composite fermions. They provide clear indication for the existence of a dominant wave vector k_F^{CF}. A direct comparison between peak position of the electron resonance with the CF resonance reveals a factor of $\sqrt{2}$ difference in their splitting.

The resonance position in B_{eff} implies $k_F^{CF} = \sqrt{2}\, k_F^e$ where $k_F^e = \sqrt{2\pi n_e}$ is the fermi wave vector of the electron system at $B = 0$. The derived value for $k_F^{CF} = \sqrt{2}\, k_F^e$ agrees with the theoretically expected value at $\nu = 1/2$. At this filling factor the density of CFs, n_{CF} equals the electron density, n_e. However, the high external (real) magnetic field, B, lifts the spin degeneracy of the CFs and, hence, their k_F grows by a factor $\sqrt{2}$, as observed. Such results are widely taken as support for the existence of a fermi surface of CFs.

Different from the electron features around $B = 0$, the CF features are not symmetric around $B_{eff} = 0$. The lower B-field ($B_{eff} < 0$) feature always exceeds the higher B-field ($B_{eff} > 0$) feature. This asymmetry is now attributed to the gradual depletion of the density around the anti-dots generating a varying B_{eff} around the anti-dots seen by the CFs and breaking the symmetry between clockwise

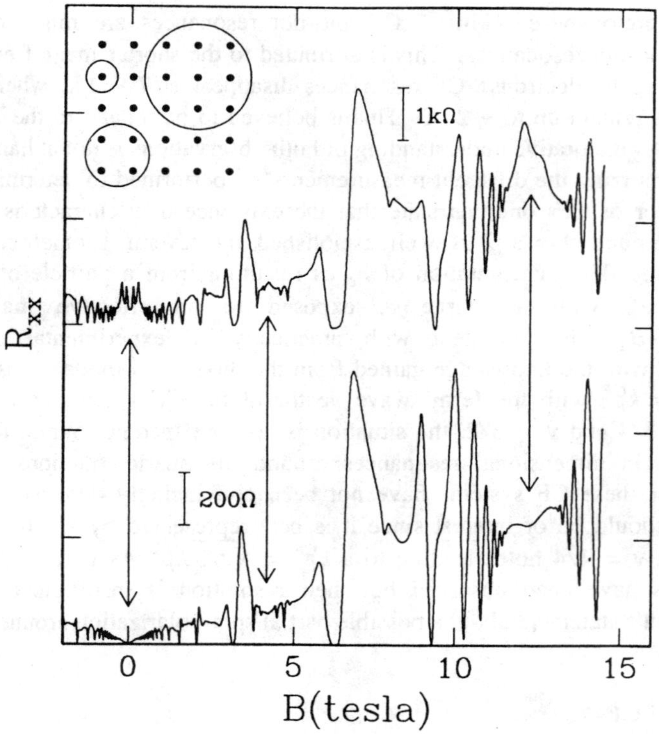

1. Comparison of the diagonal resistance R_{xx} of a bulk two-dimensional electron gas (lower trace) and R_{xx} of the d = 600 nm period anti-dot superlattice (upper trace) at $T \cong 300mK$. The fractions near the top of the figure indicate the Landau level filling factor. Inset: Schematic of commensurate orbits encircling s = 1, 4, and 9 anti-dots. Ref. 7

and counterclockwise orbit.[9,11] CF anti-dot resonances are much broader than electron anti-dot resonances. This is attributed to the shorter mean free path of CF as compared to electrons. CF resonances disappear at $T \sim 1K$, whereas electron resonances remain up to $\sim 20K$. This is believed to be related to the "stability" of the CFs. A quantitative understanding of both observations is not at hand.

Summarizing the different measurements[4-9] performed to determine the fermi wave vector of CFs one can state that the existence of a characteristic cyclotron orbit r_c around $\nu = 1/2$ is well established by several geometrical resonance experiments. The interpretation of r_c as resulting from a particle of momentum $k_F^{CF} = \sqrt{2}\,k_F^e$ with a charge e, exposed to an effective magnetic field $B_{eff} = B - B_{1/2}$ is consistent with practically all experimental observations. Combined with the knowledge gained from the theoretical models it is appropriate to identify k_F^{CF} with the fermi wave vector of the CF system at $\nu = 1/2$. For $\nu = 1/4$, $3/4$ and $\nu = 3/2$, the situation is less transparent. While features have been seen in dimensional resonances around the quarter-fractions, fermi wave vectors for these CF systems have not been convincingly deduced. Particularly $\nu = 3/4$ should be of interest since it is best represented by a full Landau level filled with $\nu = 1/4$ holes leading to a $k_F^{CF} = \sqrt{2/3}\,k_F^e$. As to $\nu = 3/2$, resolved resonances have been observed but their resolution is insufficient to make an unambiguous statement about a possible partial spin-polarization around $\nu = 3/2$.

3. Mass of CFs

In this section we limit ourselves to mass determination in a non-zero B field measuring an energy gap E_g and deducing m^* from $E_g = \hbar eB/m^*$.

Since in the CF model the features of the FQHE result from the quantization of CFs into Landau level driven by B_{eff}, measurements of E_g are synonymous with the measurement of the gaps of the FQHE. There are at least three methods to measure E_g. The most powerful is cyclotron resonance (CR). CFs are not expected to show any features in a CR experiment. Kohn's Theorem applies, which states that in a translationally invariant system, electron-electron interactions do not contribute to CR. Instead the bare carrier mass (i.e. the electron band mass) is observed. CFs, being purely of many particle origin, should have no signature in CR. Of course, any residual disorder breaks translation invariance and may provide a coupling mechanism between CR radiation and the CF cyclotron motion. So far, CR experiments have not been successful in the case of CFs. Instead, mass determination relies on other methods such as measurement of the activation energies[12-14], the Shubnikov-deHaas effect[13-17] and optical recombination measurements.[18]

As an example for such measurements we discuss briefly measurements of the activation energy.

Activation energy measurements determine the magneto-resistance R_{xx} at the minima of the FQHE as a function of temperature (T) and fit this dependence to an Arrhenius plot with activation energy E_a which is taken to be the energy gap. E_a is usually not identical to the desired E_g. A finite width Γ of the Landau levels due to carrier lifetime effects or inhomogeneous broadening often reduces the gap. Fig. 2 shows characteristic experimental results.

Activation energy measurement in the FQHE states around $\nu = 1/2$ arrive at gaps that are opening roughly linearly with increasing B_{eff} as expected from a charged particle with constant mass in the presence of a magnetic field.[12-14] This holds for positive as well as negative B_{eff} and has been measured in 2DES as well as 2D hole systems (2DHS). Practically all of the work has been performed around $\nu = 1/2$ for electron densities $n = 1-2 \times 10^{11}\, cm^2$. The so-derived masses are roughly $m^*_{CF} \sim 0.5-1m_o$ about 10 to 15 times higher than the electron band mass in GaAs, $m_b \sim 0.07m_o$. The masses, of purely many-particle origin, are consistent with theoretical estimates. It remains questionable whether the theoretically expected $\sqrt{B}$-dependence of m^*_{CF} has been observed directly in experiment (dashed line in Fig. 2). The error bars on E_a are considerable. However, an overall $\sqrt{B}$-dependence resolves the otherwise puzzling experimental observation[12] of CFs having slightly different masses for $B_{eff} > 0$ and $B_{eff} < 0$.

In analogy to the electron case around $B = 0$ the negative intercept is attributed by experimentalists to the lifetime broadening of the CF Landau levels. The deduced relaxation times are comparable to the transport scattering times derived from the mobility of CFs at $\nu = 1/2$. However, at present we have no theoretical model for CFs in the presence of disorder which would allow us to interpret the negative intercept.

Shubnikov-deHaas(SdH) measurements are another powerful tool to determine mass and scattering time in traditional metallic systems.[13-17] Since the behavior of CF around $\nu = 1/2$ resembles the behavior of regular electrons around $B = 0$, this tool has been applied to the CF system.

In general the FQHE features around $\nu = 1/2$ are well described in terms of SdH oscillations although there exist differences between experimental groups. Deduced masses are in rough agreement with masses determined from activation energy measurements as long as B_{eff} is large, i.e. as long as one is dealing with the lowest 3-4 CF Landau levels. However, even here there exists dissent. From measurement on samples varying in density by as much as a factor of 8, one group[16] derives a universal relationship for the CF mass: $m^*_{CF} = 0.51 + 0.83\ B_{eff}$. In this case, m^*_{CF} appears to be density-independent and there exists a linear dependence of B_{eff}, taken as a "non-parabolicity" of the CF mass.

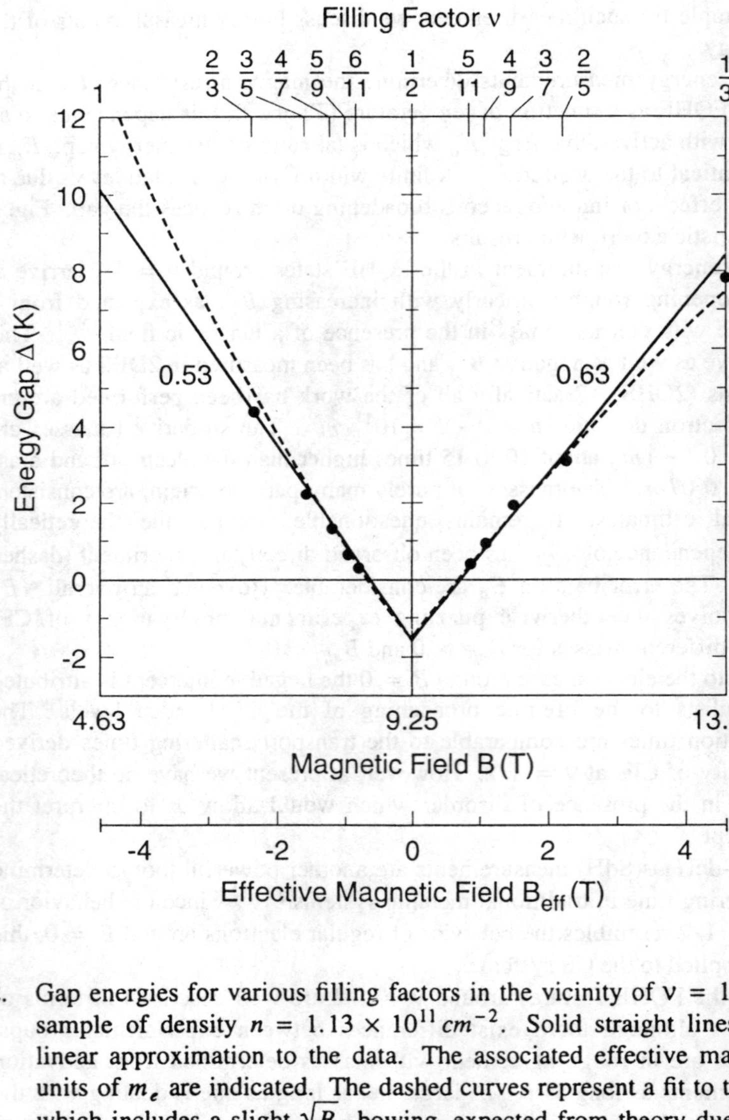

2. Gap energies for various filling factors in the vicinity of $\nu = 1/2$ in a sample of density $n = 1.13 \times 10^{11}\ cm^{-2}$. Solid straight lines are a linear approximation to the data. The associated effective masses in units of m_e are indicated. The dashed curves represent a fit to the data which includes a slight $\sqrt{B}$- bowing, expected from theory due to the requirement for electron-hole symmetry. In this case the mass at $\nu = 1/2$ is unique, $m^* = 0.57 m_e$. Ref. 14

This is astounding since m^*_{CF} for very fundamental reasons is expected to scale as $\sqrt{B}$. Being of many-particle origin the energy gap must scale[3] as $e^2/\ell_o \propto B^{-1/2}$ and, hence, $m^*_{CF} \propto \sqrt{B}$ (magnetic length $\ell_o = \sqrt{\hbar/(eB)}$). On the other hand, the $\sqrt{B}$ dependence may be modified by higher order effects such as CF Landau level interactions. At present, there is no theoretical interpretation for a linear B_{eff}-dependence of the mass. Neither seems there to be a way how the experimental data, taken over such a wide range of densities, could be consolidated with the standard CF model. Other groups do report the anticipated $\sqrt{B}$-dependence of m^*_{CF} although the range of densities over which this dependence is measured remains narrow.[13-14] On the other hand, since their sample quality allows for the measurement closer to $\nu = 1/2$, they observe a strong divergence of the SdH mass as $B_{eff} \rightarrow 0$. Activation energy measurement cannot access this region. Theoretically, due to fluctuation of the gauge field associated with CFs, their mass is expected to diverge logarithmically [3,19-21] as $B_{eff} \rightarrow 0$. The experimental divergence is much stronger, following a $\sim B_{eff}^{-4}$ or $\sim \exp(1/B_{eff})$ dependence.

In summary one may say that the determination of the CF mass remains controversial. Over the range of typical electron densities ($\sim 1 - 2 \times 10^{11}\, cm^{-2}$) masses derived away from $B_{eff} \sim 0$ are roughly $m^*_{CF} \sim 1 m_o$. Their experimentally determined B_{eff} and/or n-dependence are disputed. The mass divergence around $B_{eff} \sim 0$ seen in SdH results is not understood theoretically. Finally, almost all of the work concentrates on $\nu = 1/2$. Masses around quarter-filling fractions are hardly explored and much less understood.

4. g-factor of CFs

In trying to elucidate the attributes of CFs there arises the question as to its spin and g-factor and how such a spin would affect the CF level scheme. While the spin does not enter in the CF model for the states at Landau level filling factor $\nu = p/(2p \pm 1)$ around $\nu = 1/2$ it is known that spin effects occur in experiment around $\nu = 3/2$. Early observations date back to times before the CF model.[22,23] While the influence of the spin was deduced from such experiments there did not emerge a coherent picture. The following section will show that the FQHE states around $\nu = 3/2$ can be described in terms of CFs with a spin.[24,25]

Fig. 3 shows angular dependent magnetoresistance data at 50mK around $\nu = 3/2$ as a function of perpendicular magnetic field $B_\perp = B_{tot} \cos\theta$. The variation of the oscillations with angle is very complex. FQHE minima disappear and reappear in an opaque manner. However, a plot of the values of B_{tot} and B_{eff} at which the various FQHE states disappear results in remarkably simple diagrams.

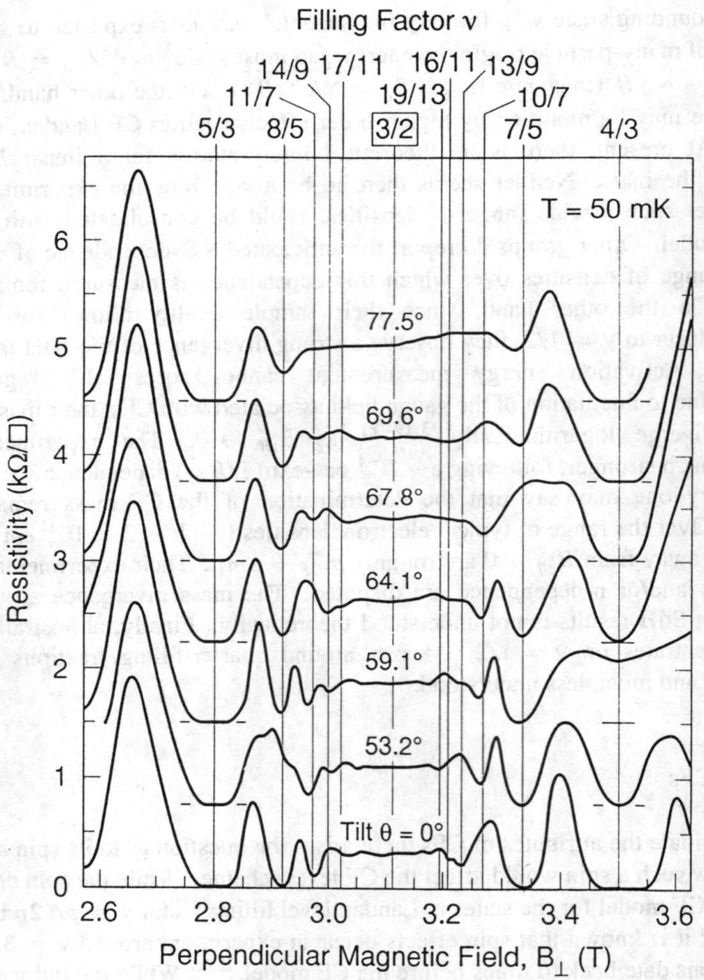

3. Magnetoresistance of a high mobility, two-dimensional electron system at 50mK around ν = 3/2 Landau level filling factor at various angles of tilt, θ, as a function of perpendicular field $B_{\perp} = B\cos\theta$. Ref. 24

Fig. 4 shows the position of all maxima (full symbol) and shoulders (open symbol) that appear at particular FQHE states in a $B_{tot} - B_{eff}$ plane. B_{eff} is the *effective* magnetic field experienced by the CF which corresponds to the external magnetic field *reduced* by the magnetic flux carried by the CFs. Following a simple counting scheme[3], it turns out to be $B_{eff} = 3(B_{\perp} - B_{\perp 3/2})$, where $B_{\perp 3/2}$ represents the perpendicular external B-field required to reach $\nu = 3/2$. Correspondingly, $\nu = 3/2$ resides at the origin of the graph. All data points lie on a fan emanating from the origin. Each line represents a fixed ratio of B_{tot}/B_{eff} and their slope follow closely a relationship of 1:2:3:4... Since B_{eff} determines the Landau level splitting for CFs ($\Delta E_{LL} = \hbar e B_{eff}/m^*$) and B_{tot} is expected to be related to the spin splitting of CFs ($\Delta E_{ss} = g^* \mu_B B_{tot}$), we may identify the lines B_{tot}/B_{eff} = const as the position of all points at which the Zeeman energy of CFs equals a multiple of their CF-cyclotron energy, $\Delta E_{ss} = j \Delta E_{LL}$. The effective mass of the CF is identified as m^* and g^* is its effective g-factor.

How do such degeneracies affect the FQHE? According to the CF-model[1-3] at any given FQHE state CFs fill an exact multiple of CF-Landau levels. The CF fermi level resides in the gap and the characteristic FQHE transport minimum occurs. If one includes the spin in such a formulation, then each Landau level is split into a spin-up and into a spin-down level similar to the electron case. However, different from the electron case this spin splitting exists already at $B_{eff} = 0$ since it is determined by B_{tot}. At specific angles, when a spin level of a CF-Landau level coincides with a spin level of another CF-Landau level, what was formerly a gap disappears, leading to the disappearance of the minimum in R_{xx}. The best way to look at the CF model around $\nu = 3/2$ is to imagine it consisting of two Landau fans both with their origin at $\nu = 3/2$: one fan describes spin-down and the other fan, shifted up by $g^* \mu_B B_{tot}$, describes spin-up. As a function of angle, both fans are shifting vertically with respect to each other, causing multiple crossings and degeneracies of CF Landau levels which are observed in the transport experiments.

The level schemes shown in Fig. 4 represent spin and Landau splitting at particular angles and FQHE states. This extraordinary simple picture explains all of the complex FQHE features in Fig. 3. From Fig. 4 and knowing the CF mass from independent measurements,[24] we can determine the g-factor of the CF. In a region around $\nu = 3/2$ it can be described as $g^* = 0.61 + 0.083(T^{-1}) B_{eff}$. The g-factor in this region does not differ very much from the bare g-factor of GaAs $g^* \approx 0.44$. It is surprising that such a simple overall picture holds for features of the FQHE around $\nu = 3/2$. Why is the CF spin not observed around $\nu = 1/2$? At typical electron densities the $\nu = 1/2$ state occurs at very high magnetic field and the CF spin splitting by far exceeds the CF Landau level splitting. Therefore only the spin-down CF Landau fan is occupied and no degeneracies occur. However, for low density samples, the CF spin will also affect the FQHE states around $\nu = 1/2$. The

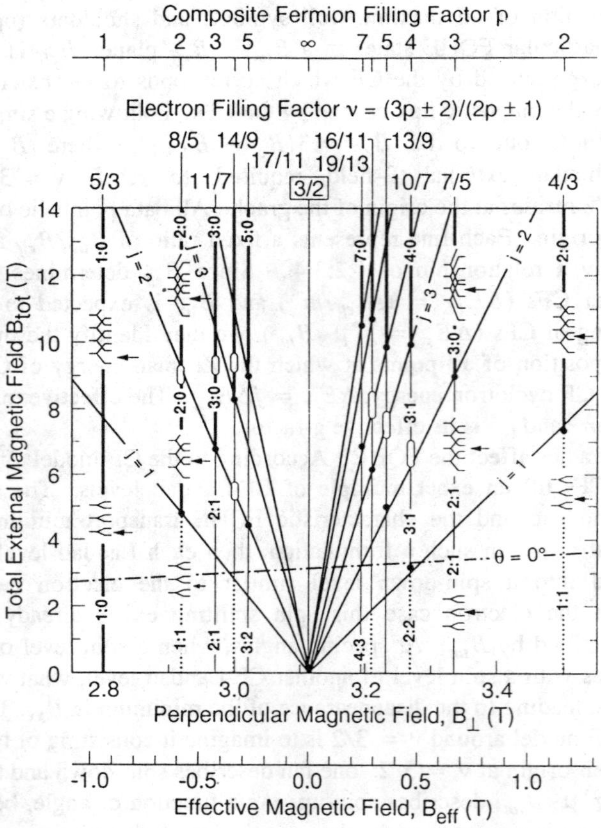

4. Maxima and shoulders found in the resistivity at the FQHE states around $\nu = 3/2$ plotted in a $B_{tot} - B_{eff}$ plane. Full (open) circles represent strong (weak) features. Inclined dash-dotted line $\theta = 0°$ indicates the motion of these traces through the graph taken at this angle. Vertical heavy, (dotted), [thin] lines indicate regions of fully spin-polarized, (fully spin-unpolarized), [partially spin-polarized] CF states, respectively. The ratios along some of these lines indicate the ratios of up-spins to down-spins which can be readily generalized. The level schemes and arrows serve as an illustration of the CF level structure and fermi level, respectively, in these particular regions. Ref. 24

impact of the spin on states such as $\nu = 2/3$ and $3/5$ has been documented.[26,27]

4. Conclusion

The CF model has been very successful in interpreting many aspects of the experimental data of the FQHE around $\nu = 1/2$ and $\nu = 3/2$. Superficially one can think of the FQHE features as resulting from Landau and spin quantization of new particles. In detail there remain many mysteries such as the apparent strong divergence of the effective mass as $\nu \rightarrow 1/2$, the unusual field dependence of the transport life time, the apparently polarization-independent g-factor around $\nu = 3/2$ and the discrepancies between different determinations of the effective mass even away from $\nu = 1/2$.

Finally, while we have learned much about CFs consisting of an electron and two flux quanta, we know very little about those with four or more flux quanta. Sample quality has to improve still further to investigate CFs of such a more fragile nature.

Acknowledgements

I would like to thank my colleagues K. W. Baldwin, R. R. Du, H. W. Jiang, W. Kang, L. N. Pfeiffer, D. C. Tsui, K. W. West, and A. S. Yeh who collaborated on most of the experiments presented in this paper.

REFERENCES

1. J. K. Jain, *Phys. Rev. Lett.* **63,** 199 (1989); Phys. Rev. **B40,** 8079 (1989); **41,** 7653 (1990), Science **266**, 1199 (1994).
2. V. Kalmeyer and S. C. Zhang, *Phys. Rev.* **B46,** 9889 (1992).
3. B. I. Halperin, P. A. Lee, and N. Read, *Phys. Rev.* **B47,** 7312 (1993) and references therein.
4. R. L. Willett, R. R. Ruel, K. W. West, and L. N. Pfeiffer, *Phys. Rev. Lett.* **71,** 3846 (1993).
5. R. L. Willett and L. N. Pfeiffer, *Surface Science* (to be published).
6. R. L. Willett, K. W. West and L. N. Pfeiffer, *Phys. Rev. Lett.* **75,** 2988 (1995).
7. W. Kang, H. L. Stormer, L. N. Pfeiffer, K. W. Baldwin and K. W. West, *Phys. Rev. Lett.*, **71,** 3850 (1993).

8. V. J. Goldman, B. Su and J. K. Jain, *Phys. Rev. Lett.* **72,** 2065 (1994).
9. J. H. Smet, D. Weiss, R. H. Blick, G. Lütjering, K. v.Klitzing, R. Fleischman, R. Ketzmerick, T. Geisel and G. Weiman (unpublished).
10. D. Weiss, M. L. Roukes, A. Menschig, P. Grambow, K. v.Klitzing, and G. Weimann, Phys. Rev. Lett., **66,** 2790 (1991).
11. D. Chklovski, private communication.
12. R. R. Du, H. L. Stormer, D. C. Tsui, L. N. Pfeiffer and K. W. West, *Phys. Rev. Lett* **70,** 2944 (1993).
13. H. C. Manoharan, M. Shayegan and S. J. Klepper, *Phys. Rev. Lett.,* **73,** 3270 (1994).
14. R. R. Du, H. L. Stormer, D. C. Tsui, A. S. Yeh, L. N. Pfeiffer, and K. W. West, *Phys. Rev. Lett.* **73,** 3274 (1994).
15. R. R. Du, H. L. Stormer, D. C. Tsui, L. N. Pfeiffer, and K. W. West, *Solid State Communication,* **90,** 71 (1994).
16. D. R. Leadley, R. N. Nicholas, C. T. Foxon, and J. J. Harris, *Phys. Rev. Lett.* **72,** 1906 (1994).
17. P. T. Coleridge, Z. W. Wasilewski, P. Zawadzki, A. S. Sachrajda, and H. A. Carmona, *Phys. Rev.* **B52,** 11603 (1995).
18. V. Kukushkin, R. J. Haug, K. v.Klitzing, and K. Ploor, *Phys. Rev. Lett.* **72,** 736 (1994).
19. B. L. Altshuler, L. B. Ioffe, A. J. Mills, *Phys. Rev.* **B50,** 14048 (1994).
20. Y. B. Kim, P. A. Lee, and X. G. Wen, *Phys. Rev.* **B52,** 17275 (1995).
21. A. Stern and B. I. Halperin *Phys. Rev.* **B52,** 5890 (1995).
22. J. P. Eisenstein, H. L. Stormer, L. N. Pfeiffer and K. W. West, *Phys. Rev. Lett.* **62,** 1540 (1989).
23. R. G. Clark, S. R. Haynes, A. M. Suckling, J. R. Mallett, P. A. Wright, J. J. Harris, and C. T. Foxon, *Phys. Rev. Lett.* **62,** 1536 (1989).
24. R. R. Du, A. S. Yeh, H. L. Stormer, D. C. Tsui, L. N. Pfeiffer and K. W. West, *Phys. Rev. Lett.* **75** 3926 (1995).
25. R. J. Nicholas, Adriatico Research Conference on "The Electron Quantum Liquid in Systems of Reduced Dimensions", Trieste, Italy, July 2-5, 1996 (unpublished).
26. J. P. Eisenstein, H. L. Stormer, L. N. Pfeiffer and K. W. West, *Phys. Rev.* **B41,** 7910 (1990).
27. L. W. Engel, S. W. Hwang, T. Sajoto, D. C. Tsui and M. Shayegan, *Phys. Rev.* **45,** 3418 (1992).

ON THE EVOLUTION TOWARDS A FRACTIONAL QUANTUM HALL STATE

R.K. HILLS, K.A. BENEDICT
Department of Physics, University of Nottingham, Nottingham U.K., NG7 2RD

A phenomenological theory for the fractional quantum Hall effect based on composite particles is proposed. An order parameter is introduced to measure the degree of attachment of magnetic flux to electrons in a two-dimensional electron system. Restricting attention to a homogeneous system in space, and assuming an underlying energy function with competing minima, an evolution equation for the order parameter is both proposed and investigated. Solutions to this differential equation clearly demonstrate the two processes of attachment and detachment of magnetic flux, and the system possesses a Hall plateau. The rate of convergence to the Hall state depends upon the applied magnetic field.

The discovery by Tsui, Störmer & Gossard[1] of the fractional quantum Hall effect (FQHE) in a two-dimensional electron system at low temperatures was something of a surprise and suggested the existence of a fundamental physical principle. These authors observed that for certain values of the applied magnetic field the longitudinal resistivity beomes zero. The current picture is that the FQHE state occurs when two-dimensional electrons and strong magnetic fields combine to form a new collective state of matter. In a non-FQHE state, the magnetic field consists of a uniform distribution of vortices. But, in the FQHE state, these vortices become attached to the electrons in a manner characterised by the rational filling factor m/n (n odd) denoting n vortices per m electrons. The transition is neither sharp nor instantaneous. If n_H denotes the relative density of vortices to electrons in the Hall state, the measured Hall resistance changes significantly for a relative density of vortices n_B in a "bandwidth" of n_H. The characteristic time of the change is small but finite. On the microscopic level, near filling factor $1/n_H$, each electron is attached to n_H vortices but, for $n_B \neq n_H$, this attachment is only possible by the introduction of a compensating number of positive or negative vortices so as to keep the applied magnetic field unchanged. Generally, by this process of attachment, the system can lower its energy. However, when the density of free vortices is too large, the configuration becomes unstable and the system reverts to the unattached topology.

Although the explanation of the FQHE must ultimately lie within microscopic physics (see, for example, Zhang[2] for a review of the Landau-Ginsburg theory), nevertheless there are macroscopic manifestations of the phenomenon and a clear need for a phenomenological theory that is capable of modelling not only transport processes but also interactions with, for example, sound propagation. This note lays the foundations of such a macroscopic theory and in this preliminary study, we concentrate on the spatially homogeneous problem with the system held in an isothermal state. Suppose that the relative density of attached vortices is n. Then for a quantitative macroscopic description we introduce an order parameter $\chi = n/n_H$, which essentially defines the extent of the deviation of the system from either the normal ($\chi = 0$) or Hall ($\chi = 1$) states. Thus the variable χ represents the proportion of electrons that have acquired n_H vortices and the intermediate states, $0 < \chi < 1$, signify an incomplete FQHE attachment of vortices. The density of free vortices (positive or negative) is given by $n_B - \chi n_H$.

The cornerstone of our model is the free energy density, E, which, in view of the isothermal assumption, we shall simply assume to be a function of χ and n_B, so that

$$E = E(\chi, n_B). \tag{1}$$

For definiteness we specialize our discussion to a single Hall plateau with a bandwidth symmetric about n_H. The methodology is clearly extendable to any number of such states and non-symmetric bandwidths. The nature of the n_B-dependence in (1) was postulated by Halperin[3] for the extreme cases $\chi = 0$ and $\chi = 1$. It is found that for the normal state, $\chi = 0$, E is a monotonic function of n_B and for the Hall state, $\chi = 1$, E has a characteristic saw-tooth form (see Figure I below).

The equilibrium values of χ for a stable or meta-stable phase are found by minimizing E considered as a function of χ at fixed n_B. Outside the critical bandwidth the absolute minimum of E should occur at $\chi = 0$. But for n_B within the band there should be a clear minimum at $\chi = 1$. These considerations suggest a model for the energy density made up of two competing parabolas. The first parabola relates to the normal state. Its minimum always lies on the abscissa $\chi = 0$ but the curve will move up or down with varying n_B. The second parabola, relating to the Hall state, moves in a similar fashion but is centred on $\chi = 1$. At the edge of the critical band the minima of the two parabolas are the same but on either side the preferred system state is indicated by that parabola having the lower minimum. Thus, if we were artificially to create at a specified n_B a partially formed Hall state with an intermediate value of χ, then the system will want to move to the equilibrium state appropriate to the parabola that provides the smaller ordinate value for the prescribed χ.

Perhaps the simplest competing parabola model that meets our criteria is given by

$$E_N = a\chi^2 + b, \qquad E_H = c(1-\chi)^2 + d, \tag{2}$$

where the coefficients a, b, c, d are functions of n_B. In particular, b is given by the equilibrium energy of the normal state and so, by the above, is monotonic increasing with n_B. Similarly d relates to the saw-tooth portion of the energy equilibrium curve. We must now determine the values of a, c. When attachment starts in the Hall band, there must be no jump in energy at $\chi = 0$. Thus $E_N(0) = E_H(0)$ in the Hall state, so that $c = b - d$. Similarly, for the normal state we need $E_H(1) = E_N(1)$, and $a = d - b$. These values also have an interesting property, for they eliminate any metastable states. A simple calculation shows that so long as $a, c > 0$ the curves never cross except at $\chi = 0$ or $\chi = 1$ (see Figure II for a graphical illustration). Note that the precise value of a in the Hall band and c outside is irrelevant. We can therefore define $a = c = |b - d|$ in the whole system.

To examine the model quantitatively, let us take a simple example. In the region around a filling factor of 1/3 ($n_H = 3$), the stable energy is piecewise linear to leading order so we can ignore asymmetries in the sawtooth of the energy function and define

$$b = (n_B - 2), \quad d = 4|n_B - n_H|. \tag{3}$$

The resulting energy function and its stable state can be seen in Figure I, where the familiar sawtooth is recovered.

Our discussion so far has been principally centred on equilibrium. However, for the non-equilibrium or time-dependent situation χ will not minimize the energy $E(\chi)$ but will differ by a term proportional to $d\chi/dt$. Thus the evolution equation is given by

$$\frac{d\chi}{dt} = -\frac{dE}{d\chi}, = \begin{cases} -2|b-d|\chi, & \text{in the normal region,} \\ 2|b-d|(1-\chi), & \text{in the Hall region.} \end{cases} \tag{4}$$

For our simple case of two competing parabolas, equation (4) can be solved exactly. It is easily shown that the solutions move towards their equilibrium value in an exponential fashion as shown in Figure III for the above special case given by (3). The characteristic time crucially depends upon the relative values of b, d, in other words, how far the system is from the transition point of the Hall state.

The above model is oversimplified in that we have only taken one Hall state and we have ignored spatial variations. As we have already remarked, the former causes no difficulty in principle, although the positions of the plateaux will have to be input into the theory. To account for spatial variations, an obvious starting point would be to include in the energy density a Landau–Ginsburg term of the form $|\nabla\chi|^2$. To minimize its energy, the system would be driven to make χ uniform. This approach is currently being investigated.

Acknowledgement

The work in this note was supported by an EPSRC grant

References

1. Tsui D C, Störmer H L, Gossard A C *Phys. Rev. Lett.* **48**, 1559 (1982)
2. Zhang S C I.J.M.P. (B) **6**, 25 (1992)
3. Halperin B I *Phys. Rev. Lett.* **52**, 1583 (1984)

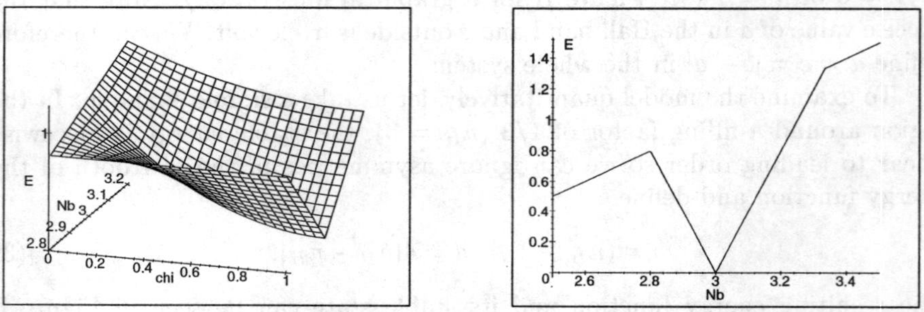

Figure I: Energy plots for the function in equations (2),(3). The figure on the right is the steady state energy value

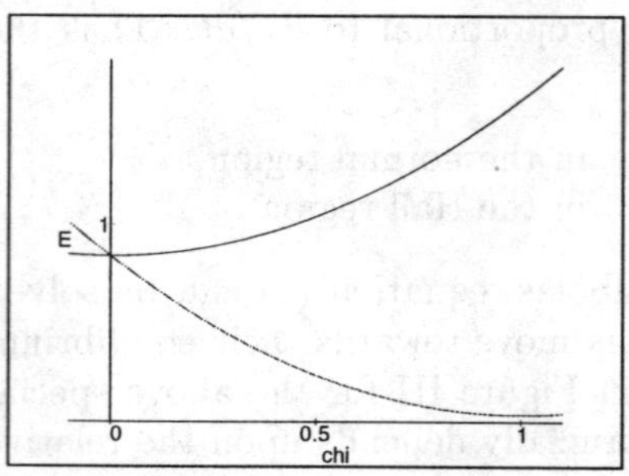

Figure II: Competing parabolas. The case shown is for the Hall band

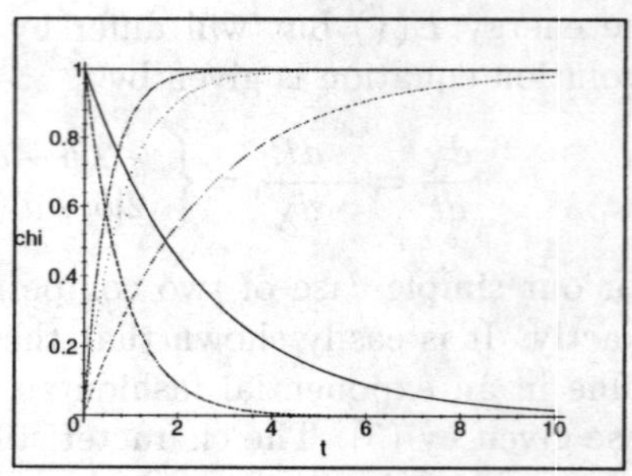

Figure III: Solution curves to (4) for $n_B = 3, 3.1, 3.2, 3.3, 3.4$, and $\chi(0) = 0, 1$

COMPOSITE FERMION (CF) ENERGY GAP VARIATION IN TILTED MAGNETIC FIELDS

P.J. GEE[1], J. SINGLETON[1], S. UJI[2], H. AOKI[2], C.T.B. FOXON[3], J.J. HARRIS[4]

[1] *Department of Physics, University of Oxford, Clarendon Laboratory, Oxford OX1 3PU, United Kingdom.*

[2] *Tsukuba National Magnet Laboratories, National Research Institute for Metals, 1-2-1, Sengen, Tsukuba, Ibaraki 305, Japan.*

[3] *Physics Department, Nottingham University, Nottingham NG7 2RD, UK.*

[4] *Department of Electronic Engineering, University College, London, UK.*

We have examined the temperature dependence (40 mK-1 K) of the oscillations in ρ_{xx} in a GaAs-(Ga,Al)As heterojunction close to filling factors $\nu = \frac{1}{2}$ and $\frac{3}{2}$ for several values of θ, the angle between the sample normal and the magnetic field. Close to $\nu = \frac{1}{2}$ the composite fermion (CF) effective mass does not follow the θ dependence expected for a 2D system; the CF energy gaps at fixed ν grow with increasing θ. Around $\nu = \frac{3}{2}$ the variation of the gaps at $\frac{7}{5}$ and $\frac{4}{3}$ with θ was fitted using a modification of the model of Du *et al.*[1] including a θ-dependent CF mass. However, the behaviour of the gaps at $\frac{5}{3}$ and $\frac{8}{5}$ is qualitatively different.

Although a number of experiments [2,3,4,5] have now demonstrated the validity of the composite fermion (CF) description [6] of the fractional quantum Hall effect (FQHE), almost all have been performed with the magnetic field B normal to the plane of the two dimensional electron system (2DES). In this work we have examined the effect of an in-plane magnetic field on the CFs. The sample was a GaAs-(Ga,Al)As heterojunction with a 120 nm spacer layer [7]; after illumination at $B = 0$ and temperature $T = 50$ mK with a red LED, the carrier density was $1.2 \times 10^{15}\text{m}^{-2}$ and the mobility 9000 $\text{m}^2\text{V}^{-1}\text{s}^{-1}$. The sample could be rotated *in situ* so that the normal to the 2DES made an angle θ to B; at each θ, ρ_{xx} and ρ_{xy} were measured at several T in the range 40 mK–1 K using a 10 nA, $\sim$ 18 Hz AC current.

Fig. 1a shows ρ_{xx} close to $\nu = \frac{1}{2}$ with $\theta = 28.2^o$; CF energy gaps were derived by fitting the amplitudes $A(T)$ of oscillations in ρ_{xx} to the function [2,3,7] $A(T) = F\frac{\chi}{\sinh(\chi)}$, where F is a function of B, $\chi = 2\pi^2 k_B T/E_{cf}$, and E_{cf} is the separation of the CF levels. We display the effect of an in-plane field [7] by deriving a CF mass which describes the gaps for $\frac{3}{5} \geq \nu \geq \frac{3}{7}$ and then evaluating how this varies with θ. We assume that E_{cf} has the form $\hbar e(B - B_{\frac{1}{2}})/m_e\mu$, where $\mu(\theta)$ is a mass (implicitly in units of m_e) to be determined and $B_{\frac{1}{2}}$ is the *total* field at $\nu = \frac{1}{2}$. Fig. 1b shows E_{cf} versus *total field* B (the sign of E_{cf} has been reversed for negative $(B - B_{\frac{1}{2}})$) at $\theta = 0.0^o$ and 39.1^o. At each θ

the points lie on a line passing through $E_{\rm cf} \approx 0$ at $\nu = \frac{1}{2}$; *i.e.* at fixed θ, to the limit of experimental accuracy the CF levels are symmetrical about $\nu = \frac{1}{2}$ and μ does not vary greatly over the small range of B covered [2,3,5]. Hence all $E_{\rm cf}$ are described using one value of μ at each θ, extracted from the gradient of lines such as those in Figure 1b.

Fig. 1c shows μ versus θ. If the CFs were purely 2D entities, one would expect $E_{\rm cf}$ to depend only on $B\cos\theta$; *i.e.* $E_{\rm cf} = \hbar e(B - B_{\frac{1}{2}})\cos\theta/(m_e\mu(0))$, yielding $\mu = \mu(0)/\cos\theta$. However, Fig. 1c shows that μ increases more slowly with θ than $1/\cos\theta$; in fact

$$\mu(\theta) \approx 0.70/[\cos(0.75\theta)], \tag{1}$$

implying that the CF gaps are growing with increasing θ. There is little theoretical work on the dynamics of CFs in tilted B. However FQHE gaps have been predicted[8] to increase due to the compression of the electronic wavefunction perpendicular to the plane of the 2DES caused by the in-plane component of B; *i.e.* the electrons become more ideally 2D. Such an effect may cause the enhancement of the CF gaps close to $\nu = \frac{1}{2}$ seen in the present work. Some studies have suggested that the CF mass is a "universal" function of effective B [2,3]; however, all of these experiments used GaAs-(Ga,Al)As heterojunctions in which the potential well shapes were rather similar. If the θ dependence of the CF mass in this work *is* due to the compressing effect of the in-plane B, it suggests that the "universal" behaviour of the CF mass will disappear when other potential well shapes (*e.g.* quantum wells) are studied.

Fig. 2 shows CF gaps at $\nu = \frac{5}{3}, \frac{8}{5}, \frac{7}{5}$ and $\frac{4}{3}$ derived[7] using the T-dependence (40 mK–1 K) of the oscillations in ρ_{xx}. Du *et al.*[1] proposed an interpretation of the variation of the gap energies with θ, in which the CF levels close to $\nu = \frac{3}{2}$ are split by a Pauli term $\pm\frac{1}{2}g_{\rm eff}\mu_{\rm B}B$ (here $g_{\rm eff}$ is the effective g-factor and $\mu_{\rm B}$ is the Bohr magneton) whereas the CF cyclotron energy depends only on the component of the effective B perpendicular to the 2DES, $3(B - B_{\frac{3}{2}})\cos\theta$. Although this approach predicts the vanishing of CF gaps[1] at certain θ, we have shown that it cannot fit the gaps shown in Fig. 2 *quantitatively* using a fixed, angle-independent CF mass[7]. The most marked failure concerns the gap at $\nu = \frac{5}{3}$; in the model[1] this is determined by spin splitting at low θ and should therefore grow as $1/\cos\theta$. In Fig. 2a this is seen not to be the case, as the gap increases much more slowly than $1/\cos\theta$ up to $\theta \sim 65^o$ and then decreases. In fact the gap at $\nu = \frac{5}{3}$ is fitted quite well by the function $3\hbar e|(B - B_{\frac{3}{2}})|/(\mu m_e)$ with μ taking the values given by Eqn. 1 (Fig. 2a).

In an attempt to fit the gaps at $\nu = \frac{8}{5}, \frac{7}{5}$ and $\frac{4}{3}$, we have derived [7] a

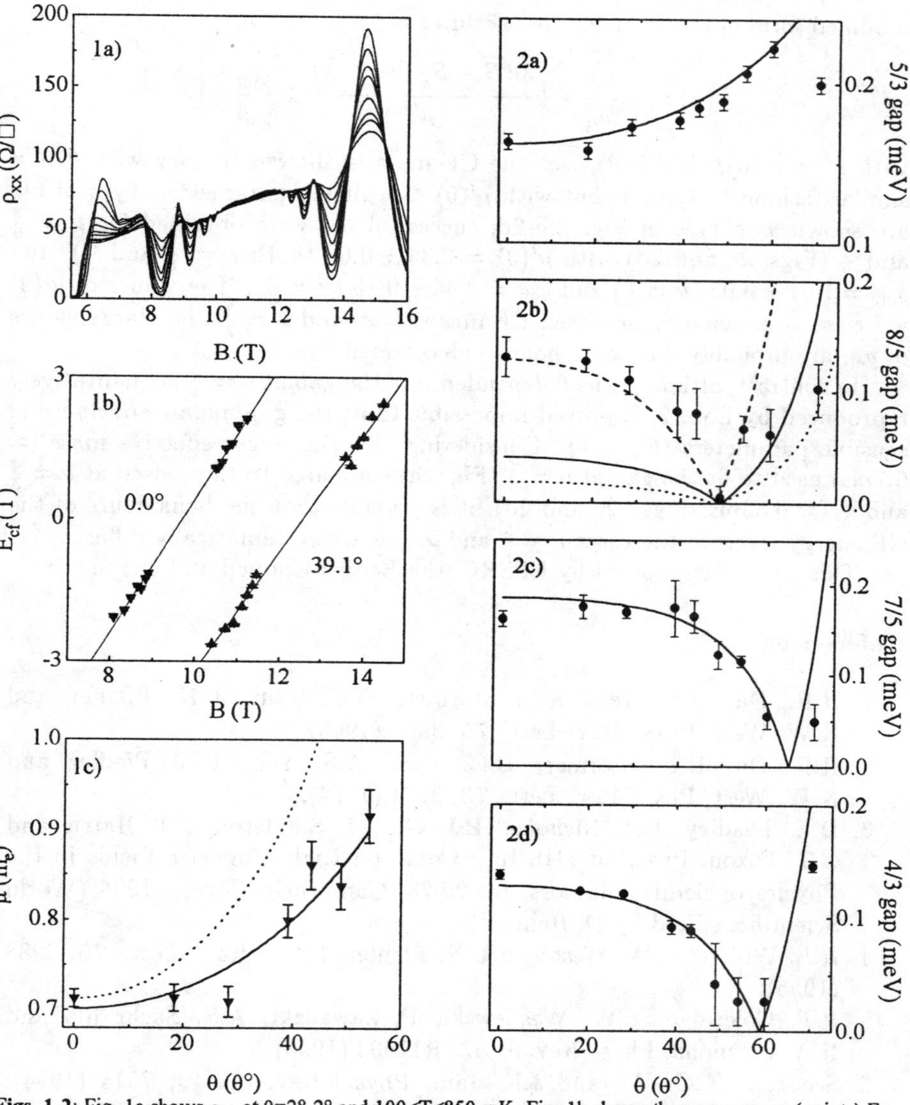

Figs. 1-2: Fig. 1a shows ρ_{xx} at θ=28.2° and 100<T<850 mK. Fig. 1b shows the energy gaps (points) E_{cf} (in Tesla) versus total field at θ=0.0° and 39.1°; lines are fits. Fig. 1c shows the values of the average cf mass μ derived from fits like those in Fig. 1b versus θ (points); the dotted line is $\mu=\mu(0)/\cos\theta$; the solid line is Eqn. 1. Fig. 2a-2d are gaps (points) for ρ_{xx} minima around ν=3/2 versus θ; curves in Figs. 2b-2d are from Eqn 2. Fig. 2a; ν=5/3 gap; line $3he|(B-B_{3/2})|/(2\pi\mu m_e)$ and μ from Eqn. 1. Fig. 2b; ν=8/5 gap; $\mu'(0)=0.1m_e$, $g_{eff}=3.2$ (dashed); $\mu'(0)=0.33m_e$, $g_{eff}=0.91$ (solid); $\mu'(0)=0.7m_e$, $g_{eff}=0.46$ (dotted). Fig. 2c; ν=7/5 gap; $\mu'(0)=0.33m_e$ and $g_{eff}=1.01$ (line). Fig. 2d; ν=4/3 gap; $\mu'(0)=0.33m_e$ and $g_{eff}=1.45$ (line).

modified form of the level scheme of Du *et al.*[1];

$$E(n) = -\frac{\hbar e}{m_e} \times \{\frac{3|(B - B_{\frac{3}{2}})|(n + \frac{1}{2})}{\mu'} \pm \frac{g_{\text{eff}} B}{4}\}, \quad (2)$$

with $\mu' = \mu'(0)/\cos(0.75\theta)$; *i.e.* the CF mass is allowed to vary with θ in a similar fashion to Eqn. 1, but with $\mu'(0)$ as a fitting parameter. Typical fits are shown as curves in Figs. 2b-2d; successful fits were obtained for $\nu = \frac{7}{5}$ and $\frac{4}{3}$ (Figs. 2c and 2d) with $\mu'(0) = 0.33 \pm 0.03$ (both $\nu = \frac{7}{5}$ and $\frac{4}{3}$), and $g_{\text{eff}} = 1.01 \pm 0.05$ ($\nu = \frac{7}{5}$) and $g_{\text{eff}} = 1.45 \pm 0.10$ ($\nu = \frac{4}{3}$). The values of $\mu'(0)$ are close to previously proposed CF masses[1] around $\nu = \frac{3}{2}$. The large values of g_{eff} are probably due to exchange enhancement[9].

In contrast, although the θ dependence of the gap at $\nu = \frac{8}{5}$ is qualitatively reproduced by Eqn. 2, it proved impossible to fit the gap *quantitatively* using sensible parameters (Fig. 2b). Considering also the larger effective mass ($\approx 0.7m_e$) used to fit the gap at $\nu = \frac{5}{3}$ (Fig. 2a) compared to those used at $\nu = \frac{4}{3}$ and $\frac{7}{5}$ ($\approx 0.33m_e$; Figs. 2c and 2d), it is possible that the behaviours of the CF energy levels in the cases $\nu \leq \frac{3}{2}$ and $\nu \geq \frac{3}{2}$ are systematically different[7].

This work is supported by EPSRC, the British Council and JSPS.

References

1. R.R. Du, A.S. Yeh, H.L. Stormer, D.C. Tsui, L.N. Pfeiffer and K.W. West, Phys. Rev. Lett. **75**, 3926 (1995).
2. R.R. Du, H.L. Stormer, D.C. Tsui, A.S. Yeh, L.N. Pfeiffer and K.W. West, Phys. Rev. Lett. **73**, 3274 (1994).
3. D.R. Leadley, R.J. Nicholas, P.J. Gee, J. Singleton, J.J. Harris and C.T. Foxon, Proc. of 11th Int. Conf. on High Magnetic Fields in the Physics of Semiconductors, pp 20-23, Cambridge, Mass., 1994 (World Scientific, edited by D. Heiman);
4. R.L. Willett, K.W. West and L.N. Pfeiffer, Phys. Rev. Lett. **75**, 2988 (1995);
5. P.T. Coleridge, Z.W. Wasilewski, P. Zawadzki, A.S. Sachrajda and H.A. Carmona, Phys. Rev. B **52**, R11603 (1995).
6. See *e.g.* X.G. Wu and J.K. Jain, Phys. Rev. B **49**, 7515 (1994); A. Lopez and E. Fradkin, Phys. Rev. B 44 5246 (1991); V. Kalmeyer and S.C. Zhang, Phys. Rev. B 46 9889 (1992).
7. P.J. Gee, J. Singleton, *et al.*, *J. Phys.: Condens. Matter*, submitted.
8. *e.g.* T. Chakraborty and P. Pietiläinen, Phys. Rev. B **39**, 7971 (1989).
9. R.J. Nicholas, R.J. Haug, K. von Klitzing and G. Weimann, Phys. Rev. B **37**, 1294 (1988).

EVIDENCE OF SKYRMION EXCITATIONS IN 2D ELECTRONS BY INTERBAND OPTICAL TRANSMISSION SPECTROSCOPY

B. B. GOLDBERG, M. J. MANFRA, AND E. H. AIFER
Boston University Physics Dept., Boston, Ma. 02215, USA

D. A. BROIDO
Department of Physics, Boston College, Chestnut Hill, Ma. 02167, USA

L. PFEIFFER AND K. WEST
AT&T Bell Labs, Lucent Technologies, Murray Hill, NJ 07974, USA

Interband magneto-absorption spectroscopy is used to determine the occupancy of the two lowest energy electron states in a 2D electron system in GaAs, the spin-up and spin-down states of the lowest Landau level. The changes in occupation provide a measurement of the absolute electron spin-polarization near $\nu = 1$. The rapid depolarization at both larger and smaller filling factors from a nearly spin-polarized state at $\nu = 1$ are evidence for the presence of Skyrmion excitations caused by the dominance of exchange energy over Zeeman energy. Recent temperature dependent spin-polarization results are consistent with theoretical models of the quantum Hall ferromagnet.

Due to the very small g-factor in GaAs, the bare Zeeman energy is two orders of magnitude smaller than the electronic exchange energy in GaAs two-dimensional electron systems at moderate magnetic fields. As a result, it can be energetically favorable to flip spins at the cost of Zeeman energy but with a net savings due to local exchange.[1] This dominance of exchange over Zeeman causes the response of a spin-polarized 2DES to a change of one quantum of magnetic flux not to be a single-particle spin-flip excitation, but rather a macroscopic spin object called a Skyrmion or charged spin-texture excitation (CSTE).[2,3] Skyrmions consist of a radial spin density distribution that is reversed at the center and gradually heals to the spin background over many magnetic lengths. With each particle in a nearly spin-aligned neighborhood, the ferromagnetic exchange contribution lowers the energy of the CSTE relative to a single flipped spin. Theoretical work[2,3] predicts of order 4 flipped spins per unpaired flux quanta for finite-sized Skyrmions, consistent with Barrett's original NMR experiment[4] as well as tilted-field transport[5] and our interband optical absorption measurements.[6]

Skyrmions cause a dramatic reduction in the degree of electron spin-polarization when the number of flux quanta per electron is different from one, that is, when the Fermi energy moves off the mid-point of the spin-gap of the lowest Landau level. Interband magneto-absorption spectroscopy measures the degree of electron spin polarization by distinguishing the occupancy of the two lowest electron spin-states.[7] At $\nu = 1$ the reduction in absorption to the lower energy, spin-up electron band is directly correlated with an increase in the higher energy, spin-down absorption. This indicates the spin-up state fills with electrons while the spin-down state empties, producing a large spin-polarization S_z just at $\nu = 1$. We report on recent temperature dependent measurements which are important in the understanding of quantum ferromagnets. The data are well fit by recent theoretical work, and provide additional confirmation of the spin-state occupation model of the absorption. We also discuss the role of electron – hole interactions on the energy and intensity of the absorption spectra.

Several sample each from two single-side n-modulation doped AlGaAs - GaAs 250Å single quantum well (SQW) wafers were studied. Wafer A had mobility $\mu = 3.2 \cdot 10^6 cm^2/Vs$ and 2DES carrier concentration of $N_s = 1.5 \cdot 10^{11} cm^{-2}$, and wafer B had $\mu = 2.6 \cdot 10^6 cm^2/Vs$ and $N_s = 1.8 \cdot 10^{11} cm^{-2}$. In transport these wafers exhibited strong fractional Hall minima at $\nu = 1/3$ and 2/3. The SQWs were chosen to minimize inhomogeneous broadening, and for absorption measurements were mounted strain-free and thinned to $\sim 0.5 \mu m$.[8] Incident power of $\sim 1 \mathrm{mW/cm^2}$ yielded typical signal-to-noise ratios of > 20 with line widths of 0.2 to 0.5 meV (FWHM). The absorption coefficients α^+, α^-, were calculated neglecting reflection which has been measured in similar samples to contribute only a small variation ($< 5\%$). The raw transmission spectra $I(w, B)$ were then normalized to obtain the magneto-absorption coefficient $\alpha(w, B) = -1/L_w \ln(I/I_0)$, where L_w = quantum well width.

The data and discussion focus on transitions to states in the lowest Landau level in the regime from $\nu = 0.7$ to 1.3 about the spin-gap. Representative spectra taken in LCP and RCP polarizations are displayed in the left and right side of Figure 1, respectively, with peak absorption as a function of field plotted in the inset. As described below, the final electron spin state for the lowest energy LCP (RCP) absorption is the lower (higher) energy spin-up (spin-down) state. The inter-band optical absorption is proportional to the available density of states in the final electron spin levels. The total spin-polarization is then given by the difference between the number of spin-up and spin-down states under the constraint that the sum yields the particle number, determining the total spin within $\leq 10\%$ accuracy. Figure 1 displays a marked decrease in the absorption into the lower spin-state, concomittant with an increase in absorption into the upper spin-state.This correlation reflects the redistribution of carriers in the spin-states, and thus a change in the spin-polarization of the electron system.

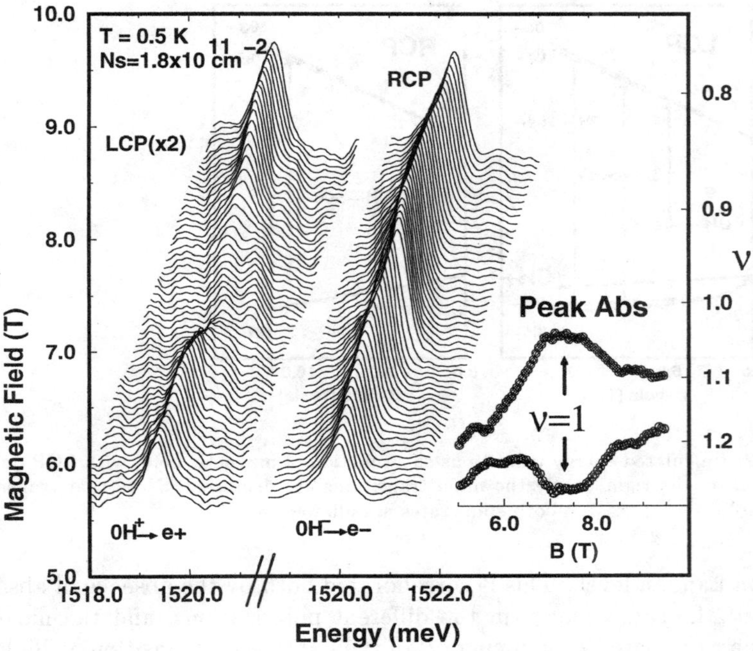

Figure 1: Absorption spectra plotted as a function of magnetic field to the left, and filling factor to the right. LCP (RCP) absorption measures the final density in the spin-up (spin-down) electron state. Inset: Peak absorption versus field about $\nu = 1$.

To determine the occupancy of the electron spin states and hence the spin-polarization, we have first calculated the inter-band optical matrix elements. Sub-band energies and wave-functions for electrons and holes were determined self-consistently within the local density approximation. The hole Landau levels were then calculated employing the Luttinger Hamiltonian[9] to take into account the valence band mixing[10]. Figure 2 shows the calculated energy levels as a function of magnetic field for the main allowed transitions in both polarizations. The lowest energy transition in RCP is from a pure heavy-hole state with $m_j = -3/2$ to the upper electron spin state $m_j = -1/2$ which we label $0H^- \rightarrow e^-$. The lowest energy state in LCP is from a mixed heavy-hole state with dominant character in the $m_j = +3/2, +1/2$ components to the lower energy spin state $m_j = +1/2$, labeled $0H^+ \rightarrow e^+$. Note that the $m_j = +1/2$ part of the hole envelope function for $0H^+ \rightarrow e^+$ is associated with a higher oscillator index and cannot cause an optical transition to the lowest

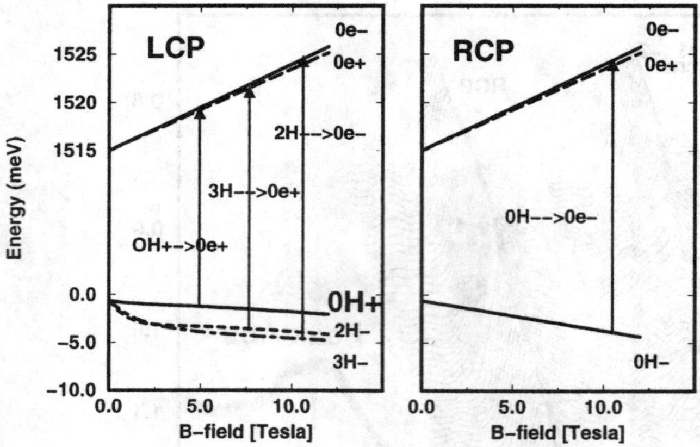

Figure 2: Calculated energy transitions as a function of magnetic field, with LCP on the left and RCP on the right. Only the main transitions are displayed. Note that transitions to both spin states are allowed in LCP.

electron Landau level.[8] This is corroborated both by the absence of absorption peaks at the same energy in the different polarizations, and the narrowness (less than the bare Zeeman energy) of the $0H^+ \rightarrow e^+$ transition at high fields. These observations confirm the validity of the matrix element calculations in the axial approximation.

We proceed to determine the spin level occupancy from the raw data as follows: The spin-polarization per particle is given by

$$S_z = \frac{N_\uparrow - N_\downarrow}{N} = \frac{N_{A_\downarrow} - N_{A_\uparrow}}{N} \tag{1}$$

where $N_{\uparrow(\downarrow)} = N_B - N_{A_{\uparrow(\downarrow)}}$. N_B is the Landau level degeneracy eB/h, and N_{A_j} is the available density of states in the $j =\uparrow (\downarrow)$ spin up (down) band of the lowest Landau level. The integrated absorption peaks are linearly related to the number of available final states N_{A_j} of the transition as

$$I_{ij} = C \cdot f_{ij} N_{A_j} \tag{2}$$

where $I_{ij} = \int \alpha_{ij} d\omega$, the oscillator strengths $f_{ij}(\omega)$ are taken to be constant over the narrow absorption peaks, and i, j label the initial and final states respectively. The constant of proportionality C may be found using the sum-rule

$$\frac{N_{A_\downarrow} + N_{A_\uparrow}}{N} = \frac{(N_B - N_\uparrow) + (N_B - N_\downarrow)}{N}$$

$$= \frac{2-\nu}{\nu} \tag{3}$$

which conserves particle number $\frac{N_\uparrow + N_\downarrow}{N} = 1$. Then

$$C(B) = \frac{\nu}{2-\nu}\left(\frac{I_{i,\uparrow}}{f_{i,\uparrow}} + \frac{I_{k,\downarrow}}{f_{k,\downarrow}}\right) \tag{4}$$

and the available densities of states are obtained from (2). Since LCP and RCP provide independent data sets, the above prescription determines both the spin polarization S_z and the constant of proportionality C – two equations in two unknowns.

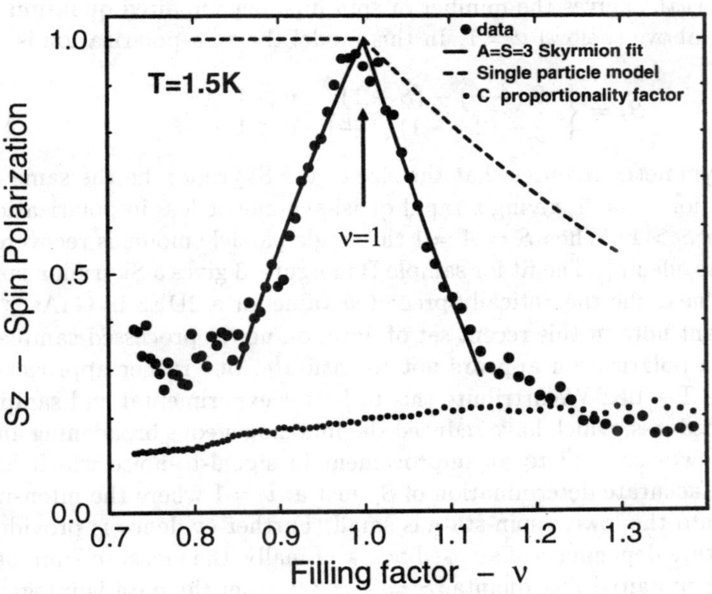

Figure 3: Spin-polarization S_z as a function of filling factor at $T = 1.5$K. The solid line represents a fit to a finite size Skyrmion model with $A = S = 3.0$ spin-flips per unpaired flux quanta. The dotted line represents the $T = 0$ single particle model with a single spin-flip per unpaired flux quanta. The dots on the lower part plot the constant of proportionality C, showing the near total absence of structure in the neighborhood of $\nu = 1$, proving how well the raw data obey the state occupation model.

In figure 3, spin-polarization versus filling factor is plotted and compared with both a single particle and a Skyrmion-based model. Previous calculations[6] have shown that a single particle model based on the exchange enhanced g-factor that modulates the overlap of the two electron spin levels[11] fails to capture the behavior of S_z. The polarization quickly saturates to unity for $\nu < 1$ and goes as $\sim \frac{2-\nu}{\nu} = N_{A_\downarrow}/N$ for $\nu > 1$ (single particle model in figure 3), in contrast to the measured polarization which decays quasi-symmetrically about $\nu = 1$ at a much more rapid rate. A spin-wave model which includes interactions [12] would provide depolarization for $\nu < 1$, though with a weaker temperature dependence than the data (see below). Recent theoretical work has also indicated that a spin-wave model is insufficient.[13]

The changes in polarization do adhere however, to a treatment which includes Skyrmion excitations (see solid line fit figure 3). In the model proposed by Barrett *et al.* [4] the one-particle available densities of states are scaled by a parameter S (A) that gives the number of spin flips per unpaired quantum of flux $|N_B - N|$, above (below) $\nu = 1$. In this model the spin-polarization is

$$S_z = \begin{cases} S\left(\frac{2-\nu}{\nu}\right) - (S-1) & \nu > 1 \\ \frac{1}{\nu} - (2A-1)\left(\frac{1-\nu}{\nu}\right) & \nu < 1 \end{cases} \tag{5}$$

Particle-hole symmetry requires that the size of the Skyrmion be the same as the Anti-Skyrmion, $S = A$, giving a rapid quasi-symmetric loss in polarization about $\nu = 1$ for $S > 1$. When $S = A = 1$ the single particle model is recovered (modulo overlap effects). The fit for sample B in figure 3 gives a Skyrmion spin of 3.0 which is near the theoretically predicted value for a 2DES in GaAs.[2,3]

Of significant note in this recent set of data, on newly processed samples, is that the spin-polarization appears not to saturate, but rather approaches unity $S_z = 1$ as T$\rightarrow$ 0K. We attribute this to better experimental and sample processing procedures, which have reduced the inhomogeneous broadening and absorption line-widths, and to an improvement in signal-to-noise which has allowed a more accurate determination of S_z just at $\nu = 1$ where the intensity of absorption into the lowest spin-state is small. Further evidence is provided in the temperature dependence discussed below. Finally, the measured number of spin-flips per unpaired flux quanta, $S = A = 3.0$, from the data is integral, as expected for the finite-sized Skyrmions.[14]

We have also recently completed a systematic measurement of the temperature dependence of the spin-polarization at $\nu = 1$ as well as at other filling factors, which may shed light on the magnetic properties of strongly correlated electron systems. Very recent theoretical work has focused on the collective-magnetization excitations of the ferromagnetic ground state.[13,15]. Figure 4 displays a set of absorption spectra taken in LCP and RCP as a function of temperature. The decrease in absorption for decreasing temperature into the

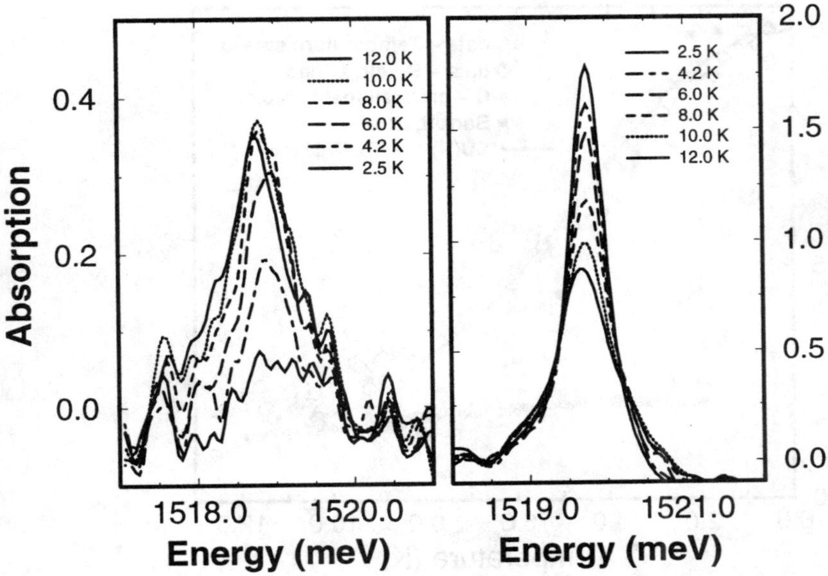

Figure 4: Absorption spectra as a function of temperature at $\nu = 1$ from 2.5 to 12.0K. The left-hand plot displays the spectra in LCP and the right-hand plot in RCP polarizations. Note that the decrease in absorption into the lower spin-state in LCP for decreasing temperature is concomitant with an increase in absorption into the upper spin-state. Also note the lack of energy shift and the narrow spectral line-widths.

lower spin-state in LCP is closely matched by the increase in absorption into the upper spin-state. From this data, S_z versus T is obtained and plotted in Fig. 5. The sets of data taken by both sweeping temperature at fixed magnetic field (large circles) and sweeping magnetic field at fixed temperature (large squares) are consistent. The scatter in the data give an indication of the uncertainty in determining the spin polarization.

The absorption data provide experimental self-consistency for our analysis in terms of spin-state occupation. First, the calculation of S_z from the $0H^- \rightarrow e^-$ and $0H^+ \rightarrow e^+$ transitions can be complimented by comparing S_z calculated from the $2H^- \rightarrow e^-$ and $0H^+ \rightarrow e^+$ transitions. Since the later set occur in a single polarization (LCP) and the $0H^- \rightarrow e^-$ and the $2H^- \rightarrow e^-$ independently monitor the occupancy of the same upper electron spin state, the similarity of the spin-polarizations calculated means that the matrix elements calculations are internally consistent with our simple sum-

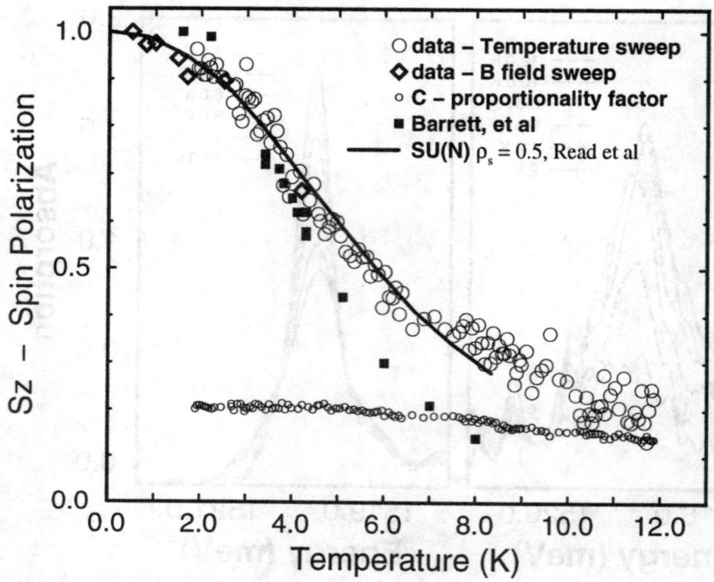

Figure 5: Spin-polarization S_z as a function of temperature at $\nu = 1$. The open circles are obtained by sweeping the temperature and correspond well to magnetic field sweeps at fixed temperature shown in the open squares. The theory of Read, et al. is displayed by the solid line, and Barrett's data is included for comparison. The small circles plot the constant of proportionality C versus temperature. From 0.5 to $\sim$ 8K, C is quite smooth and temperature independent, again showing the close adherence of the raw data to the simple sum-rule.

rule. Second, the matrix elements vary slowly over the filling factor range of interest $\nu = 0.7$ to 1.3, typically less than 20%, and hence cannot simply account for the structure observed. Nor are they particularly sensitive to the carrier density or the precise value of the zero-field splitting; these parameters have been varied with no significant change in the final spin-polarization S_z. Third, the recent set of data show a remarkable lack of structure in the constant of proportionality C, calculated for each pair of LCP and RCP spectra by requiring the sum of the density of up- and down-spins to yield the carrier density. C is displayed both as a function of magnetic field (Fig. 3) with an inverse-B dependence as expected from Fermi's golden rule, and as a function of temperature (Fig. 5), showing largely temperature independent behavior. Overall, this demonstrates that the raw data come very close to obeying the sum-rule over this field range and indicating that few higher-order processes are affecting the absorption.

Previously, a wealth of data on a seemingly similar effect, the quenching of the photoluminescence from the lowest energy transition accompanied by an increase in the emission from the higher energy transition, has been observed by one of the authors and others.[16] The total integrated emission was relatively constant, yielding an explanation based on a decrease in the re-combination rate in the lowest energy state due to localization. In the absence of significant non-radiative channels, the minority photo-excited hole must eventually emit a photon on re-combination, and hence the emission from the two electron levels must be correlated. However, not only is absorption largely unaffected by localization, but also the photons absorbed which cause transitions into the lower and higher energy spin-states are completely uncorrelated. This leads to the conclusion that the correlation observed in the absorption data is due to the changes in the occupancy and thus the total spin of the electron system.

Associated with the early work on photoluminescence in the quantum Hall regime[16] is the question of the excitonic interaction of the electron and hole and its effect on recombination. We have studied this question experimentally in the present absorption data as well, where the excitonic interaction appears in the final state, and have found the following interesting facts. Similar to earlier PL results, one of the wafers shows large energy shifts of both the LCP and RCP absorption peaks in the quantum Hall regime at $\nu = 1$ at temperatures below $T \sim 1.5$K, while in other samples where a much weaker excitonic shift is observed. The same shift appears in the photoluminescence which, unfortunately, is restricted to examining only the ground-state emission at low temperatures. The energy shifts are typically a few meV, many times the line-width, and interestingly, occur in *all* transitions, independent of either the initial or final state. Shifts are observed not only in transitions to the ground Landau level, but in a similar fashion to higher Landau levels, and even higher subband states. This is conclusive evidence that this excitonic energy shift is caused by the changes in the Coulomb hole of the final-state hole brought about by the reduced screening of the electrons at $\nu = 1$. In this wafer we observe finite-sized Skyrmion excitations as well, but in a temperature regime above the excitonic energy shift. Our conclusion is that two, somewhat independent processes are taking place, and that earlier attempts to explain the PL data within a single framework need to be revisited. We are currently studying the polarized PL and absorption simultaneously in the Skyrmion regime to further elucidate this issue.

In conclusion, interband absorption spectroscopy has proved to be an effective tool to measure the spin-polarization of a two-dimensional electron system.

Temperature and magnetic field dependent spectral data sets provide strong evidence of the validity of the spin-state occupation model used in interpreting the absorption spectra. Such analysis has led to the observation of finite-sized Skyrmion excitations near $\nu = 1$, with a measured $A = S = 3.0$ spin-flips per unpaired flux quanta. Additionally, we presently observe no saturation, with the system approaching a fully polarized state at $\nu = 1$ as $T \to 0$. Finally, the measured temperature dependence of the spin magnetization agrees well with recent theoretical work on quantum Hall ferromagnets.

Acknowledgments

This work supported by NSF DMR-9158097. We thank A. H. MacDonald and N. Read for valuable discussions.

References

1. B. I. Halperin, *Helv. Phys. Acta* **56**, 75 (1983).
2. S. L. Sondhi, A. Karlhede, S. A. Kivelson, and E. H. Rezayi, *Phys. Rev. B* **47**, 16419 (1993) and references therein.
3. H. A. Fertig, L. Brey, R. Cote, and A. H. MacDonald, *Phys. Rev. B* **50**, 11018 (1994).
4. S. E. Barrett, G. Dabbagh, L. N. Pfeiffer, K. W. West, and Z. Tycko, *Phys. Rev. Lett.* **74**, 5112 (1995).
5. A. Schmeller, J. P. Eisenstein, L. N. Pfeiffer, K. W. West, *Phys. Rev. Lett.* **75**, 4290 (1995).
6. E.H. Aifer, B. B. Goldberg, and D. A. Broido, *Phys. Rev. Lett.* **76**, 680 (1996).
7. B.B. Goldberg, D. Heiman, and A. Pinczuk *Phys. Rev. Lett.* **63**, 1102 (1989).
8. B.B. Goldberg, D. Heiman, M.J. Graf, D.A. Broido, A. Pinczuk, C.W. Tu, J.H. English, and A.C. Gossard, *Phys. Rev. B* **38**, 10131 (1988).
9. J. M. Luttinger, *Phys. Rev.* **102**, 1030 (1956).
10. The effect of the in-plane subband anisotropy on the topmost hole Landau levels is expected to be small. S.-R. E. Yang, D. A. Broido, and L. J. Sham, *Phys. Rev. B* **32**, 6630 (1985).
11. T. Ando, and Y. Uemura, *J. Phys. Soc. Japan* **37**, 1044 (1974).
12. Yu.A. Bychkov, S.V. Iordanskii and G.M. Eliashberg, *Pis'ma Zh. Eksp. Teor. Fiz.* **33**, 152 (1981) [JETP Lett. **33**, 143 (1981)]; C. Kallin and B.I. Halperin, *Phys. Rev. B.* **30**, 5655 (1984).

13. N. Read and Subir Sachdev, *Phys. Rev. Lett.* **75**, 3509 (1995).
14. A. H. MacDonald, H. A. Fertig, and Luis Brey *Phys. Rev. Lett.* **76**, 2153 (1996).
15. Marcus Kasner and A. H. MacDonald *Phys. Rev. Lett.* **76**, 3205 (1996).
16. B.B. Goldberg, D. Heiman, A. Pinczuk, L. Pfeiffer, and K. West, *Phys. Rev. Lett.* **65**, 641 (1990); A.J. Turberfield, S.R. Haynes, P.A. Wright, R.A. Ford, R.G. Clark, J.F. Ryan, J.J. Harris, and C.T. Foxon, *Phys. Rev. Lett.* **65**, 637 (1990); B.B. Goldberg, D. Heiman, M. Dahl, A. Pinczuk, L. Pfeiffer, and K. West, *Phys. Rev. B* **44**, 4006 (1991); E.M. Goldys, S.A. Brown, R.B. Dunford, A.G. Davies, R. Newbury, R.G. Clark, P.E. Simmonds, J.J. Harris, and C.T. Foxon, *Phys. Rev. B* **46**, 7957 (1992); M. Dahl, D. Heiman, A. Pinczuk, B. B. Goldberg, L. N. Pfeiffer, and K. West *Phys. Rev. B15* **45** 6957 (1992).

SKYRMIONS IN QUANTUM HALL FERROMAGNETS

A.H. MACDONALD
Department of Physics, Indiana University, Bloomington, Indiana 47405, USA

In the limit of strong magnetic fields where all electrons are confined to a single orbital Landau level, interacting two-dimensional electron systems (2DES's) exhibit a rich variety of unusual properties. The quantum Hall effect, which occurs when the chemical potential of the 2DES has a discontinuity at a density n^* which depends on magnetic field strength, is a prominent example. In the quantum Hall effect the incompressible ground state at density n^* can be a strong ferromagnet, *i.e* its total spin quantum number S can equal $N/2$ so that electronic spins are completely aligned by infinitesimal Zeeman coupling. Surprisingly, in light of the strong magnetic fields, physics associated with *spontaneous* magnetization in these systems is experimentally accessible because the Zeeman coupling is typically quite weak compared to other characteristic energy scales and can even be tuned to zero, for example by the application of hydrostatic pressure to the host semiconductor. Recently there has been considerable interest in charged excitations of the incompressible ground atate at $\nu = 1$ which has total spin quantum number $S = N/2$. As demonstrated by recent experiments the ground state spin-polarization in these systems is strongly reduced by charged excitations, increasingly so when the Zeeman coupling is weakened. This behavior can be understood qualitatively in the limit of weak Zeeman coupling, by identifying the elementary charged excitations as topologically charged Skyrmion spin-texture excitations of the non-linear sigma ($NL\sigma$) model which describes the long wavelength properties of ferromagnets. In the case of quantum Hall ferromagnets, Skyrmions carry an electrical charge as well as a topological charge, are present in the ground state for $\nu \neq 1$, and as a consequence have a more obvious influence on observable properties than in conventional ferromagnets.

This talk will review theoretical work on Skyrmions in quantum Hall ferromagnets which is based on miroscopic Hartree-Fock and generalized Hartree-Fock approximations. The objective of this work has been to develop a quantitative understanding not only of isolated Skyrmion quasiparticles but also of the dense Skyrmion systems that occur at filling factors which depart substantially from $\nu = 1$. New results on the NMR lineshape associated with the non-uniform magnetization distribution in a dense Skyrmion system will be presented.

The work described in this talk has been carried out over the last thirty months in collaboration with Luis Brey (UAM Madrid), Rene Cote (University of Sherbrooke), and H.A. Fertig (University of Kentucky).

Exact diagonalization study of interacting electrons with spin degree of freedom in the disk geometry

MARCUS KASNER
Max–Planck–Institut für Physik komplexer Systeme, Außenstelle Stuttgart, D–70569 Stuttgart, Germany

We report on an exact diagonalization study of interacting electrons in a strong magnetic field moving on a disk. We consider the case of a finite number of electrons N near filling factor one so that the number of flux quanta through the area of the system is $N_\Phi = N \pm 1$. In contrast to the case of filling factor one with its homogeneous spin density, a field theoretical model as well as Hartree–Fock calculations predict in either case ground states with a peculiar spin texture. In this work first we confirm numerically the structure of eigenstates of zero energy for a hard–core interaction that was predicted recently by MacDonald *et al.* Secondly, we identify for the more realistic Coulomb interaction even for finite N the existence of a spin textured ground state in the antiskyrmion case ($N = N_\Phi + 1$) whereas for the skyrmion case such a ground state cannot be seen for the particle numbers considered. Thirdly, we estimate the critical Zeeman coupling for which a localized quasi particle is energetically more favorable than an antiskyrmion.

1 Introduction

Interacting electrons in two dimensions that are subject to a constant perpendicular magnetic field reside at strong fields in the highly degenerate lowest orbital Landau level. Because in GaAs–heterostructures the Zeeman coupling $\Delta_z = |g\mu_B B|$ remains small in comparison with the distance between the Landau levels and the strength of the interaction even at fields of 10 Tesla, the spin degree of freedom is an important ingredient. This becomes particularly apparent at and in the vicinity of odd filling factors ν, but should also be valid at filling factors where an incompressible polarized ground state exists.
¿From a theoretical viewpoint it is known for some time that in the limit of vanishing Zeeman coupling the long wavelength physics at filling factor one can be mapped onto a nonlinear σ model [1–2] whose lowest charged excitations are known as skyrmions and antiskyrmions. These charged spin-texture excitations (CSTE) have a spin texture that can be described by a topological quantum number. At vanishing Δ_z, the neutral excitation energy of a skyrmion–antiskyrmion pair has only half of the energy of the quasihole–quasielectron pair.
However, this energetic advantage vanishes quickly if a finite Zeeman coupling is taken into account and is accompanied by a shrinking of the size of the (anti)skyrmions. In the framework of a HF–calculation, estimates were done in

order to determine the net spin at finite Δ_z compared with the spin 1/2 of the conventional quasiparticle[3]. The complete quantum mechanical treatment is actually more complicated. In order to classify the ground states of systems with one flux quantum more or less than the particle number, a Hamiltonian with a hard–core interaction is very useful[4]. Then a whole bunch of zero energy eigenstates can found whose quantum number differ in the deviation of the total spin S from the ferromagnetic value $N/2$ (for skyrmions) as well as the deviation of the total angular momentum from its maximum value. If the Zeeman coupling is switched on the degeneracy is partially lifted.
The main motivation for this work is to check these ideas for the more realistic case of Coulomb interaction in a plane geometry where all the above mentioned zero energy ground state degeneracies are lifted. The main obstacle to draw firm conclusions for the thermodynamic limit is the restricted number of particles N we are able to treat. A similar study was carried out in the spherical geometry[5]. On the experimental side, there are currently various experiments that seem to corroborate the existence of these CSTE. NMR–experiments[6], transport experiments[7] and light absorption experiments[8] are consistent with the idea that skyrmionic excitations lead to a precipitous decline of the magnetization if one deviates from filling factor one.

2 Diagonalization for hard–core and Coulomb interactions

We consider interacting electrons in the lowest Landau level moving on a disk with Zeeman coupling. The finite size version of the Hamiltonian where the one–particle basis is truncated at a maximum single particle angular momentum $m_{max} = (N_\Phi - 1)$, reads in symmetric gauge including a homogeneous positive background:

$$\begin{aligned} H &= -\frac{1}{2}\Delta_z(N_\uparrow - N_\downarrow) + \frac{1}{2}\lambda \sum_{\substack{m_1,m_2,m_3,\\ m_4=0,\ \sigma,\sigma'}}^{m_{max}} W_{m_1,m_2,m_3,m_4}\, c^\dagger_{m_1,\sigma} c^\dagger_{m_2,\sigma'} c_{m_3,\sigma'} c_{m_4,\sigma} \\ &\quad -\nu\lambda \sum_{m,m'=0,\sigma}^{m_{max}} W_{m,m',m',m}\, c^\dagger_{m,\sigma} c_{m,\sigma} + \frac{\nu^2\lambda}{2} \sum_{m,m'=0}^{m_{max}} W_{m,m',m',m}\,, \end{aligned} \tag{1}$$

where $\lambda = e^2/(4\pi\epsilon l_c)$ is the electron interaction coupling, W_{m_1,m_2,m_3,m_4} the two–particle matrix element projected onto the lowest Landau level, N_σ is the total number of electrons with spin σ and the filling factor is defined as $\nu = (N-1)/(N_\Phi - 1)$.
We did calculations for two different interactions. In the first case we chose a hard core interaction $V(\vec{r} - \vec{r}\,') = 4\pi V_0 \delta^{(2)}(\vec{r} - \vec{r}\,')$ which results in only one

pseudo potential coefficient $V_m^{(0)} = V_0\delta_{m,0}$ (m - relative angular momentum) unequal zero ($V_0 = \sqrt{\pi}/2$). In the second case we determined the spectrum for a Coulomb interaction with background.

3 Results and Discussion

The results for the hard core model whose zero–energy eigenvalues can be derived completely from the properties of the many–particle wave functions [4], are shown in Fig. 1. The degeneracy of these states has been partially lifted by a finite Zeeman coupling. The eigenvalues arranged symmetrically around zero with distance Δ_z for an angular momentum block $M - K$ ($M = N(N-1)/2 = 21$ in Fig. 1), contain multipletts with $S = |N/2 - K|$, $K = 0, 1, \ldots, 6$. Degeneracy for varying M is related to the translational invariance of the infinite system whereas the remaining degeneracy for fixed M is due to the occurence of spin multipletts labelled by K. This picture is in accordance with a recent analysis [4]. For the more realistic Coulomb interac-

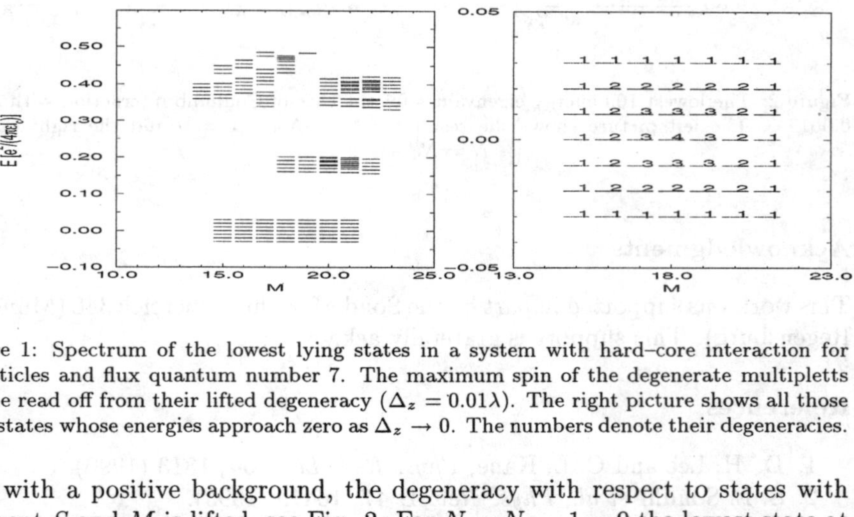

Figure 1: Spectrum of the lowest lying states in a system with hard–core interaction for 6 particles and flux quantum number 7. The maximum spin of the degenerate multipletts can be read off from their lifted degeneracy ($\Delta_z = 0.01\lambda$). The right picture shows all those eigenstates whose energies approach zero as $\Delta_z \to 0$. The numbers denote their degeneracies.

tion with a positive background, the degeneracy with respect to states with different S and M is lifted, see Fig. 2. For $N = N_\Phi - 1 = 9$ the lowest state at $M = 36$ has spin $N/2 = 9/2$. This is analogous to $\nu = 1$ with a ferromagnetic ground state. For this N we don't see a skyrmionic ground state.

In the case $N = N_\Phi + 1$, the spin of the ground state is different because $S = (N_\Phi - 1)/2 - 2$ for $N = 9, 10, 11$. These states exhibit a diminished total spin that allow a change of the ground state if the Zeeman coupling increases although the eigenfunctions remain unchanged. If one increases Δ_z mono-

tonically, subsequent level crossings occur. Eventually the Zeeman coupling reaches a value where the spin polarized state with $S = (N_\Phi - 1)/2$ is the ground state. Such a behavior was also observed for the spherical geometry which can be seen in kinks in the energy vs. Zeeman coupling plot [5].
¿From our finite size data, we determine the critical Zeeman coupling where the quasielectron state becomes the ground state. A simple quadratic extrapolation in $1/N$ yields the value $\Delta_z^{crit} = 0.041\lambda$. Probably due to the smaller systems we treated this estimate is still lower than the value 0.054λ found in [2].

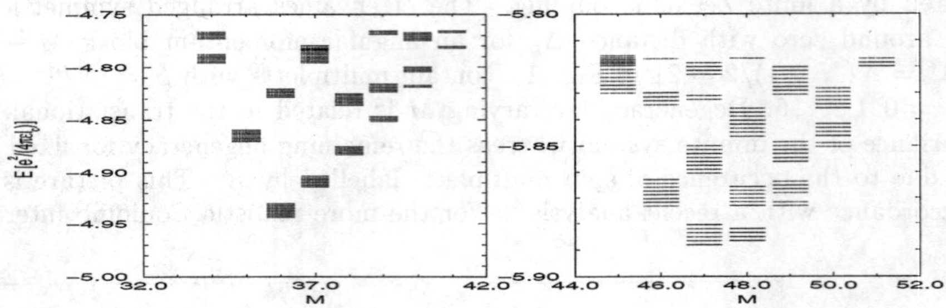

Figure 2: The lowest 100 energy eigenvalues for the case of Coulomb interaction with $\Delta_z = 0.0015\lambda$. The left picture shows the result for $N = N_\Phi - 1 = 9$ and the right one for $N = N_\Phi + 1 = 11$.

Acknowledgments

This work was supported in part by the Sonderforschungsbereich 348 (München/ Regensburg). This support is gratefully acknowledged.

References

1. D.-H. Lee and C. L. Kane, *Phys. Rev. Lett.* **64**, 1313 (1990).
2. S. L. Sondhi *et al.*, *Phys. Rev.* B **47**, 16419 (1993).
3. H. A. Fertig *et al.*, *Phys. Rev.* B **50**, 11018 (1994).
4. A. H. MacDonald *et al.*, *Phys. Rev. Lett.* **76**, 2153 (1996).
5. X. C. Xie and S. He, *Phys. Rev.* B **53**, 1046 (1996).
6. S. Barrett *et al.*, *Phys. Rev. Lett.* **74**, 5112 (1995).
7. A. Schmeller *et al.*, *Phys. Rev. Lett.* **75**, 4290 (1995).
8. E. H. Aifer *et al.*, *Phys. Rev. Lett.* **76**, 680 (1996).

LOW TEMPERATURE HEAT CAPACITY OF GaAs QUANTUM WELLS NEAR $\nu=1$

V. BAYOT, E. GRIVEI, S. MELINTE

Unité de Physico-Chimie et de Physique des Matériaux, Université Catholique de Louvain, Place Croix du Sud 1, B-1348 Louvain-la-Neuve, Belgium

M.B. SANTOS and M. SHAYEGAN

Department of Electrical Engineering, Princeton University, Princeton N.J. 08544

We report low temperature (T) heat capacity (C) data on a multiple-quantum-well GaAs/AlGaAs sample in the quantum Hall regime. Relative to its low field magnitude, C exhibits up to $\sim 10^5$-fold enhancement near ν=1 where Skyrmions are the ground state of the confined two-dimensional electrons. We attribute the large C to a Skyrmion-induced, strong coupling of the nuclear spin system to the lattice. The data are consistent with the Schottky nuclear heat capacity of Ga and As atoms in the quantum wells, except at very low T where C vs T exhibits a remarkably sharp peak suggestive of a *phase transition in the electronic system.*

In two-dimensional electron systems (2DESs), heat capacity can be a powerful probe of single- and many-body properties such as the Landau quantized density of states and the quantum Hall effect (QHE), both integral and fractional.[1,2] But in contrast to the overwhelming number of magnetotransport and magnetooptical experiments reported for 2DESs in the QHE regime,[3] very few heat capacity measurements have been reported so far.[4,5] Here we report heat capacity (C) measurements of a modulation-doped $GaAs/Al_xGa_{1-x}As$ multiple-quantum-well heterostructure down to very low temperature ($T \geq 25$ mK) and Landau level filling factor ($\nu > 0.5$).[6]

The experiments were performed on a 7×7 mm^2 multiple-quantum-well heterostructure grown by molecular-beam epitaxy and consisting of 100 GaAs quantum wells separated by $Al_{0.3}Ga_{0.7}As$ barriers δ-doped with donors (Si) near their centers.[6,7] Depending on the external time constant of the sample, three different techniques, i.e., AC or pulse in either the relaxation or quasi-adiabatic regimes, were used to measure C in a dilution refrigerator with magnetic field (B) applied either perpendicular to the 2DES plane ($\theta = 0°$) or at an angle of $\theta = 30 \pm 2°$.[6]

The data in Fig. 1(a) demonstrate the oscillatory behavior of C due to the oscillating density of states at the Fermi level of a 2DES in a perpendicular B.[2] From the position of the minima, we infer a 2DES density (per layer) of $n_e \approx 1.4 \times 10^{11}$ cm^{-2}, consistent with the magnetotransport data.[7]

Near $\nu = 1$, C reveals a completely different and unexpected behavior.

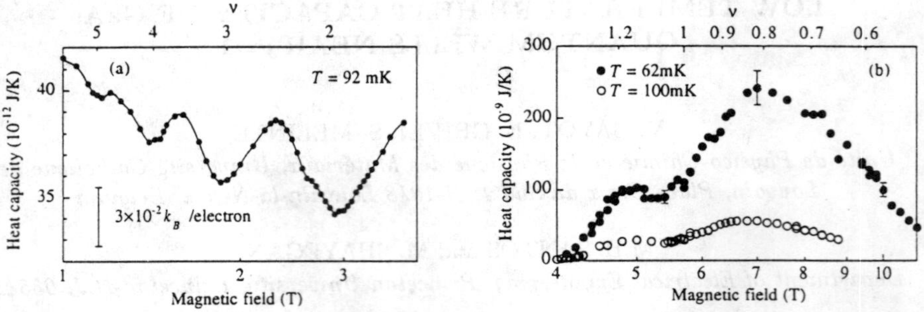

Figure 1: Heat capacity at $\theta = 0°$, showing orders of magnitude enhancement of the high-B data (b) over the low-B data (a). The line through the data points is a guide to the eye.

Figure 1(b) shows orders of magnitude enhancement of C with respect to the lower-B data of Fig. 1(a), including large maxima at $\nu \approx 0.8$ and 1.2. Moreover, (Fig. 2(a)) shows that in contrast to the low-B data where C *decreases* with decreasing T (not shown here), C first *increases* with decreasing T and then displays a very sharp peak at $T_c \approx 36$ mK before decreasing at very low T.

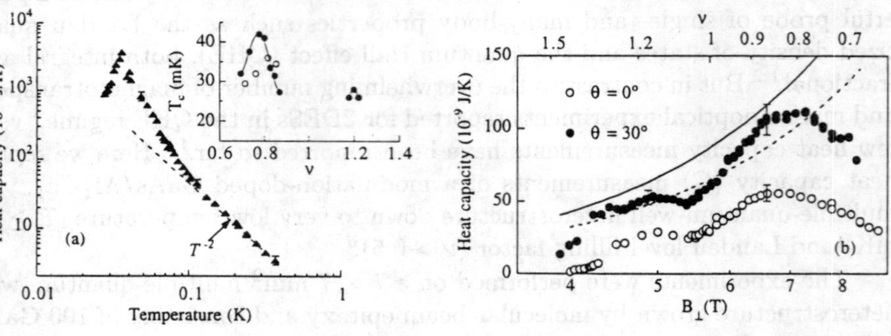

Figure 2: (a) The temperature dependence of C at B=7T and $\theta = 0°$ is shown in a log-log plot. The dashed line shows the T^{-2} dependence expected for the Schottky model. The ν-dependence of T_c at $\theta = 0°$ (∘) and 30° (•) is shown in the inset and the lines are guides to the eye. (b) C vs $B_\perp = B\cos(\theta)$ and ν at T=100mK, and at the indicated values of θ. The curves correspond to the calculated nuclear-spin heat capacity of Ga and As atoms in the quantum wells (C_N) for $\theta = 0°$ (dashed) and 30° (solid).

Both the very large magnitude of C and the T^{-2} dependence observed in the range $0.1 < T < 0.4$K point to the nuclear Schottky effect which results from the entropy reduction of the nuclear spin system (NSS) with decreasing T when the thermal energy k_BT is much larger than the spin energy level spacing Δ.[8] The observation of the Schottky effect requires good coupling of the NSS to the lattice in order to reach thermal equilibrium in the time scale of the experiment. This coupling is provided by electron spin-flip excitations and further relaxation to the lattice.[9] At first sight, because of their high purity and the low free-carrier density in the quantum wells, GaAs/AlGaAs heterostructures should not be good candidates for the observation of a nuclear Schottky effect. While this is supported by the low-B data (Fig. 1(a)), surprisingly a Schottky behavior is observed at higher B near $\nu = 1$.

Recent theoretical [10,11,12] and experimental [13,14,15,16] work on 2DESs has shown that, in the limit of a weak Zeeman coupling, electron spin textures known as *Skyrmions* are the lowest energy, charged excitations of the ferromagnetic ground state near ν=1. Skyrmions result from the dominance of the Hartree energy over the Zeeman energy, prohibiting single spin-flip excitations and favoring smooth distorsions of the spin field; the competition between the two energies determines the total spin and size of Skyrmion quasi-particles.

Moreover, nuclear magnetic resonance experiments [13,14] have provided clear evidence for a strong coupling of the NSS to the lattice through finite-size Skyrmions near $\nu = 1$. Based on these observations, we suggest that near ν=1 Skyrmions induce the strong coupling of the NSS to the lattice needed to observe the large nuclear heat capacity. The key role of Skyrmions is supported by: (1) the absence of the nuclear-spin contribution to our measured C for $\nu > 1.5$ where the Skyrmions are no longer the ground state of the 2DES,[11,12,15] and (2) our experiments in a tilted B, presented in Fig. 2(b), which clearly show that the field position of the heat capacity anomaly relates to the 2DES filling factor (rather than total B).

If we assume that only the nuclear spins of the Ga and As atoms in the quantum wells contribute to the observed C, Fig. 2(b) indicates that the calculated C_N [6,17] is semi-quantitatively consistent with both the size and the overall $\sim B^2$ dependence of the experimental data. While the decrease in C very near $\nu = 1$ can be attributed to the decreasing density of Skyrmions,[12] its decrease very far from $\nu = 1$ (i.e. $\nu > 1.2$ and $\nu < 0.85$) can be related to the 2DES approaching fillings where the Skyrmions are no longer relevant.

We observe in Fig. 2(a) that the $C \propto T^{-2}$ behavior is followed only down to ~0.1K. For $T < 0.1$K, C increases faster with decreasing T and, in a narrow range of ν, C exhibits a remarkably sharp peak at very low T. The deviation of C from the T^{-2} dependence at $T \gg \Delta/k_B$ and the shape of the observed peak are not consistent with the Schottky model which predicts a smooth maximum in C at $T \sim \Delta/2k_B$ (~2 mK in our case). Instead, the shape and sharpness of this peak are suggestive of a *phase transition.* The inset to Fig. 2(a) shows that the peak temperature T_c is maximum at $\nu \approx 0.8$ and 1.2, and decreases as $\nu \to 1$. It is tempting to associate the sharp peak in C vs T observed at low

T with the crystallization of Skyrmions, together with the associated magnetic ordering, near ν=1 which was recently proposed by Brey *et al.* [12]. The fact that our observed T_c decreases as $\nu \to 1$ is consistent with the decreasing Skyrmion density which should reduce the Skyrme-solid melting temperature. However, the details of the Skyrmion liquid-solid transition and, in particular, how it would affect the NSS are not known.

The authors are much indebted to S.M. Girvin, A.H. MacDonald and K.A. Moler for fruitful discussions and suggestions. This work has been supported by NATO grant CRG 950328 and the NSF MRSEC grant DMR-9400362. V.B. acknowledges financial support of the Belgian National Fund for Scientific Research.

References

1. For a review of the QHE see *"The Quantum Hall Effect"*, edited by R.E. Prange and S.M. Girvin (Springer-Verlag, New York, 1987).
2. For theoretical work on heat capacity of the 2DES in the QHE regime, see, e.g., W. Zawadski and R. Lassnig, Solid State Commun. **50**, 537 (1984); A.H. MacDonald, H.C.A. Oji and K.L. Liu, Phys. Rev. B **34**, 2681 (1986); D. Yoshioka, J. Phys. Soc. Jpn. **56**, 1301 (1987); Q. Li, X.C. Xie and S. Das Sarma, Phys. Rev. B **40**, 1381 (1989).
3. For recent results, see, e.g., Surf. Sci. **305** (1994).
4. E. Gornik *et al.*, Phys. Rev. Lett. **54**, 1820 (1985).
5. J.K. Wang *et al.*, Phys. Rev. B **38**, 6174 (1988); Phys. Rev. B **45**, 4384 (1992).
6. V. Bayot *et al.*, Phys. Rev. Lett. **76**, 4584 (1996).
7. V. Bayot, M.B. Santos and M. Shayegan, Phys. Rev. B **46**, 7240 (1993).
8. W. Schottky , Phys. Z **23**, 448 (1922).
9. See e.g. A. Abragam and M. Goldman, Rep. Prog. Phys. **41** 395 (1978).
10. D.-H. Lee and C.L. Kane, Phys. Rev. Lett. **64**, 1313 (1990); S.L. Sondhi *et al.*, Phys. Rev. B **47**, 16419 (1993); H.A. Fertig *et al.*, Phys. Rev. B **50**, 11018 (1994); N. Read and S. Sachdev, Phys. Rev. Lett. **75**, 3509 (1995).
11. X.-G. Wu and S.L. Sondhi, Phys. Rev. B **51**, 14725 (1995).
12. L. Brey *et al.*, Phys. Rev. Lett. **75**, 2562 (1995).
13. R. Tycko *et al.*, Science **268**, 1460 (1995).
14. S.E. Barrett *et al.*, Phys. Rev. Lett. **74**, 5112 (1995).
15. A. Schmeller *et al.*, Phys. Rev. Lett. **75**, 4290 (1995).
16. E.H. Aifer *et al.*, Phys. Rev. Lett. **76**, 680 (1996).
17. S.M. Girvin and A.H. MacDonald, private communication.

THEORY OF PHOTOLUMINESCENCE AT $\nu = 1$

N. R. COOPER* and D. B. CHKLOVSKII†
*Institut Laue-Langevin, B.P. 156, 38042 Grenoble, Cedex 9, France.
†Lyman Laboratory of Physics, Harvard University, Cambridge, MA 02138, USA.

We study the theory of intrinsic photoluminescence of two-dimensional electron systems in the vicinity of the $\nu = 1$ quantum Hall state. We identify a band of low-energy states ("excitonic states") of a photo-excited electron-hole pair in the presence of the electron system at $\nu = 1$, and focus on the photoluminescence spectra that arise from the radiative recombination of these initial states. The spectra are found to be strikingly different for the two circular polarizations. In one polarization (σ_+), radiative recombination of the valence-band hole occurs only into the $\nu = 1$ groundstate, and the photoluminescence spectrum is a narrow line. In the other polarization (σ_-), the recombination of the valence-band hole occurs into final states in which a *spin-wave* of the $\nu = 1$ groundstate has been excited. The photoluminescence spectrum in this polarization acquires a broad low-energy tail due to the "shake-up" of these spin-waves, and has a width that is closely related to the spin-wave stiffness. This polarization-dependent structure is consistent with recent observations of the photoluminescence spectra of a high-mobility GaAs quantum well at $\nu = 1$. We derive line shapes for the recombination of excitonic states in the presence of long-range disorder, which compare favourably with the experimental observations.

1 Introduction

In recent years the use of intrinsic photoluminescence (PL) to study high-mobility two-dimensional electron systems in the extreme quantum regime has attracted a great deal of experimental [1] and theoretical [2] attention. Of particular interest is the possibility of using this technique to obtain information on the strongly-correlated states in the fractional quantum Hall and Wigner crystal regimes. Due to the difficulty of describing the energy eigenstates of a photo-excited valence-band hole in the presence of the electron gas, the theory underlying PL is still not well understood and many questions remain open.

We study the theory of PL close to a filling fraction of $\nu = 1$. This presents a much simpler situation for the theorist than the fractional quantum Hall or Wigner crystal regimes. Nevertheless, we show that the PL spectra at this filling fraction contain non-trivial structure related to the spin-wave excitations and the "charged spin-textures" [3] of the $\nu = 1$ state.

In the following, we report the results we have obtained [4] for the form of the PL spectrum in GaAs quantum wells at a filling fraction of exactly $\nu = 1$. We restrict electrons to states in the lowest Landau level, but allow Landau level mixing for the valence-band hole, treating its dispersion as parabolic with an effective mass m_h. The asymmetry of the single-side doped quantum wells typically used in experiment is represented by separating the planes in which the electrons and the hole move by a distance d.

2 Excitonic Recombination at $\nu = 1$

Prior to photo-excitation, the groundstate of the electron system at $\nu = 1$ is the state $|g.s.\rangle$ in which the electrons fill all spin-$\uparrow$ states in the lowest Landau level and all spin-$\downarrow$ states are unoccupied.[3] Imagine that a (photo-excited) electron-hole pair is now introduced. There exists a set of states in which all spin-$\uparrow$ electron states remain occupied, and the additional (spin-$\downarrow$) electron binds with the valence-band hole to form an exciton. The many-body state in which this exciton has the wavefunction $\Psi_{\mathbf{P}}(\mathbf{r}_e, \mathbf{r}_h)$ is

$$|\mathbf{P}\rangle \equiv \int d^2\mathbf{r}_e d^2\mathbf{r}_h \; \Psi_{\mathbf{P}}(\mathbf{r}_e, \mathbf{r}_h)\psi^\dagger_{e\downarrow}(\mathbf{r}_e)\psi^\dagger_h(\mathbf{r}_h)|g.s.\rangle, \qquad (1)$$

where $\psi^\dagger_{e,h}$ are the electron and hole field creation operators. We refer to these states as "excitonic states". They provide a complete set of states within the subspace with maximum spin polarization.

If the electron Zeeman energy Z is large compared to all other energy scales, the excitonic states are the lowest-energy states of the system (all other states are at least Z higher in energy). Numerical calculations on finite systems suggest that they also provide a good description of the low energy states in the limit $Z \to 0$, provided the spacing d is sufficiently small ($\lesssim 1.3 l_B$).[4] It is therefore likely that excitonic states will make an important contribution to the low-temperature PL spectra. (The importance of the $\mathbf{P} = 0$ excitonic state in luminescence has been noticed previously.[5])

The radiative recombination of the excitonic states is quite different in the two different circular polarizations, which involve the recombination of electrons of opposite spin. In the σ_+ polarization, the hole recombines with the single spin-$\downarrow$ electron and can leave only the $\nu = 1$ groundstate $|g.s.\rangle$ as a final state. Momentum conservation limits this process to the $\mathbf{P} = 0$ excitonic state. In the σ_- polarization, the hole can recombine with any of the many spin-$\uparrow$ electrons. There are many possible final states, described by the complete set of single spin-wave excitations.[6] In this polarization, *all* of the excitonic states can recombine, the momentum of the initial state being transferred to the final state spin-wave. The recombination processes in both polarizations are illustrated pictorially in Fig. 1(a), and the resulting transitions are shown on the energy-level diagram in Fig. 1(b).

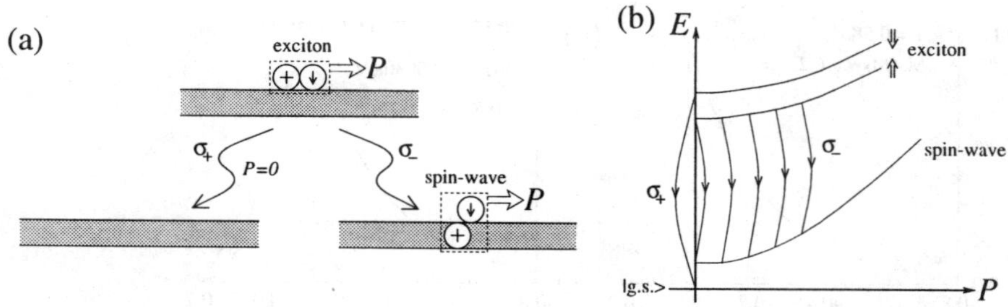

Figure 1: Schematic diagrams of (a) the radiative decay processes of an excitonic state, and (b) the resulting energy transitions (the two excitonic bands in this figure represent the two hole subband states each of which gives rise to one of the two circular polarizations).

Calculations of the exciton and spin-wave effective masses (M_X, M_{SW}) at arbitrary m_h and d show that $M_X \geq M_{SW}$ (the equality holds only for $m_h = d = 0$). For $M_X > M_{SW}$, as is the case illustrated in Fig. 1(b), the emission line in the σ_- spectrum moves lower in energy for excitonic states of increasing momentum; a population of high-momentum excitonic states will therefore produce a low-energy tail in this polarization, due to the "shake-up" of high energy spin-waves. If the excitonic states are populated according to a Boltzmann distribution at a temperature T, the line shape is

$$\exp\left[\frac{E}{k_B T(M_X/M_{SW}-1)}\right]\Theta\left(-E\right), \tag{2}$$

with energy measured relative to that of the $\mathbf{P} = 0$ excitonic state in this polarization. The linewidth depends on the spin-wave stiffness $1/M_{SW}$. In the σ_+ polarization only the zero-momentum excitonic state can recombine, so the spectrum in this polarization contains a δ-function peak.

Experimental observations [7] of the low-temperature PL spectra of a GaAs quantum well at $\nu = 1$ show qualitative behaviour similar to that which we have just described: the σ_+ spectrum is a narrow line; and the σ_- spectrum shows a broad low-energy tail. The line shape (2) with the parameters appropriate for these experiments is shown in Fig. 2(a). At low temperatures, the observed line shape becomes temperature-independent, so the form (2) cannot be correct.

The temperature-independence of the line shape suggests that disorder is important in this sample. The dominant contribution to the PL line width is thought to arise from the ionized donors lying a distance s (the spacer-layer thickness) from the quantum well. Within a theory [8] that takes account of positional correlations of these donors, we estimate the rms fluctuation in the exciton energy to be $V_{rms} = 0.093$meV. This is sufficiently large to lead to

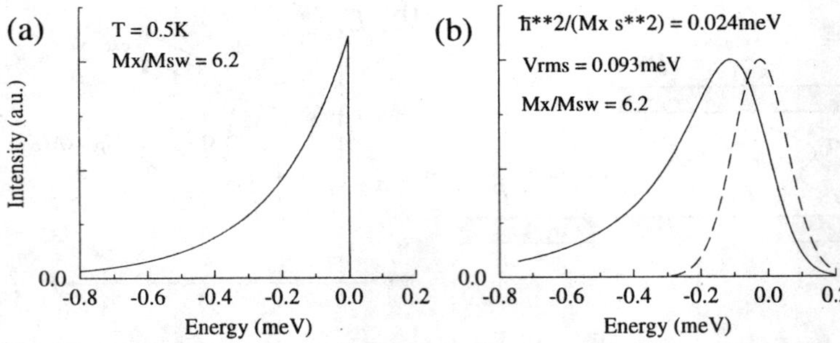

Figure 2: Spectra of recombination of the excitonic states in the σ_- (solid lines) and σ_+ (dashed line) polarizations for (a) no disorder (the σ_+ spectrum is a delta-function at zero energy and is not shown) and (b) long-range disorder arising from the remote ionized donors.

strongly localized low-energy excitonic states, which shake-up high-momentum spin-waves on recombination. We have derived line shapes for the excitonic recombination in the presence of this disorder, assuming the excitons to occupy harmonic oscillators states in all minima of the disorder potential. The results, shown in Fig. 2(b), compare favourably with the experimental line shapes.[7]

This work was partially supported by the NATO Science Fellowship Programme, by the Harvard Society of Fellows and by NSF grant DMR 94-16910.

References

1. R. G. Clark, in *Low Dimensional Electronic Systems: New Concepts*, Vol. 111 of *Springer Series in Solid-State Sciences*, edited by G. Bauer, F. Kuchar, and H. Heinrich (Springer-Verlag, Berlin, 1992), pp. 239–255.
2. A. H. MacDonald, E. H. Rezayi, and D. Keller, Phys. Rev. Lett. **68**, 1939 (1992); V. M. Apalkov, F. G. Pikus, and E. I. Rashba, Phys. Rev. B **52**, 6111 (1995); N. R. Cooper, Phys. Rev. B **53**, 10804 (1996).
3. S. L. Sondhi, A. Karlhede, S. A. Kivelson, and E. H. Rezayi, Phys. Rev. B **47**, 16419 (1993); A. MacDonald, H. Fertig, and L. Brey, Phys. Rev. Lett. **76**, 2153 (1996).
4. N. R. Cooper and D. B. Chklovskii, (unpublished).
5. B. A. Muzykantskii, Sov. Phys. JETP **74**, 897 (1992).
6. C. Kallin and B. I. Halperin, Phys. Rev. B **30**, 5655 (1984).
7. F. Plentz *et al.*, in *High Magnetic Fields in the Physics of Semiconductors*, edited by D. Heiman (World Scientific, 1995), pp. 320–323.
8. F. G. Pikus and A. L. Efros, Sov. Phys. JETP **69**, 558 (1989).

MAGNETOTRANSPORT IN COMBINED MAGNETIC AND ELECTRIC SUPERLATTICES

ROLF R. GERHARDTS
Max-Planck-Institut für Festkörperforschung, Heisenbergstraße 1,
D-70569 Stuttgart, Federal Republic of Germany

One- or two-dimensional periodic arrays of micromagnets on the surface of a GaAs-AlGaAs heterostructure containing a two-dimensional electron gas (2DEG) can impose a magnetic and a stress-induced electric lateral superlattice upon the 2DEG. We discuss combined electric and magnetic superlattices of different symmetries, which can be generated by tuning magnetization strength and direction of the surface magnets, and we calculate the resulting magnetoresistance of the 2DEG within a simple quasi-classical approach.

Introduction

Two-dimensional electron gases (2DEG) subject to a perpendicular homogeneous magnetic field B_0 and weak periodic modulation fields, show commensurability oscillations of the magnetoresistance, the so called Weiss oscillations, which have first been observed in AlGaAs-GaAs heterostructures with a weak *electrostatic* modulation[1]. They have been understood [2–4] within a quantum mechanical picture, as resulting from modulation-broadened Landau bands, with an oscillatory dependence of the bandwidth on the Landau quantum number. The dispersion of these Landau bands leads to a group velocity, and this to a "band conductivity", which both vanish when the Landau bands become flat. For a sinusoidal potential modulation $V(x) = V_0 \cos Kx$ and in the quasiclassical limit, where many Landau levels are occupied, this happens when the "electric flat-band conditions"

$$2R_c = a\,(\lambda_e - 1/4)\,, \qquad \lambda_e = 1,\,2,\,\ldots\,, \tag{1}$$

hold, where R_c is the cyclotron radius at the Fermi energy of the 2DEG and $a = 2\pi/K$ is the modulation period.

Similar oscillations (2DEG in the x-y plane) with a 1D lateral superlattice defined by a periodic *magnetic* field $B_z(x) = B_m \cos Kx$ in z direction have been predicted [5–7], with corresponding "magnetic flat-band conditions"

$$2R_c = a\,(\lambda_m + 1/4)\,, \qquad \lambda_m = 1,\,2,\,\ldots\,, \tag{2}$$

describing the minima position of these *magnetic* Weiss oscillations.

Recently magnetic Weiss oscillations were observed on samples with a surface array of metallic strips, consisting either of superconducting [8] or of ferromagnetic [9,10] material. Due to the different thermal expansion coefficients of metal and semiconductor, the strips exert, at the low temperatures of the experiment, mechanical stress on the underlying semiconductor, which leads to a periodic modulation of the conduction band edge and acts like an electrostatic modulation on the 2DEG [11,12]. For the understanding of the magnetic Weiss oscillations, it is important to be aware of the presence of this stress-induced electric modulation [9]. In the experiment with superconducting strips, the stress modulation was so dominant that it was difficult to resolve the effect of the magnetic modulation [8].

On the other hand, the deposition of dysprosium (Dy) micromagnets on the surface leads to very strong modulation effects [9]. Samples with one- and with two-dimensional periodic arrays of such micromagnets have recently been produced and investigated [9,13–15]. We will demonstrate in the following that, by tuning the strength and the magnitude of the magnetization of these surface magnets, magnetic superlattices of very different symmetries can be generated which, together with the stress-induced electric superlattice, lead to interesting interference effects on the magnetoresistance. We will treat these effects within a simple classical approach [16–18], but we will also discuss the limitations of this approach and possible improvements.

Quasi-classical approach

Resistivity formula

We treat the 2DEG (in the x-y plane) as a degenerate Fermi gas of non-interacting electrons with charge $-e$ and effective mass m, which obey classical Newton dynamics. The transport properties are determined by the electrons at the Fermi energy $E_F = \hbar^2 k_F^2/2m$, which in the homogeneous case without modulation move in the homogeneous magnetic field B_0 on circular orbits with the cyclotron radius $R_c = v_F/\omega_c$, where $v_F = \hbar k_F/m$ and $\omega_c = eB_0/mc$ are the Fermi velocity and the cyclotron frequency, respectively. Describing dissipative processes by a relaxation time τ, one obtains the Drude resistivity tensor with components $\rho_{xx}^0 = \rho_{yy}^0 = \rho_0$ and $\rho_{xy}^0 = -\rho_{yx}^0 = \omega_c\tau\rho_0$, where $\rho_0 = m/(e^2\tau n_{\rm el})$ and $n_{\rm el} = k_F^2/2\pi$ is the electron density.

We assume weak modulations, $V(\vec{r})$ of the potential energy and $B_{\rm m}(\vec{r})$ of the z component of magnetic field, which are periodic on the same square lattice and have vanishing average values, so that in their Fourier expansions $V(\vec{r}) = \sum_{\vec{q}\neq 0} V_{\vec{q}} \exp(i\vec{q}\cdot\vec{r})$ and $B_{\rm m}(\vec{r}) = \sum_{\vec{q}\neq 0} B_{\vec{q}} \exp(i\vec{q}\cdot\vec{r})$ the $\vec{q}$ sums are over $\vec{q} = (Kn_x, Kn_y, 0)$ with integers n_x and n_y which are not simultaneously zero. We assume that the forces exerted by the modulation fields on the electrons are much weaker than the Lorentz force due to B_0, so that the effect of the modulations reduces to a drift motion of the guiding centers of the (weakly disturbed) cyclotron orbits. Then one can easily calculate, from Newton's equation, the average drift velocity of the guiding centers to first order and the correction to the Drude conductivity tensor to second order in the modulation amplitudes, and, by tensor inversion, one obtains [18], to leading order in $(\omega_c\tau)^{-1} \ll 1$, the modulation correction to the Drude resistivity tensor as

$$\Delta\rho_{\mu\nu} = \rho_0 \frac{\lambda_f^2}{2E_F^2} \sum_{\vec{q}\neq 0} q_\mu q_\nu \, |\mathcal{S}_{\vec{q}}|^2 \,, \tag{3}$$

where $\lambda_f = \tau v_F$ is the mean free path of the homogeneous 2DEG and

$$\mathcal{S}_{\vec{q}} = \sigma_{B_0} V_{\vec{q}} J_0(R_c q) + (k_F/q)\hbar\omega_{\vec{q}} J_1(R_c q) \,, \tag{4}$$

with $\sigma_{B_0} = B_0/|B_0|$, $q = |\vec{q}|$, $\omega_{\vec{q}} = eB_{\vec{q}}/mc$, and $J_\nu(x)$ the Bessel functions of the first kind. It is interesting to note that the sign of the electric contribution to Eq. (4) changes if the direction of the homogeneous magnetic field $\mathbf{B_0} = B_0\mathbf{e_z}$ is reversed, while that of the magnetic contribution does not. This follows from Newton's equation of motion, and is visualized in Fig. 1 for the simple case of step modulations. For sufficiently small values of the homogeneous magnetic field, actually for $qR_c \equiv q\lambda_f/\tau\omega_c \gtrsim 2$, the Bessel functions in Eq. (4) may be replaced by their large-argument asymptotic expansions, so that

$$\frac{\mathcal{S}_{\vec{q}}}{E_F} \approx \left(\frac{2}{\pi q R_c}\right)^{1/2} \left[\sigma_{B_0} \frac{V_{\vec{q}}}{E_F} \cos\left(qR_c - \frac{\pi}{4}\right) + \frac{2\,\omega_{\vec{q}}}{q v_F} \sin\left(qR_c - \frac{\pi}{4}\right)\right] . \tag{5}$$

For a sinusoidal pure electric or pure magnetic modulation, this yields the flat-band conditions Eq. (1) and Eq. (2), respectively.

Modulation model

In order to model the magnetic stray fields of lattices of metal (Dy) strips [9] and posts [13] on the surface of AlGaAs-GaAs heterostructures, we simplify their geometrical form and assume homogeneous magnetization. We take the surface as x-y plane and the semiconductor in the halfspace $z < 0$, with the 2DEG in a plane $z = -D < 0$. For a square lattice of cylinders we assume at the lattice sites $\vec{R} = a\,(n_x, n_y, 0)$, for all integer values n_x and n_y, replica of a cylinder at the origin, with height h, $0 \le z \le h$, and radius r_0, $x^2 + y^2 \le r_0^2$. For a one-dimensional array of strips with rectangular cross sections and translational

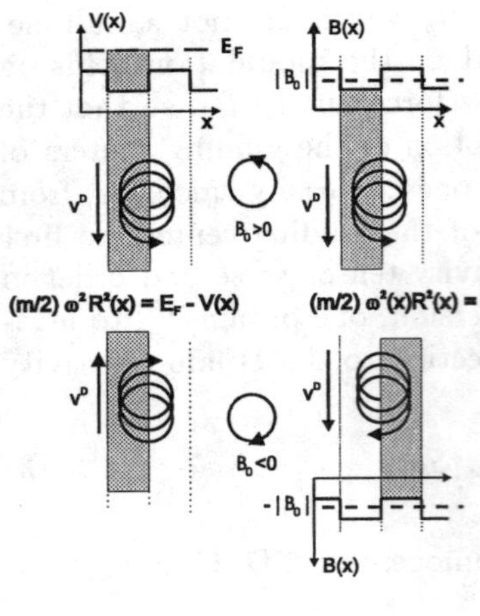

Fig. 1: Change of drift velocity $\vec{v}^D$ under inversion of B_0. Left: for pure electric modulation, electrons pass through unchanged orbits in opposite direction, $\vec{v}^D$ changes sign. Right: for pure magnetic modulation, electrons pass through changed orbits [e.g. if $B(x-x_0)$ odd, orbit reflected at $x=x_0$] in opposite direction, $\vec{v}^D$ is unchanged. The shaded area has the larger cyclotron radius $R(x)$.

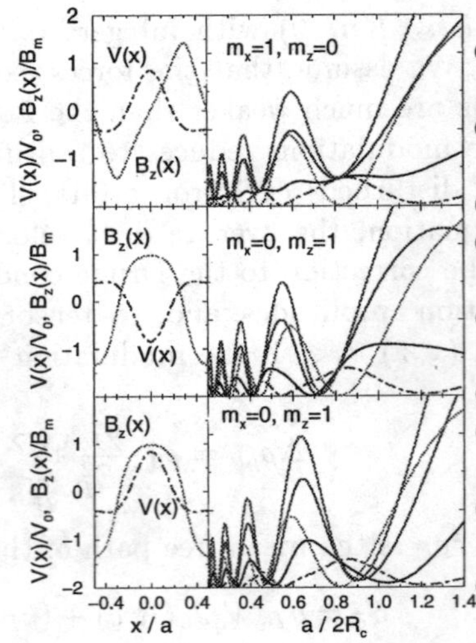

Fig. 2: Modulation fields (left panels) and resulting magnetoresistance (right panels) for a 1D lattice (period a) of magnetized strips of width $b=0.6a$ and height $h=0.4a$, and a 2DEG at $z=-0.1a$. Dash-dotted lines refer to pure electric modulation, i.e. $\mu=0$, the dotted lines to purely magnetic modulation, $\varepsilon=0$ (for definition of ε and μ see text) . Solid lines refer to mixed modulation with $\varepsilon=1$ in the top and the bottom, and $\varepsilon=-1$ in the middle panels, combined with μ =0.5, 1.0, 1.25 (from bottom to top curve near maxima) in each of the right panels. The scaling factor is $\alpha = 2a^2/(2\pi\gamma\lambda_f)^2$

invariance in y direction, we assume correspondingly a square lattice with the metal volume at the origin located in $0 \le z \le h$, $|x| \le b/2$, and $|y| \le a/2$. Then, the magnetic field $\mathbf{B}(\vec{r}, z)$ produced at an arbitrary position by the lattice of micromagnets at an arbitrary position can be calculated as a function of the magnetic moment $\mathbf{m} = \vec{m}_{\parallel} + m_z \mathbf{e_z}$ per unit cell [18].

Considering a strictly two-dimensional model for the electron gas, we neglect the x and y components of the magnetic flux density and calculate the magnetic modulation field $B_{\rm m}(\vec{r}) = B_z(\vec{r}, z=-D)$ from the Fourier coefficients

$$B_{\vec{q}}(z) = (2\pi/a^2)\left[i\vec{q}\cdot\vec{m}_{\parallel} + q\, m_z\right] F_{\vec{q}} \exp(-q|z-h/2|)\,. \qquad (6)$$

For the square lattice of cylinders and for the strip array the form factors are

$$F_{\vec{q}} = \frac{2J_1(qr_0)}{qr_0}\,\frac{\sinh(qh/2)}{qh/2} \quad \text{and} \quad F_{\vec{q}} = \delta_{q_y,0}\,\frac{\sin(q_x b/2)}{q_x b/2}\,\frac{\sinh(qh/2)}{qh/2}\,, \qquad (7)$$

respectively. The spatial orientation of magnetization of the micromagnets determines the complex phases of the Fourier coefficients. A magnetization in the z direction ($\vec{m}_{\|} = 0$) leads to real Fourier coefficients, i.e. a pure cosine expansion of the modulation magnetic field, whereas a magnetization parallel to the plane of the 2DEG ($m_z = 0$) leads to purely imaginary coefficients, i.e. a sine expansion.

In order to describe the stress effects mentioned in the introduction, we choose simple models [18] which are in accord with previous theoretical and experimental investigations [11,12]. For the 1D strip modulation we take the Fourier coefficients

$$V_{q_x} = V_0\sqrt{\pi}(\lambda/a)(\lambda q_x)^2 e^{-(\lambda q_x)^2/4} e^{-|zq_x|}\,, \qquad \lambda = b/\sqrt{6}\,, \tag{8}$$

and for the square lattice of cylinders we take the Fourier coefficients

$$V_{\vec{q}} = V_0(\pi\lambda^2/a^2)(\lambda q)^2 e^{-(\lambda q)^2/4} e^{-|zq|}\,, \qquad \lambda = \rho/\sqrt{2}\,. \tag{9}$$

Results

One-dimensional modulation

Figure 2 shows in the left panels electrostatic potentials (dash-dotted lines) and magnetic modulation fields (dotted lines) within one period of a 1D lattice defined by ferromagnetic strips on the surface, and in the right panels the magnetoresistance obtained for superpositions of these modulations with different weights. The stress-induced potential is calculated from Eq. (8) with $V_0/E_F = \gamma\varepsilon$. The strength of the magnetic modulation, calculated from Eqs. (6) and (7) with $B_{\mathrm{m}} = 2\pi|\mathbf{m}|/a^3$, is measured by $(eB_{\mathrm{m}}/mc)\,a/\pi v_F = \gamma\mu$, and the common factor γ is scaled out of the magnetoresistance curves. In the uppermost panels $\mathbf{m}$ is taken in x direction, in the other panels in z direction. In the bottom panels $V(x)$ and $B_{\mathrm{m}}(x)$ have the same phase, and in the middle panels the sign of $V(x)$ is reversed, which is equivalent with keeping $V(x)$ and $B_{\mathrm{m}}(x)$ unchanged while inverting the direction of the homogeneous field B_0. As is seen in the left panels in Fig. 2, the fields contain higher harmonics. These cause the magnetoresistance curves in the lower right panel not to go down to zero at their minima, and, for purely electrostatic modulation, they shift the minima to slightly lower B_0 values than those given by the ideal flat-band conditions, Eq. (1). The effect of higher harmonics on the dotted curves, representing the results for purely magnetic modulations, is weaker, as is to be expected from Eq. (4), and the minima occur near the magnetic flat-band conditions, Eq. (2). As the magnetic modulation is increased, the minima of $\Delta\rho_{xx}$ in the lowest panel shift from the λ_e-th electric flat-band condition towards the magnetic flat-band condition with $\lambda_m = \lambda_e$. In the middle panel they shift from the $(\lambda_e + 1)$-st electric flat-band condition towards the magnetic flat-band conditions with $\lambda_m = \lambda_e$. Comparing this result with the experimentally observed asymmetry with respect to $B_0 = 0$ under a sweep of the applied homogeneous magnetic field B_0, one finds that $V(x)$ has minima underneath the dysprosium strips ($V_0 < 0$). [9,14]

In the two lower right panels of Fig. 2, at the magnetic flat-band conditions all the curves with the same amount of electric modulation ($\varepsilon = 1$) intersect each other at the same value for the magnetoresistance. In contrast to this, the magnetoresistance curves in the upper panel do *not intersect* each other at the magnetic flat-band conditions, although they all assume the same values there. This characteristic difference between the two directions of magnetization is a direct consequence of the different complex phases of the Fourier coefficients in both cases [7,18], and has recently also been observed in experiments [14].

Square lattice of magnetized cylinders

In samples with a two-dimensional periodic array of micromagnets, an in-plane component of the magnetization leads to an anisotropic magnetoresistance tensor. In contrast to the Drude resistivity tensor $\rho^0_{\mu\nu}$, which is invariant under rotations of the coordinate system, the modulation induced correction $\Delta\rho_{\mu\nu}$ is a *symmetric* tensor, which can be diagonalized by such a rotation. Its structure is easily analyzed from Eqs. (3) and (4). Writing the magnetic moment of a cylinder as $\mathbf{m} = (m_\parallel \cos\varphi, m_\parallel \sin\varphi, m_z)$ and using Eqs. (6) and (7), we may write $k_F\hbar\omega_{\vec{q}}(z) = [i\vec{q}\cdot\vec{m}_\parallel + qm_z]\, U_q(z)$, where U depends only on $q = (q_x^2 + q_y^2)^{1/2}$. Since, according to Eq. (9), also $V_{\vec{q}} \equiv V_q$, we can exploit the square symmetry in the $\vec{q}$-sum of Eq. (3) to write

$$\Delta\rho_{\mu\nu} = \Delta\rho^\perp \delta_{\mu\nu} + \Delta\rho^\parallel_{\mu\nu}\,, \tag{10}$$

where the first, isotropic term with

$$\Delta\rho^\perp = \rho_0 \frac{2\lambda_f^2}{E_F^2} \sum_{q_x,q_y\geq 0} q_x^2 \left[\sigma_{B_0} V_q J_0(R_c q) + m_z U_q J_1(R_c q)\right]^2 \tag{11}$$

describes the interference between the electrostatic modulation and the magnetic modulation produced by the z component of the magnetization. The second, anisotropic term in Eq. (10) is independent of the electric modulation and has the components

$$\Delta\rho^\parallel_{xx} \;=\; \cos^2\varphi\, R_{xx} + \sin^2\varphi\, R_{xy}, \quad \Delta\rho^\parallel_{yy} = \sin^2\varphi\, R_{xx} + \cos^2\varphi\, R_{xy}$$

$$\Delta\rho^{\parallel}_{xy} \;=\; \Delta\rho^{\parallel}_{yx} = 2\sin\varphi\cos\varphi\, R_{xy}\,, \tag{12}$$

where

$$R_{\mu\nu} = \rho_0 \frac{2\lambda_f^2 m_\parallel^2}{E_F^2} \sum_{q_x, q_y \geq 0} q_\mu^2 q_\nu^2 \left[\frac{1}{q} U_q J_1(R_c q)\right]^2 . \tag{13}$$

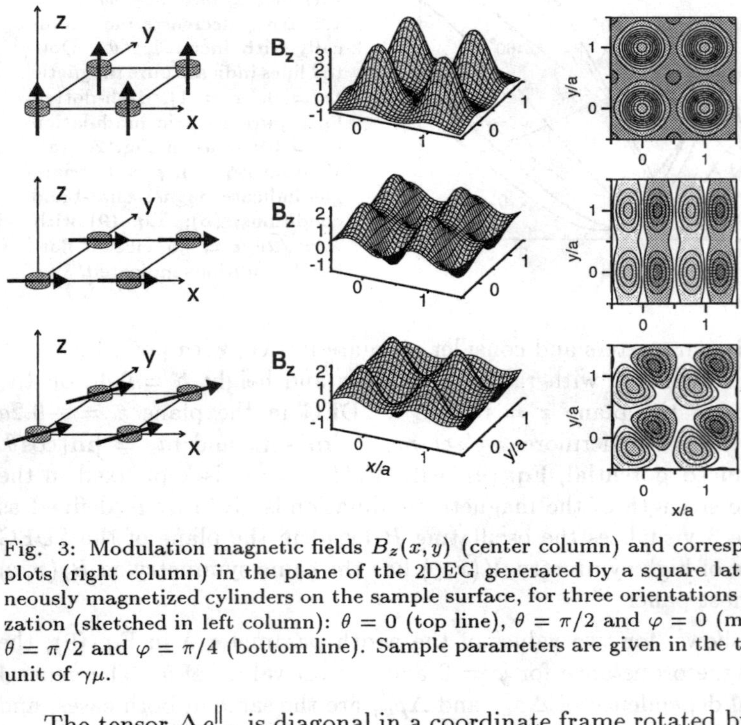

Fig. 3: Modulation magnetic fields $B_z(x,y)$ (center column) and corresponding contour plots (right column) in the plane of the 2DEG generated by a square lattice of homogeneously magnetized cylinders on the sample surface, for three orientations of the magnetization (sketched in left column): $\theta = 0$ (top line), $\theta = \pi/2$ and $\varphi = 0$ (middle line), and $\theta = \pi/2$ and $\varphi = \pi/4$ (bottom line). Sample parameters are given in the text, $B_z(x,y)$ in unit of $\gamma\mu$.

The tensor $\Delta\rho^{\parallel}_{\mu\nu}$ is diagonal in a coordinate frame rotated by the angle γ, with

$$\tan 2\gamma = -2R_{xy}\,\tan 2\varphi\,/(R_{xx} - R_{xy}) \tag{14}$$

with eigenvalues

$$\Delta\rho^{\parallel}_{\pm} = \frac{1}{2}\,(R_{xx}+R_{xy}) \pm \frac{1}{2}\left[(R_{xx}-R_{xy})^2\cos^2 2\varphi + 4R_{xy}^2\sin^2 2\varphi\right]^{\frac{1}{2}} . \tag{15}$$

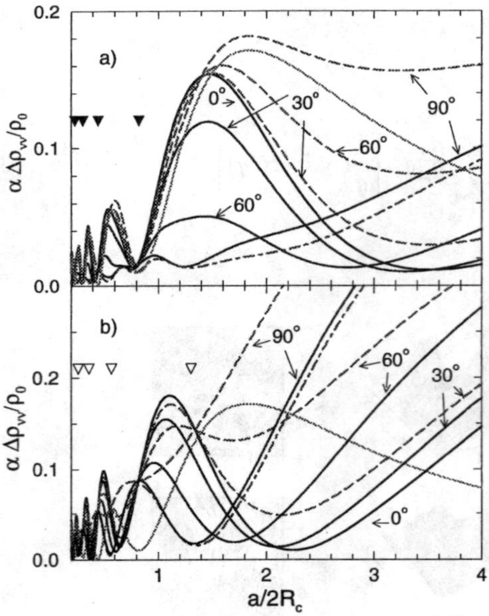

Fig. 4: Magnetoresistance curves for tilt angles $\theta = 0^o$, 30°, 60°, 90° of the magnetization, for fixed $\varphi = 0$. The upper curves (dashed) show $\Delta\rho_{xx}$, the lower ones (solid) show $\Delta\rho_{yy}$; $\Delta\rho_{xx} = \Delta\rho_{yy}$ for $\theta = 0^o$. Near $a/2R_c = 2$, $\Delta\rho_{xx}$ increased with increasing θ, near $a/2R_c = 1.2$, $\Delta\rho_{yy}$ decreases monotonically with increasing θ. Dotted lines indicate pure magnetic ($\varepsilon = 0$, $\mu = 1$), dash-dotted lines pure electric modulation ($\mu = 0$); α as in Fig. 2. (a): $V_{\vec{q}}$ from Eq. (9), $\varepsilon = 1$, triangles indicate magnetic flat-band conditions; (b): Eq. (9) with $\lambda = \sqrt{2}\rho$, $\varepsilon = 0.5$, electric flat-band conditions indicated.

We now fix parameters and consider a square lattice, with period $a = 500$ nm, of circular cylinders with radius $\rho = 0.25a$ and height $h = 0.4a$ on the sample surface in the plane $z = 0$, and a 2DEG in the plane $z = -0.2a$ below the surface. Furthermore, we set $m_{\|} = |\mathbf{m}| \sin\theta$ and $m_z = |\mathbf{m}| \cos\theta$. The stress-induced potential, Eq. (9), with $V_0/E_F = \gamma\varepsilon$, is kept fixed in the following. The strength of the magnetic modulation is given by μ, defined as before. Figure 3 visualizes the oscillating $B_z(x,y)$ in the plane of the 2DEG for three cases of high symmetry, $V(x,y)$ has the same symmetry as $B_z(x,y)$ in the uppermost panel. [18]

Figure 4 shows, for two values of the width parameter λ in Eq. (9), the calculated magnetoresistance for $\varphi = 0$ and several values of θ. The general trends of the θ-dependence of $\Delta\rho_{xx}$ and $\Delta\rho_{yy}$ are the same in both cases, and have recently also been observed in experiment [15]. Details, however, depend strongly on the model used to describe the stress-induced electric modulation.

Results for the symmetric case $m_x = m_y$, i.e. $\varphi = \pi/4$, and different tilt angles θ are discussed in Ref. [18]. For $\theta > 0$ still $\Delta\rho_{xx} = \Delta\rho_{yy}$, but the resistivity tensor is not isotropic due to the non-zero off-diagonal components $\Delta\rho_{xy} = \Delta\rho_{yx}$.

Concluding remarks

Tilting the magnetization of the periodic array of micromagnets on the surface produces in the plane of the 2DEG magnetically defined superlattices of different symmetries, which interfere with the stress-induced electrostatic superlattice and lead to characteristic features of the magnetoresistance curves. Therefore, magnetoresistance measurements may be used to probe the symmetry of the resulting superlattice.

The simple approach used here yields reasonable agreement with experimental results on samples with a 1D superlattice in the regime of Weiss oscillations (in typical AlGaAs heterostructures at $T \approx 2$ K for 0.05 T $\leq B_0 \leq$ 1 T). It can not explain the positive magnetoresistance seen in the experiments at very small B_0 values, and, of course, not the quantum effects at high magnetic fields, where the modulation leads to interesting internal structures of the Shubnikov-de Haas peaks, which recently have been discussed within a screened Hartree-Fock approximation [19]. In the classical regime, a careful treatment of Boltzmann's equation can improve upon the simple approach. If the modulation fields depend only on the x coordinate, one can show that (to all orders in $1/\omega_c\tau$) only the component ρ_{xx} of the resistivity tensor is affected by modulation [16], even if a magnetic modulation and small-angle scattering effects are included.[20] The analytical solution, which is available in the relaxation time approximation for weak modulations,[16] yields a quantitative improvement in the regime of Weiss oscillations and notably for higher B_0 values.[20] The low-field magnetoresistance is caused by electrons moving with nearly Fermi velocity in y direction, along "open" orbits within a single period in x direction. This effect, as well as the effect of small-angle scattering, has been included in a fully numerical solution of the Boltzmann equation, and very good agreement with the experiment was obtained. [20]

For the 2D modulations, the present simple approach seems to predict only the general trends concerning the symmetry of the resistivity tensor correctly. The reason for the poor agreement at $a/2R_c > 1.3$ is not yet clear. Within a classical calculation a more detailed model for the stress-induced modulation or for the magnetic modulation field may be necessary. The weak modulation approximation might also break down for large B_0 values, which unavoidably increase the strength of the magnetic modulation. The reason may, however, also be of quantum mechanical origin, since the energy spectrum in 2D mixed electric and magnetic superlattices is of the fractal Hofstadter-type [21]. For such a spectrum, the Landau bands split into narrow subbands separated by wide gaps, and, as a consequence, the band-conductivity can be drastically suppressed, as has been discussed before for the pure electric modulation. [22]

Acknowledgments

I am very grateful to D. Weiss and P.D. Ye for numerous stimulating discussions and for communicating their experimental results prior to publication.

References

1. D. Weiss *et al.*, Europhys. Lett. **8**, 179 (1989), see also in *High Magnetic Fields in Semiconductor Physics II*, edited by G. Landwehr, Springer Series in Solid-State Sciences Vol. **87** (Springer-Verlag, Berlin 1989), p. 357.
2. R. R. Gerhardts, D. Weiss, and K. v. Klitzing, Phys. Rev. Lett. **62**, 1173 (1989).
3. R. W. Winkler, J. P. Kotthaus, and K. Ploog, Phys. Rev. Lett. **62**, 1177 (1989).
4. C. Zhang and R. R. Gerhardts, Phys. Rev. B **41**, 12850 (1990).
5. P. Vasilopoulos and F. M. Peeters, Superlatt. Microstruct. **7**, 393 (1990).
6. D. P. Xue and G. Xiao, Phys. Rev. B **45**, 5986 (1992).
7. F. M. Peeters and P. Vasilopoulos, Phys. Rev. B **47**, 1466 (1993).
8. H. A. Carmona *et al.*, Phys. Rev. Lett. **74**, 3009 (1995).
9. P. D. Ye *et al.*, Phys. Rev. Lett. **74**, 3013 (1995).
10. S. Izawa *et al.*, J. Phys. Soc. Japan **64**, 706 (1995).
11. J. H. Davies and I. A. Larkin, Phys. Rev. B **49**, 4800 (1994).
12. P. D. Ye *et al.*, Semicond. Sci. Technol. **10**, 715 (1995).
13. P. D. Ye *et al.*, Appl. Phys. Lett. **67**, 1441 (1995).
14. P. D. Ye *et al.*, in *Proc. 9th Int. Winterschool on Nanostructure Physics and Technology, Mauterndorf 1996*, Semicond. Science Technol., to be published
15. D. Weiss, P. D. Ye, and R. R. Gerhardts, in *Proc. 23rd ICSP, Berlin 1996* (World Scientific, Singapore, 1997), to be published.
16. C. W. J. Beenakker, Phys. Rev. Lett. **62**, 2020 (1989).
17. R. R. Gerhardts, Phys. Rev. B **45**, 3449 (1992).
18. R. R. Gerhardts, Phys. Rev. B **53**, 11064 (1996).
19. A. Manolescu and R. R. Gerhardts, in *Proc. 12th Intern. Conf. Appl. High Magnetic Fields, Würzburg, 1996* (World Scientific, Singapore, 1997), this volume.
20. R. Menne and R. R. Gerhardts, *ibidem*; and unpublished
21. R. R. Gerhardts, D. Pfannkuche, and V. Gudmundsson, Phys. Rev. B **53**, 9591 (1996).
22. D. Pfannkuche and R. R. Gerhardts, Phys. Rev. B **46**, 12606 (1992).

QUANTUM TRANSPORT IN BALLISTIC QUANTUM DOTS

D. K. FERRY, R. AKIS, D. P. PIVIN, JR., K. M. CONNOLLY

Nanostructures Research Group, Center for System Science and Engineering, and Department of Electrical Engineering, Arizona State University, Tempe AZ 85287-5706

J. P. BIRD, K. ISHIBASHI, Y. AOYAGI, T. SUGANO

Frontier Research Program, RIKEN, 2-1 Hirosawa, Wako, Saitama 351-01, Japan

Y. OCHIAI

Materials Research Department, Chiba University, 1-33 Yayoi-cho, Inage-ku, Chiba 263 Japan

Ballistic quantum dots have found usage in a wide variety of studies ranging from single-electron charging to the universality of chaotic systems. Magneto-transport can be used to study the properties of regular (integrable) structures and the corresponding density-of-states fluctuations, in order to ascertain the quantum structure that will be useful in these dots for *open, far-from-equilibrium* device applications. Here, we discuss the observations of regular, periodic fluctuations arising from the existence of stable orbits within the dot structure in GaAs/AlGaAs and InGaAs/GaAs heterostructures. In these cases, the magneto-conductance fluctuations exhibit regular quasi-periodic structure that is dominated by a few periodic orbits giving rise to one, or a few, families of harmonically related magnetic frequencies in the Fourier spectrum. These arise from the regular trajectories which contribute to the oscillatory density-of-states in ballistic dots.

1 Introduction

Quantum dots have been studied significantly over the past few years. Most of these studies, however, have focused upon either (1) the chaotic behavior of the dots[1-8], or (2) the single-electron tunneling properties of weakly-coupled dots. We will consider only the situations, related to (1), in which the dots are *open* and strongly coupled to the environment. In most of these earlier studies, there was considerable effort expended to try to establish that the basic behavior of these ballistic quantum dots was governed by *universal* properties that were *generic* in nature and independent of the specific properties of the individual dots.

Here, however, we shall point out that the underlying physics of the ballistic quantum dots is described by characteristics that are dependent upon the individual dot under study. Nevertheless, *there is a universal behavior that is characteristic of such dots*, but it is not the generic behavior described by the pseudo-chaotic mathematics of these previous studies. Instead, the underlying properties, and the characteristic magneto-transport through the dots, is governed by the basic *regular* nature of the semi-classical orbits in the dot[9-11]. This regular nature is observed as reproducible fluctuations and quasi-periodic oscillations in the correlation functions.

In the following, we will discuss the experimental observations, in which reproducible fluctuations are observed in the magneto-transport through the dots. These fluctuations exhibit quasi-periodic oscillations, easily discernible in the correlation functions and in the Fourier transforms, which are endemic of the semi-classical regular orbits in the square quantum dots. Crucial to this result is the excitation of the ballistic dots by quantum point contacts, as the latter provide a collimated excitation of the particles within the dot. This collimation provides a specific excitation of quantum structure related to the semi-classical orbits. We discuss quantum simulations and then turn to the connection with semi-classical orbit theory, detailing how this provides the underlying basic properties of the dots themselves.

2 Experimental Overview

We consider three types of dots in this study. In all three cases, the dot geometry is square and with a size that is much smaller than the elastic mean free path, so that the underlying motion is ballistic in nature. These three structures include: (1) GaAs/AlGaAs heterostructures with surface Schottky barrier gates to define the dots, (2) GaAs/AlGaAs heterostructures with in-plane gates to define the dots, and (3) InGaAs/GaAs heterostructures with in-plane gates to define the dots. Hence, two different material systems and two distinct gate technologies are involved in the samples studied. The basic observations are similar in all three cases, and independent of the material system and gate technology.

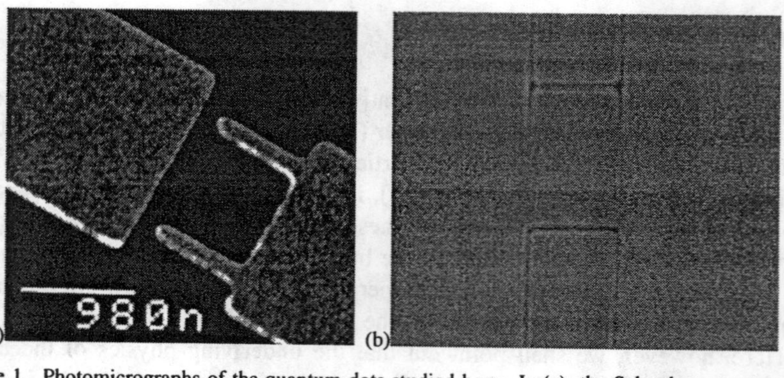

(a) (b)

Figure 1. Photomicrographs of the quantum dots studied here. In (a), the Schottky gate structure used on GaAs/AlGaAs dots is depicted. In (b), the trenches which provide the isolation for the in-plane gates are observable. These trenches are typically less than 80 nm in size.

2.1 Sample Details

The basic quantum dot studied here is a square dot with the input and output ports defined by quantum point contacts situated along one wall of the dot structure. In

Fig. 1, the two types of gates, surface Schottky barrier gates and in-plane RIE defined gates, are illustrated. In the former, the gate regions are defined by lift-off of metallic Schottky barrier metals. In the latter, the gates lie in the Q2D gas itself, with the gates isolated by trench isolation[10]. The dots are basically square in nature, and sizes ranged from 1.0 μm to 0.4 μm (the electrical sizes are somewhat smaller due to edge depletion around the gates). In nearly all cases studied, the carrier density was 4-5 × 10^{11} cm^{-2}. In the GaAs samples, mobilities are typically 40-70 m^2/V-s, while in the InGaAs, the mobility is 10 m^2/V-s. The gate design allowed electrons to be trapped in the central square cavity; note particularly the position of the openings in the quantum point contacts, as this is crucial to the excitation of regular behavior. Measurements of the GaAs samples were typically carried out at 10-30 mK base temperature in the refrigerator, while the measurements of the InGaAs samples were performed at 2 K. Source-drain excitation was kept well below the thermal voltage and lock-in techniques utilized.

2.2 *Quasi-Periodic Fluctuations*

In Fig. 2, magneto-conductance traces for the three sample types are shown for comparison. Reproducible fluctuations persist across a wide range of magnetic field. In addition, the correlation function for the fluctuations is also plotted; the latter is averaged over the range of magnetic field, and in some cases averaged over a narrow range of gate bias. It is clear that the basic nature of the reproducible fluctuations—the oscillatory nature of these fluctuations which leads to oscillatory behavior (and negative excursions) in the correlation functions—occurs in each sample, and is independent of the details of the sample material and gate technology. Instead, this behavior seems to be a basic property of the dot geometry and size, a result of the *intrinsic* dot properties and quite distinct from the generic results of chaos theory[1,2,4]. Here, the basic oscillatory properties are universal, although the specific frequency content is very dot dependent. We will discuss below how these oscillatory features of the magneto-conductance are related to the fluctuations in the density of states that arise from the regular, quasi-periodic orbits of the quantum dot itself, and are therefore a reflection of the intrinsic properties of the dots.

Although not evident in these figures, the low field magnetoresistance often exhibits a ballistic weak localization peak, which can sometimes be linear in nature. In fact, the actual lineshape of this ballistic weak localization peak is found to vary with contact opening, changing from Lorentzian to linear with bias. Lorentzian and linear lineshapes have been predicted in chaotic and regular systems, respectively[2], but it is unlikely that this transition in behavior is occurring in these dots. Lorentzian lineshapes for the ballistic weak localization peak have been observed in simple quantum point contact structures which cannot support chaotic behavior[12,13]. Thus, it is improper to assume *a priori* that the Lorentzian lineshape is connected with chaotic behavior. Indeed, phase-space filling can easily be generated by the *integrable* motion within these dots due to the magnetic field induced precession of the orbits[11]. Thus, it is not surprising that the aperiodic fluctuations observed in the magneto-conductance are quite different from those observed in chaotic dots.

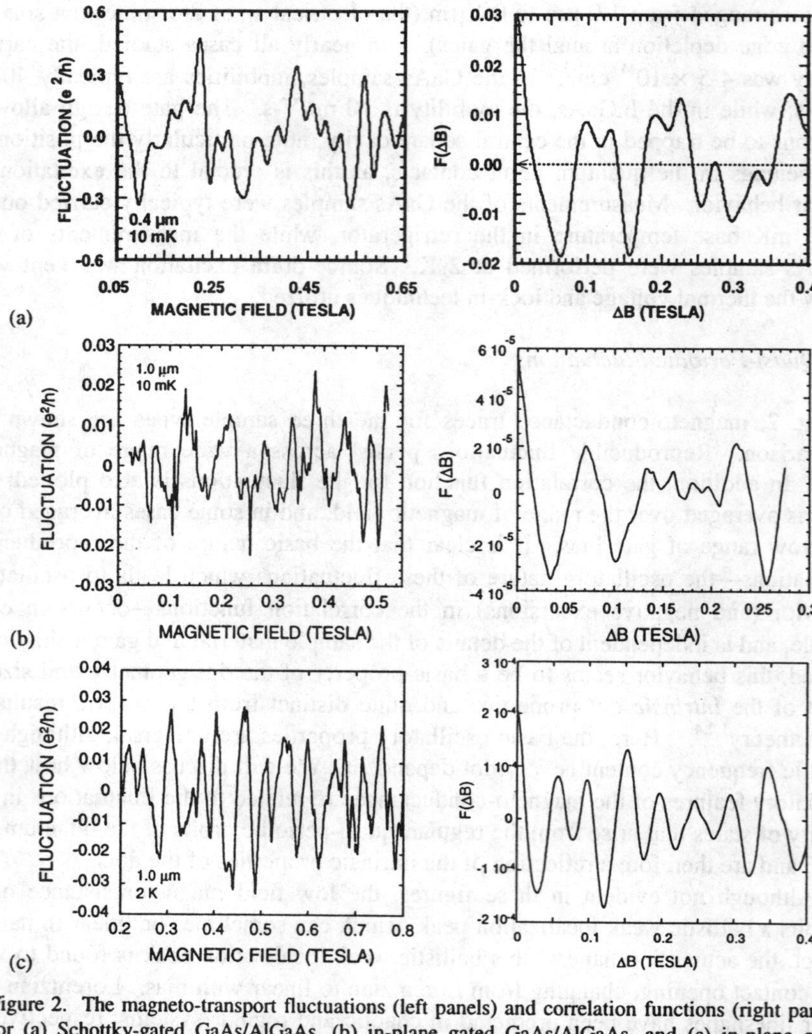

Figure 2. The magneto-transport fluctuations (left panels) and correlation functions (right panels) for (a) Schottky-gated GaAs/AlGaAs, (b) in-plane gated GaAs/AlGaAs, and (c) in-plane gated InGaAs/GaAs ballistic dots. Panels (a) and (b) are data taken at a base temperature of 10-30 mK, while (c) was obtained at 2 K.

The fluctuations and correlation functions shown in Fig. 2 exhibit considerable diffferences from one device to the next, and from one set of bias conditions to the next. In fact, the fluctuations in Fig. 2b seem to incorporate additional noise, which may arise as an intrinsic process generated by the device impurities (the mobility in this device is much less than in that of Fig. 2a), or defect states, for example. However, the resonance-like structures could be quite real, as will be discussed below. Because of these variations, it is not so easy to ascertain that in fact there is an underlying commonality to the fluctuations; one which is based upon intrinsic properties of the individual quantum dots. To illustrate that the observed features are in fact generic, we plot the results for two samples, one with Schottky gates on GaAs/AlGaAs and one with in-plane gates on InGaAs/GaAs, with the same lithographic dimension of 1.0 μm and at the same temperature of 2 K. In Fig. 3, we plot the fluctuation correlation functions for these two cases. It is clear that the same basic features are present in the correlation functions, although Fourier analysis reveals that the number of higher harmonics in the oscillations is larger in the GaAs/AlGaAs case, presumably due to the higher mobility of this latter material, which allows the carriers to undergo more traversals of a basic orbit as we discuss below. (It should be noted that the measurements were made in two different laboratories on two different continents.) This also contributes to the larger amplitude of fluctuations in the GaAs/AlGaAs case.

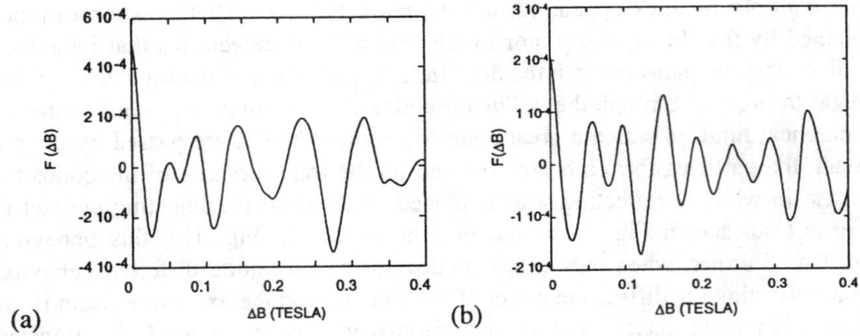

Figure 3. The correlation function for 1.0 μm dots at 2 K for the case of (a) GaAs/AlGaAs Schottky-gated dots and (b) InGaAs/GaAs in-plane gated dots.

3 Chaos and Regularity

Conductance fluctuations in a *chaotic* quantum dot are characterized by their correlation function which has a characteristic Lorentzian shape. This arises from the assumption that the ensemble of semi-classical trajectories through the quantum dot sweep out a wide range of "areas" (these areas are related to the enclosed flux winding number of the orbit, and it is intrinsically assumed that all possible exits angles for the orbits are present in the distribution) which is characterized by an exponential distribution as[1,2]

$$P(A) \sim e^{-\alpha|A|}\,, \tag{1}$$

where α is the inverse of the "average" area enclosing the flux. The correlation function of the conductance is related to this through

$$F(\Delta B) = \langle \delta G(B+\Delta B)\delta G(B)\rangle \propto \left| \int_{-\infty}^{\infty} dA e^{2\pi i \Delta BA/\phi_o} P(A) \right|^2$$
$$\propto \left[1 + \left(\frac{2\pi \Delta B}{\alpha \phi_o} \right)^2 \right]^{-2}, \tag{2}$$

where $\phi_o = h/e$ is the quantum of flux. We note that this correlation function is positive definite and exhibits no oscillatory behavior, two features which are clearly contrary to the measurements discussed here.

The measurements of conductance fluctuations in square dots, in which the classical behavior is expected to be integrable does not display the above behavior. These conductance fluctuations do not appear to be irregular but are quite periodic, and their correlation function is characterized by large amplitude oscillations. A Fourier spectrum of these correlation functions exhibits families of harmonically related peaks (Fig. 5 below); sometimes only one family appears, while at other times multiple families appear (discussed more below). These results cannot be explained by the above theory, nor by an ensemble of trajectories that includes *all* possible classical paths through the dot. Indeed, excitation of the regular behavior is thought to proceed through the *collimation* effects of the quantum point contact. In the classical limit, in which a great many waveguide modes are passed by the point contact, the particles form a beam centered on the transport axis of the contact. In the case in which a reflecting wall is placed adjacent to the injecting contact (the upper-left surface in Fig. 1a or the bottom surface in Fig. 1b), this behavior is modified. Further, when only a few modes are passed, quite different behavior is found, reflecting the diffractive effect of the contact, and the excitation beam is quite narrow. In these cases, and in our geometry, conservation of the transverse momentum causes the beam to exit the point contact at a significant angle away from the transport axis. This is shown in Fig. 4(a). Hence, excitation of the present quantum dot structures proceeds via a collimated beam, oriented away from the axis (defined by the two leads). This beam can then excite a set of regular orbits within the dot. This is further reflected by the negative excursions of the correlation function indicating anti-correlation and strong phase coherence of the particles[14].

We have carried out calculations to simulate the behavior of these dots, using a stable variant of the transfer matrix approach[11]. For example, conductance fluctuations as a function of magnetic field are studied for a 0.3 μm square dot (this size reflects the electrical size, which is less than the actual structure due to gate depletion) with 0.04 μm port openings, which allow two modes to enter and exit the dot. Instead of a random aperiodic variation with magnetic field, a series of nearly periodic oscillations is evident, as may be seen in Fig. 5. Also apparent are several resonance features. In particular, a set of resonances occurs at $B \sim 0.069$, 0.173, 0.283, and 0.397 T, in which the wave function is heavily "scarred"[15], in that the quantum mechanical amplitude appears to follow a single underlying classical orbit. Given the period for the reappearance of the diamond is $\Delta B \sim 0.11$ T, and using the criterion familiar from the Aharonov-Bohm effect, that $\Delta\phi/\phi_0 = 2\pi$ for the difference in magnetic flux, one obtains $A \sim 0.04\ \mu m^2$ for the enclosed area, which corresponds well with the enclosed area of a diamond-shaped wave function, which is depicted in Fig. 4(b), where the darker areas are points of higher wave function amplitude. This behavior, which is driven by the collimated beam, is much different than that found in resonant tunneling of closed dots[11]. In the latter case, the wavefunctions, while forming an ordered pattern, fill the entire dot. One can not make an association with single, simple periodic orbits.

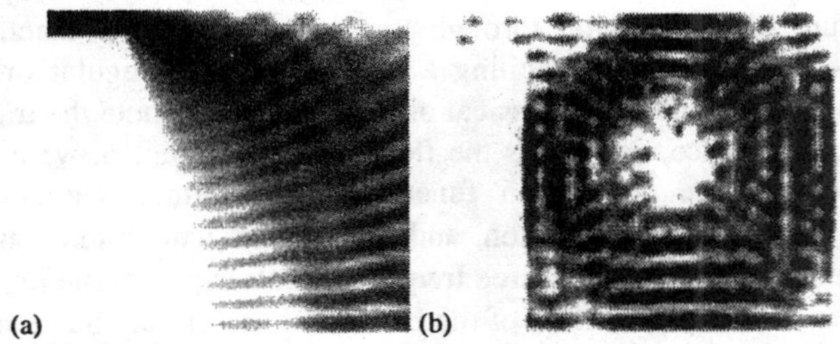

Figure 4. (a) The wave excitation from a point contact with only two modes allowed. It is diffracted away from the wall at the top. (b) $|\psi(x,y)|$ vs. x and y is plotted. Darker shading corresponds to higher amplitude. The picture correspond to the dot with 0.04 μm openings at $B = 0.173$ T.

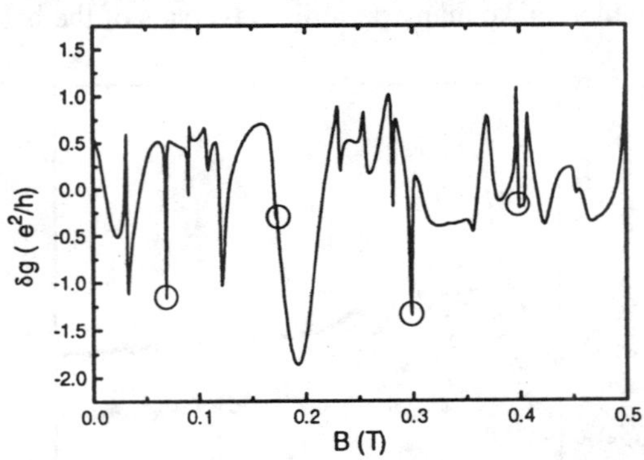

Figure 5. The computed conductance fluctuations at T = 0 K for a 0.3 μm dot. The circles indicate the periodic resonances which display the heavily scarred wave function of Fig. 4(b).

Periodic orbits have played a large theoretical role in the computation of semi-classical quantization of bound states for many years, dating to the Einstein-Brillouin-Keller view of quantization. More recent studies indicate that the "imprints" of these orbits persist up through thousands of states[16], a result that suggests the closed orbits are quite stable in regular systems and only become unstable as one passes to the ergodic regime, which seems to be replicated as one passes from the semi-classical to the quantum regime. Nevertheless, how is one to chose these semi-classical orbits? Traditionally, one examined all possible classical paths from the input to the output[1,2]. But, in such ballistic quantum dots, this produces essentially a phase-space filling and ergodic behavior of the transport, which agrees closely with chaotic dot studies. There is a clear difference between the behavior obtained when all possible trajectories are investigated and when only a limited set of trajectories are investigated[17], as is the case here with the collimation from the quantum point contacts. We focus solely on the case for a collimated beam of particles, and also have studied classical ballistics for this case. The crucial assumption in doing this is that the quantum point contact imposes a boundary condition on the particles, in which the entry angle is determined by the wave mechanical nature of propagation through the quantum point contact. It is this collimation that is important for exciting a particular set of regular orbits of the particles. From the study of these classical orbits, we can compute the transmission, and hence the conductance, as well as the fluctuations. In Fig. 6, we compare the Fourier transform of the correlation function for the three cases of (a) the experiment, (b) the quantum simulation, and (c) the classical simulation. Clearly, the strong agreement between these three frequency plots supports the importance of the regular orbits on the quantization of the ballistic dots. It may be seen that there is a strong main peak at 9 T^{-1}, which corresponds to the periodicity of the scarred diamond-shaped wave functions of Fig. 4(b). In addition, there are higher harmonics, and some other frequencies observed in the spectrum. These harmonics are related to orbits which make many periodic traversals of the basic orbit[16].

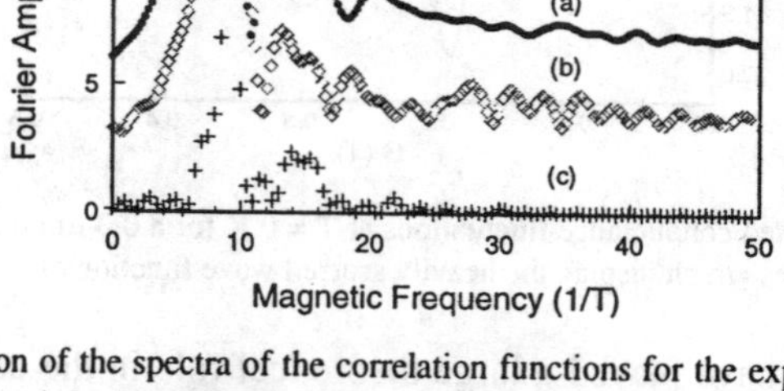

Figure 6. A comparison of the spectra of the correlation functions for the experimental dot (a), the classical billiard (b), and the quantum simulation (c), all for an effective dot size of 0.3 μm. The curves are offset for clarity.

4 Conclusions

In summary, we have presented studies of the experimental magneto-transport, the simulation of quantum mechanical transport, and the classical billiard transport through ballistic semiconductor quantum dots. These are shown to have quite similar behavior, if the classical motion is limited to a collimated set of trajectories, presumably selected by the quantum mechanical properties of the quantum point contacts by which leads are connected to the open quantum dots. The theoretical results are shown to agree substantially with the experiments performed on actual semiconductor quantum dots.

This suggests then that the collimated quantum point contacts excite a set of regular, periodic orbits within the ballistic dot that are quite stable. These trajectories couple to a large number of eigenmodes of the quantum dot, and the amplitude of the periodic orbits contributions to the oscillatory density of states is therefore heavily scarred at certain values of the magnetic field, for which there is a strong overlap with the classical trajectory[18].

The results suggest that transport in regular semiconductor quantum dots is clearly distinguished from transport in ergodic dots. The motion is heavily restricted by the beam collimation, and the fluctuation spectrum is not random, but highly oscillatory and correlated. The correlation functions for these fluctuations show regular and periodic oscillations that contain only a few, often harmonically related, frequencies, fully in keeping with the expectations of semi-classical descriptions of the fluctuations in the density of states of such structures. Thus, the fundamental mode in the underlying physics of the ballistic quantum dot is one of stable, regular orbits which provide the important connection in passing from the classical to the quantum mechanics describing transport through the dots. It is these orbits which provide the connection to the quantum modes of the dot, and for which the amplitude of the oscillatory contribution to the density of states is heavily scarred by the coupling to these eigenmodes[18], as mentioned above. The collimation of the input beam excites a particular family of repetitive orbits, which are supported by the quantum modes. The regularity of these orbits leads to periodic repetitions, and to the resulting oscillations in the density of states that are observed in the magnetotransport experiments. In this sense, the observations in these regular dots are quite different from those observed in chaotic dots, where the entire phase space of the dot is filled with orbits. In the present case, the generic behavior is one of regular, periodic orbits, rather than one of ergodic, phase space filling behavior.

Acknowledgements

The authors would like to acknowledge helpful discussions with R. P. Taylor, R. Newbury, J. R. Barker, L. Eaves, Y. Nakamura, M. Stopa, M. Tabor, and H. L. Grubin. This work was supported in part by the Office of Naval Research.

Visit http://www.eas.asu.edu/~nano/nano.html

References

1. R. A. Jalabert *et al.*, Phys. Rev. Lett. **65**, 2442 (1990).
2. H. U. Baranger *et al.*, Phys. Rev. Lett. **70**, 3876 (1993).
3. A. M. Chang *et al.*, Phys. Rev. Lett. **73**, 2111 (1994).
4. H. U. Baranger and P. A. Mello, Phys. Rev. Lett. **73**, 142 (1994).
5. C. M. Marcus *et al.*, Chaos **3**, 643 (1993).
6. I. H. Chan *et al.*, Phys. Rev. Lett. **74**, 3876 (1995).
7. R. M. Clarke *et al.*, Phys. Rev. B **52**, 2656 (1995).
8. J. P. Bird *et al.*, Phys. Rev. B **50**, R18678 (1994); **51**, R18037 (1995).
9. J. P. Bird *et al.*, submitted for publication; Superlatt. Microstruc., in press.
10. K. M. Connolly *et al.*, Superlatt. Microstruc., in press.
11. R. Akis *et al.*, Superlatt. Microstruc., in press; submitted for publication.
12. R. P. Taylor *et al.*, Phys. Rev. B **51**, 9801 (1995).
13. R. Katine *et al.*, Superlatt. Microstruc., in press.
14. The reader is reminded that the auto-correlation function of a sine wave is a cosine wave which alternates between strong correlation and strong anti-correlation each time the phase shifts by a factor of π.
15. E. J. Heller, Phys. Rev. Lett. **53**, 1515 (1984).
16. M. V. Berry, Proc. Roy. Soc. (London) A **423**, 249 (1989).
17. This has been examined for the ballistic dots of interest here in D. K. Ferry and G. Edwards, VLSI Design, to be published.
18. A. M. Ozorio de Almeida, *Hamiltonian Systems: Chaos and Quantization* (Cambridge U.P., Cambridge, 1988).

FLUX QUANTIZATION EFFECTS IN SEMICONDUCTOR-SUPERCONDUCTOR JUNCTIONS

HERBERT KROEMER, MASON THOMAS
Department of Electrical and Computer Engineering
University of California, Santa Barbara, CA 93106, USA

Semiconductor quantum wells contacted with superconducting electrodes act as superconducting weak links at temperatures sufficienty far below the critical temperature of the electrode. In the presence of a magnetic field, they exhibit periodic flux quantization effect, which persist up to temperatures at which the sample reistance, although reduced, is still far inside the non-superconducting regime.-

1 Introduction

When a superconductor and a semiconductor are brought together into atomically intimate contact, with an interface that is free from residual barriers, the electrons in the two materials can interact with each other in ways that can drastically alter the current flow across such an interface. Under favorable conditions, the resistance through a semiconductor with two closely-spaced (<1μm) superconducting electrodes becomes vanishingly small. In effect, superconductivity is induced in the semiconductor.

Superconductivity is characterized by phase coherence of the electron pair wave function over macroscopic distances. But the vector potential associated with any magnetic field causes phase shifts in the wave function, leading to a strong dependence of the transport properties on magnetic fields.

The "semi-super" junctions discussed in the present paper are all based on the generic design shown in Fig. 1.

A heavily modulation-doped InAs quantum well, typically 15nm wide, with

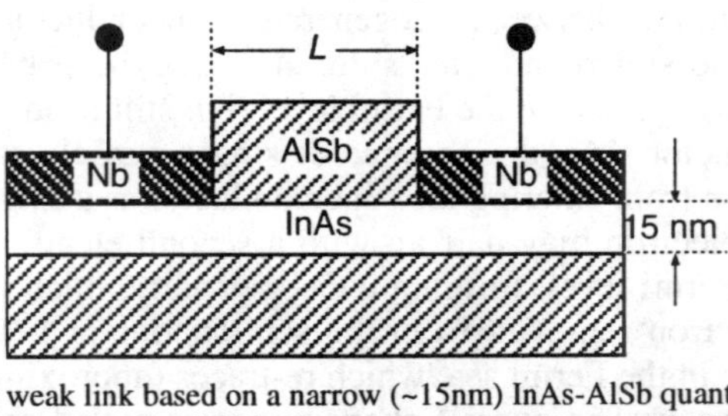

Fig.1. Superconducting weak link based on a narrow (~15nm) InAs-AlSb quantum well forming a short (<1μm) conducting link between two superconducting Nb "banks".

AlSb or (Al,Ga)Sb barriers, is contacted directly (after selective removal of one of the barriers) with superconducting Nb electrodes. The preferred semiconductor for such purposes is InAs,[1] principally because of its unique property that metal-to-InAs interfaces do not form Schottky barriers that would impede the flow of electrons. The natural barrier material is AlSb, because of its low lattice mismatch to InAs (1.3%), and a band lineup with a large conduction band offset of 1.35eV, permitting the use of modulation doping to achieve very high electron sheet concentrations (approaching $10^{13}/cm^2$), while maintaining high mobilities ($\geq 10^5 cm^2/V{\cdot}s$). Finally, Nb is the preferred superconductor, because of its high critical temperature, and its technological compatibility with InAs.

The paper is organized as follows.

Sec. 2 gives a tutorial introduction to the phenomenon of *Andreev reflections*, the central concept in the physics of electron transport across super-semi interfaces.

Sec. 3 presents data on one sample with a combination of electrode spacing (1μm), and temperature (4.2K) such that the sample resistance remains distinctly in the non-superconducting range, and merely shows a precursor of superconductivity. Nevertheless, pronounced effects of even weak magnetic fields are observed, indicating that flux quantization effects begin long before the vanishing of the resistance.

Sec. 4 presents data for *ladder arrays* of a large number of devices, each of the generic design of Fig. 1, but with shorter electrode spacings, and at lower temperatures. To the first order, these arrays behave like series-connections of “ordinary” Josephson junctions (JJ’s), but with distinctly different magnetic properties.

Sec. 5 offers some conclusions.

2. Andreev Reflections

In the weak links discussed here, the mean free path of the electrons in the semiconductor is much larger than the inter-electrode separation. The electron transport is then dominated by the phase-coherent flow of *ballistic* electrons between the banks, in contrast to the *diffusive* transport in more conventional weak links.

Consider the interface between a degenerate semiconductor and a superconductor (Fig. 2a). On the superconductor side, an energy gap has opened up. If an electron with an energy $\mathcal{E}$ *above* the Fermi level (but still inside this gap) is incident from the semiconductor side, the absence of single-particle states within the gap prevents that electron from entering the superconductor as a *single* electron.

However, the electron may pair up with a second electron at the “mirror energy” $\mathcal{E}$ *below* the Fermi level, forming a Cooper pair, which *will* enter the superconductor. The electron removed from the semiconductor below the Fermi level leaves behind a hole in the Fermi sea, which re-traces (approximately) the trajectory of the incident electron. The overall phenomenon is called an *Andreev reflection*

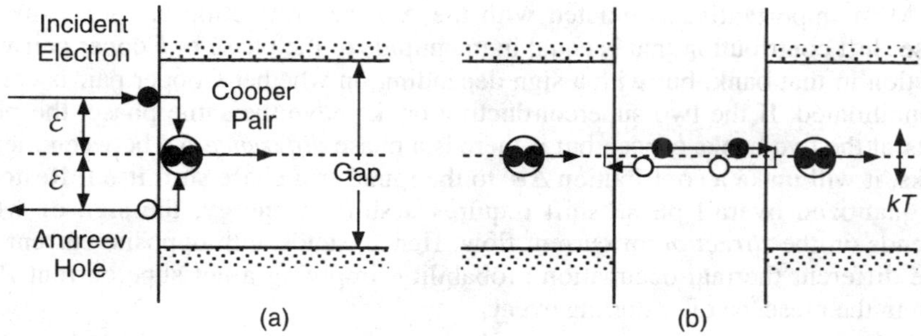

Fig. 2 Andreev reflections. (a): Basic concept. (b): Persistent current flow by multiple Andreev reflections.

(AR). Although postulated over thirty years ago,[2] AR's have received major attention only during the last few years.

A central property of AR's is that the *phase* of the Andreev hole is correlated with the phases of *both* the incident electron *and* the Cooper pairs in the superconductor. The theory shows that such a phase-correlated electron-hole pair in the semiconductor acts in many ways like a Cooper pair itself, with a pair phase equal to that of the Cooper pairs inside the superconductor. The overall effect is an extension of the superconductivity into the semiconductor, a generalized form of the superconducting *proximity effect*. If the distance over which the phase correlation exists is sufficiently large to reach the other superconducting bank, the result is a phase coupling between the two superconducting banks, that is, a superconducting *weak link*, with a Josephson current-phase relation of the form $I = I_c \cdot \sin \Delta\phi$, where $\Delta\phi$ is the phase difference between the pair wave functions in the two superconducting banks.

In terms of the *particle* picture of AR's of Fig. 3a, a supercurrent may be understood as follows. The Andreev hole will eventually reach the left superconducting electrode. In the absence of any bias across the structure, the energy of the hole is still within the superconducting gap on that side. It cannot enter the superconductor, but it can be annihilated by breaking up a Cooper pair inside the left superconductor: One of the electrons of the pair annihilates the hole, the other electron takes up the excess energy, and is injected into the semiconductor as a ballistic electron *above* the Fermi level, at *exactly* the same energy as the initial electron. This process can evidently be repeated, suggesting the possibility of a persistent-current loop.

However, such a loop can be a quantum-mechanically allowed stationary state only if the round-trip phase shift along the loop is an integer multiple of 2π, $\Delta\phi_{RT} = 2\pi n$. For every value of n, there will be two states, corresponding to opposite directions of the flow arrows in Figure 3b.

Most importantly, associated with the Andreev reflection at each bank is a phase shift contribution that is equal in magnitude to the phase of Cooper pair wave function in that bank, but with a sign depending on whether Cooper pair is created or annihilated. If the two superconducting banks have the same phase, the phase shifts at the two banks cancel, but if there is a phase *difference* $\Delta\phi$ between the two banks, it will make a contribution $\Delta\phi$ to the round-trip phase shift itself. Restoring the quantized overall phase shift requires a shift in energy, the sign of which depends on the *direction* of current flow. Hence, states with opposite current will have different thermal occupation probability, implying a net supercurrent flow, even in the presence of scattering events.

3 Superconductivity Precursors and their Magnetic Field Dependence

At temperatures above the critical temperature of Nb (7.5-9.2K, depending on purity), the current-voltage characteristics are perfectly ohmic, with no rectification effects. However, as soon as the Nb electrodes become superconducting, the *I-V* characteristics become non-linear, with a reduced differential resistance around zero bias. With decreasing temperature, that resistance decreases monotonically, and in "good" samples it eventually becomes too small to be measurable with the instrumentation at hand. The details of this temperature dependence will published elsewhere.

We show here first, in Fig. 3, the differential *conductance* of an earlier sample reported earlier, which had a fairly wide inter-electrode separation of 1μm, too wide to yield a vanishing resistance at 4.2K.[3,4] The zero-bias conductance at 4.2K is enhanced only by about a factor 7 relative to the value above the critical temperature of the Nb electrodes. This enhancement has an extraordinarily narrow width, much less than *kT/e*, a value incompatible with any model of current flow involving only uncorrelated single electrons; it requires *some* form of many-body or correlation effects. This is provided by multiple AR's.

The presence of multiple AR's is shown directtly through the pronounced oscillations in the differential resistance at finite voltages, below the voltage equivalent of the superconducting gap. Such oscillations are the "fingerprint" of multiple Andreev reflections in the presence of a non-zero bias.[5]

Fig. 4 shows the of the height of the zero-bias conductance peak of Fig. 3 as a function of magnetic field, indicating an initial precipitous drop, leveling out at about ±1μT, and falling off much more slowly at higher fields. The fields quoted are the *externally* applied magnetic fields. Because of the flux expulsion from the superconducting Nb electrodes, the flux inside the inter-electrode gap itself is much larger, by a factor we estimate to be about 50, which still leads to a very small magnetic field of about 50μT at the foot of the steep drop,.

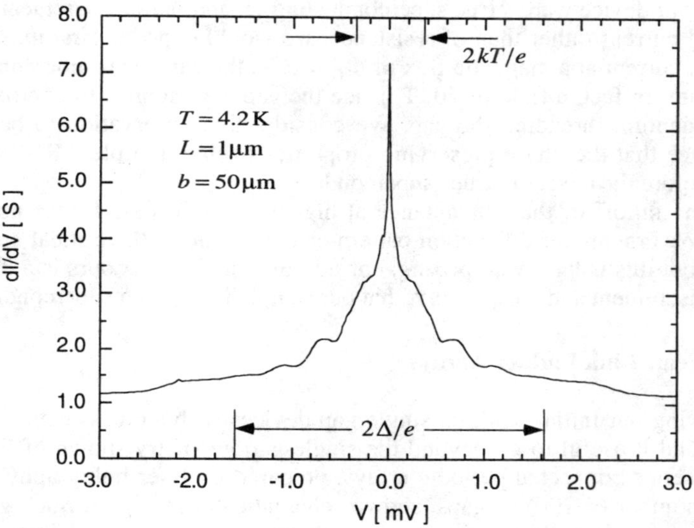

Fig. 3. Conductance peak at 4.2K around zero bias for a weak link with a 1μm electrode spacing.

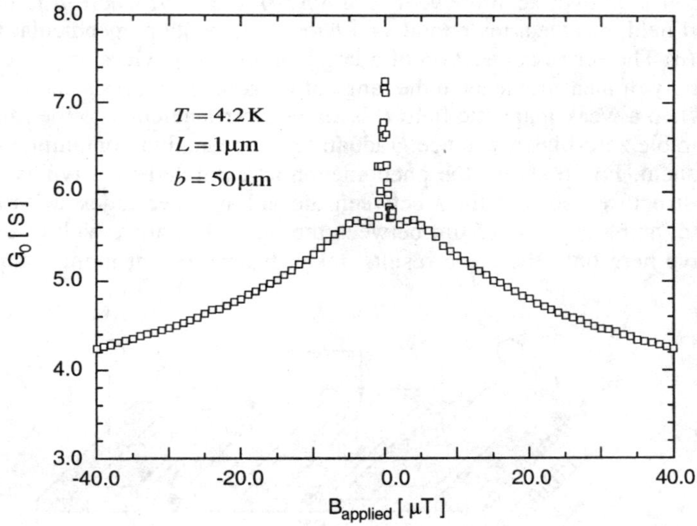

Fig. 4. Magnetic field dependence of the height of the zero-bias conductance peak of Fig. 3.

If our device were a true superconducting JJ, and our measurement were one of critical current rather than of resistance, we would expect a first minimum of the critical current at a magnetic flux of $\Phi_0 = \pi\hbar/e$, the canonical superconducting flux quantum. In fact, a field of 50μT inside the gap corresponds to approximately one flux quantum threading the gap. We consider this observation to be compelling evidence that the phase-preserving properties of the multiple AR's occur already long before the onset of "true" superconductivity.

The falloff of the conductance at higher fields in Fig. 4 does not follow the textbook Fraunhofer diffraction pattern characteristic of the critical current of true JJ's,[6] but this is hardly surprising, considering the fact that ours is a different kind of measurement and—especially, that our sample is far from superconducting.

4 Weak-Link Ladder Arrays

Following our initial work on single-gap devices with a cross-section as in Fig. 1, we found it useful to go beyond the single-gap geometry shown of Fig. 1, and to study series-connected periodic arrays, prepared by laser holography, involving a large number (≈ 300) of gaps, shown schematically in Fig. 5. Such grating structures are particularly suitable for quantitative studies of the effects of magnetic fields in the limit of vanishing resistance: (a) Except for small expulsion effects at the end of each "rung" of the array ladder, the magnetic flux "funneled" through each cell is known accurately: it is simply $\Phi = B \cdot a \cdot b$, where B is the externally applied field, a is the array period, and b the array width perpendicular to the current flow. (b) The series connection of a large number of devices improves the voltage sensitivity of measurements in the range of near-vanishing resistance.

When a weak magnetic field B is applied perpendicular to the sample plane, a measurable zero-bias resistance gradually re-appears, but containing an oscillatory component. Fig. 6 shows the phenomenon as exhibited by a typical sample. The array structure had 310 lines between the voltage electrodes, with a period a = 0.96 μm and a gap $L \approx 0.5$ μm between the lines. The array width was b = 95 μm. We give here only the basic results, taken from a recent more complete paper,[7]

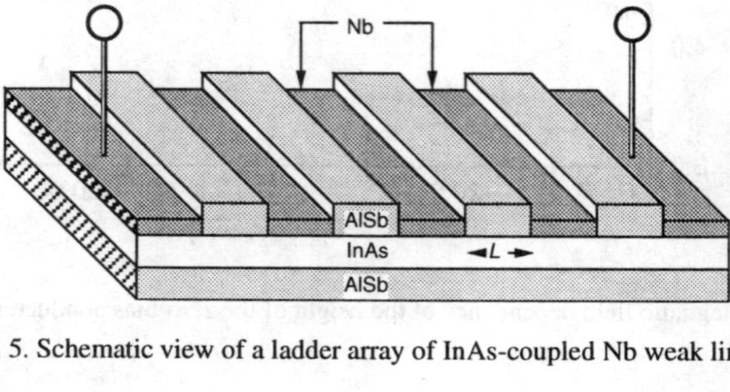

Fig. 5. Schematic view of a ladder array of InAs-coupled Nb weak links.

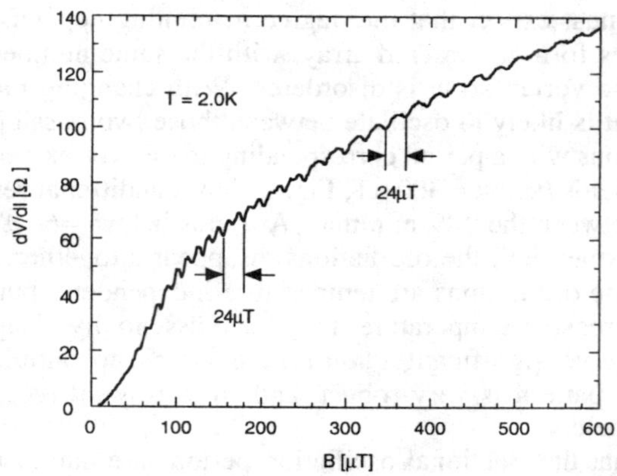

Fig. 6. Dependence of the zero-bias resistance at 2.0 K on magnetic field, for a typical sample.

where the interested reader can find more details; along with a discussion of the relation of our observations to magneto-resistance oscillations in arrays of conventional Josephson tunnel junctions.

The oscillation pattern is evidently fairly complex, except above approximately 300 μT, where the oscillations show a well-defined period ΔB of about 24 μT. If we express the period in terms of the *nominal* magnetic flux $\Phi = B{\cdot}a{\cdot}b$ through each grating cell, we obtain a flux period $\Delta\Phi = \Delta B{\cdot}a{\cdot}b = 2.2{\times}10^{-15}\mathrm{T{\cdot}m^2}$, a value close to the canonical superconducting flux quantum $\Phi_0 = 2.07{\times}10^{-15}\mathrm{T{\cdot}m^2}$. The actual flux per cell should be slightly smaller, because of lateral flux expulsion from each cell at the end of the grating lines; the agreement would be perfect if we assume that the flux expulsion is equivalent to a line shortening by about 2.6μm at each end, a correction of plausible magnitude. With increasing field, the oscillations persist for about 40 to 50 oscillation cycles, with the same period but a slowly decreasing amplitude.

This overall behavior has a very simple explanation: The magnetic flux that penetrates through the space between the Nb lines is present in the form of flux quanta, commonly referred to as *vortices*. It is well known from type-II superconductors that such vortices can move under the influence of an electric current, a dissipative process, causing a resistance. To maintain zero resistance in the presence of a current requires that the vortices are somehow pinned at defects. Because of the high structural perfection of the high-mobility InAs-AlSb quantum wells, the flux pinning in the InAs is probably weak, which should lead to a finite resistance of the structure due to transverse flux motion between the Nb "rungs". The entire structure should then behave similarly to a type-II superconductor with weak flux pinning.

Furthermore, we must expect that the degree of pinning depends on whether the vortices themselves form an ordered array, with the same number of vortices per cell, or whether the vortex array is disordered. With changing magnetic field the vortex arrangement is likely to oscillate between those two possibilities, leading to resistance oscillations with a period corresponding to one vortex per cell.

However, at fields *below* $\approx 300\,\mu T$, Fig. 6 shows additional resistance minima, exactly halfway between the "Φ_0 minima." At fields below $\approx 60\,\mu T$, corresponding to about 3 vortices per cell, the oscillations disappear altogether. Both the period and the phase of the oscillations are temperature-independent; but their amplitude decreases with increasing temperature; they have essentially disappeared at 4.2K. Although we observe significant *quantitative* sample-to-sample variations, the overall oscillation pattern is very robust, and shown by all recent samples with similar array parameters.

We believe that the fractional oscillation periods are due to the formation of *flux cell superlattice*. From the theory of vortices in long JJ's,[6] we estimate that for our sample the vortices should be relatively compact, with a width (FWHM) on the order of about 6.4μm. Our 95μm array is wide compared to this quantity, and so long as the applied magnetic fields remain small enough that the number of flux quanta per cell remains small compared to 95/6.4 ~ 15, the individual vortices should remain well-separated from each other. The separation of vortices within a cell is then also much larger than the period of the array. Because vortices repel each other, it is energetically unfavorable for vortices in adjacent cells to line up; so long as they have enough space *within* each cell, vortices in adjacent cells will tend to form a staggered array, as shown schematically in Fig. 7.

Fig. 7a shows the case where every cell contains the same number of vortices, with the predicted stagger of the vortex arrangement. This can of course only happen for specific discrete values of the magnetic field. As the field is changed, a disordered vortex arrangement must be present, but whenever the (average) number of vortices becomes a half-integer, a regular superlattice is again likely to form, in which the number of vortices in each cell alternates between an even number and an

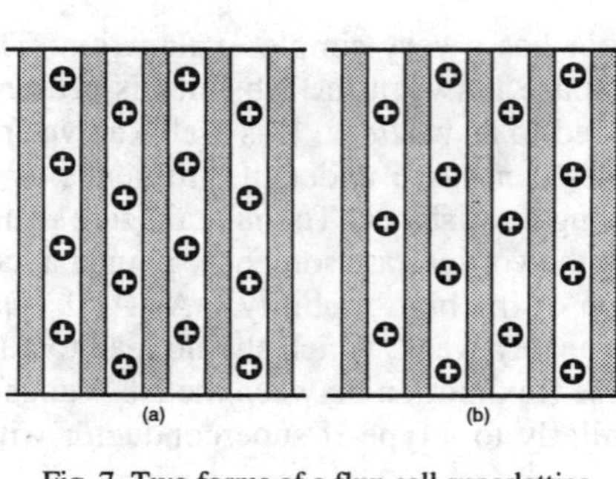

Fig. 7. Two forms of a flux cell superlattice

adjacent odd number, as shown in Fig. 7b. As stated earlier, the degree of vortex pinning, and hence the sample resistance, should be different for an ordered and a disordered vortex arrangement. Hence we must indeed expect that the resistance exhibits oscillations with a period of one-half of a flux quantum per lithographic cell.

With the oscillations being a flux quantization effect, their period should scale with the width of the array. In order to study this scaling, unencumbered by unrelated sample-to-sample variations that might obscure the width dependence, we simply decreased the width b of the existing 95μm-wide array of the sample of Fig. 6 by wet etching, first to about 38 μm, and then to 16 μm. The results, plotted in Fig. 8, in terms of Φ rather than B, show that the oscillation periods at high fields are again slightly larger than Φ_0, indicating the expected scaling with cell size, again with small end corrections for flux expulsion.

However, the frequency doubling at lower fields is much less pronounced in the 38 μm sample, and absent in the 16 μm sample. Evidently, the occurrence of fractional periods depends on the array width.

Such a dependence is readily understood: We stated earlier that a condition for the occurrence of the flux cell superlattice is an array width large compared to the width of an individual vortex, estimated as 6.4μm. With decreasing array width, this condition becomes less-well satisfied, and for sufficiently narrow arrays we must expect that the superlattice disappears altogether, with the flux distribution then showing only the lithographic periodicity of the array itself. The weakening of the half-interger oscillations at 38μm, and their absence at 16μm are plausible orders of magnitude for the phenomenon.

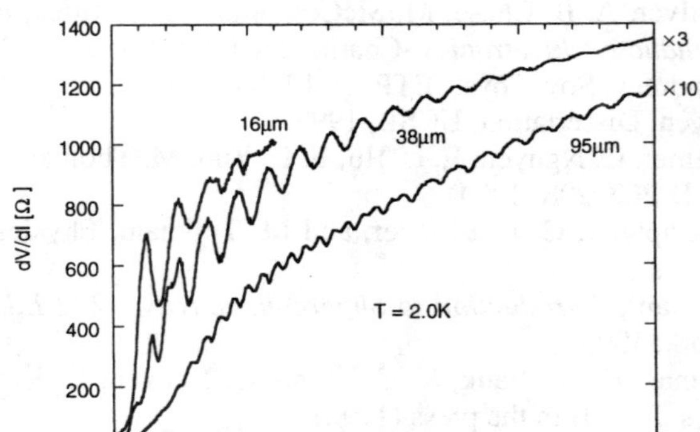

Fig. 8. Dependence of the resistance oscillations on array width.

5 Conclusions

Flux quantization effects imply phase coherence of the electron wave function over distances at least on the order of the size of the individual single-vortex flux cells. Our devices clearly have macroscopic rather than mesoscopic dimensions. In fact, it is the macroscopic size that is responsible for the weakness of the fields involved: for small mesoscopic devices the fields would be orders of magnitudes larger.

Probably the most unexpected of our observation sis that the phase coherence implied in flux quantization persists far into the regime of normal conductivity. In a recent paper, Harris et al. have come to a similar conclusion based on the observation of Shapiro steps (a.c. Josephson effect) in array structures of the kind discussed here [8]. The authors report the startling observation that Shapiro steps, and hence phase coherence could be observed up to temperatures at which the d.c. current-voltage characteristics no longer showed any significant deviations from a purely resistive behavior.

Acknowledgments

We are grateful to Prof. Evelyn Hu, Mr. Ki Wong, and Dr. Richard Blank for technical support and numerous discussions. The work was supported in part by the ONR, in part by the NSF, the latter through the NSF Science and Technology Center for Quantized Electronic Structures, Grant DMR 91-20007.

References

1 A. H. Silver, A. B. Chase, M. McColl, and M. F. Millea, *Future Trends in Superconductive Electronics*, Charlottesville, VA 1978, p. 364.
2 A. F. Andreev, Sov. Phys. JETP **19**, 1228 (1964).
3 C. Nguyen, Dissertation, UCSB, 1993.
4 H. Kroemer, C. Nguyen, E. L. Hu, E. L. Yuh, M. Thomas, and K. C. Wong, Physica B **203**, 298 (1994).
5 T. M. Klapwijk, G. E. Blonder, and M. Tinkham, Physica B+C **109/110**, 1657 (1982).
6 M. Tinkham, *Introduction to Superconductivity - 2nd Ed.* (McGraw-Hill, New York, 1996).
7 M. Thomas, H.-R. Blank, K. C. Wong, C. Nguyen, H. Kroemer, and E. L. Hu, Phys. Rev. B in the press (1996).
8 J. G. E. Harris, H. Drexler, E. L. Yuh, K. Wong, E. G. Gwinn, E. L. Hu, H. Kroemer, and S. J. Allen, Submitted to Phys. Rev. Lett. (1996).

COULOMB EFFECTS ON THE MAGNETOCONDUCTANCE OF A TWO-DIMENSIONAL ELECTRON GAS IN A LATERAL SUPERLATTICE: A SCREENED HARTREE-FOCK CALCULATION

A. MANOLESCU

Institutul de Fizica şi Tehnologia Materialelor, C.P. MG-7 Bucureşti-Măgurele, România

R. R. GERHARDTS

Max-Planck-Institut für Festkörperforschung, Heisenbergstrasse 1, D-70569 Stuttgart, Federal Republic of Germany

We calculate the magnetoconductivity tensor of a 2D electron gas in a 1D periodic potential and quantising magnetic fields. We study the internal structure of the Shubnikov-de Haas peaks and analyse recent experimental results. The electron-electron interaction is accounted for within a screened Hartree-Fock approximation (HFA), which describes both compressible strips and an exchange-enhanced spin splitting, unlike the Hartree and the standard HFA.

Recent magnetotransport experiments on GaAs-AlGaAs heterostructures with a built-in 1D lateral electric modulation, in the x direction, have focused on the internal structure of the Shubnikov-de Haas peaks of the resistivity ρ_{xx}.[1,2] This structure is determined by the density of states (DOS) of the 1D Landau bands and by their exchange-enhanced spin splitting. The interpretation of the experimental results is however difficult and still unclear.

The electron-electron interaction has important effects. The electrostatic and the exchange components act oppositely: The Hartree screening reduces the energy dispersion of the Landau bands, *increasing* the DOS, while the negative exchange energy of the occupied states broadens the bands, *decreasing* the DOS.[3] Our previous attempt to take into account such effects in a magnetotransport calculation suggested that the standard Hartree-Fock approximation (HFA) overestimates the exchange.[2] The resulting van Hove singularities (VHS) of the DOS become very sharp, and for modulation periods much larger than the magnetic length l the competition between the long-range electrostatic screening and the exchange broadening may result in unrealistic short-range oscillations of the charge density.

In the present paper we overcome such artefacts by using a screened HFA (SHFA), i.e. we include the screening influence on the exchange interaction. We show that the enhanced spin splitting can coexist with the picture of compressible/incompressible strips. Our modulation model is a simple potential

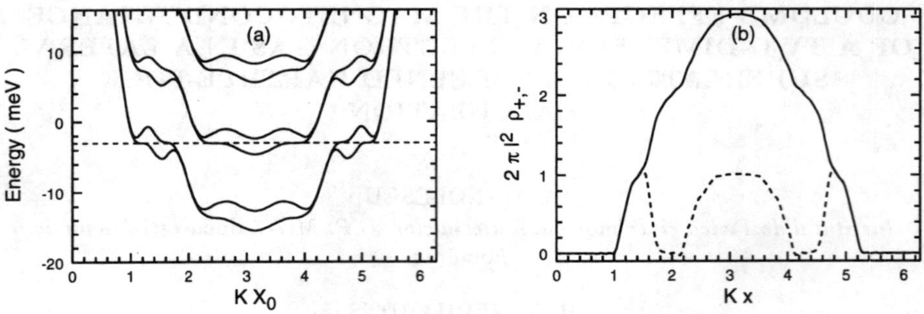

Figure 1: (a) Landau bands for isolated wires produced by a modulation with $V = 300$ meV and $a = 1000$ nm; the dashed line shows the Fermi level and X_0 is the center coordinate. (b) Particle and spin densities, $\rho_{+,-}$, full and dashed lines. $B = 6$ T and $T = 0.5$ K.

$V \cos Kx$, uniform on the y axis. The material parameters are those for GaAs.

Our one-particle Hamiltonian has the form $H = H^0 + \Sigma^{ee} + \Sigma^{ei}$, in which H^0 describes an electron in the plane $\{\mathbf{r} = (x, y)\}$, in a perpendicular magnetic field B and in the periodic potential, while Σ^{ee} and Σ^{ei} are the electron-electron and electron-impurity self-energies. We use the Hartree-Fock expression for Σ^{ee}, at a finite temperature T, and we include the polarisation loop in the interaction line of the exchange diagram by replacing the Coulomb potential $u(q) = 2\pi/q$ by $u(q)/\varepsilon(q)$. Here $\varepsilon(q)$ is the *static* dielectric function of the 2D electron gas, which we consider quasi-homogeneous, characterised by $lV/a \ll \hbar\omega_c$, $a = 2\pi/K$. We calculate $\varepsilon(q)$, for an arbitrary q, in the spirit of the random-phase approximation, using the Lindhard formula self-consistently with the eigenstates of the Hamiltonian H, within a numerical iterative scheme. The dominant screening corresponds to $ql \ll 1$ and is due to the intra-band transitions, which are determined by the DOS at the Fermi level, D_F. In our SHFA, charge-density instabilities of the homogeneous system ($V = 0$) are no longer possible, in contrast to the standard HFA.

In Fig.1 we show typical results in the limit of isolated wires, created by a long-period modulation. This picture combines the specific aspects of both Hartree [4,5] and Hartree-Fock [3,6] approximations. The former leads to strong screening, but only bare spin splitting (very small for GaAs), the latter to strong exchange enhancement, but poor screening. In the SHFA the spin splitting of the compressible edge states is accomplished by local fluctuations of the Landau bands with opposite spins, and of the spin density, but each band is individually pinned at the Fermi level. This result disagrees with the prediction of Dempsey et al., [6] that when the lateral confinement decreases, a sharp transition from spin unpolarised to spin polarised edge states, with a steep energy dispersion at the Fermi level, should occur. We believe this prediction is related to the artefacts of the HFA. Note that, due to the exchange effects we can obtain stable incompressible strips in the bulk of the wires, even without impurity broadening.

We consider the electron-impurity interaction in a phenomenological self-consistent Born approximation, taking Σ^{ei} as a c-number determined by a characteristic energy parameter $\Gamma = \gamma\sqrt{B[\text{Tesla}]}$ [meV]. We calculate the conductivities using the Kubo formalism adapted to our system, [7] in which we define the velocity operators as $\mathbf{v} = i[H, \mathbf{r}]/\hbar$. The only contribution of the self-energy to the commutator is that of the (nonlocal) exchange interaction, which is in fact the current vertex correction required by the Ward identity. [8]

The relation between the conductivities and the DOS is complicated. Both the diagonal conductivities σ_{xx} and σ_{yy} have inter-band-scattering components, proportional to $(\Gamma D_F)^2$ and sensitive to the VHS of the 1D Landau bands, when the impurity broadening Γ is small. Due to the anisotropy of the system, only σ_{yy} has an intra-band term, related to the dispersion of the Landau band, or, classically, related to the drift of the electronic orbits perpendicular to the modulated electric field. This band conductivity is, contrary to the one due to scattering, approximately proportional to $(\Gamma D_F)^{-2}$, thus vanishing near the VHS, but dominating in σ_{yy} for $\Gamma \to 0$.

For a comparison with the experiment we need to invert the conductivities into resistivities, $\rho_{xx,yy} = \sigma_{yy,xx}/(\sigma_{xx}\sigma_{yy} + \sigma_{xy}^2)$, and usually $\sigma_{xx}\sigma_{yy} \ll \sigma_{xy}^2$. For the results presented in Fig.2a we have chosen a sufficiently large impurity broadening, so that we have a small band conductivity and consequently $\sigma_{xx} \approx \sigma_{yy}$. In this case the spin splitting is not resolvable even without modulation. For a finite modulation amplitude the Shubnikov-de Haas oscillations of both longitudinal resistivities may have the double-peak structure of the DOS. Such a situation has been observed by Weiss et al. [1] for ρ_{xx}, in the second Landau band, $n = 1$, while here we get it more clearly in the third band, $n = 2$. At higher magnetic fields the stronger screening reduces the band width and the resolution of the VHS is rather poor. The spin-splitting is resolved in Fig.2b where both modulation amplitude and impurity broadening are small. For $n = 2$ the spin splitted Landau bands still partly overlap. The VHS are now observable only in ρ_{yy}, and are covered by the band conductivity in ρ_{xx}. Recent measurements have shown a more complicated, double- and triple-peak profile in ρ_{xx}, which may be attributed to a combined effect of VHS and band conductivity. [2] But as we have shown, screening effects might be very strong. The screening can be reduced for a modulation with a shorter period, comparable to l. The steeper energy dispersion may allow the resolution of the VHS when the band conductivity is suppressed by disorder, as suggested

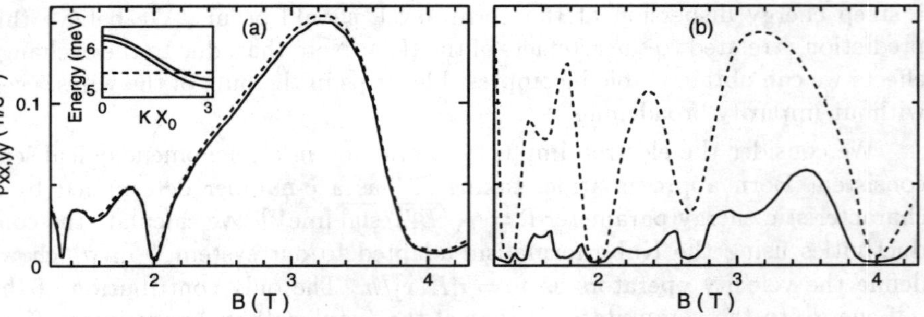

Figure 2: The resistivities ρ_{xx}, dashed lines, and ρ_{yy}, full lines, for a modulation period $a = 300$ nm. (a) $V = 12$ meV, $\gamma = 0.3$, $T = 4.2$ K. Inset: Landau bands with $n = 1$ for $B = 2.9$ T. (b) $V = 6$ meV, $\gamma = 0.05$, $T = 1$ K.

in the experiments by Sfaxi et al. [9] However, a comparative measurement of both ρ_{xx} and ρ_{yy}, in high magnetic fields, indicating the scattering and the band conductivity contributions is, to our knowledge, not available.

In conclusion, our SHFA describes both screening and exchange-enhanced spin splitting of the Landau bands, interpolating in the expected manner between the contradictory results of the Hartree and the standard Hartree-Fock approximations. The theoretical resistivities are in qualitative agreement with the available experimental data. Calculational details will be published elsewhere.

We are grateful to Gabriele Ernst, Behnam Farid, Marc Tornow and Dieter Weiss for stimulating discussions.

References

1. D. Weiss et al., *Surf. Sci.* **229**, 88 (1990).
2. A. Manolescu et al., *Surf. Sci.* , in press (Workbook EP2DS XI p.299).
3. A. Manolescu and R. R. Gerhardts, *Phys. Rev.* B **51**, 1703 (1995).
4. D. B. Chklovskii, B. I. Shklovskii and L. I. Glazman, *Phys. Rev.* B **46**, 4026 (1992).
5. K. Lier and R. R. Gerhardts, *Phys. Rev.* B **50**, 7757 (1994).
6. J. Dempsey, B. Y. Gelfand and B. I. Halperin, *Phys. Rev. Lett.* **70**, 3639 (1993).
7. C. Zhang and R. R. Gerhardts, *Phys. Rev.* B **41**, 12850 (1990).
8. R. R. Gerhardts, *Z. Phys.* B **22**, 327 (1975).
9. L. Sfaxi et al., *Surf. Sci.* , in press (Workbook EP2DS XI p.489).

Consistent Solution of Boltzmann's Equation for Two-Dimensional Electron Systems with Periodic Unidirectional Modulations of Arbitrary Strengths

R. Menne, Rolf R. Gerhardts
Max-Planck Institut für Festkörperforschung, Heisenbergstr. 1, D-70569 Stuttgart, Federal Republic of Germany

On the basis of Boltzmann's equation, and including anisotropic scattering, we investigate the effect of one-dimensional modulations on two-dimensional electron systems. In addition to modulations defined by static electric and magnetic fields, we consider a mobility modulation defined by a spatially modulated density of scattering centers. We investigate which resistivity components are affected by the different modulations and, by solving Boltzmann's equation numerically for arbitrary modulation strength, we calculate the positive magnetoresistance in weak magnetic fields and the Weiss oscillations in stronger fields within a unified approach.

Two-dimensional electron systems (2DES) in samples with a weak electric modulation [1] in x-direction reveal at low temperatures and in a very weak perpendicular applied homogeneous magnetic field B a pronounced positive magnetoresistance in the component ρ_{xx} followed at intermediate fields by the "Weiss oscillations". They are commensurability oscillations determined by the cyclotron radius R and the modulation period $a = 2\pi/q$. The resistivity component ρ_{yy} does not exhibit such a positive magnetoresistance, but shows Weiss oscillations of opposite phase, whereas no significant effect of the modulation on the Hall resistance is observed. Experiments on samples with a stronger electric modulation [2,3] and also the recent experiments on magnetically modulated systems [4] led to similar results. Beenakker [5] pointed out that the Weiss oscillations of ρ_{xx} can be understood classically in terms of a modulation induced guiding center drift of cyclotron orbits. The low-B positive magnetoresistance has been explained [3] as being due to "snake orbits" which, in x-direction, are localised within a single period of the modulation. A unified calculation of both the positive magnetoresistance and the Weiss oscillations within classical dynamics is presented here.

We consider the 2DES in the x-y plane as a degenerate Fermi gas, with Fermi energy $E_F = (m/2)v_F^2$, of (non-interacting) particles with effective mass m and charge $-e$, obeying classical dynamics in a magnetic field $\mathbf{B}(x) = (0, 0, B + B_m(x))$ and electrostatic potential $V_m(x)$. The modulation fields $V_m(x) = V_m(x + a)$ and $B_m(x) = B_m(x + a)$ are assumed to have zero average values. To describe the stationary, linear response of the modulated system to an external homogeneous electric field $\mathbf{E}^0$, we use the ansatz $f_{stat}(\mathbf{r},\mathbf{v}) = f_0 - e[\partial f_0/\partial E_F]\phi(\mathbf{r},\mathbf{v})$ for the stationary solution of the distribution function, f_0 denoting the distribution function in thermal equilibrium. With the assumption that $\phi(\mathbf{r},\mathbf{v})$ is linear in $\mathbf{E}^0$, we obtain the linearised Boltzmann equation (φ is the polar angle in velocity space)

$$\{v\cos\varphi\frac{\partial}{\partial x}+[\omega_c+\omega_m+\omega_e]\frac{\partial}{\partial\varphi}+\frac{1}{\tau}\}\phi(x,\varphi)-\frac{1}{\tau}\int\frac{d\varphi'}{2\pi}P(\varphi'-\varphi)\,\phi(x,\varphi')=\mathbf{v}\mathbf{E}^0,$$

where $\omega_c = eB/mc$ is the cyclotron frequency, and $\omega_m(x) = eB_m(x)/mc$ and $\omega_e(x,\varphi) = -\sin\varphi\, dv/dx$ are due to the magnetic and electric modulation, respectively. The energy conservation, $(m/2)\mathbf{v}^2 + V_m(x) = E_F$, fixes the magnitude of the velocity. The collision operator included in Boltzmann's equation describes elastic scattering with $\int_{-\pi}^{\pi} d\varphi P(\varphi) = 2\pi$ and $P(-\varphi) = P(\varphi)$, and τ an elastic scattering time. The special case $P(\varphi) \equiv 1$ describes *isotropic* scattering, or, equivalently, inelastic relaxation towards the *local* equilibrium distribution in the stationary state, as required by the continuity equation. Therefore inelastic scattering can be included by a suitable redefinition of $P(\varphi)$ and τ. Since, in the modulated system, the density of scatterers may also be periodically modulated, we allow in Boltzmann's equation for a x-dependent prefactor, $1/\tau(x) = (1 + r_m(x))/\tau$, where the periodic modulation of the scattering rate, $r_m(x) = r_m(x + a)$, has zero average value.

We can prove for $r_m = 0$, that the electric and magnetic modulations affect only ρ_{xx}. In order to cover the whole magnetic field range, we have solved Eq. (1) numerically, using Fourier expansions for the dependences on both x and φ. A typical result for a sinusoidal purely electric modulation or purely magnetic modulation is shown as the solid curves in Fig. 1. This result can be compared with Beenakker's approximation[5] (dash-dotted curves in Fig. 1) recovered by solving Eq. (1) in first order of the potential amplitudes, which becomes very good for B well above $B_{crit} = max\{2\pi V_m/(eav_F), B_m\}$, where only drifting orbits cover the phase space. Not seen in Beenakker's solution, but present in the exact numerical result at $B < B_{crit}$ is the positive magnetoresistance due to the influence of snake orbits. A peak occurs at $B = B_p$, where the positive magnetoresistance bends over towards the regime of the Weiss oscillations. For a sinusoidal modulations we found numerically $B_p/B_{crit} \approx 0.8$, which agrees well with the experimental observation[6] in an electrically modulated system. The dashed curve in Fig. 1 is the result of a numerical calculation including anisotropic scattering as modelled by $P(\varphi) = (8/35)(1+\cos\varphi)^4$, corresponding to a half-width of $\pi/3$, which simulates the predominant small-angle scattering by charged impurities. Fixing the amplitude of the largest Weiss oscillation by decreasing the elastic scattering time the main effect of the small-angle scattering is a rapid suppression of the Weiss oscillations and an increase of

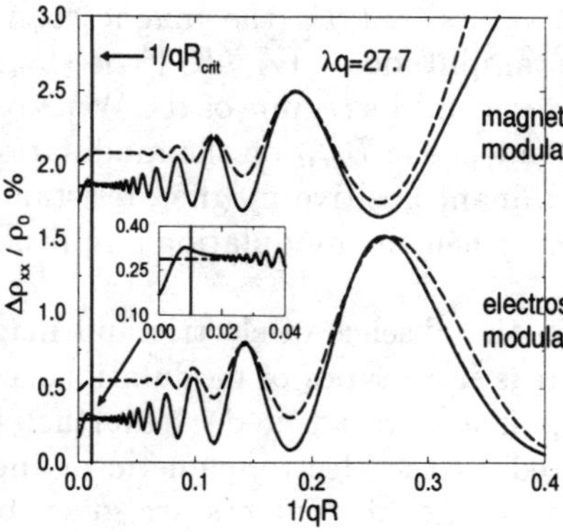

Figure 1: Magnetoresistance vs. $1/qR \propto B$ for the weak sinusoidal electric modulation and, offsetted by 1.6, the purely magnetic modulation with amplitudes V_m/E_F=0.02 and $B_m = 0.02T$, respectively; solid lines: fully numerical solution for isotropic scattering; dash-dotted line: the same for Beenakker type solution; dashed line: the same with anisotropic scattering.

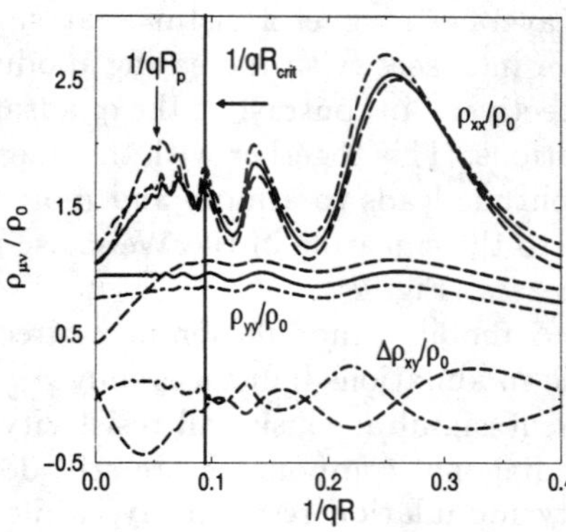

Figure 3: Magnetoresistance vs. $1/qR \propto B$ for the mixed electric and mobility modulation; solid lines: fully numerical solution for the purely modulated system with amplitudes V_m/E_F=0.2 and $r_m = 1$, respectively; dash-dotted line: the same for the system with mixed modulations; dashed line: the same for mixed modulations with amplitudes V_m/E_F=0.2 and $r_m = -1$ respectively.

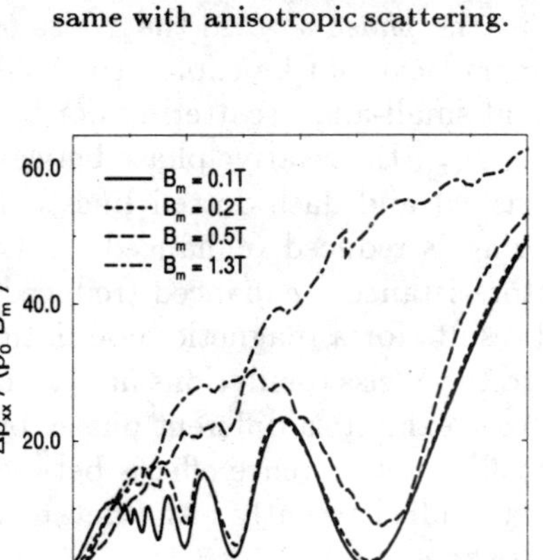

Figure 2: Fully numerical solution for the magnetoresistance vs. $1/qR \propto B$ for the strong sinusoidal purely magnetic modulations with amplitude B_m.

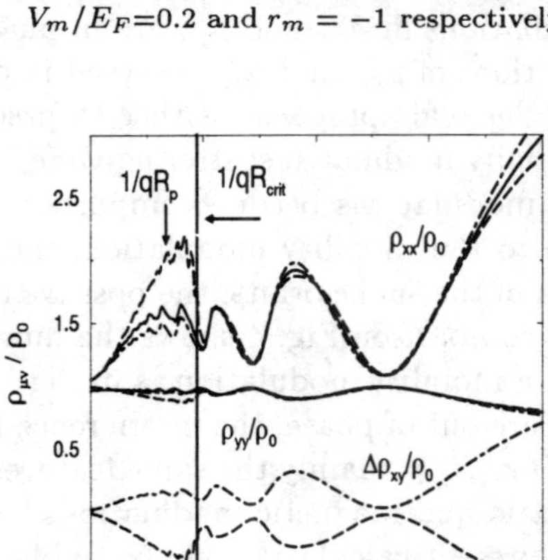

Figure 4: The same as Figure 3 for a system with mixed magnetic and mobility modulation with amplitude $B_m = 0.2T$ and $|r_m| =$ 0.9 respectively.

the average magnetoresistance at small B values. At B_p the magnetoresistance increases with increasing modulation amplitude as $|V_m/E_F|^3$ or $|B_m|^3$ respectively, in contrast to the quadratic increase in the regime of the Weiss oscillations. This together with the linear dependence of B_{crit} on the modulation amplitude leads to a more and more predominant positive magnetoresistance and to the smearing of the Weiss oscillations, when the modulation amplitude increases (Fig. 2).

A mobility modulation in x-direction in the absence of electric and magnetic modulations influences only ρ_{yy}, whereas if all types of modulations are present simultaneously, all resistivity components are affected. Nevertheless the diagonal components are still dominated by the electromagnetic or mobility modulation respectively, while the off diagonal elements are given by interferences of all modulations. In Fig. 3 we present the results of a numerical calculation for a situation with a electric modulation $V_m(x) = V_m \sin qx$ and a mobility modulation $r_m(x) = r_m \sin qx$. The solid lines refer to purely electric modulation (ρ_{xx}, $V_m = 0.2\,E_F$, $r_m = 0$, upper line) and pure mobility modulation (ρ_{yy}, $V_m = 0$, $r_m = 1$, lower line), respectively. It is seen that the effect of an extremely strong pure mobility modulation is still very weak. Especially there is no positive magnetoresistance at small B values, since there are no snake orbits in this case. If both types of modulations are present, the Weiss oscillations in ρ_{xx} and ρ_{yy} are in phase, thus, the *phase shift* in the Weiss oscillations of ρ_{xx} and ρ_{yy} observed in experiment [1] can not be explained within the classical approach, neither by predominant small-angle scattering nor by a spatially modulated scattering time. For $B < B_{crit}$ the relative phase between the modulations becomes important (see dashed and dash-dotted lines). If, due to the mobility modulation, the scattering is reduced (enhanced) in the area of the snake orbits, the positive magnetoresistance is enhanced (reduced). For comparison Fig. 4 shows the numerical results for a magnetic modulation and a mobility modulation as in Fig. 3. Here the Weiss oscillations in ρ_{xx} and ρ_{yy} are out of phase, the interference term $\Delta\rho_{yy}$ exhibits a different phase, but at $B < B_{crit}$ mainly the same features are visible. Interference effects between electric and magnetic modulations have been studied recently [7] and details of the present calculation will be published elsewhere.

1. D. Weiss *et al.*, Europhys. Lett. **8**, 179 (1989).
2. R. W. Winkler *et al.*, Phys. Rev. Lett. **62**, 1177 (1989).
3. P. H. Beton *et al.*, Phys. Rev. B **42**, 9229 (1990).
4. P. D. Ye *et al.*, Phys. Rev. Lett. **74**, 3013 (1995).
5. C. W. J. Beenakker, Phys. Rev. Lett. **62**, 2020 (1989).
6. A. Soibel, U. Meirav, D. Mahalu, and H. Shtrikman, preprint (1996).
7. R. R. Gerhardts, Phys. Rev. B **53**, 11064 (1996).

THEORY OF THE MAGNETO-TRANSPORT IN A NONPLANAR TWO DIMENSIONAL ELECTRON GAS

S.M. BADALIAN[a], I.S. IBRAHIM, and F.M. PEETERS
Department of Physics, University of Antwerp (UIA),
Universiteitsplein 1, B-2610 Antwerpen, Belgium

The magnetoresistance of a nonplanar two-dimensional electron gas (2DEG) in a magnetic field is investigated. As an example the electric field distribution of the 2DEG with a magnetic field barrier is calculated. The system satisfies the Poisson equation in which a line charge develops at the magnetic/non-magnetic field interface. The magnetoresistances across the facet and in planar regions of the 2DEG are calculated which provide an understanding of the main features of the magnetic field dependences observed in the experiment of M.L. Leadbeater *et al* [Phys. Rev. B 52, R8629 (1995)].

Recently a new approach of creating non-homogeneous magnetic structures by varying the topography of the 2DEG was proposed [1]. A non-planar 2DEG at a GaAs/AlGaAs heterojunction grown over etched facet was fabricated [2]. By applying a uniform magnetic field to this structure one obtains a spatially varying field component normal to the 2DEG. Rotating the plane of the sample allows to find different non-homogeneous magnetic structures: magnetic barriers, wells and other completely novel situations. Understanding of carrier transport properties in such non-homogeneous magnetic structures is important for the basic quantum phenomena.

The aim of the present work is to calculate the transport properties of a nonplanar 2DEG in a magnetic field. As an example we consider the magnetic field profile $B_z(y) = 0$ ($0 < y < a$ and $b < y < L$) and $B_z(y) = B$ ($a < y < b$) which is appropriate to the experimental situation of Ref. 2. Here the width ($= b - a$) of the magnetic field barrier is much larger than the magnetic length and consequently the electric field distribution is determined by classical electrostatics from which we obtain the magnetoresistance which we compare with experiment. The present situation is different from the one discussed in Ref. 3 where *quantum tunneling* through magnetic nanostructures was treated.

In the steady state the current density $\vec{J}$ and the electric field $\vec{E}$ satisfies Maxwell's equations, $\mathbf{div}\,\vec{J}(x,y) = 0$ and $\mathbf{curl}\,\vec{E}(x,y) = 0$. The system equation of charge carrier motion is determined by the Lorenz and Coulomb forces, $\vec{J}(x,y) = \sigma\,\vec{E}(x,y) - \tan\zeta(y)\,\vec{J}(x,y) \times \vec{B}(y)$. Here σ is the conductivity, the Hall angle $\zeta(y)$ is defined by $\tan\zeta(y) = \sigma\,R_B\,B(y)$, R_B is the Hall coefficient.

[a]Permanent address: Department of Radiophysics, Yerevan State University, 1 A. Manoogian St., 375049 Yerevan, Armenia, email: badalian@vxc.yerphi.am

For the considered magnetic field profile we obtain from these equations that the electric field satisfies the following equation

$$\begin{aligned}\mathbf{div}\,\vec{E}(x,y) &= \tan\zeta\,\{\sin[2\zeta(y)]\,E_y(x,y)-\cos[2\zeta(y)]E_x(x,y)\}\\ &\times[\delta(y-a)-\delta(y-b)] = 4\pi\rho_s[\delta(y-a)-\delta(y-b)],\end{aligned}\tag{1}$$

with $\delta(y)$ the Dirac-delta function. When we represent the electric field as the gradient of a potential function it satisfies the 2D Laplace equation in the *separate* planar and nonplanar regions of the 2DEG. However, this potential *does not* satisfy the Laplace equation in the whole 2DEG region. It satisfies the Poisson equation with a non-trivial right side which is determined by the field itself. Thus at the magnetic interfaces $y = a$ and $y = b$ there is an accumulation of linear charges with a charge density ρ_s.

At the boundary of the 2DEG there is no current flow out the sides ($x = -w$ and $x = w$) and no electric field parallel to the ends ($y = 0$ and $y = L$). Therefore the boundary conditions can be given by the condition that the electric field makes an angle 0 and ζ with respect to the sides of the 2DEG in the planar and nonplanar regions, respectively.

To solve such an electric potential problem we first solve Laplace equation for the whole region of the 2DEG with the same boundary conditions taking $\rho_s = 0$ in Eq. (1). Then, we include the line charges at $y = a$ and $y = b$ which implies that the electric field exhibits jumps at these points and we modify the solution of the Lalace equation such that it satisfies the Poisson equation corresponding to Eq. (1). To construct the solution of the 2D Laplace equation, the conformal mapping method[4] is used. For the field function we obtain

$$\mathbf{E}'(x,y) = \left(\frac{[sn(x,y)+sn(1,\alpha)][sn(x,y)-sn(1,\beta)]}{[sn(x,y)-sn(1,\alpha)][sn(x,y)+sn(1,\beta)]}\right)^{\zeta/\pi}. \tag{2}$$

Here $sn(x,y) \equiv sn\,(K(m)x + iK'(m)y)$ is the Jacobi elliptic function, K and K' are the complete elliptic integrals of the first kind, m is determined by the sample aspect ratio from the relation $L/K'(m) = w/K(m)$, $\alpha = a/L$ and $\beta = b/L$, $y \div (0,1)$ and $x \div (-1,1)$, $2w$ is the width of the 2DEG. For $a = 0$ and $b = L$ Eq. (2) gives the electric field distribution for an uniform magnetic field in terms of Jacobi functions which is in agreement with that obtained in Ref. 4 in the form of an expansion into hyperbolic functions.

The above field E' does not conserve current and the normal component of $\vec{J}$ jumps at the magnetic interfaces. We can remedy this as follows: in the planar regions the real electric field $\vec{E}(x,y)$ is given by Eq. (2) while in the nonplanar region, due to the jump of the y-component of the field, we have

$$E_y(x,y) = \frac{1}{\cos^2\zeta}\mathbf{Re}\,\vec{E}'(x,y) - \tan\zeta\;\mathbf{Im}\,\vec{E}'(x,y),\; E_x(x,y) = \mathbf{Im}\,\vec{E}'(x,y). \tag{3}$$

The above equation together with the auxiliary Eq. (2) is the solution of our problem and satisfies the Poisson equation (1). Now it is clear that in the whole region of the 2DEG we have $J_n(\vec{E}) = \sigma E'_n$, and since $\mathbf{div}\vec{E}' = 0$, the current is conserved at the magnetic interfaces.

The spatial distribution of the argument $\mathbf{Arg}\vec{E}(x, y)$ and the absolute value $\mathbf{Abs}\vec{E}(x, y)$ of $\vec{E}(x, y)$ in the neighbourhood of the facet is shown in Fig. 1. The field exhibits power law singularity in the diagonally opposite

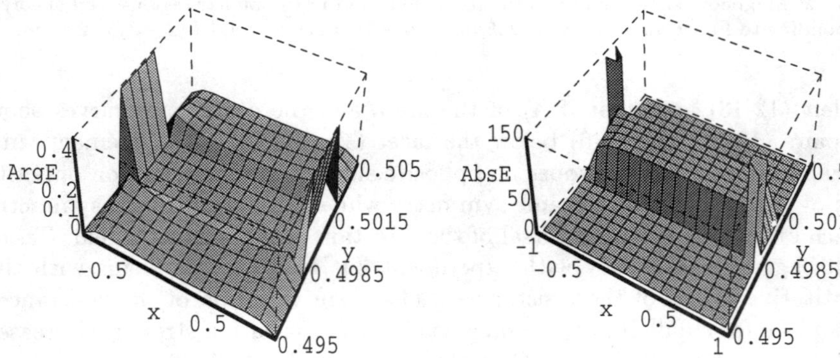

Figure 1: Spatial distribution of the argument (in units of π) and absolute value of $\vec{E}$ for the facet situated between the points 498.5 and 501.5μm. $L = 1000\mu$m, $2w = 40\mu$m, $B = 1$T.

corners of the facet while in the two other corners $\vec{E}(x, y) = 0$. Near the sides of the planar and nonplanar regions, $\mathbf{Arg}\vec{E}(x, y)$ is near to zero and ζ, respectively. $\mathbf{Arg}\vec{E}(x, y)$ has local maximum along the y-axis near the edges of the magnetic barrier and sharply drops at the magnetic interface remaining near to zero in the whole nonplanar region. The current flows between the ends of the 2DEG crossing the magnetic interfaces mainly in the singular points, i.e. it passes along the points for which $\mathbf{Arg}\vec{E}(x, y)$ is approximately zero. Such an electric field distribution differs from the one in an uniform magnetic field. The magnetic field dependence of the magnetoresistance across the facet is shown on the left in Fig. 2. In agreement with the experiment[2] the resistance increases as B^2 for small fields (due to the singularity of the electric field) and is linear in B for large B (due to the jump of the y-component of the electric field). However, the resistance increases from 13.2 to 1020.8 Ω when $B : 0 \to 10$T, which is approximately four times less than observed experimentally. The reason for such a difference could be the low ratio of the facet ($3\mu m$) to the mesa ($1000\mu m$) length. The right figure shows the magnetoresistances calculated in the planar regions. The top two curves are the resistance for probes above the facet on

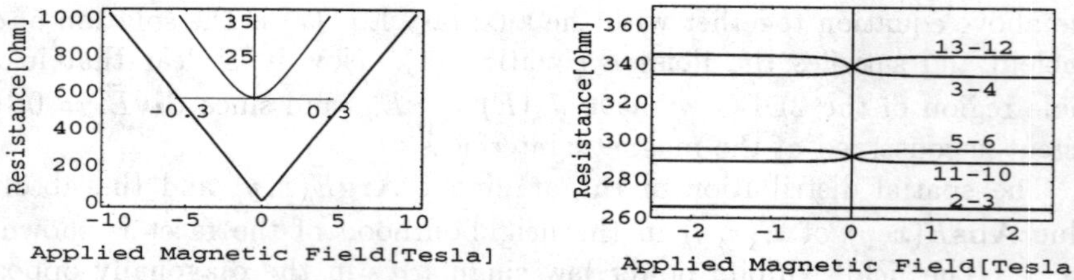

Figure 2: Magnetoresistances across the facet (left) and in the planar regions (right) corresponding to Fig. 1 for $L(3-4) = 255\mu m$, $L(5-6) = 220\mu m$, and $L(2-3) = 200\mu m$.

the left (12-13) and right (3-4) of the mesa and the center two curves show the pairs (10-11) and (5-6) below the facet (see Ref. 2). In agreement with the experiment, pairs of probes on opposite sides of the mesa and on opposite sides of the facet show the same symmetry while there is a strong asymmetry in each of the traces with reversal of the direction of the magnetic field. These results agree qualitatively with experiment but there are problems with the quantitative values of the resistances. The main variation of the resistances takes place for small B and the magnitude of the variation strongly decreases with distance from the facet. Therefore, in the scale of Fig. 2 one cannot see the resistance variation for probes (2-3) situated $300\mu m$ away from the facet.

Acknowledgments

SMB thanks for the hospitality of the Department of Physics of the University of Antwerp and the ICTP (Trieste) where this work was done. Participation of SMB in the "High Magnetic Fields in Semiconductor Physics" international conference was supported by the Deutsche Forshungsgemeinshaft (Germany). FMP is a research director with the National Science Foundation of Belgium.

References

1. C.L. Foden, M.L. Leadbeater, J.H. Burroughes, and M. Pepper, J. Phys. Condens. Matter **6**, L127 (1994).
2. M.L. Leadbeater, C.L. Foden, T.M. Burke, J.H. Burroughes, M.P. Grimshaw, D.A. Ritchie, L.L. Wang, and M. Pepper, J. Phys. Condens. Matter **7**, L307 (1995); Phys. Rev. **B 52**, R8629 (1995).
3. F.M. Peeters and A. Matulis, Phys. Rev. **B 48**, 15166 (1993); A. Matulis, F.M. Peeters, and P. Vasilopoulos, Phys. Rev. Lett. **72**, 1518 (1994).
4. R.W. Rendell and S.M. Girvin, Phys. Rev. **B 23**, 6610 (1981).

ELECTRONIC STRUCTURE OF 2D LATERAL SUPERSTRUCTURES IN STRONG MAGNETIC FIELDS

HIDEO AOKI, MASATO ANDO* and HAJIME MATSUMURA*
Department of Physics, University of Tokyo,
Hongo, Tokyo 113, Japan

Hofstadter's diagram (the energy spectrum against the magnetic field in periodic systems) is obtained for the models having flat one-electron band(s) that have originally been proposed for itinerant spin ferromagnetism. Magnetic fields let the band remain flat for the flat band arising from a topological reason. For the flat band arising from interference, the band becomes dispersive in a singular manner, which implies an anomalous orbital magnetism.

Hofstadter's butterfly, or the Landau-quantized energy spectrum against the magnetic field in periodic systems, provides an intriguing example of fractal spectrum. The quantum Hall effect in such a case has also been discussed.[1,2] Here we wish to point out that an interesting situation exists when there is *flat* (dispersionless) band(s). The flat band has been introduced in the context of the itinerant magnetism in correlated electrons by Lieb,[3] where the magnetism may also be interpreted from 'generalized Hund's coupling'.[4] The model is extended by Mielke [5] and by Tasaki [6], which introduce distant neighbour transfers to produce flat band(s) from an interference.

A class of flat bands arise from a different context of lateral superstructures, envisaged to be realized in organic π-electron materials such as the one alleged to be obtained in an attempt to fabricate fullerene.[7] A group theory [8] can classify atomic configurations into semiconducting, semimetallic and metallic classes. On top, the superperiod *enforces* the existence of flat bands in some classes [8] as a systematic realization of Lieb's case.

We consider a tight-binding system with a rectangular unit cell of $L_x \times L_y$ (twice the original unit cell for honeycomb systems). The perpendicular magnetic field, B, is characterized by $\tilde{B} \equiv BL_xL_y/\Phi_0 = p/q$, where $\Phi_0 = h/e$ is the flux quantum. The magnetic field is incorporated in the transfer energy, t_{ij}, through Peierls's phase. Details of calculation will be published elsewhere.

Figure 1 displays the Hofstadter butterfly for Lieb's, Mielke's and Tasaki's models. We can observe that the magnetic field forces the flat bands to remain flat, sandwiched between usual Hofstadter butterflies, for the Lieb's flat band arising from topology (as analytically shown). By contrast, B lifts the flatness of the bands arising from interference in Mielke's or Tasaki's models.

We can also show that the 'Wannier' states on the flat band turn into

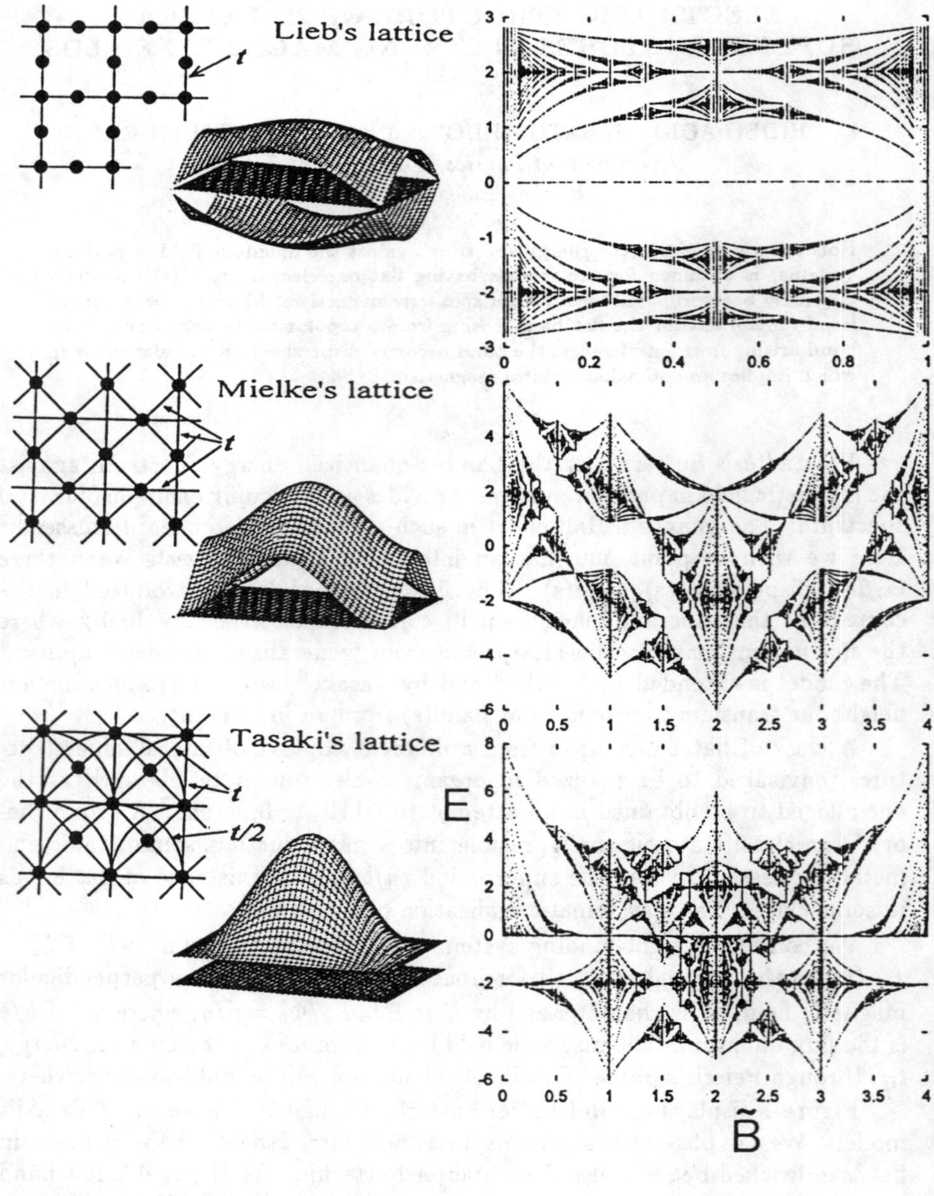

Fig.1 Hofstadter's diagram for Lieb's, Mielke's and Tasaki's models. Left panels depict the lattice structure and the band dispersion for $B = 0$.

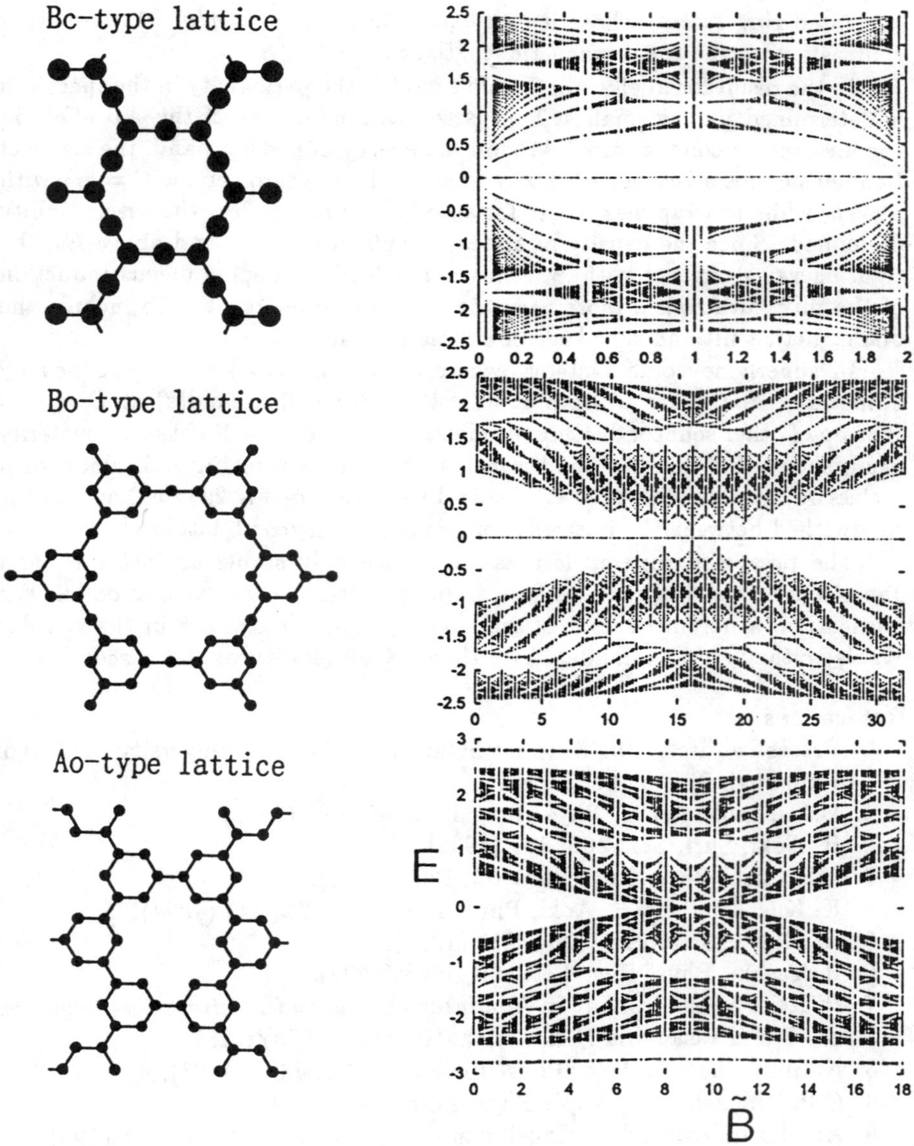

Fig.2 Hofstadter's diagram for class B_C (top), B_0 (middle) or A_0 (bottom) superhoneycomb systems. Left panels depict the lattice structure.

elongated ring states, whose length equals to p for $\tilde{B} = q/p$, which sharply contrasts with the usual magnetic length scale $\propto 1/\sqrt{B}$.

In the result for Mielke's or Tasaki's models the periodicity in the spectrum is determined by the smallest loop is Φ_0, as reminiscent of the AB effect in the mesoscopic conductance.[9] Around the energy of the flat band, the series of Landau's bands arises in such a way where we have a zero gap for $\tilde{B} = q/p$ with p even, while the gap disappears for an odd p. This renders the susceptibility ill-defined. Since the density of states spreads both *below* and above E_0, the total energy decreases with B. This might lead to a spontaneous induction of the flux, although a more accurate estimate of energy has to include the diamagnetic shift and shrinkage of atomic orbitals.[10]

In superhoneycomb systems, where a class B_0 (B_C) system defined by Shima and Aoki [8] has to have at least three (one) flat band(s) at $E = 0$ in the gap of other semiconducting (semimetallic) bands, the Hofstadter butterfly again shows that the flat band remains flat as shown in Fig.2a,b along with a class A_0 system with no flat bands. In the case of Fig.2a, the flat band is sandwiched between the butterfly for a simple honeycomb lattice [11].

The presumed superstructures are surprisingly stable against the band Jahn-Teller type distortion as seen from the total energy calculation [12]. For fabrication, a polymerization of self-aligned organic molecules in the van der Waals epitaxy [13] is a candidate. We thank Kazuhiko Kuroki for discussions.

References

* Present address: Institute for Solid State Physics, University of Tokyo, Roppongi, Tokyo 106, Japan.

1. D.J. Thouless et al, Phys. Rev. Lett. 49, 405 (1982).
2. H. Aoki, Surf. Sci. 263, 137 (1992).
3. E.H. Lieb, Phys. Rev. Lett. 62, 1201 (1989).
4. K. Kusakabe and H. Aoki, Phys. Rev. Lett. 72, 144 (1994).
5. A. Mielke, J. Phys. A 24, 3311 (1991).
6. H. Tasaki, Phys. Rev. Lett. 69, 1608 (1992).
7. H. Reiss and D.U. Kim, in *Nonlinear Optical and Electroactive Polymers*, ed. P.N. Prasad and D.R. Ulrich, (Plenum, 1988) p.281.
8. N. Shima and H. Aoki, Phys. Rev. Lett. 71, 4389 (1993).
9. C.P. Umbach et al, Appl. Phys. Lett. 50, 1289 (1987).
10. A. Alexandrov and H. Capellmann, Phys. Rev. Lett. 66, 365 (1991).
11. R. Rammal, J. Physique 46, 1345 (1985).
12. H. Aoki and N. Shima, Superlattices and Microstructures 15, 247 (1994).
13. A. Koma in *New Horizons in Low-Dimensional Electron Systems* ed. by H. Aoki et al (Kluwer, Dordrecht, 1992), p.85.

Is the Hofstadter energy spectrum observable in far-infrared absorption?

Vidar Gudmundsson
Science Institute, University of Iceland, Dunhaga 3, IS-107 Reykjavik, Iceland

Rolf R. Gerhardts
Max-Planck-Institut für Festkörperforschung, Heisenbergstraße 1, D-70569 Stuttgart, Federal Republic of Germany

The far-infrared absorption of a periodically modulated two-dimensional electron gas in a perpendicular constant magnetic field is calculated self-consistently within the Hartree approximation. For vanishing modulation the magnetoplasmon dispersion shows simple anticrossings with harmonics of the cyclotron resonance, as expected. For increasing modulation we identify intra- and intersubband magnetoplasmon modes charaterized by the number of flux quanta through each unit cell of the periodic potential.

The Hofstadter energy spectrum[1,2] indirectly determines the transport properties[3] of a two-dimensional electron gas (2DEG) with a square lattice modulation. Recently it has been claimed that even the direct effects of the splitting of the Landau bands into subbands has been observed.[4] Here we discuss the possibility to directly observe the subband splitting in optical measurements, for example, by far-infrared (FIR) absorption, or in Raman scattering. Since screening effects have been shown to be very important in the ground state[5] we shall treat the Coulomb interaction between the electrons on equal footing in the excited and the ground state by utilizing the time-dependent Hartree approximation. Results of such calculations have been published for isolated quantum dots[6,7] and for unidirectionally modulated 2DEG, but to our knowledge, corresponding results are not available for the square lattice modulation.[a]

The technical details of the ground state calculation have been published elsewhere.[5] The square lateral superlattice with period L is spanned by the lattice vectors $\mathbf{R} = m\mathbf{l}_1 + n\mathbf{l}_2$, where $\mathbf{l}_1 = L\hat{\mathbf{x}}$, and $\mathbf{l}_2 = L\hat{\mathbf{y}}$ are the primitive translations of the Bravais lattice $\mathcal{B}$; $n, m \in Z$. The reciprocal lattice $\mathcal{R}$ is spanned by $\mathbf{G} = G_1\mathbf{g}_1 + G_2\mathbf{g}_2$, with $\mathbf{g}_1 = 2\pi\hat{\mathbf{y}}/l_1$, $\mathbf{g}_2 = 2\pi\hat{\mathbf{x}}/l_2$, and $G_1, G_2 \in Z$. The ground-state properties of the interacting 2DEG in a perpendicular homogeneous magnetic field $\mathbf{B} = B\hat{\mathbf{z}}$ and the periodic potential

[a] The collective excitations of a 2DEG with a two-dimensional magnetic-field modulation have been studied by X. Wu and S. E. Ulloa in Phys. Rev. B **47**, 10028 (1993), but without considering interaction effects in the ground state.

$V(\mathbf{r}) = V\{\cos(g_1x) + \cos(g_2y)\}$ are calculated in the Hartree approximation. The Hartree single-electron states $|\alpha)$ and their energies ε_α are labelled by the quantum numbers $\{n_l, \mu, \nu\} = \alpha$, where $n_l \geq 0$ is a Landau band index, $\mu = (\theta_1 + 2\pi n_1)/p$, $\nu = (\theta_2 + 2\pi n_2)/q$, with $n_1 \in I_1 = \{0, \cdots p-1\}$, $n_2 \in I_2 = \{0, \cdots q-1\}$, $\theta_i \in [-\pi, \pi]$, and $pq \in N$ is the number of magnetic flux quanta through the lattice unit cell. The Hartree potential "felt" by each electron and caused by the total electronic charge density of the 2DEG, $-en_s(\mathbf{r})$, and the positive neutralizing background charge, $+en_b$, together with the external periodic potential $V(\mathbf{r})$ do not couple states at different points in the quasi-Brillouin zone $\theta = (\theta_1, \theta_2) \in \{[-\pi, \pi] \times [-\pi, \pi]\}$. However, states at a given point depend, in a self-consistent way, on states in the whole zone through $n_s(\mathbf{r})$. The periodic potential broadens the Landau levels into Landau bands which due to the comensurability conditions between L and the magnetic length $l = (c\hbar/eB)^{1/2}$, are split into pq subbands. The resulting energy spectrum, the Hofstadter butterfly[1,2], retains essentially it's complicated gap structure but is strongly reduced in symmetry by the electron-electron interaction.[5] The symmetry depends on the mean electron density, or, equivalently, the chemical potential.

In order to calculate the FIR absorption of the 2DEG we perturb it by a monochromatic external electric field without restricting it's dispersion relation to that of a free propagating field. The power absorption is calculated from the Joule heating of the self-consistent electric field consisting of the external and the induced field.[8] The peaks in the absorption spectra represent absorption of the collective modes of the 2DEG. Single-electron excitations are strongly damped and do not show up in the spectra.[8]

In the calculation we use GaAs parameters, $m^* = 0.067m_0$, and $\kappa = 12.4$. The power dissipation is made possible by retaining a small but finite imaginary part of the frequency in the electron susceptibility. The absorption of a homogeneous 2DEG ($V = 0$) is compared in Fig. 1 with the first-order dispersion in $\mathbf{k}$ of a magnetoplasmon, $\omega^2 = \omega_c^2 + 2\pi e^2 n_s k/(\kappa m^*)$. Kohn's theorem manifests itself by the fact that only one peak is visible for $\mathbf{k} \to 0$.[9] For finite wave vectors $\mathbf{k}$ the magnetoplasmon interacts with harmonics of the cylcotron resonance, resulting in so the called Bernstein modes.[10] To make the Hofstadter subband structure of the Landau bands discernible, we have to resort to short periods $L = 50$ nm and strong modulation, as is seen in Fig. 2. Here we use $\mathbf{k}L = k_1L = 0.2$, and the filling factor $\nu = 1/2$. For strong modulation and $pq = 3$ the dispersion of the magnetoplasmon is relatively flat for small values of $\mathbf{k}$. The structures in the absorption are very sensitive to the location of the chemical potential with respect to the subbands as can be confirmed by comparing the absorption for $\nu = 1/2$ in Fig. 2 with the absorption for $\nu = 5/6$ in Fig. 3. The inset in Fig. 2 shows the absorption for

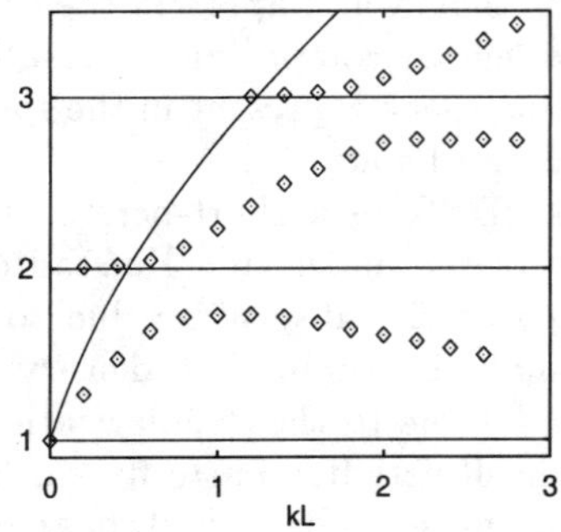

Figure 1: The dispersion of the magnetoplasmon in a homogeneous 2DEG (◇). The energy E is scaled in $\hbar\omega_c$ and the wavevector $\mathbf{k} = k_1$ is scaled with the period length $L = 200$ nm. $V = 0$, $\hbar\omega_c = 0.179$ meV, $pq = 1$, and $T = 1$ K.

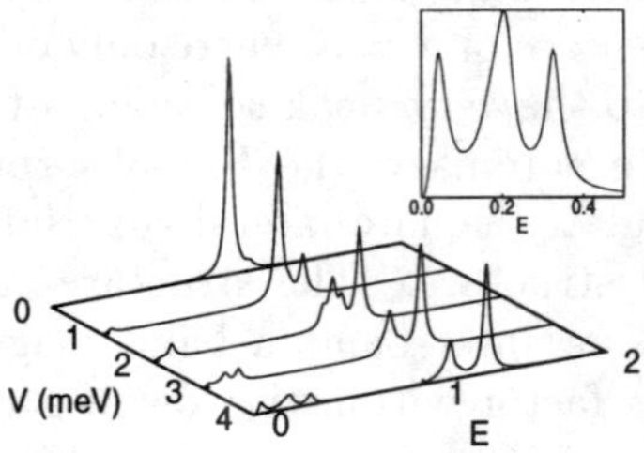

Figure 2: The absorption $P(E = \hbar\omega/\hbar\omega_c)$ in arbitrary units as a function of the modulation strength V for $\nu = 1/2$. The inset shows the intraband absorption peaks for $V = 4$ meV. $k_1 L = 0.2$, $L = 50$ nm, $\hbar\omega_c = 8.57$ meV, $pq = 3$, and $T = 1$ K.

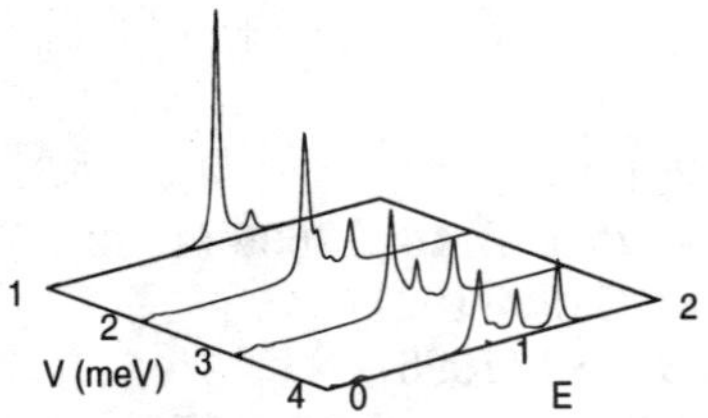

Figure 3: The absorption $P(E = \hbar\omega/\hbar\omega_c)$ in arbitrary units as a function of the modulation strength V for $\nu = 5/6$. $k_1 L = 0.2$, $L = 50$ nm, $\hbar\omega_c = 8.57$ meV, $pq = 3$, and $T = 1$ K.

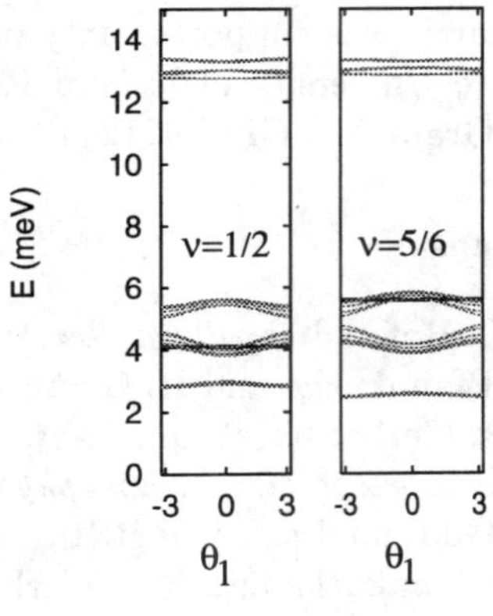

Figure 4: The Hartree dispersion of the two lowest Landau bands in the quasi-Brillouin zone (each band is split into three subbands). The chemical potential is indicated by a solid horizontal line. $L = 50$ nm, $V = 4$ meV, $\hbar\omega_c = 8.57$ meV, $pq = 3$, and $T = 1$ K.

three intra-Landau band magnetoplasmon modes. For $\nu = 1/2$ in Fig. 2 only two subbands are occupied, or partially occupied, as is illustrated in Fig. 4. We have thus *two* absorption peaks for inter-Landau band magnetoplasmons. Fig. 4 also shows that in the case of $\nu = 5/6$ we can have *three* absorption peaks for interband magnetoplasmons. The strong screening effects of partially filled subbands weakens any selections rules that would apply in the case of a noninteracting 2DEG. This is especially evident by comparing the results here to the case of $\nu = 1$, where only one absorption peak is present in the spectrum due to the very weak screening of filled Landau bands.

In summary, the FIR-absorption of a 2DEG in a short-period strongly modulated square lateral superlattice reflects the underlying Hofstadter subband structure. The structure can be seen in the absorption due to either intra- or inter-Landau band magnetoplasmons by tuning the density or the filling factor within the lowest Landau band. Due to the depolarization shift of the peaks an exact information about the Hofstadter butterfly can not be extracted from the absorption spectra. The intra-Landau band magnetoplasmon peaks in the absorption have weak oscillator strengths compared to the interband absorption and are situated in a region on the energy scale that is difficult to detect with present day FIR technology.

Acknowledgments

This research was supported in part by the Icelandic Natural Science Foundation, the University of Iceland Research Fund, and a NATO collaborative research Grant No. CRG 921204.

References

1. R.D. Hofstadter, *Phys. Rev.* B **14**, 2239 (1976).
2. D. Pfannkuche and R. Gerhardts, *Phys. Rev.* B **46**, 12606 (1992).
3. R.R. Gerhardts *et. al.*, *Phys. Rev.* B **43**, 5192 (1991).
4. T. Schloesser *et. al.*, *Europhys. Lett.* **33**, 683 (1996).
5. V. Gudmundsson and R.R. Gerhardts, *Phys. Rev.* B **52**, 16744 (1995).
6. D. Pfannkuche and R.R. Gerhardts, *Phys. Rev.* B **44**, 13132 (1991).
7. V. Gudmundsson and R.R. Gerhardts, *Phys. Rev.* B **43**, 12098 (1991).
8. V. Gudmundsson and R.R. Gerhardts, *Phys. Rev.* B, submitted (1996).
9. W. Kohn, *Phys. Rev.* **123**, 1242 (1961).
10. I.B. Bernstein, *Phys. Rev.* **109**, 10 (1958).

FRACTIONAL QUANTUM HALL EFFECT OR HOFSTADTER BUTTERFLY: INTERACTING ELECTRONS IN A LATERAL SUPERLATTICE POTENTIAL

DANIELA PFANNKUCHE[a,b] and ALLAN H. MACDONALD[a]

[a] *Department of Physics, Indiana University, Bloomington, Indiana 47405, U.S.A.*

[b] *Max-Planck-Institut für Festkörperforschung, D-50689 Stuttgart, Germany*

Competition between electron-electron interaction and a periodic external potential determines the Hall effect of a two-dimensional electron system at high perpendicular magnetic field. We investigate this system on a microscopic level and find indications for a continuous phase transition between integer and fractional Quantum Hall states at a critical strength of the superlattice potential.

The quantum Hall effect (QHE) occurs in two-dimensional electron systems which have charge-excitation gaps at densities which depend on magnetic field [1]. Two examples of the QHE which have been extensively studied experimentally are the integer QHE, in which the gap is due primarily to Landau quantization of the kinetic energy, and the fractional QHE, in which the gap, Δ, is due to electron-electron interactions. Both the integer and fractional QHE occur at fixed Landau level filling factor, i.e. at densities that are proportional to the magnetic field. The QHE will also occur for non-interacting electrons in the presence of a periodic potential [2,3]. For these systems commensurability conditions between the area containing a flux quantum Φ_0 and the area of the unit cell give rise to intricate spectra studied by Hofstadter [4] for the case of a tight-binding band. In these systems the single-particle density-of-states has many gaps which occur at densities which depend on magnetic field, and the quantum Hall effect will occur in each of these gaps. While these systems have not yet been realized experimentally in a regime which permits their quantum Hall effects to be thoroughly studied [5], they have played an important role in the development of the theory of the QHE and especially in the development of the picture in which the QHE is regarded as a topological invariant [2,6].

We have studied the competition between interactions and periodic potentials in the quantum Hall regime using notions from the topological picture of the QHE. Specifically we consider the case of a Landau level with filling factor $\nu = 1/3$ and a square lattice periodic potential $U(x, y) = U_0(\cos(2\pi x/a) + \cos(2\pi y/a))$ with $a^2 B = 3\Phi_0$. For this situation the fractional QHE gives rise to a quantized Hall conductivity with $\sigma_{xy} = (e^2/h)/3$ when $U_0 = 0$. The Hofstadter spectrum, on the other hand, leads to $\sigma_{xy} = 0$ for large U_0. The topological picture of the QHE associates the Hall conductivity, in units of

e^2/h, with topological integer 'Chern' numbers which specify the winding of the phase of the many-particle ground state wavefunction when quasiperiodic boundary conditions are applied and the Bloch wavevector is moved around the Brillouin-zone boundary. In this picture the $\nu = 1/3$ fractional QHE occurs by having a three-fold degenerate ground state whose Chern numbers sum to one. The Hall conductivity is given by the average Chern number.

We performed numerical calculations for different system sizes with up to six electrons in a lattice cell of finite area, $A = N_x a \times N_y a$ subjected to quasi-periodic boundary conditions. The low-lying excitations of the few-particle system are obtained from a diagonalization procedure.

In a system of finite size L the ground state degeneracy is lifted when the periodic potential is present. However, at small modulation strength, the numerical results confirm an exponential decrease of the ground state degeneracy splitting with increasing system size [7]. This guarantees the persistence of the fractional QHE in the thermodynamic limit at small, but finite, modulation potentials. Specifically, the calculations show that the energy difference, $W_1 = E_1 - E_0$ between the ground state and the first excited few-particle state is given by

$$W_1 \simeq nN(V/\Delta(n,N))^N \Delta(n,N)/2 \tag{1}$$

at $V = U_0 exp(-\pi/6)$ considerably smaller than the correlation gap, $\Delta(n,N) = (E_3 - E1)(U_0 = 0)$, where $n = min(N_x, N_y)$ and $N = max(N_x, N_y)$. This result is in agreement with an explicit perturbation analysis of up to fifth order in the modulation potential. At large $V \gg \Delta$ we find $W_1 = \Delta_H + \delta(n,N)$ where $\delta(n,N)$ is a small positive number and the term $\Delta_H = V(3-\sqrt{3})/2$ equals the single particle gap between the first and second subband of the Hofstadter spectrum.

In the intermediate region, i.e. $V \simeq \Delta$, analytic results [8] cannot be obtained from a microscopic treatment. We interpret our numerical results in the spirit of a second order continuous quantum phase transition, i.e. we identify the inverse gap, W_1^{-1}, with a correlation time ξ_T, $W_1^{-1} \propto \xi_T$. For modulation strength V larger than a critical value V_c the correlation time is finite and diverges as V approaches V_c from above, in accordance with a closing of the gap W_1 in the thermodynamic limit. This characterizes the phase transition from integer to fractional QH states, i.e. the transition from a non-degenerate to a (threefold-) degenerate ground state. Assuming the critical dynamical exponent to be $z = 1$, we find the critical modulation strength from the finite size scaling ansatz

$$L^z W_1 = Q(L^{\frac{1}{\nu}} \frac{V - V_c}{V_c}) \tag{2}$$

with the scaling function Q and the critical exponent ν. We perform a one-parameter scaling analysis for the relevant system size $L = \sqrt{N_x^2 + N_y^2}$ and the dimensionless modulation strength $v = V/\Delta(n, N)$, thus eliminating non-critical finite size dependencies of the correlation gap. The critical potential strength $v_c = V_c/\Delta$ is found from the fixpoint of the scaling ansatz (2).

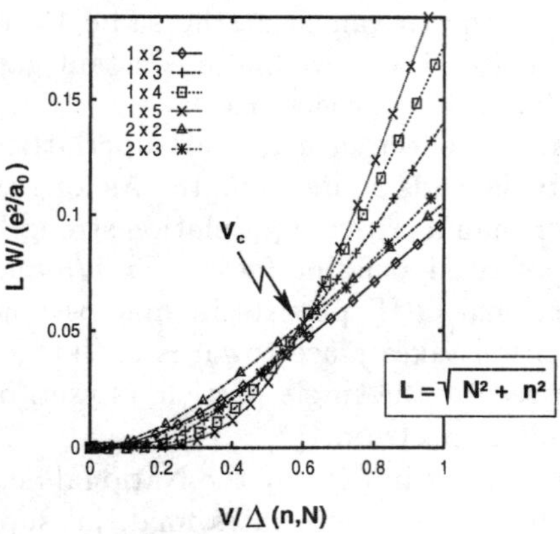

Fig. 1 Gap between the ground state and the first excited state $W \equiv W_1$ scaled by the dimensionless system size $L = \sqrt{N_x^2 + N_y^2}$ as a function of the reduced modulation amplitude $V/\Delta(n, N) \equiv v$. Different curves indicate different system sizes. The approximate crossing of these curves allow us to estimate the critical reduced modulation strength v_c.

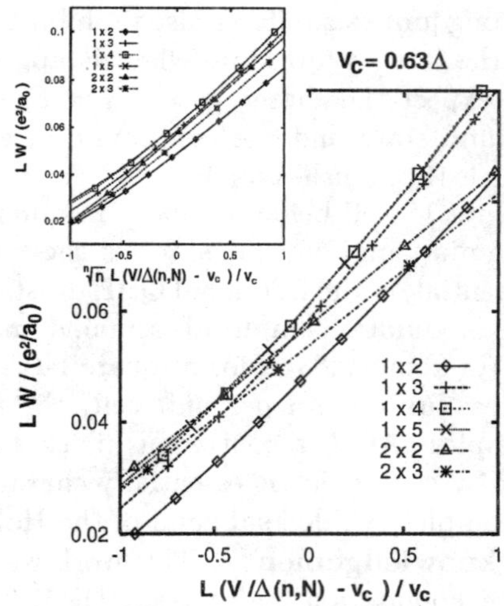

Fig. 2 Scaling behavior of the energy gap near the critical point. Curves for the $1 \times N$-system sizes are parallel, but shifted with respect to each other due to next-to-leading order finite size corrections. The $2 \times N$-systems show a different slope. Inset: Non-critical finite-size corrections in the argument of the scaling function improve the scaling behavior.

Fig. 1 shows the quantity LW_1 as a function of v for various system geometries. From this plot we deduce a critical modulation strength of $v_c \simeq 0.63$. Cross-checking with the size dependence of $W_1(V_c)$, we find $W_1(V_c) \propto L^{-1}$ in agreement with our assumption of $z = 1$.

Fig. 2, finally, shows the scaling function $LW_1 = Q$ as a function of $L(v-v_c)/v_c$, i.e. $\nu = 1$. Obviously, the curves representing the system sizes $1 \times N$ are parallel, datapoints for the system sizes 1×4 and 1×5 fall on the same curve, as it is required by the scaling ansatz. However, the curves representing the systems 2×2 and 2×3, although parallel to each other, have a different slope than the $1 \times N$ systems. We can correct for this discrepancy by an uncritical size dependence in the argument of the scaling function which is suggested by the small V behavior of W_1 (see eq. (1)). By plotting Q as a function of $n^{(1/n)}L(v - v_c)/v_c$ (inset in Fig. 2) we can force the curves for all system sizes to be parallel indicating that a one-parameter scaling might be sufficient to describe the phase transition from integer to fractional QH states. Nonetheless, a definite statement cannot be deduced from the small system sizes available.

Strong finite size effects also prohibit to study the transition from the fractional to the integer QH state when V approaches V_c from below. In this situation we expect the charge gap, $W_3 = E_3 - E_1$, between the threefold degenerate ground state and the continuum of excitations to vanish at the same V_c at which W_1 vanishes for $V \geq V_c$. For system sizes $1 \times N$ we find a constant gap W_3 for V well below V_c and a minimum for as V approaches V_c.

Summarizing, our calculations show that the effect of a lateral superlattice potential on the fractional QHE drastically depends on its strength. As long as the modulation amplitude remains smaller than a critical modulation strength V_c (of about 0.63 Δ for a square lattice potential at filling factor $\nu = 1/3$ and three flux quanta per unit cell) the fractional QHE persists in macroscopic samples. At V_c a continuous phase transition takes place towards an integer QH system which is essentially characterized by the single particle physics of a completely filled subband of the Hofstadter spectrum.

Acknowledgement - This work was supported in part by the National Science Foundation under grant DMR-9416906. D.P. appreciates financial support by the Deutsche Forschungsgemeinschaft. The authors are grateful for helpful interactions with Ulrich Zülicke and Erik Sorensen.

1. A. H. MacDonald, in *Physique Quantique Mésoscopique*, edited by E. Akkermans, G. Montambaux, J.-L. Pichard, and J. Zinn-Justin (Elsevier, Amsterdam, 1995), Chap. Session LXI.
2. D. J. Thouless, M. Kohmoto, M. P. Nightingale, and M. den Nijs, Phys. Rev. Lett. **49**, 405 (1982).
3. P. Středa, J. Phys. C: Solid State Phys. **15**, L1299 (1982).
4. D. R. Hofstadter, Phys. Rev. B **14**, 2239 (1976).
5. For recent experiments see T. Schlösser, K. Ensslin, J.P. Kotthaus, and M. Holland, Europhys. Lett. 33, 683 (1996) and R.R. Gerhardts, D. Weiss, and U. Wulf, Phys. Rev. B **43**, 5192 (1991).
6. Q. Niu, D. J. Thouless, and Y.-S. Wu, Phys. Rev. B **31**, 3372 (1985), J.E. Avron and R. Seiler, Phys. Rev. Lett. **54**, 259 (1985); D.P. Arovas, R.N. Bhatt,F.D.M. Haldane, P.B. Littlewood, and R. Rammal, Phys. Rev. Lett. **60**, 619 (1988).
7. D. J. Thouless, Phys. Rev. B **40**, 12034 (1989).
8. X.-G. Wen and Y.-S. Wu, Phys. Rev. Lett. **50**, 1501 (1993).

TRANSPORT PROPERTIES OF TWO-DIMENSIONAL ELECTRON GASES IN SHORT PERIOD LATERAL SUPERLATTICES

M. HAYNE, F. PETIT AND B. ETIENNE
Laboratoire de Microstructures et de Microélectronique (L2M), C.N.R.S., B.P. 107, 92225 Bagneux Cedex, France

Two-dimensional electron gases in a strong one-dimensional periodic potential achieved through growth on vicinal substratres exhibit a number of novel effects in magneto-transport, such as a mobility anisotropy which varies non-monotonically with temperature and electron density and large low-field magnetoresistance. The origin of these effects is demonstrated to be the modulation by their absence in a sample in which it is strongly reduced. We have also observed Weiss oscillations, but despite the fact that the system is strongly modulated, they are found to be suprisingly weak.

1 Introduction

We present transport studies of two-dimensional electron gases (2DEGs) with a strong one-dimensional potential due to the periodic modulation of an AlGaAs alloy in the channel of a modulation-doped heterojunction.[1] The lateral superlattice (LSL) of mean Al composition of 5% is obtained by a single epitaxial growth process on vicinal substrates of nominal period 32±5 nm, resulting in coupled quantum wires. When the LSL is tilted with respect to the growth direction the potential modulation is averaged out in the width of the electronic wavefunction, but the disorder remains unchanged. By comparing transport measurements on 'tilted' and 'untilted' regions of the same wafer we unambiguously identify the effects of the modulation.

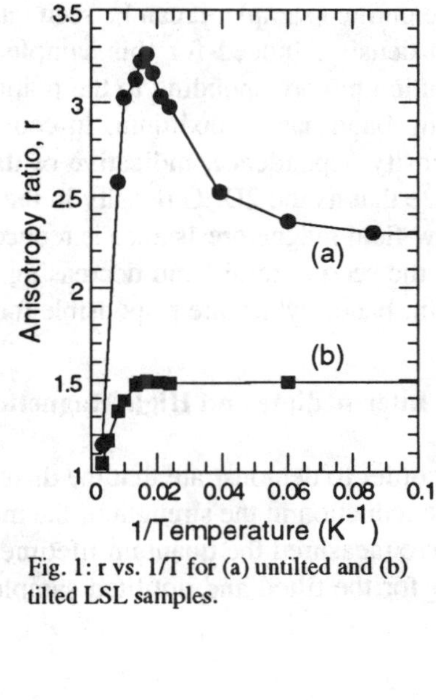

Fig. 1: r vs. 1/T for (a) untilted and (b) tilted LSL samples.

2 Experimental Details

The 2DEG density was 4.4-4.6 x10^{11} cm^{-2} in the dark (mobility 1.6 -2.1 x10^5 cm^{2}V^{-1}s^{-1} along the wires), placing the Fermi level in the second LSL mini-band. The resistivities parallel ($\rho_{//}$) and perpendicular ($\rho_{\perp}$) to the wires were probed in the same 25x25 or 50x50 μm^2 area, with the magnetic field applied normal to the plane of the 2DEG.

3 Zero and Low Magnetic Field

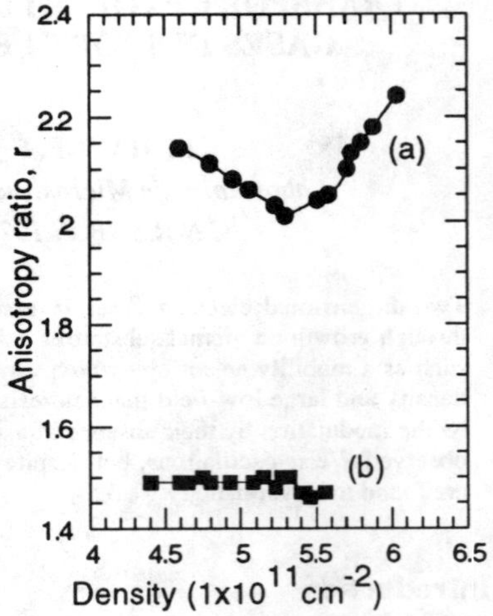

Fig. 2: Density dependence of r for (a) untilted and (b) tilted LSL samples at 1.4 K.

It has already been shown that a strong modulation potential from a short period LSL causes the opening of bandgaps in the parabolic dispersion curve of a 2DEG.[2] In transport this produces giant low-field positive magnetoresistance[2] (not shown), and a mobility anisotropy ratio, $r = \rho_{\perp}/\rho_{//}$ which shows a peak in its temperature (T) dependence (fig. 1(a)). This behaviour contrasts with that of a mesa formed in a region where the LSL is tilted, 3 mm away from the untilted LSL sample. As T is reduced a mobility anisotropy develops, but r-1 is 60% weaker than in the untilted LSL at 4K in the dark, and the non-monotonic behaviour is no longer observed (fig. 1(b)). Furthermore, we find that the low-field positive magnetoresistance is almost completely supressed.

We have further investigated the LSL mini-band structure by increasing the 2DEG density with illumination from an infra-red LED. At low temperature r is a measure of the ratio of the group velocities at the Fermi level in the directions parallel and perpendicular to the wires. Fig. 2 (a) shows r as a function of density for the untilted sample. It can be seen that there is a strong dependence of the anisotropy on density. Indeed for this sample r even shows non-monotonic behaviour, the minimum corresponding to the point where the group velocity of the second lateral mini-band has a maximum. In contrast, r for the tilted LSL shows no significant density dependence, indicative of the absence of the LSL mini-bands. Finally, we note that as the 2DEG density in the untilted sample is increased the strength of the low-field magnetoresistance is reduced. This is due to the increase in the Fermi level in the second mini-band decreasing the relative number of open orbits (in the first mini-band), which are responsible for this effect.[2]

4 Intermediate and High Magnetic Field

In order to demonstrate that the difference between the tilted and untilted samples is the reduction in the strength of the modulation and not an increase in the disorder we have measured the quantum lifetime, τ_q.[3] Fig. 3 shows the Dingle plots[3] for $\rho_{//}$ and $\rho_{\perp}$ for the tilted and untilited samples prior to illumination at 1.4 K. For the tilted

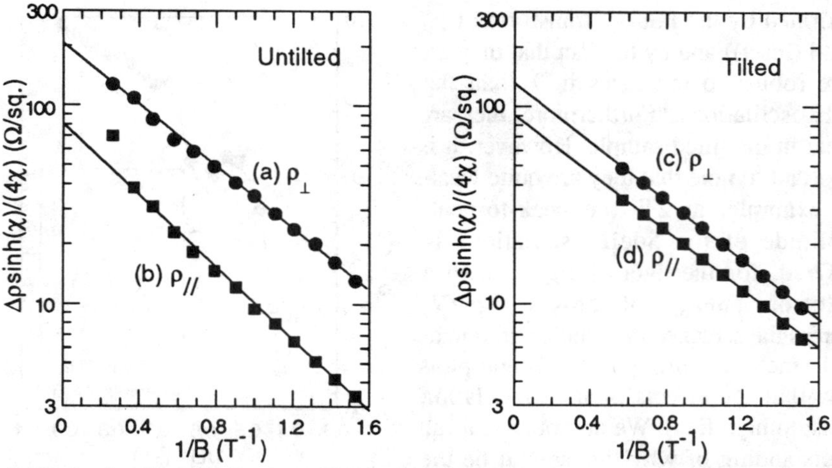

Fig. 3: Dingle plots for at 1.4 K prior to illumination. For the tilted sample the zero field resistivities are equal to the intercepts, for the untilted sample they are (a) 185 Ω/sq. and (b) 82 Ω/sq.

sample (fig. 3(c) and (d)) the Dingle plots are linear with the same slope (given by $\pi m^*/(e\tau_q)$, where m^* is the effective mass) in both directions, and have intercepts that correspond to the zero-field resistivity in each case. This implies that τ_q, which measures all scattering events equally, is independent of the direction of current flow, and that the anisotropy arises in the momentum relaxation time, which is weighted in favour of large angle scattering events.

We now turn to the untilted sample. As with the tilted sample there is an anisotropy in the intercepts with values that are close to, but not the same as, the zero field resistivities. Interestingly, we find that τ_q for $\rho_\perp$ is the same as for the tilted sample. This is consistent with the expectation that tilting the LSL does not induce any additional disorder. However, for current flow parallel to the wires (which in a magnetic field is sensitive to the perpendicular conductivity) the slope is 14% steeper. Since the equivalent slopes for the other Dingle Plots imply τ_q is invariant, we infer that the increase in slope in fig 3(b) is due to an increased m^* arising from the LSL mini-bands.

A well-known effect in LSLs is Weiss oscillations.[4] Classically, they arise as a result of the commensurability beteween the cyclotron radius and an integer multiple of the lateral period. For large period LSLs (~300 nm) they can be quite large, even if the modulation is weak.[4] In our case the short period of the LSL pushes the Weiss oscillations into the high field region (≥1 T) which is dominated by the Shubnikov-de Haas (SdeH) oscillations, making them difficult to observe. However, by exploiting the fact that the Weiss oscillations are in antiphase for $\rho_{//}$ and $\rho_\perp$[4] we can subtract out the SdeH oscillations (fig. 4). The presence of the Weiss oscillations is

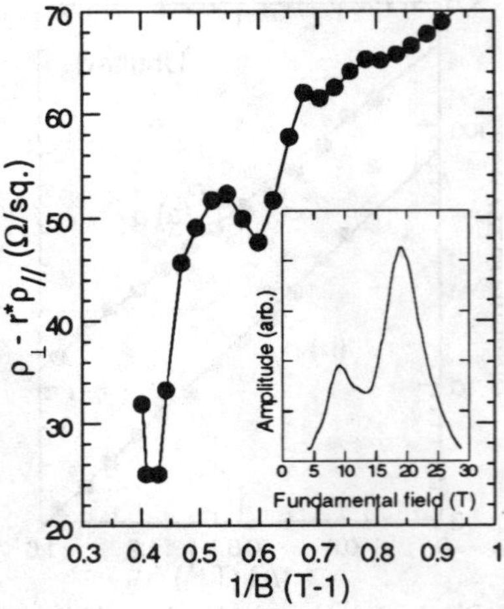

Fig 4: Weiss oscillations at T~5 K. Inset: Fourier transform of $\rho_\perp$. The small low frequency peak corresponds to the Weiss oscillations and gives a lateral period of 25 nm.

confirmed by the Fourier transform of $\rho_\perp$ (fig. 4 (inset)) and by the fact that they are more robust to increases in T than the SdeH oscillations.[4] Furthermore, they are absent in the tilted sample. However, it is important to note that they are quite weak. For example, at 2 T the peak-to-peak amplitude of the SdeH oscillations is 50 Ω/sq. for the data of fig. 4, with a cyclotron energy of only 3.5 meV whereas the screened modulation amplitude is ~10 meV. In addition, the Dingle plots show that the effects of the modulation persist to high field. We do not have a full understanding of why this should be the case, however, it may be that localised states, which are known to be present at high field,[5] inhibit the Weiss oscillations.

5 Conclusions

We have studied the transport properties of 2DEGs in a modulated potential resulting from a short period LSL grown by epitaxy on vicinal surfaces. We have reported a variety of effects which result from the formation of lateral mini-bands due to the modulation, and have demonstrated that when the LSL is tilted, severely reducing the size of the modulation, but leaving the disorder unchanged, these effects disappear. We have observed Weiss oscillations,[4] but find that they are rather weak.

References

1. B.Etienne, F. Laruelle, J. Bloch, L. Sfaxi and F. Lelarge, *J. Crystal Growth* **150** 336-340 (1995).
2. L.Sfaxi, F. Lelarge, F. Petit, A. Cavanna and B. Etienne, *Solid-State Electronics* **40**, 271 (1996).
3. P.T. Coleridge, R. Stoner and R. Fletcher, *Phys. Rev.* B **39**, 1120 (1989).
4. D. Weiss, K. von Klitzing, K. Ploog and G. Weimann, *Europhys. Lett.* **8**, 179 (1989).
5. F. Petit, M. Hayne and B. Etienne, 'The Application of High Magnetic Fields in Semiconductor Physics XII', Würzburg 29 July - 2 August 1996.

SPLIT GATE QUANTUM WIRES HAVING A FERROMAGNETIC DOT MADE BY STM FABRICATION

T. KIKUTANI, N. AOKI, A. OKI, C. U. HONG, H. HORI AND S. YAMADA
Japan Advanced Institute of Science and Technology (JAIST)-Hokuriku,
1-1 Asahidai, Tatsunokuchi, Ishikawa 923-12, Japan

We have fabricated split-gate quantum wires having a buried ferromagnetic dot interacting with a two-dimensional electron gas (2DEG). Such a structure was fabricated by successively utilizing electron-beam (EB) and two-step scanning tunneling microscope (STM) lithographies. Low temperature (0.3K) transport measurements in those wires showed some features in the conductance (G) vs gate voltage (Vg), that is, a kink = an abrupt change of dG/dVg and step structures, the differences of which are fairly from $2e^2/h$ (77μS).

1. Introduction

Semiconductor nanostructures have so far been fabricated by various sophisticated techniques utilizing EB, X-ray, focused ion-beam (FIB) etc. Although the smallest size so far attained by those technologies is less than 10nm, one promising technique in semiconductor nanofabrication is STM-based lithography[1,2] because it may offer a scheme for very narrow pattern generation with or without a mask, process as well as for ultimate atomic fabrication/manipulation.

In fact, many approaches have been proposed in an attempt to establish useful techniques for the nanofabrication of various materials. They include local heating,[3] electrochemical deposition,[4] exposure of resist materials,[5] decomposition of organometallic gases,[6] deposition of individual molecular fragments[7] and electronically induced modification[8,9]. Recently as a new approach to be classified under the last category, we reported a method for fabricating small structures (mesa or hole) on a split-gate wire surface fabricated on an AlGaAs/GaAs heterojunction[10,11]. In such samples, we have observed quantum transport phenomena, Coulomb blockades[10], and conductance telegraphic switching[11]. As a next step, to observe and analyze new kind interactions between 2DEG and ferromagnetic material (Ni in our case), we have fabricated a new structure, that is, a ferromagnetic dot buried into a semiconductor 2DEG split-gate quantum wire plane. This structure was fabricated by using two-step STM microfabrication: As a first step, a small hole, typically 200 nm upper diameter and 100 nm depth, was made in a split-gate wire surface originally made by EB lithography and then a small amount of Ni was electrically evaporated into the hole.

In such a structure, we can expect several new quantum transport phenomena. The schematic top views of a sample having such a structure are shown in Fig. 1(a) and (b). Since the 2DEG plane locates at 60nm deep from the wire surface, the typical hole made by STM lithography reached to the 2DEG plane. The ferromagnetic material buried into the hole is thus contacting to the 2DEG via depletion regions. If the split-gate wire is squeezed properly, electrons passing through the narrow channel between the ferromagnetic dot and the split-gates may have a strong interaction with spins of Ni. Moreover, when the wire becomes sufficiently narrow, 2DEG have to pass through the ferromagnet dot by tunneling the depletion regions on both side of the dot. Since usually up-spin electrons occupy in the Felmi level of bulk Ni, some kinds of spin-polarized quantum transport or tunneling could be realized. We here report the details of fabrication process of such samples and preliminary results of low temperature transport experiments.

2. Fabrication of Samples

At first, split-gate structures were fabricated on top of 2DEG wafer by using conventional EB lithography technique. This wafer consists of Si-doped GaAs (5nm), Si-doped AlGaAs(40nm), undoped

AlGaAs (15nm) and an undoped GaAs buffer layers grown successively on a semi-insulating substrate. The sheet electron density, n_s, and mobility, μ, of the 2DEG are $4.0\times10^{11}/cm^2$ and $1.7\times10^5 cm^2/V\cdot sec$, respectively at 77K. Typical length and width of the split-gate wire were 2.0 and 1.0 μm, respectively.

For STM lithography, we used an STM head installed in a scanning electron microscope (SEM), which works under a vacuum of $< 10^{-7}$ Torr. The SEM, which was tilted 45° from the normal to the sample surface, was utilized to observe the top of the STM tip and to position it to the desired location in the split-gate wire surface with an accuracy of 10nm. The method of lithography we adopted was a kind of electrical modification by using a Tungsten supertip (top curvature is less than 50nm).[10,11] In the first step, a W-tip was approached to the surface between the split-gates, and then a small hole was fabricated by applying a pulse voltage between the tip and the sample surface. The fabricated hole size was about 200nm upper diameter and 100nm depth. Corresponding fabrication conditions are as follows; Tunneling current is 0.5nA for bias voltage of 1.25V. Applied single pulse voltage is 14 V peak and 50μsec-duration.

Such a hole acts as a pin-point antidot for the 2DEG lying at a depth of 60nm from the wafer surface. In the second step, we used also a W-supertip but coated by Ni, and it was approached to the fabricated hole. Then by applying a pulse voltage between the Ni-coated tip and the sample surface, we buried electrically evaporated Ni from the tip into the small hole. The fabrication condition are as follows; Tunneling current is 0.5nA for bias voltage of 1.25V. Single pulse voltage applied is 8 V peak and 50μsec-duration. Thus, the fabrication process to make the structure of a split-gate quantum wire having a ferromagnetic dot was completed.[12] An example of SEM image of the fabricated sample are shown in Fig. 2.

3. Results of Transport Measurements and Discussion

We performed low temperature transport experiments of such a sample at 0.3 K. The measurement was carried out by alternating current 4 terminal method. The dependence of 4 terminal resistance on gate voltage is shown in Fig.3(a). As an unique feature in this result, there is a kink at about Vg = -2.3 V.

Here we denote the region for Vg > -2.3 V as region (1) and that for Vg < -2.3 V as region (2), respectively. The corresponding conductance representation is shown in Fig. 3(b). We analyzed those regions in detail by investigating the differential conductance (dG/dVg). The results for regions (1) and (2) are shown in Fig. 3(c) and (d), respectively. In Fig. 3(c) for region (1), the conductance steps with a difference of 150 - 250μS was observed. Such a difference is much bigger than $2e^2/h$ (77μS) that were observed in a simple point contact. On the other hand, in Fig. 2(d), the conductance steps having the difference of about 10μS were observed, which is fairly smaller than $2e^2/h$ (77μS).

As possible origins for the conductance behaviors mentioned above, we can consider next several causes. In region (1) where the step difference is about 200μS, we consider the possible formation of parallel transport paths near the ferromagnet dot. Actually, since it is located almost at the center line, there could be two parallel paths on both side of the dot. The step difference of about 200μS corresponds to almost three times of $2e^2/h$ (77μS). It suggests an existence of three different paths acting as parallel point-contacts when the wire is squeezed. Most probable candidate for the third path is the one connecting between 2DEG and Ni dot via the traps in the depletion region created by STM fabrication as a damage. If the number and the density of the traps are high, they could easily behave as a "wire".

In region (2) with the step difference of around 10μS, we expect that the current flows through the inside of Ni dot via the traps within the depletion region. That is, we think that the narrow path exists through the depletion layer and the Ni dot. However, when the gate voltage is decreased, the associating depletion region is difficult to become wide because it already touches to the Ni dot. This is the origin of the appearance of kink. The result that conductance step difference much smaller than $2e^2/h$ (77μS) was observed is probably due to the formation of an array of several series point contacts. Taking those results into consideration, we must find the strict condition of pulse voltage for STM microfabrication which yields relatively small damages. At present, therefore, we may not yet have clear and complete ferromagnetic dot, that is, buried Ni dot with a depletion layer in the peripheral.

4. Summary

We reported a two-step STM process for fabricating a buried ferromagnetic dot structure on the split-gate wire surface originally made on an AlGaAs/GaAs heterostructure and described preliminary results of low temperature transport experiments. From the transport behaviors of the 4 terminal conductance dependence on gate voltage, the conductance steps of about 200μS difference were observed in the region before the kink appears. In addition, the conductance steps, the difference of which was about 10μS, were observed in the region after the kink. Both results suggest the existence of irregular current flow paths inside the depletion layer and/or the Ni dot, which might be formed by the damage related trap arrays introduced at the instance of STM microfabrication.

5. References

1. As an example of review; R. Wiesendanger: *Appl. Surf. Sci.* ***54***, 271 (1992)
2. As an example of review; R. Wiesensanger: *J. Vac & Sci. Technol.* ***B12***, 515 (1994)
3. U. Staufer, R. Wiesendanger, L. Eng, L. Rosenthaler, H. R. Hidber, H. J. Guntherdot and N. Garcia: *J. Vac. & Sci. Technol.* ***A 6***, 537 (1988)
4. D. Craston, C. W. Lin and A. J. Bard: *J. Electrochem. Soc.* ***135***, 785 (1988)
5. M. A. McCord and R. F. W. Pease: *J. Vac. Sci. & Technol.* ***B6***, 293 (1988)
6. E. E. Ehrics, S. Yoon and A. L. de Lozanne: *Appl. Phys. Lett.* ***53***, 2287 (1988)
7. J. S. Foster, J. E. Frommer and P. C. Arnett: *Nature* ***331***, 324 (1988)
8. E. Hartmann, M. Enachescu, F. Koch, G. Krotz and G. Muller: *J. Non-Cryst. Solid* ***137 & 138***, 1067 (1991)
9. M. Enachescu, E. Hartmann and F. Koch: *Appl. Phys. Lett.* ***64***, 2253 (1994)
10. S. Yamada and M. Yamamoto: *Appl. Phys. Lett.* ***64***, 2253 (1994)
11. S. Yamada and M. Yamamoto: *to be published in J. Appl. Phys.*
12. N. Aoki, K. Fukuhara, T. Kikutani, A. Oki, H. Hori and S. Yamada: *to be published in Jpn. J. Appl. Phys.*

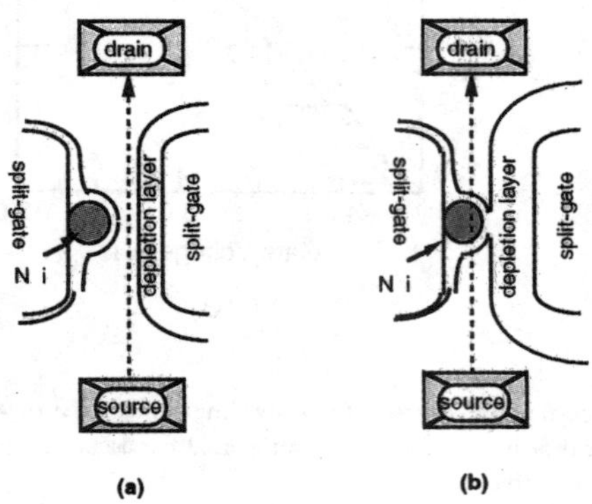

Fig. 1: The schematic top views of the sample. (a) shows point contact having a ferromagnetic dot. (b) shows double barrier structure having a ferromagnetic dot.

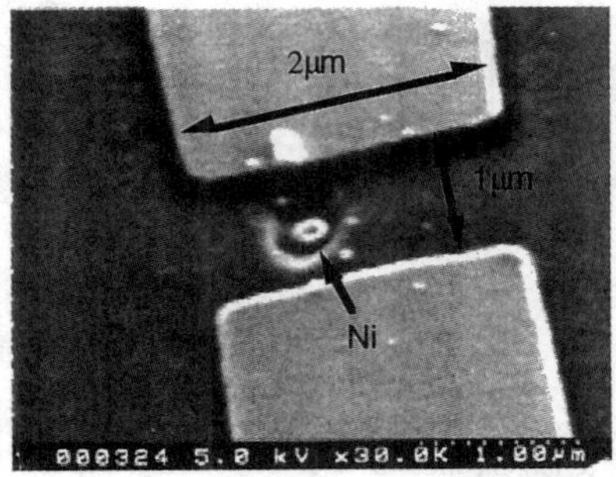

Fig. 2: An example of SEM image of the fabricated sample.

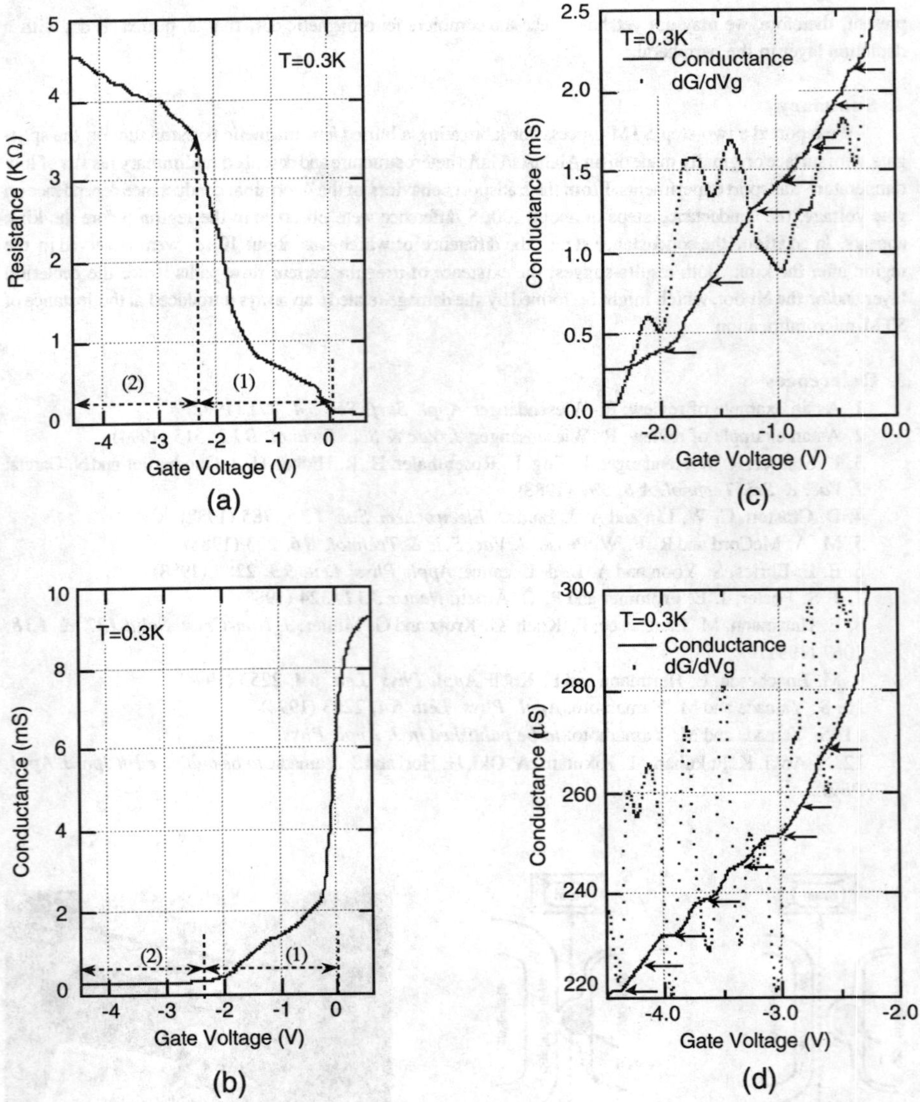

Fig. 3: (a) shows the dependence of 4 terminal resistance on gate voltage. (b) shows the dependence of 4 terminal conductance on gate voltage. (c) and (d) show the dependence of details of 4 terminal conductance and their derivative on gate voltage in the region (1) and (2), respectively.

MAGNETIC FIELD INDUCED 3D-2D TRANSITION IN InGaAs/InP - SUPERLATTICES

O. JASCHINSKI, J. SCHOENES
Technische Universität Braunschweig, Institut für Halbleiterphysik und Optik und Hochmagnetfeldanlage,
Mendelssohnstr. 3, D-38106 Braunschweig, Germany

G. NACHTWEI
Max-Planck-Institut für Festkörperforschung,
Heisenbergstr. 1, D-70569 Stuttgart, Germany

P. BÖNSCH, A. SCHLACHETZKI
Technische Universität Braunschweig, Institut für Halbleitertechnik,
Hans-Sommer-Str. 66, D-38106 Braunschweig, Germany

The magnetotransport properties of MOVPE-grown InGaAs/InP superlattices with δ-doped barriers of different doping concentrations and barrier widths have been investigated by Shubnikov-de-Haas (SdH) measurements in tilted magnetic fields up to 14 T. Depending on the barrier properties and the doping level, the angular dependence of the SdH periods shows deviations from strictly two-dimensional (2D) behaviour. The data provide evidence of a strongly anisotropic 3D-electron system within the superlattices. For smaller tilt angles, the Fermi surfaces determined experimentally can be modelled applying a Kronig-Penney model. At larger tilt angles or in-plane magnetic fields, a transition towards a 2D-behaviour is visible in the angular dependence of the cross section of the Fermi surface. This behaviour reflects the reduced overlap of the wave function in the barriers, which is due to the in-plane component of the magnetic field.

1 Introduction

Heterostructures and superlattices fabricated from InGaAs/InP are well described by a square well potential in growth direction (z-direction) [1]. For thick InP-barriers within the superlattice the sharp energy levels in the InGaAs-quantum wells represent a two-dimensional (2D) electronic system. With decreasing barrier width the potential wells become permeable for the electron wave functions. The transparency of the barriers allows an overlap of the wave functions, which introduces dispersion in z-direction. This leads to a broadening of the quantized energy levels. A very good description of the miniband structure and wave function of the electronic system is given by the Kronig-Penney model [1,2]. The penetration of the wave function into the barrier is affected by an applied magnetic field $B_{||}$ parallel to the surface of the sample, reducing the electron magnetic length $l_z = \sqrt{\hbar/eB_{||}}$ with increasing field. So the magnetic field qualitatively changes the behaviour of the anisotropic electronic system in the superlattice. For the extreme case the 3D-system will be reduced to a 2D-state.

The aim of the present work is to demonstrate experimentally such a magnetic-field-induced 3D-2D transition.

2 Experimental details

Two superlattices consisting of 20 periods of alternating InGaAs and InP layers were grown lattice matched on InP-substrates using the metal-organic vapour phase epitaxy (MOVPE) technique. Nominally the well width d_W is fixed to 6.6 nm, the barrier width d_B is either 6.6 nm or 4 nm. The barriers contain a 1 nm wide highly Si-doped layer (δ-doping layer).

To perform Shubnikov-de-Haas measurements (SdH) the superlattices were structured in the Hall-bar geometry and were mounted on a sample holder which is in the center of a superconducting coil. The sample holder allows to rotate the sample in the magnetic field. The samples were cooled down to 1.5 K. The computer aided measurements were done using a standard dc-setup.

Some additional investigations were made to get a precise knowledge about the sample parameters. Layer thicknesses were determined using X-ray diffractometry. The energy position of the minibands and the band offsets were calculated from photoluminescence data. Carrier concentrations and mobilities were obtained by SdH- and Hall measurements in perpendicular magnetic field using a Bitter coil (max. 18 T).

3 Theoretical background

3.1 Miniband structure calculations

The miniband structure of superlattices represented by a square well potential can well be described by the Kronig-Penney model. From this model we obtain the dispersion relation (assuming parabolic bands) [3]

$$cos(k_z d) = cos(k_W d_W) cosh(k_B d_B) - \frac{1}{2}\left(\epsilon - \frac{1}{\epsilon}\right) sin(k_W d_W) sinh(k_B d_B) \quad (1)$$

with $k_W = \sqrt{2m_W^* E}/\hbar$, $k_B = \sqrt{2m_B^*(\Delta E_C - E)}/\hbar$, $d = d_B + d_W$ and $\epsilon = (m_B^* k_W)/(m_W^* k_B)$, where ΔE_C is the conduction band offset and m^* is the effective mass. The indices W and B denote the well and the barrier, respectively. The different effective masses of the constituent semiconductors must be taken into account by modifying the continuity conditions of the wave functions and their derivatives at the interface. Two additional effects are considered: The nonparabolicity of the InGaAs conduction band leads to an effective electron mass depending on the well width [4]. The δ-doping layer in the barrier influences the shape of the potential well. Therefore an effective height of the potential well is used, which is smaller than the band offset of undoped material. The distortions of the well shape are neglected.

3.2 Fermi surface

Using the calculated dispersion relation we obtain surfaces of constant energy in **k**-space. For energies equal to the Fermi energy such a surface is called Fermi surface. Because of the rotation symmetry of the Fermi surface about the k_z-axis produced by the isotropy of the host semiconductor it is sufficient to take into consideration only the k_y-direction, which reduces the **k**-space to the $k_y - k_z$ plane. Superlattices exhibit an undulated and periodically repeated Fermi surface[1]. The shape depends on the curvature of the minibands and the Fermi energy level. Energies smaller than the miniband width lead to a closed surface, for greater energies one obtains open surfaces. In the latter case the Fermi surface looks like a distorted cylinder. The k_z-dependence disappears with increasing energy (or increasing doping level).

4 Data and discussion

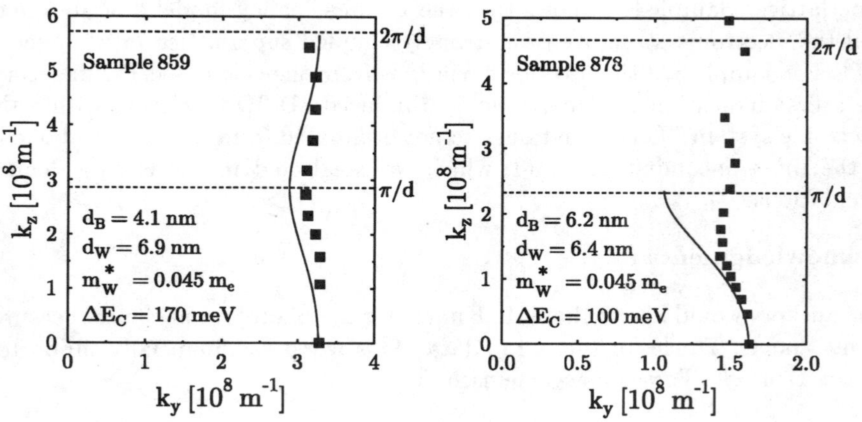

Figure 1: Experimentally determined and calculated Fermi surfaces for a weakly coupled (right side) and strongly coupled (left side) superlattice.

To examine the Fermi surface of the superlattices a detailed study of the SdH-oscillations as a function of tilt angle has been performed. All samples show only one populated miniband. For a truly 2D-electronic system the SdH-oscillation periods follow a simple cosine law. The measurements for both samples show deviations from this 2D-behaviour. This is attributed to a 3D-system within the superlattices. The Fermi surface was calculated from the

SdH-data using the corresponding Onsager relation[5]. In the perpendicular field configuration (B⊥ to the surface of the sample) the extremal cross section is a circle, which turns into an ellipse in tilted fields. In Fig. 1 the curves are calculated from Eq. (1) with the parameters given in Fig. 1. For sample 859 (electron concentration $n = 5.9 \times 10^{17}\text{cm}^{-3}$, mobility $\mu = 4600\ \text{cm}^2/\text{Vs}$) the expected Fermi surface agrees well with the experiment. This shows the suitability of the applied model. The 3D-character of this sample is evident. No influence of parallel magnetic field components is visible. The anisotropic 3D- electron system within sample 859 is not affected by magnetic field. For sample 878 ($n = 2.1 \times 10^{17}\text{cm}^{-3}$, $\mu = 4000\ \text{cm}^2/\text{Vs}$) the measurements exhibit a more distorted surface due to the smaller Fermi energy. Clear deviations from the calculated Fermi surface are visible for angles greater than 45°. This demonstrates the influence of the parallel magnetic field on the overlap of the wave functions, which reduces the dimensionality of the electronic system. For smaller angles or in-plane fields the experimental data are well described by theory.

The performed magnetotransport investigations demonstrate the influence of parallel magnetic fields on the electronic behaviour of a weakly coupled superlattice. Sample 859 shows that the Kronig-Penney model describes very well the electronic structure of a strongly coupled superlattice in a magnetic field. For sample 878 the obvious deviations from the calculated Fermi surface are a first indication of a magnetic-field-induced 3D-2D transition within the electronic system. This deviations are not originated from the nonparabolicity of the InGaAs-conduction band, which was neglected in the Kronig-Penney calculations.

Acknowledgments

The authors would like to thank L. Kuntze for the photoluminescence measurements and U. Thiele for the X-ray data. This work was financially supported by the Deutsche Forschungsgemeinschaft.

References

1. H. L. Störmer *et al.* , Phys. Rev. Lett. **56**, 85 (1986)
2. R. de L. Kronig *et al.* , Proc. Roy. Soc. (London) A **130**, 499 (1931)
3. G. Bastard, *Wave Mechanics Applied to Semiconductor Heterostructures* (Les editions de physique, Les Ulis, 1988)
4. D. Schneider *et al.* , J. Appl. Phys., **77**, 2828, (1995)
5. G. Nachtwei *et al.* , Semicond. Sci. Technol., **7**, 505 (1992)

INFLUENCE OF PARALLEL HIGH MAGNETIC FIELD ON THE BALLISTIC TRANSPORT IN TWO-DIMENSIONAL ELECTRON SYSTEM

K. OTO, S. TAKAOKA, and K. MURASE

Department of Physics, Faculty of Science, Osaka University, 1-1 Machikaneyama Toyonaka 560, Japan

We have studied the influence of the parallel magnetic field ($B_{//}$) on the ballistic scattering length (l_b) in a two dimensional electron system by the magnetic electron focusing effect. With increasing $B_{//}$, l_b decreases, where the electrons are pushed to the GaAs/AlGaAs interface by the Lorentz force from $B_{//}$. At reverse $B_{//}$, l_b increases and saturates with increasing $B_{//}$. The variation of the l_b by $B_{//}$ is attributed to the change of the electron scattering near the interface.

1 Introduction

The magnetic electron focusing effect (MEFE) is one of the ballistic transport phenomena in the two dimensional electron system (2DES). It is sensitive to electron scattering since MEFE can be observed when the ballistic electrons from the emitter enter the collector through the cyclotron trajectory. In the presence of a parallel magnetic field ($B_{//}$), the electrons moving at the Fermi velocity v_F are subjected to the Lorentz force as $-ev_F \times B_{//}$. Then, the spatial distribution of the electron wave function changes, and interface scattering, which is important in the high electron mobility system, may also be affected. The studies of the influence of $B_{//}$ on the 2DES electron transport have observed the g-factor enhancement in the quantum Hall regime[1], the spin splitting of the one-dimensional subbands[2], positive magnetoresistance[3], and the increase of the effective mass[4]. In this paper, we report the influence of $B_{//}$ on the ballistic scattering length (l_b) in the 2DES by the MEFE study.

2 Experimental Procedure

A modulation doped GaAs/$Al_xGa_{1-x}As$ (x= 0.245) heterostructure wafer was used for the sample, and shaped by electron beam lithography and an Ar-ion dry etching technique. The thicknesses of nondoped AlGaAs spacer was 250 Å, Si-doped (2.4×10^{18} cm^{-3}) AlGaAs , and the GaAs cap layer were 400, 50 Å,

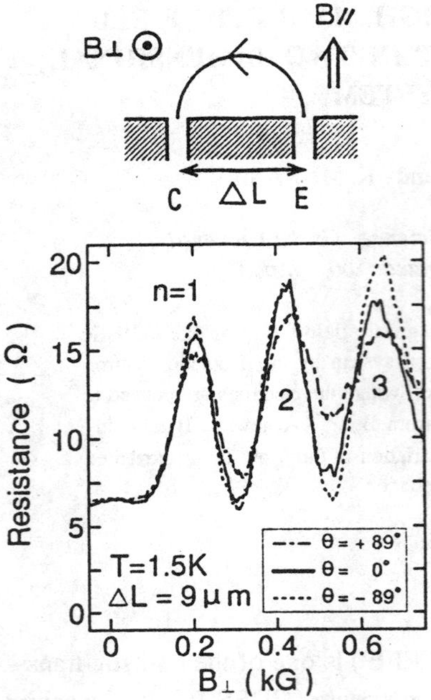

Fig. 1. Upper: Schematic view of the sample and measurement configuration. The various emitter–collector separation ΔL can be taken. The direction of $B_{//}$ and $B_{\perp}$ for $\theta > 0°$ is also shown.
Lower: Magnetoresistance with ΔL = 9μm.

respectively. The electron concentration and the mean free path of the sample were 3.5×10^{11} cm^{-2} and 10 μm at 1.5 K after brief illumination. The sample has 8 probes of 1 μm width to take various separation length (ΔL) between the emitter and collector probes, as shown in Fig. 1. The four–terminal magnetoresistance was measured by an ac lock–in technique with 300 nA at 15 Hz. The magnetic fields parallel ($B_{//}$) and perpendicular ($B_{\perp}$) to the 2DES plane can be applied to the sample by rotating the sample holder[5] in the magnet.

3 Results and Discussion

3.1. Influence of parallel magnetic fields on MEFE

The magnetoresistance in the four terminal geometry with and without $B_{//}$ are plotted as a function of $B_{\perp}$ in Fig. 1. These periodic oscillations are due to the MEFE, the peaks are observed when the ΔL is integer multiples (n, shown in Fig. 1.) of the diameter of the cyclotron motion due to $B_{\perp}$.

In this report, we concentrate on n=1 peak to avoid the effect of reflection[6] at the sample edge. However, the trend of the peaks with $B_{//}$ are almost the same for different indexed peaks. The amplitude of the oscillations are clearly affected by $B_{//}$ as seen from Fig. 1. At tilt angle θ =+89°, where the direction of the magnetic fields is schematically shown in Fig. 1, $B_{//}$ =5.7 T at $B_{\perp}$=1 kG. The amplitudes of the oscillations are reduced from those at θ =0° ($B_{//}$ =0). On the other hand, the oscillations are enhanced at θ =−89°. The temperature dependence of the influence of $B_{//}$ in the range of 4.2 K − 0.4 K was not detected. The profiles of MEFE peaks were almost the same in this temperature range, the reproducible fluctuation of the magnetoresistance was considerably enhanced by

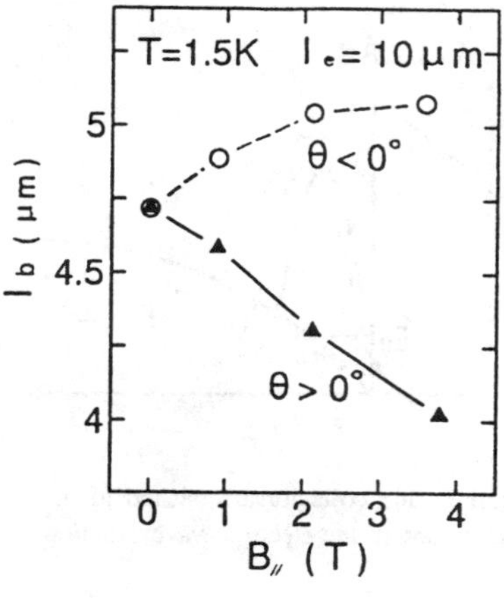

Fig. 2. Evaluated ballistic scattering length l_b vs. $B_{//}$. The l_b monotonically decreases at $\theta > 0°$, and it saturates at $\theta < 0°$ with increasing $B_{//}$.

$B_{//}$ at low temperatures.

In the geometry for $B_{//}$ rotated ±90° in the 2DES plane, the Lorentz force of $B_{//}$ is large near the emitter and collector probe. In contrast with the results in Fig. 1, the influence of $B_{//}$ has almost vanished. Due to the focusing effect of $B_{\perp}$, the electron trajectory is scarcely affected by the scattering near the probes, the MEFE peaks are thought to be almost independent of the $B_{//}$.

3.2 Ballistic scattering length

To evaluate the influence of $B_{//}$ on the MEFE peaks more quantitatively, we measured the MEFE peak heights for the same $B_{//}$ with several separation lengths ΔL between the emitter and the collector. It was found that the peak heights ΔR follow well the relation[7] :

$$\Delta R = R_0 \exp\left(-\frac{\pi\,\Delta L}{2\,l_b}\right), \qquad (1)$$

where R_0 is a constant, and the l_b is the ballistic scattering length. We evaluate the $B_{//}$ dependence of the l_b as shown in Fig. 2 by fitting the ΔL dependence with Eq. 1. The electron mean free path (l_e) estimated from the Hall and resistance measurement is 10 μm, while the l_b is 4.7 μm at $B_{//}=0$, showing the ballistic scattering length is much affected by the small angle scattering[8]. With increasing $B_{//}$ at $\theta > 0°$, where the ballistic electrons moving at the Fermi velocity v_F are pushed to the GaAs/AlGaAs interface by the Lorentz force $-ev_F \times B_{//}$, as shown in Fig. 3, the l_b monotonically decreases. At $B_{//}$ = 4 T, about 15% reduction of l_b is observed. In the reverse $B_{//}$ ($\theta < 0°$), the l_b increases, and it almost saturates at higher $B_{//}$. The displacement of the 2DES wave function in a triangle potential of the applied $B_{//}$ was roughly estimated by using values from conventional GaAs/AlGaAs heterostructures. At $B_{//}$ =4 T, it was only a few angstroms. Since the doped AlGaAs layer is separated 250 Å from 2DES by the spacer layer,

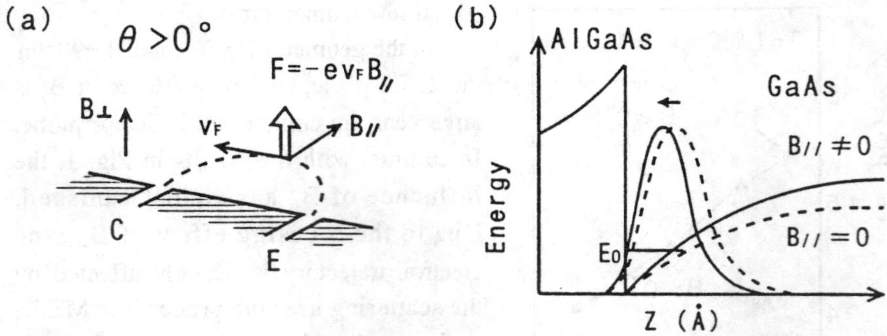

Fig. 3. (a) At $\theta > 0°$, the electrons moving between emitter and collector are pushed to the interface by the Lorentz force ($-ev_F \times B_{//}$). (b) Schematic view of the electron wave function distorted by $B_{//}$.

the change of the contribution from remote impurity scattering is small. The contribution of residual impurities, which are homogeneously contained in the epilayers, does not change with the displacement of the 2DES wave function. The variation of l_b with $B_{//}$ is attributed to the change of electron scattering near the AlGaAs/ GaAs interface, such as roughness scattering and alloy scattering.

Acknowledgements

We would like to thank H. Okada of Sumitomo Electric Industries for providing the GaAs/AlGaAs wafers. We acknowledge Prof. Y. Iye for his valuable comments. We also thank S. G. Inoue and K. Ohtsuka for their experimental support. This work is supported by the Grant–in–Aid for Scientific Research on Priority Area, from the Ministry of Education, Science and Culture. We (K.O. and S.T.) acknowledge the support of the Ogasawara Foundation for Promotion of Science and Engineering.

References

1. Th. Englert *et al*, *Surf. Sci.* **113**, 295 (1982).
2. N.K. Patel *et al*, *Phys. Rev.* **B44**, 10973 (1991).
3. P. Streda, P. Vasek, and M.Cukr, *Phys. Rev.* **B51**, 11144 (1995).
4. L. Smrcka *et al*, *Phys. Rev.* **B51**, 18011 (1995).
5. Y. Iye, and R. Yagi, *Solid State Phys.* **26**, 892 (1991) (in Japanese).
6. K. Tsukagoshi *et al*, *Appl. Phys. Lett.* **62**, 1609 (1993).
7. J. Spector *et al*, *Surf. Sci.* **228**, 283 (1990).
8. S. Takaoka *et al*, *Solid State Commun.* **83**, 775 (1992).

ANTI-CROSSING OF DIAMAGNETIC ENERGY LEVELS IN Si δ-DOPED GaAs

A.F.W. VAN DE STADT, P.M. KOENRAAD, J.H. WOLTER
COBRA, Interuniversity Research Institute, Eindhoven University of Technology,
P.O. 513, 5600 MB Eindhoven, The Netherlands

R. BOGAERTS, A. DE KEYSER, F. HERLACH
Laboratorium voor Vaste-Stoffysica en Magnetisme, Katholieke Universiteit Leuven,
B-3001 Leuven, Belgium

We have performed diamagnetic Shubnikov-de Haas measurements in pulsed magnetic fields up to 50T on GaAs layers containing single δ-layers with varying electron density and doping widths. For thin δ-layers (20Å) the measurements show a strong anisotropy of the depopulation of the n=1 subband in the transverse and longitudinal magnetoresistance. We have performed selfconsistent calculations of the diamagnetic energy levels in order to determine the influence of the layer thickness on this anisotropic behaviour. The calculations show that for thick layers (80Å) the n=1 and n=3 subband depopulations shift to higher magnetic field values. This is directly related to the higher relative occupation of these subbands in thick layers compared to thin layers. Furthermore we observe changes in the density-of-states of the n=1 and n=2 subbands. We show that this is related to the anti-crossing behaviour between the n=0 and n=1 subband and we indicate how this is related to the observed anisotropy.

1 Introduction

Although the diamagnetic Shubnikov-de Haas (SdH) effect has been widely used in the past to reveal the multiple subband structure in inversion layers [1] and δ-layers [2], the influence of current direction with respect to the in-plane magnetic field direction on magnetic depopulation is not fully understood yet. Furthermore depopulation of the n=1 subband in two-dimensional systems with high electron densities is scarcely studied because this depopulation occurs above 20T. In this paper we focus on these aspects and we test the validity of the universal depopulation curves of Reisinger and Koch [3] when applied to δ-layers having a finite width.

2 Experimental

We have performed magnetotransport measurements in pulsed parallel magnetic fields up to 50T on MBE grown Si δ-doped GaAs structures with electron densities ranging from $1.8\times10^{12}cm^{-2}$ to $7.6\times10^{12}cm^{-2}$ and layer widths ranging from 20Å to 110Å [4]. The effective layer widths were obtained indirectly by fitting the relative occupations, determined from SdH-oscillations in perpendicular fields, to selfconsistent calculations.

We have shown before [4] that for thin single δ-layers (≤20Å) the depopulation fields for the n=2 and n=3 subbands are in agreement with the depopulation curves of Reisinger and Koch. The n=1 depopulation however shows a strong anisotropy in the transverse (ρ_{yy}, B⊥I) and longitudinal (ρ_{xx}, B∥I) magnetoresistance (fig. 1). This anisotropy manifests itself as a shift towards lower magnetic fields for the n=1 depopulation in ρ_{yy} when compared to the depopulation in ρ_{xx} (fig. 2c). As a consequence the n=2 depopulation appears superimposed on the flank of the n=1 depopulation. Furthermore we observed that this shift increases with doping concentration. For thick layers (≥80Å) we still see some anisotropy but the shift is strongly reduced. This suggest that the detailed subband structure is very important. In order to understand the influence of the layer width on the diamagnetic energy levels we have extended the work of Reisinger and Koch for mathematical δ-layers to δ-layers having a finite width.

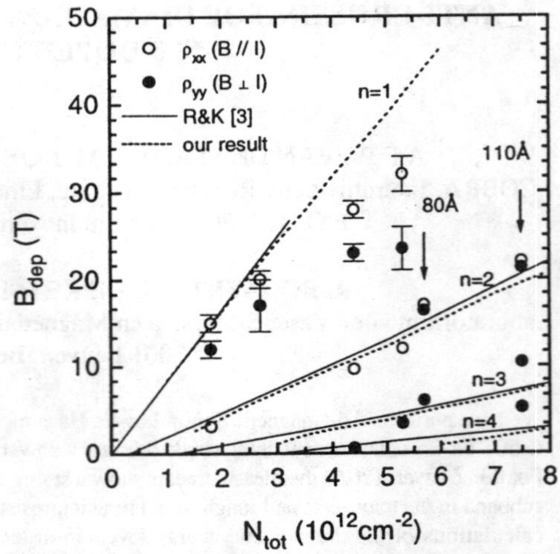

Figure 1: experimentally determined depopulation fields for single δ-layers with a width of 20Å, unless indicated otherwise. Solid lines represent the depopulation curves of Reisinger and Koch, the dashed lines are our results.

3 Calculation model

We have performed selfconsistent calculations to obtain solutions of the coupled Schrödinger and Poisson equation in the case of an in-plane magnetic field [3]. If we take a vector potential (0,-Bz,0) the wavefunctions and diamagnetic energy levels are determined by the Schrödinger equation:

$$[-\frac{\hbar^2}{2m^*}\frac{\partial^2}{\partial z^2}-\frac{e\hbar}{m^*}k_yB_\parallel z+\frac{e^2}{2m^*}B_\parallel^2z^2+U(z)]\varphi_{i,k_y} = E_{i,k_y}\varphi_{i,k_y} \qquad (1)$$

The $k_yB_\parallel z$ term couples the motion of the electrons with the lateral coordinate and represents the Lorentz-force. This term leads to a distortion of the bands in the k_y direction and thus to a change in the Fermi-contours. The term quadratic in $B_\parallel$ describes the magnetic confinement. If this confinement is strong enough a level can be pushed above the Fermi-level and becomes depopulated.

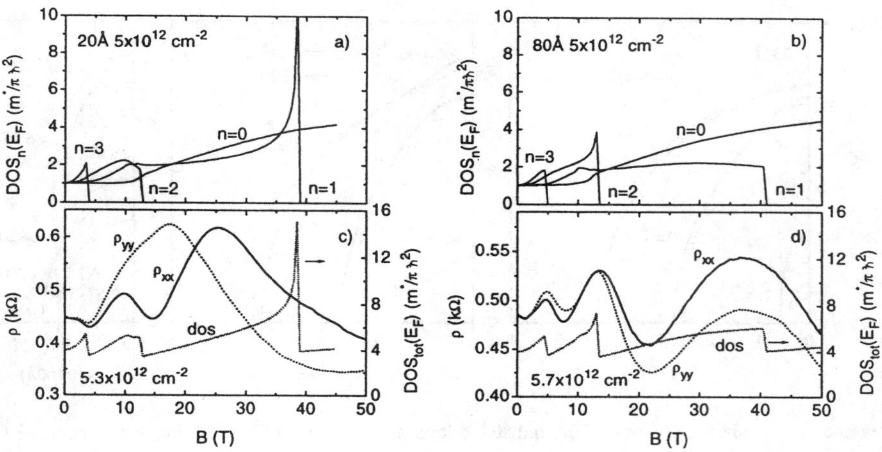

Figure 2: **Calculated individual subband density-of-states at the Fermi-level for a 20Å (a) and 80Å (b) δ-layer with an electron density of $5\times10^{12}cm^{-2}$. The total density-of-states is compared with measurements on layers with comparable layer widths and electron densities in c) and d).**

In figure 1 we compare our depopulation field strengths for thin layers (20Å, acceptor background concentration $1\times10^{14}cm^{-3}$) with the depopulation curves of Reisinger and Koch applied to δ-doping layers in GaAs [2]. For the n=2 and n=3 depopulations we find similar values. However in our calculation for layers with a high doping concentration we find that the magnetic field strengths for the depopulation of the n=1 subband are smaller. This is in agreement with our experimental results.

In figure 2a and b the calculated density-of-states at the Fermi-level of the individual subbands ($DOS_n(E_F)$) has been plotted for a 20Å and 80Å δ-layer with a doping concentration of $5.0\times10^{12}cm^{-2}$. If we compare the depopulation field strengths for these layers we see that there is a small increase for the n=3 and n=1 depopulations of the 80Å δ-layer. This increase is directly related to the higher relative occupation of the n=1 and n=3 subbands at zero magnetic field in thick layers compared to the relative occupation of these subbands in thin layers. The increase of the depopulation field of the n=3 subband is most clearly seen in de measurements on the structure having a 110Å thick δ-layer (fig. 1).

The influence of the layer width on the $DOS_n(E_F)$ of the n=1 and n=2 subbands however is more drastic: for the narrow δ-layer the high $DOS_n(E_F)$ peak just before depopulation of the n=1 subband completely disappears and changes into a broad maximum at lower magnetic fields for thick layers. Simultaneously the $DOS_n(E_F)$ just before depopulation of the n=2 subband becomes much higher. Furthermore we observe for both thin and thick layers a local maximum in the $DOS_n(E_F)$ of the n=1 depopulation near 10T. These effects on the $DOS_1(E_F)$ can be explained by studying

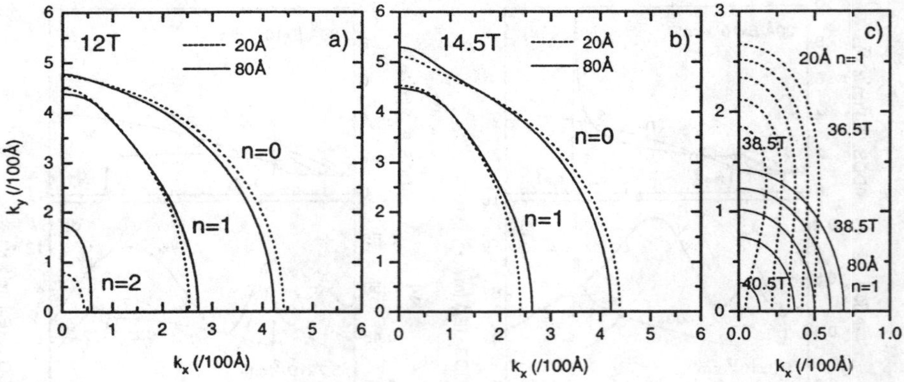

Figure 3: Fermi-contours for the 20Å and 80Å δ-layer at a) 12T, b) 14.5T and c) just before depopulation of the n=1 subband.

the Fermi-contours in figure 3a and b. Due to the parallel magnetic field the n=1 contour is distorted in the k_y direction and the $DOS_1(E_F)$ increases. At 12T the n=1 contour reaches the n=0 contour and they anti-cross. The distortion of the Fermi-surface is taken over by the n=0 subband as can be seen in the $DOS_0(E_F)$. Due to a combined effect of the difference in relative occupation and energy level separation of the 20Å and 80Å δ-layers the anti-crossing behaviour changes and we observe different Fermi-contours for the n=1 depopulation (fig. 3c). The high $DOS_n(E_F)$ just before depopulation is directly related to the strong anisotropy of the Fermi-contours.

If we simply compare the calculated total density-of-states at the Fermi-level with ρ_{xx} measurements [2] on layers with a comparable doping density and δ-layer width (fig. 2c and d), we observe that the dependence of the relative oscillation amplitude of the n=1 and n=2 depopulations on layer thickness is in qualitative agreement with the measurements. The calculations also suggest that the shift of the n=1 depopulation in ρ_{yy}, compared to the depopulation in ρ_{xx} is related to the strong anisotropy of the Fermi-contours in narrow doping layers.

References

[1] M. Kroeker, U Merkt, *Solid State Comm.* **69**, 699 (1989)

[2] A. Zrenner, H. Reisinger, F. Koch, K. Ploog, J.C. Maan, *Phys. Rev.* B **33**, 5607 (1986)

[3] H. Reisinger and F. Koch, *Surf. Sci.* **170**, 397 (1986)

[4] A.F.W. van de Stadt, R. Bogaerts, P.M. Koenraad, H. Leopold, F. Herlach, J.H. Wolter, *Physica* B **211**, 458 (1995)

EFFECTIVE MASSES AND G-FACTORS OF X_Z AND X_{XY}-ELECTRONS IN AlAs QUANTUM WELLS

A.F.W. VAN DE STADT, P.M. KOENRAAD, J.H. WOLTER
COBRA, Interuniversity Research Institute, Eindhoven University of Technology, P.O. 513, 5600 MB Eindhoven, The Netherlands

J.A.A.J. PERENBOOM
High Field Magnetic Laboratory and Research Institute for Materials, University of Nijmegen, Toernooiveld, 6525 ED Nijmegen, The Netherlands

We have performed magnetoresistance measurements up to 20T on high mobility two dimensional electron gasses (2DEG) with densities of $\sim 1\times10^{12}cm^{-2}$ in single AlAs quantum wells with well widths ranging from 40Å to 80Å. We observe a transition from an X_z-occupied subband into a twofold valley degenerate X_{xy}-occupied subband between well widths of 50Å and 60Å. Effective masses have been determined from the temperature dependence of the Shubnikov-de Haas (SdH) oscillations. We find effective masses for the X_z and X_{xy}-electrons of respectively $0.26m_0$ and $0.62m_0$. Using selfconsistent calculations we can explain this X_z-X_{xy} transition around 55Å by taking into account a strain splitting energy between the X_z and X_{xy}-levels of ~19meV. By measuring the magnetoresistance in tilted magnetic fields we have found effective g-factors of 2.9 for both type of electrons.

1 Introduction

The electrical and optical properties of AlAs differ a lot from those of GaAs because the lowest conduction minimum is at the X-point of the Brillouin-zone instead of the Γ-minimum. This X-minimum is strongly anisotropic in k-space. In two-dimensional systems the threefold degeneracy of the X-minimum is lifted in a twofold degenerate X_{xy}-state, bound by the transverse effective mass (m_t), and an X_z-state bound by the longitudinal effective mass (m_l). Because $m_l>m_t$ it is expected that the lowest bound state has X_z-character. However both optical [1] and electrical measurements [2] have shown that in type-II superlattices (SL) with wide AlAs wells (80Å) the X_{xy}-state is the lowest state. This apparent level reversal is attributed to the small amount of strain in the AlAs/GaAs system which shifts the X_z-level upward in energy with respect to the X_{xy}-level.

We were able to confine the X-electrons in a single AlAs quantum well by using a $GaAs_8/AlAs_4$ SL as barrier material. Carriers were provided by a Si-δ-doping layer in one of the GaAs layers in the SL on each side of the AlAs well. Details of this structure have been presented elsewhere [3]. Using MBE we have grown structures with well widths of 40Å, 50Å, 60Å and 80Å, with δ-doping layers of $1\times10^{12}cm^{-2}$ in either the second or third GaAs layer of the SLs.

2 Experimental

2.1 Effective mass measurements

A typical magnetoresistance measurement on a 2DEG at the X_z-minimum is shown in figure 1a. The high mobility of the sample is shown by the observation of pronounced quantum Hall plateaus. Strong magnetoresistance minima are found for every integer filling factor between 2 and 6. This can be understood if we assume that the spin to Landau-splitting ratio, $\alpha=g^*\mu_b B/\hbar\omega_c=g^*m^*/2m_0$, is approximately 0.5. In wide wells containing X_{xy}-electrons (fig. 1b) we find strong minima at filling factors ν=2,6 and 10. This can be explained by a twofold valley degeneracy and $\alpha\sim 1$.

Density-of-states (DOS) effective masses have been determined from the temperature dependence of the SdH-oscillations at low magnetic fields (B<4T), where spin and valley splittings are negligible. We used a magnetic field modulation technique to resolve the weak SdH-oscillations in this magnetic field range. For the X_z and X_{xy}-electrons we expect DOS effective masses of respectively m_t and $(m_t m_l)^{½}$. Our results are presented in table 1.

According to the valley degeneracy (n_{Hall}/n_{Sdh}) and the different effective masses we conclude that only the X_z-subband is occupied in the 40Å and 50Å wells and that only the X_{xy}-subband is occupied in the 80Å well. For the 60Å well the Hall-density is not exactly twice the SdH-density. This suggests that a slightly occupied X_z-subband, which is not resolved in the Fourier spectrum of the SdH-oscillations, is also present.

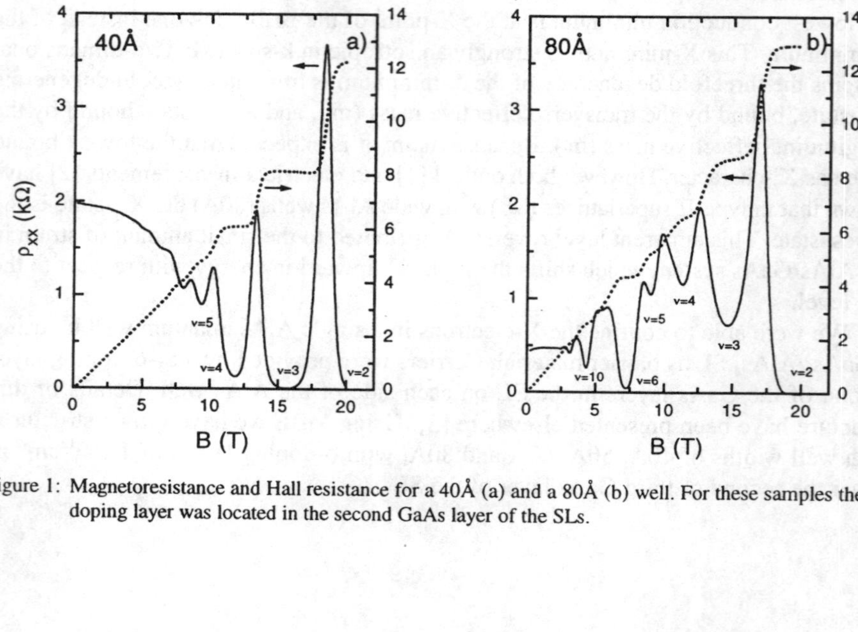

Figure 1: Magnetoresistance and Hall resistance for a 40Å (a) and a 80Å (b) well. For these samples the δ-doping layer was located in the second GaAs layer of the SLs.

Table 1: results of measurements on samples with δ-doping layers in the third period of the SLs.

well (Å)	n_{Hall} ($10^{12}cm^2$)	n_{SdH}($10^{12}cm^2$)	μ_{Hall}(cm^2/Vs)	μ_q(cm^2/Vs)	DOS m^* (m_0)
40	0.93	0.93	7900	1700	0.26±0.01
50	1.02	1.02	21600	1800	0.30±0.05
60	1.11	0.45	16300	6500	0.36±0.04
80	1.03	0.52	7100	3000	0.62±0.01

In our opinion the low effective mass value of $0.36m_0$ found in the 60Å well is due to this population of the X_z-subband. When we determine the X_{xy}-effective mass from simulated SdH-oscillations of a 2DEG with both X_z and X_{xy}-subbands occupied, we find a value much lower than $0.62m_0$.

Taking the effective mass values of the X_z and X_{xy}-electrons in the 40Å and 80Å well we obtain m_t=$0.26m_0$ and m_l=$1.48m_0$, which is consistent with the values reported in literature [2]. By solving Schrödinger and Poisson equation selfconsistently for this structure we can explain the X_z-X_{xy} transition using an X_z-X_{xy} energy splitting of ~19meV. This energy splitting is in agreement with values obtained from optical measurements [1].

2.2 *effective g-factor measurements*

By rotating a 2DEG in a magnetic field, the spin to Landau-splitting ratio becomes $\alpha = g^* m^* / 2m_0 \cos\theta$, where θ corresponds to the angle between magnetic field direction and surface normal. From the consecutive appearing and disappearing of strong magnetoresistance minima at the same filling factor during rotation of the sample we can determine α and give an estimate of the effective g-factor. We have rotated a 40Å and 80Å sample with δ-doping layers in the second SL period. The electron densities and effective masses for these structures differ slightly from those in table 1 [3]. The rotation angles were determined by the low field Hall voltage with an accuracy of a few percent.

In figure 2a the magnetoresistance traces of the X_z-2DEG is shown for various angles. The θ=0° trace shows clear minima at filling factors ν=4,5 and 6 (ν=8 and 10 are present but less clear). At intermediate angles the minima at ν=6,8 and 10 disappear and the minima at ν=5,7 and 9 appear or become stronger. For a rotation angle of 67° the minima at ν=7 and 9 have their maximum SdH-amplitude as a function of rotation angle. At this angle the spin up and down levels of adjacent Landau levels coincide, resulting in α=1. For an effective mass of $0.27m_0$, as we have found previously [3], this

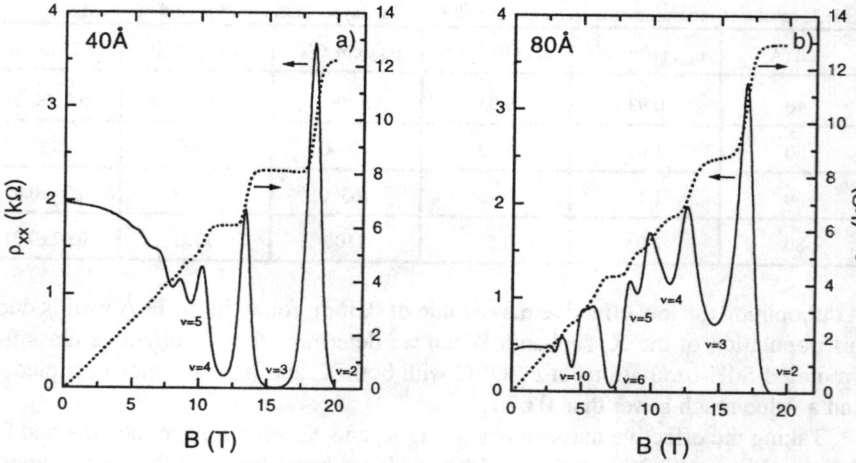

Figure 2: Magnetoresistance of a) the 40Å and b) the 80Å sample as a function of the normal component of the magnetic field for various rotation angles.

leads to an effective g-factor of 2.9.

For X_{xy}-electrons we observe a totally different behaviour (fig. 2b). At small rotation angles strong minima in ρ_{xx} are present at filling factors ν=6,10 and 14 ($\alpha\sim 1$). At $\theta=63°$ the minima at ν=8 and 12 have their maximum SdH-amplitude, corresponding to $\alpha=2$. For an effective mass of $0.63m_0$ [3] this also results in a g-factor of 2.9, which is comparable with the value found by Smith [4]. Using the effective masses and g-factors obtained from our magnetoresistance measurements we find that $\alpha_{Xz}=0.4$ and $\alpha_{Xxy}=0.9$ at $\theta=0°$. This is in agreement with our assumptions made in §2.1. The effective g-factor values show an enhancement, probably due to many body interactions similar to Γ-electron gasses [5].

References

[1] H.W. van Kesteren, E.C. Cosman, P. Dawson, K.J. Moore and C.T. Foxon, *Phys. Rev.* B **39**, 12426 (1989)

[2] S. Yamada, K. Maezawa, W.T. Yuen and R.A. Stradling. *Phys. Rev.* B **49**, 2189 (1994)

[3] A.F.W. v.d. Stadt, P.M. Koenraad, J.A.A.J. Perenboom, J.H. Wolter, accepted for publication in *Surf. Sci.*

[4] T.P. Smith III, W.I. Wang, F.F. Fang, L.L. Chang, *Phys. Rev.* B **35**, 9349 (1987)

[5] T. Ando, A.B. Stern and F. Stern, *Rev. Mod. Phys.* **52**, 437 (1982)

RING EXCHANGE INTERACTION AMONG STRONGLY LOCALIZED ELECTRONS IN Si MOSFETS

Tohru Okamoto and Shinji Kawaji
Gakushuin University, Mejiro, Tokyo 171, Japan

The activation energy in the insulating phase in two dimensional electron systems (2DES) in Si MOSFETs is studied in magnetic fields applied for various angles. Results are explained on the assumption of a 3-electron ring exchange interaction among neighboring localized electrons.

1 Introduction

In the insulating region at very low temperatures, a Si MOSFET shows a giant oscillation of the diagonal resistivity ρ_{xx} as a function of the magnetic field applied normal to the 2D plane with two minima at $eB / hN_s \sim 0.5$ and ~ 1.[1,2] Though eB / hN_s is equal to the inverse of the Landau level filling factor $1 / \nu$, it is hard to imagine that the Landau levels are separated from each other in the insulating phase where the diagonal and Hall conductivities approach to zero with decreasing temperature and no extended state exists below the Fermi level.

In this work, we performed systematic measurements of the activation energy E_A as a function of the electron concentration N_s and the magnetic field applied for various angles θ to the 2D plane. The dependencies of E_A on the strength of the magnetic field $|\boldsymbol{B}|$ and the normal component $B_\perp$ are explained on the assumption of a 3-electron ring exchange process among neighboring electrons located in the triangular lattice. In this case, an oscillation of E_A as a function of $eB_\perp / hN_s$ is attributed to the Aharanov-Bohm effect on the ring exchange interaction.

2 Experimental

The sample used has a peak electron mobility of $\mu_{peak} = 1.26\ \mathrm{m}^2 / \mathrm{V\ sec}$ at 2.6 K and a Hall bar geometry (0.6 mm x 0.1 mm). Temperature dependence of the longitudinal resistance was measured in the range $T = 0.14\ \mathrm{K} \sim 1.4\ \mathrm{K}$ and $R_{xx} = 0.2 \sim 100\ \mathrm{M}\Omega$ where the diagonal resistivity exceeds the Hall resistivity. The activation energy E_A was obtained as a function of B and N_s according to $R_{xx} = R_o \exp(E_A / 2T)$ within an accuracy of 5 % except for E_A below 1 K.[3]

3 Results

Dependence of E_A on $\boldsymbol{B}$ at fixed N_s is shown in Fig. 1. In parallel magnetic field ($\theta = 0°$), E_A increases linearly with $|\boldsymbol{B}|$ in a low field region. The slope $\partial E_A / \partial|\boldsymbol{B}|$ is close to $2\mu_B / k_B$, where μ_B is the Bohr magneton and k_B the Boltzmann

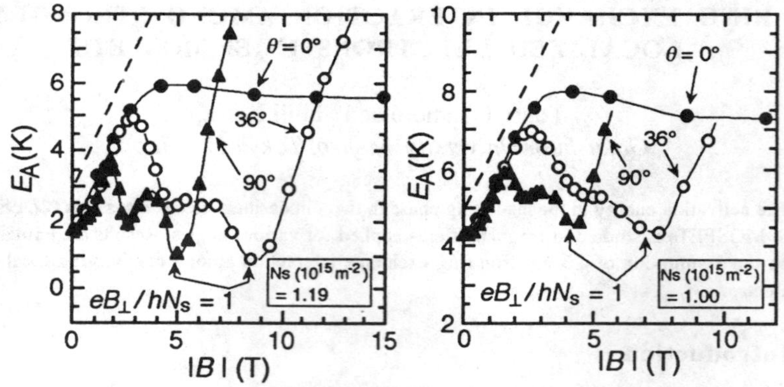

Fig. 1. Activation energy for two electron concentrations. Dashed lines represent the slope of $\partial E_A / \partial |B| = 2\,\mu_B / k_B$.

constant. In higher fields, however, the slope goes to zero and E_A takes a constant value. On the other hand, a drastic change is induced by the normal component $B_\perp$. In tilted magnetic fields, E_A deviates from that at $\theta = 0°$ and oscillates with two minima at $eB_\perp / hN_s \sim 0.5$ and ~ 1. The field dependence of E_A does not change qualitatively with N_s while the amplitude of the oscillation increases with N_s.

4 Discussion

4. 1 3-electron ring exchange interaction

In order to explain the observed field dependence of E_A, we assume two types of the exchange interactions among electrons. One is a 3-electron ring exchange interaction among localized electrons. Although our previous work demonstrated an important role of the random potential fluctuations in samples on the metal-insulator transition,[2] the Coulomb repulsive force might form a triangular lattice *at least in a short range of the spatial distribution of electrons* even in the absence of the magnetic field. In this case, various types of ring exchange processes among neighboring electrons may occur due to the quantum tunneling. Many experimental and theoretical works on the ring exchange interaction have been performed in studies of the magnetism of solid ^{3}He where competition among several types of the ring exchange interactions makes interesting magnetic ordered phases at ultralow temperatures $T \leq 1$ mK.[4,5] The 3-particle ring exchange process, which produces a ferromagnetic interaction between spin-1/2 atoms, dominates in the triangular lattice formed in 2D ^{3}He adsorbed on Grafoil[6] and in hcp solid ^{3}He.[7] In the following, for the present 2DES, we consider the 3-electron ring exchange interaction among the

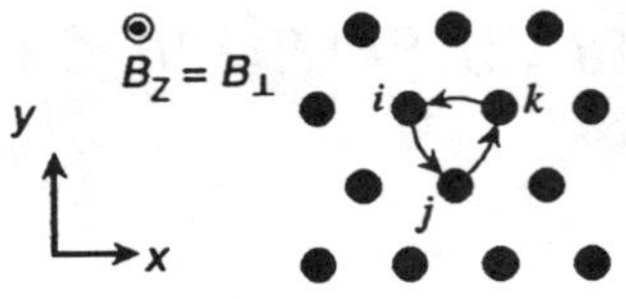

Fig. 2. 3-electron ring exchange interaction.

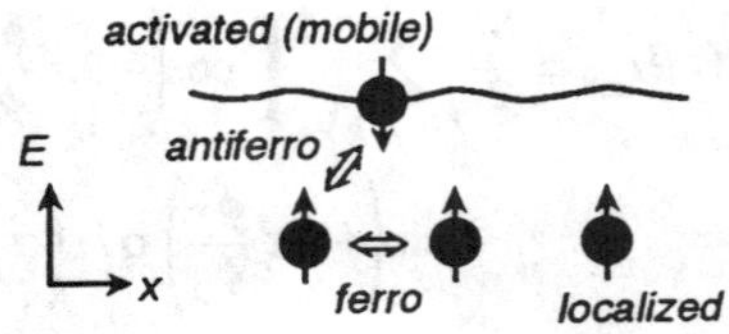

Fig. 3. Two types of exchange interactions.

nearest neighbors only (Fig. 2) although the competitions with antiferromagnetic 2- and/or 4-electron exchange interactions are expected.

4. 2 Exchange interaction with an activated electron

The other exchange interaction assumed here is an antiferromagnetic interaction between an activated electron and localized electrons as an s-d interaction in ferromagnetic metals. Here, we consider activation process as creation of a pair of mobile electron and hole. The activated electron receives a mean field by the localized electrons. If the localized electrons are ferromagnetically aligned and their spins turn toward a weak magnetic field, the activated electron takes an antiparallel spin direction against the localized electrons as shown in Fig. 3.

4. 3 Activation energy in parallel magnetic field

The derivative of the activation energy with respect to magnetic field should be considered as a change in magnetization due to one activation process, *i.e.*, $\delta M = -\partial E_A / \partial |\boldsymbol{B}|$. Thus, the observed slope in low field region indicates $\delta M = -2\mu_B$ and the spin flip from $+\mu_B$ to $-\mu_B$ caused by the antiferromagnetic exchange interaction between an activated electron and the ferromagnetically localized electrons. When the magnetic field overcomes the mean field by the localized electrons, the spin-flip process does not occur, $\delta M = 0$, and E_A takes a constant value.

4. 4 Ground state energy in tilted magnetic fields

The Aharanov-Bohm effect on the ring exchange interaction is expected in the 2DES in tilted fields with $B_\perp \neq 0$. The spin Hamiltonian established in the study of solid ^{3}He[4] should be modified. For the 3-electron ring exchange interaction among the nearest neighbors, it should be written as[3]

$$H_{ex} = \frac{J_T}{2} \sum_{\{i,\,j,\,k\}} \left[\cos\left(+ 2\pi \frac{\phi_T}{\phi_o} \right) \left(1 + \sigma_i \cdot \sigma_j + \sigma_j \cdot \sigma_k + \sigma_k \cdot \sigma_i \right) \right.$$
$$\left. - \sin\left(+ 2\pi \frac{\phi_T}{\phi_o} \right) \sigma_i \cdot \left(\sigma_j \times \sigma_k \right) \right].$$

Here, the exchange interaction constant J_T is defined to be negative and the sums are taken over all distinct 3-spin cycles $\{i, j, k\}$. The ratio of the flux ϕ_T which penetrates into the area S_T surrounded by the exchange path to the flux quantum ϕ_o plays an important role. The spin Hamiltonian leads to an antiferromagnetic ground state (AGS) where $\phi_T = (1/2)\phi_o$ while it leads to the ferromagnetic ground state (FGS) where $\phi_T = 0$ or ϕ_o. Further consideration should be needed for comparing the experimental data of E_A with the theoretical model. However, if one attributes the minima in E_A to the magnetic transition points between FGS and AGS[3] and assumes the midpoint at $eB_\perp / hN_s \sim 0.8$ as the point of $\phi_T = (1/2)\phi_o$, S_T ($= \phi_T / B_\perp$) is calculated to be $\sim 0.6\ N_s^{-1}$. This value is larger than the area of the triangle $S_\Delta = 0.5\ N_s^{-1}$. The exchange path may be expanded by the repulsive forces among the exchanging electrons.

Acknowledgments

The authors wish to thank Dr. A. Yagi (NPC Ltd.) for providing us with Si MOSFET samples, and Profs. A. Kawabata and H. Tasaki for their helpful discussions. This work is supported in part by Grants-in-Aid for Scientific Research from the Ministry of Education, Science and Culture, Japan.

References

1. M. D'Iorio, V. M. Pudalov, and S. G. Semenchinsky, Phys. Rev. B **46**, 15992 (1992).
2. T. Okamoto, Y. Shinohara, and S. Kawaji, Phys. Rev. B **52**, 11109 (1995).
3. Experimental results and theoretical procedures will be presented elsewhere in more detail.
4. M. Roger, J. H. Hetherington, J. M. Delrieu, Rev. Mod. Phys. **55**, 1, (1983).
5. H. Fukuyama, T. Fukuda, T. Okamoto, H. Akimoto, H. Ishimoto, and S. Ogawa, Physica (Amsterdam) **169B**, 197 (1991); and references therein.
6. P. Schiffer, M. T. O'Keefe, D. D. Osheroff, and H. Fukuyama, Phys. Rev. Lett. **71**, 1403 (1993).
7. T. Okamoto, H. Fukuyama, H. Akimoto, H. Ishimoto, and S. Ogawa, Phys. Rev. Lett. **72**, 868 (1994).

PHONON EMISSION FROM A 2DEG: EVIDENCE OF TRANSITION TO THE HYDRODYNAMIC REGIME

EDMOND CHOW, H.P. WEI, S.M. GIRVIN
Dept. of Physics, Swain Hall West, Indiana University, Bloomington, IN 47405

M. SHAYEGAN
Dept. of Electrical Engineering, Princeton University, Princeton, NJ 08544

Using as a thermometer the temperature dependent magneto-transport of a two-dimensional electron gas, we find that the effective temperature scales with current as $T_e \sim I^a$, where $a = 0.4 \pm 2\%$ in the *Shubnikov-de Haas* regime, and $0.53 \pm 2\%$ in between adjacent quantum Hall plateaus, both *integer and fractional*. This implies that the phonon energy emission rate changes from the expected $P \sim T^5$ to $P \sim T^4$. We explain this, as well as the dramatic enhancement in phonon emission efficiency using a hydrodynamic model.

The past three years has seen a revived interests in the inelastic scattering due to electron-phonon interactions in a 2DEG,[1] which was suggested to be affected by the quantum Hall effect (QHE). We present in this work a systematic study in the Shubnikov de-Haas (SdH) and the QHE regime to address this issue by combining current (I) and temperature (T) dependent transport measurements.

The sample used in this work is a GaAs/AlGaAs heterostructure with an electron density $n = 0.65 \times 10^{11}\,\mathrm{cm}^{-2}$, and a mobility $\mu \sim 500{,}000\,\mathrm{cm}^2/\mathrm{Vs}$ at 0.1K. The high mobility was chosen to allow the SdH oscillations to be seen at low magnetic fields (B). The donor spacer thickness is 1800Å, and the Hall bar pattern on the sample has a channel width of $300\,\mu$m. The large device is used to avoid the influence of edge effects,[2] which are known to be more pronounced in small devices,[3] and to avoid problems with heat being carried out of the sample directly by electron diffusion.[4]

The inset of Fig.1 shows the longitudinal resistivity ρ_{xx} as a function of B for $0.06\,\mathrm{T} < B < 0.16\,\mathrm{T}$ (which corresponds to spin-unresolved Landau level filling factors, $18 \leq \nu \leq 38$) at 3 different I's at a fixed bath temperature $T_b = 100\,\mathrm{mK}$. Since this I dependent evolution is similar to the T dependent evolution, we assign an effective electron temperature (T_e) for each I. We choose to plot, in Fig.1, T_e vs. I for the two oscillations pointed to by arrows in the inset of Fig.1, because their amplitudes are measurable in a wider range of I. The closed and open symbols represent the data taken from the peak at $B = 0.139\,\mathrm{T}$ and the dip at $B = 0.133\,\mathrm{T}$ respectively. Different symbols represent data taken at different T_b's which are labeled by the side of the data set. For clarity, the open symbol data set has been offset by scaling by a factor of 1/3. For each T_b, there is a current I_0, below which T_e remains constant, and above which T_e merges into a single power law:

$$T_e \sim I^a$$

over about one and a half decades in I with a close to 0.4. The solid line under each data set has a slope of 0.4 and is drawn for reference. For comparison, we also draw a dashed line of slope 0.5. The experimental value of a is obtained by first collecting all data points which have $T_e > 2T_b$ for each fixed T_b, and then performing a linear least-squares fit to the resulting data points. We obtain $a = 0.4 \pm 2\%$, where 2% sets the statistical 68% confidence limit.

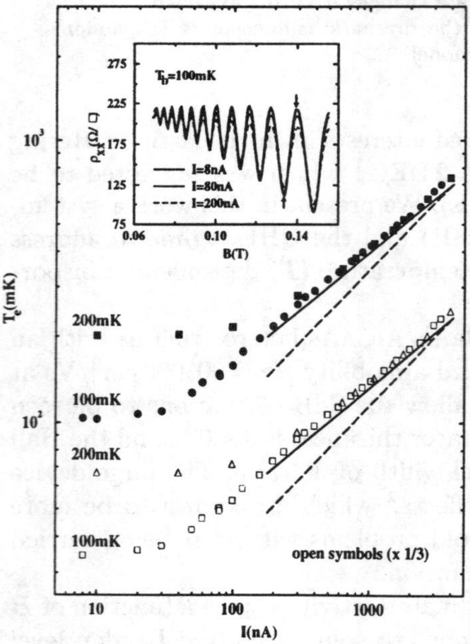

Figure 1: Electron effective temperature T_e vs. applied current I at T_b= 100 and 200 mK, for B = 0.139 T (closed symbols) and B = 0.133 T (open symbols). The solid line under each data set has slope 0.4, the dashed line 0.5. Inset: ρ_{xx} vs. B at T_b = 100 mK for three different I's.

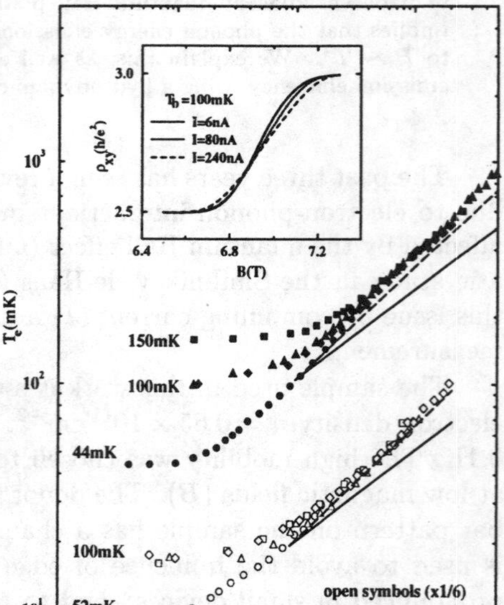

Figure 2: Electron effective temperature T_e vs. applied current I in the IQHE for $5 < \nu < 6$ (closed symbols), and in the FQHE for $1/3 < \nu < 2/5$ (open symbols). Each curve is labeled by the bath temperature. The dashed reference line under each data set has a slope of 0.5, and the solid line 0.4. Inset: ρ_{xy} vs. B at T_b = 100 mK for three different I's in the FQHE for $1/3 < \nu < 2/5$.

In the same sample, we also study the effect of I on the Hall resistivity ρ_{xy} between two adjacent FQHE plateaus. The inset of Fig.2 shows ρ_{xy} as function of B at 3 different I's with fixed $T_b = 100\,\text{mK}$, between well developed FQHE states at $\nu = \frac{1}{3}$ and $\frac{2}{5}$. By using $(d\rho_{xy}/dB)^{max}$ as a thermometer, we characterize the effect of I by assigning a T_e at each I. Here, $(d\rho_{xy}/dB)^{max}$ is the maximum slope of ρ_{xy} vs. B for $\frac{1}{3} < \nu < \frac{2}{5}$ measured with $I = 3\,\text{nA}$. In Fig.2, we plot T_e vs. I in this region using open symbols, scaled by a factor of 1/6. Different symbols represent data taken at different T_b's or upon different cooldowns of the sample. We also study plateau transitions in the IQHE for Landau level index $N = 2\downarrow$ $(5 < \nu < 6)$ and $2\uparrow$ $(4 < \nu < 5)$, where $\uparrow$ and $\downarrow$ represent spin direction. We do not study the IQHE transition for $N = 1\downarrow$ and $1\uparrow$ because there appears to have FQHE structures in ρ_{xy}. The result for $N = 2\downarrow$ is plotted in Fig.2 by closed symbols. There are no data points for $I > 2\,\mu\text{A}$ for the FQHE data because $(d\rho_{xy}/dB)^{max}$ is saturated to the classical Hall resistance. However, it is still clear that in both the IQHE and FQHE regimes T_e behaves the same and there is a single power law, $T_e \sim I^a$, over more than one decade in I with $a \sim 0.5$. The solid line of slope 0.4 and the dashed line of slope 0.5 are drawn for reference. We obtain the best fit values $a = 0.53 \pm 2\%$ in the FQHE regime and $a = 0.54 \pm 2\%$ and $0.53 \pm 4\%$ for $N = 2\downarrow$ and $2\uparrow$ respectively. This is *significantly different* from that $(0.40 \pm 2\%)$ in the SdH regime. Similar results have also been obtained in a second sample. This change in a can be viewed as a change in the cooling rate from $P \sim I^2\rho_{xx} \sim T^5$ in the SdH to $P \sim T^4$ in the QHE, since ρ_{xx} has no strong T dependence.

To understand our results, one can express P by the standard Fermi Golden Rule for the rate of phonon emission.[5] One obtains an expression which is proportional to the real part of the inverse of the conductivity (σ_{xx}). In the presence of strong disorder or high magnetic field, the reduced mean free path (and magnetic length) can put the system into the hydrodynamic regime, where typical phonon wave vectors and frequencies are negligible compared to characteristic scales over which the conductivity varies, so that σ_{xx} is a real constant. Simple power counting in the expression of P then yields T^4. The physical interpretation is the following. In this limit, momentum conservation is lost because of frequent collisions of the electrons with the disorder potential. Moreover, the static screening approximation fails. It is not the static compressibility of the electron gas that counts, but rather its limited ability to move charge quickly. The smaller the conductivity is, the poorer the screening is and hence the larger the emitted power. The rate energy is absorbed from a phonon mode depends only on the square of its electric field and is independent of wave vector and frequency. With the piezo-electric coupling, the 2DEG hence acts like a black (more precisely a "gray") body and emits phonons with the usual T^4 spectrum. Moreover, our theory without adjustable parameters gives

$$T_e = 24.9(\sigma_{xx}\rho_{xx})^{1/4}(J/A/m)^{1/2}K,$$

where J is the current density. This expression is $\sim 4\%$ below the observed values, and explains why the curves in Figure 2 for the fractional and integer transitions are nearly identical. The factor $(\sigma_{xx}\rho_{xx})^{1/4}$ (which is unity at $B = 0$) is a weak function of B in the sense that it is nearly equal for the two plateau transitions studied.[6] Thus the increased Joule heating is precisely compensated by increased electron-phonon coupling due to reduced screening.

In the clean limit, the power counting is different since the conductivity is non-local: $\mathrm{Re}(1/\sigma_{xx}) \approx q$, where q is the wave vector. This changes the temperature exponent to $P \sim T^5$. Our expression for P also correctly reproduces the usual Thomas-Fermi static screening result for the prefactor.[5] The phonon emission rate $(1/\tau_{in})$ is thus calculated to be $2.92 \times 10^8 T^3\ \mathrm{sec}^{-1}\mathrm{K}^{-3}$. Experimentally, we find $1/\tau_{in} \sim 7.3 \times 10^8 T^3\ \mathrm{sec}^{-1}\mathrm{K}^{-3}$ at $B = 0.13\,\mathrm{T}$.

Acknowledgments

The work at Indiana University is supported by IU Foundation and NSF grants DMR-9311091 and DMR-9416906. MS at Princeton is supported by NSF DMR-9222418.

References

1. H.L. Zhao, Shechao Feng, *Phys. Rev. Lett.* **70**, 4134 (1993); T. Brandes *et al*, *ibid.* **72**, 3582 (1994); T. Brandes, *Phys. Rev.* **B 52**, 8391 (1995).
2. H.Z. Zheng *et al*, *Phys. Rev. Lett.* **55**, 1144 (1985).
3. P.L. McEuen *et al*, *Phys. Rev. Lett.* **64**, 2062 (1990).
4. A. Mittal, to appear in *Advanced Nato Proceedings on 'Quantum Transport in Submicron Structures'*, (Kluwer Academic Press, 1996).
5. P.J. Price, *J. Appl. Phys.* **53**, 6863 (1982), (We have corrected a factor of 2 error in Price's result. See Mittal[4]); T. Kawamura and S. Das Sarma, *Phys. Rev.* **B 45**, 3612 (1992).
6. On the Hall plateaus this quantity is a strong function of temperature and frequency and the present analysis would have to be modified for that case.

MAGNETO-PHOTON-PHONON RESONANCES IN TERAHERTZ-DRIVEN 2DEGS IN HIGH MAGNETIC FIELDS

W. XU and C. ZHANG

Department of Physics, University of Wollongong, NSW 2522, Australia

We propose a photon-induced quantum resonance effect: *magneto-photon-phonon resonance* observed in terahertz (THz) driven two-dimensional electron gases (2DEGs) in strong magnetic fields. This effect is a consequence of electronic transitions among different Landau-levels accompanied by the resonant scattering via LO-phonon emission and optical absorption.

1. Introduction

Recent experimental measurements[1] showed that using state-of-the-art techniques such as free-electron-lasers, it has become realistic to investigate the electronic properties of a two-dimensional electron gas (2DEG) under terahertz (THz) electromagnetic (EM) radiations. Now we enter a regime with different competing energies (frequencies). For a GaAs-based 2DEG driven by THz EM fields in strong magnetic fields, the LO-phonon frequency ω_{LO}, photon frequency ω and cyclotron frequency ω_c are on the THz scale so that one can observe photon-induced novel quantum resonance effects. In this paper, we are interested in a consequence of electronic transitions among different Landau-levels (LLs) via electron interactions with photons and LO-phonons. In such a case, we predict that when a condition: $M\omega_c + \omega = \omega_{LO}$, with M an index difference between two LLs, is satisfied, resonant scattering via optical absorption and LO-phonon emission will occur between two LLs so that the d.c. resistivity and optical absorption coefficient will be enhanced. Consequently, these quantities are expected to oscillate with magnetic field and/or with radiation frequency. This resonance effect, called *magneto-photon-phonon resonance* (MPPR), is electrically analogous to the magneto-phonon resonance[2] and electro-phonon resonance[3] which are observed in the absence of the EM radiation.

2. Theoretical model

To observe the MPPR effect theoretically, in this paper we calculate frequency-dependent magnetoresistivities in a configuration: i) the magnetic field is applied perpendicular to the interface of the 2DEG, i.e., $\mathbf{B} = (0, 0, B)$; ii) the EM radiations are polarized along the x-direction; and iii) the effect of an EM field can be represented by an a.c. electric field with a frequency ω. Thus, the longitudinal and transverse magnetoresistivities can be derived using the Kubo formula[4,5] in the linear response approximation:

$$\rho_{xx} = \frac{B}{n_e e} \frac{1 - i\omega\tau}{\omega_c \tau} \quad \text{and} \quad \rho_{xy} = \frac{B}{n_e e}, \tag{1}$$

where n_e is the electron density of the 2DEG and τ is the frequency-dependent relaxation time which is connected to the memory function induced by electron interactions with impurities, phonons, etc. Here, we have accounted a fact that for a 2DEG driven by THz EM field, the real part of the memory function caused by electron–LO-phonon interactions is much smaller than the radiation frequency ω. This results in that τ can be determined by the imaginary part of the memory function. Further, it has been implied in Eq. (1) that the strength of the a.c. field provided by, e.g., free-electron-laser radiation, is much larger than the d.c. driving field.[1] In this paper, we study transport and optical properties of THz-driven 2DEGs in the linear response regime. The nonlinear effects induced by intense EM radiations can be modelled by introducing electron temperature T_e into the memory function. Recent experimental results[1] have shown that varying intensity of radiation will mainly lead to a variation in T_e. From Eq. (1), we note that in the presence of EM radiations, $\mathrm{Re}\sigma_{xy}$ ($\mathrm{Re}\rho_{xy}$) deviates from (still remains as) the conventional Hall effect.

In the present study, we consider a relatively high temperature regime ($T > 50$ K) where electron interactions with LO-phonons are the main processes for electron energy relaxation accompanied by optical absorption. We have applied the Fröhlich Hamiltonian for electron–LO-phonon interaction to calculate the electron current-current correlation function, from which the memory function has been obtained after ignoring the effects of electron-electron interactions.[5] We note that for a 2DEG in THz EM fields, $1/\tau$ is on the scale of the radiation frequency.[6]

3. Results and discussions

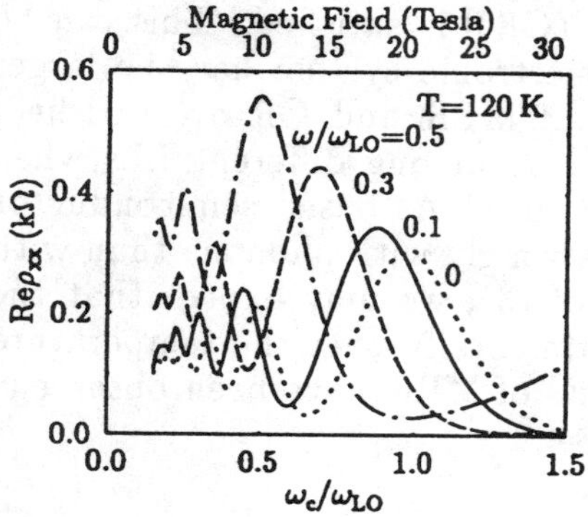

Figure 1: Reρ_{xx} as a function of magnetic field for different radiation frequencies. $\omega_{LO} = 55.6$ THz for GaAs.

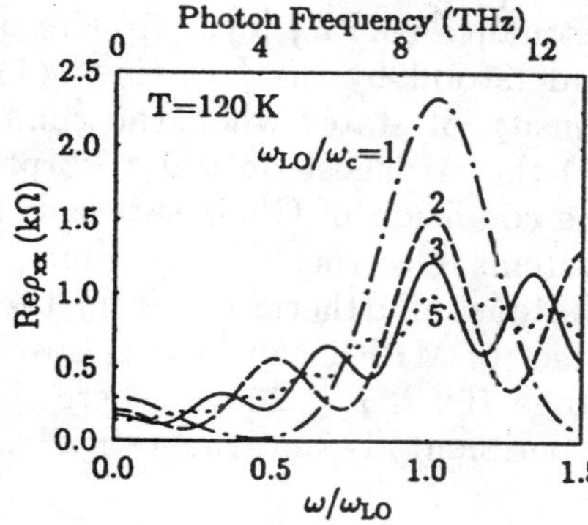

Figure 2: Reρ_{xx} as a function of radiation frequency for different magnetic fields. On the top, the photon frequency is shown by $f = \omega/2\pi$.

The results of this paper pertain to AlGaAs/GaAs heterojunctions with sample parameters: $n_e = 2 \times 10^{15}$ m^{-2} and $\mu_0 = 2$ m^2/Vs. The material parameters are taken from those for GaAs. From Eq. (1) we see that: i) $\rho_{xy} = B/n_e e$; ii) Im$\rho_{xx} = -m^*\omega/n_e e$; and iii) Re$\rho_{xx} = (B/n_e e)(\omega_c \tau)^{-1}$. Therefore, only Re$\rho_{xx}$ is of interest in studying the frequency-dependent properties.

Reρ_{xx} as a function of magnetic field (photon frequency) is shown in Fig. 1 (Fig. 2) at a fixed temperature T=120 K for different radiation frequencies (magnetic fields). We see that the MPPR effects may be observed by measuring the dependence of the d.c. resistivity and the optical absorption coefficient on B and ω. The peaks of Reρ_{xx} appear at the MPPR conditions. We note that: 1) with increasing ω, the corresponding MPPR peaks shift to the lower-B field. This implies that the MPPR effect can be observed at relatively low-B fields; 2) the number of MPPR peaks decreases with increasing radiation frequency, and the oscillation cannot be seen when $\omega > \omega_{LO}$ because the possibility of resonant scattering via photon-absorption and LO-phonon emission is suppressed; and 3) the amplitude of the MPPR peaks increases with ω. An important conclusion we draw from Fig. 1 is that a stronger resonant-absorption of higher-frequency EM radiation may occur at a lower magnetic field under the MPPR condition.

The results shown in Fig. 2 and obtained from further studies indicate that a stronger MPPR oscillation can be observed at magnetic fields [at radiation frequencies] where the condition of magneto-phonon resonance (MPR) [cyclotron resonance (CR)] is satisfied. This can be understood by the fact that: (1) the electronic system has the larger density of states when the conditions of MPR and CR are satisfied; (2) the strongest optical absorption occurs among different LLs when the condition of CR is satisfied; and (3) in GaAs-based semiconductor systems, electrons interact much more strongly with phonons than with photons. Furthermore, from the fact of (3), we may expect that the effect of MPPR can be measured experimentally over the temperature range $100 < T < 200$ K where the effects of MPR have been observed[2] experimentally and theoretically.

4. Summary

Optical and transport properties of THz-driven 2DEGs are a novel field in basic physics and in device applications. There are some recent experimental data in this field.[1] THz EM radiations give us a possibility to observe photon-induced quantum resonance effects. On the other hand, the investigation of electron (magneto-) transport in THz-driven electronic systems would be very beneficial to the design and application of novel electronic, photonic and optoelectronic devices. For example, the MPPR effect may be applied in frequency tunable FIR detectors and in the determination of the frequency of THz EM radiations. Finally, the effect proposed in this work needs experimental verification.

References

1. N.G. Asmar *et al.*, *Phys. Rev.* **B 51**, 18041 (1995); *Appl. Phys. Lett.* **68**, 829 (1996).
2. D.R. Leadley *et al.*, *Phys. Rev. Lett.* **73**, 589 (1994); W. Xu *et al.*, *Int. J. Mod. Phys.* B **10**, 169 (1996).
3. W. Xu *et al.*, *Phys. Rev.* **B 48**, 1562 (1993).
4. C.S. Ting *et al.* *Phys. Rev.* **B 16**, 5394 (1977).
5. A detailed derivation of Eq. (1) and τ induced by electron–LO-phonon interactions will be presented elsewhere.
6. W. Xu and C. Zhang, *Appl. Phys. Lett.* **68**, 3305, (1996).

THE PHONON-INDUCED DRAG OF 2D HOLES IN MAGNETIC FIELDS

D. LEHMANN
Institut für Theoretische Physik, TU Dresden, D-01062 Dresden, Germany

A.J. KENT, R.E. STRICKLAND, K.R. STRICKLAND, M. HENINI
Department of Physics, University of Nottingham, Nottingham NG7 2RD, England

CZ. JASIUKIEWICZ, T. PASZKIEWICZ
Institute of Theoretical Physics, University of Wroclaw, 50-204 Wroclaw, Poland

The magnetic field dependence of the phonon-drag current induced by acoustic phonons has been studied in 2D hole gas (2DHG) systems, formed in GaAs/AlGaAs heterostructures. The drag effect has been measured as a function of the phonon source position for different values of magnetic field applied perpendicular to the 2DHG. A theoretical model on the basis of a generalized quantum Langevin equation approach describing the response of the 2DHG due to the phonon pulse has been developed. Corresponding numerical simulations are found to be in good agreement with the experimental patterns.

1 Introduction

It is now well established that acoustic phonons can be used as a powerful probe of two-dimensional carrier gases. Phonon drag imaging is a particularly useful technique because it gives information regarding the carier phonon coupling as a function of both the phonon wavevector direction and its magnitude. To understand the images it is necessary to develop detailed theoretical models of the coupling including the effects of acoustic anisotropy. Without such models the experimental results can only be explained intuitively. In this paper we extend our previous work [1,2] on the phonon drag of 2D holes in GaAs heterostructures to include the situation where a quantising magnetic field is applied normal to the 2DHG.

2 Experiment

The experimental sample was based on a p-type GaAs/AlGaAs heterojunction grown by MBE on a 2mm thick [001] semi-insulating GaAs wafer. The 2DHG density was $3.3 \times 10^{15}\mathrm{m}^{-2}$ and the 4.2K mobility was $5\mathrm{m}^2\mathrm{V}^{-1}\mathrm{s}^{-1}$. An active area (120$\mu$m long and 50$\mu$m wide) was defined by etching and a pair of large-area ohmic contacts fabricated at its ends. On the opposite, polished, side of the GaAs substrate a 100nm thick film of CuNi was deposited by vacuum evaporation. The sample was mounted in an optical access cryomagnetic system in which it was cooled to 1.5K and a field of up to 7T could be applied in a direction perpendicular to the 2DHG. Short, 20ns, pulses of nonequilibrium acoustic phonons with an approximately Planckian frequency spectrum were generated by thermalising the light pulses from a Q-switched Nd-YAG laser in the CuNi film. The laser beam was focussed to a spot about 25μm across and this could be raster scanned over the metal film using computer controlled mirrors. The phonon drag current was detected using a low input impedance amplifier and was recorded as a function of phonon source (laser spot) position.

3 Theory

The response of the 2DHG due to the phonon pulse - the drag current $\vec{J}(t)$ - is proportional to the velocity of the center of mass of the holes. To describe this motion we consider a (quasi-) 2DHG (N holes with effective mass m^*) interacting with phonons and impurities in the presence of a perpendicular magnetic field. The Hamiltonion in the center-of-mass and relative hole description is

$$\begin{aligned} H &= \frac{1}{2Nm^*}[\vec{P} - Ne\vec{A}(\vec{R})]^2 + V(R_z) \\ &+ \sum_{m,n,k} \varepsilon_{mn} c^+_{mnk} c_{mnk} + \frac{1}{2}\sum_{\vec{q}} v(\vec{q})(\rho_{-\vec{q}}\rho_{\vec{q}} - N) + \sum_{Q} \hbar\omega_Q b^+_Q b_Q \\ &+ \sum_{Q} \frac{1}{\sqrt{V}} h(Q) e^{i\vec{q}\vec{R}}(b_Q + b^+_{-Q})\rho_{-\vec{q}} + \sum_{\vec{q}} u(\vec{q}) e^{i\vec{q}\vec{R}} \rho_{-\vec{q}} \, , \end{aligned} \tag{1}$$

where $\vec{P}$ and $\vec{R}$ are the center-of-mass momentum and coordinate, respectively, $V(R_z)$ is the confinement potential of the holes perpendicular to the 2DHG, $\vec{A}(\vec{R})$ is the vector potential. The terms in the second line represent the motion of the holes relative to the center of mass coordinate (including hole-hole interaction) and the phonons, where c^+_{mnk} and c_{mnk} are creation and annihilation operators for relative holes in the m-th sub-band due to confinement in z-direction and in the Landau level (n, k) due to magnetic quantization, ε_{mn} is the Landau level energy, $\rho_{\vec{q}}$ is the hole-density operator and $v(\vec{q})$ is the Fourier transform of the hole-hole Coulomb interaction. b^+_Q and b_Q are the phonon operators. Q stands for the phonon wave vector $\vec{q}$ and the polarization λ, ω_Q is the phonon frequency. The only coupling between the center of mass and the relative holes and phonons is due to the hole-impurity (h-im) and hole-phonon (h-ph) interaction, where $u(\vec{q})$ is the h-im potential and $h(Q)$ the h-ph matrix, which includes both deformation potential and piezoelectric coupling.

The equation of motion for the center of mass follows from (1)

$$N\left(m^*\ddot{\vec{R}}-e(\dot{\vec{R}}\times\vec{B})\right)=-i\sum_{\vec{q},j}\vec{q}e^{i\vec{q}\vec{R}}\left\{\frac{h(Q)}{\sqrt{V}}(b_Q+b^+_{-Q})\rho_{-\vec{q}}+u(\vec{q})\rho_{-\vec{q}}\right\}\ . \tag{2}$$

The terms on the right-hand side of (2) are the forces exerted on the center of mass due to h-ph and h-im interaction, repectively. Using a procedure of Hu and Connel [3] to find explicit expressions for the relative hole and phonon operators we can recast (2) into a quantum Langevin equation form which seperates the drag forces $\vec{F}$ and the frictional forces ($\sim\dot{\vec{R}}$). Thus the r.h.s of (2) can be written in the form $\vec{F}_{ph}(t)+\vec{F}_{im}(t)-\int_{-\infty}^{t}M(t-t')\dot{\vec{R}}(t')dt'$. Taking the averages over impurities F_{im} will vanish and only the absorption part of F_{ph} remains. Finally we get for the time integrated current

$$\int_{-\infty}^{\infty}dt\vec{J}(t)=\frac{\mu}{1+(\mu B)^2}\left\{\left\langle\vec{\mathcal{F}}^{abs}_{ph}(0)\right\rangle+\mu\left(\left\langle\vec{\mathcal{F}}^{abs}_{ph}(0)\right\rangle\times\vec{B}\right)\right\}\quad , \tag{3}$$

where $\mu=eN/<\mathcal{M}(0)>$ is the mobility and $<\vec{\mathcal{F}}(\omega)>$ and $<\mathcal{M}(\omega)>$ are the zero temperature averages of the Fourier transforms of $\vec{F}(t)$ and $M(t)$, respectively. We calculated the drag force $<\vec{\mathcal{F}}^{abs}_{ph}(0)>$ in RPA neglecting all higher order phonon processes and all mixed phonon-impurity processes

$$\left\langle\vec{\mathcal{F}}^{abs}_{ph}(0)\right\rangle=-i\frac{1}{V}\sum_{\vec{q},j}\vec{q}\mathcal{N}(0)\mid h(Q)\mid^2\ \frac{\chi^{(0)}(\vec{q},-\omega_Q)}{1-\sum_{\vec{q}',j'}v(q')\chi^{(0)}(\vec{q}',-\omega_{Q'})}\ .$$

$\mathcal{N}(\omega)$ is the Fourier transform of the solution of the kinetic equation for the phonon distribution function [2] and $\chi^{(0)}(\vec{q},\omega_Q)$ is the multi-subband polarizability for the quasi-2DHG in the presence of a perpendicular magnetic field taking into account all the intra and inter Landau level transitions.

4 Results and conclusions

In figure (1) the experimental and theoretical drag current images at $B=0$; 0.25; and 4.71T are compared. The important result is the apparent anticlockwise "rotation" of the axis about which the drag signal reverses polarity in the experimental and theoretical pictures. To see how this comes about one must remember that the pattern obtained in a field is the superposition of two components: The longitudinal one due to phonon momentum along the line joining the contacts and the transverse one due to phonon momentum perpendicular to the same line. Their ratio is given by the factor μB. At $B=0.25$ T,

$\mu B \approx 1$ and both components contribute equally giving a "rotation" of about $45°$ in both patterns. At $B = 4.71$T, $\mu B \gg 1$ and the transverse component dominates giving a theoretical rotation of $90°$. Strong evidence of this is seen in the experimental pattern for 4.71T, however, it is severely affected by noise. This is because the drag signal is proportional to the pre-factor $[1 + (\mu B)^2]^{-1}$.

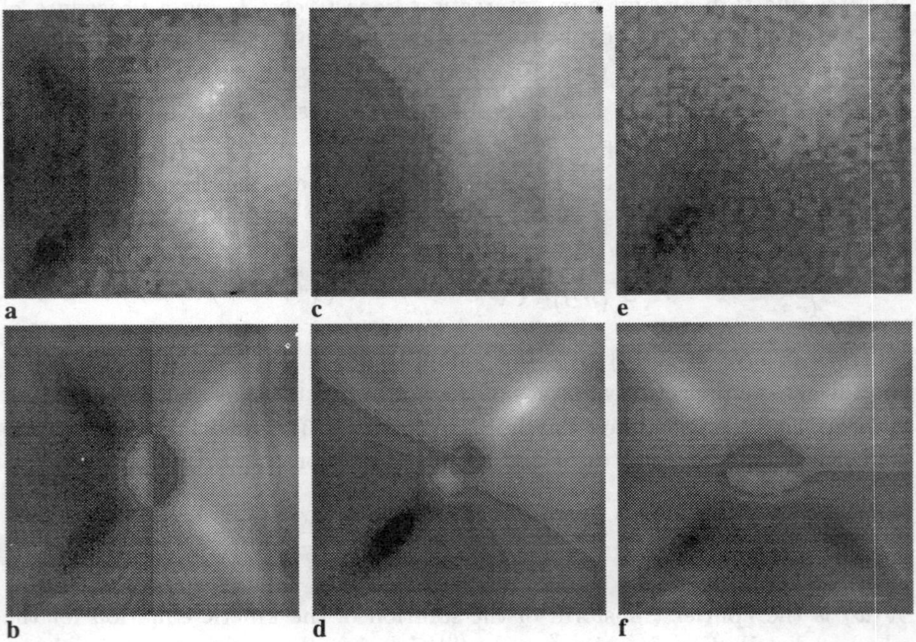

Figure 1: Comparison of experimentally obtained patterns (a,c,e) with the calculated ones (b, d, e) for different values of magnetic field: B=0T (a,b), B=0.25T (c,d), B=4.71T (e,f). The black and white regions correspond to opposite polarities of the drag current.

We would like to thank EC for supporting this work under contract nos. CHRX-CT94-0548 and CIPA CT93-0070.

References

1. D. Lehmann,Cz. Jasiukiewicz, R.E. Strickland, K.R. Strickland, A.J. Kent and T. Paszkiewicz, *Physica* B **219&220**, 25 (1996).
2. W.M. Gancza, Cz. Jasiukiewicz, A.J. Kent, D. Lehmann, T. Paszkiewicz, K.R. Strickland, R.E. Strickland, *Semic. Sci. Technol.* **11**, 1030 (1996).
3. G.Y. Hu and R.F. O'Connell, *Physica* A **151**, 33 (1988).

MODIFICATION OF THE 2D ELECTRONIC PROPERTIES IN Si-δ-DOPED InSb DUE TO SURFACE EFFECTS

A. DE KEYSER, R. BOGAERTS, L. VAN BOCKSTAL, F. HERLACH
Laboratorium voor Vaste-Stoffysica en Magnetisme, Katholieke Universiteit Leuven, B-3001 Leuven, Belgium

V.C. KARAVOLAS, F.M. PEETERS
Dept. Natuurkunde, Universitaire Instelling Antwerpen, B-2610 Antwerpen, Belgium.

W. VAN DE GRAAF, G. BORGHS
Interuniversitair Micro-Electronica Centrum, B-3001 Leuven, Belgium

A magnetotransport study of single Si-δ-layers in InSb has been performed in order to investigate the characteristics of the two-dimensional conduction channel in these structures as a function of the distance between the δ-layer and the sample surface. When the δ-layer is positioned close to the surface, the transport properties show little sensitivity to the doping characteristics. The transport can be described in terms of a two-dimensional electron gas (2DEG) in the delta-layer with a small contribution from bulk electrons. If the δ-layer is placed further away from the surface, the transport properties change drastically. Plateau-like features in ρ_{xx} are accompanied by very strong oscillatory behaviour in ρ_{xy}. This shows a strong resemblance to the effects observed in nearly compensated GaSb-InAs quantum wells where two-dimensional electrons and holes form parallel conduction channels.

1 Introduction

In recent years, considerable effort has been directed at the development of semiconductor structures in which the composition and doping density is controlled on an atomic scale. Among these, δ-doped III-V semiconductors are of particular interest for practical applications. At IMEC, Si-δ-layers in InSb with a typical doping width of 50 Å were grown heteroepitaxially on GaAs. This gives rise to a 2DEG which may be accompanied by residual bulk conduction [1]. Due to the low effective mass of the electrons, the confining potential well is sharp at the bottom, spreading out rather wide ($\sim$ 0.5 μm) at the top. To reduce the influence of defects arising from the GaAs-InSb interface, a 3 μm InSb buffer layer was grown underneath. The δ-layer was covered by a 40 nm, 120 nm and 800 nm InSb capping layer for samples G1566, G1567 and G1568, respectively. This system has been studied by means of magnetotransport experiments in pulsed fields up to 50 T at temperatures from 4.2 K to 1.4 K.

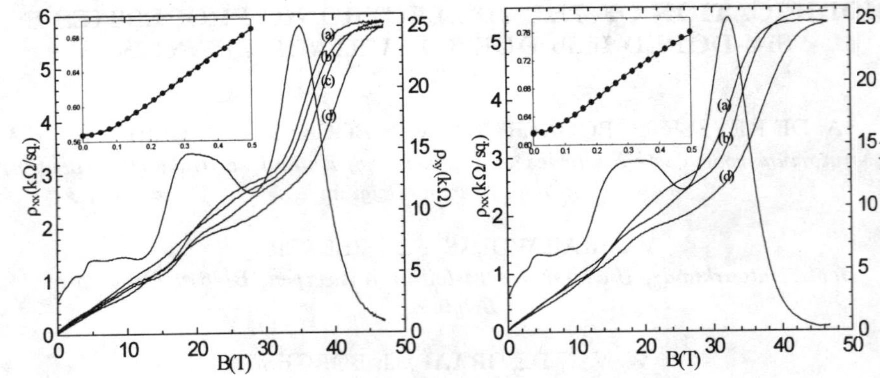

Figure 1: Magnetoresistance and Hall effect at 4.2 K for samples G1566 and G1567. The Hall curve is shown for various angles : (a) $\theta = 0^o$, (b) $\theta = 15^o$, (c) $\theta = 25^o$, (d) $\theta = 35^o$; ρ_{xx} is given for $\theta = 0^o$. The inset shows the low-field Drude fit (filled circles) to ρ_{xx}.

2 Experimental results and discussion

When the δ-layer is located near the sample surface, the transport coefficients show 2D Shubnikov-de Haas oscillations at magnetic fields below 20 T. From the slope of the low-field Hall effect, a total sheet carrier density of $n_{tot} = (1.7 \pm 0.1) \cdot 10^{12}$ cm^{-2} can be deduced for sample G1566. This is in agreement with the densities obtained from a simple 2-carrier Drude fit to the low-field magnetoresistance :

$$\rho_{xx} = \left(\frac{n_\delta e \mu_\delta}{1 + (\mu_\delta B)^2} + \frac{n_{3D} e \mu_{3D}}{1 + (\mu_{3D} B)^2}\right)^{-1}, \tag{1}$$

taking into account 2D electrons from the δ-layer with carrier density n_δ and mobility μ_δ along with 3D electrons from the bulk (n_{3D}, μ_{3D}), which yields $n_\delta = (1.54 \pm 0.06) \cdot 10^{12}$ cm^{-2}, $\mu_\delta = (0.62 \pm 0.03)$ m^2/Vs, $n_{3D} = (3.3 \pm 0.6) \cdot 10^{10}$ cm^{-2}, $\mu_{3D} = (4.1 \pm 0.3)$ m^2/Vs. The mobility of the 3D electrons is comparable to the mobility deduced from FIR cyclotron resonance experiments [2], whereas the 2D electrons have a somewhat lower mobility due to ionised impurity scattering at the Si donor atoms.

At high magnetic fields, a quantum Hall plateau at filling factor $\nu = 1$ is observed along with its corresponding ρ_{xx} minimum near 50 T. As the plateau is not completely resolved and the ρ_{xx} minimum is situated above the available field range, only a rough estimate of the 2D carrier density can be made from the assumption that it is centred around 55 T, which yields $n_\delta \approx 1.2 \cdot 10^{12}$ cm^{-2}. In the intermediate field range, a weakly resolved $\nu = 2$ quantum Hall

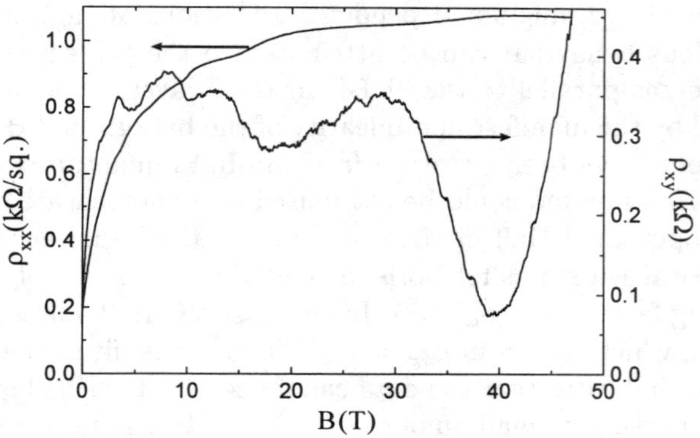

Figure 2: Magnetoresistance and Hall effect at 4.2 K for sample G1568.

plateau appears around 28 T in the $\theta = 0^o$ configuration. As the angle is increased, this feature gradually disappears while a $\nu = 3$ plateau is clearly developing. This is a result of the degeneracy of Landau levels originating from different subbands at the Fermi level, which masks the $\nu = 3$ plateau in the $\theta = 0$ configuration. When the magnetic field is rotated, the in-plane component of the field causes a diamagnetic shift of the subband energy levels, lifting this degeneracy. By the same token, the $\nu = 2$ plateau is suppressed.

For sample G1567, the situation is almost identical. The total sheet carrier density of $(1.6 \pm 0.1)\cdot 10^{12}$ cm^{-2}, derived from the low field Hall effect, is slightly lower than for G1566 and in agreement with the Drude fit: $n_\delta = (1.48 \pm 0.07)\cdot 10^{12}$ cm^{-2}, $\mu_\delta = (0.57 \pm 0.04)$ m^2/Vs, $n_{3D} = (3.7 \pm 0.5\ )\cdot 10^{10}$ cm^{-2}, $\mu_{3D} = (4.2 \pm 0.3)$ m^2/Vs. At high magnetic fields, this structure also shows a quantum Hall plateau at filling factor $\nu = 1$. The estimated centre of the plateau is near 50 T, corresponding to a total 2D density of $n_\delta \approx 1.1\cdot 10^{12}$ cm^{-2}. The reproducible discrepancy between the carrier densities obtained in low and in high magnetic fields can be attributed to magnetic freeze-out of most of the 3D and a limited fraction of the 2D electrons. Again the angular dependence of the Hall effect shows a transition from a $\nu = 2$ plateau near 27 T at $\theta = 0^o$ to a $\nu = 3$ plateau if θ is increased. The 2D properties of samples G1566 and G1567 are quite similar and are comparable to previously investigated single and multiple δ-layers. This confirms that at small distances between the δ-layer and the sample surface, the doping density and profile do not seem to have a strong effect on the observed carrier densities and mobilities.

For sample G1568, with the δ-layer at 800 nm from the surface, the data look drastically different. Hardly any quantum structure is observable in ρ_{xx}, but ρ_{xy} shows large, angular dependent oscillations at high magnetic fields. This anomalous behaviour can be attributed to the presence of an independent 2D hole gas parallel to the 2DEG in the δ-layer, an assumption that is corroborated by the manifest non-linearity of the background Hall curve. Similar phenomena have been observed in GaSb-InAs quantum wells, where the pronounced Hall minima could be attributed to a compensated quantum Hall effect [3]. Compensated Hall minima occur when the Fermi energy E_F is situated in the localised states for both the electron and hole system, yielding an effective filling factor $(\nu_e - \nu_h) = 0$. In this case the Hall voltage is completely compensated, which results in $\rho_{xy} = \sigma_{xy} = 0$. A Drude fit to the low field magnetoresistance indicates that the data can indeed be described qualitatively by a system comprising a small amount of bulk electrons and comparable densities of 2D holes and 2D electrons. In this structure, the 2D hole gas is quite likely to be situated at the sample surface, since the δ-layer is buried rather deep within the sample and epitaxial InSb has been reported to form a 2DHG at the surface [4]. For systems with a δ-layer near the surface, recombination of these p-type states with electrons furnished by the δ-layer is suggested as the mechanism that averages the effective 2D properties for structures with different doping characteristics.

Acknowledgments

ADK is supported by the Flemish Institute for Scientific and Technological Research in Industry (IWT). RB and LVB are Senior Research Assistants and FMP is a Research Director of the Flemish National Fund for Scientific Research (NFWO) which also provided funding for this research project. VCK is supported by a TMR-fellowship of the European Union.

References

1. A. De Keyser, R. Bogaerts, V.C. Karavolas, L. Van Bockstal, F. Herlach, F.M. Peeters, W. van de Graaf and G. Borghs, *Solid State Electronics* **40**, 1-8, 395-398 (1996).
2. L. Van Bockstal, M. Mahy, A. De Keyser, W. Hoeks, F. Herlach, F. M. Peeters and G. Borghs, *Physica B* **211**, 466-469 (1995).
3. K.S.H. Dalton, R. Bogaerts, R.W. Martin, M. Lakrimi, D.M. Symons, R.J. Warburton, R.J. Nicholas, F. Herlach, N.J. Mason and P.J. Walker, *Physica B* **184**, 202-205 (1993).
4. J. Wagner, A.-L. Alvarez, J. Schmitz, J.D. Ralston and P. Koidl, *Appl. Phys. Lett.* **63** (3), 349-351 (1993).

APPLICATION OF MAGNETO-CAPACITANCE TO THE STUDY OF NARROW GAP SEMICONDUCTORS

JUNHAO CHU, KUN LIU, D. Y. TANG

National Laboratory for Infrared Physics,
Shanghai Institute of Technical Physics, Academia Sinica,
420 Zhong Shan Bei Yi Road, Shanghai 200083, China

ABSTRACT

Magnetic field swept flat-band capacitance measurement and non-flat-band magneto-capacitance measurement are introduced to study the properties of narrow gap semiconductors. Typical experimental results of the non-flat-band and flat-band magneto-capacitance spectra for an n-type InSb sample have been presented and discussed which include electron Landau levels, hole quantization in the surface layer, and SdH oscillation, resonant defect states, carrier freeze-out, metal-insulator transition, etc. in the bulk.

1. Introduction

Since the non-flat-band magneto-capacitance measurement[1] was first used to observe electron quantization in surface layer this method has been widely used to study fundamental physics in low-dimensional electron systems[1-4]. Studies have shown that non-flat-band magnetocapacitance is very powerful in studies of the properties of low dimensional systems. Recently, magnetic field swept flat-band capacitance measurements[5] have been performed on narrow gap semiconductors, and it has been found that the flat-band magneto-capacitance method can be used to study magneto-transport properties of bulk semiconductors. In this paper both the non-flat-band and flat-band magneto-capacitance methods are applied to narrow gap semiconductors (NGS). In section 2 electron Landau levels and hole quantization in the surface layer of an n-type InSb sample were measured using the non-flat-band magneto-capacitance method. In section 3 magneto-transport properties of bulk of an n-type InSb sample were measured using the flat-band magneto-capacitance method.

2 Application of the Non-Flat-Band Magnetocapacitance

2.1 Electron Landau Levels

When an external magnetic field is applied perpendicular to the channel (configuration of 0°) of a gate-biased metal-insulator-semiconductor (MIS) sample, each subband in the channel is split into Landau levels. According to the resistive plate capacitor model and the DOS model[6-7], the capacitance of the MIS system reaches zero when the Fermi level lies in the centeral region between two Landau levels where the DOS is low and the states are localized. When the Fermi level lies at Landau levels where the DOS has a peak value and the states are extended, the capacitance of the system reaches a peak value. Therefore, oscilla-tions can be expected in the magneto-capacitance spectra of the MIS system when Landau levels are lifted over the Fermi level one by one by changing the gate bias or magnetic field.

Fig.1 is the magneto-capacitance measured for an n-type InSb MIS sample (N_{AD}(4.2K)=1.5x10^{14}cm^{-3}) in the configuration of 0° and different fields from 0 to

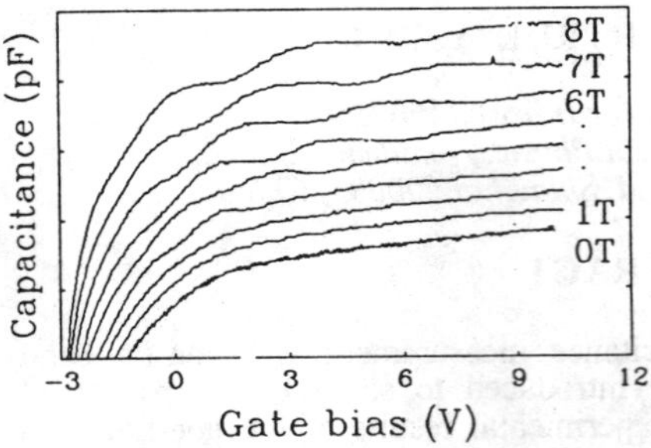

Fig.1 Magneto-capacitance for an n-type InSb MIS sample in accumulation layer at 1.2K

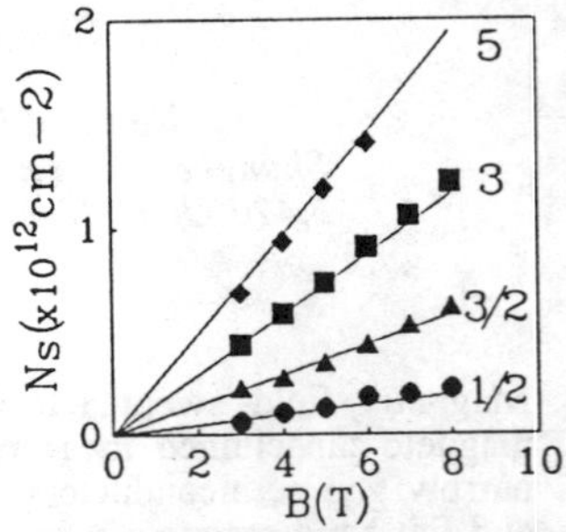

Fig.2 Electron density in the accumulation layer of the InSb sample.

8T. Capacitance oscillations have been clearly measured though they are step-like. The step-like oscillations may be attributed to non-uniform properties and edge states in the system[9]. The configuration dependence of the oscillation also has been clearly measured. This meant the capacitance oscillations arose from the Landau levels formed in the accumulation layer of the MIS sample. The onset voltage of the accumulation layer is obtained at about -3.2V from the fan chart of peak capacitance voltage. The plots in Fig.2 are the magnetic-field dependent electron density in the channel obtained from the magneto-capacitance, and the lines are theoretical results. With the quantum numbers of 1/2, 3/2, 3, 5, etc., the experimental data agreed quite well with theory. The quantum numbers of 1,2,4...have not been measured in experiments. The reason for this may be that there are resonant defect states (RDS) in the conduction band. Interaction of electrons at different Landau levels also may be responsible for the observation. The Landau level fan chart can be obtained by calculating the band-bending of the MIS sample at different peak voltages. The calculated results show that the Landau levels extrapolute to 20meV in zero field, which was equal to the ground energy of electron subbands in the accumulation layer.

2.2 Observation of Hole Quantization in Hole Channel

Fig. 3 is the magneto-capacitance of the InSb MIS sample in the p-type inversion region, which was measured in the configuration of 0° and different magnetic fields from 0 to 8T at 1.2K. With increasing magnetic field the inversion onset voltage moved to a relatively negative gate bias, which was attributed to the increase of the hole quantization energy with increasing magnetic field. By calculating band-bending of the system hole

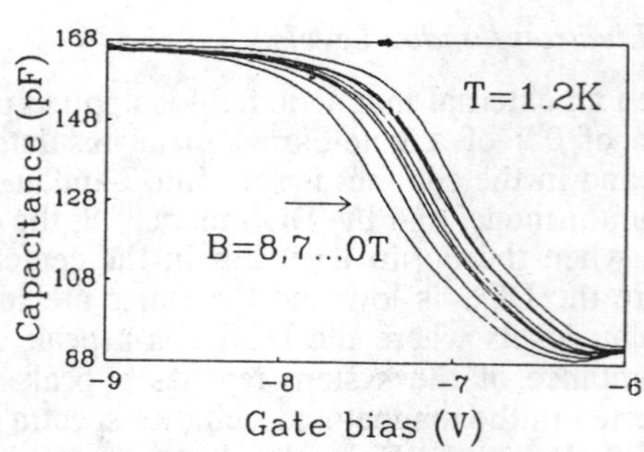

Fig.3 Manneto-capacitance of the InSb sample in p-type inversion layer at 1.2K.

quantisation can be studied.

3 Magneto-Transport Properties From Magneto-capacitance Spectroscopy

When Hall measurements are used to study magneto-transport properties of NGS, the influences of measuring circuit shortened by surface conducting layer, the field dependent carrier mobility and the geometry correction factor [10-12] should be considered. However, these influences can be avoided in the magneto-capacitance measurement. As we know the surface flat-band capacitance of a semiconductor, C_{FB}^s, is related to the free carrier concentration N_e in the semiconductor as described by following equations:

$$C_{FB}^{S} = \sqrt{2}\varepsilon_0\varepsilon_S / L_D \qquad (1)$$

$$L_D = \sqrt{2\varepsilon_0\varepsilon_s / e^2 N_e} \qquad (2)$$

Where L_D is the Debye length. Thus, the magnetic-field-dependent free carrier concentration $N_e(B)$ of a NGS sample can be obtained directly from the flat-band magneto-capacitance C_{FB}^s (B) without considering the above influences, and magneto-transport properties can be studied using proper theories.

The result shows a picture of the well-known magnetic-field-induced carrier freeze-out in semiconductors. In addition, new characteristics also have been measured that include SdH oscillations in the low field region, an unstable structure at B_R (=1.76T), the decrease of N_e in a few consecutive regions of CD, DE and EF, the unchangable behavior of N_e in FG region, etc..

The oscillations at B_0 and B_1 in the low field region can be attributed to the SdH oscillations in the bulk of the InSb sample. From the oscillation cycle the free electron concentration in the sample was measured at about $2.9\times10^{15}cm^{-3}$, which was in fair agreement with those obtained from the Hall measurement and the magneto-capacitance (see Fig.4). The unstable structure at B_R can be attributed to a donor type resonant defect state (RDS) in the conduction band in zero field[13]. With magnetic field increased to B_R, the RDS level was lowered into the forbidden gap and the RDS became a non-intrinsic shallow donor state. As a result, the transfer of electrons occured among the conduction band, the RDS-born non-intrinsic shallow donor states, and intrinsic shallow donor states. So, the unstable structure was measured. By calculating the movement of the conduction band edge with magnetic field the RDS was found lying at about 17meV above the band edge in zero field.

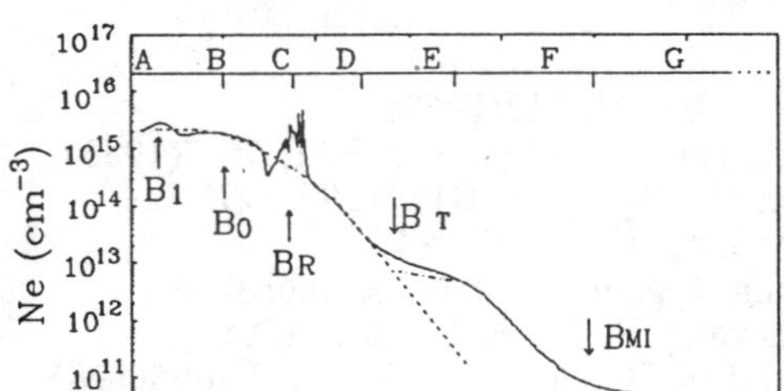

Fig.4 Magnetic field dependent electron concentra-tion in an n-type InSb sample measured from the flat-band magneto-capacitance

In the region with field larger than B_0, the InSb sample entered the quantum limit where only the last Landau subband was filled with electrons and the intrinsic shallow donor level was well defined in the forbidden gap. With increasing magnetic field the free electrons in the conduction band recombined from the lower Landau subband to the shallow donors due to both the decrease of donor radius and the increase of donor activation energy. Thus, the magnetic-field-induced carrier freeze-out was measured in the CF region.

In the region with field lower than B_0, the InSb sample stayed metallic and the conduction process was dominated by free electrons in the conduction band. In the CD and EF regions the conduction process was mainly dominated by the electron activation from the RDS-born non-intrinsic and intrinsic shallow donors to the conduction band, respectively. In the DE region the conduction process was associated with both the activation processes. The DE region was a transitional region from one conduction process to the other one, B_T was the transition field. Activation energies of the RDS-born non-intrinsic and intrinsic shallow donors in the CD and EF regions have been calculated, the results show that the activation energy of intrinsic shallow donors obtained in this work has a better agreement with the YKA theory[14] in higher field than in lower field. One reason for this is that shallow donors can be treated as hydrogen ones in higher field than in lower field. The screening effect of donor centers and the energy spreading of shallow donor levels also may be responsible for the results.

In the FG region N_e was so the small that InSb sample was insulating, donors were localized in the sample and the conduction occured by hopping. Calculation has shown that the hops were correlated. The metal-insulator transition field B_{MI} from the CF to FG region was measured as 5.0T. At B_{MI} the characteristic sizes of electronic wave-functions of shallow donors parallel and perpendicular to the applied field, $a_{//}$ and $a_{\perp}$, in the sample were about 176Å and 219Å, respectively. Thus, a value of about 0.309 was obtained for the Mott criterion using $n_0 a_{//} a_{\perp}^2 = \delta^3$ [10]. This value was in fair agreement with the theoretical one(0.30) and experimental ones(0.301)[15].

References:

1. M. Kaplit and J. N. Zemel, Phys. Rev. Lett. 21, 212(1968).
2. R. K. Goodall, R. J. Higgins, and J. P. Harrang, Phys. Rev. B31, 6597(1985).
3. S. V. Kravchenko, J. M. Caulifield et al., Phys. Rev. B47, 1296(1993-I).
4. S. Takaoka, K. Oto, et al., Phys. Rev. Lett. 72, 3080(1994).
5. Kun Liu, Junhao Chu, Dingyuan Tang, J. Appl. Phys. (to be published).
6. R.K.Goodall,R.J.Higgins, and J.P.Harrang, Phys.Rev B31,6597(1985);
7. T.P.Smith,B.B.Goldberg,P.J.Stiles,and M. Heiblum,Phys.Rev.B32,2696(1985)
9. S. Takaoka, K.Oto, et al., Phys. Rev. Lett. 72(19), 3080(1994)
10. M. Shayegan, et al, Phys. Rev. B38, 5585(1988-I)
11. S. N. Song, J. B. Ketterson, et al., Appl. Phys. Lett. 63(7), 964(1993)
12. J. Gebhardt, G. Nimtz, Solid State Commun. 63, 573(1987).
13. A. Raymond, et al., Lecture Notes in Physics 152, 387(1981).
14. Y. Yafet, R. W. Keyes, E. N. Adams, J. Phys. Chem. Solids 1, 137(1956).
15. U. Zeitler, A. G. M. Jansen et al., , J. Phys.: Condens. Matter 6, 4289(1994).

LARGE NEGATIVE MAGNETORESISTANCE IN THE INSULATING REGIME OF A TWO DIMENSIONAL ELECTRON GAS WITH AN ARTIFICIAL POTENTIAL

G.M. Gusev,[1] X. Kleber,[2,3] U. Gennser,[4] D.K. Maude,[2] J.C. Portal,[2,3]
D.I. Lubyshev,[1] P. Basmaji,[1] M. P.A. da Silva,[1] J.C. Rossi,[5]
and Yu.V. Nastaushev [6]

[1]Instituto de Fìsica de São Carlos, 13560-970, Universidade de São Paulo, SP, Brazil
[2]CNRS-LCMI, F-38042, Grenoble, France
[3]INSA-Toulouse, 31077, France
[4]Paul Scherrer Institute, CH-5232 Villigen-PSI, Switzerland
[5]Universidade Federal de São Carlos, Brazil
[6]Institute of Semiconductor Physics, Novosibirsk, Russia

The large negative magnetoresistance in a two-dimensional electron gas (2-DEG) with a periodic lattice of antidots has been studied in the regime, when the 2-DEG breaks up into separate lakes (dots), and transport is dominated by thermally activated tunnelling between dots through the constriction between antidots. The results are compared with the theory of Raikh and Glazman, who considered the increase in the transmission coefficient through the barrier due to the change of the phase factor of the electron wave function localized in the dot.

When the fluctuations of the smooth, random impurity potential are sufficiently strong, a two-dimensional electron gas (2 DEG) will break up into separate lakes (or dots) each with many electrons. In this regime the electron transport occurs via tunnelling or thermal activated hopping of electrons through the saddle points of the barriers separating adjacent dots. Recently a large negative magnetoresistance was predicted in this regime [1]. It was argued, that the transmission coefficient through the barrier increases with magnetic field due to a change in the phase factor of the electron wave function localised in the dot. A negative magnetoresistance in the hopping regime has been observed in GaAs/AlGaAs gated heterostructures [2] and δ-doped GaAs layers [3]. However, because the parameters of the smooth potential in these structures are not known, it is difficult to interpret the results as a clear manifestation of the mechanism proposed by Raikh and Glazman. The lattices of potential hills, or antidots, superimposed on a 2 DEG, presents an alternative system for studying the magnetoresistance in the insulating regime. The smooth random impurity potential is replaced by the regular antidots, which offers the possibility to control the electron tunnelling by applying a gate voltage to the antidots. In such system the regions between neighbouring antidots can be squeezed, regulating the barrier between antidots.

When this barrier is higher than the Fermi level in electron puddle, transport occurs via the hopping of electrons between adjacent lakes. Another advantage is that the geometric size of dots and parameters of saddle potential are well known, or can be recalculated from the conductivity at zero magnetic field.

The investigated structures consist of GaAs/AlGaAs double quantum wells, when only lowest symmetric level is occupied. The 2 DEG has a density of 4 x $10^{11} cm^{-2}$ and a mobility of 70- $100x10^3$ cm^2/Vs. A lateral superlattice (periodicity 0.4 - sample A, and 0.5 μm - sample B) containing a macroscopic number of antidots (10^5) was fabricated in the centre of the device between the potentiometric probes, using electron-beam lithography to pattern the PMMA resist, which was then covered by a gold gate. The physical diameter of the antidots is 0.1μm. The ballistic mean free path, 0.6-0.8μm, is larger than lattice periodicity. By applying a negative voltage, it is possible to create a periodical lattice of antidots without a significant decrease of the electron density in the centre regions in between each group of four antidots.

Fig. 1a shows the low field part of the magnetoresistance curves at high negative gate voltage (1.1-1.26 V). At stronger magnetic field the resistance reveals an exponential increase with B, which is a characteristic of a Hall insulator (not shown) [2, 3]. At low field the resistance decreases with increasing magnetic field up to 2 - 4 T, so the negative magnetoresistance is clearly seen. Starting from $|V_g|$ >1.15 V the Shubnikov oscillations disappear, and all curves have a minimum at 2 T. A large negative magnetoresistance is observed, the resistance decreases by a factor of ~ 4 with increasing B.

We have previously reported [4] the observation at $|V_g| < 1.1$ V of a transition from an insulator to a metallic (quantum Hall) state, also observed for GaAs gated heterostructures [2, 3]. At higher gate voltage an insulating behaviour was found in all regions of magnetic field. Fig. 1b shows the temperature dependence of the magnetoresistance curve at $V_g > -1.25$ V. With increasing temperature the resistance increases without significant change in the shape. This means, that relative change in the resistance does not depend on the temperature. The temperature dependence can be described by the equation $\sigma=\sigma_0 \exp(-T_0/T)^k$, where k=1/6 (fig 1c), rather than k=1/3, implied by the Mott hopping formula. A similar, k=1/6 law has also been reported for high mobility gated GaAs/AlGas heterostructures (without antidots), and as here a constant activation energy was not obtained [5]. Several theoretical models consider negative magneto-resistance in a 2 DEG. Weak localization theory predicts, that a magnetic field suppresses the quantum corrections to the conductivity, which leads to the negative magnetoresistance [6]. This magnetoresistance is clearly seen at low gate voltage and small (below 0.1 T) magnetic field (fig. 1a). We have compared it with theory [6] to obtain the phase coherent length L_φ=1.2μm at T=50mK. However, the large negative magnetoresistance at $V_g > 1.15$ V cannot be described by weak localization theory, even using a coherence length smaller than that deduced from the low voltage data. Another possibility for negative magnetoresistance is that of a point contact [7].

Our antidot lattice is fabricated between potentiometric probes, as in the point contact geometry. The 4-terminal resistance of a point contact is determined by

$R_L = h/2e^2\ (N\text{-}M)/NM$ (1),

where M is the number of channels transmitted through the point contact, and N is the number of channels in the bulk. In a magnetic field N is simply number of bulk Landau levels, which decreases with increasing magnetic field. Therefore, if M is constant, i.e. does not depends on B (for narrow point contact), the magnetoresistance is negative. However, because the regions between neighbouring antidots are squeezed, only a small number of the one dimensional channels passes through the barrier, so that N>>M. From eq. (1) is seen, that in that case the magnetic dependence of the resistance is very weak, in contrast to the observations (fig 1a, b). Recently it was demonstrated, that hopping conductivity should increase drastically with magnetic field, because the overlap between the wave functions on either side of barrier increases with B [1]. This overlap is very sensitive to the phase factor acquired in a weak magnetic field. The magnetoresistance between two lakes is given by

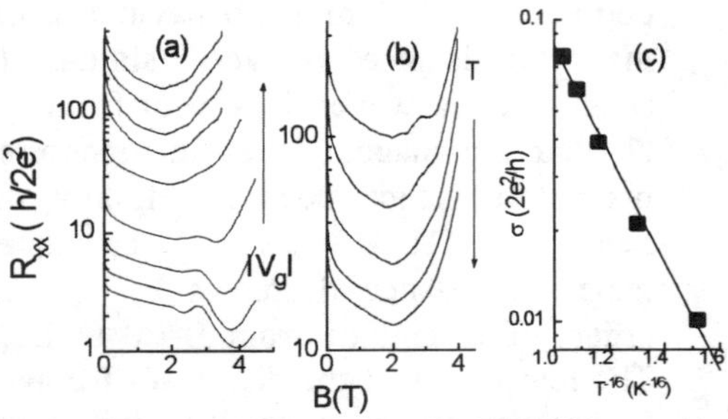

Fig. 1. Magneto-resistance of a 2 DEG with an antidot lattice for different values of the gate voltage (a) (T= 50mK, |Vg| = 1.1 - 1.26 V), and temperature (b) (|Vg| = 1.25 V, T = 80 - 800 mK) in the insulating regime. (c) The temperature dependence of the conductivity at B= 2T.

$$R(B)/R(0)= \exp(B^2/B_0^2)/\cosh^2(B/B_1) \qquad (2),$$

where the characteristic value $B_0 \approx \Phi_0 /(2\pi L\lambda)$, L is the dot radius, $\lambda=(\hbar/m\omega_{x,y})^{1/2}$, ω_x, ω_y are characteristics of the saddle potential, m is effective mass, $B_1 \approx \alpha\Phi_0 /(2\pi L^2)$, and α is a coefficient which depends on the difference in the geometry of the two lakes (see eq.(19) in reference [1]). From eq.(2) it follows, that the resistance falls off exponentially with B, reaches a minimum at $B_{min}=B_0^2/B_1$ (for $B_{min} >> B_1$), and increase sharply at higher field. The temperature independent behaviour of the minima agrees with our observations (fig. 1), and for reasonable values of the parameters B_0 and B_1 we find a good agreement with the experimentally found B_{min}. The area of dots can be obtained from the periodicity of the lattice and the size of antidots, the frequencies ω_x, ω_y from the experiments with single point contacts [7] ($\hbar\omega_x$=4.3 meV), and we assume, that the geometrical size can fluctuate by 30% from one dot to another, and, therefore $\alpha\approx 3$. Fig. 2a shows a number of curves, calculated from eq. (2) with small variation of the parameters B_0 and B_1. We see, that minimum of the curves is coincident with minimum of the magnetoresistance measured in the experiments.

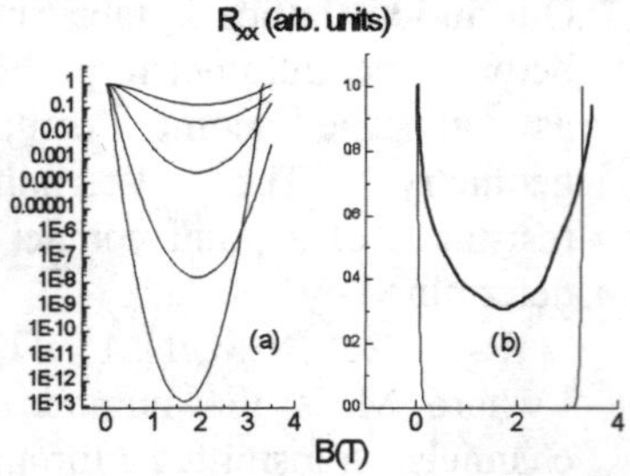

Fig. 2 (a). Theoretical curves of the magnetoresistance of a 2 DEG with an antidot lattice in the insulating regime (log-scale). Calculations were performed using eq. (2) for the following parameters (from top to bottom): B_0 (T): 1.1, 0.9, 0.62, 0.44, 0.3; B_1 (T): 0.6, 0.4, 0.2, 0.1, 0.054. (b). Experimental (thick solid line) and theoretical curve of the magnetoresistance (linear scale), for parameters realized in the experimental samples (see text): $B_0 = 0.3$ T, $B_1 = 0.054$ T.

We also find, that for the sample with periodicity d=0.5μm, the behaviour of the magnetoresistance is very similar to sample A. However, it is clear from fig. 2b that the magnitude of the measured magnetoresistance does not follow the model [1]. The calculated magnetoresistance decreases by 7 - 13 orders of magnitude from B= 0 to B_{min}. The reason is that the diameter of dots is larger than the characteristic size of the saddle point, and the resistance depends exponentially on this size. The theory, predicts an exponential sensitivity of the magnetoresistance to the dot size (fig. 2a). However, we believe that such large change in the resistance is somewhat unphysical, and other mechanisms must play an important role, reducing this exponential sensitivity. One notes, that the sample with a V_g adjusted to have a zero-field resistance of 10 MOhm would in this model still end up with a resistance much smaller than $h/2e^2$ at B_{min}, i.e. outside the hopping transport regime. and the regime of electron transport is already metallic rather than hopping. A more realistic case must be considered theoretically and would have to include the modification of the localized wave function in dots, with increasing overlap with the wave function in the adjacent dot.

Acknowledgements: We acknowledge M.E.Raikh and L.I.Glazman for helpful discussions. This work was supported by CNRS (France), CNPq, FAPESP (Brazil) and COFECUB. G.M.G. is supported by CNPq.

REFERENCES

1. M.E. Raikh and L.I. Glazman, Phys. Rev. Lett. 75, 128 (1995).
2. H.W. Jiang *et al.*, Phys. Rev. Lett. 71, 1439 (1993); T. Wang *et al.*, Phys. Rev. Lett. 72, 709 (1994).
3. R.J.F. Hughes *et al.*, J. Phys. Cond. Matter 6, 4763 (1994).
4. G.M. Gusev *et al.*, submitted to Sol. St. Commun. (1996).
5. G. Lutjering *et al.*, p.435, 11th ICEP2DS (Nottingham, 1995).
6. P.A. Lee and T.V. Ramakrishnan, Rev. Mod. Phys. 57, 287 (1985).
7. C.W.J. Beenakker and H. van Houten, in: Quantum transport in Semiconductor Nanostructures, Solid State Physics, v. 44 (Academic Press, 1991).
8. B.L. Al'tshuler and B.I. Shklovskii, Sov. Phys. JETP 64, 127 (1986).

MAGNETORESISTANCE OF FeSi IN THE REGION OF VARIABLE-RANGE HOPPING

J. BROTO, E.K. ARUSHANOV, H. RAKOTO, M. RESPAUD, J. LEOTIN
Service National des Champs Magnétiques Pulsés, UMS 819:CNRS-UPS-INSA, Complexe Scientifique de Ranguel, 31077 Toulouse-Cedex, France
E. BUCHER, CH. KLOC
Faculty of Physics, University of Konstanz, P.O. Box 5560, D-78434 Konstanz, Germany
K.G. LISUNOV
Institute of Applied Physics, Academy of Sciences of Moldova, 277028 Kishinev, Moldova

Investigations of the conductivity and the magnetoresistance (MR) have been performed in FeSi at temperatures 1.4 - 300 K in the magnetic field up to 35 T. Between ~4 - 16 K the Mott-type variable-range hopping conductivity is observed. MR exhibits a complex magnetic field and temperature dependence suggesting the positive and the negative contributions. To analyze the experimental MR data both the intra-site correlation and the spin-flip processes are taken into account in the positive component, and the Zeeman splitting of the mobility threshold in the negative contribution to MR.

The iron monosilicide, or FeSi, belongs to the hybridization-gap semiconductors, or Kondo insulators, and attracts considerable attention in recent time due to its unusual electric and magnetic properties, on one hand, and possible applications of the transition-metal silicides in microelectronic devices within existing silicon technology, on the other hand. The anomalous properties of FeSi at comparatively high temperatures, including strong increase of magnetic susceptibility above ~ 100 K and a semiconductor-to metal transition at $T \approx 200$ K, are explained by strong temperature dependence of the spin-density fluctuations.[1] However, the low-temperature properties of FeSi are less studied and still are subjects of controversy. Among them are the role of the intra-site correlation effects in the conductivity[2] and the origin of the Curie-Weiss-like susceptibility tails observed below $T \approx 70$ - 100 K.[2, 3]

In this paper the conductivity, σ, and magnetoresistance (MR) of FeSi are investigated between 1.4 - 300 K in the magnetic field up to 35 T. The experimental data are analyzed using a modified model of the density of localized states (DOS) in the forbidden band. The results of the MR measurements are discussed with different models including those which take into account the intra-site interaction between the localized electrons.

At relatively high temperatures, between $T \approx 100$ - 170 K, the conductivity of FeSi satisfies the Arrhenius law. This behavior is attributed to the intrinsic conductivity.[2] The value of the gap, $E_g = 50 \pm 5$ meV, is close to that found in Ref. 2. Below $T_v \approx 16$ K and

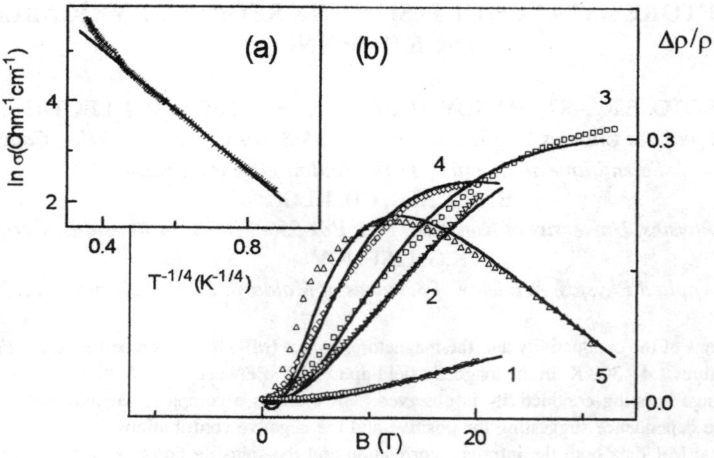

Figure 1. ln σ vs. $T^{1/4}$ (a) and Δρ/ρ vs. B at (1) 30 K, (2) 12 K, (3) 4.2 K, (4) 2.5 K and (5) 1.75 K (b). The solid lines are the fits to the experimental data as described in text.

down to $T_v^* \approx 4$ K the dependence $\ln\sigma \sim -(T_0/T)^{1/4}$ is observed (Fig. 1 (a)) suggesting the Mott-type variable-range hopping (VRH) conductivity with $T_0 =$ 1090 K, but with further decreasing of the temperature $\sigma(T)$ deviates from the Mott's law. Such behavior can be interpreted by modifying the DOS model.[2] We suppose that DOS is constant inside the strip $\Delta\varepsilon$, but decays rapidly outside this strip which is localized at the middle of the gap. Next, the Fermi energy, E_F, is close to the mobility threshold, E_c. Hence, the Mott's law takes place until the chemical potential, ζ, is inside the interval $\Delta\varepsilon$. With lowering the temperature ζ leaves the energy interval where DOS is constant and tends to E_F what explains deviation from the law $\ln\sigma \sim T^{-1/4}$ at $T < T_v^*$. The width of $\Delta\varepsilon \approx 7.5$ meV is estimated to be much smaller than that of E_g. When ζ is inside $\Delta\varepsilon$ (the H-region) the localization radius, ξ, is independent of the energy. Outside $\Delta\varepsilon$ (the L-region) it depends on the energy according to the law $\xi(\varepsilon) \sim (E_c - \varepsilon)^{-\nu}$, where ν is the critical exponent.

As seen from Fig. 1 (b) , the relative MR, $\Delta\rho(B)/\rho(0)$, exhibits complex dependence on the temperature and the magnetic field, suggesting both the positive and the negative contribution. We have achieved a consistent description of MR in FeSi when the correlation and the spin-flip processes are taken into account in the positive component, $(\Delta\rho/\rho)_p$ (the Osaka model [4]) and the Zeeman splitting of E_c (the model of Kamimura and

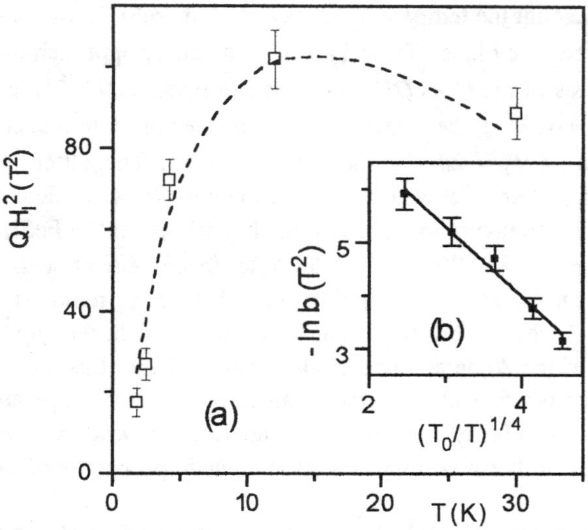

Figure 2. The experimental (□) and the calculated (dashed line) dependencies of QH_I^2 on T (a) and the dependence of ln b on $(T_0/T)^{1/4}$ (b). The solid line is the linear fit of the data (■).

Kurobe[5]) in the negative contribution to MR, $(\Delta\rho/\rho)_n$. Therefore, MR satisfies the equation

$$\Delta\rho/\rho = (\Delta\rho/\rho)_p + (\Delta\rho/\rho)_n\,, \qquad (1)$$

where

$$(\Delta\rho/\rho)_p = a[1 - \ln(1 + bH^2)/(bH^2)\,] \qquad (2)$$

and

$$(\Delta\rho/\rho)_n = \exp\{(T_0/T)^{1/4}[(1 - \gamma H)^{3\nu} - 1]\} - 1. \qquad (3)$$

In Eq. (2) $a = Q\nu_I^2/\nu^2(1)$ and $b = \nu_H^2/[H^2\nu^2(1)]$, where $\nu_H = g\mu_B H/\hbar$, $\nu_I = g\mu_B H_I/\hbar$ and $\nu(1) = \nu_0\exp[-n(T_0/T)^{1/4}\,]$. Here $Q = 2(T/T_0)^{1/4} + 5(T/T_0)^{1/2}\exp(U/k_BT)$, where U is the correlation energy, ν_0 is a pre-exponent of the hopping frequency, H_I is the internal field and $n = 1$ (Osaka [4]) or 0.53 - 0.65 (Movaghar and Schweitzer [6]). In Eq. (3) $\gamma = \mu_B/\Delta E$. In the H-region $\Delta E = E_g/2$ and $\nu = 0.5$ and in the L-region $\Delta E = E_c - \zeta$ and $\nu \approx 0.75$.

The experimental dependencies of MR shown in Fig. 1 (b) were fitted with Eq. (1). The values of the fitting parameters, a, b and γ, and those of ΔE and $QH_I^2 = a/b$ are collected in Table 1. At T = 12 - 30 K (the H-region) ΔE does not depend on T and exceeds slightly the half width of the gap (~ 25 - 30 meV) what is in agreement with our model taking into account the temperature decay of E_g in FeSi. Between 4.2 - 1.75 K (the L-region) the decrease of ΔE (Table 1) is determined by approaching of ζ to $E_F \sim E_c$. From the analysis of the plots QH_I^2 vs. T and ln b vs. $(T_0/T)^{1/4}$ (Fig. 2) are obtained the value of $H_I \approx 60$ kOe, the maximum value of the correlation energy, $U_m \approx 17$ K, and the values of $\nu_0 \approx 2.8\times10^{12}$ s^{-1} and $n \approx 0.65$. The latter is in agreement with that of Movaghar and Schweitzer.[6] Additionally, H_I was calculated for the Anderson-localized regime taking into account the hyperfine field in FeSi, $H_0 \approx 78$ - 82 kOe.[2] We obtained $H_I \approx 0.70H_0 \approx 55$ - 58 kOe, in agreement with the experimental result. The value of U_m much smaller than that of E_g removes the contradictions of Refs.2 and 3 where it has been found comparable with the gap [2] or much higher than the width of the Anderson band.[3] On the other hand, this value of the correlation energy cannot provide the observed values of the low-temperature Curie constant.[2, 3] Therefore, to clarify the nature of the low-temperature Curie-Weiss-like susceptibility tails further investigations into magnetic properties of FeSi are required.

To conclude, the conductivity and MR have been investigated in the VRH region. The values of the hopping frequency, the correlation energy and the internal field are obtained.

Table 1. The values of the parameters a, b, g, DE = m_B/g and $QH_I^2 = a/b$.

T (K)	a	$b\times10^2$ (T^{-2})	$\gamma\times10^3$ (T^{-1})	ΔE (meV)	QH_I^2 (T^2)
1.75	0.72	4.20	10	5.8	17
2.5	0.62	2.30	5.0	12	27
4.2	0.65	0.90	3.1	19	72
12	0.56	0.55	1.8	32	100
30	0.24	0.27	1.8	32	89

References

1. Y. Takahashi and T. Morya, J. Phys. Soc. Japan **46**, 1451 (1979).
2. S. Takagi, H. Yasuoka, S. Ogawa and J.N. Wernick, J. Phys. Soc. Japan **50**, 2539(1981).
3. M.B. Hunt, M.A. Chernikov, E. Felder, H.R. Ott, Z. Fisk and P. Canfield, Phys. Rev. **B 50**, 14933 (1994).
4. Y. Osaka, J. Phys. Soc. Japan **47**, 729 (1979).
5. A. Kurobe and H. Kamimura, J. Phys. Soc. Japan **51**, 1904 (1982).
6. B. Movaghar and L. Schweitzer, Phys. Status Solidi (b) **80**, 491 (1977).

INFLUENCE OF THE MAGNETIC PROPERTIES OF Mn ON SHUBNIKOV-DE HAAS (SdH)-OSCILLATIONS AND QUANTUM HALL EFFECT (QHE) IN Cd(MnMg)Te/Cd(Mn)Te QUANTUM WELLS

J. GERSCHÜTZ, S. SCHOLL AND G. LANDWEHR

Physikalisches Institut der Universität Würzburg, Am Hubland, D-97074 Würzburg

We investigated the low temperature magnetotransport properties of three modulation doped $Cd_{1-x-y}Mn_xMg_yTe/Cd_{1-x'}Mn_{x'}Te$ quantum wells. In the samples with CdTe well ($x' = 0\%$) we observed QHE and SdH-oscillations similar to those found in III-V materials, whereas the sample containing Mn directly in the well ($x' \approx 3\%$) exhibits some unusual behaviour: In QHE plateaus corresponding to certain filling factors were missing; the amplitudes of the SdH-oscillations showed beating-like modulations, the position of the maxima and minima of the oscillations was temperature dependent in the low field range. We modeled the measured ρ_{xx} and ρ_{xy} by taking into account the modification of the effective g-factor through the magnetic properties of the Mn in the well which results in a magnetic field and temperature dependence of g_{eff}.

1 Introduction and experimental setup

Due to the relatively small difference in the lattice constants of CdTe and MgTe - a semiconductor with a band gap of about 3.5 eV - $Cd_{1-y}Mg_yTe/CdTe$ quantum well structures of high quality can be grown by molecular beam epitaxy (MBE). Although the AlGaAs system provides samples of higher quality concerning mobility, dopability, gating etc. at this time, one advantage of II-VI materials is the possibility to investigate the influence of the magnetic properties of built in Mn on the two dimensional electron gas (2DEG) in the well by adding Mn in the barriers and/or in the well.

We investigated the low temperature magnetotransport properties of three $Cd_{1-x-y}Mn_xMg_yTe/Cd_{1-x'}Mn_{x'}Te$ single quantum wells. The samples were grown on undoped CdZnTe(100) substrate and modulation doped from one side using iodine as dopant. A spacer between the well and the doped layer was included to seperate the free carriers in the quantum well from the donors. Table 1 shows the growth parameters, details of the sample structures and the measurement equipment can be found elsewhere[1].

A Hall bar was prepared from all samples by standard photolithographic techniques and chemical etching. The contacts were attached by a thermobonder using indium as solder. All measurements were done under illumination with a red LED mounted close to the sample and kept at the same temperature.

2 Experimental Results

In the quantum well structure CT1047 we observed QHE and spin-split SdH-oscillations equivalent to those found in III-V materials[1]. Similar SdH-oscillations and Hall plateaus were observed in the second structure CT1103. No

	CT1047	CT1103	CT1106
x_{Mn} / %	0	≈ 3	≈ 3
x'_{Mn} / %	0	0	≈ 3
y_{Mg} / %	≈ 28	≈ 25	≈ 25
d_{Well} / Å	100	60	60
d_{Spacer} / Å	50	60	50
n_{Hall} / 10^{11} cm^2	8.0	9.2	8.5
n_{SdH} / 10^{11} cm^2	8.1	9.4	
μ_{Hall} / 10^3 cm^2/Vs	41	19	8.5

Table 1: Growth parameters and experimental results for the samples under investigation.

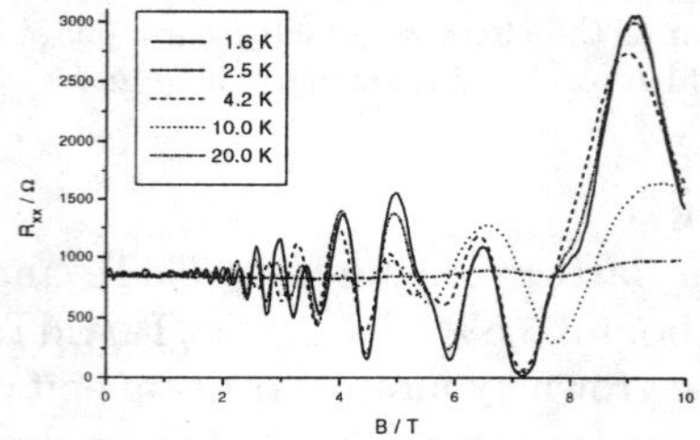

Figure 1: Magnetoresistance of the sample CT1106. Note the beating like behaviour of the SdH amplitudes and the temperature dependence of the position of the minima and maxima in the low field range.

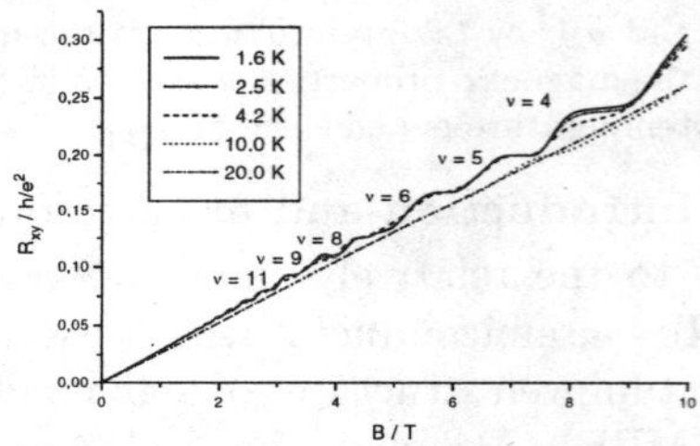

Figure 2: Hall effect of the sample CT1106. Certain filling factors (e.g. $\nu = 7, 10$) are missing in irregular series.

indication of a possible influence of the magnetic properties of the Mn in the barriers on the 2DEG in the CdTe well was found.

The third structure CT1106 also showed spin-resolved quantized Hall effect and SdH-oscillations (see figures 1 and 2), but the plateaus corresponding to certain filling factors were missing. Moreover, the amplitudes of the SdH-oscillations showed beating-like behaviour, the position of the maxima and minima of the oscillations is temperature dependent in the low field range. Since there is no significant difference between the structures and electron concentrations of this quantum well and the first two investigated, this beating-like modulation cannot be easily attributed to the occupation of higher subbands in the well.

The sample CT1106 is the only specimen containing Mn inside the well. The unusual behaviour of this sample can be modeled by taking into account the enhanced spin splitting known to exist in semimagnetic semiconductors like $Cd_{1-x}Mn_xTe$: The energy of an electron in the quantum well consisting of $Cd_{1-x}Mn_xTe$ as a function of the applied magnetic field B is given by the standard Landau quantization, but the effective g-factor g_{eff} for the spin splitting has to be modified as[2]:

$$g_{eff} = g^* + \frac{N_0\alpha \cdot x \cdot \langle S_z \rangle}{\mu_B \cdot B}; \quad \langle S_z \rangle = S_0 \cdot B_S\left(\frac{S g_{Mn} \mu_B B}{k_B \cdot (T+T_0)}\right) \qquad (1)$$

The parameters in this description are: the (intrinsic) effective g-factor $g^* = -1.77$[3], the sp-d exchange constant $N_0\alpha = 220\,meV$, the Mn content x, the average spin per Mn site $\langle S_z \rangle$, the Bohr magneton μ_B, the magnetic field B, the effective Mn spin S_0, the Mn spin $S = 5/2$, the Mn g-factor $g_{Mn} = 2$, the Boltzmann constant k_B, the temperature T and an effective temperature $T + T_0$. $B_s(q)$ denotes the Brillouin function.

Because of the opposite sign of g^* and $N_0\alpha\,\langle S_z \rangle$ and the magnetic field dependence of $\langle S_z \rangle$ through the Brillouin function, the modified g_{eff} leads to a crossover of Landau levels of different spin states with varying magnetic field. The temperature dependence of the Brillouin function modifies the Landau levels through g_{eff} for different temperatures. Figure 3 shows the calculated Landau fan for a temperature of 1.6 K, figure 4 the magnetic field and temperature dependence of g_{eff}.

Following the calculations in [4] we assumed a Gaussian broadening of the density of states (DOS) of the Landau levels with broadening parameter $\Gamma \sim \sqrt{B}$ and calculated the Fermi energy through the condition of conservation of the electron concentration. For the calculation of σ_{xx} and σ_{xy} and ρ_{xx} and ρ_{xy} as a function of the magnetic field we used a phenomenological model[5] which takes into account the tails of the DOS which do not contribute to conduction by using a Gaussian DOS of extended states with a width $\lambda < \Gamma$.

The figures 5 and 6 show the calculated resistivities ρ_{xx} and ρ_{xy}. The Parameters used for the calculations are: $n = 8.6 \cdot 10^{11}\,cm^{-2}$, $\mu = 8500\,cm^2/Vs$, $x'_{Mn} = 3.3\%$, $T_0 = 1.78K$, $S_0 = 1.73$, $m_{eff} = 0.095 \cdot m_0$, $\Gamma = 1.1 \cdot \sqrt{B}\,meV$, $\lambda = 0.6 \cdot \Gamma$.

The calculated resistivities are in good agreement with experimental data. Especially ρ_{xx} shows a comparable modulation of the amplitude of the SdH-oscillations. On the other hand the calculated ρ_{xy} indicates that an irregluar series of filling factors due to the crossing of Landau levels from different spin states can take place.

Therefore we conclude that the unusual behaviour in SdH-oscillations and QHE observed in the $Cd_{1-x-y}Mn_xMg_yTe/Cd_{1-x}Mn_xTe$ quantum well sample can be explained with our model that uses the modification of the effective

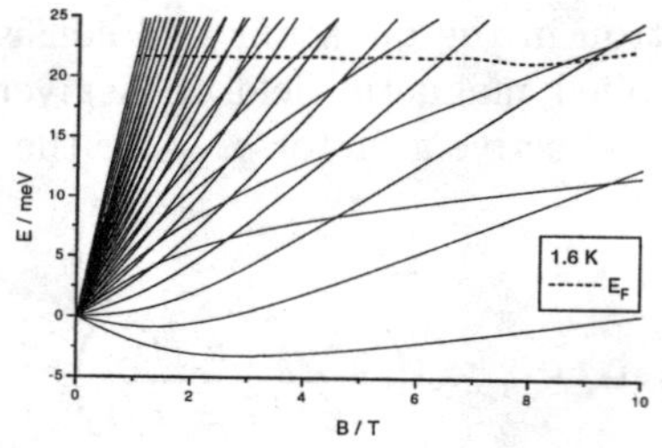

Figure 3: Calculated Landau fan for a temperature of 1.6 K.

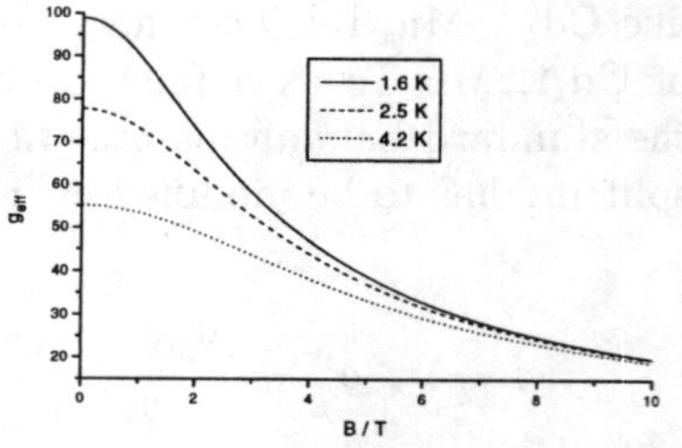

Figure 4: g_{eff} as a function of magnetic field for different temperatures

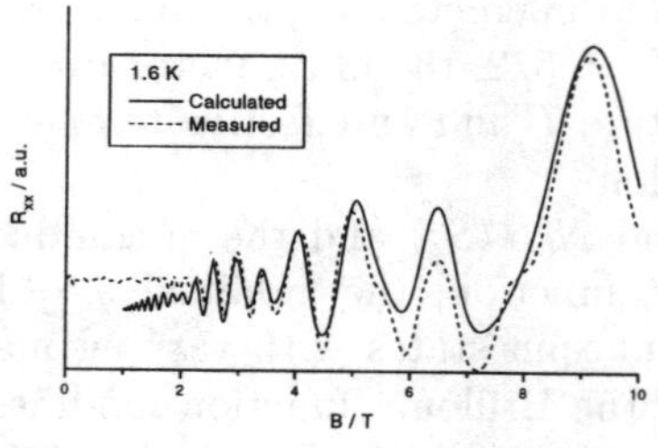

Figure 5: Calculated and measured magnetoresistance at 1.6 K.

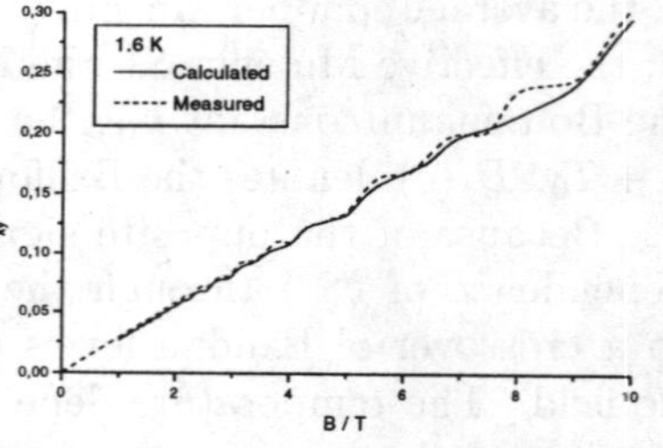

Figure 6: Calculated and measured Hall resistance at 1.6 K

g-factor through the Mn in the well. This is - to our knowledge - the first time, that the influence of the magnetic properties of Mn on a two dimensional electron gas in a quantum well could be established.

References

1. S. Scholl, J. Gerschütz, F. Fischer, A. Waag, K. Schüll, H. Schäfer, G. Landwehr, *Solid State Commun.* **94**, 935 (1995)
2. J.K. Furdyna, *J. Appl. Phys.* **64**, R29 (1988)
3. E. Molva, J.P. Chamonal, J.L. Pautrat, *Phys. Stat. Sol. (b)***109**, 635 (1982)
4. M. van der Burgt, V.C. Karavolas, F.M. Peeters, J. Singleton, R.J. Nicholas, F. Herlach, J.J. Harris, M. Van Hove, G. Borghs, *Phys. Rev.* **B52**, 12218 (1995)
5. Th. Englert, in *Application of High Magnetic Fields in Semiconductor Physics II*, ed. by G. Landwehr (Springer, Berlin, 1983), p. 165

HIGH MAGNETIC FIELD STUDY OF METAL-INSULATOR TRANSITION IN PEROVSKITE TYPE MANGANESE OXIDE $R_{1-x}Ca_xMnO_3$ (R =Pr, Nd)

M. TOKUNAGA and N. MIURA
Institute for Solid State Physics, University of Tokyo
Roppongi, Minato-ku, Tokyo 106, Japan

Y. TOMIOKA[A] and Y. TOKURA[A,B]
[A] *Joint Research Center for Atom Technology (JRCAT), Tsukuba 305, Japan*
[B] *Department of Applied Physics, University of Tokyo, Tokyo 113, Japan*

We have observed magnetic field induced phase transitions in $R_{1-x}Ca_xMnO_3$ (R=Pr, Nd) by the measurements of magnetization, magnetoresistance and magnetostriction in pulsed magnetic fields up to 40 T. We found that the transition process consists of two stages. In the lower field stage, alignment of localized moment enhances charge transfer. Then a structural phase transition accompanied with colossal magneto-resistance is induced in the higher field stage. The phase diagrams for $Pr_{1-x}Ca_xMnO_3$ (x=0.50, 0.45) and $Nd_{0.50}Ca_{0.50}MnO_3$ were constructed. It was found that the anti-ferromagnetic ordering of the localized moments stabilizes the charge ordered state.

1 Introduction

Since the discovery of high-T_C superconductors, metal-insulator transitions in $3d$ transition metal oxides have attracted much attention. The phase transitions are caused by competition between Coulomb interaction and kinetic energy of charge carriers.

Perovskite-type Mn oxide $RMnO_3$ (where R denotes trivalent rare-earth ions) contains Mn^{3+} ions which have $t_{2g}^3e_g^1$ (S=2) configuration and is a kind of Mott insulator. Substitution of trivalent rare-earth ions with divalent alkaline-earth ions produces holes in the e_g orbitals. In $La_{1-x}Sr_xMnO_3$, for example, substitution with Sr causes insulator-metal transition[1]. Electronic properties of these hole doped Mn oxides have been explained in terms of the double exchange interaction. One of the remarkable features in such hole doped Mn oxides is that the transfer of charge carriers can be controlled by external magnetic fields, and a giant negative magnetoresistance takes place around Curie temperature.

In $Pr_{1-x}Ca_xMnO_3$, however, no metallic phase shows up even in heavily doped samples[2]. In these narrow band systems, strong electron-electron correlation significantly modifies their electronic properties and gives rise to real space charge ordering at low temperatures. Modification of the charge transfer by external magnetic fields causes magnetic phase transitions from 'charge crystal' to 'charge liquid'.

In order to clarify the mechanism of the phase transition, we have carried out magnetization, magnetoresistance and magnetostriction measurements in $Pr_{1-x}Ca_xMnO_3$ and $Nd_{1-x}Ca_xMnO_3$ single crystals around x=0.50 in pulsed magnetic fields up to 40 T.

2 Experimental

Single crystals of $R_{1-x}Ca_xMnO_3$ were synthesized by floating zone method. Ca contents x of the samples for R=Pr and R=Nd were 0.45, 0.50 and 0.50, respectively. Magnetic fields were generated by using a non-destructive long pulse magnet and a 200 kJ capacitor bank which produced magnetic fields up to 40 T with a duration time of 11 ms. For magnetization measurements, we used induction method with a pair of co-axial pick-up coils. Transport measurements were carried out by the 4 probe method with dc current flowing parallel along the magnetic fields. Magnetostriction along the direction of magnetic fields was measured by the capacitance method. Those measurements were performed at 4 K to 270 K.

3 Results and Discussion

Figure 1 shows a magnetization curve of $Pr_{1-x}Ca_xMnO_3$ single crystal with x=0.50 at 1.4 K. In this temperature range, the Mn moments are aligned antiferromagnetically in zero field. With increasing magnetic fields, magnetization shows an abrupt jump at 25 T. The saturation moment is estimated to be 3.4 μ_B/Mn site by extrapolating to zero field. This value is close to the averaged spin moment of a Mn ion (3.5 μ_B/Mn site). So, this high magnetization state can be regarded as a ferromagnetic phase. In the down sweep process, the magnetization starts decreasing at 20 T, exhibiting a hysteresis which is characteristic of a first order phase transition. The transition fields H_S^+ and H_S^-, in the up-sweep and down-sweep are defined where the magnetization reaches or leaves from the higher value. Upon this transition, the magnetization shows a stepwise change as seen more distinctly in the dM/dH curve shown in Fig. 1(a).

A typical magnetoresistance trace measured at 117 K for $Pr_{1-x}Ca_xMnO_3$ with x =0.50 is shown in Fig. 1 (b). The magnetization curve at 123 K is also shown for comparison. Below 20.5 T, the resistivity decreases gradually as the Mn spins are aligned. This negative magnetoresistence originates from the same mechanism as in metallic $La_{1-x}Sr_xMnO_3$[3]. Above this field, a drastic decrease of the magnetoresistance takes place by more than two orders of magnitude, in the higher field state of the transition process. In the low magnetic field region below the transition, the temperature dependence of the resistivity shows a semiconducting behavior over the whole temperature range. In the high field region, however, the semiconducting temperature dependence changes to a metallic one below about 130 K.

The phase diagram of $Pr_{1-x}Ca_xMnO_3$ (x=0.50) determined by magnetization and transport measurements is demonstrated in Fig. 2. A metallic phase exists in the low temperature and high magnetic field regime. Below about 160 K, charge-ordered phase extends to higher field. In this area, localized moments order anti-ferromagnetically, which stabilizes the charge ordered state.

The magnetization curve for a sample of x=0.45 is compared with that of x=0.50 in Fig. 3. The onset field of the phase transition decreases with decreasing x as the hole number is not commensurate with the lattice periodicity. However, the saturation field of the magnetization does not change significantly.

The magnetostriction has also been measured for a single crystal of x=0.45. Figure 3 shows the magnetostriction at 8 K. The sample length along the magnetic field increases drastically in the order of $\Delta L/L$=10^{-3}. This result clearly demonstrates a structural phase transition. We found that the structural phase transition occurs in the higher field states of the transition process.

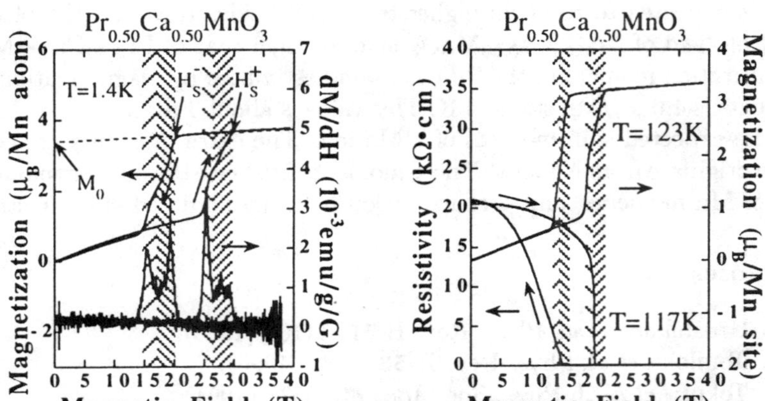

Figure 1. Magnetization and magneto-resistance of $Pr_{1-x}Ca_xMnO_3$ with x=0.50. (a) Magnetization (upper) and dM/dH (lower) at 1.4K. (b) Resistivity at 117K and magnetization curve at 123K are shown as a function of magnetic field. The transition process in the up and down-sweeps, consists of two stages.

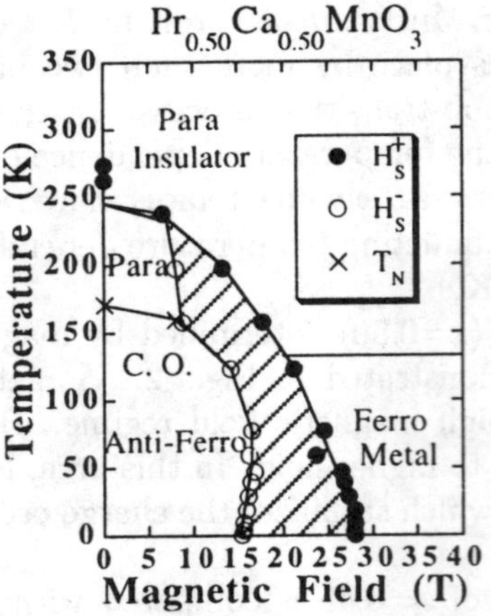

Figure 2. Phase diagram of $Pr_{1-x}Ca_xMnO_3$ with x=0.50. There is a metallic phase in the low temperature and high magnetic field region. Below Néel temperature, charge ordered state extends to higher fields.

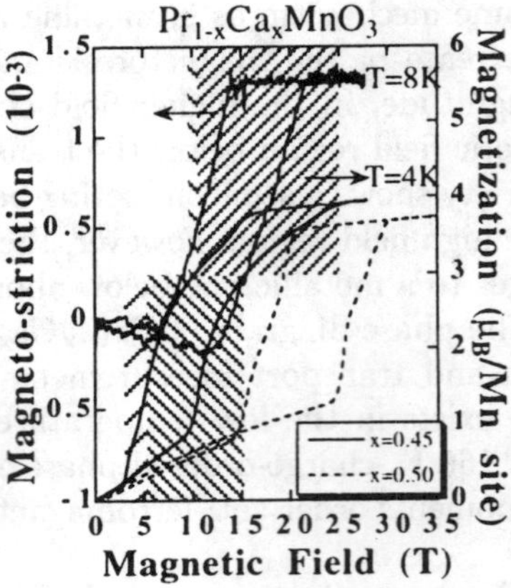

Figure 3. Magnetization at 4K and magneto-striction at 8K in $Pr_{1-x}Ca_xMnO_3$ with x=0.45. Magnetization curve of x=0.50 is also given by dotted line for comparison.

The phase diagram of $Nd_{1-x}Ca_xMnO_3$ x=0.50 was also determined by magnetization measurements. As for the charge-ordered Mn oxides, a smaller ionic radius of the *A*-site would lead to a reduction in the double-exchange interaction and hence, higher transition fields. Therefore, the substitution of Nd for Pr is expected to result in higher transition fields. However, the observed transition field of $Nd_{0.50}Ca_{0.50}MnO_3$ is lower than that of $Pr_{0.50}Ca_{0.50}MnO_3$. The saturation moment of the Nd compound estimated by magnetization measurements is 4.6 μ_B/Mn site at 4 K. This value is about 1 μ_B larger per Mn site than the expected spin moment of a Mn ion. The extra moments are considered to originate from Nd ions. These moments frustrate the anti-ferromagnetic order of Mn moments, and perhaps make the charge-ordered state unstable.

References

1. . Urushibara *et al*, Phys. Rev. B **51**, 14103 (1995).
2. . Tomioka *et al*, Phys. Rev. B **53**, 1689 (1996).
3. . Tokura *et al*, J. Phys. Soc. Jpn. **63**, 3931 (1994).

NONPARABOLICITY AND ANISOTROPY OF CYCLOTRON MASSES IN HIGHLY DOPED n-GERMANIUM

D. SCHNEIDER[1] and U. STRAUSS[1,2]

[1] *Institut für Technische Physik und Hochmagnetfeldanlage, Techn. Universität Braunschweig, Mendelssohnstr. 2, D-38106 Braunschweig, Germany*

[2] *now: Siemens AG, D-93009 Regensburg, Germany*

We used Shubnikov-de Haas- and Hall-measurements in n-Ge with carrier concentration $n = 8{\cdot}10^{17}$ up to $10^{19}cm^{-3}$ and derived the angle dependence of transversal cyclotron masses m_c for $n = 7.7 \cdot 10^{18}cm^{-3}$ in the (110) plane by Fourier analysis. The results are compared with calculations from the band model. Fermi energy W_F and the effective masses in dependence on carrier concentration are also extracted from the experiments between temperatures T=1.6 K and 35 K in magnetic fields up to B=15 T. The transversal cyclotron mass of the L-electrons show small deviations from parabolicity in the range below $W_F = 30\ meV$.

1 Introduction

Although different methods of Landau level spectroscopy have been used [1] to investigate the bandstructure of bulk Germanium (Ge), little Shubnikov-de Haas (SdH)- work has been carried out [2]: e.g. in n-Ge with magnetic fields parallel only to [110] and [111] crystal directions. Also experimental results to the problem of nonparabolicity are rare [3,4]. Our results are important as a reference for heterostructures and nanostructures with Ge and Si-Ge superlattices, where the mobility of electrons may be very enlarged which promotes SdH-oscillations.

We investigated four different doped bulk samples (1x1x6 mm with their length in [110] direction) as listed in Tab. 1. Ohmic contacts were achieved by thermocompression of Ni- or bonding of Au-wires. Transversal magnetoresistance and Hall effect measurements were performed with magnetic field B parallel to [-111]. Additionally sample 1.5 has been investigated rotating the field direction within the (110) plane in 75 steps of $\Delta\Theta$ =4.8 degrees from $\Theta = 0°$ (|| [-111]) to 360°. The experimental set up for our measurements is described in [5]. We observed the Fermi surface of the lowest conduction band L_6^- in Ge consisting of a set of four equivalent ellipsoids with rotational symmetry around [111], [-111], etc., located at the L points in k-space [6]. The axes are related to transverse and longitudinal mass parameters m_t respectively m_l.

Table 1: Experimental results for n-Ge as obtained from Halleffect (n: electron concentration, μ: mobility) at T=4.2 K and SdH-measurements (m_t: transversal electron mass, W_F: Fermi energy) between T=1.6 and 35 K. Magnetic field || [-111]; electrical current || [110]. Samples 3.5 and 1.5 grown at *SGM company*, Belgium, and the other ones at *Unterharzer Berg- und Hüttenwerke*, Germany.

Sample	dopant	$n[10^{17}\ cm^{-3}]$	$\mu[cm^2/Vs]$	$m_t[10^{-3}\ m_0]$	W_F[meV]
3.5	Sb	110	520	83±3	29.0±1.1
1.5	P	77	940	80±4	22.8±1.3
5.4	Sb	9.5	990	77±4	5.9±0.4
2.6	Sb	8.1	660	70±10	4.7±0.7

2 Results and discussion

The observed transversal magnetoresistance is modulated by SdH-oscillations with the period P, as shown in Fig. 1. An analysis gives SdH-peroid P and effective cyclotron masses m_c as described in [2,5]. We got the Fermi energy W_F by relation $W_F = e\hbar/m_t P$, including corrections[7] for oscillating W_F. Our configuration B||[-111] creates SdH-oscillations from the minimum cross-section of an ellipsoid elongated in [-111] direction, i.e. we observe the cyclotron mass $m_{c1} = m_t$. The other ellipsoids may give contributions, if the conditions for

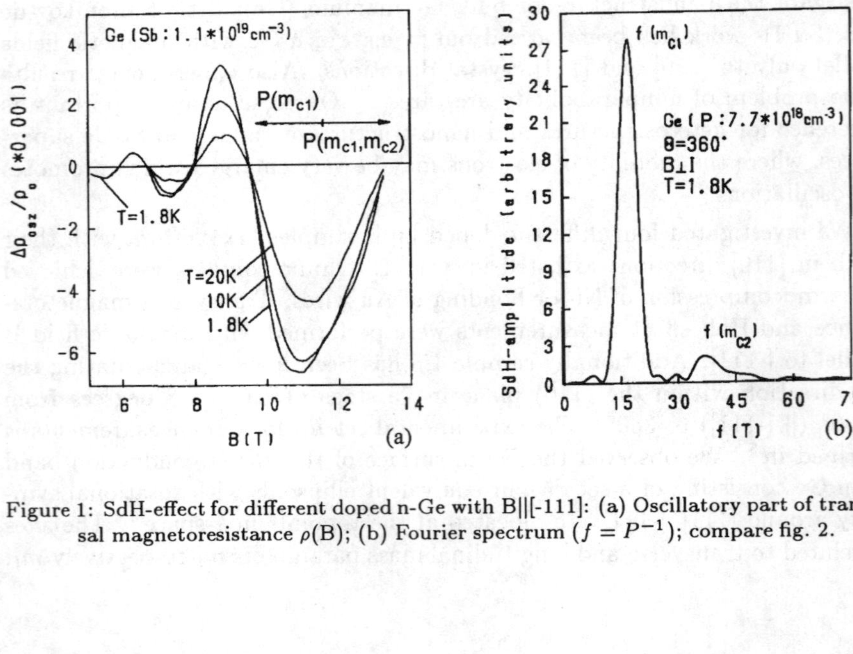

Figure 1: SdH-effect for different doped n-Ge with B||[-111]: (a) Oscillatory part of transversal magnetoresistance ρ(B); (b) Fourier spectrum ($f = P^{-1}$); compare fig. 2.

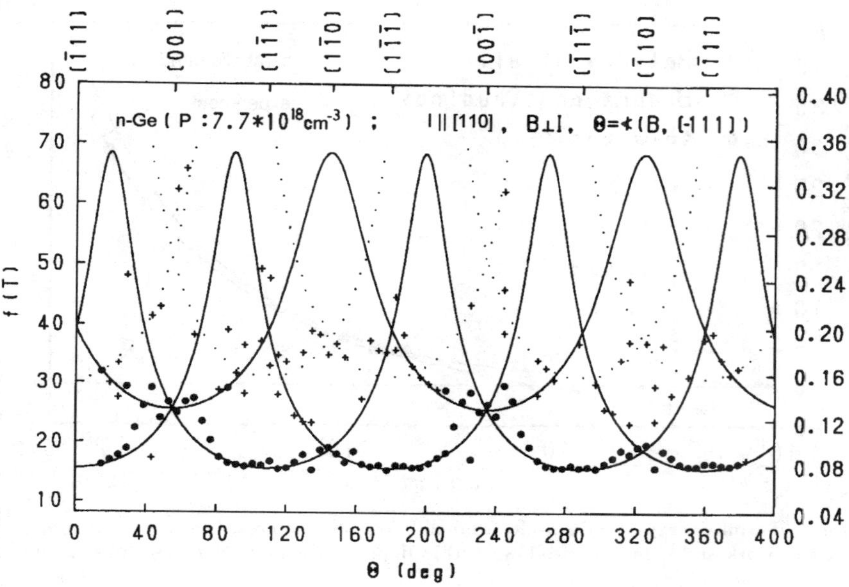

Figure 2: Anisotropy of SdH-frequency f and cyclotron mass m_c in n-Ge in a (110) plane from Fourier analysis of SdH-oscillations at T=4.2 K; Fourier peaks related to the maximum peak: • > 20 %, + > 2 %; — fitted to ellipsoid model; · · · first harmonic.

SdH-oscillations[2] are fulfilled. Fig. 1 contains one case of overlap where Fourier analysis was used. The results are presented in table 1. The angle dependence of m_t has been investigated in the sample 1.5, where the SdH-oscillations are most pronounced. When the magnetic field forms an angle Θ with the longitudinal axis of the ellipsoid, the effective mass determining the SdH-oscillations is[6] $(1/m_c)^2 = (cos\Theta/m_t)^2 + (sin^2\Theta/m_t m_l)$. Fig. 2 gives a plot of the result from Fourier analysis together with fitted curves. Since SdH-oscillations near the largest cross-section are not observable, we used[8] m_l=1.556 m_0 for the fit. Near the intersection point of three curves the resolution is reduced, because only some few SdH-oscillations give contributions. We conclude, that the relative change of effective mass in n-Ge determined by SdH-oscillations (Fig. 2) is in good accordance with cyclotron data (Fig. 5 in [6]). In Fig. 3 we show the observed Fermi energy W_F related to m_t in dependence on n, compared with other results and the case of a parabolic band. Below about n=$1.5 \cdot 10^{17}$ cm^{-3} the metal-isolator transition occurs [7] and above n=$5 \cdot 10^{17}$ cm^{-3} deviation from parabolicity starts, as observed in [7]. All results are extracted from magne-

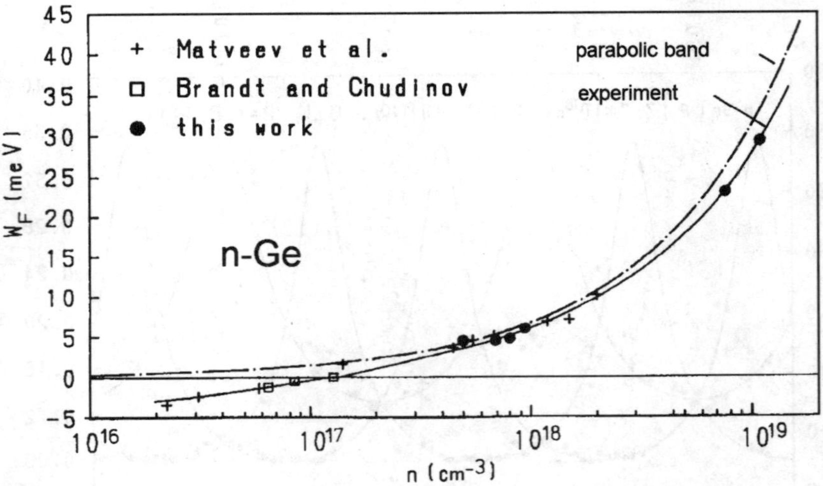

Figure 3: Fermi energy W_F in n-Ge from different magnetoresistance measurements [7,9] (• this work and [10] for the two lowest n) in dependence on carrier concentration n.

toresistance measurements and do not reflect the precision of cyclotron resonance measurements, as discussed in[4] for Fermi energies between 50 and above 100 meV. However our values are near the results of magnetopiezotransmission studies[3] below W_F=30 meV.

References

1. G. Landwehr and E.I. Rashba (eds.): Landau Level Spectroscopy, Amsterdam, Elsevier (1991).
2. D.G. Seiler, A.E. Stephans in[1] (for Ge see p. 1066/67 and references).
3. R.L. Aggarwal, M.D. Zuteck, B. Lax *Phys. Rev.* **180**, 800 (1969).
4. J.C. Ousset, J. Leotin, S. Askenazy, M.S. Skolnick, R.A. Stradling *J.Phys.C: Solid State Phys.* **9**, 2803 (1976) and references pp. 2806-08.
5. D. Schneider, D. Rürup, A. Plichta, H.-U. Grubert, A. Schlachetzki, K. Hansen *Z. Phys.* **B95**, 281 (1994).
6. G. Dresselhaus, A.F. Kip, C.Kittel *Phys. Rev.* **98**, 368 (1955).
7. G.A. Matveev, A.T. Lonchakov *Sov. Phys. Semicond.* **18**, 367 (1984).
8. B.W. Levinger, D.R. Frankl *J. Phys. Chem. Solids* **26**, 911 (1965).
9. N.B. Brandt, N.F. Chudinov *Electrons in Metals* (in Russian), Moskau 1978 (after[7]).
10. D. Schneider, J. Goerigk, unpublished.

SUBBAND LOCKING AS THE PRECURSOR TO HYBRID DOUBLE-SINGLE-LAYER BEHAVIOUR IN A DOUBLE QUANTUM WELL STRUCTURE

A. G. Davies, C. H. W. Barnes, K. R. Zolleis, J. T. Nicholls, M. Y. Simmons, D. A. Ritchie
Cavendish Laboratory, Madingley Road, Cambridge CB3 0HE, UK

Magnetoresistivity measurements of a narrow-barrier double quantum well reveal that diagonal resistivity maxima oscillate in magnetic field as the electron density is changed, giving the overall appearance of a single-layer system even when two subbands are occupied. We demonstrate that this system exhibits a hybrid behaviour between a single quantum well and two uncoupled quantum wells.

1 Introduction

Double-layer electron systems (DLESs) comprising two, high-quality, two-dimensional electron layers in close proximity are the natural extension to the single two-dimensional electron system (2DES) and a further step towards fully three-dimensionally engineered quantum structures. Investigations of the quantum Hall effects, electron tunnelling, electronic drag, and 2DES compressibility in DLESs have revealed a variety of new physical phenomena.[1]

We identify a novel hybridization of single and double-layer behaviour in a DLES formed in a narrow-barrier double quantum well. Although individual magneto-resistivity sweeps display manifestly double-layer behaviour demonstrating that two electron subbands are occupied, positions of diagonal resistivity maxima oscillate with magnetic field as the electron density is changed to give the overall appearance of a single-layer system.

2 Experimental

Our DLES comprises two 150Å-wide GaAs quantum wells separated by a 25Å $Al_{0.33}Ga_{0.67}As$ barrier. The DLES was processed into a Hall bar with a Schottky front-gate and electrical contacts were made simultaneously to both wells. The sample was mounted in a dilution refrigerator and magnetoresistivity measurements were performed using four-terminal, low frequency ac techniques.

Fig. 1(a) shows diagonal and Hall magnetoresistivity (ρ_{xx}, ρ_{xy}) traces for three gate biases (V_g) at 100mK. Insets I, II and III show the corresponding calculated conduction band-edges and probability densities. The two lowest electron subbands E_1, E_2 ($E_1 < E_2$) and the subband energy separation $\Delta(V_g)=E_2-E_1$ are identified in the schematic left inset. For V_g=+0.10V (trace I, inset I), the wells are out of balance and the subband wavefunctions are predominantly confined within separate wells. As the system is progressively brought into balance (when the electron densities associated with each well are the same), the E_1 and E_2 wavefunctions are increasingly shared between the wells. At the balance point (V_g=–0.072V, trace II), E_1 and E_2 are the symmetric and antisymmetric eigenstates which extend equally across both wells, the subband separation is just the symmetric-antisymmetric gap ($\Delta=\Delta_{SAS}\approx$20K) and the Shubnikov-de Haas (SdH) oscillations display minimized beating. At balance, the total density is $n=2.4\times10^{11}cm^{-2}$ and the mobility is $1.6\times10^6 cm^2/Vs$. If V_g is decreased

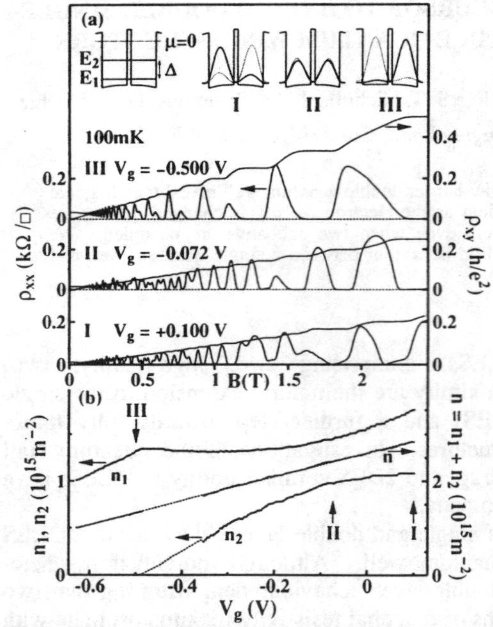

Fig. 1 (a) ρ_{xx} and ρ_{xy} vs B with corresponding Hartree band-edges and probability densities: (I) off-balance, (II) on-balance, (III) front well depleted. (b) Experimental n_1, n_2 vs V_g with total density $n=n_1+n_2$.

further, the front well depletes at $V_g \approx -0.49$V and single-layer SdH oscillations are observed from the back well (trace III). The electron densities associated with the two subbands (n_1, n_2) obtained from low-field SdH oscillations are shown as a function of V_g in Fig. 1(b) with the total electron density $n=n_1+n_2$ which varies linearly with V_g.

Fig. 2(a) shows ρ_{xx} as a function of perpendicular magnetic field (B) and V_g. White regions represent large ρ_{xx} (SdH maxima). For $V_g < -0.49$V, the front well is depleted and the ρ_{xx} maxima trajectories of the back well map out the characteristic single-layer Landau level fan. As V_g is increased, the front well starts to populate. The contributions from the two subbands become increasingly distorted as B is increased and although both wells are occupied, the trajectories of the single-layer Landau fan at the top of Fig. 2(a) are preserved across the entire V_g range, defining a constant total filling factor $\nu = nh/eB$ ($\propto V_g h/eB$). The ρ_{xx} maxima trajectories associated with the two subbands distort in sympathy to give $\rho_{xx}(B, V_g)$ the appearance of a single-layer. Circular features are formed in the regions where Landau levels from the subbands intersect.

3 Hartree Calculations

We performed Hartree calculations[2] using appropriate functional forms for the Landau extended and total density of states.[3] Exchange and correlation effects were deliberately omitted. We calculate ρ_{xx} via the empirical relation $\rho_{xx} \approx B\ \Sigma_{i,\sigma} \wp_{i,\sigma}$ where $\wp_{i,\sigma}$ is the extended density of states at the chemical potential (μ) for spin σ and subband i. The calculated ρ_{xx} is presented in Fig. 2(b) and the striking comparison with Fig. 2(a) demonstrates that this calculation contains the essential physics. Lines of constant total filling factor comprise alternating contributions from the two subbands. The ρ_{xx} maxima oscillate forwards and backwards in field, reminiscent of alternating electron and hole transport.

4 Discussion

We consider our DLES in terms of two limits: (i) two independent quantum wells in

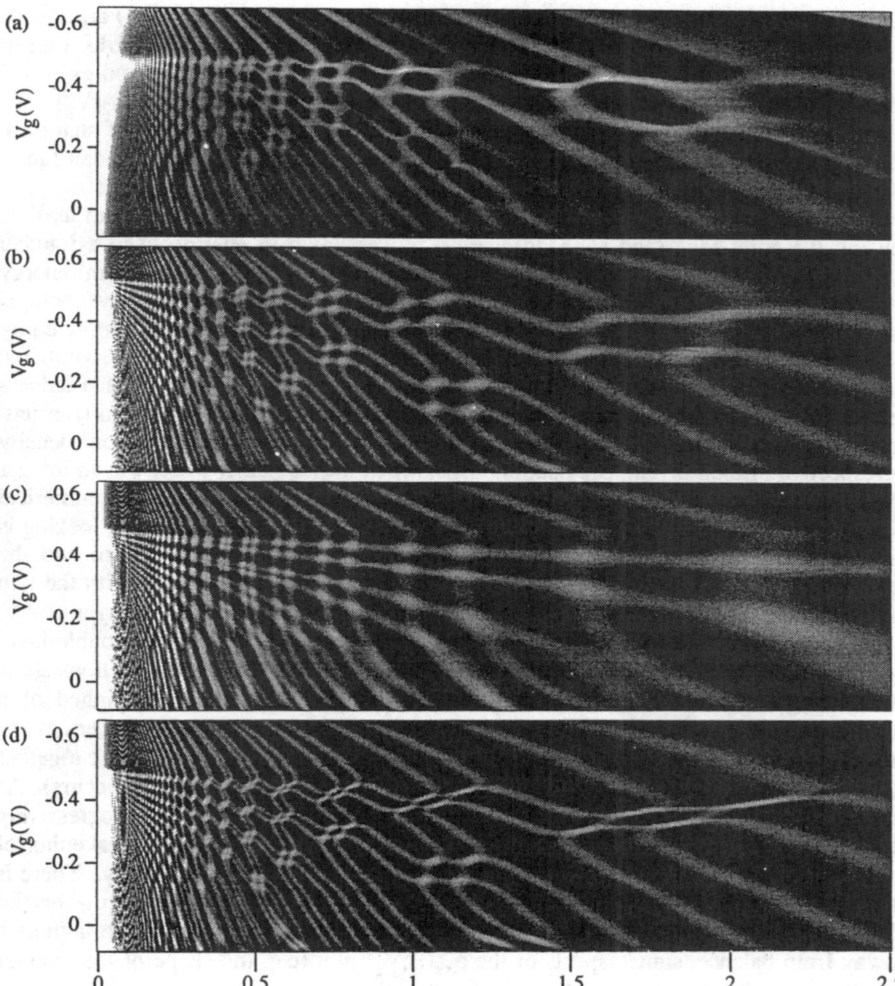

Fig. 2 $\rho_{xx}(B,V_g)$ – white represents large ρ_{xx}. (a) Experimental data. (b) Hartree calculation. (c) Independent quantum wells calcu!ation. (d) Perfect subband locking calculation. In this figure ρ_{xx} is scaled by B for maximum contrast.

which the subband wavefunctions are localized and n_1, n_2 are individually invariant and, (ii) a doubly-occupied single quantum well. Limit (ii) applies to our DLES at the balance point since the subband wavefunctions are delocalized equally across both wells. Moving away from balance, the subband wavefunctions become progressively localized within the two wells and the DLES moves towards limit (i).

In limit (i), Landau levels from each subband appear independently at μ at fields determined by n_1 and n_2 and so the ρ_{xx} maxima mapped out as a function of V_g follow

$B_{\rho xx\text{-}max}(V_g)=n_{1,2}(V_g)h/e(i+½)$ where i is an integer and $n_1(V_g)$ and $n_2(V_g)$ are shown in Fig. 1(b). The exact numerical solution for $\rho_{xx}(B,V_g)$ in the presence of disorder in this limit is shown in Fig. 2(c), calculated using the same Landau level and extended density of states as the Hartree calculation and the experimental $n_1(V_g)$, $n_2(V_g)$. [In fact, this behaviour is observed in Fig. 2(a) for B<0.5T since the density of states are only weakly modulated at low B]. The trajectories of the single-layer Landau fan at the top of Fig. 2(c) are not preserved when the upper quantum well populates.

For the electron density to remain constant in a two-dimensional system as B is swept, the band-edge (and subband energy) moves down in energy when a Landau level is pinned at μ (the Fermi energy increases) and up in energy (the Fermi energy decreases) when μ is between Landau levels.[3] In limit (ii), the subband wavefunctions are delocalized across the DLES and so both subband energies are equally affected by the changing Landau level density of states at μ. Consequently, as B is swept, the subbands oscillate together, locked with fixed separation Δ. The two Landau ladders effectively form a single ladder with only one level pinned at μ at any field (for zero disorder). Since all filled levels contribute eB/h to the total electron density irrespective of their origin, ρ_{xx} maxima occur when n=(i+½)eB/h where i is an integer and so in this limit $\rho_{xx}(B,V_g)$ is *identical* to that of a single-layer. The exact numerical solution for $\rho_{xx}(B,V_g)$ in the presence of disorder assuming perfect subband-locking is shown in Fig. 2(d), calculated using the same density of states as before and the experimental $n(V_g)$ and $\Delta(V_g)$. The ρ_{xx} maxima trajectories associated with the two subbands distort in sympathy to give $\rho_{xx}(B,V_g)$ the appearance of a single-layer.

For arbitrary V_g the DLES behaves as a hybrid between an uncoupled double-layer system (limit (i)) and a single-layer system (limit (ii)). Our Hartree calculations show that charge transfers out of the subband which has a Landau level pinned at μ (although n is invariant, n_1 and n_2 may change). This charge transfer moderates the relative motion of the two subbands producing a tendency for them to lock together (this occurs maximally in limit (ii) since no spatial redistribution of charge occurs). As the DLES is driven away from balance, the wavefunctions become progressively localized in the wells resulting in a reduction in charge transfer and a weakening of subband locking, and the two quantum wells increasingly act independently. There is sufficient wavefunction delocalization between the two wells in our narrow barrier DLES for single-layer behaviour to be observed over the entire V_g range, although away from balance, some aspects of the $\rho_{xx}(B,V_g)$ plot (e.g. the shape of the circular features) are better represented by the independent wells calculation.

AGD and CHWB thank the Royal Society and Isaac Newton Trust. KRZ and DAR acknowledge TCRC. We thank F. Stern and C. J. B. Ford for discussions. This work was funded by the EPSRC (UK).

References

1. J. P. Eisenstein, Proc. 11th. SEMIMAG, ed. D. Heiman, p78 (World Scientific, Singapore, 1995) *and references therein.*
2. A. Kumar *et al.*, Phys. Rev. **B42**, 5166 (1990).
3. C. H. W. Barnes *et al.*, Proc. 11th. EP2DS (to be published in Surf. Sci.) *and references therein.*

GaAs-SINGLE QUANTUM WELLS WITH EXTREMELY HIGH CONDUCTIVITIES

K.- J. FRIEDLAND, R. HEY, H. KOSTIAL, K. PLOOG
Paul-Drude-Institut für Festkörperelektronik, Hausvogteiplatz 5-7, D-10117 Berlin, Germany

We present a new concept to reduce impurity scattering in remotely doped GaAs single quantum wells by using heavy-mass X-electrons in the short-period AlAs/GaAs superlattice barriers to smooth the potential fluctuations of the ionized Si dopants. Electron mobilities as high as 120 m^2/Vs at electron densities of $1.6 \cdot 10^{16} m^{-2}$ are obtained in 10nm GaAs SQW in the one-subband conductivity mode without any parallel conductance. In addition to magnetotransport we present self-consistent calculations to demonstrate the applicability of our concept.

Two-dimensional electron gases (2DEG) with ultrahigh conductivity are important for both fundamental research and for applications in low-noise and high-frequency devices [1]. However, the increase of conductivity by simply increasing the dopant density is limited by the occurrence of parallel conductivity in the doping channel or by the population of excited subbands. Carriers causing parallel conductance have low mobility μ and can not increase the 2DEG conductivity, while the population of the higher subbands results in additional scattering processes. The other important prerequisite for high conductivity is high electron mobility. The remote impurity scattering can be suppressed effectively by large distances of the dopants from the conducting channel resulting, however, in low carrier densities.

We propose a new concept for both the enhancement of the carrier concentration n and the reduction of remote impurity scattering in GaAs single quantum wells (SQW), thereby significantly increasin the $n\times\mu$- product, i.e. the conductivity. The barrier for the GaAs SQW is formed by AlAs/GaAs short-period superlattices (SPSL). In AlAs/GaAs SPSL the type-I-to-type-II crossover occurs with the lowest-energy X-like conduction-band state [2] (X-electron). Due to their considerably higher effective mass, the X-electrons exhibit low mobility and contribute less to the parallel conductance. On the other hand, the heavy carriers smooth the potential fluctuations caused by the ionized impurities. We present magnetotransport experiments to confirm the applicability of this concept. Self-consistent calculations show that the lowest-energy X-like conduction-band states are occupied in the AlAs/GaAs superlattice close to the doping layer.

The structures under investigation were grown by solid-source molecular beam epitaxy on GaAs (001) substrates. The free carriers in the 10nm GaAs SQW are provided by remotely δ-doping with Si at a distance of 14nm (sample S1) and 10nm (sample S2). The barriers consist of 60 periods of 4ML AlAs /

8ML GaAs-SPSL. Single Si δ-doping sheets with a dopant concentration of $N^{2D}=2.5\cdot10^{16}m^{-2}$ were placed on both sides of the SQW into a GaAs layer of the SPSL. The low-temperature magnetotransport properties we studied on samples with Hall-bar-geometry covered by a Ti/Au gate electrode to change the electron density. The insert of Fig.1 shows the dependence of the parallel component of the resistivity tensor ρ_{xx} on the magnetic field. A very high mobility at surprisingly high electron densities is observed. The one-subband occupation is manifested by the one frequency Shubnikov-de-Haas (SdH) oscillation. The electron densities n_{SDH} derived from the SdH frequency equal n_H from the low field Hall-effect. Together with the exact value $\rho^{\nu}_{xy} = h/(e^2\nu)$ of the quantized resistivity in the quantum Hall regime for the filling factor ν these results demonstrate that the conductance is determined exclusively by electrons in the lowest 2DEG- subband. Additionally, ρ^{ν}_{xx} is zero at the Hall-plateaus with an integer filling factor ν, indicating the absence of any parallel conductance in the doping region. The dependence of the mobility μ_H on the electron density n_H is very unusual (see Fig.1). While we observe a power-like dependence $\mu_H \sim (n_H)^{\kappa}$ with $1.5<k<2$ at lower densities (for example $n_H<0.95\cdot10^{16}m^{-2}$ in sample S1) the mobility increases dramatically up to 120 m^2/Vs for $n_H>0.95\cdot10^{16}m^{-2}$ within a very narrow electron density region. The power-like behaviour of μ_H at lower densities is usually related to the scattering at remote ionized impurities randomly distributed in the doping plane [3,4]. To explain the dramatic increase of the electron mobility at higher densities we must take into account additional less movable carriers in the SPSL very close to the doping layers which are able to smooth the potential fluctuations effectively.

To clarify the nature of the additional carriers we show in Fig.2 the results of model calculations of the potential and charge distribution in our structures [5]. These calculations account for two different kinds of electrons. First, we consider the usual Γ-electrons with the isotropic effective mass $m_\Gamma/m_e = 0.07$ which form the lowest subband 2DEG in the SQW with the high electron mobility. The higher SQW-subbands as well as the Γ-electron states in the SPSL are more than 170 meV above the Fermi-energy and can, therefore, not be occupied at low temperatures. This is the reason for the possible enhancement of the electron density in the SQW up to $1.6\cdot10^{16}$ m^{-2} without causing any parallel conductance. The maximum density for single subband conductivity is not yet known. Our model calculations show this value to be much higher than $2\cdot10^{16}m^{-2}$. Next we have to include the higher conduction band minima of the AlAs/GaAs-SPSL system into the calculation. The superlattice period is chosen short enough to get the X-like conduction-band states the lowest energy states. The calculations show that states with the lower in-plane effective mass

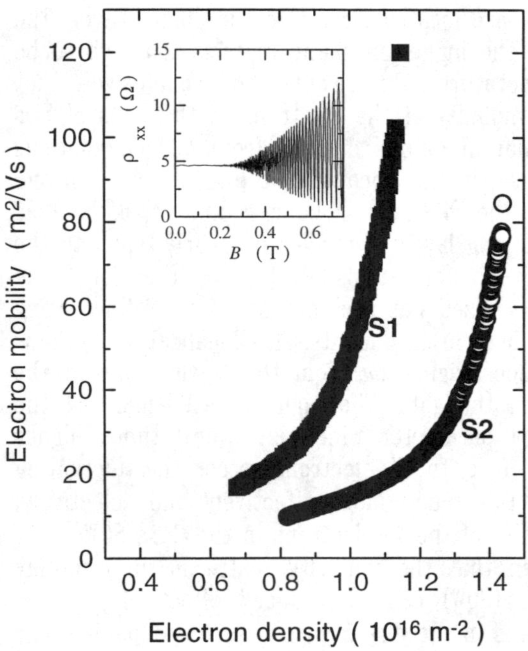

Fig.1. Dependence of the electron mobility μ_H on the electron density n_H obtained at T=0.33K and B=0.1T from samples with gate. The single points of maximum mobility in both curves were measured on the same samples without gate.
The insert shows the dependence of ρ_{xx} on the magnetic field for the sample S1 without gate.

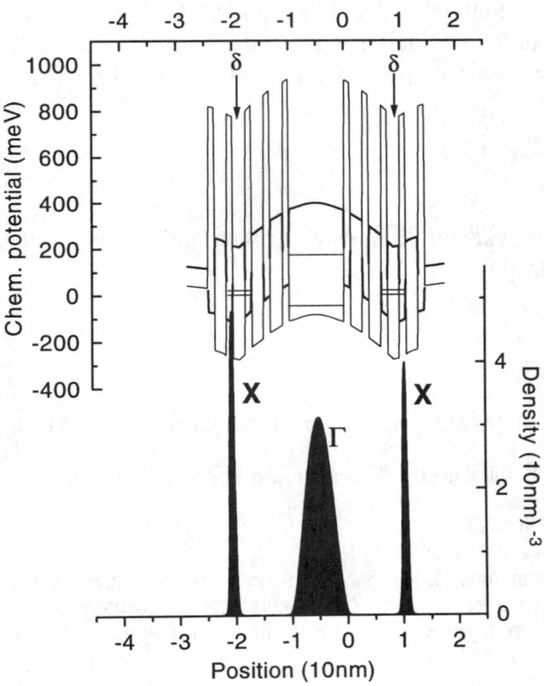

Fig.2. Calculated distribution of the potential and of the X-and Γ-carriers. Our simple calculations do not account for the stress and intermixing effects as well as for the lateral potential relief in the SPSL but show quantitatively that in the system the X-electrons appear in the AlAs -'well' closest to the δ-doping GaAs -'well'. Their quantity and spatial distribution, however, depends sensitively on the

component are occupied first. They are located closest to the doping layer. The sublevel energy of the states with the higher in-plane mass is too high to be occupied by electrons at low temperatures. Our results show conclusively that with X-electrons in the SPSL the mobility of the electrons in the GaAs SQW is considerably increased. The explanation follows directly from our calculations which show in Fig. 2 that the X-electrons are located in the AlAs- 'well' on one side of the doped GaAs- 'well' in the SPSL. Their Bohr-radius a^X_B as well as their nominal distance from the doping layer is smaller or nearly equal to the average distance between the Si-dopant atoms $d_D=(N_D)^{-0.5}$. Therefore, the X-electrons can be very easily localized at the minima of the fluctuating potential caused by the randomly distributed dopants. This localization explains the absence of parallel conductance originating from the X-electrons in the magnetotransport experiments with the gate. The high effective mass of the X-electrons results in a screening parameter which is several times higher than that of the Γ-electrons. Therefore, the X-electrons screen the fluctuating potential of the ionized Si impurities much more effectively and selectively, leading to an increase of the mobility of the Γ-electrons in the GaAs SQW.

In conclusion, we have shown that the impurity scattering in remotely doped GaAs single quantum wells (SQW) can be reduced effectively by the presence of heavy-mass X-electrons in the direct vicinity of the dopant atoms in the barrier. These X-electrons exhibit an extremely high screening capability and are able to smooth the potential fluctuations, caused by the random distribution of the dopants. They are localized and do not contribute to the conductivity.

ACKNOWLEDGEMENTS

The authors wish to thank M. Hoericke for MBE wafer growth and A. Riedel for assistance with the sample preparation.

REFERENCES

1. G. Tränkle, H. Rothfritz, R. Müller and R. Weimann, Journ. of Crystal Growth 120, No. (1-4), 240 (1992).
2. M. Holtz, R. Cingolani, K. Reimann, R. Muralidharan, K. Syassen and K. Ploog, Phys. Rev. B41 (6), 3641 (1990).
3. A. Gold, Phys. Rev. B38 (15), 10798 (1988).
4. K. Hirakawa and H. Sakaki, Phys. Rev. B33 (12), 8291 (1986).
5. K.-J. Friedland and R. Zimmermann unpublished. The calculations are performed by the one-dimensional self-consistent solution of the Schrödinger and Poisson equations in the effective mass aproximation, including exchange and correlation interaction energy.

EFFECT OF THE UNIAXIAL STRESS ON THE QUANTUM TRANSPORT OF TWO-DIMENSIONAL HOLES IN THE Si-MOSFETs

S.I. DOROZHKIN, V.E. BISTI
Institute of Solid State Physics RAS, Chernogolovka, Moscow district, 142432, Russia

G. LANDWEHR
Physikalisches Institut der Universität Würzburg, Am Hubland 97074 Würzburg, Germany

We have investigated effect of an uniaxial stress of both signs (compression and tension) on the Shubnikov-de Haas oscillations (SdHOs) and the anomalous magnetoresistance (AMR) in Si (110) MOSFETs. The stress was found to result in a drastic shift of a node in the oscillation beating pattern but to have very small influence on the occupation of two subbands. The latter result disagrees qualitatively with all available calculations.

1 Introduction

Methods of the quantum transport provide a powerful tool for investigations of the energy spectrum of electronic systems. In particular case of two-dimensional electronic systems, SdHOs are widely used to determine occupation of different subbands and an effective mass of carriers. Beating of SdHOs allows mesurements of the zero-magnetic-field spin-splitting [1]. This effect appears in Si (110) MOSFETs [2,3] due to the spin-orbit interaction in the presence of an asymmetrical quantum well at the surface. AMR provides information about spin-orbit relaxation rate from which the splitting value can be also obtained. Here we report on the effect of an uniaxial stress on this splitting and on relative population of two subbands in the Si (110) MOSFETs. We also compare our results with calculations available for this system (see [4] and references therein).

2 Experiment

We have measured five samples of metal-oxide-semiconductor field-effect transistors (MOSFETs) produced on a Si (110) surface. We used bending of samples to apply an uniaxial stress of both signs along [001] crystal axis. All the results were absolutely reproducible both for many coolings down to liquid helium temperatures and from sample to sample. The main effect of the stress σ is a shift of the node position in the SdHO beating pattern (see Fig.1 where

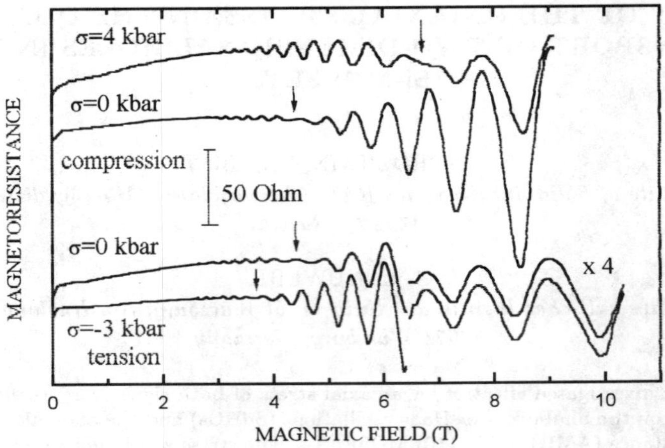

Figure 1: Magnetoresistance for stress of two signs. $n_s = 2.7 \times 10^{12}$ cm $^{-2}$, $T = 1.4$ K.

node positions are marked by arrows). Note, that the positions of the SdHO extrema are not changed by the stress and the oscillations between two shifted positions of the node are inverted. The latter clearly demonstrates that the phase of the SdHOs changes by π at the node [3]. It is important that the oscillation amplitudes far from the node (lower pair of the curves) are changed only slightly by the stress. This, in particular, shows that the sample quality is not influenced by the bending. Experimental results on the stress dependence of the node are shown in Fig.2. Note, that ν_0 varies approximately linearly with the stress and can be changed up to a factor of three. While ν_0 strongly increases with n_s its derivative $d\nu_0/d\sigma$ demonstrates only weak dependence on n_s, if any. Typical experimental data at hole densities $n_s > 2.8 \times 10^{12}$ cm^{-2} where two subbands are occupied are shown in Fig.3. There are SdHOs of two frequencies in Fig.3: the higher frequency corresponds to the first (ground) subband and the lower one to the second subband. Corresponding carrier densities n_{s1} and n_{s2} determined from these two frequencies are shown in the figure. Note, that the stress of both signs shifts slightly the positions of the high-frequency oscillations to lower magnetic fields (the shift is about 1% at 4 kbar). This reflects redistribution of holes between the subbands. Small value of the effect is a strong evidence that both subbands originate from the heavy holes, the second subband being an excited one. The suppression of the low-frequency oscillations at $\sigma = -1.3$ kbar is caused by the shift of the node in the beating effect for these oscillations.

Another effect of the deformation is a variation of AMR existing in weak

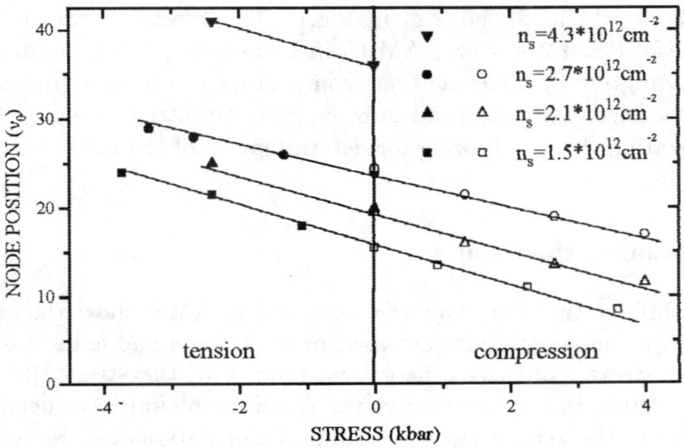

Figure 2: Dependencies of the node positions, characterized by corresponding filling factor ν_0 of the magnetic levels, on the stress for different n_s. Straight lines are guides for eyes.

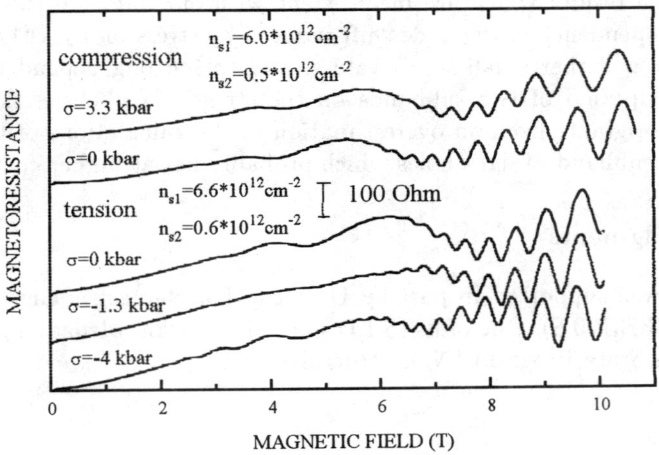

Figure 3: Magnetoresistance oscillations in the case of two occupied subbands.

magnetic fields which can be seen in Fig.1. The effect becomes much more pronounced at $T = 4.2$ K where AMR can be positive, nonmonotonic or negative depending on n_s. The tension (compression) decreases (increases) the magnetoresistance. The AMR data can be perfectly fitted by equations of the weak localization theory allowing for determination of the zero-magnetic-field spin-splitting.

3 Discussion of the results

Both, the shift of the node and the variation of AMR show that the zero-magnetic-field spin-splitting is decreased under tension and is increased under compression stress. Intensive theoretical studies of the stress effect on the beating of SdHOs and occupation of the second subband have been recently published [4]. At the zero stress the results of calculations can reasonably describe dependence of the node on n_s but only for the choice of Lattinger parameters different from the usually used Hensel's set. Note, that in contrast to all previous calculations this choice of the parameters leads to the result that the second subband is an excited subband of heavy holes, which is consistent with our observations. At a finite stress results [4] agree only qualitatively with our experimental data, giving correct sign of the node shift induced by deformation. The results which are inconsistent with our data are the following: (i) strong dependence of the node shift induced by stress on n_s, (ii) saturation of the shift at rather small stress values, and (iii) strong dependence of the relative occupation of two subbands on the stress. We believe that all the differences originate from an overestimation in the calculations of the role of the ground subband of light holes which probably lies at higher energy.

Acknowledgments

This work was supported in part by Deutsche Forschungsgemeinshaft (grant 436 RUS 113/37/0 S). One of us (S.I.D.) gratefully acknowledges a grant from the Russian State Program "Nanostructures".

References

1. S.I.Dorozhkin, *Phys. Rev.* B **41**, 3235 (1990).
2. K. von Klitzing, G. Landwehr, and G. Dorda, *Solid State Commun.* **14**, 387 (1974).
3. S.I. Dorozhkin and E.B. Ol'shanetskii, *JETP Lett.* **46**, 502 (1988).
4. W.O.G. Schmitt, *Phys. Rev.* B **50**, 15239 (1994).

MAGNETOTUNNELLING SPECTROSCOPY OF QUANTUM DOTS FORMED BY SELF-ORGANISED EPITAXIAL GROWTH

L. EAVES, I. E. ITSKEVICH*, T. IHN, A. THORNTON,
P. H. BETON, P. C. MAIN, T. J. FOSTER AND M. HENINI
Department of Physics, University of Nottingham, Nottingham NG7 2RD, UK

We investigate low temperature electron tunnelling through a single barrier GaAs/AlAs/GaAs heterostructure in which Stranski-Krastanov-grown InAs quantum dots are incorporated in the centre of the AlAs barrier. The current-voltage curves exhibit a series of pronounced peaks which are absent in a control sample grown without InAs in the barrier. For magnetic field, *B*, applied parallel to the current, each peak splits into a set of sharp components at fields as low as 0.4 T. They correspond to Landau level quantisation of the occupied electron states in the emitter accumulation layer. This proves that each peak arises from tunneling into a *single* discrete zero-dimensional state in the tunnel barrier. For increasing *B* perpendicular to the current, a quadratic shift in voltage of the peak positions and a decrease in amplitude of the peaks are observed. We thereby estimate the spatial extent of these states to be 7-10 nm, consistent with the size of InAs islands measured by scanning tunnelling microscopy.

1. Introduction

The structural and optical properties of self-assembled semiconductor quantum dots have been intensively investigated in recent years. The system which has received most attention to date consists of InAs dots grown on the GaAs or (AlGa)As surface[1-12]. The dots are formed in the Stranski-Krastanov (SK) mode of hetero-epitaxial growth. This occurs when a thin epilayer, a few atoms thick, is grown on a surface of different chemical composition and lattice constant. At a critical layer thickness, hetero-epitaxial islanding is energetically preferred to planar growth and a planar array of quantum dots (QDs) is formed. The dots can be capped by continuing growth of lattice matched layers.

Optical spectroscopy[2,3,5,7,10-12] provides precise information about the distribution of dot sizes as it is directly related to the quantum confinement

---

*On leave from Institute of Solid State Physics, Russian Academy of Sciences, Chernogolovka, Moscow District, 142432, Russia.

energies of electrons and holes. Optical spectra of individual dots have been obtained using near-field microscopy, and reveal extremely sharp intrinsic linewidths[2,3,7]. Information about the electron and hole energy levels associated with the dots have also been obtained by means of capacitance spectroscopy[5,6].

In this paper, we describe an alternative method of probing the electronic states of InAs quantum dots based on resonant tunnelling spectroscopy. The dots are embedded in a thin tunnel barrier of AlAs, sandwiched by two GaAs layers. We show that it is possible to tune the applied voltage so that resonant tunnelling occurs through an individual dot. As in the case of tunnelling studies on other types of laterally-confined heterostructure[13], a magnetic field applied either parallel or perpendicular to the direction of tunnelling has a strong effect on the tunnel current through the dots, and allows us to study the initial and final states of the tunnelling transition and estimate the spatial extent of the confined electron wavefunction in the dot.

2. Device Structure

A schematic diagram of our device is shown in Figure 1. A 1 μm-thick heavily doped Si buffer layer was grown on a (100) n^+GaAs substrate, followed by 100 nm of undoped GaAs. The InAs dots are incorporated in an undoped 10 nm AlAs barrier of thickness 10 nm. They were formed by the islanding of a 1.8 monolayer-thick InAs layer in the centre of the barrier. The barrier was followed by a 100 nm undoped GaAs layer, and capped by 1 μm of n^+GaAs. A control tunnel device, lacking the InAs layer but with other parameters identical, was also prepared. Optical lithography was used to produce circular mesas of varying diameters up to 100 μm. AuGe was alloyed into the n^+GaAs layers to form ohmic contacts.

To characterise the device, scanning electron and tunnelling microscopy (SEM and STM) and photoluminescence spectroscopy were employed. SEM and STM imaging were performed on layers with similar design, but in which growth terminated after deposition of the InAs layers. These structures had a dot density of $\sim 2 \times 10^{11}$ cm^{-2} with a typical dot size of $\sim(10 \times 10)$ nm^2.

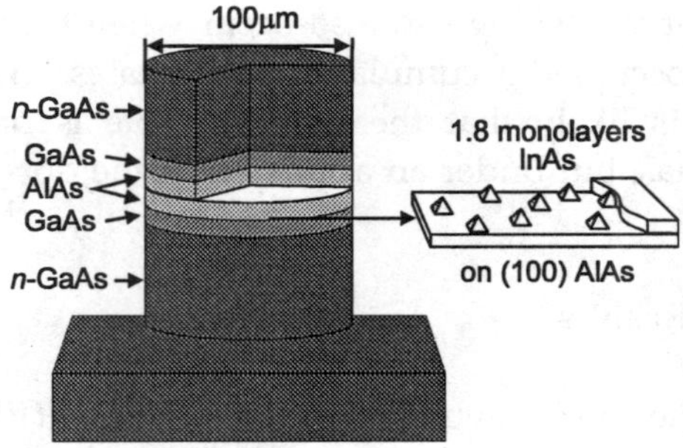

Figure 1 A schematic diagram of our device used for tunnelling spectroscopy of InAs dots.

A PL spectrum of our tunnel structure incorporating the InAs dots was obtained at 4 K using He-Ne laser excitation ($\lambda = 633$ nm), as shown in the inset of Fig. 2. The line corresponds to the emission from the dot ensemble, and is similar to that reported by other groups for similar layers[2,11,12].

The main part of Figure 2 shows a schematic band diagram of our device under an applied bias voltage. A two-dimensional electron gas (2DEG) forms an accumulation layer in the undoped GaAs emitter region adjacent to the tunnel barrier, and a tunnel current can flow through the barrier. If discrete states exist in the barrier, due for example to impurities or to the

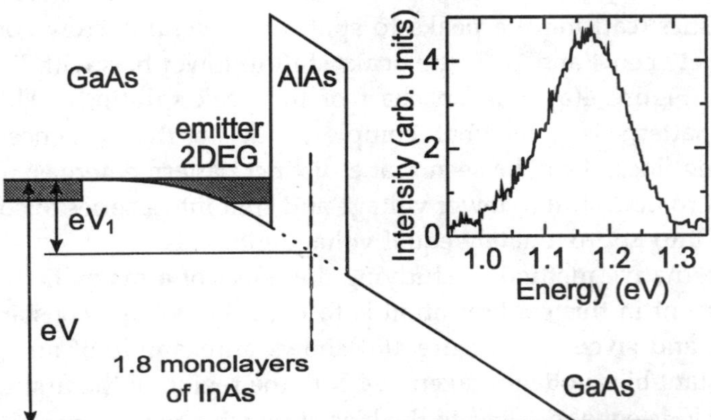

Figure 2 A schematic energy band diagram of the device under an applied voltage, V. Discrete energy levels in the tunnel barrier are associated with quantum dots. The inset shows the photoluminescence spectrum from this sample at 4 K and zero applied bias.

InAs dots, resonant tunnelling can also occur when their energy levels are resonant with the occupied accumulation layer states. As will be discussed in detail later, it is likely that the dot ensemble is partially filled with electrons at zero bias, but under an applied bias the dots discharge into the collector region.

3. Experimental Results

Figure 3 shows the current-voltage characteristics, *I(V)*, in forward and reverse bias, which correspond respectively to electron flow from and to the substrate. An *I(V)* curve of the control sample is also shown for comparison. The impedance of both devices is extremely high, $\sim 10^{13}$ Ω, at low bias voltages up to about 100 mV. In forward bias at 4 K, a series of fairly sharp features are observed with peak currents of a few picoamps (see $B = 0$ trace, upper box). These peaks sharpen further and the maximum current increases on lowering the temperature to 300 mK[14]. For reverse bias, the current increases in a series of steps of similar amplitude which become sharply defined at 350 mK (curve *c*), suggestive of charge build-up. In contrast, the *I(V)* curve of the control tunnel device (curve *a*) shows neither peaks nor steps, but a current which increases monotonically with increasing voltage.

Figure 3 also shows a series of 4.2 K current-voltage curves over a range of magnetic field, B, applied perpendicular to the plane of the tunnel barrier ($B \parallel I$). Remarkably, a magnetic field as small as 0.4 T has an effect on the *I(V)* characteristics, causing the peaks to split into several narrow components. At fields between 4 and 10 T, the peaks shift to lower bias with little change in shape. Figure 4(b) is a fan chart of the peak splitting. Although the splitting pattern is somewhat complex, due to the presence of many overlapping lines, it can be seen that a distinct pattern emerges: most of the peaks at zero field shift to lower voltage and split into a series of components separated into approximately equal voltage intervals.

An alternative method of studying the effect of a magnetic field on the tunnel current in this configuration is to hold the voltage constant at some value, V_c and sweep *B*. Figure 4(a) shows an example of an *I(B)* plot at three constant bias voltages taken at 4 K in the region of the first sharp peak in *I(V)*. If V_c is equal or close to the bias at which a peak occurs, we observe pronounced oscillations in *I(B)*. Their maxima and minima shift to smaller *B* with increasing V_c. The *I(B)* curves exhibit no structure at V_c just below, or just above, a peak in *I(V)*. The *I(B)* curves show pronounced sharpening

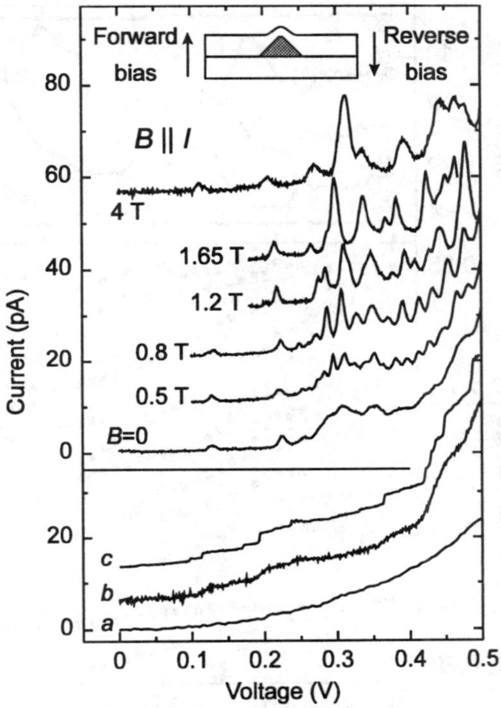

Figure 3 The *I(V)* characteristics of a 100 μm diameter mesa. Lower: (a) control sample; (b) for $B = 0$ in *reverse* bias at 4.2 K and (c) at 0.35 K. Upper: *I(V)* in *forward* bias at 4.2 K at various $B \| I$. Curves are offset.

of resonant structure at temperatures below 300 mK. The evolution of this structure with bias and magnetic field is considered in detail in reference 14.

We have also investigated the *I(V)* characteristics for magnetic field applied in the plane of the tunnel barrier ($B \perp I$), as shown in Figure 5(a). In this configuration, no clear peak splitting is observed at 4 K and there is no qualitative change in the *I(V)* curves. However, with increasing *B* the voltage position of each peak shifts quadratically to lower bias, as shown in Figure 5(b). This is accompanied by a decrease of the peak amplitude. For example the peak at around 0.1 V falls to ~50% of its zero field amplitude at 10 T.

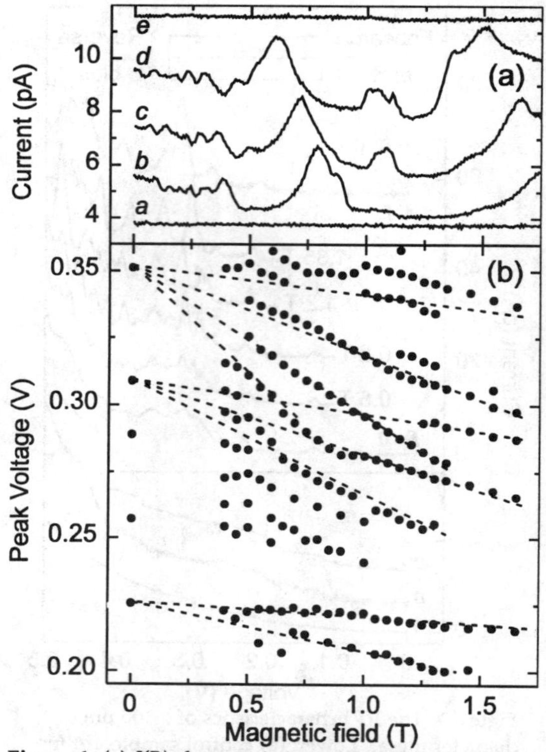

Figure 4 (a) *I(B)* characteristics at various *V*: *a* - 105 mV, *b* - 114 mV, *c* - 115 mV, *d* - 116 mV, *e* - 130 mV. Curves are offset. (b) Fan chart of the peaks in *I(V)* versus *B*‖*J*.

4. Discussion

In this section, we consider the origin of the resonant peaks in *I(V)* and argue that they arise from tunnelling through discrete, zero dimensional electron states of the InAs quantum dots in the barrier. We attribute the peaks to the resonant alignment of the occupied emitter states in the accumulation layer with zero-dimensional electronic states of individual quantum dots in the tunnel barrier. In a magnetic field normal to the sample plane, the energy spectrum of the 2DEG in the emitter accumulation layer evolves into a set of discrete Landau states. This explains the peak splitting and divergence with increasing *B*, and the shift of the peaks to lower *V*, shown in Figure 4(b).

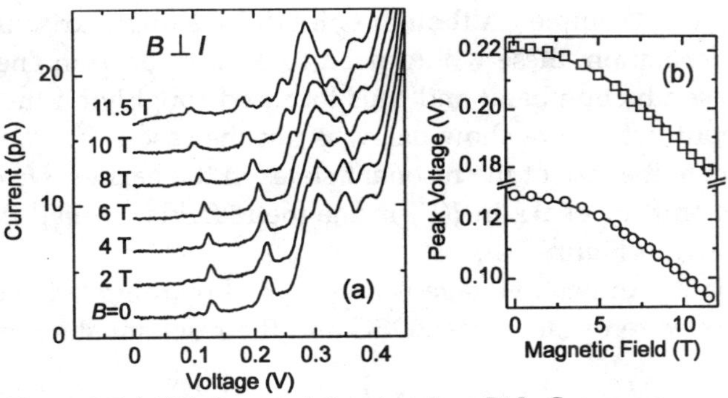

Figure 5 (a) *I(V)* characteristics at various *B⊥I*. Curves are offset. (b) Diamagnetic shift of lower-voltage peaks in *I(V)*. Solid lines are parabolic fits.

Despite the complexity of the picture due to many overlapping lines, a distinct pattern emerges: the peaks shift to lower voltage, and the bias positions of several sets of adjacent peaks diverge with *B*. The dashed lines are guides for the eye corresponding to possible Landau level fans. The splitting (corresponding to the cyclotron energy) is different for each set due to change in the electrostatic leverage with voltage. In our case, the maxima in *I(B)* occur when the magnetic field brings an occupied Landau level in the 2DEG into resonance with the energy levels in the barrier. Since the Landau quantisation energy of a given emitter state *increases* with *B*, the peaks move to *lower* bias. In effect then, due to the discrete nature of the zero dimensional states in the dot, both *I(V)* and *I(B)* probe the density-of-states (DOS) of the 2DEG in the emitter accumulation layer.

Note that the magneto-oscillations in *I(B)* shown in Figure 4(a) behave quite differently from those previously reported in single barrier tunnelling devices[15], for which the maxima generally shift to *higher B* with increasing V_c due to a Landau level depopulation effect and the increasing sheet density of the emitter 2DEG with increasing bias.

Our *I(V)* characteristics resolve the discrete DOS of 2D electrons at *B* values as low as 0.4 T, where the cyclotron splitting is less than 1 meV for Γ-valley electrons in GaAs. Such resolution is possible only if the tunnelling occurs through *zero-dimensional* states in the barrier with *discrete* energy levels. Possible zero dimensional states in the barrier might be due to donors, point defects or incorporated QDs. We reject the possibility of residual donor-assisted tunnelling, because there are no sharp features in the

$I(V)$ of the control sample. Although point defects might arise in the barrier during dot formation, these would need to be very deep in energy relative to the AlAs conduction band and thus localised roughly on the scale of the lattice constant, which we show below is not the case.

To estimate the size of the resonant states in the barrier, we consider the diamagnetic shift of peaks in $I(V)$ in magnetic fields applied *normal* to the current, shown in Figure 5(b).

The variation in peak voltage is determined by the relative diamagnetic shifts of the electron state in the 2DEG and the resonant state in the barrier:

$$e\Delta V = f\frac{e^2B^2}{2}\cdot\left(\frac{\langle z_e^2\rangle}{m^*} - \gamma\frac{\langle z_d^2\rangle}{m_d}\right).$$

Here $\sqrt{\langle z_e^2\rangle} \approx 8$ nm is the spatial extent of the electron wave function in the emitter, estimated using self-consistent numerical calculations; $\sqrt{\langle z_d^2\rangle}$ is the spatial extent of the barrier state; m^* and m_d are electron effective masses for the 2DEG and for the barrier state; γ is a geometrical factor arising from the zero dimensional confinement ($\gamma = 0.5$ for a spherically symmetric state and $= 1.0$ for a planar 2DEG, as in the case of the emitter); f is the electrostatic leverage factor. We expect conduction band non-parabolicity in InAs to increase m_d from the band-edge value. Hence, using the GaAs Γ-valley effective mass, 0.067 m_e, both for m^* and m_d, $\gamma \approx 0.5$, and $f \approx 10$, consistent with the sample geometry, we estimate the size of the resonant barrier state $\Delta z_d \approx 2\sqrt{\langle z_e^2\rangle} \approx (10 \pm 5)$ nm. The estimate is rather rough, because of uncertainty in the values of parameters used. Nevertheless, the estimate is *inconsistent* with strongly localised deep point defect states in the AlAs barrier. Conversely, it is fully consistent with the size of InAs QDs measured by SEM. Hence we conclude that the peaks in $I(V)$ originate from the tunnelling through the electron states of individual self-organised InAs dots incorporated in the barrier.

The observed decrease of peak amplitude with increasing $B \perp I$ in Figure 5(a) is qualitatively similar to that reported recently for magnetotunnelling into donor states[16]. In the experiments on donors, the decrease of peak amplitude with increasing B was used to map out the spatial extent of the donor wavefunction in **k**-space: the magnetic field causes particle tunnelling over a length Δs to acquire an additional in-plane wavevector $\Delta k = eB\Delta s/\hbar$. Since the tunnel current through the dots falls by about a half in a field of $B_{1/2} = 10$ T, this provides a second estimate of the spatial extent of dot wavefunction $\Delta z_d \simeq 2/\Delta k_d = 2\hbar/eB_{1/2}\Delta s$. Using $\Delta s \simeq 15$ nm we obtain $\Delta z_d \sim 9$ nm. This provides a separate estimate for the spatial extent of the dot wavefunction which is consistent with the value obtained from the diamagnetic shift. However, the current is not exponentially quenched at sufficiently high fields, as was observed in the case of similar experiments on resonant tunnelling through shallow donor impurities[16].

Now we discuss briefly the asymmetry of the *I(V)* characteristics in *forward* and *reverse* bias, shown in Figure 3. This results from the asymmetry of dot positions in the AlAs barrier (see Figure 3, inset). The dots are grown on the centre plane of the barrier, but the effective thickness of the AlAs layer covering the dots is reduced due to the size and shape of the dots, and thus this layer has a higher tunnelling probability than the AlAs layer on the substrate side. Thus for *reverse* bias the tunnelling rate out of the dots is much smaller than the rate of tunnelling into the dots from the accumulation layer. This leads to the charge build-up, giving the distinct steps in *I(V)* curve at very low temperatures . Conversely, for *forward* bias the tunnelling rate out of the dots is higher than the tunnelling-in rate, leading to the possibility of using the resonant tunnelling as a probe of single-particle energy levels in the dot.

We have already stated that the dot ensemble has a wide distribution of electron state energies. On the other hand, the probing of the 2DEG DOS for $B \parallel I$ shows a resolution of better than 1 meV at 4.2 K. Therefore the peaks in *I(V)* correspond to tunnelling through electron states of *individual* dots. This conclusion is also supported by the peak current values. The tunnel current through a *single* dot can be written as $I = e\nu \exp(-2\kappa d)$, where ν is the attempt frequency, $d = 5$ nm is the barrier half-thickness, $\hbar\kappa = \sqrt{(2m^*\Delta E)}$, and ΔE is the height of the barrier. Using the effective mass $m^* = 0.0067m_e$ and $\Delta E \approx 0.8$ eV, expected for this heterostructure system, we obtain peak current values of a few picoamps, consistent with the experiment.

An important question is: why do we observe tunnelling through only a few single dots rather than the ensemble of about 10^7 dots in a typical mesa? To understand this, note that the PL spectrum from the sample (inset of Figure 2) suggests that for the majority of dots the electron ground energy level is *below* the conduction band edge E_c, in agreement with capacitance spectroscopy studies[5,6]. These levels are unavailable for energy-conserving tunnelling processes. We observe tunnelling through extremal dots with electron level energies above E_c. Such dots can arise due to fluctuations in size (30-40%), shape and strain or possible Al alloying of dots. Our picture implies that a fraction of dots should be charged at zero bias, in order to align the chemical potentials of the dot ensemble and the collector and emitter n-doped contact layers. The accumulation of the 2DEG is accompanied by the dot discharge, which contributes strongly to the dependence of the leverage factor on voltage.

In conclusion, we have observed resonant tunnelling through the electron states of self-organised InAs quantum dots incorporated in an AlAs matrix. Whereas the optical spectra are dominated by the dot ensemble, the tunnel current is controlled by individual dots. This behaviour is consistent with energy-conserving tunnelling transitions from a series of occupied Landau levels in the emitter-accumulation layer to a distribution of discrete energy levels in the tunnel barrier. We associate these discrete levels with individual InAs quantum dots, since control samples grown under similar conditions but lacking the InAs layer in the tunnel barrier, show only a monotonically-varying *I(V)* characteristic, with no sharp peaks.

Acknowledgements: We are grateful to J. R. Middleton, M. Steer and M. S. Skolnick. This work is supported by EPSRC (UK). IEI & LE thank the Royal Society and EPSRC for research fellowships.

References

1. J.M. Moison *et al*, *Appl. Phys. Lett.* **64**, 196 (1994).
2. J.-Y. Marzin *et al*, *Phys. Rev. Lett.* **73**, 716 (1994).
3. S. Fafard *et al*, *Phys. Rev. B* **50**, 8086 (1994).
4. D. Leonard *et al*, *Phys. Rev. B* **50**, 11687 (1994).
5. H. Drexler *et al*, *Phys. Rev. Lett.* **73**, 2252 (1994).
6. G. Medeiros-Ribeiro *et al*, *Appl. Phys. Lett.* **66**, 1767 (1995).
7. M. Grundmann *et al*, *Phys. Rev. Lett.* **74**, 4043 (1995).
8. S. Ruvimov *et al*, *Phys. Rev. B* **51**, 14766 (1995).
9. M. Grundmann *et al*, *Phys. Rev. B* **52**, 11969 (1995).
10. D. I. Lubyshev *et al*, *Appl. Phys. Lett.* **68**, 205 (1996).
11. R. Heitz *et al*, *Appl. Phys. Lett.* **68**, 361 (1996).
12. M. Grundmann *et al*, *Appl. Phys. Lett.* **68**, 979 (1996).
13. P. H. Beton *et al*, *Phys. Rev. Lett.* **75**, 1996 (1995).
    N. Mori *et al*, *Phys. Rev. B* **52**, 1504 (1995).
14. A. Thornton *et al*, this conference proceedings.
15. F.W. Sheard *et al Phys. Scr.* **T19**, 179 (1987).
16. J.-W. Sakai *et al*, *Phys. Rev. B* **48**, 5664 (1993).

PRECURSORS AND TRANSITION TO CHAOS IN RESONANT TUNNELING

G.S. BOEBINGER
Bell Laboratories, Lucent Technologies, 700 Mountain Ave.,
Murray Hill, N.J. 07974, USA

Despite great theoretical interest, there are relatively few experimental studies of simple quantum systems whose classical counterparts exhibit an experimentally tuneable transition from integrable to chaotic dynamics. Recently, using resonant tunneling spectroscopy, Fromhold, et al. [1] have demonstrated that the wide quantum well exhibits chaotic dynamics in a sufficiently intense magnetic field which is tilted away from perpendicular to the quantum well. More recently, we have discovered a distinct transition from integrable to chaotic electron dynamics in this system [2]. The evolution of the chaos transition is mapped systematically by varying bias voltage U, magnetic field B and tilt anlge Θ.

As the tilt angle is increased, the system becomes increasingly chaotic. For a perpendicular magnetic field ($\Theta = 0°$), peaks in the tunneling current correspond to the quantum well subbands. At small tilt angles, ($10° < \Theta < 30°$) the resonant tunneling spectra show the quantum well subbands, but distinct peak doubling regions have emerged. At larger tilt angles, the quantum well subbands make a transition to a peak tripling region. Poincare section calculations identify these peak doubling and tripling regions as the precursors of the chaos transition. They result from the bifurcation and trifurcation of the stable period orbit which corresponds to the quantum well subbands.

Roughly parabolic boundaries, $U \propto B^2$, are found to border the peak doubling and tripling regions. These boundaries result from the invariance of the classical dynamics along such parabolic boundaries. Data from samples of four different well widths d find additional classical scaling, $B \propto 1/d$.

A recent periodic orbit theory [3] of the quantum well in a tilted magnetic field has found that the bifurcations and trifurcations which account for the observed peak doubling and tripling regions also nucleate an organized sequence of tangent bifurcations. The tangent bifurcations give rise to new, topologically-distinct periodic orbits. This accounts for the wide variety of periodic orbits which have been identified as playing a role in the resonant tunneling spectra [1, 2].

This work is in collaboration with G. Müller, H. Mathur, L.N. Pfeiffer, K.W. West, and A. Passner (Bell Laboratories, Lucent Technolgies), A.D. Stone and E. Narimanov (Yale University).

References

[1] T.M. Fromhold, L. Eaves, F.W. Sheard, M.L. Leadbeater, T.J. Foster, and P.C. Main, Phys. Rev. Lett. 77, 2608 (1994).

[2] G. Müller, G.S. Boebinger, H. Mathur, L.N. Pfeiffer, and K.W. West, Phys. Rev. Lett. 75, 2875 (1995).

[3] E. Narimanov and A.D. Stone, to be published.

UNIVERSAL CONDUCTANCE FLUCTUATIONS AND TELEGRAPH NOISE SPECTROSCOPY TO INVESTIGATE THE MAGNETISATION OF SELF ORGANISED ErAs QUANTUM DOTS.

F. COPPINGER[†], J. GENOE[‡], L.B. RIGAL, D.K. MAUDE, U. GENNSER[§]
AND J. C. PORTAL
Grenoble High Magnetic Field Laboratory, MPI-CNRS, 38042 Grenoble and INSA, 31077 Toulouse, France

K. E. SINGER, P. RUTTER, T. TASKIN, A.C.WRIGHT AND A. R. PEAKER
UMIST, Sackville Street, Manchester,M60 1QD, UK

Using low temperature magnetotransport techniques we demonstrate that the electrical transport is phase coherent in a GaAs matrix containing self organising ErAs quantum dots and wires of the order of 3 nm diameter. Using telegraph noise spectroscopy we are able to measure the switching of the extremely small magnetic moment (~10-50 μ_B) of an isolated antiferromagnetic cluster of ErAs. Above a critical temperature (T_c~500 mK) the switching is thermally activated. Below this temperature the switching rate saturates and it is the tunnelling of the magnetisation which dominates. The magnetic field dependence of the cross-over temperature (T_c) from thermally activated to tunnelling regimes is in good agreement with simple theory for macroscopic quantum tunnelling.

1 Introduction

We have recently shown[1] that electronic transport techniques can be used to probe the extremely small magnetic moment (~50μ_B) of an isolated antiferromagnetic ErAs cluster in a GaAs matrix. The switching of the magnetic moment of the cluster between two possible orientations leads to two level fluctuations in the resistance. Using the technique of telegraph noise spectroscopy it is possible to monitor the orientation of the magnetic moment of the cluster in real time and thus to probe the dynamics of the system. At high temperatures the switching is thermally activated but below a critical temperature ≈500 mK the switching rate saturates which is interpreted as a tunnelling of the magnetisation. This

---

[†] Current address: Electrical Engineering Dept. UCLA, CA 90024-1594, USA.

[‡] Current address : ETRO, VUB, Pleinlaan 2, B-1050 Brussel, Belgium

[§] Current address : Paul Scherrer Institut, Villigen, Switzerland.

is referred to a macroscopic quantum tunnelling (MQT) since it is a macroscopic variable (the giant spin of the cluster) which tunnels through the barrier separating the two states.

In this work we first of all present a more detailed study of the transport mechanism through the ErAs doped region. In lightly Er doped samples transport occurs via the shallow Se donor levels which have diffused from the n^+ contacts. With increasing Er concentration the rare earth getters the Se impurities and transport is dominated by a deep Er related level. For all samples tunnelling is the dominant transport mechanism at low temperatures. In the final section of the paper, the tunnelling of the magnetisation is probed using telegraph noise spectroscopy. The magnetic field dependence of the crossover temperature from the thermally activated to the tunnelling regime is compared with the WKB theory for tunnelling and a good agreement is found.

2 Samples

For the investigation n-i-n mesa structures consisting of a narrow Er-doped GaAs region sandwiched between 1μm thick 1×10^{18} cm^{-3} Se doped n+ contacts were prepared using molecular beam epitaxy (MBE). All samples have a 25nm thick layer co-doped with Er and Se on either side of the Er-doped 'intrinsic' region. The samples with different widths and Er concentrations in the intrinsic region are summarised in Table 1. Above a certain doping threshold ~ 7×10^{17} cm^{-2} the Er precipitates out to form either quantum dots or quantum wires[2] with a typical diameter ~3nm. For the 3×10^{19} cm^{-3} Er doped sample transmission electron microscope images (Figure 1) reveal that co-doping with Se modifies the growth conditions, leading to the formation of dots rather than the wires observed in the intrinsic Er doped layer. Typically the dots and wires have a diameter of ~3nm. For the

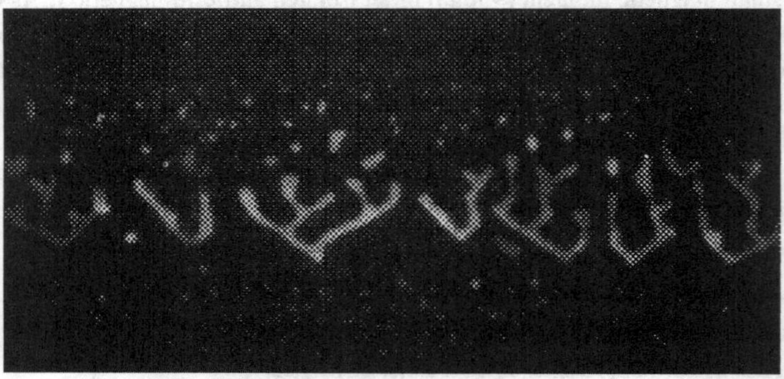

Figure 1 : Cross sectional transmission electron micrograph of sample 2077. The total height of the image is 200nm and the growth direction is upwards.

Table 1 : Summary of sample properties including the activation energy determined from the temperature dependence (see section 3).

Sample	Er (cm^{-3})	Thickness	E (meV)	Precipitates
2077	3×10^{19}	50 nm	150 ± 10	wires + dots
2112	2×10^{18}	50 nm	140 ± 5	dots
2140	2×10^{17}	100 nm	metallic	non
2141	2×10^{17}	200 nm	7 ± 2	non
2142	2×10^{17}	400 nm	> 35 meV	non

vertical transport measurements mesas of dimensions $10 \times 15\ \mu m^2$ were fabricated using standard photolithographic and etching techniques. Ohmic contacts were formed using alloyed Au-Ge-Ni. Resistance measurements were performed using classical lock-in techniques at a frequency of 670Hz with a current of 10-1000 nA in the temperature range 0.03-300K.

3 Magnetotransport

The magnetoresistance of sample 2077 with the magnetic field (H) parallel to the current at a temperature of 300mK is shown in Figure 2. A large negative magnetoresistance is observed, the temperature dependence of which is qualitatively similar to that reported in thin ErAs films and attributed to spin disorder scattering.[3] The observation of a large negative magnetoresistance suggests that the ErAs precipitates and in particular their magnetic properties play an important role in determining the electrical transport properties in our samples. ErAs is antiferromagnetic at low temperatures and low magnetic fields but an antiferromagnetic-paramagnetic phase transition is induced upon increasing temperature or magnetic field.[3] Superimposed on the magnetoresistance background are universal conductance fluctuations (UCF) of typical amplitude ~ $0.1e^2/h$, which arise from the interference of the partial electron wavefunctions as they traverse the structure. A

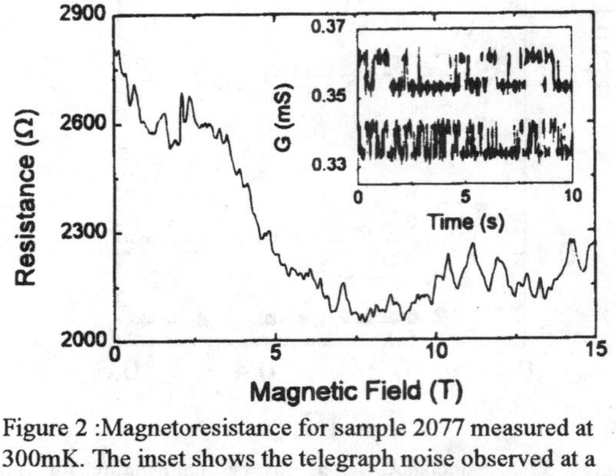

Figure 2 :Magnetoresistance for sample 2077 measured at 300mK. The inset shows the telegraph noise observed at a fixed magnetic field of 0 T and at temperatures of 1.72 (top) and 1.90 K (bottom).

phase coherence length $L_\Phi \sim 200$nm can be deduced from the typical period $B_c \sim 0.1$ T of the UCF at low temperatures. It is clear that the observed UCF cannot arise from transport through individual wires. A value of $B_c \sim 0.1$T corresponds to an interference loop equivalent to a circle of diameter 200nm which is much larger than the cross sectional surface of the wires ($\phi\sim$3nm). It therefore seems likely that electrical transport occurs via the GaAs matrix, a hypothesis which is supported by the observation of UCF in samples doped with Er but at a concentration below the precipitation limit.

More information concerning the transport mechanism can be obtained from the temperature dependence of the conductance with a magnetic field applied perpendicular (B⊥J) and parallel (B||J) to the current, shown in Figures 3 and 4 for samples 2141 and 2077. For the 200nm thick lightly Er doped sample (2141) the conductance is thermally activated at high temperature and saturates for temperatures below ~5K. Similar results have been reported for GaAs n-i-n structures by Main and co-workers.[4] The temperature independent conductance is attributed to impurity assisted tunnelling via shallow donor states which dominates over hopping processes at low temperature. The application of a magnetic field shrinks the donor wavefunction reducing the overlap and hence transmission probability in the plane perpendicular to the magnetic field and leads to a significant

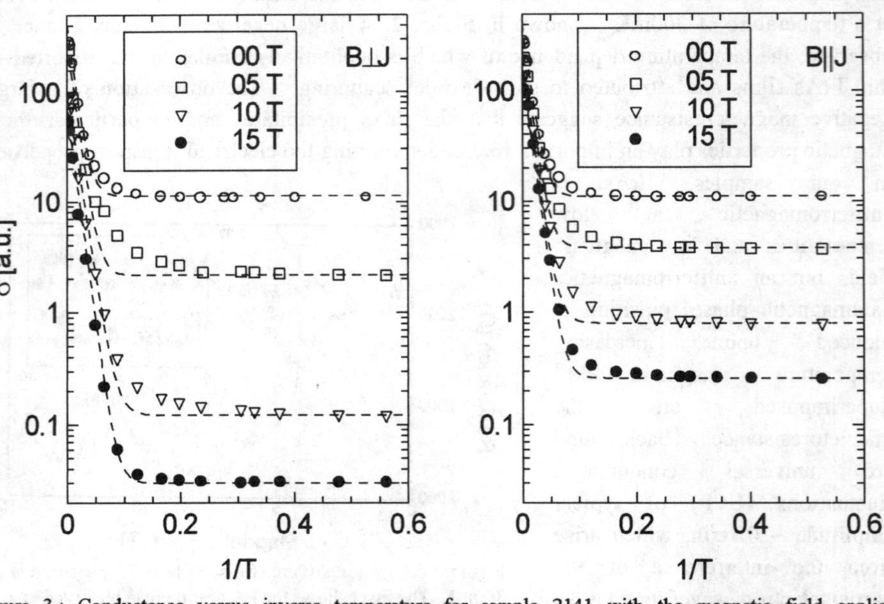

Figure 3 : Conductance versus inverse temperature for sample 2141 with the magnetic field applied perpendicular and parallel to the current. The dashed lines are fits assuming an activated behaviour with E = 7 meV as described in the text.

decrease in the tunnelling conductance. The large anisotropy observed between the B⊥J and B||J evident in Figure 3 and also observed by Main and co-workers[4] lends strong support to the impurity assisted tunnelling model. Such a large anisotropy is not expected for a hopping process since the hopping mechanism is essentially percolative and most of the anisotropy is averaged out due to the large number of sideways hops involved in traversing the sample. The dashed lines in Figure 3 indicate fits to the temperature dependent conductance using the formula $\sigma = \sigma_T + \sigma_o \exp(-E / KT)$ where σ_T is the magnetic field dependent tunnelling conductance and σ_o is a temperature independent conductance. The activation energy of 7 ± 2 meV is in good agreement with the shallow donor binding energy ~ 6meV in GaAs especially when allowance is made for the expected small magnetic field induced increase in the donor binding energy. Thus we conclude that transport in the lightly Er-doped (2×10^{17} cm^{-3}) samples is dominated by Se which has diffused from the heavily doped n^+ contacts. Note that the depart from the fit at intermediate temperatures and at low magnetic fields is due to thermally activated hopping which has an activation energy ~ 1.5 meV, much less than the shallow donor binding energy.

A somewhat similar temperature dependence is observed for the samples containing ErAs precipitates. In figure 4 we see that for sample containing dots and wires (2077) activated transport occurs at high temperature but below ~ 100K the conductance has the temperature independent signature of tunnelling. The deduced activation energy of 150±10 meV is significantly larger suggesting and Er related rather than shallow donor level is responsible. The sample containing only dots (2112) has an identical behaviour with an activation energy of 140 ± 5 meV. DLTS studies[5] of MBE grown Er-doped GaAs have revealed the presence of a large concentration of hole traps with an energy of 0.1-0.15 eV, the appearance of which coincides with the presence of ErAs precipitates and a quenching of the photoluminescence intensity, also observed in our material whenever precipitates are present.[6] Furthermore, the magnetic field has almost no effect on the conductance in the tunnel regime. If tunnelling occurs via the Er-related impurities this can be explained as being due to the much smaller Bohr radius of the deep ~ 0.14eV level. The magnetic field will significantly shrink the wavefunction when the magnetic length $(\hbar / eB)^{1/2}$ (~7 nm at 15T) is comparable to the Bohr radius ($a_B = e^2 / 8\pi\varepsilon_o\varepsilon_r E$). This is the case for a shallow donor in GaAs for which a_B~10 nm while for the Er related level (E=0.14eV) the Bohr radius a_B~0.4nm is significantly smaller than the magnetic length. For samples with precipitates it is also possible that due to the relatively thin active region a direct tunnelling process occurs which would also lead to a tunnelling conductance which is magnetic field independent.

Sample 2140 shows a metallic behaviour probably due to diffusion of Se impurities form the n^+ contacts which for the narrow 100nm thick active region leads to a dopant concentration above the Mott transition. Note this is not the case for the samples

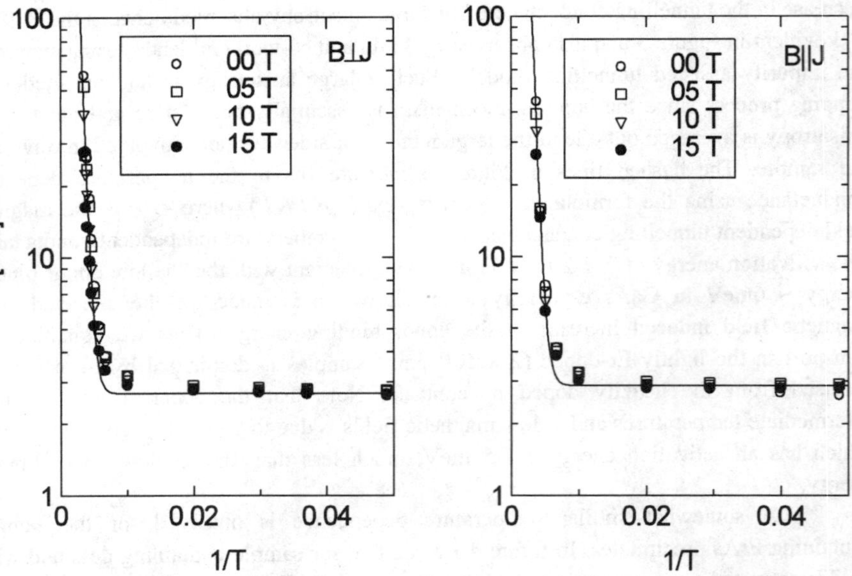

Figure 4 : Conductance versus inverse temperature for sample 2077 with the magnetic field applied perpendicular and parallel to the current. The solid lines are fits assuming an activated behaviour with E = 0.15 eV as described in the text.

with precipitates (2077,2112), despite the even narrower 50nm active region, since the concentration of Er is much higher and therefore, it is the deep Er level which dominates. In addition, rare earths have a tendency to getter shallow donors[7] so that the electrically active Se concentration will depend critically on the Er concentration. This is indeed the case for sample 2142, with a 400nm thick active region, the effective Se diffusion is reduced and it is an Er related level which dominates. As for the other samples the conductance is constant at low temperatures and thermally activated at high temperatures. The activation energy (>35meV) is clearly much larger than that associated with a shallow donor but the conductance is thermally activated over less than one order of magnitude and therefore the activation energy cannot be reliably determined.

The picture which emerges is that for narrow lightly Er-doped GaAs samples transport properties are dominated by Se which has diffused in from the contacts. With increasing active layer thickness or increasing Er concentration it is an Er related deep level which dominates. The samples can be considered as a tunnel barrier sandwiched between two n^+ contacts, the barrier height being determined by the binding energy of the dominant impurity which controls the Fermi level in the barrier. For all samples transport through the active region at low temperatures is controlled by tunnelling.

4. Macroscopic quantum tunnelling

For the heavily Er-doped sample containing both wires and dots (2077), superimposed on the UCF at certain magnetic fields are instabilities (sudden jumps) in the resistance with an amplitude similar to that of the UCF indicating that the transport through the structure is controlled by a few phase coherent sub-system of size L_Φ in parallel (see inset Figure 2). We have previously shown that such two level fluctuations (TLF) are due to the switching of the magnetic moment ($\sim 50\mu_B$) of an isolated ErAs cluster close to a critical current path, between two possible orientations. Using the technique of telegraph noise spectroscopy, essentially measuring the average time spent in each state as a function of temperature and magnetic field, it is possible to determine the energetic barrier separating the two states and magnetic moment carried by each state projected onto the magnetic field axis. The difference in energy between the two states varies as $E=\Delta m.H$ where Δm is the difference in the magnetic moment carried by each state projected onto the magnetic field axis. A particular TLF is observable over a limited field range for which the two states are almost aligned. Furthermore, the orientation of the magnetic moment can be monitored in real time simply by measuring the resistance of the sample.

This technique is of particular interest since it can be used to investigate the quantum tunnelling of magnetisation.[1] In small single domain magnetic particles such a process is referred to as macroscopic quantum tunnelling (MQT). The magnetic moment of a single domain ferromagnetic particle has been measured using a dc SQUID microsusceptometer capable of measuring as few as 10^6 electron spins.[8] However, in ferromagnetic particles the tunnelling probability is low because the barrier between the two macroscopic spin configurations is large. Antiferromagnetic particles, which have a nonzero magnetic moment due to the uncompensated spins at the surface, are better candidates for the observation of MQT because of the smaller energy barrier separating the two spin configurations.[9]

Figure 5 shows, for a typical TLF, the average time spent in one of the states as a function of inverse temperature for three different magnetic fields. At high temperatures the average time is thermally activated according to the relation,

$$\tau^{-1} = \tau_o^{-1} \exp(-\Delta E / KT)$$

where ΔE is the barrier height separating the two states. Below a critical temperature Tc ~500mK the switching rate saturates and it is the tunnelling of the magnetisation which dominates. For a magnetic field of 1.56T the two states are perfectly aligned and the switching rate below Tc is temperature independent. In contrast, at 1.61 T, the tunnelling process is thermally activated, with a slope of 1.4K corresponding to the energy difference

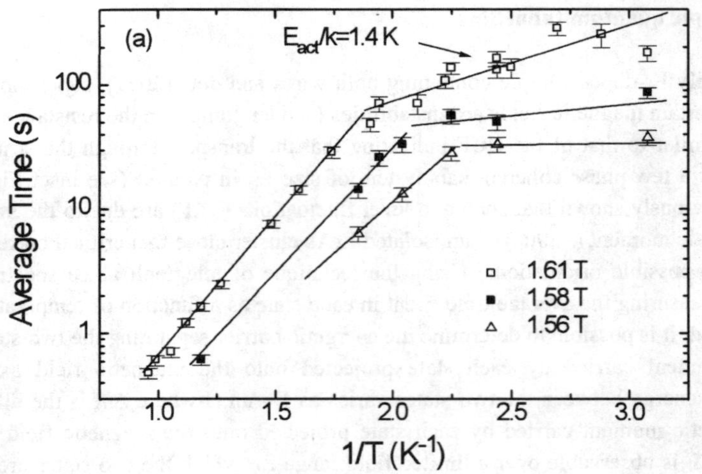

Figure 5 : Average time spent in the low resistance state as a function of inverse temperature for a typical TLF

between the two states. It is interesting to compare the magnetic field dependence of the crossover temperature (T_c) between the thermally activated behaviour and the tunnel regime, with the predictions of WKB theory.[10] By analogy with an electron tunnelling through a single barrier, the average dwell time in the tunnelling regime can be expressed as,

$$\tau^{-1} = \tau_o^{-1}\exp(-\beta)\,,$$

where the WKB coefficient $\beta \approx c(m\,\Delta E)^{1/2}\,x_c$, m can be regarded as a magnetic mass and ΔE and x_c are the barrier height and thickness respectively. The barrier height can be written as $\Delta E = E_o + M_B\Delta H$ where Eo is the barrier height when a magnetic field H_o is applied to exactly align the two states and M_B is the difference in the magnetic moment of the state and the virtual state in the barrier, projected onto the magnetic field axis. Thus the WKB coefficient can be written as

$$\beta = \beta_o(1 + M_B\Delta H / E_o)^{1/2}$$

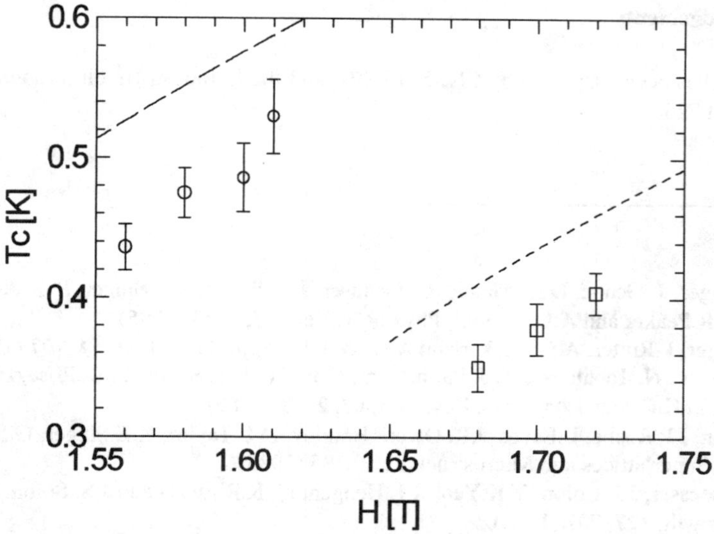

Figure 6 : Crossover temperature form the thermally activated to the tunnelling regime as a function of the applied magnetic field for two TLF. The lines are the calculated variation using E_o=5.24 K, M_B=38.5μ_B β_o=9.95 (long dash) and E_o=4.3 K, M_B=50 μ_B β_o=11.64 (short dash) deduced for each TLF .

where $\beta_o = \ln[\tau(H_o)/\tau_o]$ can be estimated from measured tunnelling rate when the states are aligned. Thus, equating the thermal and tunnelling rates the crossover temperature can be expressed as

$$T_c(\Delta H) = (E_o / K\beta_o)[1 + M_B \Delta H / E_o]^{1/2} \quad (1)$$

The magnetic moment and the barrier height can be deduced from the thermally activated regime[1] and the predictions of equation 1 are compared with the observed T_c for two different TLF in Figure 6. The good agreement, notably concerning the variation with magnetic field is surprising considering the simplicity of the model adopted. A comparison with the more detailed theory of Leggett[11] or of Chudnovsky and Gunther[12] would require the anisotropy constants for the magnetic potential, which are to our knowledge unknown for ErAs clusters. Nevertheless, the simple WKB model presented would appear to explain the data well and lends strong support for the tunnelling of the magnetisation.

Acknowledgements

This work has been supported by CNRS, EPSRC and the European Union through ESPRIT III contract 7260 .

References

[1] F.Coppinger, J. Genoe, D.K. Maude, U. Gennser, J.C. Portal, K.E.Singer, P. Rutter, T. Taskin, A.R.Peaker and A.C. Wright, Phys. Rev. Lett. **75**, 3513 (1995).
[2] K.E. Singer, P.Rutter, A.R. Peaker and A.C. Wright, Appl. Phys. Lett. **64**, 707 (1994).
[3] S.J.Allen, Jr., N. Tabatabaie, C.J. Palmstrom, G.W. Hull, T. Sands, F. DeRosa, H.L. Gilchrist and K.C. Garrison, Phys. Rev. Lett. **62**, 2309 (1989).
[4] P.C. Main, I.P.Roche, L.Eaves, J.R. Owers-Bradley, D.C.Taylor, K.E.Singer, G.Hill and M.A.Pate, Superlattices and Microstrucures, **2**, 385 (1986)
[5] D.W. Elsaesser, J.E.Colon, Y.K.Yeo, R.L.Hengehold, K.R.Evans and J.S. Solomon, J. Crystal Growth, **127**, 707 (1993).
[6] A.R.Peaker private communication
[7] J. Raczynska, K.Fronc, J.M.Langer, A.Lemanska and A.Stapor, Appl. Phys. Lett. **53**, 761 (1988).
[8] D.D. Awschalom, J.R.Rozen, M.B.Ketchen, W.J.Gallagher, A.W.Kleinsasser, R.L.Sandstrom and B.Bumble, Appl. Phys. Lett. **53**, 2108 (1988).
[9] B.Barbara and E.M.Chudnovsky, Phys. Lett. A **145**, 205 (1990).
[10] see for example L.I.Schiff, Quantum Mechanics 3rd ed. Publ. McGraw Hill (1968).
[11] A.J.Leggett et al., Rev. Mod. Phys. **59**, 1 (1987).
[12] E.M.Chudnovsky and L. Gunther, Phts. Rev. Lett. **60**, 661 (1988).

TUNNELLING BETWEEN PARALLEL 2DEGS WITH AN IN-PLANE MAGNETIC FIELD IN THE 'QUANTUM TRANSPORT REGIME'

M.J. GOMPERTZ, T. IHN, H.A. CARMONA, A. NOGARET, P.C. MAIN, L. EAVES, M. HENINI
Department of Physics, University of Nottingham, Nottingham, NG7 2RD, UK

S.P. BEAUMONT
Department of Electronics and Electrical Engineering, University of Glasgow, Glasgow, G12 8QQ, UK

We report electrical transport measurements on a system of parallel 2D electron gases confined on opposite sides of a single quantum well. By applying a magnetic field in the plane of the quantum well, the tunnelling between the 2DEGs is 'tuned' and we observe a low-field resonance in the magnetoresistance for Hall-bars. We find that for mesoscopic wires the extra disorder quenches the resonance, and we observe Universal Conductance Fluctuations in the parallel field.

In recent work on systems incorporating two spatially separate 2D electron gases (2DEGs), authors have reported resistance resonance (RR) effects in double quantum well (DQW) structures, as a function of gate voltage[1,2] and parallel magnetic field.[2] In this paper the ungated wide quantum well of Ihn *et al.*,[3] which shows a RR at non-zero magnetic field, is studied further.

The samples are MBE-grown, modulation doped structures which have a 407Å-wide GaAs QW between $Al_{0.3}Ga_{0.7}As$ barriers. The barriers are undoped up to 76Å from the well interfaces, and are elsewhere doped to a concentration of $N_D = 1.33 \times 10^{18}\text{cm}^{-3}$. Three QW subbands ($E_0$, E_1, E_2) are occupied at zero field,[3] with sheet densities $n_0 = 1 \times 10^{12}\text{cm}^{-2}$, $n_1 = 0.5 \times 10^{12}\text{cm}^{-2}$ and $n_2 = 0.11 \times 10^{12}\text{cm}^{-2}$; the measurable mobilities are $\mu_0 = 1.0\text{m}^2/\text{Vs}$ and $\mu_1 = 20.4\text{m}^2/\text{Vs}$, with $\mu_2 < \mu_1$. The two lowest subbands form the two parallel 2DEGs, separated by a distance $d = 32$nm, as obtained from the distance between the wavefunction peaks, calculated self-consistently.[3]

Figure 1(a) shows experimental magnetoresistance curves for a macroscopic Hall-bar with B applied in the plane of the 2DEGs. Two features are noteworthy: first, there is a resistance resonance at $B = 1.6$T; second, at $B = 2$T the resistance is independent of the angle between B and I.

The RR occurs when the Fermi circles of the E_0 and E_1 subbands first touch. At this field, a large number of the states at the Fermi level are delocalised across the well due to the enhanced, momentum-conserving tunnel coupling between the two lowest subbands. As a result of the large difference

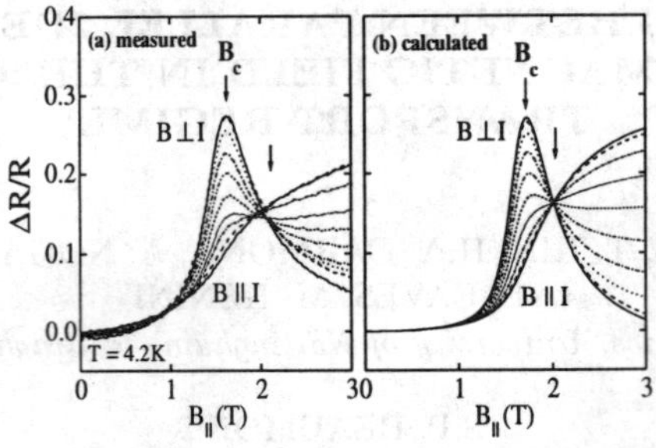

Figure 1: (a) Measured magnetoresistance with angles varied in steps of 10° from $B \perp I$ to $B \| I$. The first arrow marks the position of the peak, B_c; the second arrow marks an angle-independent point. (b) Calculated magnetoresistance for the same angles as in (a).

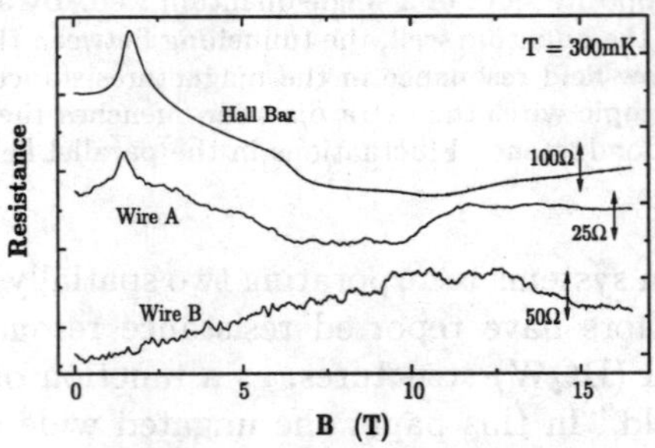

Figure 2: Diagram showing magnetoresistance of three devices. B is applied in the plane of the 2DEGs and perpendicular to I in each case.

in mobilities between the two 2DEGs, the longitudinal resistance is increased.[3]

The conductivity tensor σ is diagonal in the reference frame where B defines the x-axis, because the $\epsilon(k)$ dispersion of the system in this case is symmetric with respect to the y-axis. If the current is not aligned in either the x- or y-directions then the conductivity is no longer diagonal, but is given by the rotated tensor with component $\sigma_{x'x'}(\alpha) = \sigma_{xx}(B)\cos^2\alpha + \sigma_{yy}(B)\sin^2\alpha$, where α is the angle between B and I, and the x'-axis is the direction of the current. Calculations of $\Delta R/R$ for several angles α are shown in Figure 1(b). The angle-independent point occurs when $\sigma_{xx}(B) = \sigma_{yy}(B)$.

Recently, Vasko *et al.*[4] have pointed out that parallel 2DEGs, separated by a potential barrier, cannot be described by the semi-classical Boltzmann transport theory but require a quantum treatment of the conductivity if $\Delta_{\mathrm{SAS}} \lesssim \hbar/\tilde{\tau}$. (Here Δ_{SAS} is the symmetric-antisymmetric splitting and $\tilde{\tau}$ is a characteristic scattering time). In our Hall-bar sample $\Delta_{\mathrm{SAS}} \sim 1.1\hbar/\tilde{\tau}$ at the RR maximum.

We have attempted to study the 'quantum transport regime' further by fabricating mesoscopic wires from the layer described above. Figure 2 shows magnetoresistance measurements from devices of three different sizes at $T =$ 300mK. The uppermost trace is from the macroscopic Hall-bar (width $W =$ 125μm), whilst the traces beneath it are from wires ('A' and 'B') with lithographically defined widths $W = 800$nm and 400nm, respectively.

Notice the quenching of the resistance resonance — partially in Wire A and completely in Wire B. To begin to explain this observation we note that the diffusive boundary scattering at the edges of the deep-etched wire mesa is a source of increased disorder over that in the Hall-bar sample. The mean free path, l, of a subband can be expected to be much reduced when $W \lesssim l$. The mobility of the more disordered E_0 subband, with $l = 165$nm, is not affected, but the E_1 subband, with $l = 2.3\mu$m, is expected to undergo a reduction in its mobility. From the zero-field resistivities of Wires A and B we estimate the mean free paths in the E_1-subbands to be $l = 600$nm and $l = 150$nm respectively. Their elastic scattering times are $\tau_1 = 2$ps and $\tau_1 = 0.5$ps.

After Vasko *et al.*,[5] we note that two main factors determine the relative amplitude, $\Delta R/R$, of the resonance. First, there must be a finite scattering time asymmetry, $\beta = (\tau_1 - \tau_0)/(\tau_1 + \tau_0)$, for there to be a RR. The asymmetry gives rise to a RR of amplitude $\Delta R/R = \beta^2/(1-\beta^2) = (\tau_1 - \tau_0)^2/4\tau_1\tau_0$. In our case this simple relation must be replaced with a more complex one which takes into account that $n_0 \neq n_1$.[6] Using determined values for $\tau_{0,1}$, we are able to reproduce the experimentally observed ratios of $\Delta R/R = 11.5\%$ and $\Delta R/R \sim 0\%$ for Wires A and B.

Second, the disorder in the sample gives rise to potential fluctuations which cause electronic states to localise in space.[5] This diminishes the tunnelling between interfaces, and thus the height of the RR is reduced. The relevant quantitative parameter is the ratio of the tunnelling gap to the scattering broadening, $\Omega_T = \Delta_{SAS}\tilde{\tau}/\hbar$, where $\tilde{\tau} = 2\tau_1\tau_0/(\tau_1 + \tau_0)$.[5] We find that, for our samples, Ω_T is between 1.1 (for the Hall-bar) and 1 (Wire B), ie. the increased boundary scattering does not push the samples much further into the quantum transport regime. The reduction in amplitude of the RR is therefore attributable mainly to the strongly diminished scattering-time asymmetry, rather than to an impairment of tunnelling by disorder.

The second thing to notice in Figure 2 is the occurrence of reproducible Universal Conductance Fluctuations (UCF) in the two wire samples. For *independent* 2DEGs in the quantum limit, no UCF would be expected in this in-plane field geometry because the electron trajectories would enclose no flux.

Magnetoresistance measurements, made on Wire B, are displayed in Fig. 3

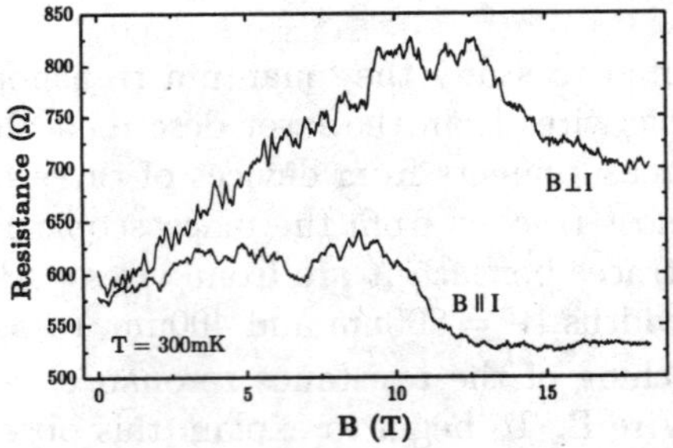

Figure 3: Two magnetoresistance curves from Wire B. In both, B is applied parallel to the plane of the 2DEGs, but the orientation of B relative to I is (a) $B\|I$ and (b) $B\perp I$.

for the orientations $B\|I$ and $B\perp I$. Again the magnetic field is applied in the plane of the 2DEGs. A notable result of the comparison of these two traces is the difference in the quasi-periods of the UCF, ΔB. The period of the fluctuations in $R(B)_{B\|I}$ is 1.3 times larger than in $R(B)_{B\perp I}$. This increase in period arises from a decrease in the area between partial electron-scattering paths perpendicular to the field direction. In the $B\|I$ orientation, the area available is $W \times d$, whereas with $B\perp I$ the area is $l_\phi \times d$, where $l_\phi \sim 500$nm is the phase-coherence length.

When B is applied *perpendicular* to the 2DEGs, the typical ΔB is 4 times smaller than in the $B\|I$ configuration. In this case the relevant area is $l_\phi \times W$. Amplitudes of the UCF are the same ($\Delta G = 0.3e^2/h$) for all three B directions. The appearance of UCF in a parallel field in this system is not surprising since the tunnelling time is short compared with the inelastic scattering time. However, it is remarkable that the ΔB's scale in the same way as for a 3D wire.[7] Further studies of UCF where the tunnelling and scattering rates are comparable should yield information about the quantum transport regime.

References

1. A. Palevski *et al.*, *Phys. Rev. Lett.* **65**, 1929 (1990).
2. Y. Berk *et al.*, *Phys. Rev.* B **51**, 2604 (1995)
3. T. Ihn *et al.*, *Phys. Rev.* B, to be published.
4. F.T. Vasko and O.E. Raichev, *Phys. Rev.* B **52**, 16349 (1995).
5. F.T. Vasko and O.E. Raichev, Zh. Éksp. Teor. Fiz. **107**, 951 (1995) [Sov. Phys. JETP **80**, 539 (1995)].
6. T. Ihn, unpublished.
7. P.A. Lee, A.D. Stone and H. Fukuyama, *Phys. Rev.* B **35**, 1039 (1987).

CURRENT-VOLTAGE CHARACTERISTICS OF QUANTUM CHAOTIC TUNNEL STRUCTURES IN HIGH MAGNETIC FIELDS

P B WILKINSON, T M FROMHOLD and F W SHEARD

Department of Physics, University of Nottingham, NG7 2RD, UK

We have calculated the current-voltage characteristics of a resonant tunnelling diode containing a wide quantum well, in which the electron dynamics is classically chaotic when a strong tilted magnetic field is applied. Our calculations enable us to identify the effects of spectral fluctuations and eigenfunction "scarring" associated with periodic orbits in the well, on the observed electrical properties. Recent magnetotunnelling experients exhibiting resonant peak doubling and tripling are discussed. It is shown that the stable-orbit trifurcations invoked by Müller *et al* are not observable in tunnelling experiments.

The classical dynamics of electrons in wide quantum wells (QWs) changes from stable to chaotic when a high magnetic field is tilted relative to the well walls. Within the chaotic sea the electron trajectories are extremely unstable. Nevertheless isolated periodic orbits exist and provide an important link between the classical and quantum features of the motion. Recent theoretical studies have shown that the shortest periodic orbits give rise to quasiperiodic clustering of the energy levels and scarring (concentrations of probability along closed orbits) of the eigenfunctions[1].

These features have been studied experimentally by incorporating the QW into a resonant tunnelling diode (RTD) since the current-voltage characteristic behaves as an energy spectrometer[2,3]. When a bias voltage V is applied to the RTD a two-dimensional electron gas (2DEG) accumulates in the undoped spacer layer adjacent to the emitter barrier. A current I flows when the electrons tunnel from the 2DEG into the quantised energy levels of the QW. The tunnelling rates into each QW state are calculated using a Bardeen transfer-Hamiltonian formalism[1,4]. Conduction band nonparabolicity is included in the calculations by using a voltage-dependent effective mass[4] $m^*(V) = 0.067m_e\{1+\alpha T(V)\}$, where $T(V)$ is the mean kinetic energy of the electron and $\alpha = 2$ eV^{-1}. The potential distribution in the RTD is modelled[2] using experimental data on 2DEG sheet density as a function of V.

We first consider measurements of the Nottingham group[2] for a RTD with 120 nm wide QW and constant magnetic field $B = 11.4$ T oriented over a range of tilt angles θ relative to the growth direction. As shown in Fig.1 the theoretical calulations reproduce very well the changes in periodicity and amplitude of the oscillations in the d^2I/dV^2 v. V plots. For $\theta = 0^\circ$ the magnetic field does not affect the tunnelling transitions which take place from the bound emitter state into the electric subbands of the well. The classical motion is regular and from the Correspondence Principle the voltage interval ΔV betwen successive resonances is given by $fe\Delta V = h/T$, where f is an electrostatic leverage factor[2] and T the periodic time for one dimensional motion between the QW barriers.

For $15^\circ < \theta < 50^\circ$, the classical motion is strongly chaotic and two distinct regions of resonant peaks, labelled t and s in Fig. 1, are observed in the tunnelling characteristics. This was attributed[2] to modulations in the density of levels with period $\Delta E = h/T_p$, associated with unstable closed orbits of period T_p, accessible to the tunnelling electrons. At the changeover from t to s peaks, ΔV is roughly halved (peak doubling). The t- or s-orbits are called type-1 or type-2 because an electron makes 1 or 2 successive impacts on the collector barrier for each impact on the emitter barrier, which accounts for the period doubling from t to s. Typical orbits are shown in Fig. 2.

Individual energy eigenfunctions may also be strongly scarred by an unstable periodic orbit at energy intervals $\Delta E = h/T_p$[2,4]. Our calculations show that at low voltages (< 150 mV) and high tilt angles (> 35°) the tunnel current is dominated by transitions into discrete strongly scarred states. This arises from the strong spatial overlap between the localised emitter state and scarred state whose probability density is concentrated along the classical orbit. Scarring thus plays the role of a quasi-selection rule. The scar patterns are better defined at low voltages when the average density of states, at the tunnel injection energy, is small[5]. It is also found that scar patterns are weakened when the orbital segment between successive bounces on the collector barrier overlaps the segment directly connecting both barriers. This is shown for $\theta = 20^\circ$, Fig. 2(b), when the two orbital segments cross in the middle of the well and produce a complicated interference structure which diffuses the scar pattern. Owing to these effects the s-resonances

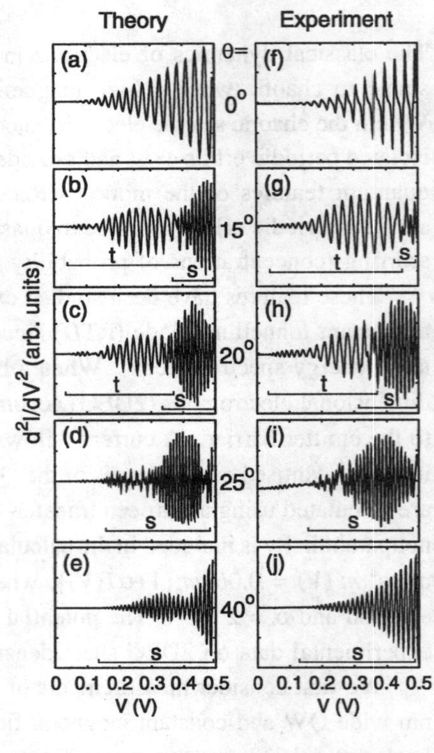

Figure 1 Comparison of theoretical and experimental tunnelling characteristics for 120 nm wide QW, $B = 11.4$ T.

in Fig. 1 originate primarily from transitions into clusters of levels (Gutzwiller fluctuations) when $\theta = 20^\circ$ but from transitions into individual scarred states when $\theta = 40^\circ$ and $V < 150$ mV. The t-resonances originate mainly from tunnelling into individual scarred states. We note that the experiments have insufficient resolution to distinguish between transitions into a cluster of states or one strongly scarred state.

Müller et al[3] have also studied the transition to chaos in a 120 nm wide QW by measuring the RTD $I(V)$ characteristics as a function of B for several tilt angles. In particular, for $\theta = 38^\circ$ a narrow island in B-V space was found, within which the spacing of adjacent resonant peaks in $I(V)$ fell by a factor 3 (peak tripling). In Fig. 3(a) the positions of the resonant peaks in the measured $I(V)$ traces are shown for a range of B and V values. Solid (open) circles denote those peaks inside (outside) the peak tripling region. Fig. 3(a) also shows our calculated $I(V)$ characteristics (with background current subtracted) displaced at 0.1 T intervals.

For the two lowest field values shown, the calculated $I(V)$ plots reveal only weak resonances which originate from tunnelling into stable traversing (type-1) orbits. The localised emitter state is only weakly coupled to the QW states here because few of the injected electrons have sufficient lateral momentum to execute type-1 orbits[6]. At higher fields the calculated $I(V)$ traces exhibit clear peak tripling within the shaded region. As shown the calculated and observed peak tripling regions occur in very similar regions of B-V space and the predicted voltage spacings are close to experimental values ($\cong$ 12 mV). Analysis of the tunnelling rates in the peak-tripling region shows that the dominant transitions involve a subset of states, regularly spaced by $\Delta E = h/T_3$, and strongly scarred by a stable type-3 orbit of period T_3 $\cong$ 1 ps (3 bounces on collector barrier for each impact on emitter barrier) as shown in Fig. 3(b). In phase space these orbits are surrounded by strongly chaotic regions[6]. Müller *et al*[3] attributed the peak tripling to a trifurcation of the basic t-orbit (type-1) as V or B is increased. We have identified states corresponding to the stable period-tripled orbit produced by this trifurcation process. As shown in Fig. 3(c) the wave

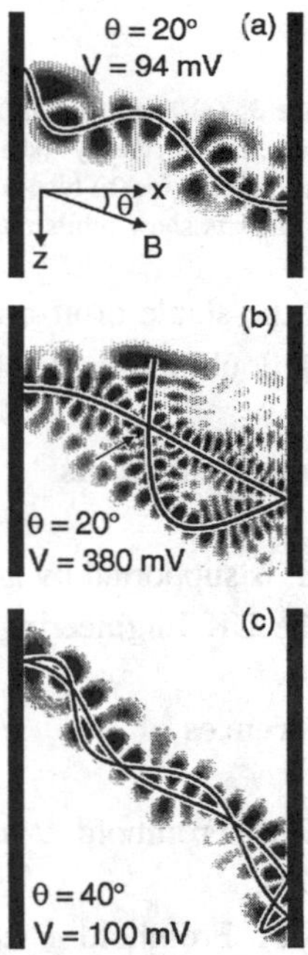

Figure 2 Scarred eigenfunctions and associated orbits for levels close to injection energy; (a) t-orbit,(b) and (c) s-orbits. Emitter barrier on left.

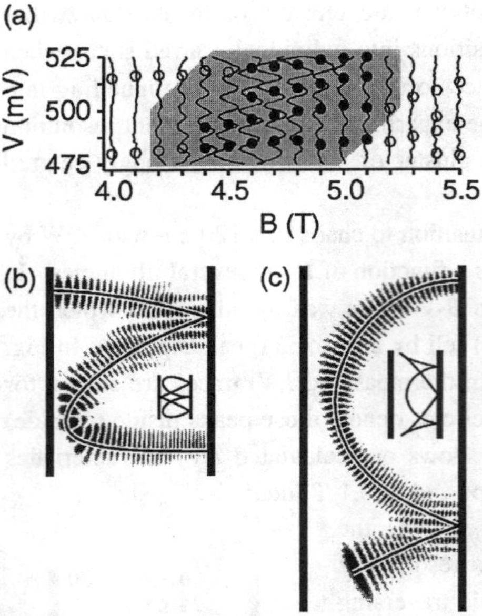

Figure 3 (a) Peak tripling region from experiment (solid circles), theory (shaded region). Scarred eigen-function for (b) type-3 orbit and (c) trifurcated orbit. Inserts show orbits rotated through 90°.

function is exponentially small at the emitter barrier since the underlying orbit does not reach this barrier. Thus these states contribute negligible tunnel current.

For $V \cong 500$ mV there is a higher field range $5.0\ \text{T} < B < 5.3\ \text{T}$ where the trifurcation process produces stable orbits which do touch the emitter barrier. But electrons from the 2DEG have insufficient lateral momentum to tunnel into these states which do not therefore contribute to the tunnel current. Instead the observed peak tripling island is well explained by the stable type-3 orbits shown in Fig. 3(b) which produce, when account is taken of band nonparabolicity, strong resonant peaks in precisely the observed region of *B*-*V* space. Calculations for other tilt angles show that the stable-orbit bifurcations and trifurcations cannot account for the observed peak doubling and tripling[3], except for the doubling found at $\theta = 11°$ and $B < 9$ T[6].

Acknowledgments

TMF is supported by an Advanced Fellowship and PBW by a Research Studentship of the UK Engineering and Physical Sciences Research Council.

References

1. T.M. Fromhold *et al*, *Phys. Rev. Lett.* **75**, 1142 (1995); *Phys. Rev.* **B51**, 18029 (1995).
2. T.M. Fromhold *et al*, *Phys. Rev. Lett.* **72**, 2608 (1994).
3. G. Müller *et al*, *Phys. Rev. Lett.* **75**, 2875 (1995).
4. P.B. Wilkinson *et al*, *Nature* **380**, 608 (1996).
5. O. Agam and S. Fishman, *Phys. Rev. Lett.* **73**, 806 (1994).
6. T.M. Fromhold *et al*, Comment submitted to *Phys. Rev. Lett.*

MAGNETIC FIELD STUDIES OF PHONON ASSISTED TUNNELLING IN A δ-DOPED DOUBLE BARRIER RESONANT TUNNELLING DEVICE

E S MOSKALENKO*, D N HILL, F F OUALI, L J CHALLIS, F W SHEARD and M HENINI

Department of Physics, University of Nottingham, NG7 2RD, UK
* *On leave from the Ioffe Physical-Technical Institute, 26 Polytechnicheskaya str, 194021 St.Petersburg, Russia*

We report measurements of phonon-assisted tunnelling through donor states in a δ-doped double barrier resonant tunnelling device. A peak additional to the 2 peaks (Stokes and anti-Stokes) resulting from the donor ground state increases strongly in amplitude with magnetic field and is tentatively attributed to phonon-assisted tunnelling through the D^- state.

1 Introduction

In recent experiments[1,2], we have made the first observations of the change in the current induced by non-equilibrium phonon pulses incident normal to double barrier resonant tunnelling devices (DBRTDs). The measurements focused on tunnelling through the ground state of the shallow donors in the well present as a result of Si diffusion from the n^+ regions during the growth process[3]. The change in tunnel current was attributed to phonon-assisted tunnelling from the emitter to donor states in the well resulting in an absorption (anti-Stokes) peak and a stimulated emission (Stokes) peak. The absorption peak occurs at a bias voltage which varies linearly with phonon frequency showing that the device has potential as a phonon spectrometer. The donor tunnel current was superimposed on a much larger (×10) background current due to tunnelling through the bound state in the well and to reduce this contribution, the present measurements were made in a high quality sample in which Si donors are introduced by δ-doping the centre of the well.

2 Experimental arrangement

The sample used was grown at 550°C and consists of two 5.6nm thick $Al_{0.4}Ga_{0.6}As$ barriers separated by a 9 nm well, the centre of which was δ-doped with Si donors to a concentration of 4×10^9 cm^{-2}. Each barrier is separated from the n^+ doped region by a 20 nm thick GaAs spacer layer to minimize diffusion of impurities during growth. The doping of the contact layers is graded from 2×10^{16} cm^{-3} to 1×10^{18} cm^{-3}.

The layers were etched down to a 50×50 μm^2 mesa. The back of the 400 μm substrate was polished and a 60×600 μm^2 constantan heater evaporated opposite the device. The bath temperature is 3.9K. Non-equilibrium phonons, which travel ballistically across the substrate, are generated by applying 200 ns current pulses to the heater[1]. The resulting transient change in tunnel current is measured as a function of the applied bias *V* in zero field and in fields up to 6T from a split-coil magnet applied parallel ($B_\parallel$) or perpendicular ($B_\perp$) to the direction of the tunnelling current.

3 Results and discussion

The current-voltage characteristic *I(V)* is shown in fig.1. The broad peak at ~0.07V is attributed, as in [3], to resonant tunnelling through the 1s ground state of the donors of energy E_d ~12.8 meV[4] below the ground state subband and as expected is appreciably larger than that observed previously[1,2]. The oscillatory period in the current with $B_\parallel$ suggests that the 2D emitter concentration is very low.

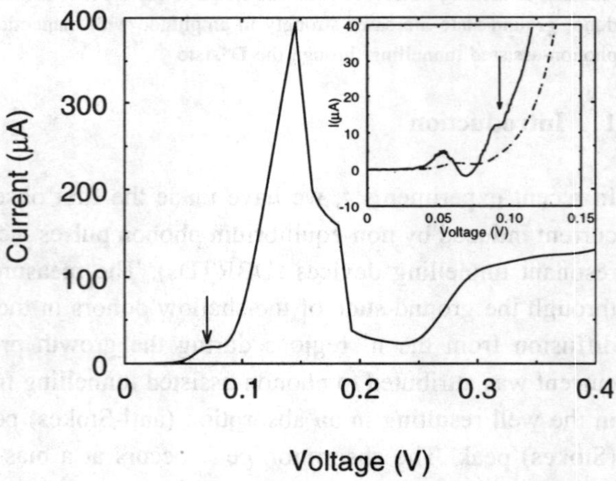

Fig 1: The I(V) characteristic at T=3.9K and B=0 showing the impurity peak at ~0.07V. The inset shows a shoulder in dI/dV indicating a peak in $I_d(V)$ at V~0.095V for B=6T.

Fig. 2 shows the change $\Delta I(V)$ induced by phonons at B=0 for heater temperature T_h=12.5K plus the donor peak $I_d(V)$ extracted from *I(V)*. Three peaks are observed in $\Delta I(V)$. Peaks **1** and **3** are attributed[1,2] to phonon-assisted tunnelling through the 1s donor ground state as a result of phonon absorption and stimulated phonon emission respectively. ΔI is proportional to the tunnelling probability into a donor times the number of (empty) donors n_{ed}. For phonons of energy $\hbar\omega$, the maximum probability is displaced from the $I_d(V)$ peak by $e\Delta V = \pm\hbar\omega$, but, since n_{ed} is least at the $I_d(V)$ peak, the maximum in phonon signal is further displaced from this peak as seen in fig. 2.[2] The negative peak, **2**, coincides with the maximum in $I_d(V)$

showing that resonant tunnelling is reduced by the incident phonons. We tentatively attribute this to a reduction in the number of emitter electrons at energies ~E_d caused by phonon absorption. The weak shoulder (arrowed) in $\Delta I(V)$ at V=0.107V is believed to be associated with an additional feature in $I(V)$ shown in the inset to Fig. 1; this becomes more apparent as B increases and the inset shows data at $B_\perp$=6T.

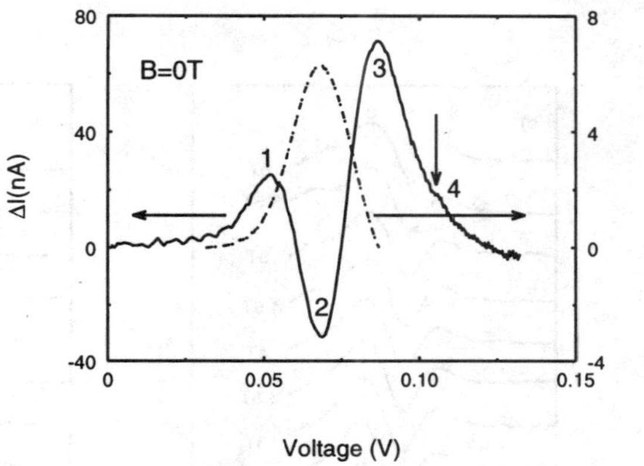

Fig 2: The variation of ΔI with bias for T_h=*12.5*K at B=0T. The dashed line shows the donor current I_d.

Fig. 3 shows that peaks **1** and **2** decrease in size and shift to lower voltages with magnetic fields $B_\parallel$ and $B_\perp$. The shift in *V* and so increase in donor binding energy with B is approximately independent of direction but the decrease in peak size is appreciably greater (~2.5 times for B=6T) for $B_\perp$ than for $B_\parallel$ (~20%). These effects are similar to those observed in $I_d(V)$ and, for example for $B_\perp$, the decrease in peak size can be explained[3,5] by the shift in the in-plane momentum of the occupied emitter states with B leading to a reduction in the overlap integral between the emitter and donor wavefunctions. As the field increases, the shoulder at V=0.107V for B=0 develops into a broad peak, **4,** with again $B_\perp$ producing the larger effect. We attribute peak **4** to phonon-assisted tunnelling through the higher energy donor state responsible for the small resonant tunnelling peak in $I_d(V)$ at V=0.095V and, since it occurs above the peak in $I_d(V)$, it represents a stimulated emission process; the corresponding absorption process presumably being unresolved from the emission process (peak **3**) associated with the donor ground state. A similar field dependent peak in $I_d(V)$ at low $B_\parallel$ fields has recently been reported by Lok et al[6] who made measurements for fields $B_\parallel$ up to 19T in comparable δ-doped DBRTDs. From the variation of its position and magnitude with $B_\parallel$, the peak was attributed to tunnelling through the D^- ground state (D^-= 2 bound electrons) so that it seems likely that the additional phonon-assisted tunnelling effects seen here are also to be attributed to this state, although tunnelling to an excited state (2s, 2p±..) cannot be excluded at this stage without further measurements at higher magnetic fields and lower temperatures.

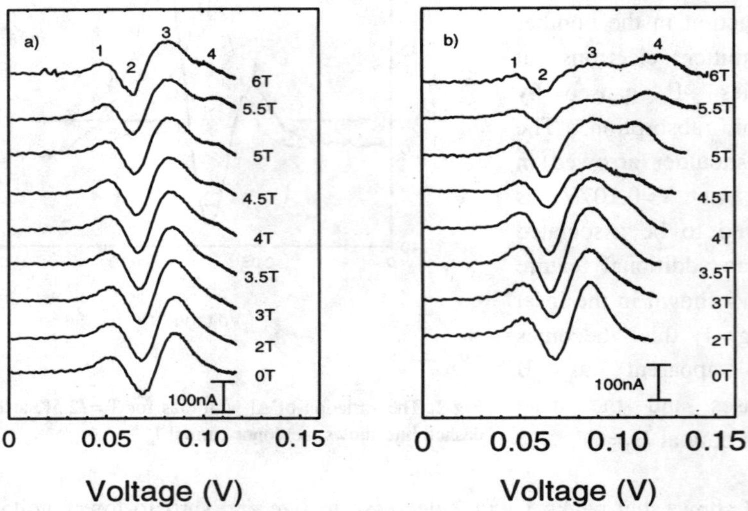

Fig 3: The effect on ΔI of a magnetic field **a)** parallel and **b)** perpendicular to the current.

References:

[1] F F Ouali, N N Zinov'ev, L J Challis, F W Sheard, M Henini, D P Steenson and K R Strickland, Phys Rev Lett, **75**, 308 (1995)

[2] F.F. Ouali, N.N. Zinov'ev, L.J. Challis, A.V. Akimov, D.N. Hill, F.W. Sheard, and M. Henini, Surf. Sci., in press

[3] J W Sakai, T M Fromhold, P H Beton, L Eaves, M Henini, P C Main F W Sheard and G Hill, Phys. Rev. **B48**, 5664 (1993)

[4] R L Green and K K Bajaj, Solid State Commun. **45**, 825 (1983)

[5] J W Sakai, T M Fromhold, P H Beton, M Henini, L Eaves, P C Main, F W Sheard and G Hill, Physica **B184**, 241 (1993)

[6] J G S Lok, A K Geim, J C Maan, I Marmorkos, F M Peeters, N Mori, L Eaves, T J Foster, P C Main, J W Sakai and M Henini, Surf. Sci., in press; see also Phys. Rev. **B53**, 9554 (1996)

INTER-LANDAU LEVEL TRANSITIONS AT THE THRESHOLD OF RESONANT TUNNELING IN A MAGNETIC FIELD

S. A. BROWN and L. D. MACKS
School of Physics, University of New South Wales, Sydney, NSW 2052, Australia

P. E. SIMMONDS
Department of Physics, University of Wollongong, Wollongong, NSW 2500, Australia

T. A. FISHER
Department of Physics, University of Sheffield, Sheffield, S3 7RH, United Kingdom

M. EMENY
Defence Research Agency, Malvern, WR14 3PS, United Kingdom

We present a magneto-transport study of a GaAs/AlGaAs double barrier resonant tunneling structure with an InGaAs prewell close to the emitter barrier. At the threshold of elastic tunneling we identify transitions from occupied Landau levels in the emitter prewell to unoccupied Landau levels of lower index in the quantum well.

1 Introduction

The application of a magnetic field parallel to the current flow in a double barrier resonant tunneling structure (DBRTS) causes Landau levels (LLs) of energy $(n+1/2)\hbar\omega_c$ to form in the emitter and quantum well (where n is the LL index and $\omega_c=eB/m^*$ the cyclotron frequency with m^* the electronic effective mass). 2D-2D tunneling from the emitter to the well obeys a $\Delta n=0$ selection rule (where $\Delta n=n_{well}-n_{emitter}$) unless there are symmetry breaking fluctuations in the structure. Inter-LL tunneling processes have been widely studied with many reports of $\Delta n>0$ inelastic and elastic transitions (see for example Refs. 1-3). Whilst $\Delta n<0$ inelastic processes have been reported[3], the absence of $\Delta n<0$ elastic transitions in samples which show $\Delta n>0$ elastic transitions is an outstanding problem.[4] To our knowledge $\Delta n<0$ elastic tunneling has only been reported in a sample with a 0D well state,[5] or where tunneling was enhanced by a component of momentum perpendicular to the barriers in a 3D emitter accumulation layer.[6]

We have studied I(V) and differential conductance in a GaAs/AlGaAs DBRTS with an InGaAs emitter prewell. We show that the tunneling threshold in a magnetic field is associated with $\Delta n<0$ elastic tunneling from an occupied LL in the emitter prewell to an unoccupied LL in the well. Elastic and inelastic $\Delta n>0$ inter-LL transitions observed at higher bias voltages in I(V) curves support our indentification of the $\Delta n<0$ transitions.

2 Sample Description

The sample comprises the following layers grown on an n^+ GaAs substrate: n^+ emitter - 0.75 μm GaAs (Si doped: $1.0\times10^{18}cm^{-3}$), 500 Å GaAs ($1.0\times10^{17}cm^{-3}$), 500 Å GaAs ($1.0\times10^{16}cm^{-3}$), and 50 Å GaAs; prewell - 200 Å $In_{0.05}Ga_{0.95}As$ and 10 Å GaAs; emitter barrier - 83 Å $Al_{0.4}Ga_{0.6}As$; well - 62 Å GaAs; collector barrier - 110 Å $Al_{0.4}Ga_{0.6}As$; n^+ collector - 50 Å GaAs, 500 Å GaAs ($1.0\times10^{16}cm^{-3}$), 500 Å GaAs ($1.0\times10^{17}cm^{-3}$), 1 μm GaAs ($1.0\times10^{18}cm^{-3}$). Measurements were performed on a 5 μm diameter circular mesa with the sample mounted in a dilution refrigerator/15 T superconducting magnet system.

3 I(V) Measurements

Fig. 1 (a) shows the 0 T I(V) curve in forward bias (electron flow from emitter to collector). Measurements on almost identical structures[7,8] have shown that there is significant charge accumulation in the well throughout the resonance regime. The bistable region observed for $820<V<980$ mV is intrinsically related to this charge accumulation.[8,9] As the magnetic field is increased, additional structure is observed in the forward bias I(V) curve, as shown at 14 T in Fig. 1 (c). The positions of strong and weak peaks in I(V) for increasing voltage sweeps are shown in Fig. 1 (b) (closed and open circles respectively). At high bias, the three lines which extrapolate to 680 mV at 0 T are elastic transitions with $\Delta n=3$ to 5. The three lines which extrapolate to 950 mV are inelastic GaAs longitudinal optic phonon emission assisted transitions with $\Delta n=2$ to 4. The slopes of these lines are given by $\alpha\Delta n\hbar\omega_c$ where $\alpha=7.5$ mV/meV represents the relative energy shift of the lowest bound states in the well and emitter under bias.

Between 6 T and 8 T weak structure is observed in I(V) at low biases (as arrowed in the inset to Fig. 1 (a) at 6 T). These data correlate with strong features observed in differential conductance measurements (positions marked as crosses in Fig. 1 (b)). The origin of this structure will be discussed in more detail elsewhere.[10]

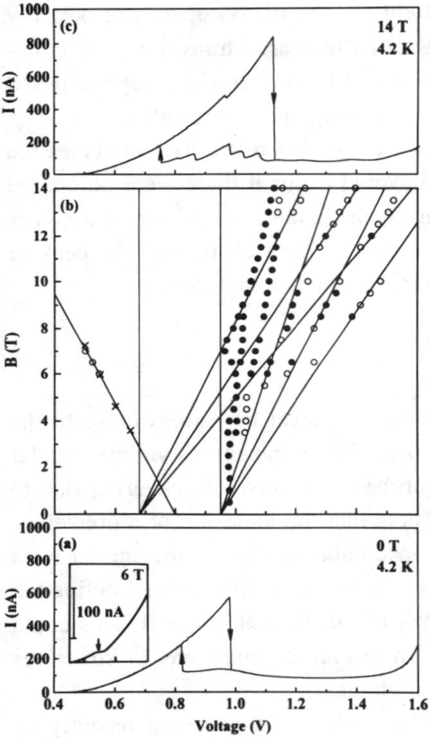

Figure 1: I(V) curves in forward bias at 4.2 K and (a) B=0, (c) B=14T. (b) Positions of strong and weak peaks (closed and open circles) in increasing voltage sweeps vs magnetic field.

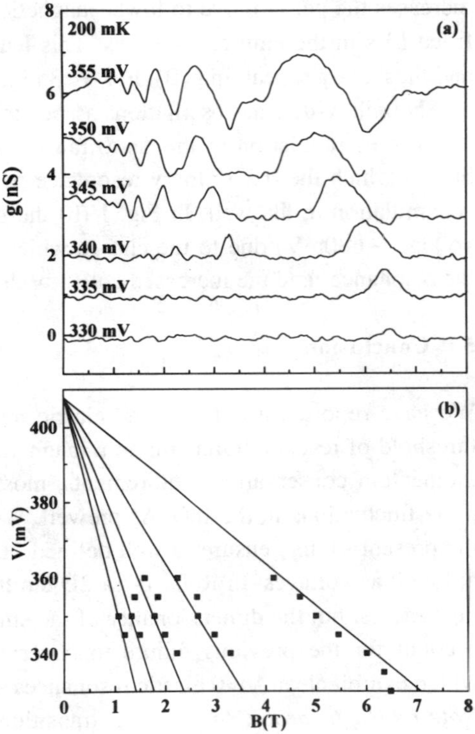

Figure 2: (a) Differential conductance (g) as a function of magnetic field for constant bias voltages at 200 mK. Adjacent traces are offset by 1 nS. (b) Magnetic field positions of maxima in g at low bias voltages. See text.

4 Differential Conductance Measurements

Differential conductance (g=dI/dV) was measured as a function of magnetic field at constant biases using AC lock-in techniques. Fig. 2 (a) shows g(B) in the forward bias tunneling threshold regime (330≤V≤355 mV). The magnetic field positions of maxima in g(B) in this regime are shown in Fig. 2 (b). The marked lines indicate elastic transitions from occupied LLs in the emitter prewell to unoccupied LLs in the well with Δn=-1 to -5. The slopes of these lines are given by $\alpha\Delta n\hbar\omega_c$ where α=6.4 mV/meV, consistent with Δn>0 data in Fig. 1 (b) (the small increase in the α at higher biases is due to the increased collector depletion length). As the bias increases the peaks move to lower magnetic fields and so the occupancy of partially filled LLs in the emitter increases. This leads to an increased transition probability and thus a larger peak in g(B). For V≥355 mV inter-LL transitions are superimposed on Shubnikov-de Haas oscillations associated with charge in the prewell.

The extrapolation of the Δn<0 fan in Fig. 2 (b) to 408 mV at 0 T indicates the bias at which the 0 T primary resonance peak would occur if there were no charge accumulation in the well. In Fig. 1 (b) the Δn>0 elastic fan extrapolates to a higher voltage (~680mV) due to the charge which remains in the well beyond the peak of the resonance[7] and the increased collector depletion length at high biases.

5 Conclusion

We have reported a fan of Δn<0 elastic inter-Landau level transitions close to the threshold of resonant tunneling in a magnetic field. These transitions violate angular momentum conservation requirements, most probably because of scattering due to alloy fluctuations in the InGaAs prewell. We note that the inclusion of a prewell in the present sample ensures a well defined 2D accumulation layer in the emitter even at low bias voltages. In Refs. 1-3 a 2D emitter accumulation layer is well defined at high biases, but the dimensionality of the emitter is less clear at low bias. This could account for the previous failure to observe Δn<0 elastic transitions in structures which exhibit clear Δn>0 elastic resonances at high bias.
Note added in proof: Δn = -1, -2 transitions have also been observed recently by B. R. A. Neves *et al*, *Phys. Rev.* B **52**, 4666 (1995).

References

1. V. J. Goldman *et al*, *Phys. Rev.* B **36**, 7635 (1987); M. L. Leadbeater *et al*, *Phys. Rev.* B **39**, 3438 (1989); C. H. Yang *et al*, *Phys. Rev.* B **40**, 6272 (1989); G. S. Boebinger *Phys. Rev.* B **47**, 16608 (1993).
2. L. D. Macks *et al*, *Phys. Rev.* B , in press (1996).
3. G. S. Boebinger *et al*, *Phys. Rev. Lett.* **65**, 235 (1990); J. G. Chen *et al*, *Phys. Rev.* B **43**, 4531 (1991).
4. C. Zhang and D. J. Fisher, *Phys. Rev.* B **52**, 14775 (1995).
5. H. Asahi *et al*, *Appl. Phys. Lett.* **59**, 803 (1991).
6. E. E. Mendez *et al*, *Phys. Rev.* B **33**, 2893 (1986); C. E. T. Goncalves da Silva and E. E. Mendez, *Phys. Rev.* B **38**, 3994 (1988).
7. T. A. Fisher *et al*, *Phys. Rev.* B **50**, 18469 (1994).
8. L. Eaves *et al*, *Solid State Elec.* **32**, 1101 (1989).
9. V. J. Goldman *et al*, *Phys. Rev. Lett.* **58**, 1256 (1987); T. C. L. G. Sollner, *Phys. Rev. Lett.* **59**, 1622 (1987); F. W. Sheard and G. A. Toombs, *Appl. Phys. Lett.* **52**, 1228 (1988); T. J. Foster *et al*, *Phys. Rev.* B **39**, 6205 (1989).
10. S. A. Brown *et al*, *to be published*.

MAGNETORESISTANCE AND ZERO-BIAS ANOMALY IN TUNNEL SCHOTTKY-BARRIER JUNCTIONS AT QUANTIZING MAGNETIC FIELDS

A.YA. SHUL'MAN, I.N. KOTEL'NIKOV

Institute of Radioengineering & Electronics of RAS, 103907 Moscow, Russia

E-mail address: ash@cplire.ru

The dependence of tunnel resistance $R = dV/dI$ on the bias V and magnetic field B have been measured and compared with the expression for $R(B)$ derived under Landau quantization (LQ) conditions. The agreement of calculated curves $R(B)$ with the measured ones allows to explain the origin of longitudinal tunnel magnetoresistance (MR) and confirms that the impurity scattering of electrons do not essentially affect the electron tunneling. The observed separation-of-variables character of $R(V, B)$ function was used to subtract uniquely the background curve in the region of zero-bias anomaly (ZBA) and LO-phonon peaks. As a result, the Lorentzian shape of these peaks and the dependence of its parameters on B were determined for the first time. The findings are discussed from the point of view of the theoretical model that explains the ZBA as a result of the discontinuity of the effective potential barrier near the Fermi level due to exchange-correlation interaction of electrons.

1 Magnetoresistance of tunnel junctions at Landau quantization

The tunnel junctions were fabricated on $n - GaAs$ wafers [the (100) surface] in a vacuum of the order of 10^{-10}torr (see [1] and [2] for details). The bulk electron density was determined using Shubnikov-de Haas (Sh-dH) effect and is equal $5.85 \cdot 10^{18} cm^{-3}$. The differential resistance $R = dV/dI$ of the tunnel $I - V$ characteristics were measured at temperature $4.2K$ and values of $B \leq 22T$. It has been observed that the dependence of R of Schottky-barrier junctions on the bias V and magnetic field B can be represented in multiplicative form $R(V, B) = f(V) * g(B)$ excluding the features in the range of tunnel spectra related to the contributions of many-particle interactions (exchange-correlation and polaron interactions) [3]. In Fig.1 the measured dependence of MR $(R(0, B) - R(0, 0))/R(0, 0)$ on B at $V = 0$ is shown. It is seen that this dependence is very close to B^2 law over a range of 10^3 times.

The longitudinal MR (LMR) turned out to be not too low in comparison with transversal one (TMR). While the TMR in the case $B \perp I$ can be associated with Lorentz force acting on a tunneling electron and described by the addition of the respective potential to the Hamiltonian, the LMR of tunnel junction can arise only due to LQ of the electron energy spectrum in GaAs. However, the measurements of the bulk Sh-dH oscillations carried out on the

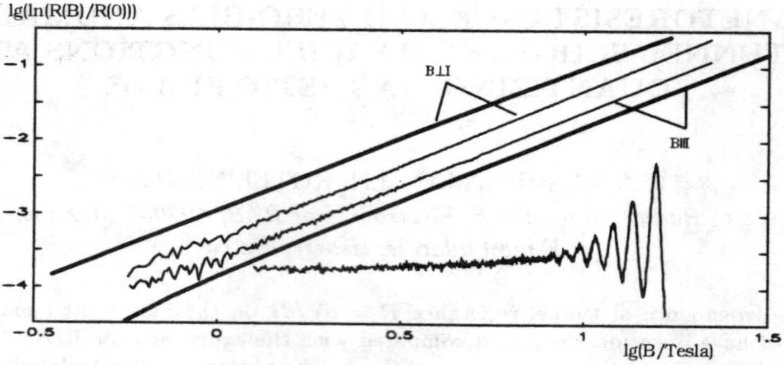

Figure 1: Tunnel resistance R versus B at $V = 0$. Both measured and calculated curves are shown. Bulk Shubnikov-de Haas oscillations are also presented.

same substrate show that LQ in the bulk starts at $B \geq 8T$, i.e. when the cyclotron frequency ω_c and the relaxation time τ calculated from the electron mobility satisfy the conditions $\omega_c\tau \geq 1$ (see Fig. 1). At weaker fields, the electrons in the bulk are thus not quantized. Nevertheless, no differences in the behavior of tunnel $R(B)$ are seen at $B < 8T$, including the case $B||I$.

The theoretical curves $\Delta R(B)$ represented also in Fig. 1 were calculated on the base of the expressions derived in [3]. For the case $B||I$ the LQ was taken into account. Calculated curves differ from experimental data less than in two times. Such an agreement proves the presence of LQ in the barrier region for $B \parallel I$ up to very low $B \leq 1T$ and indicates that electron-impurity collisions do not affect essentially the wave functions of tunneling electrons.

2 Zero-bias anomaly

The theoretical curves in Fig. 1 has been calculated without fitting parameters. The remaining quantitative divergence testifies that the change in the shape of the self-consistent Schottky barrier due to the influence of LQ on the electron screening has to be taken into account [3]. However, the self-consistent nature of the Schottky barrier implicates not only quantitative corrections.

For $B||I$ the ZBA is strongly increased (up to 5 times at $B = 22T$ in comparison with $B = 0$, see [2]). In Fig. 2a (right Y-axes) the curve $ln(\sigma(V,B)/\sigma(V,0))$ at $B = 22T$ is plotted as a function of V. Here $\sigma = 1/R$ is the tunnel conductance of the junction. The ratio to be analyzed depends weakly on V outside of ZBA and LO-phonon regions due to the above-mentioned feature of $R(V,B)$. This weak V-dependence of smooth background

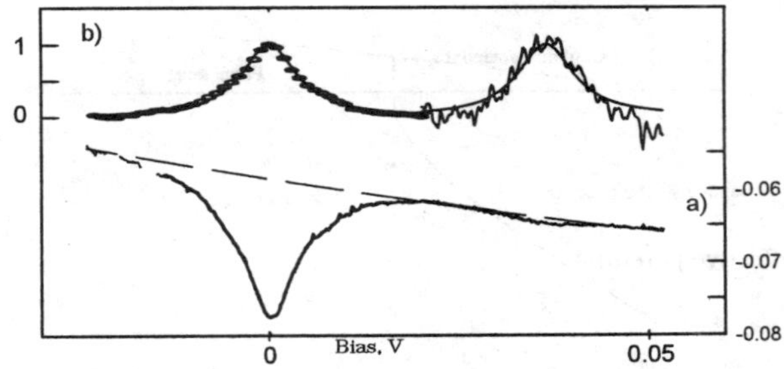

Figure 2: In Fig. 2a the function ln $[\sigma(V,B)/\sigma(V,0)]$ at $B = 22T$ is plotted for $B \parallel I$. The dashed line represents the background. In Fig. 2b the result of background subtraction is shown as well as two Lorentzian (solid lines) that fit the data in ZBA and LO-phonon region.

may be removed by the second-order polynomial fitting (dashed curve in 2a). The result of the background subtraction is plotted in Fig. 2b.

It turned out that the conductance dip in ZBA region and the LO-phonon peak can be well fitted by the Lorentzian as a function of V (solid curves in Fig. 2b). The increment of the peak amplitude grows with B as B^2 while the peak width W increases much slower, approximately as $ln(B)$.

The width at half maximum of the Lorentzian is of order of 2 meV at $B = 0$. This value is in a good agreement with the theoretical expression $\Delta = .3Ry^*$(=1.8 meV for $n-GaAs$) obtained in[4] for the jump of the exchange-correlation potential at the boundary of the electron Fermi-liquid in the Schottky barrier (see Fig. 3). Similar jump is used in the Thomas-Fermi-Dirac (TDF) model of many-electron atoms.

A number of details of ZBA and the deviations of the $I-V$ characteristics of metal/semiconductor junctions from simple models of the Schottky barrier at $B = 0$ were earlier explained on that ground[5]. The observed behavior of ZBA with B at $B||I$ under LQ conditions is likely explained by an enhancement of the reflection of the electron waves at the discontinuity of the effective potential of the Schottky barrier near the Fermi level. This enhancement is due to the transition from $3D$ motion of electrons to the one-dimensional one as B increases. The slow increasing of W similar to $ln(B)$ can be understood as a consequence of the slow increase in the magnitude of the jump. An analogous effect was calculated for TDF model in quantized magnetic field[6].

When $B \perp I$, the small but noticeable changes of many-particle features were observed. In this case the mesoscopic-like magnetooscillations of ZBA appears in the magnetic field[2].

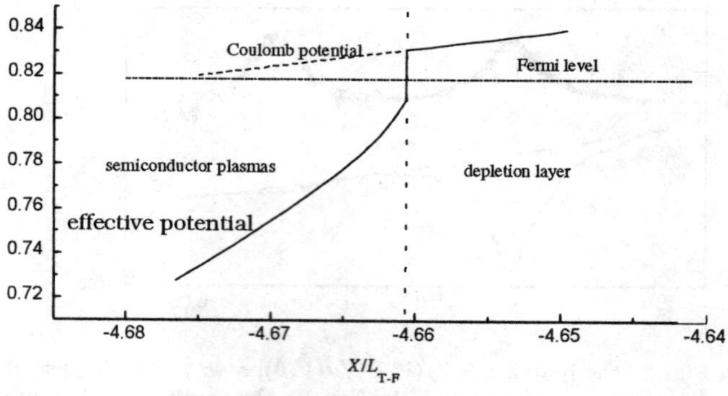

Figure 3: The effective potential $\Psi(x) = \Phi(x)+V_{xc}(n(x))$ near the region of the discontinuity. The discontinuity of Ψ is due to the cut-off of electron density that yields the disappearence of the exchange-correlation contribution to Ψ. X and Y axeses are scaled by Tomas-Fermi screening length L_{TF} and by Fermi energy of non-interacted electrons, respectively.

Acknowledgments

This work was partly carried out in Grenoble High Magnetic Field Laboratory in cooperation with J.-C. Portal and D.K. Maude. The financial support from Russian Foundation for Basic Research (grant 94-02-05712-a), International Science Foundation (grants MMR000 and MMR300) and NATO Linkage Program (HTECH.LG 960308) are gratefully acknowledged.

References

1. I.N. Kotel'nikov, I.L. Beinikhes, and A.Ya. Shul'man, *Sov. Phys. Sol. St.* **27**, 246 (1985).
2. I.N. Kotel'nikov, A.S. Rylik, and A.Ya. Shul'man, *JETP Lett.* **58**, 779 (1993).
3. I.N. Kotel'nikov, A.Ya. Shul'man, D.K. Maude, J.-C. Portal, *JETP Lett.* **60**, 863 (1994).
4. A.Ya. Shul'man and V.V. Zaitsev, *Sol. State Comm.* **18**, 1623 (1976).
5. I.N. Kotel'nikov and A.Ya. Shul'man, *Proc. XIX ICPS,* ed. by W.Zavadzki (Warsaw, Poland, 1988), Vol. 1, p. 681.
6. I. Fushiki, E.H. Gudmundsson, C.J. Pethick, and J. Yngvason, *Ann. Phys. (N.Y.)* **216**, 29 (1992)

QUANTUM–INTERFERENCE OSCILLATIONS OF THE RESONANT–TUNNELLING CURRENT IN A TRANSVERSE MAGNETIC FIELD

T. WOSIŃSKI, T. FIGIELSKI, A. MĄKOSA, W. DOBROWOLSKI
Institute of Physics, Polish Academy of Sciences,
Al. Lotników 32/46, 02-668 Warszawa, Poland

S.A. VITUSEVICH and A.E. BELYAEV
Institute of Semiconductor Physics, National Academy of Sciences
of Ukraine, 252028 Kiev, Ukraine

Quantum-interference oscillations of the resonant-tunnelling current in double-barrier AlAs/GaAs/AlAs heterostructures have been investigated in a magnetic field transverse to the current. A phase shift of the oscillation cycles proportional to the square of the magnetic field and a rapid fall-off of the oscillation amplitude with increasing magnetic field, followed by an oscillatory-type tail, have been revealed. To account for the observed oscillations a simple quantitative model is proposed which consistently explains all their features in terms of the interference of ballistic electrons tunnelled throughout the double-barrier structure.

We investigated double-barrier resonant tunnelling devices (DBRTDs) grown by molecular-beam epitaxy on n^+-GaAs substrates. Their active part consisted of a 4 nm wide undoped GaAs quantum well sandwiched between two undoped 2 nm AlAs barriers. A 100 nm buffer layer of n^+-GaAs and a 100 nm spacer layer of undoped GaAs were incorporated adjacent to the barrier on the substrate side and a 100 nm cup layer of n^+-GaAs was grown on the top of the structure. Measurements were performed on square 16×16 [μm]^2 mesa diodes fabricated from the wafers.

Two subbands of the quantum well, expected in these heterostructures, resulted in two resonance peaks displayed in the static current-voltage characteristics of the devices; Fig. 1. Each peak ended in a current bistability. Recently, we have discovered fine oscillatory structure (FOS) of the tunnel current on the increasing slope of these resonance peaks.[1] The structure, revealed while measuring the differential conductance of the devices at low temperatures, displays surprisingly regular sinusoidal shape with a period in the range 28 - 40 mV for different resonance peaks; Fig. 2a. FOS amplitude decreased with increasing temperature and the oscillations disappeared above ~100 K. Period of FOS was independent of temperature but it varied slightly in dependence on the position of device on the wafer area. In each device the period was essentially larger for the second resonance peak (corresponding to higher bias voltage) than for the first one.

In this paper we report two striking properties of FOS revealed in a magnetic field, B, transverse with respect to the current:

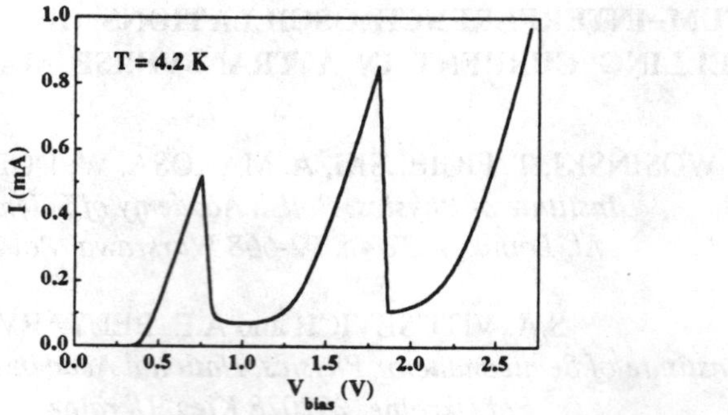

Figure 1: Current-voltage characteristic of investigated DBRTD measured at positively biased substrate.

(i) while the positions and shapes of the resonance peaks remain unchanged for $B \leq 2$ T, the FOS phase shifts on the bias-voltage scale proportionally to the square of the magnetic field induction; Fig. 2.

(ii) with increasing magnetic field, the FOS amplitude rapidly falls down to zero, reaching it at $B \approx 1.5$ T, but further goes up and falls down again, thus pointing out its oscillatory character; Fig. 3. This is in contrast to the case of magnetic field applied parallel to the tunnel current, when the FOS amplitude slowly decreases with increasing B but is still observable at 13 T.[1]

In the following we propose a simple quantitative model which can explain consistently the both properties. The model assumes that a high-energetic ballistic electron which escaped from the quantum well is partially reflected at some potential step on the collector side of device, situated at a distance L from the back

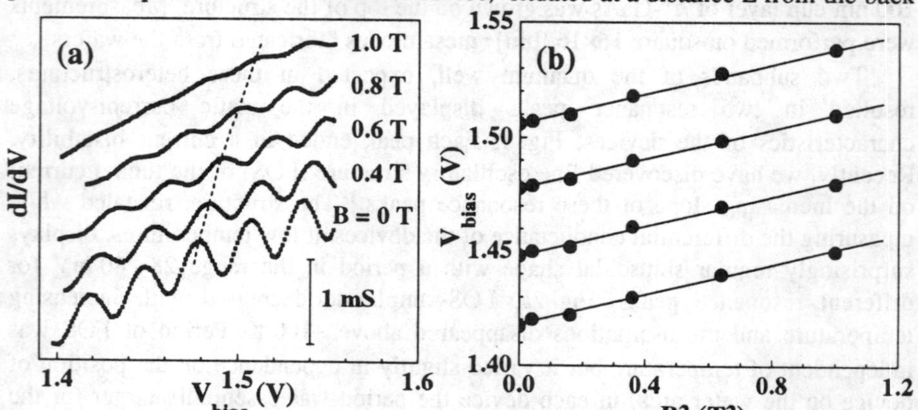

Figure 2: Effect of transverse magnetic field on the fine oscillatory structure, measured on the second resonance peak at a temperature of 2 K. (a) differential conductance vs. bias voltage in different magnetic fields. Dashed line indicates a phase shift of the oscillation cycles. (b) bias voltages corresponding to the maxima of several cycles of oscillations as a function of the square of magnetic field.

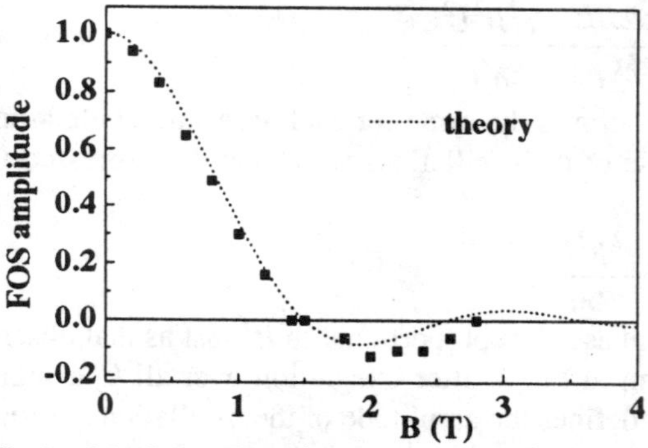

Figure 3: Relative amplitude of the fine oscillatory structure as a function of transverse magnetic field. Dotted line represents the theoretical curve, described by Eq. 4, fitted to the experimental data.

boundary of the collector barrier. Such a step can in principle arise at each interface between two technologically different epilayers. The back-reflected de Broglie wave interferes with the wave running in forward direction (along the x-axis) yielding a standing-wave component. Consequently, the squared modulus of the wave function between the collector barrier and potential step contains a part oscillating with the distance, $|\psi_{osc}(x)|^2$. Making a simplified assumption that the kinetic energy of an electron outgoing from the well is constant and equal $U_b \equiv eV_{bias}$, one gets: $|\psi_{osc}(x)|^2 \propto \cos(2k_x x)$.

For a given $x = L$, this quantity oscillates also with the variation of applied bias. Each minimum of $|\psi_{osc}(L)|^2$ corresponds to the alignment of self-energies for a "cavity" $[0, L]$ with the energy of the outgoing electron. So, the total transmittance through the considered system oscillates versus bias voltage similarly as $|\psi_{osc}(L)|^2$ does. This observation allows us to readily derive an expression for the approximate period of the oscillations, ΔU_{osc}:

$$\Delta U_{osc} = \frac{\pi \hbar}{L}\sqrt{\frac{2U_b}{m}}, \qquad (1)$$

where m is the effective mass of the electron, and the distance L is treated here as a free parameter of the model.

We can now discuss the effect of transverse magnetic field (applied along the z-axis) on the oscillatory structure. The magnetic field exerts the Lorentz force on the x- and y-component of the electron motion. In classical approximation of weak magnetic field, an electron of the wave vector $\bar{k} = [k_x, k_y, k_z]$ after travelling a distance L gains under the action of this force an additional momentum along the x-axis. It is given by the sum of a linear- and quadratic-versus B terms:

$$\Delta k_x = -\frac{k_y eBL}{\hbar k_x} - \frac{e^2B^2L^2}{2\hbar^2 k_x}. \tag{2}$$

The quadratic term is the same for each transmitted electron. It gives rise to a change in the phase of the oscillations which can be expressed in units of the bias voltage as:

$$\Delta U(B) = \frac{e^2B^2L^2}{2m}. \tag{3}$$

Eq. 3 predicts the phase shift proportional to B^2 just as that observed for FOS.

The linear term in Eq. 2, after integration over all k_y available in the electron gas of the emitter, defines the amplitude of the oscillations. Assuming the gas to be three-dimensional we obtain:

$$\left|\psi_{osc}(L)\right|^2 \propto 3\left[\frac{\sin(\alpha B)}{(\alpha B)^3} - \frac{\cos(\alpha B)}{(\alpha B)^2}\right]. \tag{4}$$

This is an oscillatory function of B, which describes well the behaviour of the observed FOS amplitude (dotted line in Fig. 3). Here $\alpha = eL^2k_F\sqrt{2/mU_b}$, and k_F is the Fermi wave vector of the electron gas in the emitter.

One may combine Eqs. 1 and 3 to eliminate uncertain effective mass from these equations to obtain more reliable values for the parameter L. All such obtained values for oscillatory structures measured on different resonance peaks and at both bias polarities lie between 110 nm, for the first peak at negatively biased substrate, and 170 nm, for the second peak at positively biased substrate. The determined values of L coincide quite well with the widths of technologically different epilayers beyond the collector barrier when assuming that the de Broglie wave of electrons outgoing from the quantum well is reflected at the metallic contact on the top of device, in the first case, and at the interface between the substrate and the epilayer, in the second case. Then, we have also to accept that in the heavily doped layer the phase-coherence length of the de Broglie wave may be as large as 100 nm at low temperature. Such large lengths in heavily doped GaAs, although surprising, have been indeed reported in the literature.[2]

This work was supported by the Committee for Scientific Research of Poland under Grant No 2 P03B 035 08.

References

1. T. Figielski, S.A. Vitusevich, A. Mąkosa, W. Dobrowolski, A.E. Belyaev, T. Wosiński, R.V. Konakova and L.N. Kravchenko, *Solid State Commun.* **94**, 93 (1995).
2. M. Heiblum, M.V. Fischetti, W.P. Dumke, D.J. Frank, I.M. Anderson, C.M. Knoedler and L. Osterling, *Phys. Rev. Lett.* **58**, 816 (1987).

DYNAMICS OF ELECTRIC FIELD DOMAINS IN GaInAs/AlInAs p-n MULITPLE QUANTUM WELLS UNDER HIGH MAGNETIC FIELDS

S. YAMADA, Y. KAWAI* AND K. YAMAMOTO
Japan Advanced Insitute of Science and Technology (JAIST)-Hokuriku,
1-1 Asahidai, Tatsunokuchi, Ishikawa 923-12, Japan

H. ASAI
NTT Optoelekctronis Laboratories,
3-1 Wakamiya, Morinosato, Atughi, Kanagawa 243-01, Japan

Y. KAWAMURA
Institute of Advanced Sciene and Technology, Osaka Prefecture University
1-2 Gakuencho, Sakai, Osaka 593, Japan

T. SASAKI and T. FUKASE
Institute for Materials Research, Tohoku University
2-1-1, Katahira, Aokaku, Sendai 980-77 Japan

We have applied the magnetotunneling method to two types of p-n GaInAs/AlInAs MQWs to analyze the dynamics of electric field domains appearing in the first negative differential resistance region. It is found that a characteristic field, Bc, deduced from the current oscillation against the magnetic field, oscillates associated with the saw-tooth current oscillations in the current-voltage curves. The oscillatory behavior shows a clear contrast depending on the electrode material (GaInAs or AlInAs).

1 Introduction

In the vertical current (I) - voltage (V) characteristics of multiple quantum wells (MQWs) or superlattices, saw-tooth current oscillations in the negative differential resistance (NDR) regions are often observed [1,2]. It is believed that the origin of the oscillation is an electric field domain associated with an electron accumulation, the peak of which moves from emitter to collector, from one well to the next [3,4]. However, the dynamics of the domain, especially the quantitative aspects, are not sufficiently clear yet. In order to solve the problem, we applied two magnetotunneling techniques [5,6] to GaInAs/ AlInAs p-n MQWs. It is well known that the magnetotunneling technique can give quantitative information about electrons accumulated in the center well of a double barrier (DB) structure. There seems to be sufficient similarity between accumulated electrons in a DB structure and the electron accumulation layers in MQWs which are the origin of the electric field domain. Therefore it can be expected that information about the dynamics of the electric field domain can be obtained by applying the technique to MQWs.

2 Sample and Measurements

We have prepared and measured two samples having different structures: one has GaInAs p-n electrodes (#1) and another has AlInAs ones (#2). Both samples include some ten periods of GaInAs (6.7 nm) / AlInAs (6.7 nm) alternating layers. The difference of the electrons will lead to the differences in the peak to valley current ratio and hence to different ratios between resonant and nonresonant terms [7]. The details of experiment are as follows: In the first one current oscillations area recorded as a function of magnetic fields up to 10 T at a fixed bias voltage. In the second one the current voltage characteristics is measured at a fixed magnetic field of up to 23 T, which was provided by the hybrid magnet of Tohoku University. In this work, we focused on the first NDR region of the I - V characteristics. This has also the purpose to confirm the first experimental approach. The data were taken at 4.2 or 1.5 K.

3 Experimental Results and Discussions

In Fig. 1 (a) and (b) the I - V characteristics for sample #1 and #2 are shown. Reflecting the difference of the electrode material, the voltage (Vp1 = 1.4 (V)) at which the current reaches its first peak in #1 sample is much smaller than that in at 2.05 V in #2. Although Vp1 should equal to the built-in voltage in each of the p-n MQWs and hence should be almost equal to the GaInAs (#1) or AlInAs (#2) band gap, the actual values are somewhat larger in both samples. For the same reason, the peak to valley ratio for the first structure is 4.7 in #1, whereas it is 2.7 in #2. Moreover, the saw-tooth oscillation is clearer in #1 than in #2. These results indicate that the resonant term is more dominant in #1 than in #2. In the figures we have attached temporal numbers to the saw-tooth structures (terraces), proceeding from low to high voltages. These numbers are believed to specify the well when counted from the emitter, where the peak of the electron accumulation is located. In addition we have measured current oscillations as a function of the magnetic field for a fixed voltage in the region of the first current peak and the subsequent NDR region. In figure 2 (a) and (b), the examples of the current oscillation are shown in their derivative form, dI/dB. The curves in Fig. 2 (a) are related to terrace 7 in Fig. 1 (a) and clearly show an increase of the main oscillation frequency as the voltage increases. In contrast, in Fig. 2 (b), where the curves correspond to the terraces of 3 and 4 in Fig. 1 (b), the decrease of the frequency in each terrace can be seen. If we carry out a systematic Fourier analysis of the frequency, that is, if we define a characteristic field (Bc) within the voltage range of interest, we can make plots of Bc against voltage in Figs. 1 (a) and (b) (right hand axis). Although the terraces analyzed are restricted, one can clearly recognize that the characteristic fields are also "oscillating" together with the current. The results indicate that the electron accumulation layers in each well grow and then shift to the next well. This behavior occurs successively in the NDR region. The reason that an increase of Bc is observed in sample #1, while a decrease is recorded in #2, is probably due to a difference in the background [8]. At present, an accurate determination of the nature of the background is difficult. Similarly it is difficult to discuss the physical meaning of the absolute values of the critical field within a particular terrace. As a simple model we chose a capacitor with parallel plates; the distance between the plates is equal to the distance emitter-well, in which the electron accumulation layer occurs. Finally, we measured I - V characteristics of the two samples under high magnetic fields up to 23 T. It is found within the range of magnetic fields used, that no anomalies occured, except that the terraces of the oscillations split at high currents. The width of the splitting becomes larger as the terrace number increases. However for the lower numbers the splitting sometimes disappears. This can probably be attributed to a change of the spatial distribution of the electron accumulation layer. However, the interpretation of the observed data is not yet clear.

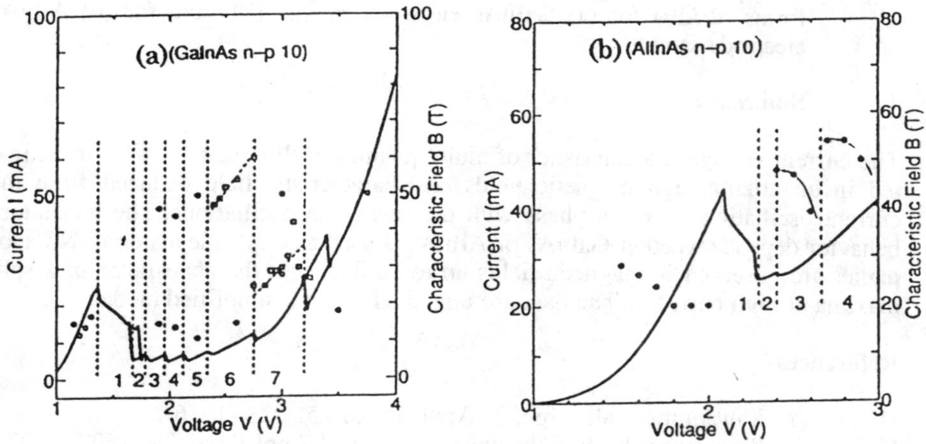

Fig. 1 Typical I - V curves together with corresponding characteristic field deduced from magnetotunneling oscillations for (a) GaInAs electrode sample (#1) and for (b) AlInAs electrode sample (#2)

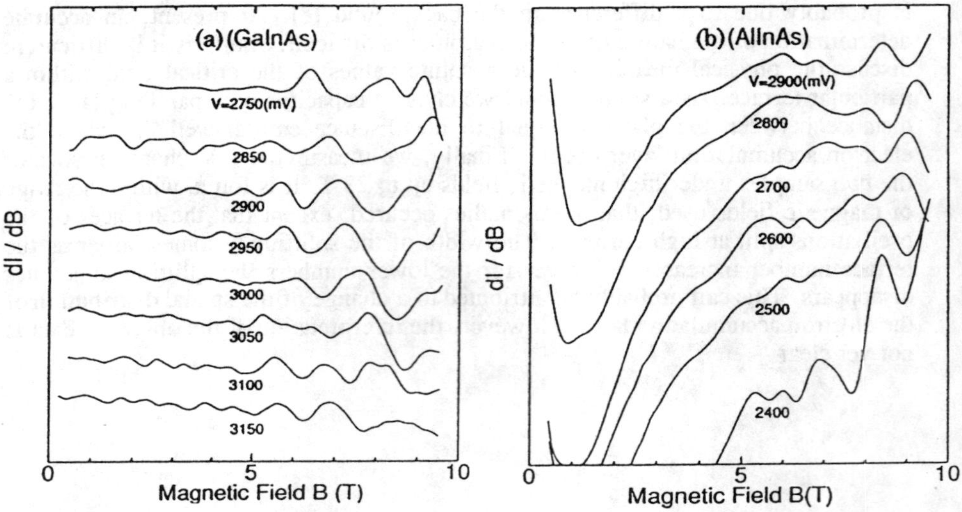

Fig. 2 Examples for magnetotunneling current oscillations in their derivative forms (dI/dB) for (a) GaInAs electrode sample (#1) and for (b) AlInAs electrode sample

4. Summary

The current voltage characteristics of multi quantum well structures were investigated in quantizing high magnetic fields. A characteristic field deduced from the current oscillation varies in phase with the saw-tooth oscillations. The oscillatory behavior depends whether GaInAs or AlInAs is used as electrode material. No anomalies are observed in magnetic fields up to 23 T except the appearance of a split maxima at high currents. The data are tentatively with a simplified model.

References

[1] Y. Kawamura et al., Jpn. J. Appl. Phys. 25, 928 (1986).

[2] Y. Kawamura, K. Wakita and K. Oe, Jpn.J.Appl.Phys. 26, 1603 (1987).

[3] F. Prengle, A. Wacker and E. Scholl, Phys. Rev. B50, 1705 (1994).

[4] S.H. Kwok, R. Merlin, H.T. Grahn et al., Phys.Rev.B50, 2007 (1994).

[5] E.E. Mendez, L. Esaki and W.I. Wang, Phys. Rev. B33, 2893 (1986).

[6] C.A. Payling et al., Surface Science, 196, 404 (1988).

[7] L. Eaves et al., Proc. Int. Conf. The Application of High Magnetic Fields in Semiconductor Physics, Springer Verlag, Vol. 87, 324 (1988).

[8] S. Ben Amor et al., Appl. Phys. Lett. 54, 1908 (1989).

SINGLE ELECTRON TUNNELLING THROUGH SELF-ASSEMBLED INAS QUANTUM DOTS AS SPECTROSCOPIC PROBE OF A 2DEG

A. Thornton, T. Ihn, I.E. Itskevich,[a] P.C. Main, L. Eaves, and M. Henini
Department of Physics, University of Nottingham, NG7 2RD, UK

Tunnelling through an individual quantum dot (QD) incorporated in the AlAs barrier of a *n-i-n* single-barrier tunnelling device can be used to probe the properties of the emitter two-dimensional electron gas (2DEG). At the magnetic field $B = 0$T single electron tunnelling through an InAs QD manifests itself as an isolated peak in the current-voltage [$I(V)$] trace. At finite $B \parallel I$ the Landau-fan and the chemical potential oscillations are mapped. Additional fine structure within the lowest Landau-level appears at low temperature as a consequence of potential fluctuations in the emitter 2DEG. The manifestation of Coulomb interaction between the tunnelling electron and the emitter 2DEG is demonstrated.

Linear response techniques such as conductivity measurements are sensitive to the Fermi-energy properties of a two-dimensional system in a magnetic field but no information can be obtained about states below the Fermi-energy, E_F. In recent years vertical tunnelling through individual localised states incorporated in a tunnelling barrier has been employed to probe the density of states (DOS) of the emitter.[1,2] The localised states have been produced, for example, by using donors in resonant tunnelling structures. Isolated resonant peaks in the $I(V)$ characteristics have been observed at zero magnetic field, but the fan charts obtained from $B \parallel I$ are very complicated and do not reflect the Landau-level (LL) structure of the emitter two-dimensional electron gas (2DEG).[1] In this paper we describe a probe of the 2DEG DOS using the single electron tunnelling current through an individual InAs quantum dot (QD) incorporated in the tunnelling barrier. A detailed description of the sample, its characterisation and single electron tunnelling through an individual QD can be found in Ref. 3.

The inset of Fig. 1a shows a schematic conduction band profile of a device under bias. At low enough carrier density in the 2DEG the non-resonant tunnelling current is much smaller than the resonant single electron tunnelling current through one QD state leading to a resonance peak in the $I(V)$ curve.[3] Figure 1a shows $I(V)$ traces recorded at $T = 4.2$K, 300mK and 30mK. An isolated peaked feature can be seen between 100mV and 150mV at all temperatures due to single-electron tunnelling through an individual QD.[3] At low T the resonance peak develops a kT-broadened sharp onset at the low voltage edge

[a]On leave from Inst. of Solid State Physics, Chernogolovka, Moscow reg., 142432, Russia.

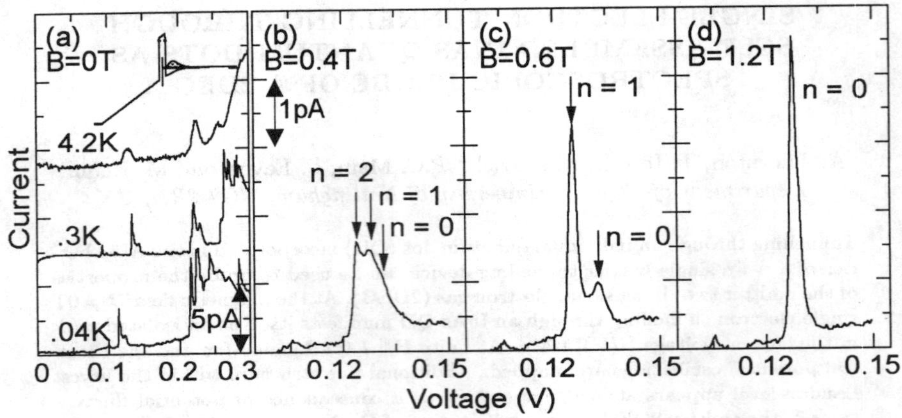

Figure 1: (a) $I(V)$ characteristics at different temperatures. (b-d) $I(V)$ characteristics of the current peak at 0.12V at different magnetic fields B showing the LL splitting.

V_F, i.e. when the energy of the localised state coincides with the Fermi-energy in the 2DEG. The high-voltage cut-off of the peak at V_0 indicates alignment of the localised state with the subband bottom of the 2DEG. Figure 1b–d shows examples of recorded $I(V)$ traces at different constant $B \parallel I$, showing three, two and one components indicated by arrows, respectively. It can be seen that the LL splitting is clearly observed but the tunnel current does not directly map out the single-particle DOS since the peak height decreases strongly as V is swept beyond V_F. The reason for that may be similar to the origin of the Fermi-edge singularity at $B = 0$T.[4] The tunnelling electron undergoes a Coulomb interaction with the 2DEG and can thereby deposit energy by creating a 2DEG excitation, e.g. an electron-hole pair. This process is most efficient close to E_F since occupied initial and unoccupied final states are required.

In order to map the oscillations of the chemical potential and the LLs in the emitter over a range of magnetic fields B we recorded a set of $I(V)$ curves at constant B and $I(B)$ curves at constant V values between 100mV and 150mV. These results are summarised in the fan-chart plot in fig. 2. The black line follows the onset of the current peak $V_F(B)$ obtained from the $I(V)$ sweeps, which is effectively the chemical potential in the emitter. Therefore as we sweep B, the variation of the current onset V_F is the variation in E_F showing $1/B$-periodic oscillations. However, we can go further than a linear transport experiment, because, above V_F (i.e. below E_F) current peaks in $I(B)$-curves occur at magnetic field values where the peak of the DOS of a certain LL aligns

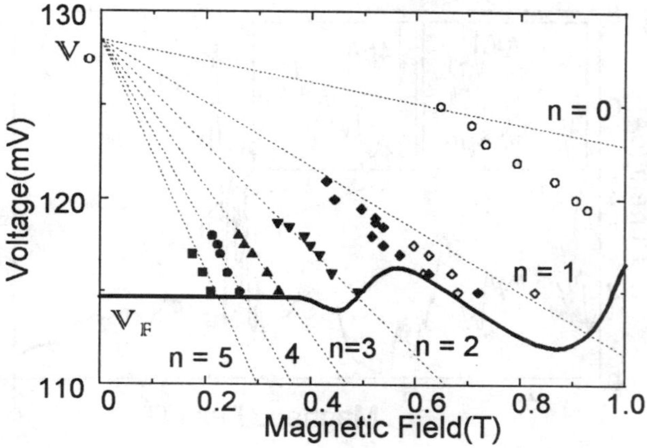

Figure 2: Fan diagram of peaks in $I(B)$ data taken at constant V around the resonance. The oscillating line represents the measured V_F, which essentially is the 2DEG chemical potential.

with the QD energy level for a given voltage V. The symbols represent these points in the B-V-plane. The straight lines labelled $n = 0, 1, \ldots, 5$ represent a Landau-fan obtained from $V_n = V_0 - f \cdot \hbar eB/m^* \cdot (n + 1/2)$, where $f = 6.5$ is the electrostatic leverage and m^* is the GaAs conduction band effective mass. We estimate the Fermi energy of the 2DEG at $B = 0$T to be $E_F \approx 2$meV and the corresponding electron density $n_s \approx 6 \times 10^{10}\text{cm}^{-2}$ at $B = 0$T.

Potential fluctuations ΔE_c in the 2DEG close to the QD can be probed either in the limit of small B as a deviation of the position of current peaks from the $1/B$-periodicity or at large B around filling factor $\nu = 1$, where $\hbar\omega_c \gg \Delta E_c$. Figure 3 shows an $I(B)$ trace recorded at constant V close to V_F. The insets are a magnification of the low field region between -0.3T and +0.3T, and of the peak of the $n = 0 \uparrow$ LL between 2T and 2.6T. In contrast to the main oscillations this fine structure disappears completely if the temperature is raised from 0.3K to 1.2K indicating that energy spacings of less than 100μeV are resolved. We interpret the fine structure as being due to local potential variations in the 2DEG. At small B they are lifting the degeneracy of momentum-eigenstates and we are looking at mesoscopic properties of the 2DEG. At large B the 2DEG states tend to be more localised in potential minima due to magnetic freeze-out and the LL degeneracy is locally lifted. Again, current peaks arise in both cases when the increasing magnetic field sweeps local 2DEG energy levels through the QD energy. A magnetic field increase $\delta B \approx 0.1$T between two current peaks is necessary to add one

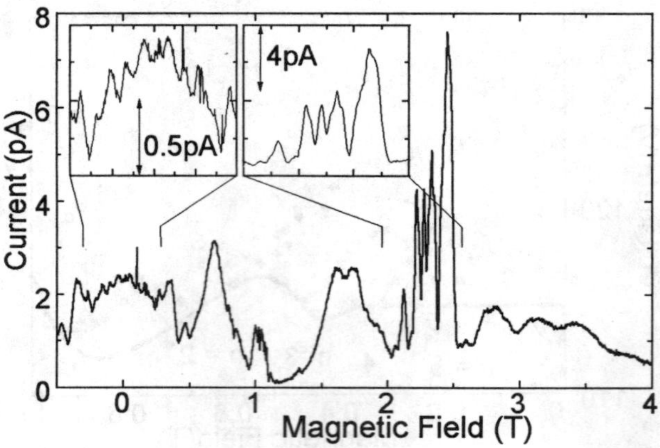

Figure 3: Fine structure in $I(B)$ curves caused by local potential fluctuations close to the probing QD state recorded at 300mK.

flux quantum to a state localised close to the probing QD. This allows us to estimate the extent of the states we are probing from the relation $\delta B \cdot A = \phi_0$, where A is the typical area and ϕ_0 is the flux quantum. We find a typical diameter of 100nm, much larger than the QD size and even than the mean QD separation. From the overall width of the Landau-band $\Delta B = 0.5T$ (see Fig. 3) we estimate $\Delta E_c \approx \hbar e \Delta B/m^* = 0.4$meV.

In conclusion, we have shown that we can probe the properties of a 2DEG using single-electron tunnelling through a self-assembled InAs quantum dot. The method is sensitive to the Landau-level structure of the 2DEG in a magnetic field applied parallel to the tunnelling current and to local potential fluctuations in the 2DEG.

This work is supported by EPSRC (UK). I.E.I. and L.E. acknowledge the Royal Society and EPSRC for research fellowships.

References

1. P.C. Main *et al.*, Physica B **189**, 125 (1993); P.H. Beton *et al.*, Phys. Rev. B **49**, 2261 (1994); Sakai *et al.*, Solid-State Electronics **37**, 965 (1994); P.J. McDonnell *et al.* Physica B **211**, 433 (1995).
2. T. Schmidt *et al.*, Solid-State Electronics **40**, 15 (1996).
3. L. Eaves *et al*, this conference proceedings.
4. A. Geim *et al.*, Phys. Rev. Lett. **72**, 2061 (1994).

MAGNETO-TUNNELING STUDY OF ZERO-DIMENSIONAL ELECTRONIC STATES IN SELF-ORGANIZED InAs QUANTUM DOT

M. NARIHIRO, G. YUSA, Y. NAKAMURA, T. NODA, AND H. SAKAKI

Research Center for Advanced Science and Technology, University of Tokyo, 4-6-1 Komaba, Meguro-ku, Tokyo 153, Japan, Quantum Transition Project, JRDC, 4-7-6 Komaba, Meguro-ku, Tokyo 153, Japan and Institute of Industrial Science, University of Tokyo, 7-22-1 Roppongi, Minato-ku, Tokyo 106, Japan

Resonant tunneling of electrons via the ground 1s states of an InAs quantum dot was studied at 4.2 K in a tunnel diode with InAs quantum dots embedded in a pair of thin AlAs barriers. Effects of inplane magnetic fields ($B \perp J$) on resonant current were evaluated and analyzed to show that spatial extent of the lowest 1s state was 18-20 nm. Effects of vertical magnetic fields ($B//J$) were also investigated to show that lateral qunatization occured not only in InAs quantum dots but also in the emitter.

Self-organized quantum dot (QD) structures (InAs QDs grown on GaAs, for example) are very attractive as a new class of quantum system. However transport study of InAs QDs has been rather scarce. Quite recently, a few groups including us independently started the study of the resonant tunneling of electrons via InAs QDs [2,3]. Here we report our work on the resonant tunneling diodes (RTDs) with InAs QDs buried in AlAs barrier and discuss the effect of magnetic fields on the tunneling through a zero dimensional state.

First we examine how the energy gap E_g in InAs QDs varies with their surrounding materials. For this purpose, we prepared by molecular beam epitaxy five different samples, in which InAs QDs were sandwiched by both sides with a 2 monolayer (ML) of GaAs and 7ML of $Al_xGa_{1-x}As$ barrier of various Al compositions x. By photoluminescence studies at 15 K, we found the average E_g to be 1.37, 1.41, 1.43, 1.47 and 1.58 eV for x =0, 0.29, 0.45, 0.61 and 1.0, respectively. Hence the ground states of electrons in InAs QDs embedded in AlAs barriers via 2 ML-thick GaAs is found to lie above the conduction band edge of GaAs.

Our RTDs with InAs QDs embedded in AlAs barriers were grown on n^+(100) GaAs substrates. We grew at temperature T_s of 540 °C a 650 nm n-GaAs buffer layer with graded Si concentration (1.3×10^{18}~6.5×10^{16} cm^{-3}), a 20 nm undoped GaAs spacer layer, a 6 ML undoped AlAs barrier and a 2 ML undoped GaAs. Then we reduced T_s to 450 °C, where InAs QDs were formed by depositing 2.2 ML InAs. On top of QDs, we deposited a 2 ML undoped GaAs, an 8 ML AlAs upper barrier and a 2.5 nm undoped GaAs. At this point we increased T_s to 540 °C and grew further a 17.5 nm undoped GaAs spacer layer, a 650 nm n-GaAs layer of graded doping (6.5×10^{16}~1.3×10^{18} cm^{-3}) to complete the growth. We also prepared a reference RTD in which only a 1.1 ML thick InAs wetting layer was embedded in a place of QDs. Using these wafers, mesa shaped RTDs of 50 μm in diameter with AuGe/Ni/Au ohmic contacts were produced.

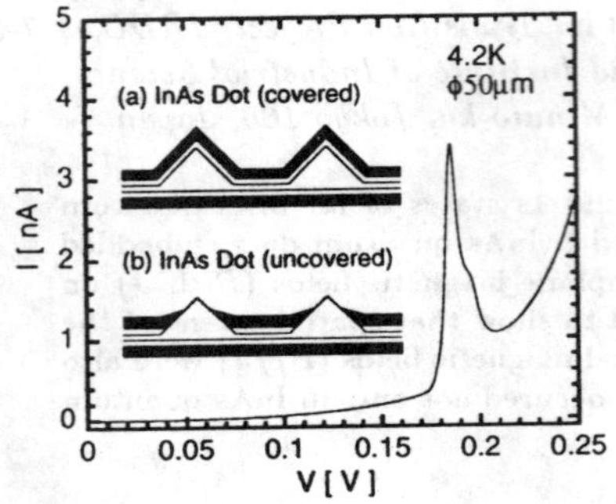

Figure 1: $I(V)$ measured at 4.2 K. The inset shows two types of InAs dots schematically.

Figure 1 shows current-voltage characteristics $I(V)$ of a RTD with InAs QDs measured at 4.2 K. A sharp current peak is seen at $V \simeq 186$ mV and a weak shoulder at $V \simeq 195$ mV. At V=0.3~0.4 V a clear oscillation in conductance-voltage characteristics $dI/dV(V)$ is observed. A RTD with only an InAs wetting layer showed no signs of structures. Hence all features in Fig.1 must originate from the tunneling through InAs QDs. The full width at half maximum ($\simeq$ 3meV) of the current peak is quite narrow. Since the current density of a RTD with 6ML and 8 ML AlAs barriers is roughly estimated to be 1 $kAcm^{-2}$, we expect that the effective area of diode necessary to support 5nA is only 22 nm × 22 nm, which is close to the area of a single InAs QD. Therefore it is likely that the current peak results from the tunneling through a single InAs QD. The reason why the current peak is supported by only a single QD is to be discussed later.

Since InAs QDs are thin as compared with their lateral size, the ground states of electrons in InAs QDs are disk-like 1s states (E_1) with the degeneracy of two and the first excited states have the degeneracy of four. Since the current peak and shoulder at V = 0.18~0.19 V are the first structures appearing in $I(V)$, these structures are most likely to result from the resonant tunneling through the ground states in which degeneracy is lifted by charging effect (E_1^+, E_1^-). Similarly the four conductance oscillation at higher bias region is ascribed to the resonant tunneling through the degeneracy-lifted excited states.

To clarify the origin of the composite peak, we studied the effect of inplane magnetic field ($B \perp J$). Figure 2 shows the result from 0 T to 9.0 T. From the behavior of the current peak at $V \simeq 186$ mV, following the framework of Sakai et al.[4], we found the spatial extent of the ground state E_1^- to be 18-20 nm which was close to the average diameter of InAs QDs. Since E_1^- state and the higher 1s state E_1^+ result from the spin degenerate states E_1 and the

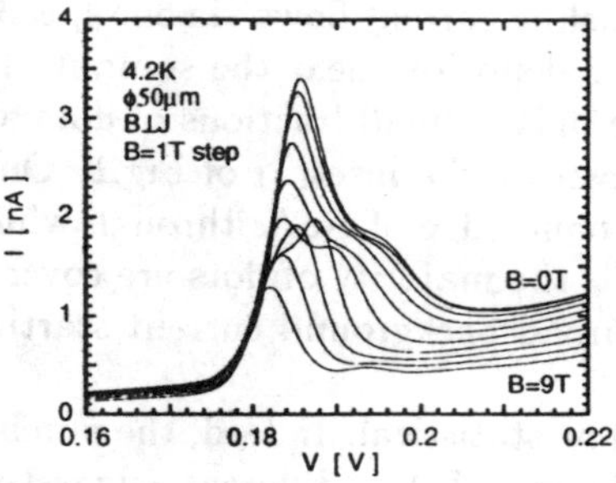

Figure 2: The tunneling current of the first peak and the shoulder measured at 4.2 K in the presence of magnetic fields B normal to the current.

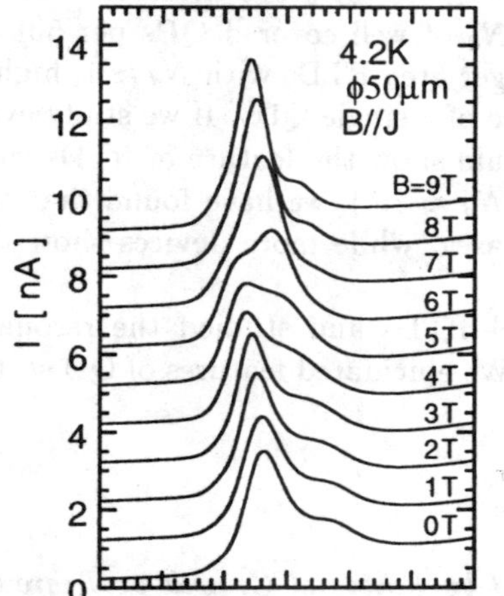

Figure 3: The tunneling current of the first peak and the shoulder measured at 4.2 K in the presence of magnetic fields B parallel to the current. Curves are offset by 1 nA.

degeneracy is lifted by electron-electron interaction, it is probable that the wavefunction of E_1^+ state is ring-like form and that the square of its Fourier transform $|F_1^+(k_x)|$ takes a camel's back form which has a peak at a finite k_{max}. Since the peak current under B is proportional to $|F_1^+(k_0)|^2$ (where $k_0 = eB\Delta s/\hbar$ and Δs is the separation between the QD and emitter), we expect that current via E_1^+ state increases with B and reaches its maximum at $B_{max} = \hbar k_{max}/e\Delta s$. Our data, where the current peaks at B =6.5 T and then decreases support this interpretation.

Next we investigated effects of vertical magnetic field ($B//J$). Figure 3 shows the result for B up to 9 T. Although the oscillation of resonant tunneling current is observed in conventional RTDs when a Landau level in the emitter and quantum well crosses with the Fermi energy, our device shows no such oscillation, implying that the Landau level formation is suppressed. Note also that the current peak at $V \simeq$186 mV shifts when B is applied but its shift is much smaller than the value expected for the Landau level in the emitter. We notice also that the amplitude of the current peak at $V \simeq$186 mV is almost unaffected by B up to 3 T and decreases with B. In contrast, a weak shoulder at $V \simeq$195 mV develops into a current peak at $B \simeq 5$ T and rapidly increases with B. The fact that the resonant tunneling current is almost independent on B in the low field region indicates that the lateral quantization occurs not only in InAs QDs but also in the emitter. Emitter states can be quantized by the strain which InAs QDs induce.

Finally we discuss why the resonant tunneling current flows via one specific InAs QD. It is known that Ga and Al atoms deposited near the summit of a InAs QD migrate to its foot region [5]. Hence only a small fractions of dots are well covered by the top AlAs barrier as shown by the inset(a) of Fig.1. Only these QDs have the bound states above the band edge of GaAs through which resonant tunneling current flows. In contrast, the majority of dots are covered insufficiently as in the inset(b) and give a broad background current starting at a low voltage region of the bias.

This selection mechanism of QDs should be statistical. Indeed, the number of structures appearing in $I(V)$ curves of different RTDs is different, suggesting that each RTD contains a different number of well covered QDs. Among our 50 μm-diameter RTDs, 20 % of them exhibit $I(V)$ curves qualitatively similar to Fig.1, while 70 % of them have no such structures, indicating the absence of well covered QDs. The remaining 10 % of RTDs show $I(V)$ curves with more crowded structures, suggesting that two or more QDs are involved. These findings suggest that the average number N_d of well covered QDs per 50μm-diameter RTD is about 0.4. If we study larger area RTDs with $N_d \approx 1$, higher percentage of RTDs should show the feature of a single QDs. If we study even larger RTDs with $N_d \gg 1$, most RTDs should show the feature of RTDs with multiple QDs. In 100μm-diameter RTDs ($N_d \approx 1.6$), we have found that the percentage of RTDs with no feature decreases, while more devices show the feature of a single QD.

In summary, we have fabricated novel RTDs and studied the resonant tunneling via discrete states in InAs QDs. We elucidated features of QD states and the lateral quantization in the emitter.

References

1. *Low Dimensional Structures prepared by Epitaxial Growth or Regrowth on Patterned Substrates, Vol. 298 of NATO Advanced Study Institute, Series E: Applied Science*, edited by K. Eberl, P. M. Petroff, and P. Demeester (Kluwer Academic Publishers, Dordrecht, 1995).
2. T. Suzuki, K. Nomoto, K. Taira and I. Hase, unpublished.
3. L. Eaves; A. Thornton, T. Ihn, I. E. Itskevich, P. C. Main, L. Eaves, and M. Henini, this conference proceedings.
4. J. W. Sakai, T. M. Fromhold, P. H. Beton, L. Eaves, M. Henini, P. C. Main, F. W. Sheard, and G. Hill, *Phys. Rev.* B **48**, 5564 (1993).
5. Q. Xie, P. Chen and A. Madhukar, *Appl. phys. Lett.* **65**, 2051 (1994).

MAGNETOTUNNELING STUDY OF A TRAPEZOIDAL SI QUANTUM WELL

J. LINDOLF, U. KUNZE
Lehrstuhl für Werkstoffe der Elektrotechnik, Ruhr-Universität Bochum, D-44780 Bochum, Germany

W. KIUNKE, I. EISELE
Fakultät für Elektrotechnik, Universität der Bundeswehr München, D-85577 Neubiberg, Germany

Using tunneling spectroscopy we have studied the subband structure and diamagnetic energy shift in a delta-doped electron surface layer with a trapezoidal confining potential. The Mg-SiO_2-Si junctions are prepared on p^+ substrates covered with an n-type delta doping layer and 2-16 nm undoped Si. At large undoped layer thickness the subband separation increases with energy, which indicates square-well like behaviour. The diamagnetic energy shift in a parallel magnetic field shows a characteristic dependence on the subband index which is approximately described by the wave-function parameters of a self-consistent calculation.

The subband structure in a two-dimensional electron gas (2DEG) in Si has been probed by tunneling in metal-oxide-semiconductor (MOS) junctions [1] and Schottky-barrier junctions on delta-doped structures [2-4]. In these experiments the confining potential is triangular, shaped, which results in a level separation which decreases with increasing energy, unlike in a square potential well where the separation increases. In the present work we investigate a 2DEG in a trapezoidal potential well at the interface of an MOS structure containing a buried δn doping layer. The oxide layer forms the abrupt first barrier, and the second barrier arises from a δn-p^+ junction separated from the Si-SiO_2-interface by a nominally undoped Si layer. This layer of thickness d ranging from 2 to 16 nm represents the flat part of the potential well.

The layers are grown by molecular beam epitaxy on degenerately p-type doped Si(100) substrate as described in an earlier publication [4]. Samples with different top layer thickness d are prepared by removing up to 14 nm by anodic oxidation and subsequent etching in HF. The SiO_2 barrier is formed by thermal oxidation at 600 °C. The cleaning procedure and oxidation consume another 2 nm Si. Metallization with Mg and Al completes the MOS structure. The tunneling junctions are defined by patterning the metal layer into 50 μm square dots. The remaining metal area of several mm^2, separated from these contacts by a 5 μm trench, provides a low-resistive contact to the electron layer. The subband structure is obtained from measurements of the

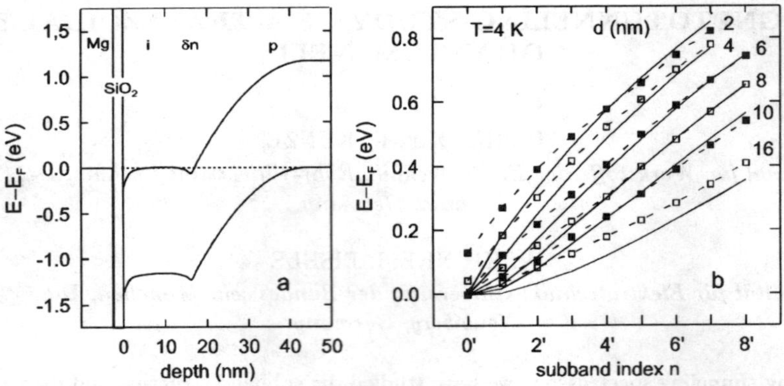

Figure 1: a) Self-consistently calculated band structure of an Mg-SiO_2-Si tunneling junction with an intrinsic layer thickness d = 16 nm . b) Subband energy $E_n - E_F$ of the primed subband series as a function of subband index n for samples with different layer thickness d measured at T = 4 K. Broken lines are guides for the eye. The solid lines represent the result of the numerical fit.

second derivative d^2I/dV^2-vs-V of the current(I)-voltage(V) characteristic at T= 4.2 K.

Figure 1a shows the band structure of a tunneling junction of d = 16 nm as obtained from a self-consistent calculation. The width d is known from two different measurements. An initial width of 20±3 nm has been determined by transmission electron microscopy. The reliability of the removing process by anodic oxidation has been established by reflection interferometry and scanning force microscopy. We estimate the error in d due to this process as ±1 nm. The other pertinent parameters of the calculation, the bulk doping N_A and the δ-doping n_D are gained from a fit to the measured subband energies. Since the spread of the δ layer has little effect on the subband structure we used a homogeneous distribution of 1 nm width [2]. An improvement of the fit to the experimental results has been achieved by taking into account a negatively charged layer of vacancies, which arise due to the high doping concentration of the δ layer [5]. We assume an areal density of 30% of n_D spread homogenously over 3 nm [5]. The resulting fit parameters $N_A = 2 \times 10^{18}$ cm^{-3}, and $n_D = 1.4 \times 10^{13}$cm^{-2} should be compared with those obtained from Schottky-barrier tunneling spectroscopy of a 2DEG created by the δ-doping layer alone, where the junctions were prepared on the same wafer, denoted in [4] as wafer B. The present data differ significantly from the earlier $N_A = 1.8 \times 10^{19}$ cm^{-3} and $n_D = 2.3 \times 10^{13}$cm^{-2}, moreover, the latter did not include the partial compensation by the charged vacancies. We believe that the MOS tunneling result is more reliable because the low work function of the metal electrode induces a high electron density at the $Si\text{-}SiO_2$ interface and thus causes an effective screening of the surface electric field. As a result, the shift of the subband energies due to the applied bias is much smaller than in Schottky barrier junctions. Additionally, in the present junctions the structures in d^2I/dV^2 can be directly attributed to the corresponding subbands, while in the Schottky-barrier junction several low-index subbands could not be detected due to the high barrier resistance at low reverse bias. Therefore in the type B samples of the earlier work the assignment of the subbands might be incorrect. Finally, unlike the Schottky-barrier junctions the series resistance is negligible in the MOS junctions. Therefore we can directly read the subband energy from the bias position V_n of the n-th subband-edge induced dip in d^2I/dV^2, i.e. $E_n - E_F = -eV_n$. We are interested in the primed subband series because of their larger signal strength [1]. A plot of the measured subband energy as a function of subband index (Fig. 1b) resembles that of a square well at an undoped layer width d >8 nm and of a triangular well at smaller d. The calculated energies reasonably describe the experimental results, except for subbands $n \leq 2'$ and the curve for $d = 16$ nm. However, the real doping profile probably differs from that expected from the growth parameters, because the Sb distribution may be asymmetric due to a spread of the dopant into the nominally undoped layer.

The shape of the surface potential well should also significantly influence the diamagnetic effect in a parallel magnetic field B, which can easily detected by tunneling [6]. Up to B =14 T has been applied along the [110] direction parallel to the surface, where the two valley pairs of the primed subband ladder are equivalent. As expected from first-order perturbation theory we observe a shift ΔV of the structures in d^2I/dV^2 proportional to B^2. However, ΔV reflects the shift of the subband with respect to E_F, i.e. $\Delta(E_n - E_F)$. ΔE_n represents the diamagnetic rise, which is proportional to the square of the spread of the wave function, $\delta_n^2 = \langle z^2 \rangle - \langle z \rangle^2$, at $B = 0$. The change ΔE_F arises from the diamagnetic shift and from the change of the density of states of the ground subband [6,7]. The latter one causes the observed shift $\Delta(E_0 - E_F) \propto \delta_0^2 B^2$, which is used to calculate the change of the Fermi level as a function of B^2. In this way we obtain the spread δ_n from a least-square fit to the experimental data ΔV vs. B^2. Figure 2 shows a plot of δ_n as a function of the subband index n for the $d = 16$ nm junction. Within experimental error δ_n shows little variation in the range $2 \leq n \leq 6$ as expected for a square well. It is interesting to note, that the theoretical curve of the self-consistent model as well as the triangular model with an electric field equal to the average field of the potential well appropriately fits to the experimental points. Obviously

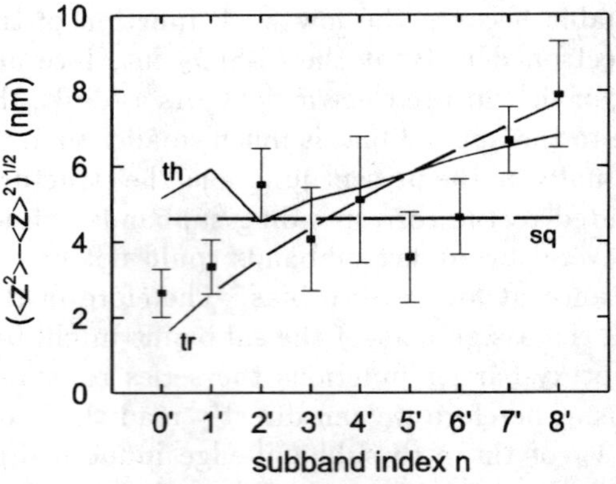

Figure 2: Spread $\delta_n = (\langle z^2 \rangle - \langle z \rangle^2)^{1/2}$ as a function of subband index n of a sample with 16 nm undoped layer. The solid line represents theoretical results obtained from the self-consistent calculation used in figure 1. The dashed lines display the spread for an ideal 16 nm wide square well (sq) and a triangular (tr) well with an electric field of 2×10^7 V/m.

further studies are required to get an understanding of the real shape of the potential well.

Acknowledgments

This work has been supported by the Deutsche Forschungsgemeinschaft.

References

1. U. Kunze, *J. Phys. C* **17**, 5677 (1984).
2. H.P. Zeindl, T. Wegehaupt, I. Eisele, H. Oppolzer, H. Reisinger, G. Tempel, and F. Koch, *App. Phys. Lett.* **50**, 1164 (1987).
3. H.-M. Lin, K.-F. Berggren, W.-X. Ni, B.E. Sernelius, M. Willander, and G.V. Hansson, *J. Appl. Phys.* **67**, 1962 (1990).
4. J. Lindolf, B. Klehn, U. Kunze, W. Kiunke, and I. Eisele, *Solid-State Electron.* **37**, 969 (1994).
5. G. Tempel, N. Schwarz, F. Müller, F. Koch, H. P. Zeindl, and I. Eisele, *Thin Solid Films* **184**, 171 (1990).
6. U. Kunze, in *High Magnetic Fields in Semiconductor Physics, Springer Series in Solid-State Sciences* **71**, ed. G. Landwehr, (Springer, Berlin 1987), p. 344.
7. U. Kunze, *Phys. Rev.* B **35**, 9168 (1987).

Resonant tunneling in ZnSe/BeTe–Structures under High Magnetic Fields

M.Keim, U.Lunz, G.Reuscher, A.Waag and G.Landwehr
Physikalisches Institut der Universität Würzburg, Am Hubland, D-97074 Würzburg, Germany

We investigated the effect of a high magnetic field on the resonant tunnel current in double barrier, single quantum–well ZnSe/BeTe–structures. This novel material combination has conduction band offsets of the two Γ–points of 2.3 eV. For magnetic fields applied parallel to the tunnel current pronounced magneto–oscillations due to Landau–splitting in a 2D–accumulation layer in front of the barrier structure could be observed. For magnetic fields perpendicular to the tunnel current the effect is reduced to a shift in position and shape of the current maxima in the I-V characteristics which enabled us to estimate the energy dispersion–relation E(**k**) in the ZnSe quantum well.

1 Introduction

Vertical electrical transport through semiconductor heterostructures has been subject to intense basic as well as device–oriented research since the pioneering work of Chang et al.[1] in 1974. Resonant tunneling under an applied high magnetic field offers the unique possibility to study electron transport through totally quantized levels and thereby to investigate the energy and momentum transfer mechanisms. In the resonant regime one can deduce from the magneto–oscillations important information such as the effective mass m^*, the dimensionality of the emitter and charge buildup in the well.

So far the work on resonant tunneling has been almost exclusively confined to III–V materials [2]. In II–VI wide–gap semiconductors Reuscher et al.[3] realized resonant tunneling in CdTe/CdMgTe double barrier structures and only recently Lunz et al.[4] presented measurements in the new material system ZnSe/BeTe. This semiconductor system shows very promising properties especially for an application in blue light emitting devices such as LEDs and LDs [5]. For basic research the ZnSe/BeTe system is of great interest due to its high conduction band offset of $\Delta E_C \approx 2.3$ eV (see e.g. GaAs/AlAs: $\Delta E_C \approx 1.0$ eV) which yields clearly separated bound states within the quantum–well. Furthermore magnetic ions like manganese can be incorporated in II–VI heterostructures to form semimagnetic semiconductors in order to study magnetic effects.

The ZnSe/BeTe resonant tunneling structures were grown by molecular beam epitaxy on GaAs (100) substrates. Details about surface preparation, growth of the heterostructures and mesa–fabrication have been described elsewhere[4]. We examined electrical transport properties of ZnSe/BeTe double–barrier, single quantum–well structures in high magnetic fields up to 10 T at 1.6 K.

2 Results

The I–V characteristics of a sample with 5 nm barrier width and 6 nm well width revealed 4 resonances in a voltage range between 0 and 12 V, whereas theoretical calculations predicted as many as 7 bound states inside the well for zero applied bias due to the high barrier energies.

Magnetotunneling experiments performed at T= 1.6 K showed no significant changes of the I–V characteristics at different magnetic fluxes **B** with **B** || I. One problem that still hampers an exact evaluation of the acquired data are the relatively high series resistances. For the above mentioned sample and the first resonance only 60% of the applied bias voltage drop across the barrier–structure and the remaining 40% are attributed to series resistances[4].

In a different set of experiments the tunnel current was measured as a function of an external magnetic field (**B** || I) at different bias voltages. Pronounced magneto–oscillations could be observed starting at around 3 T. Figure 1 shows Landau–oscillations in the tunnel current of sample CB150 (ZnSe/BeTe–DBS) at different bias voltages. These oscillations can be explained by a quantization of the electronic energy component perpendicular to the magnetic field vector **B**. By plotting the position of the extrema of the oscillations versus 1/B one can determine the Landau quantum numbers (see inset of Fig.1), the slope of the linear fit gives the oscillation frequencies B_f. A subsequent tilt of the sample in the magnetic field proved a cosine-dependence of the oscillatory frequency relative to the angle between the magnetic field **B** and the direction of the tunnel current I. We thus identified the oscillations as being of two–dimensional nature.

One possible explanation is based on the formation of a quasi–triangular potential well in the undoped spacer region in front of the barrier–structure due to carrier accumulation under an applied external voltage. It is well known that boundary quantization results in electric subbands. An external high magnetic field applied perpendicular to the plane of a quasi–2D electron gas yields Landau–splitting of the subbands and therefore causes Shubnikov–de Haas–like oscillations in the tunnel current. Using this model one can determine a 2D electron sheet–density of the accumulation layer from the oscillation frequency[6]. The electron concentrations vary between $1.0 \cdot 10^{12}$ and $2.5 \cdot 10^{12}$ cm^{-2} which is well within the expected range. A further evaluation will have to proof the origin of the sudden change in the oscillation frequencies at about 3 V bias (see the two different slopes in the inset of Fig.1).

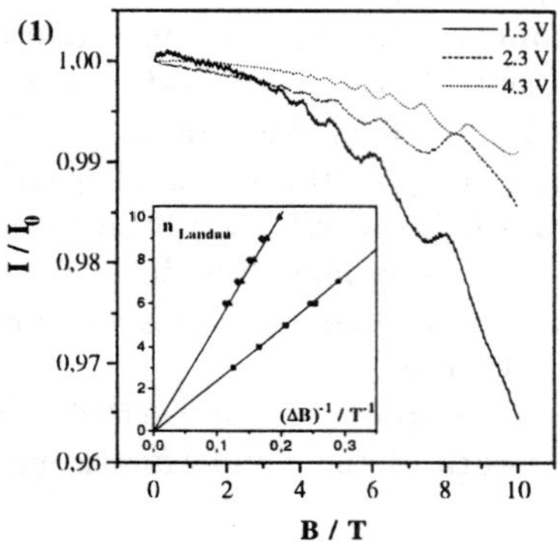

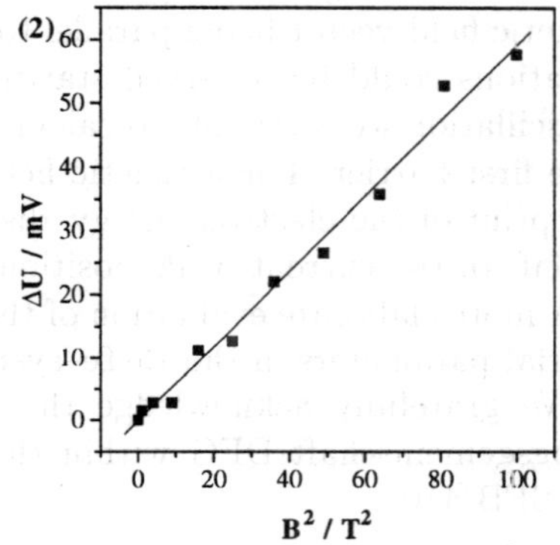

*Fig.1: Landau-oscillations (***B** *$||I$); Fig.2: Shift of the peak -current (***B** *$\perp I$).*

The second part of this paper concerns resonant tunneling with a magnetic field applied perpendicular to the growth direction of the structure (**B** $\perp$ I). As expected there was no Landau–splitting observable. However, there is a considerable effect of the magnetic field on the I-V characteristics: The position of the peak shifts to higher voltages, the peak current density is reduced with increasing magnetic field, the valley current has risen and therefore the peak–to–valley ratio is reduced. These observations are consistent with those made by Eaves et al. in the GaAs/AlAs system [7].

The effect is classically interpreted as being caused by a Lorentz–force acting on the electron on its way from the emitter–region into the center of quantum well [7]. The resulting shift in the wave vector together with a conservation of energy in the resonant tunneling regime allows to deduce the E(**k**) relation for the quantum well states. Under a magnetic field applied perpendicular to the tunnel current the modified equation for resonant tunneling reads

$$eV/2 = E_S + E\left(k_{y_0}(B)\right), \tag{1}$$

with the subband energy E_S and the Fermi energy in the emitter $E_F = 0$. Fig. 2 shows the shift of the peak position (1st resonance) as a function of the magnetic field. The resulting curve shows the expected parabolic behavior. From the slope of this graph one can in principle deduce the effective electron–mass of the well–material [7], but due to series resistances the voltage drop ΔU cannot be determined with sufficient precision at present. Therefore we did not try to interpret the data quantitatively.

3 Conclusion

In conclusion, we have investigated I–V and I–B characteristics of ZnSe/BeTe resonant tunneling structures under an applied high magnetic field. For the magnetic field vector being parallel to the tunnel current pronounced Landau–oscillations could be observed starting at about 3 T. From the frequency of the oscillation we were able to calculate the 2D electron sheet–density in front of the first barrier. For magnetic fields perpendicular to the tunnel current a fingerprint of the electron energy–dispersion relation could be obtained using the shift of the current peak–position with increasing magnetic field.

A more elaborate evaluation of the data can be done as soon as a few more material parameters of the BeTe–system are known.

We gratefully acknowledge the financial support of the Deutsche Forschungsgemeinschaft DFG within the framework of the Sonderforschungsbereich SFB 410.

References

1. L.L.Chang, L.Esaki, R.Tsu; Appl.Phys.Lett. **24**, 593 (1974)
2. E.E.Mendez, W.I.Wang, B.Ricco, L.Esaki; Appl.Phys.Lett. **47**, 415 (1985)
3. G. Reuscher, M. Keim, F. Fischer, A. Waag, G. Landwehr; Phys.Rev.**B 53**, 16414 (1996)
4. U. Lunz, M. Keim, G. Reuscher, F. Fischer, K. Schüll, A. Waag, G. Landwehr; submitted to publication
5. A. Waag, F. Fischer, H.J. Lugauer, Th. Litz, J. Laubender, U. Lunz, U. Zehnder, W. Ossau, T. Gerhard, M. Moeller, G. Landwehr; J.Appl.Phys. **80**, 792 (1996)
6. M.L.Leadbeater, E.S.Alves, L.Eaves, M.Henini, O.H.Hughes, A.Celeste, J.C.Portal, G.Hill, M.A.Pate; Phys.Rev. B **39**, 3438 (1989)
7. L.Eaves, R.K. Hayden, D.K.Maude, M.L.Leadbeater, E.C.Valadares, M.Henini, O.H. Hughes, J.C. Portal, L.Cury, G.Hill, M.A.Pate; *High Magnetic Fields in Semiconductor Physics III*, ed. G.Landwehr (Springer Series in Solid State Sciences **101**, Springer, Berlin, 1992)

RESONANT BROADENING OF TWO-DIMENSIONAL STATES IN SEMICONDUCTORS WITH INVERTED BAND STRUCTURE

A.V.GERMANENKO, G.M.MINKOV, V.A.LARIONOVA, O.E.RUT
Institute of Physics and Applied Mathematics, Ural State University
Ekaterinburg 620083, RUSSIA

Abstract

Results of tunneling studies of MOS-structures based on $Hg_{1-x}Cd_xTe$ with an inverted band structure are reported. It has been found that the 2D states have revealed themselves at negative energies, when they are in resonance with the heavy hole valence band of bulk material. To interpret the experimental data the energy spectrum and broadening of the 2D states in a surface quantum well has been theoretically investigated in the framework of the Kane model, the finite value of the heavy hole effective mass has been taken into account.

1 Introduction

The broadening of 2D states which results from the resonance with the valence band states in the volume of semiconductor has been discussed for narrow gap semiconductors.[1] In these materials the 2D electron states at energies less than the energy of the top of the valence band are separated from the valence band by a barrier, which parameters are proportional to a forbidden gap. In this case the tunneling to heavy hole states is ineffective due to the large value of their effective mass, and the broadening of 2D states is caused by tunneling to the light hole state. A specific feature of a semiconductor with inverted energy band structure is the absence of an energy gap between conduction and valence bands. Therefore, the effect of resonance broadening due to tunneling to heavy hole states should be stronger in inverted semiconductors. This work is devoted to investigations of this effect.

2 Experimental results

The experimental method used here was tunneling spectroscopy in a quantizing magnetic field. The second derivative of tunneling current with respect to voltage d^2j/dV^2 as a function of bias and magnetic field in metal-insulator-$Hg_{1-x}Cd_xTe$ ($0.095 < x < 0.105$) structures was investigated in magnetic fields up to 6 T at a temperature of 4.2 K. Tunnel junctions were fabricated on single crystal p-$Hg_{1-x}Cd_xTe$ with concentration of uncompensated acceptors

$N_A - N_D = (5-8) \times 10^{17}$ cm^{-3} by a method described in [2]. Typical magnetic field dependences of d^2j/dV^2 at different biases are presented in Fig.2 for $\mathbf{B} \parallel \mathbf{n}$, where $\mathbf{n}$ is the normal to the tunnel junction plane. Fourier transformation (FT) was carried out to analyze the oscillation curves.

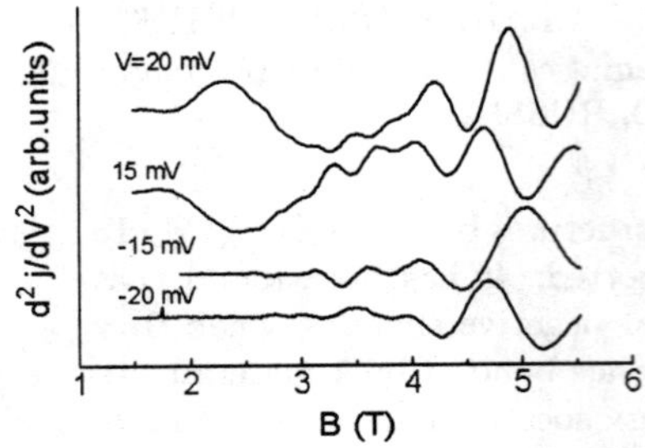

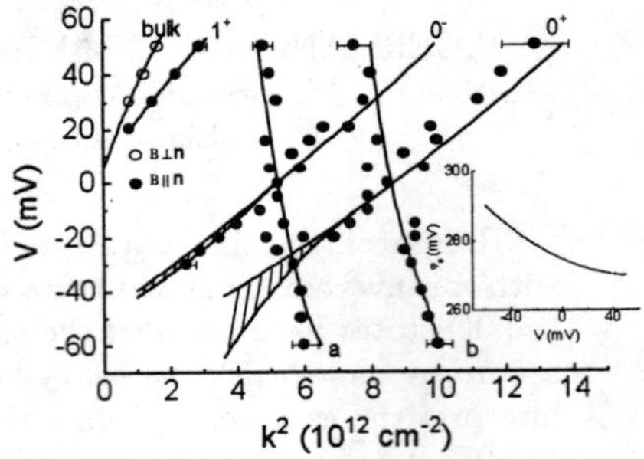

Figure 1: $\partial^2 j/\partial V^2$ **versus** B **curves at different biases.**
Figure 2: The bias dependences of k^2. **Symbols show the experimental data. The dotted curve is the dispersion law of the bulk electrons, calculated with parameters in the text. The solid curves are the theoretical bias dependences of** $k^2(E_F + eV)$ **(curves** $1^+, 0^+, 0^-$**) and** $k^2(E_F)$**(curves** a**, and** b**).**

The FT-results as a bias dependence of quasimomentum, which is connected with fundamental field B_f of Fourier spectra by the relationship $k^2 = 2(eB_f/c\hbar)$, is presented in Fig.2. An inspection of Fig.2 shows that two types of oscillation are observed: k^2 increases for branches 0^+, 0^-, 1^+ and decreases for branches a, b when the bias is increased. Typical angular dependences of B_f indicate that both types are connected with 2D states. The 1^+, 0^- and 0^+ branches correspond to so called spectroscopic oscillations arising whenever the Landau levels of the 2D states coincide with the energy $E_F + eV$, where E_F is the Fermi energy in semiconductor, and it varies in the range -5..-10 meV for different samples. The branches a and b relate to oscillations which arise whenever Landau levels of 2D states coincide with the energy E_F. So, the branches a and b give the concentration of 2D electrons, which decreases with increasing bias, as seen from the figure. It is significant that the spectroscopic oscillations are observed at negative biases down to $V = -30$ mV. At negative biases the 2D states, which are in resonance with the valence band, are tested. This means that the resonance broadening of the 2D states is not so large to destroy the oscillation picture. To our knowledge this is the first experimental observation of 2D states which lie significantly below the top of the valence band.

3 Theoretical model and discussion

To describe the energy spectrum of 2D states in a surface quantum well in an inverted semiconductor we use the Kane model. To analyze the resonance broadening of 2D states due to tunneling to heavy hole states we have to take into account the finite value of heavy hole effective mass by including in the Hamiltonian additional terms with parameters γ_1 and γ, which describe the interaction with remote bands. It has been supposed in our theoretical model that the band structure of an insulator is similar to that of the semiconductor.[1] The only difference is the value of the energy gap, which is much greater for the insulator, than that for the semiconductor (Fig. 3). Two additional parameters D_c and D_v (see Fig.3) appear in this model, but the net result mainly depends on the ratio D_c/D_v.[2] The techniques of direct numerical integration has been used to solve the eigenvalue problem.[3] A behaviour of the potential $\varphi(z)$ in a space charge region of the semiconductor has been found selfconsistently. Calculations have been carried out using the following parameter values $E_g = -110$ meV, $P = 8 \times 10^{-8}$ eV·cm, $N_A - N_D = 8 \times 10^{17}$ cm^{-3}, $\kappa = 20$, $\gamma_1 = 2$, $\gamma = 0$, $D_c = 2$ eV and $D_v = 1$ eV. Two types of solutions, which correspond to non-resonant ($E > 0$) and resonant ($E < 0$) 2D states, are presented in Fig. 4. It is clearly seen, that at $E < 0$ the wave function does not decay in the volume of semiconductor. This means that the charge carrier has a non-zero probability to tunnel from the space charge region to the volume. The Levinson theorem [1,2] has been used to find the 2D energy level position and its broadening.

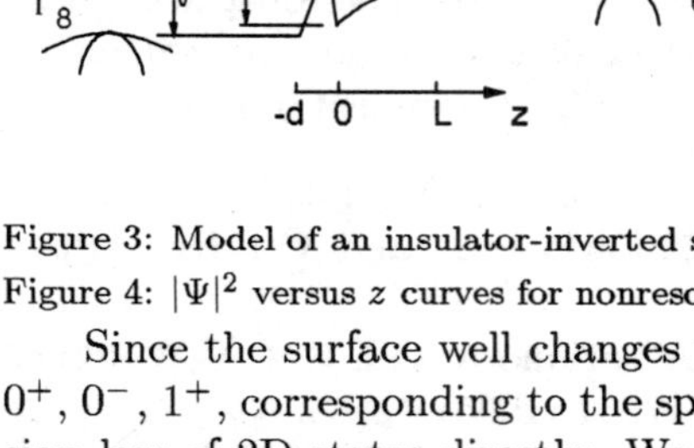

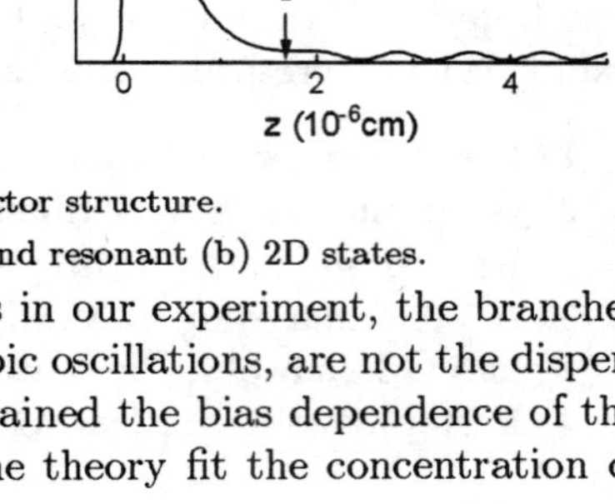

Figure 3: Model of an insulator-inverted semiconductor structure.

Figure 4: $|\Psi|^2$ versus z curves for nonresonant (a) and resonant (b) 2D states.

Since the surface well changes with bias in our experiment, the branches 0^+, 0^-, 1^+, corresponding to the spectroscopic oscillations, are not the dispersion law of 2D states directly. We have obtained the bias dependence of the surface potential (inset in Fig.2) making the theory fit the concentration of

2D electrons at different biases, which is given by quasimomentum values at $E = E_F$ (branches a and b in Fig.2). Then, using this dependence we have calculated the quasimomenta of 2D states at $E = E_F + eV$ over the entire energy range (branches 0^+, 0^-, 1^+). A comparison between theory and experiment shows that branches 0^+ and 0^- are nothing but two branches of the ground 2D subband split by the spin-orbit interaction in the asymmetric quantum well, and the 1^+ branch is the lower spin branch of the first excited subband.

Hatched regions in Fig.2 show the resonant broadening of the 2D energy levels calculated in the above model. It strongly depends on quasimomentum value. Being equal to zero at large quasimomentum values, when the 2D states lie above the top of the valence band ($V > 5$ mV), the broadening increases with decreasing quasimomentum. The increase is stronger for the 0^+ branch than that for the 0^- branch. Large broadening of 0^+ states is a possible reason that we are able to detect a maximum in the Fourier spectra, related to the 0^+ branch, down to $k^2 \simeq 6 \times 10^{12}$ cm^{-2} only, while a 0^- maximum is detected down to $k^2 \simeq 2.5 \times 10^{12}$ cm^{-2}. For both branches, maxima in Fourier spectra can be detected when the broadening does not exceed 5 meV, i.e. it is less than the cyclotron energy at the magnetic fields, where oscillations have been observed ($B > 1$ T). The broadening of 2D states seemingly increases when the quasimomentum tends to zero (Fig.2). However, the calculations show that for small quasimomentum values the broadening decreases down to zero at $k = 0$ for both branches. This agrees with an analytic solution, which can be derived for $k = 0$.[4] It should be mentioned that the net results of this calculation are not sensitive to the values of γ- parameters presented in literature.

4 Conclusion

The investigations of 2D systems in an inverted semiconductor have shown that in contrast to ordinary semiconductors the tunneling of 2D carriers to the heavy hole states of the valence band can be effective. The broadening of 2D energy levels caused by this process is different for 2D states corresponding to different spin branches of the energy spectrum, and strongly depends on qusimomentum values. The theoretical results are consistent with experimental data.

1. Pawel Sobkowicz, Semicond. Sci. Technol. **5**, 183 (1990).
2. G.M.Minkov, A.V.Germanenko, V.A.Larionova, and O.E.Rut, Phys. Rev. B **54**, 1841 (1996).
3. A.V.Germanenko, G.M.Minkov, V. A.Larionova, O.E.Rut, C.R.Becker, and G.Landwehr, Phys. Rev. B **52**, 17254 (1995).
4. R.Stepniewski, J. Phys. C: Solid State Phys. **17**, L853 (1984).

SPIN-DEPENDENT TUNNELING IN METAL-INSULATOR-NARROW GAP SEMICONDUCTOR STRUCTURES IN A MAGNETIC FIELD

G.M.MINKOV, O.E.RUT, A.V.GERMANENKO, V.A. LARIONOVA
Inst. of Phys. & Appl. Math., Ural University, Ekaterinburg, 620083, Russia

The investigations of the tunneling conductivity oscillations of the structures HgCdTe-insulator-metal in a magnetic field show that the amplitudes of the maxima associated with tunneling to different spin states are significantly distinct at magnetic field perpendicular to the normal of the tunneling contact plane. To understand the reason of this difference the contributions of the tunneling to *a* and *b* spin sublevels were calculated in framework of Kane model at this orientation of magnetic field .

The spin-dependent effects have revealed themselves in semiconductors with inverted spectrum. For the first time it has been found experimentally [1] that in bulk samples of such materials at magnetic field parallel to the current (B||j) the amplitudes of Shubnikov-de-Haas oscillations connected with the different spin states are distinguished significantly . Analogous difference in the tunneling conductivity for tunneling into the bulk states has been observed for the structures based on inverted semiconductors at B⊥n (n is the normal to the tunneling contact plane) [2]. The fact that electrons in inverted semiconductors are the carriers of fourfold degenerate at $k=0$ Γ_8 band was essential for explanation of observed peculiarities [2,3].

The surprising thing is that in narrow gap semiconductors where the electrons are the carriers of twofold degenerate Γ_6 band, the similar peculiarities of the tunneling into different spin states at B⊥n are observed too.

We have studied the tunneling conductivity oscillations in structures Al-oxide-HgCdTe with $10<E_g<100$ meV in a magnetic field. The energy diagram of the tunneling contact is shown in the Fig.1a. The bias V shifts the metal Fermi level relative to the semiconductor one and the tunneling conductivity is proportional to the density of the states in the semiconductor at the energy E_F+eV. The magnetic field quantizes the energy spectrum of the semiconductor and the tunneling conductivity becomes the oscillatory function of the magnetic field and bias. The oscillations picture for one of the structures investigated is shown in the Fig.1b. It is seen that the oscillation maxima associated with

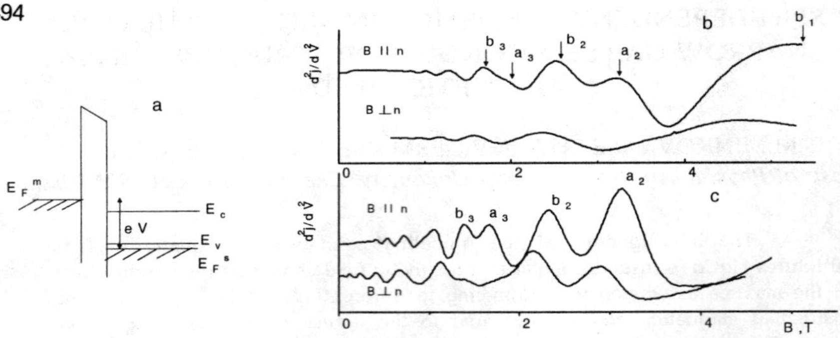

Fig.1 The energy diagram of the tunnel contact (a).The oscillations of the tunneling conductivity for structure 1 (b) and structure 2 (c) based on $Hg_{0.8}Cd_{0.2}Te$, $E_g = 80$ meV, at V= 130 mV, T= 4.2K.

tunneling into *a* and *b* sublevels of the Landau levels depend significantly on orientation of magnetic field. These amplitudes of the oscillations are about the same at **B**||**n**, but at **B**⊥**n** the amplitudes of the maxima associated with the tunneling into *b* sublevels are significantly larger and only tunneling into *b* sublevels is observable at this orientation of magnetic field. It is surprising thing that for other structures the amplitudes of maxima associated with tunneling into *a* and *b* sublevels are closely allied at **B**||**n** and at **B**⊥**n**(Fig.1b).

For both type of structures the oscillations are due to the tunneling into the bulk Landau levels. It is clear from the equality of the oscillation periods at both orientations of the magnetic field and from the fact that the maxima positions do not practically depend on the magnetic field orientation.

To understand what is the reason of such peculiarities of the tunneling in the various tunneling structures at **B**⊥**n**, let us consider the tunneling current at bias V and T=0. At this orientation of magnetic field the Landau eigenenergies are functions of distance of the orbit's center from the barrier (x_0) and the quasimomentum in the magnetic field direction (k_H). The tunneling current is the sum over the all states from E_F to E_F+eV of the tunneling probability which is proportional to the velocity squared $(v\psi)^2$ at the boundary and it is also a function of x_0 and k_H

$$j(V) \propto \sum (\hat{V}\Psi)^2 = \iint dx_0 dk_H (\hat{V}\Psi)^2$$

Thus, to calculate the tunneling current at B⊥n it is needed to calculate the distance dependencies of the velocity and energy. Going to integration over energy one can obtain for conductivity

$$\sigma(V) \equiv \frac{dj}{dV} = \int_0^{k_H^{max}} (\hat{V}\Psi)^2 \left(\frac{dE}{dx_0}\right)^{-1} dk_H \Bigg|_{E=E_F+eV}$$

This problem has been solved in the framework of the Kane model. We have supposed that the insulator has the same energy structure as the semiconductor and in this case one can write the exact wave functions in the semiconductor and in the insulator using the parabolic cylinder functions. Matching these wave functions at the boundary one can obtain the energy of the states and the velocity at the boundary at any x_0. For a first approximation we supposed that the distance dependences of the energy and velocity are similar for any k_H and have calculated them at k_H=0. Such curves for a_1 and b_1 sublevels are plotted in Fig.2,3. It is seen that the energy of the b sublevels rises slower when x_0

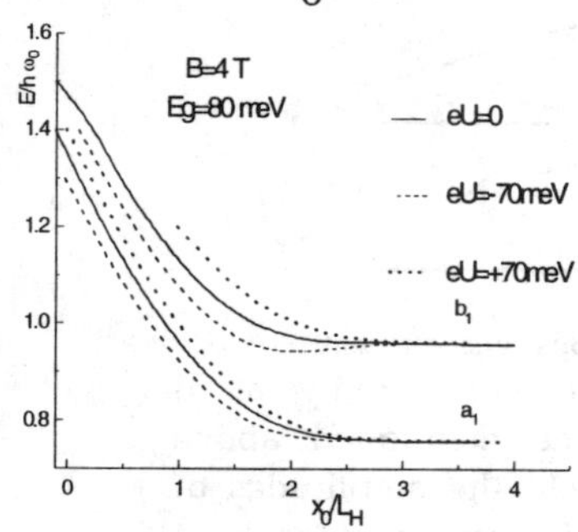

Fig.2 The distance dependencies of the energy for a_1 and $b_1$sublevels.

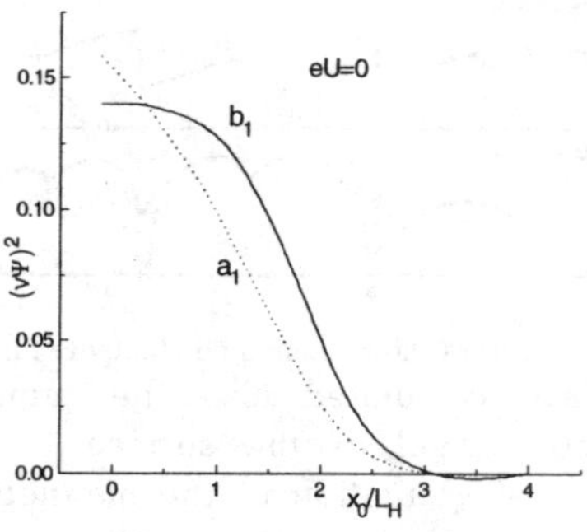

Fig.3 The distance dependences of $(v\Psi)^2$.

decreases and the velocity is larger than that for a sublevels. Such distinctions lead to different contributions in the tunneling to a and b sublevels (Fig.4) and therefore after separation of the oscillatory part of the tunneling conductivity the amplitudes of the maxima connected with tunneling into b sublevels will be larger than the amplitudes of the maxima for a sublevels (Fig.5).

In the framework of the described model we have calculated the oscillatory part of the tunneling conductivity in the structures with

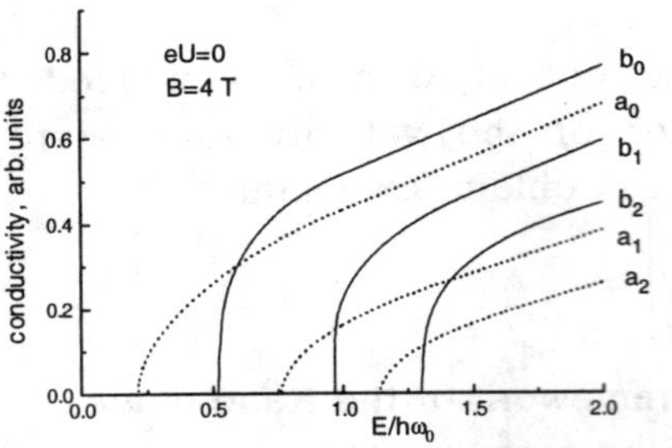

Fig.4 The contributions to the tunneling conductivity of various Landau levels.

the surface potential. To obtain the quantitative results two cases was analysed: the small rectangular well which cannot hold the 2D states and barrier . One can see that the difference between oscillatory maxima amplitudes connected with tunneling to the *a* and *b* sublevels becomes very large when the attractive potential exists near the boundary and only tunneling into *b* sublevels is visible in this case. At the repulsive potential near the surface the amplitudes of the maxima connected with tunneling into *a* and *b* sublevels become closely allied

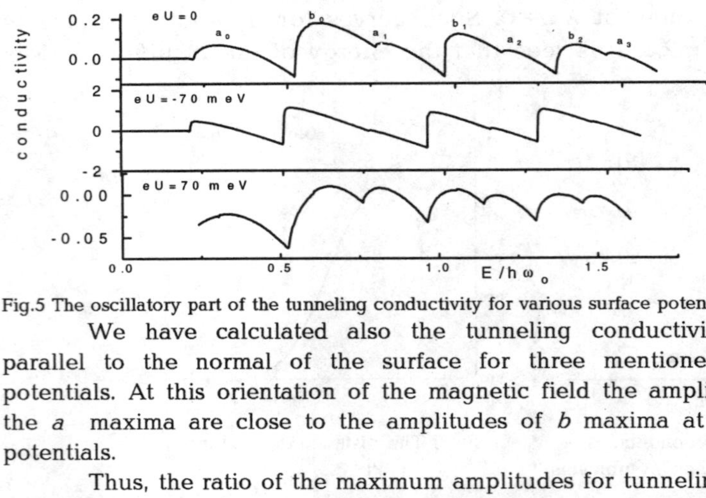

Fig.5 The oscillatory part of the tunneling conductivity for various surface potential.

We have calculated also the tunneling conductivity at B parallel to the normal of the surface for three mentioned above potentials. At this orientation of the magnetic field the amplitudes of the *a* maxima are close to the amplitudes of *b* maxima at each of potentials.

Thus, the ratio of the maximum amplitudes for tunneling into *a* and *b* sublevels at **B⊥n** depends on the surface potential and provides a way for estimation of the value of the surface potential at those conditions when it is small and cannot localize the 2D states.

References

1. Suizu K. and Narita S., Phys. Letters, 43A, 353 (1973).

2. Germanenko A.V.,Kruzhaev V.V., Minkov G.M., Rumyantsev E.L. and Rut O.E., Semicond. Sci. Technol, 8, 383 (1993).

3. Germanenko A.V., Minkov G.M., Rumyantsev E.L. and Rut O.E., Semicond. Sci. Technol, 8, 388 (1993).

INFLUENCE OF s–d EXCHANGE INTERACTION ON UNIVERSAL CONDUCTANCE FLUCTUATIONS

J. Jaroszyński and T. Dietl
Institute of Physics, Polish Academy of Sciences
al. Lotników 32/46, PL–02668 Warszawa, Poland

Mesoscopic phenomena in nanostructures of diluted magnetic semiconductors may exhibit a number of novel features driven by spin interactions between mobile electrons and localized spins. Millikelvin studies of diffusive charge transport, which have been carried out for submicron wires of n^+-CdTe and n^+-$Cd_{0.99}Mn_{0.01}Te$, are reviewed. These studies have provided information on the magnitudes of inelastic, spin-orbit, and spin-disorder scattering rates. Moreover, they put into the evidence a new driving mechanism of the magnetoconductance fluctuations – the redistribution of the electrons between energy levels of the system, induced by the giant s-d exchange spin-splitting. Important implications of these findings for previous interpretations of spin effects in semiconductor and metal nanostructures are discussed.

1 Introduction

Diluted magnetic semiconductors,[1] for which the magnitude of the exchange coupling between the magnetic ions and the electrons is well known from magnetooptical experiments, offer a worthwhile opportunity to examine quantitatively the influence of spin effects upon quantum transport phenomena.

We review here recent studies [2] of millikelvin magnetoconductance in submicron wires of a diluted magnetic semiconductor $Cd_{0.99}Mn_{0.01}Te$:In on the metallic side of the metal-to-insulator transition.[3] For a comparison, we present results of similar measurements [4] carried out on thin films and wires of a non-magnetic CdTe:In. The carrier and Mn concentrations in the studied samples were low enough to preclude the appearance of either spin-glass freezing or the Kondo effect, even at the lowest temperature of interest here, $T = 30$ mK.[5] The accumulate results [2,4] confirm a dramatic influence of the s-d coupling on the magnetoresistance at the localization boundary.[1,6,7] Surprisingly, in contrast to previous findings, [8] and theoretical suggestions, [9,10] no effect of the localized spins upon the mean *amplitude* of universal conductance fluctuations (UCF) [11] was observed. At the same time, however, a striking behavior of the *correlation field* H_c of the magnetoconductance fluctuations in the wires with Mn was detected. The data suggest that a new driving mechanism of the magnetoconductance fluctuations exists in quantum structures which contain magnetic impurities. This mechanism is associated with a redistribution of the electrons between energy levels, induced by the giant s–d exchange spin-splitting. The

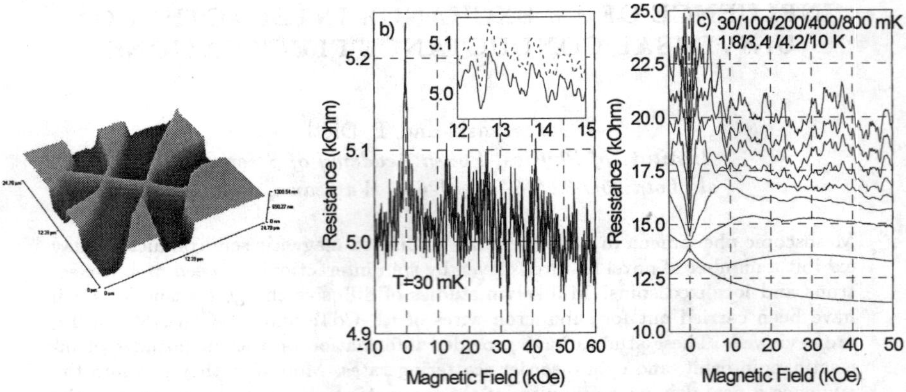

Figure 1: Atomic force micrograph of a n^+-CdTe wire (a). Resistance as a function of the magnetic field for such a wire of n^+-CdTe (b) and n^+-$Cd_{0.99}Mn_{0.01}Te$ (c). The reproducibility of the resistance fluctuations is clearly seen.

model is supported by the fact that H_c scales with the d-spin magnetization. Furthermore, the results discussed here have important implications for the interpretation of recent studies of UCF in semiconductor [12] and spin-glass [13] systems.

2 Samples

$Cd_{1-x}Mn_xTe$:In films with $x = 0$ or $x = 1 \pm 0.1\%$, thickness d down to 0.1 μm, and electron concentrations up to $n = 2 \times 10^{18}$ cm^{-3} were grown by MBE onto SI GaAs(001) substrates with 3 μm CdTe undoped buffer layers.[14]

Free standing wires with the linewidth of 0.3 μm were fabricated by means of 30 keV electron-beam lithography, followed by wet etching in 0.5% solution of Br_2 in ethylene glycol through PMMA mask. The actual shape and dimensions of the structures were visualized by an atomic-force microscope (AFM), as shown in Figs. 1 and 2.

3 Experimental Results

3.1 Magnetoresistance

Figures 1b and 1c present resistance changes as a function of the magnetic field perpendicular to the wires of CdTe:In and $Cd_{0.99}Mn_{0.01}Te$:In with the electron concentration 9.2×10^{17} and 8×10^{17} cm^{-3}, respectively. Weak-field

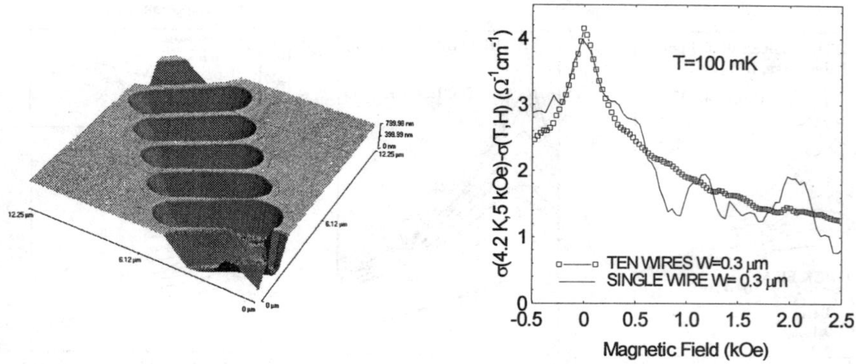

Figure 2: Atomic force micrograph of the nanostructure consisting of ten n^+-CdTe wires (a). The conductivity as a function of the magnetic field for a single n^+-CdTe wire and the structure consisting of ten such wires at $T = 100$ mK (b). A suppression of the conductance fluctuations is clearly seen in the latter sample.

magnetoresistance and reproducible resistance fluctuations were detected in both materials. According to the results presented in Figs. 1–4, the magnetoresistance of CdTe is negative in the weak magnetic fields. In order to reduce the perturbing effect of the fluctuations upon the magnetoresistance, both thin films and structures with ten wires in parallel were also studied.[4] As shown in Fig. 2, UCF are averaged out in such structures, making quantitative analysis of the magnetoresistance possible.

The negative magnetoresistance is characteristic for doped semiconductors, in which the cyclotron energy is much greater than the spin-splitting.[3,6] Its magnitude is controlled by spin-orbit scattering and phase breaking mechanisms, such as spin-disorder and inelastic scattering. Thus, the measurements for n-CdTe probe directly spin-orbit τ_{so}^{-1}, spin-disorder τ_s^{-1}, and inelastic τ_{in}^{-1} scattering rates. At high temperatures, $T > 3$ K, all samples are well described by 3D theory. This is true down to 50 mK for the sample of the thickest film, so that it can be treated as a reference 3D system. For the remaining samples the discrepancy between the measured resistance and 3D model increases below ~ 3 K. This indicates a temperature-induced crossover from a behavior specific to 3D, towards 2D or 1D, respectively. Such a crossover occurs when the phase breaking length L_φ becomes longer than the linear dimensions of the given sample, where $L_\varphi = \sqrt{D\tau_\varphi}$ and $\tau_\varphi^{-1} = \tau_{in}^{-1} + \frac{2}{3}\tau_s^{-1}$.

Phase breaking rates τ_φ^{-1} as determined experimentally for the studies structures are collected in the inset to Fig. 3. They were found to be well described by $\tau_\varphi^{-1} = aT^b$, where $a = (1.4 \pm 0.2) \times 10^{10}$ s^{-1} and $b = 1.45 \pm 0.1$

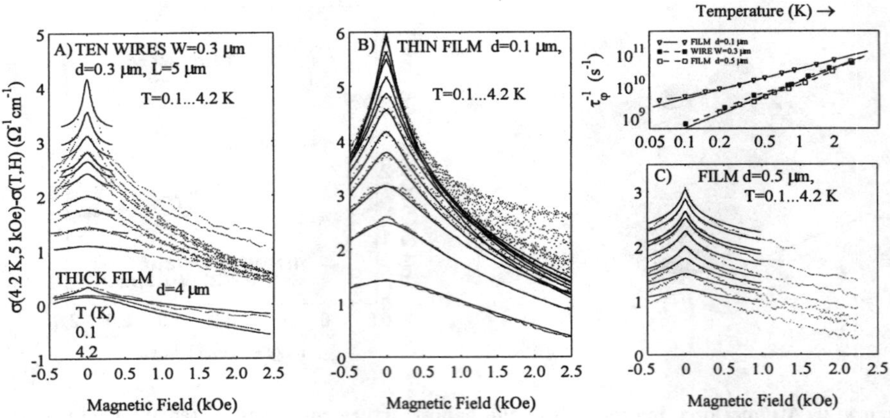

Figure 3: Negative corrections to the conductivity for n-CdTe:In as a function of the magnetic field for ten wires connected in parallel (a), and films of the thickness of 0.1 μm (b), 0.5 μm (c) and 4 μm (a). Dots depict experimental points. Solid lines show magnetoconductance calculated according to 1D (a), 2D (b,c) and 3D [bottom part of (a)] weak localization theory,[15,16,6] respectively. Inset: phase breaking rate as a function of temperature for three samples of n^+-CdTe. Solid lines represent fits to the formula $\tau^{-1} = aT^b$.

(for T in K) for both the 0.5 μm film and 0.3 $\times$ 0.3 μm wire. In contrast, for the thinnest film, $d = 0.1$ μm, $a = (3.6 \pm 0.3) \times 10^{10}$ s^{-1} and $b = 0.94 \pm 0.1$. Such temperature dependences indicate that the dominant phase-breaking mechanism comes from inelastic processes of electron scattering by thermal fluctuations of 3D or 2D electron liquid, respectively.[17] Hence, although the thicker samples have reduced dimensionality with respect to localization effects, they are still 3D regarding to the electron-electron interaction. Indeed, down to 100 mK thermal diffusion length, $L_T = 0.1/\sqrt{T}$ μm$/\sqrt{K}$, is shorter that all linear dimensions of these samples. Below 1 K, however, L_T becomes longer than the thickness of the thinnest sample. This is the reason for 2D behavior of $\tau_{in}(T)$ in this sample. Similar temperature dependences of $\tau_{in}^{-1}(T)$ were also observed in GaAs/AlGaAs heterostructures and Si MOSFETs.[18]

Spin-orbit scattering turns to be relatively weak in the studied samples. The scattering rate was estimated to be $\tau_{so}^{-1} = 5 \times 10^8$s^{-1} for the samples with the electron concentrations $n = 9.2 \times 10^{17}$ cm^{-3}, whereas $\tau_{so}^{-1} = 9.4 \times 10^8$s^{-1} for $n = 1.8 \times 10^{18}$ cm^{-3}. This strong dependence on the electron concentration ($\sim n^2$) can be explained by the Dyakonov–Perel mechanism[19] which assumes randomization of spin polarization by a fluctuating magnetic field brought about by elastic scattering and the electric field of the Cd–Te dipoles. The corresponding parameter of the spin-orbit coupling was inferred to be $a_{42} =$

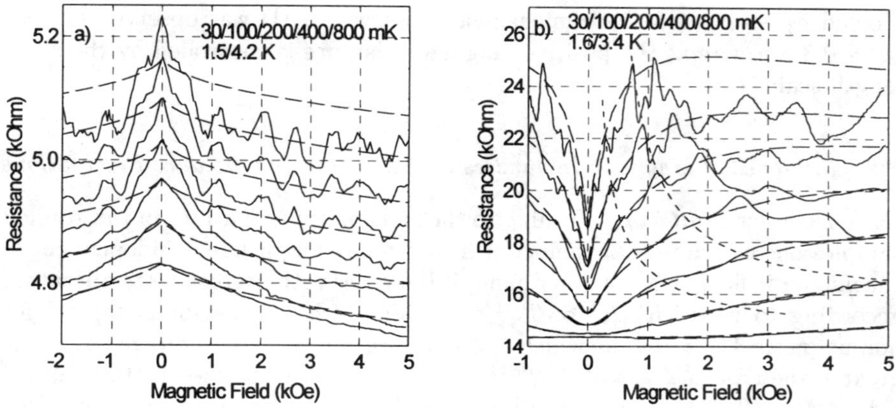

Figure 4: Resistance changes as a function of the magnetic field for the wires of n-$Cd_{1-x}Mn_xTe$ with $x = 0$ (a) and $x = 1\%$ (b) at various temperatures. Dashed lines represent magnetoresistance calculated in the framework of 3D weak-localization theory.[6] Dotted lines visualize a strong temperature dependence of the resistance features in CdMnTe (b).

9^{+3}_{-1} eVÅ^3, while band structure calculation for CdTe [20] gave $a_{42} = 11$ eVÅ^3. This good agreement together with the previously discussed dependence of τ_φ, imply that spin-disorder scattering, say by residual magnetic impurities, is totally negligible in the nonmagnetic semiconductor under consideration.

A striking influence of the magnetic impurities upon the magnetoresistance and UCF is shown in Fig. 4b, where data for n^+-$Cd_{0.99}Mn_{0.01}Te$ are shown together with the results of a theoretical computation.[3,6] It is seen that at $T > 4.2$ K the magnetoresistance is negative, as observed for the nonmagnetic CdTe. However, it changes sign when temperature decreases below 4.2 K, and the electron spin-splitting becomes greater than the cyclotron energy. As demonstrated previously,[6] the positive magnetoresistance is caused by the effect of the giant exchange spin-splitting upon the electron-electron interaction. This spin-splitting depends on the temperature and the magnetic field according to [1,21]

$$\hbar\omega_s \equiv \tilde{g}\mu_B H = g^\star \mu_B H + \alpha M(T,H)/(g\mu_B), \qquad (1)$$

where $\alpha N_o = 0.22$ eV is the s–d exchange energy, $M(T,H)$ is the magnetization of the Mn spins given by a modified Brillouin function, $M(T,H) = \bar{x}N_o g\mu_B S B_S(T+T_o,H)$, where $N_o = 1.48 \cdot 10^{22}$ cm^{-3}, $g = 2.0$, $S = 5/2$ and, for $\bar{x} = 1\%$, $T_o \approx 80$ mK.[5,21] These parameters imply $\tilde{g}\mu_B H = 5.5$ meV for $Sg\mu_B H \gg k_B(T+T_o)$ and $\tilde{g} = 150$ K$/(T+T_o)$ in the opposite limit. In that way, the Coulomb amplitude was the only adjustable parameter and its fitting yielded $g_3 + g_4 = 1.3$. No dimensional crossover to 1D was observed for this $0.3 \times 0.3\ \mu$m wire as the positive magnetoresistance is controlled by the short length scale, L_T.

3.2 Influence of Magnetic Impurities on Universal Conductance Fluctuations

Spin–Disorder Scattering. Turning to the resistance fluctuations in the studied samples one can note that their root mean square amplitude is independent of the magnetic field. On the other hand, it increases with decreasing temperature according to $\text{rms}(\Delta R)/R^2 \propto T^{-1/2}$. While such a behavior is typical for nonmagnetic 1D wires, in which the distance between the voltage probes is greater than both L_T and L_φ,[10,11,18] it comes as a surprise in the case of $Cd_{0.99}Mn_{0.01}Te$. Indeed, in the latter $\text{rms}(\Delta R)$ is expected to be controlled by τ_s which for $T > T_0$ is independent of temperature but increases with the magnetic field.[8,9] However, because of the low density of states specific to semiconductors, we obtain $\tau_s^{-1} = \alpha^2 m^* k_F \bar{x} N_o S(S+1)/(4\pi\hbar^3)$ to be as low as 10^{10} s^{-1} at $H = 0$. This estimate explains the minor importance of spin-disorder scattering in the studied system. At the same time the data put into a question a suggestion[12] that spin-disorder scattering by etching-induced defects can account for the absence of the resistance fluctuations in the region of weak magnetic fields in n^+-GaAs wires at $T \geq 1.3$ K.[12] Also, in view of the results for both n-CdTe and n-CdMnTe, the low-temperature saturation of L_φ deduced from magnetoresistance measurement on n^+-GaAs wires[22] and AlGaAs/GaAs heterostructures[23] can hardly be explained by the presence of residual magnetic impurities. If not by external noise, this saturation is most probably caused by spin-orbit coupling, which is twice stronger in GaAs[24] than in CdTe.

Spin–Splitting. Another important aspect of the data depicted in Fig. 4 concerns an unusual behavior of the correlation field H_c of the resistance fluctuations in $Cd_{0.99}Mn_{0.01}Te$. As shown by dotted lines, the fields H_i corresponding to characteristic points of the fluctuation pattern tend to increase with either temperature or the magnetic field, a behavior not observed in nonmagnetic wires, including those of n^+-CdTe. This new effect is visible for both the perpendicular and the parallel orientation of the magnetic field with respect to the wires.

It should be noted that field-induced changes of spin configurations have been proposed as the mechanism driving magnetoconductance fluctuations in spin-glass Cu:Mn wires.[13] The data for n-$Cd_{0.99}Mn_{0.01}Te$ discussed here have led to suggest[2] the existence of another spin effect operating also in the paramagnetic phase. This effect has its origin in the spin-splitting-induced redis-

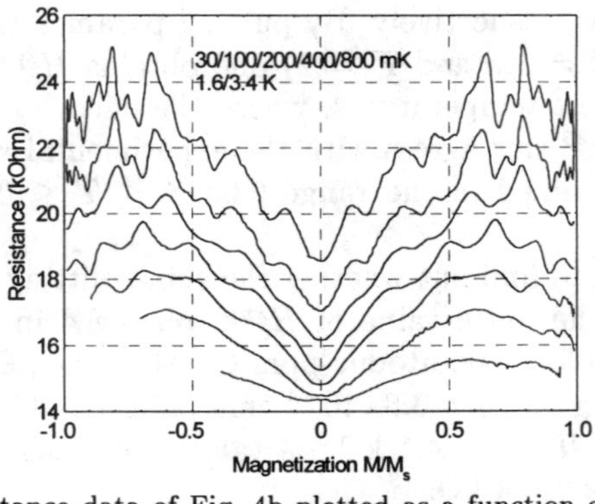

Figure 5: Magnetoresistance data of Fig. 4b plotted as a function of magnetization in the units of $M_s = g\mu_B S N_o \bar{x}$ (magnetization parameters are given below Eq. 1).

tribution of carriers between the spin subbands. The redistribution, and the corresponding shift of the Fermi energy ε_F with respect to the bottom of the spin-down and spin-up subbands, result in a gradual change of energy levels of the system that contribute to the conductance. Since, according to Eq. 1, the spin-splitting is proportional to the magnetization, the positions of the characteristic points of the fluctuation pattern should be temperature independent if the resistance would be plotted as a function of the magnetization M, not of the magnetic field H. That this is indeed the case is shown in Fig. 5.

In order to evaluate the correlation field related to the spin-splitting H_c^{spin} one can observe that, if $\varepsilon_F \gg \hbar\omega_s$, an increase of the magnetic field by ΔH leads to the shift of the Fermi energy $\Delta\varepsilon_F = \pm\frac{1}{2}\Delta H \partial\hbar\omega_s/\partial H$. Hence, for $\hbar\omega_s > \mu_c$, where [10] $\mu_c \approx \max(k_B T, h/\tau_\varphi)$ is the energy correlation range, we obtain H_c^{spin} in the form $H_c^{spin} = \sqrt{2}\mu_c(\partial\hbar\omega_s/\partial H)^{-1}$, where the factor $1/\sqrt{2}$ appears because the fluctuations result from a superposition of two independent contributions associated with two different spin subbands. One can presume that in magnetic materials phase breaking time $\tau_\varphi \equiv h/\mu_c$ is equivalent to spin-relaxation time T_2 of the itinerant electrons. According to detailed experimental and theoretical studies of T_2 in paramagnetic n-$Cd_{0.95}Mn_{0.05}Se$ near MIT [25] one can get in the limit of weak magnetic fields

$$H_c^{spin} = \frac{2\sqrt{2}k_B T \alpha m^\star k_F}{\hbar^2 g\mu_B}\left[1 + \frac{3\sqrt{3}}{\pi(k_F\ell)^2}\right]\left(1 + \frac{F}{2}\right). \tag{2}$$

The terms in the square bracket and the parenthesis represent enhancement factors of the s–d scattering caused by nonmagnetic impurities and electron-

electron interactions, respectively. By putting parameters suitable for studied CdMnTe wire ($k_F\ell = 1.5$ and $F = 2$) one obtains $H_c^{spin} = 410 \times T$[K] Oe. Except for the lowest temperatures, where the effect of bound magnetic polaron may appear,[6,25] this agrees with the experimental values, $H_c^{exp} = 360 \times T$[K]+36 Oe, determined in the range 0.03 K $\leq T \leq$ 0.8 K and $0 \leq H \leq k_B(T+T_o)/(g\mu_B)$.

It is interesting to find out whether the spin-splitting effect discussed here could account for the finite value of H_c^{spin} observed in Cu:Mn.[13] By taking parameters suitable for 1000-atomic-ppm Cu:Mn,[13] *i.e.*, $\ell = 200$ Å, $L_\varphi = 0.35$ μm, $m^\star = m_o$, $\chi = \chi_C(T_g) = 3.8 \cdot 10^{-4}$ emu, and $|\alpha N_o| = 0.5$ eV, one obtains from Eqs. 1 and 2, $H_c^{spin} = 4.1$ kOe, a value quite close to the experimental finding, $H_c^{spin} = 4.2 - 6.4$ kOe.[13]

4 Summary

We have described results of the low-temperature magnetoresistance studies[2,4] on nanostructures, for which the incorporation of magnetic impurities could be controlled during the growth process, while the details of magnetic properties and the value of the exchange coupling between the spins and the electrons could be monitored by means of a quantitative study of magnetooptical effects and weak-localization magnetoresistance. The presented data reemphasize a relatively small importance of spin-disorder scattering in semiconductors. At the same time, they demonstrate the significance of the spin-splitting-induced redistribution of the carriers between the spin subbands in quantum transport phenomena. It was so far noted that the redistribution may affect the conductivity in two ways. First, by its influence on the mean free path that appears for $\hbar\omega_s \geq \varepsilon_F$.[26] Second, by changing the distance of ε_F to the mobility edge ε_m,[27] an effect operating if $\hbar\omega_s \geq |\varepsilon_F - \varepsilon_m|$. The findings discussed here show that yet another, third, mechanism can play a role, namely that in mesoscopic systems the redistribution between the nanostructure energy levels is relevant, resulting in that the scales of characteristic energies and the fields are rather small, $\hbar\omega_s \geq \max(k_BT, h/\tau_\varphi)$.

Acknowledgments

The authors express their thanks to G. Karczewski, T. Wojtowicz and J. Kossut for growing, J. Wróbel, M. Sawicki, T. Skośkiewicz, E. Kamińska, E. Papis and A. Piotrowska for nanostructurization and R. Duś and R. Nowakowski for AFM visualization of the samples and valuable discussions. Also, we would like to thank W. Plesiewicz for cryogenic expertise as well as J.A. Gaj and M.

Grynberg for their encouragement and stimulation in the course of this work and Polish KBN for financial support under Contracts No. 2–P03B–06411 and 8–T11B–02108.

References

1. see, e.g., T. Dietl, in *Handbook on Semiconductors*, edited by T.S. Moss, (North-Holland, Amsterdam, 1994) vol. 3b, p. 1251.
2. J. Jaroszyński, J. Wróbel, M. Sawicki, E. Kamińska, T. Skośkiewicz, G. Karczewski, T. Wojtowicz, A. Piotrowska, J. Kossut and T. Dietl, *Phys. Rev. Lett.* **75**, 3170 (1995).
3. see, B.L. Altshuler and A.G. Aronov, in *Electron-Electron Interaction in Disordered Systems*, edited by A. L. Efros and M. Pollak (North-Holland, Amsterdam, 1985), p. 1; H. Fukuyama, *ibid*, p. 155; P.A. Lee and T.V. Ramakrishnan, *Rev. Mod. Phys.* **57**, 287 (1985).
4. J. Jaroszyński, J. Wróbel, M. Sawicki, T. Skośkiewicz, G. Karczewski, T. Wojtowicz, J. Kossut, T. Dietl, E. Kamińska, E. Papis, A. Piotrowska, R. Duś and R. Nowakowski, *Acta Phys. Pol.*, in press.
5. M.A. Novak, O.G. Symko, D.J. Zheng, and S. Oseroff, *J. Appl. Phys.* **57**, 3418 (1985).
6. M. Sawicki, T. Dietl, J. Kossut, J. Igalson, T. Wojtowicz and W. Plesiewicz, *Phys. Rev. Lett.* **56** 508 (1986).
7. Y. Shapira, N.F. Oliveira Jr., P. Becla, and T.Q. Vu, *Phys. Rev. B* **41**, 5931 (1990).
8. A. Benoit, S. Washburn, R.A. Webb, D. Mailly and L. Dumoulin, in *Anderson Localization*, edited by T. Ando and H. Fukuyama (Berlin, Springer, 1988), p. 346; C. Van Haesendonck, H. Vloeberghs, Y. Bruynseraede and R. Jonckheere, in *Nanostructures Physics and Fabrication*, edited by M.A. Reed and W.P. Kirk (Academic Press, Boston, 1989), p. 467; A. Benoit, D. Mailly, P. Perrier and P. Nedellec, *Superlattices and Microstructures* **11**, 313 (1992); T. Dietl, G. Grabecki and J. Jaroszyński, *Semicond. Sci. Technol.* **8**, S141 (1993); G. Grabecki, T. Dietl, W. Plesiewicz, A. Lenard, T. Skośkiewicz, E. Kamińska and A. Piotrowska, Physica B **194-196**, 1107 (1994).
9. A.A. Bobkov, V.I. Falko and D.E. Khmelnitskii, *Zh. Eksp. Teor. Fiz.* **98**, 703 (1990) [*Sov. Phys.: JETP* **71**, 393 (1990)]; V. Chandrasekhar, P. Santhanam and D.E. Prober, Phys. Rev. B **42**, 6823 (1990); V.I. Falko, *J. Phys.: Conden. Matter* **4**, 3943 (1992).
10. B.L. Altshuler and B.I. Shklovskii, *Zh. Eksp. Teor. Fiz.* **91**, 220, (1986) [*Sov. Phys.: JETP* **64**, 127 (1986)]; P.A. Lee, A.D. Stone and H.

Fukuyama, *Phys. Rev. B* **35**, 1039 (1987).
11. for a review on UCF, see, *Mesoscopic Phenomena in Solids*, edited by B.L. Altshuler, P.A. Lee and R.A. Webb (Amsterdam, Elsevier, 1991).
12. A.K. Geim, S.V. Dubonos and I.Yu. Antonova, *Pisma Zh. Eksp. Teor. Fiz.* **52**, 873 (1990) [*JETP Lett.* **52**, 247 (1990)].
13. P.G.N. de Vegvar and T.A. Fulton, Phys. Rev. Lett. **71**, 3537 (1993).
14. G. Karczewski, A. Zakrzewski, L. Dobaczewski, W. Dobrowolski, E. Grodzicka, J. Jaroszyński, T. Wojtowicz and J. Kossut *Thin Solid Films*, **267**, 79 (1995).
15. B.L. Altshuler and A.G. Aronov, *Pisma Zh. Eksp. Teor. Fiz.* **33**, 515 (1981) [*JEPT Lett.* **33**, 499 (1981)].
16. S. Hikami, A.I. Larkin and Y. Nagaoka, *Prog. Theor. Phys.* **63**, 707 (1980).
17. B.L. Altshuler, A.G. Aronov and D.E. Khmelnitskii, *J. Phys. C* **15**, 7367 (1982).
18. see, C.V.J. Beenakker and H. Van Houten, *Solid State Phys.* **44**, 1 (1991).
19. B.L. Altshuler, A.G. Aronov, D.E. Khmelnitskii and A.I. Larkin, in *Quantum Theory of Solids* p. 137, Ed. by I.M. Lifshits, MIR Publishers, Moscow 1982.
20. M. Cardona, G. Fasol and N.E. Christensen, *Phys. Rev. B* **38**, 1806 (1988).
21. see, J.A. Gaj, W. Grieshaber, C. Bondin-Deshayes, J. Cibert, G. Feuillet, Y. Merle d'Aubigné and A. Wasiela, *Phys. Rev. B* **50**, 5512 (1994); B. Kuhn–Heinrich, W. Ossau, T. Litz, A. Waag and G. Landwehr, *J. Appl. Phys.* **75**, 8046 (1994).
22. A. Ramon, M. Heiblum and H. Shtrikman, *Semicond. Sci. Technol* **8**, 2176 (1993).
23. T. Ikoma, T. Odagiri and K. Hirakawa, in *Quantum Effect Physics, Electronics and Applications* p. 157, Ed. by K. Ismail, IOP, Bristol 1992.
24. P.D. Dresselhaus, C.M.A. Papavassiliou, R.G. Wheeler and R.N. Sacks, *Phys. Rev. Lett.* **68**, 106 (1992).
25. T.Dietl, M. Sawicki, E.D. Isaacs, M. Dahl, D. Heiman, M.J. Graf, S.I. Gubariev and D.L. Alov, *Phys. Rev. B* **43**, 3154 (1991).
26. Y. Shapira and R.L. Kautz, Phys. Rev. B **10**, 4781 (1974).
27. H. Fukuyama and Y. Yosida, *Physica* (Amsterdam) **105 B+C**, 132 (1981); T. Wojtowicz, T. Dietl, M. Sawicki, W. Plesiewicz and J. Jaroszyński, *Phys. Rev. Lett.* **56**, 2419 (1986); B.A. Aronzon, N.K. Chumakov, T. Dietl and J. Wróbel, *Zh. Eksp. Teor. Fiz.* **105**, 405 (1994) [*Sov. Phys. JETP* **78**, 216 (1994)].

MAGNETISATION AND PERSISTENT CURRENTS IN TWO-DIMENSIONAL RINGS

W.-C. TAN and J. C. INKSON
Department of Physics, University of Exeter, Exeter EX4 4QL, UK

Using an exactly soluble model we study magnetisation and persistent currents in two dimensional mesoscopic rings over a large range of magnetic field strength. In the weak magnetic field regime the persistent current and the magnetisation are simply proportional to each other, both show rapid Aharanov-Bohm (AB) type oscillations. In the strong magnetic field regime, however, they have very different behaviours. While the persistent currents only show AB-type oscillations, the magnetisation is dominated by de Haas-van Alphen (dHvA) type oscillations with the much weaker AB-type oscillation superimposed on them.

The recent experimental observation of persistent currents in mesoscopic rings[1,2] has received considerable theoretical attention.[3,4] One of the most important goals has been to go beyond the one dimensional picture of Büttiker *et al*[5] and to study the persistent current in a ring of a finite width. Because of the difficulty in dealing with an annular geometry, most theoretical studies have been performed using simple models, in which a ring with a finite width is approximated by more manageable models, such as two dimensional (2D) or three dimensional (3D) *straight wires* with periodic boundary conditions, and the uniform magnetic field is replaced by a thin magnetic flux tube confined to the hole region of the ring. Within such simple models, the magnetisation and the persistent current in a ring are simply proportional to each other. However, for a real annular ring in a uniform magnetic field the penetration of magnetic field into the conducting region will not only result in aperiodic oscillations in the persistent currents but also break down the simple linear relation between the persistent current and the magnetisation. At present measuring the magnetisation is the only way to experimentally probe the persistent currents in mesoscopic rings, therefore it is highly desirable to gain a thorough understanding of the persistent currents, magnetisation and their relation for a 2D ring over the whole range of magnetic field strength. In this paper, we present just such a study using an exactly soluble ring model.[6] The model has analytical solutions for both the electron energy spectrum and the wavefunctions. This allows us to calculate various physical properties for rings and dots containing a large number, say 10^4 electrons with little computational requirement, and therefore direct comparison between throretical and experimental results can be made.

We consider non interacting spinless electrons in a 2D ring defined by a radial potential:[6]

$$V(r)=\frac{a_1}{r^2}+a_2r^2-2\sqrt{a_1a_2} \tag{1}$$

The average radius of the ring is $r_0 = (a_1/a_2)^{1/4}$. The width of the ring at Fermi energy E_f is $\Delta r = \sqrt{8E_f/\mu\omega_0^2}$. For r near r_0, the potential of the ring is parabolic, $V(r) \cong \frac{1}{2}\mu\omega_0^2(r-r_0)^2$, where $\omega_0 = \sqrt{8a_2/\mu}$ and μ is the electron effective mass.

In the presence of a uniform perpendicular magnetic field B and an infinite thin magnetic flux $\Phi = 1\phi_0$ (where $\phi_0 = h/e$) piercing through the ring centre, the eigen energies are given by[6]

$$E_{n,m} = \left(n + \frac{1}{2} + \frac{M}{2}\right)\hbar\omega - \frac{m-1}{2}\hbar\omega_c - \frac{\mu}{4}\omega_0^2 r_0^2, \tag{2}$$

where

$$\omega_c = \frac{eB}{\mu}, \quad \omega = \sqrt{\omega_c^2 + \omega_0^2}, \quad \lambda = \sqrt{\frac{\hbar}{\mu\omega}}, \quad M = \sqrt{(m-1)^2 + \frac{2a_1\mu}{\hbar^2}}. \tag{3}$$

At zero temperature, the magnetisation of a system containing a fixed number (N) of electron is

$$M(B) = -\left.\frac{\partial U(B)}{\partial B}\right|_N \tag{4}$$

where

$$U(B) = \sum_{n,m} E_{n,m}(B) \tag{5}$$

is the total energy. The summation is taken over the N occupied states.

Using either the current density operator or the Byers-Yang relation,[7] one can show the circular current carried by the (n,m)th state is

$$I_{n,m} = \int j_{n,m}(r)dr = \frac{e\omega}{4\pi}\left(\frac{m}{M} - \frac{\omega_c}{\omega}\right) \tag{6}$$

The total persistent current is then given by

$$I = \sum_{n,m} I_{n,m} \tag{7}$$

We have calculated the magnetisation and persistent current as a function of magnetic field for two GaAs rings: a narrow ring (r_0=1350nm, $\hbar\omega_0 = 1.97$meV) and a wide ring (r_0=300nm, $\hbar\omega_0 = 0.63$meV). Each of the two devices contains N=1400 electrons, which gives a Fermi energy $E_f \approx 11$meV in both devices. At the Fermi energy, the widths of the two rings are 160nm and 500nm respectively. It should be noted that the parameters of the narrow ring is very similar to those of the 2D GaAs ring used in the experiment of Mailly *et al.*[2]

Fig. 1 shows that, in the weak magnetic field region, the magnetisation is dominated by strong mesoscopic oscillations (or AB type oscillations) in both rings. However, the device geometry has significant effect on the oscillation patterns: The

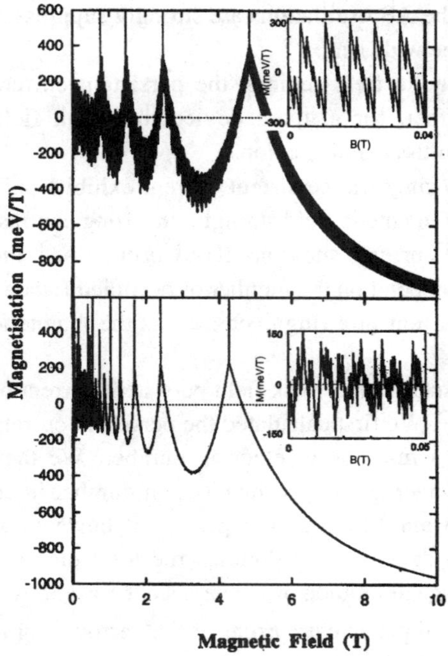

Figure 1 The magnetisation of the narrow ring (upper panel) and the wide ring (lower panel) as a function of magnetic field strength. The insets show the weak field behaviour of the magnetisation.

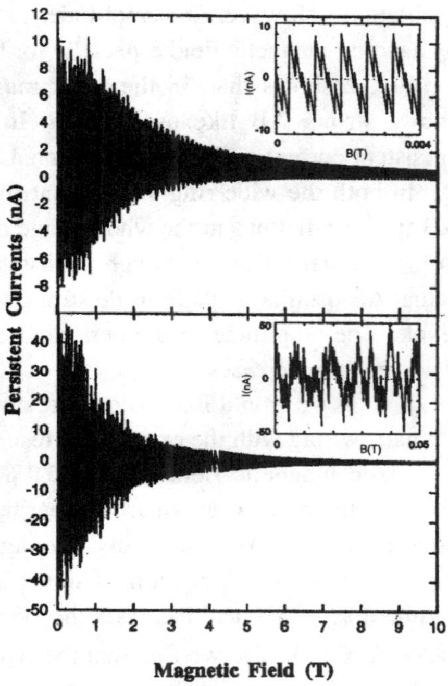

Figure 2 The persistent currents of the narrow ring (upper panel) and the wide ring (lower panel) as a function of magnetic field strength. The insets show the weak field behaviour of the persistent currents.

narrow ring exhibits periodic oscillations with the period of h/e, while the oscillation pattern of the wide ring is clearly aperiodic. The reason for the aperiodic oscillations of the wide ring is that the electron states in different subbands have very different radii and therefore lead to different AB frequencies.[6]

On the large magnetic field scale, the magnetisation is dominated by strong de Haas-van Alphen (dHvA) type oscillations, which correspond to intersections of the Fermi energy by a Landau subband. Even in the strong magnetic field regime, we can still see the AB type mesoscopic oscillations that are superimposed on the dHvA oscillations. However, the amplitudes of the AB oscillations are strongly suppressed by a strong magnetic field especially for the wide ring.

Fig. 2 shows that, in the weak magnetic field regime, the persistent current shows strong AB like oscillations. In fact, for a given device the weak field persistent current is simply proportional to the magnetisation.

In both the wide ring and the narrow ring, the persistent current exhibits only AB type oscillations in the whole range of magnetic field strength, in strong contrast to the magnetisation. Although persistent currents show no dHvA oscillations, the actual AB oscillation patterns do strongly depend on the number of occupied Landau levels. The amplitude of the persistent current in a ring decreases as the magnetic field strength increases.

Finally we would like to compare our calculated weak field persistent current of the narrow ring with the experiment result[2]. We first calculated the persistent current at a fixed magnetic field B=0.0005T as a function of electron number. We then obtained the root-mean-square current by averaging over the electron number in an interval of 100. The calculated average amplitude of the persistent currents at N=1400 is 3.2nA. Inclusion of the spin degeneration doubles the total electron number to N=2800 and increases the average amplitude of the persistent current by a factor of $\sqrt{2}$. Hence we find that the typical persistent current of the narrow ring is 4.5 nA, which is in very good agreement with the experimental value 4±2nA.[2]

Acknowledgments: This work was supported by the EPSRC (UK).

References

1. L. P. Levy et al, Phys. Rev. Lett. **64**, 2074 (1990); V. Chandrasekar et al, Phys. Rev. Lett. **67**, 3578(1991)
2. D. Mailly, C. Chapelier and A. Benoit, Phys. Rev. Lett. **70**, 2020 (1993)
3. Y. Avishai and M. Kohomoto, Phys. Rev. Lett. **71**, 279 (1993);Y. Avishai, Y. Hatsugai and M. Kohomoto, Phys. Rev. B **47**, 9501(1993)
4. G. Bouzerar, D. Poilbolanc and G. Motambaux, Phys. Rev. B **49**, 8258 (1994); H. F. Cheung, E. K. Riedel and Y. Gefen, Phys. Rev. Lett. **62**, 587 (1989) G. Kirzenow, Superlattices and Microstructures, **14**, 237 (1995); . A. Jalabert *et al* (to be published)
5. M. Büttiker, Y. Imry, and R. Landauer, Phys. Lett. **96A**, 365 (1983)
6. W.-C. Tan and J. C. Inkson, Phys. Rev. B **53**, 6947 (1996)
7. N. Byers and C. N. Yang, Phys. Rev. Lett. **7**, 46 (1961)

BOUNDARY ELEMENT METHOD FOR CALCULATING ELECTRONIC STATES OF QUANTUM DOT STRUCTURES IN MAGNETIC FIELDS

P. A. KNIPP
Christopher Newport University, Newport News, VA 23606-2998, USA

T. L. REINECKE
Technische Physik, Universität Würzburg, D-97074 Würzburg, Germany
Naval Research Laboratory, Washington, DC 20375-5347, USA

A boundary element method is developed for calculating the electronic properties of quantum dots and related nanostructures in the presence of magnetic fields. It provides an efficient method to treat systems having non-separable geometries. Calculations to illustrate this method are given for several systems.

1 Introduction

The effects of magnetic fields on the electronic properties of low dimensional nanostructures such as quantum wires, quantum dots, and related structures are currently of interest for interpreting magneto-luminescence data for their electronic states and also for understanding transport properties such as those involving quantum beats, edge states in the Hall effect, and the Bohm-Aharanov effect. Schrödinger's equation, however, is separable only for certain highly symmetrical nanostructure shapes such as cylindrical wires and dots in the shape of disks or rings. Therefore for most systems of experimental interest, numerical techniques must be used to obtain accurate results for the single particle properties. Numerically intensive techniques such as finite differences methods, finite elements approaches, or functional expansions[1] have been used to study such problems.

In previous work[2] we developed an efficient technique called the Boundary Element Method (BEM) as an alternative, and often more efficient, approach for the electronic properties of realistic structures within the effective mass approximation. This method is especially effective for nanostructures composed of regions with constant potentials, which are called "piecewise uniform." Here we extend this formalism to treat the case of systems in magnetic fields.

In the present work we address quantum dots and structures of more complex geometries formed by lithographic lateral patterning or by the gating of quantum wells or heterojunctions. Typically their lateral size considerably exceeds their vertical extent. Thus the motion in the vertical direction decouples from the lateral motion, and the problem then depends on the two variables in the lateral direction. We consider the magnetic field to be uniform and oriented in the growth direction of these structures.

2 Formalism

Using the effective mass approximation, polar coordinates (ρ and ϕ), and the symmetric gauge for the vector potential, one obtains Schrödinger's equation for this two-dimensional system:

$$\left[-\frac{\hbar^2\nabla^2}{2m} - \frac{i\hbar^2 b}{m}\frac{\partial}{\partial\phi} + \frac{\hbar^2 b^2 \rho^2}{2m} + V(\overline{\rho})\right]\psi(\overline{\rho}) = E\psi(\overline{\rho}) \tag{1}$$

where $\overline{\rho}$ is a two-component vector in the lateral plane, the band offset $V(\overline{\rho})$ is piecewise uniform, and the "reduced" magnetic field strength is defined by $b \equiv eB/(2\hbar)$. Within a particular area A of the structure through which V is uniform, the equation governing the Green's function $G(\overline{\rho}, \overline{\rho}'; E)$ is

$$(-\nabla^2 + b^2\rho^2 - k^2)\, G(\overline{\rho}, \overline{\rho}'; E) = \delta(\overline{\rho} - \overline{\rho}') \tag{2}$$

where $\hbar^2 k^2 \equiv 2m(V - E)$. For $b = 0$ the solution for G is given by a cylindrical Bessel function, and for $b \neq 0$ the solution for G can be related to the Kummer function,[3] which can be evaluated straightforwardly.[4]

By combining Eqs. (1) and (2), integrating over $\overline{\rho}$ through an area A in which V is uniform, and using Green's theorem we obtain

$$\begin{aligned}\psi(\overline{\rho}') = \oint_C ds\, \{\phi(s) G(\overline{\rho}_s, \overline{\rho}'; E) \\ -\psi(s)\left[\partial_s G(\overline{\rho}_s, \overline{\rho}'; E) + ib\rho\tfrac{\partial|\overline{\rho}_s - \overline{\rho}'|^2}{\partial s} G(\overline{\rho}_s, \overline{\rho}'; E)\right]\}\end{aligned} \tag{3}$$

where C is the contour bounding A, s indicates a point $\overline{\rho}_s$ on C, $\phi(s) \equiv \partial_s \psi(\overline{\rho})$, $\partial_s \equiv \hat{\mathbf{n}}_s \cdot \overline{\nabla}|_s$, and $\hat{\mathbf{n}}_s$ is the unit vector normal to C at the point s. Eq. (3) is an integral relation which gives the wave function ψ in all space in terms of quantities [$\psi(s)$ and $\phi(s)$] on the interfaces. In order to calculate these quantities on the interface we evaluate Eq. (3) at a point $\overline{\rho}'$ within A but arbitrarily close to $\overline{\rho}_{s'}$ on C to obtain the homogeneous integral equation:

$$\oint_C ds' [P(s, s'; E)\phi(s') + Q(s, s'; E)\psi(s')] = 0 \tag{4}$$

where the kernels P and Q are related simply to Kummer's function. Similar equations hold for the several other homogeneous regions of the nanostructure. These equations give a homogeneous system of coupled linear integral equations which are solved by discretizing the integrals in them, calculating the determinant of the resulting matrix, and finding the energies for which the determinant vanishes. These energies correspond to the system's stationary states. For nonzero magnetic fields the wave functions are complex-valued.

3 Examples

In this section we give results of numerically implementing the formalism of Sec. 2 for systems having barriers with infinite band offsets, in which case only the first term in the integrands of Eqs. (3) and (4) is nonzero. In Table I and Fig. 1 results are given for the states of a "stadium" shape of size 25 nm × 50 nm, an outline of which is indicated by the dashed line in Fig. 1. This shape has been studied extensively[5] and is of current interest in the study of quantum chaos and related topics. From the results in the Table it can be seen that the energies of these states converge well with a modest number of boundary elements N.

Table 1: Convergence of the BEM-calculated energies (in meV) of the lowest three states of a 25 nm × 50 nm stadium-shaped dot for which $m = 0.0665$ and $B = 10$ T as a function of N, the number of boundary elements.

N	E_0	E_1	E_2
8	13.43	18.99	29.59
16	13.76	19.71	29.22
32	13.73	19.77	29.44
64	13.70	19.75	29.44
128	13.70	19.75	29.43

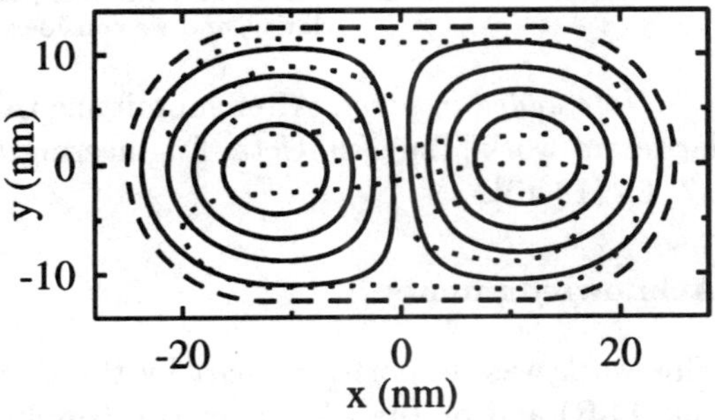

Figure 1: Contour plot of the first excited state (E_1) of stadium shape, whose outline is indicated by the dashed line. Solid (dotted) lines are the real (imaginary) parts of the wave function. Here we consider $B = 10$ T.

The BEM can also be used to calculate states which are excluded from particular regions, such as is the case for "antidots." When $B = 0$ such states are delocalized, and their spectrum is continuous.[2] For nonzero B in these structures all states are localized, and the energies of most states are near $(2n+1)\frac{\hbar e B}{2m}$ for n =0, 1, 2, ... However some states have energies which differ significantly from this owing to their localization near the antidot. A contour plot of such a state is seen in Fig. 2(a) for the case of a square antidot.

A shape of interest in the study of quantum chaos is a "Sinai billiard,"[6] an outline of which is shown by the dashed lines in Fig. 2(b). Here the electron is confined within the rectangle and excluded from the circle. A contour plot of the ground state for $B = 10$ T is shown in this figure. The fact that the ground state wave function has many nodes results from the term in the Hamiltonian [Eq. (1)] proportional to $-ib\partial/\partial\phi$.

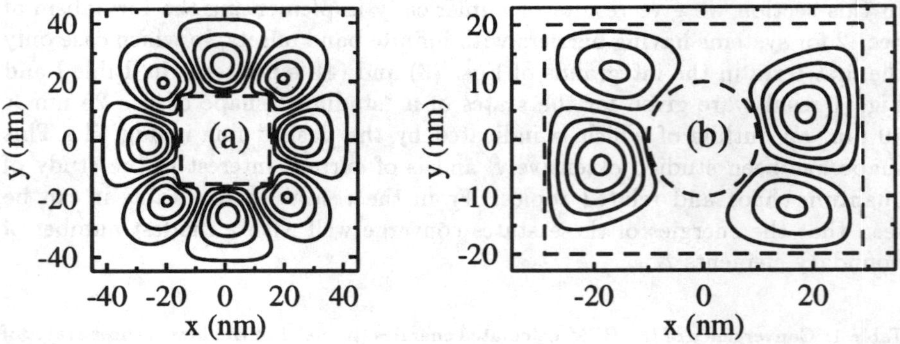

Figure 2: Contour plot of the real part of the wavefunction of (a) a square antidot and (b) the ground state for a quantum dot shaped like a "Sinai billiard." In both cases the dashed lines indicate the dots' boundaries, only the real parts of the wave functions are shown for clarity, and we consider $B = 10$ T.

Note added in proof. After submitting this paper we became aware of similar recent work [Tsuyoshi Ueta, Engineering Analysis with Boundary Elements **17**, 69 (1996)].

Acknowledgments

This work was supported in part by the U. S. Office of Naval Research (PAK and TLR) and by the Alexander von Humboldt Foundation (TLR).

References

1. J. E. Müller, Phys. Rev. Lett. **72**, 2616 (1994).
2. P. A. Knipp and T. L. Reinecke, Phys. Rev. B **54** (15 July 1996 issue).
3. Milton Abramowitz and Irene A. Stegun, *Handbook of Mathematical Functions* (Dover, New York, 1965).
4. E. Badralexe *et al.*, Comput. Phys. Commun. **71**, 47 (1992).
5. See, e.g., E. J. Heller *et al.*, Nature **369**, 464 (1994).
6. See, e.g., A. Kudrolli *et al.*, Phys. Rev. E **49**, R11 (1994).

STUDY OF PHASE BREAKING OF MESOSCOPIC TRANSPORT IN QUANTUM WIRE AND DOT

Y. OCHIAI, Y. OHKUBO, A. W. WIDJAJA, K. YAMAMOTO
Department of Materials Science, Chiba University,
1-33 Yayoi-cho, Inage-ku, Chiba 263, Japan

J. P. BIRD, K. ISHIBASHI, Y. AOYAGI, T. SUGANO
Nanoelectronics Materials Laboratory, RIKEN,
2-1 Hirosawa, Wako, Saitama 351-01, Japan

D. K. FERRY
Department of Electrical Engineering, Arizona State University,
Tempe, Arizona, 85287-5706, USA

We analyze the magnetic field dependence of the correlation field, B_c, of conductance fluctuations (CF), observed in a GaAs/AlGaAs quantum wires and dots. In both devices at high magnetic fields, B_c is found to increase linearly with magnetic field and the phase breaking time is given by the slope of the linear dependence. In addition, the amplitude of the CF shows a monotonic decrease with decreasing lead width. Using the latter dependence, we are able to obtain another estimate for the phase breaking rate near zero magnetic field using a recent prediction from random matrix theory.

1 Introduction

Chaotic orbital motion has been studied in the conductance fluctuations (CF) of the low temperature magnetotransport in semiconductor mesoscopic systems.[1,2] These have analyzed the low temperature transport by focusing on the phase breaking mechanism of ballistic electron waves in small cavities fabricated in the high mobility 2-dimensional electron gas (2DEG) system. Our interest is in the effect of the magnetic field on the chaotic transport. It is well known that Anderson localization and phase coherency in diffusive systems are strongly affected by applying a magnetic field. Here, it is very important to investigate how the field affects weak localization in ballistic systems. In our previous studies, the phase breaking time has been discussed by means of a correlation field analysis but can also be determined by the field dependence of the correlation field B_c.[3] We discuss the origin of an observed discrepancy between the two analyses by considering the mechanisms of phase breaking at high fields, prior to edge state formation.

2 Experiments

Split gate dots and corrugation wires were fabricated in GaAs/AlGaAs heterojunctions using lithographic techniques. The 2DEG carrier density and mobility of the dots and wires were 4.1 and 4.0×10^{15} m^{-2}, and 20 and 52 m^2/Vs, respectively. The detailed sample characteristics were reported previously.[4,5)]

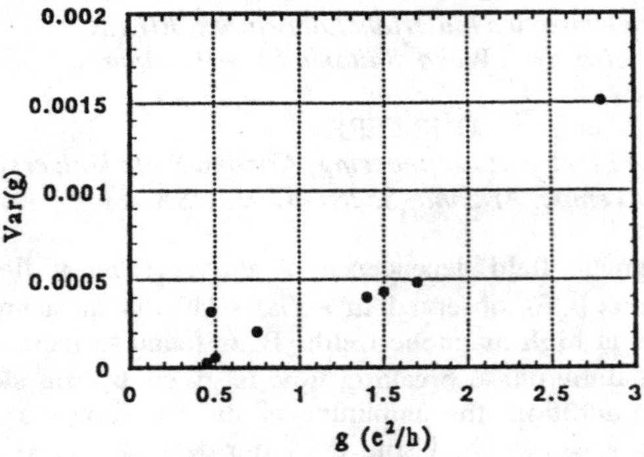

Fig. 1. The amplitude of the conductance fluctuations δg vs. the channel conductance g in a single dot at 10 mK.

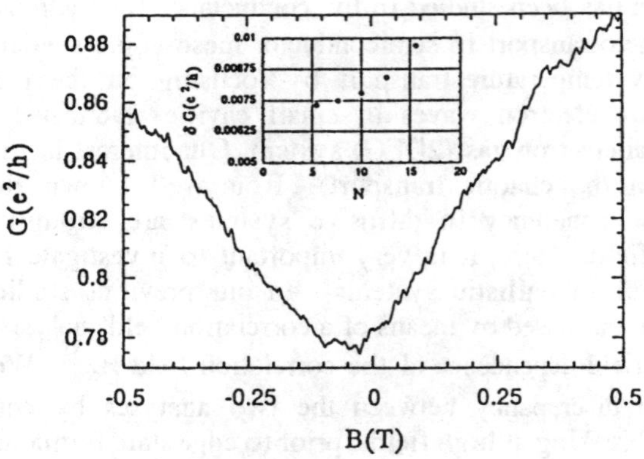

Fig. 2. Magneto-conductance in a corrugation wire just before pinch-off at 0.5 K. The inset shows the amplitude of the conductance fluctuations.

3 Results and Discussion

At low magnetic fields, we observe CF due to coherent interference of electron waves trapped in the single dot. We apply a correlation field analysis to obtain the field dependence of the B_c.[4)] From the slope of the field dependence, we can estimate the phase breaking time τ_ϕ, as we have shown in previous studies.[3,4)] The phase breaking time at higher field is 30 ps. However, at low fields B_c is almost field independent so that we can not apply the correlation field analysis for the determination of the low field τ_ϕ. The weak localization peak can be fit with a Lorentzian line shape in this dot and is thought to be well explained by the chaotic motion of the electron waves in the dot.[4)] We analyzed the amplitude of the CF by applying the results of random matrix theory (RMT)[6)],

$$\delta g = N/((4N^2-1)^{1/2} + N_\phi) \qquad (1)$$

where N is the channel number in the point contact leads and N_ϕ is an effective channel number in the dot. The channel number dependence of the amplitude is shown in Fig.1. Considering the characteristic parameters of the single dot, the phase breaking time τ_ϕ is given by

$$\tau_\phi = h/(N_\phi \Delta) \qquad (2)$$

where $\Delta = h^2/(2\pi m^* A)$ and A is the area of the dot. We thus obtain $\tau_\phi = 5$ ps using a single fitting parameter of $N_\phi = 65$. This value is small and there exits an unexpected difference between high and low field values. The smallness of τ_ϕ is considered to arise from a straight path in the single dot. As the field increases, the effect of the straight path should disappear when skipping orbits are formed in the dot.

Recently, we studied ballistic transport in the corrugation wires and analyzed the transport with a model of successive coherent resonance scattering. The large negative magnetoresistance is considered to come from such a multiple scattering.[7)] Conductance fluctuations are observed in the low field magnetoconductance in the corrugation gate wire as the gates are narrowed and this is shown in Fig.2. At high fields, we have no fluctuations in the magneto-conductance but can observe only SdH oscillations. We analyze the fluctuations using the correlation field, B_c,which should give some measure of A. A is found to be about 0.4 μm^2 and decreases as the gate voltage increases. In this regime, we can image an array of small dots. An electron wave is trapped and scattered randomly while coherently wandering in the array with multiple resonant scattering. The amplitude of the CF is plotted as a function of the entrance channel number of the corrugation unit as shown in the inset of Fig.2. If we apply a similar fit as discussed for the dot, the effective channels N_ϕ is several hundred. On the other hand, τ_ϕ is of ps order from considering the value of A of the wire. There is a large difference between the two since N_ϕ is very large. Even if we take account of a thermal smearing effect on the amplitude of the CF, the number N_ϕ will be reduced very little because our experiment has been performed at 0.5 K. Therefore, in order to explain fully with a chaos model, we must consider the exact peak shape of the negative magnetoresistance arising from chaotic orbits in a dot array.

4 Summary

We analyze the magnetic field dependence of the correlation field, B_c, of the CF, observed in a GaAs/AlGaAs quantum wire and dot. We have discussed the motion of the ballistic electron waves in single dots and dot arrays in mesoscopic systems. The amplitude of the CF shows a monotonic decrease with decreasing lead width. From this dependence, we are able to obtain a rough estimate for the phase breaking rate near zero magnetic field using a recent prediction from RMT. However, this result does not agree with estimates of τ_ϕ made at higher magnetic field. In the dot case, it is found that the magnetic field breaks the chaotic behavior and leads to regular motion of the electron. In order to fit to the RMT, it is necessary to use a value N_ϕ that is too large; e.g., it corresponds to a "lead" much larger than the size of the dot or wire. While the overall appearance of the CF seems to agree with RMT, the actual values for parameters such as τ_ϕ are not in agreement with other measurements and suggests an inadequacy of RMT in semiconductor dots.

Acknowledgements
We are grateful to the NEDO, *Nanostructures and Electron Waves Project*, for their financial support. Also, the financial support of Grant-inAid for Scientific Research on Priority Areas supported by the Ministry of Education, Science and Culture, *Quantum Coherent Electronics*, is gratefully acknowledged.

References

1.C.M.Marcus *et al, Phys.Rev.* B, **48**, 2460 (1993).
2.A.M.Chang, *et al, Phys.Rev.Lett.* **69**, 506 (1994).
3.Y.Ochiai, *et al, Jpn.J.Appl.Phys.* **34**, 4345 (1995).
4.N.Ohkubo, *et al*, submitted for publication.
5.A.W.Widjaja, *et al*, to be published in *Superlattices and Microstructures*.
6.H.U.Baranger, and P.A.Mello, *Phys. Rev.* B,**51**, 4703 (1995).
7.A.Grincwajg and D.K.Ferry, to be published in *Superlattices and Microstructures*.

MAGNETOTRANSPORT PROPERTIES OF A NEW-TYPE STAR-LIKE SINAI BILLIARD

M. V. BUDANTSEV, Z. D. KVON, A. G. POGOSOV,
A. I. TOROPOV, N. T. MOSHEGOV, A. E. PLOTNIKOV
Institute of Semiconductor Physics, Novosibirsk, Russia

The star-shaped solid state Sinai billiard was created and investigated for the first time. Its classical magnetotransport is characterized by the commensurability features connected with skipping and pinned orbits, previously observed in the antidot lattices. The Fourier power spectrum of the measured mesoscopic conductance fluctuations is similar to that of chaotic cavities, but in contrast with them it has a maximum, corresponding to the magnetic flux quantization through the area, enclosed by semiclassical skipping trajectories.

In the recent few years the transport properties of a variety of two-dimensional (2D) electron billiard systems have been widely investigated. Among this number are periodical lattices of artificial scatters, characterized by the classical magnetic field transport anomalies connected with commensurability of the diameter of the electron cyclotron orbit and lattice period [1,2], as well as single cavities of different shapes with characteristic sizes lower than the mean free path l_p of the electron in the initial material [3–6]. If the electron phase-coherence length l_φ is greater than the period or the size of the devices, they exhibit random magnetoresistance fluctuations similar to universal conductance fluctuations in disordered systems, caused by the quantum interference effects. Theoretical study shows that quantum transmission probability, which determines the conductance of a microstructure, directly reflects the classical behavior.

In this paper we report the experimental study of both classical and quantum transport properties of a new type of electron Sinai billiard of the form of a star-shaped cavity schematically shown in the inset of Fig.1. The samples were prepared on the basis of a two-dimensional electron gas (2DEG) in GaAs/AlGaAs heterojunction with electron mobility μ=600000cm^2/Vs and concentration n_s=3.510^{11}cm^{-2}. The cavity connected with 6 probes by constrictions was introduced into 2DEG by means of electron beam lithography and subsequent reactive ion etching. The cavity was shaped by 6 arc segments, thus forming a regular symmetrical star. The characteristic size of the device was about 1.6μm. The distance between neighboring arc centers was 0.8μm. The lithographic diameter of arcs was 0.3μm. We used the infrared LED to change the electron concentration of 2DEG, thus changing the width of the depletion layers from 0.25μm (almost closed state with effective arc diameter approaching to 0.8μm) to 0.1μm (opened state).

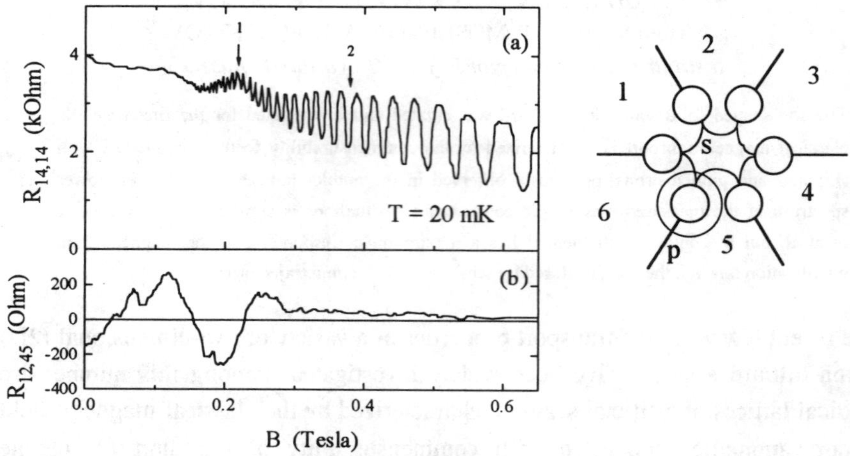

Figure 1: Two-terminal $R_{14,14}$ and four-terminal $R_{12,45}$ magnetoresistance traces after illuminating and a schematic view of the star-shaped billiard .

The magnetoresistance measurements were carried out in the temperature range 20mK–4.2K under magnetic fields up to 10T. The two and four probe measurements of different geometries were used, when the pairs of current and potential probes were chosen among 6 possible probes of the star. All magnetoresistance curves $R_{ij,kl}(B)=U_{kl}(B)/I_{ij}$ (where I_{ij} is the value of current flowing from probe i to j, U_{kl} is the voltage drop between probes k,l) shows striking anomalies in correlated magnetic fields. Each anomaly can be attributed to the appearance under certain magnetic field of the regular stable electron orbits localized inside the star against the background of the chaotic ones. The typical magnetoresistance trace $R_{14,14}(B)$ measured in the open state of the device is presented in Fig.1a. One can see Shubnikov–de Haas (SdH) oscillations, the envelope of which displays two distinguished peaks marked by the arrows in the figure. The first one corresponds to the cyclotron orbit (p) of the radius r_c=0.4μm analogous to the pinned orbit in the antidot lattices [1]. The second one (r_c=0.23μm) originates from the stable skipping orbit (s) which is also present in the antidot lattices [2]. Fig.1b shows the sign-changing $R_{12,45}(B)$ magnetoresistance analogous to the band resistance in the cross-junction devices [3]. Both curves remain practically unchanged at higher temperature T=4.2K except for the usual decrease of the amplitude of SdH oscillations. This supports that the mentioned above features are of classical nature.

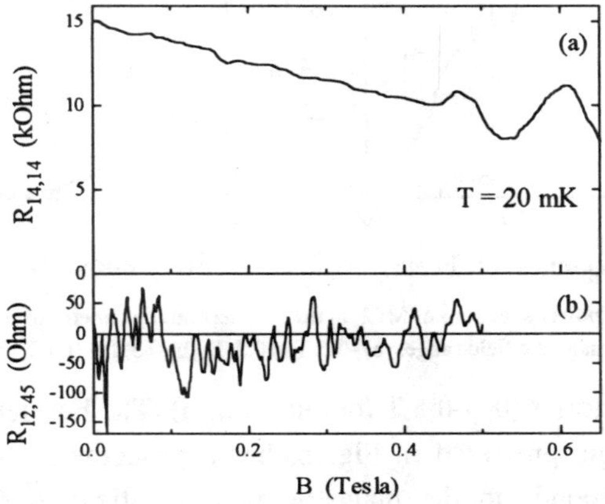

Figure 2: Two-terminal $R_{14,14}$ and four-terminal $R_{12,45}$ magnetoresistance traces before illuminating.

In the almost closed state of the device the mentioned above anomalies were accompanied by the mesoscopic conductance fluctuations (MCF). The corresponding magnetoresistance curves are presented in Fig.2 for the two geometries of measurements ($R_{14,14}$ – two-terminal) and ($R_{12,45}$ – four-terminal). The amplitude of these oscillations is much higher for the four terminal geometry (such as the oscillations becomes even sign reversing) then in the two-terminal one, thus indicating that they are connected primarily with the interference of electron paths existing inside the star. The Fourier power spectrum (FPS) of the oscillations is shown in Fig.3a. It exhibits the decrease with frequency according to [3] $S(f)=S(0)\,(1+2\pi\alpha\phi_0 f)\,\exp(-2\pi\alpha\phi_0 f)$, where f is the magnetic frequency in cycles/T, α is the characteristic inverse area of semiclassical coherent trajectories and ϕ_0 is the magnetic flux quantum. It means that the system investigated is rather chaotic. A least squares fit to the data (the dotted line in Fig.3a) yields $\alpha=2.1\mu m^{-2}$. The mentioned behavior of the FPS was predicted and observed in the chaotic cavities such as stadium [4] and cross-junction [3].

But in contrast with these systems, the MCF of the star-shaped billiard has the characteristic feature for the magnetic field range, corresponding to the appearance

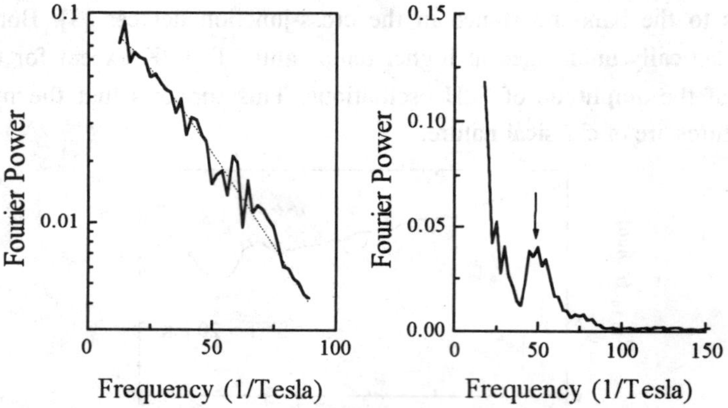

Figure 3: Fourier power spectrum of MCF in the four-terminal geometry of measurements in magnetic field range: (a) -0.5 T–+0.5 T, (b) 0.3 T – 0.5 T.

of the skipping trajectories (0.3–0.5 T for our billiard). The Fourier Power spectrum of MCF in this range, presented in Fig.3b, has a pronounced peak at frequency f=50T^{-1}, that corresponds to the magnetic flux quantization through the area 0.22μm^2. This area is much less than that of the device, which is about 0.67μm^2 taking into account the depletion layers, and can be attributed to the area enclosed by regular skipping trajectories inside the star.

Acknowledgments

This work was supported by the Russian Fund for Fundamental Research (Grant 95-02-04583a)

References

1. D.Weiss, M.L.Roukes, A.Menschig, P.Grambow, K.von Klitzing, G.Weimann, *Phys. Rev. Lett.* **66**, 2790 (1991).
2. E.M.Baskin, G.M.Gusev, Z.D.Kvon, A.G.Pogosov, M.V.Entin, *JETP Lett.* **55**, 678 (1992).
3. R.A.Jalabert, H.U.Baranger, A.D.Stone, *Phys. Rev. Lett.* **65**, 2442 (1990).
4. A.M.Chang, H.U.Baranger, L.N.Pfeiffer, K.W.West, *Phys. Rev. Lett.* **73**, 2111 (1994)
5. I.H.Chan, R.M.Clarke, C.M.Markus, K.Campman, A.C.Gossard *Phys. Rev. Lett.* **74**, 3876 (1995)
6. J.P.Bird, K.Ishibashi, D.K.Ferry, Y.Ochiachi, Y.Aoyagi, T.Sugano, *Phys. Rev.* **B 52**, 8295 (1995)

THE ASYMMETRY OF MESOSCOPIC PHOTOVOLTAIC EFFECT IN MAGNETIC FIELD IN BALLISTIC AND ADIABATIC TRANSPORT REGIMES

A.A. BYKOV, L.V. LITVIN, E.B. OLSHANETSKII, N.T. MOSHEGOV, A.I. TOROPOV
Institute of Semiconductor Physics, Russian Academy of Sciences, Siberian Branch, Novosibirsk, 630090, RUSSIA

The mesoscopic E.M.F. versus magnetic field dependencies in GaAs/AlGaAs rings for $\omega_c\tau>1$ have been found to be asymmetric with respect to changing the sign of magnetic field. This is explained by asymmetry of geometrical backscattering of carriers arising in ballistic and adiabatic regimes in microstructures whose geometry lacks central symmetry.

1 Introduction

The symmetry of transport coefficients with respect to changing the sign of magnetic field follows directly from the Onsager relations, according to which the two-terminal resistance of a mesoscopic conductor with no magnetic impurities should obey the equation R(B)=R(-B) [1]. This is also true for microwave mesoscopic E.M.F. in such conductors if the transport is diffusive: E.M.F.(B)=E.M.F.(-B) [2-4]. On the other hand when the transport is ballistic or adiabatic the effect of scattering impurities on transport coefficients is much diminished. In this case the asymmetry of coherent scattering, crucial for the generation of mesoscopic E.M.F, would be determined not so much by the random potential of impurities and defects as by the confining potential of a mesoscopic sample. The lack of central symmetry in the geometry of a mesoscopic sample under condition $\omega_c\tau>1$ should lead to the asymmetry of geometrical backscattering of nonequilibrium carriers and, therefore, give rise to a photovoltaic effect for which E.M.F.(+B)≠E.M.F.(−B).

2 Samples and experiment

In this work we have studied in detail the behavior of two submicron AlGaAs/GaAs rings. The asymmetrical geometry of these rings was not created on purpose but resulted from imperfection inherent to the fabrication technology. The two rings were made each on a different type of AlGaAs/ GaAs heterolayer. These heterolayers had the following parameters at T=4.2K: sample N1 – the electron density $n_s=1.3\times10^{12}$ cm^{-2} and the electron mobility $\mu=3\times10^4$ cm^2/Vs; sample N2 – $n_s=5\times10^{11}$ cm^{-2} and $\mu=8\times10^5$ cm^2/Vs. First, macroscopic samples were fabricated using photolithography to define a Hall bar 50 μm in width and with voltage probes placed 100 μm apart. Then, the microstructures were produced using electron lithography to define a submicron ring in the center of the current carrying arm of a Hall bar. The zero magnetic field resistance of the interferometers at 4.2K was 2-20 kΩ. The effective radius of the rings was calculated from the h/e-oscillations period and was found to be $r_{eff} \approx 0.15$ μm for N1 and $r_{eff} \approx 0.3$ μm for N2.

The measurements were carried out at T=50mk-4.2K in the microwave radiation frequency range 1GHz-145GHz in magnetic fields up to 13T. The E.M.F. signal was measured between ohmic current contacts to the rings at the frequency of the microwave field amplitude modulation. The electron gas heating (ΔT) by microwave radiation was estimated from the reduction of the Aharonov- Bohm and Shubnikov- de Haas oscillations amplitude and was kept much lower than the experimental temperature, ΔT<<T.

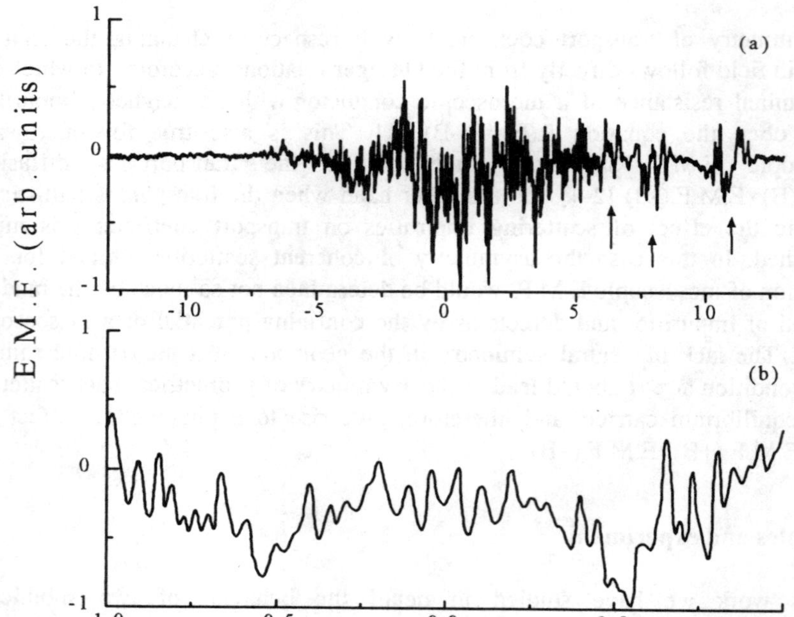

Figure 1: E.M.F. versus B for sample N1: a - over entire B range; b - around B=0. T=50 mK, microwave frequency 1.6 GHz, $R_0 \approx 15$ kΩ. Arrows indicate the regions where $R_{xx}=0$ in the Hall bar.

3 Experimental results and discussion

Fig.1 shows the microwave E.M.F. versus magnetic field dependencies for sample N1. The $\omega_c\tau>1$ condition for this sample is satisfied at B≈0.3T. Over the entire magnetic field range studied the two-terminal resistance of the sample was found to be symmetric with respect to changing the sign of magnetic field: R(B)=R(-B). In contrast to this the microwave E.M.F. at $\omega_c\tau>1$ exhibits asymmetric behavior: E.M.F.(B)≠E.M.F.(-B). This asymmetry is most pronounced at B>5T i.e. where there should be quantum Hall regime in the Hall bar and edge current states coupled by tunneling in the ring. A more detailed view of E.M.F.(B) in weak magnetic fields $B<B_c$ shows that in this region the symmetry of E.M.F. is preserved: E.M.F.(B)=E.M.F.(-B). For sample N2 the $\omega_c\tau>1$ condition is satisfied at B_c≈0.015T and, as one can see in Fig.2, throughout the entire microwave frequency range studied the E.M.F. signal for $B>B_c$ is asymmetric with respect to B=0.

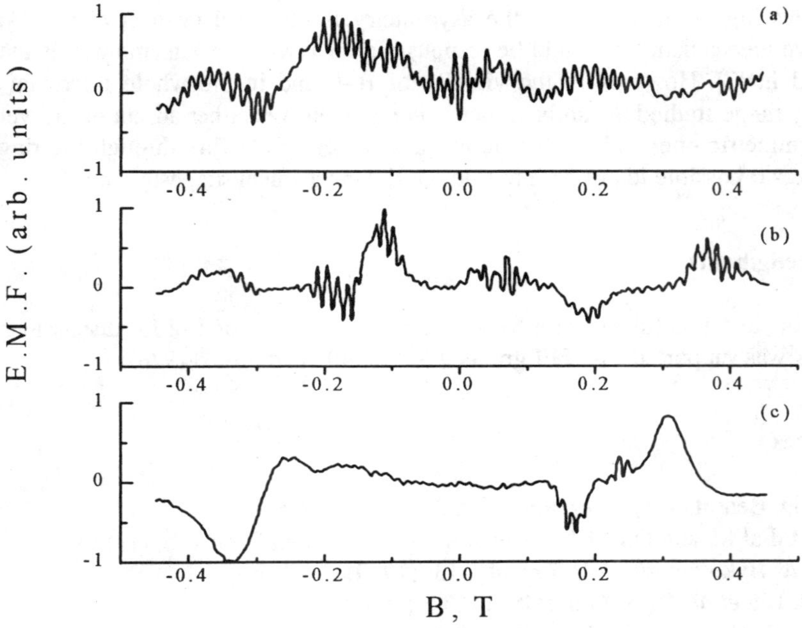

Figure 2: E.M.F. versus B for sample N2 for different microwave frequencies: a - 9 GHz, b - 74 GHz, c - 135 GHz, T=4.2 K, R_0≈3.5 kΩ.

So, it is experimentally established that the E.M.F.(±B) asymmetry arises under the condition $\omega_c\tau>1$, i.e. when the nonequilibrium carriers are induced by the Lorentz force either to perform a circular motion or to skip in a cycloid trajectory along the edge of a 2D gas. In these circumstances there should arise the effects of cyclotron trapping in the rings [5] and of backscattering at the sites where the ring is joined to the Hall bar [6]. In this case the photovoltaic effect is most probably caused by the asymmetry in backscattering arising from the geometrical difference in the two junctures of the ring to the Hall bar. As evident from Fig.2, in the case of a ballistic ring there are resonances both in the E.M.F. versus microwave frequency dependence and in the E.M.F. versus magnetic field dependence. It is known [2] that random potential asymmetry results in a smooth E.M.F. versus microwave frequency dependence. On the other hand a geometrical asymmetry would lead to E.M.F. resonances as the Fermi energy is varied. This proves that it is the latter that is the main cause of photovoltaic effect in ballistic rings. One of the possible reasons for the observed E.M.F.(±B) asymmetry is a nonzero elliptisity of the microwave radiation resulting in a circular photovoltaic effect. In our experimental conditions this is further supported by the fact that at frequencies higher than 50 GHz the asymmetry is observed even at $\omega_c\tau<1$. An alternative mechanism that might be responsible for E.M.F.(±B) asymmetry is that suggested in [7]. However in the vicinity of B=0 and in the whole microwave frequency range studied the antisymmetric component was either absent or, as well as the symmetric one, had the h/e period in magnetic field flux through the ring. This makes it possible in our case to rule out the mechanism suggested in [7].

Acknowledgments

The authors are grateful to Entin M.V. and Pogosov A.G. for fruitful discussions. The work was supported by RFFI grants 95-02-04409 and 96-02-19262.

References

1. A.D. Benoit *et al, Phys. Rev. Lett.* **57**, 1765, (1986).
2. V.I. Fal'ko and D.E. Khmel'nitskii, *Sov. Phys. JETP* **68**, 186, (1989).
3. A.A. Bykov *et al, JETP Lett.* **42**, 13, (1989).
4. J.J. Lin *et al, Phys. Rev.* **B45**, 14231, (1982).
5. H. van Houten *et al, Phys. Rev.* **B37**, 8534, (1988).
6. J.Liu *et al, Phys. Rev.* **B47**, 13039, (1993).
7. A.G. Aronov and V.E. Kravtsov, *Phys. Rev.* **B47**, 13409, (1993).